AF694556

INTERNATIONAL JOURNAL OF IMPACT ENGINEERING

Editor-in-Chief: Professor W. Johnson
Ridge Hall, Chapel-en-le-Frith, via Stockport SK12 6UD, U.K.

Editor: Professor Norman Jones
Department of Mechanical Engineering, University of Liverpool, P.O. Box 147,
Liverpool L69 3BX, U.K.

Associate Editor: Professor S. R. Reid
Applied Mechanics Division, Department of Mechanical Engineering, UMIST,
Manchester M60 1QD, U.K.

EDITORIAL ADVISORY BOARD

C. E. Anderson
Southwest Research Institute
San Antonio
TX 78284, U.S.A.

Y. L. Bai
Institute of Mechanics
Academia Sinica
Peking, People's Republic of China

V. L. Biderman
Bauman Institute
Moscow, U.S.S.R.

S. R. Bodner
Faculty of Mechanical Engineering
Technion-Israel Institute of Technology
Haifa, Israel

N. Cristescu
Department of Mechanics
University of Bucharest
Bucharest 70109, Romania

A. A. Deribas
U.S.S.R. Academy of Sciences
Novosibirsk 90, U.S.S.R.

W. Goldsmith
Department of Applied Mechanics
University of California, Berkeley
CA 94720, U.S.A.

N. K. Gupta
Department of Applied Mechanics
Indian Institute of Technology
Delhi, Hauz Khas
New Delhi-110029, India

K. Kawata
Science University of Tokyo
Faculty of Science and Technology
Department of Mechanical
Engineering, 2641 Yamazaki
Noda 278, Japan

H. Lippmann
Lehrstuhl für Mechanik
Technische Universitat München
8 München, Postfach 20 24 20,
F.R.G.

J. B. Martin
Department of Civil Engineering
University of Cape Town
Private Bag, Rondebosch
Cape Town, South Africa

Paul S. Symonds
Division of Engineering
Brown University
Providence
RI 02912, U.S.A.

P. S. Theocaris
Department of Mechanics
Athens National Technical University
Athens, Greece

R. L. Woodward
Department of Defence
Materials Research Laboratories
Ascot Vale
Victoria 3032, Australia

J. Zukas
Computational Mechanics Associates
Baltimore
MD 21239, U.S.A.

Publishing, Subscription and Advertising Offices:
Headington Hill Hall, Oxford OX3 0BW, U.K. (*Phone* Oxford 64881, *telex* 83177)

Annual institutional subscription rate (1987) DM 450.00. Two year institutional rate (1987/88) DM 855.00. Personal subscription rate for those whose library subscribes at the regular rate (1987) DM 143.00. Prices subject to change without notice. Subscription enquiries from customers in North America should be sent to: Pergamon Journals Inc., Maxwell House, Fairview Park, Elmsford, NY 10523, U.S.A., and for the remainder of the world to: Pergamon Journals Ltd, Headington Hill Hall, Oxford OX3 0BW, U.K.

Microform Subscriptions and Back Issues

Back issues of all previously published volumes are available in the regular editions and on microfilm and microfiche. Current subscriptions are available on microfiche simultaneously with the paper edition, and on microfilm on completion of the annual index at the end of the subscription year.

Copyright © 1987 Pergamon Journals Ltd. Published quarterly

It is a condition of publication that manuscripts submitted to this journal have not been published and will not be simultaneously submitted or published elsewhere. By submitting a manuscript, the authors agree that the copyright for their article is transferred to the publisher if and when the article is accepted for publication. However, assignment of copyright is not required from authors who work for organizations which do not permit such assignment. The copyright covers the exclusive rights to reproduce and distribute the article, including reprints, photographic reproductions, microform or any other reproductions of similar nature and translations. No part of this publication may be reproduced, stored in a retrieval system or transmitted in any form or by any means, electronic, electrostatic, magnetic tape, mechanical, photocopying, recording or otherwise, without permission in writing from the copyright holder.

Photocopying Information for Users in the U.S.A. The Item-fee Code for this publication indicates that authorization to photocopy items for internal or personal use is granted by the copyright holder for libraries and other users registered with the Copyright Clearance Center (CCC) Transactional Reporting Service provided the stated fee for copying, beyond that permitted by Section 107 or 108 of the United States Copyright Law, is paid. The appropriate remittance of $3.00 per copy per article is paid directly to the Copyright Clearance Center Inc., 27 Congress Street, Salem, MA 01970.

Permission for other use. The copyright owner's consent does not extend to copying for general distribution, for promotion, for creating new works, or for resale. Specific written permission must be obtained from the publisher for such copying.

The Item-fee Code for this publication is: 0734-743X/87 $3.00 + 0.00

HYPERVELOCITY IMPACT

PROCEEDINGS OF THE 1986 SYMPOSIUM

San Antonio, Texas

21–24 October 1986

Guest Editor
CHARLES E. ANDERSON, JR

PERGAMON PRESS
OXFORD · NEW YORK · BEIJING · FRANKFURT · SÃO PAULO
SYDNEY · TOKYO · TORONTO

U.K.	Pergamon Press, Headington Hill Hall, Oxford OX3 0BW, England
U.S.A.	Pergamon Press, Maxwell House, Fairview Park, Elmsford, New York 10523, U.S.A.
PEOPLE'S REPUBLIC OF CHINA	Pergamon Press, Qianmen Hotel, Beijing, People's Republic of China
FEDERAL REPUBLIC OF GERMANY	Pergamon Press, Hammerweg 6, D-6242 Kronberg, Federal Republic of Germany
BRAZIL	Pergamon Editora, Rua Eça de Queiros, 346, CEP 04011, São Paulo, Brazil
AUSTRALIA	Pergamon Press Australia, P.O. Box 544, Potts Point, N.S.W. 2011, Australia
JAPAN	Pergamon Press, 8th Floor, Matsuoka Central Building, 1-7-1 Nishishinjuku, Shinjuku-ku, Tokyo 160, Japan
CANADA	Pergamon Press Canada, Suite 104, 150 Consumers Road, Willowdale, Ontario M2J 1P9, Canada

Copyright © 1987 Pergamon Journals Ltd.
A member of the BPCC PLC Group of Companies

All Rights Reserved. No part of this publication may be reproduced, stored in a retrieval system or transmitted in any form or by any means: electronic, electrostatic, magnetic tape, mechanical, photocopying, recording or otherwise, without permission in writing from the publishers.

ISBN 0-08-0358993

First edition 1987

Published as Volume 5 of the *International Journal of Impact Engineering* and supplied to subscribers as part of their subscription. Also available to non-subscribers.

Papers reproduced by direct photographic process and authors remain fully responsible for the quality and content of the text.

Printed in Great Britain by Aberdeen University Press, Aberdeen

HYPERVELOCITY IMPACT
PROCEEDINGS OF THE 1986 SYMPOSIUM

Contents

PREFACE

The phenomenology of hypervelocity impact has been of interest since the mid-1950's, though interest waned somewhat in the late 1960's and most of the 1970's. Studies in the United States during the late 1950's and early 1960's centered around the meteorite hazard to NASA spacecraft and the feasibility of using solid projectiles to defend against incoming reentry vehicles. In the late 1970's, interest was revitalized with an active program in a nonnuclear kill (NNK) defense system against reentry vehicles. With the advent of spacecraft visits to comets, NASA's interest in the space station, the Strategic Defense Initiative, and the theories of large meteoroid or cometary impact against the earth, other planets, moons, and asteroids, the interest in hypervelocity impact has grown considerably.

Hypervelocity research began in the mid-1950's. In an attempt to understand and assess the state of the art in hypervelocity impact phenomenology, the Office of Scientific Research of the Air Research and Development Command commissioned the Rand Corporation to conduct a symposium in 1955. A strong interest in hypervelocity impact phenomenology continued for the next ten years. Six other hypervelocity impact symposia were conducted between 1955 and 1964. Although the workshops were normally limited to a small group of researchers in certain areas, these seven symposia were very successful in informing the community of the relevant work being conducted in hypervelocity impact research.

The third through seventh symposia were conducted under the direction of a tri-service committee (Army-Navy-Air Force) in the United States. All three members of the chairing committee, Atkins (Navy), Dittrich (Air Force), and Eichelberger (Army), who were responsible for the meetings, had been actively involved in the Hypervelocity Kill Mechanisms Program sponsored by the Advanced Research Projects Agency (ARPA) under Project Defender. The four areas normally considered in these symposia were: experimental techniques, theory, experiments, and applications. Only specific papers in the applications sessions were classified. Project Defender, which studied the NNK of reentry vehicles, was closed out in 1965 and NNK studies were subsequently deemphasized. Thus, the tri-service sponsorship of these symposia was discontinued, although hypervelocity impact phenomena were far from being understood. NASA still had a strong interest in hypervelocity impact because of the Apollo program. Under their auspices, another symposium was sponsored by the American Institute of Aeronautics and Astronautics (AIAA) in 1969, but interest was down dramatically and plans for further meetings were dropped.

Southwest Research Institute (SwRI) completed a survey of hypervelocity impact as it relates to the mission of the Ballistic Missile Defense (BMD) in 1982, and then updated the report in 1984. This survey, entitled "BMD Impact Phenomenology Data Base: Bibliography," by Wilbeck and Anderson, compiled the known literature in the area of hypervelocity impact which might be of interest to BMD personnel or contractors. Besides the relevant work in impacts against reentry vehicles, the bibliography includes work on projectile studies, target studies, impact conditions (velocity, yaw, obliquity), impact effects, failure modes, numerical techniques and studies, unique experimental considerations, and analytical techniques. While the bibliography is very useful, two facts became evident during its compilation. First, the bibliography only contains published reports, and thus is "dated" with respect to active and on-going research areas. Second, some of the relevant technologies were found to be fragmented, with related efforts being sponsored by a number of organizations.

NASA was beginning a new thrust in space debris, and planning for the space station. At the same time, all three of the U.S. armed services were expressing considerable interest in the technology and phenomenology of hypervelocity impact because of the high closing velocities of aircraft, and the research into advanced armors and faster projectiles. In many instances these various communities were only partially aware of what was occurring outside their own narrow field of interest. There was no concerted effort to coordinate research between the various groups studying hypervelocity impact. A review of current symposia and conferences also showed that none of these meetings offered a forum conducive to the presentation of research in the overall area of hypervelocity impact. Thus, the 1986 Hypervelocity Impact Symposium (HVIS) was born.

An Organizing Committee was formed in 1985 with Dr. Harry Fair of the Defense Advanced Research Projects Agency (DARPA) as Chairman. Other committee members are Charles E. Anderson, Jr. (Southwest Research Institute), Burton G. Cour-Palais (NASA-Johnson Space Center), Walter Herrmann (Sandia National Laboratories), William M. Isbell (General Research Corporation), William Johnson (Editor-in-Chief, **International Journal of Impact Engineering**), and James S. Wilbeck (Winzen International). In addition, the following Session Chairmen contributed significantly in selecting the technical content of the 1986 HVIS and ensuring its quality: Thomas Ahrens (California Institute of Technology), James Asay (Sandia National Laboratories),

Robert Bjork (Pacifica Technology), Price Boyd (U.S. Army Strategic Defense Command), Alex Charters (General Research Corporation), Gordon Johnson (Honeywell), Dennis Orphal (California Research and Technology), and William Snowden (Office of the Undersecretary of Defense for Research and Engineering).

The Organizing Committee reviewed the previous hypervelocity impact symposia, and found the Rand Symposium to be an excellent role model for the 1986 HVIS. The objectives of the 1986 HVIS were twofold. The first objective was to promote an understanding of the basics physics of the hypervelocity impact event. The second objective was to provide a comprehensive historical review of the research that had been conducted in the various hypervelocity impact disciplines.

There was a concerted effort to bring together the researchers in the many diverse disciplines with an interest in hypervelocity impact, both to promote the technical exchange of information, and to promote synergism through the interaction of these researchers. It has been approximately 20 years since the last symposium. Significant research was conducted during the years of the previous symposia, and much of it is still relevant. The experimental work, particularly at the GM Defense Laboratories in Santa Barbara, still stands as some of the best and most definitive data collected at hypervelocities. In addition, a whole new generation of researchers are now working in the field. In an attempt to prevent duplication of effort, a number of papers were solicited which critically review specific technologies, summarize the state of the art, and discuss what the future may hold. Other papers were selected which describe the results of current research investigations in hypervelocity impact.

This volume constitutes the papers presented at the 1986 Hypervelocity Impact Symposium, held in San Antonio, Texas during October 21-24. Authors are to be commended on their keeping to the schedule for providing their papers for peer review, making corrections, and preparation of their final manuscript. The first paper is a keynote by Dr. Harry Fair; it gives a history of the rise, fall, and then again the rise of hypervelocity impact research. It was primarily through the support of Dr. Fair that the 1986 HVIS was organized and held.

There are many ways to organize a symposium proceedings. Typically, they are organized by the discipline or topical category under which the paper falls, e.g., experimental techniques, numerical techniques, material response, etc. In this volume, papers are listed in the Table of Contents by disciplines/topical areas, but are arranged in the body of the volume alphabetically by the last name of the first author of each paper. The exception is Dr. Fair's symposium keynote, which is presented first. Hopefully this arrangement will make it easy to use the volume to locate either specific papers or topical groups of papers. We think that this volume, because of its technical content, and because it is archival literature, will be an oft quoted reference.

Charles E. Anderson, Jr.
Neil W. Blaylock
James S. Wilbeck

San Antonio, TX
December 1986

EDITORIAL

This special issue of the *International Journal of Impact Engineering** contains the Proceedings of the 1986 Hypervelocity Impact Symposium which was held in San Antonio, Texas, 21–24 October 1986. A glance through the contents page reveals that a distinguished group of authors attended and contributed articles on a wide range of theoretical, experimental and numerical aspects of hypervelocity impact which we are pleased to publish in this special issue of the Journal.

The editors wish to point out that authors in the field of hypervelocity impact are encouraged to consider submitting their work for publication in the regular quarterly issues of the *International Journal of Impact Engineering.*

We wish to take this opportunity to thank Charles E. Anderson, Neil W. Blaylock and James S. Wilbeck for their considerable assistance in preparing this special issue of the *International Journal of Impact Engineering.* Thanks are also due to the Organizing Committee of the Hypervelocity Impact Symposium:

Dr Harry Fair — DARPA,

Dr Charles Anderson — Southwest Research Institute,

Mr Burton Cour-Palais — NASA,

Mr Walt Herrmann — Sandia National Laboratories,

Mr William Isbell — General Research Corporation,

Professor William Johnson — Editor-in-Chief, *International Journal of Impact Engineering,*

Dr James Wilbeck — Winzen International.

The next Hypervelocity Impact Symposium will be held in the spring of 1989. Further details will be published in this journal as they become available.

W. Johnson
N. Jones
S. R. Reid

*The regular issues of this journal are now typeset, as remarked in an Editorial in the Volume 4, No. 1, issue.

Slow, slow, quick quick, slow.

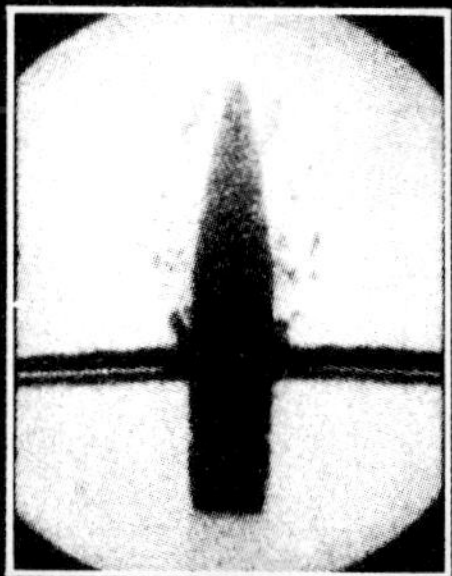

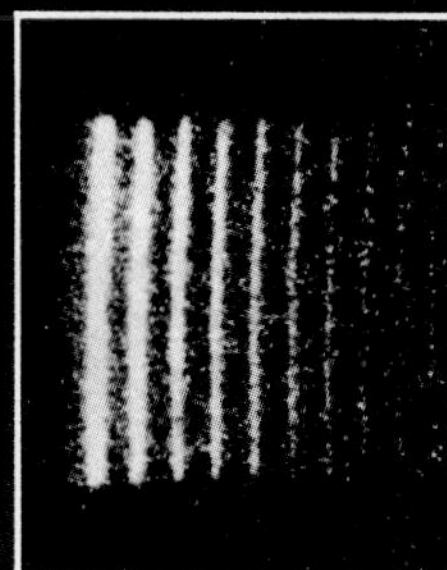
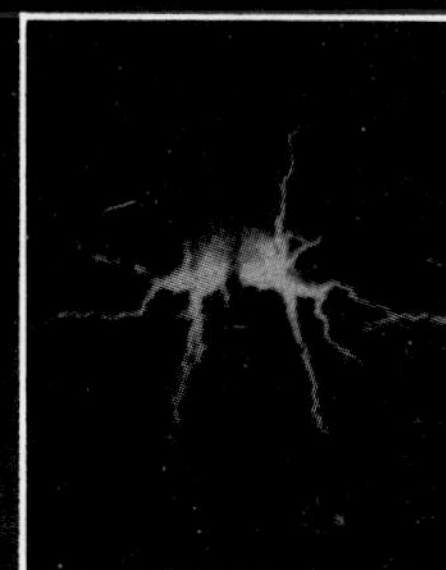
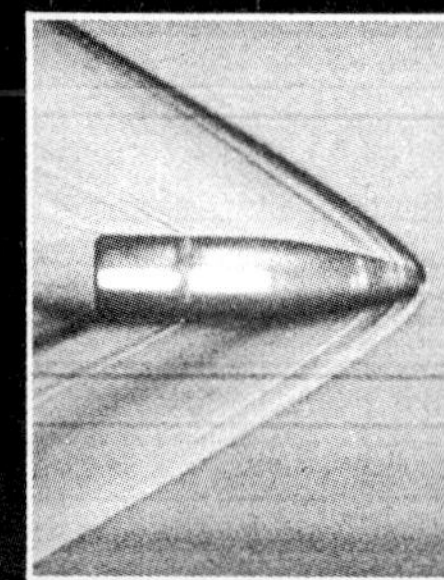

Whether you're measuring the penetration of a ballistic round in a metal target; analysing stress and crack propogation in limestone; recording a series of laser pulses at 20 picosecond intervals; simulating a lightening strike on a supersonic aircraft wing; or evaluating the shock waves of an in-flight projectile......Hadland Photonics have the High Speed Camera System to keep pace with your demands.

For over 20 years Hadland have led the way in Image Converter Camera technology, and the Imacon Division offers a range of 'ultra-fast' cameras capable of framing 2-dimensional images at rates of up to *600 million frames per second,* and 'streak' cameras with resolutions of *one picosecond!*

To maximise on information received from streak cameras, the Hadland 2D Image Analysis System will acquire, store and analyse the streak image.

Send us the coupon, or use the Reader Enquiry Service for full information and literature on the Hadland Photonics range of High Speed Cameras and the 2D Image Analysis System.

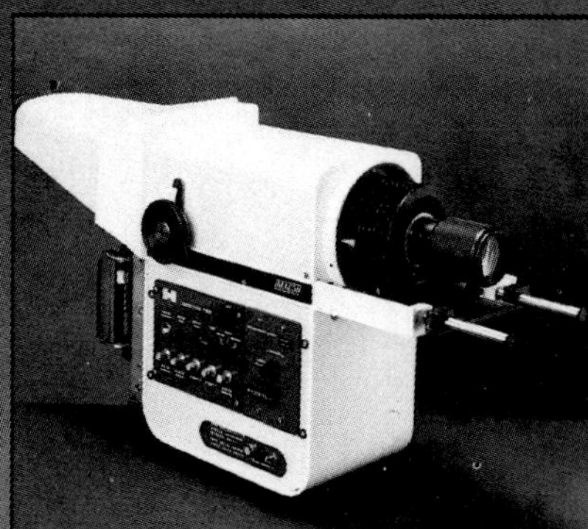

Imacon 500 Streak Camera

Major Features:

- Guaranteed sensitivity on delivery and after one full year.
- Single pulse or synchroscan recording.
- Sub-picosecond resolution up to 1.5μm wavelength.

Imacon 792 Framing and Streak Camera

Major Features:

- Framing up to 600 million fps with 6-20 frames per sequence.
- Streak up to 1 nanosecond per mm.
- Fibre Optic image tube for maximum dynamic range and definition.
- Wide spectral sensitivity of 180 – 850nm.
- Compatible with X-Ray Systems.

Imacon 2D Image Analysis System

Major Features:

- Combines latest CCD technology with 68000 based microcomputer system.
- Fully compatible with all Imacon 500 Series, Hamamatsu, and other streak cameras.
- Fast image display and processing.
- 16-bit frame store.

HADLAND PHOTONICS

Much more than meets the eye!

Hadland Photonics Limited, Newhouse Road, Bovingdon, Hemel Hempstead, Hertfordshire HP3 0EL United Kingdom.
Telephone: 0442 832525 Telex: 82344

Hadland Photonics GmbH, Robert-Bosch-Str 6, D-5030, Huerth-Efferen, West Germany.
Telephone: (0) 2233-64022
Telex: 88 9994/888 2420

Hadland Photonics Inc., 600 West Cummings Park, 2500 Woburn, MA 011801 United States of America.
Telephone: 617 933 7860 Telex 949373

Please send me further information and literature on the following:
☐ Streak Cameras ☐ Framing/Streak Cameras ☐ 2D Image Analysis

Name ____________________

Company ____________________

Address ____________________

____

INTRODUCTION TO IMPACT ENGINEERING

M. A. Macaulay,
Senior Research Fellow, Brunel University

Impact is an area of expanding scientific and engineering research. Impacts occur too quickly for us to follow what is happening and the evidence they leave behind is often ambiguous. However, recent developments in high-speed instrumentation and the use of computers have increased our understanding, although a multi-disciplinary approach is not yet established and information is scattered and expressed in a variety of jargons.

This book provides an introduction to the subject for the non-specialist but in sufficient detail to enable an interested reader to move on to more specialist published work. The text is written in an easy-to-read style and it will appeal to students and researchers in mechanical, structural and materials engineering as well as those working in the medical field.

Contents: Preface; Linear elasticity; Non-linear and time-dependent elasticity; Plasticity; Fracture; Rigid body motion; Undamped linear vibrations; Deformable bodies; Energy Absorbers; Structures; Impact injury; Testing; Biblography; References; Index.

January 1987 Hb 0 412 28930 X
276pp £19.95

CHAPMAN HALL
11 New Fetter Lane, London EC4P 4EE

Back Issues of Pergamon International Research Journals

Back issues of all Pergamon journals are available in hard copy or microform. New subscribers to a journal may purchase back issues of that publication in hard copy or microform editions at a 25% discount. Stocks of back issues are maintained at all Pergamon offices worldwide, and orders may be placed with confidence at the location most convenient to you. Should any issue of a volume be temporarily out of stock at the time of ordering, a photoduplicated copy will be supplied at no extra charge to complete your order.

Save up to 25% by Purchasing Complete Sets (Hard Copy)

Subscribers wishing to purchase complete sets of back volumes can do so at a saving of 25%.

The Microform Alternative

A very competitive price alternative to the hard copy is microform which represents a substantial saving of a further 50% on microfilm and 20% on microfiche prices.

Microform at 50% Special Trade-in Price for Back Issues

A hard copy back issue volume of any Pergamon journal, or the immediately preceding volume in hard copy, may be traded in against the same volume or volumes in microform. The trade-in allowance is 50% off the listed price of the microfilm or microfiche, and applies to both separate reels and complete back issue sets.

Back Issues Price List

Full details of the rates for back issues of all Pergamon journals can be found in our Back Issues Price List. Please contact your nearest Pergamon office for a copy.

Pergamon Journals Ltd.,
Headington Hill Hall, Oxford OX3 0BW, England

Pergamon Journals Inc.,
Fairview Park, Elmsford, New York 10523, USA

Int. J. Impact Engng Vol. 5, pp. 1–11, 1987
Printed in Great Britain

0734-743X/87 $3.00 + 0.00
Pergamon Journals Ltd.

HYPERVELOCITY THEN AND NOW

Harry Fair

Defense Advanced Research Projects Agency
Washington, DC

ABSTRACT

A history of hypervelocity impact research, particularly in the United States, is discussed, emphasizing the technical areas and objectives, and to some extent, the funding sources. The reasons for a hiatus of 10 to 15 years, followed by the current resurgence of interest also are given.

INTRODUCTION

The ability to accelerate macroscopic objects to hypervelocities can potentially have an impact on our society equivalent to the development of nuclear power or directed energy weapons such as lasers. The achievement of hypervelocity impact and our study of its phenomena and physics will have broad consequences including the transport of vehicles to and through space, the survivability of space structures, the survivability and lethality of military systems, the development of new forms of energy such as impact fusion, and understanding of the origins and evolution of the Earth and solar system.

"Hypervelocity impact" is a term which encompasses events in which impact-generated pressures are well in excess of the projectile and target strength. In contrast, at what we normally term ordnance velocities, particularly below 1.5 km/s, the shock phase of an impact event quickly decays. Consequently, ordnance impact pressures are on the order of or below target and projectile material strengths. In fact, though the shock phase can be included in the analysis of ordnance velocity impacts, excellent analytical results have been obtained in modelling such penetration dynamics by ignoring the shock phase completely.

However, as impact velocities increase to values in excess of 3 km/s, not only do shock dynamics become important, they can dominate the overall response of projectile and target. For some materials, such as lexan, the hypervelocity impact regime begins at 1.5 km/s. For a very high speed of sound material such as beryllium, it is well above 3 km/s. But for practical purposes, for most metals, the lower velocity limit for hypervelocity impact is on the order of 3 km/s. In this paper we will use the term hypervelocity to designate those velocities in excess of 3 km/s.

The ability to accelerate objects to hypervelocities, and the phenomenology of hypervelocity impact, received considerable interest in the mid-1950's. Unfortunately, this interest declined significantly in the following decades. In fact, the most recent United States (U.S.) hypervelocity conference was held 17 years ago in 1969. Prior to that, from 1955 through 1964, hypervelocity impact conferences were held approximately every 18 months. These had strong participation by industry, university, and government scientists and engineers.

There is clearly a story behind this significant rise and fall and current renaissance of interest in hypervelocity.

HISTORY OF HYPERVELOCITY RESEARCH

There are several histories of hypervelocity and hypervelocity impact contained in this story. One is the history of that for which hypervelocity impact is needed. Another is the history of how it can be achieved. The heyday of hypervelocity was pulled along by a desire for spacecraft survivability and missile defense--that for which it was needed. But today's renaissance has been pushed as much by what we can now achieve--the coming of age of electromagnetic propulsion--as it has been pulled by the renewal of interest in strategic defense.

Strategic Defense: A Road Once Traveled

In the mid-1950's, the dominant driver of hypervelocity impact research was the potential development of long range ballistic missiles which could reach continental ranges (ICBMS) or even launch payloads to space. This potential attracted the attention of those military and space scientists who were concerned with the development of long range weapons, with the defense against these missiles, and with the exploration of space.

The first Rand Symposium on High-Speed Impact, held 31 January - 2 February 1955, arose out of an attempt to "assess the vulnerability of the intercontinental ballistic missile to damage by high-speed fragments." The sparkplug behind the symposium, General Bernard Schriever, prophesied in his introductory talk, "These missiles (ICBMs) are certain to become available within the next few years."

The participants, stimulated by Fred Whipple and George Gamov, recommended that the U.S. gather data on meteoroid impacts using sounding rockets, and develop three laboratory tools to achieve hypervelocity: the light gas gun, shaped charge jets, and electromagnetic guns. The first paper at the symposium, by W. W. Salisbury, argued presciently that if sufficiently large magnetic fields could be achieved, a reasonably compact and lightweight electromagnetic gun might be feasible. Others indicated that light gas guns might be capable of routinely accelerating small masses to several km/s.

Then, on 4 October 1957, the first of the Sputniks was launched. The Soviets demonstrated an impressive throw weight. For example, Sputnik 3 ,launched on 15 May 1958, attained orbit at 2,926 lbs, compared to the American record satellite weight, at the time, of 32 lbs (Explorer II, launched 5 March 1958).

In the excitement, President Dwight Eisenhower upgraded the National Advisory Committee on Aeronautics (NACA) into the National Aeronautics and Space Administration (NASA), formed the Advanced Research Projects Agency (ARPA)--what we now know as DARPA--and gave ARPA its first charter: Project Defender. The goal of Defender was to make ICBM-delivered nuclear weapons impotent and obsolete. But unlike President Ronald Reagan's Strategic Defense Initiative, Defender was carried on with little fanfare. To this day, even the existence of the Defender crash program is little known outside the circle of those who worked on it.

Although Defender considered lasers, particle beams and even directed plasmas from nuclear explosives, hypervelocity kinetic energy warheads emerged as the strongest contender. Much of this activity was simply known by the name of "ARPA funding order 149." The kinetic energy approach was most highly developed in the Bambi program. This system employed the Space Patrol Active Defense (SPAD) warhead, a spinning spiderweb studded with small pellets.

Defender also motivated the development of armor against hypervelocity impact. Strategists expected the Soviets to try to counter space-based portions of a ballistic missile defense with kinetic weapons such as clouds of pellets.

Thus, as a consequence of the central role of hypervelocity impact in Project Defender, hypervelocity impact research entered its boom years. Symposia were held at approximately eighteen month intervals (see Table 1) to enable researchers to share their results in this rapidly developing competitive field.

Hypervelocity in the Laboratory

One of the major hopes and disappointments described in the early hypervelocity symposia concerned electromagnetic guns. The electromagnetic gun expertise developed by the Japanese and Germans (who had approached hypervelocity by accelerating 10 gm projectiles to 1.1 km/s) had been largely lost or forgotten after World War II.

Authors such as Arthur C. Clark and Robert A. Heinlein foresaw electromagnetic guns for space transportation and weapons. But engineers were frustrated by the lack of data from earlier research, the critical weakness of supporting technologies, and a generally fragmented research community.

Several participants at the first hypervelocity symposium described a variety of alternate electromagnetic guns and launchers. Many of these showed considerable promise for hypervelocity laboratory experiments and for such military applications as hypervelocity tank guns. However, during this critical period, these technologies actually lost ground. By the second symposium,

Table 1

HYPERVELOCITY IMPACT SYMPOSIA

Symposium	Date	Location
The Rand Symposium on High-Speed Impact Sponsor: The Rand Corporation and Office of Scientific Research of the Air R&D Command Host: The Rand Corporation	31 Jan-2 Feb 1955	Santa Monica, CA
The Second Hypervelocity and Impact Effects Symposium Sponsor: Naval Research Laboratory and the Air Research and Development Command Host: Naval Research Laboratory	22-24 May 1957	Washington, DC
The Third Symposium on Hypervelocity Sponsor: Tri-Service Host: Armor Research Foundation of ITT	7-9 Oct 1958	Chicago, IL
Hypervelocity Impact Fourth Symposium Sponsor: Tri-Service Host: Air Research and Development Command	26-28 Apr 1960	Eglin AFB, FL
Fifth Symposium on Hypervelocity Impact Sponsor: Tri-Service Host: Colorado School of Mines	30 Oct-1 Nov 1961	Denver, CO
Sixth Symposium on Hypervelocity Impact Sponsor: Tri-Service Host: The Firestone Tire & Rubber Co.	30 Apr-2 May 1963	Cleveland, OH
Seventh Hypervelocity Impact Symposium Sponsor: Tri-Service Host: Martin Company (Orlando)	17-19 Nov 1964	Orlando, FL
AIAA Hypervelocity Impact Conference Sponsor: American Institute of Aeronautics and Astronautics Host: AIAA	30 Apr-2 May 1969	Cincinnati, OH
1986 Hypervelocity Impact Symposium Sponsor: Defense Advanced Research Projects Agency Host: Southwest Research Institute	21-24 Oct 1986	San Antonio, TX

one of the larger electromagnetic gun efforts had failed to reproduce the earlier German accomplishments. The proceedings reported that "it is not likely that electromagnetic gun techniques will be successful in the near future for attainment of hypervelocity." This unfortunate but correct assessment of the specific experiments reported at the second symposium had far-reaching and long-term influence on funding for efforts to develop practical electromagnetic guns and launchers.

No project ever dies quite that easily. At the third symposium, K. W. Miller and R. M. Bergslien of the Armor Research Foundation at least granted that "electrical guns give promise of being an effective tool for achievement of hypervelocities of small masses for research purposes." However, they discounted their value as a potential weapon. But electromagnetic launch research was definitively laid in its grave, as far as any Defender funding was concerned, by the general conclusion at the second symposium that the technology of electromagnetic guns, and their necessary pulsed power supplies, were not, at that time, sufficiently mature to warrant their development.

During this period in which electromagnetic gun research was spectacularly unsuccessful, a reliable laboratory device emerged with the capability of accelerating objects to hypervelocities. The light gas gun was invented by W. D. Crozier and W. Hume of the New Mexico School of Mines, and came into initial operation as a laboratory device in the early 1950s. The light gas gun is able to accelerate materials to velocities higher that those attainable from conventional guns by using a gas of low atomic weight as the working fluid to propel the projectile. As a consequence, more of the gun kinetic energy is partitioned into the projectile and less into the reaction of the gas itself. Conventional guns can achieve projectile velocities on the order of 1.5 to 2 km/s. Light gas guns, however, can readily achieve velocities up to 10 km/s. Light gas guns also have a cost and compactness advantage over rocket sleds and sounding rockets in cases where a small mass projectile is acceptable.

Much of the credit for developing the light gas gun into a reliable laboratory device goes to Alex Charters, working at NASA-Ames Research Center. At the second hypervelocity symposium, Charters reported on the successful launch of 1-5 gm projectiles to over 4-5 km/s.

Once the light gas gun reached the laboratory, it opened up a new area of physics and chemistry. William Isbell, in a series of experiments from 1964 through 1971, replicated conditions in the center of the Earth by impact shocking a mixture of iron, nickel and sulfur at over 6 megabars. From 1960 to 1970 researchers at NASA-Ames experimentally studied hypervelocity crater formation in lunar-like material to aid in distinguishing impact from volcanic craters on the Moon. Light gas gun experimentation has also led to the development of equations of state to describe the behavior of matter under the conditions of high pressure and temperature found in hypervelocity impact.

Attempts to develop variants of light gas guns for weapons systems were never successful since the guns generally proved to be too heavy and bulky. But they have served as the most reliable launch mechanism for laboratory hypervelocity investigations.

Decline of Hypervelocity Research

Although in the 1950's and 1960's defense was the primary source of funding for hypervelocity impact research, another significant factor was the drive to explore and exploit space. Well before the rise of Project Defender, NASA's predecessor, NACA, began funding research into hypervelocity. The concern was meteors. How abundant were they? What would they do to a thin space structure when they hit? Could meteorite bumpers be developed for protection?

The U.S. Pegasus 1 (16 February 1965), 2 (25 May 1965) and 3 (30 July 1965) micrometeoroid satellites proved the threat to be worthy of concern, but to have been exaggerated. However, a new care arose. With President John Kennedy's call to land a man on the Moon by the end of the decade, some astronomers argued that billions of years of meteoritic bombardment of the lunar surface had created seas of dust hundreds to thousands of feet thick. Any place smooth enough to serve as an Apollo landing site would surely swallow its visitors. Again, a spacecraft--the Soviet Luna 9 (launched 31 January 1966), which landed in a basin and discovered luna firma--showed the fear to be baseless. NASA lost interest in sponsoring basic research on hypervelocity impact.

In the early to mid 1960's a number of major political and economic events coalesced to spell the ultimate destruction of Project Defender. The first blow to Defender was the deployment of the U2 reconnaissance aircraft and U.S. spy satellites. The first of these satellites, the Air Force Samos 2, was orbited 31 January 1961, and was followed by Samos 3 (9 September 1961) and the CIA Keyhole-4 CORONA (7 March 1962). These assets revealed that the missile gap which had most motivated Defender was not as bad as we had feared, or Nikita Khrushchev had led us to believe. In fact, those fearsome Soviet ICBMs based on the gigantic boosters which had lofted Sputnik 3 had no internal guidance, requiring radio command guidance all the way. They took hours of preparation to launch, and were based naked on the surface of the ground, vulnerable to any nuclear blast within miles.

With the race to land a man on the Moon, and then the Vietnam War, the United States had more pressing matters on hand than to combat a missile threat that might never materialize. Later in the decade, as that threat did develop, Defender faced the most implacable foe of all. Defense Secretary Robert McNamara began a campaign to base our nation's defense on his doctrine of Mutual Assured Destruction, MAD.

The Soviets were initially appalled at the concept of trading a defensive shield against ICBMs for an assurance that in the event of nuclear hostilities, aggressor and defender alike would remain unprotected and be mutually destroyed. "I believe that the defensive systems which prevent attack are not the cause of the arms race....Maybe an antimissile system is more expensive than an offensive system, but it is designed not to kill people but to preserve human lives." This argument was provided at a 1967 press conference by then Soviet Premier Aleksey N. Kosygin.

But the Johnson Administration took its own rhetoric seriously. It appeared that missile defenses would cost more than the ICBMs they were designed to block. McNamara's cost effectiveness calculations persuaded the U.S. that this inexorably meant that more missile defenses would simply inspire yet more nuclear-armed ICBMs. Beginning in 1965, Defender's budgets were slashed. Hypervelocity impact conferences lost their DOD sponsors. In 1969, the faltering Defender program was terminated, and the American Institute of Aeronautics and Astronautics hypervelocity conference held that year was the last of its kind until 1986. From then, until President Reagan's Strategic Defense Initiative, the U.S. ballistic missile defense was focused solely on preserving our retaliatory capacity. This was underlined by the fact that although the 1972 Antiballistic Missile (ABM) Treaty allowed two ABM sites, the U.S. only deployed one Sprint site, to protect our retaliatory missiles near Concrete, North Dakota. When we realized how readily an attacker could blind the Sprint radars, we closed even that. In this climate of overt hostility toward strategic defense, hypervelocity impact research necessarily languished.

The Hypervelocity Renaissance

Several factors emerged during the 1970's and early 1980's to bring new life to the science and technology of hypervelocity impact. These arose from such diverse interests as space colonization, electromagnetic launch, strategic defense, nonnuclear strategic offense, armor/antiarmor, planetary science, paleontology, and fusion research.

Space Colonization and the Rebirth of Electromagnetic Launch

Hypervelocity impact research received an indirect and totally unanticipated boost from space colonization enthusiasts, most significantly Princeton physicist Gerard O'Neill. Publishing in the September 1974 issue of Physics Today, O'Neill proposed using a coaxial linear accelerator electromagnetic gun--the "mass driver"--to launch lunar ore to the L-5 orbit for use in building cities, farms and industry in space. NASA funded O'Neill from 1975 to 1978 to build concept designs of these large space structures and the mass driver.

In 1977, O'Neill spent his sabbatical at MIT, where he began collaboration with Professor Henry Kolm of the Francis Bitter National Magnet Laboratory and a group of students who had a knack for building sophisticated physics experiments from surplus parts. It was by this group that the first mass driver was built and tested. The significance of this technology is that it for the first time demonstrated the capability of accelerating larger masses than possible with the rail gun which Hansler had used to achieve his WWII successes.

Meanwhile, Richard Marshall and co-workers at the Australian National University in Canberra were developing unique solid metal brushes to withstand extremely high pulsed current densities. Using the world's largest homopolar generator (550 MJ) to power a five-meter-long rail gun, they accelerated intact a 3 gm Lexan projectile to 5.9 km/s. This was a world record launch velocity for electromagnetic guns, and demonstrated that electromagnetic guns could, in fact, enter the realm of hypervelocity. Marshall's goal was to ultimately accelerate projects to sufficiently high velocities for the initiation of a thermonuclear fusion reaction in a fuel pellet.

At the University of Texas, William Weldon and Herbert Woodson were inadvertently developing the critical component technology necessary to make electromagnetic guns practical with their work on pulsed power, using rotating electrical machinery for welding and other metal forming applications.

Thus it was through the motivation of such diverse interests as space colonization, impact fusion, and welding and metal forming, that the key technology components for electromagnetic propulsion were brought into place. This is where Peter Kemmey, Thadeus Gora and I came into the picture. At the time, I was director of the Propulsion Laboratory at the U.S. Army Armament Research and Development Command. Frustrated by the difficulty of squeezing the last ounce of higher performance from an extremely mature solid propellant artillery propulsion technology, we decided to break out with a project we called "Nutty Ideas."

In keeping with the tenor of the study, Kemmey and Gora often did their best work at the local greasy spoon, and accordingly, briefed me on their initial electromagnetic launch concepts there. Working from first principles--we at the time were unaware of the sudden progress emerging from the laboratories of MIT, Texas and Australia--or even of Hansler's work--we concluded there might be something to the particular nutty idea of using electromagnetic energy rather than chemical energy for gun propulsion.

We were attracted to the electromagnetic gun concept because, if it could be made to work, it offered incredible advantages over chemically propelled munitions:

- the possibility of precise automatic control of projectile energy;
- the physics limit of velocity of light as opposed to velocity of sound for chemical propellants;
- elimination of the conventional gun tube, with its requirement for an axially symmetrical projectile and problem of barrel erosion;
- rapid fire;
- low vulnerability;
- instantaneous ignition;

- o ease of supply;
- o reduced cost.

In 1978, I was invited to present this concept to the Technology Trends Colloquium, a meeting organized by the Office of the Secretary of Defense (OSD) to explore new or emerging technologies and their applications. The OSD tasked the Army Propulsion Laboratory to assess the state of the art and potential military uses of electromagnetic propulsion. To capitalize on the recent extensive work in accelerators, directed energy, and pulsed power, we selected a National Advisory Panel consisting of Henry Kolm of MIT; Peter Mark of Princeton University; James Powell of Brookhaven National Laboratory; William Weldon of the University of Texas; Richard Marshall of the Australian National Laboratory; Philip Thullen of Los Alamos National Laboratory; and Peter Kemmey, Thaddeus Gora and myself of the U.S. Army R&D Command.

The initial meeting of the advisory panel was held at MIT. I outlined our intention to exploit the advances in pulsed power, directed energy, and electromagnetic accelerators, to possibly develop an electromagnetic gun program. Had the component technologies advanced during the past decade to now enable the development of practical electromagnetic gun systems? The general consensus of the panel was that the component technologies were indeed close at hand. The next step was to organize a technical workshop to further discuss the advances in component technologies and possible applications. This workshop was held at the U.S. Naval Academy in Annapolis in December 1978. Concepts were presented for numerous applications:

- o armor penetrators;
- o indirect fire (mortar, howitzer, etc.);
- o incoming round interceptor;
- o air defense;
- o direct rapid fire;
- o battlefield supply;
- o aircraft launcher;
- o integrated electric self-propelled weapon system;
- o anti-ballistic missile system;
- o orbit-to-orbit space transportation;
- o Earth-to-space (and Moon-to-space) launch;
- o impact fusion/equation of state.

Since then, significant progress has been made in armor penetrators, direct and indirect fire, aircraft launchers, space propulsion, impact fusion, and, of course, rail guns for ballistic missile defense.

In 1980 we held the First Conference on Electromagnetic Guns and Launchers in San Diego. The focus was on basic technology rather than applications. The highlights included reports on:

- o initiation of the design of a compact 5MJ homopolar generator;
- o rail gun experiments using explosive magnetocumulative generators;
- o proposal to use rotating bed nuclear reactor for power in space;
- o construction of EMACK rail launcher;
- o space propulsion and impact fusion analyses.

One problem was how to develop a body of expertise in electromagnetic guns that would be sufficient to get the technology beyond the laboratory and into a working weapon system. You cannot just throw money at contractors. You first need a corps of scientists and engineers who thoroughly understand the problem. Back in 1978, when we were first reviving these technologies, there were not that many people who were experts in electromagnetic propulsion. We tackled that problem right at the start by ensuring much of our funding (provided by DARPA and the Army R&D Command) went to support graduate students.

Subsequent developments were fast paced, often exceeding expectations. DARPA became the major sponsor for university and industry research efforts. In addition to sponsoring the First Symposium on Electromagnetic Guns and Launchers in San Diego, DARPA held a second symposium in Boston in October 1983. At the 1983 symposium, the first Peter Mark Medal "for outstanding contributions in electromagnetic launch technology" was awarded, and plans to hold the symposium every two years were announced. At the second symposium, the first results were reported on EMACK, a railgun accelerator built by Westinghouse which accelerated a 300 gm projectile to 4.2 km/s. Compact power sources were reported by the University of Texas. They had built a compact 5 MJ homopolar generator with a goal of delivering 0.75 MA, but which actually achieved a current of 1 MA.

These advances in the technology were extremely rapid, and several military tactical applications were selected as focus projects. As a consequence of these applications, the general technology goals were to accelerate projectiles through the atmosphere at velocities up to 3 km/s. However, a major redirection of the technology was to occur as the result of President Reagan's 23 March 1983 speech to the American people.

Making Nuclear Weapons Obsolete

It was on 23 March 1983 that President Reagan challenged the science and engineering communities to explore the possibility of developing a defensive shield against nuclear ballistic missiles, to "render these nuclear weapons impotent and obsolete." In response, the Office of the Secretary of Defense empaneled the Defensive Technologies Study Team (DTST), headed by Dr. James Fletcher, to investigate a broad range of technologies including chemical rocket interceptors, directed energy weapons, and hypervelocity electromagnetic guns. When the study started, directed energy weapons and chemical rocket interceptors were the favored technologies. But because of the strong potential effect against all phases of ballistic missiles in flight, the DTST recommended a major enhancement in the efforts to develop the technology for space-based electromagnetic guns.

With the initial formation of the Strategic Defense Initiative Organization (SDIO), DARPA continued to play a leading role in developing and directing the technology programs, but with funds provided primarily by the SDIO. This shift in technology emphasis led to significant new technology opportunities and challenges; for example, the need for higher velocity launchers (10-20 km/s), lightweight guided exoatmospheric projectiles, and larger energy density pulsed power sources, than had been envisioned previously.

An even higher velocity regime derived from the DTST, and is called the GEDI (Guided Electromagnetic Defensive Interceptor) program. This could stand as well for "Goldilocks," as you will see in a minute.

During the DTST, I became intrigued by the anomaly in energy and velocity for all of the weapons systems considered for SDI. The directed energy weapons traveled at or close to the speed of light, but did not have much energy per pulse, and as a consequence, were of marginal lethality against hard targets. On the other hand, the kinetic energy weapons had assured lethality since they contained 10's to 100's of megajoules of energy per pulse (from their impact), but were slow (approximately 5-7 km/s for chemical interceptors, and 10-20 km/s for electromagnetic guns). Having raised three sons and being quite a fan of the story of Goldilocks and the 3 bears, I believed we needed a velocity which was fast enough, and projectiles energetic enough, that they were "just right."

GEDI seeks to achieve velocities of at least 50-100 km/s with gram-sized particles. My initial guidance scheme was to fire the projectile through a laser gun tube in an annular laser beam which would hold the projectile in its interior by ablation of the projectile whenever it strays into the laser gun tube wall. Alternative approaches include the use of a focused beam to ablate the sides of the projectile as required, but at the cost of a tougher control problem.

Construction of the GEDI experimental facilities is underway at the Center for Electromechanics at the University of Texas at Austin, and GT Devices of Alexandria, VA. At UT, the power supply will be their 60 MJ, 9-MA homopolar generator. Acceleration will be achieved by a two-stage railgun. The first stage will be two meters long and will accelerate the projectile up to 5-10 km/s for injection into the second stage. The program will initially develop a 5 m second stage, with a later rail gun of 25 m to achieve GEDI velocity. The GT Devices approach is to use a combination electrothermal/rail gun in which they use a plasma created electrically to provide the initial projectile velocity. The guidance schemes and projectiles are being developed by Los Alamos and Spectra Technologies.

Clearly, if these accelerations can be achieved with an intact projectile, the GEDI test facility will enable breakthroughs in hypervelocity impact research by providing a manyfold increase in currently achievable velocities.

Armor/Antiarmor

The most recent military impetus for hypervelocity impact research was underscored in the fall of 1985, by the report on armor/antiarmor authored by the Defense Science Board. Conventional antiarmor rounds travel no faster than approximately 1.5 km/s. However, recent advances such as reactive armor, which explodes when impacted, shattering an antitank round before it penetrates, and certain ceramics, which are, counterintuitively, far harder to penetrate than steel, have currently raised serious concerns for future kinetic energy antiarmor systems.

The conventional terminal ballistics wisdom is that the penetration of armor reaches a plateau or even diminishes at hypervelocities. As a strong proponent of hypervelocity tank guns, I finally recognized that I had been asking the wrong questions. Given that an antiarmor gun can provide 2-4 km/s velocity to a projectile, how do you design this projectile to exploit this capability? After noticing promising hydrocode data on the penetration of armor with separated or broken shaped charge jets, DARPA funded several parallel analytical and experimental efforts to resolve this issue.

Experimental work by Alex Charters (General Research Corporation), Steve Bless (University of Dayton Research Institute) and Clay Miller (AVCO), and analyses by Dennis Orphal (CRT), and E. Blaise and B. Yoon (RDA), show that the use of segmented penetrators appears especially promising. While segmentation appears to give little advantage at conventional velocities, above 1.5 km/s it begins to be impressive. At 5 km/s penetration of a device with 15 segments is twice as effective as the same length of penetrator in a single rod, and with increasing velocity the advantage of the segmented device continues to grow. Another approach, emphasized by Orphal, is hollow rod penetrators. While some hollow rods penetrate less efficiently than solid rods, other hollow geometries have demonstrated significantly improved penetration. What geometric relationships enable improvement? How far can the improvement be carried?

The Secretary of Defense has recently characterized the challenge of Warsaw Pact armor as one approaching a matter of "national urgency" and directed the Army and DARPA to establish a major new initiative in armor/antiarmor to resolve this technology imbalance. The DARPA Tactical Technology Office and U.S. Army have responded to this directive. A key element of this initiative is the electromagnetic gun weapon system program in which a mobile hypervelocity electromagnetic gun weapon system will demonstrate the ability to accelerate the projectile to 2-4 km/s. The prime Army player in the new program is the Armament Research, Development and Engineering Center (ARDEC). The Defense Nuclear Agency is stressing capacitor-based power supplies for the demonstrator gun as an alternative to the compulsator and homopolar generator power supplies now dominating the field. Funding for projectiles is provided by the SDIO.

Four tasks have been identified for the program. Task A calls for design and fabrication of a mobile system compatible with the Army's AFV (armored family of vehicles) concepts for future weapons. The gun will fire a projectile with 15 megajoules of kinetic energy. (By comparison, the high-performance 120-mm gun of the German Leopard 2 tank and the U.S. M1 tank achieve essentially 9 MJ.). The minimum muzzle velocity required for the gun is 2.5 km/s, although we hope to get into the 4.0 km/s bracket. The minimum rate of fire will be 3-5 rounds per minute with a burst capability preferred. Synergisms with advanced vehicle concepts--e.g., all-electric vehicles--are to be explored as well. While the gun will be designed for anti-tank warfare at extended ranges, it also should be useful for tactical air and missile defense.

Under Task B, bidders are to build and operate a 9 MJ, one-shot laboratory gun over a 15-month period. This 9-MJ gun will use projectiles developed under a Hypervelocity Projectile Development Program. While initial tests will use rounds developed by the contractor, the facility must also be made available to related projectile programs. Awards for Task B have gone to the University of Texas at Austin, and to Maxwell Laboratories of San Diego, Calif.

Task C features the fabrication and demonstration of a multi-shot 9-MJ gun under a 30-month Phase 1 program. A maximum gun weight of 20 tons is permitted, while vehicle engineers will develop an "integrated prime power and pulsed power supply" for the gun. Minimum firing frequency for the gun is set at three rounds per minute for a three-minute period, using either EM or ET launch. A lifetime of 100 rounds is required. The weapon will include a projectile and loader, and a gun capable of elevation. Only primitive boresight fire control is required. The system will be able to operate as a stand-alone mounted on a skid and will be readily transportable.

Three awards were released under Task C. One contract went to a team comprising Maxwell Laboratories and FMC Corporation's Defense Systems Group in Santa Clara, Calif. FMC's Northern Ordnance Division in Minneapolis will design the gun and its associated equipment, including an innovative ammunition autoloader and a preaccelerator. FMC's Ordnance Division in San Jose will manage the system design and integrate the gun with a Bradley Fighting Vehicle chassis. A capacitor-type power supply will be developed by Maxwell, which is working with the Defense Nuclear Agency to develop capacitors with very high energy densities.

Other Task C awards went to the University of Texas/Austin and to Kaman Aerospace Corporation (KAC) of Bloomfield, Connecticut. KAC will use two subsidaries--Kaman Sciences Corporation in Colorado Springs and EML Research, Inc. The latter firm, headed by Dr. Kolm, was acquired by Kaman in 1985.

Ford Aerospace Corporation in Newport Beach, California has won a contract under Task D1 to develop unguided projectiles that will be used in the 9-MJ guns of Tasks B and C. Physics International (PI) will be a subcontractor to Ford for the projectile program. PI has specialized in electromagnetic pulse phenomena and will participate in the modeling of projectile/barrel interactions.

Fire-control systems for antiarmor, tactical air defense, and tactical missile defense, we believe, would require ranges beyond 5 kilometers. A 60-month program will lead to their demonstration on the 15-MJ gun system. The rounds should offer precision terminal accuracy, high resistance to countermeasures, and performance in adverse weather. We are also asking contractors to develop a proof-of-principle fire control system.

The electromagnetic guns being developed under these efforts will likely also be useful as laboratory instruments for hypervelocity research. The 9 MJ device may become available for research around early 1988. While it will not offer the extremely high acceleration promised by the GEDI research facility, it will handle projectiles in the size range we are accustomed to using with the light gas gun.

Planetary Science

Another motivation behind the renaissance in hypervelocity impact research has been recent discoveries in the fields of planetary science, geology and paleontology. Impact phenomena have been discovered to be central to understanding these disciplines. Highlights include:

- The discovery that nearly all solar system bodies are heavily cratered.
- Calculations by H.J. Melosh of the Lunar and Planetary Laboratory of the University of Arizona have indicated that the major differences in composition of the Moon versus Earth may be explainable by a collision of an Earth-size with Mars-size body resulting in coalescence of today's Earth/Moon system.
- Evidence, first reported by Luis Alvarez and associates, has indicated that the Cretaceous extinction (i.e., demise of the dinosaurs) was caused by impact on the Earth of a major extraterrestrial body.
- Discovery, by Eleanor Helin of the Jet Propulsion Laboratory, Eugene Shoemaker of the U.S. Geological Survey and others, of a populous class of large, near-Earth asteroids with high probability of impacting Earth.
- Accretion of planets and moons and the evolution of their atmospheres has been shown to be heavily influenced by impact phenomena.

Further exploration of these lines of inquiry cries out for better laboratory tools to quantify behavior of hypervelocity impact.

Impact Fusion

Friedward Winterberg of the University of Nevada was the first to propose, in 1964, a means to achieve hypervelocity impact fusion. He calculated that for impact to economically catalyze a fusion reaction, it appears necessary for it to generate plasma temperatures of over 5 kilovolts, plasma pressures on the order of 1000 megabars, and ion densities of about 10^{22} ions/cc.

In 1979, the Department of Energy began to take impact fusion seriously and held a workshop organized by R. N. Kostoff of the University of Washington in July at Los Alamos. The study concluded that impact fusion promises several advantages over conventional confinement fusion concepts. Injection of a hypervelocity particle into the fusion chamber is likely to be more readily commercialized than laser beam insertion because there are no optical surfaces to be either protected or burned out with each power pulse. Impact fusion, they reported, appears likely to be able to provide the greater than 30 gain predicted to be necessary for economical power production if velocities can be achieved on the order of 100 km/s; laser implosion experiments suggest these gains will be difficult to attain under any conditions. They also concluded that electromagnetic propulsion also appears to be inherently more efficient than lasers, and the technology development closer to the power levels--the delivery of 10-100 MJ to a target; this energy is believed to be the requirement for fusion.

Indeed, before the end of this decade, the GEDI accelerator may be able to achieve velocities required for impact fusion tests.

FUTURE OF HYPERVELOCITY IMPACT

Where is hypervelocity impact R&D headed? From our vantage point today, its greatest near-term potential appears in the military arena. However, as we have seen above, the electromagnetic guns being developed by the Strategic Defense and Armor/Antiarmor Initiatives will also become invaluable laboratory tools. As has happened from the invention of the compass and sextant to the development of spaceflight, military motivations may foot most of the bill, but in the long term, scientific and commercial enterprises may reap most of the benefit.

The impending new generation of hypervelocity laboratory tools promises to take impact physics research well beyond the plateau of the era of the light gas gun. Within the next decade, experiments should become feasible that can replicate conditions and enable researchers to solve for the equations of state in the depths of the giant gas planets and the nuclear furnaces of stellar interiors. Impact fusion may finally be able to escape the realm of pure theory and be blessed with an experimental approach which may, in fact, ignite thermonuclear reactions in the laboratory--and which has the potential for scaling to become a viable commercial means to generate power.

The contents of the 1986 Hypervelocity Impact Symposium provide an excellent window to the research topic which will benefit and thrive from the renewed interest in and improved laboratory facilities for hypervelocity research.

The prime motivator of the first conference, ballistic missile defense, is again a strong national goal and is well represented. Antiarmor may be the dominant near-term military application, although at this symposium, hypervelocity impact relevant to air defense has also surfaced as a serious topic. Air defense applications have highlighted a relatively new area for hypervelocity research: the chemistry of high shock-induced reactions.

Space scientists are well represented at this symposium (and would have been even better represented if we were not competing with a Halley's Comet meeting in Heidelburg). Ann Vickery and her husband Jay Melosh each have a paper on how hypervelocity impacts on the Moon and Mars may have accelerated ejecta to escape velocities. Peter Schultz has suggestions on impact experiments we can perform to answer questions of planetary cratering shock physics; crater scaling; planetary accretion/disruption; variables affecting cratering phenomena; and the effects of cratering on surface and atmospheric evolution.

An early topic of the hypervelocity impact symposia is also reappearing: impact damage of spacecraft. Encounters with hypervelocity space junk have recently scarred both the Salyut space station and the U.S. shuttle. Shrapnel from Soviet and U.S. ASAT tests, and space junk from more peaceful activities, is generating a low Earth orbit meteoroid belt which is in danger of growing geometrically as collisions among the "junk" generate yet more debris. Earth may yet join the ringed planet club.

You will probably agree, however, that the most dramatic paper to be presented here is the one by Dave Roddy of the U.S. Geological Survey and his associates with California Research and Technology. They have modeled a Cretaceous extinction-type event, the "impact of a 10 km diameter asteroid into a terrestrial ocean." They show a computer simulation of formation of a 200 km wide ejecta blanket, a hole as deep as five Mount Everests below sea level, and a rim four Everests above sea level.

With the advances we anticipate in the laboratory, we will also become able to develop improved analytic tools. A close reading of the phenomenology papers uncovers results which appear to conflict both with other experiments and with theory. Sessions on numerical techniques and analytical theory will attempt to wrestle with this.

New Military Applications

Hearkening back to the 1978 Army R&D Command study of possible applications of electromagnetic launch, we see that several promising applications have as yet received little attention: air defense, orbit-to orbit spacecraft propulsion, and interception of incoming rounds. In addition, there clearly are numerous applications beyond those initially suggested by the National Advisory Panel.

Lethal mechanisms exploiting kinetic kill against long range fixed or relocatable area targets is an important emerging area of military interest. The Army Space Initiative, headquartered at Ft. Leavenworth, Kansas, is currently investigating the use of long range, highly accurate kinetic kill against second echelon soft targets. The Gladiator program, headquartered at Eglin Air Force Base, is addressing similar possibilities.

For long range weapon delivery, it is advantageous, in order to reduce response time, to use long range artillery such as electromagnetic or electrothermal guns or missiles for a ballistic delivery, or else a modification of such a trajectory called "boost-glide." A boost-glide type vehicle saves kinetic energy by using its high lift to drag ratio to convert its upward velocity into horizontal velocity, thus extending its range. It glides at very high altitude, where drag is negligible, until near the target, at which point it can dive, make wide cross range maneuvers, or even spiral in, or reverse direction. There can be significant kinetic energy remaining, making these potential delivery systems good candidates for hypervelocity weapons.

Endoatmospheric guidance and control of hypervelocity vehicles will be difficult due to the plasma sheathe they create in flight. Solutions are being sought by a recently initiated program run by Dr. Jasper Lupo at DARPA/TTO. This problem is nontrival. It has been wrestled with by reentry vehicle designers for decades, but if solved, clearly opens the door to a host of new military applications including very long range artillery, antiarmor, air defense and terminal phase tactical and ballistic missile defense.

Strategic Implications: The Abolition

But there may be more at stake than a merely quantitative improvement in mature defense activities. Highly accurate, very long range hypervelocity weapons are in the process of turning today's strategic orthodoxies upside down. Most significantly, hypervelocity antiarmor guns provide the potential to break out of the conventional model-after-model evolutionary increase in armor penetration.

In 1982, New Yorker staffer Jonathan Schell electrified the nation with his book, The Fate of the Earth. Schell argued that nuclear explosives "were built as 'weapons' for 'war,' but their significance greatly transcends war and all its causes and outcomes. They grew out of history, yet they threaten to end history. They were made by man, yet they threaten to annihilate man. They are a pit into which the whole world can fall--a nemesis of all human intentions, actions and hopes."

Schell's book galvanized the Freeze movement, a political effort which tried to harness the fears Schell had aroused with a frankly uninspiring proposal to merely freeze nuclear arms at their current level. In 1983, after coming within 2 votes of passing the U.S. House of Representatives, the Freeze fizzled.

President Reagan accomplished much to defuse the Freeze with his morally appealing Strategic Defense shield. Consequently, Schell has considered how to ensure that the President's initiative would lead not simply to strengthened deterrence but all the way to the abolition of nuclear weapons. In 1984 he published The Abolition, which proposed a strategic defense-aided process of reductions in nuclear arms. Schell's latest book did not spark another mass movement. His erstwhile arms control allies volunteered scathing book reviews for the national press, and the antiestablishmentarians who had marched for a Freeze did not feel comfortable about demonstrating in the streets in support of the President's Strategic Defense Initiative. However, recent polls show the American people now overwhelmingly back SDI.

The big problem critics of SDI hang their arguments upon is that, as long as our defenses are not 100% leak proof, we will never abolish nuclear weapons because we will need them for deterrence. However, hypervelocity kinetic energy weapons can provide alternatives to nuclear retaliation.

Is it necessary to remain confined to these nuclear weapons of mass destruction? Can advances in precision long range hypervelocity kinetic energy weapons provide the nonnuclear retaliation required for world stability? Substitution of precision nonnuclear for nuclear strategic weapons could easily shape up to be one of the more interesting debates of the next decade.

CONCLUSION

Hypervelocity impact research clearly has broad societal impacts. These include fusion energy, space transportation, understanding the physics of stellar interiors, the evolution of our solar system and life on Earth, revolutionary approaches to conventional warfare, but perhaps most significantly, nonnuclear alternatives for strategic defense. The supporting technologies have now matured to the point where vastly higher velocities of macroscopic particles will be achievable in the laboratory, and soon, as practical commercial devices on and off the battlefield. We may, in fact, be on the threshold of a quantum technological leap whose full implications we can as little predict as the domestication of fire.

Acknowledgements

I would like to thank Carolyn Meinel of Analytic Decisions, Inc.; Charles Anderson of Southwest Research Institute; and John Bosma of Cardinal Communications Corp. for assistance in developing this paper.

Int. J. Impact Engng Vol. 5, pp. 13–32, 1987
Printed in Great Britain

0734-743X/87 $3.00 + 0.00
Pergamon Journals Ltd.

IMPACT ON THE EARTH, OCEAN AND ATMOSPHERE

Thomas J. Ahrens and John D. O'Keefe

Seismological Laboratory 252-21
California Institute of Technology
Pasadena, California 91125

ABSTRACT

Several hundred impact craters produced historically and at times as early as 1.9 x 10^9 years ago with diameters in the range 10^{-2} to 10^2 km are observed on the surface of the earth. Earth-based and spacecraft observations of the surfaces of all the terrestrial planets and their satellites, as well as many of the icy satellites of the outer planets, indicated that impact cratering was a dominant process on planetary surfaces during the early history of the solar system. Moreover, the recent observation of a circumstellar disk around the nearby star, β-Pictoris, appears to be similar to our own hypothesized protosolar disk. A disk of material around our sun has been hypothesized to have been the source of the solid planetesimals from which the earth and the other planets accreted by infall and capture. Thus it appears that the earth and the other terrestrial planets formed as a result of infall and impact of planetesimals. Although the present planets grew rapidly via accretion to their present size (in $\sim 10^7$ years), meteorite impacts continue to occur on the earth and other planets.

Until recently meteorite impact has been considered to be a process that was important on the earth and the other planets only early in the history of the solar system. This is no longer true. The Alvarez hypothesis suggests that the extinction of some 90% of all species, including 17 classes of dinosaurs, is associated with the 1 to 150 cm thick layer of noble-element rich dust which is found all over the earth exactly at the Cretaceous-Tertiary boundary. The enrichment of noble elements in this dust is in meteorite-like proportions. This dust is thought to represent the fine impact ejecta from a ~10 km diameter asteroid interacting with the solid earth. The Alvarez hypothesis associates the extinction with the physics of a giant impact on the earth.

Using finite-difference techniques, cratering flow calculations are used to obtain the spatial attenuation of shock pressure with radius, r, along the impact axis for the impact of silicate rock and iron impactors on a silicate half-space at speeds of 5 to 45 km/sec. Stress wave attenuation is found to be represented by two regimes, if the peak pressure, P, is fitted to expressions of the form $P \propto r^a$. At distances from 2.2 to 5.6 projectile radii into a silicate target, the constant, a, is on the order of -0.2. This low-attenuation rate impedance matching regime extends further into the target at the slower impact velocities. This occurs because of the slightly divergent flow associated with the penetration of a spherical projectile. For the near-field impact regime, an impact at 5 km/sec of an iron object with silicate surface will induce complete melting for silicate; the iron will remain solid. At 15 km/sec, partial vaporization occurs for both silicate and iron whereas at 45 km/sec, complete vaporization occurs in both materials. Similar calculations were conducted for a silicate meteoroid striking a silicate surface at velocities ranging from 5 to 45 km/sec. At greater radii in the far-field regime, the exponent, a, varies systematically from -1.45 to -2.15 for impacts of silicate onto silicate as the impact velocity is increased from 5 to 45 km/sec. For an iron projectile impacting at speeds of 5 to 45 km/sec, the exponent, a, varies from -1.67 to -2.95.

Upon impact of a 10 to 30 km diameter silicate or water object onto a 5 km deep ocean overlying a silicate half-space planet at 30 km/sec, we find that from 12 to 15% of the incident energy is coupled into the water. In the gravity field of the earth, some 10 to 30 times the impactor mass of water is launched on trajectories which can achieve stratospheric heights. The amount of ejecta launched to stratospheric altitudes is similar to the 10^1 to 10^2 projectile masses which result from impact of objects on an ocean-free silicate half-space (land). In the case of impact directly onto a silicate-half-space, only ejecta launched on trajectories which would carry it to stratospheric heights, has an impactor to target mass ratio which matches the fraction (10^{-2} to 10^{-1}) of extraterrestrial material found in the platinum-metal-rich Cretaceous-Tertiary boundary layer.

Oceanic impact results in impulsive-like giant tsunamis initially having amplitudes of ~4 km, representing the solitary waterwave stability limit in the deep ocean, and containing 10^{-2} to 10^{-1} of the energy of the impact.

Calculation of the interaction of a ~10 km bolide with the atmosphere indicates that only some 8% of the energy is imparted to the air during initial passage through the atmosphere. However, upon impact with the earth $\sim 10^1$ to 10^2 times the bolide mass of water or rock is ejected into the stratosphere, although, only ~0.1 bolide masses are in $<1 \mu m$ particles. The vaporized, melted, and (<1 mm) solid ejecta transfer up to ~40% of their energy to the atmosphere. Using the results of a similarity solution for the flow of gas as a result of an explosion in exponential atmosphere it is found that the atmosphere above such a large energy source is entirely ejected at speeds exceeding the escape velocity of the earth. Using the similarity solution we have calculated the mass of atmosphere lost due to impacts of 1 to 5 km radius projectiles. No atmosphere is lost for surface sources with energies less than 10^{27} ergs. Impact of objects in the energy range 10^{27} to 10^{30} ergs causes gas losses of 10^{11} to 10^{14} kg or 10^{-8} to 10^{-5} of the total present atmospheric mass. Impact energies of greater than 10^{30} ergs cause little increase in atmospheric loss.

INTRODUCTION

Impact cratering on planetary surfaces began to be intensely studied as part of our present exploration of space after 1959 when the Soviet spacecraft Luna 3 obtained the first indistinct images showing that the back side of the moon was as intensely cratered as the front side. The controversy as to whether the major craters on the moon were volcanic, a view championed by J. Green (e.g. Green, 1971), or, resulted from impact as championed by Shoemaker (e.g. Shoemaker, 1962) raged from the early 1960's for nearly a decade. In the years that followed the United States' Ranger, Surveyor, and finally, the Apollo lunar exploration projects demonstrated that craters on the Moon were impact craters. Previously, the American geologist, G. K. Gilbert (1893), and later others, developed a theory to explain impact craters, whereas the famous geophysicist Mohorovicic (1928) and others believed that the craters visible on the Moon were volcanic in origin. Subsequent photos of the ancient terranes of Mars and Mercury by the Mariner IV and X spacecrafts and of the large impact craters on Venus by the orbital radar imaging spacecraft, Venera 13 and 14, supported the concept of the importance of impact cratering in planetary surface evolution. Between 1979 and 1986 the Voyager spacecraft have shown us that the icy satellites of the major planets Jupiter, Saturn, and Uranus are distinctly different, and many display impact craters.

Within the Earth-Moon system several important correlations have been established between the observed cratering rates and potential impactors observed telescopically in Earth crossing orbits. For cratering on terranes younger than 0.5 x 10^9 years, the rate observed on the Earth on the North American and European Pre-cambrian platforms for craters of diameter greater than 20 km, is 0.35 $\pm$ 0.13 x 10^{-14} and 0.33 $\pm$ .2 x 10^{-14} craters/km^2/year (Figs. 1-2). These values are close to those observed on the now dated Mare surfaces of the Moon which indicate a rate of 0.37 x 10^{-14} craters/km^2/year (for craters having a diameter $>$ 20 km).

Finally, if we look at the distribution of asteroids which are now on earth-crossing orbits and thus will eventually impact the earth, the so-called Apollo bodies, their distribution with diameter (D) (Fig. 3) indicates that the asteroid number, N_c, will be of the form $N_c \propto D^{-1.6}$. This, when taken with crater scaling laws, yields crater dstributions close to those observed (Fig. 2) for craters with diameters $>$ 22.6 km. Moreover, using the absolute number of observed objects in Earth crossing orbits infers a terrestrial crater rate of 0.55 x 10^{-14} craters/km^2/year which is slightly greater than the observed rate (Grieve and Dence, 1979).

Thus using the cratering record of terranes younger than 0.5 x 10^9 years on the earth or younger than 3.4 x 10^9 years on the Moon, the impact rate on the Earth-Moon system appears to be nearly constant. Also, based on their total number and size-number distribution, objects currently in earth-crossing orbits are predicted to impact the earth at slightly greater than the observed rate.

Impact processes thus undoubtedly occurred over a long period on the earth. Moreover, hypotheses that the terrestrial planets accumulated from accretion processes in which many planetesimals were originally in orbit around the sun and coalesced by impact of the smaller planetesimals onto the largest growing planetesimals, as pictured in Fig. 4, have recently received support from visual observations of material around nearby sun-like stars such as β-Pictoris (Smith and Terrile, 1984) (Fig. 5). We infer that planetary systems are now forming by impact accretion like our own system did 4.5 x 10^9 years ago. The impact of large projectiles (toward the end of the $\sim 10^7$ year accretion period) undoubtedly gave rise to a wide range of angles of inclination of the spin axes of the planets, relative to the normal to the ecliptic (Fig. 6).

In 1980, Alvarez et al. (1980) proposed that the 1 to 150 cm thick worldwide layer of siderophile-element rich dust exactly at the 65 Ma, Cretaceous-Tertiary boundary resulted from impact of a large meteorite on the earth. Sediments containing the dust layer were formed in deep sea, shallow marine, and lacustrine environments. Alvarez et al. suggested that large quantities of impact ejecta were lofted into the stratosphere. This gave rise to a sudden climatic change, which in turn caused extinction of approximately 90% of all existing fauna 65 Ma ago. The physical evidence of the impact-induced layer is now found in 66 localities world-wide, by many other workers. The layer has a concentration of noble metals such as iridium, osmium, gold, platinum, rhenium, ruthenium, and palladium, in addition to nickel and cobalt, which is larger by factors of 5 to 1000 over crustal abundances and occurs in meteoritic proportions (Fig. 7). Although the issue of whether this impact caused the sudden demise of the dinosaurs is still widely debated (Sloan et al., 1986) there seems to be little controversy that there was a sudden and pronounced extinction of life at the Cretaceous-Tertiary boundary which had long been recognized and puzzled over by paleontologists.

Although the Alvarez hypothesis as to the cause of mechanism of sudden extinctions is still controversial, the present authors believe that the impact mechanics support his model (Ahrens and O'Keefe, 1982a,b). We conclude that not only is impact accretion a process which is vital to the formation of the earth and the terrestrial planets, but that the evolution of the surfaces with time is affected by impacts and that large impacts appear to have a pronounced effect on the evolution of life. In the present paper we review our previous calculations of the effect of impact on the rock surface of the earth, on the ocean, and the atmosphere, and present some new work on mechanism of impact-induced atmospheric escape.

IMPACT ON ROCK

When an impact occurs on a planetary half space, shock waves are driven into both the target and projectile. The upper propagating shock is reflected at the rear surface of the projectile and subsequently a rarefaction wave is driven forward into the shock projectile and target. The downward traveling shock decays and deposits energy in the form of irreversible work into the target (Fig. 8). The locus of shock states achieved in either the projectile or target is described in terms of the equation of state which specifies the locus of both shock and release wave states in stress, volume, and energy space. Stress conditions must also be specified so as to describe dynamic yielding. Also specified in the equation of state are the conditions required for incipient melting (IM), complete melting (CM), incipient vaporization (IV), and complete vaporization (CV). These are prescribed by entropy values, internal energies, and specific volumes along the Hugoniot (Ahrens and O'Keefe, 1972, 1985).

An often-used formulation for the equation of state is that first proposed by Tillotson (1962). In the compressed region this is given by

$$P = \left[a + \frac{b}{\left(\frac{E}{E_0 \eta^2} + 1 \right)} \right] \frac{E}{V} + A\mu + B\mu^2 \tag{1}$$

where a = 0.5 is the polytropic constant minus 1, at high temperature. The constant, b, is defined such that (a+b) is the STP Gruneisen parameter, $\gamma \equiv V(\partial P/\partial E)_V$, and A is the bulk modulus. Here $\eta = V_0/V$ and $\mu = \eta - 1$, where V_0 is the STP specific volume. The parameters b, B, and E_0 are obtained by fitting to Thomas-Fermi calculations of the equation-of-state in the 10^2 Mbar range, and Hugoniot data in the 10^{-1} to 10^0 Mbar range.

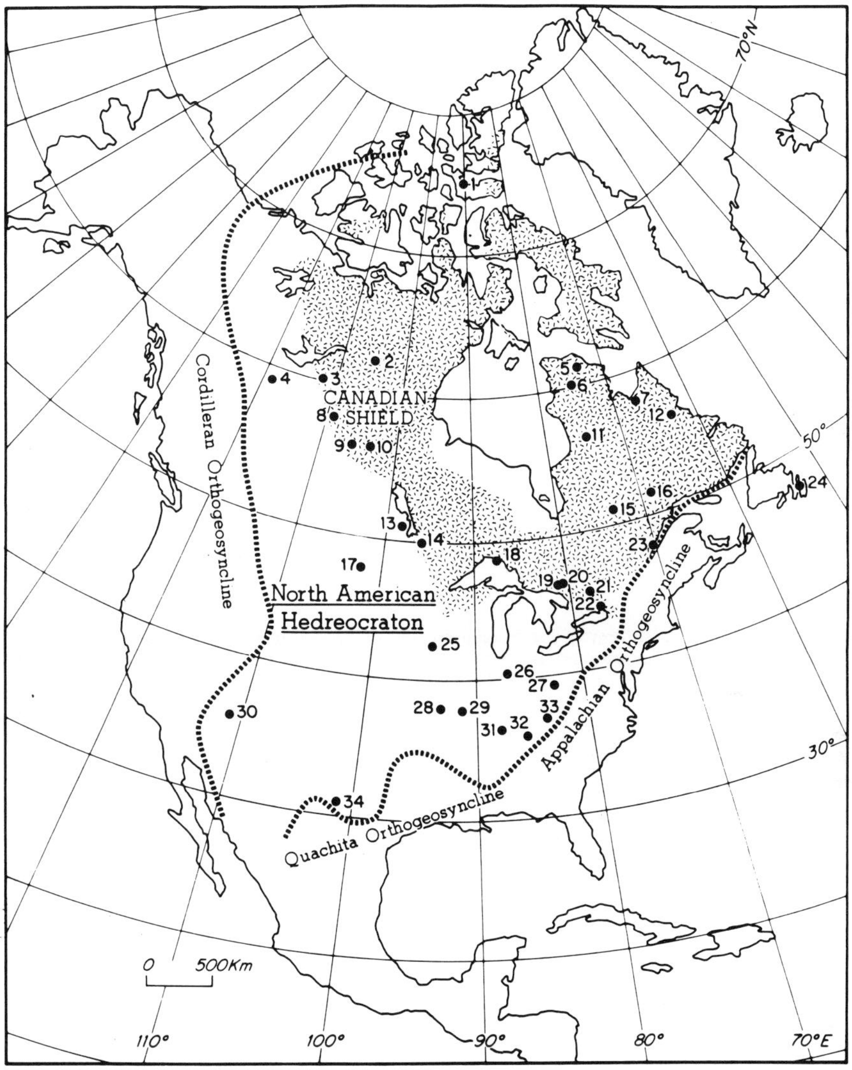

Fig. 1. Location of currently known impact structures > 1 km in diameter in North America. 1, Houghton Dome; 2, Nicholson Lake; 3, Pilot Lake; 4, Steen River; 5, New Quebec; 6, Lac Couture; 7, Lac La Moinerie; 8, Carswell; 9, Gow Lake; 10, Deep Bay; 11, Clearwater (East and West); 12, Mistastin; 13, Lake St. Martin; 14, West Hawk Lake; 15, Ile Rouleau; 16, Manicouagan; 17, Red Wing; 18, Slate Islands; 19, Sudbury; 20, Wanapitei; 21, Brent; 22, Holleford; 23, Charlevoix; 24, Conception Bay; 25, Manson; 26, Kentland; 27, Serpent Mound; 28, Decaturville; 29, Crooked Creek; 30, Barringer; 31, Wells Creek; 32, Flynn Creek; 33, Middelsboro; 34, Sierra Madera (after Grieve & Dence, 1979).

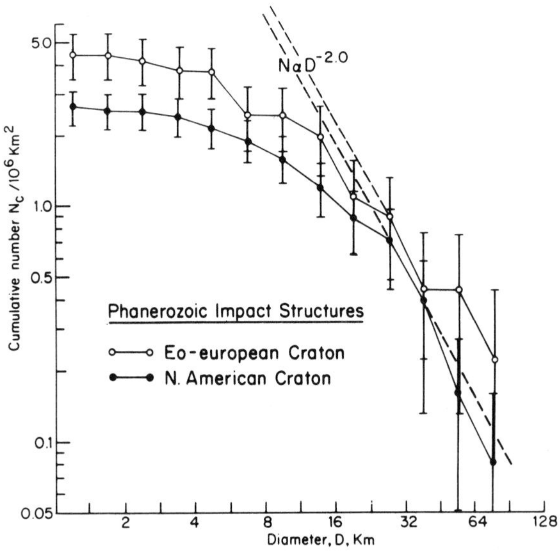

Fig. 2. Log (cumulative number/10^6 km^{-2}) -log (diameter) plot of Phanerozoic impact structures > 1 km in diameter on North America and Eo-european cratons. Bars indicate 1σ confidence interval (after Grieve and Dence, 1979).

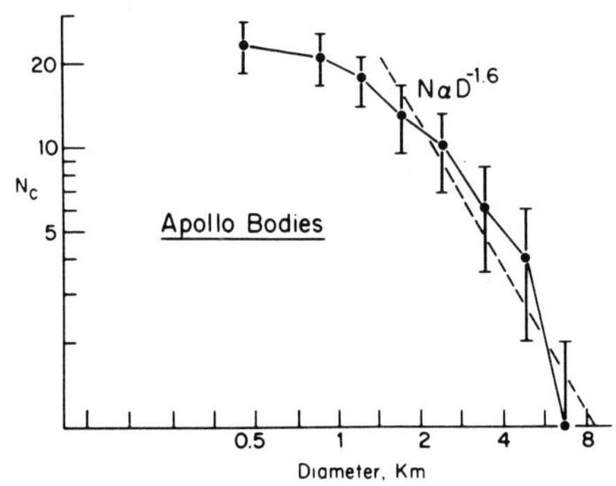

Fig. 3. Log (cumulative number) versus log (diameter) of Apollo earth-crossing asteroids (after Grieve and Dence, 1979).

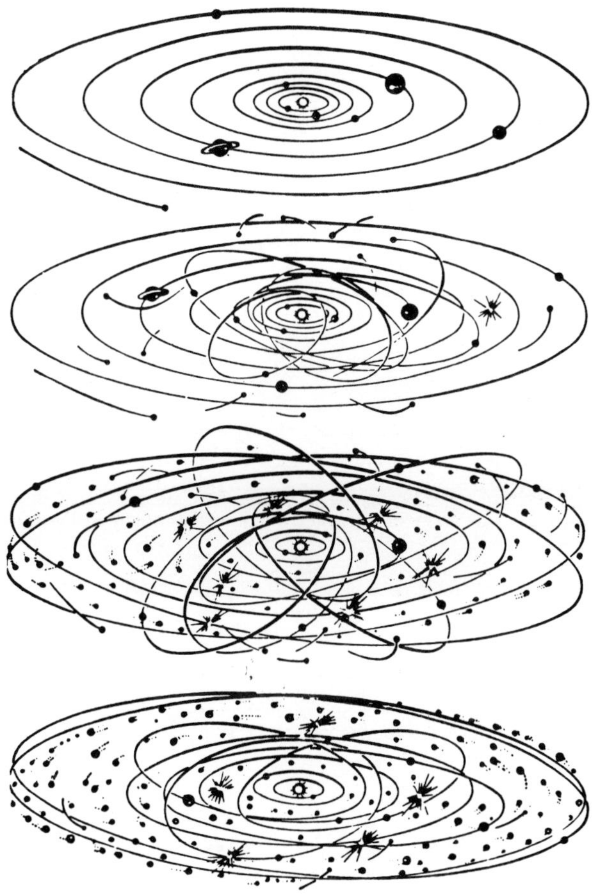

Fig 4. Accretion of the planets from the proto-planetesimal swarm in orbit about the sun (after Levin, 1972).

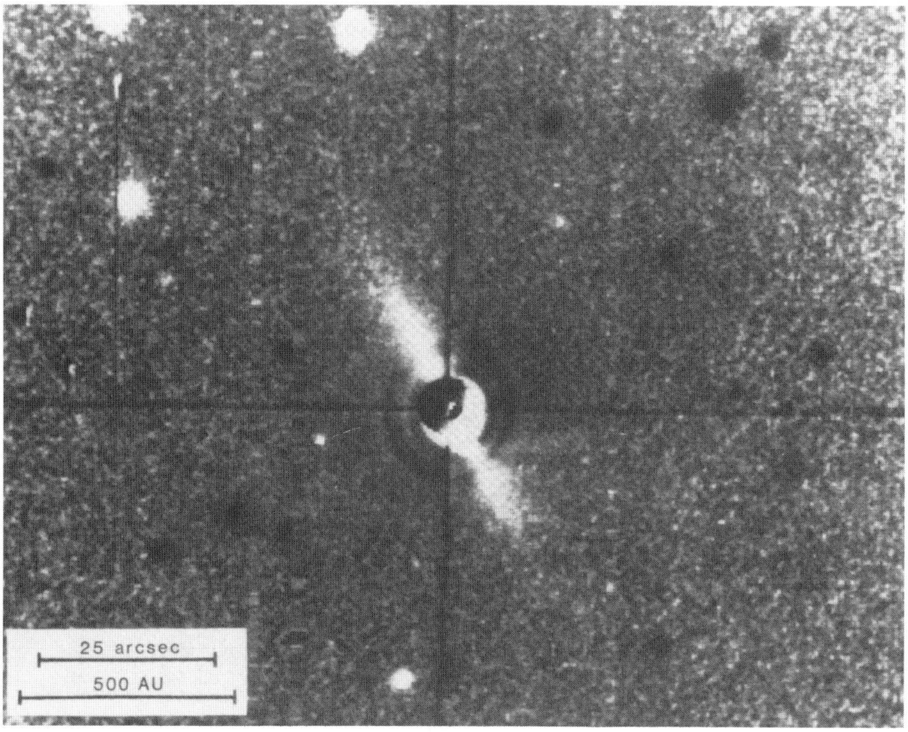

Fig. 5. Enhanced image of the star β-Pictoris shows what may be another solar system. The disk of material surrounding β-Pictoris extends 60 x 10^9 km from the star, which is located behind a circular occulting mask in the center of the picture. This material is probably composed of ices, carbonaceous chondrite-like organic substances, and silicates. These are the materials from which the comets, asteroids, and planets of our own solar system are thought to have formed (after Smith and Terrile, 1984).

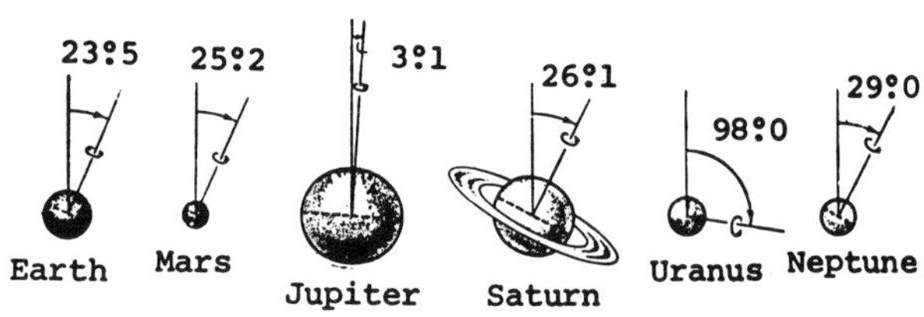

Fig. 6. Obliquity of planetary skin axes relative to the ecliptic (after Zharkov and Trubitsyn, 1978).

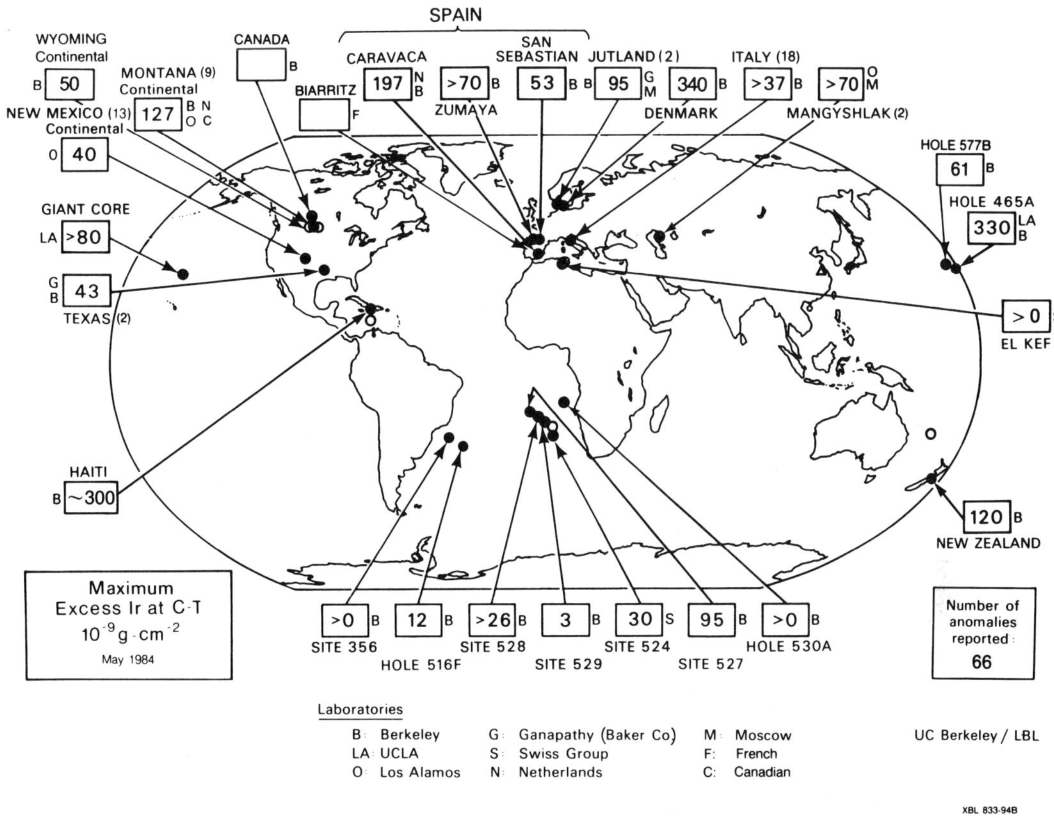

Fig. 7. Location map of 66 noble-metal rich Cretaceous-Tertiary boundary layer samples which have been analyzed (at the 9 laboratories listed). For each locality, the Ir concentration in microgram/cm^2 at the boundary is shown in the box. The laboratory which conducted the siderophile elemental analysis (primarily via neutron activation) is indicated. Most sites of the Ir anomaly are in marine rocks; however, the Wyoming, New Mexico, and Montana sites, as indicated, occur in continental rocks. Multiple localities are indicated by parentheses. Five marine rock localities which are expected on the basis of micro-fossils to be at the Cretaceous-Tertiary boundary and contain siderophile element concentrations, but in fact are barren, are indicated by open circles. (H. Michel, private communication.)

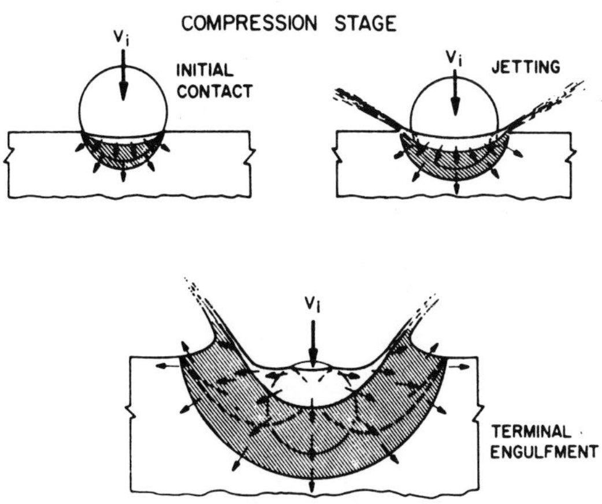

Fig. 8. The compression stage of impact mechanics after Gault et al. (1968). Shock wave propagation begins at initial contact between the projectile and the target. Shocked projectile and target material is stippled and lined, respectively. Arrows denote particle velocities. Rarefaction "waves" (dashed lines) unload the shocked zone from available free surfaces, causing jetting. Compression ends when the backward propagating shock reaches the rear of the meteorite.

For silicate rock, Ahrens and O'Keefe (1977) give the complete Hugoniot parameters which result in incipient melting (43 GPa), complete melting (52 GPa), incipient vaporization (102 GPa), and complete vaporization (590 GPa) upon isentropic release to ambient pressure. The energy along the Hugoniot is given as

$$E = P(V_0\text{-}V)/2 \tag{2}$$

solution of (1) and (2) yields the Hugoniot pressure.

$$a' P_H^2 + b'P_H + c'=0 \tag{3}$$

where

$$a' = \frac{s}{k}\left[\frac{as}{V} - 1\right] \tag{4}$$

$$b' = (a + b)s/V + l\ s/k - 1 \tag{5}$$

$$c' = l \tag{6}$$

and where

$$s=(V_0\text{-}V)/2,\ k=E_0\eta^2,\ \text{and}\ l=\mu A+\mu^2 B \tag{7}$$

The relation of the critical isentropes to the Hugoniot is shown in Fig. 9. The pertinent thermodynamics which specify the critical isentropes for solid and porous silicates, and ice is given in Ahrens and O'Keefe (1972, 1977, 1985). The mass of material melted and vaporized inferred from a series of calculations is summarized in Fig. 10.

Using finite difference techniques, we have examined the energetics of impact cratering on planetary surfaces and addressed such issues as the partitioning of energy (O'Keefe and Ahrens, 1977a) the amount of melt production as a result of impact and distribution of ejecta traveling at different velocities (O'Keefe and Ahrens, 1977b, 1982a). Recently we have examined the effect of planetary gravity and impact size on maximum crater depth (Fig. 11).

The peak shock pressure as a function of radius beneath the impact crater displays three regimes. Regime 1 is the impedance match regime, Regime 2 is the pressure decay regime, and Regime 3 is the elastic regime. Regime 1 extends to several projectile radii (R_o) into the target where the peak shock pressure is given by the planar impedance match pressure (McQueen et al., 1970). In Regime 2 the pressure decay is given by a relationship of the form (Fig. 12).

$$\log_{10}P = n \log_{10}(R/R_o) + b \tag{8}$$

We have found that the exponent of pressure decay in this regime is given as a function of impact velocity, V_I, for a silicate impacting a silicate halfspace by (Fig. 13).

$$n \cong -0.625 \log_{10}[V_I(\text{km/sec})] - 1.25 \tag{9}$$

Finally, when the shock decays to a value which does not exceed the strength of the target (the Hugoniot elastic limit), further wave decay occurs primarily as geometrical spreading. For elastic wave decay, $P \propto R^{-3}$ in the elastic regime.

In practice, much of the target material which is in Regime 1 is thrown out of the resulting crater. Thus Regime 2 becomes the zone of shock energy deposition as depicted in Fig. 14. Regime 3 begins at a radius beyond the Hugoniot elastic limit (assumed to be 3.5 GPa) (Fig. 14).

A few, relatively unshocked, samples of apparently the moon and possibly Mars are delivered to the earth as part of the terrestrial meteorite collection. How these samples could achieve escape velocity from these objects and not be heavily shocked, if not melted, has presented an enigma. Melosh (1984) has pointed out that a region near the target surface is never subjected to the peak pressure expected (from expression of the form of Eq. 8). Lunar and possibly Martian meteorites might be launched, via impact, from this region. If the shock pressure pulse emanating from the impact is describable as a pressure pulse which has a rise time, τ, centered at the depth of maximum projectile penetration, d (Fig. 14) this occurs at an approximate depth of 4 projectile radii. The free surface zone from which surface spalls may launch samples of planets upon impact is defined in terms of the coordinates of Fig. 14 as

$$0 = (x^2/2 + d)^{1/2} + [z^2+(s/2)^2]^{1/2} - [(z-d)^2+x^2]^{1/2} - \tau c \tag{10}$$

where c is the zero pressure sound speed. Using the approximation that $\tau = 2R_o/U$ where U is the shock speed in the projectile, the free surface zone sketched in Fig. 14 is obtained.

IMPACT INTO THE OCEAN

In order to gain some insights into large body impact into the oceans it is useful to examine first the partitioning of energy upon impact without a water layer, as for example, that of a silicate impactor on a silicate halfspace. The partitioning of energy from a 10 km diameter silicate impactor impacting at a velocity of 30 km/sec with a kinetic energy of 7 x 10^{30} ergs versus time is shown in Fig. 15. In the figure, the total initial energy, 7 x 10^{30} ergs, is unity and the energy budgets in the four quantities (defined in the captions) is plotted versus dimensionless time, τ. Dimensionless time is defined in terms of the travel-time of the projectile across its own diameter prior to impact, which is 0.33 sec in the present case. Figure 15 shows that because of shock heating at 2-3 τ, the internal energy of the projectile peaks at a value of 0.18 of the total energy, and at a similar value of τ, the kinetic energy of the projectile is only at 0.2 of its initial unity value and has imparted to the target a peak of 0.35 of the total energy. As time progresses the internal energy of the projectile and the kinetic energy of the target both decrease, whereas the internal energy of the target increases to a value of 0.85 of the total energy. The relative final energy budget upon impact onto the solid planet for different density impactors is given in Fig. 16.

In the case of impact into an ocean, the particle velocity flow fields produced from a 10 km diameter silicate projectile impacting a 5 km thick ocean overlying a silicate halfspace at 30 km/sec is shown in Fig. 17. Notably, the water plus silicate target demonstrates a transient cavity some 2 1/2 projectile diameters deep and a steam and ejecta bubble extending to a height of ~25 km. A water wave (Fig. 18) which extends to some 4 km in height is produced in the impact

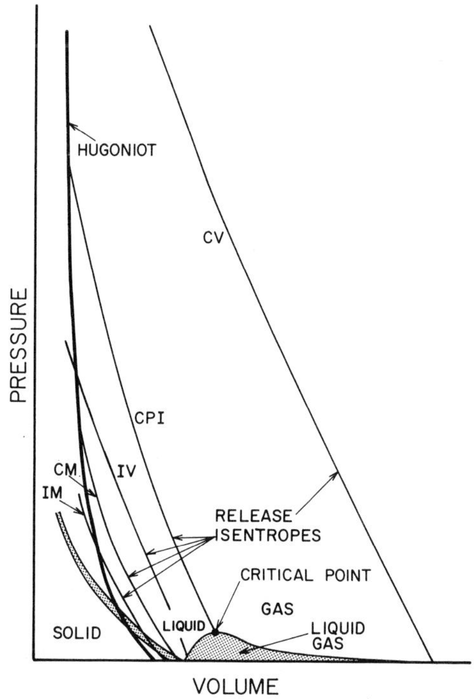

Fig. 9. Diagrammatic sketch of critical isentropes, which are defined as isentropes passing through point on the Hugoniot at onset or completion of phase change upon release to ambient pressure. IM, CM, IV, and CV, indicates isentropes resulting in incipient melting, complete melting, incipient vaporization and complete vaporization, respectively. CPI indicates isentrope passing through the liquid-vapor (gas) critical point (discussed by Zel'dovich and Raizer, 1967).

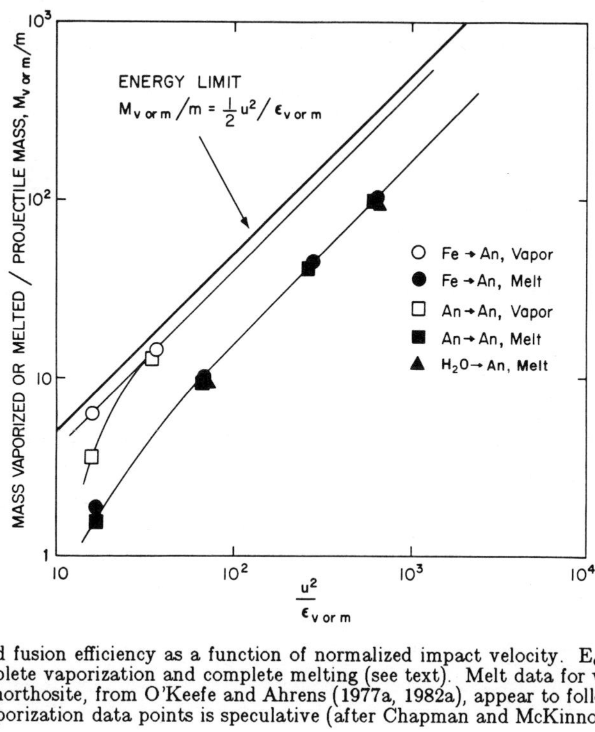

Fig. 10. Vaporization and fusion efficiency as a function of normalized impact velocity. E_{cv} and E_{cm} are the internal energies required for complete vaporization and complete melting (see text). Melt data for water ice, anorthosite, and iron projectiles striking anorthosite, from O'Keefe and Ahrens (1977a, 1982a), appear to follow energy scaling. Extension of curves through vaporization data points is speculative (after Chapman and McKinnon, 1986).

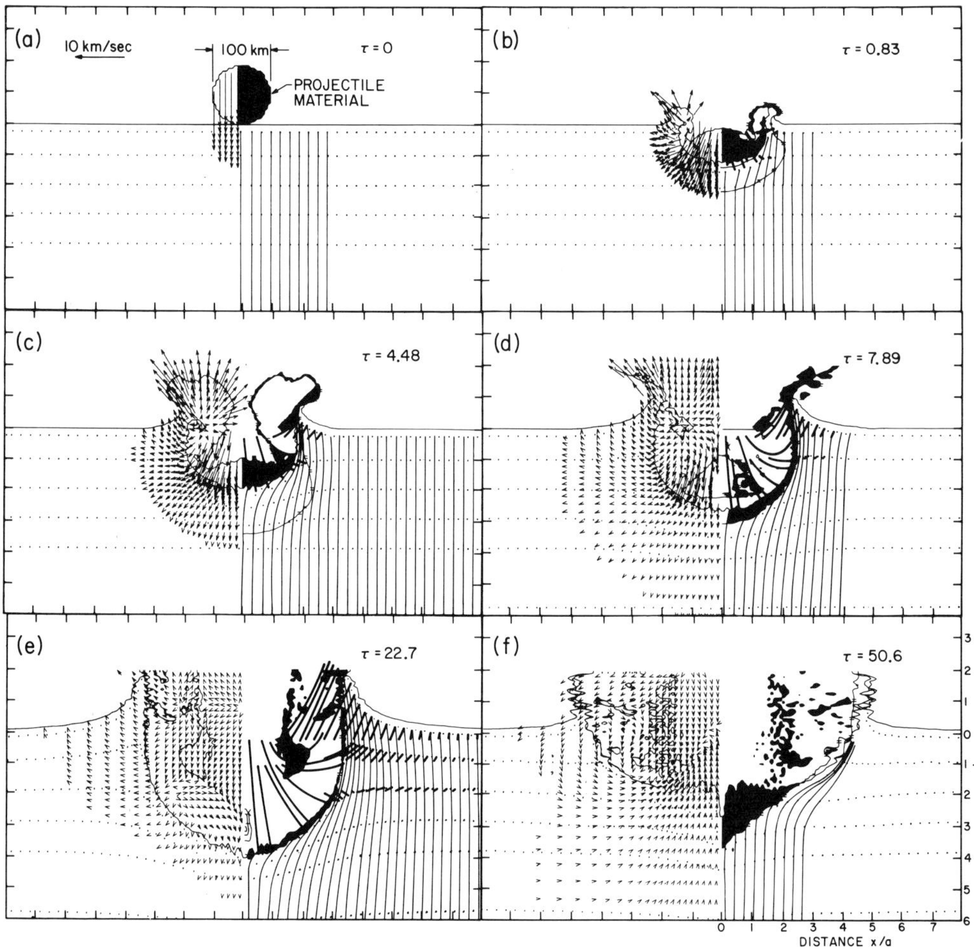

Fig. 11. Flow field upon impact of 100 km diameter silicate impactor onto silicate halfspace with earth gravity at 12 km/sec. Panels (a), (b), (c), (d), (e), and (f) show flow at $\tau = 0$, .83, 4.48, 7.89, 22.7, and 50.6. Particle velocity arrows in left panels indicate particle velocity motion scaled to 10 km/sec vector of (a). Heavy curve in right hand panels of (b), (c), (d), and (e) indicate trajectories of particle motion in transient cavity. Light vertical lines in right hand panel, and horizontal aligned dots in both right and left hand panels are fixed to material in target and demonstrate total displacement. Note gravity driven rebound motion of floor of transient cavity between (e) and (f) resulting in flat floored crater and reduction in displacement of vertical lines and horizontal dots.

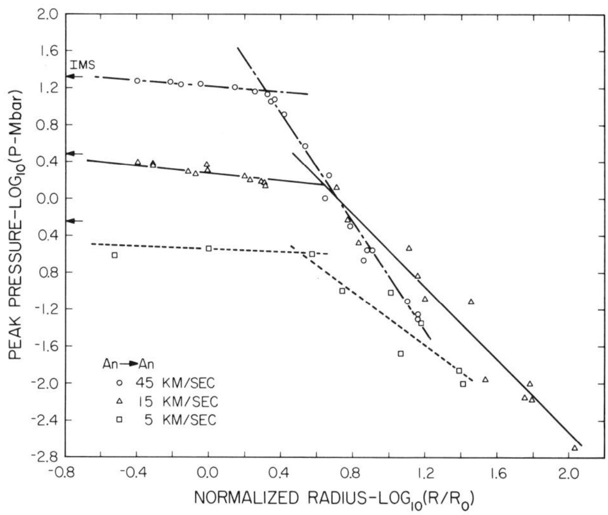

Fig. 12. Log_{10} (peak shock pressure) versus log_{10} (normalized radius) at various impact velocities for gabbroic anorthosite impactors. Arrows indicate one-dimensional impedance match pressures.

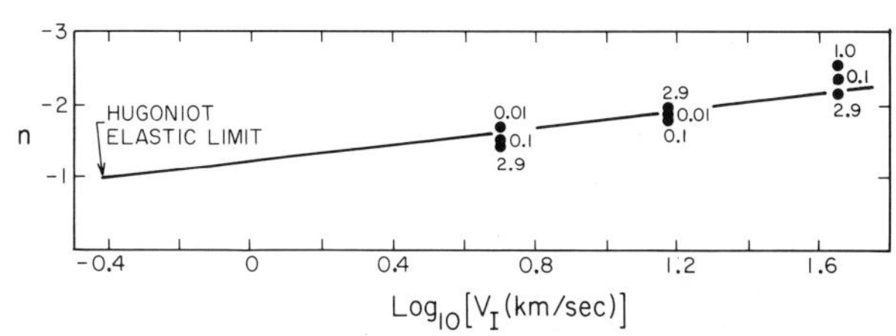

Fig. 13. Pressure decay exponent (Regime 2) versus log_{10} (impact velocity, km/sec) for various porous and solid silicate and ice projectiles impacting a 2.9 g/cm^3 density, silicate-halfspace. Density of projectiles, g/cm^3, is indicated by values shown next to the points.

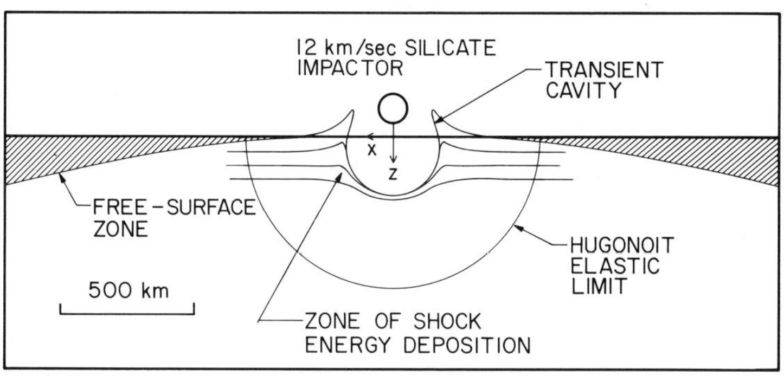

Fig. 14. Crater configuration 74 seconds after impact of 10^2 km diameter, 12 km/sec silicate projectile on a silicate planet with a gravitational acceleration of 1g. The range to the Hugoniot elastic limit (3.5 GPa) calculated from Eq. 8 and the extent of the free-surface zone obtained from Eq. 10 is indicated.

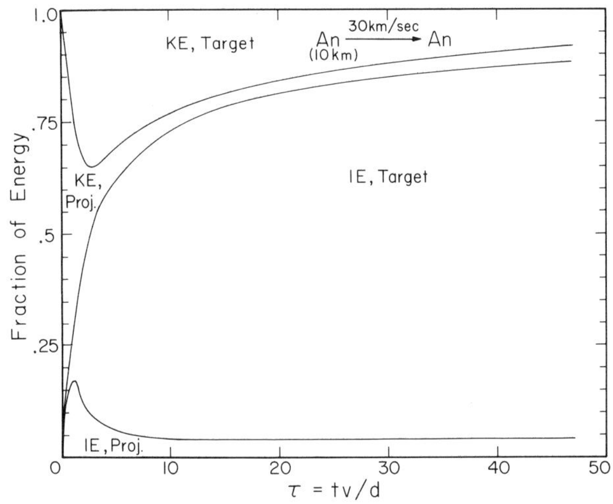

Fig. 15. Fraction of impact energy partitioned into kinetic energy, KE, of projectile (Proj.) and target and internal energy, IE of target versus dimensionless time, τ for impact of a 30 km/s silicate bolide onto a silicate Earth.

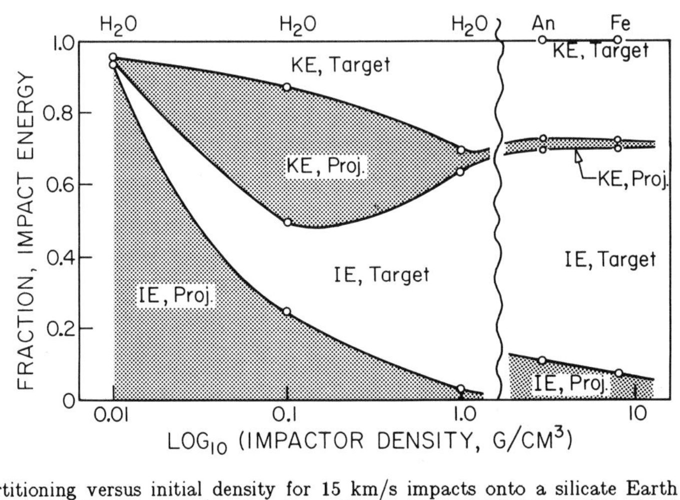

Fig. 16. Energy partitioning versus initial density for 15 km/s impacts onto a silicate Earth by meteoroid swarm objects (H_2O) and silicate (An) and iron (Fe) meteorites. KE, Target and KE, Proj. indicate final kinetic energy partitioned into planet and projectile material, respectively. IE, Target, and IE, Proj. indicate final internal energy partitioned into planet and projectile material, respectively (after O'Keefe and Ahrens, 1982a).

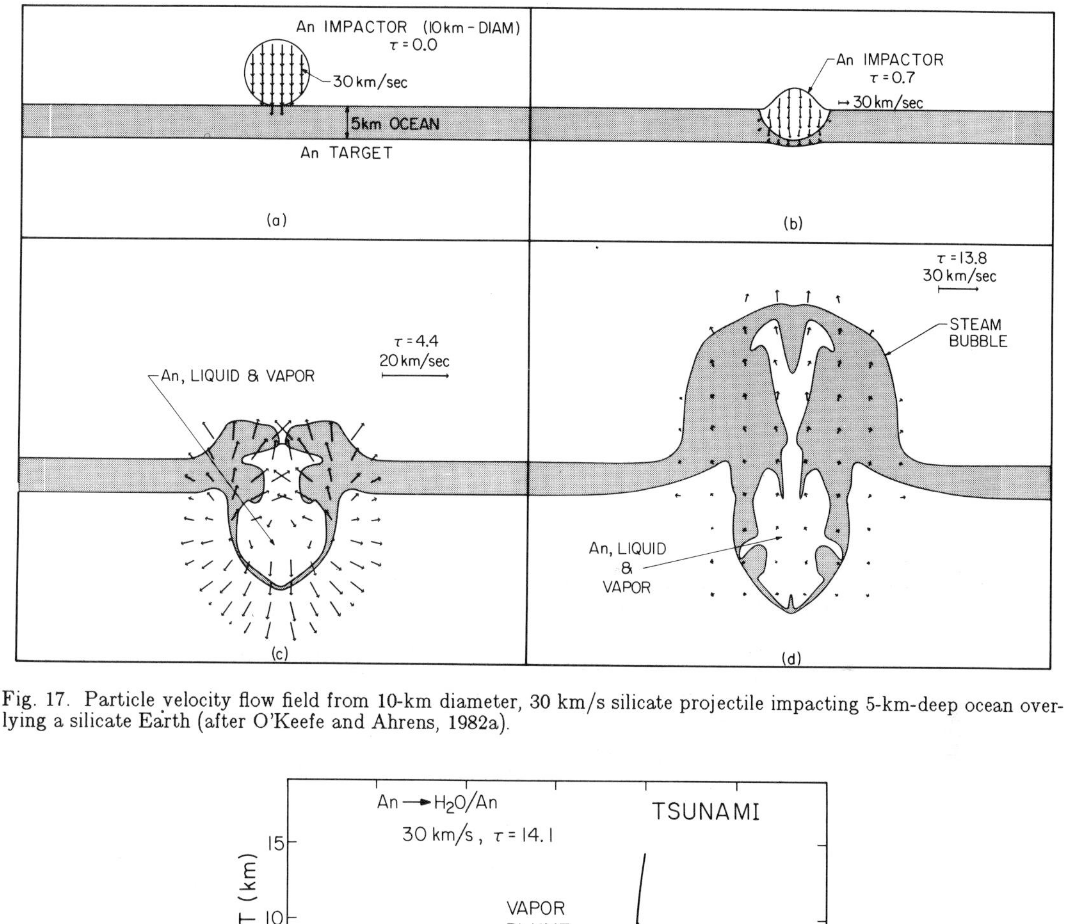

Fig. 17. Particle velocity flow field from 10-km diameter, 30 km/s silicate projectile impacting 5-km-deep ocean overlying a silicate Earth (after O'Keefe and Ahrens, 1982a).

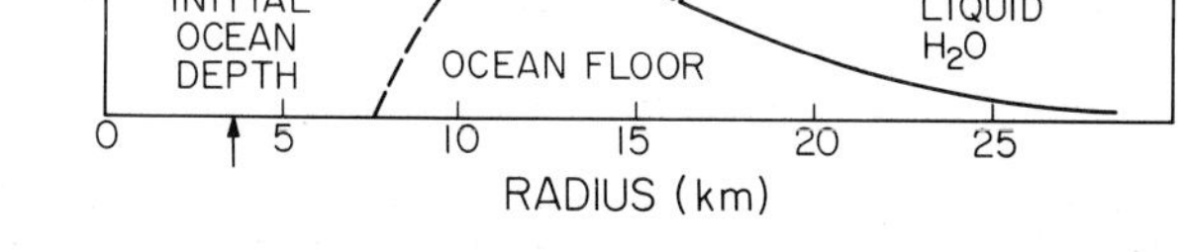

Fig. 18. Initial configuration of transient crater and water-wave for the impact flow depicted in Figure 16 (after Ahrens and O'Keefe, 1983).

region. Both the large quantity of water ejected into the stratosphere and the giant tsunamis are expected to have extreme environmental effects as discussed in the case of the Cretaceous-Tertiary extinction bolide by O'Keefe and Ahrens (1982a), and Ahrens and O'Keefe (1983). Although impact into a water-bearing layer produces dramatic effects in the impact regime, the partitioning of energy plot (Fig. 19) demonstrates that for the impacts of Figs. 17 and 18, only ~ 12% of the total energy of the impactor eventually resides in the ocean layer, whose thickness is 0.5 the diameter of the projectile. Parametric studies (O'Keefe and Ahrens, 1982a) demonstrate that upon impact in the ocean, some ~ 10 projectile diameters of water, will take up nearly all the impact energy of the silicate projectile. Thus a projectile less than ~0.5 km in diameter will be nearly stopped by 5 km of water.

IMPACT INTO THE AIR

Because of the strong mechanical impedance mismatch between the air and silicate, the calculation of the impact interaction of a projectile with an atmosphere, and the description of the ejecta interaction are difficult. Earlier (O'Keefe and Ahrens, 1982a,b) studied the interaction of a 5 km/sec, 10 km diameter silicate impactor interacting with a silicate halfspace covered with a 10 km thick layer of air. Partitioning of energy (Fig. 20) and the resulting flow field (Fig. 21) provide considerable insight into the physical processes which operate. As demonstrated in Fig. 20 the total fraction of energy initially partitioned into the atmospheric layer is only ~ 0.08 of the bolide energy. However, most of the kinetic energy of the ejecta, 0.09 of the total energy, is also later transferred to the air by air-drag. Moreover, some 0.27 of the resulting earth and projectile ejecta internal energy is transferred to the atmosphere via radiation and conduction processes. Thus O'Keefe and Ahrens (1982a) estimated some 0.39 fraction of the total energy of a very large bolide (~10^{28} ergs) would be transferred to the atmosphere. Depending on density, smaller (~ 10^2 m diameter) objects transfer virtually all of their energy into the atmosphere. Thus impactors with energies $>10^{28}$ ergs impact the earth and virtually all interactions with the atmosphere occur very near the surface of the earth.

Using the approximation that the energy of the impactor is totally delivered to the atmosphere at the surface (expected to be closely valid for projectile energies $>10^{28}$ erg) we have examined the shock-induced flow from such impacts on the earth for an exponential atmosphere. This interesting problem has been studied by Zel'dovich and Raizer (1967) and is often called the Zel'dovich flow. Previous studies of the Zel'dovich flow were carried out by Kompaneets (1960), Troutman and Davis (1965), Grover and Hardy (1966), Hayes (1968), Lutzky and Lehto (1968), Laumbach and Probstein (1969), and Bach et al. (1975). These works demonstrate that in an exponential atmosphere the downward propagating shock (from a finite height explosion) slows down and the upward propagating shock accelerates in both shock velocity and particle velocity behind the shock front. The general idea of the Zeldovich flow is demonstrated in Fig. 22 in which the shock front is shown at various non-dimensional times, T. T is related to real time, t, by

$$T = t \left[\frac{E_o}{4\pi\Delta^5 \rho_c J_o} \right]^{1/2} \equiv At \tag{11}$$

where E_o is the energy of the explosion, Δ is the atmospheric scale height (7 km for the earth), ρ_c is the density at the height of the explosion (taken here to be 10^{-3}g/cm^3) and J_o = 0.4233 is a constant which is a function of the polytropic exponent, γ, of the atmosphere (γ is taken to be 1.4). Thus one can see that as a shock propagates upward it accelerates. The strong shock conditions yield density and particle velocity at the shock front, in terms of the density, ρ_a, at a given height

$$\rho_a = \rho_c \exp(-R \cos\theta/\Delta) \tag{12}$$

where R is the radius to the shock front and θ is defined in Fig. 22. At the shock front, for strong shocks:

$$\rho = \frac{2\,\rho_a}{\gamma + 1} \tag{13}$$

also the shock velocity is given by

$$U = 4\Delta\, A/[\xi^{3/2}(\exp(\lambda_1\omega/2)+\omega^2)\lambda_2] \tag{14}$$

where

$$\xi \equiv \omega/\cos\theta \tag{15}$$

and

$$\omega \equiv d\ln \rho_a/d\ln \xi \tag{16}$$

Here λ_1 = -0.9085 and λ_2=0.02902 are additional constants which apply for γ = 1.4 case. The particle velocity behind the shock front is

$$u = 2U/(\gamma+1) \tag{17}$$

For altitudes which are a factor of 10 greater than the atmospheric scale height, we assume that when the outward particle velocity is greater than the earth's escape velocity which is 11 km/sec. Thus, within the continuum approximation, the gas in these regions escape the earth.

Some typical calculated particle velocity profiles along the (θ= 0° and θ = 80° directions for two times are shown in Figs. 23 and 24. In the case of θ = 0°, virtually all the gas above an elevation of 40 km, escapes the earth, whereas at θ = 80°, at a much later time, no gas has sufficient velocity to escape the earth.

For the values of Δ = 7 km, γ = 1.4, ρ_a = 10^{-3} g/cm^3, and an earth with a surface area of 5.1x10^{18}cm^2 a total hemispherical mass of atmosphere of 1.8 x 10^{21} g is calculated upon integrating Eq. 12. This compares to the actual earth value of 2.5 x 10^{21} g. Within the framework of the present theory of atmospheric escape for an atmosphere lying above a halfspace (no earth curvature) the maximum mass of gas which can escape from a single explosion or impact will be the mass of gas <u>above</u> a plane tangent to the spherical earth. For the above atmospheric parameters, this mass of gas is 2.6 x 10^{18} g and compares to a mass of 1.6 x 10^{16} g which is the largest mass which we calculate can escape from a single impact (Table 1, Fig. 25). Thus we conclude that within the framework of the Zel'dovich flow problem, only some 5 x 10^{-6} of the total atmosphere of the earth can be lost by a single impact. According to Table 1, the minimum impact energy for this maximal atmospheric loss event is in the range of 0.5 x 10^{30} to 1 x 10^{30} ergs. This corresponds to a 2 g/cm^3 density, 3-6

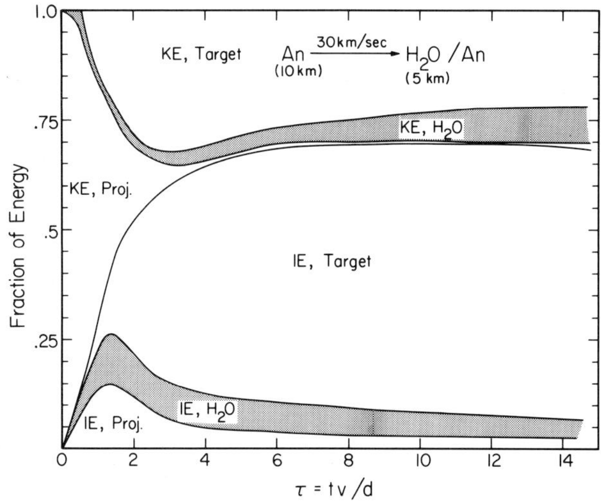

Fig. 19. Fraction of impact energy partitioned into kinetic and internal energy of projectile and target (silicate and water) materials versus dimensionless time for a 10-km-diameter silicate bolide impacting a 5 km ocean overlying a silicate Earth at 30 km/s (after O'Keefe and Ahrens, 1982).

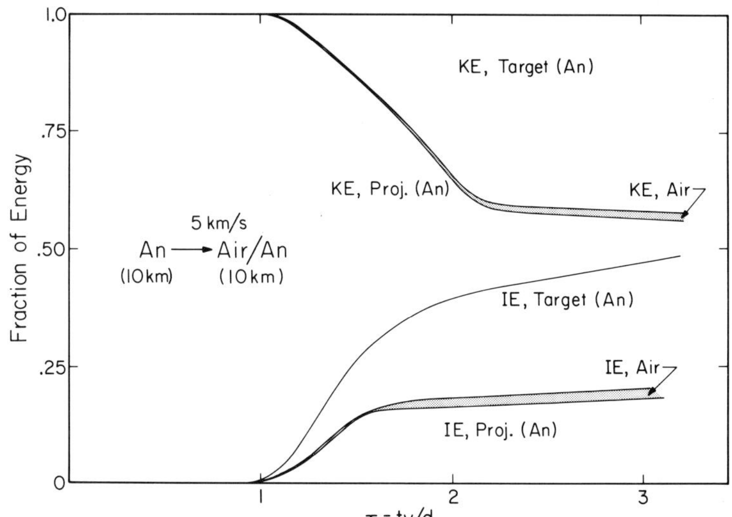

Fig. 20. Fraction of impact energy partitioned into kinetic and internal energy of projectile and target (air and silicate) versus dimensionless time for a 10-km-diameter silicate projectile impacting a 10-km-thick, 0.001 g/cm^3 atmosphere overlying a silicate Earth (after O'Keefe and Ahrens, 1982a,b).

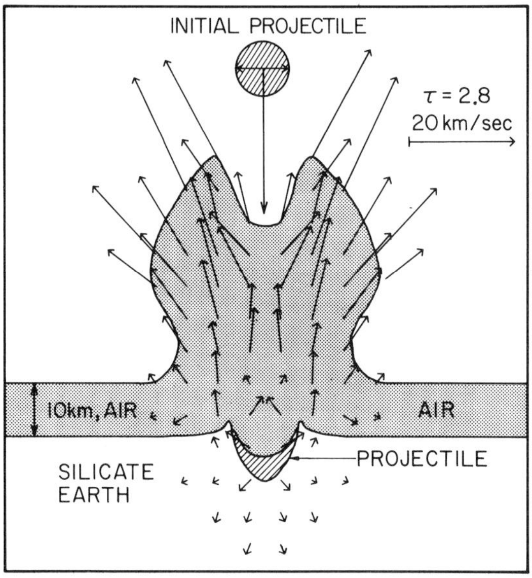

Fig. 21. Particle velocity flow field from a silicate projectile impacting a uniform 0.001 g/cm^3 atmosphere overlying a silicate Earth at 5 km/s at $\tau = 2.8$ when projectile has delivered ~90% of its energy to the target(s) (after O'Keefe and Ahrens, 1982a,b).

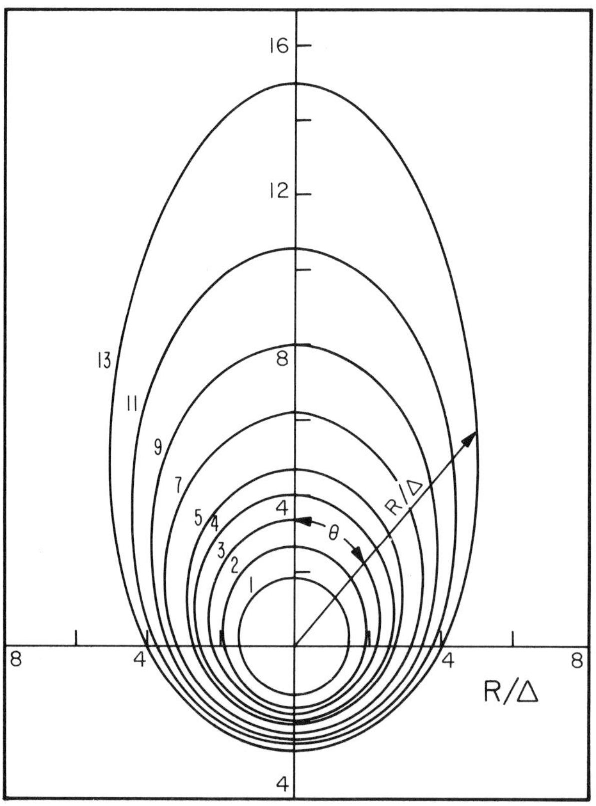

Fig. 22. Shock envelopes for various values of the non-dimensional time scale $T(\gamma = 1.4)$ (after Bach et al., 1975).

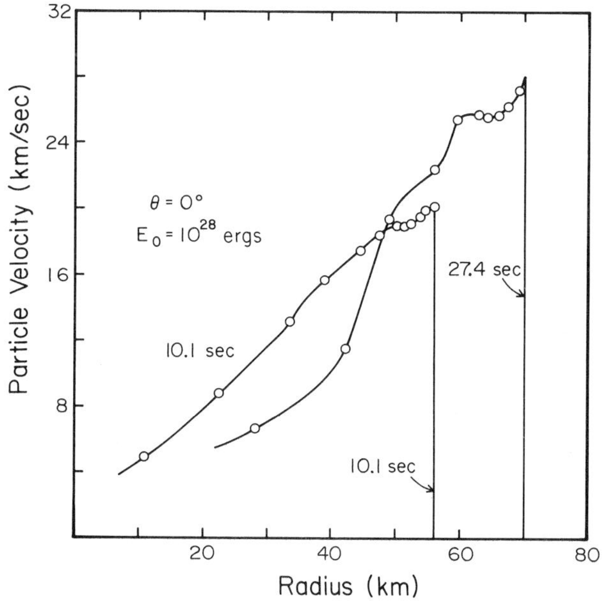

Fig. 23. Particle velocity profile at two different times along $\theta = 0°$ for impacts of 10^{28} erg.

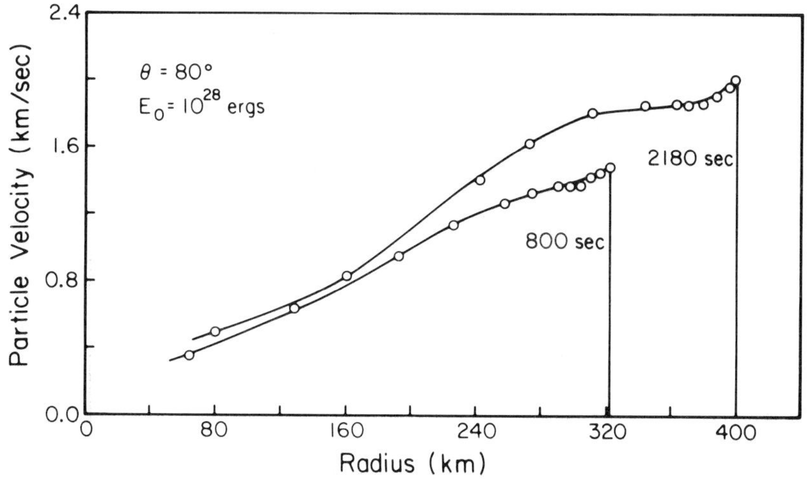

Fig. 24. Particle velocity profile at two different times along $\theta = 80°$ for impacts of 10^{28} erg.

Table 1. Mass Fraction of Atmospheric Hemisphere Lofted to Escape vs. Planetary Escape Velocity

Energy (ergs)	Velocity (km/sec) 1	2	3	4	5	6	7	8	9	10	11	30	50
1.0×10^{26}	4.4-08*	7.0-09	0.0	0.0	0.0	0.0	0.0	0.0	0.0	0.0	0.0	0.0	0.0
0.5×10^{27}	4.4-07	1.1-07	2.2-08	8.8-09	3.5-09	5.6-11	0.0	0.0	0.0	0.0	0.0	0.0	0.0
1.0×10^{27}	5.6-07	1.1-07	4.4-08	4.4-08	1.8-08	8.8-09	3.5-09	2.2-10	0.0	0.0	0.0	0.0	0.0
0.5×10^{28}	8.8-06	5.6-07	4.4-07	1.1-07	1.1-07	1.1-07	4.4-08	4.4-08	4.4-08	1.8-08	1.8-08	0.0	0.0
1.0×10^{28}	8.8-08	5.6-07	5.6-07	4.4-07	4.4-07	1.1-07	1.1-07	1.1-07	1.1-07	4.4-08	4.4-06	0.0	0.0
0.5×10^{29}	8.8-06	8.8-06	8.8-06	5.6-07	5.6-07	5.6-07	5.6-07	5.6-07	4.4-07	4.4-07	4.4-07	2.2-08	3.5-09
1.0×10^{29}	8.8-06	8.8-06	8.8-06	8.8-06	8.8-06	1.1-06	5.6-07	5.6-07	5.6-07	5.6-07	5.6-07	4.4-08	4.8-08
0.5×10^{30}	8.8-06	8.8-06	8.8-06	8.8-06	8.8-06	8.8-06	8.8-06	8.8-06	8.8-06	8.8-06	1.8-06	4.4-07	1.1-07
1.0×10^{30}	8.8-06	8.8-06	8.8-06	8.8-06	8.8-06	8.8-06	8.8-06	8.8-06	8.8-06	8.8-06	8.8-06	5.6-07	4.4-07

$*4.4 \times 10^{-8}$

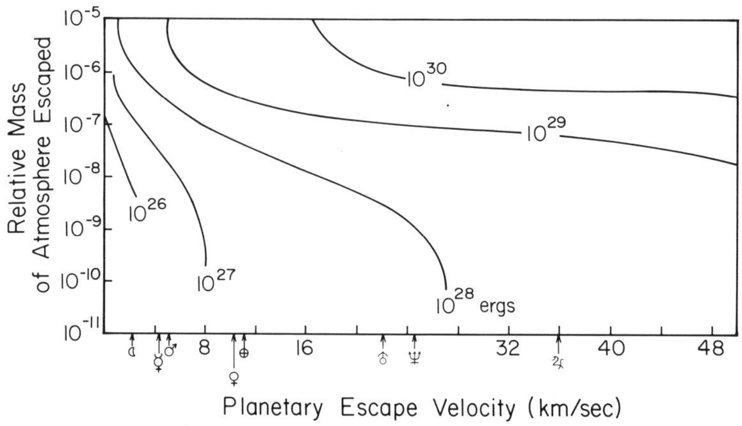

Fig. 25. Relative mass of a hemisphere of atmosphere, which will escape for an exponential atmosphere with $\gamma = 1.4$, $\rho_a = 10^{-3}$ g/cm^3, and $\Delta = 7$ km, versus, planetary escape velocity. For comparison escape velocity for moon, Mercury, Mars, Venus, Earth, Uranus, Neptune, and Saturn is indicated.

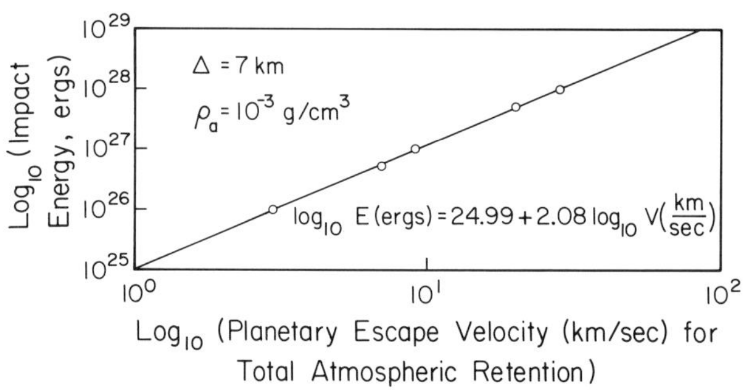

Fig. 26. Log_{10} (impact energy, ergs) versus $\log_{10}$ (planetary escape velocity, V (km/sec) for total atmospheric retention). Linear fit to present calculations indicated by straight line.

km diameter, projectile infalling at 11 km/sec into the present atmosphere. This maximal atmospheric loss projectile is only a factor of ~10 lower in energy than that inferred by Alvarez et al. (1980), O'Keefe and Ahrens (1982b) for the Cretaceous-Tertiary extinction bolide.

Finally, we note that actually a very narrow energy regime range exists between the energy for maximal loss (10^{30} ergs) versus that for impacts (or explosion) for which the result is total atmospheric retention which occurs for an escape velocity of 11 km/sec. According to the equation of Fig. 26 for escape velocity of 11 km/sec, the total atmospheric retention occurs at 10^{27} ergs.

ACKNOWLEDGMENTS:

Research supported under NASA grants NSG-7129 and NGL-05-002-105. Contribution number 4341, Division of Geological and Planetary Sciences, California Institute of Technology. We appreciate receiving Fig. 7 from Dr. Helen Michel of Lawrence Berkeley Laboratory and the helpful comments of Y. Meroz and two reviewers on the manuscript.

REFERENCES

Ahrens, T. J. and J. D. O'Keefe (1972). Shock melting and vaporization of lunar rocks and minerals. *The Moon*, **4**, , 214-249.

Ahrens, T. J. and J. D. O'Keefe (1977). Equation of state and impact-induced shock-wave attenuation on the moon. In D. J. Roddy, R. O. Pepin, and R. B. Merrill (Eds.), *Impact and Explosion Cratering*, Pergamon Press, pp. 539-656.

Ahrens, T. J. and J. D. O'Keefe (1983). Impact of an asteroid or comet in the ocean and the extinction of terrestrial life, part 2. *Proc. 13th Lunar and Planetary Science Conf. Supplement, J. Geophys. Res.*, **88**, A799-A806.

Ahrens, T. J. and J. D. O'Keefe (1985). Shock vaporization and the accretion of the icy satellites of the Jupiter and Saturn. In P. Klinger, A. Dollfus, and R. Smoluchowski (Eds.), *Ices in the Solar System*, Reidel, pp. 631-654.

Alvarez, L. E., W. Alvarez, F. Asaro, and H. V. Michel (1980). Extraterrestrial cause for the Cretaceous-Tertiary extinction. *Science*, **208**, , 1095-1108.

Bach, G. G., A. L. Kuhl, and A. K. Oppenheim (1975). On blast waves in exponential atmospheres. *J. Fluid Mech.*, **71**, part 1, 105-122.

Chapman, C. R. and W. B. McKinnon (1986). Cratering of planetary satellites. In F. Burns and D. Morrison (Eds.), *Natural Satellites*, in press.

Gault, D. E., W. L. Quaide, and V. R. Oberbeck (1968). Impact cratering mechanics and structures. In B. M. French and N. M. Short (Eds.) *Shock Metamorphism*, Mono Press, pp. 87-100.

Gilbert, G. K. (1893). The Moon's face: A study of the origin of its features. *Bull. Phil. Soc. Wash.*, **12**, , 241-292.

Green, J. (1971). Copernicus as a lunar caldera. *J. Geophys. Res.*, **76**, 5719-5731.

Grieve, R. A. F. and M. R. Dence (1979). The terrestrial cratering record II: The crater production rate. *Icarus*, **38**, , 230-242.

Grover, R. and J. W. Hardy (1966). The propagation of shocks in exponentially decreasing atmospheres, *Astrophys. J.*, **148**, , 48-60.

Hayes, W. D. (1968). The propagation upward of the shock wave from a strong explosion in the atmosphere. *J. Fluid Mech.*, **32**, 317-331.

Kompaneets, A. S. (1960). A point explosion in an inhomogeneous atmosphere. *Soviet Phys., Doklady*, **5**, 46-48.

Laumbach, D. D. and R. F. Probstein (1968). A point explosion in a cold exponential atmosphere. *J. Fluid Mech.*, **32**, 53-75.

Levin, B. J. (1972). Origin of the earth. In A. R. Ritsema (Ed.), *The Upper Mantle*, Elsevier, pp. 7-30.

Lutzky, M. and D. L. Lehto (1968). Shock propagation in spherically symmetric exponential atmospheres. *Phys. Fluid*, **11**, 1466-1472.

McQueen, R. G., S. P. Marsh, J. W. Taylor, J. N. Fritz, and W. J. Carter (1970). The equation of state of solids from shock wave studies. In Kinslow (Ed.), *High Velocity Impact Phenomena*, Academic, New York, pp. 294-491.

Melosh, H. J. (1984). Impact ejection spallation, and the origin of the meteorites. *Icarus*, **59**, 234-260.

Mohorovicic, S. (1928). Experimentalle untersuchungen über die entstehung der Mondkrater: eineuer Beitrag zur. Explosionshypothese. *Archiv za Hemija i Farmacizu Zagreb*, **2**, 66-76.

O'Keefe, J. D. and T. J. Ahrens (1977a). Impact induced energy partitioning, melting, and vaporization on terrestrial planets. *Proc. Lunar Sci. Conf., 8th*, 3357-3374.

O'Keefe, J. D. and T. J. Ahrens (1977b). Meteorite impact ejecta: Dependence of mass and energy lost on planetary escape velocity. *Science*, **198**, 1249-1251.

O'Keefe, J. D. and T. J. Ahrens (1982a). The interaction of the Cretaceous-Tertiary extinction bolide with the atmosphere, ocean, and solid earth. *Proc. Conf. on Large Body Impacts and Terrestrial Evolution, Sp. Paper #190 Geol. Soc. Am.*, 103-120.

O'Keefe, J. D. and T. J. Ahrens (1982b). Impact mechanics of the Cretaceous-Tertiary extinction bolide. *Nature,* **298,** 123-127.

O'Keefe, J. D. and T. J. Ahrens (1986). Effect of planetary gravity and impact size on crater maximum penetration and excavation depth. *Abstract, Lunar and Planetary Science,* **XVII,** 632-633, Lunar and Planetary Science Institute, Houston, TX.

Shoemaker, E. M. (1962). Interpretation of lunar craters. In Z. Kopal (Ed.), *Physics and Astronomy of the Moon,* Academic Press, pp. 283-375.

Sloan, R. E., J. K. Rigby, Jr., L. M. VanValen, and D. Gabriel (1986). Gradual dinosaur extinction and simultaneous ungulate radiation in the Hell Creek formation. *Science,* **232,** 629-633.

Smith, B. A. and R. J. Terrile (1984). A circumstellar disk around β-Pictoris. *Science,* **226,** 1421-1424.

Tillotson, J. H. (1962). Metallic equations of state for hypervelocity impact. *General Atomic Report GA 3216.*

Troutman, W. W. and C. W. Davis (1965). The two-dimensional behavior of shocks in the atmosphere. Air Force Weapons Lab. Report, AFWL-TR-65-151.

Zel'dovich, Ya. B. and Yu. P. Raizer (1967). *Physics of Shock Waves and High-Temperature Hydrodynamic Phenomena,* **Vol. II,** Academic Press, NY.

Zharkov, V. N. and V. P. Trubitsyn (1978). In W. B. Hubbard (Ed./transl.), *Physics of Planetary Interiors,* Pachart Press, Tucson.

Int. J. Impact Engng Vol. 5, pp. 33–59, 1987
Printed in Great Britain

0734-743X/87 $3.00 + 0.00
Pergamon Journals Ltd.

AN OVERVIEW OF THE THEORY OF HYDROCODES

Charles E. Anderson, Jr.

Southwest Research Institute
San Antonio, TX

ABSTRACT

Hydrocodes are large computer programs that can be used to simulate numerically highly dynamic events, particularly those which include shocks. Lagrangian and Eulerian descriptions are reviewed, and advantages and disadvantages are summarized. The question of how to best represent the continuum equations on a finite computer is answered by summarizing the topics of accuracy and stability. The concept of artificial viscosity is introduced to permit the continuum code to deal with the discontinuities of shocks. Finally, a review of the treatment of materials, i.e., equation of state and constitutive response, including failure, is presented.

INTRODUCTION

Classical continuum mechanics attempts to describe the dynamics of a continuous media with a set of differential equations established through the application of the principles of conservation of mass, momentum, and energy from a macroscopic point of view. An equation of state relates the density (or volume) and internal energy (or temperature) of the material with pressure. A constitutive relation describes the particular nature of the material by relating the stress in the material with the amount of distortion (strain) required to produce this stress. The constitutive relation may include strain rate effects, work hardening, thermal softening, etc.

The differential equations relate material density ρ, velocity v_i, specific total energy e, the stress tensor σ_{ij}, and external body forces per unit mass f_i, where the subscripts represent standard tensorial notation. Two fundamental descriptions of the kinematic deformation of continuous media exist: the Eulerian (spatial) and Lagrangian (material) description. The conservation equations for the two descriptions are:

	Lagrangian	Eulerian	
Conservation of Mass	$\frac{D\rho}{Dt} + \rho \frac{\partial v_i}{\partial x_i} = 0$	$\frac{\partial \rho}{\partial t} + \frac{\partial}{\partial x_i}(\rho v_i) = 0$	(1)
Conservation of Momentum	$\frac{Dv_i}{Dt} = f_i + \frac{1}{\rho}\frac{\partial \sigma_{ji}}{\partial x_j}$	$\frac{\partial v_i}{\partial t} + v_j \frac{\partial v_i}{\partial x_j} = f_i + \frac{1}{\rho}\frac{\partial \sigma_{ji}}{\partial x_j}$	(2)
Conservation of Energy	$\frac{De}{Dt} = f_i v_i + \frac{1}{\rho}\frac{\partial}{\partial x_j}(\sigma_{ij} v_i)$	$\frac{\partial e}{\partial t} + v_i \frac{\partial e}{\partial x_i} = f_i v_i + \frac{1}{\rho}\frac{\partial}{\partial x_j}(\sigma_{ij} v_i)$	(3)

The summation convention is implied by the repeated indices. In Eqs. (1-3), the "ordinary" mechanical effects have been considered, but electromagnetic, chemical, and heat conduction effects have been ignored.

The differences between the two sets of equations are inherent in the definition of the derivative D/Dt:

$$\frac{D}{Dt} = \frac{\partial}{\partial t} + v_i \frac{\partial}{\partial x_i} \qquad (4)$$

which goes by the various names of material derivative, substantial derivative, or total time derivative. The significance of this derivative will be discussed in the next section.

The total specific energy is the sum of kinetic energy and specific internal energy E:

$$e = \frac{1}{2} v_i v_i + E \tag{5}$$

Equation (3) is often rewritten in terms of the specific internal energy:

Lagrangian **Eulerian**

$$\frac{DE}{Dt} = \frac{P}{\rho^2}\frac{D\rho}{Dt} + \frac{1}{\rho} s_{ij}\,\dot{\varepsilon}_{ij} \qquad \frac{\partial E}{\partial t} + v_i \frac{\partial E}{\partial x_i} = \frac{P}{\rho^2}\left(\frac{\partial \rho}{\partial t} + v_i \frac{\partial \rho}{\partial x_i}\right) + \frac{1}{\rho} s_{ij}\,\dot{\varepsilon}_{ij} \tag{6}$$

where s_{ij} and $\dot{\varepsilon}_{ij}$ are the stress deviators and strain rates, respectively, and P is the hydrostatic pressure.

Two more expressions are required to complete the set of equations, an equation of state, and a constitutive model:

$$P = P(\rho,E) \qquad \sigma_{ij} = f(\varepsilon_{ij}, \dot{\varepsilon}_{ij}, E, K) \tag{7}$$

The equation of state accounts for compressibility effects (changes in density) and irreversible thermodynamic processes such as shock heating. The constitutive model, in general, permits the stress to be a function of strain ε_{ij}, strain rate effects $\dot{\varepsilon}_{ij}$ (both in loading and unloading, i.e., stress relaxation), internal energy E (thermal softening), and damage K. Of course, before a solution can be obtained, appropriate boundary and initial conditions must also be prescribed.

The set of coupled, highly nonlinear equations represented by Eqs. (1-7) has been solved exactly for only a dozen or so problems where certain simplifying restrictions have been invoked (e.g., incompressible flow, inviscid flow, irrotational flow, potential flow). Approximate or asymptotic solutions have been obtained by discarding certain terms, and these solutions have practical utility in certain physical circumstances, for example, boundary layer theory. Perturbation theory and singular perturbation theory have been used to obtain solutions over a restricted range where the assumptions imposed are valid. In general, however, analytical solutions rely on linearity, symmetry, and a small number of variables. But, there exist many problems of practical interest whose solutions are not amenable to these techniques; the full set of coupled equations must be solved simultaneously. With the advent of digital computers, thousands of mathematical operations can be performed quickly; thus, numerical techniques provide a method of obtaining solutions.

Computer programs which handle the propagation of shock waves and compute velocities, strains, stresses, etc., as a function of time and position are called "hydrocodes." Early formulations did not include strength effects. Thus, metals were treated as a fluid, with no viscosity, and the expression, "hydrodynamic computer code" came into being; with time, this was shortened to hydrocodes. Hydrodynamic computer codes can compute the approximate response of a material object providing the pressures are sufficiently high that strength effects can be ignored. However, in developing better predictive capabilities, material modeling has become more sophisticated.

These computer programs can predict the response of a penetrator and target from first principles. They are powerful tools to the analyst; they can provide insight and help explain phenomena observed in experiments. They can be used to test hypotheses concerning fundamental mechanisms, and be used in a systematic application to obtain the influence of variables. They can be used to extend experimental data by performing parametric variation of variables; they can be used as an alternative to experiments; and, they may be the only recourse in the extreme high velocity regimes since current launch techniques do not permit the velocities such as may be encountered outside the earth's atmosphere. In one sense, hydrocodes are the best instrumented experiment. By simulating the experiment numerically, strains, strain rates, velocities, displacements, stresses, etc., as a function of time and space are obtained.

There are several good articles which review and compare one, two, and three dimensional hydrocodes (Mescall, 1974; Belytschko, 1975; Herrmann, 1975; Jonas and Zukas, 1978; Zukas, 1980; Zukas et al., 1981; Zukas, et al., 1982). The reader is referred to these articles for further information. The remainder of this paper is concerned with the numerical issues of hydrocodes.

EULERIAN AND LAGRANGIAN DESCRIPTIONS

The Eulerian description is a spatial description; the Lagrangian description is a material description. From a computational point of view, a few less derivatives need to be computed in the Lagrangian description, e.g., Eqs. (1-3). Fundamentally, however, there are significant differences between the two mathematical descriptions and their numerical solutions. In the numerical representation of the differential equations, it becomes very easy to visualize the distinction between these two descriptions, and understand inherent advantages and disadvantages of each.

Solution of Eqs. (1-7) first requires that the differential equations be discretized. A lattice of points, or grid, is generated to approximate the geometry of interest. If adjacent points are connected by lines, the area or volume enclosed by connecting adjacent points in the grid is referred to as cells.

To obtain an Eulerian or spatial description, one would measure (or compute), as time progresses, velocity, pressure, density, temperature, etc., at the fixed points of the grid. All grid points, and consequently cell boundaries, remain spatially fixed with time. Mass, momentum, and energy flow across cell boundaries. The quantities of flow into and out of a cell are used to compute the new mass, pressure, velocity, energy, etc., of that cell. In this formulation, the cell volume is obviously an invariant since points remain fixed in space.

In the Lagrangian description, however, the grid points are attached or fixed to the material and move with the local material velocity. Velocity, pressure, density, temperature, etc., are computed as time progresses for each discrete point in the continuum. Adjacent points can be stretched or come closer together, as in a spring, depending upon the forces applied. Mass, momentum, and energy are transported by material flow--the grid points move relative to a fixed spatial coordinate system. In mathematical terminology, the grid points follow particle paths of points in the material. In this formulation, mass within a cell is invariant, but the volume of the cell may change with time because of expansion or compression of the material.

Again note the distinction between the two mathematical descriptions. For example, in computing the density of a cell, $\rho = M/V$, where M is the mass within the cell of volume V, the Eulerian description has a constant volume, V, and the mass changes as more or less mass accumulates in the cell due to mass flow across the boundaries of the cell. The density would decrease if more mass flowed out of a cell during a time step, Δt, then flowed into the cell. In the Lagrangian description, the mass M is a constant, and the volume of the cell changes due to the movement of the boundaries firmly attached to the moving material. For example, the density would increase as the cell is compressed.

In Lagrangian calculation, the mesh is generated to approximate the materials of interest with points (Lagrangian lines) placed along exterior boundaries (free surfaces) and material boundaries. Boundary conditions at free surfaces and contact surfaces between different materials are straightforward. The positions of free surfaces and contact surfaces are automatically determined in the computations, and, thus, are one of the features which makes the Lagrangian formulation so useful. For example, Fig. 1 depicts the normal impact of a steel cylindrical projectile against an aluminum plate using a Lagrangian grid. Obviously, the free surfaces are well-defined, as well as the ability to clearly distinguish steel and aluminum cells. Likewise, if the plate (or projectile) were made of layers of materials, the different material interfaces could be easily distinguished, since each grid cell is uniquely a specific material.

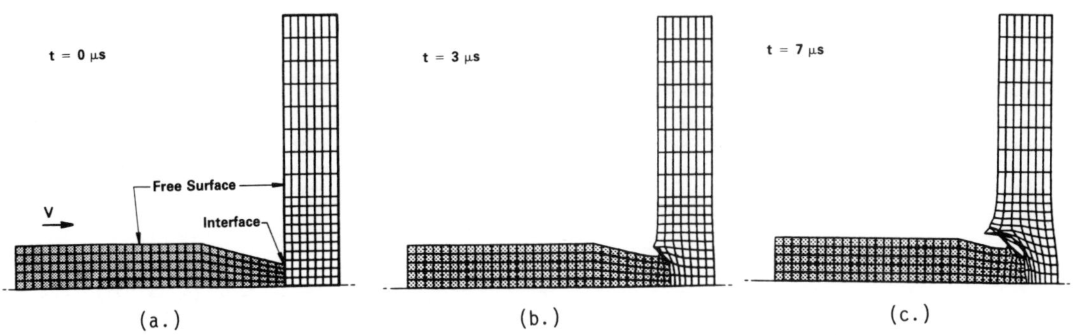

Fig. 1. Lagrangian grid.

Behavior at material interfaces, such as the opening and closing of voids, can be computed by a Lagrangian calculation by providing spatial slidelines. Slidelines are generally used when large transverse motion along a material interface is expected (Wilkins, 1964; Swegle, 1978). Opposite sides of the sliding interface are often referred to as the "master" and "slave" surfaces. The master-slave pair is called a slideline. The master surface is advanced using its own force field and the force field of the slave surface. The slave surface is then brought back into contact with the master. The normal components of velocity are assumed to be continuous across the slideline, but the tangential components may be discontinuous. Points may be tied initially and allowed to break if some criterion is met. Friction effects can be included for motion along slidelines. Of course, the code user must understand the problem well enough to know where slidelines would naturally occur; otherwise, these effects will not be modeled accurately.

Another important advantage of the Lagrangian description is in its ability to follow the history of a material particle. A material whose properties depend upon its previous history can be modeled. Thus, the Lagrangian description allows an excellent treatment of material models (constitutive relations) and failure models, e.g., strain hardening and total plastic work.

Lagrangian codes can have severe difficulties, however, where large grid distortions occur, such as turbulent flow and high velocity impacts. In the Lagrangian grid, the mesh points follow the motion of particles of mass. For example, in an impact problem, the computational mesh becomes severely distorted near the projectile/target interface, Fig. 1b and 1c. Sometimes a zone may flip back on itself resulting in a computed negative mass. Even with a slideline at the projectile/target interface, the numerical calculations bog down during the first few microseconds. The time step becomes extremely small due to compression and distortion, and the accuracy of the calculations in the distorted area is highly suspect because of possible large errors associated with the finite difference approximations in the extremely deformed regions.

From the Lagrangian formulation, the only way to obtain answers at later times for problems experiencing extreme grid compression and distortion is to rezone the computational grid. Basically, a new, undistorted grid is overlayed on the old, distorted mesh. Once the geometric overlay is completed, the physical properties of the new cells are computed from the partial zone properties of the old lattice, assuring that mass, momentum, and energy are conserved. Care must be taken to obtain the new mesh. Essentially, the procedure requires deciding which points need to be moved to make the grid more regular, and then moving these points to new locations. For large problems, even with semi-automatic help from the computer, this can be very tedious and time-consuming. Material regions should not overlap in a new cell (or else the material boundary is destroyed); cells can only exist in material regions; i.e., a cell cannot pass over a free surface, otherwise, for example, conservation of mass would be violated if the material density is held constant. And, usually, the physics which caused the distortion is still manifested in the problem--the newly rezoned grid will become distorted and must be rezoned again. To advance the problem in time may thus require that the rezoning process be performed again and again.

The problems which can be encountered in the Lagrangian calculations because of compression or severe distortion of the grid are alleviated in the Eulerian description because of its fixed grid system. In principle, all hydrodynamic problems could be solved numerically by using a multi-material Eulerian calculation which computes the mass, momentum, and energy flows across the fixed cell boundaries. Turbulent flow, rotational flow, high velocity impacts, highly compressed states, etc., can be computed using a Eulerian computer code. Diffusion and the mixing of gases can be handled by an Eulerian description. (Because each cell in a Lagrangian description is uniquely one material, and flow across cell boundaries is not allowed, diffusion and mixing problems cannot be analyzed with a Lagrangian code.) As already stated, a mesh must first be generated; this mesh must be large enough to contain all of the physical space in which action may occur. Cells must be present in the Eulerian grid to catch the flow of mass, momentum, and energy into initially unoccupied regions.

Most of the current codes, while allowing as many as nine distinct materials in the computational grid, restrict a particular cell to a maximum of two materials, though recent work has expanded this to four or more materials (W. Johnson, 1986; Thompson, 1986). (This constraint was the result of bookkeeping on the part of the code, and was a reflection of the needs of users--there had been no significant demand, until recently, for requiring three or more different materials within the same cell.) The collapse and formation of a shaped charge, and the dynamics of high velocity impacts, are typically treated with an Eulerian code.

However, because the Eulerian description describes only what is crossing grid boundaries, the position of lines which approximate material interfaces and exterior boundaries is difficult to determine. The curves which approximate the boundary and material interfaces must move with the material, and consequently will become irregular, time-dependent lines in a fixed Eulerian mesh. Typically, they are known only within the dimensions of a Eulerian cell. In a typical Eulerian hydrocode, interfaces and boundaries are designated by mixed cells, i.e., cells with two materials. Figure 2 depicts an impact between two materials in Eulerian coordinates. At some time $t > 0$, the placement of materials might look like Fig. 2b; mixed cells (whether they be void and material, or materials 1 and 2) are denoted by the "M". The finer the mesh, the more accurately the boundary is represented, but at the price of more computational zones. The

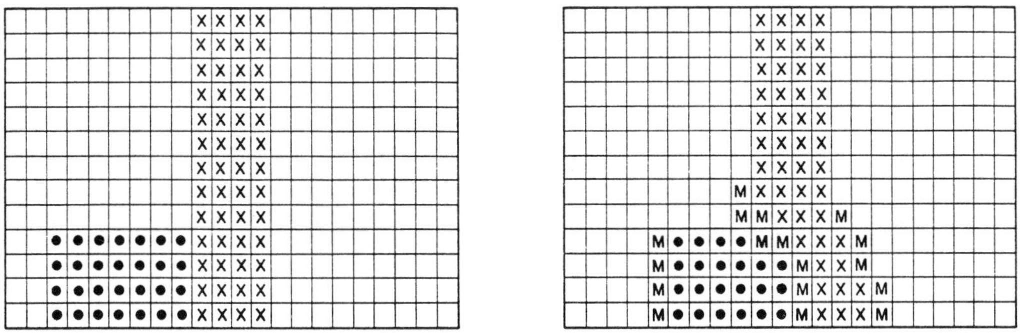

(a.) $t = 0$ (b.) $t > 0$

Fig. 2. Eulerian grid with two materials denoted by dots and crosses; "M" denotes mixed cells.

numerical approximation of the differential equations in these irregular boundary cells are not to the same degree of accuracy as interior cells of the mesh. Thus, the application of boundary conditions is generally more difficult and less accurate than for the Lagrangian description.

There is a tendency in the numerical Eulerian formulation for the materials to diffuse across material interfaces at nonphysical rates (because once a material has entered a zone, it is considered to be everywhere in that zone.) This is particularly true when the problem is being carried to long times to find the ultimate configuration or state of the system after most or all external forces have been removed. A finer spaced grid, that is, more zones, can diminish this numerically induced material diffusion, but again at the expense of more computation time. This numerical diffusion problem can be quite severe. The method most commonly used to limit numerical diffusion is one which preferentially transports material out of a cell according to the material contained in neighboring cells. The fraction of each species in a zone is known, and species "a" will be given preference to move into a neighboring cell if the neighboring cell consists mostly of species "a," etc. A procedure for limiting numerical diffusion is required or else the number of mixed zones continues to grow, thereby smearing the interface between the materials. Also, in Eulerian codes, materials are in pressure equilibrium in mixed zones, which requires some sort of iterative procedure between the equations of state of the materials. Thus, the more mixed zones, the more the computational burden.

The Eulerian description describes the instantaneous state of the material in the zone, but has no history of this material. Because the Eulerian description tracks only what crosses cell boundaries, it cannot compute the time history of a particular material particle, which can be important in formulating realistic constitutive, fracture, and failure relations. At the expense of more computational time, "internal state" variables can be devised to transport history dependent variables, though mixing of these variables diminishes their accuracy.

The advantages and disadvantages of the two mathematical schemes can now be summarized. In a Lagrangian calculation, the grid mesh is required only when there is material. The Lagrangian description allows the mesh to be placed such that maximum resolution is obtained for a specific number of zones. Figure 3 depicts a Eulerian grid for the same impact problem of Fig. 1. Figure 3a depicts the grid geometry with approximately the same number zones as shown in Fig. 1, and Fig. 3b depicts the same problem, but with the same spatial resolution (i.e., same zone size). In addition, the Lagrangian grid clearly defines material interfaces and boundaries--their treatment is straightforward. Material and failure modeling can account for rate and history effects. But, as demonstrated in Fig. 1c, the Lagrangian mesh can become severely compressed and distorted which causes the time step to become extremely small, effectively terminating the problem. Also, the numerical differencing schemes become very inaccurate in regions of extreme grid distortion, thus yielding computed quantities that may have little to do with reality. The Lagrangian grid can be rezoned, but in the area where large distortions have occurred, rezoning is akin to an Eulerian calculation.

SPACE DISCRETIZING, TIME INTEGRATION, ACCURACY, AND STABILITY

Although Eqs. (1-7) describe a continuum, the computer is finite. Somehow the problem must be discretized, and then the computer solves these discretized equations. So, the question becomes, how can we best describe or approximate the differential equations describing a continuum with a

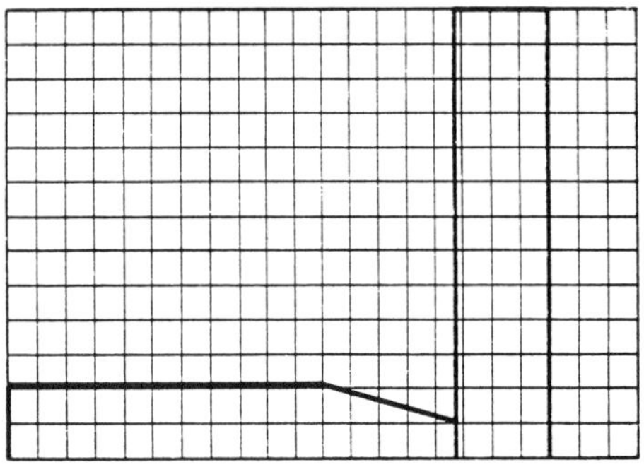

286 Zones

a. Same spatial resolution as Fig. 1.

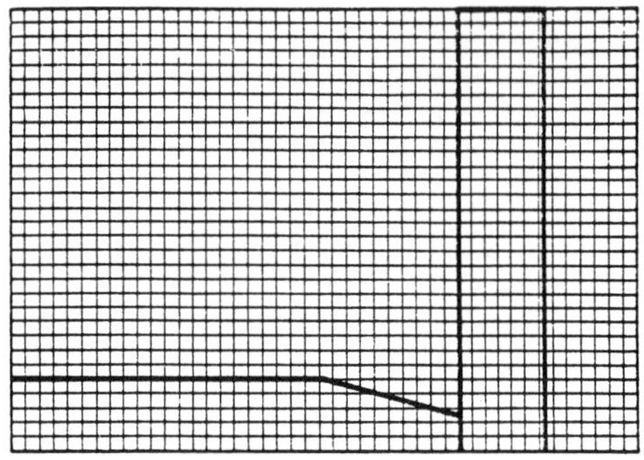

1395 Zones

b. Same number of zones as Fig. 1.

Fig. 3. Comparison of mesh efficiency.

finite and discrete number of points in a computer? A brief overview of the issues in answering this question gives considerable insight to the general features of all hydrocodes, and reasons for the specific algorithms used.

Space Discretization

Two methods for solving continuum problems on the computer are used, finite difference and finite element. Finite element methodology will be discussed later. In the finite difference representation, a lattice of points, or grid, is generated to approximate the geometry of interest. The spatial derivatives in the differential equations are replaced by difference equations, e.g., for some functional F, the partial derivative $\partial F/\partial x$ becomes $\Delta F/\Delta x$ where the differences are computed from values at grid points. The finite differencing is a pointwise approximation. The values computed at a point are then taken to represent the physical parameter over some finite region of space, e.g., a grid cell.

Figure 4a depicts the continuous function F as a function of the spatial coordinate x. By dividing the x-internal into discrete intervals of Δx, the function F at the point $x_n = n\Delta x$ is represented by F_n. The first derivative of F at x_n can be represented by a variety of difference formulas, e.g.,

$$\left.\frac{\partial F}{\partial x}\right|_{x_n} = \frac{F_{n+1} - F_n}{\Delta x} \; ; \qquad \left.\frac{\partial F}{\partial x}\right|_{x_n} = \frac{F_n - F_{n-1}}{\Delta x} \; ; \qquad \left.\frac{\partial F}{\partial x}\right|_{x_n} = \frac{F_{n+1} - F_{n-1}}{2(\Delta x)} \tag{8}$$

which corresponds, respectively, to forward, backward, and central difference equations. Which of these difference formulas, or alternative difference formulas, is the best approximation of the derivative is not as straightforward as one might suppose.

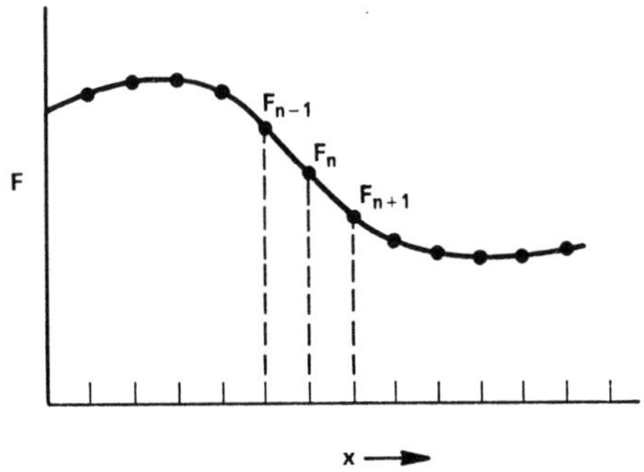

a. Smooth function.

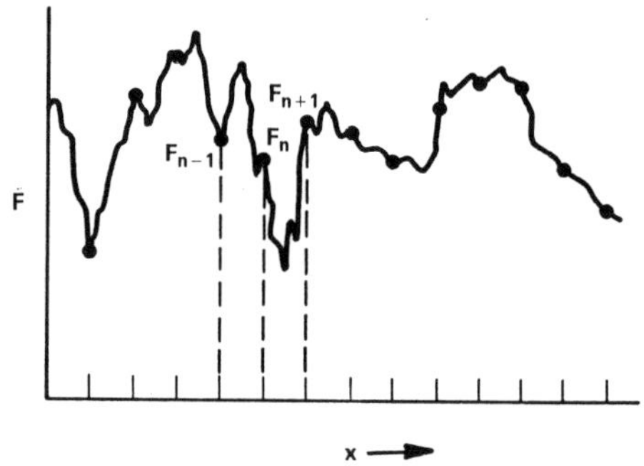

b. Rapidly varying function.

Fig. 4. Space descritization

Time Integration

Several alternatives exist for advancing the solution in time, and just as there is a concern for accuracy in the spatial discretization, the same concern also applies to time integration. The time derivative can be replaced by a difference equation; however, as the solution is advanced in time, the terminology generally used is to refer to the time integration of the differential equations.

Starting at time t^{ℓ}, the solution is advanced by an increment of time Δt to $t^{\ell+1}$ $(t^{\ell+1} - t^{\ell} = \Delta t)$.
Equations (1-3, 6) can be written in the general form:

$$\frac{\partial U}{\partial t} = - \frac{\partial}{\partial x} F(U) \equiv G(U) \tag{9}$$

The Mean Value Theorem states that over the interval $[t^{\ell}, t^{\ell+1}]$, there exists some mean value or average G, called $\bar{G}$, which is given by (Hildebrand, 1962):

$$\bar{G} = \frac{1}{\Delta t} \int_{t^{\ell}}^{t^{\ell+1}} G(\tau)d\tau \tag{10}$$

$\bar{G}$ can be computed from known values of G at times t^{ℓ} and $t^{\ell+1}$:

$$\bar{G} = \varepsilon\, G^{\ell+1} + (1-\varepsilon)G^{\ell} \tag{11}$$

where ε has some value between 0 and 1. Integrating Eq. (9) and using the Mean Value Theorem gives:

$$U^{\ell+1} = U^{\ell} + \Delta t\left(\varepsilon G^{\ell+1} + (1-\varepsilon)G^{\ell}\right) \tag{12}$$

Equation (12) is called an implicit integration scheme for any $\varepsilon \neq 0$, meaning that the computed value depends upon both the new and current values of the parameters. As there are N grid points, there are N equations for the N unknowns of each variable. As the equations are nonlinear, an iterative procedure is required to determine the solution at the new time.

By choosing $\varepsilon=0$, the variables at time $(\ell+1)$ are found directly from parameters already known at time ℓ. The procedure whereby functions at the new time step are determined from the known functions at a previous time step is called an explicit integration scheme. Explicit time integration is used in hydrocodes; the reason for this will be given later.

Criteria for Discretization

There are four properties to be asked of a particular difference representation in order to ascertain the "best" approximation of the differential equations: consistency, accuracy, stability and efficiency. Each of these will be discussed.

Consistency: The first property to be demanded of a difference technique is that the difference equations in some manner approximate the differential equations they are replacing. Formally, it is required that the difference scheme reduce to the differential equation in the limit as Δt and Δx go to zero; e.g., $\lim (\Delta F/\Delta x)\ \Delta x \to 0\ \partial F/\Delta x$. This requirement is fundamental, otherwise the difference scheme does not simulate the differential equation (e.g., Lax and Richtmyer, 1956). However, the use of finite spatial and time steps produces errors and makes the solution deviate from the solution to the differential equation; thus, the difference solution must be examined in more detail.

Accuracy: Two types of errors occur which impair the accuracy of a difference scheme, round-off errors and truncation errors. Round-off errors are associated with the precision of the computer, that is, the number of significant digits carried by the computer to represent a variable. Scientific computers represent each number with 60 to 64 bits called a word, which permits the retention of approximately fourteen to eighteen "significant" digits. Variables, such as pressure, will change by many orders of magnitude in a dynamic shock event simulated by hydrocodes. Additionally, a large number of cycles are executed to advance the solution in time. These two factors require that a large number of digits be carried by the computer. These "significant" digits are not carried to increase the accuracy of an answer by giving more significant digits per se (often, the input parameters may not be known to more than two or three significant digits, so the final answer cannot be more accurate than the input), but rather, to minimize round-off errors. Even with the large number of computer digits, occasions do arise where round-off errors create difficulties. Sometimes, special algorithms are necessary to handle certain types of round-off errors.

A second type of error is the result of the approximation scheme used, and is called truncation error. The essence of this error arises from representing a continuous variable with a discrete number of points, and the magnitude of the errors depends on the mesh intervals in time and space. Taylor series expansion provides a means to estimate the truncation errors in representing derivatives with difference equations. For example, the backward difference and forward difference schemes for $\partial F/\partial x$ ignored terms of the order (Δx) and higher; whereas, the central difference scheme ignored terms of the order of $(\Delta x)^2$. These difference schemes are referred to as first-order accurate, and second-order accurate, respectively. Since any spatial interval can be normalized to run from x=0 to x=1, it is clear that difference schemes with truncation errors of $(\Delta x)^2$ are more accurate than first-order accurate difference equations. However, because of stability issues, an alternate method for estimating accuracy will be discussed.

J. von Neumann developed an analysis procedure for analyzing the frequency response of finite difference equations with respect to both accuracy and stability (von Neumann and Richtmyer, 1950; Richtmyer and Morton, 1967). Consider a problem in one-dimension whose solution is given by a continuous function F(x) on the interval $0 \leq x \leq L$. The solution can be written as a Fourier series:

$$F(x) = \sum_{k=-\infty}^{\infty} g_k e^{i\pi kx/L} \qquad g_k = \frac{1}{L} \int_0^L F(x)\, e^{-i\pi kx/L} dx \tag{13}$$

On the computer, a discrete number of points N is used to represent the function. Thus, the infinite sum given in Eq. (13) is replaced by a finite series. The spacing between grid points (assumed here to be constant) is Δx; thus, the total length of the interval is given by $L = N(\Delta x)$, and x by $n(\Delta x)$. Hence, the function at the n-th grid point is given by:

$$F_n = \sum_{k=1}^{N} g_k\, e^{ik\pi n/N} \qquad g_k = \frac{1}{N} \sum_{k=1}^{N} F_n\, e^{-i\pi kn/N} \tag{14}$$

Wavelengths go from a minimum of $2(\Delta x)$ to a maximum of $2N(\Delta x)$. No wavelengths shorter than $2(\Delta x)$ can be defined on the spatial mesh. Thus, if the function "wiggles" in an interval shorter than Δx, there is no way to resolve these "wiggles." This is depicted in Fig. 4b. In order to resolve a rapidly varying function, more points are required on the interval. The greater number of points implies a larger N and smaller Δx. The effect in terms of wavelength is that by decreasing Δx, the spatial grid has an improved frequency response.

In using these Fourier modes, an estimate of the accuracy of the finite-difference scheme can be made. For example, the central difference scheme, Eq. (8c) gives the following (Potter, 1973):

$$\Delta_n' = \left(1 - \frac{k^2(\Delta x)^2}{6} + O[(\Delta x)^4]\right) \frac{\partial}{\partial x} \tag{15}$$

where the Δ'_n is the difference operator. Thus, the difference operator is equal to the partial derivative within a factor of $(\Delta x)^2$, i.e., second-order accurate. However, the real importance of using Fourier modes is in the investigation of numerical stability.

<u>Lax-Wendroff Method</u>: From 1960 through the mid 70's, a number of investigators searched for the "best" difference representation of the continuum equations. Much of this research is documented in the <u>Journal of Computational Physics</u>, though considerable "testing" and comparison of the various schemes were performed at different Government laboratories and universities. In this search for an accurate simulation of the difference equations, Lax and Wendroff (1960) devised an integration scheme that used a difference representation developed by Lax (1954). Richtmyer and Morton (1967) simplified the scheme somewhat by rewriting the original Lax-Wendroff method as a two-step method.

Consider the general advection equation, Eq. (9). By computing variables at a time $t^{\ell+\frac{1}{2}}$ and half-way between grid points, $n+\frac{1}{2}$, the difference equations become second-order accurate in both time and space.

This scheme is depicted in a space-time difference mesh in Fig. 5. In essence, the two-step Lax-Wendroff method is a predictor-corrector method. The first step predicts the values of U at the half-time, half-space points:

$$U_{n+\frac{1}{2}}^{\ell+\frac{1}{2}} = \tfrac{1}{2}(U_n^{\ell} + U_{n+1}^{\ell}) - \frac{\Delta t}{2(\Delta x)} (F_{n+1}^{\ell} - F_n^{\ell}) \tag{16}$$

and then these values of U define the fluxes at the intermediate time and space points:

$$F^{\ell+\frac{1}{2}}_{n+\frac{1}{2}} = F\left(U^{\ell+\frac{1}{2}}_{n+\frac{1}{2}}\right) \qquad (17)$$

The corrector step computes U at the final time for each spatial grid point:

$$U^{\ell+1}_{n} = U^{\ell}_{n} - \frac{\Delta t}{\Delta x}\left(F^{\ell+\frac{1}{2}}_{n+\frac{1}{2}} - F^{\ell+\frac{1}{2}}_{n-\frac{1}{2}}\right) \qquad (18)$$

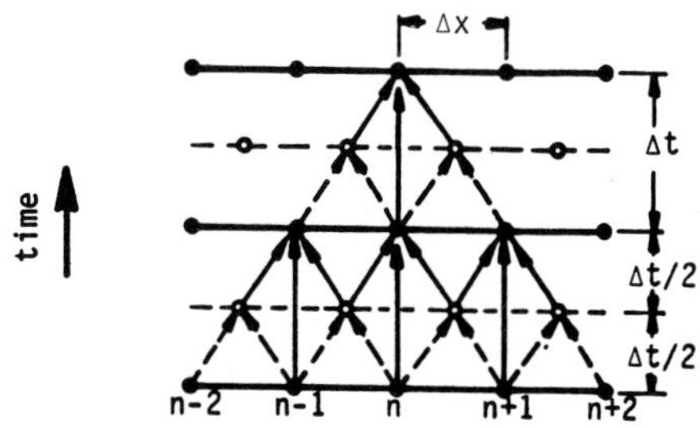

Fig. 5. Lax-Wendroff method.

While no hydrocode may explicitly use the Lax-Wendroff method, the essence of the method is used by most of the codes. In particular, by defining cell-centered quantities for the stresses (and pressures), densities, and internal energies, the velocities at each grid point are computed at $t^{\ell+\frac{1}{2}}$. Then these velocities are used to advance the cell-centered variables to $t^{\ell+1}$.

Conservation on a Difference Mesh: The system of partial differential equations is nonlinear, and in defining a particular difference method, there are a variety of ways of differencing the nonlinear terms. When nonlinear terms are present, the difference methods do not have the same accuracy for all terms. The effect is to have local sources and sinks of mass, momentum and energy (Emery, 1967). These anomalies are most often manifested in the presence of strong pressure gradients (shocks). Since Eqs. (1-3) were derived from application of conservation of mass, momentum and energy, it would seem natural to ask if the difference equations also could be conservative. Lax and Wendroff (1960) showed that if the hydrodynamic equations could be put in the form (Richtmyer and Morton, 1967; Roache, 1972):

$$\frac{\partial U}{\partial t} + \frac{\partial}{\partial x} F(U) = 0 \qquad U = \begin{pmatrix} \rho \\ \rho U \\ e \end{pmatrix} \qquad F(U) = \begin{pmatrix} \rho U \\ \rho U^2 + P \\ (e+P)U \end{pmatrix} \qquad (19)$$

then the difference equations are conservative. Both the Eulerian and Lagrangian descriptions can be put in a conservative form (Richtmyer and Morton, 1967). The implication is that the difference scheme, independent of truncation errors, will identically conserve mass, momentum and energy (except for round-off errors).

Stability: Unfortunately, answering the question of accuracy is not sufficient--certain difference schemes lead to numerical instabilities (for example, see Walsh, 1972). If the difference scheme produces a solution which is not bounded, the scheme is said to be numerically unstable. A numerical method is stable if a small error at any stage produces a smaller cumulative error. If an error is amplified from time step to time step, the error will quickly swamp the solution and the result will be meaningless. For example, assume that an error ε exists in the parameter U at time step ℓ. The calculation of the U at time step $(\ell+1)$ will have an error associated with it, and both $U^{\ell+1}$ and $\varepsilon^{\ell+1}$ will be functions of U and ε at the previous time steps:

$$U^{\ell+1}_{n} + \varepsilon^{\ell+1}_{n} = f\left(U^{\ell}_{n} + \varepsilon^{\ell}_{n}\right) \qquad (20)$$

The amplification of the error at time step $(\ell+1)$ is given by:

$$\varepsilon^{\ell+1} = g\,\varepsilon^{\ell} \qquad (21)$$

where "g" is termed the amplification factor. This amplification factor is related to truncation errors and the integration schemes used, but is not associated with round-off errors. For stability, the requirement is that the error at time $(\ell+1)$ be less than or equal to the error at time ℓ, otherwise the error will grow as time increases. This leads to the very important conclusion: the absolute magnitude of the amplification factor must be less than or equal to one for a numerical scheme to be stable:

$$|g| \leq 1 \qquad \text{for stability} \qquad (22)$$

In general, a whole system of difference equations exists (at least one for each grid point); the amplification factor is a tensor, and the errors can be represented as components of a multi-element vector (Richtmyer and Morton, 1967). However, the conclusion of Eq. (22) remains the same except that the condition for stability must be applied separately to the amplitude of each error eigenvector. In general, the eigenvalues of the amplification matrix may be complex, in which case the magnitude of the amplitude is given by $[g^*g]^{1/2}$, where g^* is the complex conjugate of g.

The errors in Eq. (20) also satisfy the general partial differential equations. Separation of variables is used to separate the time component of ε from the spatial component:

$$\varepsilon = \varepsilon^*(t)e^{ikx} \tag{23}$$

where $\varepsilon^*(t)$ gives the amplitude of the Fourier mode. As any waveform can be approximated by the superposition of the various wavelengths in a Fourier series, then the procedure is to demand stability for an arbitrary wavelength, represented by the wavenumber k. Application of the von Neumann stability methodology to the Lax-Wendroff method yields for the amplification matrix (Richtmyer and Morton, 1967):

$$g = 1 - i\alpha\sin(k\Delta x) - \alpha^2[1 - \cos(k\Delta x)] \qquad \alpha = \frac{c\Delta t}{\Delta x} \tag{24}$$

$$g^*g = 1 - \alpha^2 (1 - \alpha^2) [1 - \cos(k\Delta x)]^2 \tag{25}$$

where c is the wave speed. The requirement that $[g^*g]^{1/2}$ be less than or equal to one, for all k, gives the famous Courant-Friedrichs-Lewy (CFL) stability condition (Courant, et al., 1928):

$$\alpha \leq 1 \qquad \text{i.e.,} \quad \frac{c\Delta t}{\Delta x} \leq 1 \quad \text{or} \quad \Delta t \leq \frac{\Delta x}{c} \tag{26}$$

Equation (26) is for Lagrangian coordinates; in Eulerian coordinates, the sound speed is replaced by the sum of the particle velocity and sound speed:

$$\Delta t \leq \frac{\Delta x}{|v| + c} \tag{27}$$

The stability criterion is essentially a statement of casualty, i.e., no signal (information) can be allowed to propagate across the shortest dimension of a zone in a time Δt. Though the CFL criterion was found here for the Lax-Wendroff method, it applies equally to other stable difference representations. (Some difference representations lead to an amplification matrix which gives $|g| > 1$ for any value of Δt.)

It is worth remarking here that for small wave numbers ($k\Delta x$ small), the cosine in Eq. (25) can be expanded in a Taylor series to yield:

$$g^*g = 1 - \alpha^2(1 - \alpha^2) \frac{k^4(\Delta x)^4}{4} + O(k^6(\Delta x)^6) \tag{28}$$

Thus, numerical diffusion, i.e., the smearing of a variable over space, occurs only to the fourth order in the wave number (Potter, 1973). Long wavelengths are only minimally affected, while the effect of fourth-order diffusion tends to smooth short wavelength discontinuities on the difference mesh (see the section on artificial viscosity in this paper). Additionally, because there is no imaginary component in Eq. (28), no unwanted oscillations appear in the solution. These features, in addition to being second-order accurate in the spatial derivative and time integration, have lead to wide application of the Lax-Wendroff method.

Efficiency: The last requirement for a numerical technique is efficiency. A computer is finite; only so much memory is available, and the calculation time required to get a numerical solution must be reasonable. For many computer systems, the costs of running a problem are related to central processing time, memory requirements, and input/output operations. The efficiency of a particular numerical scheme can be defined as the total number of arithmetical, logical, and storage operations performed by the computer to obtain a solution over a unit time-length of the problem. Efficiency decreases with greater complexity of the numerical technique; however, the accuracy of the numerical solution is generally increased with increasing complexity. In a hydrocode, several thousands of variables typically are computed for each time step. It is important that the numerical differencing and integration technique be as efficient as possible; but at the same time, for the solution to be of any value, the numerical techniques must be accurate. A compromise must be reached between efficiency and accuracy. It has been found that the small grid sizes necessary to resolve important shock interactions give quite accurate results using second-order accurate methods (Herrmann, 1977). The increased complexity and extra computations necessary to go to higher order methods, even with some attendant increase in grid size, appear to be computationally inefficient.

Finite Element Discretization

Whereas the finite difference technique is a pointwise discretization of the continuum, finite element techniques envision the solution region as being composed of many small interconnected sub-regions or elements. Thus, a piecewise approximation of the differential equations is made. Nodes are assigned to elements and then an interpolating function, typically a polynomial (a spline function is another possibility), is used to represent the variation of the variable over the element.

The degree of the polynomial is related to the number of nodes and degrees of freedom of the nodes in the element. The displacements at these nodal points are the basic unknowns of the problem. These displacement functions define a state of strain and strain rate within each element; with the constitutive equation, the state of stress is determined. The forces, concentrated at the nodes, relate internal loads, external loads, and the nodal displacements. The total potential energy is found, then variational calculus is used to minimize the potential with respect to the coefficients of the polynomials, resulting in a system of algebraic equations for the unknown coefficients.

Traditionally, finite element formulations have inverted the so-called stiffness matrix. An iterative scheme is then required to obtain the solution. For dynamic, time-dependent problems, Belytschko recognized that the stiffness matrix was in fact unnecessary for the explicitly integrated finite element methods (Chiapetta, Belytschko and Rouse, 1973; Belytschko, Chiapetta and Bartel, 1976; Belytschko, 1976). Thus, the finite element hydrocodes are integrated directly. The first adaptation of an explicit time integration scheme to a large deformation finite element code which also could handle shocks was HONDO (Key, 1974; Key, 1974b; Key, et al., 1978). Independent work by Johnson (1976, 1977, 1977b, 1977c, 1982), and by Hallquist (1976; 1976b; 1978; Goudreau and Hallquist, 1982) has led to the EPIC and DYNA families of codes, respectively. These last two families of codes are used extensively to solve shock propagation problems.

In summary, Gordon Johnson has described the difference between the two techniques as, "Finite difference techniques are an approximate solution to an exact problem, and finite element techniques are an exact solution of an approximate problem." There is no basic mathematical difference between the two methods; Belytschko (1976) has shown that the discrete forms of the equations of motion of the finite element are equivalent to those of the finite difference method for a number of cases. The differences are in the implementation.

Finite element codes, by the nature of their formulation, automatically incorporate certain types of boundary conditions. The equations of motion are formulated through nodal forces for each element, and do not depend on the shape of the neighboring mesh. Equations of motion in the finite difference method, on the other hand, are expressed directly in terms of the stress gradients of the neighboring grid.

Historically, a criticism of the finite difference technique was that it could not represent an irregular surface; finite elements were noted for their ability to handle complicated geometries. Conversely, finite element techniques have been criticized for their poor constitutive capabilities and the lack of capabilities seen in some finite difference codes, e.g., slidelines, rezoning, etc., (Belytschko, 1976). However, these criticisms are no longer applicable. Mesh generator schemes in finite difference codes have become fairly sophisticated; and slidelines, and material libraries with attendant constitutive models have become very much a part of finite element hydrocodes. The large difference between the finite element and finite difference techniques has closed considerably, and perhaps the best way of summarizing is to say that the two techniques provide different algorithms for obtaining solutions to the same differential equations.

The four properties of the discretization process (consistency, accuracy, stability, and efficiency) apply equally to the finite element process, though the explicit implementation might vary in comparison with finite difference techniques. For example, consistency is automatically implied by the minimization technique. Accuracy is just as relevant to finite elements as finite difference. Large zone sizes can be accomplished by going to higher order elements, i.e., using quadratic and cubic polynomials (akin to going to higher order difference approximations). In finite element hydrocodes, the complexity of using higher order elements versus resolution of wave interactions is not warranted (efficiency), and constant stress elements are typically employed. Stability criteria, which result from separating the spatial and time discretization process, apply equally as well to finite element. The CFL condition holds for constant stress elements, and perhaps a more severe restriction holds if higher order elements are used. And, finally, since the mesh size in both finite difference and finite elements are considerably larger than the thickness of the shock, both techniques require artificial viscosity so that the same equations will handle the propagation of shocks. This is the topic of the next section.

TREATMENT OF SHOCKS

The discussion to this point has been concerned with the field equations where it has been assumed that the physical variables of interest vary smoothly enough that they can be resolved on a finite difference or finite element grid. We already have discussed that a complex waveform can be reconstructed from an infinite number of terms in a Fourier series. However, wavelengths smaller than the grid spacing cannot be described, thus limiting the grid's ability to resolve physical phenomena which may be occurring very rapidly. Shocks, on the other hand, can be described by an instantaneous jump in the velocity, pressure, and density. At first glance, it would appear that a finite difference representation of the continuum equations would not permit a calculation involving the propagation of shocks. However, two approaches exist for the calculation of problems containing shocks.

One approach is to use the Rankine-Hugoniot jump conditions. This approach generally is combined with the mathematical technique called method of characteristics. In a sense, information flows along the characteristics which move at the wave speed. This method is excellent for understanding the detailed wave motion; however, if the ultimate state of the material is desired, a finite-difference code is generally superior (Karpp, 1972).

The second approach is to use the discretization procedure on the differential equations. But the equations must be modified slightly in order to obtain any meaningful numerical results. Before discussing the modifications, it is instructive to look at the numerical solution of a shock propagating into steel at two different times, Fig. 6, using the Lax-Wendroff method. The dots on the figure denote the values of the pressure at the grid points. The dashed line represents the true solution. Large amplitude oscillations are observed in the numerical solution. These oscillations are not instabilities; a calculation can be continued indefinitely and the amplitude of the oscillations will remain approximately the same. We note here that a Lax-Wendroff scheme was used in Fig. 6, and that if a different finite-difference scheme were used, then the solution might well appear as Fig. 7. (The reason for the marked difference in Figs. 6 and 7 is the result of a dissipation term inherent in the Lax-Wendroff method, which dampens the short wavelength oscillations away from the shock front.) The large oscillations in Figs. 6 and 7 have little to do with the solution to the exact differential equations, and, furthermore, can swamp the true solution.

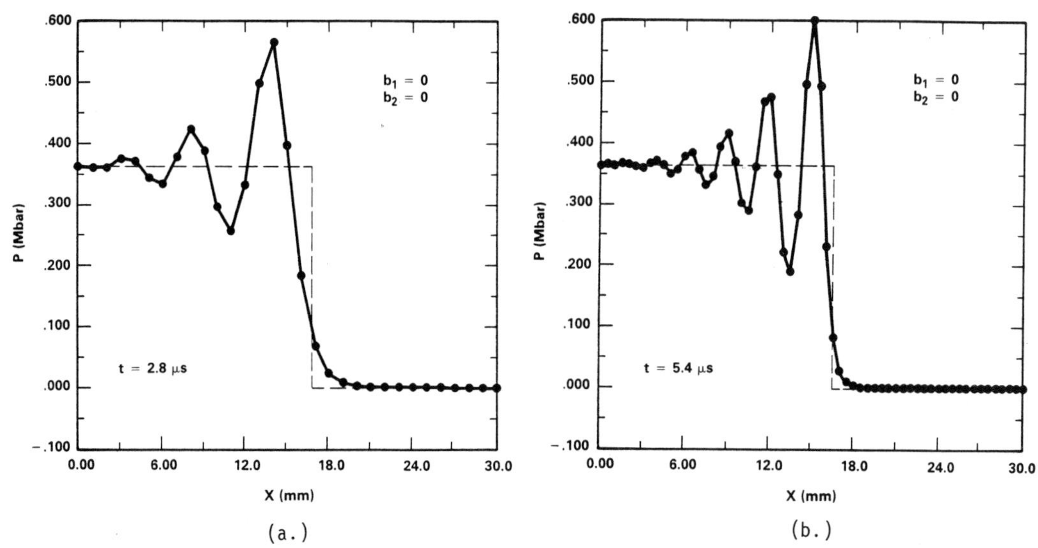

Fig. 6. Shock propagation into steel bar.

Several interpretations have been used to describe physically the oscillations in Figs. 6 and 7. One interpretation by Potter (1973), is that the energy of the large amplitude disturbance tends to aggregate in the shortest wavelengths on a difference mesh (that is, the high frequency components of the Fourier series representation contain the energy). Consequently, the difference solution will induce very large oscillations between the variables on adjacent mesh points. Other interpretations also have been given (Richtmyer and Morton, 1967). Regardless of the interpretation, the "excess" energy manifested in the large amplitude variations needs to be converted to a true internal energy.

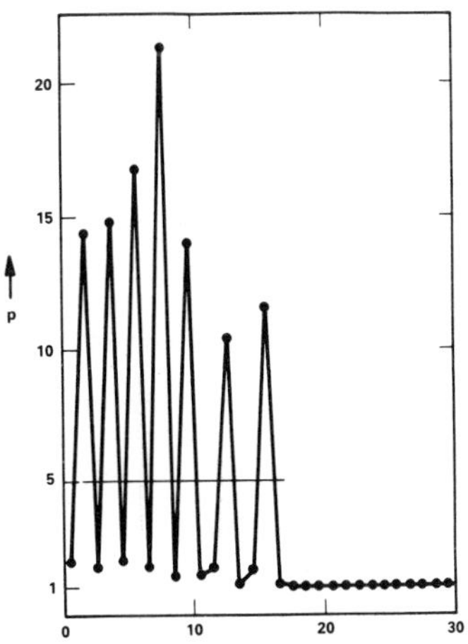

Fig. 7. An alternate differencing scheme (from Richtmyer and Morton, 1967).

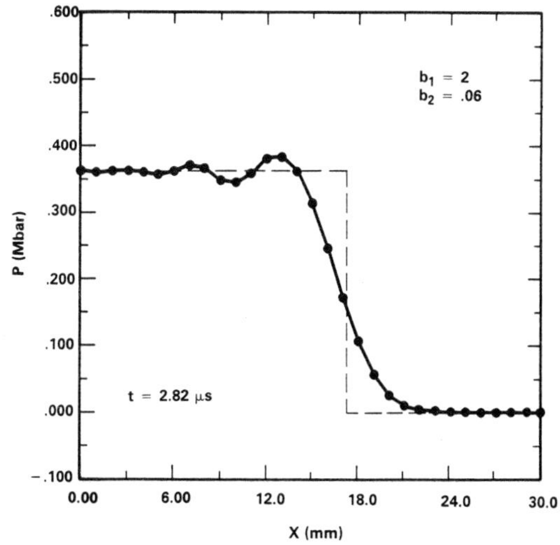

Fig. 8. Shock propagating into steel bar, with artificial viscosity.

Artificial Viscosity

Von Neumann and Richtmyer (1950) devised a method for handling shocks by introducing a purely artificial dissipative mechanism. Physical insight suggested this type of formulation. Becker (1922) had studied the effect of heat conduction and viscosity on shocks. As reported in Richtmyer and Morton (1967):

> He [Becker] showed that, when heat conduction is allowed for, the temperature varies smoothly through the shock layer and that for shocks of strength less than a critical value, the pressure and density vary smoothly also; whereas, for stronger shocks, the transition of pressure and density from their initial to their final values is partly by a smooth variation and partly by a discontinuous jump. When viscosity is allowed for, however, all quantities vary smoothly through the shock region for a shock of any strength. In either case the thickness of the transition zone is proportional to the coefficient of the dissipative mechanism, so that in the limit of no heat conduction and no viscosity the variations approach the discontinuous ones of the Hugoniot theory.

The idea then was to introduce a dissipative mechanism that would result in the shock being spread smoothly over several mesh intervals. Viscosity is a mechanism which has such a dissipative effect, and certainly appears to be more appropriate than a heat conduction term. The procedure thus required that some artificial viscosity term be devised that had the right strength to spread the shock over several computational zones, and could be incorporated into the difference equations. In traditional fluid mechanics, a viscosity term appears in the one-dimensional momentum equation as:

$$\text{viscosity term: } \frac{\partial}{\partial x}\left(\mu \frac{\partial v}{\partial x}\right) \qquad (29)$$

where μ is a coefficient of viscosity. This would suggest that an artificial viscosity term q could be added directly to the pressure in the differential equations, e.g., $\partial(p+q)/\partial x$, where q is called the artificial viscosity and would be of the form $\mu \partial v/\partial x$.

Becker had shown that with ordinary viscosity, in which the stress is proportional to the rate of shear and is represented by a linear term in the velocity, the thickness of the transition layer varies with the shock strength, approaching zero for very strong shocks and infinity for very weak shocks. In their original formulations, von Neumann and Richtmyer were concerned that the thickness of the shock should be independent of the shock strength since this was to be a numerical artifact. Indeed, physical reasoning implied that the strength of the artificial viscosity term would be orders of magnitude larger than real viscous effects, since normally, real viscous effects can be ignored in most treatments of shocks. The form of the artificial viscosity term was chosen as:

$$q_1 = -a_1^2 \rho \left| \frac{\partial v}{\partial x} \right| \frac{\partial v}{\partial x} \qquad (30)$$

where a_1 has the dimensions of length. It was later found that a more appropriate expression was:

$$q_1 = \begin{cases} a_1^2 \rho (\partial v/\partial x)^2 & \partial v/\partial x < 0 \\ 0 & \partial v/\partial x \geq 0 \end{cases} \qquad (31)$$

that is, the artificial viscosity term need only be present during material compressions. For most problems, either formula is satisfactory because q is extremely small except at the shock front. However, when extreme expansion (as of a free surface) can take place, the second expression is superior (Richtmyer and Morton, 1967). The adjustable constant a_1 can be written as $b_1^2(\Delta x)^2$, thereby making the difference representation of Eq. (31) independent of zone size.

Adding a linear artificial viscosity term, with a_2 a constant with the dimensions of length, and c the sound speed:

$$q_2 = a_2 c \rho \frac{\partial v}{\partial x} \qquad (32)$$

has the advantage of further smoothing the oscillations not totally dampened by the quadratic artificial viscosity term. The constant a_2 can be defined as $b_2(\Delta x)$ to keep the difference representation independent of mesh size. Greater care must be exercised with the magnitude of b_2 since there is a much greater chance of distorting the solution away from the shock.

Figure 8 shows the numerical shock profile at the same time as Fig. 6b, but with both the quadratic and linear artificial viscosity terms present. The choice of the constants b_1 and b_2 determine the number of zones over which the shock front is smeared, versus the oscillations behind the front. In Figure 8, the shock front is smeared over approximately eight zones; the front could be steepened if finer zoning was used.

Shock Viscosity

Notwithstanding that code results are often in excellent agreement with experiments, researchers were not necessarily happy with the concept of artificial viscosity, but rationalized its use as a practical substitute for real viscous processes. But a very significant finding by Swegle and Grady (1985) has altered the above objection. Figure 9 is a plot of the Hugoniot stress σ_H, versus the maximum strain rate $\dot{\varepsilon}_m$, for a number of different materials. Note that all materials have essentially the same slope. From Fig. 9, the maximum strain rate can be written as:

$$\dot{\varepsilon}_m = A \sigma_H^4 \qquad (33)$$

where A is the intercept, a material dependent parameter. Whereas the artificial viscosity is proportional to sum of the strain rate and the square of the strain rate, the actual viscosity in the shock front is proportional to the strain rate to the one-fourth power.

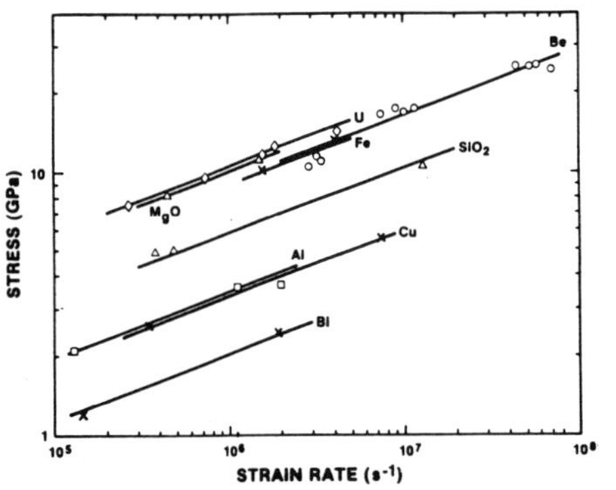

Fig. 9. Maximum shock stress versus strain rate (from Swegle and Grady, 1985).

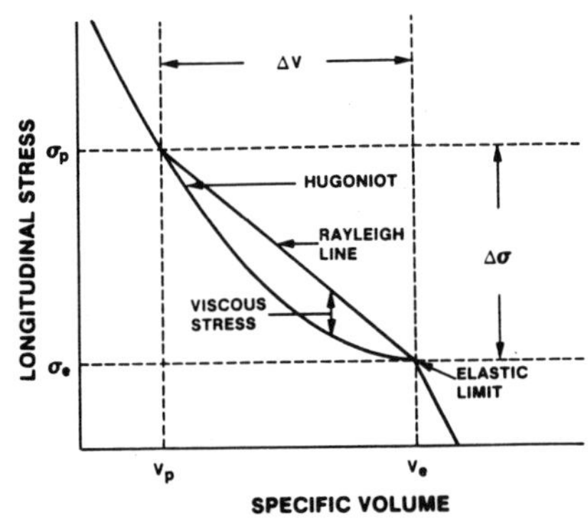

Fig. 10. Schematic for viscous stress (from Swegle and Grady, 1985).

It is assumed that the viscous stress is a purely deviatoric stress (there is no bulk viscosity), and that this viscous stress, τ_v, can be identified as the stress on the Rayleigh line minus the equilibrium stress, taken to be the Hugoniot, Fig. 10. Swegle and Grady show that the differential change in the octahedral shear stress τ, is a function of both total strain rate $\dot{\varepsilon}$, and the viscous stress τ_v:

$$d\tau = 2G\dot{\varepsilon}dt - 2GA'\tau_v^2 dt \qquad A' = 8\sqrt{2}\,\rho_e^2 c_o^4 A/3s^2 \tag{34}$$

where G is the shear modulus, and A' is a constant which incorporates the material properties of Fig. 9; ρ_e is the density at the Hugoniot elastic limit, c_o is the bulk sound speed, s is the slope in the linear shock-particle velocity relationship, and A the intercept from Fig. 9.

Normally, for an elastic, perfectly plastic material, the material state point is calculated assuming that the current strain increment is strictly elastic (Wilkins, 1964). If the resulting state point lies outside the von Mises yield surface, then the stresses are adjusted radially to bring the state point back to the yield surface. But now, the difference in the stress at the "elastic" state point, computed from the first term of Eq. (34a), and the yield surface is assumed to be the viscous stress. The stress is then relaxed by the second term in Eq. (34a), which is proportional to the square of the viscosity. Subcycling on the timestep is required to insure that the finite difference expression for Eq. (34a) is an accurate solution of the differential equation (otherwise, the stress setback could exceed the first term, and the stress state would lie inside the yield surface).

With this technique, shocks have a finite rise time and thus a finite thickness. The thickness of the shock front is given by the material parameter A' of Eq. (34b). Of course, the computational zone size must be sufficiently small to resolve the details of the shock front. Figure 11 shows comparison between experiment and the hydrocode calculations for copper at two different shock amplitudes. These hydrocode calculations were performed without any artificial viscosity. A zone size of 10^{-5} to 10^{-4} cm, depending upon the shock strength, is required to resolve the shock front.[1] Swegle and Grady report that if the zoning is too coarse for the given problem, the solution becomes very noisy, i.e., if the grid size cannot resolve the shock, then standard artificial viscosity methods must be used to spread the shock over several computational zones.

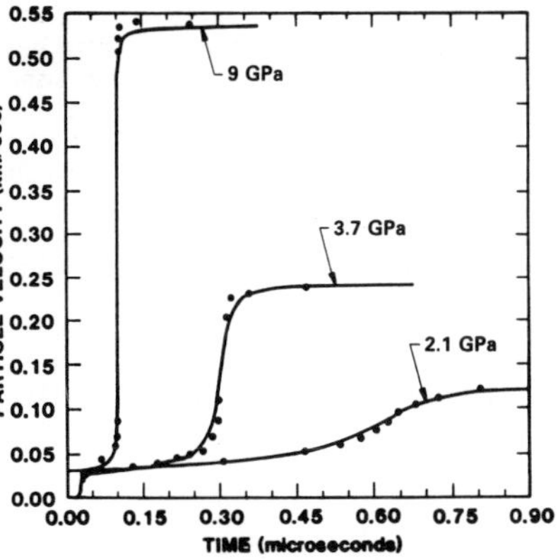

Fig. 11. Comparison of experiment (dotted line) and calculation (solid line) wave profiles for aluminum (Swegle and Grady, 1986).

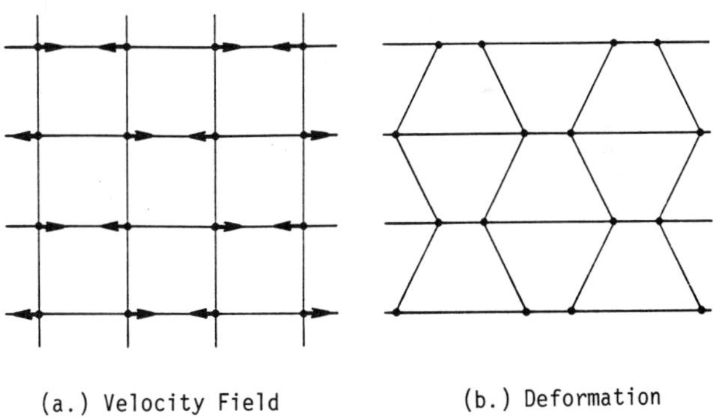

Fig. 12. Keystone velocity field and deformation.

[1] Walsh and Tillotson (1963), using a qualitative argument, point out that data (Eichelberger and Gehring, 1962) for crater depth show simple linear scaling. The validity of simple linear scaling follows directly from Rankine-Hugoniot and continuous hydrodynamic flow equations which are homogeneous in the space and time variables. This is true only if thermal conduction and viscous effects are negligible; otherwise, second derivatives enter the equations and the scaling no longer applies. Since the experimental data used projectiles with dimensions from 10^{-4} cm up to 10 cm, the conclusion was that it was appropriate to neglect viscosity effects for dimensions larger than 10^{-4} cm. It is interesting to speculate that if Eichelberger and Gehring had been able to fire slightly smaller projectiles (which already were on the order of 10^{-11} gm), and taken measurements, that the effects of shock front viscosity might have been evident.

Obviously, this very, very small zone size is unacceptable in many applications; nevertheless, the model permits the calculation of the correct strain rates in the shock front. The development of micromechanical models in which the evolution of parameters is controlled by the strain rate, such as adiabatic shear instabilities, can now be investigated at a more fundamental level. If the details of the shock front are not important, or are of secondary importance, then the usual artificial viscosity methods should be used.

Keystone Viscosity

A common numerical phenomena observed in dynamically loading, e.g., impact, is the situation where the velocities of the mesh points on opposite corners of the mesh are equal, Fig. 12a. This velocity field produces no strain or volume change in the mesh since the finite difference equations for the components of the strain rate tensor involve velocity differencing of the grid points. The stress and viscosity components are all zero and there is no resistance to the deformation caused by this type of velocity field. In finite element theory, the difficulty arises because of the assumption of a homogeneous strain field which is incompatible with a four sided element (Belytschko, 1974). The grid will take on the characteristic "keystone" appearance shown in Fig. 12b. To control this type of deformation, another type of viscosity is introduced to give artificial rigidity to the mesh, called keystone viscosity or hourglass viscosity. The reader is referred to Swegle (1978) or Belytschko (1984) for appropriate expressions. Care must be exercised with keystone viscosity since it invokes nonphysical restoring forces and thus may produce nonphysical solutions. Typically, the coefficient for keystone viscosity is zero or very small unless the user encounters a situation where the grid is distorting into the keystone appearance.

Stability and the Time Step

Stability has been discussed in some detail earlier in this paper. However, a modification to the CFL stability condition deduced from the simple transport problem, Eq. (26) or (27), has to be made because of artificial viscosity.

Addition of the artificial viscosity to dampen the oscillations resulting from the passage of shocks also limits the time step. When the von Neumann stability method is used on a diffusion-type differential equation (i.e., a parabolic differential equation, whereas the wave equation is a hyperbolic equation), the stability criterion is found to be dependent on the diffusion coefficient. The mathematical treatment is still not understood (Richtmyer and Morton, 1967) and a form of the stability requirement must rely on empiricism; the method employed in the Sandia codes has been found to work well (Kipp and Lawrence 1982; and Swegle 1978):

$$\Delta t \le \frac{\Delta x}{A + [A^2 + c^2]^{\frac{1}{2}}} \qquad A = 2b_1^2\,(\Delta x)\,|\partial v/\partial x| + b_2 c \tag{35}$$

Note that the time step must decrease as the "diffusion" term (related to the artificial viscosity) grows larger. Wilkins (1964), and Richtmyer and Morton (1967) give slightly different expressions. Equation (35) reduces to the familiar CFL condition if the artificial viscosity coefficients are set to zero (as they are during an expansion phase); but the diffusion term controls the time step when it is large, i.e., when shocks are present. For multiple dimensions, the definition of A changes slightly as the quadratic artificial viscosity term is extended to two or three dimensions.

In the section on time integrations, an implicit integration scheme was discussed. If the von Neumann stability analysis is applied to Eq. (12), $\varepsilon \neq 0$, the amplification factor is found to be less than or equal to one for all wave numbers, implying that the implicit integration scheme is unconditionally stable. Thus, the difference equation is numerically stable regardless of the size of the time step, though in practice the time step might be limited by the ability of the convergence scheme to converge to a solution. One might then ask, "Why not use an implicit integration scheme for hydrocodes?" Because shocks travel distances of millimeters per microsecond, the time step must be small to resolve the temporal details. Since the time steps are necessarily limited by the desire to resolve shock wave interactions, the added complexity of an implicit integration scheme, with its attendant number of interactions (and often convergence problems) is not warranted (also, see Herrmann, 1977). In structural response calculations, or heat transfer calculations, significant changes occur over timeframes which are orders of magnitude longer than microseconds, thereby making implicit schemes useful in those problems. Thus, for hydrocode calculations, the time step is limited by the modified Courant condition, Eq. (35).

The time step is dramatically affected by changing the grid size. In one dimension, halving the grid size quadruples the number of computations, since twice as many zones are now involved in the mesh, and by Eq. (26c), the time step has been halved. In two dimensions, doubling the number of zones in the two dimensions increases the run time by at least a factor of 8 (to reach the same time), while in three dimensions, doubling the number of zones in all directions requires at least 16 times more computational steps.

MATERIAL MODELING IN HYDROCODES

The preceding discussions have dealt solely with how the continuum equations are solved on the computer. Specific material properties govern the response of materials, and result in different behavior for nominally the same impact conditions. Material modeling can be divided into three areas: volumetric response, or resistance to compressibility (equation of state), the resistance to distortion (constitutive); and the reduction in ability to carry stress as damage accumulates (failure). The importance in understanding and modeling material response is reflected in the papers by Asay and Kerley (1986), Holian and Burkett (1986), Kerley (1986), Steinberg (1986), Drumheller (1986), and Trucano and Asay (1986). An excellent monograph by the National Academy of Sciences (1980) reviews the state of understanding of the response of metals to ultra-high loading rates, though with particular application to armor. The following paragraphs summarize material modeling in hydrocodes.

Equation of State

Whether a material has strength or not, some expression is required to express the relationship between pressure, volume, and internal energy. This relationship is called the equation of state. Thermodynamic theory allows a description of a material to be given by three thermodynamic variables, such that one variable can be expressed in terms of the other two, e.g., the pressure in terms of the internal energy E, and specific volume V ($\rho = 1/V$):

$$P = f(E, V) \tag{36}$$

Equation (36) can be thought of as a surface in pressure-energy-volume space. As such, a change in the pressure dP can be written in terms of the changes in the other two variables:

$$dP = \left(\frac{\partial P}{\partial V}\right)_E dV + \left(\frac{\partial P}{\partial E}\right)_V dE \tag{37}$$

where the subscript E and V on the parentheses refer to the derivative being performed with one variable while holding the other constant. Integration of Eq. (37) allows the pressure to be expressed in terms of the volume V and energy E relative to the pressure at a reference energy and volume, E_R and V_R.

The Gruneisen parameter is defined as:

$$\Gamma = V \left(\frac{\partial P}{\partial E}\right)_V \tag{38}$$

and if it assumed that Γ is pressure independent, then it can be shown that the integration of Eq. (37) gives (Rice, et al., 1958; McQueen, et al., 1970):

$$P = P_H(V) + \frac{\Gamma}{V}(E - E_H) \tag{39}$$

where the reference state is taken as the Hugoniot. Equation (39) is referred to as the Mie-Gruneisen equation of state, and it permits the evaluation of the pressure at some volume and energy off the Hugoniot. The Gruneisen parameter is found from evaluating Eq. (38) using various thermodynamic identities (Zel'dovich and Raiser, 1967): $\Gamma = \beta\kappa/\rho C_p$, where β is the coefficient of volumetric expansion, κ is the adiabatic bulk modulus, ρ is the density, and C_p the specific heat at constant pressure. Normally, the dynamic Gruneisen parameter is determined from $\Gamma\rho = \Gamma_o\rho_o$ where the subscript refers to the room-temperature, zero-pressure values (McQueen, et al., 1970).

Equation (39) has certainly been shown to be valid for solids, and in particular, metals, as well as liquids. But at very high pressures, and at distended states (e.g., vaporization), Eq. (39) must be replaced by other formulas. Porous materials need special treatment (Herrmann, 1969; Carroll and Holt, 1972; 1972b). The Tillotson equation of state (Tillotson, 1962) has been used with success for impacts at hypervelocities (Sedgwick, et al., 1978). Trucano and Asay (1986) show that the form of the equation of state is very important in predicting the dynamics of the debris bubble after hypervelocity impact of a bumper shield. A conclusion of Holian and Burkett (1986) is that the ability to predict the dynamics of the debris bubble not only is dependent upon the equation of state, but the ability of the code to treat correctly two-phase flow. The reader is referred to the excellent review article by Asay and Kerley (1986) for further information on equations of state.

Constitutive Modeling

Elastic-Plastic Formulation: Figure 13a displays the pressure-volume relation (Hugoniot) if the material has no strength (hydrodynamic). However, if the material has strength, then it will load elastically until yield, at which point it will begin to permanently deform plastically. An elastic, perfectly plastic material response is shown in Fig. 13b, both for compression and tension. The "superposition" of the curves of Figs. 13a and 13b is shown in Fig. 13c for the condition of uniaxial strain. The material is seen to load elastically to the Hugoniot elastic limit, and then flow plastically parallel to the Hugoniot. Wilkins (1964, 1969) developed the methodology for the inclusion of strength effects which is used in today's finite difference and finite element hydrocodes. For the condition of uniaxial strain, the principal stresses, t_i, are given by:

$$t_1 = -P + \frac{4}{3}\tau_m \qquad t_2 = t_3 = -P - \frac{2}{3}\tau_m \qquad P = \frac{1}{3}(t_1 + t_2 + t_3) \tag{40}$$

where τ_m is the maximum shear stress, which, at plastic flow, is equal to the flow (yield) stress $Y°/2$. The hydrostatic pressure is simply the average of the principal stresses. Thus, an applied load is carried by the hydrostatic pressure and in part by its resistance to shear (Davison and Graham, 1979). From Eqs. (40), it can be seen that if the hydrostatic pressure is very large, it is reasonable to ignore material strength, i.e., the hydrodynamic approximation.

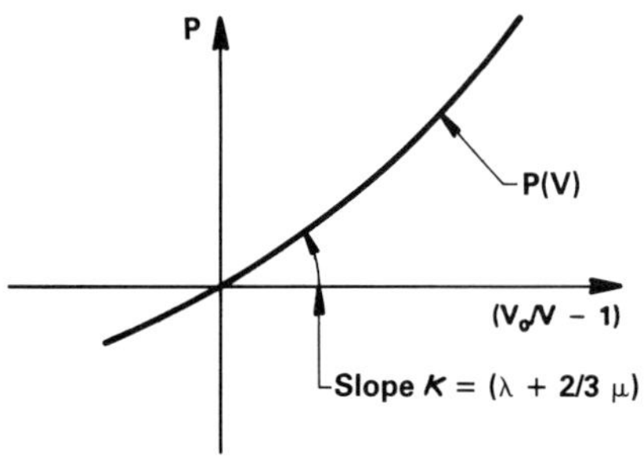

a. Hydrodynamic response.

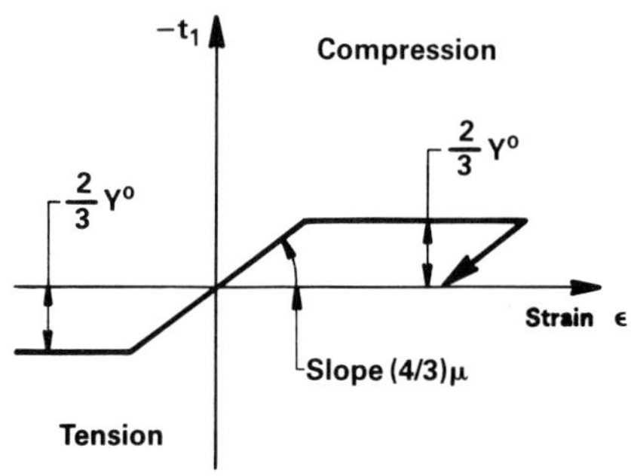

b. Elastic-perfectly plastic response.

In general, the stress can be divided into its volumetric and deviatoric parts:

$$\sigma_{ij} = -P\,\delta_{ij} + s_{ij} \tag{41}$$

The first term accounts for volumetric changes, and is obtained from the equation of state. The stress deviators, s_{ij}, are related to distortion changes, and in particular, the time rate of change of the deviators is computed from the strain rates:

$$\dot{s}_{ij} = 2G\dot{\varepsilon}_{ij} + \dot{\Delta}_{ij} \tag{42}$$

where G is the shear modulus, and $\dot{\Delta}_{ij}$ is a correction for rigid body rotation.

To compute elastic-plastic flow, it is assumed that the total strains can be separated into elastic ε^e, and plastic ε^p components, giving for the strain rate tensor:

$$\dot{\varepsilon}_{ij} = \dot{\varepsilon}^e_{ij} + \dot{\varepsilon}^p_{ij} \qquad \dot{\varepsilon}^p_{ii} = 0 \qquad \dot{\varepsilon}^p_{ij} = \beta s_{ij} \tag{43}$$

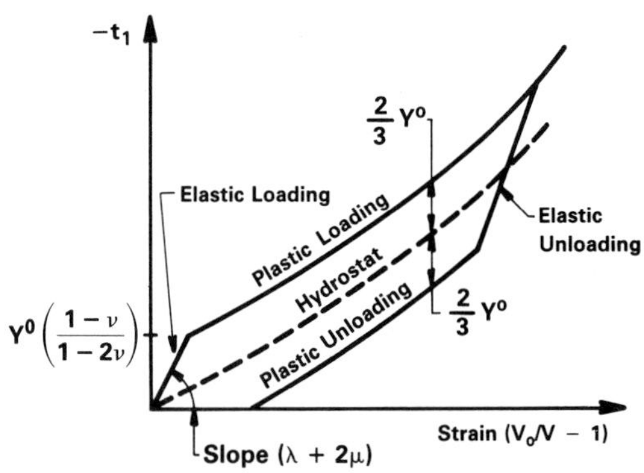

c. Combination of a. and b.

Fig. 13. Stress-strain relations in one dimension for a material with strength.

The second equation, Eq. (43b), is a statement of incompressibility of the plastic part of the strain rate. The third equation (45c) states the assumption that the plastic strain rate is proportional to the stress deviator (Wilkins, 1964):

An equivalent stress σ_{eq} can be defined from the second invariant of the deviatoric stress tensor:

$$\sigma_{eq} = (3\ s_{ij}s_{ij}/2)^{\frac{1}{2}} \qquad \sigma_{eq} \le Y^o \qquad (44)$$

Equation (44b), the von Mises yield criterion, states that the equivalent stress must be less than or equal to the yield stress. In other words, for complex stress states, the von Mises yield criterion provides the determination if a material has flow plastically.

For computational purposes, it is first assumed that $\beta = 0$ in Eq. (43c). The stress deviators are computed from integrating Eqs. (42), and then σ_{eq} is compared to $Y°$. If σ_{eq} is less than $Y°$, the material is elastic; if greater than $Y°$, the material has flowed plastically and the stresses must be allowed to relax back to the yield surface. This is done by redefining $\beta = Y^o/\sqrt{3s_{ij}s_{ij}/2}$. By adjusting the stress deviators in this manner, perpendicular to the yield surface, only the plastic component of the stresses are affected (Wilkins, 1964). Once the stress deviators are adjusted, the material state point will lie on the yield surface.

General Constitutive Equations in Hydrocodes: The original formulation of strength effects considered that materials were elastic, perfectly plastic. But the elastic-plastic constitutive formulation summarized in the last section is sufficiently general to permit work hardening and other effects which are very important for many materials.

Most hydrocodes have various material options. The first several options are variations of an elastic, perfectly plastic constitutive equation:

- o If the yield stress $Y°$ is set to zero, the material is treated hydrodynamically.
- o If $Y°$ is taken as a non-zero positive constant, then the material is treated as elastic, perfectly plastic, undergoing plastic deformation whenever $\sigma_{eq} \ge Y^o$.
- o If $Y°$ is taken as infinity (or a very large number), the material is treated as perfectly elastic since it can never yield.

Note that three different material responses are obtained by modifying "Y°". In general, this is how most material strength effects can be modeled. For example, the effects of thermal softening sometimes are approximated by

$$Y = Y^o\ (1 + Y_1\ \eta)\ (1 - \frac{E}{Y_2}) \qquad (45)$$

where Y_1 gives an increase in the yield strength with compression and a decrease with expansion ($\eta = \rho/\rho_o - 1$), and the Y_2 term accounts for the decrease in the flow stress with increasing internal energy, with Y_2 generally taken to be the melt or sublimation energy.

Strain hardening, strain rate dependence, and temperature effects can all be incorporated into the codes by modifying the flow stress. In visco-plastic material response, the state of the material may exceed its equilibrium value due to strain-rate effects, and the stress will relax back to the yield surface (see the discussion on viscous effects in the section on shock viscosity). Strain-rate dependent stress relaxation is treated similarly to other effects by modifying the flow stress, except that the stress relaxation process requires subcycling (Herrmann, Lawrence and Mason, 1970).

Discussions of these various types of constitutive relations are given by Swegle (1978), Kipp and Lawrence (1982), and Johnson and Cook (1983). Anisotropic effects also have been considered, such as the Bauschinger effect (Kipp and Lawrence, 1982) and orthotropic materials (Johnson, Vavrick and Colby, 1980). However, there are certain limitations on constitutive models for the hydrocodes. Johnson and Cook (1983) state:

> It is recognized that more complicated models may indeed give more accurate descriptions of material behavior. Similarly, various models may give better descriptions for various materials. In many instances, however, the computational user cannot readily incorporate complicated and diverse models. The result is that a constant "dynamic flow stress" is often used.

Certainly the model must use parameters computed by the hydrocode. Davison (1984) discusses that a continuum description of phenomena must be consistent with macroscopic thermodynamic principles, and reflect the microscope basis of the phenomena to be represented. He discusses the role of internal state variables in constitutive modeling. Constitutive modeling is discussed further in the review paper by Asay and Kerley (1986), and Steinberg (1986).

Failure Modeling

Material failure modeling, within the context of the hydrodynamic codes, has evolved over the years. In recent years, it has reached levels of sophistication and complexity to predict accurately certain specific failure modes. It was recognized very early in code development that while materials can and do support high pressures in compression, it is physically untenable for materials to support large (negative) pressures or stresses in tension. The very first criterion applied in the hydrocodes was a tensile pressure or stress cutoff which "clamped" the material pressure or stress at some physically reasonable, but maximum negative value, generally referred to in the codes as σ_{min} or P_{min}. The material retained no knowledge of former "fracturing"; that is, the material was still treated as a continuum and all elements or grid points representing the material still could support tensile and shear stresses. The next level of sophistication was to set the stress to zero in a computational zone that reached a prescribed tensile fracture stress. Memory of the fracture was retained and the computational zone could no longer support tensile or shear stresses. This procedure has been applied successfully to predict and model spall (Mescall and Papirno, 1974; Shockey, Curran and De Carli, 1975; Bertholf, et al., 1975). In some Eulerian codes such as CSQII (Thompson, 1979), interior failure in tension introduces a void which results in the contraction of the surrounding matter which relieves tensions. The void can grow in size in subsequent cycles until the zone no longer exceeds the failure criterion. For computational purposes, these voids are treated as an additional "material." Voids can grow to neighboring zones and subsequently look like cracks which can coalesce and form spall (Thompson, 1979).

A variety of instantaneous "critical value" variables has been used to model instantaneous material failure, including a fracture stress, maximum principal stress, volumetric strain, maximum principal strain, shear strain, equivalent effective strain (second invariant of the strain deviator), and total plastic work. The success of each of these failure criteria, and why one might be more suitable than another, generally depends upon the material being modeled and the type of failure to be expected.

But damage typically grows with time and applied stress, and except when the intensity of the applied load is so severe that damage occurs in negligibly short time, it is unreasonable to expect satisfactory results from an instantaneous failure criteria. Zukas (1980) summarizes experimental observations that lead to the following conclusions about general features of material failure:

- o A range of damage is possible;
- o Damage grows as a function of time and the applied stress;
- o As the level of damage increases, the material is weakened and its stiffness reduced.

More complicated failure criteria have been used to account for progressive development of failures. For example, one of the earliest time-dependent models was due to Tuler and Butcher (1968) to account for the observed dependence of spall failure on stress pulse duration. Failure is assumed to occur instantaneously when a critical value of the damage, K, defined by the time integral is reached:

$$K = \int_0^t (\sigma - \sigma_o)^\lambda \, dt \qquad (46)$$

where $\sigma(t)$ is the tensile stress pulse of arbitrary shape, and σ_o is a threshold stress level below which no significant damage will occur regardless of stress duration. The material constant, λ, is used to fit experimental data.

A more sophisticated fracture model was used by Bertholf, et al., (1977, 1979) to model long rod penetration into spaced armor. Briefly, the model is a biaxial fracture-stress model where the threshold plastic work, W_p, and threshold tensile stresses must be exceeded before a crack will form normal to the maximum tensile stress. The angle of fracture is retained and the normal stress decays to zero over a few computational time cycles. The orthogonal direction can also fracture.

Note that Eq. (46), when differentiated, implies that the rate of damage accumulation is a function of stress:

$$\dot{K} = (\sigma - \sigma_o)^\lambda \qquad (47)$$

Criteria in which the damage accumulation is also a function of the current extent of damage, strain, and temperature have been devised by Davison, and co-workers (1973, 1977)

$$\dot{K} = f(K, \varepsilon, T) \qquad (48)$$

to describe more accurately the response of an elastic-viscoplastic deformation with spall damage accumulation. The work of Davison and Stevens (1973) presents a theory of spallation for the case where damage takes the form of small, penny-shaped cracks, such as those which occur in relatively brittle materials. Later work (Davison, et al., 1977, 1978, and Kipp and Davison, 1981) considered the case in which the material is very ductile and spall damage takes the form of a diffuse distribution of small spherical voids. These material descriptions have not been used extensively, but are suitable for incorporation into the numerical hydrocodes. Johnson (1982, 1985) has performed recent work with EPIC where an accumulative damage model for fracture takes into account strain, strain rate, temperature, and a dimensionless pressure which is the pressure divided by the yield stress (the yield stress is a function of strain, strain rate, and temperature).

Micromechanical behavior has been incorporated in a continuum damage model by researchers at SRI International which attempts to account explicitly for the nucleation and growth (NAG) of voids, cracks, and shear bands in ductile, brittle, and shear failure, respectively (Seaman, Curran and Shockey, 1976; Hageman and Herrmann, 1978; Erlich, Seaman, Shockey and Curran, 1980). Ductile failure damage is initiated when the average stress exceeds a tensile-pressure criterion. Brittle failure is initiated when the maximum normal stress exceeds a tensile threshold. Shear banding begins when the plastic shear strain exceeds a critical value. After initiation, voids, cracks, or shear bands nucleate and grow according to experimentally determined rate equations. As damage accumulates, the stresses are relaxed by amounts and in direction governed by the damage distribution functions (the damage distribution functions, e.g., crack, void or shear band size and orientation are treated as internal state variables). The ductile and brittle fragmentation (FRAG) and shear banding (SNAG) have been used to predict fragment size distributions. However, as shown by the work of Mock and Holt (1982, 1983), considerable time and effort are required to determine the dynamic fracture parameters required by the models. Indeed, except for very few materials, the NAG/FRAG/SNAG models are not used because the dynamic fracture parameters are not known, and to obtain them requires a major research effort.

Lagrangian Hydrocode Modeling of Deep Penetration: Over the last several years, considerable effort has been focused on improving the ability of Lagrangian hydrocodes to compute the entire history of a penetrating projectile. Generally, the realm of penetration and perforation has been studied with Eulerian codes because of the severe grid distortion of Lagrangian calculations. But the advantages inherent in a Lagrangian code, as discussed earlier, has provided the impetus for finding ways to perform penetration calculations with a Lagrangian code. The first attempt to extend the time solution of an impact event with Lagrangian hydrocodes utilized rezoning procedures. This approach has been successfully applied to impact problems (Bertholf, et al., 1977, 1979), but can be manhour and computer intensive.

The first approach to allow deep penetration into thick targets using a Lagrangian code by modifying the slideline algorithms was performed by G. Johnson using EPIC-3 (Johnson, et al.; 1980; Johnson, 1982). The target material was allowed to fail completely and the master surface redefined to include the next layer of grid points. The failure had to be along a predetermined region. A number of approximations and assumptions had been made which were not particularly satisfying; thus, it was not used extensively.

In a very real sense, the ability to model deep penetration is dependent on physically realistic failure modeling. In EPIC (2 or 3), material fracture has been handled by failing elements in one of two ways. Partial failure occurs when an element is predicted to fracture by a cumulative damage model (Johnson and Cook, 1985). When this type of failure occurs, no tensile or shear stresses are allowed to develop in the failed element, but the element can support hydrostatic compression. A second option for treating failure is to set all stresses (including the pressure) to zero. This is called total failure (or erosion). The element essentially dissappears except mass is retained at the nodes (Johnson and Stryk, 1986). Normally, sliding surfaces are used at the projectile/target interface. Techniques have been developed to handle dynamic relocation of sliding surfaces to model two types of failure which can occur with penetration--adiabatic shear, and erosion--depending upon the velocity of the projectile and material strengths. Each will be discussed below.

Adiabatic Shear: During a penetration event, as target material is being "pushed" by the projectile, the work of plastic deformation is converted almost entirely into heat. Because of the high deformation rates, this heat is unable to conduct away from the plastic zone, resulting in an increase in temperature in the plastic zone. This leads to additional local plastic flow, concentrating the local plastic strain even further (Trucano and Grady, 1984). It is usually envisioned that plastic shear instabilities are initiated at sites of stress concentration in an otherwise uniformly straining solid (Burns, et al., 1981). The result is the propagation of a narrow band of intense plastic strain through the material along planes of maximum shear stress or minimum strength until unloading occurs or the material fractures. Called adiabatic shear, this mechanism is present in most impact scenarios, though other physical processes can dominate. For striking velocities near the ballistic limit (perforation) velocity of a "thick" target, an intact plug will result (Wilkins, 1978; Zukas, 1980).

Ringers (1983) at BRL has modified EPIC-2 (Johnson, 1978b) to allow for deep penetration when the failure mode is target plugging. Essentially, a new node is generated by splitting a node, and the master slide surface is effectively redefined. This redefinition of the slide surface is referred to as dynamic relocation. Certain criteria establish whether a node can split. Considerable work was required to establish the criteria for the splitting of a node (which must be on the master surface), and the direction of the split. A number of approaches were found to be intractable; at the same time, a definite attempt was made to keep the criteria physical and reasonable. The criterion adopted was that a split would occur when the equivalent strain reached a specified critical level and the shear stress was greater than the axial or radial deviator stresses. The direction of the split was determined by the strain of the element that had met the criteria for splitting. If the magnitude of the axial strain was greater than the magnitude of the radial strain, splitting was to the nearer radial node; otherwise, splitting was to the nearer axial node.

To account for "sliding" between the two sides of the split which develops in the target, another set of master and slave surfaces was established between the developing plug and remaining target material. This second master/slave surface allows for the plug material elements to "slide" past the remaining target material--this is readily visible in Fig. 14.

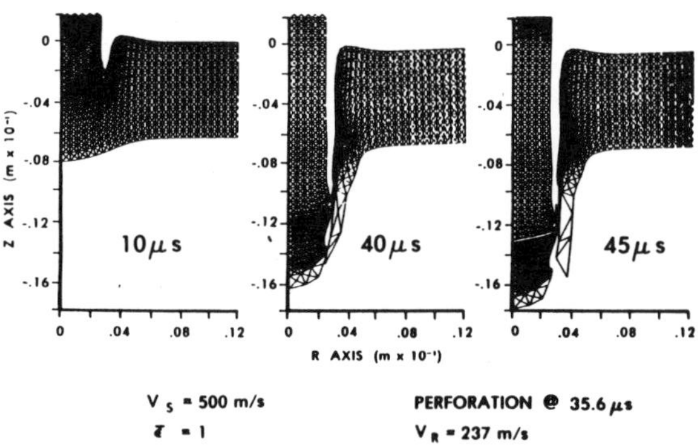

Fig. 14. Deep Penetration by a Lagrangian calculation--adiabatic shear (from Ringers, 1983).

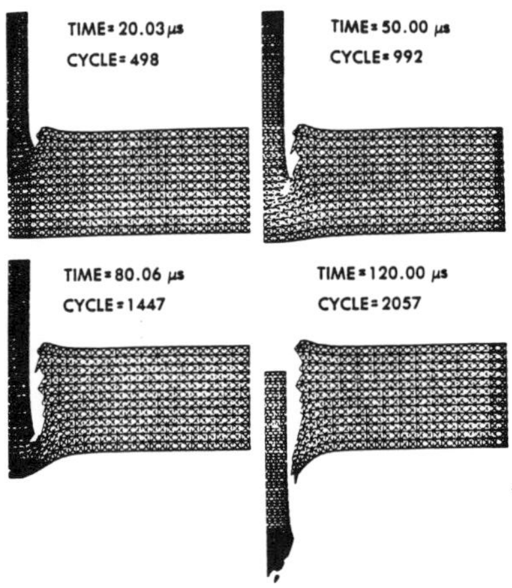

Fig. 15. Plate perforation by a lagrangian calculation--erosion (courtesy of Kimsey and Zukas).

<u>Erosion</u>: The failure mode is significantly different for hypervelocity impact where the impact pressures substantially exceed material strength. In hypervelocity impact, the principal factor in characterizing material behavior is the high pressure equation of state. Material response is characterized by large strains and total failure generally referred to as erosion (both for the target and the penetrator). The resulting material flows are similar to colliding water jets, hence the term "hydrodynamic."

The fundamental difference, from the computational viewpoint of modeling failure by erosion as opposed to adiabatic shear, is the treatment of the slidelines. In the plugging mode, the slideline is redefined to follow the failure surface, e.g., the separating of elements by the splitting of nodes. However, with erosion, the sliding surfaces are redefined in the presence of total element failure. These two approaches tend to mimic the physical processes one might heuristically expect for the two failure modes.

Kimsey and Zukas (1986) developed an algorithm for EPIC-2 to allow redefinition of the master surface and a symmetric treatment of the master-slave interface. The procedure is based on discrete master segments, removing the criteria for a continuous master surface. Failure of both the target and penetrator is modeled. Figure 15 shows the results obtained for penetration of a steel plate by a 65 gram, hemispherically nosed steel rod with a striking velocity of 1103 m/s. A plot of the number of cycles versus time (microseconds) is approximately linear, implying that the computational time step is essentially a constant. Kimsey and Zukas also conducted the numerical simulation of a long rod penetrator impacting a semi-infinite target at 3.1 km/s. They obtained excellent agreement with experiment (Kimsey and Zukas, 1986).

Belytschko and Lin (1985, 1986) used a different approach for erosion using EPIC-3. Their algorithm requires no definition or tracking of sliding interfaces. The interaction is handled by operations on slave nodes and master elements. Hexahedral elements were introduced to simplify the interaction algorithm, and hourglass or keystone viscosity was required to stop spurious deformation modes caused by one point-integration.

Strecher and Johnson (1984) modified EPIC-2 to have an eroding interface. The normal impact of a 5 km/s tungsten projectile into a semi-infinite steel target is shown in Figure 16. Johnson and Styrk (1986) also have modified EPIC-3 to account for total material failure in the projectile and target. Their algorithm redefines the sliding interface.

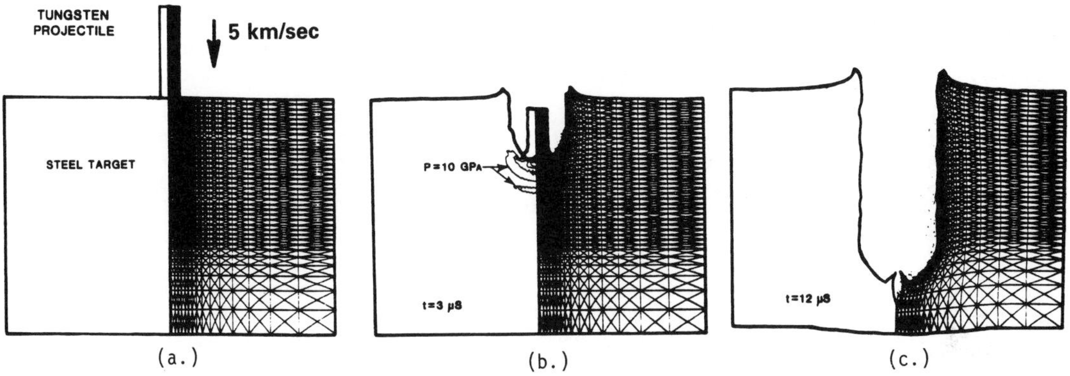

Fig. 16. Deep penetration into a semi-infinite target--erosion (courtesy of G. Johnson).

Summary of Deep Penetration Modeling: In conclusion, note two important considerations with respect to these two failures, shear failure and erosion. First, the dominant mode of failure has been assumed. Work to date has been to understand and realistically model a specific failure mechanism; the transition from one failure mode to another is generally beyond current predictive capabilities. Second, adiabatic shear failure is more complex to model, requiring failure criteria (some reasonable criteria did not work), a procedure for extrapolating the direction of failure, and separation of material along the failure plane. Erosion is characterized by a much simpler failure mechanism, consistent with the previous subsection discussion that when damage levels are high, failure is easier to characterize. However, though one type of failure may be more complex to model, implementation of the failure model into the computer program can be as difficult or even more difficult for the simpler model, particularly where element or node removal may be desired, or relocation of slidelines is required. Certainly implementation of either of these failure models required a substantial effort and detailed knowledge of the computer program logic.

Summary of Material Failure Modeling

In summary, the most used models to predict material failure in hydrocode simulations require some critical stress, strain, plastic work or other instantaneous criteria, mostly because some appropriate critical level can be estimated from known material properties. As long as the conditions for failure are far above the threshold, damage occurs in negligible times and these single-criterion models are generally adequate. However, near threshold, a time-dependent description of failure growth is required (with a threshold below which no damage occurs); these models realistically predict increasing damage rates as the threshold is exceeded. If the threshold is exceeded by a sufficiently large quantity, the damage rate saturates and the model effectively can be replaced by a single criterion, sudden damage model. Realistic numerical modeling of failure near thresholds also requires a fine grid mesh, and attendant smaller time step, to resolve accurately the physical state of the material such as the stresses (e.g., reference Bertholf, et al. (1975) who found that a high degree of numerical resolution was required to predict the back-surface spallation observed in an impact experiment--coarse zoning resulted in a solution which predicted no back-surface spallation). Thus, the numerical analyst must exercise judgement to obtain an acceptable answer. The detail necessary, and the complexity of the failure model, is failure-mode dependent. Also, the numerical analyst must foresee the dominate failure mode expected since state of the art in failure modeling *a priori* assumes a failure mode and then tries to predict that failure mode from basic principles. These last two points are vividly demonstrated in the last section where two different failure mechanisms encountered in penetration and perforation are modeled explicitly in a hydrocode.

SUMMARY

Hydrocodes are large computer programs used to simulate numerically highly dynamic events, particularly those involving shocks, by solving the conservation equations coupled with material descriptions. The equations are discretized, using either the finite difference or finite element technique. The conservation equations can be written in one of two mathematical descriptions, Eulerian (spatial) or Lagrangian (material), and each of these descriptions has advantages and disadvantages. In particular, Lagrangian codes provide good spatial resolution relative to the number of grid cells needed to describe the problem; free surfaces and material interfaces are well defined. Finally, because they are written in material coordinates, Lagrangian codes can provide excellent material constitutive modeling. However, severe distortion of the grid sometimes limits the usefulness of the Lagrangian codes to early times. Eulerian codes, because of their fixed coordinate systems, can handle problems where large distortions take place, at the expense of the attractive features of Lagrangian codes (grid economy, interfaces, material descriptions).

The formulation of the numerical algorithms to solve the conservation equations require specific treatments to ensure numerical stability (a limiting integration time step), and unwanted large oscillations at shock discontinuities (artificial viscosity). A trade-off in efficiency versus accuracy has led to the codes as they are today, and also has led to the development of specific algorithms to take advantage of Eulerian-like and Lagrangian-like attributes (Johnson and Anderson, 1986).

The hydrocode codes have been very useful in understanding and predicting transient phenomena which occur on the order of microseconds and tens of microseconds. Two-dimensional codes have been in existence since the early-to-mid 60's and their sophistication has continued to grow during the 70's and early 80's. The advent of the larger (more memory) and faster computers have allowed three-dimensional versions of the codes to be formulated and solutions obtained (Johnson and Anderson, 1986). Since the development of a specific code usually has grown out of a particular need, some of the codes have enhanced features in certain areas; for example, the coupling of radiation effects in the codes developed by Sandia Laboratories. Zukas (1980) has described the major physical weakness of the codes to be their material descriptions. It is true that the numerical simulation can only be as accurate as the material models. Traditional material parameters consisted of static or slow strain-rate parameters measured in the laboratory. However, an active research area has been the measurement and understanding of high strain-rate phenomena, and our understanding of material response has improved considerably. But obviously, there are still challenges.

In recent years, considerable efforts have been spent on post-processing, particularly in the graphical display of data. The output from the hydrocodes can be voluminous, and the only hope for assimilation is by an appropriate graphical display of the data. Indeed, these graphic routines can become "monster" programs themselves. An undesirable outgrowth of improved pre- and post-processing packages is that many people who do not (or do not want to) understand the "physics" can now use the codes. This is leading to an emphasis on impressive graphics, movies, etc., opposed to analysis and understanding of the numerical results.

Hydrocodes can be very powerful tools in research if one recognizes their limitations and understands their operations. They can give detailed understanding of physical processes and can be used to perform analytical experiments. These computational experiments can be cheaper than laboratory experiments, and, in some cases, because the actual experiment is extremely difficult or impossible to perform, offer the only viable way to obtain information.

REFERENCES

_____________ (1980). "Materials Response to Ultra-High Loading Rates," U.S. National Materials Advisory Board, Washington, DC.

Asay, J. R., and G. I. Kerley (1986). "The Response of Materials to Dynamic Loading," International Journal of Impact Engineering, Vol. 5.

Becker, R. (1922). "Stosswelle and Detonation," Z. Physik, Vol. 8, p. 321.

Belytschko, T. (1974). "Finite Element Approach to Hydrodynamics and Mesh Stabilization," in Computational Methods in Nonlinear Mechanics, edited by J. T. Oden et al., Texas Institute for Computational Mechanics, pp. 231-238.

Belytschko, T. (1975). "Nonlinear Analysis--Descriptions and Numerical Stability," in Computer Programs in Shock and Vibration, edited by W. and B. Pilkey, Shock and Vibration Information Center, Washington, DC., pp. 537-562.

Belytschko, T. (1976). "Computer Methods in Shock and Wave Propagation Analysis," in Computing in Applied Mechanics, AMD-Vol. 18, American Society of Mechanical Engineers, pp. 139-161.

Belytschko, T., R. L. Chiapetta, and H. D. Bartel (1976). "Efficient Large Scale Nonlinear Transient Analysis by Finite Elements," International Journal for Numerical Methods in Engineering, Vol. 10, pp. 579-596.
Belytschko, T., and J. I. Lin (1985). " A New Interaction Algorithm with Erosion for EPIC-3," BRL-CR-540, USA Ballistic Research Laboratory, Aberdeen Proving Ground, MD.
Belytschko, T., and J. I. Lin (1986). "A Three Dimensional Impact-Penetration Algorithm with Erosion," International Journal of Impact Engineering, Vol. 5.
Belytschko, T., W. K. Liu, J. M. Kennedy and S. J. Ong (1984). "Hourglass Control in Linear and Nonlinear Problems," Computer Methods in Applied Mechanics and Engineering, Vol. 43, pp. 251-276.
Bertholf, L. D., L. D. Buxton, B. J. Thorne, R. K. Byers, A. L. Stevens, and S. L. Thompson (1975). "Damage in Steel Plates from Hypervelocity Impact II. Numerical Results and Spall Measurement," Journal of Applied Physics, Vol. 46, No. 9, pp. 3776-3783.
Bertholf, L. D., M. E. Kipp, and W. T. Brown (1977). "Two-Dimensional Calculations for the Oblique Impact of Kinetic Energy Projectiles with a Multi-Layered Target," BRL-CR-333, USA Ballistic Research Laboratory, Aberdeen Proving Ground, Maryland.
Bertholf, L. D., M. E. Kipp, W. T. Brown, and A. L. Stevens (1979). "Kinetic Energy Projectile Impact on Multi-Layered Targets: Two-Dimensional Stress Wave Calculations," ARBRL-CR-00391, USA Ballistic Research Laboratory, Aberdeen Proving Ground, Maryland.
Burns, T. J., D. E. Grady, and L. S. Costin (1981). "On a Criterion for Thermo-Plastic Shear Instability," in Shock Waves in Condensed Matter--1981, edited by W. J. Nesslis, L. Seaman and R. A. Graham, American Institute of Physics, pp. 372-375.
Carroll, M. M. and A. C. Holt (1972). "Suggested Modifications of the P-alpha Model for Porous Materials," Journal of Applied Physics, Vol. 43, No. 2, pp. 759-761.
Carroll, M. M., and A. C. Holt (1972). "Static and Dynamic Pore Collapse Relations for Ductile Porous Materials," Journal of Applied Physics, Vol. 43, No. 4, pp. 1626-1636.
Chiapetta, R., T. Belytschko, and J. Rouse (1973). "A Computer Code for Dynamic Stress Analysis of Media-Structure Problem with Nonlinearities (SAMSON)," Report AFWL-TR-72-104, Air Force Weapons Laboratory, Kirtland AFB, NM.
Courant, R., K. O. Friedrichs, and H. Lewy (1928). "Uber die partiellen Differenzengleichungen der mathematischen Physik," Mathematische Annalen, Vol. 100, pp. 32-74.
Davison, L. (1984). "Numerical Modeling of Dynamic Material Response," in Shock Waves in Condensed Matter--1983, edited by J. R. Asay, R. A. Graham, and G. K. Straub, Elsevier Science Publishers, pp. 181-186.
Davison, L. W., and R. A. Graham (1979). "Shock Compression of Solids," Physics Reports, Vol. 55, No. 4, pp. 257-379.
Davison, L., and M. E. Kipp (1978). "Calculation of Spall Damage in Ductile Metals," in High Velocity Deformation of Solids, K. Kawata and J. Shioiri, eds., Springer-Verlag, West Berlin, Germany.
Davison, L. W., and A. L. Stevens (1973). "Thermomechanical Constitution of Spalling Elastic Bodies," Journal of Applied Physics, Vol. 44, No. 2, pp. 668-674.
Davison, L. W., A. L. Stevens, and M. E. Kipp (1977). "Theory of Spall Damage Accumulation in Ductile Metals," J. Mech. Phys. Solids, Vol. 25, pp. 11-28.
Drumheller, D. S. (1986). "Hypervelocity Impact of Mixtures," International Journal of Impact Engineering, Vol. 5.
Emery, A. F. (1967). "An Evaluation of Several Differencing Methods for Inviscid Fluid-Flow Problems," SCL-DC-66-78, Sandia National Laboratories, Albuquerque, NM.
Erlich, D. C., L. Seaman, D. A. Shockey, and D. R. Curran (1980). "Development and Application of a Computational Shear Band Model," ARBRL-CR-00416, USA Ballistic Research Laboratory, Aberdeen Proving Ground, Maryland.
Goudreau, G. L. and J. O. Halquist (1982). "Recent Developments in Large-Scale Finite Element Lagrangian Hydrocode Technology," Computer Methods in Applied Mechanics and Engineering, Vol. 33, pp. 725-757.
Hageman, L. J., and R. G. Herrmann (1978). "Incorporation of the NAG-FRAG Model for Ductile and Brittle Fracture into HELP, a 2D Multimaterial Eulerian Program," ARBRL-CR-00380, USA Ballistic Research Laboratory, Aberdeen Proving Ground, Maryland.
Halquist, J. O. (1976). "A Procedure for the Solution of Finite-Deformation Contact-Impact Problems by the Finite Element Method," Report UCRL-52066, Lawrence Livermore National Laboratory, Livermore, CA.
Halquist, J. O. (1976b). "Preliminary User's Manual for DYNA3D and DYNAP--Nonlinear Dynamic Analysis of Solids in Three Dimensions," UCID-17268, Lawrence Livermore National Laboratory, Livermore, CA.
Halquist, J. O. (1978). "DYNA2D--An Explicit Finite Element and Finite Difference Code for Axisymmetric and Plane Strain Calculations (User's Guide)," UCRL-52429, Lawrence Livermore National Laboratory, Livermore, CA.
Herrmann, W., (1969). "Constitutive Equation for the Dynamic Compaction of Ductile Porous Materials," Journal of Applied Physics, Vol. 40, No. 6, pp. 2490-2499.

Herrmann, W. (1975). "Nonlinear Transient Response of Solids," in Shock and Vibration Computer Programs, Reviews and Summaries, edited by W. and B. Pilkey, Shock and Vibration Information Center, Washington, D.C, pp. 151-172.

Herrmann, W. (1977). "Current Problems in the Finite Difference Solution of Stress Waves," in Proceedings of the Workshop on Nonlinear Waves in Solids, edited by T.C.T. Ting, R. J. Clifton and T. Belytschko, sponsored by US Army Research Office and National Science Foundation, pp. 31-52.

Herrmann, W., R. J. Lawrence, and D. S. Mason (1970). "Strain Hardening and Strain Rate in One-Dimensional Wave Propagation Calculations," SC-RR-70-471, Sandia National Laboratories, Albuquerque, NM.

Hildebrand, F. B., (1962). Advanced Calculus for Applications, Prentice-Hall, Inc.

Holian, K. S., and M. W. Burkett (1986). "Sensitivity of Hypervelocity Impact Simulations to Equation of State," International Journal of Impact Engineering, Vol. 5.

Johnson, G. R. (1976). "Analysis of Elastic-Plastic Impact Involving Severe Distortions," Journal of Applied Mechanics, Vol. 43, No. 3, pp. 439-444.

Johnson, G. R. (1977). "A New Computational Technique for Intense Impulsive Loads," Proceedings of the Third International Symposium on Ballistics, Karlsruhe, Germany.

Johnson, G. R. (1977b). "High Velocity Impact Calculations in Three Dimensions," Journal of Applied Mechanics, Vol. 99, No. 1, pp. 95-100.

Johnson, G. R. (1977c). "EPIC-3, A Computer Program for Elastic-Plastic Impact Calculations in 3 Dimensions," Contract Report No. 343, USA Ballistic Research Laboratory, Aberdeen Proving Ground, MD.

Johnson, G. R. (1978b). "EPIC-2, A Computer Program for Elastic-Plastic Impact Calculations in 2 Dimensions Plus Spin," ARBRL-CR-00373, US Army Ballistic Research Laboratory, Aberdeen Proving Ground, MD.

Johnson, G. R. (1982). "Status of the EPIC Codes, Material Characterization and New Computing Concepts at Honeywell," in Computational Aspects of Penetration Mechanics, Springer-Verlag.

Johnson, G. R., D. D. Colby, and D. J. Vavrick (1979). "Three Dimensional Computer Code for Dynamic Response of Solids to Intense Impulsive Loads," International Journal for Numerical Methods in Engineering, Vol. 14, pp. 1865-1871.

Johnson, G. R., and W. H. Cook (1983), "A Constitutive Model and Data for Metals Subjected to Large Strains, High Strain Rates and High Temperatures," Proceedings of the Seventh International Symposium on Ballistics.

Johnson, G. R., and W. H. Cook (1985). "Fracture Characteristics of Three Metals Subjected to Various Strains, Strain Rates, Temperatures and Pressures," Engineering Fracture Mechanics, Vol. 21, No. 1, pp. 31-48.

Johnson, G. R., and R. A. Stryk (1986). "Eroding Interface and Improved Tetrahedral Element Algorithms for High Velocity Impact Computations in Three Dimensions," International Journal of Impact Engineering, Vol. 5.

Johnson, G. R., D. J. Vavrick, and D. D. Colby (1980). "Further Development of EPIC-3 for Anisotropy, Sliding Surfaces, Plotting and Material Models," Contract Report ARBRL-CR-00429, USA Ballistic Research Laboratory, Aberdeen Proving Ground, Maryland.

Johnson, W. E. (1986). Personal communication.

Johnson, W. E. and C. E. Anderson (1986). "History and Application of Hydrocodes to Hypervelocity Impact," International Journal of Impact Engineering, Vol. 5.

Jonas, G. H., and J. A. Zukas (1978). "Mechanics of Penetration: Analysis and Experiment," International Journal of Engineering Science, Vol. 16, No. 11, pp. 879-904.

Karpp, R. R. (1972). "The Method of Characteristics," in Dynamic Response of Material to Intense Impulsive Loading, P. C. Chou and A. K. Hopkins, eds., Air Force Materials Laboratory, Ohio.

Kerley, G. I. (1986). "Theoretical Equation of State for Aluminum," International Journal of Impact Engineering, Vol. 5.

Key, S. W. (1974). "HONDO - A Finite Element Computer Program for the Large Deformation Dynamic Response of Axisymmetric Solids," SLA-74-0039, Sandia National Laboratories, Albuquerque, NM.

Key, S. W. (1974b). "A Finite Element Procedure for Large Deformation Dynamic Response of Axisymmetric Solids," Computer Methods of Applied Mechanics and Engineering, Vol. 4, pp. 195-218.

Key, S. W., Z. E. Beisinger and R. D. Krieg (1978). "HONDO II - A Finite Element Computer Program for Large Deformation Dynamic Response of Axisymmetric Solids," SLA-78-0422, Sandia National Laboratories, Albuquerque, NM.

Kimsey, K. D., and J. A. Zukas (1986). "Contact Surface Erosion for Hypervelocity Problems," BRL-MR-3495, USA Ballistic Research Laboratory, Aberdeen Proving Ground, MD.

Kipp, M. E., and L. Davison (1981), "Analysis of Ductile Flow and Fracture in Two Dimensions," SAND 81-0070C, Sandia National Laboratories, Albuquerque, New Mexico.

Kipp, M. E., and R. J. Lawrence (1982). "WONDY V--A One-Dimensional Finite-Difference Wave Propagation Code," SAND 81-0930, Sandia National Laboratories, Albuquerque, New Mexico

Lax, P. D. (1954). "Weak Solutions of Nonlinear Hyperbolic Equations and Their Numerical Computation," Communications on Pure and Applied Mathematics, Vol. 7, pp. 159-193.

Lax, P. D., and R. D. Richtmyer (1956). "Survey of the Stability on Linear Finite Difference Equations," Communications on Pure and Applied Mathematics, Vol. 9, pp. 267-293.

Lax, P. D., and B. Wendroff (1960). "Systems of Conservation Laws," Communications on Pure and Applied Mathematics, Vol. 13, pp. 217-237.

McQueen, R. G., S. P. Marsh, J. W. Taylor, and J. N. Fritz (1970). "The Equation of State of Solids from Shock Wave Studies," in High Velocity Impact, edited by R. Kinslow, Academic Pr.

Mescall, J. F. (1974). "Shock Wave Propagation in Solids," in Structural Mechanics Computer Programs, W. Pikley, S. Saczalski and H. Schaeffer, editors, University of Virginia Press, VA.

Mescall, J. F. and Papirno, R. P., (1974). "Spallation in Cylinder-Plate Impact," Experimental Mechanics, Vol. 14, No. 7, p. 257-266.

Mock, W. Jr., and W. H. Holt (1982). "Determination of Dynamic Fracture Parameters for HF-1 Steel," Journal of Applied Physics, Vol. 53, No. 8, pp. 5660-5668.

Mock, W. Jr., and W. H. Holt (1983), "Fragmentation Behavior of Armo Iron and HF-1 Steel Explosive-Filled Cylinders," Journal of Applied Physics, Vol. 54, No. 5, pp. 2344-2351.

Potter, D. (1973). Computational Physics, John Wiley & Sons.

Rice, M. H., R. G. McQueen, and Walsh (1958). "Compression of Solids by Strong Shock Waves," in Solid State Physics, Vol. 6, edited by Seitz and Turnball, Academic Press.

Richtmyer, R. D., and K. W. Morton (1967). Difference Methods for Initial-Value Problems, 2nd edition, Interscience Publishers, New York.

Ringers, B. E. (1983). "New Sliding Surface Techniques Enable the Simulation of Target Plugging Failure," ARBRL-TR-02541, USA Ballistic Research Laboratory, Aberdeen Proving Ground, MD.

Roache, P. J. (1972). Computational Fluid Mechanics, Hermosa Publishers, Albuquerque, NM.

Seaman, L., D. R. Curran, and D. A. Shockey (1976). "Computational Models for Ductile and Brittle Fracture," Journal of Applied Physics, Vol. 47, No. 11, pp. 4814-4826.

Sedgwick, R. T., L. J. Hageman, R. G. Herrmann and J. L. Waddell (1978). "Numerical Investigations in Penetration Mechanics," International Journal of Engineering Science, Vol. 16, No. 11, pp. 859-870.

Shockey, D. A., D. R. Curran, and P. S. De Carli (1975). "Damage in Steel Plates from Hypervelocity Impact. I. Physical Changes and Effects of Projectile Material," Journal of Applied Physics, Vol. 46, No. 9, pp. 3766-3775.

Stecher, F. P. and G. R. Johnson (1984). "Lagrangian Computations for Projectile Penetration into Thick Plates," in Computers in Engineering 1984, edited by W. A. Graver, AMSE, NY.

Steinberg, D. J. (1986). "Constitutive Models Used in Computer Simulation of Time-Resolved, Shock-Wave Data," International Journal of Impact Engineering, Vol. 5.

Swegle, J. W. (1978). "TOODY IV--A Computer Program for Two-Dimensional Wave Propagation," SAND-78-0552, Sandia National Laboratories, Albuquerque, New Mexico.

Swegle, J. W., and D. E. Grady (1985). Shock Viscosity and the Prediction of Shock Wave Rise Times," Journal of Applied Physics, Vol. 58, No. 2, pp. 692-701.

Thompson, S. L. (1979). "CSQII--An Eulerian Finite Difference Program for Two-Dimensional Material Response - Part 1. Material Sections," SAND77-1339, Sandia National Laboratories, Albuquerque, New Mexico.

Thompson, S. L. (1986). Personal communication.

Tillotson, J. H. (1962). "Metallic Equations of State for Hypervelocity Impact," Report GA-3216, General Atomic, San Diego, CA.

Trucano, T. G., and J. R. Asay (1986). "Effects of Vaporization on Debris Cloud Dynamics," International Journal of Impact Engineering, Vol. 5.

Trucano, T. G., and D. E. Grady (1984). "Intermediate Velocity Penetration of Steel Spheres into Deep Aluminum Targets," in Shock Waves in Condensed Matter--1983, edited by J. R. Asay, R. A. Graham, and G. K. Straub, Elsevier Science Publishers, pp. 187-190.

Tuler, F. R., and B. M. Butcher (1968). "A Criterion for the Time Dependence of Dynamic Fracture," International Journal of Fracture Mechanics, Vol. 4, pp. 431-437.

von Neumann, J., and R. D. Richtmyer (1950). "A Method for the Numerical Calculation of Hydrodynamic Shocks," Journal of Applied Physics, Vol. 21, No. 3, pp. 232-237.

Walsh, R. T. (1972). "Finite-Difference Methods," in Dynamic Response of Materials to Intense Impulsive Loading, P. C. Chou and A. K. Hopkins, eds., Air Force Materials Laboratory, Ohio.

Wilkins, M. L. (1964). "Calculations of Elastic-Plastic Flow," in Methods of Computational Physics, Vol. 3, edited by B. Adler, S. Fernback, and M. Rotenberg, Academic Press, New York.

Wilkins, M. L. (1969). "Calculation of Elastic-Plastic Flow," UCRL-7322, Rev. 1, Lawrence Livermore National Laboratory, Livermore, CA.

Wilkins, M. L. (1978). "Mechanics of Penetration and Perforation," International Journal of Engineering Science, Vol. 16, No. 11, pp. 793-808.

Zel'dovich, Y. B., and Ya. P. Raiser (1967). Physics of Shock Waves and High-Temperature Hydrodynamic Phenomena, Vol. II, edited by W. D. Hayes and R. F. Probstein, Academic Press.

Zukas, J. A. (1980). "Impact Dynamics: Theory and Experiment," ARBRL-TR-02271, USA Ballistics Research Laboratory, Aberdeen Proving Ground, MD.

Zukas, J. A., G. H. Jonas, K. D. Kimsey, J. J. Misey, and T. M. Sherrick (1981). "Three-Dimensional Impact Simulations: Resources and Results," in Computer Analysis of Large-Scale Structures, edited by K. C. Parck and R. F. Jones, Jr., AMD, Vol. 49, American Society of Mechanical Engineers.

Zukas, J. A., T. Nicholas, H. F. Swift, L. B. Greszczuk, and D. R. Curran (1982). Impact Dynamics, John Wiley and Sons.

Int. J. Impact Engng Vol. 5, pp. 61–67, 1987
Printed in Great Britain

0734-743X/87 $3.00 + 0.00
Pergamon Journals Ltd.

SCALING FLOW FIELDS FROM THE IMPACT OF THIN PLATES

Salvo E. Arione and Michael D. Bjorkman

The Boeing Aerospace Company
P. O. Box 3999
M/S 87-68
Seattle, WA 98124

ABSTRACT

A dimensional analysis is performed to obtain velocity scaling relationships for the perforation of thin plates. The approach used is an extension of Dienes and Walsh's "late-stage equivalence" and Holsapple and Schmidt's "coupling parameter" concepts, used to simplify velocity scaling of impact phenomena. The coupling parameter C for plate perforation, is shown to have the form $C=dU^{\mu}\delta^{\nu}$ for the perforation of thick plates and the form $C=dU^{\mu}\delta^{\nu} f(t/d)$ for the perforation of thin plates (d is the projectile diameter, t is the plate thickness, U is the impact velocity and δ is the projectile density). It is shown that $\mu=1/2$ for momentum scaling and $\mu=1$ for energy scaling, however, from scaled hydrocode output it is found that, for aluminum impacting aluminum, the value of μ is equal to 0.83 ± 0.03, which is neither energy nor momentum scaling. It is also shown that velocity scaling of thick plate perforation, using the same materials in the model and prototype and the same t/d, is not possible. An example of velocity scaling hydrocode output is given where the radial particle velocity wave profiles from the model calculation at U=55.6 km/s and t/d=0.675 are similar to those from the prototype calculation with U=100 km/s and t/d=1.08.

INTRODUCTION

Design of the space station common modules will require demonstrating the survivability of the design to possible impact by orbital debris and meteoroids. At low earth orbit, orbital debris closing velocities with respect to the space station will vary between 0 and 16 km/s. In addition, meteoroids could impact the space station at velocities up to 70 km/s, with the most probable closing velocity at 20 km/s.

Present laboratory test techniques are incapable of launching projectiles at velocities typical of meteoroids or even the faster orbital debris particles. Current two-stage light gas gun experiments are limited to collisions in the 0 to 7 km/s range while faster techniques, such as electromagnetic and explosive launchers, do not launch well characterized projectiles and thus are not well suited to scaling experiments. Thus, at least for the next few years, velocity scaling of laboratory tests will be necessary for demonstrating the survivability of the space station and other space structures.

The plate perforation velocity scaling methodology developed at Boeing has built upon two sources: the late-stage equivalence concept developed by Dienes and Walsh (1970) and the coupling parameter concept developed by Holsapple and Schmidt (1986). Dienes and Walsh were able to show with hydrocode calculations that the cratering motions produced by an impact approach asymptotic limits in the intermediate times of the flow beyond which the motions are similar for impacts with the same dU^{μ}, where d is projectile diameter and U is projectile velocity. Holsapple and Schmidt have elucidated the effect of target constitutive relation on the coupling parameter dU^{μ} and determined the connection between the intermediate times of the flow at which the coupling of the projectile's momentum and energy is complete with the late times of the flow during which the final crater forms.

In this paper the coupling parameter method is extended to thin plate perforating impacts. A methodology for velocity scaling the response of a plate to perforating impact is developed and demonstrated by velocity scaling hydrocode output.

COUPLING PARAMETER FOR SEMI-INFINITE TARGETS

The cratering process can be divided into two parts: early-times during which the length scale of the motions is comparable to d, and late-times during which the length scale of the far field motions is large relative to d. In some cases the far field solution is equivalent to the point source limit where d goes to zero (a more detailed discussion of this case can be found in Holsapple and Schmidt (1986)). If the point source limit is to produce the same motions in the far field as a nonzero diameter source, then U will have to increase as d decreases to preclude the motions going to zero as d vanishes. Therefore the point source limit exists provided that there exists a function of the parameters of the projectile

$$C = f\,(d, U, \delta) \tag{1}$$

that remains finite and constant as d approaches zero and U approaches infinity while the projectile density δ remains constant. Thus, if the point source solution for the wave motion is known, then Eq. (1) can be used to find all finite source impacts which produce similar motions to the point source solution.

The functional dependence of the coupling parameter on d and U for impacts of semi-infinite targets can be determined by performing a dimensional analysis on Eq. (1). There are four parameters in Eq. (1) and three fundamental units (mass, length and time), thus by the pi theorem of dimensional analysis, there exists a single pi group

$$C = d\,U^{\mu}\,\delta^{\nu} \tag{2}$$

where μ and ν are constants not determined by the dimensional analysis. Holsapple and Schmidt (1982) have shown for the cratering of semi-infinite targets that μ is bounded between 1/3 and 2/3; the momentum and energy scaling limits. Furthermore, Holsapple and Schmidt (1986) have given several examples showing μ for many metals is not 1/3 or 2/3, but a value slightly less than energy scaling, μ=0.58, which is also the value reported by Dienes and Walsh (1970). The constancy of this result for many metals was examined by Bjorkman and Holsapple (1983) with a series of one dimensional hydrocode calculations and was found to be a result of the high pressure limit of the constitutive relation for metals. For porous geologic materials Holsapple and Schmidt (1986) have found that μ is closer to momentum scaling, with a value of μ=0.47 typical of Ottawa sand and they have been able to relate the near momentum scaling result to the porosity of the target material.

COUPLING PARAMETER FOR PERFORATING IMPACTS

The results from the previous section for impacts of semi-infinite targets require modification at two points for application to the plate perforation impact case. First, the coupling parameter velocity exponent has a different range of values for the perforation case, and second, the plate thickness t enters the dimensional analysis so consideration must be given to whether t is a late or an early time variable.

The first modification of the point source limit is necessary when a plate is perforated in a much shorter time than that required for the motions to expand far from the penetration. For this case the energy of the impact is coupled to the target along a roughly cylindrical volume. In the limit of small projectile diameter and large impact velocity the source approaches a line and not a point, which we shall refer to as the "line source limit".

In the line source limit the values of the coupling parameter velocity exponent μ are also restricted between momentum and energy scaling. However, because the source is now a line source and not a point source the limits on μ are now energy and momentum per unit length.

The range of values μ may take in the line source limit can be estimated from a dimensional argument. The momentum per unit length has the same units as $d^2 U\delta$ and the energy per unit length has the same units as $d^2 U^2 \delta$. Taking the square root of the above two dimensional relations yields limits on μ between 1/2 and 1, for momentum and energy scaling, respectively.

The second modification of the point source case is necessary because plate thickness is now a parameter of the problem. When a thick target plate is perforated a significant portion of the early time motions produced in the target plate will be unaffected by the free surfaces of the target. Thus for a thick target, the target plate thickness t will be a late time variable only and will not occur in the coupling parameter and Eq. (2) will apply. However, for thin plate perforation, the early time motions of the target plate will be affected by the target's free surfaces. Thus for a thin target, t will be an early time variable and will occur in the coupling parameter, as well as a late time variable. Thus, for thin plate perforation the coupling parameter will be,

$$C = d\, U^{\mu}\, \delta^{\nu}\quad f(t/d) \tag{3}$$

The above definitions for thin and thick plate targets are not quantitative. The most straightforward method for determining when a target is thick and when it is thin is to perform hydrocode calculations and determine which coupling parameter, Eq. (2) or (3), governs the scaling. With a series of calculations it should prove possible to map out the boundary separating the two types of scaling.

STRONG SHOCK SOLUTIONS

Before hydrocode results can be velocity scaled the value of the coupling parameter velocity exponent μ for the particular constitutive equation of interest must be determined. The simplest method for determining μ follows from scaling the strong shock solutions from hydrocode calculations. The portion of the strong shock wave profile selected for scaling is the radial particle velocity shock amplitude.

The general functional form of the particle velocity field can be written as

$$v = f\,(d, U, \delta, \rho, c, t, r, z, \theta) \tag{4}$$

where d is the projectile diameter, U is the impact velocity, δ is the projectile mass density, ρ is the target mass density, c is the target sound speed, t is the target thickness, r is the radial coordinate, z is the axial coordinate, and θ is the polar angle. (Here, and throughout this paper, the function symbol f is used to denote the relationship between the parameters of the problem. Even though the same symbol f is used for all the relationships in this paper it is not to be construed that the functional relation is the same for all equations.)

Combining Eqs. (1) (or (2)) and (4) gives:

$$v = f\,(C, \rho, c, t, r, z, \theta). \tag{5}$$

In the strong shock limit $v \gg c$, thus c is not a parameter of the problem, so Eq. (5) may be written as

$$v = f\,(C, \rho, t, r, z, \theta). \tag{6}$$

Equation (6) is a function of seven variables with 3 fundamental units which by the pi theorem of dimensional analysis reduces to a relation among four dimensionless groups of variables.

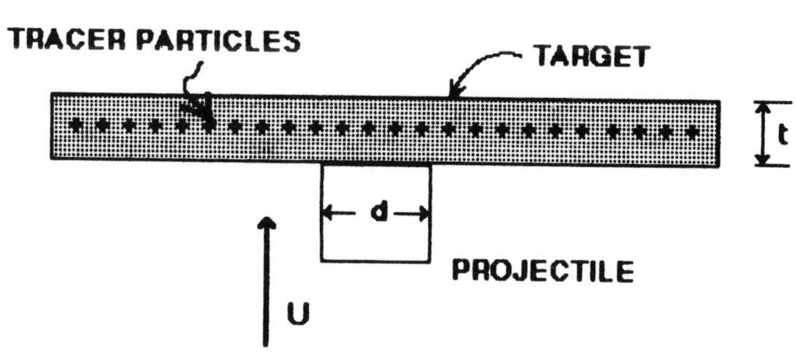

Fig. 1. Hydrocode impact setup.

If t is a late time variable only, then Eq. (6) may be written as,

$$r v^{\mu} \rho^{\nu}/C = f\,(r/t,\ z/t,\ \theta). \qquad v >> c \tag{7}$$

If t is an early time variable as well as a late time variable then, Eq. (7) remains unchanged, only C now contains a function of t/d.

If t is a late time variable only, then Eq. (7) can be used to obtain the velocity exponent μ. Two hydrocode calculations are performed in which the peak amplitude of the particle velocity is recorded at the same scaled positions, z/t, and θ. Then Eq. (7) may be rewritten as,

$$v/U \quad [r/d]^{1/\mu} = f\,(r/t). \qquad z/t,\ \theta \text{ const.} \tag{8}$$

Thus for fixed z/t (e.g. the center line of the target plate) it should be possible to use Eq. (8) to superimpose all strong shock solutions, when t is a late time variable only.

If t is an early time variable, as well as a late time variable, then Eq. (7) reduces to,

$$v/U = [r/d]^{-1/\mu} \quad f\,(r/t,\ t/d). \tag{9}$$

Note, when t/d is held constant Eq. (9) reduces to,

$$v/U = f\,(r/d), \qquad t/d,\ z/t,\ \theta \text{ const} \tag{10}$$

when t is a early time variable.

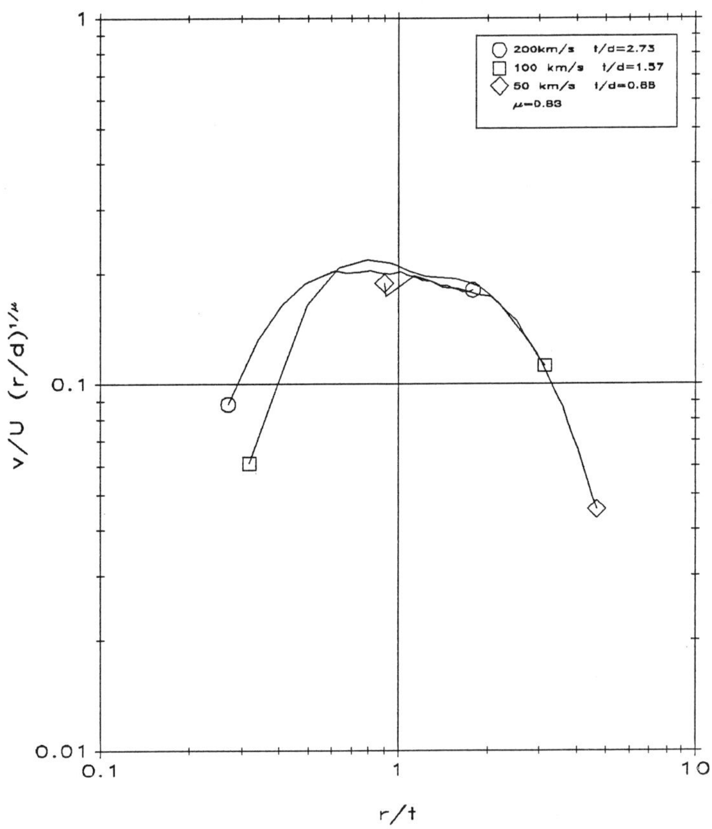

Fig. 2. Plot of three hydrocode calculations scaled with Eq. (8).

Strong shock solutions were generated with the two-dimensional hydrocode CSQ II (Thompson, 1979). The response of the plate was monitored with tracer particles positioned for all runs at fixed z/t of 1/2 as illustrated in Fig. 1, however, the spacing between the tracer particles was varied for some of the calculations. Both the projectile and the target constitutive relations were ANEOS library aluminum equations of state. ANEOS is a complete solid-liquid-vapor equation of state developed for CSQ by Thompson and Lauson (1972). The projectile was a cylinder with a diameter of 1 cm and length to diameter ratio of one. All calculations were for impacts at normal angles of incidence.

To test whether t is a late time variable only a plot of three hydrocode calculations scaled with Eq. (8) is shown in Fig. 2. A variety of values of μ between 1/2 and 1 were tried, with reasonable superposition of the curves obtained for $0.80 < \mu < 0.85$.

Calculations were performed to test whether t is also an early time variable and the results are shown in Fig. 3. Note that good superposition is not possible indicating that t is a late time variable for $t/d > 0.5$.

It is important to note that the values for the coupling parameter velocity exponent μ obtained from Fig. 2 lie within the energy and the momentum scaling limits identified in the previous section. Also note that energy scaling (μ = 1) is not obtained just as in the cratering of semi-infinite targets (Holsapple and Schmidt, 1982).

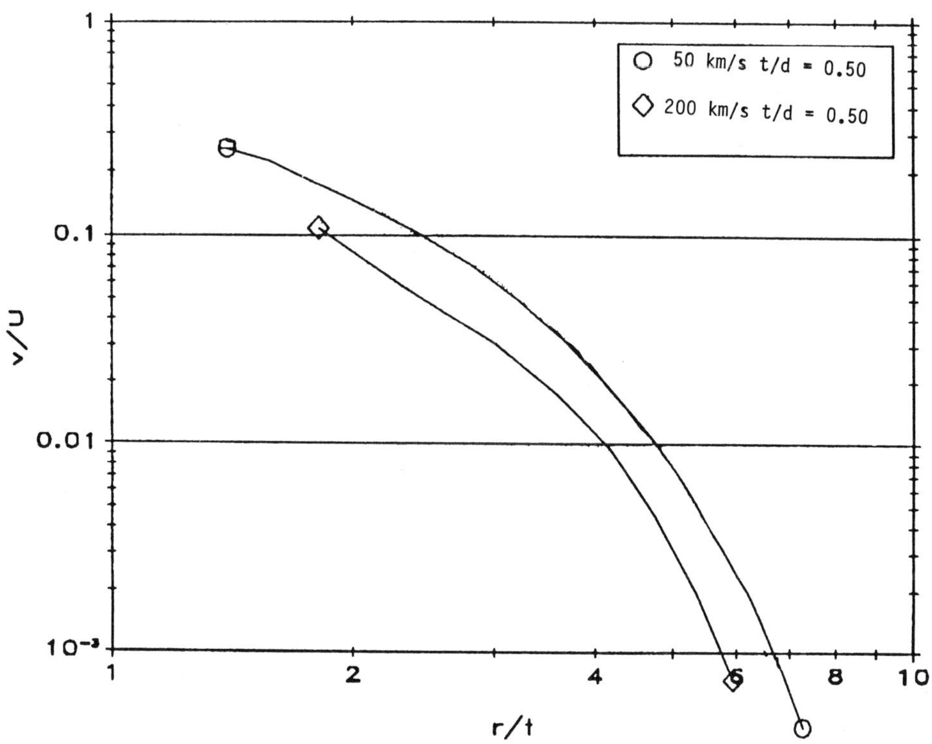

Fig. 3. Calculations to show that t is a late time variable.

VELOCITY SCALING WAVE PROFILES FROM IMPACTS OF THICK PLATES.

With the identification μ = 0.80 to 0.85 for perforating impacts of fully consolidated metals it is now possible to demonstrate the velocity scaling of flow fields from impacts of thick plates ($t/d > 0.5$) as calculated with hydrocodes.

The scaling relationships for the radial particle velocity profiles below the strong shock region are developed from

$$v = f\,(C, \rho, c, t, r, z, \theta, T), \tag{11}$$

where T is time. There are nine variables in Eq. (11) and three fundamental units thus there are six independent groups of variables. One possible set of pi-groups is

$$v/c = f\,(tc^{\mu}\,\rho^{\nu}/C,\ r/t,\ z/t,\ \theta,\ Tc^{\mu+1}\,\rho^{\nu}\ /C). \tag{12}$$

Note, from Eq. (12), that velocity scaling the perforation of thick plates cannot be accomplished with the prototype materials and the same t/d. In order to velocity scale the perforation of thick plates the target material must be changed or the t/d must be changed. Instead of changing the model material it was decided to alter the model's t/d using the first pi group on the right hand side of Eq. (12).

Two calculations were performed, the prototype at 100 km/s impact velocity and the model at 55.6 km/s. The parameters of the two calculations are listed in Table 1.

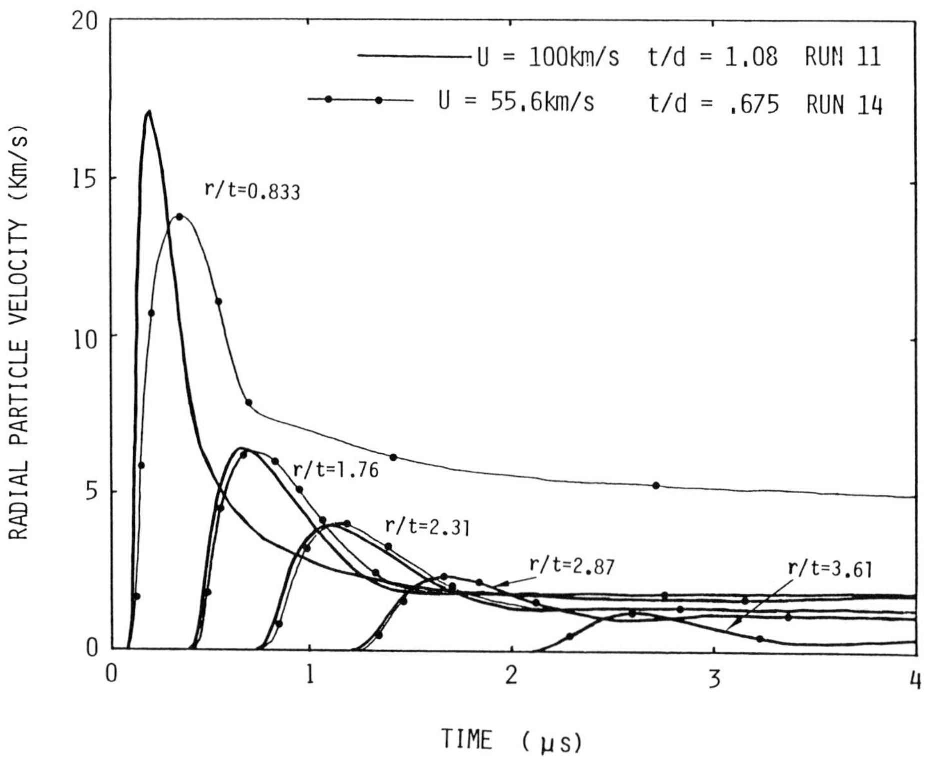

Fig. 4. Velocity wave profiles for two scaled runs.

TABLE 1 Hydrocode Calculation Parameters

Calculation	Impact Velocity (km/s)	t/d	z/t	Mesh size (cm)
Prototype	100	1.08	0.5	0.1
Model	55.6	0.675	0.5	0.625

The model calculation mesh size was also scaled in order for the model calculation to have the same scaled numerical accuracy as the prototype calculation.

The scaled particle velocity wave profiles at five Lagrangian positions for both prototype and model calculations are shown in Fig. 4. The wave profiles are labeled with the scaled radial Lagrangian position r/t. The wave profiles labeled r/t=0.833 in Fig. 4. are in the early time regime before coupling is complete and thus before the finite impactor solution has approached the line source solution. Consequently the wave profiles are very different near the source. The remaining radial particle velocity profiles at Lagrangian positions, labeled r/t = 1.76, 2.31, 2.87, 3.61 in Fig. 4, show improved matching of the scaled wave profiles as the distance from the impactor increases and the peak amplitude has decayed to small values. The wave profiles at the greatest distance superimpose indicating that the motions away from the source have been correctly velocity scaled.

CONCLUSIONS

The line source limit concept was utilized to establish similitude between prototype and model calculated radial particle velocity flow fields. Similar radial particle velocity shock amplitudes were demonstrated in the strong and weak shock regimes and similar radial particle velocity wave profiles were in the weak shock regime. The coupling parameter velocity exponent μ was computed to have a value between 0.80 and 0.85. The plate thickness is a late time variable only for impact of plates with $t/d > 0.5$.

ACKNOWLEDGMENT

This work was supported by Boeing Independent Research and Development funds.

REFERENCES

Bjorkman, M.D. and K.A. Holsapple (1984). in Shock Waves in Condensed Matter -1983, ed. by J.R. Asay, R.A. Graham, and G.K. Straub, Elsevier Science Publishers B.V., pp. 195-8.

Dienes, J.K. and J.M. Walsh (1970). in High-Velocity Impact Phenomena, ed. by R. Kinslow, Academic Press, New York, pp. 46-104.

Holsapple K.A. and R.M. Schmidt (1982), J. Geophys. Res. 87(B3) pp. 1849-70.

Holsapple K.A. and R.M. Schmidt (1986), submitted to J. Geophys. Res.

Thompson, S.L. (1977). CSQ II - An Eulerian Finite Difference Program for Two-Dimensional Material Response -Part 1. Material Sections. SAND 77-1339. Sandia National Laboratories, Albuquerque, NM.

Thompson, S.L. and H.S. Lauson (1972). Improvements in the Chart D Radiation-Hydrodynamic Code III: Revised Analytic Equations of State. SC-RR-71 0714, Sandia National Laboratories, Albuquerque NM.

Int. J. Impact Engng Vol. 5, pp. 69–99, 1987
Printed in Great Britain

0734-743X/87 $3.00 + 0.00
Pergamon Journals Ltd.

THE RESPONSE OF MATERIALS TO DYNAMIC LOADING*

J. R. Asay and G. I. Kerley

Sandia National Laboratories
Albuquerque, New Mexico

ABSTRACT

The response of materials to hypervelocity impact spans a wide region of material behavior, ranging from high impact pressures and temperatures, where thermodynamic effects prevail, to low pressures where mechanical properties are important. This paper discusses thermomechanical and physical processes important to hypervelocity impact events, presents a perspective of theoretical foundations, and provides an overview of current equation-of-state and constitutive modeling capabilities.

INTRODUCTION

An understanding of material response to dynamic loading is needed in many hypervelocity impact problems and other applications. The fundamental importance of material models in computer simulations of hypervelocity impact phenomena is discussed in the review article by Anderson and Johnson (1986). In order to describe the phenomena produced by the impact of a high-velocity particle onto a target, it is necessary to know physical, mechanical, and thermodynamic properties for various types of materials over a wide range of densities, temperatures, and deformation conditions. The purpose of this paper is to review current knowledge of these topics, particularly those relevant to such problems as spacecraft shielding (Agureikin, *et al.*, 1985; Rosenblatt, *et al.*, 1968; Swift and Hopkins, 1970; Swift, *et al.*, 1970; Swift, *et al.*, 1982), armor penetration (Zukas in Zukas, *et al.*, 1982), and satellite protection against hypervelocity projectiles (Spillman, 1986).

A hypervelocity impact normally produces states of both high compression and expansion. The highest pressures are generated by a shock wave in the area of direct impact, while progressively lower pressures are created as the shock propagates through the material and is attenuated. Hence, a wide range of compressed states is produced, so that purely elastic behavior, plastic flow, melting and other phase transitions, and changes in chemical or electronic structure can all occur, in different regions, during a single impact event. As a specific example, the impact of a 10-km/s tungsten projectile on an aluminum target produces peak pressures of about 300 GPa, compressions of 50% in the aluminum, and temperatures ranging from 20000 K in the shocked state to 3000 K after release.

Expansion follows shock compression when release waves arrive from the free surfaces. The expansion behavior is very dependent upon the state produced by the initial shock. The material may undergo tensile fracture on release and be further fragmented into small particles. Rosenblatt and colleagues (1968) have used two-dimensional code calculations to analyze the behavior of the expanded projectile and target for the impact of a 7.5-km/s aluminum projectile onto an aluminum target. Perforation of the target results in the formation of a hole and ejection of debris consisting of solid fragments and liquid droplets. Analysis of target response, the subsequent motion of target debris, and the destructive potential of the debris when interacting with secondary structures, are matters of considerable interest (Swift and Hopkins, 1970; Howard, 1969; Lipkin and Kipp, 1976).

In very high velocity impacts, sufficient energy can be dissipated by the shock process to vaporize some of the material upon release. Experimental studies by Swift and coworkers have shown that vaporization markedly increases the divergence of an expanding debris cloud; momentum and energy are spread over a large region, thereby producing a decrease in destructive potential. Hopkins and coworkers (1972) studied the thickness of rear plates necessary to stop debris produced by impacts of aluminum or cadmium spheres into target plates for velocities up to 7 km/s. They showed that penetration depth is a complicated function of impact velocity; peak penetrations were obtained at certain velocities. This behavior has been attributed to changes in the mechanical and physical states of the debris cloud (Swift, *et al.*, 1970; Swift and Hopkins, 1970).

Many studies of hypervelocity impacts make use of computer calculations with numerical continuum mechanics codes to simulate and analyze experiments or to evaluate the consequences of damage in the design of systems (Gehring, 1970). The material properties required for these calculations depend upon the sophistication of the code and of the constitutive models used. The basic information needed by most codes includes the equation of state, yield strength, shear modulus, and tensile strength for each material in the problem. For example, Bertholf and coworkers (1975) compared calculations using both a Lagrangian and an Eulerian code with experimental data for the impact of a nylon sphere onto a steel plate. They showed that the predictions were sensitive to the material properties; it was necessary to include a treatment of the liquid-vapor transition in nylon and the polymorphic phase transition in steel and to use the correct compressive and tensile strengths in order to agree with the measurements.

A full understanding of hypervelocity impacts may require a description of other aspects of material behavior not usually treated in most codes. For example, the size distribution of fragments will affect the destructive potential of the debris cloud. In addition, continuum mechanics mixture models are needed to improve calculations of the expansion of a multiphase cloud consisting of solid fragments, liquid droplets, and vapor.

The remainder of this paper is organized into four sections. The first section is a brief discussion of some basic theoretical concepts that underly the development of material models. The second section gives an overview of experimental techniques for measuring material response. The third section, entitled "Equations of State," describes current capabilities for constructing models for the equilibrium hydrostatic response of materials. The last section, entitled "Constitutive Response," is a review of nonhydrostatic and nonequilibrium behavior, including material strength, nonhomogeneity (composite materials), porosity, failure, and fragmentation. A complete discussion of all these subjects is beyond the scope of this article. Our objective is to present the foundations and the principal ideas in a qualitative manner. In addition to citations of original work, the references include a number of review papers that can be consulted for additional information.

THEORETICAL FOUNDATIONS

Macroscopic Point of View - Continuum Mechanics

The theoretical description of a hypervelocity impact or other dynamic event is normally developed from the principles of continuum mechanics. From a macroscopic point of view, the three basic equations that describe the motion can be derived from the balance laws for mass, momentum, and energy, by regarding matter as a continuous medium described by field variables. The momentum equation, or equation of motion, is

$$\partial(\rho\mathbf{v})/\partial t + \nabla\cdot(\rho\mathbf{v}\mathbf{v}) = \nabla\cdot\sigma \quad , \tag{1}$$

where ρ, $\mathbf{v}$, and σ are the density, the velocity vector, and the stress tensor, respectively. Constitutive equations are also required to describe the dependence of σ on ρ, the internal energy E, and other dynamical variables. If σ is assumed to depend only on ρ and E, it immediately follows that the stress tensor reduces to a scalar quantity, and that the constitutive behavior is completely described by the equilibrium hydrostatic equation of state. This result suggests that the principles of thermodynamics should provide the basis for developing a more general constitutive model (Herrmann, 1982; Wallace, 1984). Although these arguments establish the existence of the stress tensor and some of its properties, they do not provide a theoretical foundation for calculating it from first principles. Therefore, the macroscopic approach to constitutive modeling is inherently phenomenological. The equations are chosen to describe observed trends in material behavior; they may or may not be based upon explicit assumptions about the microscopic processes that are involved. Parameters in the model are normally determined from experimental data.

Molecular Point of View - Statistical Mechanics

In addition to their intrinsic interest, theories based on first principles help to understand the important mechanisms in material response and are necessary for making predictions in the absence of experimental data. In order to develop such theories, it is necessary to begin with the submicroscopic view of matter as a system of interacting particles. Instead of actually solving the equations of motion for any particular system, the techniques of statistical mechanics are used to average the equations over a collection of systems (an ensemble), each of which has the same macroscopic properties (Balescu, 1975). The result is similar to what would be obtained by averaging over many experiments. The equations of continuum mechanics were first derived from statistical mechanics by Irving and Kirkwood (1950) and more recently by Frölich (1967). They showed that the stress tensor and other variables of the continuum theory can be expressed as averages of dynamical quantities over the coordinates and momenta of individual atomic particles. The derivation is valid for both the equilibrium and nonequilibrium cases.

The calculation of material response parameters using statistical mechanics requires knowledge of the probability distribution function used to average over all possible configurations of the system. By definition, a system is at equilibrium when the probability is independent of time. For that case, the distribution function is well known (Balescu, 1975); it is found that a system having a given density ρ and internal energy E has evolved to the highest possible entropy state and that it has no memory of its history. Therefore, equilibrium statistical mechanical theories describe only hydrostatic behavior, where the stress σ depends only on ρ and E.

In the description of dynamic problems, deviations from equilibrium hydrostatic behavior are very important, particularly for solids. For more general theories, new variables must be used in addition to ρ and E. For example, material strength can be described by making the stress tensor a function of shear strain, which accounts for the fact that an elastic solid retains a memory of its initial configuration. Additional variables are needed to describe inelastic behavior. In principle, such problems could also be attacked using statistical mechanics. However, calculations would require theories for describing nonequilibrium distribution functions that presently do not exist. Therefore, it is necessary to rely upon phenomenological methods in order to describe deviations from equilibrium hydrostatic material response.

Current Capabilities

The steps taken for constructing a model of dynamic material response are outlined in Fig. 1. We have divided the topics to be discussed into two main areas. The term "equation of state" (EOS) is used when referring to equilibrium hydrostatic behavior, while everything else is called "constitutive response."

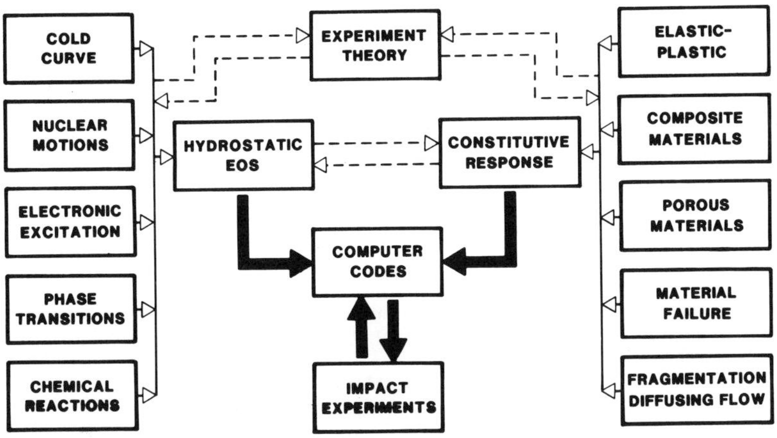

Fig. 1. Steps in constructing a model of dynamic material response.

EOS models are usually developed using a combination of statistical mechanical theory and experimental data. Rigorous calculations using statistical mechanics can be made only for highly idealized problems. However, the past 20 years have seen enormous progress in the development of predictive theories that are applicable to all aspects of EOS modeling. As a result, the role of experiments is gradually changing; instead of using data simply as input to a fitting process, the measurements can be used to test theoretical assumptions and to determine where they need to be improved. Application of the more fundamental methods should also increase confidence in the EOS in regimes inaccessible to experiment.

The situation is very different when it comes to the treatment of constitutive response. Because non-equilibrium statistical mechanical theories are extremely primitive, constitutive modeling typically starts by assuming a known equilibrium EOS and adding phenomenological corrections to it. Despite this limitation, enormous progress has been made in recent years; experimental work has led to improved understanding of the microscopic aspects of material response, and theories have been developed that treat microscopic structural parameters explicitly (Davison, 1984).

EXPERIMENTAL METHODS

Experimental data provide a necessary foundation for evaluating basic theories and phenomenological models of dynamic response. Traditionally, both static and dynamic methods have been used for this purpose. Early experimental EOS measurements relied heavily on static means to determine pressure-volume response. Because of the need for EOS measurements at very high pressures and limitations in high-pressure technology, shock-wave methods have played an increasingly important role in acquiring EOS data in regimes inaccessible by other methods. Early EOS applications of this technology date to the mid-1940s. In the past 20 years, these techniques have progressed to the point where a wide variety of physical, mechanical, and chemical properties can be studied with shock-wave techniques.

Static Methods

The earliest techniques for determining the isothermal pressure-density relationship involved the use of hydrostatic compression methods (Bridgman, 1931; Crossland and Spain, 1983). Hydrostatic loading is generally limited to pressures below a few tens of GPa, because typical fluids used for the pressure transmitting medium freeze and are not hydrostatic at extremely high pressures. Recent advances in diamond cell technology have extended these measurements to a few hundred

GPa; pressures as high as 550 GPa have been recently reported (Xu, *et al.*, 1986; Jayaraman, 1983), although the accuracy of measurements in this regime has not yet been established, and the stress state is not truly hydrostatic.

Static pressure measurements have provided and continue to provide an important source of information regarding the isothermal hydrostatic behavior of materials at high pressure. However, there are both low- and high-pressure limits to the use of these data. At very high pressures, major problems are the accuracy of pressure-volume measurements, non-hydrostatic loading and non-uniform stress states . At low pressures, the strength of materials must be considered explicitly when using hydrostatic data for constructing dynamic equations of state (Fowles, 1961).

Shock-Wave Techniques

Hugoniot measurements. Shock-wave techniques provide an even more important source of EOS data. In these approaches, plane shock compression is produced with various methods, including explosive loading, projectile impact, and energy deposition (Rice, McQueen and Walsh, 1958; McQueen, *et al.*, 1970; Graham and Asay, 1978). Typical plate-impact methods are illustrated in Fig. 2. Usually, two kinematic properties of a steady shock wave are measured, such as the shock velocity and the final stress or mass velocity. Assuming that the shock transition occurs between equilibrium states, the conservation equations of mass, momentum and energy can be used to determine the stress-volume-energy state:

$$\begin{aligned} V(U - u_o) &= V_o(U - u) \ , \\ \sigma - \sigma_o &= (U - u_o)(u - u_o)/V_o \ , \\ E - E_o &= \tfrac{1}{2}(\sigma + \sigma_o)(V_o - V) \ , \end{aligned} \tag{2}$$

where V is specific volume, σ is longitudinal stress, E is internal energy, U is shock velocity, u is particle velocity, and subscript o refers to initial conditions ahead of the shock (Rice, *et al.*, 1958).

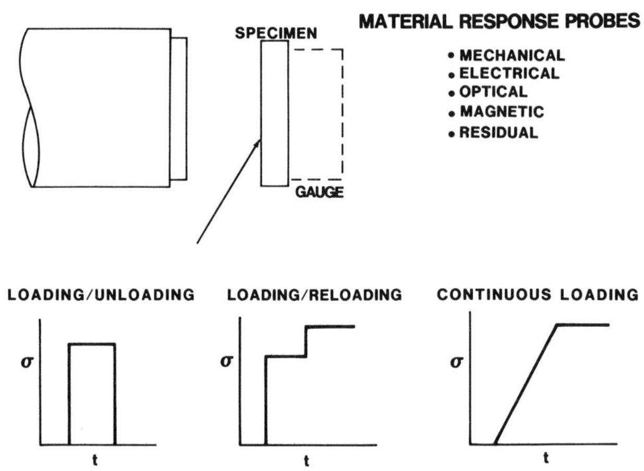

Fig. 2. Plate-impact techniques for measuring the dynamic properties of materials.

The locus of end states produced by shock compression is referred to as the Hugoniot curve. The relationship between a Hugoniot, isotherm, and isentrope passing through the same initial state is depicted in Fig. 3. The Hugoniot lies above the other two loading curves because of irreversible heating. The Hugoniot curve does not explicitly give the dependence of pressure on entropy or temperature and must be supplemented with other experimental data or theoretical calculations to develop a complete EOS.

Extensive Hugoniot measurements have been performed with various shock-wave techniques since the mid-1940s, which have been refined in recent years, so that data of high precision can now be easily acquired. Major new developments involve the refining of gas-gun and electrical pin techniques for performing highly accurate shock velocity measurements; pressure-volume data can routinely be determined to accuracies of 1% (Mitchell and Nellis, 1981; Nellis, 1982). At present, the most precise Hugoniot measurements, using two-stage gas guns, can be made for impact velocities up to about 8 km/s, corresponding to pressures of a few hundred GPa in metals. However, a number of special techniques have been developed for shock wave studies at much higher pressures. Al'tshuler (1985) has reported high-explosive driver systems that can launch iron plates to 15–18 km/s for EOS measurements. Nuclear explosives (Ragan, 1982) have been used to obtain Hugoniot data for several materials at pressures up to about 10 TPa (100 Mbar), relative to a molybdenum standard. Recently, Vladimirov and colleagues (1984) have reported Hugoniot data on aluminum to pressures of about 400 TPa (4000 Mbar), relative to an iron standard. Measurements using laser-driven shocks have also been reported for aluminum (Ng, *et al.*, 1985). Unfortunately, many of the ultrahigh pressure data have large error bars. Extensive Hugoniot data are tabulated for a wide variety of materials (Rice, *et al.*, 1958; McQueen, *et al.*, 1970; van Thiel, 1977; Marsh, 1980).

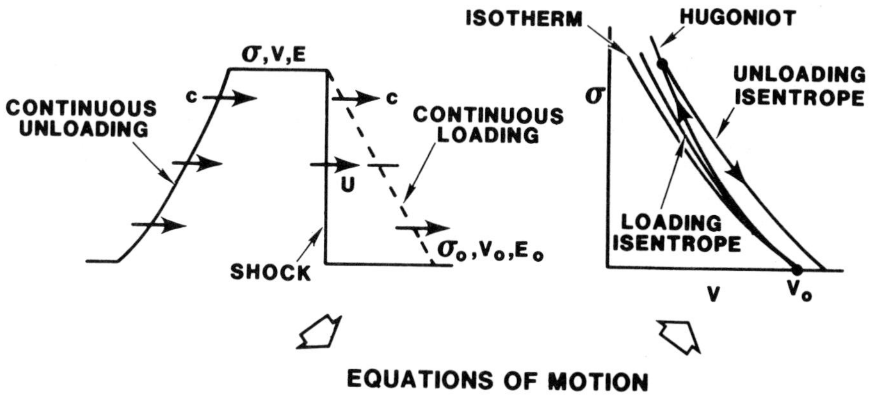

Fig. 3. Reduction of shock and continuous loading data to obtain Hugoniot and off-Hugoniot data.

Off-Hugoniot Measurements. Recent advances in obtaining high-pressure EOS data include development of plate-impact techniques for measuring off-Hugoniot material response. A variety of methods are available for studying these states, including unloading and reloading from the shocked state, reverberating wave techniques, use of porous materials, and wave-profile measurements, as reviewed by Graham and Asay (1978) and more recently by Chhabildas (1986).

Time-resolved shock-wave instrumentation using a variety of detectors, such as piezoelectric, piezoresistive, and interferometric gauges, have played an important role in these studies. In this approach, wave structures are measured at different points in their evolution. Since the non-linear response of materials normally produces a spreading of unloading waves from the shocked state and a steepening of compression waves (Graham and Asay, 1978; Grady, 1980; Asay, 1985), measurements of wave evolution combined with the differential conservation equations allow determination of the continuous stress-volume response, if rate effects are negligible (Fowles and Williams, 1970; Grady, 1980), as follows:

$$
\begin{aligned}
dV &= -V_o du/c \ , \\
d\sigma &= c du/V_o \ , \\
dE &= -\sigma dV \ ,
\end{aligned}
\tag{3}
$$

where c is Lagrangian wave velocity. Wave-profile measurements can presently be made to shock pressures of several hundred GPa using a variety of techniques. An example of the use of interferometric methods (Barker and Hollenbach, 1972) to obtain wave profiles at high shock pressures is given in Wise and Chhabildas (1986).

Wave-profile measurements have been useful in determining decompression curves (which are nearly isentropic) from high-pressure shocked states, as illustrated in Fig. 3; recent applications to aluminum are given by Asay and coworkers (1986). In addition to evaluating the quasi-isentropic response of materials, these data have been useful for determining a wide variety of mechanical and physical properties in the shocked state. A recent advance in this area concerns the development of plate-impact techniques for producing continuous, nearly isentropic compression of materials (Barker and Hollenbach, 1970; Barker, 1984), including graded-density impactors for plate-impact experiments. These impactors produce a small amplitude shock upon impact, followed by continuous compression to high pressure (Barker, 1984). Time-resolved instrumentation techniques are used to measure the evolution of the compressive wave structure, leading to the determination of continuous loading paths. The advantage of this method is that continuous loading can be produced to very high pressures, while keeping the specimen in a relatively low temperature state. The method has been used to obtain continuous loading of tungsten to pressure of about 100 GPa (Chhabildas and Barker, 1986). Another technique for generating isentropic compression in extremely compressible materials, such as gaseous fluids, involves plate-impact techniques, wherein a gas cushion between the specimen and projectile produces continuous loading. This technique is presently being used to obtain the quasi-isentropic response of hydrogen for pressures to 400 GPa (Barker, *et al.*, 1985).

Wave-profile techniques have also been used to study dynamic phase transitions occurring under shock compression, including detection of melting and polymorphic phase transitions. Specifically, Barker and Hollenbach (1974) have used these techniques to investigate the kinetics of the 13-GPa polymorphic transition that occurs under impact loading in iron; a complete review of shock-induced phase transitions is given by Duvall and Graham (1977).

More recent applications of wave-profile techniques concern studies of shock-induced vaporization. Trucano and Asay (1986) have shown how measurements of the stagnation pressure produced by expanding vapors with time-resolved gauges can be used to evaluate theories of material response in hot, expanded regions.

Temperature measurements. Although many instrumentation techniques are available for studying the mechanical response of materials to high pressures (Graham and Asay, 1978), there is an important need to develop techniques for measuring thermal properties under shock compression. Traditionally, shock luminescence methods have been used to measure temperature, as illustrated by the classic work of Kormer and coworkers (1965) on alkali halides. In that work, shock temperatures were determined by fitting a black-body spectrum to the light emitted at the shock front in transparent materials; this procedure allows direct determination of temperature, thereby producing a sensitive measure of melting under shock compression. Recent refinements of the method (Ahrens, *et al.*, 1982; Lyzenga, *et al.*, 1983; Nellis, *et al.*, 1984a; Boslough, 1986) include the use of multiple-wavelength pyrometry to improve the accuracy of temperature measurement. It is presently possible to determine shock temperatures in transparent materials to accuracies of a few percent. The use of these techniques to study opaque materials is presently limited, although there have been attempts to measure the interface temperature between metallic surfaces and a transparent window (Urtiew and Grover, 1977).

Expansion measurements. EOS data in the vapor region are presently extremely limited. Soviet investigators have conducted shock-wave experiments on lead for initial pressures sufficiently high to produce states in the liquid-vapor coexistence region after unloading from the shocked state (Fortov, *et al.*, 1974; Al'tshuler, *et al.*, 1977; Al'tshuler, *et al.*, 1980), using the impedance-matching technique. A lower impedance "buffer" material was placed behind the sample to be studied. Shock-wave arrival at the interface between materials produced a low-pressure release state in the sample and an initial shock in the buffer. Measurement of shock velocity in the buffer determined the unloaded pressure-particle velocity state in the unloaded sample; a variety of lower impedance materials were used to establish the piecewise unloading curve of the specimen. This technique has been useful for evaluating theories of hot, expanded material response. One limitation of the method is that it does not provide direct information about the kinetics of the vaporization process.

Isobaric Measurements

Another technique for studying hot, expanded states is the isobaric heating technique, in which a thin-wire specimen is resistively heated to high temperatures at constant pressure (Shaner, *et al.*, 1976; Gathers, 1983). The volume of the specimen is determined by photographic methods, the temperature is estimated with pyrometric techniques, and the enthalpy from resistive energy deposition. Shaner obtained thermophysical data on several metals with this technique which was compatible with Soviet shock loading/unloading measurements. Recently, a laser-generated pulse technique has been used to measure sound velocity in liquid lead (Hixson, *et al.*, 1986).

EQUATION OF STATE

Overview

The development of EOS for use in hypervelocity impact and other dynamic phenomena has always relied heavily on shock wave experiments. Nearly 30 years have passed since Rice, McQueen and Walsh wrote their classic review (1958) of work in this field. Their analysis, like much of the early work on EOS, was based upon the Mie-Grüneisen (MG) approximation,

$$P(\rho, E) = P_r(\rho) + \rho\gamma(\rho)[E - E_r(\rho)] \quad , \tag{4}$$

where P_r and E_r are the pressure and internal energy on some "reference curve" (usually the Hugoniot), and γ is the Grüneisen parameter, assumed to be a function only of the density ρ. This expression has proven to be a powerful tool for describing EOS because of its simplicity and the fact that it can be fit directly to experimental data without the use of theoretical models. Although the MG formula continues to be a useful EOS option in many problems, it has a number of serious limitations. For example, it gives poor results for the expanded liquid and the vapor regions. If the Hugoniot is used as the reference curve, the formula also fails at high densities and cannot be used to describe some multiply shocked or isentropically compressed states. The dependence of γ on temperature becomes important when chemical dissociation and electronic excitation occur. Finally, and perhaps most importantly, the MG formula provides little insight into how to construct the EOS in regions where there are no experimental data.

In order to describe the correct asymptotic behavior in both the high-temperature and the low-density extremes, Tillotson (1962) developed a formula that uses separate expressions in different regions of the EOS surface. For the compressed region, he generalized Eq. (4) by allowing γ to decrease with increasing energy E, in order to account for the effects of thermal electronic excitation; the analytical expression and constants were chosen to fit Thomas-Fermi results. For expanded states, he selected formulas that approach the ideal gas result at low densities. For energies above the sublimation energy, the transition between the compressed and expanded regions is smooth and continuous; at lower energies, the formulas represent the mixed phase vaporization region in an approximate way. No attempt was made to describe melting or polymorphic phase transitions. The Tillotson EOS is an improvement to the MG formula that has been useful for hypervelocity impact calculations. However, it is oversimplified and does not give a good description of many of the phenomena that are discussed in this paper. For example, it has been shown to give poor agreement with porous Hugoniot measurements for aluminum (Bjork, 1986) and with isobaric expansion data for lead (Holian and Burkett, 1986).

In order to construct an EOS that is accurate over a wide range of densities and temperatures, it is necessary to develop a theoretical model that includes a treatment of many physical phenomena, as are illustrated in Fig. 4. At low temperatures in the solid region, the EOS is determined principally by the ground state electronic energy and by lattice vibrations; changes in thermodynamic properties due to polymorphic phase transitions may also be important features to be included. Melting occurs with increased temperatures, while vaporization requires both heating and expansion; both transitions can affect the way energy is converted upon impact. In the case of chemical compounds, dissociation can have significant effects on the EOS. At very high temperatures, thermal excitation and ionization of the electrons become the problem. In the past 20 years, a vast amount of experimental and theoretical work has been devoted to the study of these topics. Several recent review articles are available that discuss this work in greater detail than is given in this paper (Davison and Graham, 1979; More, 1981; Godwal, *et al.*, 1983; Bushman and Fortov, 1983; Ross, 1985).

Separation of Degrees of Freedom

The first step in constructing an EOS model is to find a suitable division of the problem into separate and tractable parts. In the following discussion, we will consider the main topics shown in Fig. 1. The simplest problem is one in which there are no phase transitions or chemical reactions; for a single phase and a single chemical component, it is common to assume that the thermodynamic functions can be written as the sum of several terms that are approximately independent of one another. The Helmholtz free energy is

$$F(\rho,T) = E_c(\rho) + F_n(\rho,T) + F_{vr}(\rho,T) + F_e(\rho,T) \quad . \tag{5}$$

The pressure P and internal energy E can be written in a simular fashion. Definitions of the terms in Eq. (5) are most easily given for a solid. E_c is the internal energy on the zero-Kelvin isotherm, for the electrons in the ground state and the nuclei in a perfect crystal lattice; this term accounts for both the cohesive forces that give rise to condensed phases and the repulsive forces that determine most of the response to compression. The terms F_n and F_{vr} arise from nuclear motions. F_n is the contribution from the vibrations of atoms or molecules about their equilibrium lattice positions. For molecular substances, F_{vr} is the contribution from internal vibrational and rotational degrees of freedom. F_e, the contribution due to thermal excitation and ionization of the electrons, becomes important at high temperatures. For substances that have either phase transitions or chemical reactions, it is not a good approximation to write the thermodynamic functions as simple sums of terms, as in Eq. (5). Such systems are best regarded as *mixtures* of several phases and chemical species. Methods for treating these problems are discussed in the last two sections on equations of state.

Zero-Kelvin Isotherm

The zero-Kelvin isotherm, or "cold curve," is the largest term in the EOS over much of the density-temperature region of interest. It is also used to compute the nuclear contributions to the EOS in some approximate models of solids and liquids. No one way to construct the zero-Kelvin curve works well for every material. It is possible to make sophisticated calculations in some cases, while very approximate schemes must suffice in others.

Experimental data. Experimental information about the solid isotherm can be obtained using static methods, as discussed earlier. In addition, the solid isotherm can be determined by correcting shock Hugoniot data for irreversible thermal heating (Rice, *et al.*, 1958). This procedure is straightforward and accurate if the thermal corrections can be described using the Mie-Grüneisen EOS and the Grüneisen parameter has a simple volume dependence. Comparisons between static data and isotherms computed from shock data have been found to agree very well in many cases. However, the results are less accurate if melting occurs, or if thermal electronic contributions to the EOS become significant. The thermal corrections are much smaller in the dynamic isentropic loading experiments discussed earlier.

Analytical formulas for the cold curve. The idea of a "universal" analytic expression, one that accurately describes the zero-Kelvin isotherm using only a few parameters, is appealing. However, no truly universal formula is known; the fact that high pressures induce changes in electronic structure and compressibility suggests that no simple relationship can be found. Nevertheless, many useful formulas have been proposed for fitting experimental data over a limited range of densities and for extrapolating low-pressure results into regions where no measurements are available. For example, the formula recently proposed by Rose and colleagues (1984) appears to give good agreement with experimental and theoretical results for pressures up to several hundred GPa in many materials. A number of other analytic expressions have been reviewed by Godwal and colleagues (1983). It should be noted that some of the popular formulas do not go to the correct high-density limit. At high densities, one can switch smoothly to an interpolation formula based on Thomas-Fermi theory, thereby covering the entire range in a reasonable way (Kerley, 1981).

Band-theoretical calculations. In recent years there have been significant advances in understanding the electronic structure of condensed matter, using band theory, and in the development of numerical methods and computer codes for calculating EOS. Because this subject has been discussed in some detail in other review articles (Godwal, *et al.*, 1983; Bushman and Fortov, 1983; Ross, 1985), we will limit our remarks to a few main points.

In band theory, the electrons in a solid are treated as independent, in the sense that the motion of each depends only on the average behavior of the others. Hence, the allowed states of the electrons are described by one-electron wave equations and by wave functions that extend throughout the solid. The allowed energies fall into bands of levels that roughly correspond to the atomic orbitals. For simple crystal structures, the calculations are similar to those for an isolated atom, except for the boundary conditions on the wave functions.

In the one-electron model, the average potential seen by an electron in a solid consists of electrostatic interactions with the nuclei and the other electrons and also an exchange term that arises from the Pauli exclusion principle. In a systematic derivation, leading to the Hartree-Fock (HF) scheme, the exchange term has a "non-local" character that increases the difficulty of solving the equations. In most band calculations, the problem is simplified by using a form of the exchange potential that depends only on the local electron density. However, the local density formulation should not be regarded as an approximation to the HF scheme, because it gives significantly different results; in fact, it is found to be better than the HF method for metals. Corrections to the one-electron model (correlations) are also included in the local density formulation. At present, there is no definitive argument for choosing the exchange-correlation potential from among several options. Variations in the treatment of this term can cause differences of up to 5% in density, 10% in compressibility, and 30% in cohesive energy. However, these differences usually become less important at high pressures.

Band theory plays an important role in EOS modeling. Numerous calculations have shown that the theory gives quantitative agreement with both static and dynamic data and provides the most reliable predictions for cases where no data exist. It is also a valuable tool for unraveling problems where the understanding of experiments is complicated by melting, solid-solid transitions, and chemical reactions, so that the cold curve is not readily extracted from the measurements. Many recent studies, too numerous to mention here, have succeeded in explaining unusual or anomalous phenomena arising from electronic phase transitions, metallization, and other effects of pressure on electronic structure (Ross, 1985).

On the other hand, most band theory codes are limited to calculations for relatively simple crystal lattices. In order to model polymers, composites, and various other complicated materials that arise in certain hypervelocity impact problems, it is still necessary to employ much more approximate models and empirical methods. Recent progress in fast linear methods offers hope that calculations for these difficult problems will become available in the future.

Other theories. The pseudopotential method is an approximate form of band theory that can be used to calculate both the zero-Kelvin isotherm and an effective interatomic potential that is needed in models for the nuclear degrees of freedom (Godwal, *et. al.*, 1983). In its most common and basic form, the theory is applicable only to "simple metals," in which the valence orbitals have a nearly free-electron character and the interaction between core orbitals on different atoms is weak. Extensions of the method to more complicated materials have been discussed by Yin and Cohen (1982) and by Moriarty (1982). Less accurate estimates of the cold curve can be obtained using average atom theories, such as the quantum-corrected Thomas-Fermi model (More, 1981) and the INFERNO model (Liberman, 1979); these methods are also useful for calculating thermal electronic contributions to the EOS, as discussed below. A third model, developed by Gordon and coworkers, can be used to compute both the cold curve and the intermolecular potential from wavefunctions for isolated molecules; this method has been shown to give very good results for closed shell systems such as the rare gases and molecular solids (LeSar and Gordon, 1983).

Nuclear Degrees of Freedom

A theoretical model for the nuclear degrees of freedom requires the solution of two problems – determination of the forces between molecules, and calculation of thermodynamic properties from these forces. At the present time, the second problem is far better understood than the first. Once the potentials have been defined, accurate models are available for computing the properties of solids, liquids, and gases. However, the potentials are extremely difficult to measure or to calculate, except for the simplest materials. In most cases of practical interest, quantities that depend upon the intermolecular forces must be treated in an *ad hoc* or empirical manner.

Mie-Grüneisen models. The usual starting point for computing the contributions from nuclear degrees of freedom is to model a solid as a collection of coupled harmonic oscillators. The thermodynamic properties can be determined by summing over the vibrational levels computed using the theory of lattice dynamics. However, such calculations are often avoided by using the popular Debye model that describes the principal features of the nuclear term with a few simple parameters. This theory uses a simplified description of the lattice vibrational spectrum, obtained by treating the solid as a continuous medium and imposing a frequency cutoff that yields the correct degrees of freedom. In the simplest form of the theory, the nuclear contribution to the internal energy is found to be (Zharkov and Kalinin, 1971).

$$E_n(\rho, T) = Nk[9\Theta_D/8 + 3TD(\Theta_D/T)] \ , \tag{6}$$

where D is the Debye function and Θ_D is the Debye temperature. Making the usual assumption that Θ_D depends only on density, the pressure term is

$$P_n(\rho, T) = \rho\gamma(\rho)E_n \ , \tag{7}$$

where γ is the Grüneisen function,

$$\gamma(\rho) = d \ln \Theta_D / d \ln \rho \ . \tag{8}$$

For the case where only the cold curve and the nuclear term contribute to the EOS, it is easy to show that these relations reduce to the Mie-Grüneisen approximation, Eq. (4).

The above formulas are normally used in an empirical manner. At ambient solid density, Θ_D can be determined by fitting experimental heat capacity data or from Slater's formula (1939), using the sound speed and Poisson's ratio; γ can be computed from thermodynamic data, using Grüneisen's formula. The principal problem in calculating the nuclear term is that of specifying the density dependence of the Grüneisen parameter. A wide variety of formulas for $\gamma(\rho)$ are in use (Godwal, *et al.*, 1983); for example, the Dugdale-MacDonald expression that relates γ to derivatives of the zero-Kelvin isotherm is often used. For many practical applications, almost any of the common expressions will give reasonable results because γ is only a weak function of density (unless there is a phase transition or change in electronic structure). Even so, evidence is mounting that none of the standard approximations are completely satisfactory. γ can also be measured with shock-wave techniques (Godwal, *et al.*, 1983). Data available at the present time are limited and have large error bars; deviations of up to 25% from the usual formulas were observed for beryllium and aluminum (Neal, 1976, 1979). Errors of this magnitude are particularly serious when making predictions of melting and other phase transitions that are sensitive to relatively small errors in the thermodynamic functions.

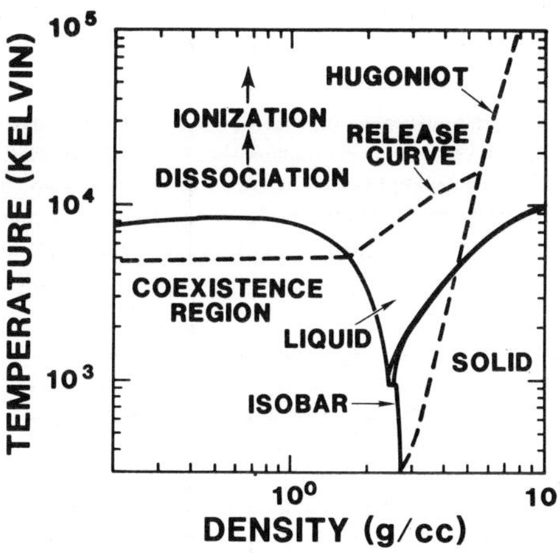

Fig. 4. Phenomena important to EOS modeling. Curves are shown for a fictitious material.

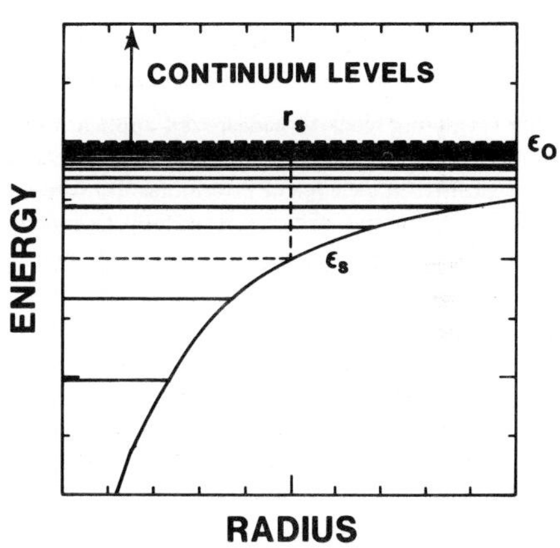

Fig. 5. Potential and energy levels for an electron in an isolated ion. At a density corresponding to radius r_s, all levels with energy $\epsilon > \epsilon_s$ are in the continuum.

Lattice dynamical models. If the intermolecular forces are known (or assumed), theories of lattice dynamics can be used to compute thermodynamic properties of crystalline solids. Most EOS work has relied upon the quasiharmonic approximation, (Born and Huang, 1954), in which the dynamical equations are decoupled by expressing the degrees of freedom in terms of collective coordinates, the phonons. This method was used to calculate Grüneisen parameters for Be and Al, the results were in much better agreement with experimental data than those obtained from the standard approximations (Godwal, *et al.*, 1983).

Fluids and gases – phenomenological models. For satisfactory results, models of the solid cannot be extrapolated very far into the fluid range without some changes. Realistic theories of liquids and dense gases have become available only recently and still have certain limitations; therefore, many EOS models for practical applications continue to rely on phenomenological treatments of the fluid regime. The simplest approach is to use an "interpolation formula," in which the solid equations are modified to approach the ideal gas formula in the limit of high temperatures and/or low densities. For example, the ANEOS code (Thompson and Lauson, 1972) uses such an expression. Many tables in the Sesame library use formulas of Cowan (Cranfill and More, 1978) and of Barnes (1975). These methods can be improved by building in discrete changes in density and energy at melting. In the GRAY code (Royce, 1972), formulas developed by Grover (1971, 1977) are used to scale the fluid EOS from that of the solid. A similar approach is used in the ANEOS code. Certain parameters in these models are determined from experimental data.

A wide variety of analytic equations are available for the limited purpose of fitting the EOS in the region of the vapor-liquid coexistence boundary. For example, the "soft-sphere" expression of Young (1977), that contains five parameters adjusted to fit experimental density and enthalpy data, has been found to give good results for liquid metals. To construct a global EOS, this formula can be used in the coexistence region, while other models are used for higher densities and temperatures.

Theories of liquids and dense gases. The development of accurate models to describe EOS in the fluid regime has always been hampered by two problems. First, the structure of a fluid is qualitatively different from that of a solid. The physical picture of a lattice, consisting of atoms vibrating about equilibrium positions, is not a good starting point for constructing a fluid model, and there is no fundamental basis for separating the EOS into a cold curve and a nuclear term. However, a useful alternative to the solidlike picture did not emerge until recently. Second, experimental data for testing theories are much more limited in the fluid regime. For example, melting usually causes only a small change in the slope of the Hugoniot curve.

A significant advance in the theory of liquids and dense gases has been the development of perturbation theories (Barker and Henderson, 1976; Mansoori and Canfield, 1969; Anderson, *et al.*, 1976; Kerley, 1983). It has been shown that the structure of a dense fluid is determined principally by repulsive forces and that an idealized fluid, like the hard-sphere system, is a useful starting point for èxpanding the free energy in powers of the intermolecular potential. The terms above first-order can be made quite small when the hard-sphere diameter is chosen in an optimum way. For example, minimization of the free energy with respect to the hard-sphere diameter gives a very accurate *a priori* calculation of the thermodynamic properties from a given intermolecular potential. (The theories discussed here should be distinguished from phenomenological models, in which the parameters are determined from experimental data and *ad hoc* assumptions.) These theories have been used extensively to compute EOS for noble gases and simple molecular fluids, using pair potentials constructed partly from theory and partly from experimental data (Ross, 1980a; Ross and McMahan, 1980; Young, *et al.*, 1981; Ross, *et al.*, 1983). They have also been applied to simple liquid metals, using potentials calculated from pseudopotential theory (Jones, 1973; Ashcroft and Stroud, 1978; Moriarty, *et al.*, 1984).

A major stumbling block to widespread application of these models is that the intermolecular forces cannot be calculated or measured with the necessary accuracy. For this reason, Kerley (1983) has developed a perturbation theory (the CRIS model) in which explicit knowledge of intermolecular forces is not required. The key concepts of liquid structure are retained, but terms in the expansion are computed approximately, using only the zero-Kelvin isotherm of the solid. Hence the model can be used in practical applications without the need to perform direct EOS measurements to calibrate parameters. The method has been shown to give good results for liquid metals as well as for noble gases and molecular fluids (Kerley, 1983).

Internal degrees of freedom. The equations of state for molecular fluids not only have basic scientific interest but are also needed in theories for calculating the properties of substances that undergo chemical reactions when compressed or heated (see below). For small molecules, it is usually assumed that the motions of the molecular centers of mass are not strongly coupled to internal degrees of freedom and that the translational degrees of freedom, F_n, can be calculated by treating the molecules as spherically symmetric. The contributions from internal vibrations and rotations, F_{vr}, are computed from well known formulas, using the vibrational levels and constants of the isolated molecule (Herzberg, 1945). The internal degrees of freedom are important because they act as energy "sinks" - they contribute to the heat capacity but not to the pressure.

Thermal Electronic Excitation

Because current experimental techniques provide very little information about the contributions from thermal electronic excitation, F_e, almost all models rely on theories of electronic structure to construct this part of the EOS. This term *dominates* the EOS at high temperatures, but it is relatively small over most of the range presently accessible to experiment. Hugoniot data and isobaric expansion measurements at high temperatures provide information that can be used to test theories, although they do not uniquely separate effects of thermal electronic excitation from the other terms in the EOS.

In order to construct a model for the thermal electronic degrees of freedom, it is necessary to address two problems - determination of the energy for particular electronic configurations of the system, and calculation of a statistical mechanical average over all configurations. A general solution to this problem lies so far beyond existing computational capabilities that drastic approximations must be made. One of the principal difficulties is that of describing the pressure ionization of electronic levels that occurs with increasing density. The effective potential and the energy levels for an electron in an isolated ion are depicted schematically in Fig. 5. The discrete levels, with energies $\epsilon < \epsilon_0$, are bound by the potential well and hence are localized about the nucleus. The continuum levels, with $\epsilon > \epsilon_0$, have sufficient energy to escape the potential of the ion and hence are delocalized, or free to move about the system. The changes in electronic structure that occur with increasing density can be described heuristically by placing the ion at the center of a sphere with radius r_s, determined from the average volume per atom,

$$4\pi r_s^3/3 = V/N \ . \tag{9}$$

As seen in Fig. 5, those bound levels for which $\epsilon > \epsilon_s$ have sufficient energy to escape the ion sphere and so have become free states under compression. In other words, an increase in density lowers the bottom of the continuum and causes the electronic levels to change from localized to delocalized character. (The energy levels also shift with compression, but this fact does not affect the qualitative picture given here.) At low temperatures, pressure ionization of the highest occupied (valence) level is accompanied by a change from insulating to metallic properties; for substances that are metallic at room temperature and pressure, this insulator-metal transition normally occurs near the vapor-liquid coexistence region.

In constructing a theoretical model, it is difficult to treat the localized and delocalized states in a consistent fashion; different approximations are normally used in the two cases. Most models are based on one of two approaches, each of which has certain limitations. Ionization equilibrium theory, which treats the localization of atomic levels correctly, is most rigorous and easy to formulate at low densities. The other approach is to compute the electronic structure for an average configuration; this approximation vastly simplifies the computational problem at high densities, but it does not treat the localized states correctly. Attempts to merge these two theories have had only limited success, because neither approach gives satisfactory results at intermediate densities. Therefore, attempts to include pressure ionization in a model may lead to spurious predictions (Bushman and Fortov, 1983).

Ionization equilibrium theory. In the simplest form of the theory, (Saha, 1920), the system is treated as a weakly interacting mixture of free electrons and ions. The contributions of the ions to the EOS are expressed in terms of the partition functions (sums over the bound levels), and the free electrons are treated as an ideal gas. The ionization state is a function of density and temperature and is determined by balancing the ionization and recombination rates or, more generally, by minimizing the free energy, the two procedures giving identical results (More, 1981). The ANEOS code (Thompson and Lauson, 1972) uses an approximate form of the Saha theory (Zel'dovich and Raizer, 1966) to compute the thermal electronic contribution to the EOS. To eliminate divergence of the partition functions, it is necessary to cut off the sum over all bound states. However, the obvious procedure, simply dropping a state when it crosses into the continuum, is not acceptable because it predicts discontinuous changes in the EOS as a function of density. In order to avoid such difficulties, the ANEOS model sets all ion partition functions to unity and does not include any treatment of pressure ionization. This approximation is not accurate at low temperatures, but it should give reasonable results at high temperatures, where the thermal electronic terms are most important.

Efforts to extend the Saha theory to high densities incorporate a number of effects, including a systematic treatment of continuum lowering and pressure ionization, energy level shifts due to interactions between the bound and free electrons, electron degeneracy, and coupling between the electronic and nuclear degrees of freedom, due to the long range of the Coulomb interaction. Important contributions in this area include a number of papers by Rouse (Rouse, 1971; Godwal, *et al.*, 1981), the chemical model (Graboske, *et al.*, 1971; Gryaznov, *et al.*, 1980), and the activity expansion method (Rogers, 1981). Computer codes employing these models are used to construct parts of an EOS table in the density and temperature regime where they are deemed to be valid; they are then merged with the results of other models to define a complete EOS surface (Long, 1982). Still another approach is to treat the electrons as classical interacting particles, building the effects of the bound states into the interaction potential (Bushman and Fortov, 1983).

Average configuration models. At high densities, the isolated atom becomes less useful as a starting point for developing theoretical models, and it is desirable to make *a priori* calculations that incorporate the effects of density in a self-consistent manner. Because it is not feasible to make separate calculations for each electronic configuration, the common practice is to determine the structure and energy for only an *average* configuration and to compute the entropy from the Fermi statistics of noninteracting particles. This procedure is not satisfactory for describing excitations among the localized states and is found to give systematic errors in the entropy at low densities. However, it is a good approximation when the average electron-electron interaction is nearly independent of the configuration, as it is in the metallic regime at high densities.

The Thomas-Fermi (TF) "statistical" model, (More, 1981; Godwal, *et al.*, 1983), is the oldest and most widely used of the average atom theories; for example, it was used to compute many of the Sesame EOS tables (Holian, 1984). The TF model employs a simplified expression for the kinetic energy that is derived from the uniform electron gas problem and is asymptotically correct in the very high density limit. A consequence of the approximation is that the model does not predict any discrete bound states and so smears out structure in the EOS found in more sophisticated models. Because it does not predict an energy gap for electronic excitation or ionization, it overestimates the thermal electronic contribution at low temperatures in many cases. However, comparisons with both experiment and other theories show that the TF model does give reasonable results in an average sense and is a useful approach. A number of methods for introducing bound states and other improvements to the TF model have been proposed (More, 1981; Godwal, *et al.*, 1983).

Recently, interesting and encouraging results have been obtained using the INFERNO model of Liberman (1979). The potential for an average atom is defined by surrounding the ion sphere with an electron gas and a uniform positive charge that simulate the environment of neighboring atoms. The Dirac equation is solved to obtain wave functions and energies for both the discrete bound states and the continuum levels, with the usual requirement that the screened potential and the electron charge distribution be self-consistent. INFERNO calculations exhibit shell structure effects at low densities but agree with TF results in the high-density limit (Kerley, 1983). When a complete EOS was constructed for xenon and compared with Hugoniot data, it was found that the results at high pressures were rather sensitive to the model used for the thermal electronic term, and that the predictions using INFERNO were significantly better than those using TF theory (Kerley, 1983). A notable achievement of the model is that it treats bound and continuum states in a consistent way and predicts a reasonable transition from one to the other. However, because it uses the oversimplified expression for the entropy (which treats the bound states as delocalized), it has been found to predict spurious behavior near the metal-insulator transition. Therefore, the model has not been widely used to construct EOS for practical applications.

Generalizations of band theory to finite temperatures have also been used to compute contributions from thermal electronic excitation. These models also use the average configuration approach and calculate the entropy from the approximate formula used in the average atom models discussed above. Calculations have been reported for the metallization of iodine (McMahan, *et al.*, 1977) and xenon (Ross and McMahan, 1980) and for shock heating of the alkali metals (Ross, 1980b).

Another model proven to be useful, especially for inertial fusion research, is that employed in the XSNQ code (Zimmerman and More, 1980) which uses similar expressions for the thermodynamic functions as in other average atom theories, but computes energy levels from simple hydrogenic formulas fit to various data instead of from electronic structure calculations. Advantages of the model are that it is computationally fast enough to run in-line in radiation-hydrodynamic codes, that it provides data for mixtures of any elemental composition, and that it can be used for non-equilibrium problems.

Phase Transitions

In the previous discussion we explained how an EOS for a single phase is constructed by combining contributions from the electronic and nuclear degrees of freedom. If a substance can exist in more than one phase, the stable phase at a given pressure and temperature is the one having the lowest Gibbs free energy. The problem of including phase transitions in an equilibrium EOS requires only the application of thermodynamic principles. In the case of the vaporization transition, it is found that a single EOS describes both the vapor and condensed phases, as discussed below. To describe melting and solid-solid transitions, separate equations-of-state including the free energy must be generated for each phase. A number of EOS models now include a free-energy treatment of melting, but polymorphic phase transitions are normally handled more crudely. Computer programs have been developed for locating phase boundaries and constructing EOS tables that include an arbitrary number of phases (Cooper, 1979). The main difficulty encountered with these methods is that extra effort must be taken in constructing the EOS for the different phases, because the predicted phase lines are sensitive to small errors in free energy.

The problem of describing nonequilibrium phase transitions is more complicated. A rate equation must be formulated to describe the time evolution of the composition of the system. Because each phase has different dynamical properties, this equation is coupled to the hydrodynamic flow equations and must be solved simultaneously. A general approach to this problem has been described by Hayes (1975).

Solid-solid transitions. Although polymorphic phase transitions frequently exhibit slow transformation rates under static loading conditions, they usually occur very rapidly in dynamic experiments. The most common explanation given for this fact is that stress waves generate enormous numbers of defects that serve as nucleation sites for the process. Because of these rapid transition rates, an equilibrium model of the EOS surface is often found to be adequate for most purposes in describing dynamic phenomena. Classic triumphs of shock wave methods include discoveries of the 13 GPa $\alpha \rightarrow \epsilon$ transition in iron (Bancroft, *et al.*, 1956) and of the $\alpha\epsilon\gamma$ triple point (Johnson, 1962), that were later confirmed by static measurements. Duvall and Graham (1977) give numerous other examples, together with a discussion of how phase transitions affect the propagation of shock waves.

If two or more solid phases can exist in addition to the liquid, determination of the phase boundaries becomes much more complicated, and none of the commonly used EOS models solve this problem in a general way. A simple approximation is to use the Mie-Grüneisen formula with a Hugoniot reference curve that includes the measured changes in density and compressibility at the transition. The same results can be obtained by building the transition into the zero Kelvin isotherm, without any change in the thermal terms. These methods may not be satisfactory if the transition densities are strongly dependent on temperature. Grover's theoretical work on carbon (1979) illustrates a more general approach. He constructed separate EOS models for graphite, diamond, a metallic solid, and the liquid, and located the phase boundaries by matching free energies. An EOS computed in this fashion is usually put into tabular form for use in hydrodynamic codes.

Melting. An understanding of melting in dynamic experiments is needed not only for describing the EOS but also for determining the loss of material strength and elastic shear modulus. Static measurements of melting are presently limited to pressures below about 10 GPa (Cannon, 1974); melting is difficult to observe in standard Hugoniot experiments, but it can be studied using double shock-wave techniques. Temperature measurements give striking results, as illustrated by the classic work of Kormer and coworkers (1965) shown in Fig. 6. The data clearly exhibit discontinuities as the Hugoniot crosses and follows the phase line. Unfortunately, reliable temperature data can be obtained only for transparent materials. The second technique is to measure the release wave velocity in the shocked state (Asay, 1975; McQueen, *et al.*, 1984;

Brown and Shaner, 1984). Upon melting, there is a rapid drop in velocity from the longitudinal to the bulk value. Most of the existing data are consistent with the assumption that shock-induced melting is a very rapid process and that the general features of the process can be described using an equilibrium EOS. However, some experiments have shown that melting is a heterogeneous phenomenon in some solids, such as quartz (Grady, *et al.*, 1978); a description of such processes would require corrections to the simple equilibrium picture.

A number of semi-empirical expressions have been developed to estimate the pressure dependence of the melting temperature without explicit treatment of the liquid phase. One popular formula, derived from the Lindemann criterion, expresses the melting curve in terms of the Debye temperature for the solid (Urtiew and Grover, 1977). A natural extension of these ideas is to build the assumed melting law into the liquid EOS by "scaling" the solid properties, as discussed above, and using the free energies to determine the phase boundary (Grover, 1971; Thompson and Lauson, 1972). This approach is found to give reasonable results for many materials, and equations of state constructed for practical applications seldom go beyond this level of sophistication. However, many materials have complicated nonmonotonic melting curves, due to transitions in the solid and to changes in electronic structure that are not well described by these simple models (Duvall and Graham, 1977; Bushman and Fortov, 1983). A realistic treatment in such cases requires application of the free energy method, using carefully constructed equations of state for both the solid and liquid phases.

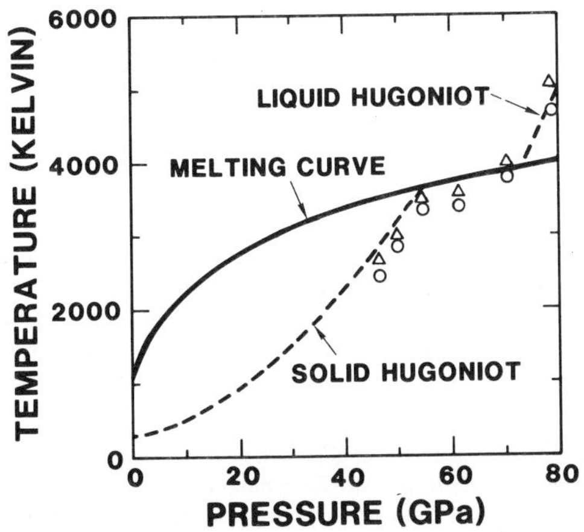

Fig. 6. Shock melting in NaCl (Kormer, *et al.*, 1965). Triangles and circles are temperature measurements at two different wavelengths. The curves are calculated, as discussed in the original paper.

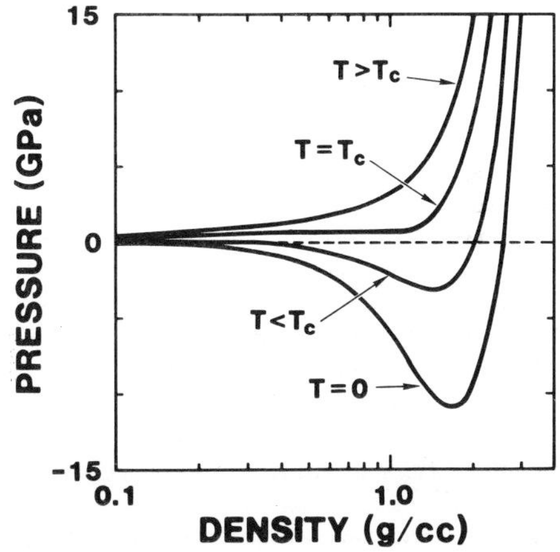

Fig. 7. Liquid-vapor coexistence region. Isotherms are shown for a fictitious material.

Vaporization. A realistic treatment of the vapor-liquid coexistence region and of expanded fluid states is necessary for describing the release behavior that follows shock loading in many hypervelocity impact problems. The most important features of vaporization, first explained by van der Waals and Maxwell (Hirschfelder, *et al.*, 1964), are illustrated in Fig. 7. In the gaslike regime at low densities, the pressure is determined by thermal motion of the atoms; in the liquid regime at high densities, it is determined by the intermolecular forces. Any attempt to describe the change from one regime to another by a single smooth EOS always predicts an unstable region, $(\partial P/\partial \rho)_T < 0$, at temperatures below the critical temperature T_c. These "van der Waals loops" do not describe the equilibrium behavior, because the real system can lower its free energy by separating into coexisting vapor and liquid phases. The properties of the two phases are determined by matching pressures and free energies (the "Maxwell construction"), leading to a coexistence region in which the composition of the vapor-liquid mixture varies and the pressure is constant. Some features of the transition are also present for temperatures above T_c, although there is no phase separation.

Even the simplest fluid models are found to give a rough description of vaporization, but quantitative results can seldom be obtained without adjusting parameters in the model to fit experimental data. These data are very limited in the case of most metals and many other materials that have critical temperatures well above the regime accessible to current experimental techniques. Static measurements of vapor pressure and coexistence curves have been made up to the critical point for a few low boiling metals (Hensel, 1981) and exhibit behavior similar to that observed in simple molecular fluids. Most theoretical models are based on the assumption that these same features will persist in all materials, even at much higher temperatures. For describing hypervelocity impact phenomena, the details of the vaporization region are not likely to be important unless the temperature on release is well above the boiling point, so that substantial momentum and energy can be imparted to a vapor cloud; this regime is likely to be important at the extremely high velocities associated with some applications. The very high temperature expanded fluid region has been studied experimentally using three dynamic techniques that were discussed earlier. Expansion isentropes for strongly shocked porous Pb and Cu were measured using the impedance-matching technique (Al'tshuler, *et al.*, 1980). Isobaric expansion measurements have been made to temperatures as high as 8000 K for several metals (Shaner, *et. al.*, 1976; Gathers, 1983). These two methods provide information about the expanded liquid but are not very sensitive to the vaporization process. A newer method, measurement of the stagnation pressure produced by shock-induced vaporization, provides data that should help to understand the expansion of a debris cloud produced by hypervelocity impact (Trucano and Asay, 1986).

Plasma phase transitions. Experimental data and theoretical arguments both suggest that vaporization and the properties of hot, expanded fluids are further complicated by thermal electronic excitation. Because of its low critical temperature, mercury has been studied extensively, using static methods (Hensel, 1981); the data show the existence of a metal-insulator transition and other effects of changes in electronic structure that are not well understood. Recent measurements for cesium and rubidium have revealed subtle differences in the coexistence curves from those of insulating fluids (Jungst, *et al.*, 1985). It is reasonable to expect that these anomalies become more important at higher temperatures where the degree of electronic excitation increases. Kerley (1986a) has made a theoretical study of the metal-nonmetal transition in aluminum, using both ionization equilibrium theory and the INFERNO code to compute the thermal electronic contributions. His model predicts the existence of two liquid phases at temperatures in the range of 4000–7000 K, a phenomenon not observed in simple fluids. Experimental data to test this prediction are not yet available.

Mixtures and the Effects of Chemical Reactions

In the previous discussion, we assumed that the EOS could be separated into additive contributions from different degrees of freedom. This approximation can be very poor for materials that vary in chemical composition with changes in temperature and pressure. Chemical reactions cause particularly dramatic effects in explosives, where the release of chemical energy and subsequent expansion of the reaction products can produce a detonation (Mader, 1979). Considerable effort has been devoted to the measurement and calculation of detonation-product EOS, which are crucial to the description of detonation waves and their interactions with other materials. Although explosives have received more attention, plastics, polymers, and many other "inert" compounds also undergo chemical reactions when subjected to high temperatures and pressures and have EOS similar to those for detonation products. For example, most hydrocarbons display a break or change in slope in the Hugoniot curve at about 20 GPa that is due to decomposition into elemental carbon and small molecules (Nellis, *et al.*, 1984a). Samples recovered from shocked teflon show that decomposition into carbon and gaseous fluorocarbons has occurred (Morris, *et al.*, 1984). Hugoniot measurements for liquid nitrogen and oxygen show a dramatic softening above 40 GPa that is due to molecular dissociation (Nellis, *et al.*, 1984b).

Semi-empirical calculations. Despite recent progress, all current theories of chemical effects have a number of difficulties and limitations, so that an explicit treatment of the problem is not available in any of the commonly used EOS models. In most cases, it is necessary to ignore the above objections and to use one of the simple models, varying the zero Kelvin isotherm and Grüneisen parameter to force agreement with Hugoniot data. Some effects due to chemical reactions can be reproduced by putting a break or change in the slope of the isotherm, as is done to treat phase transitions. Furthermore, most models are formulated to give the correct limit at high temperatures, where the material has completely dissociated to elemental species. The principal weakness of this methodology, when applied to substances that undergo chemical reactions, is lack of predictive power. Because the model is built upon assumptions that are inconsistent with the behavior of the material, the EOS can give poor results outside the range of experimental data.

Theory of chemical equilibrium. A realistic model of dissociation is formulated by treating the material as a homogeneous mixture (or solution) of several chemical species, having a composition that can vary as a function of density and temperature. For simplicity, it is almost always assumed that the system is at equilibrium, so that the composition can be determined by minimizing the free energy, as in treatments of ionization equilibrium. Conclusions about the general validity of the equilibrium approximation must await further study, but most of the work to date suggests that it is at least a reasonable starting point for developing EOS theories. With this assumption, the principal tasks are to develop a model for the EOS of the mixture and to determine the important chemical components and their EOS parameters.

Several theories are available for calculating the properties of mixtures. In the "one fluid" approximation, the intermolecular potential parameters are averaged over all species, and the system is treated as if it consisted of a single constituent, except that an entropy of mixing term is added to the free energy. This model has been used successfully in the BKW EOS (Mader, 1979) and in the CHEQ code (Ree, 1984), for calculating detonation properties of explosives. A second method, that treats the multicomponent problem explicitly, is based upon generalization of liquid perturbation theory. This formulation is computationally more expensive, but Chirat and Pittion-Rossillon (1981) have found a simplified version that gives good results for detonation products. A third approach, "ideal mixing," treats the system as a heterogeneous mixture with all species at the same temperature and pressure, except that the entropy of mixing term is included in the free energy. The advantage of this model is that the properties of the mixture are computed directly from the EOS of the individual species, without the use of intermolecular pair potentials; therefore, it can be applied to mixtures where complex species, such as liquid metallic carbon, are present. This theory has given good results when applied to detonation products, molecular fluids, and hydrocarbons (Kerley, 1985, 1986b; Kerley and Switendick, 1986).

Nonequilibrium chemical reactions. As in the case of phase transitions, the description of nonequilibrium chemical processes requires the formulation of a rate equation that is coupled to the hydrodynamic flow equations. An important application of these methods is in the treatment of initiation in explosives (Mader, 1979; Nunziato, *et al.*, 1982).

CONSTITUTIVE RESPONSE

Previous discussions have centered on thermodynamic descriptions of the hydrostatic response of materials, which describes only fluidlike behavior. Shock loading of solids produces a three-dimensional stress state that is usually described in terms of an average stress, assumed to be the hydrostatic pressure, and a deviatoric term that is related to the shear stress in plane shock propagation. The shear stress results in a wide variety of mechanical states important to the compressive and tensile yielding of materials during dynamic loading. Description of these effects is referred to as constitutive modeling and requires not only the EOS, but also the yield behavior and the shear modulus of materials. In this section we review the constitutive descriptions of high-rate material response necessary to describe hypervelocity impact phenomena. Our intent is to discuss the current material models used to describe high-rate deformation, to illustrate how dynamic material properties can be obtained from experimental data, and to identify recent developments in this area.

Plane shock loading plays a central role in obtaining dynamic properties to construct constitutive relations for use in continuum mechanics codes. For this reason, the following discussion will emphasize material response corresponding to uniaxial strain. Generalized three-dimensional descriptions are given in a number of works, including (Herrmann, 1970; Dienes and Walsh, 1970; Moss and Glenn, 1984).

Elastic-Plastic Response

Overview. Many materials of engineering interest can be appropriately described in terms of elastic-plastic response. In this model, the initial response is elastic, or reversible, until a threshold stress is reached, where the response becomes plastic, or irreversible. For use in large computer codes, the following features of an elastic-plastic model are desirable: (1) the model should allow implementation of von Mises and possibly more general yield surfaces; (2) it should be capable of describing large strain, strain-rate, pressure and temperature effects on dynamic yield strength; (3) it should allow independent evaluation of model parameters from a variety of experimental testing methods; (4) it should be compatible with damage-accumulation models for predicting tensile failure; (5) it should properly account for any history dependence of strength properties, including degradation of properties due to shock loading and recovery of strength after rapid deformation; and (6) it should be compatible with computational requirements on storage and computing time.

The theoretical basis for describing elastic-plastic compression in dynamic experiments was established several years ago; key early studies can be found in the works of Taylor (1942), Pack, Evans and James (1948), Wood (1952), Morland (1959), and Fowles (1961). More recent theoretical investigations of elastic-plastic behavior are reported by Herrmann (1970), Herrmann, *et al.* (1970), Moss and Glenn (1984), Wallace (1984), and Bodner and Rubin (1985), although considerable controversy exists regarding theoretical modeling of plasticity at large strains and strain rates.

Stress states. For use in computer models the generalized stress state, σ_{ij}, can be written as

$$\sigma_{ij} = (P + Q)\,\delta_{ij} + S_{ij} \quad , \tag{10}$$

where P is the hydrostatic pressure determined by the EOS theories discussed earlier, S_{ij} are deviator stresses that can be related to the yield stress through the flow function, and Q is an artificial viscosity used for numerical stability. For plane shock loading in an isotropic material, the principal longitudinal stress in the direction of motion, σ_x, and the principal lateral stresses, $\sigma_y(= \sigma_z)$, are related to the maximum resolved shear stress as $\sigma_x - \sigma_y = 2\tau$. Using the von Mises yield condition, the longitudinal stress, is given by the hydrostatic pressure and the deviatoric stress as (Fowles, 1961; Herrmann, 1976; Duvall, 1971; Chartagnac, 1982)

$$\sigma_x = P + \frac{4}{3}\tau \quad . \tag{11}$$

The cyclic response of an elastic-perfectly-plastic material in uniaxial strain is illustrated in Fig. 8a. The material response is elastic to state A, the Hugoniot elastic limit (HEL), followed by plastic compression to state B. During unloading, the response is again elastic to C, with plastic decompression from C to zero stress. The corresponding stress-wave structure for plane loading is illustrated in Fig. 8b. At low shock stresses (a few GPa), an elastic precursor is produced during compression, followed by a plastic wave. At sufficiently high stresses (on the order of several tens of GPa), the elastic wave is overdriven, resulting in a single elastic-plastic shock wave. For unloading, a continuous elastic wave is produced to state C, followed by a continuous plastic rarefaction wave. The solid wave profile in Fig. 8b corresponds to ideal, rate-independent behavior, although strain-rate effects, viscoplastic response, strain hardening and Bauschinger effects produce deviations from the ideal behavior, as illustrated by the dashed profile in the figure.

To utilize the general capabilities of the elastic-plastic model, it is desirable to describe the dependence of the yield function on strain, strain-rate, pressure, temperature, and loading history. Clearly, this is a formidable task that is often not practical to accomplish, because of the lack of property data and also from limitations in computing capability.

Although yield parameters can be estimated from shock-wave data alone, a common approach is to employ several different experimental methods for estimating the appropriate parameters. Quasi-static uniaxial stress tests, Taylor tests, Hopkinson bar experiments, and torsion testing have been used to define the effects of plastic strain, strain-rate and temperature on yielding (Herrmann and colleagues, 1970; Cochran and Guinan, 1976; Johnson, 1978; Johnson and Cook, 1983; Steinberg, *et al.*, 1973, 1980, 1984; Moss and Glenn, 1984; Steinberg, 1986). A particularly interesting technique for investigating high-strain-rate and strength effects involves the pressure-shear loading of thin specimens developed by Clifton (1984).

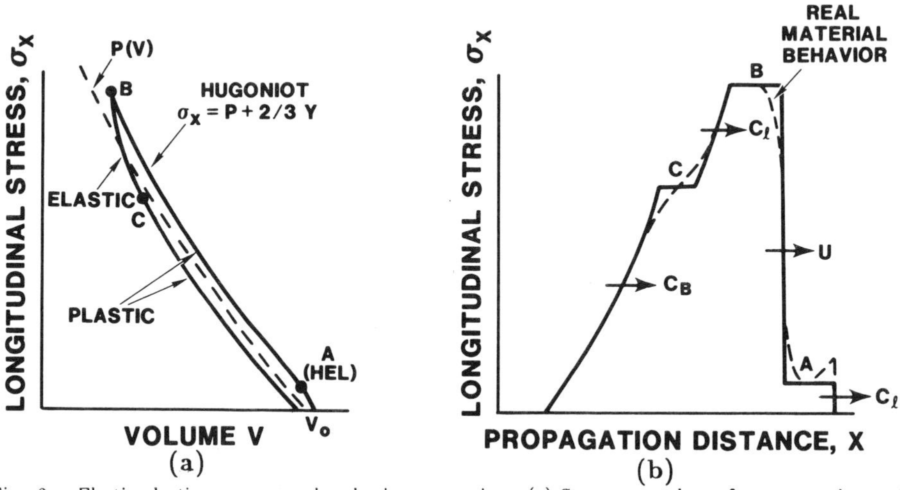

Fig. 8. Elastic-plastic response under shock compression. (a) Stress vs. volume for compression and release under uniaxial strain. (b) Shock and rarefaction waves in an elastic-plastic material.

Determination of Strength Parameters. The compressive yield strength after shock loading is important to several aspects of dynamic material response, including shock attenuation (Curran, 1963; Herrmann and Lawrence, 1978) and impact crater growth and final size (Rosenblatt, 1970). Since the high loading rates achieved by shock loading result in unique defect and strength states, measurements of yield behavior at low-to-intermediate strain rates and stresses achieveable in quasi-static or Hopkinson-bar experiments often cannot be reliably extrapolated to the ultra-high shock-loading stresses and rates, although some success in correlating the two regimes is being achieved (Steinberg, 1986; Follansbee, 1986). In addition, present theories of dynamic response are not adequate for predicting material strength from first principles, so that yield behavior under dynamic loading is presently determined from a combination of low, intermediate and high strain-rate data (Steinberg, 1986).

Steinberg and coworkers (1980, 1986) and Johnson and Cook (1983) have developed generalized strength models which incorporate the effects of strain hardening, strain-rate, temperature, and pressure effects on material strength for use in wavecodes, such as HEMP (Wilkins, 1969), WONDY (Kipp and Lawrence, 1982), and EPIC (Johnson, 1978). Strength models have also been extended to describe the effects of shock-induced melting on material strength (Steinberg and Guinan, 1973; Steinberg, 1986) and anisotropic strain hardening (Herrmann, *et al.*, 1970; Steinberg and coworkers, 1973, 1980, 1984; Bammann and Johnson, 1986). Currently, such models have the capability for describing the essential features of dynamic loading and unloading, although they are essentially limited to the experimental data ranges.

The initial dynamic yield strength can be determined from the amplitude of the HEL (Fowles, 1961), as suggested in Fig. 8, or estimated from low strain-rate data if rate effects are negligible (Steinberg, 1986); Jones and Graham (1971) have compiled Hugoniot elastic limits obtained from wave structure measurements for a large number of materials. This approach only determines the initial value of Y, and does not provide information about material strength in the shocked state.

Asay and Lipkin (1978), and Asay and Chhabildas (1981) have shown that unloading and reloading wave profile measurements can be used to determine material strength in the shocked state. These data are also useful for determining elastic moduli in the compressed state, for estimating the high-pressure sound speed, for evaluating Bauschinger effects during unloading, and for determining strain-rate effects. Alternatively, these effects can be predicted from low strain-rate data, as discussed by Steinberg (1986), although extrapolation of data over large ranges of strain rate should be carefully evaluated. Fig. 9 gives the dependence of material strength on shock stress for several materials, which illustrates that order-of-magnitude changes are often obtained after shock compression to high stresses. In many cases, strength changes are minimal until a critical shock stress is reached, where a significant change in hardening is observed. The abrupt increase in shock hardening observed for these cases is not presently understood, but may be due to heterogeneous yielding effects (Grady, *et al.*, 1983; Swegle and Grady, 1986). After reaching a maximum, the gradual decrease in strength with Hugoniot stress shown in Fig. 9 for several materials results from the shock-induced temperature approaching the pressure-dependent melt boundary. Loss of both shear modulus and shear strength has been observed in several materials at stresses sufficient to produce shock-induced melting. Examples include aluminum (Asay and Hayes, 1975; McQueen, *et al.*, 1984) and tantalum (Brown and Shaner, 1984).

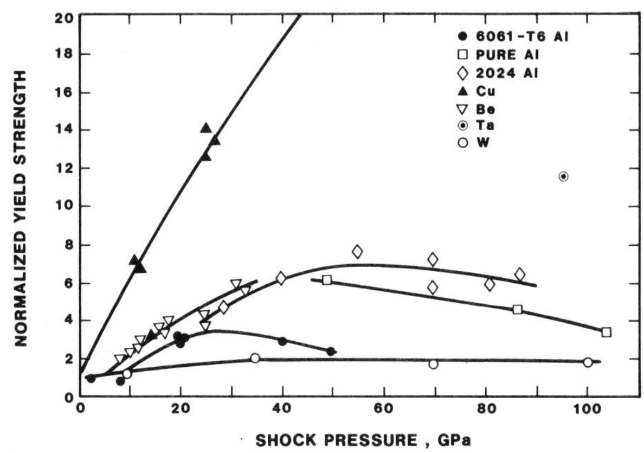

Fig. 9. Dynamic yield strength in the shocked state vs. shock stress for several metals. High pressure strengths have been normalized to initial values for all materials. 6061-T6Al (Asay and Chhabildas, 1981); pure Al (Asay, *et al.*, 1986); 2024 Al (Morris, *et al.*, 1982); Be (Chhabildas, *et al.*, 1982); Ta (Asay and Chhabildas, 1986); Tungsten (Chhabildas and Barker, 1986); Cu (Chhabildas and Asay, 1982).

Strain-rate effects. Strain-rate effects occurring during initial loading in crystalline materials are manifested by dynamic elastic over-stresses which exceed static values, and by viscous stresses developed during plastic deformation. The importance of strain-rate effects is not clearly apparent in all dynamic applications. Viscous stresses arising during shock compression control the loading rate achieved in steady compression waves, which is important in several applications of dynamic response. On the other hand, the large strain-rate effects of yielding observed at low-to-intermediate rates are often not apparent in shock-wave experiments. Therefore, these effects have to be carefully evaluated for each material. Although the use of rate-independent material models gives good results for many practical engineering problems, inclusion of rate-dependence is necessary to achieve accurate numerical solutions in several dynamic problems, such as fracture and fragmentation and in calculations of stress-wave structure.

Viscoplastic response is often described in terms of a Maxwell-type relation (Herrmann, 1976), where deviations from elastic response are described by a relaxation function. Stress relaxation occurs on a time scale determined from either micromechanical models of inelastic deformation (Johnson and Barker, 1969; Davison and Graham, 1979), or from empirical theories based on an overstress from equilibrium (Herrmann, *et al.*, 1970; Steinberg and Sharp, 1980). These models generally produce a decaying elastic wave with stress relaxation. In fact, study of the elastic precursor wave structure can be used to deduce mechanical and physical properties controlling dynamic yielding; a large number of investigations have been performed on various aspects of rate-dependent yielding, as discussed by Davison and Graham (1979). Clifton (1984) has shown that the pressure-shear loading technique can be used directly to evaluate the dependence of shear stress on strain rate in the high-rate regime. This method provides an alternate technique for directly determining rate-dependent effects, which is extremely valuable for constitutive model development.

Another consequence of strain-rate effects is apparent in the plastic shock-wave structure. The plastic shock wave is not discontinuous, as illustrated by the ideal (solid lines) behavior depicted in Fig. 8b, but has a finite risetime. Viscoplastic response results in steady plastic-wave propagation for materials with positive stress-volume curvature, and is a consequence of the balance between non-linear material response tending to steepen the wave and dissipative effects that cause dispersion. Studies of steady shock compression have been utilized for the purpose of studying viscous properties; for example, see Johnson and Barker (1969), Herrmann (1970), Grady (1980), Wallace (1984), Swegle and Grady (1985).

Since viscous properties are usually determined experimentally and interpreted in terms of empirical or micromechanical models of limited applicability, they cannot usually be extrapolated outside the experimental measurement range with confidence. Recently, however, Grady (1981) has made progress in understanding the basic mechanisms of shock viscosity that offers promise for accurately extrapolating empirical measurements over a broad range of shock stresses and loading rates. The approach is based on the observation that the risetime of steady plastic shock waves varies as the inverse fourth power of shock amplitude for a wide variety of materials in the intermediate stress range. The fourth-power dependence is illustrated in Fig. 10, and implies that the viscous shear stress varies as the inverse square root of strain rate. Although the underlying mechanisms for this observation are not known, Moss (1985) has proposed a possible explanation based on the effective stress field of dislocations. This observation has been used by Swegle and Grady (1985) to develop a constitutive model that includes both viscous behavior and strain hardening in a three-dimensional constitutive framework suitable for implementation in wavecodes, which has been successful in reproducing the steady-wave features of propagating waves. The use of real material viscosities in numerical solutions is potentially useful for engineering problems, since the requirement for numerical stability is satisfied simultaneously with the calculation of realistic wave risetimes.

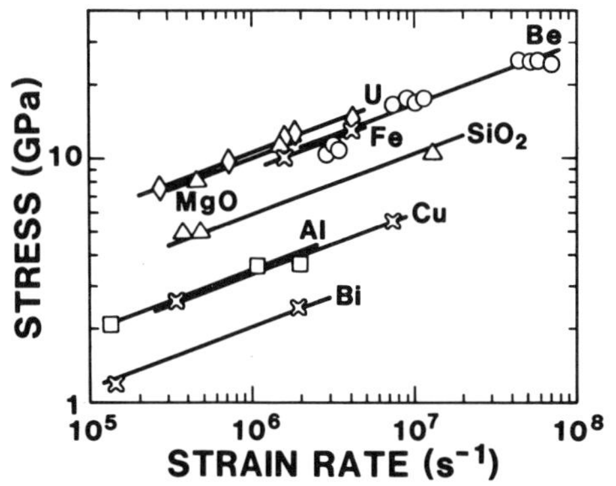

Fig. 10. Experimental measurements of peak strain rate in steady waves versus Hugoniot stress.

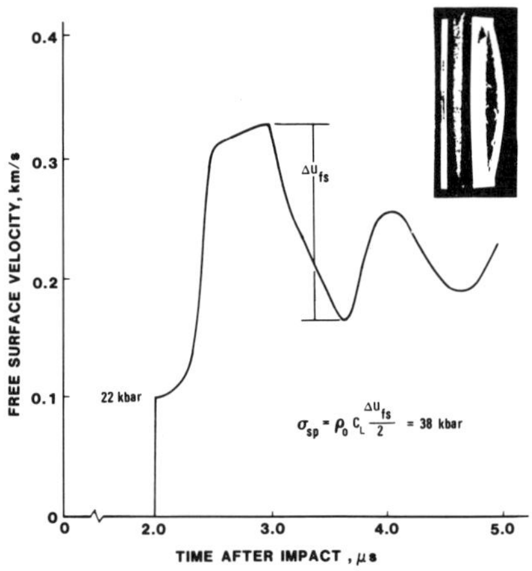

Fig. 11. Experimental measurement of the dynamic spall strength of steel using velocity interferometery. The inset shows a post-shock polished section of the target and spalled piece (Berthoff, *et al.*, 1975).

Heterogeneous Deformation. Recent studies by Asay and Chhabildas (1981) and Grady and coworkers (1983) indicate major discrepancies with the elastic-plastic model at high strain rates achieved in steady shock compression. These results show that homogeneous deformation is unstable at the high rates of deformation, resulting in microscopic shear bands produced in the shock-induced microstructure. At sufficiently high deformation rates, the heat produced by "inelastic" shock compression is deposited in the bands, resulting in temporary loss of shear strength, which is later recovered as the bands cool. The consequence of this shock-induced softening mechanism is that the stress state lies within the yield surface immediately after shock compression. Additional compression from the shocked state results in large elastic stresses, in contrast to predictions of the elastic-plastic theory. Additional discussions of shock-induced microstructure on material properties can be found in Murr, Staudhammer, and Meyers (1986).

Recent theoretical studies (Swegle and Grady, 1986) indicate that such complex material response can be treated within the context of continuum models in a self-consistent theory appropriate for use in wavecodes. An elastic-plastic model is proposed that treats microscopic shear bands and matrix material as separate phases in a continuum mixture. The theory allows for thermal nonequilibrium during shock dissipation, combined with thermal transport between phases, to produce a homogeneous temperature state after shock passage. The yield stress is taken to be a function of the shear-band temperature, so that thermally softened states are produced immediately after shock compression, followed by hardening as the bands cool. The model is an important advance in describing complex material behavior at high rates and represents a class of next-generation models needed for improved plasticity descriptions. Other viscoplastic theories have also been developed that show promise for describing similar history-dependent effects (Bammann, 1986; Shockey, *et al.*, 1985). These developments are important because they offer the possibility for basing model construction on physical mechanisms appropriate to shock compression in order to provide better extrapolations outside experimental measurement ranges. An important point is that these mechanisms are unique to high-rate processes and therefore cannot be extrapolated from low-strain-rate data. For this reason, the use of lower-rate strength data should be viewed with caution when determining the constitutive response of materials at extremely high rates.

It is also important to note that advanced elastic-plastic models are typically implemented in Lagrangian codes, although Eulerian codes remain the primary computational tools for hypervelocity impact problems. Thus, limitations in the use of advanced models are often imposed by the complexity of the problem, which places restrictions on the type of computer code needed. An excellent summary of numerical capabilities for treating hypervelocity phenomena is given in the review article by Anderson and Johnson (1986).

Composite Materials

Composite materials are playing an increasingly important role in hypervelocity applications because of their high strength-to-weight structural bearing capabilities. The dynamic response of these materials, which involves the interaction of different constituents, is an area of material response that is presently not as well understood as elastic-plastic behavior under hypervelocity impact conditions. In constructing material models for composites, the major considerations are (1) "equations of state" that are valid from low to high pressures, (2) descriptions of compressive and tensile failure, (3) non-equilibrium heating and thermal transport between phases, and (4) decomposition of constituents at high energies.

Composite materials fall in the general class of mixture materials, which have been recently reviewed by several authors (Bedford, *et al.*, 1976; Bedford and Drumheller, 1983; Passman and Batra, 1984). The two general classes consist of high-quality regular arrays of constituent materials, such as laminated carbon-carbon composites, and engineering composites, which are typically random arrays of constituent materials, such as particulate-loaded polymers.

There are several approaches for modeling the dynamic response of mixtures. The older techniques employ equivalent modulus models, in which the composite is assumed to be homogeneous and volume-averaged properties are determined under the assumption of equal pressures. McQueen and colleagues (1970) used this approach to determine the high-pressure Hugoniots of several metal mixtures, *e.g.*, alloys, over the pressure range from a few tens of GPa to a few hundred GPa with good success. They used ideal mixing rules to calculate a zero-Kelvin isotherm for the mixture, corrected for the different temperatures occurring in the constituents during shock compression, by using the M-G equation, weighted specific heats and average Grüneisen parameters. This approach gave good results for the ultrasonic velocities at ambient pressure; data were not available to evaluate high-pressure unloading wave velocities.

An important class of composites includes laminated composites which play an important role in engineering applications. Although the equal-pressure assumption should apply to the dynamic response at high pressure, the behavior of composites is more complex at low pressures, where shock-wave dispersion and strength effects prevail. Drumheller and Sutherland (1973) developed equivalent lattice models for laminated composites that account for the experimentally observed dispersive behavior of shock propagation (Drumheller and Lundergan, 1975). Similar models have been developed by Hegemier and Bache (1973). Munson and Schuler (1971a) neglected strength effects and developed effective moduli relations for describing the low-pressure Hugoniot response of carbon-carbon composites that adequately predicted pressure-volume states from constituent properties. The approach was extended by Barker (1971) who used a Maxwell relation to describe the relaxation from instantaneous to equilibrium behavior in laminated composites. The theory has been extended to both regular and random polymeric composites with good success (Munson and Schuler, 1971a; Munson and Schuler, 1971b; Munson, 1973). This model is easy to implement, using a limited number of material properties of each constituent and a single shock-wave profile measurement to establish the relaxation time; it gives reasonably good predictions of wave dispersion in composites at low pressures. A limitation is that relaxation times cannot be reliably determined *a priori* from properties of the individual layer thicknesses so that it is necessary to experimentally measure these parameters.

Drumheller (1982) has recently developed an improved three-dimensional model for granular composites, based on immiscible mixture theory, that properly accounts for the shear states produced under dynamic loading. The model has been used to successfully describe both longitudinal and transverse stress fields for shock loading of particulate-loaded polymeric materials (Chhabildas and Swegle, 1982) and can also be used to account for porosity in a composite (Drumheller, 1986).

An issue that has not yet received adequate attention concerns the thermodynamic response of composites at very high shock pressures. A major question concerns the relative portion of energy deposited in each constituent during steady-wave propagation. Although the total energy dissipated by the shock is properly accounted for by the conservation equations, it is uncertain whether this energy is equally shared by the constituents during initial shock compression or deposited preferrentially in one. The dynamic material response can be substantially influenced by the energy deposition process, with possible phase transitions resulting from high local temperatures. Recent work by Drumheller (1986) is beginning to illuminate this question.

Porous Materials

Solids in distended or porous states play an exceedingly important role in a variety of applications. Some of the advantages of these materials accrue from the fact that they absorb energy upon impact, thus mitigating the consequences of shock loading. Shock loading of distended materials usually results in compaction to a partially or fully compacted state, with attendant heating associated with pore compaction. In describing the dynamic response of these materials, it is generally necessary to describe both the compaction behavior at low pressures and the effects of heating at high pressures.

Historically, the development of theoretical models for describing shock compaction of porous materials evolved from the assumption of rigid locking during the compaction response (this model is usually referred to as the "snow plow" model). In this approximation, compression from the initial distended state to the final state, assumed to be fully compacted, occurs at infinitesimally small loading pressures (Zel'dovich and Raizer, 1966; McQueen, *et al.*, 1970). The resulting thermodynamic state of the compacted solid is hotter than that obtained by shock compression of initially fully dense material because of the mechanical work performed during compaction. The MG theory, Eq. (4), has been traditionally used to describe the Hugoniot response of distended materials at high shock pressures (above a few tens of Gpa), since it allows use of the known Hugoniot for the fully dense material; however, any of the equations of state discussed earlier can also be used, in principle. The functional form of the resulting Hugoniot is particularly useful when the product of the density and the Grüneisen parameter is assumed constant, and a liner relationship for the shock velocity – particle velocity relation applies. The Hugoniot equation can be used for analytic calculations or employed in a variety of Lagrangian and Eulerian computer codes for numerical solution. When used in a complete thermodynamic theory, this approach should give realistic estimations of temperature states achieved during shock compression to full compaction. A limitation is that the model does not account for states of partial compaction important at low pressures (a few GPa).

To overcome this limitation, Herrmann and colleagues developed a set of phenomenogical equations, commonly referred to as the P-α model, that accurately reproduce the experimentally observed low-pressure compaction behavior of porous materials (Herrmann, 1969a). The model requires, in addition to the Hugoniot of the solid, specification of the initial elastic response in the porous state, a function which describes the compaction behavior, and the pressure at which full compaction occurs. The P-α model generally provides realistic representations of experimental data and has been used for a large variety of materials (Butcher, 1966; Lee, *et al.*, 1970; Butcher, 1971; Butcher, *et al.*, 1974). It has also been used to describe initially heated materials (Herrmann, 1969b; Perry, 1973).

In its simplest form, the P-α model does not account for rate effects associated with pore collapse. Carroll and Holt (1972) developed a theory of rate-dependent pore collapse that has been used to describe the risetime of compressive shock waves in aluminum with good accuracy. However, rate effects are predominant primarily at lower impact pressures, and generally have negligible consequence on compaction behavior in practical applications.

In some applications, it is desirable to describe the dynamic response of porous materials that contain another constituent, such as water. Various mixture theories have been developed for describing such effects (Bedford and Drumheller, 1983; Passman and Batra, 1984; Drumheller, 1985; Drumheller, 1986).

Dynamic Fracture and Fragmentation

Overview. To describe the dynamic debris clouds generated by hypervelocity impact, it is necessary to determine the size and velocity distribution of individual fragments or particles, as well as the thermodynamic state of the cloud constituents. For numerical simulations, it is also desirable to simulate the relative motions of particles and other debris products in the mixture. The interaction of high-amplitude compressive stress waves produced during hypervelocity impact with free surfaces on the target structure can produce high residual temperatures, as well as tensile loading resulting in material fracture.

Dynamic fracture proceeds by the nucleation and growth of cracks or voids under tensile loading; these defects ultimately coalesce, producing material failure. To provide realistic computer simulations of the fragmented state, the following features of a fracture model are desirable: (1) the model should describe material failure resulting from ductile or brittle fracture, including strain-rate effects; (2) it should be cast in a properly invariant three-dimensional form and provide an accurate representation of material response in tension; (3) it should be compatible with the compressive yield behavior, allowing for preconditioning by the stress wave and degradation of material properties due to partial states of damage; (4) it should accurately describe the tensile pressure, deviatoric stress, and temperature dependence of material failure, including the effects of phase transitions; (5) it should be amenable to predicting the sizes and velocities of fragments resulting from material failure; and (6) it should be possible to construct the model from easily obtained experimental data.

Spallation. The interaction of rarefaction waves in a body can produce tensile states of stress and strain, often resulting in fracture of the body. A common form of dynamic fracture is referred to as spallation, which is defined as the complete separation or partial separation of a material resulting from the production of tensile states within the interior. Spallation was first reported by Hopkinson in 1914 and occurs in a wide range of applications, such as explosive loading, flyer plate impact, and electron beam or laser deposition. Reviews of spallation phenomena are given in Butcher and coworkers (1964), Curran and coworkers (1977), Novikov (1981), and Davison and Graham (1979).

Spallation occurs as the consequence of the nucleation and growth of cracks or voids produced under the action of a tensile stress. A region of damage develops, ultimately resulting in complete material separation. The tensile stress necessary to cause a given level of damage (usually complete separation) is referred to as the spall strength of the material and, in general, is a function of the pulse duration or tensile loading rate, and possibly the loading prehistory. Spallation experiments under well controlled planar impact conditions provide a means of studying the dynamic fracture processes.

Figure 11 illustrates an experimental free surface velocity profile obtained on a steel specimen undergoing spallation, obtained from an interferometer measurement of a specimen (Berthoff, *et al.*, 1975) impacted with a thin plate. The inset illustrates the resulting failure induced in a specimen for this loading process. The pull-back in free surface velocity can be used to infer the spall strength of the material for the imposed loading condition, whereas the structure of the free surface velocity in the vicinity of the pullback can be used to study damage evolution processes leading to material failure (Shockey, *et al.*, 1973a; Davison, *et al.*, 1977; Cochran and Banner, 1977; Grady and Kipp, 1980).

Threshold Damage Theories. A generalized description of dynamic fracture requires definition of a threshold stress for material failure. Recent models of dynamic failure are based on the use of internal state variable theories to describe the degree of material degradation, which is usually characterized by a scalar or tensor quantity referred to as material damage. The damage parameter can be used to specify void or crack intensity in both ductile and brittle fracture; an evolution equation is usually developed to describe the dependence of material damage on loading history.

The simplest description of material failure is embodied in the critical tensile stress model, in which fracture is specified by a critical mean tensile stress threshold necessary to cause complete material separation. This failure criterion, which is not a damage model, is used in the majority of impact problems because of its simplicity, the ease of characterizing spall strength from experimental data, and the fact that it can be used in most Lagrangian and Eulerian wavecodes. Its limitations are that it does not describe effects of pulse duration or fracture-time dependence on failure, the degradation of material properties induced by partial states of damage, or a description of stress reduction resulting from material damage.

The Butcher-Tuler model is representative of a class of failure models allowing the next degree of sophistication in describing dynamic failure (Tuler and Butcher, 1968). A similar approach has been used by Soviet investigators (Zlatin and Ioffe, 1973). In these models, material damage is described as the integral of a stress function over time. The model provides for a fracture-time dependence to spall and has been successfully used in wavecodes (Tuler and Butcher, 1968; Rajendran and Bless, 1985) to describe the effects of pulse duration on spallation in plate-impact experiments. Although developed for one-dimensional strain loading, it has been applied to more general loading conditions (Wilkins, 1977; Rajendran and Bless, 1985). A limitation is that it does not account for material degradation due to partial damage.

A more general damage accumulation model, developed by Johnson and Cook (1983), includes a dependence on homologous temperature and equivalent strain rate, as well as equivalent tensile stress. The model requires five constants, which can be evaluated from a combination of torsional, Hopkinson bar, quasi-static tension and dynamic plate-impact tests, and has been used in various impact calculations, including Taylor-impact tests.

Nucleation and Growth Models. One of the most general capabilities currently available for modeling dynamic fracture damage is embodied in the nucleation and growth (NAG) theories, which attempt to link material damage occurring at the microscopic level with material response at the continuum level; the main objective is to describe resulting material damage produced by the nucleation of cracks or voids at imperfections, and subsequent growth of individual microcracks and voids with time. The models evolved as a consequence of the inability of simpler models to extrapolate to loading conditions outside experimental domains and to provide a means of evaluating the effects of partial damage.

The development of NAG models for describing dynamic fracture dates to the early 1970s. Several dynamic fracture models have been developed to account for either the nucleation and growth of microcracks under dynamic tensile loading to describe brittle failure, or the nucleation and growth of voids in tensile states to describe ductile failure. The simplest of this class of models is represented by the work of Cochran and Banner (1977), who developed a spallation model with two parameters to account for degradation in material properties with damage accumulation. The model utilizes a damage parameter proportional to the volume of microcracks and a constant spall strength, which is related to a Griffth-like energy criterion in fracture mechanics. A central feature of the model is a constitutive relationship that incorporates material damage for states of partial fracture, which allows determination of gradual stress reduction resulting from tensile damage. The model is useful in wavecode calculations because of its simplicity and the fact that only two parameters are necessary for implementation. However, limitations are that it does not account for fracture time-dependence or describe microdamage and the resulting residual material properties. It has been implemented into wavecodes and successfully used to describe planar fracture damage in several applications (Cochran and Banner, 1977; Steinberg, 1986).

The group at SRI International has been instrumental in developing more general NAG models for describing ductile and brittle failure under dynamic loading. These models incorporate realistic physical processes for the nucleation and growth of either voids or fractures and have been used to describe material failure resulting from microdamage in a variety of materials, including metals, porous materials and composites (Barbee, *et al.*, 1972; Seaman, *et al.*, 1976; Curran, *et al.*, 1977; Shockey, *et al.*, 1985; Seaman, *et al.*, 1974). The NAG models have been incorporated in two-dimensional Lagrangian codes (Seaman and Shockey, 1975) and two-dimensional Eulerian codes (Hageman and Herrmann, 1978) for describing dynamic tensile failure. An improved version for describing ductile tensile failure has been recently developed (Shockey, *et al.*, 1985), which accounts for the kinetic microscopic processes associated with shear band and void nucleation. The models are capable of describing fracture-time dependence of material damage, damage-induced stress reduction, changes in residual properties resulting from microdamage and damage anisotropy under generalized loading conditions.

Because generic models incorporating micromechanical mechanisms of damage are physically based, they have the potential for providing more accurate solutions to dynamic problems, and in general good agreement with experimental data has been achieved with the NAG models. Limitations of the models are that fracture parameters are obscure and not easily related to conventional parameters, such as fracture toughness. The current models are complicated and difficult to implement routinely in computer solutions because a large number of parameters, typically 5–10, is required to characterize the nucleation and growth functions. A need exists to simplify the models for use in practical applications and to improve computing capabilities to allow implementation of the models.

Other damage theories have been developed for use in computer codes (Davison, *et al.*, 1977; Kipp and Stevens, 1976, Grady and Kipp, 1980) and applied to plate impact and explosive loading problems.

The physical mechanisms of dynamic fracture are complex and difficult to establish *a priori*. Although it is desirable to use low-strain-rate data to define the damage functions, physical phenomena unique to high-rate deformation, such as microscopic shear banding and thermal localization, may be controlling factors to dynamic fracture. For this reason, wave profile data such as that shown in Fig. 11, coupled with post-shock analyses on recovered specimens, provide an important means for evaluating damage theories.

Fragmentation. As indicated earlier, it is necessary to calculate the distribution of fragments resulting from the dynamic fracture of materials during hypervelocity impact in order to assess impact lethality. For calculations of debris cloud dynamics, it is desirable to (1) describe the size, velocity distribution and divergence of fragments, (2) evaluate the thermomechanical state of the fragments, (3) specify the thermodynamic state of material in the clouds, and (4) calculate the relative flow of fragments and vapor in the debris cloud. To accomplish these objectives, it is necessary to determine the influence of material properties and loading rates on fragment generation.

A desirable objective of a fracture model is to calculate the fragment size distribution in conjunction with calculation of microdamage distributions. Theories of dynamic fracture are presently limited in their capabilities for predicting fragment distributions under arbitrary loading. There are presently three general approaches for predicting fragment distributions, comprising of the nucleation and growth theories, energy balance considerations, and statistical concepts.

The nucleation and growth models have been used to determine fragment size through calculation of the characteristic spacing of microfractures and determination of the coalescence of cracks to form isolated fragments (Seaman, *et al.*, 1976; Shockey, *et al.*, 1985). Fragmentation is assumed to occur when the damage attains a critical value, corresponding to inability of the material to transmit stress. The NAG models have been successful in calculating fragment sizes of armor steel under various loading conditions (Shockey, *et al.*, 1973b). A similar damage model for brittle failure developed by Grady and Kipp (1980) relates fragment size to the damage accumulated by penny-shaped crack, which gave reasonable fragment size distributions occurring during the explosive rubblization of oil shale. This work also illustrated the importance of loading rate on fragment size.

Grady and colleagues have developed theories of fragment size based on the balance of kinetic energy available for fragmentation with the energy dissipated during fragment formation, either in the form of surface energy of liquid fragments or the plastic work performed during fragmentation of ductile materials (Grady and coauthors, 1982a, 1982b, 1982c, 1984, 1985). This approach holds considerable promise for general use because of its simplicity and its general applicability to a wide range of fragmentation problems. The model has been used to describe dynamic fragmentation occurring in brittle and ductile solids with reasonable success, and is applicable to fluids as well. In liquids, the surface energy associated with fragment formation is equated to excess kinetic energy; for solids the plastic work performed is related to the excess kinetic energy; and in brittle solids the fracture toughness is related to the excess energy. Similar approaches have been developed to describe the breakup of thin sheets and jets, such as occur in impact-induced debris plumes and shaped-charge jets (Grady, 1986).

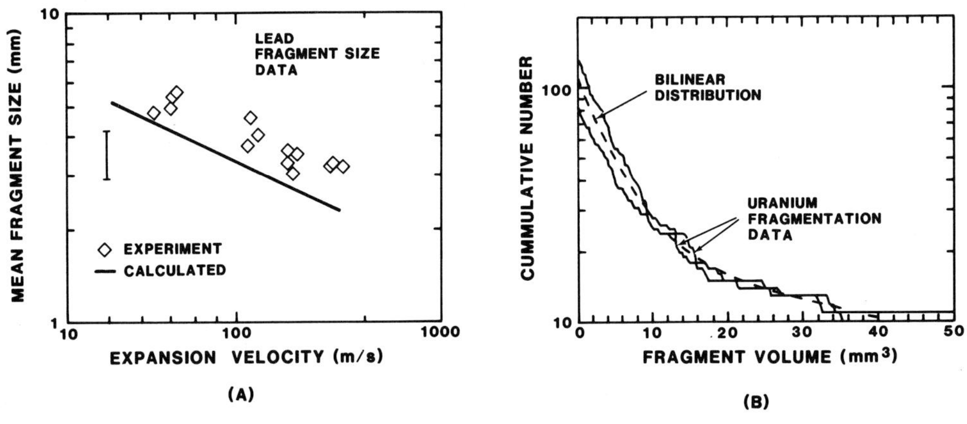

Fig. 12. Dynamic fragmentation experiments and predictions. (a) Experimental and calculated mean fragment sizes in lead projectile-on-lead plate experiments (Grady, *et al.*, 1985). (b) Cumulative fragment number distribution from two uranium projectile impacts on uranium plates and comparison with a statistical representation (Grady, *et al.*, 1985).

Grady has applied the energy-balance approach in a series of experiments on various materials and for different loading conditions with reasonably good success. Fig. 12a illustrates the mean fragment size obtained in lead projectile impacts on lead plates, in comparison with theoretical predictions of the average size. The predictions are based on the local volumetric strain rate at corresponding material points where fracture into a particle occurs, which strictly applies to simple expansion. Recently, Grady and colleagues (1985) have developed a more general relation for the fragmentation expansion rate, involving the velocity gradient tensor, which is more applicable to general problems of hypervelocity impact. Generalizations of the model have also been made by Glenn and Chudnovsky (1985) to include the effects of strain energy on dynamic fragmentation.

Dynamic fragmentation is not completely characterized by an average fragment size, because it results in a spectrum of particle sizes, due to the statistical nature of the fracture process. Material and loading parameters that influence fragment distributions are currently not well understood, leading to the use of various theories for describing particle distributions. Many of the current statistical treatments are based on the early work of Mott and Linfoot (1943) concerning the fragmentation of detonating shells. Grady and coworkers (1985) have proposed a modification to Mott's approach that has shown relatively good agreement with experimental data from projectile and expanding ring experiments. Fig. 12b illustrates the comparison for uranium projectile/uranium plate impact experiments, assuming a bimodal distribution resulting from a strain-rate inhomogeneity in the impact problem. The bimodal response corresponds to material failure at the back-surface peripheral crater and in the highly-stressed main body of ejected material.

The material and kinematic properties of importance to dynamic fragmentation are only beginning to emerge; considerable work is needed to develop broader-based theories for general use. In particular, there is a need to (1) develop fracture models which are self-consistent in predictions of fragmentation distributions, (2) relate fragmentation processes to underlying microscopic deformation mechanisms in order to develop physically-based theories that account for pre-tensile loading history, and (3) incorporate the proper thermodynamic and thermomechanical properties. Experimental data on fragmentation play a crucial role in the evaluation of statistical predictions of fragment distributions. In this respect, expanding ring data (Grady and Benson, 1983) and projectile-impact experiments (Grady, *et al.*, 1985) have been useful.

SUMMARY

In this paper, we have discussed material modeling and experimental capabilities important to hypervelocity impact problems. Our approach has been to discuss current modeling capabilities based on modern experimental and theoretical investigations, identify technical needs for improving the understanding of hypervelocity impact phenomena, and to provide an extensive bibliography for those interested in more details.

Methods for constructing equilibrium hydrostatic equations of state from theoretical models and experimental data were discussed in some detail. We showed how the problem can be subdivided into several parts. Starting with a single phase and a single chemical species, an EOS can be written as the sum of a zero-Kelvin isotherm, contributions from nuclear degrees of freedom, and contributions from thermally excited electronic degrees of freedom. We discussed the various models that have been developed to compute each term, ranging from those that are based mainly upon experimental data to those that are based upon first principles. Next we described how phase transitions can be treated by constructing a separate EOS for each phase and locating the phase boundaries using the free energy; applications to polymorphic transitions, melting, and vaporization were reviewed. Finally, we discussed the importance of chemical reactions in EOS modeling and some ways of treating this problem.

In addition to discussions of the EOS, techniques for evaluating the constitutive response of materials under dynamic loading conditions were presented. These discussions centered on the deviations from purely hydrostatic response produced in most impact problems, and the use of traditional and advanced material models for describing such effects. The section included a discussion of the elastic-plastic behavior of materials, modeling techniques for composite materials, and models to describe the low- and high-pressure response of porous materials. A discussion of dynamic fracture and fragmentation was also given that identifies present modeling capabilities for describing material failure and the resulting fragmentation, which provides the starting point for a more complete description of debris cloud dynamics.

Throughout this discussion, we have given examples of how experimental data can be used, sometimes as direct input to a model and sometimes as a general verification of the theory; we also noted some of the regimes where experimental data are not available.

ACKNOWLEDGEMENTS

We would like to express our appreciation to the following people for critical comments on the paper: L. C. Chhabildas, D. S. Drumheller, D. E. Grady, M. E. Kipp, R. J. Lawrence, S. L. Passman, S. L. Thompson, and T. G. Trucano. We also thank Mrs. T. G. Romero for her painstaking efforts in preparing this manuscript.

REFERENCES

*This work performed at Sandia National Laboratories supported by the U.S. Department of Energy, under contract DE-ACO4-76DP00789.

Agureikin, V. A., S. I. Anisimov, A. V. Bushman, G. I. Kanel', V. P. Karyagin, A. B. Konstantinov, B. P. Kryukov, V. F. Minin, S. V. Razorenov, R. Z. Sagdeev, S. C. Sugak, and V. E. Fortov (1985). High Temperature, **22**, 761-778.

Ahrens, T. J., G. A. Lyzenga, and A. C. Mitchell (1982). Temp. Induced by Shock Waves in Minerals: Appl. to Geophys. In S. Akimato and M. H. Manghnani (eds.), Advances in Earth and Planetary Sciences, **12**, 579–594.

Al'tshuler, L. V., A. A. Bakanova, A. V. Bushman, I. P. Dudoladov, and V. N. Zubarev (1977). Sov. Phys. JETP, **46** (5), 980–983.

Al'tshuler, L. V., A. V. Bushman, M. V. Zhernokletov, V. N. Zubarev, A. A. Leont'ev, and V. E. Fortov (1980). Sov. Phys., JETP, **51**, 373-383.

Al'tshuler, L. V. (1985). Results and Prospects for Experimental Studies of External States of Matter. Sov. Phys. Usp., **28** (9), 847–850.

Anderson, C. E., Jr. and W. Johnson (1986). An Overview of Hydrocodes: Theory and Application. Int'l J. of Impact Engr., **5**.

Anderson, H. C., D. Chandler, and J. D. Weeks (1979). Adv. Chem. Phys., **34**, 105.

Asay, J. R. and D. B. Hayes (1975). J. Appl. Phys., **46**, 4789-4800.

Asay, J. R. and J. Lipkin (1978). J. Appl. Phys., **49**, 4242-4247.

Asay, J. R. and L. C. Chhabildas (1981). Determination of the Shear Strength of Shock Compressed 6061-T6 Aluminum. In M. A. Meyers and L. E. Murr (eds.), Shock Waves and High-Strain-Rate Phenomena in Metals, Plenum Publ. Corp., NY.

Asay, J. R. (1985). Measurement of Material Properties to Mbar Pressures Using Time-Resolved Shock Wave Techniques. Sandia National Laboratories Report SAND84-2657, unpublished.

Asay, J. R., L. C. Chhabildas, G. I. Kerley, and T. G. Trucano (1986). High Pressure Strength of Shocked Aluminum. In Y. M. Gupta (ed.), Shock Waves in Condensed Matter–1985, Plenum Publ. Corp.

Asay, J. R. and L. C. Chhabildas (1986). To be published.

Ashcroft, N. W. and D. Stroud (1978). Solid State Phys., **33**, 1–81.

Balescu, R. (1975). Equilibrium and Nonequilibrium Statistical Mechanics. Wiley, New York.

Bammann, D. (1986). Sandia National Laboratories, private communication.

Bammann, D. J. and G. C. Johnson (1986). On the Kinematics of Finite-Deformation Plasticity. Sandia National Laboratories Report SAND86-8609, unpublished.

Bancroft, D., E. L. Peterson, and F. S. Minshall (1956). J. Appl. Phys., **27**, 291–298.

Barbee, T. W., Jr., L. Seaman, R. Crewdson, and D. Curran (1972). J. Materials, JMLSA, **7**, 393-401.

Barker, J. A. and D. Henderson (1976). Rev. Mod. Phys., **48**, 587.

Barker, L. M., J. R. Asay, T. G. Trucano, and J. L. Wise (1985). Preliminary Considerations for the Gasdynamic Precompression of Solid Molecular Hydrogen. Sandia National Laboratories Report SAND85-0111.

Barker, L. M. and R. E. Hollenbach (1970). J. Appl. Phys., **41** (10), 4208–4226.

Barker, L. M. (1971). J. Compos. Mater., **5**, 140–162.

Barker, L. M. and R. E. Hollenbach (1972). J. Appl. Phys., **43** (11), 4669–4675.

Barker, L. M. and R. E. Hollenbach (1974). J. Appl. Phys., **45** (11), 4872–4887.

Barker, L. M. (1984). High-Pressure Quasi-Isentropic Impact Experiments. In J. R. Asay, R. A. Graham, and G. K. Straub (eds.), Shock Waves in Condensed Matter–1983. Elsevier Sci. Publ., B.V.

Barnes, J. F. (1975). An Equation of State for Sodium Over an Extended Temperature and Density Range. In Thermodynamics of Nuclear Materials – 1984, International Atomic Energy Agency, Vienna. Vol. I, 327-339.

Bedford, A., D. S. Drumheller, and H. J. Sutherland (1976). Mechanics Today, **3**, 1–54.

Bedford, A. and D. S. Drumheller (1983). Int. J. Eng. Sci., **21** (8), 863–960.

Bertholf, L. D., L. D. Buxton, B. J. Thorne, R. K. Byers, A. L. Stevens, and S. L. Thompson (1975). J. Appl. Phys., **46**, 3776-3783.

Bodner, S. R. and M. B. Rubin (1985). A Unified Elastic-Viscoplastic Theory with Large Deformations. Proc. of Jean Mandel Memorial Sym., Physical Basis and Modeling of Finite Deformation of Aggregates, Paris, France, 9/30–10/2/85.

Bjork, R. L. (1986). The Physics of Hypervelocity Impact. Int'l. J. Impact Engr., **5**.

Born, M. and K. Huang (1954). Dynamical Theory of Crystal Lattices. Oxford University Press, Oxford.

Boslough, M. B. (1986). Application of Fast Radiation Pyrometry to Time-Resolved Studies of Shock-Induced Chemical Reactions in Solids. Int'l. J. of Impact Engr., **5**.

Bridgman, P. W. (1931). The Physics of High Pressures, Dover, NY.

Brown, J. M. and J. W. Shaner (1984). Rarefaction Velocities in Shocked Tantalum and the High Pressure Melting Point. In J. R. Asay, R. A. Graham, and G. K. Straub (eds.), Shock Waves in Condensed Matter 1983. Elsevier Science Publ., B.V.

Bushman, A. V. and V. E. Fortov (1983). Sov. Phys. Usp., **26**, 465-496.

Butcher, B. M., L. M. Barker, D. E. Munson, and C. D. Lundergan (1964). AIAA Jour., **2**, 977-990.

Butcher, B. M. (1966). Porous Carbon Materials. Sandia Corporation Development Report SC-RR-66-438, unpublished.

Butcher, B. M. (1971). Numerical Techniques for One-Dimensional Rate-Dependent Porous Material Compaction Calculations. Sandia Corporation Research Report SC-RR-710112, unpublished.

Butcher, B. M., M. M. Carroll, and A. C. Holt (1974). J. Appl. Phys., **45** (9), 3864–3875.

Cannon, J. F. (1974). J. Phys. Chem. Ref. Data, **3**, 781.

Carroll, M. M. and A. C. Holt (1972). J. Appl. Phys., **43** (4), 1626–1636.

Chartagnac, P. F. (1982). J. Appl. Phys., **53**, 948–953.

Chhabildas, L. C. and J. W. Swegle (1982). J. Appl. Phys., **53** (2), 954–956.

Chhabildas, L. C. and J. R. Asay (1982). Time-Resolved Wave Profile Measurements in Copper to Mbar Pressures. In C. M. Backman, T. Johanisson, and T. Legner (eds.), High Pressure in Research and Industry, 8th AIRAPT Conference.

Chhabildas, L. C., J. L. Wise, and J. R. Asay (1982). Reshock and Release Behavior in Beryllium. In W. J. Nellis, L. Seaman, and R. A. Graham (eds.), Shock Waves in Condensed Matter – 1981, Amer. Inst. of Phys.

Chhabildas, L. C. and L. M. Barker (1986). Dynamic Isentropic Compression Techniques with Application to Aluminum and Tungsten. Sandia Laboratories Report SAND86-1888, unpublished.

Chhabildas, L. C. (1986). Survey of Diagnostic Tools Used in Hypervelocity Impact Studies. Int'l. J. of Impact Engr., **5**.

Chirat, R. and G. Pittion-Rossillon (1981). J. Chem. Phys., **74**, 4634-4642.

Clifton, R. J. (1984). Pressure-Shear Impact and the Dynamic Plastic Response of Metals. In J. R. Asay, R. A. Graham, and G. K. Straub (eds.), Shock Waves in Condensed Matter–1983, Elsevier Science Publ., B.V.

Cochran, S. G. and M. W. Guinan (1976). Bauschinger Effect in Uranium. Lawrence Livermore Laboratory Report UCID-17105, unpublished.

Cochran, S. and D. Banner (1977). J. Appl. Phys., **48**, 2729-2737.

Cooper, N. G. (1979). An Invitation to Participate in the LASL Equation of State Library. Los Alamos Scientific Laboratory report LASL-79-62, 16-18, unpublished.

Crossland, B. and I. L. Spain (1983). High Pressure Generation and Containment. In G. N. Peggs (ed.), High Pressure Measurement Techniques, Appl. Sci. Publ., London.

Cranfill, C. W. and R. More (1978). IONEOS: A Fast, Analytic, Ion Equation-of-State Routine. Los Alamos Scientific Laboratory Report LA-7313-MS,

Curran, D. R. (1963). J. Appl. Phys., **34**, 2677-2685.

Curran, D. R., L. Seaman, and D. A. Shockey (1977). Physics Today, **31**, 46-55.

Davison, L., A. L. Stevens, and M. E. Kipp (1977). J. Mech. Phys. Solids, **25**, 11-28.

Davison, L. and R. A. Graham (1979). Phys. Reports, **55**, 255-379.

Davison, L. (1984). Modeling of Dynamic Material Reponse. In J. R. Asay, R. A. Graham, and G. K. Straub (eds.), Shock Waves in Condensed Matter–1983. Elsevier Science Publ., B.V.

Dienes, J. K. and J. M. Walsh (1970). Theory of Impact: Some General Principles and the Method of Eulerian Codes. In R. Kinslow (ed.), High-Velocity Impact Phenomena, Academic Press, New York.

Drumheller, D. S. and H. J. Sutherland (1973). J. Appl. Mech., March, 149–154.

Drumheller, D. S. and C. D. Lundergan (1975). Int. J. Solids and Struc., **11**, 75–87.

Drumheller, D. S. (1978). Int. J. Solids and Struct., **14**, 441-456.

Drumheller, D. S. (1982). J. Appl. Phys., **53** (2), 957–969.

Drumheller, D. S. (1985). A Theory for Dynamic Compaction of Wet Porous Solids. Sandia National Laboratories Report SAND85-1128, unpublished.

Drumheller, D. S. (1986). Hypervelocity Impact of Mixtures. Int'l. J. of Impact Engr., **5**.

Duvall, G. E. (1971). Shock Waves in Condensed Media. In P. Calderola and H. Knoepfel (eds.), Physics of High Energy Density, Academic Press, NY.

Duvall, G. E. and R. A. Graham (1977). Rev. Mod. Phys., **49** (3), 523–579.

Follansbee, P. S. (1986). High-Strain-Rate Deformation of FCC Metals and Alloys. In L. E. Murr, K. P. Staudhammer, and M. A. Meyers (eds.), Metallurgical Applications of Shock-Waves and High-Strain-Rate Phenomena, Marcel Dekker, Inc., NY.

Fortov, V. E., A. A. Leouit'en, A. N. Dremin, and S. V. Preshin (1974). JETP Lett., **20** (1), 13–14.

Fowles, G. R. (1961). J. Appl. Phys., **32**, 1475-1487.

Fowles, F. R. and R. Williams (1970). J. Appl. Phys., **41**, 360-363.

Frölich, H. (1967). Physica, **37**, 215-226.

Gathers, G. R. (1983). Int. J. Thermophys., **4**, 209.

Gehring, J. W., Jr. (1970). Engineering Considerations in Hypervelocity Impact. In R. Kinslow (ed.), High-Velocity Impact Phenomena, Academic Press, New York.

Glenn, L. A. and A. Chudnovsky (1985). Strain Energy Effects on Dynamic Fragmentation. Lawrence Livermore National Laboratory Report UCRL-93289, unpublished.

Godwal, B. K., S. K. Sikka, and R. Chidambaram (1981). Phys. Rev. Lett., **47**, 1144-1147.

Godwal, B. K., S. K. Sikka, and R. Chidambaram (1983). Phys. Rep., **102**, 121-197.

Graboske, H. C., D. J. Harwood, and H. E. Dewitt (1971). Phys Rev., **A3**, 1419-1431.

Grady, D. E., W. J. Murri, and P. S. DeCarli (1978). J. Geophys. Res., **80** (35), 4857–486l.

Grady, D. E. and M. E. Kipp (1980). Int. J. Rock Mech., Min. Sci., and Geomech. Abstr., **17**, 147-157.

Grady, D. E. (1980). J. Geophys. Res., **85**, (B2), 913-924.

Grady, D. E. (1981). Appl. Phys. Lett., **38**, 825–826.

Grady, D. E. (1982a). Fragment Size Prediction in Dynamic Fragmentation. In W. J. Nellis, L. Seaman, and R. A. Graham (eds.), Shock Waves in Condensed Matter–1981. American Institute of Phys.

Grady, D. E. (1982b). J. Appl. Phys., **53** (1), 322-325.

Grady, D. E. (1982c). Analysis of Prompt Fragmentation in Explosively Loaded Uranium Cylindrical Shells. Sandia National Laboratories Report SAND82-0140, unpublished.

Grady, D. E., J. R. Asay, R. W. Rhode, and J. L. Wise (1983). Microstructure and Mechanical Properties of Precipitation Hardened Aluminum under High Rate Loading. In J. Mescall and V. Weiss (eds.), Metal Behavior under High Stress and Ultrahigh Loading Rates, Plenum Press, NY.

Grady, D. E. and D. A. Benson (1983). Exp. Mech., **23** (4), 393-400.

Grady, D. E. and M. E. Kipp (1984). Mechanisms of Dynamic Fragmentation: Factors Governing Fragment Size. In Workshop in Inelastic Deformation and Failure Models, Northwestern University, unpublished.

Grady, D. E., T. K. Bergstresser, and J. M. Taylor (1985). Impact Fragmentation of Lead and Uranium Plates. Sandia National Laboratories Report SAND84-1545, unpublished.

Grady, D. E. (1986). Fragmentation of Rapidly Expanding Jets and Sheets. This conference.

Graham, R. A. and J. R. Asay (1978). High Temp.-High Press., **10**, 355-390.

Grover, R. (1971). J. Chem. Phys., **55**, 3435-3441.

Grover, R. (1977). In A. Cezairliyan, (ed.), Proceedings of the Seventh Symposium on Thermophysical Properties, 67-74.

Grover, R. (1979). J. Chem. Phys., **71**, 3824-3829.

Gryaznov, V. K., M. V. Zhernokletov, V. N. Zubarev, I. L. Iosilevskii, and V. E. Fortov (1980). Sov. Phys. JETP, **51**, 288-295.

Hageman, L. J. and R. G. Herrmann (1978). Incorporation of the NAG-FRAG Model for Ductile and Brittle Fracture into HELP. A 2D Multimaterial Eulerian Program. Systems, Science, and Software Corp. Report ARBRL-CR-00380, unpublished.

Hayes, D. B. (1975). J. Appl. Phys., **46**, 3438-3443.

Hegemier, G. A. and T. C. Bache (1973). J. Elasticity, **3**, 125–140.

Hensel, F. (1981). In J. V. Sengers (ed.), Proceedings of the Eighth Symposium on Thermophysical Properties. National Bureau of Standards, Gaithersburg, MD., 151-158.

Herrmann, W. (1969a). J. Appl. Phys., **40** (6), 2490–2499.

Herrmann, W. (1969b). On the Dynamic Compaction of Initially Heated Porous Materials. Sandia Laboratories Development Report SC-DR-68-865, unpublished.

Herrmann, W. (1970). Nonlinear Stress Waves in Metals. Sandia Laboratories Report SC-R-70-4323, unpublished.

Herrmann, W., R. J. Lawrence, and D. S. Mason (1970). Strain Hardening and Strain Rate in One-Dimensional Wave Propagation Calculations. Sandia National Laboratories Report SC-RR-70-471, unpublished.

Herrmann, W. (1976). Some Recent Results in Elastic-Plastic Wave Propagation. In E. Varley (ed.), Propagation of Shock Waves in Solids, ASME.

Herrmann, W. and R. J. Lawrence (1978). J. Engr. Materials and Technology, **100**, 84-95.

Herrmann, W. (1982). On Constitutive Modeling for the Shock Physicist. In W. J. Nellis, L. Seaman, and R. A. Graham (eds.), Shock Waves in Condensed Matter–1981. American Institute of Phys.

Herzberg, G. (1945). Molecular Spectra and Molecular Structure II. Infrared and Raman Spectra of Polyatomic Molecules. Van Nostrand Reinhold: New York. Chap. 5, 501-537.

Hirschfelder, J. O., C. F. Curtiss, and R. B. Bird (1964). Molecular Theory of Gases and Liquids. Wiley, NY, 363.

Hixson, R. S., M. A. Winkler, and J. W. Shaner (1986). Physica **139** and **140B**, 893–896.

Holian, K. S. (1984). T-4 Handbook of Material Properties Data Bases. Los Alamos Scientific Laboratory Report LA-10160-MS, unpublished.

Holian, K. S. and M. W. Burkett (1986). Sensitivity of Hypervelocity Impact Simulations to Equations of State. Int'l. J. Impact Engr., **5**.

Hopkins, A. K., T. W. Lee, and H. F. Swift (1972). J. Spacecraft, **9**, 342-345.

Howard, J. R. (1969). Development of the Mariner Mars 1971 Meteroid Shield. In Proc. of the AIAA Hypervelocity Impact Conf., Cincinnati, OH.

Irving, J. H. and J. G. Kirkwood (1950). J. Chem. Phys., **18**, 817.

Jayaraman, A. (1983). Rev. Mod. Phys., **55**, 65.

Johnson, P. C., B. A. Stein, and R. S. Davis (1962). J. Appl. Phys., **33**, 557–56l.

Johnson, J. N. and L. M. Barker (1969). J. Appl. Phys., **40**, 4321-4334.

Johnson, G. R. (1978). EPIC-2. A Computer Program for Elastic-Plastic Impact Computations in 2 Dimensions Plus Spin. BRL Contract DAAD05-77-C-0730, unpublished.

Johnson, G. R. and W. H. Cook (1983). A Constitutive Model and Data for Metals Subjected to Large Strains, High Strain Rates and High Temperatures. Proc. 7th Int. Symp. on Ballistics, The Hague, The Netherlands, 541-547.

Johnson, G. R. (1985). Engr. Frac. Mech., **21**, 31-48.

Jones, O. E. and R. A. Graham (1971). Shear Strength Effects on Phase Transition "Pressures" Determined from Shock-Compression Experiments. In E. C. Lloyd (ed.), Accurate Characterization of the High Pressure Environment, National Bureau of Standards.

Jones, H. (1973). Phys. Rev., **A8**, 3215.

Jungst, S., B. Knuth, and F. Hensel (1985). Phys. Rev. Lett., **55**, 2160.

Kerley, G. I. (1981). User's Manual for PANDA: A Computer Code for Calculating Equations of State. Los Alamos National Laboratory Report LA-8833-M, unpublished.

Kerley, G. I. (1983). In J. M. Haile and G. A. Mansoori (eds.), Molecular Based Study of Fluids, Am. Chem. Soc., Washington, DC., 107-138.

Kerley, G. I. (1985). In Proc. of the Eighth Int'l Symposium on Detonation, Albuquerque, NM, to be published.

Kerley, G. I. and A. C. Switendick (1986). In Y. M. Gupta (ed.), Shock Waves in Condensed Matter - 1985, Plenum Publ. Corp.

Kerley, G. I. (1986a). Theoretical Equation of State for Aluminum. Int'l. J. of Impact Engr., **5**.

Kerley, G. I. (1986b). Unpublished work.

Kipp, M. E. and A. L. Stevens (1976). Numerical Integration of a Spall-Damage Viscoplastic Constitutive Model in a One-Dimensional Wave Propagation Code. Sandia National Laboratories Report SAND76-0061, unpublished.

Kipp, M. E. and R. J. Lawrence (1982). WONDY V-4 One-Dimensional Finite Difference Wave Propagation Code. Sandia National Laboratories Report, SAND81-0930, unpublished.

Kormer, S. B., M. V. Sinitsyn, G. A. Kirillov, and V. D. Urlin (1965). Sov. Phys. JETP, **21**, 689-700.

Lee, L. M, R. P. May, and T. R. Guess (1970). AIAA J., **8** (8), 1421–1428.

LeSar, R. and R. G. Gordon (1983). J. Chem. Phys., **78**, 4991–4996.

Liberman, D. A. (1979). Phys. Rev., **B20**, 4891.

Lipkin, J. and M. E. Kipp (1976). J. Appl. Phys., **47** (5), 1979-1986.

Long, K. S. (1982). A New Full-Range Equation of State for Copper. In W. J. Nellis, L. Seaman, and R. A. Graham (eds.), Shock Waves in Condensed Matter–1981. American Inistitute of Phys.

Lyzenga, G. A., T. J. Ahrens, and A. C. Mitchell (1983). J. Geophys. Res., **88** (B3), 2431–2444.

Mader, C. L. (1979). Numerical Modeling of Detonations. University of California Press, Berkeley.

Mansoori, G. A. and F. B. Canfield (1969). J. Chem. Phys., **51**, 4958.

Marsh, S. P. (1980). LASL Shock Hugoniot Data. Univ. of California Press, Berkeley.

McMahan, A. K., B. L. Hord, and M. Ross (1977). Phys. Rev., **B15**, 726-737.

McQueen, R. G., S. P. Marsh, J. W. Taylor, J. N. Fritz, and W. J. Carter (1970). The Equations of State of Solids from Shock Wave Studies. In R. Kinslow (ed.), High-Velocity Impact Phenomena, Academic Press, NY.

McQueen, R. G., J. N. Fritz, and C. E. Morris (1984). The Velocity of Sound Behind Strong Shock Waves in 2024 Al. In J. R. Asay, R. A. Graham, and G. K. Straub (eds.), Shock Waves in Condensed Matter–1983, Elsevier Science Publ., B.V.

Mitchell, A. C. and W. J. Nellis (1981). Rev. Sci. Instrum., **52**, 347.

More, R. M. (1981). Atomic Physics in Inertial Confinement Fusion. Lawrence Livermore Laboratory Report UCRL-84991, unpublished.

Moriarty, J. A. (1982). Phys. Rev. **B26**, 1754–1780.

Moriarty, J. A., D. A. Young, and M. Ross (1984). Rev. Phys., **B30**, 578.

Morland, L. W. (1959). Phil. Trans. Royal Soc., **A251**, 341-383.

Morris, C. E., J. N. Fritz, and B. L. Holian (1982). In W. J. Nellis, L. Seaman, and R. A. Graham (eds.), <u>Shock Waves in Condensed Matter – 1981</u>. American Institute of Physics.

Morris, C. E., J. N. Fritz, and R. G. McQueen (1984). J. Chem. Phys., **80**, 5203-5218.

Moss, W. C. (1985). Viscosity and Steady Waves. Lawrence Livermore National Laboratory Report, unpublished.

Moss, W. C. and L. A. Glenn (1984). A Bauschinger Effect Model Suitable for Use in Large Computer Codes. In J. R. Asay, R. A. Graham and G. K. Straub (eds.), <u>Shock Waves in Condensed Matter–1983</u>, Elsevier Science Publ., B.V.

Mott, N. F. and E. H. Linfoot (1943). A Theory of Fragmentation. Extra-Mural Research No. F72/80, Ministry of Supply (England), unpublished.

Munson, D. E. and K. W. Schuler (1971a). Steady Wave Analysis of Wave Propagation in Laminates Using Effective Moduli. Proc. of Sagamore Army Materials Research Conf., Raquette Lake, NY, 9/1–4/70.

Munson, D. E. and K. W. Schuler (1971b). J. Compos. Mater., **5**, 286–304.

Munson, D. E. (1973). Anisotropic Hugoniots of Cloth-Laminate Quartz Phenolic (0, 45, and 90 Degree Lay-up Angles). Sandia Laboratories Development Report SLA-73-0505, unpublished.

Murr, L. E., K. P. Staudhammer, and M. C. Meyers (1986). <u>Metallurgical Applications of Shock-Wave and High-Strain-Rate Phenomena</u>, Marcel Dekker, Inc. NY.

Neal, T. (1976). Phys. Rev., **B14**, 5172.

Neal, T. (1979). Determination of the Grüneisen γ for Beryllium at 1.2 to 1.9 Times Standard Density. In K. D. Timmerhaus and M. S. Barber (eds.), <u>High Pressure Science and Technology</u>, Vol 1, Plenum Press, New York, 80-87.

Nellis, W. J. (1982). Properties of Condensed Matter to Ultrahigh Dynamic Pressures. In G. N. Peggs (ed.), <u>High Pressure Measurement Techniques</u>, Appl. Sci. Publ., London.

Nellis, W. J., F. H. Ree, R. J. Trainor, A. C. Mitchell, and M. B. Boslough (1984a). J. Chem. Phys., **80**, 2789-2799.

Nellis, W. J., N. C. Holmes, A. C. Mitchell, and M. van Thiel (1984b). Phys. Rev. Lett., **53**, 1661.

Ng, A., D. Parfeniuk, and L. DaSilva (1985). Phys. Rev. Lett., **54**, 2604–2607.

Novikov, S. A. (1981). J. Appl. Mech. & Tech. Phys., **11**, (3), 385-394.

Nunziato, J. W., M. E. Kipp, R. E. Setchell, and E. K. Walsh (1982). Shock Initiation in Heterogeneous Explosives. Sandia National Laboratories Report SAND81-2173, unpublished.

Pack, D. C., W. M. Evans, and H. J. James (1948). Proc. of the Physical Society, **60**, 1-8.

Passman, S. L. and R. C. Batra (1984). Archive for Rational Mech. and Anal., **87** (1), 11-33.

Perry, F. C. (1973). J. Appl. Phys., **44** (11), 4975–4981.

Ragan, C. E. (1982). Phys. Rev., **A25**, 3360–3375.

Rajendran, A. M. and S. J. Bless (1985). High Strain Rate Material Behavior. Air Force Material Laboratory Report AFML-TR-85-4009, unpublished.

Ree, F. H. (1984). J. Chem. Phys., **81**, 1251-1263.

Rice, M. H., R. G. McQueen, and H. M. Walsh (1958). Compression of Solids by Strong Shock Waves. In F. Seitz and D. Turnbull (ed.), Solid State Phys., **16**, 1-63.

Rogers, F. J. (1981). Phys. Rev., **A24**, 1531-1543.

Rose, J. H., J. R. Smith, F. Guinea, and J. Ferrante (1984). Phys. Rev., **B29**, 2963-2969.

Rosenblatt, M., K. N. Kreyenhagen, and W. D. Romine (1968). Analytical Study of Debris Clouds Formed by Hypervelocity Impacts on Thin Plates. Air Force Materials Laboratory Report AFML-TR-68-266, unpublished.

Rosenblatt, M. (1970). Numerical Calculations of Hypervelocity Impact Crater Formation in Hard and Soft Aluminum Alloys. Air Force Materials Laboratory Report AFML-TR-70-254, unpublished.

Ross, M. (1980a). J. Chem. Phys., **73**, 4445.

Ross, M. (1980b). Phys. Rev., **B21**, 3140.

Ross, M. and A. K. McMahan (1980). Phys. Rev., **21**, 1658.

Ross, M., F. H. Ree, and D. A. Young (1983). J. Chem. Phys., **79**, 1487-1494.

Ross, M. (1985). Rep. Prog. Phys., **48**, 1-51.

Rouse, C. A. (1971). Prog. High Temp. Phys. Chem., **4**, 139.

Royce, E. B., (1972). GRAY, A Three-Phase Equation of State for Metals. Lawrence Livermore Laboratories Report UCRL 51121, unpublished.

Saha, M. N. (1920). Phil Mag., **40**, 472.

Seaman L. and D. A. Shockey (1975). Models for Ductile and Brittle Fracture for Two-Dimensional Wave Propagation Calculations. Stanford Research Institute Report AD-A008 992, unpublished.

Seaman, L., R. Tokheim, and D. Curran (1974). Computational Representation of Constitutive Relations for Porous Material. Stanford Research Report DNA 3412F, unpublished.

Seaman, L., D. R. Curran, and D. A. Shockey (1976). J. Appl. Phys., **47**, 4814-4826.

Shaner, J. W., G. R. Gathers, and C. Minichino (1976). High-Temp.-High-Press., **8**, 425.

Shockey, D. A., L. Seaman, D. R. Curran, P. S. DeCarli, M. Austin, and J. P. Welhelm (1973a). A Computational Model for Fragmentation of Armor Under Ballistic Impact. Stanford Research Institute Report, unpublished.

Shockey, D. A., L. Seaman, D. R. Curran, P. S. DeCarli, M. Austin, and J. P. Wilhelm (1973b). The Influence of Microstructural Features on Dynamic Fracture. Stanford Research Institute Report, unpublished.

Shockey, D. A., D. R. Curran, and L. Seaman (1985). Development of Improved Dynamic Failure Models. SRI International Report DAAG-29-81-K-0123, unpublished.

Slater, J. C. (1939). Introduction to Chemical Physics, McGraw-Hill, NY.

Spillman, G. R. (1986). Hypervelocity Impact Bumpers as Countermeasures. Los Alamos National Laboratory Report LA-10629-MS, unpublished.

Steinberg, D. J. and M. W. Guinan (1973). Constitutive Relations for the KOSPALL Code. Lawrence Livermore Laboratory Report UCID-16226, unpublished.

Steinberg, D. J. (1984). The Effect of Strain Hardening in Dynamic Deformation of Beryllium. Lawrence Livermore National Laboratory Report UCID-91891, unpublished.

Steinberg, D. J. (1986). Hydrodynamic Computer Simulation of Time Resolved Shock Wave Data. Int'l. J. of Impact Engr., **5**.

Steinberg, D. J. and R. W. Sharp, Jr. (1980). An Elastic-Viscoelastic Constitutive Model for Metals. Lawrence Livermore Laboratory Report UCRL-18729, unpublished.

Swegle, J. W. and D. E. Grady (1986). Calculation of Thermal Trapping in Shear Bands. In L. E. Murr, K. P. Staudhammer, and M. A. Meyers (eds.), Metallurgical Applications of Shock-Wave Data and High-Strain-Rate Phenomena, Marcel Dekker, Inc., NY.

Swegle, J. W. and D. E. Grady (1985). J. Appl. Phys., **58**, 692-701.

Swift, H. F. and A. F. Hopkins (1970). J. Spacecraft, **17**, 73-77.

Swift, H. F., D. D. Preonas, P. W. Dueweke, and R. S. Bertke (1970). Response of Materials to Impulsive Loading. Air Force Materials Laboratory Report AFML-TR-70-135, unpublished.

Swift, H. F., R. Bamford, and R. Chen (1982). Designing Dual-Plate Meteroid Shields–A New Analysis. Jet Propulsion Laboratory Report JPL Pub. 82-39, unpublished.

Taylor, G. I. (1942). The Plastic Waves in a Wire Extended by an Impact Load. British Official Report RC329, unpublished.

Thompson, S. L. and H. S. Lauson (1972). Improvements in the Chart D Radiation-Hydrodynamic CODE III: Revised Analytic Equations of State. Sandia Laboratories Report SC-RR-710714, unpublished.

Tillotson, J. H. (1962). Metallic Equations of State for Hypervelocity Impact, General Atomic Division of General Dynamics Report GA-3216, unpublished.

Tuler, F. R. and B. M. Butcher (1968). Int'l Jour. of Frac. Mech., **4**, 431-437.

Trucano, T. G. and J. R. Asay (1986). Effects of Vaporization on Debris Cloud Dynamics. Int'l. J. of Impact Engr., **5**.

Urtiew, P. A. and R. Grover (1977). J. Appl. Phys., **48** (3), 1122–1126.

van Thiel, M., J. Shaner, and E. Salinas (1977). Compendium of Shock Wave Data. Lawrence Livermore National Laboratory Report UCRL-50108, unpublished.

Vladimirov, A. S., N. P. Voloshen, V. N. Nogin, A. V. Petrovtsev, and V. A. Simonenko (1984). JETP Lett., **39**, 82-85.

Wallace, D. C. (1984). Thermoelastic-Plastic Flow in Solids. Los Alamos National Laboratory Report X-4: 84-114U, unpublished.

Wilkins, M. (1969). Lawrence Livermore Laboratory Report UCRL 7322-Rev A, unpublished.

Wilkins, M. L. (1977). Fracture Studies with Two- and Three-Dimensional Computer Simulation Programs. Lawrence Livermore National Laboratory Report UCRL-78376, unpublished.

Wise, J. L. and L. C. Chhabildas (1986). Laser Interferometer Measurements of Refractive Index in Shock-Compressed Materials. In Y. M. Gupta (ed.), Shock Waves in Condensed Matter – 1985, Plenum Publ. Corp.

Wood, D. S. (1952). Trans ASME J. Appl. Mech., **19**, 521-525.

Xu, J. A., H. K. Mao, and P. M. Bell (1986). Science, **232**, 1404–1406.

Yin, M. T. and M. L. Cohen (1982). Phys. Rev., **B26**, 5668-5687.

Young, D. A. (1977). A Soft Sphere Model for Liquid Metals. Lawrence Livermore Laboratories Report UCRL 52352, unpublished.

Young, D. A., A. K. McMahan, and M. Ross (1981). Phys. Rev., **B24**, 5119.

Zel'dovich, Y. B. and Y. P. Raizer (1966). In W. D. Hayes and R. F. Probstein (eds.), Physics of Shock Waves and High-Temperature Hydrodynamic Phenomena, Academic Press, New York.

Zimmerman, G. B. and R. M. More (1980). J. Quant. Spectrosc. Radiat. Transfer, **23**, 517.

Zharkov, V. N. and V. A. Kalinin (1971). Equations of State for Solids at High Press. and Temp. Consult. Bureau, NY.

Zlatin, N. A. and B. S. Ioffe (1973). Sov. Phys.–Tech. Phys., **17**, 1390–1393.

Zukas, J. A. (1982). Penetration and Perforation of Solids. In J. A. Zukas, T. Nicholas, H. F. Swift, L. B. Greszczuk, and D. R. Curran (eds.), Impact Dynamics, John Wiley and Sons, New York.

Int. J. Impact Engng Vol. 5, pp. 101–110, 1987
Printed in Great Britain

0734-743X/87 $3.00 + 0.00
Pergamon Journals Ltd.

HYPERVELOCITY PENETRATION OF PLATE TARGETS BY ROD AND ROD-LIKE PROJECTILES

J. Baker and A. Williams
U.S. Naval Research Laboratory, Washington, D.C.

INTRODUCTION

The ability of an elongated projectile to obtain deep penetration in the hypervelocity regime is what makes it particularly interesting. This paper summarizes the results of research performed at NRL on the impact of hypervelocity rod and rod-like projectiles over the six years from 1964 to 1969. Semi-empirical correlations are reported which relate the projectile and target characteristics and the impact configuration to the resultant impact damage.

ROD PROJECTILES

Single Plate Perforation

Rod length loss for like-material impact. Baker (1969) reported the following equation for the rod length lost, Δl, as a function of the impact parameters:

$$\frac{\Delta l}{d} = \frac{t}{d} + \frac{\Delta t}{d} + \frac{v\tau}{d} \{ 1 - \exp [- (\frac{\Delta l/d + t/d + \Delta t/d}{fv\tau/d})] \} \tag{1}$$

where all lengths have been normalized by the rod diameter, d, t is the target thickness, v is the impact velocity, and f, τ, and $\Delta t/d$ are empirical constants. Note that Eq.(1) must be solved iteratively since the unknown parameter $\Delta l/d$ appears on both sides of the equation.

The first term in Eq.(1), t/d, is simply the length loss due to steady state hydrodynamic flow. The second term, $\Delta t/d$, is a small additional loss due to a non-steady state process that is possibly associated with the initial shock phase of the impact process. For most cases of practical interest ($t/d>0.05$), this term can be taken as a constant, however, as noted below, this is not strictly true. The third term, involving $v\tau/d$, is the primary non-steady state loss associated with the final breakout phase of the perforation process. The value of $v\tau/d$ can be computed from the equation

$$v\tau/d = 1.98\ v\ /\ \sqrt{[c^2 - (u_s - v/2)^2]} \tag{2}$$

where c is the sound speed in the shocked rod material and u_s is the shock velocity. These quantities can be obtained using either published equation-of-state data or coefficients for the approximation that the shock velocity is a linear function of the particle velocity.

It should be noted that for thick target plates ($t/d>2$) the rod loss approaches a linear dependence on t/d given by $\Delta l/d = t/d + \Delta t/d + v\tau/d$. For very thin target plates ($t/d<0.1$) the term $\Delta t/d$ in Eq.(1) should be replaced by the expression $(\Delta t/d)[1-\exp(-\xi t/d)]$ so that the rod loss approaches zero as t/d approaches zero. The effect of this modification is however negligible for $t/d>0.05$.

Rod length loss for unlike-material impact. For an impact configuration in which the projectile and target plate are of different materials, the rod loss can be still be computed using a modification of Eq.(1). By analogy with the theory of steady state, incompressible fluid flow which predicts that for unlike fluids, the penetration varies as the square root of the density ratio of the two fluids, so too, does the length of rod lost depend on the rod and target materials, i.e.

$$(\Delta l/d)_{1\to 2} = \sqrt{(\rho_2/\rho_1)}\ (\Delta l/d)_{1\to 1} \tag{3}$$

Figure 1 shows the result of applying Eq.(3) to Eq.(1) for a wide range of projectile/target material combinations.

Rod length loss for oblique impact. For oblique impacts in which the angle between the rod axis and the target surface (Θ) is less than 90°, the impact process behaves as if the target thickness is greater than its nominal value. This additional thickness arises because of the greater distance which is traversed by the rod in passing through the target and is given by a factor of $1/\sin(\Theta)$. Also, because of the finite diameter of the rod, the point of contact between the rod and the target (and therefore the effective thickness of the target) is extended by a distance $d/2\tan(\Theta)$. Therefore Eq.(1) can still be applied to oblique impacts if the target thickness, t/d, is replaced by its effective thickness given by

$$(t/d)_{eff} = (t/d)/\sin(\Theta) + 1/[2\tan(\Theta)] \tag{4}$$

A yawed rod is one in which there is an angle between the velocity vector of the rod and its axis. For the most part, the NRL data was limited to yaw angles below 10° and it is within that range that the results given here are valid and independent of yaw angle. The NRL experiments also showed that, for yaw angles greater than 15°, the rod was completely destroyed. See Bertke et al (1973) for additional information on yawed rod cratering.

Hole diameter for normal impact. The target plate hole diameter approaches the diameter of the rod as the target plate becomes very thin, and, conversely, the diameter of the hole approaches the diameter of a crater in a semi-infinite target as the target plate becomes very thick. The following analytical expression provides such a dependence:

$$D/d = 1 + (D_{\infty}/d - 1)\ [1 - \exp(-kt/d)] \tag{5}$$

where k is a constant that depends on the rod and target materials and D_{∞}/d, the corresponding crater diameter, has been shown by Baker (1969) to be, very nearly, a linear function of velocity, $D_{\infty}/d = 1 + \sigma_{\infty}v$, so that Eq.(5) becomes

$$D/d = 1 + \sigma_{\infty}v\ [1 - \exp(-kt/d)] \tag{6}$$

where σ_{∞} is a constant which depends on the rod and target materials. Figure 2 shows Eq.(6) for an aluminum rod and a variety of target plate materials.

More recently, Baker (1974) analyzed some crater diameter data which was better fitted by an expression of the form

$$D_{\infty}/d = 1 + \sigma_{o}v\ \{1 - (v_{o}/v)[1 - \exp(-v/v_{o})]\} \tag{7}$$

where σ_{o} and v_{o} are constants for a given combination of projectile and target materials. This alternate equation for crater diameter is especially appropriate when there is a large difference in the densities of the target and projectile materials.

Hole diameter for oblique impact. For an oblique impact, the hole in the target plate is distorted into an approximately elliptical shape in which the major diameter lies along the trajectory of the rod. The equation for the major diameter is given by purely geometrical considerations while the minor diameter was shown experimentally to have a weak dependence on the impact angle, Θ.

$$\frac{D_{major}}{d} = \frac{D/d}{\sin(\Theta)} \qquad \frac{D_{minor}}{d} = \frac{D/d}{[\sin(\Theta)]^{1/5}} \tag{8}$$

Multiple Plate Perforation

Rod length loss for spaced, multiple plate targets. For a target consisting of a series of single, spaced plates, the rod length loss and target plate hole diameter equations can simply be applied in a sequential manner to each plate individually using the residual rod from the preceding plate. Note that, because of a lack of sufficient experimental data to establish an anlytical correlation, the velocity of the rod is assumed to remain undimished throughout the impact process.

Rod length loss for multilayer targets. In 1970 Baker and Zalesak extended the single plate analysis to a multilayer target using the concept of "corresponding layers." This concept asserts that the rod length lost in a particular layer of a multilayer target is given by the incremental loss that would occur, according to Eq.(1), in passing through the corresponding portion (i.e., that portion located in the same relative position) in a target which has a thickness equal to that of the layer being calculated plus all the preceding layers and which is composed entirely of the same material as the layer being calculated. This approximation assumes that the rod penetrates each layer entirely under steady state conditions which will be nearly true as long as each subsequent layer is of lesser density than the preceding one.

When a subsequent layer has a higher density than the preceding one, then the pressure at the rod-target interface must undergo a transition from one steady state pressure in the first layer to a new, higher pressure in the second. The strength of this effect will depend on the relative magnitudes of the two steady state pressures. The greater the difference in pressures, the more nearly the rod loss in the subsequent layer will approach that of a single plate rather than that of a corresponding layer. The following equation expresses this relationship:

$$\Delta l/d = (\Delta l/d)_{cl} + [(\Delta l/d)_{sp} - (\Delta l/d)_{cl}]\ (1 - P_{s12}/P_{s13}) \tag{9}$$

where $(\Delta l/d)_{cl}$ is the corresponding layer rod loss, $(\Delta l/d)_{sp}$ is the single plate rod loss, and P_{snm} is the incompressible steady state stagnation pressure f material n penetrating into material m and m=2 refers to the preceding layer and m=3 refers to the current layer being calculated. The steady state pressure is given by

$$P_{snm} = \tfrac{1}{2}\rho_m v^2 \ /\ [1 + \surd(\rho_m/\rho_n)]^2 \tag{10}$$

where ρ is the material density.

In addition, if the preceding layer is relatively thin ($t/d<0.75$) then the rod-target interface may not yet have reached the steady state condition when the more dense subsequent layer is encountered. In that case the change in conditions between layers will not be as great as otherwise and therefore the amount of additional rod loss will be lessened. Assuming that this effect has the same dependence on target thickness as the single plate rod loss expression, then the second term in Eq.(9) should be multiplied by the factor

$$\Lambda = 1 - \exp\{ - [\ (\Delta l/d)_e + (t/d)_e + (\Delta t/d)\ ] \ /\ (fv\tau/d)\ \} \tag{11}$$

where $(t/d)_e$, the effective thickness of the target, is the distance from the current layer to the location of the last strong shock (either the initial impact point or a preceding interface with a large density change, nominally a factor of two or more, whichever distance is less) and $(\Delta l/d)_e$ is the rod loss for a single plate of thickness $(t/d)_e$ of the same material as the rod.

Denoting the steady state factor by Π_{123}, then the rod loss for each layer of a multilayer target is given by

$$\Delta l/d = (\Delta l/d)_{cl} + [(\Delta l/d)_{sp} - (\Delta l/d)_{cl}]\ \Pi_{123}\Lambda \qquad \text{for } \rho_3 > \rho_2 \tag{12}$$

and

$$\Delta l/d = (\Delta l/d)_{cl} \qquad \text{for } \rho_3 \le \rho_2 \tag{13}$$

The total rod loss is given by summing the losses for each layer.

Rod length loss for oblique impact into multilayer targets. As for single plates, an oblique impact into multilayer target should use Eqs.(12) and (13) with an effective target thickness similar to that of Eq.(4). It must be recognized, however, that the second term in Eq.(4), $1/[2\tan(\Theta)]$, is independent of target thickness and arises from the geometry of impacting the front surface of the target. Therefore this term is modified so that it enters into the calculation only to the degree that the transition between layers is like a new impact as expressed by the steady state factor Π_{123}, and is furthermore modified by the factor Λ for similar reasons. The equation for the effective thickness of an oblique, multilayer target is therefore given by

$$(t/d)_{eff} = (t/d)/\sin(\Theta) + \Pi_{123}\Lambda \ /\ [2\tan(\Theta)] \qquad \text{for } \rho_3 > \rho_2 \tag{14}$$

and

$$(t/d)_{eff} = (t/d)/\sin(\Theta) \qquad \text{for } \rho_3 \le \rho_2 \tag{15}$$

Hole diameter for multilayer targets. Lacking sufficient experimental data to determine the effect of adjacent layers of a multilayer target on the hole diameter, the single plate hole diameter equations are used as an approximate value for the hole diameter in each layer of the multilayer target. There is some evidence that the multilayer hole diameters tend to be larger than the single plate values, perhaps because of the petalling and other structural effects that occur in the multilayer target. The single plate hole diameter should therefore be considered to be a minimum estimate of hole size.

Single Plate Cratering

An impact will create a crater in the target when the rod projectile either comes to rest or is totally consumed before it perforates the target. The impact analysis is then primarily concerned with the depth and diameter of the crater and only secondarily with any residual rod length.

Steady state cratering model. Tate (1967 and 1969) proposed a model of rod cratering in which he modified Bernoulli's equation for incompressible, steady state flow by adding, a dynamic strength, σ, for each material to both sides of the equation, i.e., $\frac{1}{2}\rho_p(v-u)^2+\sigma_p = \frac{1}{2}\rho_t u^2+\sigma_t$, where v is the projectile velocity, u is the velocity of the interface, ρ_p is the projectile density, and ρ_t is the target density. The final target penetration, p, consists of a hydrodynamic phase (if the impact velocity is high enough) which is calculated by integrating the term u/(u-v) over the initial length of the projectile, l (in general this must be done numerically) plus a rigid projectile phase (if the dynamic strength of the rod is greater than that of the target).

The Tate model shows good agreement with experimental data below about 2 km/s, however in the hypervelocity regime, the crater depth of the model becomes independent of velocity and asymptotes to the steady state limit given by the initial rod length times the square root of the density ratio. This does not agree with the hypervelocity data presented by Christman and Gehring (1965) in which the penetration continues to rise above the steady state limit with increasing impact velocity. This divergence from Tate's one-dimensional, steady state model is most likely due to the two-dimensional, non-steady effects which dominate the final phase of the crater formation.

Modification of the Tate model. Christman and Gehring (1966) proposed an empirical model of rod cratering in which the penetration occurs hydrodynamically for all but the last diameter of rod length and that that last diameter of length creates a hemispherical crater charateristic of a chunky fragment impact. Their model of crater penetration depth is given by

$$p/d = \sqrt{(\rho_p/\rho_t)}(l/d - 1) + \tfrac{1}{2}\, D_\infty/d. \qquad (16)$$

The factor ½ implies that the chunky fragment portion of the crater is always hemispherical, but that is only true if the projectile and target are of the same material. In general, the final portion of the crater will be either greater than or less than hemispherical depending on the relative densities of the target and projectile. Therefore the factor ½ should be replaced by an expression such as $\frac{1}{2}(\rho_p/\rho_t)^m$. If it is assumed that this dependence on the density ratio also goes as the square root (m=½) then Eq.(16) for rod penetration depth becomes

$$p/d = \sqrt{(\rho_p/\rho_t)}(l/d - 1 + \tfrac{1}{2}\, D_\infty/d) \qquad (17)$$

Baker (1974) observed that the difference between the simple empirical model of Eq.(17) and the hydrodynamic limit to which the Tate model asymptotes is the second factor in parentheses which multiplies the square root of the density ratio. He therefore suggested that the Tate model could be modified to have the correct dependence at high velocity by simply multiplying the square root of the material density ratio, $\sqrt{(\rho_p/\rho_t)}$, by the factor $\beta = 1 - \frac{1}{2}\,(1-\sigma_\infty v)/(l/d)$.

When this modification is introduced into Tate's analysis, it not only has the desired effect of having the penetration continue to increase at the higher velocities to values above the steady state limit but also leaves the lower velocity behavior and the residual rod length virtually unchanged. In addition, the modification introduces a dependence on the initial rod length such that the shorter the initial rod, the greater will be the deviation of the penetration from the original Tate model at high velocities. Figure 3 compares the penetration (normalized by the inital rod length) versus velocity calculated from the original Tate model with that calculated from the empirically modified model. The values of the dynamic strengths of the rod and target materials differ for each combination off materials and have been selected so that the modified model best fits the experimental data points (also shown in the figure) which are taken from Christman and Gehring (1965) for a C1015 steel rod with an l/d of three impacting an 1100-0 aluminum target.

ROD-LIKE PROJECTILES

Having developed a reasonably good understanding of the penetration process of a rod projectile, the next step was to investigate the possibility of designing a projectile which would be an even more efficient penetrator than a rod. The five designs which were examined are illustrated in Fig.4 and are designated as TYPES 1 through 5. The baseline rod projectile is designated as TYPE 0. Since the shapes of the rod-like projectiles are all very different, the common measure of their penetrating ability was taken to be the amount of mass (rather than length) lost by each projectile in perforating the same target.

Application of Rod Analysis to Rod-like Projectiles

The rod analysis can be applied to the rod-like projectile designs by using the following procedure: (1) Calculate the steady state projectile length loss for penetrating a plate of thickness, t, $\Delta l_{ss} = t\sqrt{(\rho_t/\rho_p)}$; (2) Calculate the steady state mass loss, Δm_{ss}, corresponding to the above length loss, which is a simple geometric problem that depends on the projectile design; (3) Calculate the diameter at the front of the residual projectile, d_r, after removing the amount of projectile lost by steady state; (4) Use t/d_r in Eq.(1) without the t/d term to determine the non-steady state mass loss, Δm_{nss}, that a rod projectile of diameter d_r would experience. To do this, calculate the total length loss for a rod of diameter d_r impacting a plate of thickness t and then subtract the steady state length loss obtained in step 1, i.e., $\Delta m_{nss} = (\Delta l - \Delta l_{ss})\pi\rho_p d_r^2/4$; (5) The total mass loss is then $\Delta m = \Delta m_{ss} + \Delta m_{nss}$; (6) Calculate the total length loss corresponding to the total mass loss and then compare it to the measured experimental length loss.

The key to this procedure is the assumption in step 4 that the diameter of the projectile after perforating the target determines the amount of continued plastic flow which produces the additional projectile mass loss. This is a reasonable assumption and a comparison of the test results with the length loss calculations shows good agreement.

Discussion of Each Rod-like Projectile Design

TYPE 1 projectile. The TYPE 1 projectile design, a truncated cone, is subdivided into two categories designated as TYPE 1A with small values of the nose diameter and TYPE 1B with larger values of the nose diameter.

TYPE 1A projectile. The projectile-target material combinations were aluminum-aluminum, steel-aluminum, steel-titanium, and tungsten carbide-steel. In general, only for the steel-aluminum tests with small yaw angles did the projectile survive the severe lateral forces of the impact.

A plot of the calculated fractional mass loss versus target plate thickness for both the TYPE 1A and rod projectiles is shown in Fig.5. The ratio of cone mass loss to rod mass loss is obviously small for values of t/d less than about two and the rate of mass loss for the cone remains less than that for the rod for values of t/d up to approximately seven, above which the rate of loss for both are approximately equal.

For a spaced target configuration, the TYPE 1A projectile becomes a TYPE 1B after impacting the first target plate. The TYPE 1A projectile would therefore be particularly effective against a target configuration in which the first plate is thick and made of a low density material.

TYPE 1B projectile. Two of the impacts which failed with the TYPE 1A projectile were successful with the TYPE 1B projectile which has a larger nose diameter. If the target hole diameter is greater than the diameter of the projectile base then there should be no side contact between the projectile and the target. A plot of the calculated fractional mass loss versus target plate thickness for a steel TYPE 1B and rod projectile impacting a titanium target is shown in Fig.6.

TYPE 2 projectile. The primary advantage of the TYPE 2 projectile, a pointed rod, is its simplicity. This design was intended for like-material impact in the expectation that the pointed nose reduce the effect of the impact shock which would more than compensate for the side cavitation mass loss. A plot of the calculated fractional mass loss versus target plate thickness for steel rod and TYPE 2 projectiles with three cone tip angles, $\beta = 15^\circ, 30^\circ, 60^\circ$, impacting a steel target is shown in Fig.7. The calculation of the steady state mass loss for the pointed rod will change from a cone volume evaluation to include a rod volume evaluation if the steady state length loss exceeds the length of the conical tip. This change in mass loss rate is the knee in the graph of Fig.7.

TYPE 3 projectile. The concept of the TYPE 3 projectile is that the leading smaller diameter rod portion (snout rod) will produce a hole in the target plate sufficient to pass the remaining larger diameter rod without any additional projectile loss. Two snout rod configurations were used. One had a snout diameter that was half of the main rod diameter and a snout length that was four diameters long. The second configuration had a snout diameter that was two-thirds of the main rod diameter and a snout length that was four and a half diameters long. A plot of the calculated fractional mass loss for a steel rod and the two TYPE 3 projectiles impacting a steel target is shown in Fig.8.

TYPE 4 projectile. The TYPE 4 projectile is a combination of a high strength TYPE 1A truncated cone projectile and a low density sleeve. The five successful tests had a 4340 steel cone hardened to Rockwell "C" 55 with a Zelux (polycarbonate plastic) sleeve with a density of 1.19 g/cc. The ratio of the steel cone nose to base diameters was 0.08 while the ratios of Zelux sleeve diameter to cone base diameter were 1.0, 0.66, and 0.58.

The impact effect of this combination projectile is calculated as the sum of the impact of each part. The sleeve is modeled as a short rod with a diameter and a length such that the mass of the short rod is equivalent to the sleeve mass. The desired effect of the sleeve is the creation of a hole in the target with sufficient diameter and depth to prevent contact between the hole wall and the conical projectile base.

A plot of the calculated fractional mass loss for a steel rod and two TYPE 4 projectiles impacting a steel target is shown in Fig.9. Due to the low strength of the sleeve material, the calculations assume that the entire sleeve is destroyed by the impact, even for very thin targets. The steel cone loses mass according to the procedure for the TYPE 1A projectile. The mass loss for the TYPE 4 projectile is therefore the sleeve mass plus the cone mass loss.

TYPE 5 projectile. The TYPE 5 projectile consists of two masses: a leading "bumper" and a long following rod which are connected by a smaller diameter rod, or stem. This projectile configuration is an efficient one for perforating a t/d=2 target of like-material. The impact characteristics of the bumper (l/d~1) are essentially the same as that of a sphere of equal mass.

For very thick targets, the target damage due to bumper impact is the same as crater formation in a semi-inifinite target. The bumper will be destroyed by the impact shock in all cases except for low projectile velocities or very thin targets (t/d~0.01). The fragile connecting stem is also likely to be destroyed in most impact situations. The mass loss of the TYPE 5 projectile is therefore the sum of the bumper and stem masses plus the mass loss of the following rod.

A plot of the calculated fractional mass loss for a steel rod and TYPE 5 projectiles impacting a steel target is shown in Fig.10. Note the greater effectiveness of the bumper rod with the larger diameter of the connecting stem.

CONCLUSIONS

Rod Projectiles

The terminal ballistic results of a hypervelocity rod impacting a single target plate has been successfully quantified with a few, relatively simple, analytic expressions from which one can calculate the length of the residual rod and the diameter of the hole in the target plate given the value of a few empirically determined constants which depend on the rod and target materials. These analytic correlations have been shown to be valid for material combinations ranging from a very dense uranium rod impacting a low density aluminum target plate to a low density aluminum rod impacting a high density uranium target plate. The effect of target plate obliquity has also been quantified. The single plate equations can be readily applied, in a sequential manner, to a target consisting of a series of spaced plates.

Furthermore, the single target plate analysis has been extended to a multilayered target using a concept called "corresponding layers" along with some easily computed factors which control the extent to which the penetration process diverges from that for a single plate and approaches that of a "corresponding layer."

The rod projectile analysis concludes with an analytical model of crater penetration depth and residual rod length. The proposed model slightly modifies that of A. Tate by intoducing a factor which multiplies the density ratio. This factor, which is velocity dependent, permits the calculated crater penetration depth to rise above the simple hydrodynamic limit at very high velocities in agreement with the experimental data of Christman and Gehring.

Rod-like Projectiles

The five rod-like projectile designs, illustrated in Fig.4, were chosen to provide a significant reduction in the projectile mass loss, compared to that of a rod projectile, for particular impact configurations. This improved performance is due to one or more of the following mechanisms:

(1) a smaller nose diameter having less mass per unit length than a rod but which still creates a large enough clearance hole;

(2) a projectile of two independent sections with the leading, lower mass one sacrificed to create a clearance hole for the other;

(3) a projectile which is strong enough to create a hole in the target without suffering any damage to itself.

The most effective of the new projectile designs is the TYPE 1A sharp steel cone. For all but very thick targets the projectile mass losses are negligible. However, the impact configuration is somewhat limiting as it requires a weak, low density target and a high strength projectile.

The TYPE 1B truncated cone and the TYPE 3 snout rod projectiles should be particularly effective against thin targets since the mass loss would be a function of a small projectile nose diameter.

When the target thickness is such that the steady state length loss is greater than the length of the conical point, the reduction in mass loss for the TYPE 2 projectile over that for the rod will simply be twice the mass of the conical point. For thinner targets the TYPE 2 projectile suffers an unknown amount of side erosion to the conical point during target penetration and the modified rod loss theory will therefore predict low values of mass loss.

The target penetration efficiency of the TYPE 5 bumper rod projectile is due to the decoupling of the shock and plastic flow in the bumper from the following rod. The connecting stem must provide a margin of time to allow for sufficient flow of the target debris such that the following rod does not encounter any significant amount of target material. The connecting stem may also act like the snout of a TYPE 3 projectile, punching a clearance hole in any remaining target thickness.

The TYPE 4 sleeve/cone projectile combines the penetration mechanisms of the TYPE 1 and TYPE 5 projectiles. For thin targets the low density sleeve simply punches a clearance hole in the target while for thick targets the sleeve creates a crater of sufficient diameter and depth to give hole clearance for the projectile base and to allow the steel cone to operate as a TYPE 1B projectile which penetrates the remaining target thickness. The steel cone must be strong enough to withstand the shock waves generated by the impact of the low density sleeve.

Summary

This paper has summarized the results of experimental tests and analytical studies of the hypervelocity impact of rod and rod-like projectiles which were conducted at the Naval Research Laboratory. The results presented here provide relatively simple analytic expressions from which one can calculate the results of a hypervelocity impact of a rod or rod-like projectile even into complex targets under most impact configurations of interest. The methodology does require a knowledge of certain empirical constants which depend on the projectile and target materials. For those cases where the values of these constants have not been provided, they can easily be determined by performing a relatively few experimental impacts.

REFERENCES

Baker, J. (1969). See author for complete reference.

Baker, J. and Zalesak, S. (1970). See author for complete reference.

Baker, J. (1974). See author for complete reference.

Bertke, R., Heyda, J., Swift, H. and Lehman, M. (1974). A Study of Yawed Rod Impacts. University of Dayton Research Institute, U.S. Army Ballistic Research Laboratory Report No. BRL-CR-182.

Christman, D. and Gehring, J. (1966). Analysis of High-Velocity Projectile Penetration Mechanics. J. Appl. Phys., 37, 1579-1587.

Tate, A. (1967). A Theory for the Deleceration of Long Rods after Impact. J. Mech. Phys. Solids., 15, 387-399.

Tate, A. (1969). Further Results in the Theory of Long Rod Penetration. J. Mech. Phys. Solids., 17, 141-150.

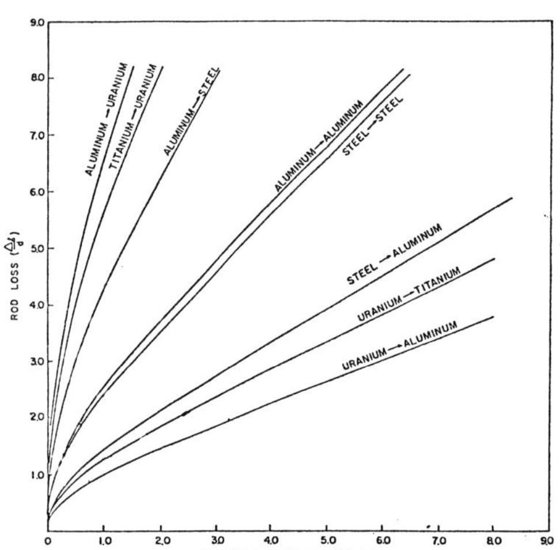

Fig.1 - Rod loss vs target plate thickness for selected materials with 5 < l/d < 10 and 3.1 < v < 4.6 km/s.

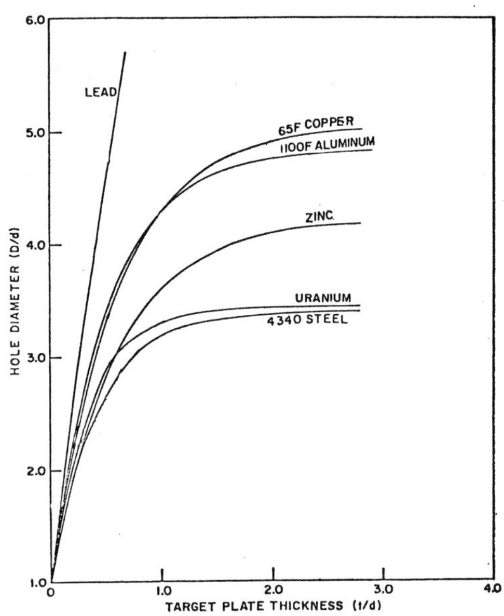

Fig.2 - Hole diameter vs target plate thickness for an aluminum rod with 5 < l/d < 10 impacting at 4.6 km/s.

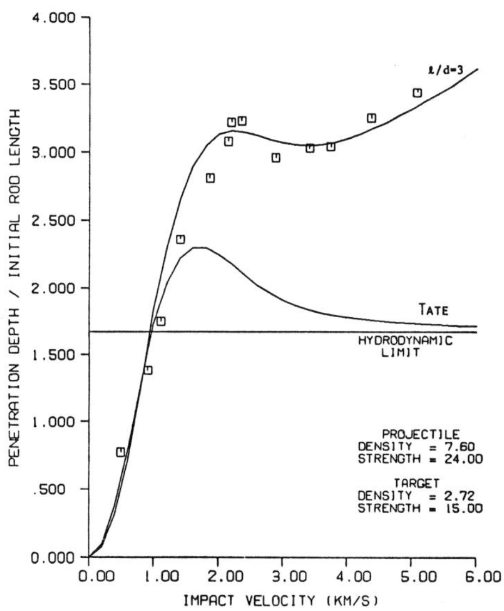

Fig.3 - Crater depth vs impact velocity for the Tate theory and the modified Tate theory with experimental data for a C1015 steel rod impacting an 1100-O aluminum target.

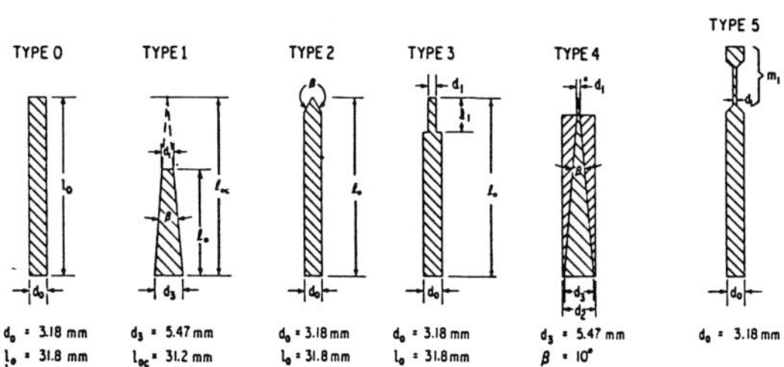

Fig.4 - Rod and rod-like projectile designs.

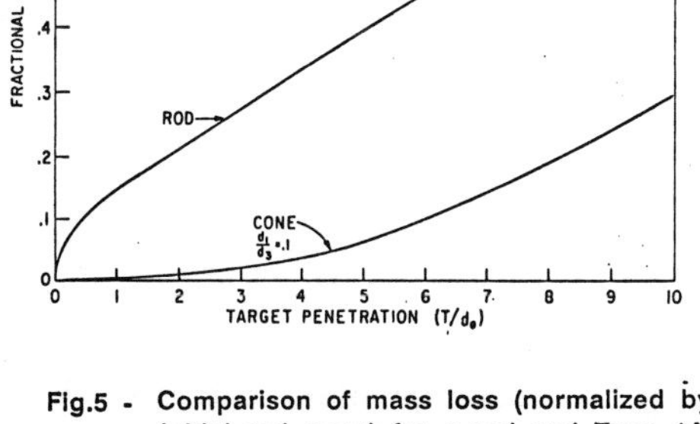

Fig.5 - Comparison of mass loss (normalized by initial rod mass) for a rod and Type 1A 4340 steel projectiles impacting a 2024-T4 aluminum plate at normal incidence.

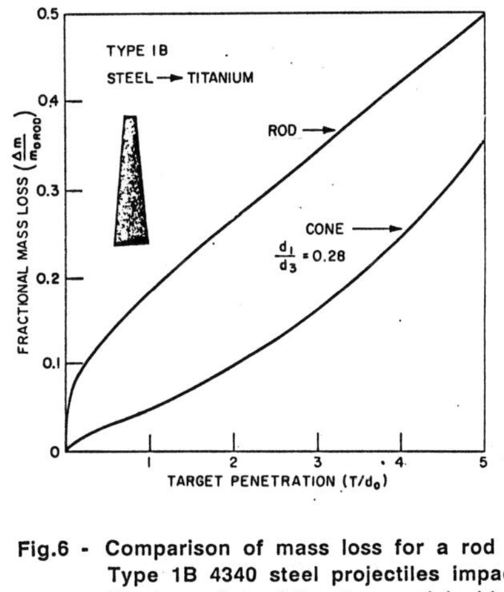

Fig.6 - Comparison of mass loss for a rod and Type 1B 4340 steel projectiles impacting a titanium alloy plate at normal incidence.

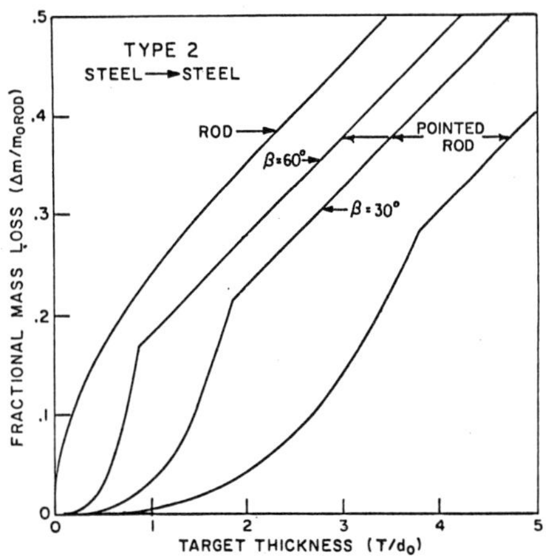

Fig.7 - Comparison of mass loss for a rod and Type 2 (conical tip angles of 15, 30 and 60 degrees) 4340 steel projectiles impacting a 4340 steel plate at normal incidence with velocities of 4.6 to 4.8 km/s.

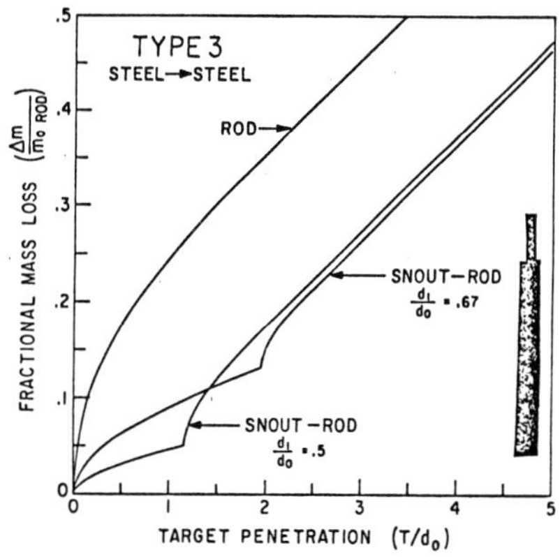

Fig.8 - Comparison of mass loss for a rod and Type 3 4340 steel projectiles impacting a 4340 steel plate at normal incidence with 4.7 < v < 5.2 km/s.

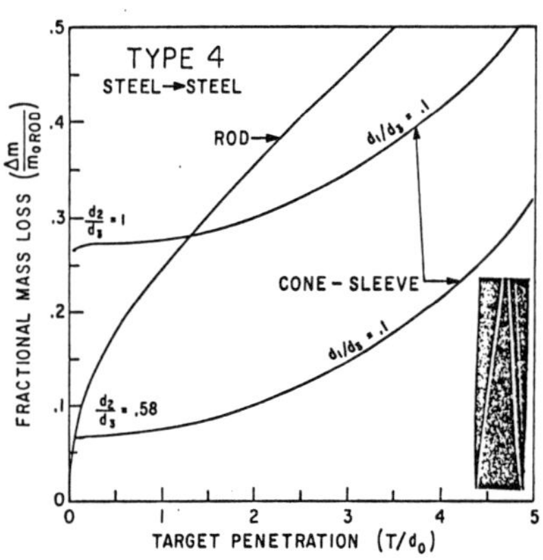

Fig.9 - Comparison of mass loss for a rod and Type 4 4340 steel projectiles impacting a 4340 steel plate at normal incidence with 4.6 < v < 4.8 km/s.

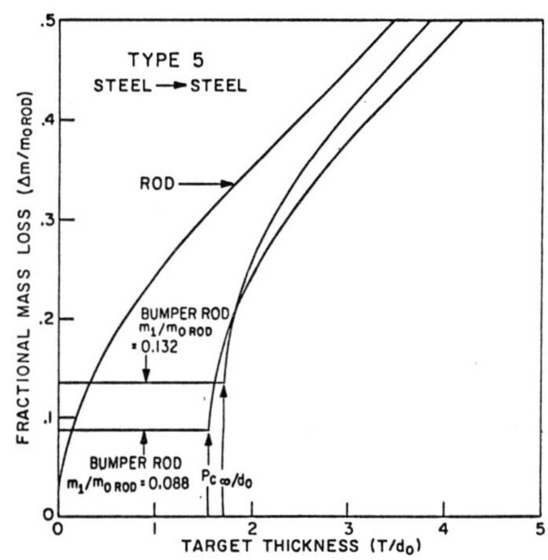

Fig.10 - Comparison of mass loss for a rod and Type 5 4340 steel projectiles impacting a 4340 steel plate at normal incidence with an impact velocity of 4.5 km/s.

Int. J. Impact Engng Vol. 5, pp. 111–127, 1987
Printed in Great Britain

0734-743X/87 $3.00 + 0.00
Pergamon Journals Ltd.

A THREE-DIMENSIONAL IMPACT-PENETRATION ALGORITHM WITH EROSION

Ted Belytschko and Jerry I. Lin

Department of Civil Engineering, Northwestern University
Evanston, Illinois 60201

ABSTRACT

A new interaction algorithm for treating three-dimensional impact-penetration simulations with erosion is presented. This algorithm is based on interactions between slave nodes (projectile) and master elements (target) and no tracking of the sliding interfaces is needed. It employs an assembly of normals to identify the interface surface adaptively so that it can handle eroding elements in both the target and the penetrator. In the interaction algorithm, momentum is exchanged between interacting nodes so that the total momentum is conserved. Examples of three dimensional calculations are also given.

INTRODUCTION

In impact-penetration problems, interface algorithms based on interactions between master nodes and slave nodes (Johnson 1977,1982; Johnson et al., 1978; Herrmann and Bertholf, 1983) are often used. When elements on the contact surfaces are eroded, these algorithms require redefinition or tracking of the sliding interfaces, which is quite complicated in three dimensions.

A new algorithm which allows arbitrary erosion in the target or projectile is developed. This algorithm handles the interaction by operations on slave nodes (projectile) and master elements (target) and requires no definition of the contact surfaces. The interaction procedure is designed in such a manner that total linear momentum is always conserved. This may result in some loss of energy, but it is not considered harmful if the interaction nodes contain only a small portion of the total mass.

An isoparametric hexahedral element with 8 nodes and 24 degrees of freedom is incorporated with the new algorithm. The stresses and strains are evaluated only at a single point in the element (Flanagan and Belytschko, 1981; Goudreau and Hallquist, 1981). This can result in certain physically unrealistic modes of deformation, known as hourglass modes. These are avoided by the use of hourglass control (Belytschko et al., 1984).

Several other important features of this algorithm are:

1. A cell structure which subdivides interaction domain into smaller zones to localize the detailed penetration detection.
2. A two-step check procedure, consisting of a rough-but-quick check and a precise check, to detect the penetration.
3. A standard finite element assembly procedure to assign normal directions for the nodes on the sliding surfaces so that the outside interacting surfaces of both the target and penetrator are readily identified regardless of any erosion that takes place on either body.

OVERVIEW OF THE ALGORITHM

The basic purpose of this algorithm is to treat the interaction of two bodies with eroding elements. Eroding elements are elements which are destroyed during the course of the computation because of very high strains, which implies that they represent elements of the target or penetrator which have ceased to play a significant role in the physics of the problem. When an element is eroded all its stress components are set equal to zero, but its mass and momentum are retained at the nodes associated with that element.

Algorithms with eroding elements in three dimensions are very difficult to treat within the conventional framework of master and slave contact surfaces. It is difficult to design algorithms to redefine the contact surfaces when elements are destroyed, particularly in three dimensions. Moreover, the corners which are created in the surfaces by the erosion of elements present severe algorithmic difficulties.

Therefore, this new algorithm uses a concept of slave nodes and master elements. One of the two bodies, usually the projectile, is defined by the nodes, hereafter called slave nodes, the second body is defined by elements, called master elements. If a slave node is found to be inside a master element, it is then brought back to an outside surface. In order to determine the outside surfaces, a set of normal vectors is assembled for all master elements as described in Section 5. Because of the character of the assembly process, a nonzero normal vector will only result on outside surfaces and will provide an effective average normal to the surface.

The mechanics of the interaction of the two bodies is executed completely through the interaction of the slave nodes with the master elements. The rules of this interaction are as follows:

i) Slave nodes are not permitted to penetrate master elements.

ii) Whenever penetration of a slave node into a master element is detected, the slave node is returned by a projection to the surface of the element it has penetrated and the associated loss of momentum is transferred to the appropriate nodes of the master element. If a check shows that this projection has not moved the slave node to an exterior surface, the node is moved to the appropriate edge.

The efficacy of this procedure hinges strongly on the use of explicit time integration, since stability requirements then restrict the time step so that the master element which is penetrated effectively represents the zone of the interaction; it should be noted that in cases of high-velocity impact the time step of nodes on the interface must be restricted so that node does not move more than 10-20% of a zone size during a time step. Because of the large number of slave nodes and master elements involved in three dimensional simulations, special techniques are needed to quickly identify the slave nodes and master elements between which interaction is possible. This is accomplished by using a cell structure which is fixed in space. Cells are considered to be substantially larger than elements and so may include many master elements and slave nodes. The basis of the procedure is to identify all slave nodes in a cell, and then in treating an element to identify the cell number in which it is located and to check the slave nodes which occupy the same cells in the interaction algorithm.

The steps of the interaction procedures within the structure of the complete algorithm is shown in Table 1. As can be seen, the locations of the slave nodes are determined during the time integration of the velocities. Penetration of master elements by slave nodes is then determined in a loop over all master elements. This loop must precede the element loop in which new nodal forces are determined because the interaction algorithm modifies the velocities of the slave nodes and master element nodes.

The methods for treating the frictional force and examples are given in the last two sections.

Table 1. The interaction procedures

1. initial conditions: velocities and positions of all nodes
2. integrate velocities to obtain new displacements
3.* determine the cell locations of all slave nodes
4.* for each master element:
    4a. compute surface normal vectors and assemble into global array
    4b. determine cells in which element is located
    4c. by checking all slave nodes in these cells, determine if any slave nodes are in the element
    4d. if a slave node is in the element, move it back to an outside surface and transfer the momentum to the element nodes, which modifies its nodal velocities

5. for each element:
   - 5a. compute strain-rates from the nodal velocities and stress-rates from the constitutive equations
   - 5b. integrate stress-rates to obtain new stresses and compute nodal forces
   - 5c. assemble nodal forces into global array
6. find nodal accelerations from equations of motion
7. integrate accelerations to find new velocities; go to 2

* denotes steps which pertain to the interaction algorithm

CELL STRUCTURE AND LOCATION OF NODES

The cell structure which is used to identify the slave nodes and master elements for which interaction is possible is shown in Fig. 1. The cells are uniform in size in all directions and their edges are parallel to the x, y and z coordinates.

The cell domain is described by the following:

$$(x_{min},\ x_{max},\ NCX) \tag{3.1a}$$

$$(y_{min},\ y_{max},\ NCY) \tag{3.1b}$$

$$(z_{min},\ z_{max},\ NCZ) \tag{3.1c}$$

Here x_{min} and x_{max} are the x-coordinates of the extreme points of the cell domain and NCX is the number of cells in the x-direction; the definitions of the terms in (3.1b) and (3.1c) are analogous.

The cell location of a node with coordinates (x, y, z) is computed in two steps. First the integer cell numbers in the x, y and z directions are computed by

$$IX = NCX*\ (x - x_{min})\ /\ (x_{max} - x_{min}) + 1 \tag{3.2a}$$

$$IY = NCY*\ (y - y_{min})\ /\ (y_{max} - y_{min}) + 1 \tag{3.2b}$$

$$IZ = NCZ*\ (z - z_{min})\ /\ (z_{max} - z_{min}) + 1 \tag{3.2c}$$

These numbers are then used to compute the cell number of the node, ICELL, as follows. If the node is outside the cell domain, the cell number ICELL is set to zero, i.e.

$$\begin{aligned} ICELL = 0 \quad \text{if} \quad & IX < 1 \quad \text{or} \quad IX > NCX \\ \text{or} \quad & IY < 1 \quad \text{or} \quad IY > NCY \\ \text{or} \quad & IZ < 1 \quad \text{or} \quad IZ > NCZ \end{aligned} \tag{3.3}$$

Otherwise, ICELL is computed by

$$ICELL = (IZ-1)*NCX*NCY + (IY-1)*NCX + IX \tag{3.4}$$

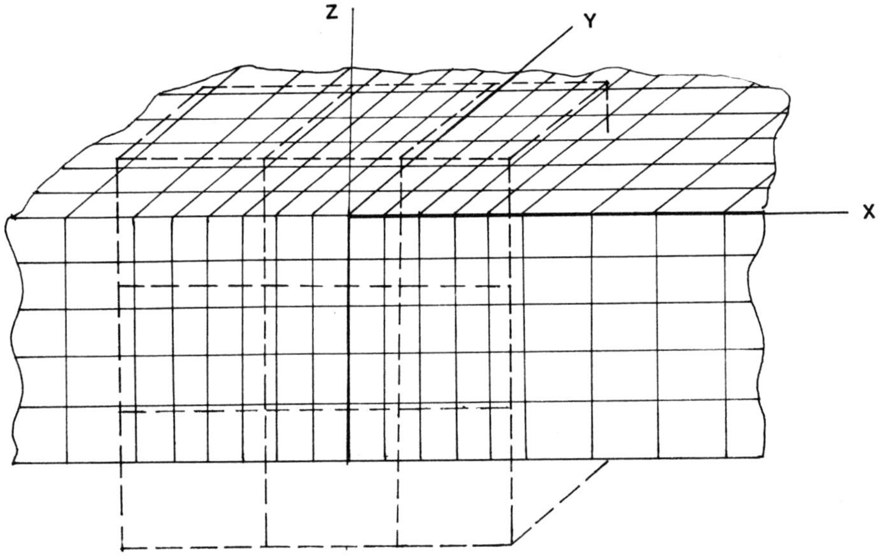

Fig. 1. A 3x1x3 cell structure.

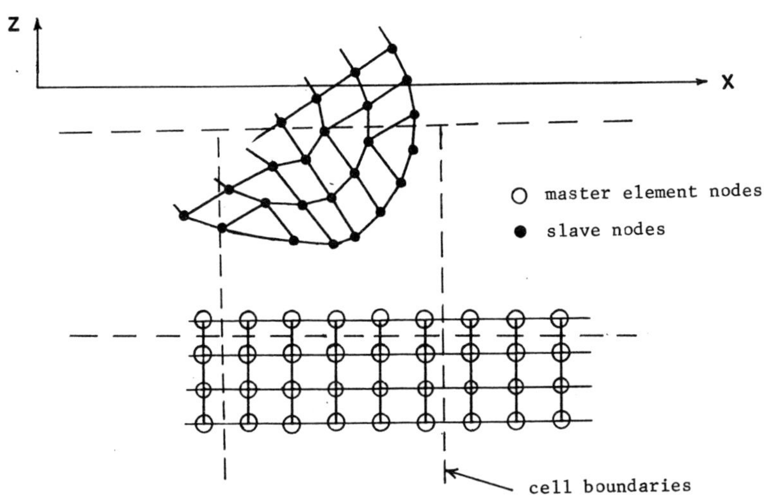

Fig. 2. Two dimensional view of cell structure showing elements in more than one cell.

The locations of all slave nodes are stored in an array K=LOCSLA(IC,J) where IC is the cell number and K is the node number of the Jth slave node in cell IC. IF LOCSLA(IC,1)=0, then no slave nodes are located within cell IC.

The identification of the cell locations of slave nodes is made during the time integration loop. As the new positions of slave nodes are calculated (prior to any adjustment for interaction with master elements), the cell number is obtained by the procedures (3.2) through (3.4).

For purposes of efficiency, the cell structure should extend only over the domain in which interaction is expected. This includes the domain occupied by the master elements in the undeformed configuration and any part of the domain into which master elements are anticipated to move as a result of the interaction. Although the optimal relation between cell size and master element size has to be determined, a cell should span at least two element lengths in each direction to reduce the number of elements which occupy more than one cell. A cell structure consisting of 3 x 3 x 1 cells in the x, y and z directions as shown in Fig. 1 has been found quite effective in the calculations performed so far.

ELEMENT PENETRATION DETECTION

Any slave nodes which penetrate an element are treated during a loop through the elements; this loop is separate from the element loop in which the nodal forces are updated, and considers only master elements. The element level steps consist of the following for each element e:

1. determine whether any slave node has penetrated element e;
2. if a slave node has penetrated element e, place the node outside the element and transfer the momentum loss to the element nodes.

Note that this procedure is carried out only for master elements so the subdomain of the target that is designated to consist of master elements must be sufficiently large so that no slave node can penetrate the target without penetrating a master element.

The algorithm for step 1 must be streamlined as much as possible in order to insure that computer time is not wasted. The cell scheme is used to minimize as much as possible the number of computations required. The details of the procedure are as follows:

The cell number of each node of the element is determined. As shown in Fig. 2, it is possible for the element to be in several cells. The cell numbers of the nodes are determined in the same manner that is used for determining the slave node cell location. If ICELL = 0, the node is not in the interaction zone, so it need not be considered. The cell numbers of all nodes of the element are checked against those of the other nodes so that no duplicate cell numbers are considered in checking for slave node penetration of the element.

To determine which slave nodes are in an element, all slave nodes in the same cell as the element are checked. First a rough check is made. For this purpose, the element confines are defined by

$$\begin{aligned} x^e_{max} &= \max_{I\varepsilon e} (x_I) \\ x^e_{min} &= \min_{I\varepsilon e} (x_I) \\ y^e_{max} &= \max_{I\varepsilon e} (y_I) \\ y^e_{min} &= \min_{I\varepsilon e} (y_I) \\ z^e_{max} &= \max_{I\varepsilon e} (z_I) \\ z^e_{min} &= \min_{I\varepsilon e} (z_I) \end{aligned} \tag{4.1}$$

where x_I, y_I and z_I are the coordinates of node I. A node J is within the confines of element e if

$$
\begin{aligned}
x^e_{min} &< x_J < x^e_{max} \\
y^e_{min} &< y_J < y^e_{max} \\
z^e_{min} &< z_J < z^e_{max}
\end{aligned} \tag{4.2}
$$

All slave nodes which are located in cells which are occupied by nodes of the master element are processed to check whether they are in the element. If an element node occurs in a cell which has a leading zero in LOCSLA, then the cell contains no slave nodes, so no checks need to be made. Otherwise, the process of determining whether a slave node occurs in the element begins. In order to reduce computations, the slave nodes which are sufficiently far from the master element are first eliminated from consideration of the element confines.

If a slave node is within the confines of the master element, the more exact and time consuming checks are made on the slave node to see whether it is within the element. If the slave node is within the element, the position of the slave node is adjusted and the corresponding momentum is transferred to the appropriate nodes of the master element. In addition, the slave node number is deleted from the LOCSLA list so that this slave node is not checked against other elements. The details of this are given in Section 6.

Once a slave node has passed the simple checks for possible penetration, a definitive check is made as to whether the node is within the element. This is accomplished by constructing six pentahedra, each consisting of a side of the hexahedron and the slave node, as shown in Fig. 3. If the volumes of all six pentahedra are positive, the slave node must be within the element.The volumes of the pentahedra are computed by the formula for hexahedron volume (Flanagan and Belytschko, 1981) with nodes 5 to 8 considered coincident. This gives the following formula for the volume of a pentahedron

$$V = \sum_{I=1}^{5} x_I \, B_I^{(5)} \tag{4.3}$$

where

$$
\begin{aligned}
B_1^{(5)} &= \frac{1}{12} [(2y_5 - y_3) \, z_{42} + y_2 \, (z_{53} + z_{54}) + y_4 \, (-z_{53} - z_{52})] \\
B_2^{(5)} &= \frac{1}{12} [(y_4 - 2y_5) \, z_{31} + y_3 \, (z_{54} + z_{51}) + y_1 \, (-z_{54} - z_{53})] \\
B_3^{(5)} &= \frac{1}{12} [(y_1 - 2y_5) \, z_{42} + y_4 \, (z_{51} + z_{52}) + y_2 \, (-z_{51} - z_{54})] \\
B_4^{(5)} &= \frac{1}{12} [(2y_5 - y_2) \, z_{31} + y_1 \, (z_{52} + z_{53}) + y_3 \, (-z_{52} - z_{51})] \\
B_5^{(5)} &= \frac{1}{12} [(y_{54} - y_{52}) z_{31} - (y_{53} - y_{51}) z_{42} - (z_{51} - z_{53}) y_{42} + (z_{52} - z_{54}) y_{31}]
\end{aligned} \tag{4.4}
$$

In using these formulas, nodes 1 to 4 must define a side of an element and must be numbered so that they are counterclockwise when viewed from a point inside the element.

As soon as any volume is found to be negative, the slave node can be considered to definitely lie outside the element. All 6 pentahedral volumes must be positive if the nodes are inside the element.

NORMAL DIRECTIONS

An important ingredient in defining the interaction of the slave nodes and master elements is that any transfer of momentum which occurs between the target and penetrator (other than that due to friction, which is not considered here) should be in directions normal to the interface. For this purpose, the normal vectors must be available when the positions of the slave nodes are adjusted. Since an interaction surface is never defined, it is necessary to construct normals in an alternative manner.

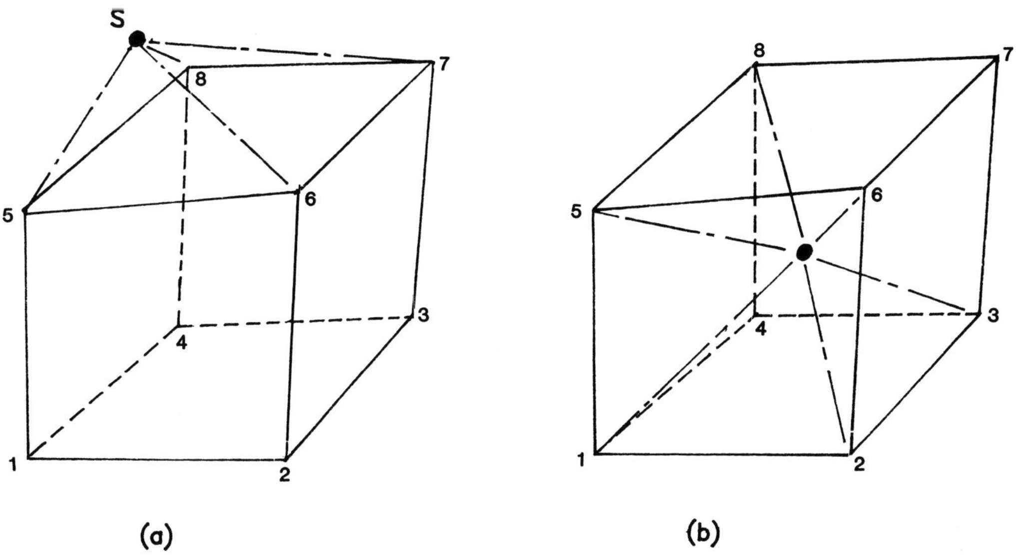

Fig. 3. Penetration check showing the pentahedral volumes which are computed. In (a) the volume 5 - 8 - 7 - 6 - S is negative, in (b) all 6 pentahedra have positive volumes.

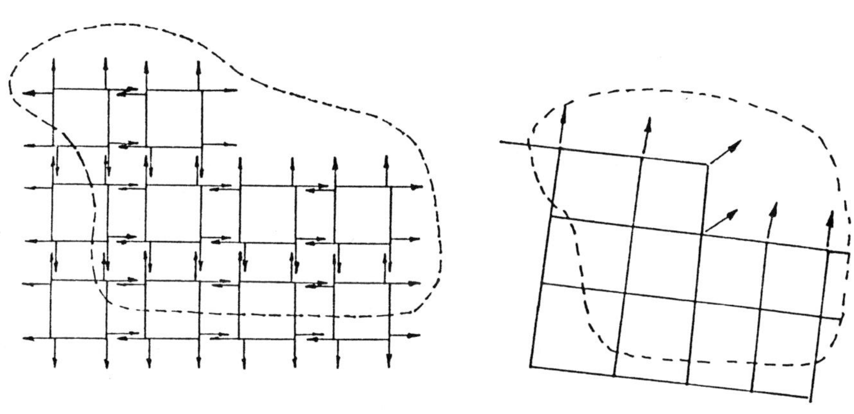

Fig. 4. Assembly of normals from master elements to determine outside surfaces.

The assembly procedure of the finite element method provides a very natural and concise way of computing these normal vectors. We will first present the procedure in a two dimensional setting so that its ingredients may be understood and visualized more readily and then present the three dimensional equations.

The basic idea of this method is that the normals for each side of the element are computed and added component by component into a normal vector array according to node numbers. The procedure is illustrated in Fig. 4. Note that on interior nodes the assembled normal vectors essentially cancel, so their components are very small or zero. On exterior nodes, the normal vectors point out from the domain with a direction which reasonably approximates a normal to a surface on the edge of the domain. In particular, the normal at a corner takes an average direction of the surfaces that meet at the corner.

In three dimensions, the procedure consists of the following: for each side of the element with local nodes 1, 2, 3, 4, a vector normal to the side is approximately computed by

$$\underset{\sim}{n} = \underset{\sim}{x}_{42} \times \underset{\sim}{x}_{31} \tag{5.1}$$

where (x) designates a vector cross-product. This vector $\underset{\sim}{n}$ is normalized and then assembled into the global arrays of the normals of 4 nodes by adding each component of the vector $\underset{\sim}{n}$ to the existing vector in the nodal array. When the contribution of each element has been added to the nodal arrays, the procedure is complete. The procedure can be summarized as follows:

1. Initialization at beginning of time step: zero the vectors $\underset{\sim}{n}_I$ for all nodes I.
2. For each element e in the set of master elements:
   - 2.1 For each side of the element, compute $\underset{\sim}{n}$;
   - 2.2 add $\underset{\sim}{n}$ into the arrays $\underset{\sim}{n}_I$ of the four nodes which define the side.

Remarks:

1. In order to avoid difficulties with master elements at the edge of the master element domain, the master elements should be defined to extend sufficiently far so that the edge master elements are not penetrated from the sides.
2. For nodes of master elements which lie on a plane of symmetry, the components of the normal vector which do not lie in the plane of symmetry are set to zero.
3. For master elements which lie on the bottom of the target (opposite to the original position of the projectile), the normals to the bottom surface are omitted to avoid driving the slave nodes in the wrong direction.

ADJUSTMENT OF SLAVE NODE POSITIONS

Once it is determined that a slave node has penetrated a master element, it is necessary to adjust its position consistent with the fact that its normal momentum has been transferred to the target. This procedure involves three steps:

1. using the normal vectors associated with the element, determine in which direction the position of the slave node has to be adjusted;
2. displace the slave node to an outside surface in the direction of the normal $\bar{\underset{\sim}{n}}$;
3. if the surface to which a slave node is brought is not an outside surface, bring the slave node back to an edge of the surface.

The procedure is implemented as follows. The average normal of the element is found by

$$\bar{\underset{\sim}{n}} = \Big(\sum_{I=1}^{8} \underset{\sim}{n}_I \Big) \, / \, \| \quad \| \tag{6.1}$$

where the division by "$\| \quad \|$" designates normalization of the vector in the numerator. Let the current coordinates of the slave node be $\underset{\sim}{x}_0$. Then the node is displaced by the procedure

$$\underset{\sim}{x} = \underset{\sim}{x}_0 + \eta \, \underset{\sim}{\bar{n}} \tag{6.2}$$

where η is an undetermined parameter $\eta > 0$.

The magnitude of η is determined by checking which of the sides of the hexahedron is intersected by the line of Eq. (6.2). This is accomplished as follows. Each side is subdivided into 2 triangular surfaces (note that this is only an approximation to the surface of an isoparametric hexahedron but it simplifies computations enormously). By taking 3 nodes of the surface in turn, the surface is defined by

$$x_i = x_{iI} \, \xi_I \qquad I = 1 \text{ to } 3 \tag{6.3}$$

$$\xi_1 + \xi_2 + \xi_3 = 1 \tag{6.4}$$

The intersection of the line defined by Eq. (6.2) and the plane is determined by solving the 4 equations in 4 unknowns represented by Eqs. (6.3) to (6.4). The solution is given by

$$\eta = \frac{-1}{D} \det \begin{bmatrix} x_{13} & x_{23} & x_{03} \\ y_{13} & y_{23} & y_{03} \\ z_{13} & z_{23} & z_{03} \end{bmatrix} \tag{6.5a}$$

$$\xi_1 = \frac{1}{D} \det \begin{bmatrix} x_{03} & x_{23} & \bar{n}_x \\ y_{03} & y_{23} & \bar{n}_y \\ z_{03} & z_{23} & \bar{n}_z \end{bmatrix} \tag{6.5b}$$

$$\xi_2 = \frac{1}{D} \det \begin{bmatrix} x_{13} & x_{03} & \bar{n}_x \\ y_{13} & y_{03} & \bar{n}_y \\ z_{13} & z_{03} & \bar{n}_z \end{bmatrix} \tag{6.5c}$$

$$\xi_3 = 1 - \xi_1 - \xi_2 \tag{6.5d}$$

where

$$D = \det \begin{bmatrix} x_{13} & x_{23} & \bar{n}_x \\ y_{13} & y_{23} & \bar{n}_y \\ z_{13} & z_{23} & \bar{n}_z \end{bmatrix} \tag{6.6}$$

A particular triangular surface is intersected by the parametric ray of Eq. (6.2) if and only if

$$\eta > 0 \tag{6.7}$$

$$0 < \xi_I < 1 \qquad \text{for} \quad I = 1 \text{ to } 3 \tag{6.8}$$

Once the surface on which the slave node is projected is determined, the surface is checked to ascertain whether it is an outside surface. This is done by checking whether the 4 normals of the nodes of the surface are non zero. If this check fails, the node is projected to an edge of the surface as shown in Fig. 5.

The equations which govern this realignment of the slave node are the following. Let $\underset{\sim}{x}_n$ be given by

$$\underset{\sim}{x}_n = \underset{\sim}{x}_o + \Delta t\ \underset{\sim}{v}^{old} \tag{6.9}$$

where $\underset{\sim}{x}_o$ is the position of the slave node at the beginning of the time step as shown in Fig. 5. The edges of the side are then considered in turn and its nodes are generically identified as 1 and 2, with the vector connecting them denoted by $\underset{\sim}{x}_{21}$. The previously computed new position of the node is denoted by $\underset{\sim}{x}_m$ where

$$\underset{\sim}{x}_m = \underset{\sim}{x}_n + \Delta\ \underset{\sim}{r} \tag{6.10}$$

The node is then repositioned on the intersection of the line $\underset{\sim}{x}_{12}$ with plane defined by the vectors $\underset{\sim}{x}_{mn}$ and $\underset{\sim}{x}_{no}$. The equations to be solved are

$$\underset{\sim}{x}_{1o} + \underset{\sim}{x}_{21}\ \eta_1 - \underset{\sim}{x}_{on}\ \eta_2 - \underset{\sim}{x}_{mn}\ \eta_3 = \underset{\sim}{0} \tag{6.11}$$

where η_i, $i = 1$ to 3, are the unknowns. If the edge $\underset{\sim}{x}_{21}$ is the correct one η_1 must satisfy $0 < \eta_1 < 1$. Otherwise, another edge of the surface is considered.

The final position is denoted by $\underset{\sim}{x}_f$ which is given by

$$\underset{\sim}{x}_f = \underset{\sim}{x}_1 + \eta_1\ \underset{\sim}{x}_{21} \tag{6.12a}$$

where η_1 is the above solution. The reposition vector is then redefined by

$$\Delta\underset{\sim}{r} = \underset{\sim}{x}_{fn} = \underset{\sim}{x}_f - \underset{\sim}{x}_n \tag{6.12b}$$

If a slave node is on a plane of symmetry, the component of $\Delta\underset{\sim}{r}$ normal to the plane of symmetry is now set to zero.

One situation which can foil even this check and allow a slave node to remain within the target is shown in Fig. 6. Here the slave node has penetrated element with nodes numbered 1 to 8. The repositioning will place the node on the surface defined by nodes 1 to 4, and because a single intact element, defined by nodes 1 to 4 and 9 to 12 lies above the intact plane, the normals will be nonzero on each of the four nodes, 1 to 4. Therefore, the node will remain on this surface. However, if all projectile nodes are slave nodes, a subsequent slave node should engage the element defined by nodes 1 to 4 and 9 to 12.

Once the new position of the slave node is determined the change in its velocity is computed by

$$\Delta\underset{\sim}{v} = \Delta\underset{\sim}{r}/\Delta t \tag{6.13a}$$

where $\Delta\underset{\sim}{r}$ is the total displacement of the slave node needed. The velocity of the slave node is then modified by

$$\underset{\sim}{v}^{new} = \underset{\sim}{v}^{old} + \Delta\underset{\sim}{v} \tag{6.13b}$$

In unusual circumstances it is possible for the normal of an element to form a wide angle with the slave node velocity. This situation is illustrated in Fig. 5, for node B, where α is the angle between the velocity vector and the normal. If the previously described procedures are

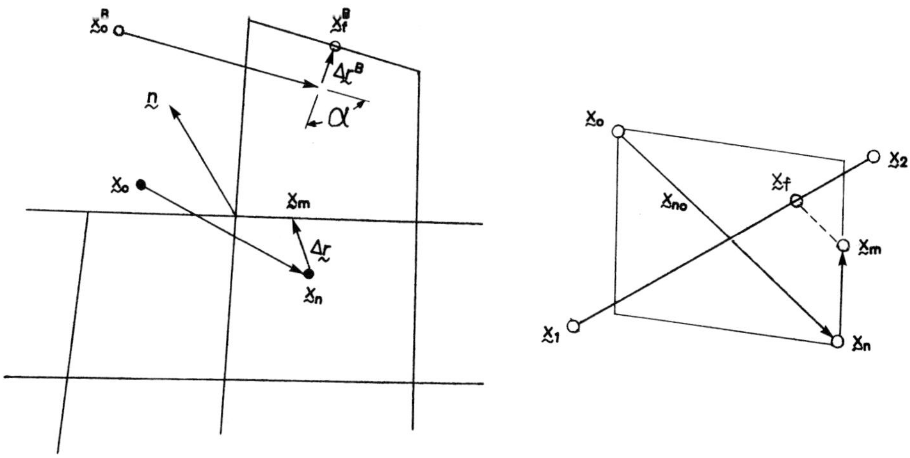

Fig. 5. Procedure for repositioning a slave node when the normal projection leaves it inside the target; node B illustrates a case where velocity is increased by original repositioning.

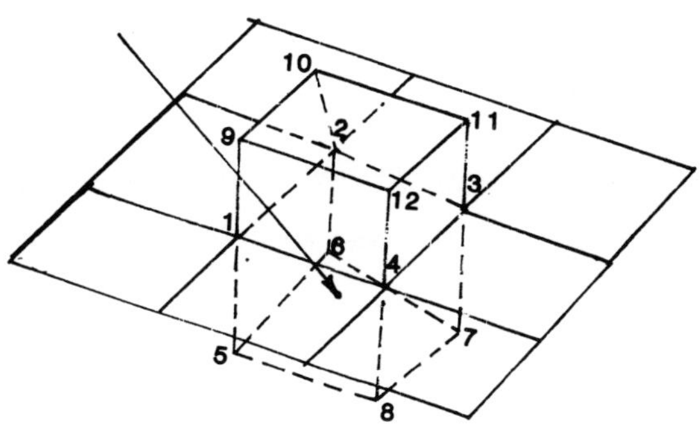

Fig. 6. Depiction of a situation where checking for nonzero normals is not sufficient to detect an interior surface.

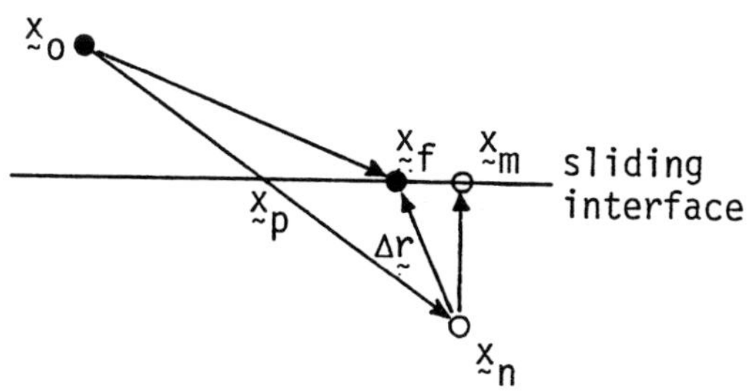

Fig. 7. Procedure for repositioning a slave node to account for the frictional effect on the sliding interface.

used, the repositioning of the slave node will increase its velocity and its kinetic energy as shown; note the new velocity vector is longer than the original velocity vector. This repositioning is of course completely contrary to physical laws, since it increases the kinetic energy of the system.

In order to avoid this, the new velocity $\underset{\sim}{v}^{new}$ is compared with the original velocity, $\underset{\sim}{v}^{old}$. It is required that the following inequality be satisfied by these velocities

$$\| \underset{\sim}{v}^{new} \| < \| \underset{\sim}{v}^{old} \| \tag{6.14}$$

If this is not satisfied, the node is moved back along the vector $\underset{\sim}{v}^{new}$ until its velocity satisfies the following equation

$$\| \underset{\sim}{v}^{new} \|^2 + \| \Delta \underset{\sim}{r}/\Delta t \|^2 = \| \underset{\sim}{v}^{old} \|^2 \tag{6.15}$$

This implies that the angle between Δr and $\underset{\sim}{v}^{new}$ will be a right angle. Although this procedure will usually position the slave node outside of the target, it insures that energy is not generated by the procedure. In subsequent time steps the slave node will again penetrate a master element so the procedure is not harmful.

The momentum loss associated with this adjustment is $M\Delta\underset{\sim}{v}$, where M is the mass of the slave node. This momentum is now transferred to the nodes of the penetrated surface. The formula used for node J of the surface is

$$\Delta \underset{\sim}{v}_J = \frac{M}{rM_J} \underset{\sim}{n}_J^T \bar{\underset{\sim}{n}} \Delta \underset{\sim}{v} \qquad \text{(no sum on J)} \tag{6.16}$$

where M_J is its mass and

$$r = \sum_{I=1}^{4} \underset{\sim}{n}_I^T \bar{\underset{\sim}{n}} \tag{6.17}$$

This formula apportions the momentum to the nodes according to how strongly their vectors point in the direction of the interface normal $\bar{\underset{\sim}{n}}$.

FRICTIONAL FORCE ON THE INTERFACE

Frictional forces between the two bodies sometimes play an important role in the impact-penetation simulation. Firction usually restraints the projectile from slipping along the surface, so the smaller angles of incidence can result in penetration.

In this interface algorithm, the frictional effect on the interface is taken into account by repositioning the slave node. The procedure is best demonstrated by Fig. 7. Suppose a slave node travels from $\underset{\sim}{x}_o$ to $\underset{\sim}{x}_n$ in a time step, it would be returned to $\underset{\sim}{x}_m$ in a frictionless case. With the presence of friction, the slave node is moved to $\underset{\sim}{x}_f$ instead of $\underset{\sim}{x}_m$. Here $\underset{\sim}{x}_f$ is determined by

$$\underset{\sim}{x}_f = a \underset{\sim}{s} + \underset{\sim}{x}_m \tag{7.1}$$

where

$$\underset{\sim}{s} = \frac{\underset{\sim}{x}_p - \underset{\sim}{x}_m}{\| \underset{\sim}{x}_p - \underset{\sim}{x}_m \|} \tag{7.2a}$$

$$a = \begin{cases} \beta \| \underset{\sim}{x}_m - \underset{\sim}{x}_n \| & \text{if} \quad \beta \| \underset{\sim}{x}_m - \underset{\sim}{x}_n \| < \| \underset{\sim}{x}_p - \underset{\sim}{x}_m \| \\ \| \underset{\sim}{x}_p - \underset{\sim}{x}_m \| & \text{if} \quad \beta \| \underset{\sim}{x}_m - \underset{\sim}{x}_n \| > \| \underset{\sim}{x}_p - \underset{\sim}{x}_m \| \end{cases} \qquad (7.2b)$$

β = coefficient of kinetic friction between the projectile and the target (7.2c)

$\underset{\sim}{x}_p$ = the intersection of the initial velocity of the slave node and the sliding interface. (7.2c)

The momentum transfer between the slave node and the master element is accomplished by following Eq. (6.13) to (6.17).

NUMERICAL EXAMPLES

The sample problems include a thick and a thin target penetration problem. In each case, a table of key problem parameters and some computer graphics of the evolution of the response are given. All computations are performed by the Harris-800 computer system.

Example 1.

The first example is a problem with a thick target object, involving 104 elements in the projectile and 1040 elements in the target. Table 2 gives the problem parameters. The evolution of the problem is shown in Fig. 8

Table 2. Parameters for Example 1

Projectile		
shape	:	rod with a round nose
dimensions	:	4.04 in (0.103 m) long, 0.201 in (0.005 m) radius
density	:	0.00073 lb-sec^2/in^4 (7805 kg/m^3)
bulk modulus	:	23,810,000 psi (1.633x10^5 MPa)
shear modulus	:	11,630,000 psi (8.019x10^4 MPa)
yield stress	:	250,000 psi (1.724x10^3 MPa)
ultimate stress	:	310,000 psi (2.137x10^3 MPa)
initial velocity	:	x-component 56155.0 in/sec (1426 m/sec) z-component -32421.0 in/sec (-823.5 m/sec)
Target		
shape	:	plate
dimensions	:	5.6 in x 0.6 in x 1 in (0.142 m x 0.015 m x 0.025 m) (half plate)
density	:	0.00073 lb-sec^2/in^4 (7805 kg/m^3)
bulk modulus	:	27,780,000 psi (1.915 x 10^5 MPa)
shear modulus	:	11,360,000 psi (7.833 x 10^4 MPa)
yield stress	:	160,000 psi (1.103x10^3 MPa)
ultimate stress	:	185,000 psi (1.276x10^3 MPa)
initial velocity	:	0

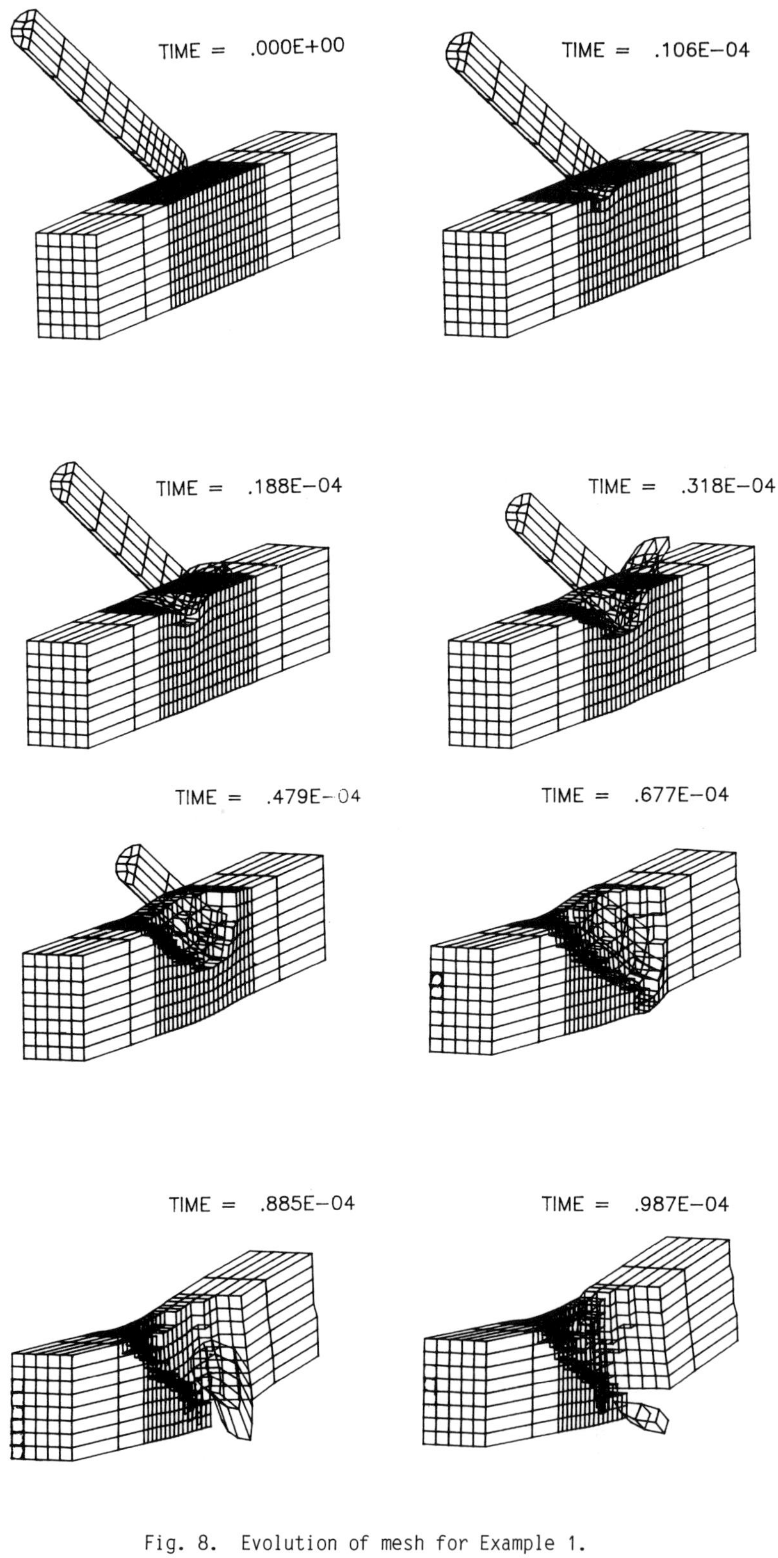

Fig. 8. Evolution of mesh for Example 1.

Example 2

Example 2 is taken from the EPIC-3 manual (Johnson et al., 1978). It consists of a copper rod striking a steel plate at 6562 ft/second. The projectile is modeled with 414 elements, the target with 1014 elements. The copper is assigned a erosion strain of $\bar{\varepsilon}_p = 2.0$. A complete listing of material and problem parameters is given in Table 3. Note that arbitrary erosion can occur in both the target and projectile.

Table 3. Parameters for Example 2

Projectile

shape	:	rod with a round nose
dimensions	:	4.9 in (0.124 m) long, 0.5 in (0.0127 m) radius
density	:	0.000831 lb-sec^2/in^4 (8884 kg/m^3)
bulk modulus	:	20,739,000 psi (1.430x10^5 MPa)
shear modulus	:	6,380,000 psi (4.40x10^4 MPa)
yield stress	:	20,300 psi (1.400x10^2 MPa)
ultimate stress	:	65,300 psi (4.502x10^2 MPa)
initial velocity	:	x-component 55660.0 in/sec (1414 m/sec) z-component -55660.0 in/sec (-1414 m/sec)

Target

shape	:	plate
dimensions	:	7.9 in x 3.95 in x 0.375 in (0.2 m x 0.1 m x 0.0095 m) (half plate)
density	:	0.000734 lb-sec^2/in^4 (7847 kg/m^3)
bulk modulus	:	24,200,000 psi (1.669x10^5 MPa)
shear modulus	:	9,300,000 psi (6.412x10^4 MPa)
yield stress	:	160,000 psi (1.103x10^3 MPa)
ultimate stress	:	203,000 psi (1.400x10^3 MPa)
initial velocity	:	0

The simulation is shown in Fig. 9. Note that the projectile starts jetting in the positive x-direction early in the simulation. These large shears result in rapid erosion of the projectile. Subsequently, large deformations in the target result in shear failure in the target. Erosion takes place in both the target and projectile.

The projectile velocity after the penetrating process, the projectile exit velocity, is of interest to many researchers in this area. Four test problems were run to study the effect of initial velocity, angle of incidence, mesh refinement and friction on the exit velocity.

The geometry and the material properties are identical for the four problems, as listed in Table 2. ϕ_{inc}, v_i and v_e are the angle of incidence, initial and exit velocity for the projectile respectively. The results are summarized in Table 4. Example 3 and 5 are identical except that the frictional force is taken into acocunt with a friction coefficient of 0.18 in Example 3.

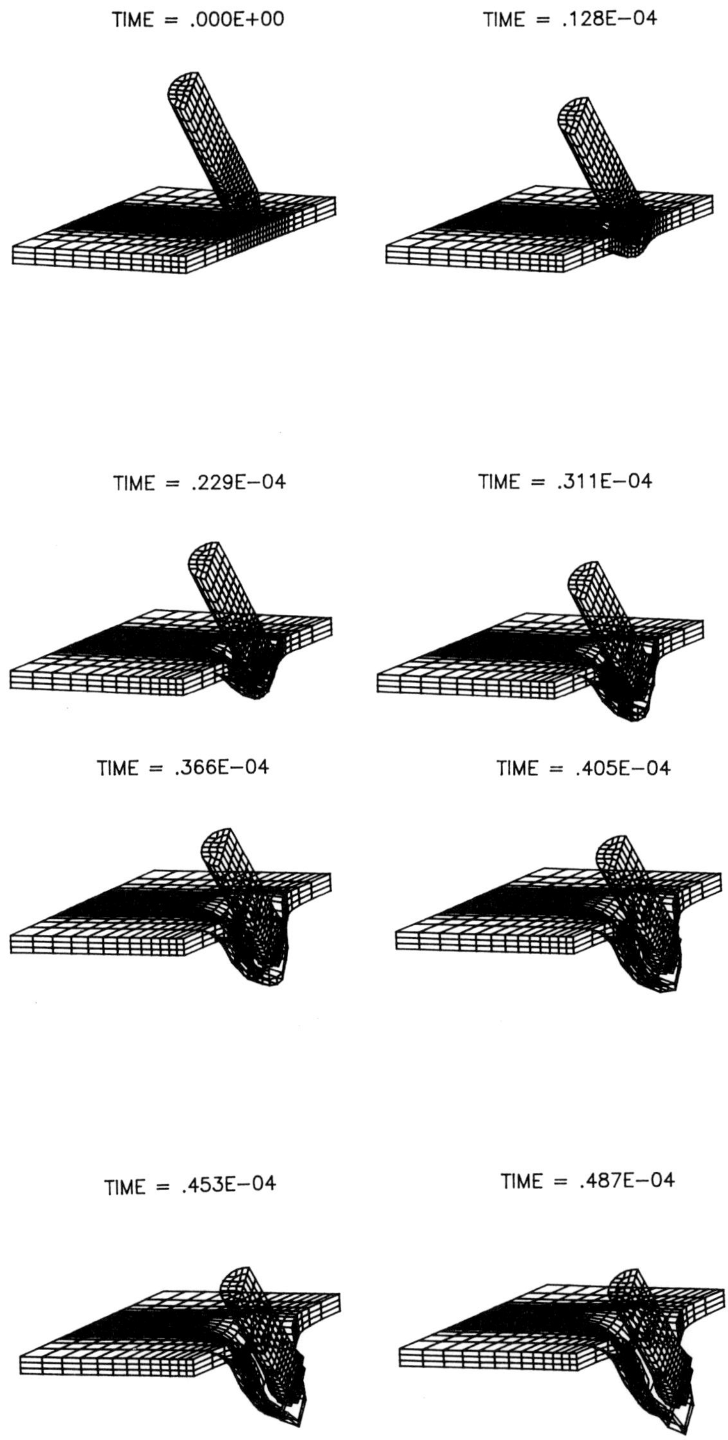

Fig. 9. Evolution of mesh for Example 2.

Table 4. Summary of projectile velocity ratio

example	nodes (elements) projectile	target	ϕ_{inc}	v_i(m/sec)	v_e(m/sec)	v_e/v_i
1	210 (104)	1458 (1040)	60°	1647	654	0.397
3	784 (486)	2310 (1700)	45°	1647	1097	0.66
4	784 (486)	2310 (1700)	45°	3294	2578	0.783
5	784 (486)	2310 (1700)	45°	1647	714	0.434

The combination of a sharper projectile angle and a finer mesh cause the higher v_e/v_i ratio in Example 3 than in Example 1.

The increase of the ratio from Example 3 to Example 4 may be attributed mainly to the higher initial velocity of the projectile. Comparison between Examples 3 and 5 indicates that the addition of friction results in a higher exit velocity; this is apparently caused by a penetration path which is less oblique. In the absence of friction, the penetrator slides and then penetrates at a rather shallow angle and consequently has to remove more material in passing through the target.

ACKNOWLEDGEMENT

We gratefully acknowledge support of the U.S. Army Research Office in Contract DAAG29-84-K-0057 and the advice of K. Kimsey, H. Jonas and J. Zukas of U.S. Ballistic Research Laboratory, Aberdeen Proving Grounds.

REFERENCES

Belytschko, T., W.K. Liu, J.M. Kennedy and S.J. Ong (1984). Hourglass control in linear and nonlinear problems. Compt. Meth. Appl. Mech. Engrg., 43, 251-276.

Flanagan, D.P. and T. Belytschko (1981). A uniform strain hexahedron and quadrilateral with orthogonal hourglass control. Intl. J. Num. Meth. Engrg., 17 679-706.

Goudreau, G.L. and J.O. Hallquist (1981). Recent development in large scale finite element Lagrangian hydrocode technology. Report UCRL 86460 Lawrence Livermore Laboratory.

Hermann, W. and L.D. Bertholf (1983). Explicit Lagrangian finite difference methods. In T. Belytschko and T.J.R. Hughes (Ed.), Computational Methods for Transient Analysis. North-Holland, Amsterdam, pp. 316-417.

Johnson, G.R. (1977). EPIC-3, a computer program for elastic-plastic impact calculations in 3-dimensions. Report BRL 343 Honywell Inc., Defense Systems Division.

Johnson, G.R., D.D. Colby and D.J. Vavrick (1978). Further development of the EPIC-3 computer program for three-dimensional analysis of intense impulsive loading. Report AFATL-TR-78-81.

Johnson, G.R. (1982). Status of the EPIC codes, material characterization and new computing concepts at Honeywell. In J. Chandra and J.E. Flaherty (Ed.), Computational Aspects of Penetration Mechanics. Springer-Verlag, New York, pp. 24-35.

Int. J. Impact Engng Vol. 5, pp. 129–154, 1987
Printed in Great Britain

0734-743X/87 $3.00 + 0.00
Pergamon Journals Ltd.

THE PHYSICS OF HYPERVELOCITY LETHALITY

R. L. Bjork

Pacifica Technology, Box 148, Del Mar, CA 92014

ABSTRACT

The lethal effects of a high speed impact on a space structure are considered. The prediction of the lethal results of such an impact is complicated by the large number of parameters which influence the physical phenomenology, including shock-induced melting and vaporization. This paper looks into the theory of hypervelocity impact to address the question of whether melting and vaporization are beneficial or harmful to lethality. The first half of the paper reviews theoretical research that has been conducted in the complex two-phase region of metals, and on the supporting experimental data base. Attention is focused on aluminum and lead; aluminum because of its importance as an aerospace structural material, and lead because it may be shock vaporized at impact speeds available in the laboratory. The necessity for better experimental data in the two-phase region is made clear. The second half of the paper reports on experimental-calculational comparisons of selected impacts of interest to the hypervelocity lethality problem.

INTRODUCTION

This paper will discuss the damage imparted to spacecraft by high velocity particle impacts. A prediction of the damage and lethal effects of such an impact is complicated by the rich variety of concommitant physical phenomena, and the parameter space in which each comes into play. The phenomena include varying degrees of shattering of the projectile and the material it strikes, shock-induced melting and vaporization of these materials, and the effects of the debris so produced on various components of the spacecraft.

The materials for which these phenomena must be studied are diverse. Spacecraft incorporate a wide spectrum of materials, and the particles which may strike them are also diverse. Among the spacecraft materials, aluminum is by far the most prevalent material, but important components are also made of other lightweight metals, steel, impregnated composites, and plastics. In addition, the impacting particles may be naturally-occurring meteoroids, whose materials range from very porous stone to solid iron, or may be any of a wide variety of man-made materials.

Several different types of damage have been identified, including cratering by individual particles, erosion and scouring by the concerted action of many small particles, and general structural failure or disintegration, caused by pressure-area loading due both to vapor and to clouds of small particles. As will be discussed later, each of these damage mechanisms is affected differently by melting and vaporization, some favorably and some detrimentally. To understand and predict lethality, the physical processes which influence the efficacy of each damage type must be addressed.

The velocities of interest range from 2 or 3 km/s to roughly 80 km/s, corresponding to the range of observed meteoroid velocities relative to earth, augmented by the vector addition of the spacecraft velocity relative to earth.

It has long been recognized (see McQueen & Marsh, 1960) that impacts within this velocity range will result in varying degrees of melting and vaporization because of shock effects. To exemplify these effects, Table 1 tabulates some impact conditions which lead to incipient vaporization and 30% vaporization of the common structural material, aluminum.

TABLE 1 Impact Conditions to Vaporize Aluminum Targets

INCIPIENT VAPORIZATION

Projectile Material:	Uranium	Steel	Aluminum
Impact speed (km/s):	8.3	9.4	12.4

30% VAPORIZATION BY MASS

Projectile Material:	Uranium	Steel	Aluminum
Impact speed (km/s):	13.5	15.3	20.2

In many important cases, the interaction begins when a particle strikes a relatively thin outer wall or skin, perforates it easily, and proceeds into the target's interior. Often, the hole produced in the skin is of small, or secondary importance, whereas the ensuing damage is of primary importance. The skin impact is very efficient in shattering the particle, and the material which enters the target is in the form of a debris cloud composed of both projectile and skin material. Depending on the impact parameters and materials involved, the cloud may be composed of large fragments, finely divided fragments, or even a vapor cloud.

The effects of melting and vaporization on the lethality of the debris cloud are substantial. It will be demonstrated that the average particle size in a debris cloud decreases with increasing impact speed; at low speeds through the effects of mechanical breakup of solid and liquid fragments, and at high speeds through vaporization, which renders the material into atoms or molecules. In the completely vaporized state, the debris cloud will impose a simple pressure-area-time loading on internal components. Vaporization changes the geometry of the debris cloud in a manner which reduces the peak imposed stress by a large factor compared with an unvaporized cloud. However, it should be noted that the total momentum imparted by the debris cloud is not substantially altered by the processes of melting and vaporization.

The question arises: Are the effects of melting and vaporization beneficial or harmful to lethality? The answer to the question depends on the type of damage mechanism under consideration. If a vulnerable component has the character of a small, hard shell which must resist penetration in order to survive, melting and vaporization are demonstrated herein to be detrimental to lethality. The reasons are that they lead to smaller particle sizes, which reduces the likelihood of lethality due to particle penetration, and they reduce the peak stress on the component, which in most cases raises the impact velocity required to induce a structural failure of the component.

If the vulnerable component has the character of thin shell encompassing a broad area, which must undergo complete structural failure to void the spacecraft's mission, then melting and vaporization are in many cases helpful. As an example consider the reinforced skin of a spacecraft, whose function is to provide the structural framework to support all the other components. Such a skin can tolerate a large number of isolated holes and still preserve the majority of its structural strength. Thus a debris cloud composed of many isolated high energy fragments which simply produces many isolated holes in the skin, is not very damaging. The impacts which create the holes impart only a small fraction of the particles' momentum to the skin. Most of the momentum passes harmlessly through the holes, and is lost for lethal purposes. On the other hand, a vaporized debris cloud having the same mass and momentum, may in certain cases impart all of its momentum to the shell, and produce the massive structural failure sought.

Finally, there are intermediate cases, such as small, vital components whose survival is mandatory to the spacecraft mission. For these, the effects melting and vaporization are not obvious a priori, because they reduce the probability of killing the component (given a hit on it), while enhancing the probability of obtaining a hit.

The general lesson to be learned from these brief examples is that the lethal processes are complex and dependent on the specific nature of the target and the associated damage mechanisms. Therefore the quantitative assessment of impact lethality, and defenses against those impacts, will require both the identification of the important damage mechanisms, and a quantitative evaluation of how they are influenced by the physical phenomena just discussed. Shock-induced melting and vaporization will play a central role in these considerations in the velocity regimes where they come into play.

To illustrate the shock vaporization of metals, we consider the EOS of lead. Figure 1 shows several isentropes based on the SESAME 3201 EOS. The values of entropy are not specified by SESAME, so the isentropes are indexed by the pressure at which they intersect the Hugoniot. The leftmost curve in the figure is the principal Hugoniot, which represents the family of states which may be obtained by a single shock acting on material in the standard reference state (zero pressure and normal density).

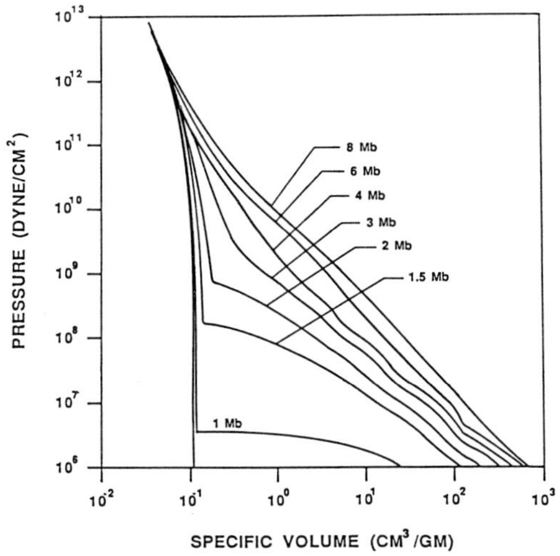

Fig. 1. Lead isentropes based on SESAME 3201 EOS.

A diagram such as Fig. 1 allows one to trace out the paths traversed in the P-V plane (P = pressure, V = specific volume) by particles which are involved in a high velocity impact or explosive process. In such events, the first signal to reach a particle is a shock, because strong shocks move supersonically with respect to the medium in which they propagate. Until the shock arrives, the particle "has no knowledge" that an impact or explosion has occurred, hence it remains in its initial P-V state until the shock arrives. The shock is characterized by an effectively discontinuous jump in pressure, density, and other thermodynamic variables. On shock arrival, the particle is suddenly brought to some point, A, on the Hugoniot, the precise point being determined by the strength of the shock (magnitude of the pressure jump). Besides increasing the pressure and density of the particle, the shock also increases the particle's tropy. The stronger the shock, the greater the entropy increase. If the pressure on the particle decreases with time after the shock passage, as is usual in impacts and explosions, the particle's states traverse the isentrope passing through point A. This family of states is called the "release isentrope."

It is instructive to discuss two examples of the paths traversed by particles in two different processes. In the first, suppose that the particle is first reached by a shock of 800 GPa (8 x 10^{12} dynes/cm^2 = 8 Mb). It is suddenly brought to the point on the Hugoniot corresponding to P = 8 x 10^{12} dynes/cm^2. The release isentrope from that point has a gas-like character along its entire length, that is, its slope on this plot of log P vs. log V is between -1 and -2, corresponding to gases having values of γ between 1 and 2. When it has expanded to 0.1 MPa (10^6 dynes/cm^2 = 1 bar), it has attained a volume of 700 to 800 cc/gm, which corresponds to that of superheated lead vapor. In other words, on release it behaves as though it had been instantly vaporized by the initial shock.

As a second example, consider a particle first reached by a 100 GPa (= 1 x 10^{12} dynes/cm^2 = 1 Mbar) shock. Its initial expansion closely follows the Hugoniot, which means that it retains densities characteristic of the liquid throughout a substantial pressure drop. When it crosses the saturated liquid line at about.4 MPa, the slope of the isentrope is sharply reduced because of the onset of vaporization. On release to 0.1 MPa, it attains a specific volume of about 20 cc/gm. Material in such a state is a mixture of liquid droplets and vapor in equilibrium at the normal boiling point of lead, namely 2024° K. Essentially all of the volume is provided by the vapor, a fact which allows one to estimate quite precisely the fraction of mass which is vaporized.

The velocities acquired by material elements as they follow a release isentrope depends on the shape of the isentrope in the P-V plane. For example, in a one-dimensional rarefaction fan which adjoins a region of constant state, the change in velocity, du, on expanding through a specific volume increment, dV, is given by

$$du = \pm c\, dV/V \tag{1}$$

where c is the sound speed defined through $c^2 = V^2\left(-\frac{\partial P}{\partial V}\right)_s$. The sign is determined by the direction which the rarefaction fan faces. The relation is more complex in a multidimensional rarefaction process, and the shape of the isentrope influences both the increments of forward and lateral velocity. In summary, the shape of the isentropes influences the distribution of momentum in the spray pattern. In turn, the isentrope's shape is influenced by the presence of the two-phase region.

INVESTIGATING THE SHOCK VAPORIZATION OF METALS

Requirements and Background

If one desires to compute only the fraction of material vaporized by a shock-release process, it is sufficient to set entropy on the Hugoniot, that is, to specify the increase in entropy imparted to the material by shocking it to a given pressure. The material will retain that entropy during isentropic release to any final release pressure, no matter what the shape of the release isentrope in the P-V plane. The computation is completed by identifying the material state which has that entropy at the desired release pressure, for example by recourse to thermodynamic tables such as Stull and Prophet (1971) or Stull and Sinke (1956).

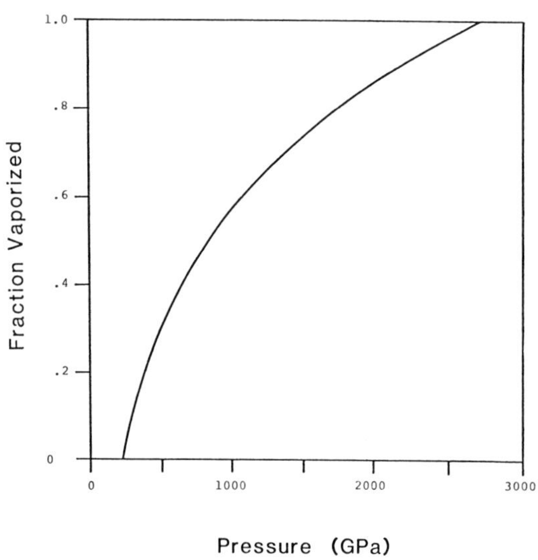

Fig. 2. Fraction of aluminum vaporized as a function of shock pressure.

Hugoniot measurements usually provide measurements of the shock speed, D, and particle velocity, U, from which one may construe pressure, density, and specific internal energy. For many materials sufficient Hugoniot data have been taken to provide an accurate, continuous prescription of P, ρ, and e along the Hugoniot. Unfortunately, these data give no information as to the temperature or entropy on the Hugoniot. To set entropy, one must provide additional independent data which determines either the temperature along the Hugoniot, or a specific heat. The specific heat at either constant volume or pressure will do. Elementary applications of thermodynamics shows that setting any one of these three quantities will set all the others, as well as setting the specific entropy. In general, for materials of interest, there is insufficient experimental data to set any of these three quantities on the Hugoniot over the required pressure range. Therefore, they must be estimated from an EOS theory. This procedure was used to compute the fraction of aluminum vaporized on release to one atmosphere after being shocked to an arbitrary pressure. The results are presented in Fig. 2, where the abscissa is the shock pressure.

To conduct a hydrocode computation of an impact process, it is necessary to know the P-V-e EOS everywhere in the P-V plane covered by the release isentropes. This forms a roughly triangular region bounded on the left by the Hugoniot, below by the smallest release pressure which must be considered, and on the upper right by the release isentrope which intersects the Hugoniot at the highest shock pressure which must be considered. In most cases, this data by itself is sufficient to conduct an adequate computation of the kinematic response. For example, the EOS of the two-phase liquid-vapor region is adequately set in these terms for the hydrocode, which does not require a specification of the temperature.

From such P-V-e data, one may compute the shape of the release isentropes in the P-V plane, and one can set temperature on each isentrope up to a multiplicative factor. If temperature is then known along some reference curve in the P-V plane, temperature, entropy, and both specific heats are determined throughout the region formed by the intersection of the region over which the EOS is valid and the region between the isentropes passing through the termini of the reference curve. Such information is usually a useful adjunct in theories which estimate the EOS.

Conditions in and around the two-phase region are very difficult to investigate theoretically. Although a large body of theoretical research has been devoted to analyzing such regions, no complete theoretical model has been identified which holds promise of reliably predicting data in them. Instead, numerous approximate theories have been developed which require experimental data both to establish the validity of the theory's approximations, and to determine the parameters of the theory. A brief summary of more widely-used EOS theories is given in what follows.

On the experimental side, there is a scarcity of data in and above the two-phase regions of metals of interest. Metals heated to this degree become exceedingly corrosive and difficult to confine under controlled conditions in the laboratory, and it is very difficult to maintain samples in a constant state on which diagnostics can be conducted. A brief summary of experimental data available for aluminum is given in what follows.

Soviet Research

The Soviet scientific literature has reported the results of a very extensive experimental and theoretical program to generate data on metals (and a few other materials) which have been strongly shocked, and to formulate equations of state to interpret that data. Experimental shock measurements to very high pressures have been conducted for several metals. Shock-release experiments have been conducted for lead in regimes which produce substantial vaporization. The EOS formulations which have resulted fit both their data base and that compiled by U. S. scientists. In addition, important contributions have been made to first-principle theories which cover states not experimentally accessible. In general, the research has emphasized the importance of the semi-empirical approach, in which theory provides the analytical form of the EOS fit, but includes parameters which may be fit by recourse to appropriate experimental data. A brief, but not comprehensive, review follows, which contains references to enable the interested reader to delve more deeply into the results.

Because extensive use is made of a particular form of the Gruneisen model, it will be summarized here to prevent repetition in the following text. The EOS results from the assumptions that 1) There is a temperature-independent contribution to the specific internal energy, $e_c(V)$, corresponding to the energy of the atoms in their equilibrium lattice positions. 2) There is a harmonic potential for each atom about its equilibrium position, which results in a spectrum of normal modes of lattice vibration. 3) The normal mode frequencies depend on volume only, and that all frequencies have the same volume dependence, as given by the Gruneisen postulate,

$$\gamma(V) = -\frac{d \ln \nu}{d \ln V} \tag{2}$$

where V is the specific volume, and $\gamma(V)$ is a dimensionless quantity known variously as the Gruneisen ratio, coefficient, parameter, or even constant (although it is volume-dependent!). By writing the partition function of the system, it may then be deduced from statistical mechanical considerations that the EOS has the form:

$$\begin{aligned} P &= -\frac{de_c(V)}{dV} + \frac{\gamma(V)}{V} e_n = P_c(V) + \frac{\gamma(V)}{V} e_n \\ e &= e_c(V) + e_n(V,T) \end{aligned} \tag{3}$$

The quantity, $e_c(V)$, is known as the cold curve, inasmuch as it is the lattice energy at 0° K (aside from the ground states of the lattice vibrations). To comply with thermodynamic self-consistency, pressure on the cold curve must be $P_c(V) = -de_c/dV$. The quantity, e_n, is called the nuclear contribution to energy, inasmuch as it arises from the kinetic and potential energy of the atoms, whose mass resides almost entirely in their nuclei. In this case, the dynamics gives rise to lattice vibrations, whose energy is a function of both T and V.

Given the form of Eq. (3), it is easily shown that if P and e are given parametrically as $P = P_r(v)$, $e = e_r(v)$ along some reference curve, then, over the volume range spanned by the reference curve, the P-V-e EOS is given by

$$P - P_r(V) = \frac{\gamma(V)}{V} [e - e_r(V)] \tag{4}$$

The Hugoniot is an example of a reference curve suitable for this purpose.

It is clear from the form of Eq. (4) that

$$\frac{\gamma(V)}{V} = \left.\frac{\partial P}{\partial e}\right|_V \tag{5}$$

Various relations have been postulated which give $\gamma(V)$ in terms of $P_c(V)$ and its first two derivatives. Vashchenko and Zubarev (1963) derived a relationship between γ and P_c based on the free volume model of Lennard-Jones and Devonshire (1937). They also worked out an expression for the contribution of the longitudinal and transverse vibrational modes for cases wherein the Poisson ratio exhibited a certain weak dependence on volume. This comprised an extension of Slater's (1939) model, which assumed a constant Poisson ratio. They identified a general form which described all their new results in addition to the Slater (1939) and Dugdale-MacDonald (1953) models, to wit:

$$\gamma = -\frac{4-3n}{6} - \frac{V}{2}\,\frac{\frac{d^2}{dV^2}(P_c V^n)}{\frac{d}{dV}(P_c V^n)} \tag{6}$$

where $n = 0 \rightarrow$ Slater (1939)
$n = 2/3 \rightarrow$ Dugdale-MacDonald (1953)
$n = 4/3 \rightarrow$ Free volume model of Vashchenko & Zubarev (1963)
$n < 0 \rightarrow$ Poisson ratio decreases with compression.

The EOS model just described will be called herein the Mie-Gruneisen model, after Gruneisen (1926). Modern usage has broadened the name to encompass any EOS having the form of Eq. (4). The use of a cold curve is almost universal, both in the early EOS efforts and in modern usage. Also, both early and recent formulations have made extensive use of the form to prescribe nuclear contributions, even in state regimes wherein the concept of lattice vibrations ceases to have meaning. For a further discussion of this model, see Gilvarry (1956a, 1956b). It has also been shown that at high temperatures important contributions to the pressure and energy are made by he electrons, which can be accommodated by adding terms to Eq (3). When this is done, the EOS assumes the form

$$\begin{aligned} P &= P_c(V) + P_n(V,T) + P_e(V,T) \\ e &= e_c(V) + e_n(V,T) + e_e(V,T) \end{aligned} \tag{7}$$

which has been very widely used. This form, sometimes with modest variations, is still widely used today.

Al'tshuler and colleagues (1958a) reported on Hugoniot measurements on iron in the pressure range from 40 to 487 GPa. They found that their results could be fit over the entire range investigated by a linear relation between shock velocity, D, and particle velocity, U; namely D = 3.80 + 1.58U, where D and U are in km/s. They also reported experiments on porous iron, using initial specific volumes of .127 and .181 cc/gm. Their results were interpreted using an EOS of the form of Eq. (7), wherein P_e and e_e were zero, $e_n = C_V T$, and $P_n = BT/v$. This is equivalent to setting $\gamma = B/C_V$, a constant. These prescriptions lead to the following relation between the Hugoniot and cold curve:

$$V\frac{\partial e_c(V)}{\partial V} + \gamma e_c(V) = -\gamma\frac{P_H(V)}{2}V\left(1 + \frac{2}{\gamma} - \frac{V_o}{V}\right) \tag{8}$$

where $P_H(V)$ is the experimentally-determined Hugoniot curve. This differential equation was solved for $E_c(V)$. Extrapolations of the cold curve were graphically compared with the 0° K isotherms predicted by the Thomas-Fermi model (TFM) and Thomas-Fermi-Dirac (TFD) model as computed by Gombas (1949). At pressures less than 40 TPa, their extrapolated cold curve was found to have lower pressures at a given compression than TFD, which in turn fell beneath the TFM results.

Al'tshuler and colleagues (1958b) reported Hugoniot measurements on copper, zinc, silver, cadmium, tin, gold, lead, and bismuth in the pressure range from 35 to 513 GPa, using the reflection method with iron as a standard material. For all the metals except tin, the results could be fit by a linear relation between D and U. A correlation was noted between shock compression and the initial atomic volume of the metals, namely, those with high initial atomic volumes underwent the greatest compressions.

Al'tshuler and colleagues (1960a) reported experimental Hugoniot measurements on aluminum to 200 GPa, copper to 419 GPa, lead to 417 GPa, and iron to 402 GPa. The paper stresses the difficulty of obtaining the EOS from purely theoretical considerations. Instead a semi-empirical approach is used, in which plausible analytical expressions for the various contributions are deduced from theory, and the parameters in the expressions are fit by comparison with experimental data. The EOS they proposed had the form of Eq. (7).

The nuclear contributions were assumed to be given by[1]:

$$e_n = C_v(T - T_o + e_o/C_v) \tag{9}$$
$$P_n = \frac{\gamma(V)}{V} C_v(T - T_o + e_o/C_v)$$

C_v, the specific heat at constant volume, was assumed to be constant. The Slater (1939) and Dugdale-MacDonald (1953) relations between the cold curve and Gruneisen coefficient were cited. From their data, they concluded that γ exhibited a general trend to decrease with increasing compression, but also underwent oscillations about this trend.

By using the Thomas-Fermi solutions of Gilvarry (1956a,b) and Latter (1955), the electronic contributions were found to be suitably expressed by

$$P_e = \frac{1}{4}\beta_o(V_o/V)^{1/2}T^2 \tag{10}$$
$$e_e = \frac{1}{2}\beta_o(V/V_o)^{1/2}T^2$$

The temperatures on the Hugoniots of lead, aluminum and copper, calculated on the basis of the above assumptions, were computed and are tabulated in the paper. The method of Walsh and colleagues (1957) was used to find $P_c(v)$ from the Hugoniot data and other assumptions.

Kormer and Urlin (1960), undertook to analyze the Hugoniot data taken previously on metals. The EOS used had the form of Eq. (7).

The cold curve was assumed to be expressed by

$$P_c = \sum_{i=1}^{6} a_i \delta^{1+i/3}, \text{ where } \delta = V_o/V \tag{11}$$

The six coefficients, a_i, were determined by matching the density, sublimation energy, bulk modulus, and Gruneisen parameter at zero temperature and pressure, and the pressure and its volumetric derivative at a high pressure where the TFD model was thought to be accurate. Slater's (1939) model was used for the Gruneisen parameter. The criterion to establish where the TFD model was sufficiently accurate was $(P_{TF} - P_{TFD})/P_{TF} < 0.2$.

The form of the nuclear and electronic contributions were identical with those used by Al'tshuler and colleagues (1960). The electronic specific heat, β_o, was assumed to be that found at very low temperatures. The Hugoniots predicted from this purely theoretical method were compared with those found experimentally for aluminum, nickel, copper, lead, and tin. The theoretical predictions were stated to agree to within 5-8% over the entire experimental pressure range. For silver, zinc, and gold, somewhat larger discrepancies were found. It was concluded that the method was adequate to estimate the cold curves and Hugoniots for metals where shock data was lacking.

Kormer (1962a), used the same EOS model as Kormer (1960) to predict the Hugoniots of 13 metals on a purely theoretical basis. However, he used two different inputs to fit the six values of a_i. First, the TFC model was used to prescribe pressure and its volumetric derivative at a sufficiently high pressure. The criterion for sufficiently high pressure was that $(P_{TF} - P_{TFC})/P_{TF} =$ .25-.30. Second, a small correction was made to γ_o to account for the electronic contribution. The discrepancy with Hugoniot data at 50 GPa was quoted as being less than 4% for all metals except silver and tantalum. In the range from 50 to 200 GPa, the discrepancy from the data of McQueen and Marsh (1960) was less than 5% for tungsten, copper, nickel, zinc, and molybdenum; 10-15% for lead, tin, silver, and gold. At pressures of about 400 GPa, the discrepancies were about 8% for aluminum, copper, tin, and lead; and about 15% for zinc, silver, and gold.

Even better agreement was obtained by adding a seventh term to the cold curve, and fitting it from a single shock measurement. Using this method for aluminum and copper, whose Hugoniots were the most extensively measured, reproduced all Hugoniot pressure measurements to within ±3%. In

---

[1]Note that these relations are thermodynamically inconsistent unless $e_o = C_vT_o$, in which case the correct relations are $P_n = \gamma(V)\, C_vT/V$, $e_n = f(T)$, an arbitrary function of T, which may have the form, C_vT + constant.

addition, the theoretically predicted sound speeds at high pressure on the Hugoniot reproduced the measurements of Al'tshuler & colleagues (1960b) to within ±3%.

Several conclusions were drawn based on these good comparisons. It was felt that this method gave a more accurate determination of the volume-dependence of the Gruneisen parameter than previous methods. With Kormer's method, the Gruneisen coefficient was found to exhibit a uniform decrease with compression, in contrast to the oscillatory behavior found previously by other means. The use of Slater's (1939) relation to determine the volume dependence of the Gruneisen ratio was favored over the Dugdale-MacDonald (1953) relation. The TFC model was favored over the TFD model.

Further background on the electronic contribution was given. Ya. B. Zel'dovich and A. S. Kompaneets were credited with being the first to recognize that electronic contributions to the pressure and energy must be included at high temperatures and pressures. This conclusion was cited to be reached as a result of their computations on the Thomas-Fermi-Dirac model at non-zero temperatures. Although Kormer cites the date 1947 in his attribution, he gives no other data on the reference. Credit was claimed for suggesting the form of the electronic contributions to pressure and energy used in the EOS calculations of Al'tshuler (1960a). It was noted that their formulation relied heavily on solutions to the Thomas-Fermi equation by Gilvarry (1954, 1955, 1956) and on the Thomas-Fermi calculations of Latter (1955)[2].

Kormer (1962a) went on to note that the temperatures on the lead Hugoniot as computed by Al'tshuler (1960a) were in substantial disagreement with those computed by McQueen and Marsh (1960). He noted specifically that at a compression of 1.70 on the Hugoniot, temperatures of 16,500° K and 8500° K were obtained by McQueen and Al'tshuler, respectively. In addition, he noted discrepancies between his computed cold curves (which agreed closely with Al'tshuler & colleagues, 1960a) and those of McQueen & Marsh (1960). He attributed both disagreements to the inclusion of the electronic terms on his and Al'tshuler's part, and their omission on McQueen and Marsh's part.

Kormer (1962b) reported results of shock compression measurements using iron driver plates at impact speeds up to 15.45 km/s. The experiments utilized porous samples of aluminum, copper, tungsten and nickel at pressures up to 1000 GPa. These results gave important new information on thermodynamic states at high pressure not on the principal Hugoniot. It may be shown in general that a given driver acting on a porous sample produces a higher specific volume and energy at a given shock pressure than the same driver acting a sample of normal density. It was noted that the shape of the porous Hugoniot curves in the P-V plane provided not only further support for the concept of electronic contributions to pressure and energy, but also provided data on the volume dependence of the electronic analogue of the Gruneisen ratio, $\partial P_e/\partial e_e|_V$.

A modified form of the nuclear contributions was proposed, namely

$$
\begin{aligned}
P_n &= \frac{3\gamma + z}{1 + z}\,\rho R(T-\hat{T}) \\
e_n &= \frac{2 + z}{1 + z}\,\frac{3}{2}\,R(T-\hat{T})
\end{aligned}
\qquad (12)
$$

where γ is given by Slater's (1939) relation, R is the universal gas constant divided by the average molecular weight, and $z = \ell RT/c_c^2$, $c_c^2 = dP_c/d\rho$, and ℓ is a constant to be determined. At low temperature, z approaches 0, and the expressions reduce to the Mie-Gruneisen form, with $C_V = 3R$. At high temperatures, the nuclear contribution assumes the form

$$P = \rho RT, \quad C_V = \frac{3}{2}R \qquad (13)$$

which prescribes the pressure and energy of an ideal monatomic gas of the correct atomic weight. This provides a smooth transition for the nuclear contributions, from lattice vibrations at low temperatures to an ideal gas at high temperatures. In a sense, this approach may be said to introduce a temperature dependence to the specific heat and Gruneisen coefficient.

---

[2]Latter noted that the contribution of exchange terms to the electronic specific heat at low temperatures is almost exactly cancelled by correlation effects, and that exchange effects become negligible at high temperatures. Hence he concluded that the overall behavior of the atom was better described by omitting exchange terms altogether, if correlation effects were not also included.

The electronic contributions were described by a fit to the results of Latter (1955) and Gilvarry (1954a,b) and Gilvarry and Peebles (1955)[3]:

$$e_e = (b^2/\beta)\ \ln \cosh\ (\beta T/b)$$
$$P_e = g\rho(b^2/\beta)\ \ln \cosh\ (\beta T/b) \qquad (14)$$
$$g = -\frac{d\ \ln \beta}{d\ \ln \rho}$$

Al'tshuler & colleagues (1962) compare the cold curves as deduced from their shock experiments with the zero temperature isotherm computed by Kalitkin (1960) on the TFC model. For nickel, cadmium, zinc, and lead, good agreement was obtained between the TFC computations and reasonable extrapolations of the cold curves. To obtain agreement for the more compressible metals, it was necessary to use a new form for the cold curves, and to account for the anharmonicity of the lattice vibration potential. For this purpose the "free volume" approximation of Lennard-Jones & Devonshire (1937) was used. As a result, a temperature-dependent Gruneisen coefficient was computed.

Al'tshuler and colleagues (1968) used the reflection method with lead as the standard material to determine the Hugoniot of iron to 3.1 TPa. The Hugoniot of lead was based on shock measurements to 1 TPa, and on the Thomas-Fermi calculations of Latter (1955), which were deemed to be valid above 30 TPa. The nuclear contribution was the same as Kormer and colleagues (1962b), while the cold curve used the model of Al'tshuler and colleagues (1962). The lead Hugoniot between 1 and 30 TPa was determined by graphical interpolation.

The experimental results of Kormer (1962b) are still used as a standard of comparison for many EOS formulations (See Trainor, 1983).

Al'tshuler (1965) gives a good review of research in shock physics prior to that time. Many results are summarized and an extensive set of references is provided.

Trunin and colleagues (1969) employed the reflection method, using lead as a standard material to measure Hugoniot points for copper and cadmium to pressures of about 1.5 TPa. Trunin and colleagues (1972) used the same methods to measure Hugoniot points for iron, copper, and cadmium to 5.1, 3.8, and 3.3 TPa, respectively. They constructed curves in the P-ρ plane which passed through their new high pressure data and earlier data at lower pressures, and compared their graphical extrapolation with the Thomas-Fermi calculations of Latter (1955). The comparison resulted in satisfactory merges with the TF calculations for all materials in the region of 15 to 20 TPa. However, it was noted that the TFC method had been used to construct the EOS of the standard material, lead.[4]

Kirzhnits, Lozovik, and Shpatakovskaya (1975) gives a good review of the modifications made to the Thomas-Fermi model (TFM, also called TF elsewhere in this paper) and the limits of its application. The effects of exchange, correlation, quantum effects and atomic shells, and the corrections to the TFM which may be made to account for them are discussed. For compressed states, the corrections lead to a smaller pressure at a given compression. A good set of references is given, both to Kirzhnitz's extensive research and that of others.

Al'tshuler and colleagues (1977a) compared experimental Hugoniot data for quartz, aluminum, iron, copper, cadmium, and lead with the theoretical Hugoniots computed on the TF and TFC models. It was found that the TFC data could be adequately expressed in the form

$$D = a + bU + cU^2. \qquad (15)$$

---

[3]Bjork and colleagues (1967) noted that the data of Kormer and colleagues (1962b) could not be well fit by a Tillotson EOS, and also was in substantial disagreement with other EOS in use at that time. Hence, they undertook to incorporate Kormer's data, and that of Anderson (1965) in a global EOS for aluminum which also incorporated Latter's (1955) Thomas-Fermi data. To facilitate use in hyrdocodes, the EOS was put into tabular form, and was used to compute impacts on aluminum and steel sheets by meteoroids having a wide range of densities and velocities.

[4]Ragan and colleagues (1977) reported an absolute measurement of the molybdenum Hugoniot at 2.0 TPa. Ragan and colleagues (1984) used the reflection method with molybdenum as a standard to obtain Hugoniot measurements for aluminum, carbon, copper, lithium hydride, iron, porous molybdenum, and quartz in the 1.2 to 6.0 TPa range. Pressure in the molybdenum base was cited to be 6.43 TPa.

Values of (a,b,c) were reported for large number of solids.

As a criterion for the applicability of the TFC model, a 5% error in density at a given T and V was arbitrarily chosen. At zero temperature, this criterion was cited to be met for P > 30 TPa (Kalitkin, 1975). For the interpolation Hugoniots studied, the criterion was met at about 5 TPa, where the temperature is on the order of 10 eV. This was cited as experimental confirmation that the TFC model becomes more accurate at high temperatures.

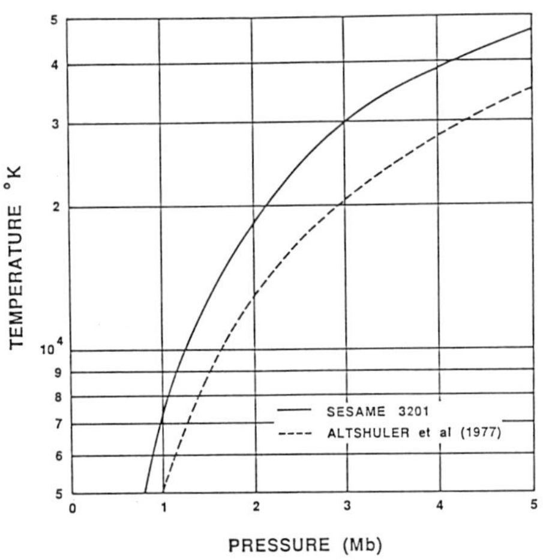

Fig. 3. Temperature as a function of pressure on the lead Hugoniot based on two different EOS formulations.

Al'tshuler and colleagues (1977b) conducted flier plate impacts on lead samples in the ambient atmosphere, and measured the parameters of the air shock driven by the expanding lead. Interpretation of the data identified intersections of the release isentropes with both the saturated liquid and saturated vapor curves. The pressure and particle velocity of an intersection with the saturated liquid curve is given, and a similar point can be construed for the intersection with the saturated vapor curve from the data given. It was claimed that the experiments verified that thermodynamic equilibrium was maintained during release through the liquid-vapor region. Inasmuch as the experiments were conducted on a microsecond time scale, it was concluded that evaporation and condensation in the two-phase region had a relaxation time of much less than a microsecond. This result was cited to support the theoretical study of Fortov and Leont'ev (1976) on the relaxation times of lead and aluminum. Based on the new data, computations of temperature, entropy, specific heat and Gruneisen coefficient were made and reported. In general, the temperature and Gruneisen coefficient were higher, and the entropy and specific heat were lower than given by previous EOS formulations. It was noted that the derivation of a thermodynamically complete EOS of lead in a wide region was the subject of a separate investigation. Figure 3 compares Al'tshuler's temperatures with those given by SESAME 3201, and it is seen that the Soviet data is indeed some 40% to 50% lower. If the P-V Hugoniot values of the two EOS's are comparable, as they should be, this indeed implies that the specific heat is higher, and entropy lower than the SESAME values. As a result, the Soviet EOS will predict a greater degree of shock vaporization than SESAME for the same impact conditions. The discrepancy between Hugoniot temperatures as given by these two EOS models, and others, is an outstanding question at the present time. To resolve this issue will probably require new experimental data.

Al'tshuler and colleagues (1980), report an analytical EOS which presumably fits the data of Al'tshuler and colleagues (1977b). It specifies the specific Helmholtz free energy as a function of temperature and specific volume, so the thermodynamic variables derived from it will be thermodynamically consistent. The model has the form of Eq. (7), except that two different expressions for the nuclear contributions are used, the first for the solid phase, and the second for the melted and vaporized phases. The high temperature form accounts for anharmonicity of the lattice where appropriate, and goes to an ideal atomic gas in the high temperature limit. The electronic contributions describe a degenerate electron gas at low temperatures, ionization in the plasma region, and go to an ideal electron gas in the high temperature limit. Adjustable constants were included in the model, which enabled the EOS to reproduce the results of static and dynamic experiments. Previous wide-range EOS research efforts were criticized as having omitted this important consideration.

A complete set of coefficients were reported for lead and copper. In summary, this research culminated in a wide-range EOS which reproduces the Soviet shock vaporization experiments conducted on copper and lead, their extensive Hugoniot data on both solid and porous samples, and their quantum statistical models at high temperature and pressure.

Vladimirov and colleagues (1984) reported on Hugoniot measurements on aluminum to 400 TPa. Only a crude determination of density was made. However, the data is useful for EOS validations.

Volkov and colleagues (1980), and Simonenko and colleagues (1985) report the absolute measurement of the aluminum Hugoniot at about 1 TPa, obtained by simultaneous measurement of the shock and particle velocities. To measure the particle velocity, thin reference sheets of Eu_2O_3 were embedded in the aluminum sample. The driving device provided a flux of 1 MeV neutrons which were moderated by the aluminum, and captured by the Eu_2O_3, which produced a strong flux of γ rays from the thin reference sheets. γ-ray detectors viewed the side of the stack through a collimator and recorded the time of passage of each reference layer. The pressures measured by this experiment were higher than those previously predicted by extrapolation curves based on data taken using the reflection method.

Other Modern EOS Formulations

The SESAME library. The SESAME library (Cooper, 1980; Bennett and colleagues, 1978; Abdallah and colleagues, 1980) is a general purpose system for storing and accessing EOS and other material properties. The most important file type contains entries of pressure, specific internal energy, and (sometimes) Helmholtz free energy on a rectangular grid of density and temperature. For tables of this type, routines are provided which prescribe derivatives with respect to density and temperature. A program called HYDSES is available which forms tables of pressure and temperature on rectangular grids of density and specific internal energy, as well as routines for determining derivatives with respect to energy and density. Such tables are particularly appropriate for hydrocodes. The data in the tables has been computed using a wide variety of models, and new tables are added from time to time. In particular, the tables have been revised to agree with the solid and porous Hugoniot data taken by Ragan and colleagues (1984), and Vladimirov and colleagues (1984).

PANDA. PANDA (Kerley, 1981) is a family of codes, each valid in a certain equation of state region. PANDA utilizes the CRIS model to prescribe the translational degrees of freedom over the entire fluid range, which encompasses the liquid, vapor and supercritical regions. (Kerley, 1980a, 1980b, 1980c). The model has routines which specify the vapor pressure and properties of the liquid and vapor on the liquid-vapor coexistence curves as a function of temperature. It also identifies the spinodal points on isotherms. The model also provides estimates of the structure factor, the radial distribution function, the bulk and shear viscosities, and the diffusivity.

The thermodynamic properties are computed from an expansion about a hard-sphere fluid reference system, in which the hard sphere diameter is chosen by a variational principle. The zero temperature isotherm of the solid is used to define an expression of a molecule in the field of its neighbors. A summary of the computations is given in Kerley (1981), together with a description of the inputs required to formulate an equation of state.

In the CRIS model, each molecule (atom) is assumed to be located in a spherical cage formed by its neighbors. The radius of the cage, R, and coordination number, ν, vary from molecule to molecule. The coordination number is the average number of nearest neighbors. The coordination number varies with R in such a way that the volume per molecule is constant throughout the fluid and equal to the macroscopic value. The energy of the molecule in a cage of radius R is equal to the cold curve of the solid times a correction factor for the coordination number. At low densities, when the electronic wave functions are well localized, the correction is nearly proportional to the macroscopic density. However, the electrons tend to become delocalized at high densities, and a further correction is required.

PANDA contains a dimerization model to treat fluids that are monatomic in the condensed fluid phase, but which form some diatomic molecules in the vapor. The diatomic molecules are treated as hard spheres, rigid rotators, and harmonic oscillators.

A new theoretical EOS of aluminum has been reported by Kerley (1986). His approach was to determine the Helmholtz free energy in the form

$$F_{solid}(\rho,T) = E_c(\rho) + F_{n_{solid}}(\rho,T) + F_e(\rho,T) \tag{16}$$

for the solid, and

$$F_{fluid}(\rho,T) = F_{n_{fluid}}(\rho,T) + F_d(\rho,T) \tag{17}$$

for the fluid and vapor regions.

The functions were evaluated using PANDA, augmented by other theories and computational codes. He cites modern approaches to determine the cold curve and Gruneisen parameter. An important new result was obtained for the electronic contributions for expanded states below T = 1 eV. Using two different models for high ($\rho > .64$ gm/cc) and low ($\rho < .25$ gm/cc) densities, he identified the regions in which liquid aluminum behaves as an insulator (electrons localized) and as a metal (electrons non-localized), and a coexistence region for the metallic and non-metallic liquids. The region is disposed in such a way that an isobaric experiment must traverse it to access the liquid-vapor coexistence region. (See Section titled Isobaric Experiments.) Including the electronic contributions improved the agreement with the isobaric data of Gathers (1983). The new model could provide the critical pressure of aluminum, but this was not specifically reported in the paper. Based on the graphical results presented, the critical pressure appears to be 0.3 GPa, or less.

The LLNL formulation. Scientists at the Lawrence Livermore National Laboratory (LLNL) utilize a family of six codes, each valid in a certain equation of state region, together with interpolative routines to provide smooth merges between the results of the various regions. The codes and their region of applicability are discussed in Trainor (1983). The results of all the codes are generally summarized in an equation of state table, which consists of pressure and specific internal energy entries on a rectangular grid in the temperature-density plane.

The EOS region needed to conduct impact response calculations is treated by the APW, GRAY, TFNUC, and Soft Sphere models. The cold curve is set by APW, using the augmented plane wave models of McMahan (1979) and McMahan and Ross (1979). TFNUC embodies the TFC model. GRAY is a semi-empirical model which specifies the nuclear and electronic contributions at small compressions and temperatures below 10 eV. The expanded region at temperatures less than 1 eV. is treated by the Soft Sphere model. This region includes the important and complex liquid-vapor coexistence region.

The soft sphere model. Hoover and colleagues (Alder, Hoover & Young, 1968; Hoover & Ree, 1968; Hoover and colleagues, 1970; Hansen, 1970; Hoover, Gray & Johnson, 1971; Hutchinson & Conkie, 1971, 1972) have conducted an extensive series of Monte Carlo calculations for particles interacting with pair potentials of the form $\phi(r) = \varepsilon(\sigma/r)^n$. The calculations suggest a configurational heat capacity term of the form

$$E_{config} = \frac{NkT}{6}(n+4)\,\rho^{n/9}\,(\varepsilon/kT)^{1/3} \tag{18}$$

The form is derived by correcting the computed Monte Carlo energy for the kinetic energy, $\frac{3}{2}NkT$, and a static lattice constant term $C_n\rho^{n/3}$ where C_n is the f.c.c Madelung constant for the pair potential and ρ is the dimensionless density $\rho = N\sigma^3/\sqrt{2}V$. N is the number of molecules in the system, and V is the volume they occupy.

Young (1977) adopted Hoover's soft sphere model with two modifications. First, a cohesive term of the form $-N\varepsilon\rho^m$ was added to the energy, and second, the configurational term was multiplied by Q, a constant whose value is set empirically. Values of Q greater than 1.0 provide a first order correction for the electronic contribution to energy, which is assumed to have the same form as the configurational term.

The equations which result are

$$E = NkT\left\{\frac{3}{2} + C_n\rho^{n/3}(\varepsilon/kT) + \frac{Q}{6}(n+4)\rho^{n/9}(\varepsilon/kT)^{1/3} - \rho^m(\varepsilon/kT)\right\} + E_{coh} \tag{19}$$

$$P = \frac{NkT}{V}\left\{1 + \frac{n}{3}C_n\rho^{n/3}(\varepsilon/kT) + \frac{Qn}{18}(n+4)\,\rho^{n/9}(\varepsilon/kT)^{1/3} - m\rho^m(\varepsilon/kT)\right\} \tag{20}$$

The form of the pressure equation is derived from the expression for E by using the equation of thermodynamic consistency. Note that each equation has two temperature-independent terms which define the cold curve, and the remaining terms provide the nuclear and (through Q) the electronic contributions. In the P-V plane, the isotherms exhibit a van der Waals-like behavior in the liquid-vapor coexistence region. The pressure and boundaries of the coexistence region are computed from a Maxwell construction on the isotherms.

Eqs. 19 and 20 contain 5 adjustable parameters, namely ε, σ, n, m, and Q, which Young determined by fitting experimental data. For the case of lead, the data consisted of 300 MPa isobaric data of specific enthalpy and specific volumes versus temperature up to about 5000 K, obtained by personal communication from Hodgson, and sound speed versus temperature data taken at 0.1 MPa by Markov (1975).

The values obtained are:

$$
\begin{aligned}
n &= 11 \\
m &= 1.05 \\
Q &= .85 \\
N_A\, \sigma^3/\sqrt{2} &= 2.68262 \times 10^5\ m^3/kg \\
N_A \varepsilon &= 4.74657 \times 10^6\ J/kg \\
C_n &= 6.08756
\end{aligned}
$$

Using these parameters, the model gives the following critical point data for lead:

$$
\begin{aligned}
T_c &= 5158°\ K \\
V_c/V_o &= 3.70 \\
P_c &= 226\ MPa \\
Z_c &= 0.36
\end{aligned}
$$

For the case of aluminum, the experimental data used are: a. Enthalpy versus temperature at one atmosphere for temperatures up to 1800 deg K (Hultgren, and colleagues 1973). b. Density versus temperature at one atmosphere for temperatures up to 1800 deg K (Gol'tsova, 1965, and Levin, 1967). c. Sound speed versus temperature at one atmosphere for temperatures up to 1200 deg K (Stremousov and Tekuchev, 1976).

Using this data, the following critical parameters of aluminum were found: P_c = 0.182 GPa, T_c = 5726 K, V_c/V_o = 6.37, Z_c = 0.24. The critical pressure is substantially smaller than the values given by previous research. Some earlier predictions are (in GPa) 0.616 by Thompson and Lauson (1972), 0.546 by Young and Alder (1971), and 0.447 by Fortov, Dremin, and Leont'ev (1975).

A major source of inaccuracy is the absence of experimental data in the important state regions where the model is to be applied. The data used to fit the constants was all taken at one atmosphere, whereas the model is used to predict pressures in the several thousand atmosphere range. A second source of inaccuracy arises because the model does not fit the data base precisely, and the selection of the best fit requires judgement on the part of the user.

Isobaric Experiments

The requirement for experimental data on expanded states at high pressures has been discussed. One of the few ways to obtain such data is through the so-called isobaric experiments. These have been used to obtain data in the melted region between the Hugoniot and saturated liquid curve. In these experiments, a wire made of the material to be investigated is heated by a sudden burst of electrical current. Although the heating is rapid, it can be shown that controlled experiments lead to isobaric expansion.

This technique has been used on metallic wires in a pressurized argon atmosphere. For references to this work, see Gathers (1983). Physical considerations provide both upper and lower limits on the rate of electrical heating. As a result most experiments are conducted on a time scale of a few tens of microseconds. The time duration of heating is long enough to permit several transits of a sound wave across the specimen diameter, hence the loading is isobaric in character. Confining pressures up to 8 kbars may be used. Ohmic heating is used, and the current required depends on the material's electrical resistivity. The materials best characterized by this technique are those of modest resistivity, such as tantalum, lead, niobium, molybdenum, platinum, iridium, and vanadium.

An optical measurement is made of the wire diameter as a function of time, which provides the specific volume as a function of energy deposited. Pyrometric techniques are used to measure temperature. Once the wire is heated, it may be loaded on one side by a laser. The loading produces a sound wave which traverses the wire diameter and produces a recordable response on the other side. The measurement of the transit time provides a measurement of the sound speed in the heated material.

The technique cannot be extended into the vaporized region, because vaporization leads to excessive deformation of the sample. It has provided useful information on the melted region between the Hugoniot and the saturated liquid curve.

Gathers (1983) reports isobaric data on copper and aluminum taken at a pressure of 0.3 GPa over a substantial temperature range. The highest temperatures at which measurements could be made were 4000° K and 4500° K for aluminum and copper, respectively. At higher temperatures, it was observed that the volume and electrical resistance rose sharply, which precluded accurate measurements. Theoretical estimates of the saturated liquid curves were examined, with the conclusion that the data did not extend across them for either metal, but were limited to the liquid region. We note that this behavior is consistent with the findings of Kerley (1986), the implication being that the sudden rises in volume and resistance are associated with the formation of the non-metallic liquid phase.

COMPUTATIONAL/EXPERIMENTAL APPROACH TO LETHALITY

The EOS/hydro Committee Program

The EOS-HYDRO committee, chaired by Maj. D. Apo of the Defense Nuclear Agency, formulated a program whose primary objective was to identify a combination of computer code and EOS which could predict the parameters of debris cloud produced in very high velocity impacts. The program consisted of high speed impact experiments which produced debris clouds, combined with pre- and post-shot calculations to predict and interpret the cloud parameters. Inasmuch as shock-vaporization will occur for all materials at sufficiently high speeds, it was important to model the effects of vaporization on the cloud parameters. For this purpose some of the experiments used lead projectiles and breaker plates. As a control, parallel experiments using the refractory material, molybdenum, were conducted.

Lead and molybdenum have similar densities, namely 11.34 and 10.22 gm/cc. The parallel experiments used spherical projectiles weighing 20 g., and breaker plates 6.35 mm thick. The impact speeds were to be identical, namely 7.3 km/s. Hence the parallel experiments had the same momentum and kinetic energy, and employed breaker plates of nearly the same areal density. Because the kinematic parameters were so close, differences in the results could be attributed almost entirely to the difference in the constitutive relations of the two materials. (The EOS is a subset of the constitutive relations.) It was anticipated that a substantial fraction of the lead would be vaporized and melted in the impact, and that no molybdenum would be vaporized or melted.

Other subobjectives were to benchmark the codes against experiment, to determine the sensitivity of the computational results to changes in the EOS, and to determine those EOS which are most suitable for predictive purposes. It was recognized that computational errors could arise from two sources, namely, those associated with solving the equations of motion, and those caused by inaccuracies in the constitutive relations. Hence, it was also desired to intercompare computations made by various codes using the same EOS.

The Experiments

The experiments to be discussed were conducted by Delco, and had the parameters given in Table 2. The instrumentation consisted of two orthogonal channels of flash X-rays which provided pictures at the times shown in Table 2. Step wedges were emplaced in the X-ray field of view, so that the density in the debris cloud could be determined by photographic densitrometry. An aluminum witness plate was emplaced 50 cm behind the breaker plates. The witness plate was 2" thick, and was a square, four feet on a side. Step wedges were in the view of each X-ray to enable the determination of the density field by photodensitometry.

TABLE 2 Summary of Shot Parameters

Shot number	4005	4006	4007
Projectile	20 gm Mo	20 gm Mo	20 gm Pb
Target plate	.660 cm Mo	0.660 cm Mo	0.635 cm Pb
Velocity (km/s)	6.60	6.58	6.58
X-ray times (μs)	20.5, 31.2	30.8, 40.1	30.6, 40.2

In the molybdenum shot, there were three distinct areas of damage to the witness platte: 1) A central region, about 8 inches in diameter, was almost completely covered by impact craters of about the same size, almost complete overlain by smaller surface craters characteristic of very small particles. 2) A ring, extending from about 8 to 15 inches, which contained larger craters than those of the central core, and spaced farther apart. This area was also covered by smaller surface craters. 3) An outer ring, extending from about 15 to 34 inches, containing impact craters substantially smaller than the inner ring or central core, decreasing in size and areal density as one goes to greater radial distances. No small surface craters were in evidence in this ring.

In the lead shots, no surface pitting at all was visible to the naked eye. The aluminum witness plate appeared to be lightly covered with lead, the plating being apparently heavier in the central region. No distortion of the plate was evident. A gauge, encased in plastic, having a total thickness of less than 0.1 mm, had been glued to the front face of the witness plate. Although it was well within the region plated with lead, most of the plastic still adhered to the plate post-shot, and was only lightly damaged. This result stands in strong contrast to the damage of the witness plate in the molybdenum shots.

Stress gauges were fielded on replicate molybdenum and lead shots. On these shots, the witness plates consisted of two aluminum plates bonded together, with the stress gauges in the middle of the sandwich. The Pb shot gave a peak stress of about 0.14 GPa and a width at half maximum of about 6 μs. The Mo shot gave a peak stress of about 1.45 GPa and about a 2 μs width at half maximum.

PacTech Calculations

Pacifica Technology (PacTech) conducted pre-and post-shot computations of the molybdenum and lead shots. The code used for these computations was PEACH, an Eulerian hydrocode which incorporates a high-accuracy transport algorithm. The transport algorithm is designed to be accurate for all densities, and contains no adjustable parameters.

In addition, computation of the expected structural response of a cylinder fielded in the debris cloud of the lead shot were conducted. For this purpose, the code STAGS was used, with surface loads prescribed as a function of position and time by the PEACH results.

Molybdenum-molybdenum calculations and comparison with experiment. A pre-shot computation of the Mo-Mo experiment was conducted using a velocity of 7.3 km/s, the velocity specified by the EOS/HYDRO Committee. A Tillotson EOS, as modified by Allen (1967) was used.

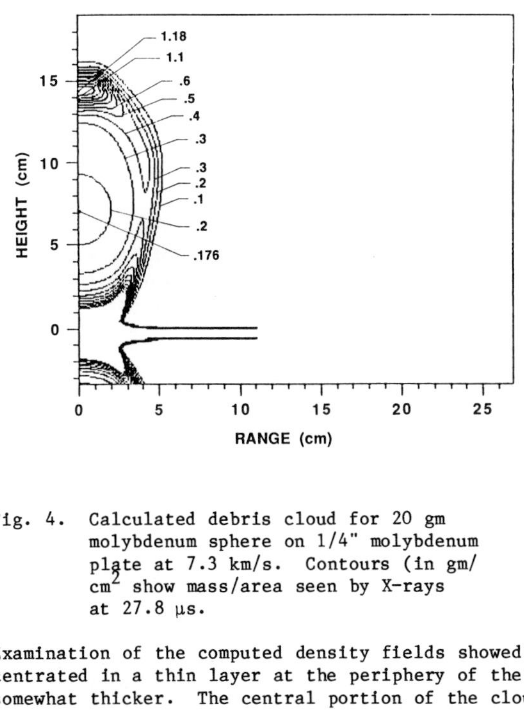

Fig. 4. Calculated debris cloud for 20 gm molybdenum sphere on 1/4" molybdenum plate at 7.3 km/s. Contours (in gm/cm^2 show mass/area seen by X-rays at 27.8 μs.

Figure 4 presents contours of the computed mass per unit area seen by a parallel flux of X-rays propagating normal to the axially-symmetric debris cloud. We will call this the "X-ray density field." It is computed by assuming a uniform, parallel flux of X-rays normal to the plane of the figure. At each point, the mass per unit area is the $\int \rho \, dx$ along the X-ray path through that point. The X-ray density is a useful quantity, because the X-ray absorption in the cloud is related to it through the relation

$$I(\nu) = I_o(\nu) \exp \{- \frac{\mu(\nu)}{\rho} \int \rho \, dx\} \quad (21)$$

where the quantities $I(\nu)$ and $I_o(\nu)$ are the spectral intensities of the incident and final X-ray flux, and μ is the spectral absorption absorption coefficient, which is well-known for lead and molybdenum. By accounting for the spectral composition of the flash X-ray source, and for the spectral responses of the X-ray screen and film, one may compute the average photographic density on the X-ray negative. This provides a direct comparison between computation and experiment.

Examination of the computed density fields showed that the mass in the debris cloud was concentrated in a thin layer at the periphery of the cloud, except near the tip, where the layer was somewhat thicker. The central portion of the cloud was empty.

The X-rays of the Mo-Mo shots strongly suggest that the cloud mass indeed resides in a thin layer near the periphery, and that the interior is empty. A major difference with computation is that the mass is subdivided into a great many particles, whereas the computation treats it as being continuously distributed. However, the data suggests that the computation gives a good indication of the bulk density of fragments, that is, the mass/volume existing in a small test volume. In addition, the velocity of each fragment is well-specified. To predict the lethal effects, it will be necessary to determine empirically the size distribution of particles, and correct for individual particle effects.

The computed and measured dimensions of the debris cloud are compared in Table 3. In conducting the pre-shot computation, it was anticipated that the velocity would be 7.3 km/s, and that flash X-rays would be taken at 17.2 and 27.8 μs. However, the shot was actually conducted at 6.6 km/s, and X-rays were taken at 30.8, and 40.1 μs. Hence, a direct comparison of the shot and pre-shot computation is not possible. However, a reasonable comparison can be made by using a simple time scaling. The computed results can be reasonably scaled to a shot at 6.6 km/s by defining a scaled time, $t_s = t(7.3/6.6)$, and then assuming that the density field computed at t are those which would be computed a time t_s for the slower shot. This is equivalent to reducing the entire velocity field by the factor of 6.6/7.3. Such scaling is clearly only valid over a small velocity range. Fortunately, as Table 3 shows, the scaled X-ray times for the computation are close to the experimental times for the two earlier X-rays. The pre-shot results were then augmented by running them forward to 36.5 μs to provide a scaled comparison with the experimental X-ray at 40.1 μs.

TABLE 3 Comparison of computed and measured Moly debris cloud dimensions

	t	t_s	t_{exp}	D_{neck}(cm)		D_{debris}(cm)		Location of L_{debris}(cm)		$(\rho t)_{min}$(cm)	
	μs	μs	μs	meas	comp	meas	comp	meas	comp	meas	comp
Moly	17.2	19.0	20.5	5.0	5.0	7.8	7.6	9.9	10		4.2
Moly	27.8	30.8	30.8	5.6	5.4	11.2	10.4	15.3	16.2		7.2
Moly	36.5	40.4	40.1	5.8	5.4	13.8	13.8	20.1	21		9.5

D_{neck} = Debris cloud diameter as it emerges from hole in target plate.

D_{debris} = Maximum debris cloud diameter.

L_{debris} = Length of debris cloud.

Computations were conducted at 7.3 km/s.

Shots were fired at ~6.6 km/s.

$$t_s = t\left(\frac{7.3}{6.6}\right)$$

On this basis, at least, the computed dimensions of the debris cloud are all reproduced quite accurately by the computation. The last column of Table 3 gives the prediction of the minimum value of X-ray mass per unit area. As yet the X-rays have not been reduced to provide the comparable experimental result.

Lead-Lead calculations and comparison with experiment. A pre-shot computation of the Pb-Pb experiment was conducted using a velocity of 7.3 km/s, the velocity specified by the EOS/HYDRO Committee. The SESAME 3200 EOS for lead was used. The computation was repeated pre-and post-shot using all the same parameters, but with two other EOS's, namely the SESAME 3201 EOS, which is a recent addition to SESAME made by Holian, and the Tillotson/Allen EOS as modified by Holian. Figure 5 shows the X-ray density field for a computation using SESAME 3201, and Fig. 6 shows the X-ray density fields using the Tillotson/Allen/Holian (T/A/H) EOS. The results are seen to be quite similar to each other, and they are also quite similar to the pre-shot calculation using SESAME 3200.

The pre-shot calculation was conducted at 7.3 km/s, and the later ones were conducted at the same speed so the results could be directly compared with each other. The experiment was conducted at about 6.6 km/s. By using the same time scaling as for the Mo-Mo shot, one may deduce that the computed result at 36.5 μs may be compared with the experimental X-ray taken at 40.2 μs. The agreement between the computed and experimental results for lead is not nearly as good for the molybdenum shot. First, the shapes of the debris cloud do not match well. The X-ray of the Pb debris cloud shows that it has an acorn shape, that is, a narrow neck where the cloud exits from the target plate hole, rapidly expanding to a maximum diameter about a third of the distance to the tip. Ahead of this is a short cylindrical section, with a nearly-hemispherical cap. No individual fragments are evident within the debris cloud, but there is a dark mass within the cloud near its tip. On the periphery, particularly near where the cloud emanates from the plate, there are filaments of material, which are suggestive of melted strings which elongate as they fly.

The computational outline, on the other hand, has a smoothly increasing diameter in front of the target plate, which attains its maximum value about two-thirds of the way to the tip. Beyond this, there is a smooth cap.

The dimensions of the experimental and computed lead debris cloud are compared in Table 4.

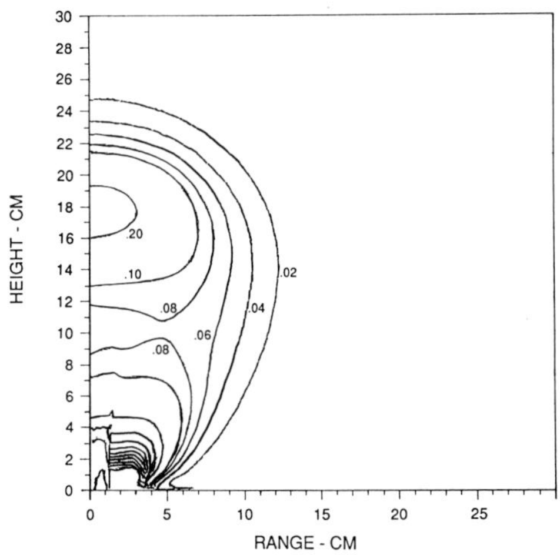

Fig. 5. Debris cloud calculated using SESAME 3201 EOS for 20 gm lead sphere on 1/4" lead plate at 7.3 km/s. Contours (in gm/cm^2) show mass/area seen by X-rays at 36.5 μs.

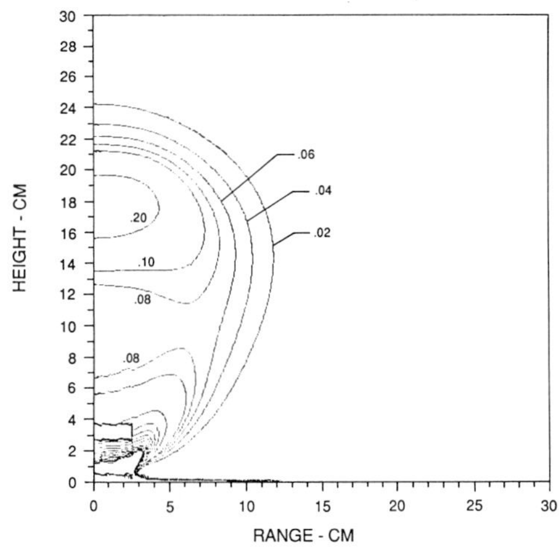

Fig. 6. Debris cloud calculated using Tillotson Allen/Holian EOS for 20 gm lead sphere on 1/4" plate at 7.3 km/s. Contours (in gm cm^2) show mass/area seen by X-rays at 36.5 μs.

TABLE 4 Comparison of Lead Debris Cloud Dimensions

Experiment time: 40.2 μs. Computation time: 36.5 μs

D_{neck} (cm)			D_{max} (cm)			L_{debris} (cm)		
measured	computed		measured	computed		measured	computed	
	3201	T/A/H		3201	T/A/H		3201	T/A/H
5.9	5.6	10.5	16.7	23.8	24.4	25.0	24.3	24.7

The conclusions to be derived from these results are:

- o The agreement between the calculation and experiment in the Mo-Mo shots validates PEACH's accuracy.
- o The lead calculations agree among themselves, but do not agree well with experiment. This suggests that the three EOS's used share a common error.

Finite element calculation. STAGS was used to compute the structural response of a cylinder loaded by the debris cloud from Shot 4007, which involved lead on lead at 7.3 km/s. The parameters of the calculation were as follows:

Cylinder:
Material: 6061 T6 aluminum
Outer diameter: 24.94 cm
Wall thickness: 0.756 cm
Cylinder height: 32.26 cm
Boundary conditions: Free

Standoff: 32 cm from target plate to nearest point of cylinder

Loading: The entire surface of the cylinder was loaded as a function of position and time by the impact of the debris cloud, as computed by PEACH. Figure 7 shows the load history at selected locations. The cylinder axis was oriented perpendicular to the shot line, and hence normal to the cloud's axis of symmetry. The coordinate denotes distance along the cylinder's axis, and theta the angular distance around the cylinder. Hence, z = 0, $_{r}$ = 0 is the intersection of the shot line and cylinder wall. As the Fig. 7(a) shows, the peak stress there is about 1.8 kbar. This may be compared with the 1.45 Kbar stress measurement at a standoff of 50 cm. The onset of the loading becomes later with increasing distance from the center, and the peak stress decreases as well. However, the load duration of about 15 μs at half max remains about the same.

Figure 8 shows the radial velocity field, whose integral is the radial displacement. The central point, (a) essentially undergoes a rapid inward velocity pulse, which peaks at about 900 m/s at a time of about 0.08 ms. The velocity then diminishes in magnitude, and becomes nearly zero at 0.8 ms. The other points undergo velocity histories with more structure, reflecting the excitation of various modes of oscillation.

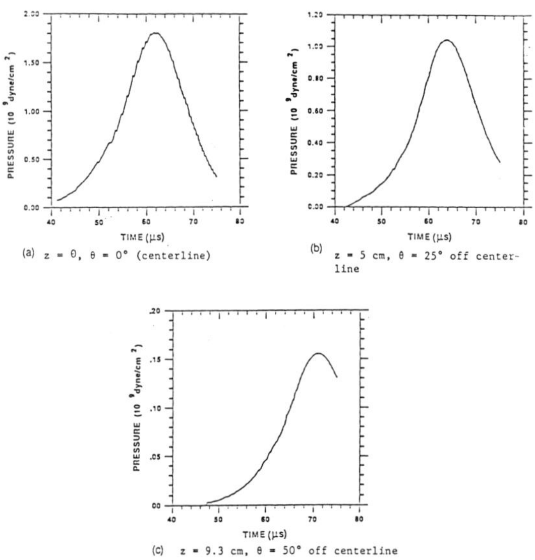

Fig. 7. Examples of surface loads applied to cylinder for finite element calculation. Actually loading over entire cylinder was made at all times using loads computed by the PEACH calculation.

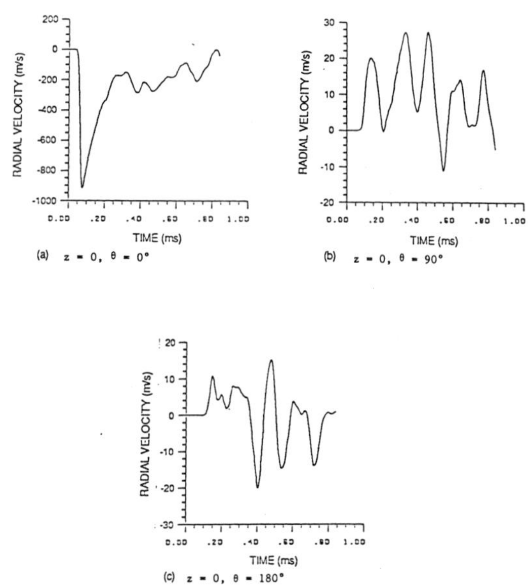

Fig. 8. Finite element calculation of radial velocity of cylinder wall at selected locations. Locations are on center plane of cylinder (z = 0). a. Front point (8 = 0 deg). b. Side point (8 = 90 deg). c. Rear point +0 = 180 deg).

Figure 9 shows the radial displacement versus time at the front, side and back of the cylinder. Displacement is positive outwards. The negative displacement of the front point therefore represents an inward deflection. The positive displacements at the back point represent outward displacements, which are away from the shot point. The code predicts a permanent set, corresponding to an ovalling mode. The diameter is reduced in the direction of impact by about 2 cm, and increased by about 1.6 cm in the perpendicular direction.

Figure 10 shows the time histories of the axial stress components which arise on the inner and outer cylinder walls. Stress, elastic strain, and plastic strain are defined at five locations through the shell thickness. The large differences which arise between the stress on the inner and outer fibers reflect the large bending moments which occur during the response.

PACTECH EXPERIMENTAL PROGRAM

Description and Results

A sequence of shots was conducted in the Pacifica Technology light gas gun to examine the effects produced on the lethality of the debris cloud by the onset of vaporization. Lead projectiles and breaker plates were used, and the impact speed was varied to encompass regions in which no vaporization would occur, and in which a great deal of vaporization would occur. The projectiles were spheres 0.96 cm in diameter. Flash X-rays were taken of each projectile in flight after launch to make sure it retained its sphericity. The breaker plates were pure lead, of thickness 0.32 cm, to provide a thickness/diameter ratio of 1/3. Each cloud impinged on an aluminum cylinder made of 6061-T6 aluminum. All cylinders had the same dimensions, namely, 4" outer diameter, 8" height, and .083" wall thickness. Flash X-rays were taken of the debris clouds of each shot. The impacts were conducted in an evacuated range, wherein the pressure was about 13 Pa.

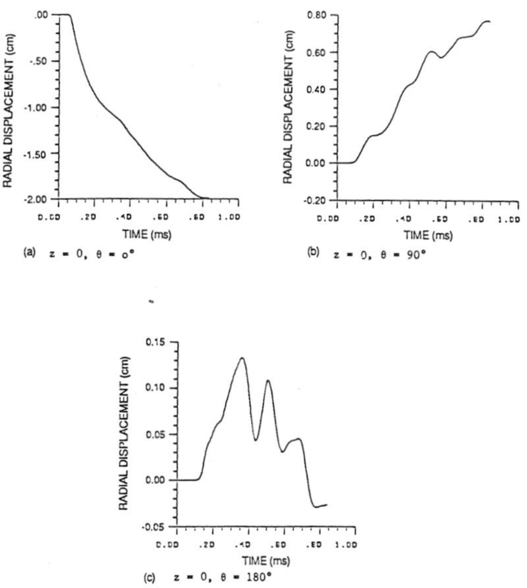

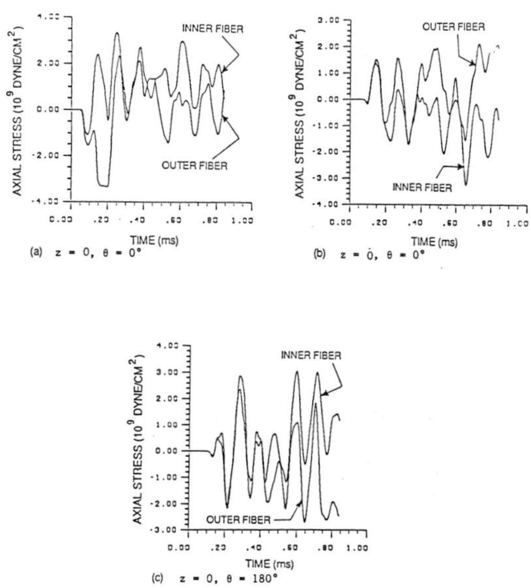

Fig. 9. Finite element calculation of radial displacement of cylinder wall at selected locations. Locations are on center plane of cylinder (z = 0). a. Front point (8 = 0 deg). b. Side point (8 = 90 deg). c. Rear point ←0 = 180 deg).

Fig.10. Finite element calculation of axial stress in cylinder wall at selected locations. Locations are on center plane of cylinder (z = 0). a. Front point (8 = 0 deg). b. Side point (8 = 90 deg). c. Rear point ←0 = = 180 deg). Stresses on both inner and outer wall fibers are shown.

Figure 11 shows the debris clouds at four impact velocities. At 1.72 km/s, the debris cloud consists of discrete particles. There is a cluster of finely divided material near the front of the cloud which contains the fragments from the projectile and the portions of the breaker plate which were directly struck by the projectile. Based on our calculations of the event, the shoulder in the profile at the periphery of the cluster roughly marks the boundary between target material which was directly in the path of the projectile, and target material outside this region which is ejected in the process of hole formation. No evidence of melting or vaporization is seen at this velocity.

At 2.93 km/s, the cloud front has the appearance of a billowy continuum, which assumed to be formed of vapor. No individual particles are in evidence. Behind the billowy region is a region comprised of filaments, which are interpreted as tendrils of melted material which have streamed from the forming hole. Further back, the material is composed of small particles, and the average particle size increases as one proceeds back toward the breaker plate. The hole in the plate is still forming at this time, and sizable particles are being ejected from its periphery. It is also clear that particles are being ejected in the rearward direction.

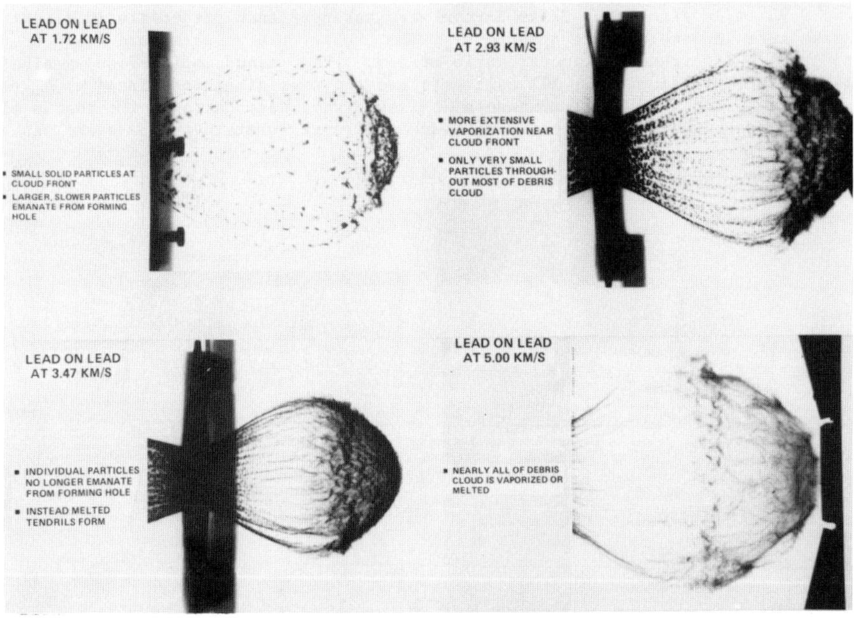

Fig. 11. Experimental debris clouds for lead spheres on lead breaker plates at various velocities. Sphere diameters were all 0.96 cm. All plates were 0.32 cm thick.

At 3.47 km/s, the same regions of vapor, melted material and particles exist. However, the vaporized region has a somewhat smoother surface, the region occupied by melted tendrils is longer, and the average particle size is smaller.

At 5.00 km/s, the cloud consists of vaporized material ahead of the shoulder, and melted tendrils behind it.

Figure 12 shows the damage produced on the cylinders by the action of the debris clouds. A surprising trend is illustrated, namely that the damage to the cylinder, as measured by the size of hole blown in it, decreases as the velocity is increased from 1.45 to 2.49 km/s. A large hole is blown in the cylinder at 1.45 km/s, but at 2.49 km/s, where the kinetic energy in the cloud is about three times as large, the cylinder survives intact.

At 1.45 km/s, a large hole is blown in the cylinder. Close examination showed that the cylinder's central region was extensively pitted by the individual lead fragments. Material ejected from the craters resulted in substantial thinning and weakening of the wall, to the extent that the wall could not withstand the momentum load on it.

At 1.92 km/s, many small pits due to individual lead particles covered the cylinder's central portion, but the individual pits were smaller than before. This is attributed to the diminution of average particle size with increasing impact speed. As a result, less thinning and weakening occurred, and a smaller hole was produced in the cylinder.

At 2.49 km/s, the size of individual pits is reduced to such an extent that the cylinder survives intact, although the momentum loading produces a significant bulge in the cylinder. The debris cloud at 2.49 km/s has nearly three times the kinetic energy and nearly twice the momentum as at

1.45 km/s. Momentum increases faster than linearly with velocity because of momentum multiplication.

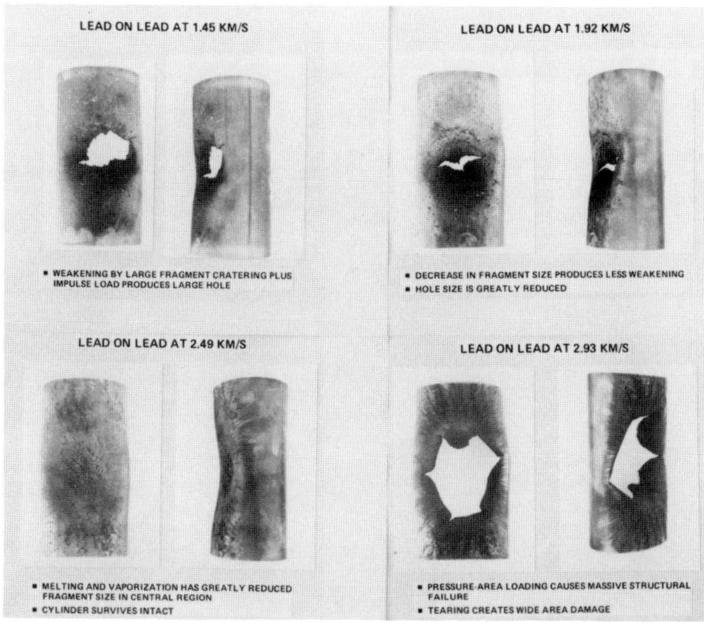

Fig. 12. Cylinder damage caused by debris clouds shown in Fig. 11. All cylinders were of 6061-T6 aluminum, 10.2 cm in diameter, 20.3 cm high, with wall thickness of 0.21 cm

At 2.93 km/s, the pits due to individual particles are smaller still, but now the cylinder succumbs structurally to the area-momentum-time loading, and undergoes a dramatic rupture and inward petalling. On the inner surface of the cylinder opposite the impact point, a few bulges are produced, and the surface is coated with a thin layer of lead. At 3.47 km/s (not shown in Fig. 11), the same qualitative behavior is observed, except that now the bulges are no longer produced on the inner surface.

This series of experiments vividly illustrated that damage mechanisms exist, such as those which produce the hole in the cylinder, whose lethal effects progressively diminish as impact velocity increases in certain velocity regimes. The argument is advanced that the behavior of lead suggests that other materials may also exhibit this phenomenon. Now, it is not intuitively obvious that the lethality of lead simulates that of other stronger, less ductile metals. To address this issue, let us examine Fig. 13, which shows the debris clouds of mo-mo at 5.1 km/s and pb-pb at 1.58 km/s. The figure illustrates that the debris clouds have very similar characteristics. Each is composed of discrete particles which appear to be solid. The general conformations are the same. Each exhibits the shoulder associated with breaker plate material directly in the path of the projectile. Near the front of the molybdenum cloud can be seen a group of very small particles, the rear of which has a nearly circular boundary. This is interpreted to be fragments of the projectile itself. The group fills the interior of the cloud. A similar group of very small particles is seen in the lead cloud, although not so many are in evidence. The central point is that the damage from both clouds is caused by the collective action of the particles. Above the transition velocity, where severe particle deformation occurs on impact, and "hypervelocity" penetration occurs, crater dimensions depend primarily on particle mass, density, and velocity, and only secondarily on strength or ductility. Hence, it is proposed that the size spectrum of particles in the molybdenum cloud will decrease in the same manner as the lead particles, and an analogous variation of lethality with velocity will result. Of course it is obvious that the molybdenum cloud, whose particles have velocities up to 5 or 6 km/s, is more lethal than the lead cloud at 1.58 km/s. The illustration pertains to the variation of lethality with velocity, rather than its absolute magnitude.

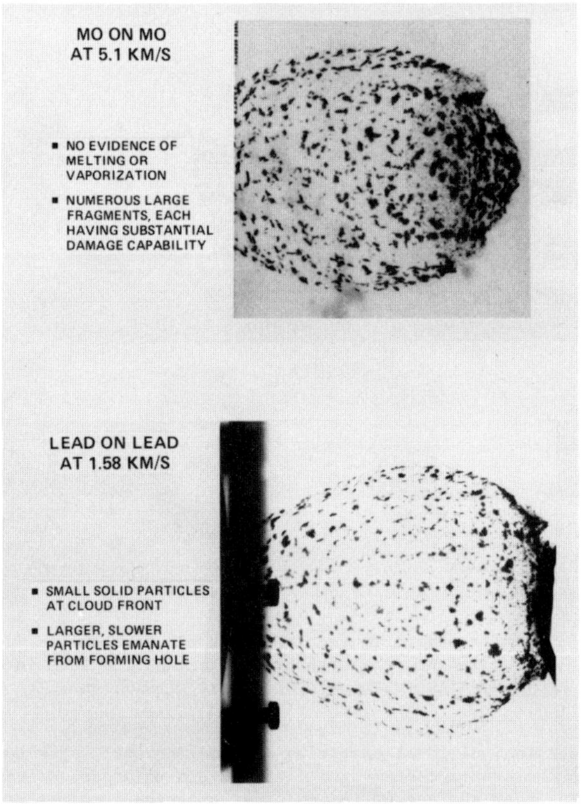

Fig. 13. Debris clouds for molybdenum on molybdenum at 5.1 km/s and lead on lead at 1.58 km/s.

Conclusions

The experiments illustrate that the damage caused by debris clouds may decrease with velocity, especially in the velocity region where the materials begin to undergo shock-vaporization. This has been illustrated by a particular damage mechanism, namely the hole size produced in a particular cylinder. However, the possibility exists that a similar behavior will be found for other damage mechanisms, and each postulated mechanism should be investigated to determine its velocity dependence. In particular, it may not be argued a priori, that a level of damage found at low impact speeds will be higher at higher speeds. The dynamic processes which create various types of damage should be investigated to ascertain their efficacy at the impact speeds of interest.

Computations using a family of computing codes have been presented which can examine the damage to be expected from a wide variety of damage mechanisms, including deformation of structural members and other components, to the hole sizes (or lack thereof) produced in spacecraft components.

REFERENCES

Abdallah, J., G. I. Kerley, B. I. Bennett, J. D. Johnson, R. C. Albers, and W. F. Huebner (1980). *HYDSES: A subroutine package for using Sesame in hydrodynamic codes.* Los Alamos Scientific Laboratory report LA-8209.

Alder, B. J., W. G. Hoover, and D. A. Young (1968). Studies in molecular dynamics. V. High density equation of state and entropy for hard disks and spheres. *J. Chem. Phys.*, 49, 3688.

Allen, R. T. (1967). *Equation of State of Rocks and Minerals*, Report GAMD-7834, General Dynamics, General Atomic Division, San Diego, California.

Al'tshuler, L. V., K. K. Krupnikov, B. N. Ledenev, V. I. Zhuchikhin, and M. I. Brazhnik (1958a). Dynamic compressibility and equation of state of iron under high pressure. *Soviet Physics - JETP*, 34, 606-614.

Al'tshuler, L. V., K. K. Krupknikov, and M. I. Brazhnik (1958b). Dynamic compressibility of metals under pressures from 400,000 to 4,000,000 atmospheres. *Soviet Physics - JETP*, 34, 614-619.

Al'tshuler, L. V., S. B. Kormer, A. A. Bakanova, and R. F. Trunin (1960a). Equation of state for aluminum, copper, and lead in the high pressure region. *Soviet Physics - JETP*, 11, 573-579.

Al'tshuler, L. V., S. B. Kormer, M. I. Brazhnik, L. A. Vladimirov, M. I. Speranskaya, and A. I. Funtikov (1960b). The isentropic compressibility of aluminum, copper, lead, and iron at high pressures. *Soviet Physics - JETP*, 11, 766-775.

Al'tshuler, L. V., A. A. Bakanova, and R. F. Trunin (1962). Shock adiabats and zero isotherms of seven metals at high pressures. *Soviet Physics - JETP*, 42, 65-74.

Al'tshuler, L. V. (1965). Use of shock waves in high-pressure physics. *Soviet Physics - Uspekhi*, 8, 52-91.

Al'tshuler, L. V., B. N. Moiseev, G. V. Simakov, and R. F. Trunin (1968). Relative compressibility of iron and lead at pressures of 31 to 34 megabar. *Soviet Physics - JETP*, 27, 420-422.

Al'tshuler, L. V., N. N. Kalitkin, L. V. Kuzmina, and B. S. Chekin (1977a). Shock adiabats for ultrahigh pressures. *Soviet Physics - JETP*, 45, 167-171.

Alt'shuler, L. V., A. A. Bakanova, A. V. Bushman, I. P. Dudoladov, and V. N. Zubarev (1977). Evaporation of shock-compressed lead in release waves. *Soviet Physics - JETP*, 46, 980-983.

Al'tshuler, L. V., A. V. Bushman, M. V. Zhernokletov, V. N. Zubarev, A. A. Leont'ev, and V. E. Fortov (1980). Unloading isentropes and the equation of state of metals at high energy densities. *Soviet Physics - JETP*, 51, 373-383.

Bjork, R. L., K. N. Kreyenhagen, and M. H. Wagner (1967). *Analytical study of impact effects as applied to the meteoroid hazard.* NASA Contractor's Report CR-757, Washington, D. C.

Bennett, B. I., J. D. Johnson, G. I. Kerley, and G. T. Rood (1978). Recent developments in the Sesame equation-of-state library. Los Alamos Scientific Laboratory report LA-7130.

Cooper, N. G. (1980). *An invitation to participate in the LASL equation of state library.* Los Alamos Scientific Laboratory brochure LASL-79-62.

Dugdale, J. S. and D. K. C. MacDonald (1953). The thermal expansion of solids. *Phys. Rev.*, 89, 832.

Fortov, V. E., and A. A. Leont'ev (1976). Kinematics of vaporization and condensation with the isentropic expansion of metals. *High Temperature*, 14, 634-639.

Fortov, V. E., A. N. Dremin, and A. A. Leont'ev (1975). Evaluation of the Parameters of the Critical Point. *Teplofizika Vysokikh Temperatur*, 13, 984.

Gathers, G. R. (1983). Thermophysical properties of liquid copper and aluminum. *International Journal of Thermophysics*, 4, 209-226.

Gilvarry, J. J. (1954a). Thermodynamics of the Thomas-Fermi atom at low temperature. *Phys. Rev.*, 96, 934-943.

Gilvarry, J. J. (1954b). Solution of the temperature-perturbed Thomas-Fermi equation. *Phys. Rev.*, 96, 943-948.

Gilvarry, J. J. and G. H. Peebles (1955). Solutions of the temperature-perturbed Thomas-Fermi equation. *Phys. Rev.*, 99, 550-555.

Gilvarry, J. J. (1956a). The Lindemann and Gruneisen laws. *Phys. Rev.*, 102, 308-316.

Gilvarry, J. J. (1956b). Gruneisen's law and the fusion curve at high pressure. *Phys. Rev.*, 102, 317-325

Gilvarry, J. J. (1956c). Equation of the fusion curve. *Phys. Rev.*, 102, 325-331.

Gilvarry, J. M. (1956d). Gruneisen parameter for a solid under finite strain. *Phys. Rev.*, 102, 331-340.

Gol'tsova, E. I. (1966). *High Temperature*, 4, 348.

Gombas, P. (1949). *Die statische Theorie des Atoms und ihre Anwendung.* Springer, Vienna.

Gruneisen, E. (1926). Zustand der festen Korper. *Handbuch der Physik*, 10, 1-52.

Hansen, J. P. (1970). Phase transition of the Lennard-Jones system. II High temperature limit. *Phys. Rev.*, A2, 221.

Hoover, D. A. and F. H. Ree (1968). Melting transition and communal entropy for hard spheres. *J. Chem. Phys.*, 49, 3609.

Hoover, W. S., M. Ross, K. W. Johnson, D. Henderson, and B. C. Brown (1970). Soft sphere equation of state. *J. Chem. Phys.*, 52, 4931.

Hoover, W. S., S. G. Gray, and K. W. Johnson (1971). Thermodynamic properties of the fluid and solid phases for inverse power potentials. *J. Chem. Phys.*, 55, 1128.

Hultgren, R. (1973). *Selected values of the thermodynamic properties of the elements.* American Society for metals, Metals Park, Ohio.

Hutchinson, P. and W. R. Conkie (1971). A functional derivative approach to thermodynamically self-consistent radial distribution functions. *Mol. Phys.*, 21, 881.

Hutchinson, P. and W. R. Conkie (1972). Thermodynamically self-consistent radial distribution functions for inverse power potentials. *Mol. Phys.*, 24, 567.

Kalitkin, N. N. (1960). The Thomas-Fermi model of the atom with quantum and exchange corrections. *Soviet Physics - JETP*, 11, 1106-1111.

Kalitkin, N. N. (1975). Doctoral dissertation, *Appl. Mech. Inst.*, USSR Acad. Sci.

Kerley, G. I. (1980a). Perturbation theory and the thermodynamic properties of fluids. I. General theory. *J. Chem. Phys.*, 73, 469-477.

Kerley, G. I. (1980b). Perturbation theory and the thermodynamic properties of fluids. II. The CRIS model. *J. Chem. Phys.*, 73, 478-486.

Kerley, G. I. (1980c) Perturbation theory and the thermodynamic properties of fluids. III. Inverse-power and 6-12 potentials. *J. Chem. Phys.*, 73, 487.

Kerley, G. I. (1986). Theoretical equation of state for aluminum. *This proceedings.*

Kirzhnitz, D. A., Yu E. Lozovik, and G. V. Shpatakovskaya (1975). Statistical model of matter. *Soviet Physics - Uspekhi*, 18, 649-672.

Kormer, S. B. and V. D. Urlin (1960). Interpolation equations of state of metals for the region of ultrahigh pressures. *Soviet Physics - Doklady*, 5, 317-320.

Kormer, S. B., V. D. Urlin, and L. T. Popova (1962a). Interpolation equation of state and its application to experimental data on impact compression of metals. *Soviet Physics - Solid State*, 3, 1547-1553.

Kormer, S. B., A. I. Funtikov, V. D. Urlin, and A. N. Kolesnikova (1962b). Dynamic compression of porous metals and the equation of state with variable specific heat at high temperatures. *Soviet Physics - JETP*, 15, 477-488.

Latter, R. (1955). Temperature behavior of the Thomas-Fermi model for atoms. *Phys. Rev.*, 99, 1854-1870.

Latter, R. (1956). Thomas-Fermi model of compressed atoms. *J. Chem. Phys.*, 24, 281-292.

Lennard-Jones, J. E. and A. F. Devonshire (1937). Critical phenomena in gases. *Proc. Roy. Soc. (London)*, A163, 53-70.

Levin, E. S., G. D. Ayushina, and P. V. Gel'd (1968). *Teplofizika Vysokikh Temperatur*, 6, 432.

Markov, B. G. (1975). The speed of ultrasound and the thermophysical properties of liquid metals Sn, Pb, Cd and of their binary alloys Pb-Sn and Pb-Cd. *High Temp. Engl. Transl.*, 13, 1027.

McQueen, R. G. and S. P. Marsh (1960). Equation of state for nineteen metallic elements from shock-wave measurements to two megabars. *J. Appl. Phys.*, 31, 1253-1269.

Ragan, C. E. III, M. G. Silbert, and B. C. Diven (1976). Shock compression of molybdenum to 2.0 TPa by means of a nuclear explosion. *J. Appl. Phys.*, 48, 2860-2870.

Ragan, C. E. III (1984). Shock-wave experiments at three-fold compression. *Phys. Rev. A*, 29, 1391-1402.

Simonenko, V. A., N. P. Voloshin, A.S. Vladimirov, A. P. Nagibin, V. N. Nogin, V. A. Popov, V. A. Vasilenko, and Yu. A. Shoidin (1985). Absolute measurements of shock compressibility of aluminum at pressures P > 1 TPa. *Soviet Physics - JETP*, 61, 869-873.

Stremousov, V. I., and V. V. Tekuchev (1976) *Russ. J. Phys. Chem. Engl. Transl.*, 50, 1418.

Stull, D. R. and H. Prophet (1971). *JANAF Thermochemical Tables*, Second Edition. United States Department of Commerce, National Bureau of Standards, Washington, D.C 20234.

Stull, D. R. and G. C. Sinke (1956). *Thermodynamic Properties of the Elements*, No. 18 in the Advances in Chemistry Series, American Chemical Society, Washington, DC.

Thompson, S. L. and H. S. Lauson (1972). *Improvements in the Chart D Radiation-Hydrodynamic CODE, III; Revised Analytic Equations of State*, Sandia Report SC-RR-0714 (Reprinted in November 1979).

Trainor, K. S. (1983). Construction of a wide-range tabular equation of state for copper. *J. Appl. Phys.* 54, 2372-2379.

Trunin, R. F., M. A. Podurets, B. N. Moiseev, G. V. Simakov, and L. V. Popov (1969). Relative compressibility of copper, cadmium, and lead at high pressures. *Soviet Physics - JETP*, 29, 630-631.

Trunin, R. F., M. A. Podurets, G. V. Simakov, L. V. Popov, and B. N. Moiseev (1972). An experimental verification of the Thomas-Fermi model for metals under high pressure. *Soviet Physics - JETP*, 35, 550-552.

Vashchenko, V. Ya. and V. N. Zubarev (1962). Concerning the Gruneisen constant. *Soviet Physics - Solid State*, 5, 653-655.

Vladimirov, A. S., N. P. Voloshin, V. N. Nogin, A. V. Petrovtsev, and V. A. Simonenko (1984). Shock compressibility of aluminum at P > 1 Gbar. *JETP Letters*, 39, 82.

Volkov, A. P., N. P. Voloshin, A. S. Vladimirov,V. N. Nogin, and V.A. Simonenko (1980). Shock compressibility of aluminum at a pressure of 10 Mbar. *JETP Letters*, 31, 588-592.

Walsh, J. M., M. H. Rice, R. G. McQueen, and F. L. Yarger (1957). Shock-wave compressions of twenty-seven metals. Equations of state of metals. *Phys. Rev.*, 108, 196-216.

Young, D. A. (1977). *A Soft Sphere Model for Liquid Metals*, UCRL-52352, Lawrence Livermore Laboratory.

Young, D. A. and B. J. Alder (1971). Critical point of metals from the van der Waals Model. *Physical Review A*, 3, No. 1, 364 ff.

Int. J. Impact Engng Vol. 5, pp. 155–163, 1987
Printed in Great Britain

0734-743X/87 $3.00 + 0.00
Pergamon Journals Ltd.

VELOCITY SCALING IMPACT MELT VOLUME

Michael D. Bjorkman* and Keith A. Holsapple**

*Boeing Aerospace Co. Seattle, WA
**Dept. of Aeronautics and Astronautics
University of Washington Seattle, WA

ABSTRACT

Velocity scaling relationships for the amount of melted target material produced by hypervelocity impacts are examined. Scaling of the melt volume with nearly the momentum of the impactor has been obtained by Gault and Heitowit using a semi-analytical energy balance approach and energy scaling has been obtained from scaled hydrocode output by O'Keefe and Ahrens. Both results are inconsistent with the point source limit of cratering motions in non-porous media which requires somewhat less than energy scaling. The nearly momentum scaling result from the energy balance approach of Gault and Heitowit has been reexamined and it is shown to be incorrect given their assumptions. A corrected derivation is given which results in energy scaling for all equations of state. The energy scaling result obtained by O'Keefe and Ahrens with hydrocodes has been tested by performing new hydrocode calculations at larger impact velocities. The inconsistency between the coupling parameter approach and O'Keefe and Ahrens hydrocode calculations has been resolved. For aluminum onto aluminum impacts at velocities less than 50 km/s, the new hydrocode calculations have demonstrated that the melt volume forms too close to the impactor for the point source solution to govern the melt volume dimensions, hence energy scaling is a possible result. However, at greater than 50 km/s impact velocity, less than energy scaling is obtained because the flow in the melt volume region approaches the point source limit.

INTRODUCTION

The volume of melt and vapor produced during planetary impacts has attracted attention in applications to various problems in planetary geology. Since these applications have impact velocities in excess of those experimentally achievable, most effort has been put into analysis. Methodologies used to estimate the volume of melted material are of two types: analytical energy balance models and numerical solutions of the fundamental balance equations. To date, it is controversial whether the volume of melted material scales with the momentum of the impact, the energy of the impact, or with some other rule.

In the first section the Gault and Heitowit (1963) energy balance model (hereafter called the GH model) is reexamined. Although apparently never made explicit in the literature the GH model requires the melt mass to be nearly proportional to the momentum of the impactor. Values for the melt mass velocity scaling exponent determined from the GH model output reported in Gault, Hortz and Hartung (1972), O'Keefe and Ahrens (1977), and Grieve and Cintala (1981) range from 1.0 to 1.2; nearly momentum scaling. Holsapple and Schmidt (1982), however, have demonstrated that models with a fixed percentage of the impactor energy coupled to the target, such as the GH model, must scale with the energy of the impactor. This inconsistency prompted a reexamination of the assumptions of the GH model which has led to a demonstration that those assumptions are contradictory and that energy scaling is the only possible result.

The second section is a discussion of hydrocode calculations of melt volume, particularly those by O'Keefe and Ahrens (1977) who obtained energy scaling for impacts of gabbroic anorthosite. A dimensional analysis performed by Bjorkman (1984) based on the coupling parameter (Holsapple, 1983; Holsapple and Schmidt, 1986) indicated that the melt volume should scale with a velocity exponent that is less than that based on the energy of the impact, but greater than that based on the momentum. Even though the energy scaling result obtained by O'Keefe and Ahrens and the nearly energy scaling result obtained by Bjorkman do not differ by much, the apparent discrepency is troubling prompting a recalculation of O'Keefe and Ahrens' melt volume results.

GAULT AND HEITOWIT ENERGY BALANCE MODEL

The Gault and Heitowit theory is based on a simple model of the cratering process, but that simplicity is achieved at the expense of inconsistency with the fundamental balance laws of physics. Due to that inconsistency, a variety of scaling laws can be obtained, depending on which laws are adopted, and which are ignored.

A New Derivation of the GH Model

The GH model first assumes that the shock wave encompasses a hemispherical shell of target material as shown by the shaded area in Fig. 1. That shell of material is assumed to be at a uniform pressure, particle velocity and density, those values being a state on the Rankine-

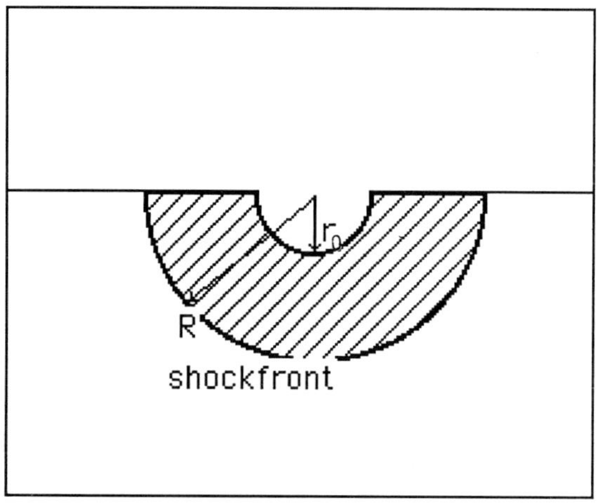

Fig. 1 Cratering flow field geometry for the GH model.

Hugoniot curve (hereafter called the R-H curve), the curve of all states that can be reached by a single shock from the reference state. The inner surface at the radius r_0 is assumed to be stress free. Therefore, the total energy E_t, (internal plus kinetic) of that target material is constant in time. From the shock jump-conditions it can be shown that the specific energy immediately behind the shock front (and hence for the GH model, throughout the wave) is the sum of a kinetic energy term equal to 1/2 u^2 and an equal internal energy term, where u is the shock particle velocity amplitude. Thus, the total energy in the wave, the sum of those two components, is,

$$E_t = M u^2 , \tag{1}$$

where M is the mass of shocked material.

Since E_t in the crater bowl is constant, the decay of the particle velocity amplitude can be obtained by rewriting (1) as,

$$u \propto M^{-1/2}, \tag{2}$$

which is equivalent to the particle velocity decaying with range as $r^{-3/2}$. In the strong-shock regime, where the pressure is proportional to the square of the velocity, the pressure would decay as r^{-3}. The mass of melt, consisting of all material that is shocked to a given pressure, would consequently scale with the energy of the impactor. These are energy-scaling results, a direct consequence of the constant energy assumption. That energy scaling result is also the least rapid shock amplitude decay physically permissible in the strong-shock regime. (See Holsapple and Schmidt, 1986).

The fundamental problem with the GH model is that the assumption of a uniform state behind the shock is only consistent with mass balance and with energy balance, but not with force (momentum) balance. This is apparent in two important ways. First, if the shocked target material is at constant pressure, then that pressure must also be applied to the inner radius in order to balance forces. Since that surface has the same particle velocity as the entire shocked region, that pressure inputs work into the region, and the energy is then not constant, but must increase in time. This invalidates the assumption of constant energy. Furthermore, even if a new decay relation is worked out based on the actual increase of energy from the force on the crater boundary at radius r_0, then it is clear that the decay relation for the particle velocity must be less steep that the result (2) above. Such a decay is not physically permissible in the strong shock regime. The second way that the uniform shocked state assumption can be seen inconsistent with force balance is by considering the shocked material. It is assumed in the GH model that a uniform velocity field in the shocked target material is decreasing with time; the velocity decay requires that there be accelerations hence there must be pressure gradients, which contradicts the uniform pressure assumption.

Gault and Heitowit go on to introduce the concept of waste heat and obtain different decay relations, apparently by assuming that it is the available energy, equal to the total energy minus the waste heat, that is constant. However, we do not understand the basis for this assumption, or the relevance of waste heat to E_t. Consider the definition of waste heat. Shown in Fig. 2 is the p-V R-H curve, the Rayleigh line which describes the p-V path of a steady

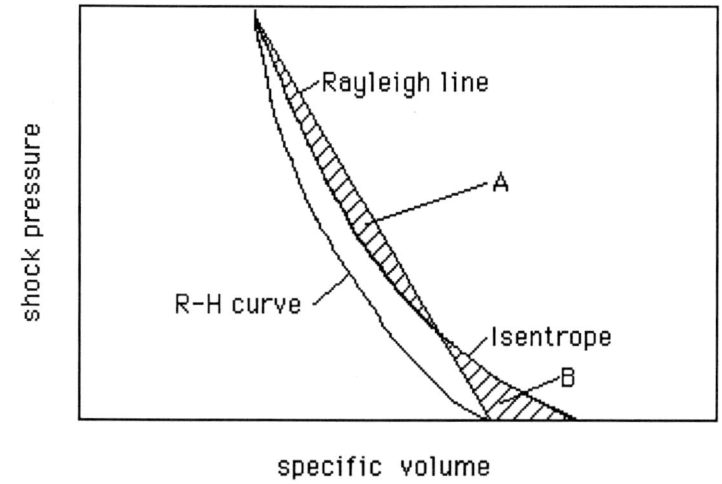

Fig. 2 Geometric interpretation of waste heat.

continuous shock front, and the release isentrope. The well known energy jump-condition for the amount of internal energy increase behind the shock can be interpreted geometrically as the area under the Rayleigh line. Integrating the quantity (-PdV) along the release isentrope gives the amount of work done by the material upon release. Area B in Fig. 2 subtracted from area A in Fig. 2 is the amount of internal energy behind the shock which is not recovered upon release (i.e. waste heat) due to the irreversible entropy increase across the shock front.

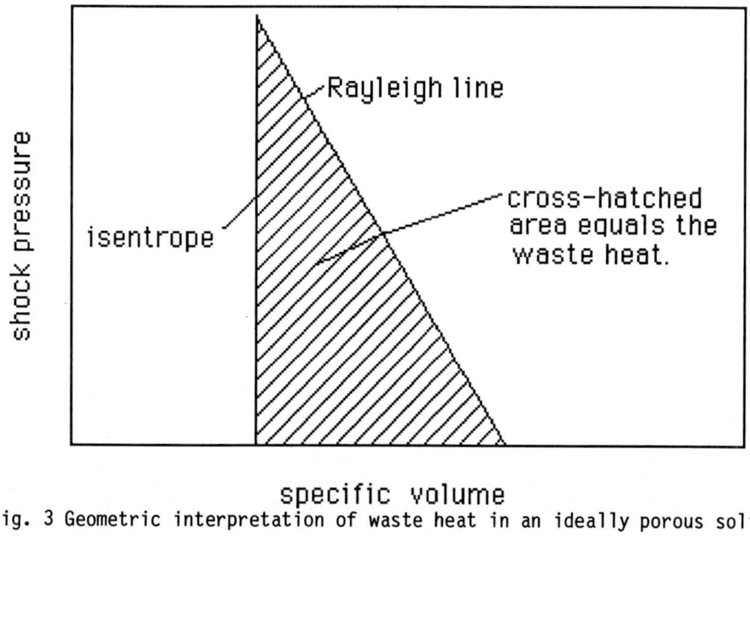

Fig. 3 Geometric interpretation of waste heat in an ideally porous solid.

Thus, though waste heat may remove energy from that available for cratering, the waste heat remains as internal energy of the released target material and the total energy E_t in the crater bowl remains constant. If the total energy in the crater bowl E_t remains constant, then (1) is still valid even though there is waste heat, and (2) still holds for the shock amplitude decay relation. Thus, the GH model should obtain energy scaling for any R-H curve and any isentropic release path. The only mechanism which can lead to other than energy scaling with the GH model is ejecta which will decrease E_t in the crater bowl by removing material which contains kinetic and internal energy.

Comparison of the GH Model with an Exact Solution

O'Keefe and Ahrens (1977) noted results from their hydrocode calculations of impact that were substantially different from the GH model, and attributed the discrepancies to the assumption in the GH model that the adiabat and the R-H curves were the same. However, it is claimed here that the most important facet of the GH model that violates physical reality is assuming the shock in the target to be a square wave and the resulting violation of balance of forces.

As an example, it is useful to consider the GH model applied to a point explosion in an ideally porous solid. This porous solid is assumed to crush up at a stress incrementally above zero, and once crushed up to be rigid. The Rayleigh line, p-V R-H curve and the p-V release path are shown plotted in Fig. 3.

Because of the simplicity of the constitutive relation for this porous material, an exact solution of the Euler equations for the wave motions exists for comparison with the GH model square wave assumption. A simple derivation given by Zeldovich and Raizer (1967) is repeated here. The density profile of the compaction shock passing through the porous solid is

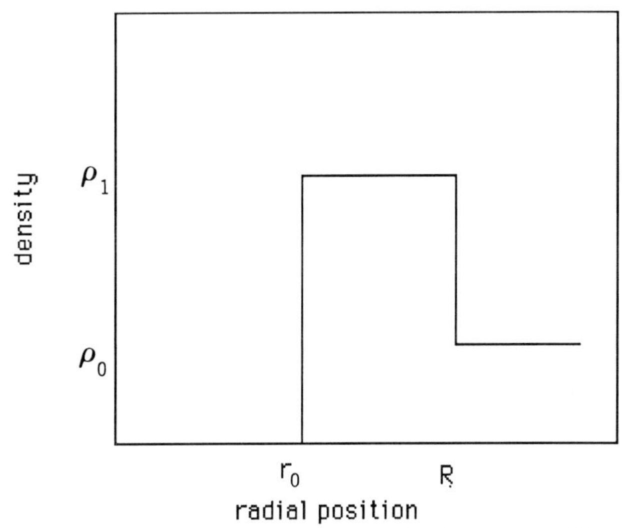

Fig. 4 Density profile for the compation shock in an ideally porous solid.

illustrated in Fig. 4. The radius of the shock front is R and the radius of the internal cavity is r_0. Therefore conservation of mass gives

$$M = (4\pi/3)R^3\rho_0 = (4\pi/3)(R^3-r_0^3)\rho_1. \quad (3)$$

Factoring the terms gives

$$r_0^3 = R^3(1-\rho_0/\rho_1) = R^3 k \quad (4)$$

where the porosity k is defined as $1-\rho_0/\rho_1$, ρ_0 is the density of the porous solid and ρ_1 is the density of the compressed material. The particle velocity wave profile is found by using the equation of continuity, which gives from the incompressibility assumption behind the shock,

$$u = u_1(R/r)^2, \quad r_0 < r < R \quad (5)$$

where u_1 is the particle velocity behind the shock front, as shown in Fig. 5.

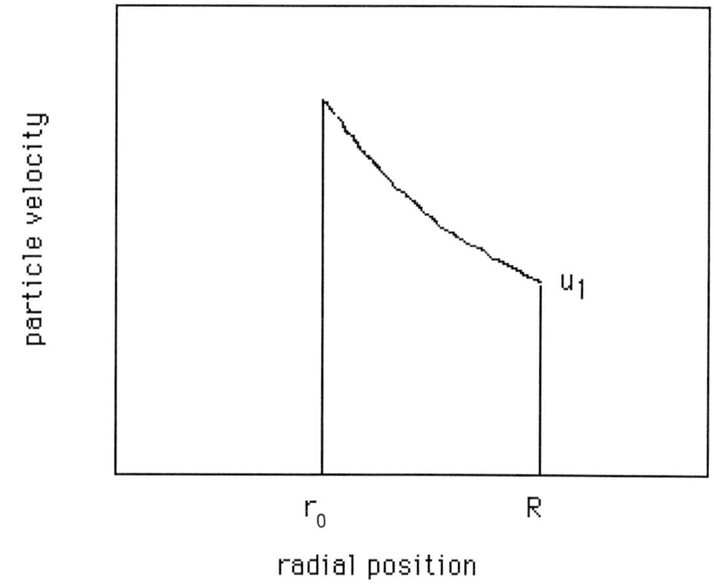

Fig. 5 Particle velocity profile for a compaction shock in an ideally porous solid.

Note that the wave profile shown in Fig. 5 is not the square wave assumed in the GH model. Furthermore, the incompressible flow assumption which leads to (5) is not so far removed from real materials as one might first think. The Z model of cratering flows (Maxwell, 1977) assumes the particle velocity field in the target after passage of the outgoing shock wave has the form,

$$u \propto R^{-Z}. \tag{6}$$

with values of Z typically between 2 to 4. None of these values for Z result in a square wave and the smaller values are close the value of 2 obtained in (5).

After solving for the wave profile (5), Zeldovich and Razier go on to solve for the shock amplitude u_1 decay as a function of engulfed mass M. They argue that it follows from the incompressibility assumption that the internal energy of each material element is fixed in time so that the waste heat is the entirety of the internal energy imparted at the shock. The "available energy" is equal to the kinetic energy. Then from the conservation of total energy the decrease in kinetic energy over the entire shocked mass M in time dt will equal the increase in internal energy of mass dM encompassed by the shock in the time dt, or,

$$-d(Mu^2/2) = -\beta d(Mu_1^2/2) = u_1^2/2 \; dM \tag{7}$$

where $u^2 = \beta u_1^2$ denotes the mean square velocity in the shocked mass M. The coefficient β is determined from the velocity profile from (4) and (5), and can be shown to be given by an explicit dependence on

$$\beta = 3/(k + k^{2/3} + k^{1/3}) \tag{8}$$

the porosity of the material. Integrating (8) gives the shock particle velocity amplitude decay relation

$$u_1^2 \propto M^{-n} \tag{9}$$

where $n=(1+\beta)/\beta$ then depends on the material porosity. Examination of the formula for n shows that (9) varies continuously from energy scaling (the GH model result for all porosities) at zero porosity (n=1) to momentum scaling (n=2) for total porosity (i.e. $\rho_o = 0$).

Suppose now the square wave assumption as used in the GH model were to be invoked in this case, rather than the actual profile calculated to be (5). If the wave is assumed square then the mean square velocity must equal the shock amplitude u_1 and the coefficient β must equal 1, independent of the material porosity. The resulting decay relation would have momentum scaling (n=2) for all porosities, clearly the incorrect result.

HYDROCODE CALCULATIONS OF MELT

Simultaneously solving the Euler equations of motion and the constitutive equation with a hydrocode allows direct calculation of the quantity of melt from a hypervelocity impact. Repeated calculation of each case of interest can be eliminated by scaling the results of a hydrocode parameter study as follows.

The volume of melted planetary material V_M is expected to be given by the form

$$V_M = f(a, U, \delta, \rho, E_M) \tag{10}$$

where a is the projectile radius, U is the impact velocity, δ is the projectile density, ρ is the target density, and the material property E_M is the specific energy of the Rankine-Hugoniot state from which isentropic release ends at the 1 atm point on the liquidus. A dimensional analysis of this relation gives the reduced form

$$V_M / a^3 = f(\delta / \rho, U^2/E_M). \tag{11}$$

Only the scaling between different impact conditions for fixed materials is of interest. Then the dependence on the density ratio can be omitted. In addition, for given materials there is a one-to-one relation between the impact velocity U and the initial one-dimensional particle velocity v in the target at the impact point, which can be found by standard shock impedance considerations. Therefore, (11) can also be written in terms of v:

$$V_M / a^3 = f\left[v2/(2EM)\right]. \tag{12}$$

The factor of 2 is inserted for convenience. Since the internal energy at the shock (equal to the kinetic energy density) is 1/2 v^2 at the point of impact, then the independent parameter in (12) is equal to unity at the minimum impact velocity at which melt first occurs and increases for all greater impact velocities.

At the minimum impact velocity for melt, that melt region in the target is immediately adjacent to the impactor. As the impact velocity increases, that melt region grows much larger: for very high impact velocities it will be many times larger than the impactor size. In this latter case a further assumption of a point-source replacing the actual impactor may be warranted.

It has been demonstrated (Holsapple and Schmidt, 1986) that a function C of the early time variables a, U, δ exists and remains finite in the point source limit ($a \rightarrow 0$, $U \rightarrow \infty$) where the effects of the early time variables a and U are not distinguishable individually. Thus, one may use the function C to relate the motions from a finite-source impact to the motions in the point source limit as follows. In the point source limit, (10) is replaced by

$$V_M = f(C, \rho, E_M) \tag{13}$$

The coupling parameter C can be shown to have the form

$$C \propto a U^{\mu} \delta^{\nu} \tag{14}$$

where the velocity exponent μ varies between 1/3 and 2/3 for all materials (Holsapple and Schmidt, 1982). Metals have been found to have μ=0.58 and porous soils to have μ=0.47 (Holsapple and Schmidt, 1986). Substituting (14) into (13), followed by a dimensional analysis gives the form

$$(V_M / a^3) \propto (E_M / U^2)^{-3\mu/2} \tag{15}$$

as a specific special case of (11) that should hold for sufficiently high impact velocities. To put both of (15) and (12) into a common form, rewrite (12) as

$$(V_M / a^3)(E_M / U^2)^{3\mu/2} = f\left[v^2/(2E_M)\right] \tag{16}$$

where the function f is zero when the variable $v^2/(2E_M)$=1 and, according to the arguments above, should approach a constant when $v^2/(2E_M)$ becomes large. This expression is the primary form that should relate the melt volume for different impactor sizes and velocities. A linear dimension such as the melt depth D_M follows a similar law given by

$$(D_M / a)(E_M / U^2)^{\mu/2} = f\left[v2/(2E_M)\right]. \tag{17}$$

The scaled depths of melted planetary material calculated with hydrocodes by Ahrens and O'Keefe (1977) and Orphal, Borden, Larson and Schultz (1980), are shown plotted in Fig. 6 scaled according to the form of (17). (Neither reference reported the melt depths. Melt depths were inferred from the plots of peak shock amplitude as a function of depth in Ahrens and O'Keefe, 1977, and from the melt contour plots in Orphal, Borden, Larson and Schultz, 1980.) The exponent μ was taken to be 0.58 and the exponent ν was taken to be 0.67, values obtained for crater depth growth scaling in Holsapple and Schmidt (1986). The curves all asymptotically approach the y axis, i.e. $v^2/(2E_M)$ equals 0, as they must, and, if the coupling parameter C is valid, then all should approach a horizontal line as the variable $v^2/(2E_M)$ increases. They do seem to have this property, confirming the predictions.

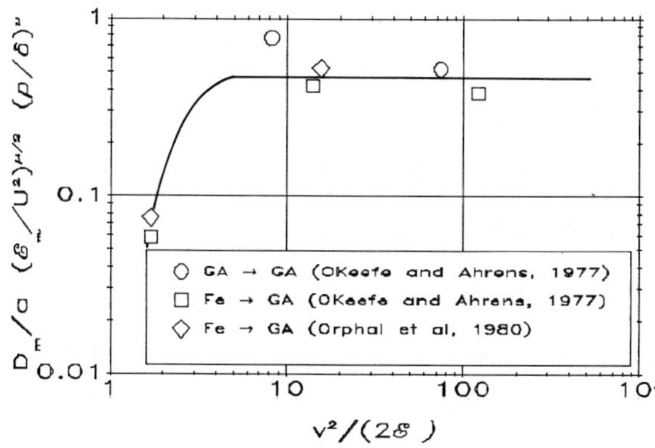

Fig. 6 Scaled melt depths.

The deviations of the Ahrens and O'Keefe scaled melt depth data from a horizontal line are large but are of the same magnitude as the oscillations of the shock amplitude as a function of range shown in Ahrens and O'Keefe (1977). Thus, the errors in determining melt depth from Ahrens and O'Keefe's shock amplitude decay as a function of depth are thought responsible for the deviations from a horizontal line in Fig. 6.

The value for E_M used in plotting Fig. 6 is 3.42 kJ/g for gabbroic anorthosite. This value was calculated from the R-H energy jump condition, the high pressure phase gabbroic anorthosite p-V R-H curve given in Ahrens and O'Keefe (1977), and from using 52 GPa as smallest pressure on the R-H curve which isentropically releases into the completely melted state at 1 atm.

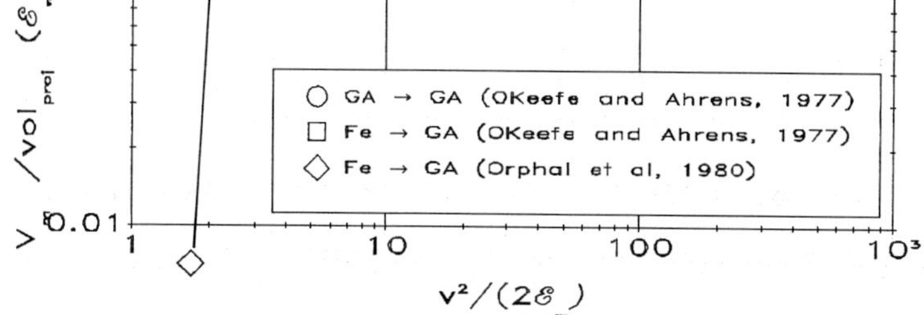

Fig. 7 Scaled melt volumes.

Shown plotted in Fig. 7 are the scaled melt volumes from O'Keefe and Ahrens (1977) and Orphal, Borden, Larson, and Schultz (1980). Note that the large impact velocity data do not fall on a horizontal line as the melt depth data do. In fact, the calculated melt volumes appear to scale with the energy of the impact (μ=2/3) and not μ=0.58. (I.e., if the value μ=2/3 were used in the ordinate, then the large impact velocity data would lie on a horizontal line). Even though the difference is not large when compared with other uncertainties, the trend is bothersome.

This trend prompted a calculation of melt volume using the 2-d finite difference hydrocode CSQ II (Thompson, 1977) as a check. Calculations were performed of a 10 cm diameter aluminum cylinder striking an aluminum semi-infinite plate at impact velocities of 15, 20, 30, 50, and 100 km/s. The ANEOS library equation of state for aluminum was used. ANEOS is an analytical complete 3-phase equation of state which can be applied to a variety of materials (Thompson and Lauson, 1972).

Table. Hydrocode Melt Mass Results

mesh	velocity	15 km/s	20 km/s	30 km/s	50 km/s	100 km/s
coarse						
upper bound			$9.43\ 10^3$ g	$2.15\ 10^4$ g	$6.98\ 10^4$ g	$2.67\ 10^5$ g
lower bound			$4.45\ 10^3$ g	$1.24\ 10^4$ g	$4.29\ 10^4$ g	$1.65\ 10^5$ g
fine						
upper bound		$4.62\ 10^3$ g	$1.11\ 10^4$ g	$3.03\ 10^4$ g	$9.40\ 10^4$ g	$3.21\ 10^5$ g
lower bound		$3.11\ 10^3$ g	$7.81\ 10^3$ g	$2.39\ 10^4$ g	$7.41\ 10^4$ g	$2.56\ 10^5$ g
very-fine						
upper bound				$3.78\ 10^4$ g	$1.17\ 10^5$ g	$4.32\ 10^5$ g
lower bound				$3.41\ 10^4$ g	$1.03\ 10^5$ g	$4.00\ 10^5$ g

The melt volumes calculated for aluminum impacting aluminum are shown plotted in Fig. 8.

The value of E_M for aluminum used in plotting Fig. 8 was 6.45 kJ/g and was calculated from the R-H energy jump condition, the aluminum R-H curve given in Thompson and Lauson (1972), and 100 GPa as the smallest pressure state on the aluminum R-H curve which releases into the completely melted state at 1 atm pressure.

The volume of melted target material was calculated from time edits of the hydrocode runs by summing the mass of projectile and target material which experienced 100 GPa pressure (which corresponds to the state of pressure on the R-H curve which releases into the completely melted state) and multiplying the melt mass by the original density of the target and projectile, 2.7 g/cm^3. Due to the descretization of the projectile and target materials the melt mass could only be determined to within the range of values shown in the Table. The upper limit was calculated by summing all zones with at least some material at 100 GPa pressure, while the lower limit was calculated by summing only those zones which were completely engulfed by 100 GPa pressure. This range in values does not represent the uncertainty in the volume of melt calculated with the hydrocode; it represents the uncertainty over determining the volume of melted material from the hydrocode edits.

In order to address the question of the accuracy of the numerical calculations, calculations were performed with three spatial resolutions: coarse resolution calculations were performed with zones 2 cm on a side, fine resolution calculations with zones 1 cm on a side, and very-fine resolution calculations with zones 0.5 cm on a side. The calculations shown in Fig. 8 show that the solution is slowly converging on the answer.

It is apparent from Fig. 8 that for values of $v^2/(2E_M)$ above 50 the melt volume scales with a coupling parameter with μ less than 2/3 (where μ = 2/3 is energy scaling). The fine resolution calculations give μ= 0.58 and the very-fine resolution calculations give μ= 0.61 for $v^2/(2E_M) > 50$. However, it is also apparent from Fig. 8 that for $8 < v^2/(2E_M) < 50$ the melt volume scales with the energy of the impactor as the scaling transitions from one-dimensional flow to the point source limit of the flow.

Hence, the resolution of the original problem with why the melt volume does not scale with less than energy scaling is that the coupling parameter does not govern all of the flow over which the final melt volume forms until the final melt/solid boundary in the target is far from the source.

It appears from Fig. 6 that the flow directly under the impactor is governed by a coupling parameter with μ = 0.58 from small impact velocities ($v^2/(2E_M) \sim 5$) and higher, while it appears from Fig. 8 that the flow near the free surface doesn't scale with a coupling parameter with μ = 2/3 until $v^2/(2E_M) > 50$.

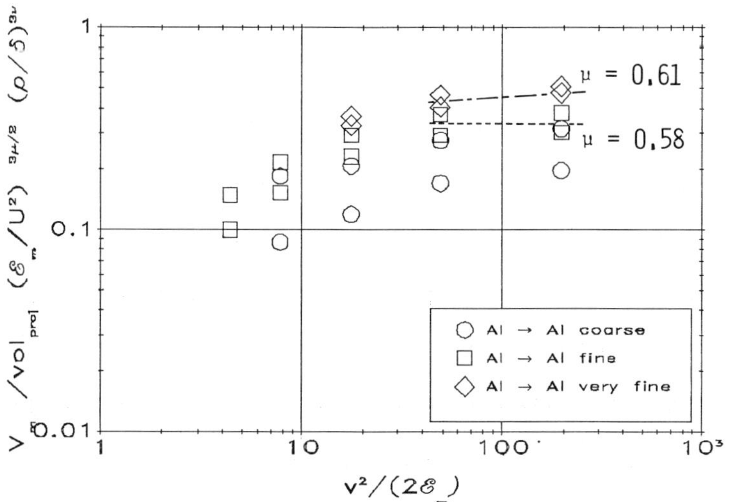

Fig. 8 Aluminum impacting aluminum melt volumes.

Even though the original difficulty has been resolved by demonstrating melt volumes scale with a coupling parameter with m < 2/3 the demonstration is of limited utility to planetary applications. The less than energy scaling result has been shown to occur for impact velocities 50 km/s and larger for aluminum onto aluminum impacts which exceeds the impact velocities of interest to most planetary applications. Thus most planetary applications will require energy scaling of the melt volume.

ACKNOWLEDGMENT

We are indebted to Mr. Kyu Choe who performed most of the calculations reported here. The research was supported by NASA grant NAGW-328 and contract NASW-3291.

REFERENCES

Ahrens, T.J., and J.D. O'Keefe (1977). Equations of state and impact-induced shock-wave attenuation on the moon. in, D.J. Roddy, R.O. Pepin, and R.B. Merill (eds.) Impact and Explosion Cratering. Pergamon Press, New York. p. 639-56

Bjorkman, M.D. (1984). Feasibility of determining impact conditions from total crater melt (abstract). Lunar and Planet. Sci. XV, 64-5.

Gault, D.E., and E.D. Heitowit (1963). The partion of energy for hypervelocity impact craters formed in rock. Proceedings of Sixth Hypervelocity Impact Symposium, Vol. II part 2, pp. 419-56.

Gault, D.E., F. Hortz, and J.B. Hartung (1972). Effects of microcratering on the lunar surface. Proc. Lunar Sci. III, 2713-34.

Grieve, R.A.F., and M.J. Cintala (1981). A method for estimating the initial impact conditions of terrestrial cratering events, exemplified by its application to Brent crater, Ontario. Proc. Lunar Planet. Sci. XII, 1607-21.

Holsapple, K.A. (1983). On the existence and implications of coupling parameters in cratering mechanics (abstract). Eos Trans. AGU, 64, 319-20.

Holsapple, K.A., and R.M. Schmidt (1982). On the scaling of crater dimensions 2. Impact processes. J. Geophys. Res., 87, 1849-70.

Holsapple, K.A., and R.M. Schmidt (1986). Point source solutions and coupling parameters in cratering mechanics. Accepted for publication, J. Geophys. Res.

O'Keefe, J.D., and T.J. Ahrens (1977). Impact-induced energy partioning, melting, and vaporization on terrestrial planets. Proc. Lunar Sci. Conf. 8th, 3354-74.

Orphal, D.L., W.F. Borden, S.A. Larson, and P.H. Schultz (1980). Impact melt generation and transport. Proc. Lunar Planet. Sci. Conf. 11th, 2309-2323.

Thompson, S.L. (1977). CSQ II - An eulerian finite difference program for two-dimensional material response - part 1. materials sections. Sandia National Laboratories, Abuquerque, NM. SAND77-1339,

Thompson, S.L., and H.S. Lauson (1972). Improvements in the CHART D radiation-hydrodynamic code III: revised analytic equations of state. Sandia Laboratories, Albuquerque, NM. SC-RR-71 0714.

Zeldovich, Ya.B., and Yu.P. Raizer (1967). Physics of Shock Waves and High-Temperature Phenomena. Academic Press, New York. Vol. 2.

Int. J. Impact Engng Vol. 5, pp. 165–171, 1987
Printed in Great Britain

0734-743X/87 $3.00 + 0.00
Pergamon Journals Ltd.

HYPERVELOCITY PENETRATION OF CERAMICS

S.J. Bless (1), Z. Rosenberg, (1,2), B. Yoon (3)

1-University of Dayton Research Institute, Dayton, Ohio 45469
2-Permanent Address Rafael Institute, Haifa, Isreal
3-R&D Associates, Arlington, VA 22209

ABSTRACT

The penetration resistance of alumina was found to decrease with velocity for armor-piercing bullets. However, it was relatively independent of velocity for rods and fragment-simulating projectiles. These results are explained in terms of compressive yielding caused by high velocity pointed projectiles.

INTRODUCTION

Ceramic armor has evolved as one of the most useful techniques for defeat of gun-launched projectiles. There are presently emerging technologies that may result in substantial escalation of achievable muzzle velocities. The question arises as to whether or not ceramics will continue to exhibit efficient performance against such future projectiles.

The performance of ceramic armor is determined by the properties of its constituents and its geometric design. This paper is concerned with the former. Specifically, we wish to determine the ability of ceramics to resist penetration at very high impact velocities.

HYPOTHESES

Consistent with published analyses of ballistic impact on ceramics (for example, Wilkins, et al. 1970), we assume that there are four possible phases to penetration of ceramic armor: (1) Shock - in which a high pressure shock wave followed by a release wave propagates through the ceramic. (2) Penetration - in which the projectile penetrates the target by compressive yielding of the target material in its path. (3) Fracture - in which a fracture front propagates into the ceramic ahead of the projectile, causing comminution. (4) Penetration of fractured material - in which comminuted material is moved out of the path of the projectile and the projectile advances through the target.

This study is focussed on penetration phenomena that are mainly dependent on ceramic properties. Target geometry effects are not addressed. Hence we are mainly concerned with the first two phases.

What are the expected influences of impact velocity on penetration of ceramics? The principal shock parameter of ceramics is the Hugoniot Elastic Limit (HEL). Shocks above the HEL induce extensive microfracture in ceramics (Bless, et al., 1985; Yaziv, 1985). In fact, even shocks in the elastic region apparently cause some microcracks, resulting in loss of spall strength (Bless, et al., 1985; Rosenberg and Yeshurun, 1985).

If the shock response plays an important role in penetration of ceramics, then there should be an observable effect as the peak shock stress exceeds the HEL; we define the velocity at which this first occurs as u_1. Representative values of u_1 are listed in Table 1. The effect of impact velocity (u_0) exceeding u_1 will be more manifest when a substantial portion of the target experiences the peak shock stress - e.g. when the projectile is flat or blunt and of diameter (D) comparable to the ceramic thickness (T). Otherwise, the effect of shock degradation of the ceramic will appear only gradually, as a larger and larger volume of the target experiences a stress above the HEL. If we use the approximations that the peak stress is proportional to u_0 and that the peak transient stress falls to D/T of its peak value at the ceramic rear surface (where D is now the diameter of the impact region), then the critical velocity for massive shock damage for a non-flat projectile will be approximately $(T/D)u_1$. Between u_1 and this velocity, the effects of shock degradation will gradually become dominant.

Table 1 Values of Critical Velocities

Projectile	u_1 km/s for AD85	u_2 km/s for AD85	u_2 km/s for B_4C
aluminum	0.62	2.02	3.28
steel	0.37	1.06	1.85
tungsten	0.30	0.79	1.31

In general, values of u_1 are well below usual ballistic impact velocities. This indicates that there is probably a region of extensive shock degradation around the impact site when conventional armor-piercing bullets strike ceramic targets. Since such bullets are relatively sharp, the total effect on the ceramic is minimal. For a 7.62 mm APM2 projectile, for example, the diameter of the tip is only about 1 mm. Thus, if shock degradation is important for this type of projectile, one should observe a gradual decrease of ceramic effectiveness as the impact velocity is increased above conventional values. However, even at hypervelocities, it will probably not be possible to completely degrade the ceramic by this mechanism.

Penetration criteria have been treated by many authors. We use the penetration criterion of Tate (1969). A projectile can penetrate a target when

$$\rho_p u_0^2/2 > R_t - Y \tag{1}$$

where Y is the projectile yield stress, R_t is the target resistance, and ρ_p is the projectile density. The parameter R_t is related to the target yield stress. A value of several times the yield stress is normally found to describe experimental data; this increase is usually attributed to effects of confinement. When the penetration criterion is exceeded, we would expect the projectile to advance into the target from the impact face.

It is not known if the Tate model (1969) is valid for ballistic-grade ceramics. Nor are the appropriate values of R_t for ceramics well defined. Unlike metals, the strength of ceramics is highly dependent on pressure. For example, the strength of AD85 alumina increases from 20 to 40 kbar as the deformation changes from one-dimensional stress to one-dimensional strain (Rosenberg and Yeshurun, 1985; Rosenberg, 1985). Moreover, unlike metals, some ceramics lose shear strength after initial failure, at least under shock loading (Rosenberg and Yeshurun, 1985; Gust and Royce, 1971). We have assumed that for ceramics R_t is equal to the HEL, since that represents failure under very high confinement. Doing so results in values for the critical velocity for ceramic penetration (u_2) given in Table 1. The values of u_2 range from slightly above conventional ordnance velocities for impact onto alumina, to hypervelocities for impacts onto boron carbide.

If this conceptual framework is correct, when u_0 exceeds u_2, the penetration resistance of ceramics will start to decrease. In analogy with metals, as u_0 approaches $2u_2$, the strength of the ceramic will be of little significance, and the projectile will penetrate the ceramic with a velocity given by

$$v = \beta u_0/(1 + \beta);\ \beta = \sqrt{\rho_p/\rho_t} \quad . \tag{2}$$

Penetration will be to a depth equal to $\sqrt{\rho_p/\rho_t}$ times the length of the penetrator. (Here ρ_t is the density of the ceramic.)

In the case of thick ceramic tiles struck by high-density projectiles at hypervelocities, there is also the possibility of secondary penetration. This process involves a second collision of projectile material with the base of the penetration cavity. Secondary penetration will not be considered in the present paper. General criteria for secondary penetration are given by Bless, et al. (1986).

EXPERIMENT DESIGN

The ideal experimental configuration would probe the impact response of ceramics independent of details of target construction. This might be accomplished, for example, by a semi-infinite block of ceramic struck by a projectile, with measurement of the depth of penetration. Unfortunately, this experiment was impossible due to technical and financial constraints.

As an alternative, we examined the performance of thin tiles. Failure of tiles in ceramic armor usually occurs by tension. In order to increase sensitivity to impact surface effects and decrease sensitivity to details of armor design, it is necessary to suppress early tensile failure and its consequences. This can be accomplished by mounting tiles on very thick substrates. Tiles are much more efficient in this configuration. For example, considering alumina/aluminum armor--if the aluminum plate of finite thickness from an armor that is just matched to a given threat is replaced with a very thick aluminum plate, then it will be found that the bullet fails to penetrate the aluminum plate at all. This is the configuration that was used in most of our targets.

The ceramic tiles used in our work were Coors AD85 alumina. This grade of alumina is widely used for ballistic protection. The aluminum used was 6061-T651. The tiles were bonded with a thin layer (approximately 0.2 mm) of epoxy. Figure 1 illustrates the usual arrangement.

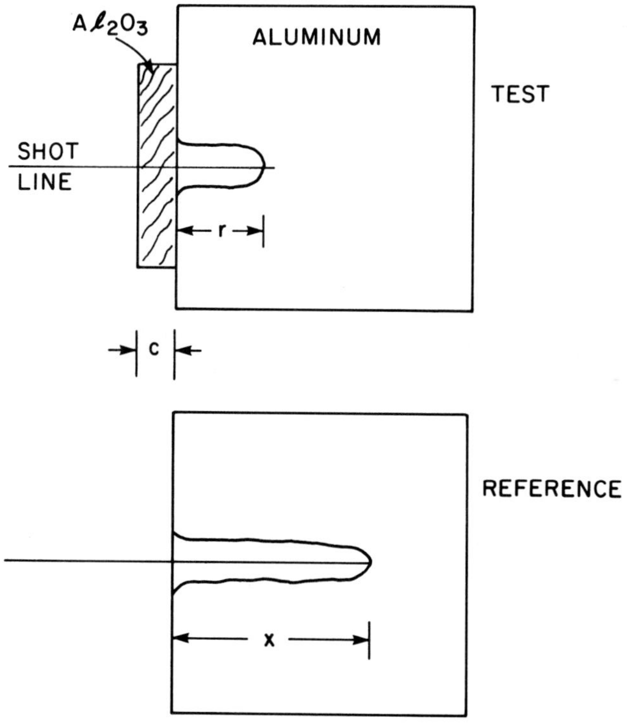

Fig. 1. Experiment design and definition of terms.

Using the target design of Fig. 1, the efficiency of the ceramic can be compared to the efficiency of aluminum. One way to do this is by the Differential Efficiency Factor, DEF (Yaziv, et al., 1986), defined as

$$DEF = \rho_0 (x - r)/\rho_t c \tag{3}$$

where r, x, and c (defined in Fig. 1) are, respectively, the depth of penetration behind the ceramic, the depth of penetration into aluminum that results when there is no ceramic, and the thickness of the ceramic. The densities ρ_t and ρ_0 refer to the ceramic and the aluminum, respectively. The DEF is equal to 1 for an efficiency equal to that for aluminum; it is greater than one for more efficient ceramics. However, it is difficult to associate a unique value of DEF with a good-performance ceramic because the DEF is very sensitive to the thickness of the tested tile.

For present purposes, it is more useful to define an efficiency parameter that varies from zero to one as the ceramic efficiency varies from equal to aluminum to the highest possible value. This is done by defining the efficiency ε as

$$\varepsilon = 1 - (\rho_0 r + \rho_t c)/\rho_0 x \tag{4}$$

Using the design of Fig. 1 and the efficiency parameter ε, it is possible to fire a given projectile at a given ceramic and determine how the performance of the ceramic varies with velocity. In particular, we can examine changes in ε associated with u_1 or u_2.

In conducting these experiments, it was desirable to use well known or standard materials. The ceramic, AD85, has been well characterized by the authors and others (Bless, et al., 1985; Yaziv, 1985; Rosenberg and Yeshurun, 1985; Rosenberg, 1985). The HEL of this material is 60 kbar. We measured the high strain rate flow stress of 6061-T6 aluminum using a split Hopkinson bar (Bless, et al., 1983); it was nearly constant at 4 kbar.

A variety of projectiles were used in this study. The steel projectiles used were standard Armor piercing bullets -- .30 caliber APM2 copper jacketed steel. This steel has a hardness of RC55. The core mass is 5.3 g, and the entire projectile mass is 10.3 g. The tungsten projectiles used were made of Kennametal W2 WNiFe materials. They were machined to the same shape and dimensions as the APM2 cores, and their mass was 10.7 g. The yield strength of W2 is 8 kbar. Aluminum projectiles were also prepared that had the same shape as the APM2 bullet. These had a mass of 11.3 g. Both 6061-T651 and 7075-T6 alloys were used. Fragment simulating projectiles (FSP) used were standard Army "chisel nose" FSPs, as described, for example, by Mascianica (1980). They were 12.7 mm diameter, of mass 13.4 g, and made from steel having a hardness RC30. They were fired unspun with separating sabots. The tantalum rods used were length to diameter ratio of five. The diameter was 4.9 mm. The flow stress was measured on a split Hopkinson bar; the peak value was 9 kbar at 4 percent strain.

The projectiles were launched on various impact ranges available at the Impact Physics Laboratory of the University of Dayton Research Institute. Depending on velocity and projectile, launchers were a 7.62 mm rifled barrel, a 20 mm smooth bore barrel, or a 50 mm/20 mm two-stage light-gas gun. Velocity was measured by interruption of light sources. Yaw was checked in high velocity shots with either flash x-rays or an orthogonal Hall camera.

CALIBRATION

In order to calculate ε, it was necessary to determine the reference penetration for the various projectiles into 6061-T651 aluminum. The data are presented in Fig. 2 for tantalum rods and Fig. 3 for FSPs. Some of the FSP data were generated with 7.62 mm diameter projectiles; hence the data are shown scaled by diameter. The reference penetration depth for the W2 projectile was 115 mm at an impact velocity of 1.06 km/s. The penetration of the 7.62 mm APM2 was 43 mm at 0.86 km/s and 73 mm at 1.63 km/s. The penetration was estimated to be 91 mm at 2.3 km/s. The penetration depth for the aluminum version of the APM2 was 30 mm at 1.45 km/s and 51.5 mm at 2.72 km/s.

EXPERIMENTAL RESULTS

The results of the experiments to date are listed in Table 2 and graphically displayed in Fig. 4. As indicated in Table 2, some of the hypervelocity shots with Ta projectiles employed aluminum cover plates over the ceramics.

The data from bullets made from steel or W2 show that the efficiency of the ceramic is near maximum at conventional ordnance velocities. However, the efficiency decreases markedly as the impact velocity exceeds u_2. This is expected according to the hypothesis of penetration by yielding of ceramic ahead of the projectile.

The data for Ta projectiles show that the ceramic is never effective against this threat. In order to determine whether or not shock damage due to the blunt end of this projectile was causing shock degradation, a Ta rod with a 45 degree cone was tested in shot 6-967. However, this projectile also easily penetrated the ceramic. At the highest velocities, $\varepsilon < 0$, indicating that the ceramic is less effective then a similar areal density layer of aluminum.

The FSPs can also be considered a blunt projectile, since the flat face of these projectiles measures approximately 5 x 13 mm, and the ceramic tile is only 9.14 mm thick. The ceramic exhibited poor efficiency against this projectile at 1.01 and 1.66 km/s.

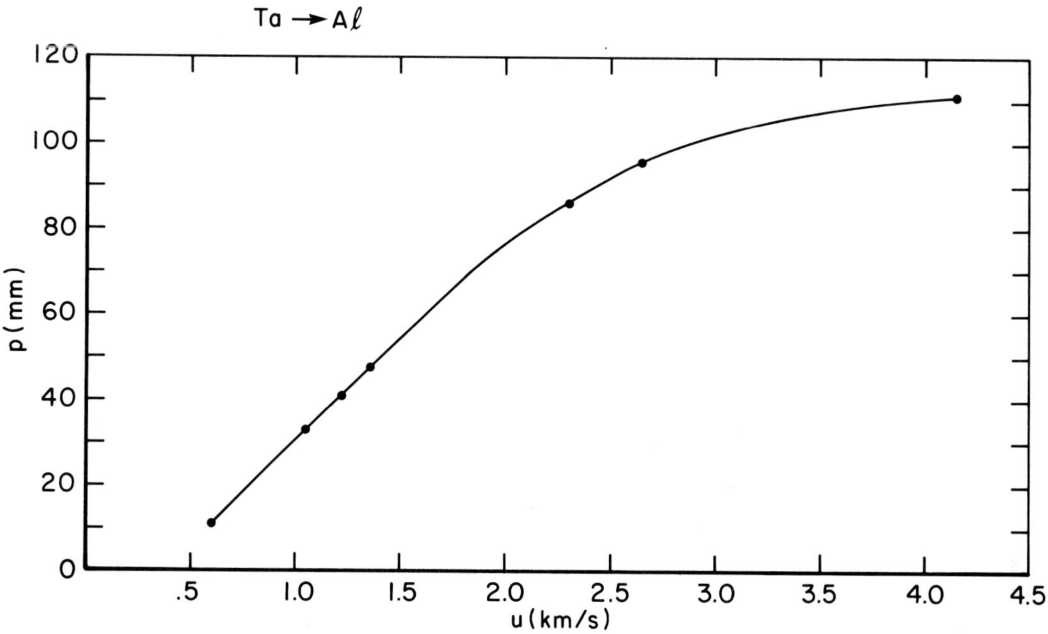

Fig. 2. Penetration of 8 g L/D = 5 Ta rods into 6061-T651 aluminum.

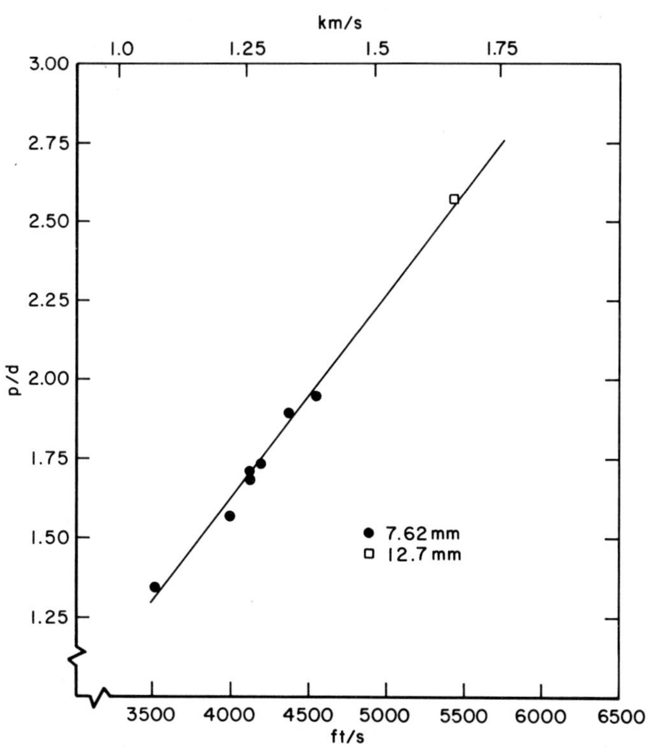

Fig. 3. Penetration scaled by projectile diameter for FSPs into 6061-T651 aluminum.

Table 2 Experimental Results

S.N.	Projectile	u_0 km/s	c mm	r mm	x mm	ε
4-324	8 g Ta rod	2.55	9.3	96*	93	-0.15
4-325	8 g Ta rod	1.96	9.3	84*	74	-0.27
6-777	8 g Ta rod	1.35	9.3	36	48	0.01
6-778	8 g Ta rod	0.61	6.3	4.8	11	-0.16
6-967	7.6 g Ta sharp rod	0.70	9.19	5.3	16	-0.05
6-958	12.7 mm FSP	1.66	12.7	14.8	33	0.07
6-969	12.7 mm FSP	1.01	9.14	4.8	17.5	0.07
4-452	7.62 mm APM2	2.30	9.14	64	91	0.17
9-843	7.62 mm APM2	0.84	9.14	1	48	0.74
10-452	7.62 mm APM2	0.78	6.35	1	48	0.81
9-1117	7.62 mm APM2	0.94	9.14	1	51	0.75
6-832	7.62 mm W2 APM2	0.88	9.14	4	115	0.87
6-966	7.62 mm APM2	1.64	9.14	30	73	0.46
4-579	6061-T6 APM2	2.70	9.14	31.3	51.5	0.17
4-596	7075-T6 APM2	2.80	9.14	39.4	53	0.05
6-1122	7075-T6 APM2	1.47	9.14	3	31	0.85

* Includes 12.7 mm cover plate

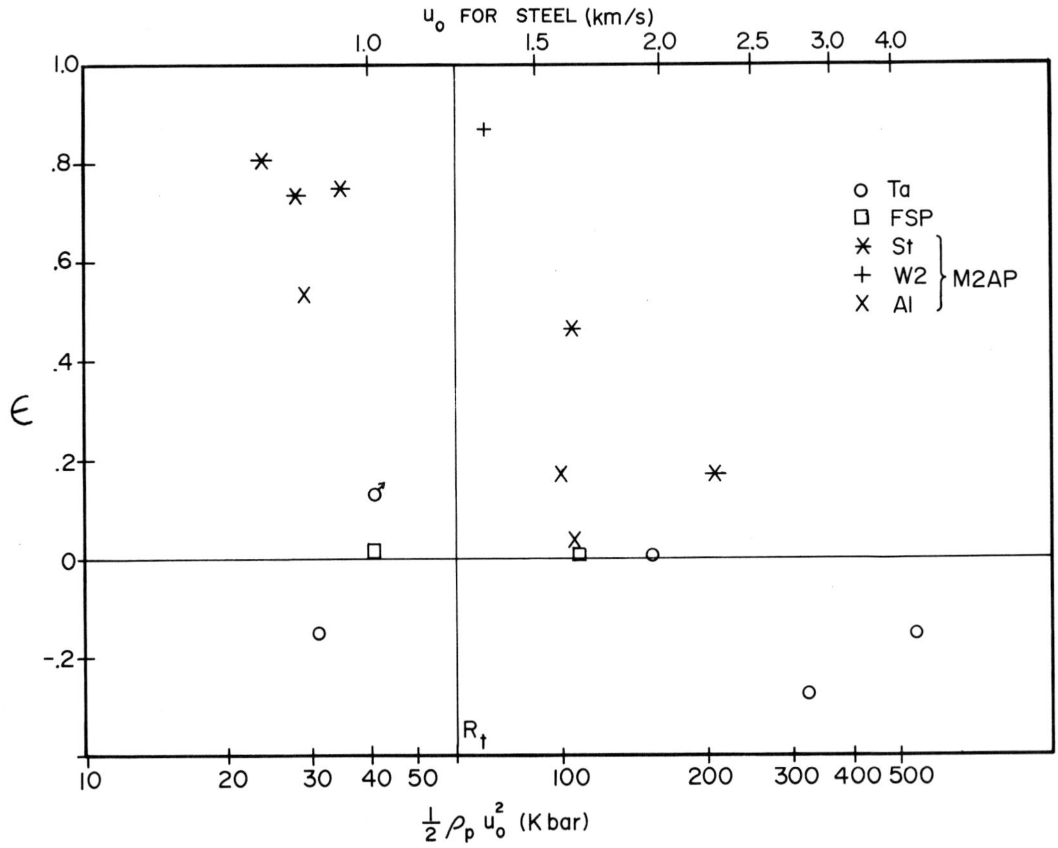

Fig. 4. Efficiency of AD85 ceramic for various projectiles. (Arrow indicates pointed rod projectile.

CONCLUSIONS

These experiments show that the penetration resistance of ceramics to conventional armor piercing bullets decreases at very high velocities. The fall off in protection is consistent with the hypothesis of yielding of the ceramic ahead of the projectile. However, the decrease in ceramic effectiveness is only gradual as the critical velocity is exceeded. If these results can be generalized, and if we ignore the strength of the projectile material, we can say that ceramics are efficient armor materials when

$$\text{HEL} > \rho u_0^2/2 \tag{5}$$

This requirement is an addition to the requirement given by Wilkins, et al. (1970), that $R_t > Y$.

By inference from Fig. 4 and by analogy with metals, we might further speculate that by the time that $u_0 = 2u_2$, penetration of the ceramic will be essentially hydrodynamic, and its value as armor may be less satisfactory. For example, steel projectiles fired against B_4C might be able to penetrate hydrodynamically by 3.7 km/s. Therefore, investigations should be continued at hypervelocities using these materials. Future studies should include the effects of confinement which have not been addressed in the current work.

The inefficiency of the ceramic in defeating the FSP could be due to shock degradation. This projectile is relatively blunt. The shot at 1 km/s was at a D/T ratio of 1.4; thus, essentially all of the ceramic experienced a shock above the HEL. Note that Wilkins et al. (1971) also found that ceramics were less effective against blunt than against sharp projectiles.

Our results also show that ceramics are ineffective against long high-density ductile rods, such as tantalum, even at relatively low velocities. Details of the mechanism by which such rods penetrate ceramic materials have not been reported.

ACKNOWLEDGEMENTS

This work was sponsored in part by the U.S. Army Armament Research and Development Command under Contract No. DAAK10-85-C-0036. The Project Monitor was Dr. Harry Fair. High strain rate tests were performed by Dr. A.M. Rajendran. A.J. Piekutowski, K. Okajima, and S.J. Hanchak aided in the ballistic tests.

REFERENCES

Bless, S.J., D. Jurick, and B. Yoon (1986). Secondary penetration of rod projectiles. 9th Int'l Ballistics Symp. April.

Bless, S.J., D. Yaziv, and Z. Rosenberg (1985). Spall zones in polycrystalline ceramics. Proceedings of. 4th APS Conference Shock Waves in Condensed Matter. Spokane, WA. 22-25 July 1985.

Bless, S.J., A.M. Rajendran, D. Dawicke, and D.L. Paisley (1983). High strain rate characterization of materials. AFWAL-TR-83-4038, August.

Gust, W.H. and E.B. Royce (1971). J. Appl. Phys., 42, 274.

Mascianica, F.S. (1980). In R. Laible (Ed.), Ballistic Materials and Penetration Mechanics. Elsevier. Ballistic testing methodology.

Rosenberg, Z. (1985). Dynamic uniaxial stress experiments on alumina with in-material manganin gauges. J. Appl. Phys., 57, 5087-5088.

Rosenberg, Z. and Y. Yeshurun (1985). Determination of the dynamic response of AD-85 alumina with in-material manganin gages. J. Appl. Phys., 58, 3077-3080.

Tate, A. (1969). Further results in the theory of long rod penetration. J. Mech. Phys. Solids, 17, 141.

Wilkins, M.L., C.F. Cline, C.A. Honodel (ca 1970). Fourth progress report of light armor program. UDRL-50694.

Wilkins, M.L., R.L. Landingham, and C.A. Honodel (1971). Fifth Progress Report on Light Armor Program. UCRL-50980.

Yaziv, D., G. Rosenberg, and Y. Partom (1986). Differential ballistic efficiency of applique armors. 9th Int'l Ballistics Symp. April.

Yaziv, D. (1985). Shock fracture and recompaction of ceramics. Ph.D. Thesis. University of Dayton.

Int. J. Impact Engng Vol. 5, pp. 173–180, 1987
Printed in Great Britain

0734-743X/87 $3.00 + 0.00
Pergamon Journals Ltd.

SHOCK-INDUCED SOLID-STATE CHEMICAL REACTIVITY STUDIES USING TIME-RESOLVED RADIATION PYROMETRY

M. B. Boslough

Sandia National Laboratories
Albuquerque, New Mexico

ABSTRACT

Time-resolved radiation pyrometry has been used to study materials which undergo solid-state chemical reactions due to shock loading. Shock-induced chemical reactivity in solids is fundamentally different than that in high explosives and other energetic materials because, if no volatiles are present, the reaction products end up in the condensed, rather than the vapor, state. Bulk property changes accompanying the solid-state reactions may therefore be too small to be observable with wave profile or shock-velocity measurements. However, some solid-state reactions, such as that between metallic nickel and aluminum, are exothermic enough to give rise to a measurable increase in temperature, so pyrometry can be used to detect the reactions. Unfortunately, these measurements are complicated by the large temperature increases generated by other sources. Possible mechanisms for generation of these high temperatures, and their effect on the chemical reaction, are suggested.

INTRODUCTION

The triggering of solid state chemical reactions by the passage of strong shock waves has been the subject of many studies. There are two main reasons for the increasing interest in shock-induced solid state reactions: 1) By studying chemical effects, our understanding of the fundamental science of the shock process should improve; 2) Shock-chemistry holds the potential to generate new or radically modified materials which might have physically interesting or technologically useful properties. Several recent studies have made use of controlled explosive loading fixtures (Graham and Webb, 1984, 1986) to shock metal and inorganic powder mixtures to well-defined pressure and mean-bulk temperature states. Recovered samples have been subjected to sophisticated analysis such as scanning electron microscopy (SEM), transmission electron microscopy (TEM), energy dispersive spectroscopy (EDS), differential thermal analysis (DTA), Mössbauer spectroscopy, x-ray diffraction and magnetometry. This post-shock analysis has demonstrated that, for example, nickel aluminides and titanium aluminides can be generated by shocking mechanical and composite mixtures of their respective component metals (Horie, Graham and Simonsen, 1985; Pak, Horie and Graham, 1986; Horie, Hoy, Simonsen, Graham, and Morosin, 1986). The issue which these studies do not address is: at what stage of the process does the chemistry occur? In order to answer this question, "real-time" measurements must be made while the sample is undergoing the shock loading and unloading process.

Radiation pyrometry is a real-time technique which lends itself to shock-induced chemical reaction studies by measuring temperature. Chemical reactions, like polymorphic phase transformations, can involve large endothermic or exothermic energies which can manifest themselves in significant, measurable temperature changes. Despite the large temperature change, the bulk properties (such as density and sound speed) of the product (or final phase) may differ little from those of the reactants (or initial phase). These bulk properties may therefore be difficult to measure with methods such as wave profile or shock-velocity experiments, and the chemical reaction or phase transformation will go undetected. This was found, for example, to be the case for the melting of stishovite by Lyzenga, Ahrens and Mitchell (1983), who observed a large temperature drop along the Hugoniot of quartz in a region with no obvious bulk property changes.

Shock induced chemical reactions do not, in general, have only a small effect on bulk properties. An extreme example of chemical reaction with large bulk property change would be high explosive detonation, in which a change of physical state (solid to vapor) accompanies the chemical reaction. This type of reaction is obviously observable with bulk property measurements. We are concerned, however, with solid state chemical reactions in which both the reactant and product are condensed matter, as in the case of the intermetallic reactions as observed in recovery experiments.

Boslough and Graham (1985) and Boslough, Graham and Webb (1986) were the first to apply radiation pyrometry to the study of the intermetallic nickel-aluminum reaction. They observed extremely large values of spectral radiance for visible and near-infrared light emitted from shocked compacts of nickel-aluminum powder mixtures. The high temperatures implied by these spectral radiance values were interpreted as being due to the large exothermic energy release of the shock-induced intermetallic reaction previously observed in recovery experiments. There were several potential problems with that interpretation, however, and the present study was intended to resolve them.

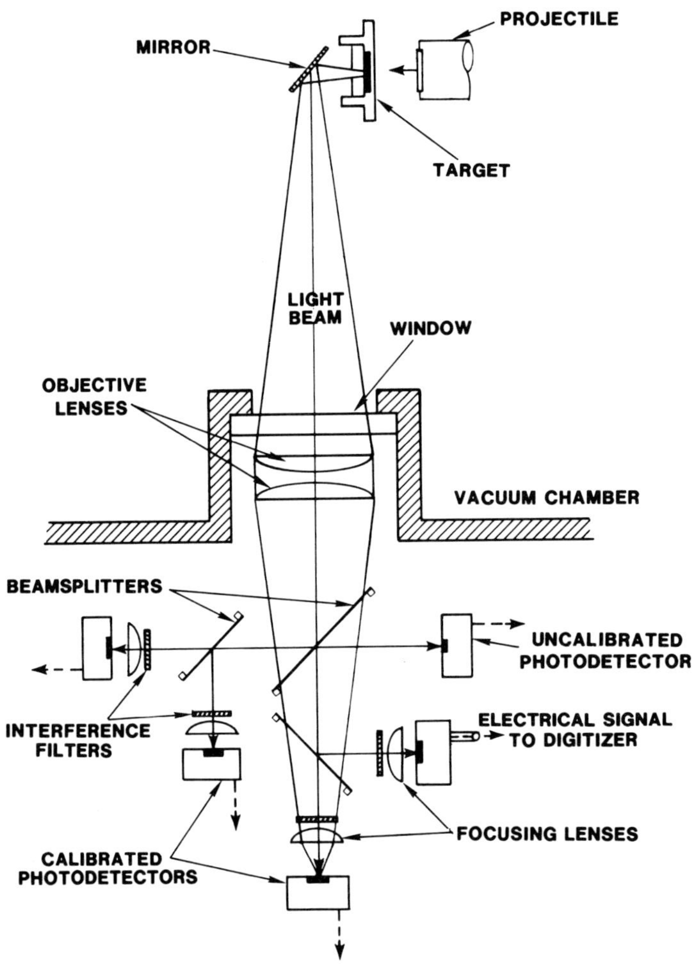

Fig. 1. Schematic diagram of radiation pyrometer and its relation to gas-gun impact chamber.

In the earlier experiments, the targets were modified explosive loading fixtures. That configuration was useful for comparing time-resolved data to the results of the recovery experiments, but the targets were not designed to constrain the observed region to one-dimensional flow as is usually the case for real-time shock wave experiments. Thus, at any given time a large range of material states (pressures and temperatures) were observed by the pyrometer. The distribution of temperatures complicates the interpretation of the data from the pyrometer, which is ideally intended to observe only one temperature at a given instant. The distribution in pressures makes it impossible to assign an observation to a particular shock state. Also, there was no way of precisely controlling the gas present in the pores of the sample compact. Spurious light from trapped air undergoing shock, or from oxidation reaction was therefore a concern. In addition, there was some evidence for light emission from the shocked sapphire window, which could lead to problems. Finally, there was no test in the earlier studies of the possibility that the high temperatures were due to physical processes, such as localized heating due to plastic deformation or pore collapse.

EXPERIMENTAL

The present experiments were carried out with a radiation pyrometer of the type described by Boslough and Ahrens (1986) and depicted schematically in Fig. 1. Absolute light intensity (spectral radiance) is simultaneously measured as a function of time at four wavelengths between .5 and 1.8 μm in the visible and near-infrared using large aperture silicon and germanium photodiodes. The signals are recorded with an array of Tektronix 7612 digitizers with a sampling rate of 5 ns. The pyrometer is calibrated prior to each experiment with a tungsten filament spectral irradiance source. By using large apertures at the target (up to 2.54 cm) and large bandwidth interference filters (up to 0.2 μm half-height full-width), the sensitivity of the system is maximized. A least squares fit to these data determines a time-resolved temperature and effective emissivity (Boslough and Ahrens, 1986). A fifth photodiode provides a faster time response to ensure that any high frequency component of light is recorded. It is a silicon avalanche photodiode with a nominal rise time of 2 ns and its signal is recorded using a Tektronix 7912 digital oscilloscope. It receives unfiltered, uncalibrated stray reflected light, and its output is not used in the temperature calculation.

In order to circumvent the problem of two-dimensional flow, a new target fixture has been designed which makes use of a very small sample-thickness to diameter ratio (Fig. 2). This

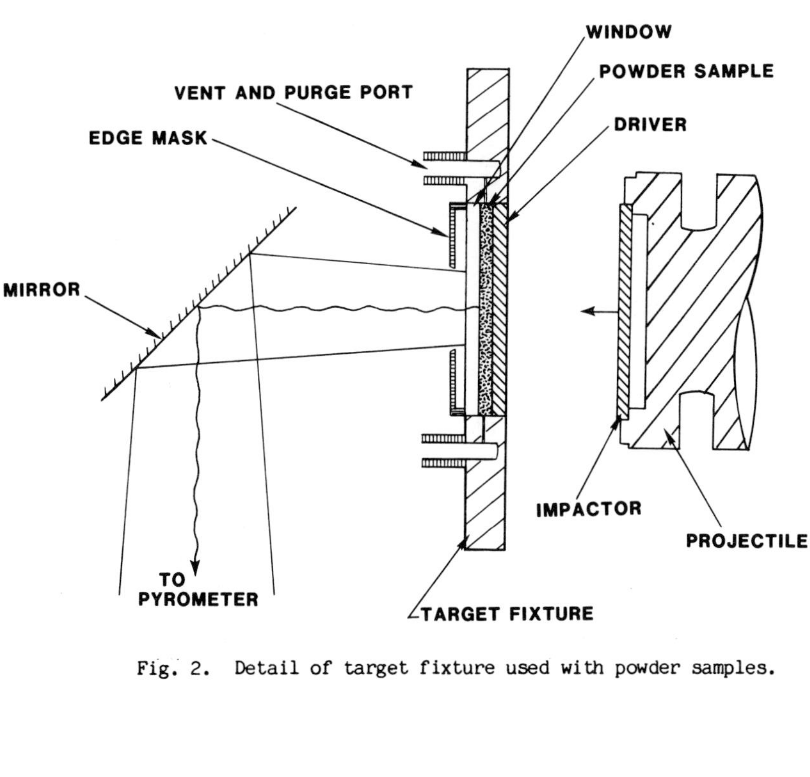

Fig. 2. Detail of target fixture used with powder samples.

fixture also has ports through which purging gas can be flowed and/or air can be evacuated. An upper bound on the amount of light emitted from the sapphire window was obtained by carrying out an experiment on alumina powder. Shock experiments were carried out on pure nickel powder in order to measure temperature increases due to sources other than chemical reactions between Ni and Al powders.

Three experiments on three different powders will be presented here. The powders consisted of: 1) a mechanical mixture of nickel and aluminum, 70% Ni and 30% Al by volume pressed to an initial density of 4.25 g/cm^3. 2) pure nickel at 4.99 g/cm^3 and 3) aluminum oxide at 2.36 g/cm^3. Powders were CERAC N-1095 nickel with particle sizes between 44 and 74 μm, CERAC A-1189 aluminum from 5 to 15 μm, and spherical alumina from 20 to 60 μm. In all three experiments, the impact conditions are the same: gas gun-launched copper flyer plates impact copper drivers at nominal velocities of 1 km/sec. The shock wave passes through the driver and into the powder sample. Peak pressures vary somewhat among the three experiments, as the powders have various shock impedances.

RESULTS

Time-resolved data were obtained at four wavelengths for all three experiments. Data for experiment 2073 (nickel-aluminum) are presented in Fig. 3. The shock wave arrives at the powder-window interface about 2.6 microseconds after the trigger signal. Light that is detected prior to this time is presumably generated within the powder and scattered by multiple reflections through the interconnected pore space ("shine-through"). Release of pressure due to the arrival of the rarefaction wave from the free surface of the window occurs about 0.5 microseconds later. The edge mask aperture insured that the powder was under uniaxial strain during this entire time interval. Air pressure in the target chamber was measured to be less than 2.6 Pa just prior to the experiment. The spectral radiances measured

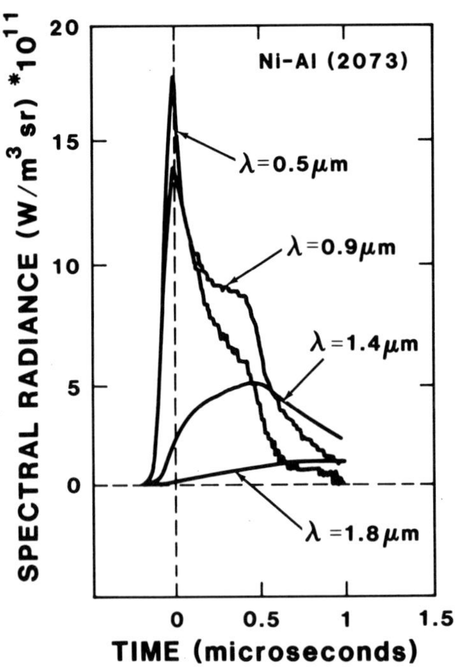

Fig. 3. Time-resolved spectral radiance data from nickel-aluminum experiment 2073 at four wavelengths. Shock wave reaches powder-window interface at time t=0.

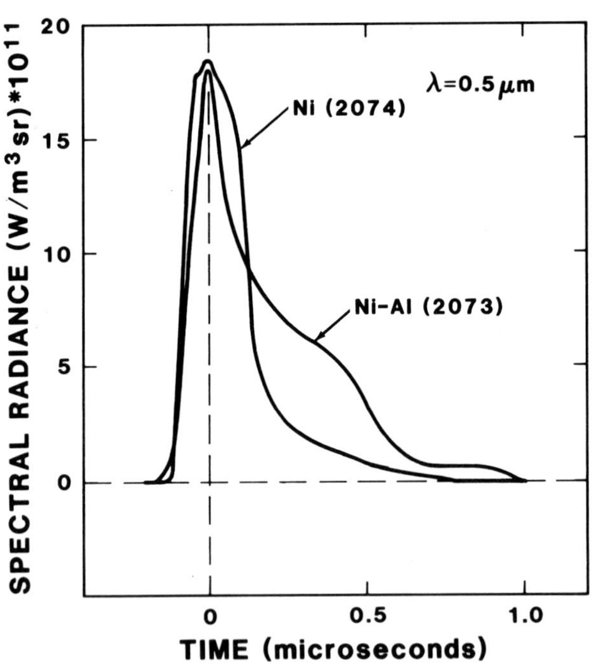

Fig. 4. Time-resolved spectral radiance data at 0.5 μm from nickel aluminum (2073) and nickel (2074) experiments.

in this experiment are similar in magnitude to these measured by Boslough and Graham (1985) in the earlier experiments on this powder mixture. However, single well-defined states are observed in the present experiment due to the uniaxial strain condition. Furthermore, the fact that the sample was in contact with the well-evacuated target chamber demonstrates that the emission of light is not due to air in the system.

Spectral radiances for 0.5 μm light radiated from various shocked powders are compared in Fig. 4. Data from the nickel-aluminum mixture are shown along with data from the shocked end-member nickel (experiment 2074). Peak spectral radiances are almost precisely the same for both nickel and nickel-aluminum at 0.5 μm. This would be expected if the early behavior is dominated by physical, rather than chemical processes, as the mixture is about 87% nickel by mass. Subsequent behavior, however, shows a difference between the powders; the brightness of the mixture decays more rapidly at early times, and more slowly at later times than the brightness of the pure nickel. Time-resolved temperatures determined from the spectral radiances are depicted in Fig. 5.

The experiment on alumina powder was intended to provide a measure of the maximum amount of light radiated from the sapphire window under the loading conditions of these experiments. Experiments on sapphire by Barker and Hollenbach (1970) suggest that its use as an optical window is questionable in this pressure range, while Cremers, Marston and Duvall (1979) measured significant light emission from sapphire. Alumina powder was used in this experiment because it has the same composition as sapphire, and its shock impedence happens to be very similar to that of nickel. Powder was used to simulate the physics of the other experiments, in which the shock arrives at the sapphire window through powder. At 0.5 μm, Fig. 6 shows that the light emission from the alumina experiment is significantly less than that from the

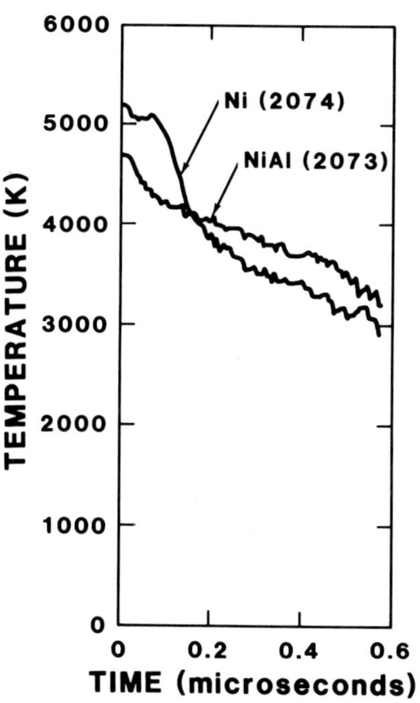

Fig. 5. Time-resolved temperatures for two experiments determined from measured spectral radiances. As implied by increasing "noisiness", error increases with time due to decreasing spectral radiance signals.

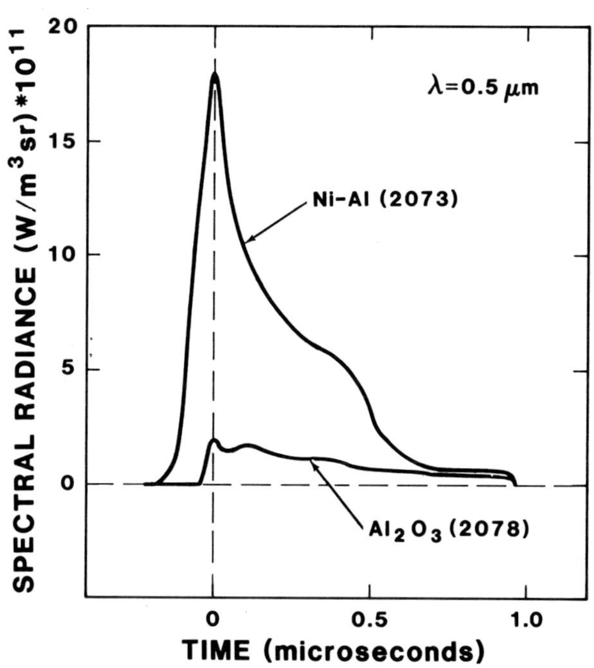

Fig. 6. Spectral radiance at 0.5 μm of alumina powder compared to that of the nickel aluminum mixture.

nickel-aluminum mixture, at least while the powders remain shock-loaded for the first 0.5 µs. At later times, light emission from the alumina appears to become relatively important, because of the drop in brightness of the nickel-aluminum. The measured spectral radiance is, however, an upper bound on that radiated from the sapphire window, as it was impossible from the geometry of the experiment to determine what fractions were radiated from the window, the powder, and the driver-powder interface. It is conceivable that the fraction radiated from the window itself is small.

DISCUSSION

Our understanding of shock-chemistry can be increased from experiments such as these in two ways: 1) something can be learned of the conditions required for chemistry to occur by carrying out real-time experiments, and 2) the reaction itself may be detectable in real time. This will allow models of shock-chemistry processes (e.g. Horie, 1986) to be well grounded in experimental evidence.

Results of the present experiments can be used in the first way, and possibly in the second. The extremely high initial temperature present in the nickel-aluminum mixture surpasses the melting points of both metals, and even their vaporization points at one atmosphere. There is no way of determining the mass fraction of hot material from present experiments. However, it can be crudely estimated by equating the thermal energy gain in a mass at the measured temperature to the Rankine-Hugoniot energy. These energy balance considerations require the mass to be small (about 17%). These small quantities of very hot material can aid the diffusion and mixing rates required for the chemistry to occur on such fast time scales.

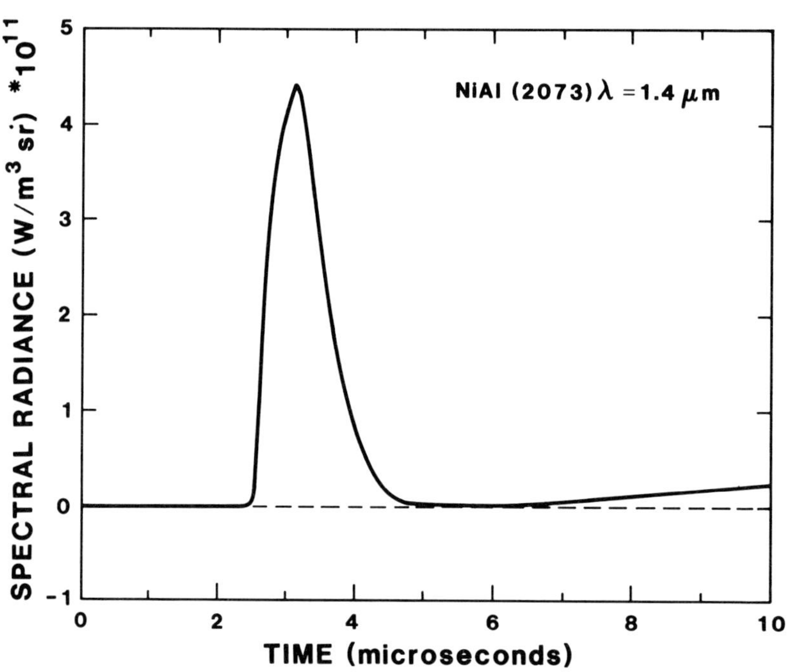

Fig. 7. Data at 1.4 µm for the nickel-aluminum experiment showing a long-term increase in spectral radiance. Absolute time scale is arbitrary.

Use of the present data as a measurement of the reaction in real time is not as straightforward. It is impossible by measuring absolute temperatures of heterogeously heated shocked materials to determine whether or not chemistry has taken place because physical processes such as pore collapse can acheive temperatures high enough to dominate the light emission. However, temperature gains due to exothermic reaction energy can be seen under certain circumstances. First, by comparing the temperature history of a reactive powder to that of an inert powder with the same physical properties, the temperature difference can be explained in terms of chemistry. The nickel-aluminum (reactive) mixture differs from the pure nickel (inert) in the present experiments only by the addition of about 13 weight percent aluminum. The initial shock temperature in the pure nickel is somewhat higher than that of the mixture, possibly because the average void is bigger in the nickel powder (aluminum grains are smaller and tend to fill the nickel voids in the mixture). The initial cooling rate is higher in the mixture. This would be expected, as aluminum has a higher thermal diffusivity than nickel. However, at later times, the temperature of the mixture is higher than that of the pure metal. This is suggestive of exothermic chemistry, but further and more precise experiments need to be undertaken to be certain.

A second way in which exothermic reaction temperatures might be detected is to make observations at longer times. If enough time has elapsed so that the temperature distribution has become uniform, the temperature of a compressed inert material becomes equal to the mean-bulk (continuum) shock temperature. On release, the temperature decreases adiabatically to a residual post-shock temperature. Both of these temperatures can be calculated in a straightforward fashion, and if two materials have the same physical properties, these temperatures should be the same. Again, any differences in temperature can be attributed to chemistry. If the exothermic reaction continues after the temperature distribution becomes homogeneous, the temperature will begin to increase. Fig. 7 shows behavior which suggests this possibility. However, this experiment was not designed for observation times longer the 1 μsec, so the state of the powder, and of the sapphire window at late times are very poorly characterized; flow is two-dimensional, fracture of the window may be taking place, and so forth. A possible method of overcoming these late-time problems would be to use much finer-grained powders. Because of smaller and more numerous pores, initial hot-spot temperatures would be lower, and would equilibrate much faster to a uniform distribution. By the same token the powders would probably react more quickly as the components would not have as far to diffuse. By using powders with sufficiently small grain sizes, the observations of interest could all be made before two dimensional effects take over.

CONCLUSIONS

Observations of light emitted from shocked metal powders demonstrate that extremely high temperatures exist for brief periods of time. It may be these high temperatures which trigger shock-induced chemical reactions in reactive powders, by providing a mechanism for mass diffusion. Subsequent temperature histories show some evidence for heating due to chemical reaction on a submicrosecond timescale. Further experiments will be carried out in an attempt to isolate and quantify this shock-induced reaction in real time.

ACKNOWLEDGEMENTS

This work was performed at Sandia National Laboratories under U.S. Department of Energy under contract #DE-AC04-DP00789

REFERENCES

Barker, L. M., and R. E. Hollenbach (1970). Shock-wave studies of PMMA, fused silica, and sapphire. J. Appl. Phys., 41, 4208-4226.
Boslough, M. B., and T. J. Ahrens (1986). A sensitive time-resolved radiation pyrometer for shock-temperature measurements. J. Phys. E., submitted.
Boslough, M. B., and R. A. Graham (1985). Submicrosecond shock-induced chemical reactions in solids: first real-time observations. Chem. Phys. Letters, 121, 446-452.
Boslough, M. B., R. A. Graham, and D. M. Webb (1986). Optical measurements of shock-induced chemical reactions in mixed aluminum-nickel powder. In Y. M. Gupta (Ed.), Shock Waves in Condensed Matter - 1985. Plenum Press, NY. pp 767-772.
Cremers, D. A., P. L. Marston, and G. E. Duvall (1980). Sources of light emission during shock experiments. High Temperatures-High Pressures, 12, 109-112.
Graham, R. A., and D. M. Webb (1984). Fixtures for controlled explosive loading and preservation of powder samples. In J. R. Asay, R. A. Graham, and G. K. Straub (Eds.), Shock Waves in Condensed Matter - 1983. North Holland Press, New York. pp 211-214.
Graham, R. A., and D. M. Webb. (1986). Shock-induced temperature distributions in powder compact recovery fixtures. In Y. M. Gupta (Ed.), Shock Waves in Condensed Matter - 1985. Plenum Press, NY. pp 831-836.
Horie, Y., (1986). Personal communication.

Horie, Y., R. A. Graham, and I. K. Simonsen (1985). Synthesis of nickel aluminides under high-pressure shock loading. Mat. Letters, 3, 354-359.
Horie, Y., D.E.P. Hoy, I. K. Simonsen, R. A. Graham, and B. Morosin (1986). Shock-wave synthesis of titanium aluminides. In Y. M. Gupta (Ed.), Shock Waves in Condensed Matter - 1985, Plenum Press, NY, 761-766.
Lyzenga, G. A., T. J. Ahrens, and A. C. Mitchell (1983). Shock temperatures of SiO_2 and their geographical implications. J. Geo. Phys. Res., 88, 2431-2444.
Pak, H. R., Y. Horie, and R. A. Graham (1986). Synthesis of nickel-aluminum alloys by shock compression of composite particles. In Y. M. Gupta (Ed.), Shock Waves in Condensed Matter - 1985, Plenum Press, NY, 749-754.

Int. J. Impact Engng Vol. 5, pp. 181–203, 1987
Printed in Great Britain

0734-743X/87 $3.00 + 0.00
Pergamon Journals Ltd.

DEVELOPMENT OF THE HIGH-VELOCITY GAS-DYNAMICS GUN

A. C. CHARTERS

General Research Corporation
Santa Barbara, California

ABSTRACT

The subjects of this paper are the historical overview and development of the high-velocity gas-dynamics gun. These are guns that derive their energy from a reservoir of compressed gas. Other guns derive their energy from electricity or from high explosive. Their historical overviews and developments are covered in papers by Mr. William Weldon and Mr. Alex Wenzel.

The gas dynamics gun is viewed first from the standpoint of modern technology. An idealized configuration, the "Reference Gun", is analysed in order to quantify the effects of gun diameter and length, projectile mass, and propellant gas pressure and composition. The analysis assumes that the propellant is an ideal gas, and formulae are derived for the base pressure and velocity of the projectile as functions of the size and loading parameters of the gun. The analysis demonstrates that the prime requirements for high velocity are a high gas pressure, a low molecular weight gas, a light projectile, and a long gun.

The history of guns is reviewed briefly from 14th century black-powder muzzle-loaders to 20th century, nitrocellulose -propellant, breech-loaded guns. The velocity limit of the modern gun is shown to be around 3 km/s, if the gun is loaded with nitrocellulose propellant and is very long (200 calibers). However, if the gun is loaded with hydrogen and the length doubled, it is shown that the velocity limit can be increased to 7 km/s, thus approaching current needs.

The problem of using hydrogen has been solved by the invention of the piston-compression light-gas gun (PCLGG). However, the limited strength of the fragile, sabot-model projectiles of experimental research has capped the maximum acceleration and has placed a demand on the gun's operating cycle to generate a constant pressure at the base of the projectile for the launching run. This second problem has been partially solved by the invention of a modification to the PCLGG known as the piston-compression, accelerated-reservoir, light-gas gun (PCARLGG). Both the PCLGG and the PCARLGG are described. The performance of the PCARLGG has been analyzed by a hydrocode developed for this purpose, and the results of the calculations are presented and compared with experiment.

The concept of a frictionless, adiabatic "Ideal Gun" is introduced in order to simplify the analysis of performance. It is shown that the performance of any ideal gun is given by a simple equation involving two dimensionless parameters that relate the projectile's velocity to its mass, its average base pressure, and the diameter and length of the gun. Based on the ideal gun equation, the maximum operating velocity of the gas-dynamics gun is estimated to be about 12 km/s.

INTRODUCTION

The development of the high velocity gun is the subject of the first session of this symposium. The choice is appropriate because the gun is central to all experiments in high velocity impact. Impact experiments are always flight tests. Either projectile or target or both must be put into motion so that the two collide at the desired velocity. There are no test techniques similar to the wind tunnel or the water tunnel that eliminate flight tests.

High velocity guns may be classified according to the source of energy that powers them; namely, mechanical, gas dynamic, electrical, and high explosive. The announced subject of this paper is "An Historical Overview of Hypervelocity Launcher Technology." Actually, the "Historical Overview" will be covered in three papers: that of electrically powered guns in William Weldon's paper "Development of Hypervelocity Electromagnetic Launchers"; that of explosively powered launchers in Alex Wenzel's paper "A Review of Explosive Accelerators for Hypervelocity Impact"; that of gas-dynamic guns in this paper.

Mechanical launchers and rocket launchers will not be covered. Mechanical launchers reached their apex in the fourteenth century with the English long bow (Churchill, 1956). This kinetic energy weapon, new to the Continent at that time, destroyed the power of the mounted, armored knight. No longer was "the terrible worm in the iron cocoon" master of the battlefield (Tuchman, 1978). In a certain sense, it is ironic that mechanical launchers reached their apex at the very time that gas dynamic launchers made their initial appearance. Cannon and firearms were shortly to supersede the long bow as a dominant weapon of medieval warfare.

High velocity rockets, particularly multi-stage rockets, have been essential to the development of modern, long-range weapons and to the realization of space flight (Burgess, 1961). They have been used as launchers for weapons and for range tests, but they have not been used in the laboratory because the path length required for the rocket to accelerate to high velocity is too long. The restriction to range tests limits their use in hypervelocity impact research, since most experiments are performed in the laboratory.

In choosing a title for this paper, a choice was made between calling a gas-powered "throwing engine" a "launcher" or a "gun." The Late Latin root of launcher is "lanceare", which means "to wield a lance" (Mish, 1984). The old English forms of gun are "gunne" and "gonne", which are derived from the Latin word "mangonel", described by Goddard, Herr and Thompson (1969) to be "a powerful, pre-gunpowder, stone-hurling engine of war." A "powerful ... stone-hurling engine" seemed a more appropriate antecedent of the modern "throwing engine" than a "lance", and the designation of "gun" won out.

DISCOVERY OF BLACK POWDER

The development of the gas-dynamics gun, begins with the invention of black powder, a mixture (by weight) of 75 parts of potassium nitrate, 15 parts of charcoal, and 10 parts of sulfur (Mauncy, 1949). The black powder mixture contains both oxidant and fuel. If ignited, it combusts in a rapid, exothermic reaction and generates a high temperature gas suitable for use as a gun propellant. Chronologically, this was its first application. The use of black powder as a blasting agent in mining came some 300 years later.

Some people have called black powder and guns "inventions of the devil." But there are other opinions. Anderson (1969), in his article on the history of gunpowder in the Encyclopaedia Britannica, begins: "Discovery of black powder and its adaption to firearms marks one of the most important events in the history of civilization. The blasting power of black powder enabled men to do work that had formerly required brute force, and its use to propel missiles gave civilized men ascendancy over barbarians."

Many people have laid claim to the discovery of black powder, including Chinese, Hindus, Greeks, Arabs, English (Roger Bacon), and Germans (Bertold Schwarz). The dates of its discovery range from the 10th to the 14th century. There is a statue in the town square of Freiburg in southern Germany honoring Bertold Schwarz for its discovery. Another significant comment on Bertold Schwarz's discovery is a cartoon in a German artillery officer's handbook (Fig. 1); a free translation of the caption is:

> "Berthold Schwarz, two seconds ago,
> Found powder strikes a mighty blow!"

Fig. 1. Berthold Schwartz, 13th century inventor of gunpowder.

The cartoon conveys the message that black powder is an energy source capable of delivering a large force in a very short time. A large force delivered in a short time is the requirement for the energy source of all guns, ancient and modern. It also is the requirement for the energy sources of all launchers, be they gas dynamic, electric, or high explosive.

A REPRESENTATIVE GUN

Our narrative shifts briefly from the 14th to the 20th century and presents a description of the gun as an instrument of modern technology. In principle, the gun is a simple device consisting of a long tube open at one end, the muzzle, and closed at the other, the breech (Fig. 2). The projectile is placed part way down the tube from the breech, leaving a volume, the chamber, which holds the propellant.

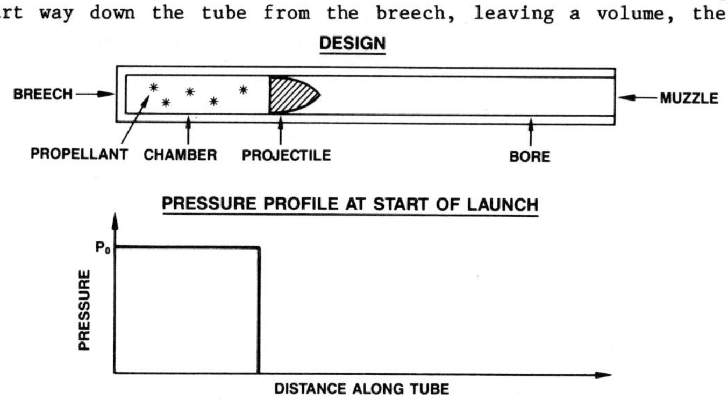

Fig. 2. Gas dynamics gun. Design of gun (sketch at top); gas pressure profile at start of launch cycle (sketch at bottom).

After ignition, the propellant combusts rapidly filling the chamber with gas at high pressure and temperature. If the combustion is very rapid so that the propellant is consumed before the projectile has moved an appreciable distance, as was the case in 14th and 15th century cannon, the chamber is filled with gas at uniform pressure at the start of launch, as shown in the bottom sketch of Fig. 2.

The gas pressure starts the projectile moving; the motion of the projectile generates an expansion wave which moves back through the gas column. The projectile accelerates moving down the bore with ever increasing velocity. The gas follows the projectile, and the pressure in the gas decreases from its value in the undisturbed reservoir to a lower value at the projectile base. The situation is shown in Fig. 3. The pressure decrease is caused in part by the increase in volume due to the projectile's movement and in part by the acceleration of the gas due to the projectile's velocity. In short, the pressure energy of the gas is expended accelerating the gas as well as the projectile.

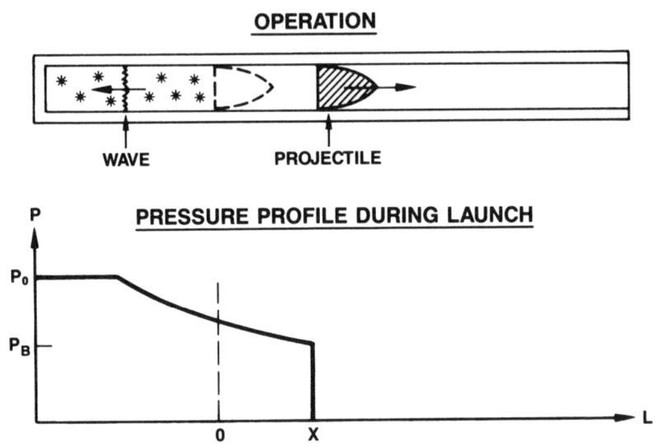

Fig. 3. Gas dynamics gun. Kinematics of gun's operation (sketch at top); gas pressure profile during launch (sketch at bottom).

In real guns, the chamber is usually larger in diameter than the bore and is short compared to the length of the gun (Hayes, 1938; Seigel, 1965). Also, the propellant is configured to burn relatively slowly in order to reduce the maximum pressure by taking advantage of the volume opened up by the motion of the projectile (Corner, 1950). The analysis of the dynamics of gas and projectile in the real gun is complex.

Fortunately, the essential features of the launch cycle are reproduced in a simple gun configuration, called herein the "Reference Gun." Its configuration is shown in the sketches of Figs. 4 and 5. Bore and chamber have the same diameter. The chamber is long enough so that the expansion wave traveling to the breech and then reflecting back through the gas column does not reach the projectile's base before the projectile is launched. Calculations for a typical case suggest that a chamber length of 1/3 the total gun length is effectively infinite, in so far as the projectile is concerned. The propellant is assumed to combust completely before the projectile moves, so that the chamber is filled with gas at uniform pressure and temperature at the start of the launch cycle. The gas is assumed to be ideal with a constant ratio of specific heats. Losses due to friction and heat transfer are neglected.

For this simple case, Seigel (1965) has developed explicit equations for the projectile's base pressure and velocity. His analysis has been recast in the present treatment in terms of the following dimensionless parameters:

$$\text{Effective projectile density} = \rho_e = M/AD \tag{1}$$

$$\text{Pressure coefficient} = C_p = P_0/[(1/2)\rho_e\, a_0^2] \tag{2}$$

All symbols used in this and subsequent analyses presented in this paper are defined in Table 1, Definition of Symbols.

The effective projectile density, ρ_e, is significant to the analysis of high-velocity guns, because many high velocity projectiles have a value of ρ_e close to unity. In addition, the pressure coefficient, C_p, as defined, reflects the conventional treatment of pressure, density, and velocity in fluid dynamic systems.

In terms of the pressure coefficient, defined by equations (1) and (2), the projectile's base pressure is given by

$$\frac{P}{P_0} = \left(1 - \frac{\dfrac{U}{2a_0}}{(\gamma - 1)} \right)^{\frac{2\gamma}{(\gamma - 1)}} \tag{3}$$

and the projectile's velocity by

$$C_p\,(X/D) = \frac{4}{(\gamma+1)}\left\{\frac{\dfrac{2}{(\gamma-1)} - \dfrac{(\gamma+1)}{(\gamma-1)}\left[1 - \dfrac{U(\gamma-1)}{2a_0}\right]}{\left[1 - \dfrac{U(\gamma-1)}{2a_0}\right]^{(\gamma+1)/(\gamma-1)}} + 1\right\} \qquad (4)$$

The ratio of specific heats of the gas is a parameter in equations (3) and (4). Equation (3) is plotted in Fig. 4 and equation (4) in Fig. 5 for $\gamma = 1.25$, approximately the value of the ratio for the gases generated by the combustion of nitrocellulose propellant. In this sense, the graphs of Figs. 4 and 5 for the Reference Gun are also representative of modern service guns.

Inspection of Fig. 5 and equation (4) shows that high velocity requires a large value of the distance parameter, and, hence, in terms of the physical variables, a high gas pressure, P_0, a low effective projectile density, ρ_e, and a long launch tube, X/D. The role of the speed of sound, a_0, is not obvious from equation (4), because it appears in the denominators of both the velocity ratio and the distance parameter, but one can infer the effect of a_0 from equation (3). For a given value of velocity, increasing the value of a_0, increases the pressure ratio and, hence, the acceleration of the projectile.

From another viewpoint, the effect of the speed of sound may be assessed by noting that this quantity is a measure of the inertia of the gas: The higher the speed of sound, the lower the inertia of the gas. The inertia of the gas affects the velocity of the projectile, because it accounts for part of the drop in pressure from the reservoir to the base of the projectile. The total pressure drop is due to (1) the additional volume opened up to the gas by the movement of the projectile and (2) the pressure gradient required to accelerate the gas. The ratio of the pressure drops caused by these two factors is a measure of the effect of the gas inertia.

A brief analysis illustrates the effect of gas inertia on the pressure drop from reservoir to projectile. The case treated is one analysed later in this paper. At this point, suffice it to note that $P_0 = 500 \times 10^6$ N/M^2 (= 5 kilobars = 72,526 psi), ρ_e = 9.5 gm/cc, and a_0 = 1,150 M/Sec, thereby giving C_p = 0.080.

As the projectile moves down the gun tube a distance X, a rarefaction wave, moving at the velocity a_0, runs back into the reservoir of gas a distance $X_0 = a_0 t$, where t is the travel time of the projectile. If there were no inertial effects, the pressure would decrease due to an adiabatic expansion from a volume of AX_0 to a volume of $A(X + X_0)$, where X is the distance traveled by the projectile. The problem is to compute t as a function of X, because equation (4) is not directly integrable in closed terms. However, the graph of Fig. 5 is closely represented by the function

$$U/a_0 = 0.845[C_p(X/D)]^{1/4} \qquad (5)$$

and equation (5) can be integrated, giving an equation for dimensionless time $\tau = ta_0/D$, namely

$$\tau = 1.58(C_p)^{-1}[C_p(X/D)]^{3/4} \qquad (6)$$

The dimensionless time is also the dimensionless distance the rarefaction wave travels into the gas reservoir; hence, equation (6) gives X_0/D in terms of the projectile's travel, X/D, and permits the calculation of the pressure drop due to increase in volume, namely

$$P/P_0 = [X_0/(X + X_0)]^{1.2} \qquad (7)$$

Consider first that the projectile has traveled a relatively short distance of 33 calibers reaching a velocity of 1.2 km/s. The ratio of reservoir pressure to projectile base pressure due to volume increase alone is 0.48. The true dynamic pressure ratio, P/P_0, is 0.25. Hence, the actual pressure drop is twice that due to volume increase alone. Consider, next that the projectile has traveled a relatively long distance of 200 calibers reaching a velocity of 2.0 km/s. Now, the pressure ratio due to volume increase alone is 0.36; the true dynamic pressure ratio is less than 0.10. Hence, the actual pressure drop is nearly 4 times that due to volume increase alone. The relative increase in pressure drop due to gas inertia shows that the kinetic energy of the gas consumes more and more of the pressure energy as the velocity increases.

Noting that the speed of sound is given by

$$a_0^2 = qRT_0/\mathfrak{m}$$

one sees that the properties of the gas needed for high velocity are high temperature and low molecular weight. In other words, the ideal propellant for a high-velocity gas dynamics launcher

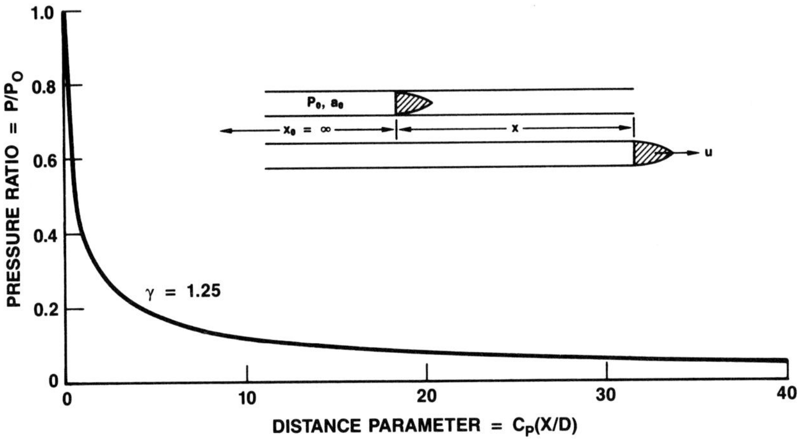

Fig. 4. Pressure profile of reference gun. Ratio of projectile base pressure to the reservoir pressure, P/P_o, as a function of the distance parameter, $C_P(X/D)$.

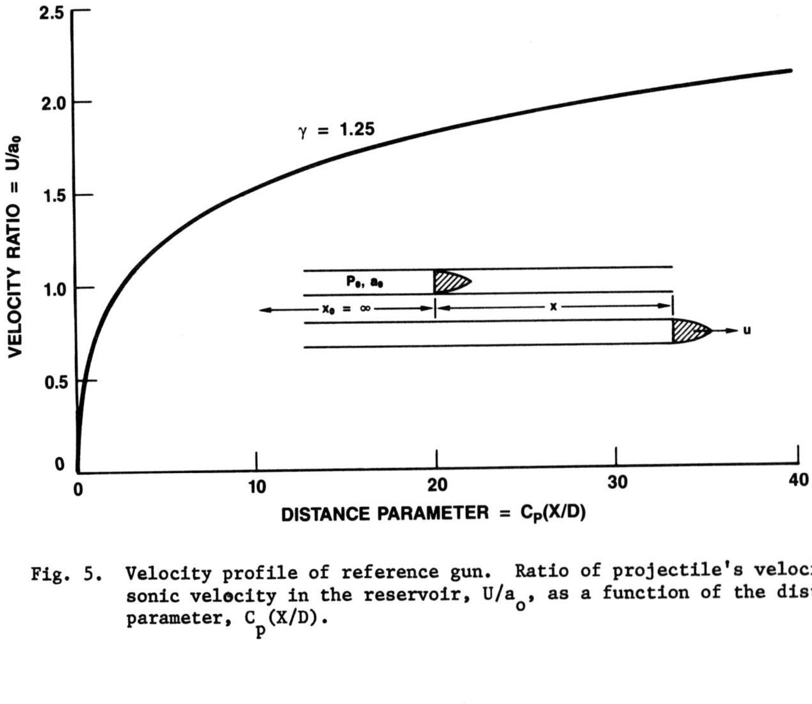

Fig. 5. Velocity profile of reference gun. Ratio of projectile's velocity to sonic velocity in the reservoir, U/a_o, as a function of the distance parameter, $C_p(X/D)$.

is a hot, light gas. Analyses similar to this have pointed the direction taken in the development of high velocity guns.

GUNS FROM 14TH TO 20TH CENTURY

Our narrative now returns to the 14th century and moves rapidly through five hundred years to the middle of the 19th century. Guns of this five hundred period were similar in design and operation, although there were many technical innovations through the years (Mauncy, 1949). The top sketch of Fig. 6 shows the basic design. The breech was permanently closed; a hole (touch-hole) was bored through the tube wall near the breech; the gun was loaded with black powder, wad, and projectile from the muzzle and the touch-hole was primed with fine-grain powder. The bore of the tube was smooth and the projectile was usually a ball or a package of shot, chain, etc. The powder charge was ignited by applying fire to the powder at the mouth of the touch-hole.

Several important advances were made during the nineteenth century (see bottom sketch of Fig. 6). Techniques for sealing a removable breech plug were developed so that guns could be loaded from the breech. The strength of gun tubes was greatly increased by the use of high-strength steel and by cold-working processes, and the tube was lengthened. The concept of spin stabilization had been discovered, the bore was rifled to impart spin to the projectile, and the projectiles were changed from round balls to the familiar shape of a pointed cylinder, four to five diameters long. Finally, nitrocellulose propellants were invented and replaced black powder with a more stable, controllable, and powerful propellant (Hayes, 1938; Goddard, Herr and Thompson, 1969).

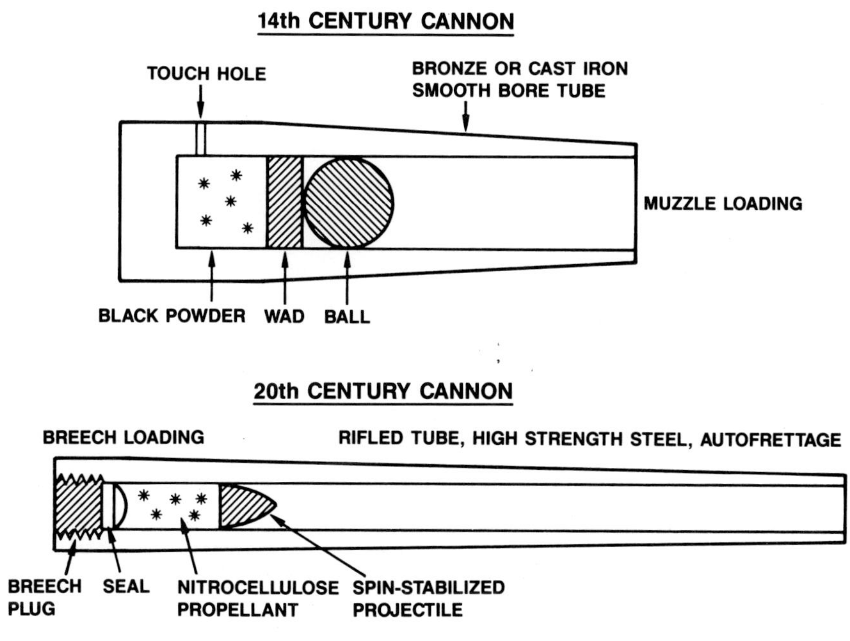

Fig. 6. Design of cannon. Top sketch illustrates cannon made during the period from the 14th to mid 19th centuries; bottom sketch illustrates he basic design of modern cannon (mid 19th through 20th century).

The performances of 16th and 20th century cannon are compared in Fig. 7. The maximum velocity increased by a factor of more than 3, the maximum throw weight by a factor 20; and the maximum kinetic energy of the projectile by a factor of 200. Three decades into the 20th century, the maximum velocity had again been increased from 0.8 km/s for military cannon to 1.4 km/s for special types of guns firing light projectiles. At that time, a maximum velocity of 1.4 km/s was considered adequate for most purposes.

CENTURY	GUN LENGTH CALIBERS	MAX PROJ wt, kg	MAX MUZZLE vel, m/sec
16th	32	50	250
20th	50	1,000	800

$(KE_{MAX}$ 20th CENT.)/(KE_{MAX} 16th CENT) = 200

Fig. 7. Comparison of characteristics of 16th and 20th century cannon.

THE DRIVE FOR HIGH VELOCITY

The advent of the high velocity, multistage rocket during and after World War II gave rise to the Inter-Continental Ballistic Missile and, shortly thereafter, to satellites and other spacecraft. These vehicles re-entered the earth's atmosphere at velocities of 7 to 8 km/s and were subject to aerodynamic heating and other phenomena caused by their flight velocity that had not been encountered before (Allen and Eggers, 1958; Allen, 1964; Martin, 1966). There was a urgent need to study these phenomena in the laboratory, and strenuous efforts were directed to improving the high velocity capabilities of both supersonic wind tunnels and aeroballistic ranges. The urgency was increased by the beginning of an anti-ICBM program employing kinetic energy weapons (KEWs) and by the realization of the meteoroid hazard to spacecraft.

An immediate result of these events was the recognition of the need to increase the velocity of laboratory launchers. Guns using nitrocellulose (NC) propellants were the launchers of that time, and their performance was similar to that of the Reference Gun (discussed previously). Analyses similar to the one presented suggested that the velocity of these guns could be increased to 2 or possibly 3 km/s but that 7 km/s was beyond their reach. The difficulty lay in the share of pressure energy required to accelerate the propellant gas. Hence, the solution appeared to be to change from the gaseous products of the combustion of NC propellants to a gas with a lower molecular weight.

Hydrogen is the gas with the lowest molecular weight. Putting aside for the moment the matter of generating a reservoir of hot, high-pressure hydrogen, one asks the question "If hydrogen were substituted for the combustion products of NC propellant, could the muzzle velocity be increased to the desired value of 7 km/s?" The following exercise with the Reference Gun provides a tentative answer.

The muzzle velocity of the Reference Gun is computed from equation (4) as a function of the equivalent projectile density for the gun loaded, first, with the combustion products of NC propellant and, second, with hydrogen. The pressures and temperatures of both gases are assumed to be the same on loading, and both gases are assumed to be "ideal." Values of the pressure and temperature on loading and of the launch tube length are chosen to be high but within practical limits. The initial values are as follows: $P_0 = 500 \times 10^6$ N/M^2 = 5 kilobars = 72,526 psi; T_0 = 3,0000K; γ = 1.25 for NC Propellant; $\mathfrak{M}$ = 23.4^0 for NC Propellant gas; γ = 1.40 for hydrogen; $\mathfrak{M}$ = 2.00 for hydrogen; L\D = 200 calibers.

The results of the calculation are shown in Fig. 8. For NC propellant, the maximum velocity of the lightest projectile (ρ_e = 1 g/cc) is just over 3 km/s. For hydrogen, the maximum velocity of the lightest projectile is 6 km/s. The dashed line above the hydrogen curve shows that increasing the length of the launch tube to 400 calibers increases the velocity to 7 km/s. Since high velocity guns of 400 caliber length have been built and operated successfully, a tentative answer to the question posed is: "Yes, a gun similar to the Reference Gun with hydrogen can probably launch a light projectile at velocity of 7 km/s."

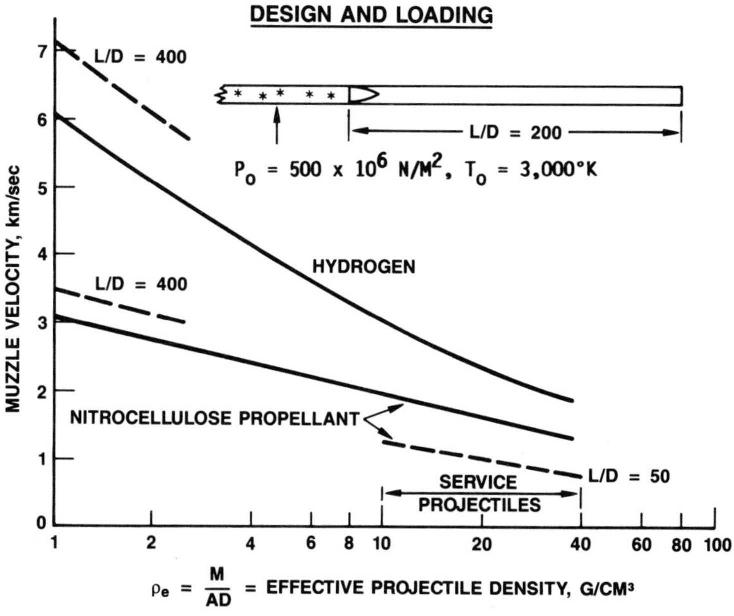

Fig. 8. Comparison of reference gun performance for the gun loaded, first, with a heavy gas gas propellant and, second, with a light gas propellant.

Figure 8 also answers two questions about the gun loaded with NC propellant. First: "Would increasing the launch tube length to 400 calibers bring about a significant increase in velocity?" The dashed line above the NC propellant curve shows that doubling the length only increases the velocity by 10%. Second: "Why do the velocities of service projectiles approach 2 km/s, while in practice they normally range between 1.0 and 1.5 km/s [the range of weights of service projectiles is marked on the abscissa (Hayes, 1938)]?" The high velocities computed for service projectiles is due to the assumed gun length of 200 calibers. Most service guns have lengths of about 50 calibers (Hayes, 1938). The velocities were recomputed for a gun length of 50 calibers and are given by the dotted curve below the NC propellant curve. The new calculations show that reducing the gun length to 50 calibers brings the velocities down to the normal range.

Two tentative conclusions may be drawn from the results of the above exercise. First, the use of hydrogen significantly increases the velocity for all projectile weights. Second, a gun similar to the Reference Gun could probably fire a light projectile at the desired velocity of 7 km/s, if a practical method could be developed for loading the gun's chamber with hot hydrogen at high pressure.

THE PISTON COMPRESSION LIGHT-GAS GUN

Hydrogen and helium are the two gases with the lowest molecular weights. To the author's knowledge, there is no compound or mixture that can generate either gas by an exothermic reaction. It seems clear that the energy for loading the gun's chamber with high-pressure, hot hydrogen or helium must come from a source external to the gas itself. The problem is to develop a technique for loading the chamber with hot, high-pressure gas.

The concept of the piston-compression light-gas gun was invented in the late 1950s by several researchers of Dr. E. J. Workman's group at the New Mexico School of Mines (NMSM) (Crozier and Hume, 1957) and has led to the design of nearly all of the light-gas guns in current use. The development of this type of light-gas gun was continued at the Ames Aeronautical Laboratory of the NACA shortly after the NMSM work terminated (Charters, Denardo and Rossow, 1957).

Figure 9 shows the Ames' design of the piston compression light-gas gun. The gun consists of two basic units, a launcher (or launch tube) and a pump. The two are connected end to end on a common axis. The port between the two is closed by a shear disk designed to open at a specified value of pressure (the actual disk is scored and ruptures by petaling, although it is called a "shear disk" from an early design). The pump is made up of a pump tube, an open-ended valve tube, and a powder chamber. The pump and valve tubes are coaxial and the powder chamber, oriented at right angles to the tube axis, joins the tubes at their junction. The diameters of the pump and valve tubes are equal and are normally greater by a factor of three or more than that of the launch tube.

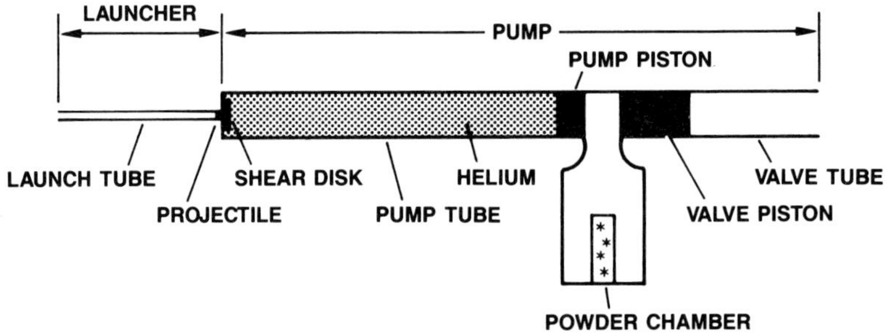

PERFORMANCE

LIGHT GAS	MAX OPER VEL
HELIUM	3 km/sec
HYDROGEN	6 km/sec

Fig. 9. Initial design of the heavy-piston, piston-compression, light-gas gun developed at the Ames Aeronautical Laboratory of the NACA. The maximum operating velocities attained with this gun using helium and hydrogen as propellants are listed in the table at the bottom.

Figure 9 also shows the gun loaded for firing. The projectile is placed at the breech of the launch tube ahead of the shear disk. Pistons are placed at the breeches of the pump and valve tubes next to the junction with the powder chamber. The pump tube is filled with a light-gas, helium or hydrogen. A charge of powder is placed in the powder chamber.

In the design presented here, pump and valve pistons are relatively heavy; that is, the powder charge and pump piston weight are adjusted so that the pump piston moves slowly enough to avoid the generation of strong shock waves but rapidly enough for the piston to have sufficient kinetic energy to compress the light-gas by its momentum alone. The compression cycle approximates an isentropic compression of the light gas, although shock waves are not completely avoided. The valve piston is made heavy enough so that the valve tube remains closed during the pump cycle.

The operating cycle is illustrated by the sketches of Fig. 10. In "Position 1", the powder has been fired, and the pistons are beginning to move. In "Position 2", the pump piston, moving down the pump tube, has compressed the light gas to the rupture pressure of the shear disk; the disk ruptures, releasing the gas into the launch tube, and starting the projectile. In "Position 3", the projectile has completed its run through the launch tube, and the pump piston has come to rest near the end of the pump tube. The valve piston has left the valve tube releasing the powder gas and clearing the way for the return of the pump piston on its rebound.

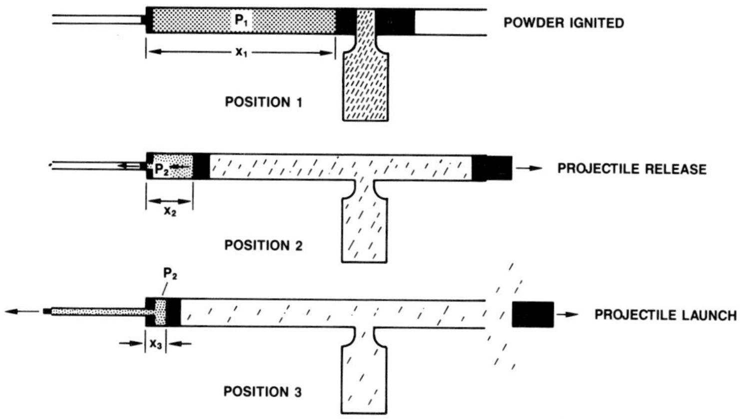

Fig. 10. Operating sequence of the initial Ames piston-compression light-gas gun. Top sketch shows the gun at the beginning of the firing cycle just after ignition of the powder in the chamber; middle sketch shows the situation when the piston has compressed the light gas to the rupture pressure of the shear disk; bottom sketch shows the situation at the time of projectile launch.

The unusual gun configuration with the powder chamber placed to one side and a open gun tube facing to the rear is a safety feature incorporated in early Ames designs. The gun was loaded so that a relatively large buffer of high-pressure light-gas remained at the front of the pump tube when the piston came to rest at the end of the pump cycle in order to prevent any possibility of the piston colliding with the end of the pump tube. The gas buffer effectively "fired" the piston back down the pump tube. The Ames design provided an escape route for the piston on its rebound. With the advent of the accelerated-reservoir light-gas gun, to be discussed shortly, the design was changed to the normal in-line powder chamber, since the piston did not rebound and an escape route was no longer needed.

The projectiles for the gun consisted of "model" and "sabot" (see Fig. 11). The "model" was a model of the vehicle being tested (sphere, cylinder, cone, etc.). The "sabot" was a structure that supported the model in the bore of the launch tube and applied the accelerating force of the propellant gas. Usually, the sabot was made from a strong plastic, such as a nylon or a polycarbonate. The sabot thus supported the model in the bore, regardless of the model's configuration, and reduced the effective projectile density, since the plastics in normal use have a density close to one g/cc.

Helium was the light gas used at first, because hydrogen was considered too dangerous. In the initial operation, saboted spheres of aluminum and steel were launched routinely at 3 km/s, twice the velocity attainable in powder guns. Thus the gun's performance was partly vindicated. But, in truth, the velocity was far from the desired value of 7 km/s, the re-entry velocity of the ICBM, and great emphasis was placed on increasing the launch velocity to at least 7 km/s and, preferably, a higher value.

The use of helium was pushed to the limit, but the velocity could not be extended much beyond 4 km/s. Serious erosion problems were encountered. The hot helium melted the bore at the breech end and deposited the molten metal along the length of the launch tube, requiring the tube to be refinished each shot. It was decided to try hydrogen as a last resort, despite the increased hazard.

Hydrogen immediately increased the gun's performance by a factor of nearly two. Saboted spheres could now be fired at 6 km/s. Simple plastic cylinders were fired at velocities of 7 km/s and above. At last, re-entry velocities were achieved by a laboratory launcher!

At the beginning of the program, "high velocity" was the prime objective. Now that high velocity had been achieved, it was discovered that high velocity was a necessary but not sufficient condition. The "sufficient condition" meant that a meaningful object must be placed in flight not only at the desired velocity but also structurally intact and oriented correctly. The meaningful object could be a KEW projectile or the model of a re-entry vehicle, depending on the test involved. Thus, new demands were made on the performance of the launcher, demands whose importance had not been recognized early on.

THE CONSTANT PRESSURE LAUNCH CYCLE

Figure 11 shows the model of a conical vehicle and its segmented sabot prior to final assembly and loading (Canning, Seiff and James, 1970). The model is representative of a projectile designed for hypervelocity aerothermodynamic tests in an aeroballistic range. It is constructed with a heavy metal nose section and a light metal base section to ballast the center of gravity forward for aerodynamic stability. The sabot is made from high-strength polycarbonate plastic. The segments of the sabot separate from the model after launch by moving laterally and are stopped in the blast chamber by terminating this section of the range with a barrier having a port on the flight line just large enough to allow the model to fly through (Charters and Curtis, 1963).

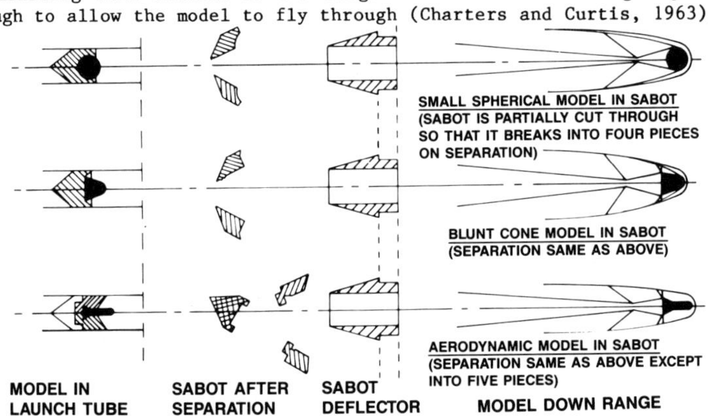

Fig. 11. Typical model-and-sabot launch and separation sequences.

The pressure of the light gas acts on the base of the sabot to accelerate the model-sabot projectile, and the inertial reactions of model and sabot generates stresses in their materials. These stresses, as well as the strength of the gun, limit the maximum base pressure (Charters, 1961). Plastics have tensile strengths running from 70 x 10^6 N/M^2 = 10,200 psi) to 170 x 10^6 N/M^2 = 24,700 psi (Department of Defense, 1975). The strength of the sabot material at critical points in its structure determines the maximum acceleration, although, in some projectiles, the strength of the model is the determining factor. The strength of the sabot material in compression is enhanced somewhat by the confinement of the gun bore, and base pressures somewhat in excess of the static limits can be tolerated without structural failure. Experience suggests that a base pressure of 150 x 10^6 N/M^2 (21,800 psi) is safe for most projectiles and that a base pressure of 200 x 10^6 N/M^2 (29,000 psi) is a good working maximum for analysis of gun performance, although some designs have tolerated pressures as high as 250 x 10^6 N/M^2 (36,300 psi).

The previous analysis of the performance of the Reference Gun assumed a maximum projectile base pressure of 500 x 10^6 N/M^2 (72,500 psi) at the start of launch, a value 2.5 times the maximum allowable working stress of the sabot material. The sabot of the projectile shown in Fig. 11 would be expected to fail if launched in the Reference Gun under the assumed loading conditions. However, if the light-gas loading pressure were lowered to 200 x 10^6 N/M^2 (29,000 psi) for a successful launch, the velocity would be reduced to 4.7 km/s, far less than the desired velocity of 7 km/s (see figure 12). The experimenter faces a dilemma: Either a damaged projectile or an unacceptably low velocity.

The dilemma is due to the pressure cycle of the launch. In the Reference Gun, the projectile's base pressure starts at the maximum value and monotonically decreases as the projectile accelerates. If the pressure cycle could be varied so that the pressure remained constant during the launch, a much lower value would suffice to produce the same launch velocity (Seigel, 1965; Charters and Curtis, 1964; Lukasiewicz, 1967).

The history of ordnance provides a cogent example of the effect of the pressure cycle. In describing the advance in artillery achieved by Gen. Thomas J. Rodman's (USA) invention of perforated grain gunpowder, Goddard, Herr and Thompson (1969) comment "... The final effect was ... to develop, in effect, a prolonged steady push against the base of the projectile rather than a sudden blow, with maximum stress much moderated." One is reminded of the aphorism of George Santayana (1954): "Those who cannot remember the past are condemned to repeat it."

Figure 12 compares the performances of the Reference Gun operated, first, in the conventional manner with a high reservoir pressure and, second, with a constant base pressure. The gun is assumed to be 200 calibers long, to be loaded with hydrogen at 3,000° K, and to fire a light projectile having an effective density of 1.0 gm/cc. The muzzle velocity is plotted as a function of the constant reservoir pressure cycle in the lower curve, and of the constant projectile base pressure cycle in the upper curve.

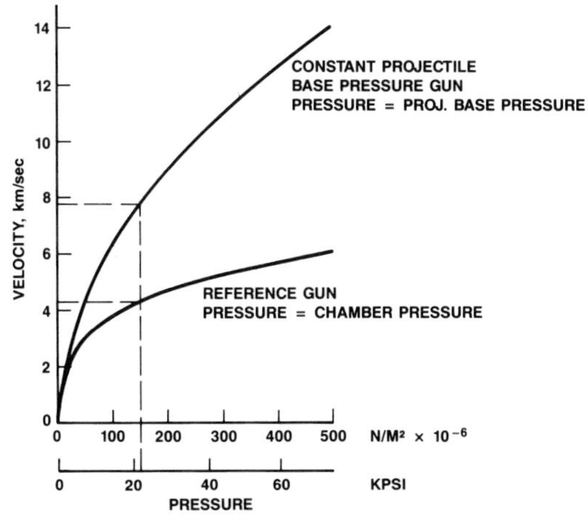

Fig. 12. Muzzle velocity as a function of (1) chamber pressure of the reference gun and (2) projectile base pressure of a constant base pressure gun. Launch tube length = L/D = 200 calibers; effective projectile density = 1 g/cm^3; conditions of reference gun propellant: gas = hydrogen; initial gas temperature = 3,000° K.

Inspection of Fig. 12 shows the advantage of the constant base pressure cycle over the constant reservoir pressure cycle. For example, if the pressure is held to the safe value of 150 x 10^6 N/M^2 (21,800 psi), the velocity of the reference gun is only 4.4 km/s; on the other hand, the velocity of the constant base pressure gun is 7.6 km/s, well into the range of re-entry velocities. The problem at this point in the development of high-velocity gas dynamic launchers was to devise a light-gas gun configuration that generated a constant base pressure cycle.

THE PISTON-COMPRESSION, ACCELERATED-RESERVOIR LIGHT-GAS GUN

Curtis (1962) modified the piston compression light-gas gun by (1) adding a gently tapered section to join the launch and pump tubes and (2) constructing the pump piston in two parts: a heavy, dense rear section and a light, deformable-plastic forward section (see Fig. 9). These modifications enabled the gun to be operated with a pressure cycle that produced an approximately constant base pressure over the first part of the launch run (Charters and Curtis, 1964). Curtis called his modified gun "An Accelerated Reservoir Light-Gas Gun", a name which has been expanded somewhat in the heading of this section (PCARLGG).

The sketches of Fig. 13 show the design and operation of the PCARLGG. Sketch A shows the gun loaded for firing; comparison of this sketch with Fig. 9, shows that the valve tube has been eliminated, the powder chamber has been brought around into line with the pump tube, the pump piston has been made of two parts, and the pump tube ended in a tapered section. Sketch B shows the configuration after the powder has been fired and the pump piston is part way down the pump tube. Sketch C shows the piston about to enter the taper section; the pressure in the light-gas at this time has risen to the rupture pressure of the break valve (another name for the scored, petaling disk described previously), which has opened releasing the hydrogen into the launch tube. Sketch D shows the front of the piston extruded into the taper section and the projectile exiting the launch tube, thus completing the launch cycle.

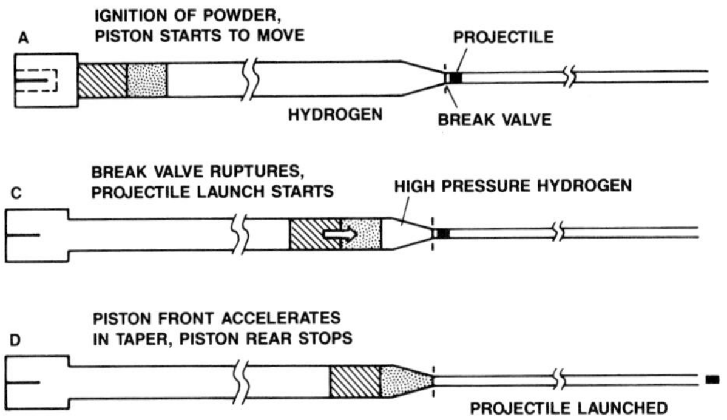

Fig. 13. Operating sequence of accelerated-reservoir light-gas gun. Sketch A: condition at start of firing cycle; sketch C: situation when pressure in pump tube reaches the rupture pressure of the break valve; sketch D: situation at time of projectile launch.

The logic behind the design of the PCARLGG is as follows. Holding the pressure constant at the base of the projectile as it accelerates down the length of the launch tube requires increasing the pressure at the breech of the launch tube at an ever increasing rate as the velocity of the projectile increases. In fact, there is a theory of the constant base pressure gun which prescribes the variation in the pressure and velocity at the entrance to the launch tube needed to hold the base pressure constant (Bixler and Seigel, 1966; Charters and Curtis, 1964).

The launch cycle starts when the break valve ruptures. The pump piston has compressed the light-gas to the rupture pressure of the break valve, which has been designed to rupture at the projectile base pressure desired. At this time, the volume of the pump tube between the piston and the projectile is filled with a moving reservoir of compressed hydrogen. After the break valve opens, the reservoir of hydrogen must be further compressed at an ever increasing rate to generate the increases in pressure and velocity at the launch tube breech needed to hold the base pressure constant.

The requirement of a continually increasing rate of compression of the hydrogen reservoir is met by the combination of taper section and two-part piston. The heavy rear section of the piston drives the front section forward by its inertia. The front section enters the taper and deforms as though it were an incompressible fluid. The velocity of the piston front increases continually as the area of the bore decreases through the taper. Thus, the piston compresses the reservoir of hydrogen at an accelerated rate during the first part of the launch cycle, thereby generating the desired pressure-time profile at the breech of the launch tube. The acceleration of the piston front moving through the taper and compressing the reservoir of hydrogen suggested the name of "accelerated reservoir light-gas gun".

Is the base pressure of a projectile launched by the PCARLGG actually held constant, as predicted? This question is best answered by the results of computations of the gun's dynamics, because it has not proven feasible to measure directly the base pressure during the launching run. The computational scheme is verified by comparison of its predictions with measurements of the launch tube's breech pressure and of the muzzle velocity.

Piacesi, Gates and Seigel (1963) developed a computer code for calculating the dynamics of the piston compression light gas guns at the Naval Ordnance Laboratory (NOL). They treated all dynamic components (powder gas, piston, light gas, and projectile) as compressible fluids with known equations of state. They assumed that all properties were constant across the bore and, hence, were functions only of distance along the bore and time (one dimensional analysis). However, changes in area were taken into account, both abrupt (powder chamber to pump tube) and gradual (pump tube through taper to launch tube). Losses due to friction and heat transfer were neglected. They then developed a computer code to integrate the difference equations based on the differential Lagrangian equations for the conservation of mass, momentum, and energy. Rapid compression of a gas produces shock waves, and the computational problem of discontinuities due to shock waves was solved by the introduction of a pseudo-viscosity pressure, a method suggested by von Neumann and Richtmyer (1932) (see also Richtmyer and Morton, 1967).

Other laboratories have used the code developed by Piacesi, Gates and Seigel and have made modifications to suit their particular purposes. An updated version of the NOL code was prepared by Charters and Sangster (1973) for the Lawrence Livermore Laboratories (LLL) in 1973 and used to compute the performance of their light-gas guns. The results of one calculation are presented in Figs. 14 and 15.

Figure 14 shows the variation of pressure with time. The break valve time is marked; prior to this time, the computed pressure is the pressure at the break valve; after this time, the computed pressure is the pressure at the projectile's base. The pressure profile shows the shock waves generated by the compression of the hydrogen. Although the fronts of these waves have been smeared out by the computing technique, the magnitude of the pressure jump is correct. The second wave breaks the valve. The third wave brings the pressure to its maximum value. After the third wave, the pressure remains high and constant to within 20% of the average value for a limited time. It then falls rapidly, as the gas expands.

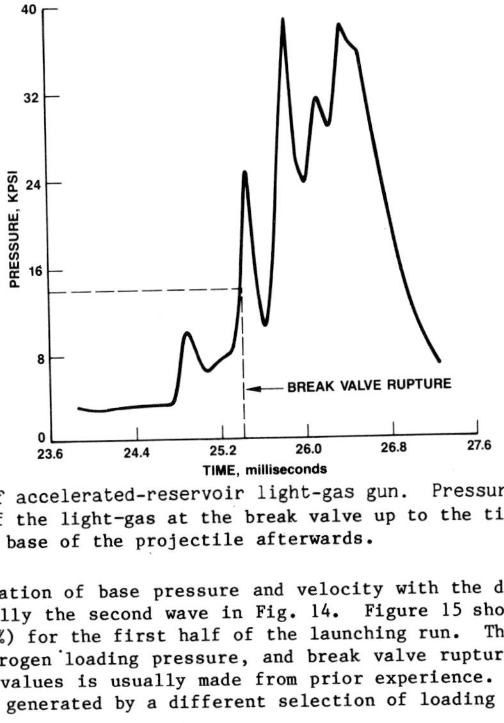

Fig. 14. Dynamics of accelerated-reservoir light-gas gun. Pressure as a function of time: pressure of the light-gas at the break valve up to the time of break valve rupture and at the base of the projectile afterwards.

Figure 15 shows the variation of base pressure and velocity with the distance. The first pressure wave in Fig. 15 is actually the second wave in Fig. 14. Figure 15 shows that the base pressure is roughly constant (+/- 10%) for the first half of the launching run. The loading parameters (powder charge, piston mass, hydrogen loading pressure, and break valve rupture pressure) are interactive, and the choice of their values is usually made from prior experience. A smoother pressure profile could probably have been generated by a different selection of loading parameters.

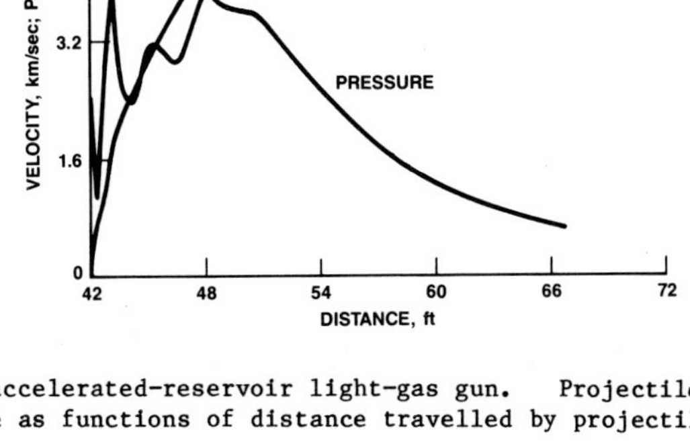

Fig. 15. Dynamics of accelerated-reservoir light-gas gun. Projectile velocity and projectile base pressure as functions of distance travelled by projectile during launch.

Figure 16 compares the measured and computed pressures at the breech of the launch tube (Collins, Charters and Christman, 1967). The original gage record is shown in the photograph at the top; the measured and computed pressures are compared in the sketch at the bottom. Computed and measured pressures are seen to be in good agreement.

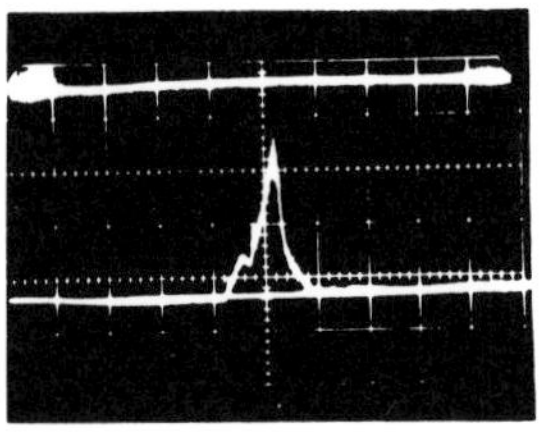

SCALE: 100,000 psi/4cm
SWEEP: 1 millisec/cm

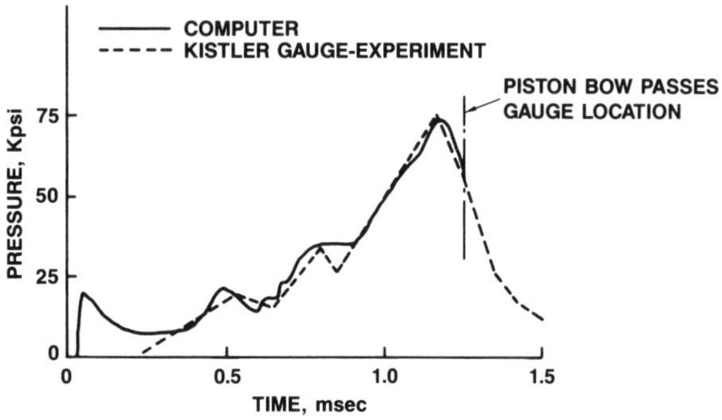

Fig. 16. Measured and computed pressures at breech of launch tube of accelerated-reservoir light-gas gun. Photo at left: oscilloscope record of measured pressure; sketch at right: comparison of measured and computed pressures.

Figure 17 presents the ratio of computed to measured muzzle velocities as a function of the measured velocity (Charters and Sangster, 1973). Although the computed velocities are about 10% high on the average, the consistency of the comparison vindicates the use of the computer program. Correcting the computed velocity by a factor of 10% brings computed and measured values into agreement to within +/-10% for all the runs of Fig. 17.

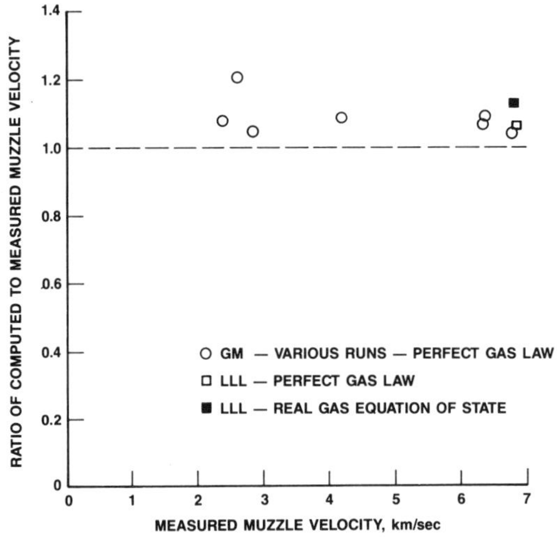

Fig. 17. Performance of accelerated-reservoir light-gas gun. Comparison of computed and measured muzzle velocities.

OTHER LIGHT-GAS GUN OPERATING CYCLES

Our narrative has followed a single line of development of the gas dynamic launcher from the black powder gun to the piston-compression, accelerated-reservoir, light-gas gun. The intent has been to present the logic of the gun's development in a continuous, uninterrupted series of concepts from the beginning to the present. This approach has omitted the development of other concepts of light-gas guns, and our narrative will digress briefly to discuss some of these concepts.

Figure 18 shows sketches of two types of hypervelocity, light-gas guns that have gone through development cycles in the past but are not in common use at the present. Both types represent different methods of heating and compressing the light-gas.

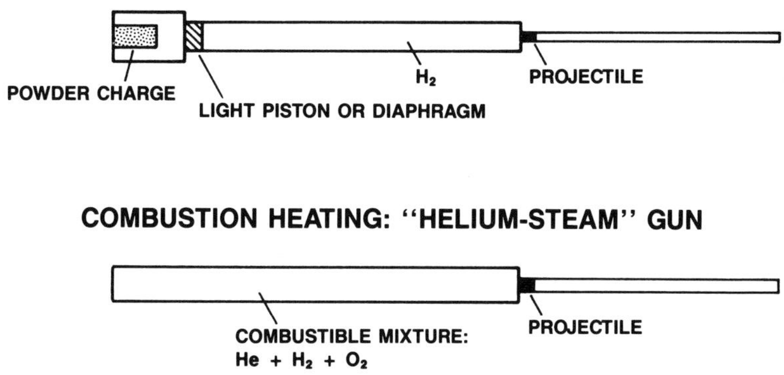

Fig. 18. High-velocity light-gas guns. Top sketch: gun designed to heat the light-gas by shock waves generated during the compression cycle; bottom sketch: "helium-steam" gun.

The bottom sketch of Fig. 18 shows a single-stage, light-gas gun known as a "steam gun" (Seigel and Slawsky, 1956). The configuration is simpler than the piston-compression light-gas gun, which is called a "two stage" gun, because the pump tube can be thought of as a first-stage gun firing the piston, and the launch tube as a second-stage gun firing the projectile. The chamber of the steam gun is filled with a combustible mixture of hydrogen and oxygen diluted by helium. The combustion of the hydrogen and oxygen produces steam which pressurizes and heats the mixture of steam and helium. The molecular weight of the propellant gas is higher than that of hydrogen alone but much less than that of the combustion products of nitrocellulose propellants.

The steam gun was operated successfully. Unfortunately, the mixture had a tendency to detonate, if not ignited properly, and the tendency increased as the caliber of the gun was increased. The problem with detonation eventually terminated the development of the steam gun.

The top sketch of Fig. 18 shows a variation of the piston-compression light-gas gun differing from the PCLGG described in this paper in the weight and velocity of the pump piston (Bioletti and Cunningham, 1960). The pump piston is comparatively light and is fired at relatively high velocity in order to produce very strong shock waves thus heating the light-gas to a higher temperature than can be reached through isentropic compression (Baker, 1964).

The light-piston concept produced high gas temperatures and correspondingly high projectile velocities. The gun operated successfully in firing projectiles strong enough to withstand the shock wave pressures, but it is not suitable for launching relatively fragile sabot-model projectiles.

The theory predicts "the higher the temperature, the lighter the gas, and the higher the velocity," as evidenced by the effect of the speed of sound of the gas in the chamber of the Reference Gun. The top sketch of Fig. 19 depicts one of the methods that was tried to increase the temperature of the compressed gas by adding energy to the gas from an external source. In the method illustrated, the pump tube was heated before firing, thus increasing the initial temperature of the light-gas and, hence, the final temperature (for a given compression ratio). Other methods of adding energy were tried: the gas was heated by an electric arc; the gas was heated in a container external to the pump tube and inserted into the pump tube during the compression cycle (Stevenson and Anderson, 1962; Swift and Baker, 1966).

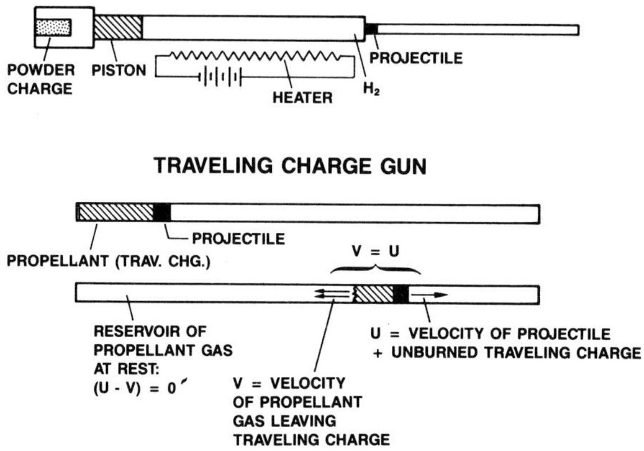

Fig. 19. High velocity light-gas guns. Top sketch: gun designed to heat the light-gas with externally applied energy; bottom sketch: travelling charge gun.

Unfortunately, none of the methods produced an increase in velocity beyond the value obtained with the unheated gas. Various reasons were advanced to explain why there was no increase in velocity with increase in temperature -- contamination of the light-gas, the effects of the equation of state of a real gas, etc., -- but no one understood the difficulty well enough to fix the problem.

The bottom sketch of Fig. 19 illustrates a different approach to the problem of sharing the pressure energy of the propellant gas between the kinetic energy of the projectile and kinetic energy of the gas itself. The traveling charge projectile operates somewhat like a rocket in that the unburned propellant is attached to and moves with the projectile (Baer, 1960). The concept is to fill the volume opened up by the movement of the projectile with propellant gas at constant pressure and zero velocity. In this way, the traveling charge projectile leaves a reservoir of gas behind it that is at rest and has no kinetic energy. The net effect is to transfer all the pressure energy of the gas to the kinetic energy of the projectile alone.

In order to fill the volume opened up by the projectile at the rate required to hold the gas at rest, the burning rate of the propellant must change with the velocity of the projectile. The burning rates required at high velocity greatly exceed the maximum burning rates of conventional nitrocellulose propellants. Some inventors have claimed to have developed propellants with the required burning rates, but a clear-cut demonstration of their claims is lacking so far.

Figure 20 shows still another type of hyper-velocity launcher, the "two-stage, explosively-driven, light-gas gun" (Moore, 1961). The implosion of a "tamped linear explosive driver" compresses the light gas in the pump tube (the first stage) generating a column of high-pressure, high-temperature gas fronted by a shock wave. When the wave reaches the launch tube (the second stage) and impinges on the projectile, an explosive lens of diluted nitromethane is detonated and the resultant shock wave follows the gas column behind the projectile, thereby maintaining the base pressure during the launch. Some success has been obtained with explosively driven light-gas guns, although these devices have not been employed to any extent in aeroballistic ranges.

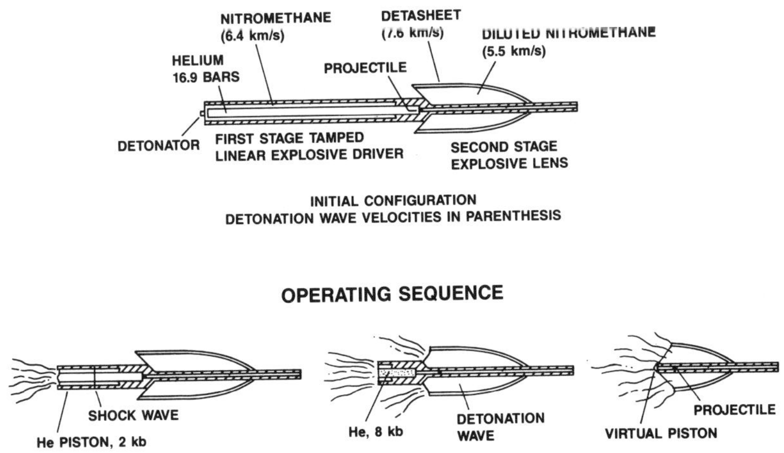

Fig. 20. High-velocity light-gas guns. Two stage explosively-driven light-gas gun: top sketch shows design of gun; bottom sketch shows operating sequence.

MAXIMUM VELOCITY: MECHANICAL LIMITATIONS

The question is often asked: "What is the maximum velocity of a gas dynamics launcher?" A direct answer depends on an analysis of the launch cycle involved, but one often finds that the answer depends on factors and intangibles that are beyond the scope of his analysis. For example, the velocity limit of the PCARLGG depends on the ultimate strength of the taper section of the pump tube, on losses due to friction and heat transfer, and on the maximum acceleration that the projectile can withstand. Each of these factors will be examined briefly in the following discussion.

The taper section of the PCARLGG deforms slightly with each hypervelocity shot; also, fissures may develop along its bore. On the first few shots, the steel cold works and its strength increases. On later shots, fissures can develop and form lines of stress concentration so that the strength may decrease. Hence, the ultimate strength of the taper section is partly a matter of the history of its use. Also, it is normally a very expensive unit, and range operators are chary about firing rounds that will work the taper section to the limit of its strength and require its early replacement.

The launch tube of the PCARLGG can be made longer, and the added length may generate some increase in velocity. However, losses due to friction and heat transfer increase with launch tube length and reduce the anticipated velocity increase. If the length is too great, the velocity may actually decrease with further increase in length, because the bore friction and the shock pressure (in the gas ahead of the projectile) exceed the pressure drive.

The strength of the projectile launched by the PCARLGG limits the maximum acceleration and, hence, the maximum base pressure. The limit depends on the design of the model and sabot, the strengths of their materials, and the increase in compressive strength due to bore confinement. Preliminary estimates of ultimate strengths can be made using handbook data taken from static tests, but these values may be low, because projectiles have been launched successfully with acceleration stresses exceeding the static limits. One depends on experience in order to predict the working values of material strengths.

VELOCITY FUNCTION OF THE IDEAL GUN

An analysis of the performance of an "ideal" gas-dynamics gun has proven useful in relating gun performance to design. The intent is to provide a first order estimate of performance, a "first cut," as it were, in forecasting velocities or roughing out gun dimensions. Accordingly, the effects of friction and heat transfer are neglected; hence, the designation of "ideal" gas-dynamic gun.

The basic physical variables of a gas dynamic gun (or an EML, for that matter) are the length and caliber of the launcher, L and D, the weight of the projectile, M, and the average base pressure during the launch, P_c ("c" for constant pressure = average pressure). These four variables determine the muzzle velocity of the ideal launcher, U. The analysis is best formulated by transforming these physical variables into dimensionless parameters so that the results are independent of gun size and projectile weight. A reference velocity is required for the transformation. The author has arbitrarily chosen a value of 340 meters per second, which is approximately the speed of sound in standard, sea-level air (15.00C, 1.013 x 10^5 N/M^2). The physical variables are transformed into the dimensionless parameters similar to those used in the analysis of the Reference Gun. An exception is the pressure coefficient, C_p, which is now based on the average pressure, P_c, and the reference velocity a_s, namely

$$C_{pc} = P_c/[(1/2)\ \rho_e a_s^2]$$

where the double subscript, pc, is used to distinguish the pressure coefficients of the Reference and Ideal Guns.

MAXIMUM VELOCITY: ESTIMATE FROM EXPERIENCE

The equation of the launch velocity is derived by integrating the equation of motion for the projectile, namely

$$U/a_s = [C_{pc}(L/D)]^{1/2} \tag{8}$$

Equation (8), called the "Ideal Gun Law", is plotted in Fig. 21. Inspection of equation (8) and Fig. 21 shows that launching at hypervelocity requires a long launch tube length, a high average projectile base pressure, and a light projectile, conclusions reached previously in the analysis of the Reference Gun.

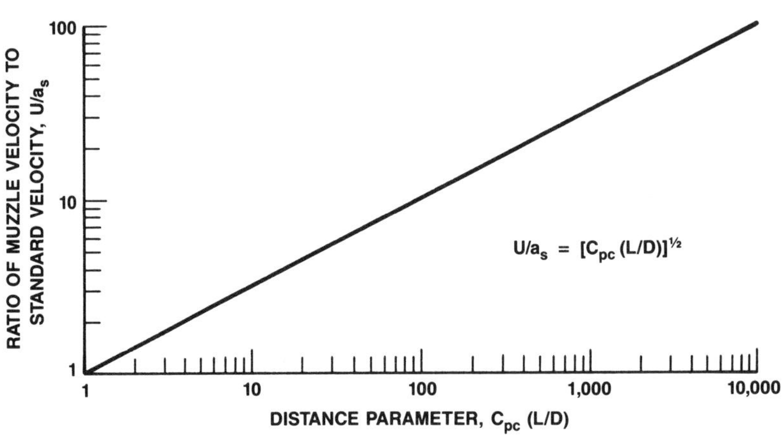

Fig. 21. Performance curve of ideal gas-dynamics gun. Ratio of muzzle velocity to standard velocity, U/a_o, as a function of the distance parameter, $C_{pc}(L/D)$.

The ideal gun law is used to estimate the maximum velocity of the gas-dynamic gun. It is assumed that the ballistic efficiency of the gun approaches 0.5 and that the estimates of the maximum or (minimum) values of the physical variables fall within the limits of current technology. It is understood that the estimates are based on experience.

The experiential basis of the analysis is both its weakness and its strength. The analysis is weak because anyone can take exception to the choice of values of the physical variables, claiming that the experience involved in their choice is too limited or is faulty. The analysis is strong because anyone can substitute the values that he believes to be correct and readily compute a revised velocity.

The maximum velocity of the PCARLGG is based on the following estimates of the values of L/D, P_c, and ρ_e: L/D = 500; P_c = 1.5 x 10^8 N/M^2 = 21,755 psi; ρ_e = 1,000 KG/M^3 = 1 gm/cc. The selection of these values is based on the following rationale. The 500 caliber length is 100 calibers longer than the longest light-gas gun in operation at the present time (400 calibers). The value of 1.5 x 10^8 N/M^2 for the average pressure is based on the observation that the maximum pressure sustained without damage by a hypervelocity sabot-model projectile is about 3.0 x 10^8 N/M^2 and that the average pressure is almost never more than one-half the maximum pressure. An effective projectile density of 1 gm/cc is the density of a bore-diameter, plastic cylinder about 0.8 caliber long, and experience suggests that 1 gm/cc is the minimum effective projectile density for a practical design.

The dimensionless parameters computed from the above values of the physical variables are C_{pg} (L/D) = 1,300 and U/a_s = 36. Hence, the estimated maximum velocity for a gas-dynamic gun is 12.2 km/s. Although the firing of a light-weight proof slug can possibly set a velocity record higher than the estimated maximum, the above study suggests that the "maximum operating velocity," defined as the highest velocity that a sabot-model projectile can be fired on demand in repetitive tests, is about 12 km/s.

In 1965, Seigel predicted "projectile velocities in excess of 60,000 ft/sec (18.3 km/s) will be obtained for projectiles weighing more than 0.1 gm by 1985." The technology has not advanced at the anticipated rate. In the author's opinion, the 18 km/s milestone will await the development of an advanced hypervelocity gun, whose technology is beyond that of the present PCARLGG, possibly an electromagnetic launcher or some hybrid gun combination.

TABLE 1 Definition of symbols

Symbol	Definition
A	Cross-section area of projectile = gun bore area
a_0	Sonic velocity in reservoir gas
a_s	Velocity standard = 340 m/s = sonic velocity in standard air
C_p	Pressure coefficient = $P_0/(1/2)\rho_e\, a_0^2$
C_{pc}	Average pressure coefficient = $P_c/(1/2)\rho_e\, a_s^2$
D	Diameter of projectile
L	Length of launch tube
M	Mass of projectile
$\mathcal{M}$	Gram molecular weight
P	Pressure
P_c	Average projectile base pressure during launching run
P_0	Pressure in reservoir
$\mathcal{R}$	Universal gas constant
t	Travel time of projectile
T_0	Temperature of gas in reservoir
X	Distance of projectile travel
X_0	Distance of wave travel in reservoir = a_0t
U	Velocity of projectile
γ	Ratio of specific heats
ρ_e	Effective projectile density = M/AD
τ	Dimensionless time = ta_0/D

REFERENCES

Anderson, B.E. (1969). Gunpowder. In Sir Wm. Haley (Ed. in Ch.), Encyclopaedia Britannica, Vol. 10, 1038-1040.

Baer, P.G. (1960). The traveling charge gun as a hypervelocity launching device. Proc. 4th Hypervel. Impact Symp, Bal. Res. Labs., Aberdeen Proving Gnd.

Baker, J.R. (1964). A comparison of shock and isentropic heating in light-gas gun compression. NRL Rept. 6063, U.S. Naval Research Lab., Wash. D. C.

Bioletti, C., and B.E. Cunningham (1960). A high velocity gun employing a shock compressed light gas. NASA TN D-307.

Bixler, D.N., and A.E. Seigel (1966). Calculation of the flow conditions in a constant base pressure launcher with a real hydrogen gas propellant. NOL TR 66-157.

Burgess, E. (1961). Long-Range Ballistic Missiles. Chapman and Hall, London.

Canning, T.N., A. Seiff, and C.S. James (1970). Ballistic-range technology. AGARDograph No. 138.

Charters, A.C. (1961). The free-flight range: A tool for research in the physics of high-speed flight. Internat. Hypersonics. Conf., Aug. 16-18, ´61, Mass. Inst. Tech. Amer. Rocket Soc., New York.

Charters, A.C., and J.S. Curtis (1963). How an accelerated -reservoir light-gas gun is used in a hypervelocity free flight range. Gen. Mtrs. Eng. J., Vol. 10, No. 2, 28-37.

Charters, A.C., and J.S. Curtis (1964). High velocity guns for free flight ranges. In W. C. Nelson (Ed.), Temperature Aspects of Hypersonic Flow, Pergamon Press. pp. 371-404.

Charters, A.C., B.P. Denardo, and V. J. Rossow (1957). Development of a piston-compressor type light-gas gun for the launching of free-flight models at high velocity. NACA TN 4143.

Charters, A.C., and D.K. Sangster (1973). Fortran computer program for the interior ballistic analysis of light-gas guns. Contr. Rept. for Lawrence Livermore Lab.

Churchill, W.S. (1956). A History of the English-Speaking Peoples, Vol. 1, The Birth of Britian. Bantam, New York.

Collins, D.J., A.C. Charters, and D.R. Christman ((1967). Parametric studies of light-gas guns. TR No. 77-01F, AC Electronics-Defense Res. Labs., Santa Barbara, Calif.

Corner, J. (1950). Theory of the Interior Ballistics of Guns. Wiley, New York.

Crozier, W.D., and W. Hume (1957). High-velocity light-gas gun. J. Apl. Phy., 28, p. 892.

Curtis, J.S. (1962). An accelerated reservoir light gas gun. NASA TN D-1144.

Department of Defense (1965). Military Standardization Handbook: Plastics. MIL-HDBK-700A. 17 MARCH 1975. U.S. Govt. Print. Off.

Goddard, C., M.B. Herr, and H.C. Thompson (1969). Artillery. In Sir W. Haley (Ed. in Ch.)l, Encyclopaedia Britannica. Vol. 2. pp. 529-534.

Hayes, T.J. (1938). Elements of Ordnance. Wiley, New York.

Lukasiewicz, J. (1967). Constant acceleration flows and applications to high-speed guns. Amer. Inst. Aero. Astro., Vol 5, No. 11, pp. 1955-1963.

Martin, J.J. (1966). Atmospheric Reentry. Prentice-Hall, New Jersey.

Mauncy, A. (1949). Artillery Through the Ages. Div. Pub., Nat. Park Serv., U.S. Dept. Int., Wash. D.C.

Mish, F.C. (1984). Webster´s Ninth New Collegiate Dictionary. Merriam-Webster, Springfield, Mass.

Moore, E.T., Jr. (1967). Explosive hypervelocity launchers. Physics. Internat. Final Rept.-0511

Piacesi R., D.F. Gates, and A.E. Seigel (1963). Computer analysis of two-stage hypervelocity model launchers. NOL TR 62-87.

Richtmyer, R.D., and K.W. Morton (1967). Difference Methods for Initial-Value Problems. Interscience, New York.

Santayana, G. (1954). The Life of Reason. Scribner.

Seigel, A.E. (1965). The theory of high speed guns. AGARDograph 91.

Seigel, A.E., and Z.I. Slawsky (1956). A hypervelocity gun using a shock-compressed steam-heated propellant. NavOrd Rept. 4345.

Stephenson, W.B., and D.E. Anderson (1962). Performance of two-stage light-gas model launchers. Inst. Aero. Sci., 13th Ann. Meet., New York.

Swift, H.F., and J.R. Baker (1966). Hypervelocity capability and impact effect program. NRL Memo.Rept. 1677.

Tuchman, B.W. (1978). A Distant Mirror. Knopf, New York.

von Neumann, I., and R.D. Richtmyer (1932). A method for the numerical calculation of hydrodynamic shocks. J. Appl. Phy., 21, p. 232.

Int. J. Impact Engng Vol. 5, pp. 205–220, 1987
Printed in Great Britain

0734-743X/87 $3.00 + 0.00
Pergamon Journals Ltd.

SURVEY OF DIAGNOSTIC TOOLS USED IN HYPERVELOCITY IMPACT STUDIES

L. C. Chhabildas

Sandia National Laboratories
Albuquerque, New Mexico

ABSTRACT

This paper surveys a variety of diagnostic tools that are available for use in studies of hypervelocity impact. Emphasis is on time-resolved methods for measuring pressure or particle velocity histories of stress waves induced under these conditions, and on the new developments of the past decade.

INTRODUCTION

In this paper, diagnostic techniques commonly used in conjunction with hypervelocity impact are reported. The concept of hypervelocity impact, however, is somewhat arbitrary, since a physical phenomenon associated with impact can be observed in different materials at different impact velocities, *i.e.*, at different stresses. As an example, shock-induced melting is observed in lead at ~ 50 GPa, and at ~ 280 GPa in tantalum. The former can be achieved at plate impact velocities of ~ 2 km/s obtained on a propellant gun, whereas the latter requires plate-impact velocities of 5.6 km/s, which can be achieved with two-stage light gas guns or through the use of explosively-driven flyer plates.

Hypervelocity impact studies entail a wide range of topics, such as constitutive equations of state for materials (Asay and Kerley, 1986), the study of shock Hugoniots (McQueen, *et al.*, 1970), strength of materials (Davison and Graham, 1979), fracture (Curran, 1982), kinetics of phase transition (Duvall and Graham, 1977), shock-induced melting and vaporization, impact cratering, and penetration mechanics (Zukas, *et al.*, 1982). Needless to say, the gauge or the experimental technique employed to study these phenomena should be sensitive enough to observe the physical effect to be determined. A wide variety of tools such as piezoelectric, piezoresistive (Murri, *et al.*, 1974; Graham and Asay, 1978), electromagnetic (Fowles, 1972), interferometric (Barker, 1983), photographic, and radiographic techniques (Swift, 1982) are available and have been used to study thermophysical and mechanical phenomena resulting from hypervelocity impact. Prior to qualifying a gauge for standardization, well controlled experiments are needed to determine the properties of the gauge, its reliability, precision, time resolution and limitations. The experimental geometry/configuration used to qualify these gauges is frequently that of one-dimensional uniaxial-strain loading obtained with smooth-bore guns or through the use of explosive loading techniques (Graham and Asay, 1977; Murri, *et al.*, 1974). In most hypervelocity impact events, however, the materials are subjected to a complex state of loading. With the recent development of well controlled one dimensional pressure-shear loading conditions, the gauge response can be determined under planar biaxial strain configurations. This is a crucial development since it is the first step towards characterizing both the material and the gauge under general states of loading. In this paper, diagnostic techniques successfully used at high stresses and strain rates will be reported. Although the emphasis will be on gauges used in uniaxial-strain configurations, the response of these gauges in biaxial strain configurations is also considered.

ELECTROMAGNETIC TECHNIQUES

Electromagnetic Particle Velocity Gauge

The electromagnetic particle velocity gauge technique is based on Faraday's law of induction. The motion of a conductor of length $\vec{l}$, when placed in a magnetic field of intensity $\vec{B}$ generates an EMF E, which is proportional to its velocity $\vec{u}$, as given by the relation

$$E = \vec{l} \cdot (\vec{u} \times \vec{B}) \tag{1}$$

In a uniaxial-strain configuration, *i.e.*, in a material shocked by a plane wave, if the active gauge element, the particle velocity direction, and the magnetic field vector are oriented orthogonal to each other, then the measurement of induced voltage $E(t)$, gives a direct measurement of the particle velocity motion $u(t)$, through the relation $u(t) = E(t)/lB$.

The accuracy of this technique will therefore depend upon the precision with which this alignment can be performed and the precision with which the measurements of magnetic field intensity and the gauge lengths are determined. In particular, the passive signal leads are oriented parallel to the applied field (or perpendicular to the active gauge element), so that there is no voltage induced in them. In principle, calibration of the gauge is not necessary, and the gauge is suited for use as an embedded gauge if the matrix material does not become conductive at high stresses. In practice, the gauge is calibrated to account for effects due to the uncertainty in the thickness of the gauge element, magnetic fringing, and difficulty in obtaining perfect orthogonality between the gauge element, the magnetic field and the particle-velocity flow field. For a nominal gauge thickness of 35 μm, the wave reverberation time when embedded in sapphire shocked to 60 GPa is determined to be ~ 30 ns (Mashimo and Sawaoka, 1981). The gauge has been successfully used to determine both loading and release states in materials such as fused silica up to 22 GPa (Sugiura, *et al.*, 1981), and rocks and minerals up to 40 GPa (See references in Grady, 1977). These investigations report an uncertainty of 2% in determining the particle velocity at these high stresses. An important feature of this technique is that gauges can be embedded at different locations in a sample to allow the use of Lagrangian wave analysis to determine the relevant stress-volume behavior (see Fig. 1).

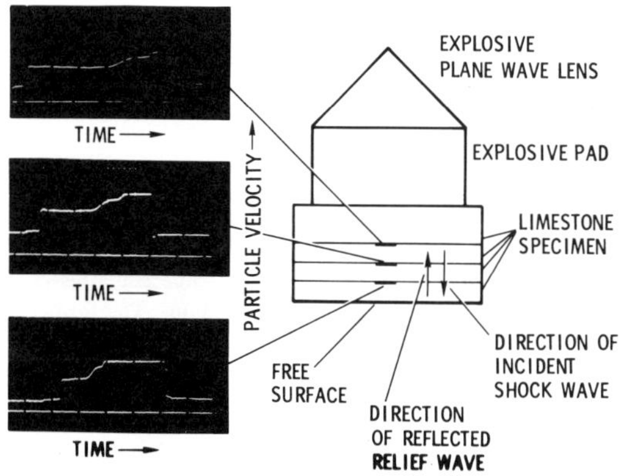

Fig. 1. Electromagnetic particle velocity profiles in limestone at 12 GPa. Profiles indicate both the arrival of the shock wave, and the release wave from the free surface (Murri, *et al.*, 1975).

This gauge is particularly attractive for use as an embedded gauge in an insulating material for biaxial-strain loading (Koller and Fowles, 1979; Young and Dubugnon, 1977). In particular, in a one-dimensional pressure-shear environment (Abou-Sayed, *et al.*, 1976; Chhabildas and Swegle, 1980; Gupta, *et al.*, 1980a), the electromagnetic particle-velocity gauge can be used to determine both the longitudinal and shear particle velocities associated with the pressure-shear waves. The technique has been described in detail by Gupta (1976) and Swegle (1978). In a pressure-shear configuration, the normal to the impact surface can be defined as the x-axis, and the direction of the shear-wave particle velocity as the y-axis; then the gauge element can be located in the impact plane (yz plane) and orthogonal to both the x- and y- axis, *i.e.*, along the z-axis. (This is not necessary, but is done for simplification.) If the magnetic field is oriented such that it is parallel to the xy plane, *i.e.*, it has components B_x and B_y, then the induced EMF, E, will be given by

$$E = l(u_x B_y - u_y B_x) \quad , \tag{2}$$

where u_x and u_y are the longitudinal and shear components of the particle velocity, respectively. A minimum of two experiments with identical impact configurations but with different B components thus allows a determination of u_x and u_y. In particular, if the magnetic field is chosen to be either parallel to, or perpendicular to, the impact interface, then each experiment would yield u_x or u_y accordingly. This gauge has been used to detect pressure-shear waves in PMMA (Gupta, 1980b) and in polycrystalline alumina (Gupta, 1983) (see Fig. 2).

OPTICAL TECHNIQUES

Optical Interferometry

The advent of lasers has made optical interferometry an attractive and useful technique to determine surface motion in shocked materials. Of all the interferometer systems available for use in shock studies (Barker, 1983; Graham and Asay, 1978; Goosman, 1983), the two most commonly used are the VISAR (Barker and Hollenbach, 1972) and the Fabry-Perot velocity interferometer (Johnson and Burgess, 1968; Durand, *et al.*, 1977). There are distinct similarities between these two systems. Both systems use Doppler-shifted light from the target, which can have a reflecting surface that is either specular or diffuse. Both systems in the diffuse mode are relatively insensitive to surface tilt. In both cases, the fringe sensitivity of the interferometer is dependent on the delay time. In the case of the VISAR, the delay time is introduced through the use of glass etalons or a lens combination (Amery, 1976) in one leg of the wide-angle Michelson interferometer system, whereas a set of (Fabry-Perot) plates separated by a distance L provides the necessary delay for the Fabry-Perot system. The time resolution of the VISAR is dependent on the delay time used in the delay leg system, whereas that of the Fabry-Perot system is determined by the photon fill time (Burgess, 1967) of the Fabry-Perot plates.

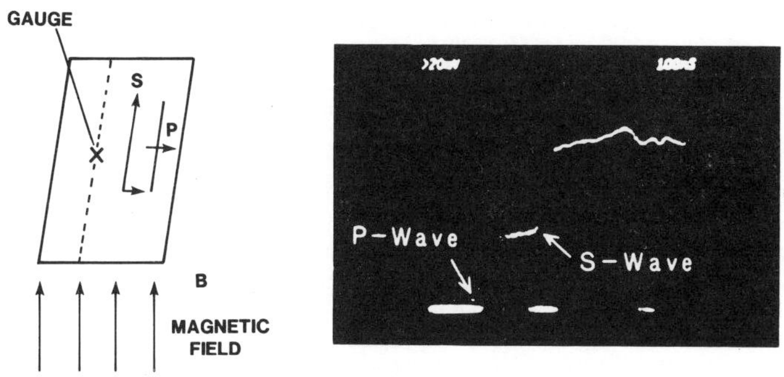

Fig. 2. Electromagnetic particle velocity profiles in alumina. Profile indicates the arrival of the longitudinal wave and the shear wave (Gupta, 1983).

VISAR System

In a standard VISAR system, the reflected beam from a diffused target is split equally into two beams to form the two different legs of a wide-angle Michelson interferometer, in which one leg is delayed in time with respect to the other. Either glass etalons or a lens systems are used to introduce a temporal delay in the delay leg of the interferometer, while maintaining the apparent optical path length of the two legs to be the same. When these two beams are superimposed, fringes $F(t)$, are produced in the interferometer and are related to the change in velocity $u(t)$ of the reflecting surface by the relation (Barker and Hollenbach, 1970; Barker and Schuler, 1974):

$$u(t - \frac{1}{2}\tau) = \frac{\lambda}{2\tau} \frac{F(t)}{(1+\delta)(1+\Delta\nu/\nu)}. \tag{3}$$

The wavelength of the light used is λ, and δ is a correction term due to the wavelength dependence of the refractive index of the etalon material ($\delta = 0.034$ at 514.5 nm when fused silica etalons are used to achieve the necessary delay, or $\delta = 0$ when a lens combination is used). The optical correction term $\Delta\nu/\nu$ results from the change in refractive index of the window material with shock stress (Barker and Hollenbach, 1974). If the measurements are made at a free surface, $\Delta\nu/\nu$ is 0. The use of window materials for laser interferometry will be discussed in a later section.

In the VISAR (Barker and Hollenbach, 1972), quadrature coding has been included to eliminate ambiguity in the sign of acceleration and to improve fringe resolution. This is accomplished by adding a quarter-wave retardation plate, and a polarizing beam splitter to separate the two fringe signals 90° out of phase. These fringe signals are recorded with the aid of photomultiplier tubes, although streak cameras have been used to record fringes (Parker and Chau, 1977; Bloomquist and Sheffield, 1982). The VISAR has an accuracy of $\sim$ 2% for peak surface velocities of $\sim$ 0.1 km/s, with much better accuracy at higher velocity since more fringes are measured. Amery (1976) has developed an air delay leg lens interferometer which permits longer delay times to be used in the interferometer, thereby allowing an improved sensitivity in measuring velocities as low as 0.02 km/s. The analysis of VISAR data is quite sophisticated and usually a computer is required for best accuracy.

A significant improvement to the original VISAR (Barker and Hollenbach, 1972) has been made (Hemsing, 1977) which has the effect of increasing the fringe signal amplitude, while eliminating time-dependent beam intensity signals due to incoherent light sources generated during the experiment. This makes it obsolete to monitor the reflected beam intensity signal as is presently done. In the VISAR system it is essential that the beam intensity be monitored separately to account for reflectivity changes (on the fringe signals) of the mirrorized surface during shock compression.

Fabry-Perot Velocity Interferometer

In the case of a Fabry-Perot interferometer, the Doppler-shifted light from the moving surface is passed through a set of flat and parallel plates (Fabry-Perot plates) separated by a distance L. A spherical lens is then used to image either the circular fringe pattern or dots at the slit of an electronic streak camera. The fringe diameter of the resulting fringes can be controlled by changing the focal length of the lens, or by adjusting the Fabry-Perot plate separation. As the frequency of the light from the moving target changes, the fringe diameter assumes a new value of D_1' from its static value D_1. The velocity V of the moving surface can be calculated using the relation

$$V = \frac{c\lambda}{4L}\left(\frac{(D_1'^2 - D_1^2)}{(D_2^2 - D_1^2)} + m\right), \tag{4}$$

where λ is the wavelength of the light used, D_2 is the static fringe diameter of the next fringe, c is the speed of light, and m is an integer. In case of shocks, the number of fringe jumps inserted is controlled by the integer m, through a prior knowledge of V, or the fringe constant $c\lambda/4L$ may be adjusted to obtain the desired velocity with minimum number of jumps. The precision of the system can be varied over the range of 0.1 to 2% and is determined by the spacing between the plates, the number of fringe jumps inserted, and the focal length of the lens used. The data reduction is simple, since only estimates of D_1' as a function of time are needed to estimate the velocity profile. Some of the applications of this interferometer system have been summarized in the SPIE Proceedings 427, Paisley (1983).

Window Materials

In principle, there is no upper bound in measurements of particle velocity (or stress) using the laser velocity interferometer. In practice, there are limitations in performing free-surface velocity measurements, due to surface instabilities such as jetting from the free surface (Asay, *et al.*, 1976) caused by the arrival of the shock front. To alleviate this, optically transparent materials are usually bonded to the sample, and particle velocity measurements are made at the sample/window interface. This has the added advantage of simulating *in-situ* conditions if the shock impedance of the window material approximates that of the sample. This necessitates that the window materials be both optically transparent and be characterized to account for refractive index changes with shock stress. Solid-state window materials that have been optically calibrated to date include polymethylmethacrylate (PMMA), fused silica, z-cut sapphire, (Barker and Hollenbach, 1970; Setchell, 1979) and lithium fluoride (Wise and Chhabildas, 1986). An aqueous solution of zinc chloride (Wise, 1984), has also been characterized for use as a window.

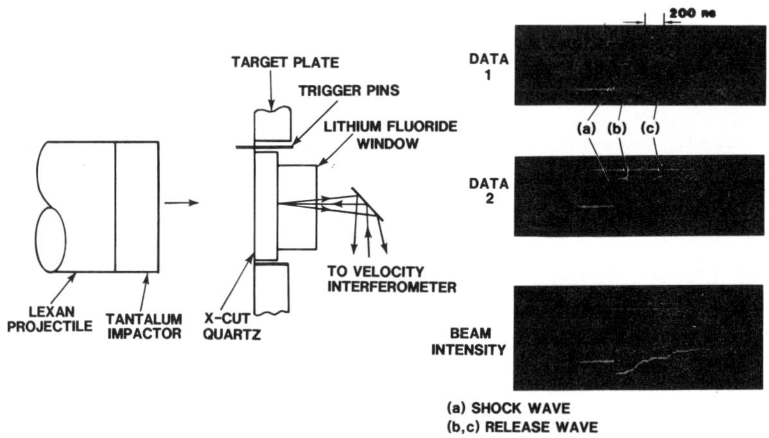

Fig. 3. VISAR fringe records at a quartz/lithium fluoride window interface at 123 GPa. The arrival of the shock wave (a), and the reverberating release waves (b) and (c) in tantalum are clearly indicated (Chhabildas, 1986).

The upper stress limit for PMMA windows has been determined to be 22 GPa (Chhabildas and Asay, 1979), after which there is a transparency loss. Although there is no reported transparency loss in fused silica, there is a significant loss in fringe contrast (Setchell, 1981) when used above its elastic limit of 8.7 GPa (Chhabildas and Grady, 1984). Z-cut sapphire has been optically calibrated up to its elastic limit of 15 GPa (Barker and Hollenbach, 1970). Shock experiments in z-cut sapphire by Lee (1976) up to 43 GPa and by Chhabildas (1986) over the stress range of 90 to 170 GPa suggest a severe loss of fringe contrast. This is also corroborated by recent studies of ramp loading of z-cut sapphire, which also suggest a loss of contrast at stresses above 30 GPa (Barker, 1986). Sapphire is apparently not suited for use as a laser interferometer window at stresses above 30 GPa. It should, however, be pointed out that sapphire has been used as a transparent window anvil in shock luminosity studies for shock stresses up to 80 GPa (Urtiew, 1974). The physical mechanism for this apparent transparency loss needs to be determined to resolve the conflicting results of these two different types of experiments. The transparency of single-crystal lithium-fluoride windows has been established up to 160 GPa (Grady, 1985) and has been optically characterized over the stress range of 1.6 to 115 GPa (Wise and Chhabildas, 1986) for use as a laser interferometer window. It would seem, therefore, that lithium fluoride is the most obvious and versatile material for use with laser interferometry at extremely high stresses (see Fig. 3). When used with high-impedance materials such as tungsten, equation-of-state studies to stresses greater than 600 GPa can be performed. The upper stress limit for the use of LiF as a laser interferometer window has not been determined. The fact that LiF is optically transparent at these high shock pressures and temperatures also suggests that it should be an insulator, and should potentially be a good candidate material when electrical or magnetical insulating properties are desired. The combination of lithium fluoride windows and laser interferometry has allowed material studies to megabar pressures such as the strength of aluminum (Asay, *et al.*, 1986), the reverse phase transformation kinetics of stishovite quartz to a low density phase (Chhabildas, 1986), the dynamic isentropic compression of tungsten up to 80 GPa (Chhabildas and Barker, 1986) and the identification of rarefaction shocks in z-cut calcite at ~ 15 GPa (Grady, 1986), which is indicated in Fig. 4.

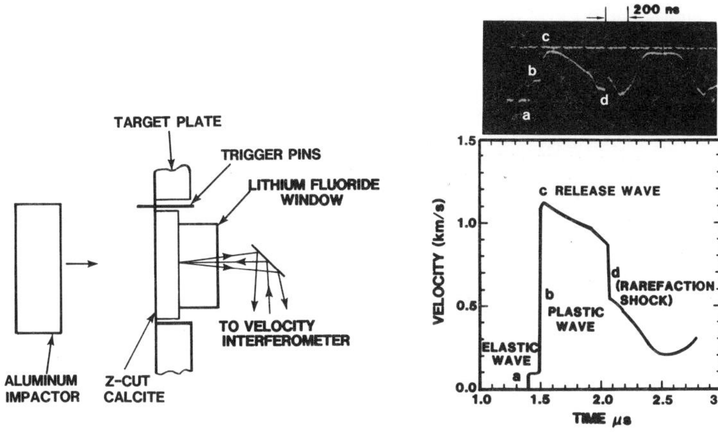

Fig. 4. VISAR fringe record and the velocity profile at the calcite/lithium fluoride interface at ~ 18 GPa. The excellent time resolution of the interferometer allows an unambiguous determination of the rarefaction shock in calcite (Grady, 1986).

Fiber Optics Interferometry

With the development of fiber optics technology, laser light is being transmitted to and received from the moving target using multimode silica/silicone type optical fibers (Durand, 1984; Gidon, *et al.*, 1984). The diameter of the fibers that have been used range from 0.2 mm to 0.6 mm. A lens is positioned at each end of the fiber a focal length away. With this geometry a large spatial area (2 mm to 6 mm) of the target is illuminated when the fiber is placed ~ 40 mm away from the moving target. Durand (1984) used the same fiber to transmit light both to and from the target, whereas Gidon and coworkers (1984) used two fibers placed side by side, one for transmitting the light signal, and the other to receive the reflected light signal from the target. In both instances, the reflected light was channeled to a Fabry-Perot interferometer. Results from both investigations indicate much better fringe sharpness when the signals are channeled through fiber optics, as compared to air transmission techniques. This is illustrated in Fig. 5. In addition, a static fringe pattern is superposed (Durand, 1984) on the dynamic fringe pattern as a result of reflections from the front end of the fiber. Needless to say, light transmission by fiber optic cables has many advantages, including enhanced operational safety, and the ability to transmit light easily in hostile or remote environments such as those caused by explosive products or by electromagnetic pulses.

Laser Interferometric Techniques to Measure Transverse Motion

In the last decade, a variety of techniques has become available for dynamic pressure-shear loading of materials (Abou-Sayed, *et al.*, 1976; Chhabildas and Swegle, 1980; Gupta, *et al.*, 1980; Mashimo, *et al.*, 1984). In addition, a number of interferometric techniques (Abou-Sayed, *et al.*, 1976; Kim, *et al.*, 1977; Chhabildas, *et al.*, 1979) have been developed to determine the transverse motion resulting from pressure-shear loading. Only the transverse displacement interferometer (TDI) (Kim, *et al.*, 1977) and the shear-wave velocity interferometer (Chhabildas, *et al.*, 1979) will be discussed here since these are more commonly used due to their fringe sensitivity, and also because both the longitudinal and transverse motions can be determined simultaneously at the same location with one normally incident laser beam.

Transverse Displacement Interferometer (TDI)

In the transverse displacement interferometer technique, a diffraction grating perpendicular to the plane of transverse wave propagation is deposited on the free-surface of the specimen. This is usually done using photoresist techniques. When a laser beam is incident normally on this diffraction grating, a reflected beam normal to the grating and diffracted beams at angles θ_n given by the relation $n\lambda = d sin\theta_n$ are generated; d is the spacing between the diffraction grating, and n is the order of diffraction and takes on positive or negative integral values. When the two diffracted beams located symmetrically from the normally incident beam are recombined, then the change in phase $\Delta\phi$ due to the transverse displacement y is given by

$$\Delta\phi = 4\pi n y(t)/d \quad , \tag{5}$$

and the fringe sensitivity, *i.e.*, the transverse displacement per fringe, is $d/2n$. Typical fringe sensitivity is less than 1 μm per fringe. The fringe sensitivity can be controlled by varying d and/or n, the order of the diffracted beam to monitor both small and large transverse displacements without exceeding the frequency response of the recording system. The normally reflected beam can be monitored directly by using one of the laser interferometric techniques discussed above to obtain the longitudinal motion due to pressure-shear loading. The transverse displacements determined using this technique are totally decoupled from the longitudinal motion, which enhances the sensitivity for detecting shear motions at large normal displacements.

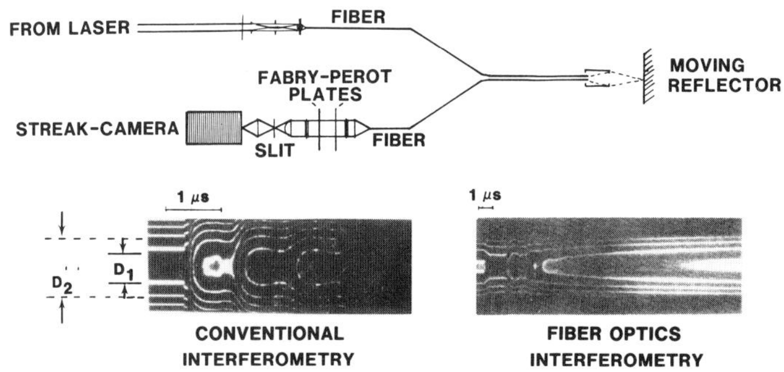

Fig. 5. Fabry-Perot fringe records of a moving reflector. Comparison of the two different records suggest superior fringe signals when optical fibers are used to transmit light signals (Gidon, *et al.*, 1984).

Shear-Wave Velocity Interferometry

Experimentally, this technique is very similar to the TDI technique described above. A laser beam is incident normally on a diffraction grating or a preferentially scratched mirror deposited on the surface to obtain the normally reflected beam and the diffracted beams as described above. Instead of recombining the two beams that are located symmetrically from the normally reflected beam, each individual beam at an angle $\pm\theta$ is monitored by a VISAR. Fringes $F_\theta^\pm$ produced in the interferometers are proportional to a linear combination of both the longitudinal $U(t)$ and shear components $V(t)$ of the free surface velocity (Chhabildas, *et al.*, 1979) and are given by the relations

$$\frac{\lambda}{2\tau}\frac{F_\theta^\pm(t)}{(1+\delta)} = \frac{1}{2}U(t)(1+cos\theta) \pm \frac{1}{2}V(t)sin\theta. \tag{6}$$

Thus, by using two VISARs and by monitoring two beams at $\pm\,\theta$ both the longitudinal velocity and the shear-wave velocity can be determined simultaneously by solving the two above equations. With an air delay leg VISAR (Amery, 1976), a precision in determining $V(t)$ to $\sim 2\%$ can be achieved. The longitudinal and transverse particle velocity profiles obtained in a study of aluminum (Chhabildas and Swegle, 1980) are indicated in Fig. 6.

Fiber-Optic Pins

The fiber optic technology has also had an impact in the development of fiber-optic pins to be used as wave arrival time indicators. In one version of the fiber optic pin (Benjamin, *et al.*, 1984) a small microballoon 0.25 mm in diameter filled with a noble gas such as argon or xenon is attached to the end of an optical fiber. A strong pressure pulse shock heats the gas in the microballoon producing a flash which is channeled to the recording system via optical fibers. The recording system could be either a combination of photo diodes, photomultiplier tubes, or a streak photography. The whole assembly is shielded by using a metal case to protect the microballoon and to eliminate stray optical signals. In an impact experiment where a flat BB pellet is made to impact the pin at $\sim$ 0.2 km/s, the rise time of the flash signal as registered by a photomultiplier tube is $\sim$ 80 ns. At higher impact velocities rise times are expected to be much shorter and the flash signals much stronger. The pin has been used in explosives to determine the detonation velocity (Benjamin, *et al.*, 1984) and also to diagnose a magnetic flux compression generator (Caird, *et al.*, 1984), which produces 20 MJ of electrical energy and currents in excess of 50 MA (see Fig. 7). Some of the useful features of this pin are (i) it is passive, *i.e.*, no external power sources are required as in electrical pins, and (ii) it is electrically and magnetically inert and is therefore not susceptible to electrical noise especially when used in a harsh electromagnetic pulse environment such as those that exist with railguns.

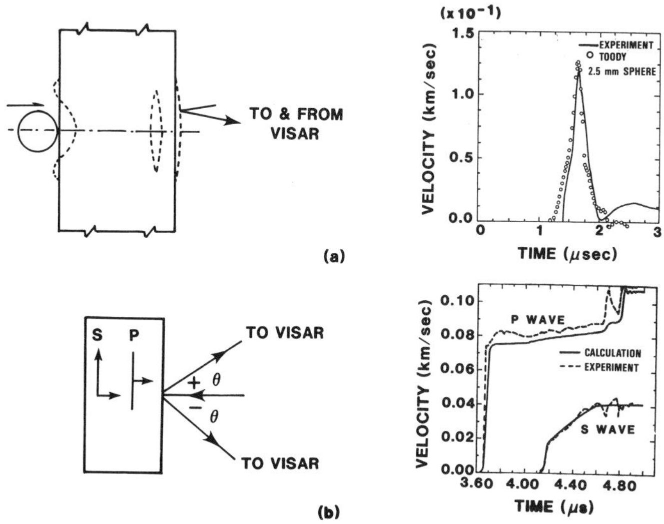

Fig. 6. Experimental configuration and velocity profiles demonstrating the use of VISAR interferometric techniques in (a) a diverging flow environment resulting from hypervelocity impact (Lipkin and Kipp, 1976) and (b) pressure-shear environment to determine the in-plane motion (Chhabildas and Swegle, 1980).

Shock Luminescence: Optical Pyrometry

Most of the optical techniques discussed above are based on the principle of laser interferometry with an external light source and are used to determine particle velocities associated with the propagation of stress waves. In addition to stress-volume change upon dynamic compression, there is also an internal energy change of the material. The technique mentioned in this section takes advantage of the copious radiation emitted by materials when shocked to high pressure from which the temperature of the shocked material is inferred. Temperature measurements allow a critical evaluation of the equation of state, and the determination of melt boundaries of a material in the shocked state. The technique was first proposed and implemented by Kormer and coworkers (1965) to determine the shock-induced temperatures of alkali halides. It was not until much later that the technique was further refined (Lyzenga and Ahrens, 1979; McQueen, *et al.*, 1982; Sugiura, 1980) both in time resolution and in spectral coverage to increase the accuracy of the temperature measurements. In the technique used by Lyzenga and Ahrens (1979), the emitted radiation is filtered (half-width $\sim$ 9 nm) into six channels over the wavelength range of 450 nm to 800 nm and recorded using silicon photodiodes and is illustrated in Fig. 8. The measured radiation is calibrated against the radiation of a tungsten ribbon lamp to determine the spectral radiance N_λ. A least squares fit to the measured values of N_λ as a function of the wavelength λ using the Planck radiation function

$$N_\lambda = \epsilon C_1 \lambda^{-5}\left(e^{C_2/\lambda T} - 1\right)^{-1} \tag{7}$$

yields both the emissivity ϵ and the temperature T of the radiating body (C_1 and C_2 are constant). McQueen and Fritz (1982) use a streak camera and a detonating wave in nitromethane as a standard to determine the temperature of the emitting body based on the assumption that the emitted radiation is a black-body source, *i.e.*, $\epsilon = 1$. The uncertainty in determining temperature measurements ranges from 2% to 5%. This technique is useful to determine the temperature of the radiating body at high stresses and over the temperature range of 2000 K to 6000 K. Temperature determinations above 6000 K are not very accurate due to the relative insensitivity of the visible spectrum shape to the temperature in this region. The technique has been used most recently to determine shock-induced temperatures in transparent materials (Lyzenga, *et al.*, 1982, 1983; Nellis, *et al.*, 1984; Boslough, *et al.*, 1984), so that the thermal radiation from the hot compressed material is observed through the unshocked portion of the transparent sample. At low stresses, however, shock luminescence resulting from inhomogeneous deformation (Brannon, *et al.*, 1984), triboluminescence (Graham and Brooks, 1971), and spall fracture (Brannon, *et al.*, 1984) becomes significant so that the radiation spectrum is not black body. As an extension of this technique, infrared pyrometry (Van Holle, 1984) has been developed for use at relatively low stresses to estimate hot-spot temperatures in explosives, as observed through a transparent window (KBr), to determine the detonation reaction zone structure. However, the infrared radiation spectrum observed is usually difficult to interpret (Van Holle, 1984) and requires supplementation through framing camera pictures of the surface to identify the source of the radiation.

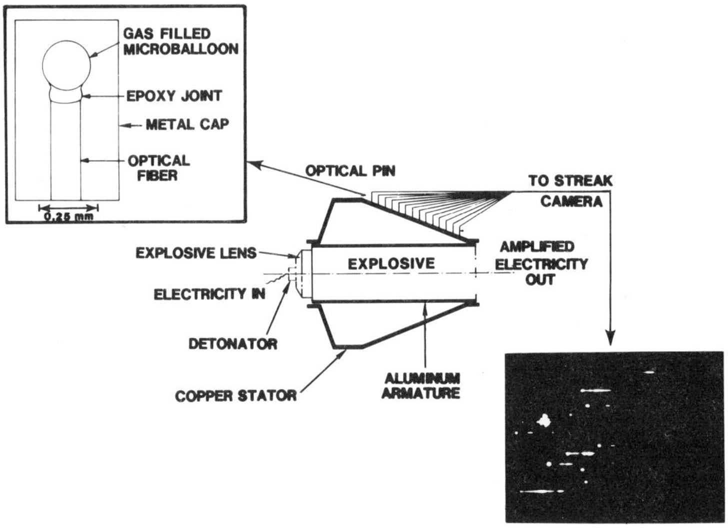

Fig. 7. Fiber optic pin (Benjamin *et al.*, 1984) used to diagnose a magnetic flux compression generator (Caird, *et al.*, 1984).

Shock Luminescence: Optical Analyzer

The technique of using an optical analyzer (McQueen, *et al.*, 1982) to measure arrival times of release waves is based on the principle that the intensity of the shock-induced luminescence observed would be sensitive, since it is proportional to the fourth power of temperature. In a flat-plate impact experiment, a shock wave travels both into the target and into the impacting plate. If the impact plate is backed by a low-impedance material, the shock in the impact plate will reflect back as a rarefaction wave. If material rigidity is important, then the leading edge of the rarefaction wave travels at the elastic wave speed of the material which is much faster than the shock speed of the shock travelling into the target. If the sample target is thin, then the dwell time of the shock at the sample/optical analyzer will be relatively long. As the sample thickness is increased, the dwell time of the shock at the interface will decrease, assuming a constant impactor thickness. A critical target dimension exists for which the shock in the target and the release wave from the back surface of the impactor arrives at the sample/analyzer interface simultaneously, *i.e.*, the dwell time of the shock at the sample/analyzer interface is zero. If x_C is the critical target dimension for which this occurs for an impactor thickness of x_I, then it can be shown that C_L the Lagrangian wave speed for the leading edge of the release wave is given by

$$C_L = U_s(R+1)/(R-1), \tag{8}$$

where U_s is the shock velocity and $R = x_C/x_I$. For simplicity, symmetric impact loading conditions are assumed. Experimentally, since explosive loading techniques were used (McQueen, *et al.*, 1982) and the dwell time Δt of the shock is determined at ten different sample/analyzer locations by registering the intensity of the shock radiation, the critical target dimension x_C is deduced very accurately (0.5%). Fiber optic cables were used to transmit the radiation signals to photomultiplier tubes whose output was recorded on scopes. Sound speed determinations in shocked material allow determination of either melt boundaries or phase boundaries that cannot be easily observed in the shock velocity - particle velocity relation, *i.e.*, on the shock Hugoniot. This technique has been used to locate the melt boundary in iron (Brown and McQueen, 1980), tantalum (Brown and Shaner, 1983) and 2024 aluminum (McQueen, *et al.*, 1983). Bromoform, fused quartz, and a high density glass were commonly used for optical analyses in the above investigations. The requirement is that the analyzers be optically transparent, emit radiation at stresses of interest like a black body. Al'tshuler and coworkers (1971) used the same technique to determine the elastic release wave speeds and the shear stress in the shocked state by measuring the free-surface velocity of the target and determining the thickness of the target at which the main shock attenuates. Velocity interferometric techniques, axisymmetric gauges (Fritz and Morgan, 1973) and embedded gauges (Grady, *et al.*, 1975) are also appropriate tools to determine the sound speed in the shocked state from which the melt boundary of a material in the shocked state could be inferred (Asay and Hayes, 1975).

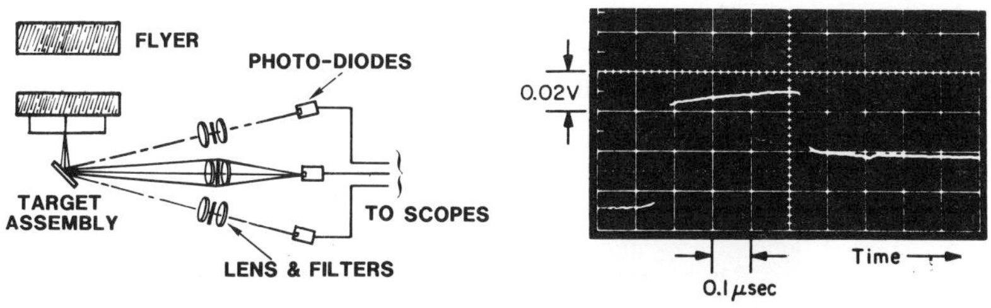

Fig. 8. Time-resolved shock luminosity record from one of the channels in shock pyrometry experiment (Lyzenga and Ahrens, 1979).

PIEZORESISTIVE TECHNIQUES

Many materials exhibit a change in resistivity as a function of both pressure and temperature. The resistance change in manganin is much more sensitive to pressure than it is to temperature and thus is a good candidate for use as a dynamic stress transducer. At low shock stresses, however, carbon (Horning and Isbell, 1975) or ytterbium (Ginsberg, *et al.*, 1973) can be used since the resistance change in these materials is larger that it is for manganin when shocked to the same stress. Experimentally, the resistance change in carbon (Perez and Chartagnac, 1980) as a function of stress is reproducible up to 2 GPa and is limited for accurate use over this low stress range. Ytterbium is restricted for use to stresses below 4 GPa, due to a phase transformation approximately at that stress. Manganin, however, has been demonstrated to be useful to stresses greater than 100 GPa (DeCarli, *et al.*, 1986) and is, therefore, the most widely used piezoresistive gauge at high stresses (see Fig. 9). The main emphasis in this section will be focused on the manganin gauge, although the discussion in general is valid for any piezoresistive gauge.

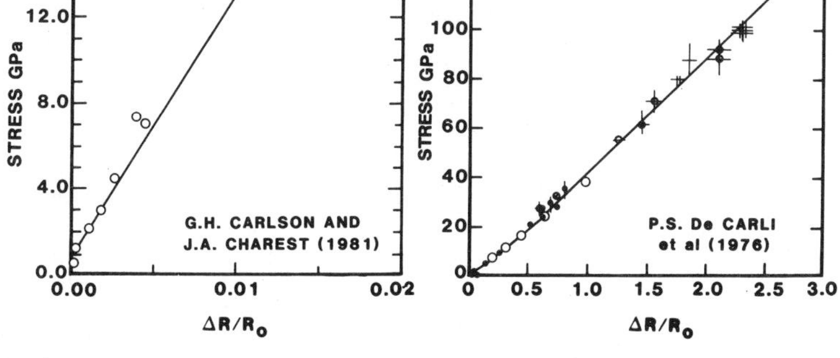

Fig. 9. Fractional change in resistance as a function of shock stress for constantan (Carlson and Charest, 1981) and manganin (P. S. DeCarli, *et al.*, 1976).

The gauge is usually calibrated in a well-controlled uniaxial strain experiment by measuring the fractional change in resistance $\Delta R/R_o$ and the results are empirically correlated to the stress σ through the relation

$$\Delta R/R_o = k\sigma \quad , \tag{9}$$

where k is a piezoresistive coefficient. In general, k is dependent on the stress (Keough and Wong, 1970), the composition of the material (Graham and Asay, 1978), and the gauge shape, *i.e.*, whether a foil gauge or a wire gauge is used (Graham and Asay, 1978; Rosenberg, *et al.*, 1980).

The gauge element in the form of a foil (50 ohms) is normally embedded in materials such that the active gauge element is normal to the stress wave propagation direction. It is usually electrically isolated from its matrix material through the use of insulating gauge package materials (such as kapton, mica, or teflon). Unfortunately, at higher stresses, most gauge package materials become partially conducting (Graham, 1979), which hampers an accurate calibration of the gauge at higher stress. This is alleviated to some extent through the use of low-impedance gauges (DeCarli, *et al.*, 1976; Vantine, *et al.*, 1980), in which the initial manganin gauge resistance used is as low as 50 milliohms. Manganin, however, is the only *in-situ* stress gauge available for use at these high stresses and can be embedded at multiple locations to determine the wave evolution. It has been used successfully (see Fig. 10) with relatively little difficulty to stresses up to 50 GPa (Grady, *et al.*, 1974a; Murri, *et al.*, 1975).

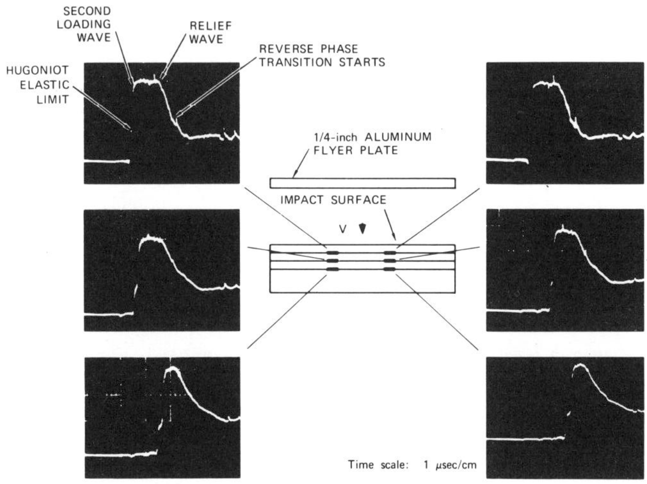

Fig. 10. Manganin stress-wave profiles for Arkansas novaculite at 25 GPa (Grady, *et al.*, 1974).

Recently, both manganin (Kanel, *et al.*, 1978; Rosenberg and Partom, 1985c; Chartagnac, 1982) and ytterbium (Gupta, *et al.*, 1980; Gupta, 1983; Gupta and Gupta, 1985) have been used to measure the lateral stress state that exists in a material under uniaxial-strain loading, by placing the gauge element parallel to the shock propagation direction. Manganin has also been used in a preliminary attempt to detect shear waves in compressed materials (Rosenberg and Bless, 1986).

There has always been a theoretical effort (Barsis, *et al.*, 1970; Grady and Ginsberg, 1977) to understand the dynamic response of manganin and ytterbium gauges, respectively. More recently, these attempts have been continued by Partom and coworkers (1981) and Gupta (1983). In their theoretical framework, they propose that the resistance change of a gauge element is made up of components due to a piezoresistive effect and a gauge plasticity effect. The gauge plasticity effect in these theoretical studies (Partom, *et al.*, 1981; Gupta, 1983) demonstrates that the stress and the strain state in the gauge is more complex and will be different than in the matrix material (Gupta, 1983; Partom, *et al.*, 1981). This analysis is quite significant because it allows an understanding of the piezoresistive gauge response when used also as a lateral stress gauge (Gupta and Gupta, 1985; Rosenberg and Partom, 1985e). The different gauge calibration coefficients observed for wire gauges or for foil gauges (Graham and Asay, 1978; Rosenberg, *et al.*, 1980) and the hysteresis effects observed in manganin upon unloading (Keough, 1968; Yaziv, *et al.*, 1980; Vantine, *et al.*, 1980) is also explained through gauge elastic-plastic effects (Rosenberg and Partom, 1985b). A finite yield strength for the gauge material, such as manganin (Rosenberg and Partom, 1985a), would cause the stress release path in the gauge to be different than the loading path (Gupta, 1983), which would cause the gauge behavior to appear hysteretic.

Strain Compensated Gauge

Constantan is unique in that the fractional change in resistance as a function of shock stress (up to 15 GPa) is negligibly small (Carlson and Charest, 1981; DeCarli, 1981; Rosenberg, *et al.*, 1984). This is indicated in Fig. 9. However, under compressive strain or tensile strain, the strain-gauge coefficient is approximately 2 (Carlson and Charest, 1981; Chen, *et al.*, 1984) (see Fig. 11). The strain gauge coefficient is defined to be the ratio of the fractional change in resistance to the fractional change in gauge length. Hence, constantan is used as a strain gauge material. Attempts have been made recently to use manganin in an axially symmetric diverging stress flow in a material to determine the axial stress history (Carlson and Charest, 1981; DeCarli, 1981; Rosenberg, *et al.*, 1984) and is illustrated in Fig. 12. This is based on the principle that the fractional change in resistance of the manganin gauge (or any piezoresistive gauge) when placed in a diverging flow is a result of both the changing axial stress history and gauge strain (stretching). Estimates for the magnitude of gauge strain (stretching) is determined through measurements of the change in resistance of a constantan gauge located in the same plane alongside the piezoresistive manganin gauge. This is an interesting concept that definitely needs to be pursued further. It is at this stage strictly based on the assumption that the strain measured by constantan is the same as that in the matrix material and in manganin. Although this assumption seems plausible, more careful study is still necessary to determine the accuracy and reliability of the resulting measurements.

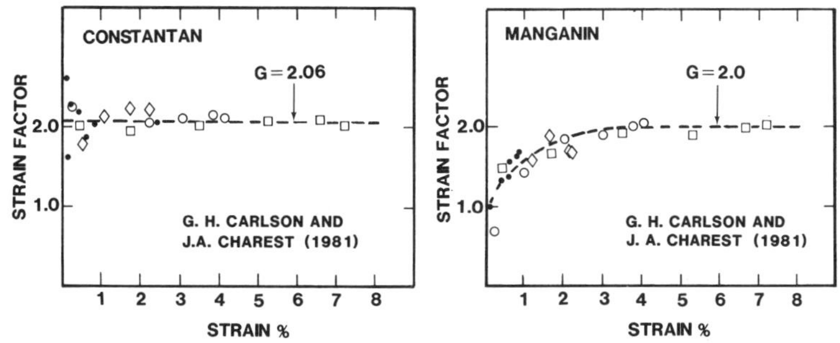

Fig. 11. Strain gauge coefficient as a function of strain for constantan and manganin (Carlson and Charest, 1981).

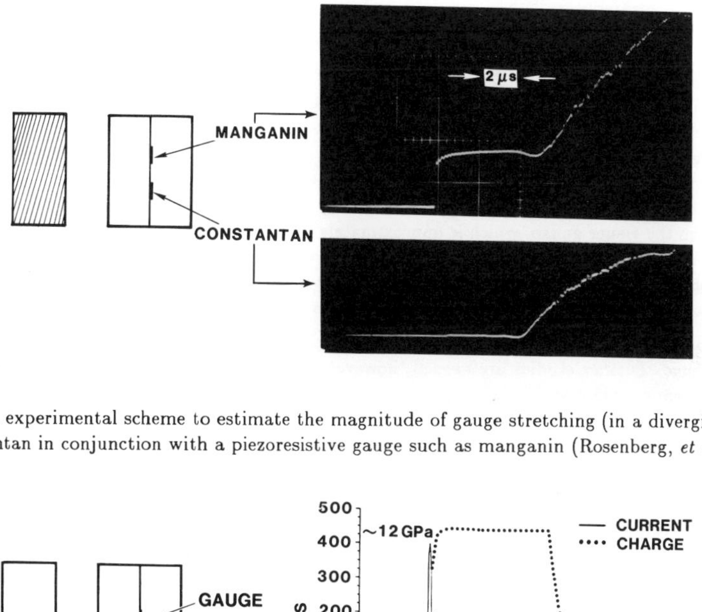

Fig. 12. An experimental scheme to estimate the magnitude of gauge stretching (in a diverging flow) by using constantan in conjunction with a piezoresistive gauge such as manganin (Rosenberg, *et al.*, 1984).

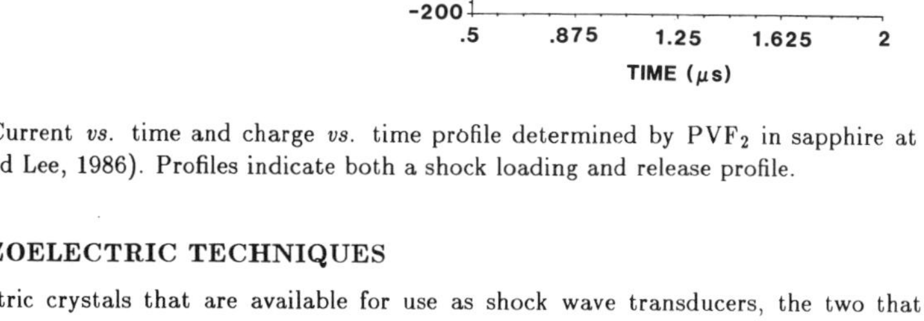

Fig. 13. Current *vs.* time and charge *vs.* time profile determined by PVF_2 in sapphire at ~ 12 GPa (Graham and Lee, 1986). Profiles indicate both a shock loading and release profile.

PIEZOELECTRIC TECHNIQUES

Of all the piezoelectric crystals that are available for use as shock wave transducers, the two that have received the most attention are x-cut quartz and lithium-niobate crystals (Graham and Reed, 1978). They are the most accurately characterized stress wave transducers available for use up to stresses of 4 GPa and 1.8 GPa, respectively. Excellent review articles describing the gauge characteristics and capabilities also exist (Graham and Asay, 1978; Murri, *et al.*, 1975; Grady,

1977; Davison and Graham, 1979) and will not be repeated here. Gupta and Murri (1982) have demonstrated that lithium-niobate crystals when cut to the proper orientation could be used also as shear-wave transducers.

Recently, mechanically-stretched polyvinylidene fluoride (PVF_2), a polymeric film which exhibits piezoelectric properties when poled in an electric field (Kawaii, 1969), has received some attention (Bauer, 1982, 1983, 1984). Since it is in the form of a 25-μm-thin flexible piezoelectric plastic, it has the potential of developing into a thin piezoelectric stress transducer to be used in materials at the impact interface or in an embedded mode. It has the advantage of being self-powered, yielding large voltage signals upon shock loading. The wave reverberation time will depend on the thickness of the gauge and the shock impedance of the material that it is embedded in, and is on the order of 25 ns in sapphire. Bauer (1982, 1983, 1984), realizing this potential, undertook a careful and systematic study of various polymeric films. He noted that the piezoelectric activity of the film depended on the polarization process which was highly irreproducible with conventional processes. He has since developed a poling process in which the thin PVF_2 film attains a remnant polarization approaching 10 $\mu C/cm^2$. This seems to be the dominant criteria to achieve the highly reproducible behavior required for shock gauges.

The current output from a thick piezoelectric gauge is proportional to the stress difference between the two faces of the gauge and the reading time depends on the thickness of the gauge. For a thin piezoelectric film, embedded in a material, the current output, $i(t)$, will be proportional to the time-averaged stress derivative which is a significant feature in that for times greater than about 20 ns, the current is a function of the stress rate. Such measurements provide a much more sensitive indication of rate-dependent wave properties. The charge, Q, which is the time integral of the measured current, will therefore be proportional to the stress, *i.e.*,

$$\sigma \propto \int i(t)dt \quad . \tag{10}$$

The gauge, henceforth referred to as the Bauer gauge, or the ISL gauge has been calibrated up to 4 GPa (Bauer, 1984) for both shock loading and release. The same Bauer gauge has been used by Lee and coworkers (1985) in shock studies and has shown a reproducible charge vs stress behavior. Recently, Graham and Lee (1986) have extended these calibration studies to $\approx$ 20 GPa, and have measured both a shock loading and release profile in sapphire at 12 GPa and is indicated in Fig. 13. Effects of gauge hysteresis and a spall signal have been observed when released from 12 GPa. Studies by Perez and Tougue (1985) on a different batch of material exhibiting a remnant polarization of $\approx$ 5.3 $\mu c/cm^2$ show a different calibration curve from the Bauer gauge, which is approximately 30% less sensitive at $\approx$ 5 GPa. This study and other work by Bauer show the dependence of material and the poling process on the piezoelectric response curve and points toward a need in establishing both material and gauge qualification procedures. The same study (Perez and Tougue, 1985) also suggests the range of practical utilization of their gauge to be $\approx$ 13 GPa.

The early studies on thin PVF_2 polymeric films are encouraging and demonstrate the usefulness of this material as a thin piezoelectric gauge. Further studies are needed, however, not only to characterize the gauge but also to develop an understanding of the gauge response.

SUMMARY

A variety of time-resolved techniques that can be used in conjunction with hypervelocity impact studies have been summarized. Manganin gauges have been developed to be used to stresses above 100 GPa, and when used jointly with constantan have the potential of yielding axial stress properties in diverging flow situations. The use of electromagnetic particle velocity gauges has been extended up to 60 GPa, but is limited to insulating materials. The transparency of lithium-fluoride windows to stresses approximating 160 GPa allows velocity interferometric techniques to be used to much higher stresses in high-impedance materials. The development of both fiber optic interferometry and fiber optic pins has a particularly significant impact on hypervelocity impact studies, since they could potentially be used in electromagnetic environments for equation-of-state studies at extremely high impact velocities. The development of thin polymeric piezoelectric PVF_2 films is also encouraging since they would be self-powered and provide large voltage signals, which is a desired property in electrically noisy environments.

ACKNOWLEDGMENTS

The author would like to acknowledge J. R. Asay, L. M. Barker, M. B. Boslough, D. E. Grady, R. A. Graham, R. E. Setchell, and J. L. Wise, Sandia National Laboratories; and L. M. Lee, Ktech Corporation, for fruitful discussions and a critical review of the manuscript. Special thanks are also due to P. S. DeCarli, Stanford Research Institute; W. M. Isbell, General Research Corporation; and Z. Rosenberg, University of Dayton Research Institute, for discussions relating to strain compensated gauges.

REFERENCES

This work performed at Sandia National Laboratories supported by the U.S. Department of Energy under contract number DE-AC04-76DP00789.

Abou-Sayed, A. S., R. J. Clifton, and L. Herman (1976). Exp. Mech., **16**, 127–132.

Al'tshuler, L. V., M. I. Brazhnick, and G. S. Telegin (1971). J. Appl. Mech. Tech. Phys., **12**, 921–926.

Amery, B. T. (1976). 6th Symposium on Detonation AC-R-221, Office of Naval Research - Department of the Navy, Arlington, VA, 673–681.

Asay J. R. and D. B. Hayes (1975). J. Appl. Phys., **46**, 4789–4800.

Asay J. R., L. P. Mix, and F. C. Perry (1976). Appl. Phys. Letters, **29**, 284–286.

Asay J. R. and G. I. Kerley (1986). The Response of Materials to Dynamic Loading. Int'l. J. of Impact Engr., **5**.

Asay, J. R., L. C. Chhabildas, T. G. Trucano, and G. I. Kerley (1986). High Pressure Strength of Shocked Aluminum. In Y. M. Gupta (ed.), Shock Waves in Condensed Matter, Plenum Publishing, 145–149.

Barker, L. M. and R. E. Hollenbach (1970). J. Appl. Phys, **41**, 4208–4226.

Barker, L. M. and R. E. Hollenbach (1972). J. Appl. Phys, **43**, 4669–4675.

Barker, L. M. and K. W. Schuler (1974). J. Appl. Phys, **45**, 3692–3693.

Barker L. M. (1983). Velocity Interferometer for Time-resolved High-velocity Measurements. In Dennis L. Paisley (ed.), High Speed Photography, Videography, and Photonics, Proc. SPIE, **427**, 116–126.

Barker, L. M. (1986). Sandia National Laboratories, private communication.

Barsis, E., E. Williams, and C. Skoog (1970). J. Appl. Phys., **41**, 5155–5162.

Bauer, F. (1982). Behavior of Ferroelectric Ceramics and PVF_2 Polymers Under Shock Loading. In W. J. Nellis, L. Seaman, and R. A. Graham (eds.), Shock Waves in Condensed Matter–1981, American Institute of Physics, 251–267.

Bauer, F. (1984). Ferroelectrics, **49**, 231–240.

Bauer, F. (1984). Piezoelectric and Electric Properties of PVF_2 Polymers Under Shock Wave Action: Application to Shock Transducers. In J. R. Asay, R. A. Graham, and G. K. Straub (eds.), Shock Waves in Condensed Matter–1983, Elsevier Science Publ., B.V., 225–228.

Benjamin, R. F., F. J. Mayer, and R. L. Maynard (1984). Microshell-tipped Optical Fibers as Sensors of High-pressure Pulses in Adverse Environments. In R. A. Greenwell (ed.), Fiber Optics in Adverse Environments 11, Proc. SPIE, **506**, 116–121.

Bloomquist, D. D. and S. A. Sheffield (1982). Optically Recording Velocity Interferometer System (ORVIS) for subnanosecond Particle Velocity Measurements in Shock Waves. In L. Endelman (ed.), 15th High Speed Photography, Proc. SPIE, **348**, 523–528.

Boslough, M. B., T. J. Ahrens, and A. C. Mitchell (1984). J. Geophys. Res., **89**, 7845–7851.

Brannon, P. J., R. W. Morris, C. H. Konrad, and J. R. Asay (1984). Shock-induced Luminescence from X-cut Quartz and Z-cut Lithium Niobate. In J. R. Asay, R. A. Graham, and G. K. Straub (eds.), Shock Waves in Condensed Matter-1983, Elsevier Science Publ., B.V., 303–306.

Brown, J. M., and R. G. McQueen (1980). Geophys. Res. letters., **7**, 533-536.

Brown, J. M. and J. W. Shaner (1984). Rarefaction Velocities in Shocked Tantalum. In J. R. Asay, R. A. Graham, and G. K. Straub (eds.), Shock Waves in Condensed Matter-1983, Elsevier Science Publ., B.V., 91–94.

Burgess, T. J. (1967). The Photon Fill Time of a Fabry-Perot Spectrometer, Lawrence Livermore National Laboratory Report UCID-15224, unpublished.

Caird, R. S., R. F. Benjamin, R. G. McQueen, and D. J. Erickson (1984). Application of Optical-fiber Pins to Explosive, Pulse-power Generators. In R. A. Greenwell (ed.), Fiber Optics in Adverse Environments II, Proc. SPIE, **506**, 122–126.

Carlson, G. H. and J. A. Charest (1981). Pressure Measurements of Nonplanar Stress Waves. Los Alamos Scientific Laboratory Report No. LAUR-812343, unpublished.

Chartagnac, P. F. (1982). J. Appl. Phys., **53**, 948–953.

Chen, D. Y., Y. M. Gupta, and M. H. Miles (1984). J. Appl. Phys., **55**, 3984–3993.

Chhabildas, L. C. and J. R. Asay (1979). J. Appl. Phys., **50**, 2749–2756.

Chhabildas, L. C., H. J. Sutherland, and J. R. Asay (1979). J. Appl. Phys., 50, 5196–5201.

Chhabildas, L. C. and J. W. Swegle (1980). J. Appl. Phys., **51**, 4799–4807.

Chhabildas, L. C. and D. E. Grady (1984). Shock Loading Behavior of Fused Quartz. In J. R. Asay, R. A. Graham, and G. K. Straub (eds.), Shock Waves in Condensed Matter-1983, Elsevier Science Publ., B. V., 175–178.

Chhabildas, L. C. (1986). Shock Loading and Release Behavior of X-Cut Quartz. In Y. M. Gupta (ed.), Shock Waves in Condensed Matter, Plenum Publishing, 601–605.

Chhabildas, L. C. (1986). Unpublished data.

Curran, D. R. (1982). Dynamic Fracture. In J. A. Zukas, T. Nicholas, H. F. Swift, L. B. Greszczuk, and D. R. Curran (eds.), Impact Dynamics, John Wiley, New York.

Davison, L. W. and R. A. Graham (1979). Phys. Rep., **55**, 257–379.

DeCarli, P. S., D. C. Erlich, L. B. Hall, R. G. Bly, A. L. Whitson, D. D. Keough, and D. Curran (1976). Stress-Gage System for the Megabar (100 GPa) Range. Defense Nuclear Agency Report No. DNA 4066F, unpublished.

DeCarli, P. S. (1981). Stanford Research Institute, private communication.

Durand, M., P. Laharrague, P. Lalle, A. Lebihan, J. Morvan, and H. Pujois (1977). Rev. Sci. Instrum., **48**, 275–278.

Durand, M. (1984). Use of Optical Fibers for Velocity Measurement by Laser Doppler Interferometry with a Fabry-Perot Interferometer. In M. Andre and M. Hugenschmidt (eds.), High Speed Photography and Photonics, Proc. SPIE, **491**, 650–656.

Duvall, G. E. and R. A. Graham (1977). Rev. Modern Phys., **49**, 523–579.

Fowles, G. R. (1972). Experimental Technique and Instrumentation. In P. C. Chou and A. K. Hopkins (eds.), Dynamic Response of Materials to Intense Impulsive Loading, 405–480.

Fritz J. N. and J. A. Morgan (1973). Rev. Sci. Instrum., **44**, 215–221.

Gidon, S., G. Garcin, and H. Behar (1984). Doppler Laser Interferometry with Light Transmission by Two Optical Fibers. In M. Andre and M. Hugenschmidt (eds.), High Speed Photography and Photonics, Proc. SPIE **491**, 894–898.

Ginsberg, M. J., D. E. Grady, P. S. DeCarli, and J. T. Rosenberg (1973). Effects of Stress on the Electrical Resistance of Ytterbium and Calibration of Ytterbium Stress Transducers. Stanford Res. Inst. Final Report, Menlo Park, CA, unpublished.

Goosman, D. R, A. M. Franck, H. H. Chau, and N. L. Parker (1983). Fabry-Perot Velocimetry Techniques: Is Doppler Shift Affected by Surface Normal Direction? In D. L. Paisley (ed.), High Speed Photography, Videography, and Photonics, Proc. SPIE, **427**, 127–135.

Grady, D. E., W. J. Murri, and G. R. Fowles (1974). J. Geophys. Res., **79**, 332-338.

Grady, D. E. and M. J. Ginsberg (1977). J. Appl. Phys., **48**, 2179–2181.

Grady, D. E. (1977). In S. Akimoto and M. H. Mangnani (eds.), High-Pressure Research, Applications in Geophysics, Academic Press, New York, 389–438.

Grady D. E., W. J. Murri, and P. DeCarli (1975). J. Geophys. Res., **80**, 4857–4861.

Grady, D. E. (1985). Sandia National Laboratories, private communication.

Grady, D. E. (1986). High-Pressure Release-Wave Measurements and Phase Transformation in $CaCO_3$. In. Y. M. Gupta (ed.), Shock Waves in Condensed Matter, Plenum Publishing, 589–594.

Graham R. A. and W. P. Brooks (1971). J. Phys. Chem. Solids, **32**, 2311.

Graham R. A. and J. R. Asay (1978). High Temperatures–High Pressures, **10**, 355–390.

Graham, R. A. and R. P. Reed (1978). Selected Papers on Piezoelectricity and Impulsive Pressure Measurement. Sandia National Laboratories Report No. SAND78-1911, unpublished.

Graham, R. A. (1979). J. Phys. Chem., **83**, 3048–3056.

Graham, R. A. and L. M. Lee (1986). Sandia National Laboratories and Ktech Corporation, private communication.

Gupta, Y. M. (1976). Appl. Phys. Lett., **29**, 694–697.

Gupta, Y. M., D. D. Keough, D. Henley, and D. F. Walter (1980). Appl. Phys. Letters, **37**, 395–397.

Gupta, Y. M., D. D. Keough, D. F. Walter, K. C. Dao, D. Henley, and A. Urweider (1980a). Rev. Sci. Instrum., **51**, 183–194.

Gupta, Y. M. (1980b). J. Appl. Phys., **51**, 5352–5361.

Gupta, Y. M. and W. J. Murri (1982). Piezoelectric Shear Stress Gage for Dynamic Loading. In W. J. Nellis, L. Seaman, and R. A. Graham (eds.), Shock Waves in Condensed Matter - 1981, American Institute of Physics, 525.

Gupta, Y. M. (1983). J. Geophys. Res., **88**, 4304–4312.

Gupta, Y. M. (1983). J. Appl. Phys., **54**, 6256–6266.

Gupta, S. C. and Y. M. Gupta (1985). J. Appl. Phys., **57**, 2464–2473.

Hemsing, W. F. (1979). Rev. Sci. Instrum., **50**, 73–78.

Horning, R. R. and W. M. Isbell (1975). Rev. Sci. Instrum., **46**, 1374–1379.

Johnson, P. M. and T. J. Burgess (1968). Rev. Sci. Instrum., **39**, 1100–1103.

Kanel, I., A. M. Moladets, and A. N. Dremin (1978). Combustion, Explosion, and Shock Waves, **13**, 772–777.

Kawaii, H (1969). Jpn. J. Appl. Phys., **8**, 975.

Keough, D. D. (1968). Procedure for Fabrication and Operation of Manganin Shock Pressure Gauges. Stanford Res. Inst. Final Report, Menlo Park, CA, unpublished.

Keough, D. D. and J. I. Wong (1970). J. Appl. Phys., **41**, 3508-3515.

Kim, K. S., R. J. Clifton and P. Kumar (1977). J. Appl. Phys., **48**, 4132–4139.

Koller, L. R. and G. R. Fowles (1979). Simultaneous Generation and Measurement of Longitudinal and Shear Waves in Shock Compressed Media. In K. D. Timmerhaus and M. S. Barber (eds.), High Pressure Science and Technology, Plenum Publishing, New York, **2**, 927–934.

Kormer, S. B., M. V. Sinitsyn, G. A. Kirillov, and V. D. Urlin (1965). Sov. Phys. JETP, **21**, 689–700.

Lee, L. M. (1976). Shock-Induced Index of Refraction Variations in PMMA, Sapphire and Lithium Fluoride. Ktech Report No. TR-76-04, unpublished.

Lee, L. M., W. D. Williams, R. A. Graham, and F. Bauer (1986). Studies of the Bauer Piezoelectric Polymer Gauge (PVF_2) Under Impact Loading. In Y. M. Gupta (ed.), Shock Waves in Condensed Matter, Plenum Publishing, 497–502.

Lipkin, J. and M. E. Kipp (1976). J. Appl. Phys, **47**, 1979-1986.

Lyzenga, G. A. and T. J. Ahrens (1979). Rev. Sci. Instrum., **50**, 1421–1424.

Lyzenga, G. A., T. J. Ahrens, W. J. Nellis and A. C. Mitchell (1982). J. Chem. Phys., **76**, 6282–6286.

Lyzenga, G. A., T. J. Ahrens, and A. C. Mitchell (1983). J. Geophys. Res., **88**, 2431–2444.

Mashimo, T., A. Sawaoka (1981). Jpn. J. Appl. Phys., **20**, 963–970.

Mashimo, T., S. Ozaki, and K. Nagayama (1984). Rev. Sci. Instrum., **55**, 226–230.

McQueen, R. G., S. P. Marsh, J. W. Taylor, J. N. Fritz, and W. J. Carter (1970). The Equation of State Of Solids from Shock Wave Studies. In R. Kinslow (ed.), High Velocity Impact Phenomena, Academic Press, NY, 293.

McQueen, R. G., J. W. Hopson, and J. N. Fritz (1980). Rev. Sci. Instrum., **53**, 245.

McQueen R. G. and J. N. Fritz (1982). Some Techniques and Results from High-pressure Shock-wave Experiments Utilizing the Radiation from Shocked Transparent Materials. In W. J. Nellis, L. Seaman, and R. A. Graham (eds.), Shock Waves in Condensed Matter - 1981, American Institute of Physics, 193–207.

Murri, W. J., D. R. Curran, C. F. Petersen, and R. C. Crewdson (1974). In R. H. Wentorf, Jr. (ed.), Response of Solids To Shock Waves in Advances in High-Pressure Research, Academic Press, NY.

Murri, W. J., D. E. Grady, and K. D. Mahrer (1975). Equation of State of Rocks. Stanford Research Institute Final Report, Menlo Park, CA, unpublished.

Nellis W. J., F. H. Ree, R. J. Trainor, A. C. Mitchell, and M. B. Boslough (1984). J. Chem. Phys., **80**, 2789–2799.

Parker, N. L. and H. H. Chau (1977). Streak Camera Recording of Interferometer Fringes, Lawrence Livermore National Laboratory Report No. UCID-17531, unpublished.

Partom, Y., D. Yaziv, and Z. Rosenberg (1981). J. Appl. Phys., **52**, 4610–4616.

Perez, M. and P. Chartagnac (1980). Rev. Sci. Instrum., **51**, 921–925.

Perez, M. and D. Tougne (1985). J. Phys. Colloq. C5, Suppl. to No. 8, **46**, 655-660.

Rosenberg, Z., D. Yaziv, and Y. Partom (1980). J. Appl. Phys., **51**, 3702-3705.

Rosenberg, Z., Y. Partom, M. Mayseless, and J. Falcovitz (1984). J. Appl. Phys., **56**, 1434–1439.

Rosenberg, Z. and Y. Partom (1985a). J. Appl. Phys., **57**, 5084–5086.

Rosenberg, Z. and Y. Partom (1985b). J. Appl. Phys., **58**, 1814–1818.

Rosenberg, Z. and Y. Partom (1985c). J. Appl. Phys., **58**, 3072–3076.

Rosenberg, Z. and S. J. Bless (1986). J. Appl. Phys., **59**, 3928-3930.

Setchell, R. E. (1979). J. Appl. Phys., **50**, 8186–8192.

Setchell, R. E. (1981). Combust. Flame, **43**, 255–264.

Sugiura, H., K. Kondo, and A. Sawaoka (1980). Rev. Sci. Instrum., **51**, 750–752.

Sugiura, H., K. Kondo, and A. Sawaoka (1981). J. Appl. Phys., **52**, 3375–3382.

Swegle, J. W. (1978). J. Appl. Phys., **49**, 4280–4281.

Swift, H. F. (1982). Image Forming Instruments in Impact Dynamics. In J. A. Zukas, T. Nicholas, H. F. Swift, L. B. Greszczuk, and D. R. Curran (eds.), Impact Dynamics, John Wiley, New York.

Urtiew, P. A. (1974). J. Appl. Phys., **45**, 3490–3493.

Vantine, H., J. Chan, L. Erickson, J. Janzen, and R. Weingart (1980). Rev. Sci. Instrum., **510**, 116-122.

Von Holle, W. G. (1984). Shock Wave Diagnostics by Time-Resolved Infrared Radiometry and Non-Linear Raman Spectroscopy. In J. R. Asay, R. A. Graham, and G. K. Straub (eds.), Shock Waves in Condensed Matter-1983, Elsevier Science Publ., B.V., 283–291.

Wise, J. L. (1984). Refractive Index and Equation of State of a Shock-Compressed Aqueous Solution of Zinc Chloride. In J. R. Asay, R. A. Graham, and G. K. Straub (eds.), Shock Waves in Condensed Matter-1983, Elsevier Science Publ., B.V., 317–320.

Wise, J. L. and L. C. Chhabildas (1986). Laser Interferometer Measurements of Refractive Index in Shock-Compressed Materials. In Y. M. Gupta (ed.), Shock Waves in Condensed Matter, Plenum Publishing, 441–454.

Yaziv, D., Z. Rosenberg, and Y. Partom (1980). J. Appl. Phys., **51**, 6055–6057.

Young, C. and O. Dubugnon (1977). Exp. Mech., **17**, 470–477.

Zukas, J. A. (1982). Penetration and Perforation of Solids. In J. A. Zukas, T. Nicholas, H. F. Swift, L. B. Greszczuk, and D. R. Curran (eds.), Impact Dynamics, John Wiley, New York.

Int. J. Impact Engng Vol. 5, pp. 221–237, 1987
Printed in Great Britain

0734-743X/87 $3.00 + 0.00
Pergamon Journals Ltd.

HYPERVELOCITY IMPACT IN METALS, GLASS AND COMPOSITES

Burton G. Cour-Palais
NASA Lyndon B. Johnson Space Center
Houston, Texas 77058

ABSTRACT

This paper is a review of hypervelocity impact research carried out during the intense activity period leading up to the Apollo lunar missions. It is intended as a historical note on the research into hypervelocity impact phenomena in metallic, glass, and composite materials and the spacecraft applications of that research. The specific areas covered include cratering and spallation in thick, semi-infinite targets, perforation and hole formation in thin, single-thickness targets, spaced dual sheet armor, impact radiation, and impact ionization. Optimum and nonoptimum dual sheet combinations are treated in some detail because of the current interest in hypervelocity impact protection for the Space Station. On the other hand, the treatment of hypervelocity impacts on composites, phenolic resins and thermosetting epoxy systems reinforced with graphite or other high strength fibers, is limited because work in this area has just begun.

INTRODUCTION

Some twenty years after the Apollo designers had to contend with the unknown consequences of meteoroid impacts on manned spacecraft there is a new concern. A manmade orbiting debris population has been and is being added to the natural meteoroid environment and poses a threat to long-duration, long-term manned missions in near-Earth space (Kessler, 1981; 1982; 1985; Kessler and Cour-Palais, 1978). A sense of complacency seems to have pervaded the planners of new space operations after Skylab survived in Earth orbit without any penetration of the habitable volume for 6 years. However, there is now an awareness that the designers of new space ventures will have to contend with the consequences of rather poorly controlled littering of the near-Earth environment with dead spacecraft or fragments of spacecraft. Now that civilian and military space planners have begun to understand the problem, greater efforts to control the unnecessary littering of space is expected in the future.

The purpose of this paper is to provide some insight into the effect of hypervelocity impacts on spacecraft, the ability to take advantage of certain physical and dynamic aspects of an encounter with very high speed "projectiles" to minimize shielding, the nature of the threat that we are facing today, and the direction of the research being conducted in the civilian space arena. This includes a selective review of the pertinent work done by a number of investigators.

The Environment

What **was** the threat to manned space operations when President Kennedy set a national goal of landing a man on the Moon's surface by 1970? Very high speed solid particles, or meteoroids, had been observed as "meteors" in the atmosphere from the beginning of recorded history. Those that survived atmospheric ablation and were recovered as "meteorites" were broadly classified as stones or irons. This meant that some of the impacting particles could have a high mass density and could be more penetrating than similar mass low-density meteoroids.

Scientific observations of meteors began in the United States in 1911 with visual observations by the American Meteor Society (McKinley, 1961). McKinley states that the modern technique of photographic observations was initiated by Whipple in the 1930's as part of the Harvard Observatory's regular night sky patrol program. The two modifications that enhanced the value of photographic records for meteor work were the rotating shutter placed in front of the lens and the mounting of two cameras 38 km apart directed at the same sky area (Whipple, 1938). These two developments were used to determine the angular velocity and the path parameters of the meteors detected on the photographic plates (McKinley, 1961).

The advent of the Super-Schmidt camera, designed by James Baker of Harvard specifically for meteor work, led to an intense study of meteors by the Harvard group in New Mexico between March, 1952, and October, 1953 (Whipple, 1952, 1954, 1958). This work resulted in the determination of meteor influx rates as a function of photographic magnitude (Hawkins and Upton, 1958) and the calculation of a meteoroid velocity distribution relative to the Earth (Hawkins and Southworth, 1958; McCrosky, 1957). Other experimenters in Europe and Canada also determined meteor influx rates from visual and telescopic observations (McKinley, 1961).

Relationships between the observed meteor magnitudes and the corresponding in-space meteoroid masses were required before a design meteoroid mass distribution could be derived. McKinley (1961) covers the attempts by Whipple, Opik, and others to derive an acceptable value for the mass of a zero magnitude meteoroid, which was a derivative of the assigned mass density. The mass density of most of the photographic meteors was assumed to be under 1 g/cm^3 due to the evidence of fragmentation high in the atmosphere (Jacchia, 1955).

It was apparent from the available mass distributions that the photographic and visual meteoroids were not a problem to the Apollo missions due to the very low probability of a damaging impact. However, it was known that meteoroids that did not leave a luminous trail in the atmosphere could be detected by radar and that these were far more numerous (McKinley, 1961). Also, the direct measurement of meteoroid impacts by satellites with various types of detectors had become a viable option (Alexander and colleagues, 1963). These results also showed that there were many more potentially damaging particles in space than had been detected by either the photographic records or the ground-based radar observations.

Under the impetus of the Apollo Program, the necessary radar and satellite measurements were made (Clifton and Naumann, 1966; Elford, 1967; Hastings, 1964; Hawkins, 1963; McKinley, 1961; O'Neal, 1968) and a design meteoroid environment model was derived for the Apollo Program in consultation with Whipple and the other experimenters at the Smithsonian Astrophysical Observatory and the NASA Centers (Cour-Palais, 1969a). This design environment model specified an Earth-centered velocity distribution which had an average of 20 km/s, a mass density of $0.5 g/cm^3$, consistent with the observations of meteoroid fragmentation in the upper atmosphere as reported by Jacchia (1955), and a cumulative distribution of masses between a gram and a nanogram. The distribution for the gram to microgram range is expressed in logarithmic form as

$$\log N = -14.37 - 1.21 \log m \tag{1}$$

where N is the integrated flux of particles passing through one side of a square meter area every second, and m is the mass of the particle in grams.

The flux of particles between a microgram and a nanogram with the same units is

$$\log N = -14.34 - 1.58 \log m - 0.06(\log m)^2 \tag{2}$$

Several other environment models of the meteoroid population in the Earth-Moon region of space have been formulated as the result of investigations of the returned lunar samples and spacecraft measurements since the end of the Apollo Program.

The orbital debris environment has had a similar history. Ground-based radar observations have catalogued Earth-orbiting objects to a limiting diameter of about 10 cm at an altitude of 1000 km (Kessler, 1981; Kessler and Cour-Palais, 1978). Taff and his colleagues (1985) report that they have been able to discriminate debris fragments as small as 1 cm with their electro-optical system, and impacts observed on spacecraft have confirmed the presence of debris with particle diameters in the micrometer to the sub-millimeter range (Kessler, 1985). Intermediate sizes have not been confirmed, but they are expected to exist based on laboratory hypervelocity tests. Sufficient information is available, however, to provide designers with a reasonably secure model of the cumulative number versus size distribution of the near-Earth orbital debris for diameters ranging from about 10 meters to a tenth of a millimeter (Kessler, 1982). The average collision velocities between a spacecraft in 30 and 60 degree inclination near-Earth orbits and orbiting debris fragments are 10 and 14 km/s, respectively (Kessler, 1982). Kessler states that these high speeds are a result of the concentration of debris at these and higher latitudes and the rapid precession of their ascending nodes (Kessler and Cour-Palais, 1978). The average mass density of debris fragments 1 cm and smaller in diameter is expected to be that of aluminum, 2.78 g/cm^3.

Impact Damage in Metallic Targets

Thick targets. The damage apparent to the block of aluminum depicted in Fig. 1 dramatically demonstrates the damage that results from a hypervelocity impact into a solid, metallic target. What is seen is a section through the crater and rear surface spall produced by a polyethylene projectile, like the one shown on the base of the stand, impacting a block of 1100-0 aluminum alloy at 7 km/s. The target was impacted at the GM Defense Research Laboratories, now Delco, at Santa Barbara, California, during the 1960's. What is seen is a nearly hemispherical, smooth crater with a raised lip on the front surface. This is typical of the hydrodynamic phase of a very high speed impact, when the impact pressure exceeds the strength of the target material. If the block had been made of a more brittle material, the lip would have been missing and the crater less smooth.

Other typical features of a hypervelocity impact on a solid target are seen at the other end of the target. The most prominent of these are the almost detached spalls from the back and top sides. Spallation is the result of the block not being thick enough to be "semi-infinite", that is, the reflected tensile stress wave from the rear free surface is still higher than the critical normal fracture stress (Rinehart and Pearson, 1963). A target can be considered to be semi-infinite when the rear surface is just on the point of separation. In this target, we see that the projectile hit off the center of the block; the top edge has started to spall, but the bottom edge has not. Another interesting feature of this target is the corner cracking at 45 degrees. This is brought about by the resultant stress caused by the meeting of two reflected tensile stresses from the free surfaces of the block. Notice that at the bottom of the block the crack can occur without the spall.

The size and impact energies of the meteoroids most likely to be encountered during the Apollo missions were simulated by launching particles ranging from 1.27 cm to 50 µm in diameter at hypervelocities. Typical two-stage light-gas gun velocities for projectiles between 0.16 cm and 1.27 cm were from about 6 to nearly 9 km/s (Denardo, Nysmith, and Summers, 1967; Goodman and Liles, 1963; Halperson, 1963). However, the 50 µm projectiles could be accelerated to between 12 and 18 km/s using an electro-thermal gun (Scully and colleagues, 1965).

According to Gehring (1970b) it was well established that crater shapes tended to be hemispherical in the hypervelocity regime and that the crater dimensions scaled linearly with the projectile diameter for any velocity. Figure 2 is a plot of the projectile diameter to crater depth ratio, p/d, as a function of the impact velocity V in km/s, for spheres ranging from 50 µm to 1.2 cm. It is apparent that the 50 µm data falls below the other projectile results. The validity of this effect was confirmed by the author of this report by plotting the projectile diameter as a function of p/d for three widely separated pyrex glass sphere sizes, 0.635 cm, 0.4 mm, and 50 µm impacting a 2024-T4 target at 7 km/s. It was found that the p/d scaled as the 19/18 power of the projectile diameter for this velocity and for these projectile and target materials. If the results of Fig. 2 are replotted with the projectile diameter raised to the 19/18 power, as shown in Fig. 3, all the data points are in much better agreement. This size scaling factor was discussed with Maiden (1964) who felt that the nonlinearity must apply to the strength dependent phase of the cratering process, as the crater to diameter ratio must be linear during the hydrodynamic phase.

Fig. 1. Sectioned hypervelocity impact crater and spallation in a 1100-0 aluminum block.

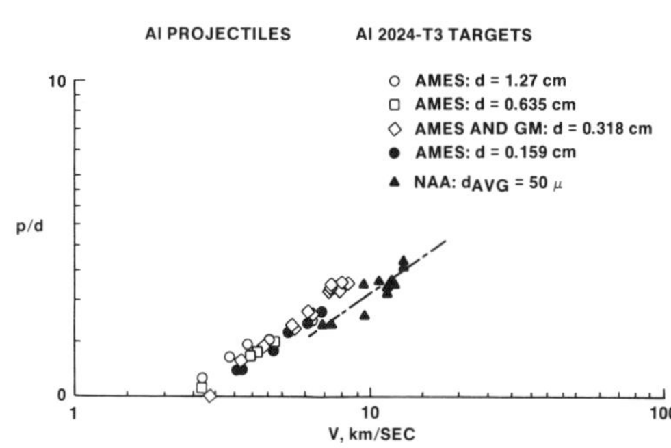

Fig. 2. Penetration depth to projectile diameter ratio as a function of velocity without diameter scaling.

While investigating the momentum transfer associated with the hypervelocity impact of aluminum spheres into thick aluminum targets at velocities up to 7.3 km/s, Denardo and Nysmith (1964), found that a correction factor using the projectile diameter raised to the 1/18 power improved the correlation of the crater depths and diameters.

Denardo, Summers, and Nysmith (1967) reported the results of a detailed experimental investigation of this phenomenon with projectiles ranging between 1.27 cm and 0.16 cm at impact velocities up to 8.5 km/s. They concluded that the agreement between macroparticle and microparticle penetration, with allowance for the 1/18 power law for diameter scaling, was remarkable for the velocity range that they tested. They further stated that in conjunction with the results obtained by Slattery and Friichtenicht (1966), with hypervelocity microparticle impacts into 1100-0 aluminum targets, the 1/18 scaling law extended over a diameter range of four orders of magnitude. This was significant because it allowed the higher speed data available for the small projectiles.to be used to extend the results obtained for the larger, relatively slower ones.

The size scale factor was later shown by Gault and Moore (1965) to be related to the rupture stress and mean deformation strength of the target material. They stated that the exponent of the projectile diameter varies between 1.2 for very brittle materials such as rock and glass to nearly one for a ductile aluminum alloy. Subsequent analytical studies (Riney and Heyda, 1964; Rosenblatt, 1970; Sedgwick, 1966) showed that this nonlinear scaling phenomenon could be accounted for by the inclusion of material strain-rate and incipient melting effects in their numerical descriptions of the impact process. Diameter scaling remains an important factor in the comparison and the extrapolation of test data over a significant size range.

The strength and ductility of the target material also influence the final dimension of the crater in a semi-infinite target. In Fig. 4 the penetration depth to projectile diameter ratio, with size scaling, is plotted as a function of velocity for two aluminum alloys, 1100-0 and 2024-T3. The 1100-0 is a very soft, ductile material with a Brinell Hardness Number of 27. The 2024-T3 is harder and less ductile, and its Brinell Hardness number is 145. A comparison of the laboratory data from two different sources in Fig. 4 shows that the difference in the depth to diameter ratio is distinct, and the reasons for this are understood (Holsapple and Schmidt, 1982; Meyers and Charest, 1964; Prater, 1970). As mentioned previously, the initial shock pressure during a hypervelocity impact is high enough to cause the target material to behave as a fluid. Crater growth during this phase is identical for the hard as well as the soft aluminum alloy, because their relative strengths are not important. As the shock pressure decays and the material properties begin to dominate again, the crater dimensions are eventually "frozen" at different times for the two alloys. To illustrate this, the 1100-0 alloy has a yield stress of $4.4 \times 10^{10}\ N/m^2$ and an ultimate strength of $8.96 \times 10^{10}\ N/m^2$. The comparable values for the 2024-T3 are $2.48 \times 10^{11}\ N/m^2$ and $3.31 \times 10^{11}\ Nm^2$, respectively. For a given impact shock pressure and similar decay characteristics, the 2024-T3 crater will stop growing before the 1100-0 crater. According to Meyers and Charest (1964) the 2024-T3 crater growth limitation begins at 21 μs after impact, whereas the 1100-0 crater continues for another 16 μs. The same authors also state that the final crater dimensions for the 1100-0 case can be as much as 15% less than the maximum achieved during the impact process. This rebound effect is probably more noticeable for the ductile materials and is a transient event that can only be detected by x-radiography.

Several empirical hypervelocity equations describing crater formation in semi-infinite metallic targets were derived during the Apollo era (Herrmann and Jones, 1961; Herrmann and Wilbeck, 1986;

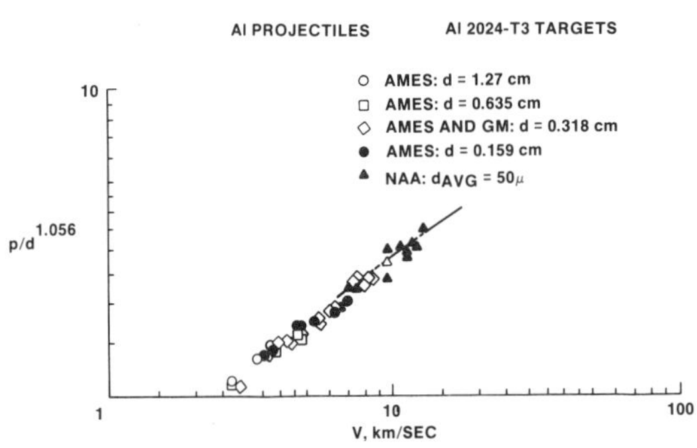

Fig. 3. Penetration depth to projectile diameter ratio as a function of velocity with diameter scaling.

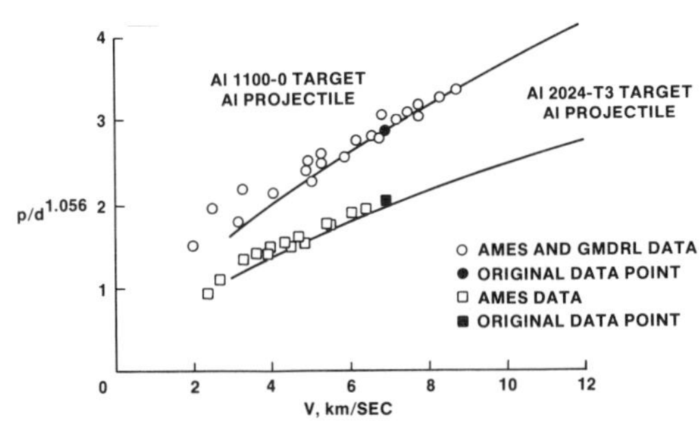

Fig. 4 Penetration depth to projectile diameter ratio as a function of velocity with diameter scaling and material strength effects.

Wenzel and Gehring, 1965). Gehring (1970b) discusses two of these equations and describes some of the fundamentals involved in deriving practical cratering equations. Empirical equations are considered valid only a few kilometers per second beyond the limits of the hypervelocity test data used for their derivation. Extrapolation to meteoroid or orbital debris velocities depended on the results obtained from numerical calculations based on an understanding of the impact process (Bjork, Kreyenhagen, and Wagner, 1967; Dienes and Walsh, 1970; Riney, 1964; Rosenblatt, 1970; Wagner, Brooks, and Bjork, 1965; Walsh and Johnson, 1965). For the average debris encounter velocity, 10 km/s, for a spacecraft in a 30 degree inclination orbit, little or no extrapolation is necessary.

Experimental results at the highest light-gas gun speeds attained in the laboratory favored a velocity scaling to the 2/3 power (Halperson, 1963; Sorensen, 1965; Wenzel and Gehring, 1965). Numerical analyses by Walsh and his colleagues (1964, 1965) showed that the crater dimensions increased as the 0.58 power of the impact velocity for velocities that were greater than about twice the target sound speed. However, Riney and Heyda (1964) concluded that the depth of penetration in a thick target varies with the impact velocity to the 2/3 power in the hypervelocity regime. Both groups agreed that there was no effect of projectile density on the crater dimensions at constant mass, provided that the target and projectile densities differed by no more than a factor of three. They also found that the target strength played an important role in limiting the crater size at all hypervelocity levels.

Light-gas gun hypervelocity test results, plotted as p/d as a function of velocity, could not differentiate between the 0.58 and 2/3 velocity scaling (Halperson, 1963; Wenzel and Gehring, 1965). Wenzel and Gehring also show comparisons between shaped charge hypervelocity test results to 16.3 km/s and the same two velocity power law dependencies, but again without clearly settling the issue. The empirical equation, given by Wenzel and Gehring (1965) in which the penetration depth in a semi-infinite target is proportional to the 1/3 power of the impact energy, correlates well with the shaped-charge data to over 16 km/s. Kineke (1965), using impact data obtained by various explosive devices to 12 km/s with steel, aluminum, and titanium projectiles and copper and lead targets, finds good correlation with velocity scaling to 0.58.

Recent investigations of the existing theoretical and empirical scaling laws for cratering phenomena by Holsapple and Schmidt (1980, 1982), originally concerned with large terrestrial and planetery craters where gravity and strength are very significant factors, have led to a better understanding of scaling in general. They show that scaling laws are more complicated than the preceding discussion might imply and that energy scaling, momentum scaling, strength scaling, and gravity scaling are points on an envelope of permissible scaling rules.

Thin or plate targets (less than semi-infinite). Up to this point, the discussion has been about semi-infinite targets. If the target is made progressively thinner and if the projectile mass and velocity are kept constant, a piece of the rear surface will spall as can be seen in Fig. 1. The size and cohesiveness of the spall depends on the material properties, as well as the strength of the shock as it reaches the rear surface of the target. Further reduction of the target thickness will result in the crater and the spall meeting, and the target is perforated. Since it is more often necessary in aerospace applications to assess the penetration resistance of targets that are much thinner than semi-infinite, "finite thickness" equations were needed. The first approach was to determine empirical finite thickness factors, applicable to varying degrees of spallation, to be applied with the crater depth equation (Gehring, 1970b). The spall factors reported by Gehring were for 2024-T3 aluminum alloy targets impacted by 0.32 cm diameter aluminum projectiles at a nearly constant velocity of 7.4 km/s. Spallation began to occur at a thickness of three times the crater depth and broke away when the thickness of the target was about 2.3 times the crater depth. Perforation resulted when the thickness was only 1.8 times the crater depth.

Although these factors were determined for a specific brittle aluminum alloy, they were used for other aluminum alloys with similar properties. The point is that finite thickness equations were not formulated early in the Apollo Program, and the procedure of applying spall factors to cratering equations was fairly standard procedure, until Fish and Summers (1965) determined the perforation limit of a number of materials used in satellite meteoroid impact detectors. These authors defined this limit as the velocity at which a given target was damaged to the point that it could not sustain a pressure differential of 1×10^5 N/m^2 without leaking. The tests were performed with 0.16 cm aluminum projectiles at hypervelocities up to 8.5 km/s, and they found that the perforation limit varied inversely with the square root of the density of the target material and the 1/18 power of its ductility. The Fish-Summers perforation formula is usually written in a more convenient form (Frost, 1970; Naumann and Jex, 1969) as follows

$$t = K \times (\rho)^{0.148} \times (m)^{0.352} \times (V)^{0.875} \qquad (3)$$

in which the units are V in km/s, m in g, ρ in g/cm3, and K is a constant for the material calculated from its elongation, expressed as a percentage, and mass density. Values for K for some of the more commonly used spacecraft and satellite penetration materials are given by Naumann and Jex (1969) and Frost (1970). For aluminum 2024 alloy in the T3 condition, K is 0.54, and for the annealed aluminum alloy 1100-0, K is given as 0.52.

Naumann and Jex (1969) calibrated the Explorer 23 and Pegasus meteoroid satellite impact detectors made from 0.0025 and 0.005 cm stainless steel, and 0.023 and 0.041 cm aluminum, respectively, using 24 to 33 μm borosilicate projectiles accelerated to 18 km/s in an electro-thermal gun facility (Scully and colleagues, 1965), and 40 to 200 μm borosilicate and plastic projectiles to 8 km/s using light-gas guns. They concluded that the Fish-Summers equation could not be extrapolated to very thin, metallic materials, because the perforation mechanisms were different than those normally seen in the thicker materials. Naumann and Jex (1969) mentioned that in several instances the crater depths exceeded the thickness of the sheet, indicating that the material had been drawn in the impact process.

Thin sheet spaced targets. Although there are many instances in which spacecraft structures can be adequately protected by a single thickness of material against hypervelocity impact, weight considerations usually prevent such protection. The principle of placing a thin sheet out in front of the main structural wall as a "bumper" was well known before the Apollo Program began, having been first suggested by the noted astrophysicist Whipple (1946, 1952). The initial tests on spaced targets confirmed that as a defense against hypervelocity impacts, a structure consisting of two thin sheets was significantly lighter and better than a single thickness of material.

During the early days of practical shielding design, it was necessary to arrive at the total thickness of a dual-sheet combination by comparing it with the equivalent single sheet or plate thickness. Hypervelocity tests had determined these equivalences for the same "ballistic limit" (Nysmith and Summers, 1962). The ballistic limit was defined by these authors as the velocity required to cause perforations resulting in a loss of pressure with a differential of 10^5 N/m^2. Nysmith and Summers (1962) tested several dual-sheet combinations at hypervelocities, varying the thicknesses, spacing, and materials, and later Nysmith (1968) updated and extended this work to 8.8 km/s. Posever and Scully (1965) investigated dual-sheet configurations of aluminum alloy 5052-0 foils, 0.013 to 0.13 mm thick, using their electro-thermal gun facility. They obtained over 2700 data points with 50 to 75 μm borosilicate projectiles impacting at 15 to 18 km/s.

A large amount of research was done on the effects of varying bumper materials and thicknesses on the damage to the second sheet of a dual-wall configuration, and the properties required for the optimum performance of the bumper were identified (Gehring, 1970a; Maiden, McMillan, and Sennett, 1965; Meyers and Charest, 1964; Swift and Hopkins, 1969). The factors involved in the formation of the hole in the bumper were also studied (Nysmith and Summers, 1962; Swift and colleagues, 1969; Turpin and Carson, 1970), and several empirical equations were developed (Carey, McDonnell, and Dixon, 1985; Gehring, 1970a; Nysmith, 1968; Nysmith and Denardo, 1969; Sawle, 1969). The hole diameter equations (Herrmann and Wilbeck, 1986) included parameters such as the impact velocity, sheet thickness, and projectile diameter, and material terms such as density, strength, and shock properties for the projectile and the target. Comparisons of some of these equations with data are given by Turpin and Carson (1970) for typical spacecraft sheet thicknesses and by Carey, McDonnell, and Dixon (1985) for foil thicknesses.

Phenomenological approaches for sizing the elements of multielement meteoroid protection, in which the elements were dealt with separately, made it possible to optimize the beneficial properties of the bumper, the spacecraft shell or second sheet, and the spacing between them (Cour-Palais, 1969; Gehring, 1970a; Maiden, McMillan, and Sennett, 1965; Richardson, 1969; Swift and colleagues, 1982; Wilkinson, 1968; Woodall, 1967). Because dual or multisheet protection for space vehicles is still the most effective passive defense against hypervelocity projectiles, a discussion of some of the principles involved follows.

The outer wall, generally known as the "bumper" or shield, is the key element, and it determines the condition of the projectile after initial impact. When a metallic target subjected to hypervelocity impact is made progressively thinner to the point at which complete penetration occurs, an expanding cloud of projectile and target material is generated. This cloud can consist of solid, liquid, or vaporized material, or a combination of all three, depending on the initial impact pressure.

Figure 5 shows the results of an aluminum-aluminum impact by a 0.95 cm projectile at nearly 7 km/s, with an initial impact pressure of approximately 10^{11} N/m^2. On the right is a view of a 0.16 cm shield with an intact projectile on it for scale. On the left is the front surface of the second plate which was spaced 30.5 cm behind the shield. The brightness of the impact zone shows that the material was in the liquid state. The surface is roughened by multiple small craters in the central zone, and their density falls off with distance from the center of impact. Also, there is a radial pattern with strings of very small craters. All of these effects are a result of the remains of the projectile and of the shield hole material, and they illustrate rather dramatically how the shield works. The impact of the debris cloud on the second or backup

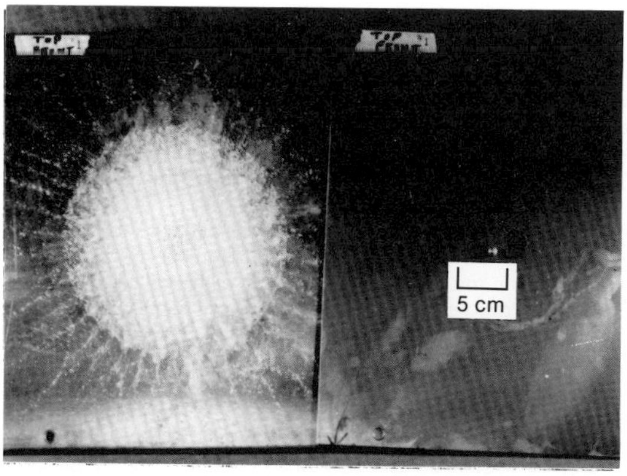

Fig. 5. Characteristic hypervelocity impact damage patterns for dual-sheet configurations: surface damage with no penetration.

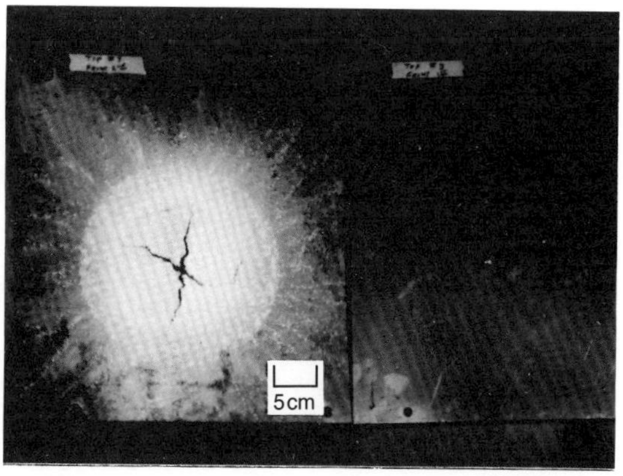

Fig. 6. Characteristic hypervelocity impact damage patterns for dual-sheet configurations: rupture, front view.

sheet can result in various types of damage. In the case shown in Fig. 5, the damage is superficial because the backup sheet is too thick to be efficient. If it had been thinner, there would have been spall off the rear surface.

The next stage in the dual-sheet interaction with a hypervelocity projectile is illustrated in Fig. 6. With the same projectile, velocity, and spacing, the backup sheet has experienced a pre-petalling or rupture failure. The wall thickness was reduced to the point at which it could not withstand the impulse generated by the debris from the shield, and it split. The rear of this same plate is seen in Fig. 7, which shows the deep drawing of the material that preceded the split. Obviously, the ultimate tensile allowable and ductility play a large part in minimizing this type of failure. Another interesting phenomenon, seen in Fig. 6, is the aluminum vapor deposit on the rear surface of the front sheet. The secondary impact of the molten debris from the bumper interaction, at 80 to 90% of the initial projectile velocity (Swift and colleagues, 1969), is sufficient to exceed the vaporization temperature of aluminum. The vapor cloud is reflected from the front of the second sheet onto the rear of the front one, coating it in the process.

Figure 8 shows yet another type of failure in which the second sheet is perforated by numerous small crater and spall combinations. The front surface is easily recognized by the familiar bright, molten splatter from the bumper. In this case, the only variation from the target shown in Fig. 7 was that the bumper thickness was tripled. Other illustrations of second-sheet failure and some excellent time-resolved photography of the shield debris and its interaction with the second sheet are given by Swift and Hopkins (1969) and Gehring (1970a).

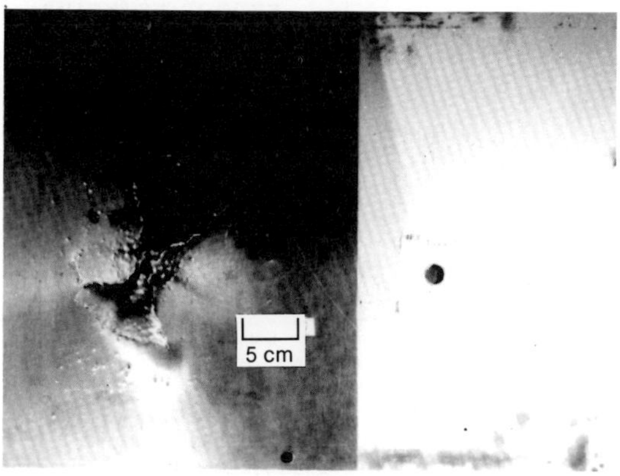

Fig. 7. Characteristic hypervelocity impact damage patterns for dual-sheet configurations: rupture, back view.

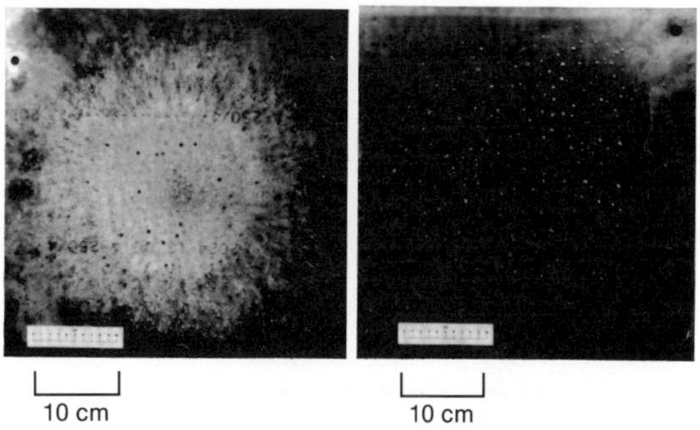

Fig. 8. Characteristic hypervelocity impact damage patterns for dual-sheet configurations: perforations.

The reaction of the second sheet of a dual-sheet configuration is, as has been mentioned, conditioned by the state of the debris from the projectile-bumper interaction. Gehring (1970a) summarized the work done by the General Motors group during the 1960's on the response of the second sheet to an optimum impact situation, which was achieved by using cadmium projectiles impacting cadmium shields at 6.5 km/s. The momentum imparted to the second sheet was measured and modelled, and an equation derived to determine the second-sheet thickness for fracture and yielding. The thickness, t_b, of the second sheet may be calculated from the equation

$$t_b = C \times (m \times V)/S^2 \qquad (4)$$

where C = 41.5+ or -14 and 8.2+ or -1.4 for the 0.2% yield and fracture criteria for a 7075-T6 aluminum, respectively, t_b is in cm, V is the initial impact velocity in km/s, and S is the space between the bumper and the second sheet in cm. The 0.2% yield equation will give a sheet thickness that should not deflect or spall. On the other hand, the fracture criterion will result in a second sheet that may fail in tension, as depicted in Fig. 6.

When the projectile and shield are not in an optimum relationship, the second sheet is subjected to hypervelocity fragments from the shield and the molten residue of the projectile. Some of the nonoptimum projectile and shield combinations that result in more serious consequences to the second sheet are shown by Cour-Palais (1969b) and Gehring (1970b). Nonoptimum shield conditions are always the case for any dual-sheet configuration designed to operate in an environment with the size distributions previously described for the meteoroid and the orbital debris populations. The reason for this is that if a bumper is designed to be optimum for a particular design particle size, it is nonoptimum for both smaller and larger particle sizes. As the smaller particles are more numerous for either of the environments mentioned, the probability of penetration is higher than it was designed to be with an optimum bumper. Nonoptimum dual-sheet combinations were investigated by Cour-Palais (1969b, 1979), Gehring (1970b), and Swift and colleagues (1969), which resulted in various approaches to solving the problem, including qualification of a configuration by hypervelocity testing. Cour-Palais (1969) derived an empirical nonoptimum equation for the thickness of the second sheet, t_b, as follows

$$t_b = C \times m^{0.33} \times V/(S \times Y)^{0.5} \qquad (5)$$

where C = 19.8 for projectiles up to 0.32 cm, m is the mass of the projectile before impact in g, V is the initial impact velocity in km/s, S is the space between the two sheets in cm, Y is the 0.2% yield stress of the second sheet material, and t_b is the required thickness in cm.

One way to depict the behavior of a dual-wall configuration subjected to hypervelocity impact is a plot such as Fig. 9 (Cour-Palais, 1969b, 1979; Gehring, 1970b; McMillan and colleagues 1966). The ratio of the sum of the shield thickness, t_s, and the backup thickness, t_b, to the projectile diameter is plotted as a function of the shield thickness to projectile diameter ratio. In Fig. 9 the data is for the aluminum alloy 7075-T6, a velocity of 7.4 km/s, a spacing of 5.08 cm, and 0.16 and 0.32 cm projectiles. The data at t_s/d zero correspond to a backup sheet without a bumper, and the value used for the calculated point is for a finite plate that is cratered and

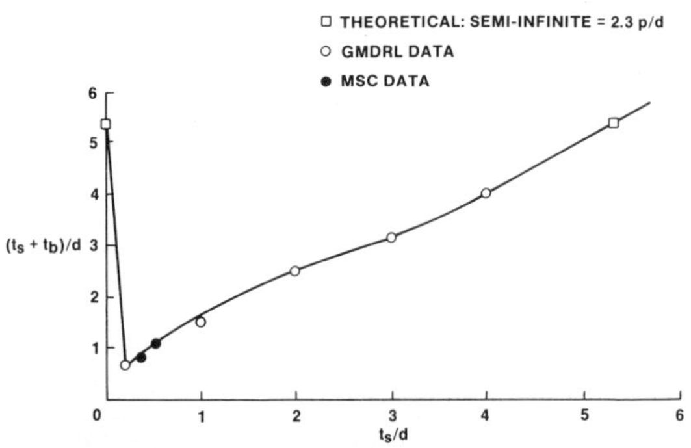

Fig. 9. Spectrum of nondimensionalized total thicknesses of spaced sheets required to prevent hypervelocity penetration, expressed as a function of shield thickness. The data refers to 7075-T6 aluminum sheets spaced 5.08 cm apart impacted at 7.4 km/s.

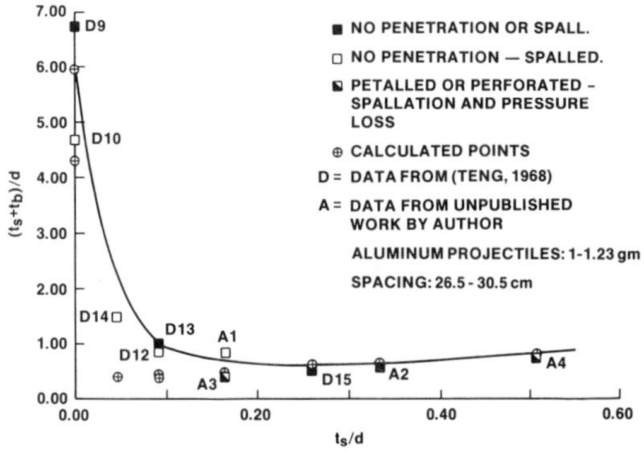

Fig. 10. Nondimensionalized total thicknesses of spaced sheets required to prevent hypervelocity penetration by 0.9 cm aluminum projectiles, for a limited range of shield thickness. The data refers to 2024-T3 aluminum sheets spaced 30 cm apart and impacted at approximately 7 km/s.

spalled but not penetrated. The thickness of the plate was calculated using a factor of 2.3 times the crater depth for the target condition described. As t_s/d increases, the total thickness decreases until it reaches a minimum at a t_s/d of about 0.15, which for this combination of projectile and target, is the optimum dual-sheet structure. Further increases in the ratio of shield thickness to projectile diameter is accompanied by an increase in the total thickness, as shown by the data points on the plot. The larger values of t_s/d, 1 through 4, are characterized by cratering and spall detachment from the rear surface of the bumper and fragments that can perforate the second wall (Gehring, 1970b). The plot in Fig. 9 is terminated at a value of t_s/d which corresponds to a shield which is thick enough by itself to prevent spall detachment, and consequently, the backup wall is unnecessary. Figure 9 illustrates the complete spectrum of effectiveness of the dual-sheet protection technique although it is only necessary to consider the minimum thickness portion of the curve such as is shown in Fig. 10. The data for Fig. 10 was obtained by the author for aluminum projectiles 0.88 and 0.95 cm diameter, impacting at approximately 7 km/s on two aluminum alloy 2024-T351 sheets spaced at 30 to 40 times the projectile diameter. The points marked A1, A3, and A4 are for the same targets shown in Figs. 5, 6, 7, and 8; A2 is an intermediate point with one small perforation and no penetration or spall. The other points marked "D" are data from an earlier series of tests (Teng, 1968) that represent other stages of the shield and backup interaction. The line through the data points divides the region between no penetration or rear surface spallation, and perforation or rupture leading to a loss of internal pressure. The points marked with crosses represent the calculated values for the borderline between the safe and the unsafe dual-wall protective structures, and although the plot is strictly valid between 4 km/s and 7 km/s, it can be used up to 10 km/s, the average debris impact velocity given previously in this paper. The calculated points correspond to the use of the nonoptimum backup thickness equation (5) with $C = 31.75$.

Figure 11 is a plot of the optimum shield thickness to projectile diameter as a function of velocity for an aluminum shield impacted by aluminum and glass projectiles. The solid line, reproduced from Maiden and his colleagues (1963, 1965), is the boundary between molten and solid projectile fragments, and the dashed line marks the melting threshold of the shield "hole". Aluminum is completely molten at an impact pressure of 0.88×10^{11} N/m^2 which occurs at 6.6 km/s for an aluminum on aluminum impact (Swift, 1982). The hole diameter equation given by Gehring (1970a) was used to calculate the amount of material to be melted, and the heat of fusion, Hf, was assumed to be 4% of the heat available in the impact. Data points in Fig. 11 were obtained by the author and are seen to agree with the projectile and shield melting thresholds. The open symbols are for aluminum projectiles, which left a bright splatter on the second sheet, and the solid symbols indicate results from glass projectiles impacting the aluminum shields. Glass projectiles were used because they exceed the melting impact pressure for aluminum at 7 km/s and allow the molten shield splatter to be clearly distinguished from the projectile. The zone between the solid and the dashed lines gives the range of t_s/d values for complete melting of the projectile and the shield. At laboratory light-gas gun speeds, and with projectiles up to 0.32 cm in diameter, the range of optimum t_s/d values is quite limited, but at the higher speeds there is a greater latitude for optimization. It is expected that the optimum zone will be much larger for the 1 cm projectiles as is noticeable in Fig. 10.

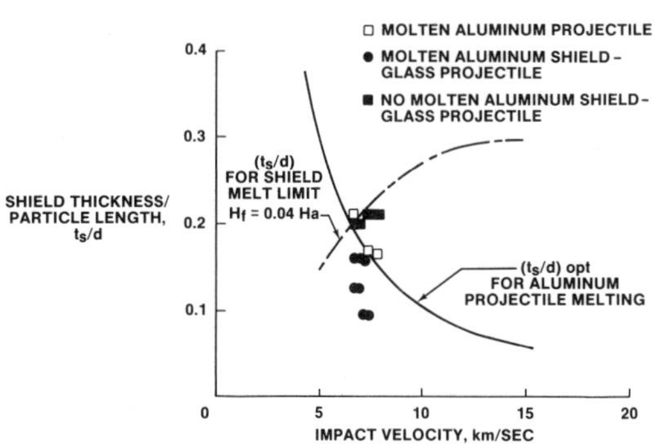

Fig. 11. Nondimensionalized limiting thickness of aluminum shields for complete melting of the projectile and shield plug material expressed as a function of impact velocity.

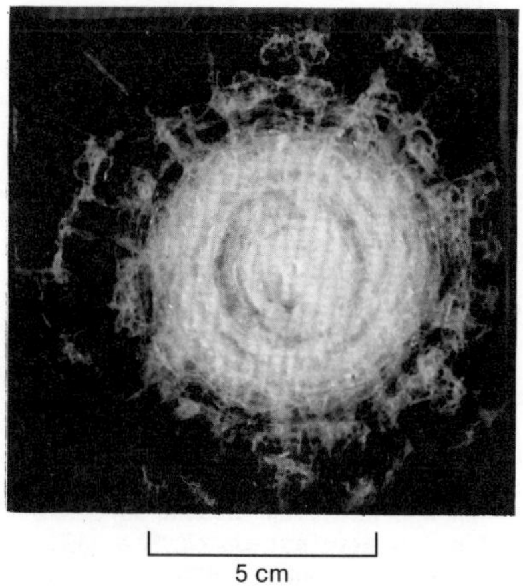

Fig. 12. Top view of hypervelocity impact damage in a fused silica (glass) target.

The previous discussion is applicable to hypervelocity impacts at right angles to the target. Most meteoroid and orbital debris impacts will be at angles between 30 and 50 degrees to the surface normal axis. McMillan and colleagues (1966) investigated oblique angle impacts into dual-wall, spaced targets at 30, 45, and 60 degrees and velocities in excess of 7 km/s. They found that unlike normal impacts at these speeds, the major source of damage was caused by fragments from the shield. The larger amount of discrete fragments can be expected because of the lower peak pressures associated with oblique impacts. McMillan and his colleagues state that the degree of fragmentation decreases as the angle from the normal is increased, with the greatest damage to the wall occurring at 45 degrees. Because the normal component of the impact velocity determines the initial impact pressure, the fragmentation problem will be lessened as projectile and target again achieve postimpact melting temperatures. This will begin to happen when the normal component of the velocity is above 7 km/s. It is expected that the zero degree impact total dual-wall thickness, as determined by a plot such as Fig. 11, will apply to the 30 to 45 degree case, if the normal component of the impact velocity is substituted.

It is very important to understand that the dual-wall configuration, capable of defeating a hypervelocity meteoroid or orbital debris particle, is vulnerable to projectiles that are not travelling fast enough to be broken up by the shield. Investigators of representative hypervelocity dual-wall combinations at speeds below 7 km/s (Gehring, 1970b; McMillan and colleagues, 1966) found that the maximum penetration effect, and consequently the thickest rear wall, occurred at about 2.5 km/s. In fact, the total thickness to prevent penetration at this velocity was the same as it would be for a 10 km/s impact for the same projectile diameter, although the failure mechanism would be different (Gehring, 1970b).

Impact Damage in Non-Metallic Targets

Glass targets. A hypervelocity impact in a thick glass target is far more dramatic than one into a block of metal. Although the initial impact pressure generated by a 7 km/s impact on glass is not much less than on aluminum, the damage is much more extensive. This is because of the brittle nature of all glasses and their low tensile strength, which allows the shock-related shear and release tensile stresses to dominate the material strength over a longer time. The damage is characterized by concentric rings of surface spallation which are progressively below the surface as the impact point is approached. This leads to a typically conical crater formation (Flaherty, 1969). The impact shown in Fig. 12 has these concentric zones well delineated in addition to other typical effects found in targets that are not semi-infinite. The concentric, discontinuous flaws are fracture zones within the glass caused by the reflected shocks off the free boundaries of the target. Notice also the radial cracks, which are most prominent at the corners.

The corner cracks are a feature of the target being too small for this particular impact, a 0.16 cm aluminum projectile at a velocity of about 7 km/s. The surface spall is between 6 and 7 cm in diameter, and the depth of the pit is 0.46 cm from the surface of the glass. The diameter to depth ratio is 13 to 15, and the surface spall is 40 to 45 projectile diameters. These are in sharp contrast to the impact damage in a brittle aluminum target, typically a pit depth of about 0.3 cm, and a surface spall ring only slightly larger than the pit diameter, which is about 4 projectile diameters. The target shown is 2 cm thick, and there is a considerable amount of rear surface spallation. A glass target must be much thicker than its aluminum counterpart to prevent rear spall. Flaherty (1969) studied the phenomenology of crater and flaw formation in glass caused by a range of impact velocities from very low speeds to hypervelocities. He also looked into the effect of subsequent thermal loads on crater morphology. Other investigators (Cour-Palais, 1982; McHugh and Richardson, 1971, 1974; Wenzel and Gehring, 1965) provided the crater formation equations and the design parameters required to prevent cracking and to prevent spallation off the rear surfaces for the Apollo Program.

The impact shown in Fig. 12 for a 0.15 cm projectile at 7 km/s has done a lot of damage to the glass target. Had this been an observation port, it would have been of very little further use. However, the projectile energy and surface impact energy can be low enough, about 1 erg and 0.3 to 0.35 × 10^{11} N/m^2, to cause a lipped pit, as was observed on micron-sized particle hypervelocity impacts in glass (Roy, Slattery and Friichtenicht, 1972). As the energy increases, the first spall is detached around the pit, and sometimes this is accompanied by the ejection of the pit lips. Eventually, the energy level is sufficient for the concentric spallation to develop and for the crater to be ejected, leaving either a blocky, fractured indentation, marked by a sharp change of slope from the surrounding spallation, or a fused remnant (Cour-Palais, 1973). An appreciation of the forces at work in a glass target impacted at hypervelocity may be obtained by an examination of Fig. 13. This is a view of a sectioned, laminated target made up of a stack of aluminum sheets bolted together to provide good surface contact without a mechanical bond. The impact was at about 7 km/s, and one can see the crater, the shock compression under it, and the rebound of the upper layers as the release wave overcame the compression and the tensile "strength" across the interface. The significance of this target is that is simulates the low tensile strength of glass and its inherent brittleness. As one looks at the successive peeled-back layers of the target depicted in Fig. 13, it is easy to see how the concentric spall zones are formed. Finally, the crater is beginning to be ejected as it is carried upward by the deeper spallation. A top view of the same target is shown in Fig. 14, and it clearly shows the spall zones and the radial tearing that is analogous to the radial cracks in the glass target.

<u>Composite targets</u>. The fiber-reinforced, cured resins and epoxy systems known as "composites" that compete on a strength-weight-cost basis with the aluminum and magnesium alloys used in spacecraft did not exist twenty years ago. These materials became commercially available in the late 1970's, and they have been used extensively for aircraft applications. Their use for spacecraft has only recently become significant, and the Space Shuttle external tank and the Orbiter cargo bay doors are examples of the acceptance of these new materials for important aerospace applications.

The use of the composites primarily for military aircraft meant that the only impact testing done until very recently was at typical ordinance velocities. Military hypervelocity impact research into the relative merits of either composites or the aluminum alloys has not been available. As a result, NASA has undertaken to investigate the hypervelocity impact properties of composite materials because of its own needs.

The NASA research in progress involves the experimental and analytical description of crater formation and delamination in various fiber and resin or epoxy combinations, using projectiles ranging in diameter from 0.2 mm to 0.32 cm, at impact velocities from 3.5 to 9 km/s. The target variables include the fiber strengths, the fiber orientation in the layups, and the total laminate thickness. Only nylon and aluminum projectiles have been used in the tests up to this point, but other materials consistent with orbiting debris will be investigated.

Preliminary results of this program are reported in these proceedings by Yew (1986), and several photographs of sections taken through the postimpact target damage areas are given by the author. Because composites are laminated materials, the reaction to the passage of a strong shock is very similar to the laminated aluminum target previously discussed and shown in Figs. 13 and 14. Like the aluminum target, the cratering process in a composite material will cause front surface delamination and peeling under the action of the shearing forces. However, the extent of the peeling in the composites is very directional and very extensive (Yew, 1986). Also, the target depicted in Fig. 13 shows the shock compression in the aluminum layers, which can be compared with the discolored area under the crater in the sectioned 96 lamina plate shown in Yew's Fig. 15 (Yew, 1986). Both the aluminum laminate and the 96 lamina composite are semi-infinite; therefore, there is no rear surface spall detachment, although the 96 lamina composite does show the onset of spallation by discoloration. Another noticeable effect of a strong shock on a laminated composite is the formation of fractures in the laminations (Yew, 1986), which could be a problem if the expected structural loads are across the fracture planes.

The initial results of hypervelocity impacts on a limited sample of composite materials show that composites are good candidates for spacecraft applications exposed to the meteoroid or debris environment, although the potential weight advantage has not been quantified.

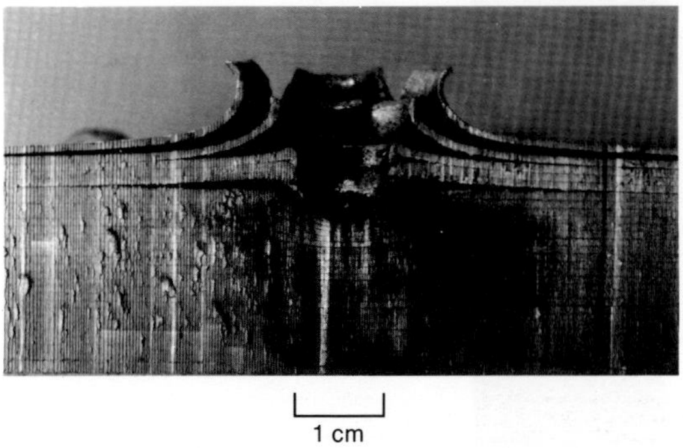

Fig. 13. Sectioned hypervelocity crater in a laminated aluminum sheet target.

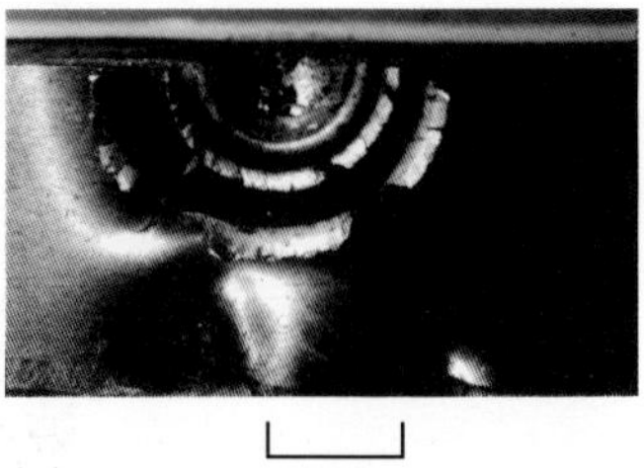

Fig. 14. Top view of sectioned crater in a laminated aluminum sheet target.

Impact Radiation and Ionization

Although the phenomenon most associated with hypervelocity impact is crater formation, there are two other effects that assumed some significance for their usefulness in the detection and identification of meteoroids. These phenomena are the flash of visible radiation and the ionization which are detectable at the instant of a high velocity particle impact (Eichhorn, 1976; Friichtenicht and Slattery, 1963; Gehring and Warnica, 1963; Jean, 1966; Jean and Rollins, 1969; MacCormack, 1963; Rosen and Scully, 1965). It may be of interest to discuss some of the experimental and analytical results of these investigations.

Jean and Rollins (1969) reported that they had isolated a narrow, fast-rising light pulse, followed by a much wider pulse which decayed slowly to zero intensity, in the impact flash. Typically, the time signature of the flash shows a spike with a half pulse width of 1 μs or smaller, and a rise time smaller than 0.2 μs followed by a slow-rising tail with a half pulse width of 3 to 5 μs. A specific example cited by these authors was for a 0.32 cm titanium projectile impact on a cadmium target at 3.5 km/s and a range pressure of 1.33×10^{-3} N/m^2. The spike height was 0.16 μs, its half pulse width was 0.5 μs, and the tail reached a maximum 3.4 μs after the onset of the spike. Jean and Rollins state that in some cases the spike and the tail merge and in others the spike is predominant and the tail very small.

Open shutter photographs such as Fig. 15 show the presence of two different light intensity levels corresponding to the two pulses that were observed. Time-resolved photographs, taken at 1 to 5.5 μs after the start of the spike, indicated that the tail of the impact flash was associated with the radiation emitted by an expanding, luminous, well-defined ring. The expansion velocity measured for the luminous ring produced by the 0.32 cm titanium sphere impacting the cadmium target at 3.5 km/s was 11.6 km/s. Other data presented by Jean and Rollins for aluminum-aluminum impacts also show that the ring expands at 2 to 3 times the projectile impact velocity.

The luminous spike was shown to be associated with the appearance of a "fast jet" of particles travelling at twice the speed of the luminous ring and detected by secondary targets at right angles to the main target. Jean and Rollins (1969) state that the fast jet and the luminous ring are produced by the same process at different pressures. These authors also show that the process that produces both the luminous spike and the tail is related to the production of a jet of material in a shaped charge. The angle between the tangent to the surface of a spherical projectile and the target at the impact interface, changes from 0 to 90 degrees during the penetration process. It passes through a critical angle for the formation of a jet which is akin to the critical angle in shaped-charge theory. The critical angle for the impact flash was measured using conical projectiles with different cone angles, and for copper-aluminum impacts, was found to range from 12.5 degrees at 1 km/s to 32 degrees at 6.3 km/s.

Jean and Rollins also conducted parametric studies of the flash to determine whether any of its characteristics could provide quantitative information about the size and velocity of the projectile. Copper spheres ranging in diameter from 0.16 to 0.64 cm were impacted into cadmium targets at 2 to 8 km/s, and only the impact spike measurements were used in the correlations. It was found that the peak intensity, I, of the radiation and the initial rate of change of the intensity, dI/dt, of the spike were strongly and consistently dependent on the diameter and

Fig. 15. Open shutter potograph of a typical impact flash.

velocity of the projectile for the chosen emission lines of cadmium: 5085, 3610, and 3261 Å. The measured peak intensity varied as the eighth power of the velocity for all three of the cadmium lines and the fourth power of the diameter at the 3610 Å transition for the copper projectiles. The rate of change of the spike intensity varied as the sixth power of the projectile velocity and approximately the square of the diameter for the same emission lines and projectile material and sizes.

Jean and Rollins also conducted spectrographic studies, using cadmium and aluminum targets and aluminum projectiles at 6 to 7 km/s with a range pressure between 1.33×10^3 to 1.33×10^4 N/m^2. They concluded that the short-duration, transient spike could be attributed to the presence of a hot, dense plasma at the impact point, while the long-duration tail was a result of the radiation of a neutral gas expanding from the impact zone.

Eichhorn (1976) continued the early research conducted by Jean and Rollins, performing his experiments with a 2 MV Van de Graaff dust accelerator. He confirmed that the mass and velocity dependence of the various parameters of the impact flash were suitable for the measurement of micrometeoroid impacts on spacecraft and gave the relationships for I, the light intensity, and E the total light energy as follows

$$I = C1 \times m \times V^{4.1} \tag{6}$$

$$E = C2 \times m \times V^{3.2} \tag{7}$$

where the units are m, the projectile mass in grams, g; V, the projectile velocity in km/s; $C1$ and $C2$ are constants depending on projectile and target material; I is in ergs/s; and E is in ergs.

Eichorn (1976) also found that impact flash measurements provided details about the crater-forming process. Using the spectral distribution of the light emission, he estimated the temperatures of the radiating gas and plasma. He was able to obtain the total thermal energy of the radiating gas and combine this with the estimated gas temperature to arrive at the total mass of the vapor produced during an impact. This led Eichorn to conclude that for iron particles impacting gold targets at 7.4 km/s, at least 1.6% of the displaced projectile and crater material was vaporized.

Spray particles at ejection angles less than 20 degrees were also detected by Eichorn (1976) for the iron particles impacting gold targets. He states that the velocity of this ejecta increased with primary velocity, but only up to 5 km/s, after which it remained practically constant. The maximum velocity measured by the author was 30 km/s for an initial particle impact at 4.8 km/s.

The impact ionization effect is the production of free charge at the site of a hypervelocity impact on a suitable target (Friichtenicht and Slattery, 1963). These authors did their research using micron-sized iron and graphite particles impacting targets of various materials at velocities up to 16 km/s in a 2 MV Van de Graaff accelerator. The impact creates temperatures high enough to vaporize the projectile and some of the target material and also to thermally ionize a fraction of the vaporized atoms. The result is a microplasma from which charge may be extracted through the application of electric fields (Becker and Slattery, 1972). Friichtenicht and Slattery (1963) state that the charge, Q, produced at impact is proportional to the mass of the projectile and the cube of the impact velocity and give the following relationship

$$Q = K \times E \times V/A \tag{8}$$

where Q is in coulombs; K is a quantity that contains target material properties; E is the particle energy in ergs; V is the particle velocity in km/s; and A is the atomic weight of the particle material.

Like the impact flash, the ionization charge signal shows two components for particle impacts under about 10 km/s. The first one follows immediately after the impact with an orgin that is close to the impact site. A second component is observed about 2 µs after the first, and its charge is collected on the target at some distance from the impact site (Becker and Slattery, 1972). These authors state that iron spheres less than 1 µm radius, impacting a solid surface at velocities less than 10 km/s, throw back small "spray" particles. The impacting particle is vaporized at this speed, and the charge produced per unit particle mass is approximately proportional to the fourth power of the impact velocity. This velocity dependence is the same reported by Jean and Rollins (1969) for the rate of change of flash intensity, as was mentioned previously in this report. The particles that impact at a velocity higher than 10 km/s are more completely vaporized, and produce either much smaller, or many fewer spray particles. Becker and Slattery (1972) feel it is these spray particles which create the delayed charge component, and they note that the earlier component of ionization is always present, but that the delayed charge component is only seen for velocities below 10 km/s. Jean and Rollins (1969) say that for the similar case involving impact flash, there is no tail, or later component, at higher velocities because the amount of material produced is not enough to be excited. In conclusion, Becker and Slattery (1972) warned that particle velocity measurements based on the character of the impact ionization signal should be treated with great care.

Cosmic dust impact detectors, which made use of the ionization produced by an impact on a known target material, were designed and successfully flown on the Helios and Heos 2 spacecraft (Dietzel and colleagues, 1973; Grun and colleagues, 1980). The Helios 1 instrument had a sensitivity threshold of 3×10^{-16} kg at an impact speed of 10 km/s, and the measured parameters allowed the particle mass, speed, electrostatic charge, and composition of the plasma produced by an impact to be determined. Impact plasma detectors similar to the Heos 2 instrument were used on the European Space Agency's Giotto mission through the tail of Comet Halley and will be used on board the Ulysses mission, formerly ISPM, to detect interplanetary dust and measure its dynamic and physical properties (Grun and colleagues, 1983). The information available in the characteristics of an impact flash was never used for flight instrumentation, and this technique remains a valuable diagnostic tool for laboratory hypervelocity impact studies.

CONCLUSIONS

A historical overview of the research into the phenomenology of hypervelocity impact, especially as it relates to protection of spacecraft from meteoroid impacts during the manned Apollo Program, has been presented. The Apollo Program was the driver for a very concerted effort to understand not only the nature of the threat facing man's first venture away from his planet but also how to ensure his survival in a hostile meteoroid environment. A long hiatus in civilian hypervelocity impact research ensued after the successful completion of the manned missions to the Moon. This hiatus has now been broken with the onset of the Space Station design, and there has been a resumption of civilian research and development impact testing. Much that had been previously learned is still applicable, but new research is particularly required to investigate the hypervelocity properties of the composite materials. A partial bibliography of the hypervelocity impact papers that were used for the design and certification of the Apollo meteoroid protection is given in Cour-Palais (1982).

REFERENCES

Alexander, W. M., C. W. McCracken, L. Secretan, and O. Berg (1963). Review of direct measurements of interplanetary dust from satellites and probes. In W. Priester (Ed.), Space Research, 3rd ed. pp. 891-917.

Becker, D. G., and J. C. Slattery (1972). Investigation of high-speed phenomena. Report Number 16623-6005-RO-00, TRW Systems Group, California.

Bjork, R. L., K. N. Kreyenhagen, and M. H. Wagner (1967). Analytical study of impact effects as applied to the meteoroid hazard. NASA CR-757.

Carey, W. C., J. A. M. McDonnell, and D. G. Dixon (1985). An empirical penetration equation for thin films used in capture cells. In R. H. Giese and P. Lamy (Ed.), Properties and Interactions of Interplanetary Dust. pp. 131-136.

Clifton, K. S., and R. J. Naumann (1966). Pegasus satellite measurements of meteoroid penetration. NASA TM X-1316.

Cour-Palais, B. G. (1969a). Meteoroid environment model-1969 (near-Earth to lunar surface). NASA SP-8013.

Cour-Palais, B. G. (1969b). Meteoroid protection by multi-wall structures. Paper Number 69-372, AIAA Hypervelocity Impact Conference, Vol. of Technical Papers.

Cour-Palais, B. G. (1973). Apollo window meteoroid experiment. In Apollo 17 Preliminary Science Report, NASA SP-330, No. 21.

Cour-Palais, B. G. (1979). Space vehicle meteoroid shielding design. In N. Longdon (Ed.), The Comet Halley Micrometeoroid Hazard, ESA SP-153. European Space Agency, Paris. pp. 85-92.

Cour-Palais, B. G. (1982). Hypervelocity impact investigations and shielding experience related to Apollo and Skylab. In Orbital Debris, NASA Conf. Publication 2360. pp. 247-275.

Denardo, B. P., and C. R. Nysmith (1964). Momentum transfer and cratering phenomena associated with the impact of aluminum spheres into aluminum targets at 24,000 ft/s. Agardograph 87, Vol. 1.

Denardo, B. P., J. L. Summers, and C. R. Nysmith (1967). Projectile size effects on hypervelocity impact on craters in aluminum. NASA TN D-4067.

Dienes, J. K., and J. M. Walsh (1970). Theory of impact: some general principles and the method of eulerian codes. In R. Kinslow (Ed.), High-velocity Impact Phenomena. pp. 45-104.

Dietzel, H., G. Eichorn, H. Fechtig, E. Grun, H. J. Hoffmann, and J. Kissel (1973). The Heos 2 and Helios micrometeoroid experiments. J. Phys. E. Sci. Instrum., 6. p. 209.

Eichorn, G. (1976). Analysis of the hypervelocity impact process from impact flash measurements. Planet. Space Sci., 24. pp. 771-781.

Elford, W. G. (1967). The incidence of meteors on the Earth derived from radio observations. In G. Hawkins (Ed.), Meteors Orbits and Dust, NASA SP-135.
Fish, R. H., and J. L. Summers (1965). The effect of material properties on threshold penetration. In Proc. Seventh Hypervelocity Impact Symposium, Vol. 6. pp. 1-25.
Flaherty, R. E. (1969). Impact characteristics in fused silica. Paper Number 69-367, AIAA Hypervelocity Impact Conference, Vol. of Technical Papers.
Friichtenicht, J. F., and J. C. Slattery (1963). Ionization associated with hypervelocity impact. In Proc. Sixth Symposium on Hypervelocity Impact, Vol. 2. pp. 591-612.
Frost, V. C. (1970). Meteoroid damage assessment. NASA SP-8042.
Gault, D. E., and H. J. Moore (1965). Scaling relationships for microscale to megascale impact craters. In Proc. Seventh Hypervelocity Impact Symposium, Vol. 6. pp. 341-351.
Gehring, J. W. (1970a). Theory of impact on thin targets and shields and correlation with experiment. In R. Kinslow (Ed.), High Velocity Impact Phenomena. Academic Press, New York. pp. 105-156.
Gehring, J. W. (1970b). Engineering considerations in hypervelocity impact. In R. Kinslow (Ed.), High Velocity Impact Phenomena. Academic Press, New York. pp. 463-514.
Gehring, J. W., and R. L. Warnica (1963). An investigation of the phenomena of impact flash and its potential use as a hit detection and target discrimination technique. In Proc. Sixth Symposium on Hypervelocity Impact, Vol. 2. pp. 627-681.
Goodman, E. H., and C. D. Liles (1963). Particle-solid impact phenomena. In Proc. Sixth Symposium on Hypervelocity Impact, Vol. 2. pp. 543-576.
Grun, E., H. Fechtig, R. Giese, J. Kissel, L. Linkert, J. McDonnell, G. Morfill, G. Schwehm, and H. Zook (1983). The ISPM dust experiment. ESA SP-1050. European Space Agency, Paris.
Grun, E., N. Pailer, H. Fechtig, and J. Kissel (1980). Orbital and physical characteristics of micrometeoroids in the inner solar system as observed by Helios 1. Planet. Space Sci., 28. pp. 333-349.
Halperson, S. M. (1963). Some phenomena associated with impacts into aluminum. In Proc. Sixth Symposium on Hypervelocity Impact, Vol. 2. pp. 525-541.
Hastings, E. C. (1964). The Explorer 16 micrometeoroid satellite. NASA TM X-949, Supplement 3.
Hawkins, G. S. (1963). The Harvard radio meteor project. In Smithsonian Contributions to Astrophysics, Vol. 7. pp. 53-62.
Hawkins, G. S., and R. B. Southworth (1958). The statistics of meteors in the Earth's atmosphere. In Smithsonian Contributions to Astrophysics, Vol. 2. pp. 349-364.
Hawkins, G. S., and E. K. Upton (1958). The influx rate of meteors in the Earth's atmosphere. Astrophysical J., 128. pp. 727-735.
Herrmann, W., and A. H. Jones (1961). Correlation of hypervelocity impact data. In Proc. Fifth Hypervelocity Impact Symposium.
Herrmann, W., and J. Wilbeck (1986). Review of hypervelocity penetration theories. In Intern. J. Impact Eng., Vol. 5.
Holsapple, K. A., and R. M. Schmidt (1980). On the scaling of crater dimensions, 1, Explosive processes. J. Geophys. Res., 85(B12). pp. 7247-7256.
Holsapple, K. A., and R. M. Schmidt (1982). On the scaling of crater dimensions, 2, Impact processes. J. Geophys. Res., 87(B3). pp. 1849-1870.
Jacchia, L. G. (1955). The physical theory of meteors No. 8: Fragmentation as a cause of the faint meteor anomaly. Astrophys. J., 121. pp. 521-527.
Jean, B. (1966). Experimental observations of optical radiation associated with hypervelocity impact. AIAA J. pp. 4-10.
Jean, B., and T. L. Rollins (1969). Radiation from hypervelocity impact generated plasma. Paper Number 69-364, AIAA Hypervelocity Impact Conference, Vol. of Technical Papers.
Kessler, D. J. (1981). Sources of orbital debris and the projected environment for future spacecraft. J. Spacecraft and Rockets, Vol. 18, 4. pp. 357-360.
Kessler, D. J. (1982). Proposed preliminary design criteria: model environment for the 1990's. In D. Kessler and S. Su (Ed.), Orbital Debris, NASA Conf. Publication 2360. pp. 79-83.
Kessler, D. J. (1985). Orbital debris issues. In Space Debris, Asteroids and Satellite Orbits, Advances in Space Research, Vol. 5, Number 2. Pergamon Press, Oxford. pp. 3-10.
Kessler, D. J., and B. G. Cour-Palais (1978). Collision frequency of artificial satellites: the creation of a debris belt. J. Geophys. Res., 83(A6). pp. 2637-2646.
Kineke, J. H. (1965). Hypervelocity impact: dependence of crater dimensions on impact velocity. Memorandum Report Number 1652. Ballistic Research Laboratories, Maryland.
MacCormack, R. W. (1963). Investigation of impact flash at low ambient pressures. In Proc. Sixth Symposium on Hypervelocity Impact, Vol. 2. pp. 613-625.
McCrosky, R. E. (1957). A rapid graphical method of meteor trail reduction. Smithsonian Contributions to Astrophysics, 1. pp. 215-224.
McHugh, A. H., and A. J. Richardson (1971). Glass equations for solar cells. J. Spacecraft and Rockets, Vol. 8, 6.
McHugh, A. H., and A. J. Richardson (1974). Hypervelocity impact damage to glass. STR 241. Rockwell International, Space Division, Downey, California.
McKinley, D. W. R. (1961). Meteor Science and Engineering. McGraw-Hill, New York.

McMillan, A. R., A. H. Jones, K. H. Meyer, B. L. Lathrop, C. J. Maiden, and R. E. Sennett (1966). Experimental investigations of simulated meteoroid damage to various spacecraft structures. TR66-67. GM Defense Research Laboratories, Santa Barbara, California.
Maiden, C. J. (1963). Experimental and theoretical results concerning the protective ability of a thin shield against hypervelocity projectiles. In Proc. Sixth Symposium on Hypervelocity Impact, Vol. 3. pp. 69-156.
Maiden, C. J. (1964). Private Communication.
Maiden, C. J., J. W. Gehring, and A. R. McMillan (1963). Investigation of fundamental mechanism of damage to thin targets by hypervelocity projectiles. TR63-225. GM Defense Research Laboratories, Santa Barbara, California.
Maiden, C. J., A. R. McMillan, and R. E. Sennett (1965). Thin sheet impact. In Proc. Seventh Hypervelocity Impact Symposium, Vol. 4. pp. 63-123.
Meyers, C. L., and J. A. Charest (1964). Research on the properties of optimum meteoroid shields. TR64-48. GM Defense Research Laboratories, Santa Barbara, California.
Naumann, R. J., and D. W. Jex (1969). Calibration of Pegasus and Explorer 23 detector panels. NASA TR R-321.
Nysmith, C. R. (1968). Penetration resistance of double-sheet structures at velocities to 8.8 km/s. NASA TN D-4568.
Nysmith, C. R., and B. P. Denardo (1969). Experimental investigation of the momentum transfer associated with impact into thin aluminum targets. NASA TN D-5492.
Nysmith, C. R., and J. L. Summers (1962). An experimental investigation of the impact resistance of double-sheet structures at velocities to 24,000 ft/s. NASA TN D-1431.
O'Neal, R. L. (1968). The Explorer 23 micrometeoroid satellite. NASA TN D-4284.
Posever, F. C., and C. N. Scully (1965). Investigation of meteoroid impacts on two-sheet configurations. AFFDL-TR-65-196. Wright-Patterson Air Force Base, Ohio.
Prater, R. F. (1970). Hypervelocity impact: material strength effects on crater formation and shock propagation in three aluminum alloys. ASML-TR-70295. Wright Patterson Air Force Base, Ohio.
Richardson, A. L. (1969). Theoretical penetration mechanics of multi-sheet structures based on discrete particle modeling. Paper Number 69-371, AIAA Hypervelocity Impact Conference, Vol. of Technical Papers.
Rinehart, J. S., and J. Pearson (1963). Explosive Working of Metals. The Macmillan Company, New York.
Riney, T. D. (1964). Theoretical hypervelocity impact calculations using the Pickwick code. R64SD13. GE Space Sciences Laboratory, Philadelphia.
Riney, T. D., and J. F. Heyda (1964). Hypervelocity impact calculations and their correlations with experiment. R64SD64. GE Space Sciences Laboratory, Philadelphia.
Rosen, F. D., and C. N. Scully (1965). Impact flash investigations to 15.4 km/s. In Proc. Seventh Hypervelocity Impact Symposium, Vol. 6. pp. 109-140.
Rosenblatt, M. (1970). Analytical study of strain rate effects in hypervelocity impacts. Final Report, Contract NAS8-20235.
Roy, N. L., J. C. Slattery, and J. F. Friichtenicht (1972). Study for Apollo meteoroid window experiment. Report Number 20921-6001-RO-00. TRW Systems, California.
Sawle, D. R. (1969). Hypervelocity impact in thin sheets and semi-infinite targets at 15 km/s. In Paper Number 69-368, AIAA Hypervelocity Impact Conference, Vol. of Technical Papers.
Scully, C. N., E. A. Escallier, F. D. Rosen, and J. D. O'Keefe (1965). The electrothermal gun for hypervelocity ballistics research. In Proc. Seventh Hypervelocity Impact Symposium, Vol. 1.
Sedgwick, R. T. (1969). Theoretical study of size scaling in cratering resulting from hypervelocity impact. Final Report, Contract NAS8-20239.
Slattery, J. C., and J. F. Friichtenicht (1966). Experimental research on hypervelocity cratering by microscopic particles. Report Number 03246-6001-RO-00, TRW Systems, California.
Sorensen, N. R. (1965). Systematic investigation of crater formation in metals. In Proc. Seventh Hypervelocity Impact Symposium, Vol. 6.
Swift, H. F. (1982). Hypervelocity impact mechanics. In Impact Dynamics, John Wiley, New York. pp. 215-230.
Swift, H. F., R. Bamford, and R. Chen (1982). Designing dual-plate shields--a new analysis. JPL Publication 82-39. Jet Propulsion Laboratory, California.
Swift, H. F., and A. K. Hopkins (1969). The effects of bumper material properties on the operation of spaced hypervelocity particle shields. Paper Number 69-379, AIAA Hypervelocity Impact Conference, Vol. of Technical Papers.
Swift, H. F., D. D. Preonas, W. C. Turpin, and J. H. Cunningham (1969). Characterization of debris clouds behind impacted meteoroid bumper plates. Paper Number 69-380, AIAA Hypervelocity Impact Conference, Vol. of Technical Papers.
Taff, L. G., D. E. Beatty, A. J. Yakutis, and P. M. Randall (1985). Low-altitude one centimeter space debris search at Lincoln Laboratory's (M.I.T.) experimental test system. In Space Debris, Asteroids and Satellite Orbits, Advances in Space Research, Vol. 5, Number 2. Pergamon Press, Oxford. pp. 34-45.
Teng, R. N. (1968). Hypervelocity impact damage in aluminum targets. DAC-59816. McDonnell-Douglas Aircraft Company, California.

Turpin, W. C., and J. M. Carson (1970). Hole growth in thin plates perforated by hypervelocity pellets. AFML-TR-70-83. Wright-Patterson Air Force Base, Ohio.

Wagner, M. H., N. B. Brooks, and R. L. Bjork (1965). Impact of a porous aluminum projectile on aluminum at 20 and 72 km/s. In Proc. Seventh Hypervelocity Impact Symposium, Vol. 3. pp. 1-54.

Walsh, J. M., and W. E. Johnson (1965). On the theory of hypervelocity impact. In Proc. Seventh Hypervelocity Impact Symposium, Vol. 2. pp. 1-75.

Walsh, J. M., W. E. Johnson, J. K. Dienes, J. H. Tillotson, and D. R. Yates (1964). Summary report on the theory of hypervelocity impact. GA-5119. General Atomic Division of General Dynamics, California.

Wenzel, A. B., and J. W. Gehring (1965). Hypervelocity impact studies against Apollo-type structures up to 16.5 km/s. TR65-56. GM Defense Research Laboratories, Santa Barbara, California.

Whipple, F. L. (1938). Photographic meteor studies: No. 1. Proc. Amer. Phil. Soc., 79 pp. 499-548.

Whipple, F. L. (1946). Possible hazards to satellite vehicle from meteorites. In The Collected Contributions of Fred L. Whipple, Vol. 2. Smithsonian Institution, Washington, D. C. pp. 1072-1079.

Whipple, F. L. (1952). Meteoritic phenomena and meteorites. In The Physics and Medicine of the Upper Atmosphere. University of New Mexico Press, Albuquerque. pp. 137-170.

Whipple, F. L. (1954). Photographic meteor orbits and their distribution in space. Astron. J., 59. pp. 201-217.

Whipple, F. L. (1958). The meteoric risk to space vehicles. In Proc. International Astronautical Congress. Springer-Verlag, Vienna. pp. 418-428.

Wilkinson, J. P. (1968). A penetration criterion for double-walled structures subject to meteoroid impact. 68SD275. GE Technical Center, Schenectady, New York.

Woodall, S. R. (1967). A numerical method for analyzing meteoroid two-sheet structural configuration impact problems. Mechanics Section Technical Memo 067-6A. GE Space Sciences Laboratory, Philadelphia.

Yew, C. H., and R. R. Kendrick (1986). A study of damage in composite panels produced by hypervelocity impact. In International J. Impact Eng., Vol. 5. Pergamon Press, Oxford.

Int. J. Impact Engng Vol. 5, pp. 239–248, 1987
Printed in Great Britain

0734-743X/87 $3.00 + 0.00
Pergamon Journals Ltd.

A UNIFIED THEORY OF PENETRATION

James Dehn

U. S. Army Laboratory Command
Ballistic Research Laboratory
Aberdeen Proving Ground, MD 21005-5066

ABSTRACT

This paper describes a single equation of motion which, in its most general form, describes penetration by a forming, stretching, eroding jet. In various specialized forms it describes penetration by pre-formed rods and compact projectiles. One specialized form reduces to the classical jet penetration formula according to which penetration into a semi-infinite target is proportional to the square root of the ratio of the jet and target densities. However, in the present theory, a correction for target hardness is automatically included. The equation is derived and certain solutions are illustrated by comparisons with experimental data.

INTRODUCTION

Because of the differences between jets and kinetic energy projectiles, separate theories were developed. Hydrodynamic theories were favored for high speed jets, while various theories were used for low-speed bullets and fragments. With the advent of high-speed long rods, amended jet theories were developed. In this paper we will develop a single theory which describes all types of penetrator from a unified viewpoint. Our aim is to be as simple as possible without omitting any important features. Since this is a symposium on hypervelocity impact, the examples we have selected to illustrate the theory will emphasize this regime.

THE EQUATION OF MOTION

The force exerted by the target opposing the motion of the penetrator is taken to be

$$F = a + b\dot{P} + c\dot{P}^2 \tag{1}$$

where $\dot{P}$ is the speed of the penetrator nose in a fixed coordinate system with origin at the target surface. This form with b = 0 is attributed to Poncelet. The constant, a, depends on the target hardness

$$a = 3Y_t A = 3C_m(BHN)A \tag{2}$$

where (BHN) is the Brinell Hardness Number and Y_t is the ultimate tensile strength. Tabor (1948 and 1951) has shown that the mean pressure exerted by a metal target on a metal penetrator is about three times the ultimate yield strength, Y_t, as determined in standard tensile tests. He also noted the proportionality between Y_t and the Brinell Hardness Number with $C_m = 0.3 \times 10^8$ (dyne/cm^2)/(kg/mm^2) in general. Here A is the cross-sectional area presented to the target by the projectile. For a rod projectile with initial cross-sectional area, A_o, for example, we will usually have $A/A_o > 1$ because of nose deformation. The constant b has the form

$$b = 6\pi R\eta = 6\sqrt{\pi A}\,\eta \tag{3}$$

where R is the radius of the deformed nose cap and η is the coefficient of target viscosity. For all cases to be treated in this paper, $b\dot{P}$ is negligible compared to the other two terms in Eq (1). The reason for this is the gap which exists between the incoming penetrator body and the deformed target and projectile material. Provided this material does not touch the penetrator, the manner in which it comes to rest is of no importance to the motion of the penetrator. What is important is the bond-breaking measured by the constant, a, and the acceleration of target material to the speed of the penetrator. The last term is

$$c\dot{P}^2 = .5\rho_t A\dot{P}^2 \tag{4}$$

where ρ_t is the target density so that $.5\rho_t\dot{P}^2$ is the energy per unit volume or force per unit area, while A is the nose area as before. In the two force expressions we will be concerned with, Eqs (2) and (4), everything is known except for $A \geq A_o$. In many cases this can be estimated from experimental information also.

Let us begin with an eroding rod penetrator of density ρ_p and mass $M = \rho_p A_o L$. During the penetration the length L is less than the initial length, L_o, but A_o remains the same over the undeformed part of the penetrator, so this is a reasonable estimate for the remaining mass. Then

$$\dot{L} = \dot{M}/(\rho_p A_o) = -\dot{M}_p/(\rho_p A_o) \tag{5}$$

where $\dot{M}_p = |\dot{M}|$ (for $\dot{M} < 0$) is the mass lost per unit time at the nose position P. Let R and $\dot{R}$ be the position and speed of the rear of the projectile. Just before impact all parts of the projectile move with the same speed. Just after impact, the target resistive force increases from zero to a large value, then starts to decrease. During this brief, transient period the nose speed decreases to a value $\dot{P}_o$ which is less than the striking speed, $\dot{S}_o$. Here S is the position of the center of mass which is halfway between the nose and rear at any time. Consequently

$$\dot{P} = \dot{S} + \dot{L}/2 = \dot{S} - \dot{M}_p/(2\rho_p A_o) \tag{6}$$

and

$$\dot{R} = \dot{S} - \dot{L}/2 = \dot{S} + \dot{M}_p/(2\rho_p A_o) \tag{7}$$

where we have used Eq (5). Addition of these equations expresses the fact that the speed of the center of mass is the average of $\dot{P}$ and $\dot{R}$. Subtraction of these equations leads to $\dot{R} > \dot{P}$, as expected. In the absence of erosion, $\dot{L} = \dot{M}_p = 0$ so $L = L_o$, $M = M_o$ and $\dot{P} = \dot{S} = \dot{R}$. Here we are ignoring the details of the transient and begin our problem just after the resistive force has peaked. The rear of the rod decelerates a bit later than the front because it takes time for pressure waves generated at the nose to reach the rear and reflect from the solid-air interface. This deceleration may occur in several steps. Here we are ignoring such details and using average values.

In the penetration problem we are describing, mass dM_p is lost at the front while the speed of the center of mass becomes $\dot{S} + d\dot{S}$. Thus

$$\left[(M-dM_p)(\dot{S}+d\dot{S}) + \dot{P}dM_p\right] - M\dot{S} = -Fdt. \tag{8}$$

Here $M\dot{S}$ is the momentum at the beginning of the time interval, dt, while the square bracket is the momentum of the system, rod and eroded mass at the end of this interval. The difference is equal to the impulse imparted, with F given by Eq (1). In the limit $dt \rightarrow 0$, we can ignore the second order product $(dM)(d\dot{S})$, divide by dt and obtain

$$Md\dot{S}/dt = (\dot{S}-\dot{P})\dot{M}_p - F = \dot{M}_p^{\,2}I_o - F \tag{9}$$

where we have used Eq (6) for $(\dot{S}-\dot{P})$ and introduced the constant $I_o = (2\rho_p A_o)^{-1}$. In all cases of interest the right side of Eq (9) is negative. That is, the erosion rate, $\dot{M}_p$, is never so large that the right side is positive. In an extreme case (mud splattering on steel) the right side may be zero with $\dot{P} = 0$ (no penetration), $F = a$ and $\dot{M}_p = \sqrt{a/I_o}$ as an upper limit. Of course $\dot{M}_p = 0$ is the lower limit. From Eq (9) it is evident that the loss of mass tends to accelerate the penetrator and counteract the target resistive force.

Now let us extend this theory to jets. Commonly, erosive penetration begins even before formation is complete. Mass is added to the rear at a rate $\dot{M}_R$ even as it is lost at the front, so the net mass change is $\dot{M} = \dot{M}_R - \dot{M}_p$. Jets in common use are also tapered, being fatter at the rear than at the front. Here we will let A_o be the average cross-sectional area at any moment such that its product with the jet density and length gives us the mass, namely, $M = \rho_p A_o L$. It is possible to make a non-stretching jet (Chanteret and Jamet, 1984), but for most fully-formed jets of interest, observations made during free-flight (no target present) reveal that the length increases ($\dot{L} > 0$) while the area decreases ($\dot{A}_o < 0$) without change in mass. Consequently, $\dot{M} = \rho_p(A_o\dot{L}+\dot{A}_oL) = 0$ or $\dot{L} = L|\dot{A}_o|/A_o$ is the contribution to the change in length due to stretching. Let us assume that this contribution can be added to the change in length due to changes in mass, so

$$\dot{L} = (\dot{M}_R-\dot{M}_p)/(\rho_p A_o) + L|\dot{A}_o|/A_o = 2I_o(\dot{M}+\dot{\mu}) \tag{10}$$

where $\dot{\mu} = \rho_p L|\dot{A}_o| = \rho_p A_o \dot{L}'$ is the stretching rate. For a pre-formed ($\dot{M}_R = 0$), non-stretching ($\dot{\mu} = 0$) penetrator, Eq (10) reduces to Eq (5).

At any instant a jet has a definite mass and length. However, because of taper, the center of mass will be located to the rear of the midpoint of its length. In a one-dimensional problem where all forces are exerted along the penetrator length, we can choose any point along its length to describe the motion. If C is the geometrical midpoint of the length, then

$$\dot{P} = \dot{C} + \dot{L}/2 = \dot{C} + (\dot{M}+\dot{\mu})I_o \tag{11}$$

and

$$\dot{R} = \dot{C} - \dot{L}/2 = \dot{C} - (\dot{M}+\dot{\mu})I_o \tag{12}$$

where we have used Eq (10). After formation ($\dot{M} \to -\dot{M}_p$) it is possible for $\dot{\mu} = \dot{M}_p$ so $\dot{P} = \dot{C} = \dot{R}$. Usually, however, $\dot{R} < \dot{P}$. Of course Eqs (11) and (12) reduce to Eqs (6) and (7) for a pre-formed, non-stretching penetrator. The analog of Eq (8) is

$$[(M+dM_R-dM_p)(\dot{C}+d\dot{C}) + \dot{R}dM_R + \dot{P}dM_p] - M\dot{C} = -Fdt \tag{13}$$

and in the limit $dt \to 0$, we neglect $(dM_R)(d\dot{C})$ and $(dM_p)(d\dot{C})$, so

$$Md\dot{C}/dt = -(\dot{C}+\dot{R})\dot{M}_R + (\dot{C}-\dot{P})\dot{M}_p - F \tag{14}$$

We can use Eqs (11) and (12) to find expressions for $(\dot{C}+\dot{R})$ and $(\dot{C}-\dot{P})$ and simplify to obtain

$$Md\dot{C}/dt = I_o\dot{M}(\dot{M}+\dot{\mu}) - (F+2\dot{C}\dot{M}_R) \tag{15}$$

This equation simplifies somewhat after formation is complete ($\dot{M}_R \to 0$, $\dot{M} \to -\dot{M}_p$) and reduces to Eq (9) in the absence of stretching ($\dot{\mu} = 0$). If, in addition, there is no erosion, then Eq (9) describes a constant mass penetrator. Eq (15) is a single equation which provides a unified viewpoint for describing the motion of a penetrator which may stretch or not, be forming or not and erode or not. From Eq (15) it is evident that stretching and erosion contribute to accelerating the penetrator, while mass addition during formation contributes both to acceleration and deceleration, the net effect depending on particular cases.

EXAMPLES

A. Compact Penetrators

Figure 1 shows penetration depth in semi-infinite targets versus striking speed for 4.76mm diameter aluminum spheres impacting hard (2014-T6) and soft (1100-F) aluminum targets. This example has been chosen because it covers the largest speed range for a given projectile/target combination ever reported (Halperson, 1963 and 1965; Halperson and Atkins, 1962; and Atkins, 1960). It also illustrates the fact that target hardness plays an undiminished role to speeds in excess of typical jet tip speeds. This has also been

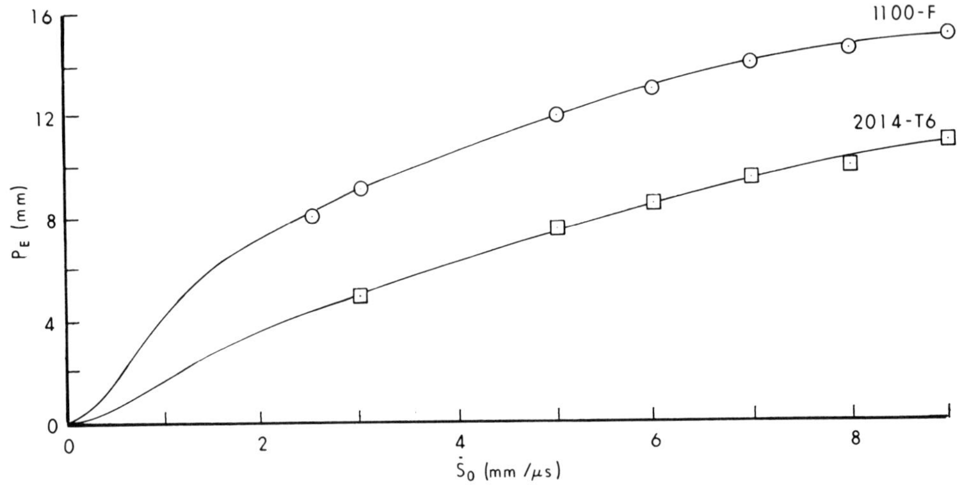

Fig. 1. Penetration depth versus striking speed for aluminium spheres impacting hard and soft aluminium targets.

noted by other workers (Sorensen, 1965; Mobley and Pond, 1965; Walsh and Johnson, 1965; and Kineke and Richards, 1963). A more extensive analysis (Dehn, 1986) shows that erosion is a negligible factor in such impacts. There was no difference between the two types (2024 and 2017) of aluminum spheres used. All spheres flattened and covered part of the crater bottom in such a way that some forward component of momentum was retained by every element of the sphere until the end of the motion. By erosion we mean some elements of the penetrator lose all forward motion at some time during the penetration. It seems to be generally true that projectiles with an aspect ratio (length to diameter ratio) less than or equal to one, do not erode in this sense. Consequently, $\dot{M}_p = 0$ in Eq (9) so $\dot{P} = \dot{S}$ in Eq (6) and $F = a + c\dot{S}^2$ in Eq (1) with $a = 3Y_t A_o (A/A_o) = a_o (A/A_o)^p$ and $c = .5\rho_t A_o (A/A_o) = c_o (A/A_o)$. Since no time-dependent data are available, a solution in the $\dot{P}, P$ plane is sufficient. The position of the projectile when it becomes embedded in the target, P_E, is found by setting $\dot{P} = 0$ in the solution, so

$$P_E = .5\left[(M_o/C_o)/(A/A_o)\right] \ln\left[1+(\dot{S}_o/\alpha)^2\right] \qquad (16)$$

where $\alpha = \sqrt{a/c} = \sqrt{6Y_t/\rho_t}$ depends only on target hardness and density. The first derivative of Eq (16) with respect to the striking speed, $\dot{S}_o$, is

$$dP_E/d\dot{S}_o = (M_o/a)\ \dot{S}_o/[1+(\dot{S}_o/\alpha)^2] \qquad (17)$$

which vanishes for both low ($\dot{S}_o \to 0$) and high ($S_o \to \infty$) striking speeds. It is easy to show that α is the inflection point of this curve. This law of diminishing returns which says that, for speeds considerably greater than the inflection point, further increases in striking speed increase the penetration only slightly, has been known for a long time (Bethe, 1941). However, increasing the mass will increase the penetration depth in a linear manner according to Eq (16). This trend is also well known.

For a sphere, $M_o = \rho_p(\pi/6)D_o^3$ where D_o is its original diameter. Since $c_o = .5\rho_t(\pi/4)D_o^2$ and $\rho_p = \rho_t$ in this example, we find $M_o/c_o = (4/3)D_o = 6.35$mm for this sphere. If we use the reported strengths of the target materials we find $\alpha = 0.5$mm/μs for the soft aluminum and $\alpha = 1.1$mm/μs for the hard aluminum aproximately. The only unknown in Eq (16) is then the flattening factor, (A/A_o). Evidence from other experiments indicate that this factor has a time average value of about 1.3 for high speed impacts such as we have in this example. This single value of (A/A_o) enables us to calculate the solid curves in reasonable agreement with experiment. If we assume that A/A_o declines slightly with increasing speed, we can improve the agreement. However, this is not worth doing in view of the nature of the approximations we have used. The main point to be made is that we can represent the experimental data closely by a simple formula derived from a general theory which connects all tyes of penetrations.

B. Rod Penetrators

After considering a number of possibilities we have found that an erosion rate proportional to the penetration speed agrees most closely with all of the available data. This takes the form

$$\dot{M} = -\mu_o\dot{P} \tag{18}$$

where μ_o is a constant. Since the initial value of P is zero in the coordinate system we have chosen, we find

$$M = M_o - \mu_o P \tag{19}$$

upon integrating Eq (18). Then $\dot{M}_p = |\dot{M}| = \mu_o\dot{P}$ in Eq (6), so

$$\dot{S} = (1+\mu_o I_o)\dot{P} = k\dot{P} \tag{20}$$

from Eq (6). Since $k \geq 1$, it is clear that $\dot{P} \leq \dot{S}$, as we expect when erosion is taking place. Since $d\dot{S}/dt = kd\dot{P}/dt$ from Eq (20) and $\dot{M}_p^{\,2} = \mu_o^{\,2}\dot{P}^2$, Eq (9) becomes

$$M_o[1-(\mu_o/M_o)P]kd\dot{P}/dt = -(a+\bar{c}\dot{P}^2) \tag{21}$$

where

$$\bar{c} = c - I_o\mu_o^{\,2} \tag{22}$$

may be called the reduced inertial coefficient. When time-dependent data is not available, as is usually the case in penetration mechanics, we can use the relation $d\dot{P}/dt = \dot{P}d\dot{P}/dP$ to transform Eq (21) to a separable form in the $\dot{P}$, P plane. The solution is

$$\dot{P} = \sqrt{(a/\bar{c})\{[1+(\bar{c}/a)(\dot{S}_o/k)^2][1-(\mu_o/M_o)P]^{1/\varepsilon} - 1\}} \tag{23}$$

where the dimensionless erosion rate is $\varepsilon = (\mu_o k)/(2\bar{c})$. When embedment occurs in a semi-infinite target, $\dot{P} = 0$ and

$$P_E = (M_o/\mu_o)\{1-[1+(\dot{S}_o/\alpha)^2]^{-\varepsilon}\} \tag{24}$$

which reduces to Eq (16) for $\mu_o \to 0$, $\bar{c} \to c$, $k \to 1$. Here $\alpha = k\sqrt{a/\bar{c}}$ reduces to the previous α when there is no erosion. However, the inflection point of Eq (24) is $k\sqrt{a/(\bar{c}+\mu_o k)}$, slightly smaller than α. We also note that Eq (24), unlike Eq (16), approaches a finite limit (M_o/μ_o) as $\dot{S}_o$ increases without limit. If sufficient data is available to very high striking speeds, this enables us to estimate μ_o in the absence of time-dependent data from which more direct estimates can be made.

In the special case $\bar{c} = 0$ in Eq (22) we find $\mu_o = \sqrt{c/I_o} = A_o\sqrt{\rho_p\rho_t/\gamma}$ with $\gamma = (A_o/A) \leq 1$. Since $M_o = \rho_p A_o L_o$, $(M_o/\mu_o) = L_o\sqrt{\gamma\rho_p/\rho_t}$. Instead of Eq (24) we have in this special case

$$P_E/L_o = \sqrt{\gamma\rho_p/\rho_t}\,\{1-\exp[-(\mu_o k/2a)(\dot{S}_o/k)^2]\} \tag{25}$$

Here $\mu_o/(2k)$ does not differ greatly from unity in many cases of interest. Then for a high speed impact, $\dot{S}_o^{\,2}$ dominates the strength factor, a, and the exponential correction term in Eq (25) is negligible. In such a case and for the maximum erosionrate ($\bar{c}$=0), Eq (25) reduces to the classical density law which was derived for a penetrator of constant length and constant speed (Birkhoff and co-workers, 1948; Eichelberger, 1956). From Eq (21) it is clear that if $\bar{c}$ turns negative because μ_o exceeds the value above, the force on the right side approaches zero, an unrealistic value. In this sense, the value of μ_o which makes $\bar{c}$ vanish is the maximum erosion rate. This can be shown to imply a non-zero residual mass in every physical case (Dehn, 1986). The same reference also shows how to use this value to estimate μ_o for various penetrator/target combinations.

Now let us consider another example from the first series of Hypervelocity Impact Symposia. Fig. 2 shows data for steel (0.6g) and aluminum (0.2g) rods (L_o/D_o = 9.52mm/3.17mm = 3) impacting semi-infinite mild steel targets over a wide range of speeds (Christman and co-workers, 1965). Soft and hard aluminum rods were used but no differences in penetration were observed. The hydrodynamic limit, $P_E = L_o\sqrt{\rho_p/\rho_t}$, is 9.5mm and 5.6mm for steel and aluminum rods respectively and is greatly exceeded for these short rods. The steel rods were made from the same mild steel as the targets with reported Brinell hardness 110kg/mm^2. From this we find $Y_t = 33 \times 10^8$ dyne/cm^2 and for A_o = .08cm^2, we find $a_o = 7.9 \times 10^8$dyne = .0079g mm/μs^2 for both projectiles. Moreover, c_o = .031g/mm for both cases. By the methods mentioned above, we find μ_o = .032g/mm for the steel rod and μ_o = .019g/mm for the aluminum rod. The steel rod has I_o = 8.0mm/g, k = 1.26, $\alpha = 1.26\sqrt{a/\bar{c}}$ mm/μs and $\varepsilon = .02/\bar{c}$ with $\bar{c}$ = c - .008g/mm, $a = a_o(A/A_o)$ and $c = c_o(A/A_o)$. From a variety of sources it can be shown that (A/A_o) is a function of striking speed and aspect ratio as shown in Fig. 3 (Dehn, 1986).

If we use these values in Eq (24) we can calculate the upper solid curve in Fig. 2 which agrees closely with the data. For the aluminum projectile, I_o = 23.4mm/g, k = 1.44, $\alpha = 1.44\sqrt{a/\bar{c}}$ and $\varepsilon = .014/\bar{c}$ with $\bar{c}$ = c - .008. A set of curves similar to that shown in Fig. 3 holds for an aluminum rod versus steel. For L_o/D_o = 3, the high speed limit of (A/A_o) is about 1.6, somewhat lower than for a similar steel rod versus steel. When the target is denser than the projectile it offers more resistance to lateral spreading and leads to a smaller average hole diameter. In this way we can use Eq (24) to calculate the lower solid curve in Fig. 2 which also agrees with experiment.

Next, let us consider a more recent example for which time-dependent data are available. Figure 4 is for a steel rod (L_o/D_o = 22mm/2mm = 11, M_o = .575g) striking a semi-infinite magnesium target at $\dot{S}_o$ = 3.27mm/μs. It shows the positions of the front and rear of the rod versus time as determined by X-ray photography (Perez, 1982). The final depth of penetration was 60mm, about 30% higher than the hydrodynamic limit even for this long rod when striking a low density target. From Fig. 4 we can find L = P - R for any time and for known ρ_p, L_o and A_o, we can use Eq (19) with $M = \rho_p A_o L$ to determine μ_o from the data. We find μ_o = .009g/mm, independent of speed. From Perez' X-rays we obtain (A/A_o) = 2.5 which together with his target density, ρ_t = 1.8g/mm and strength, $Y_t = 10^9$dyne/cm^2, enables us to calculate the parameters we need. In particular, we find the dimensionless erosion rate to be $\varepsilon = 1$, so the solution of Eq (21) in the P, t plane is

$$P = 64\text{mm}\,\{1-[1+(4-.1t)^2]/17\} \qquad (26)$$

from which we can also find $\dot{P}(t)$. Eq (26) gives us the upper solid curve in Fig. 4 which agrees closely with the data. Eq (19) gives us M versus P or t and $L = M/(\rho_p A_o)$ as well as R = P - L, the lower solid curve in Fig. 4. Expressions like Eq (26) for P in terms of a finite number of elementary functions of t can be found for any integer or half-integer value of ε. For intermediate values a simple numerical evaluation of a single integral is sufficient.

C. Jet Penetrator

For the BRL 81mm precision shaped charge device (42° copper cone, composition B explosive) a great deal of experimental information is available. This has been exploited (Dehn,

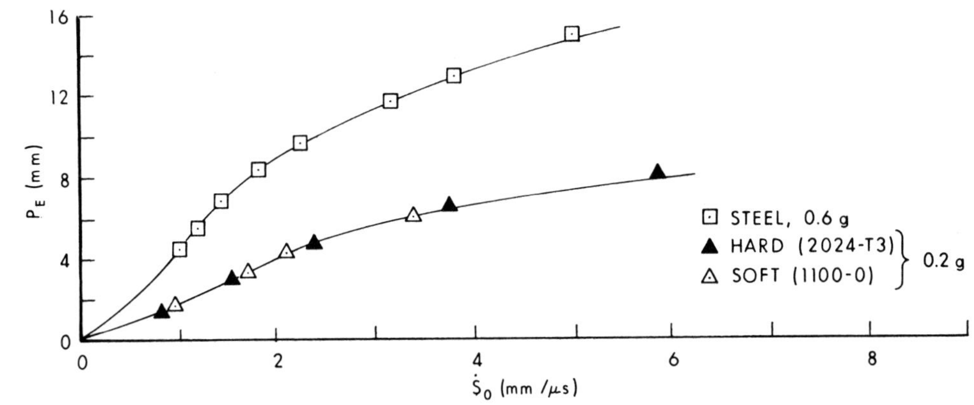

Fig. 2. Penetration depth versus striking speed for $L_0/D_0 = 3$ steel and aluminium rods impacting mild steel targets.

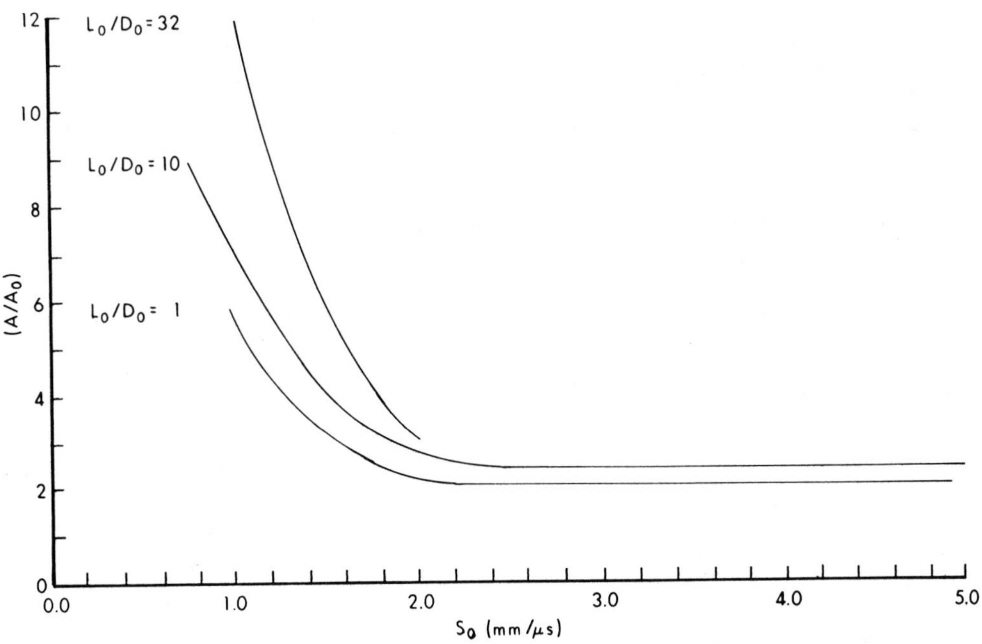

Fig. 3. Flattening factor versus striking speed for steel rods of various aspect ratios impacting steel targets.

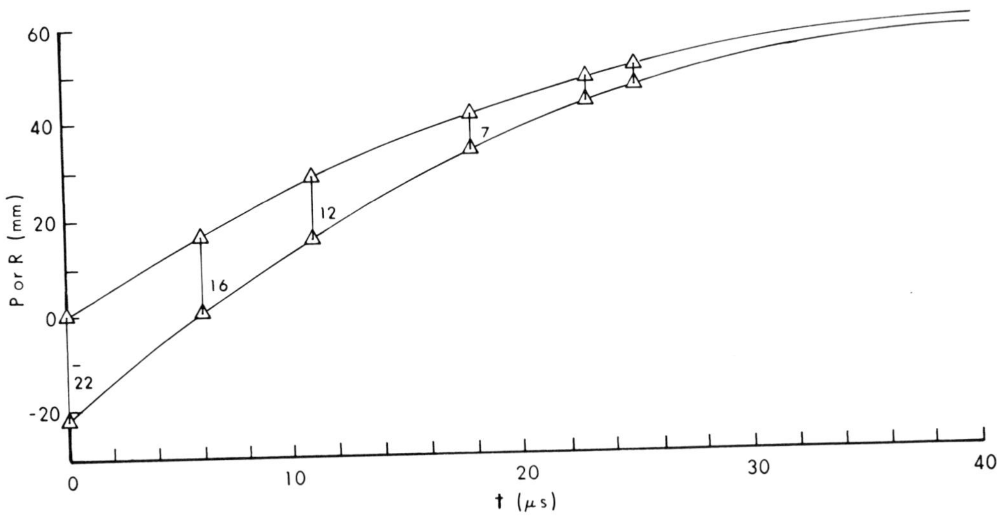

Fig. 4. Front and rear positions versus time for an $L_0/D_0 = 11$ steel rod impacting a magnesium target at 3.27mm/μs. Numbers indicate remaining length in mm.

1986) to illustrate various solutions of the equation of motion, Eq (15). In particular, penetration/standoff curves as well as penetration/time curves have been successfully calculated for a variety of targets. Since the length of this paper is limited we will give only one example taken from the last Hypervelocity Symposia series. One of the puzzles of early hydrodynamic theory was deeper penetration into lead than into steel in spite of lead's greater density, contrary to the density rule. Fig. 5 shows penetration versus time for this shaped charge at 2.44CD standoff versus semi-infinite lead and steel targets. The calculated solid curves agree quite well with the data reported by Feldman (1956) as solid curves rather than points. As is evident from the figure, penetration into lead is lower than into steel especially at early times when the speed-dependent forces dominate the strength term. However, the greater strength of steel terminates the penetration at an earlier time. Penetration into the weak lead target lasts more than 150μs longer and results in a larger final depth.

CONCLUSION

There is a great deal more that can be said about the foundations, applications and possible extensions of this theory (Dehn, 1986). For example, if T is the thickness of a finite target plate, and P = T is taken to be the perforation criterion in Eq (23), we can find the residual speed, $\dot{P}$, versus the striking speed, $\dot{S}_o$, as well as the residual mass. What has been presented here, however, should be sufficient to indicate that it is possible to construct a reasonably simple yet accurate unified theory of penetration.

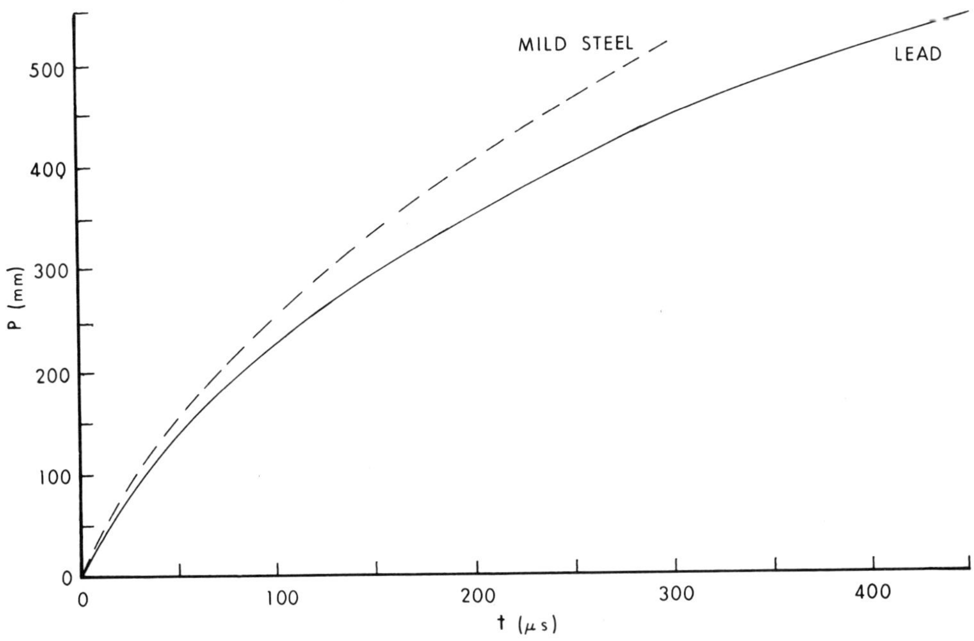

Fig. 5. Penetration (tip position) versus time for an 81mm 42° copper cone shaped charge jet impacting steel and lead targets.

REFERENCES

Atkins, W. W. (1960). Hypervelocity Penetraton Studies. Fourth Hypervelocity Impact Symposium, Vol. 1, paper 11.

Bethe, H. A. (1941). Attempt of a Theory of Armor Penetration. Frankford Arsenal R 492.

Birkhoff, G., D. P. MacDougall, E. M. Pugh and G. I. Taylor (1948). Explosives with Lined Cavities. J. Appl. Phys., 19, 563.

Chanteret, P. Y., and F. Jamet (1984). Quasi Non-Stretching Hypervelocity Jets. Eighth International Symposium on Ballistics, VI-15.

Dehn, J., (1986). A Unified Theory of Penetration. BRL report to be published.

Eichelberger, R. J. (1956). Experimental Test of the Theory of Penetration by Metallic Jets. J. Appl. Phys., 27, 63.

Feldman, J. B., Jr. (1959). Volume Energy Relation from Shaped Charge Jet Penetrations. Third Hypervelocity Impact Symposium, Vol. 1, p. 215.

Halperson, S. M. and W. W. Atkins (1962). Observations of Hypervelocity Impact. Fifth Hypervelocity Impact Symposium, Vol. 2, Part 2, p. 497.

Halperson, S. M. (1963). Some Phenomena Associated with Impacts into Aluminum. Sixth Hypervelocity Impact Symposium, Vol. 2, Part 2, p. 525.

Halperson, S. M. (1965). Comparisons between Hydrodynamic Theory and Impact Experiments. Seventh Hypervelocity Impact Symposium, Vol. 5, p. 235.

Kineke, J. H. Jr., and L. G. Richards (1963). Influence of Target Strength on Hypervelocity Crater Formation in Aluminum. Sixth Hypervelocity Impact Symposium, Vol. 2, Part 2, p. 513.

Mobley, C. E. Jr., and R. B. Pond (1965). Energy Balances for Hypervelocity targets. Seventh Hypervelocity Impact Symposium, Vol. 5, p. 73.

Perez, E. (1982). Etude Experimentale et theorique de la Penetration de Cibles Metalliques Semi-infinies par des Projectiles Metalliques de Grand Allongement et de vitesse Superieure a 2,000m/s. Sciences et Techniques de L'Armament, 56 (1), 11.

Sorensen, N. R. (1965). Systematic Investigation of Crater Formation in Metals. Seventh Hypervelocity Impact Symposium, Vol. 6, p. 281.

Tabor, D. (1948). A Simple Theory of Static and Dynamic Hardness. Proc. R. Soc. A192, 247.

Tabor, D. (1951). The Hardness of Metals. Clarendon Press, Oxford.

Walsh, J. M. and W. E. Johnson (1965). On the Theory of Hypervelocity Impact. Seventh Hypervelocity Impact Symposium, Vol. 2, p. 1.

Int. J. Impact Engng Vol. 5, pp. 249–260, 1987
Printed in Great Britain

0734-743X/87 $3.00 + 0.00
Pergamon Journals Ltd.

FRAGMENT BREAKUP

David L. Dickinson*, Jerome D. Yatteau** and Rodney F. Recht**

*Naval Surface Weapons Center
Dahlgren, Virginia 22448-5000
**Denver Research Institute
University of Denver
Denver, Colorado 80210-0127

ABSTRACT

This paper summarizes observations that were made in conducting a series of tests with compact steel fragments penetrating spaced plates at impact speeds up to about 5 km/sec. The discussion focuses on the onset of fragment breakup and a description of the particle cloud behind the target plate. The effects of plate thickness, fragment size, and fragment hardness on the impact speed required to produce breakup are presented. The effects of impact orientation and target incidence on the formation of the residual cloud are discussed.

INTRODUCTION

This paper pertains to compact fragments penetrating relatively thin plates (Plate Thickness/Fragment Diameter, T/D $\leq$ 1) in air or a vacuum with minimum plate spacings greater than several fragment diameters. At low speeds, fragments perforate thin plates without deformation or mass loss. As impact speed increases, fragment deformation and plate hole size increase. Against harder and/or heavier plates, material will be sheared from the deformed fragment as it passes through the plate. Steel cylinders (BHN 285) perforating thin mild steel plates deform enough to cause hole enlargement when impact speed exceeds about 500 m/s. Further increases in impact speed will eventually cause breakup of the fragment and plate plug. Mild steel cubes begin to fracture on impact with steel plates at speeds as low as 750 m/s. Plate plug breakup for mild steel plates under flat impacts by steel cylinders begins at about 1000 m/s [1].

If impact speeds are not much higher than the fracture threshold, the ballistic threat to spaced plates remains well represented by single fragment and plate particles. However, breakup of typical warhead fragment simulators becomes extensive at speeds between 1500-2000 m/s. Penetration and plate damage predictions for impact speeds above 2000 m/s require models to predict the transformation of single fragments into high speed multi-particle debris clouds. Most previous studies of high speed fragments have involved impact speeds above 5000 m/s. Beginning in 1978, the Naval Surface Weapons Center (NSWC) supported a series of experimental and analytical efforts to expand the available data base and to develop required new penetration methodology for impact speeds up to 5000 m/s. This paper addresses selected phenomena just above the breakup threshold.

The initial test program was conducted at the Naval Research Laboratory (NRL) and involved fragments of various sizes, shapes, materials, and hardnesses impacting at speeds between 1500- and 4500 m/s [2]. These and other test results were used by the Denver Research Institute (DRI) to develop the high speed multiple plate penetration model and associated computer program called FATE [3]. Additional experiments and analyses have been performed at NRL and DRI to expand the high speed penetration data base and to improve the penetration model. The tests have involved primarily steel cylinders and cubes. Also under sponsorship by NSWC, the Boeing Aerospace Company conducted a careful data analyses effort that has resulted in a better understanding of the variables affecting the fragment fracture threshold speed [4]. The following discussion draws upon all this work but primarily from the intial work at NRL and DRI and from a recently completed test program at DRI to determine fracture threshold speeds and debris characteristics for steel cubes.

EXPERIMENTAL SETUP

The experiments at NRL utilized a two-stage light gas gun which has been used to achieve impact speeds up to 5500 m/s. The tests at DRI were conducted with powder guns that have launched fragments at speeds up to 2700 m/s. An important aspect of the test program has been the reliable measurement of impact speed and impact orientation. Figures 1 and 2 illustrate the typical multiple and single plate target arrangements at NRL and DRI. At NRL, double exposure radiographs were used to obtain impact and residual speed and impact orientation. At DRI, a combination of foil and magnetic switches and orthogonal radiographs in front and behind the shatter plate provided redundant measurements of impact orientation. Some of the DRI tests utilized a rifled barrel to spin the cubes for stablization and achievement of flatter impacts.

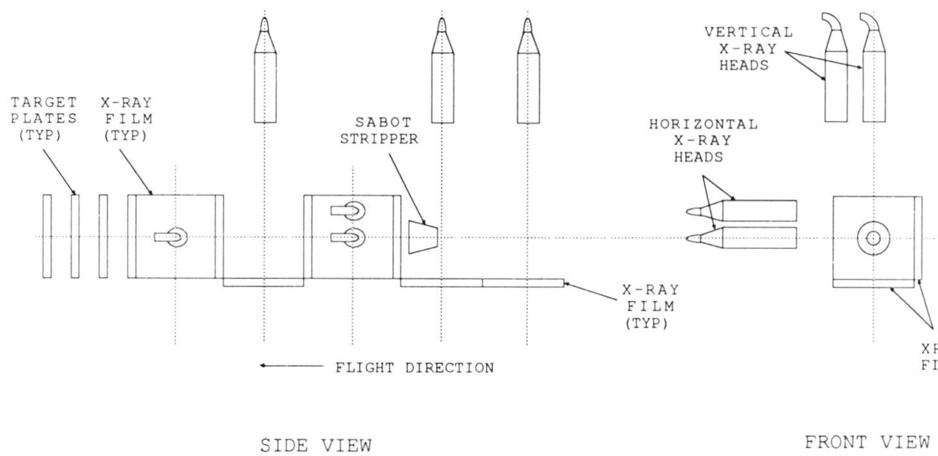

Fig. 1. NRL Multiple Plate Target Arrangement

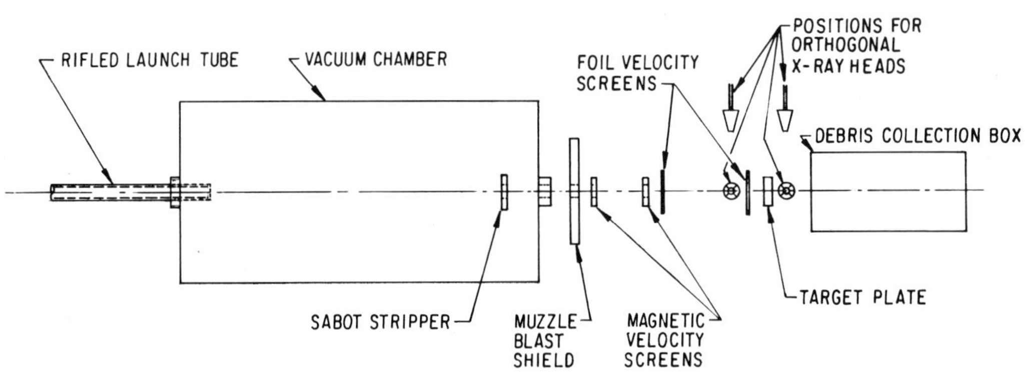

Fig. 2. DRI Single Plate Target Arrangement

Single and multiple plate tests have been conducted at both facilities. The multiple plate tests provided high speed penetration and plate damage characteristics and the single plate tests provided post-impact debris characteristics. The debris particles were captured behind the plate. The recovery medium was Verilite,™ an expanded clay product aggregate manufactured by Mapleton Development Inc. The fragment and plate particles are easily separated from the Verilite™ using a shaker table precision sieves, and a magnet. The technique enables recovery of particle masses down to 0.1 grain.

HIGH SPEED PENETRATION CHARACTERISITCS

High speed plate damage is illustrated in Fig. 3. The plate array was impacted face-on by a 240 grain steel cube at 2560 m/s. Figure 4 contains double exposure radiographs showing the dispersion of the fractured cube behind the first plate (1.6 mm 2024-T3 aluminum alloy) for the same test. The third plate in Fig. 3 is 3.2 mm mild steel; all other plates in the array are 3.2 mm 2024-T3 aluminum alloy. The cross-shaped hole pattern in the second plate is typical for fractured cubes after flat impacts. The arms of the cross emanate from the lateral faces indicating seperation along diagonal planes. Hole patterns in subsequent plates are consistent with the outer debris particles being gradually stripped away. The larger particles penetrated six of the seven plates in the array. A single large plate particle can barely be seen moving ahead of the main fragment cluster in Fig. 4. Smaller aluminum alloy plate particles are not visible.

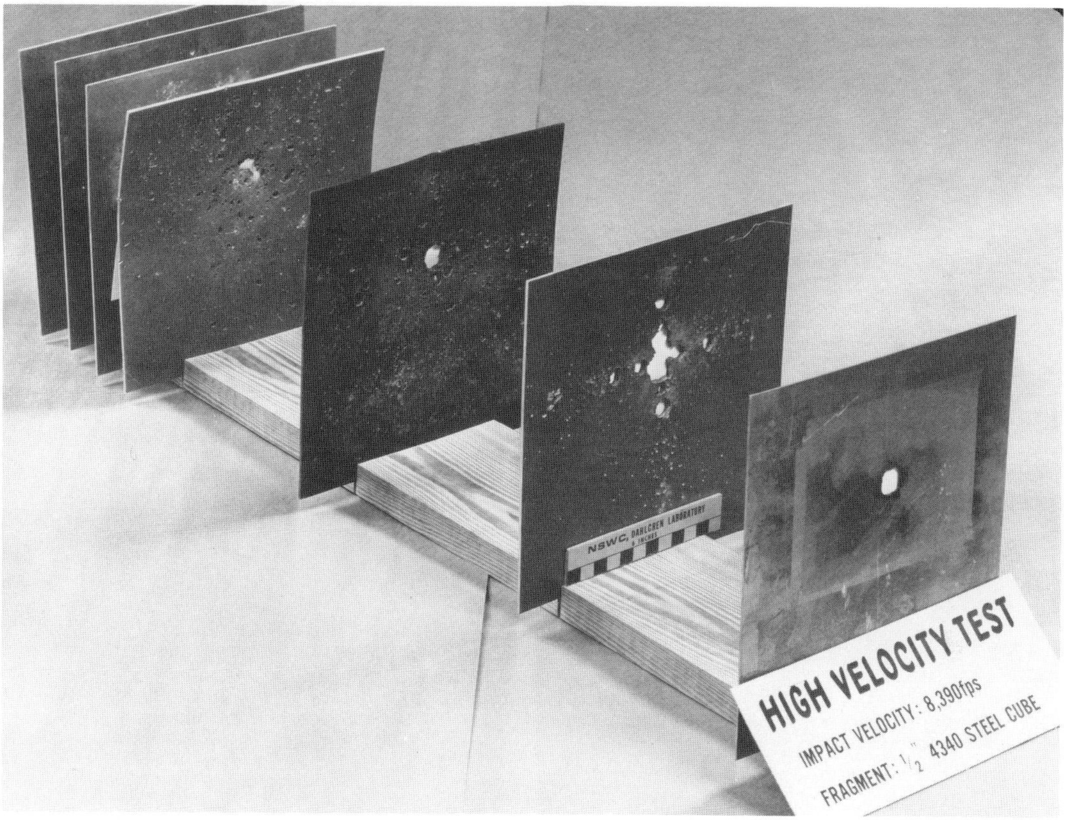

Fig. 3. High Speed Multiple Plate Penetration

The envelope of typical high velocity debris cloud can be seen more clearly in the radiograph in Fig. 5. In this case, a 1.6 mm steel plate was impacted by a 6.4 mm steel sphere at 2775 m/s. Broken plate particles emerge from the back of the plate on the surface of a hollow expanding cloud which is nearly ellipsoidal in shape. For normal impacts, the axis of the cloud is aligned with the shotline; for oblique impacts, the axis of the cloud will be rotated toward the plate normal. Fragment recovery experiments have shown that plate particle sizes vary considerably in a given cloud but there is no apparent bias with respect to trajectory inclination. These studies also show that fragment particles are usually clustered in a tighter pattern about the shotline than the plate particles [1].

Figures 6 and 7 illustrate the effects of increasing impact speed above the fracture threshold on fragment breakup and second plate damage. Each figure contains sequential (double exposed) radiographs of fragment debris particles behind a 1.6 mm aluminum alloy plate and damage to a 3.2 mm aluminum plate placed 254 mm behind the first plate. With increasing speed, fragment breakup becomes more severe; the sizes of individual debris particles decrease and the number of particles increase. Dispersion of fragment debris particles also increases with increasing impact speed.

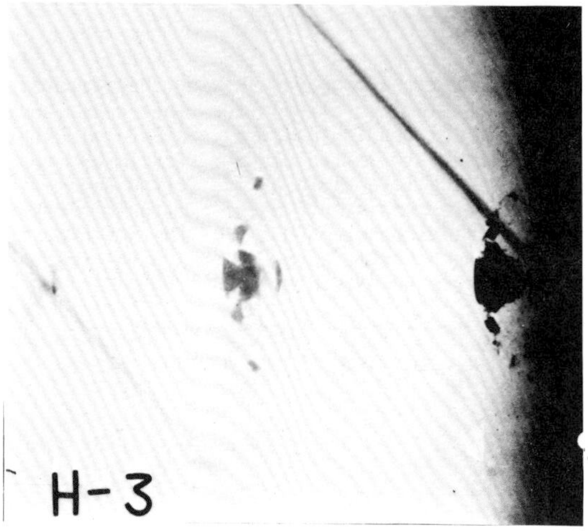

Fig. 4. Double Exposure Radiograph of Fractured Cube

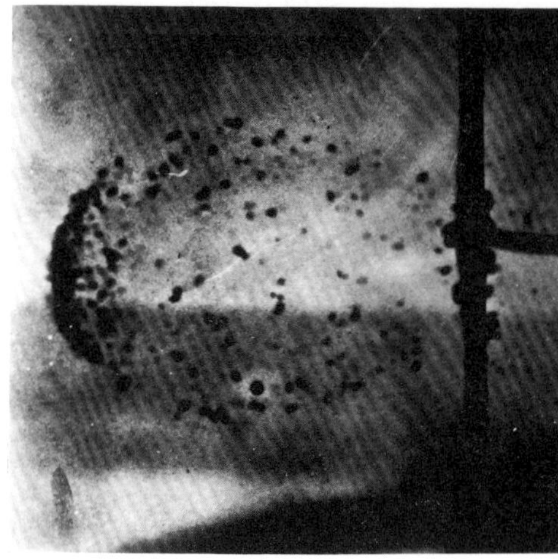

Fig. 5. Debris Cloud Behind a Steel Plate

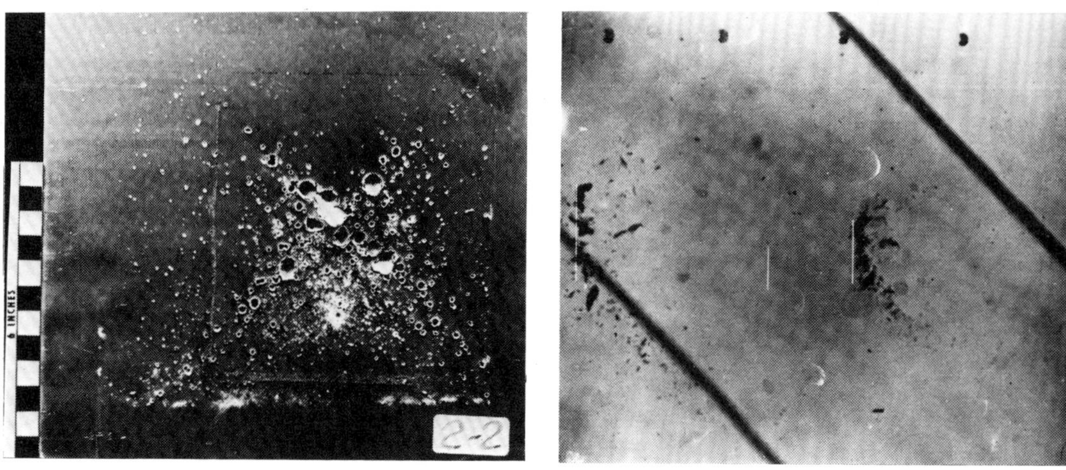

Fig. 6. Witness Plate Damage and Double Exposure Debris Radiograph for a 31 Grain Steel Cube at 3300 m/s

The number of perforations in the second plate increases with increasing impact speed. Individual hole sizes near the center of the patterns are seen to be relatively insensitive to speed. Apparently the decrease in fragment size with increasing speed is offset to some extent by an increase in hole size with speed. The hole patterns and fragment radiographs indicate that the aerial density of fragment trajectories increases towards the center of the fragment beam. As a result, individual perforations tend to overlap near the center of the pattern to form an enlarged single hole. Central hole enlargement can also occur after impact when the plate responds to the impulse transmitted by the perforating and/or cratering particles. Large plate deformations lead to tearing and petalling in thinner plates, as shown in Fig. 8, and failure due to combined shear, bending, and membrane stresses in thicker plates. The latter failure mode appears to have occurred in the aluminum plate shown in Fig. 9. Note that the material removed from the plate (inset photograph) was heavily cratered, but not perforated.

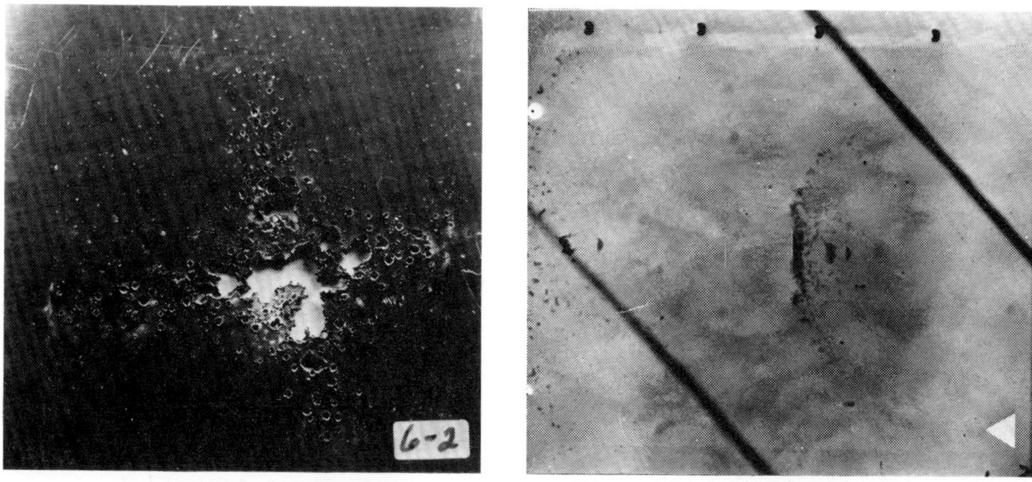

Fig. 7. Witness Plate Damage and Double Exposure Debris Radiograph for a 31 Grain Steel Cube at 4700 m/s

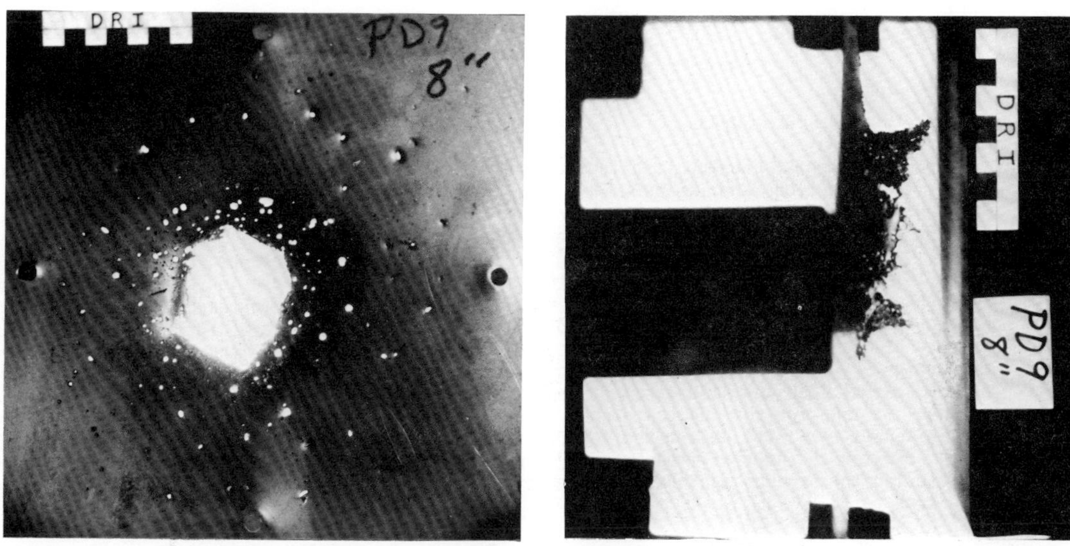

Fig. 8. Damage to a 0.8 mm Steel Witness Plate 203 mm Behind a 2.4 mm Steel Plate Impacted by a 11.9 mm Steel Sphere at 2140 m/s

The cartoon in Fig. 10 depicts the main features of high speed multiple plate penetration. Upon impact with the first plate, the fragment loses a portion of its mass on the impact side of the plate, and the remainder appears in the debris cloud behind the plate. At higher speeds, the debris cloud will consist of broken fragment and plate particles. At lower speeds, the debris will contain an intact fragment and either a number of plate particles or an intact plate plug. The particles will disperse radially in transit to the next plate, where, in general, some will perforate and some will be stopped. Radial dispersion is bounded by the cone half-angles ϕ_c and ϕ_s as shown. The perforating particles will lose mass and speed and drive new plate material out the exit side of the plate. The debris cloud behind second plate will thus have a different mass and speed distribution and generally contain penetrator fragments along with plate fragments from the first and second plates. The perforation process is then repeated at successive plates until the mass and speed of the particles are reduced to the point where none can perforate the next plate. Damage to any given plate will consist of a number of holes and craters produced by a perforating and nonperforating penetrator and plate particles. Depending on the incident fragment number, mass, and speed distributions, there may or may not be an enlarged central hole about the shotline.

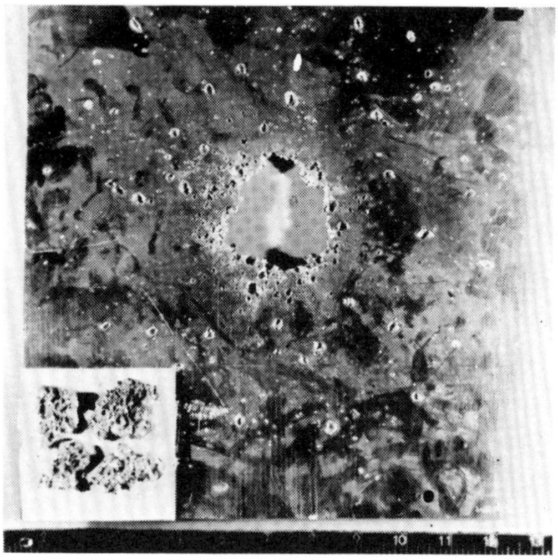

Fig. 9. Damage to a 6.4 mm Al Alloy Witness Plate 203 mm Behind a 3.2 mm Steel Plate Impacted by a 12.7 mm Steel Sphere at 2260 m/s

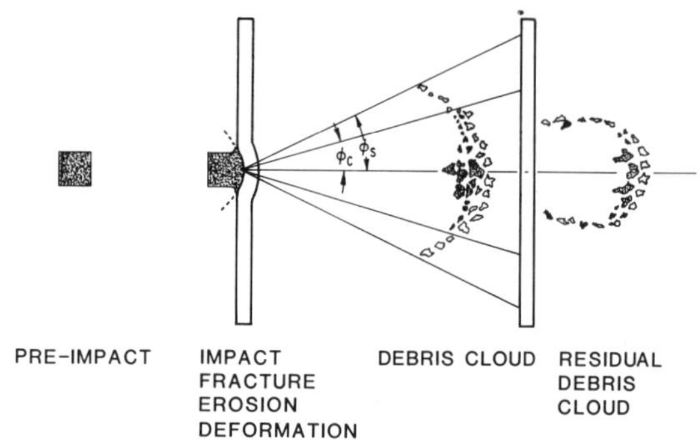

Fig. 10. Typical High Speed Multiple Plate Penetration

BREAKUP FOR STEEL CUBES

For nonspherical fragments, breakup is affected by impact orientation. Breakup data for cubes that include debris collection and measurements of impact orientation were found to be relatively scarce. A test program was therefore undertaken at DRI to develop a fairly complete data base describing the breakup of steel cubes. A systematic approach was followed. Three cube masses were selected: 31-, 104-, and 248 grains (6.3, 9.5, and 12.7 mm). The reference cube material was selected as 1018 (BHN 183) steel and the reference plate material was selected as 2024-T3 aluminum alloy. The threshold speed, Vc, for the onset of breakup was determined for each size cube against aluminum alloy plates of various thicknesses for flat impacts. These data would serve to establish the dependence of Vc on T/D and verify a possible size effect indicated by the fragment fracture study at Boeing [4]. To obtain values of Vc for other plate materials, the reference cubes were fired face-on against 1010 and 4130 steel plates at normal obliquity (incidence). Additional data were obtained at NRL for the 104 grain cube fired against a 6Al-4V titanium plate. Values of Vc for 4130 steel cubes were also determined for flat impact against 2024-T3 aluminum and 1010 steel plates. In addition to the flat impact tests, a number of experiments were performed to verify an existing criterion for breakup against oblique plates by yawed cubes. Impact orientation was measured in each test.

The fragment and plate debris particles were collected and analyzed at NSWC to determine the effect of increasing impact speed on the number and size of debris particles.

THRESHOLD FRACTURE SPEED FOR FLAT IMPACTS

The cubes were sabot-launched from a rifled barrel to consistently obtain reasonably flat impacts. Fragment breakup was determined from a combination of post-impact radiographs and recovered debris analysis. Fig. 11-13 contain composite radiographs showing the effect of increasing impact speed on fragment breakup for the 104 grain cubes impacting flat against three plate thicknesses. The impact speed for each test is shown on the right of the figure. Also shown on the right is the ratio of the mass of the largest recovered cube particle, Mr, to the original mass of the cube, M. Figure 14 is a plot of the normalized residual mass of the largest recovered cube fragment (Mr/M) vs. impact speed for two of the plate thicknesses. The solid symbols denote fractured cubes, the open symbols denote intact cubes, and the half-filled symbols denote barely fractured cubes. The threshold breakup speed, Vc, was determined by the intersection of a line (Log-Log plot) fitted to the fractured cube data points near the point where the cube started to lose mass with the horizontal line where no mass loss occurred. Only data with impact angles, μ (between cube face and plate), less than 10 deg and usually less than 5 deg were used in determining Vc. All of the Vc values for the aluminum alloy plates are plotted vs. T/D in Fig. 15.

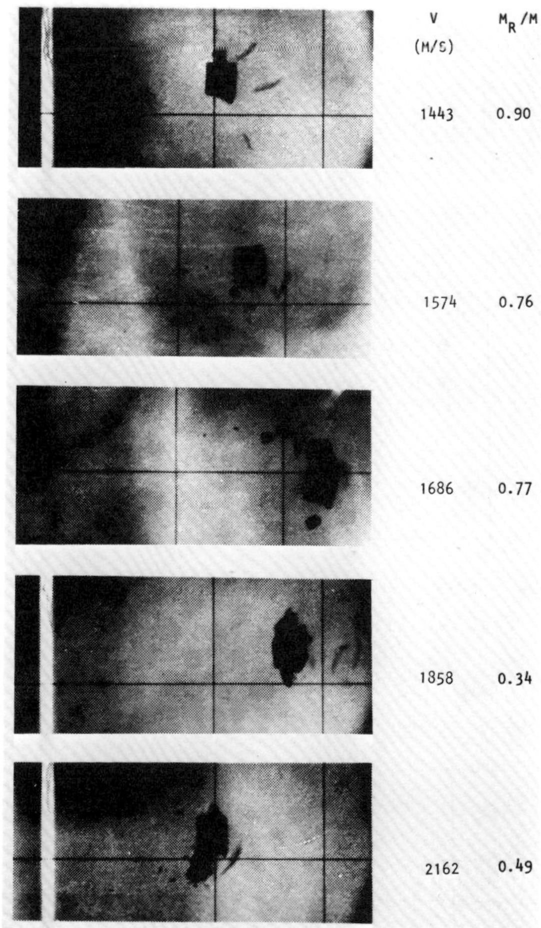

Fig. 11. 104 Grain Mild Steel Cubes vs. 0.6 mm Al Alloy Plates

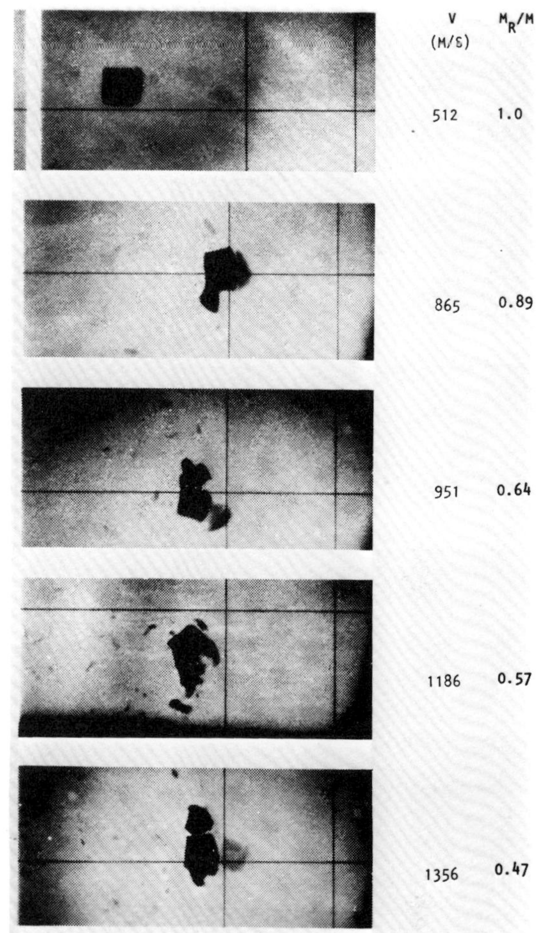

Fig. 12. 104 Grain Mild Steel Cubes vs. 2.3 mm Al Alloy Plates

A detailed analysis of the fractured specimens has not yet been made. The radiographs, the cross-shaped hole patterns in witness plates, and the occasional "apple core" shape of fractured specimens suggest that near the threshold speed fracture probably results from tensile stresses across the diagonals associated with the interaction of relief waves from adjacent sides of the cube. The arrival of the relief wave from the rear of the plate could determine lateral spall depth in the cube. The decrease in Vc for thin plates with increasing T/D is consistent with a lower required fracture for longer pulse durations. The gradual increase of Vc with T/D values above 0.25 may be due to increased lateral confinement from deeper penetration before arrival of the rear plate surface relief wave. The test results in Fig. 15 for T/D values above 0.25 indicate the size effect that, for consistent T/D, Vc decreases with increasing cube size. A similar size effect was determined for steel spheres against plates with T/D less than 0.5 [4]. The size effect cannot be ascertained from the cube data for T/D values less than 0.25.

CUBE HARDNESS EFFECTS ON CRITICAL SPEED

The effect of cube hardness was examined by firing cubes with three different hardnesses against an aluminum alloy plate to determine the critical impact speed for breakup. The results are plotted in Fig. 16. The graph shows that the threshold breakup speed for 4130 steel cubes varied linearly with cube hardness over the values for the 1018 steel cubes. Use of the cube hardness to correlate the test results presumes that, for the same impact geometry and impact pressure, resistance to fracture is determined by dynamic strength of the cube.

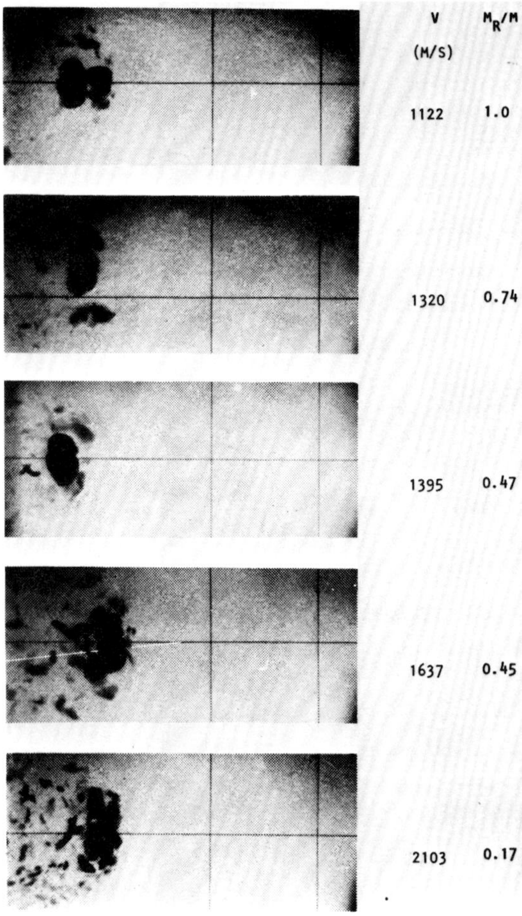

Fig. 13. 104 Grain Mild Steel Cubes vs. 9.5 mm Al Alloy Plates

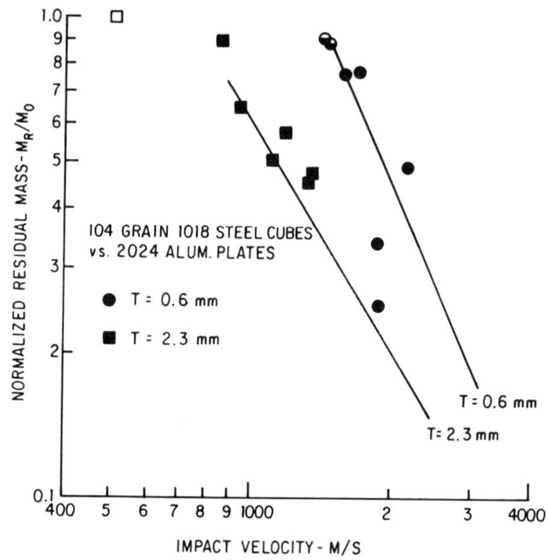

Fig. 14. Primary Fragment Residual Mass

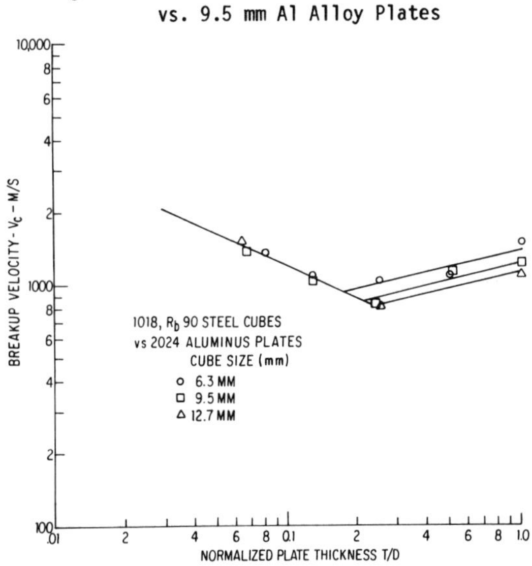

Fig. 15. Threshold Fracture Speed For Mild Steel Cubes Impacting Flat Against Normal Al Alloy Plates

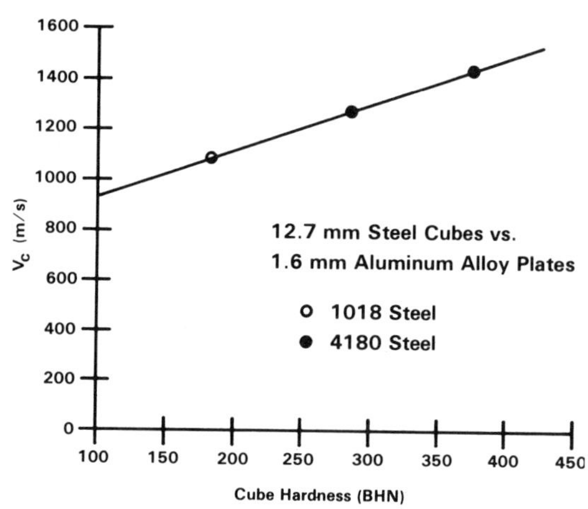

Fig. 16. Threshold Fracture Speed for Three Hardnesses of Steel Cubes

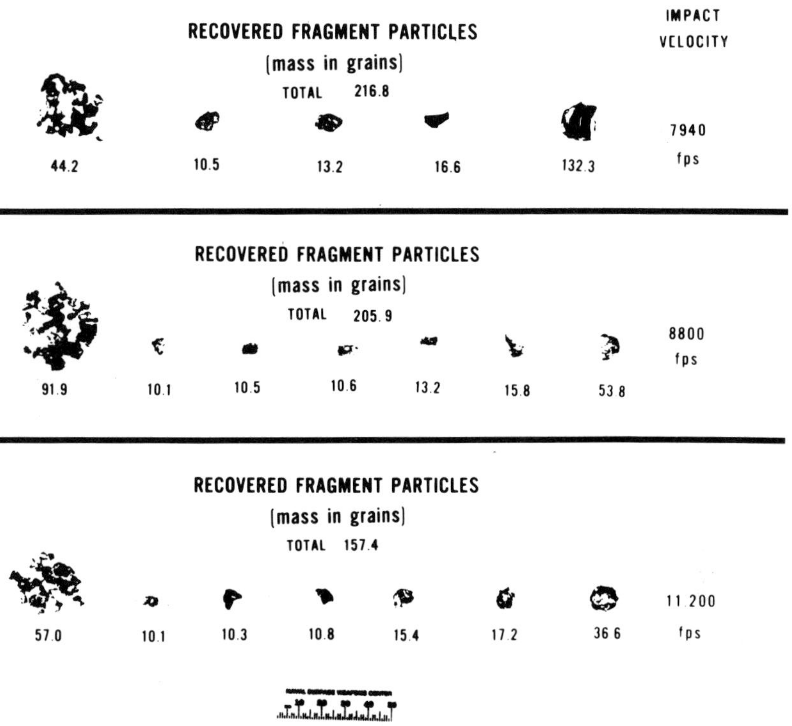

Fig. 17. Residual Fragment Population for 12.5 mm Diameter Cylinders (248 Grain) Striking 1.6 mm Al Alloy Plates

RESIDUAL FRAGMENT CLOUD POPULATION

As impact speed increases the largest residual particle decreases in mass, Fig. 14. Figure 17 shows the recovered fragment population for three different impact speeds. Fragments less than 10 grains are grouped together. The recovery technique limits the minimum mass of the recovered particles to 0.1 grains. As speed increases, the size of the particles in the middle of the fragment population remains relatively the same. The total recovered mass decreases with increasing speed.

NUMBER OF SECONDARY FRAGMENT PARTICLES FOR FLAT IMPACTS

For fixed impact geometry, the number of secondary fragment particles, Nc, generated above the threshold speed is also presumed to vary with some power of V/Vc. Fragment particle number data are illustrated in Fig. 18 for flat impacting 104 grain cubes.

EFFECTS OF IMPACT ORIENTATION AND PLATE OBLIQUITY

Figure 19 depicts a yawed steel cube perforating an oblique aluminum alloy plate. The speed of the cube plate intersection, Vi, along the cube face depends on the impact speed, V; the plate obliquity, θ ; and the impact angle, μ ; as indicated below the figure. It has been previously proposed that the ratio of the intersection speed to the sonic speed in the cube, C, be used as a speed dependent measure of the flatness of an impact as follows [1]. If $Vi \geq C$, pressure relief cannot propagate from the free surface ahead of the intersection and the impact is considered to be "flat." If $Vi < C$, then breakup cannot occur.

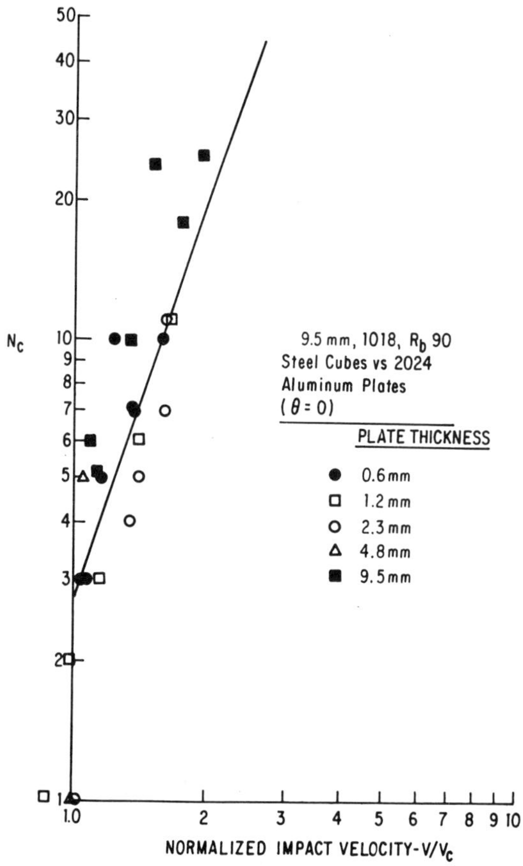

Fig. 18. Number of Fragment Particles vs. Normalized Impact Speed

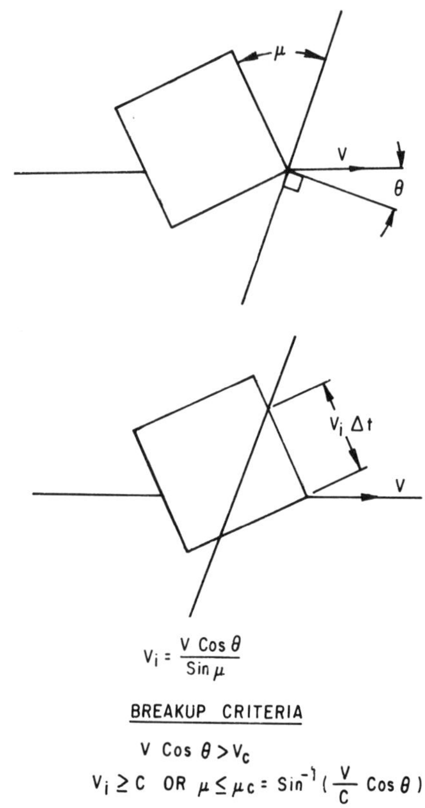

Fig. 19. Cube Impact Geometry

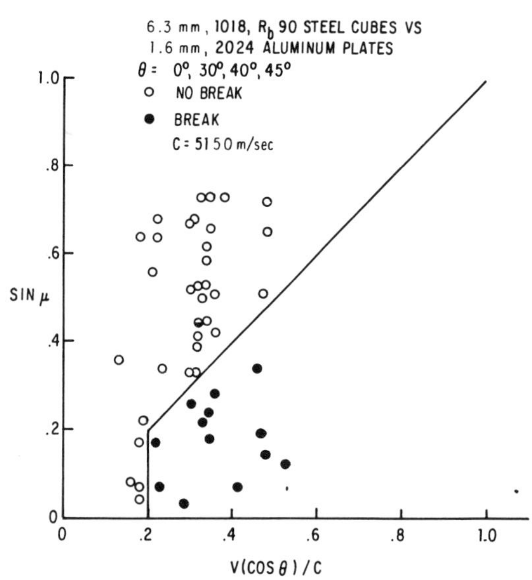

Fig. 20. Effects of Impact Angle and Plate Obliquity on Cube Fracture

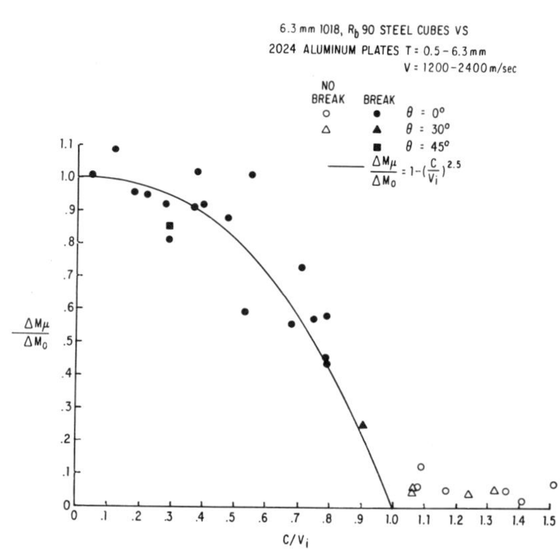

Fig. 21. Effects of Impact Angle and Plate Obliquity on Fracture Mode Mass Loss

The tests involving deliberately yawed cubes and oblique plates were designed to verify the breakup criteria. Figure 20 depicts the breakup results for the 31 grain, 1018 steel cubes. The two line segments in the graph correspond to the breakup criteria. Impact conditions lying below and to the right should produce fracture and those falling above and to the left should not produce fracture. The data in Fig. 20 provide convincing verification of the breakup criteria. It should be noted that the requirement on the normal speed component (first relationship in Fig. 19) does not apply at high obliquities (θ >60 deg). At high obliquity the velocity component parallel to the plate has a significant effect on breakup.

Finally, it was observed that although the yawed cubes fractured against oblique plates according to the above criteria, the extent of fracturing was less than for normal, unyawed impacts. The ratio of the sonic speed to intersection speed, C/Vi, appears to correlate the test results regarding the effects of μ and θ on fracture for impacts satisfying the breakup criteria. Figure 21 contains mass loss data for the 31 grain cubes impacting aluminum plates above the fracture speed for flat-normal impacts. The ordinate is the ratio of measured mass loss to the mass loss for flat normal impacts. A value of zero for C/Vi corresponds to a perfectly flat impact and the mass loss ratio therefore approaches unity as C/Vi approaches zero. Also, fracture does not occur when C/Vi exceeds unity according to the breakup criteria, and hence the mass loss ratio (due to fracture) should be zero for C/Vi values above unity. The nonzero mass loss ratios for C/Vi > 1 in Fig. 20 are due to a small amount of extrusion/shear mass loss.

Figure 22 illustrates the effect of changing impact orientation at fixed speed. The top radiograph of the figure shows a fractured cube, the middle radiograph shows a less severe fracture, and the bottom radiograph illustrates an intact residual fragment.

The test results presented in this paper were obtained by Mr. Andrew Williams of the NRL and Messrs. Edward Wittrock, James Dunn, and Steven Lightsey of the DRI. This work was funded by NSWC.

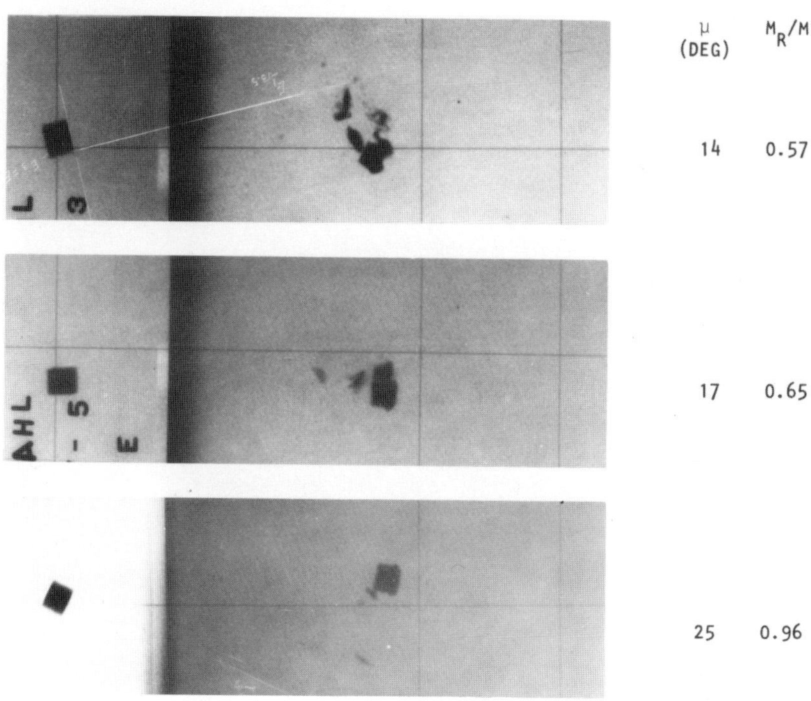

Fig. 22. Effect of Changing Impact Orientation on Fragment Breakup

REFERENCES

[1] Recht, R. F. , T. W. Ipson, and E. P. Wittrock (1969). Transformation of Terminal Ballistic Threat Definitions into Vital Component Malfunction Predictions. Denver Research Institute, Naval Weapons Center, China Lake, California, (NWC TP 4871).

[2] Dickinson, D. L. (1979). Investigations of High Velocity Fragments Impacting Plate Arrays. Naval Surface Weapons Center, Dahlgren, Virginia, (NSWC TR 79-66).

[3] Yatteau, J. D. (1982). High Velocity Multiple Plate Penetration Model. Denver Research Institute, Naval Surface Weapons Center, Dahlgren, Virginia, (NSWC TR 82-123).

[4] Bjorkman, M., T. O'Brian, E. Wilhelm, A. Zimmershied (1984). Fragment Fracture Study, Vol. I. Boeing Co., (Boeing Document No. D180-28210-1).

[5] Rolens, D. K., R. F. Recht, and E. S. Grubin (1976). Effects of 'G' Loading on Aircraft Vulnerability, A Users Guide to Structural Residual Determinations. Denver Research Institute, Ballistics Research Laboratories, Aberdeen, Maryland, (BRL CR 293).

Int. J. Impact Engng Vol. 5, pp. 261–268, 1987
Printed in Great Britain

0734-743X/87 $3.00 + 0.00
Pergamon Journals Ltd.

HYPERVELOCITY IMPACT OF MIXTURES

D. S. Drumheller

Thermomechanical and Physical Division
Sandia National Laboratories
Albuquerque, New Mexico

ABSTRACT

The study of the behavior of mixtures during shock loading entails several special problems which are both physical and computational in nature. On the physical level, many mixtures of interest such as engineering composites and water-saturated geological materials have constituents which are both soft and porous. Thus hypervelocity impact produces enormous heating. The distribution of this heating between the constituents of the mixture must be understood before accurate predictions of Hugoniot states and release paths can be achieved. On the computational level, numerical solutions within a wavecode framework require simultaneous solutions of an equation of state and conservation of energy equation for each constituent of the mixture. At present, to achieve these solutions a numerical subcycling scheme is required.

In this paper these problems are discussed in detail. A formulation of the theory of mixtures will be presented which is both complex enough to encompass the essential physics of many problems and simple enough to be incorporated into many wave propagation codes. Results of calculations will be compared to impact and release experiments on a mixture of water and powdered calcite.

INTRODUCTION

Many materials, both processed and naturally occurring, are heterogeneous mixtures or composites. They are composed of two or more constituents intermixed in random or regular patterns. Examples of these materials include both water-saturated geological formations and polymer based composites containing either regular arrays of reinforcing fibers or random distributions of particles. Many complex phenomena associated with strength and failure mechanisms have been identified with stress wave propagation through these media. The propagation of high-amplitude shock waves produced by hypervelocity impact results in still another phenomenon. Because of the compliant and porous structure of many of these materials, a large amount of heating is generated by the shock wave, and in general the distribution of this heat between the constituents of the material is not known. Since the thermal characteristics of the constituents may differ greatly, it is of primary importance to model this phenomenon. This paper presents a model which can be used for this purpose. The study of a water-calcite system is used as an illustration of the method.

The propagation of shock and release waves in heterogeneous mixtures poses several problems which are best understood by first reviewing a few basic points about the theory of shock wave propagation in homogeneous materials. In a homogeneous isotropic material, a shock wave propagates not as a discontinuity in stress and strain, but rather as a structured wave. It has a finite, albeit very small, thickness, and the values of stress and strain vary smoothly from their initial to their final values. For materials which can support shock waves, two important properties of the material contribute to the formation of these types of waves. First, the mechanical response is nonlinear. In particular, the stiffness of the material increases with increasing load. Thus highly dispersed wavefronts tend to steepen and form shock waves because the trailing high-pressure portion of these waves travels more rapidly than the leading low-pressure portion. This steepening continues until the second important property of the solid, dissipation produced by rate-sensitive behavior, takes effect. The final profile assumed by the shock wave is formed by a balance between the steepening effects of the nonlinear response and the dispersive effects of the rate-sensitive behavior.

The analysis of these types of waves is straightforward. It is outlined here as a prelude to the ensuing discussion of large amplitude waves in mixtures.[1] To begin, the kinematics of this problem are described by a vector X_i which gives the reference position of an element of the material. The current position of this element is denoted by x_i, and if t denotes time, then a function χ_i exists such that

$$x_i = \chi_i(X_j, t) \ . \tag{1}$$

The velocity and acceleration of this element of the material are

$$v_i = \frac{\partial}{\partial t}\chi_i(X_j, t) \tag{2}$$

and

$$a_i = \frac{\partial^2}{\partial t^2}\chi_i(X_j, t) \ . \tag{3}$$

The behavior of stress waves within the material is controlled by three conservation laws. The conservation of mass equation

$$\dot{\rho} + \rho\frac{\partial v_i}{\partial x_i} = 0 \tag{4}$$

provides a link between the motion of the material and its mass density ρ. Here $(\dot{\ }) = \partial/\partial t(\ ) + v_i\partial/\partial x_i(\ )$. The conservation of momentum equation

$$\rho a_i = -\frac{\partial}{\partial x_i}(P + Q) + \rho f_i \tag{5}$$

links the motion to the equilibrium pressure P and the dissipative pressure Q. Here f_i is a specific external body force. Because high amplitude waves are being considered, strength effects are ignored, and the stresses within the body are represented by pressure terms alone. It is also assumed that the total pressure within the body can be represented by additive terms representing equilibrium and dissipative contributions. The conservation of energy equation

$$\rho\dot{E} = (P + Q)\frac{\dot{\rho}}{\rho} + \rho r \tag{6}$$

links the mechanical work to stored strain and thermal energy. Here E is the specific internal energy, and r is the specific external heat supply. The contribution to this relation from heat flux has been ignored.

These relations are accompanied by constitutive relations for P and Q. For example, a function for P can be specified so that

$$P = \hat{P}(\rho, E) \ . \tag{7}$$

This is often called an equation of state. Correspondingly, Q can be specified by

$$Q = \hat{Q}(\rho, \dot{\rho}, E) \tag{8}$$

with the additional requirement that $Q \to 0$ as $\dot{\rho} \to 0$.

Plane shock waves which do not interact with other compression or release waves form steady structured waves. When the above relations are applied to a steady wave problem, Equations (4), (5), and (6) can be integrated analytically. To achieve this, it is assumed that only motion in the x_1 direction exists, and all variables are functions of the term $(x_1 - Ut)$ where U is the speed of propagation of the steady wave. The results are

$$\frac{v_1}{U} = 1 - \frac{\rho_0}{\rho} \ , \tag{9}$$

$$P + Q = \rho_0 U v_1 \ , \tag{10}$$

and

$$E = \frac{1}{2}v_1^2 \ . \tag{11}$$

The initial conditions in front of the wave are given by $\rho = \rho_0$ and null values for P, Q, and v_1. Normally these relations are viewed as jump conditions which connect the initial equilibrium state in front of the wave to the final equilibrium state behind the wave. The final state is also called the Hugoniot state. At both of these states $\dot{\rho} = 0$, and therefore $Q = 0$. If the value of U or the values of either v_1, ρ, P, or E behind the wave are specified, then Equations (9), (10), (11), and (7) can be solved for the remaining variables. These integral conditions can then be used to determine the solution at any point along the wavefront provided that the constitutive equation (8) is known. For a specific problem both ρ_0 and U are constants. Thus, Equation (10) gives a straight line relationship between the total pressure $P + Q$ and the particle velocity v_1 at any point along the wavefront. This is called the Rayleigh-line loading path. It is important to note that the Hugoniot state does not depend upon the constitutive equation for the dissipative pressure; however, the profile of the steady propagating wave does depend on this relation.

This observation has both experimental and computational implications. Measurements of Hugoniot states are considerably easier to achieve than measurements of shock wave profiles. Thus, the decoupling of the influence of the constitutive relation (8) upon the Hugoniot state simplifies the experimental techniques which are required to study Hugoniot states. This decoupling also allows relevant numerical computations of these same experiments using only artifical constructs for (8). However, as will be shown in the next section, when one turns from problems involving single-constituent materials to problems involving mixtures, these advantages disappear.

MIXTURE THEORY

In this section a simple theory of mixtures is presented. The assumptions behind the development of this theory are motivated by a desire to study the behavior of very high amplitude waves in a heterogeneous agglomeration of two or more constituents. Only time scales associated with the passage of a pressure wave are considered. On these short time scales it is assumed that no relative motion exists between the constituents. Thus the position, velocity, and acceleration of the mixture are given by Equations (1), (2), and (3).

Each constituent of the mixture has its own intrinsic mass density which is denoted by $\bar{\rho}_\xi$. Here the subscript ξ denotes a particular constituent such as water, plastic, or void. Each constituent occupies only a fraction of the total volume of the mixture at any given point within the body. This is specified by the volume fraction ϕ_ξ, a variable which assumes values between 0 and 1. The sum of the volume fractions over all constituents of the mixture is subject to the following constraint:

$$\sum_\xi \phi_\xi = 1 . \tag{12}$$

If void is present in the mixture, it is treated as one of the ξ constituents. Consequently, Equation (12) does not preclude the presence of porosity.

The partial mass density of a constituent is denoted by ρ_ξ where

$$\rho_\xi = \phi_\xi \bar{\rho}_\xi . \tag{13}$$

When the partial mass densities are summed, the mass density of the mixture is obtained; that is,

$$\rho = \sum_\xi \rho_\xi . \tag{14}$$

By using this relation, the conservation of mass equation (4) can be applied to this problem. It is also easy to show that the mass fraction

$$\gamma_\xi = \frac{\rho_\xi}{\rho} \tag{15}$$

is a constant (Drumheller and Bedford, 1980). Therefore, if ρ is known, ρ_ξ can be determined from Equation (15).

Each constituent has an equilibrium and dissipative pressure denoted by P_ξ and Q_ξ.

$$P_\xi = \hat{P}_\xi(\bar{\rho}_\xi, E_\xi) , \tag{16}$$

and

$$Q_\xi = \hat{Q}_\xi(\bar{\rho}_\xi, \dot{\bar{\rho}}_\xi, E_\xi) , \tag{17}$$

where E_ξ is the specific internal energy for the ξth constituent. If a relationship (called a mixing rule) is established between these terms and the mixture counterparts P and Q, the conservation of momentum equation (5) can be applied to this problem. Since strength effects have been ignored, the mixture can be viewed as an agglomeration of fluids. An equal-pressure assumption then seems a reasonable approximation for a mechanical mixing rule; that is,

$$P_\xi + Q_\xi = P + Q \, . \tag{18}$$

It should be noted that the product $\phi_\xi \, (P_\xi + Q_\xi)$ is commonly called the partial pressure. Thus (18) is equivalent to using Dalton's Law wherein the total pressure is equal to the sum of the partial pressures. The conservation of energy equation (6) must also be satisfied. This relation can be used to determine the energy of the mixture which is related to the constituent energies by

$$\rho E = \sum_\xi \rho_\xi E_\xi \, . \tag{19}$$

The additional relations which are required to solve the problem come from another mixing rule.

One assumption that can be made is that the temperatures of each constituent, T_ξ, are equal to the mixture temperature; that is,

$$T_\xi = T \, . \tag{20}$$

This is called the isothermal mixing rule, and it assumes that thermal conduction and radiation between neighboring constituents is large even on stress wave time scales.

At the other extreme, it can be assumed that these thermal transport mechanisms are very weak, and no thermal energy is conducted between neighboring constituents. This is called the adiabatic mixing rule. In this case the following energy equations apply for each of the constituents (Drumheller and Bedford, 1980):

$$\rho_\xi \dot{E}_\xi = \phi_\xi (P_\xi + Q_\xi) \frac{\dot{\bar{\rho}}_\xi}{\bar{\rho}_\xi} + \rho_\xi r_\xi \tag{21}$$

where r_ξ is the specific external heat supply for the ξth constituent. Given the definition that

$$\rho r = \sum_\xi \rho_\xi r_\xi \, , \tag{22}$$

Equation (21) can be summed to obtain Equation (6).

When applied to steady wave problems, examination of these field equations shows that the Hugoniot state achieved by the mixture is not dependent upon the constitutive equations for Q_ξ, provided the isothermal mixing rule is used. This is not the case when the adiabatic mixing rule is used.

To illustrate this latter point, consider an example of a mixture of a fluid and a solid. First assume that the dissipative pressure in the solid is always negligible, $Q_s = 0$. The equations necessary to solve this problem are (4), (5), (6), (16), and the solid equation from (21). These equations have been chosen so that the dissipative pressure appears only in the conservation of momentum and energy for the mixture. A steady wave analysis will result in the integral relations (9), (10), and (11). An integral of the energy equation for the solid constituent cannot be obtained in closed form; however, it is obvious that the integral does not explicitly contain any expression for the dissipative pressure. The discussion presented in the previous section can then be used to illustrate that the Hugoniot state can now be determined without any additional assumptions about the dissipative pressure which resides solely within the fluid.

If this problem is solved again using the adiabatic assumption but with the dissipative pressure in the fluid (rather than the solid) set to zero, the energy equation for the fluid must be used in place of the energy equation for the solid. Different integral relations are obtained, and a different Hugoniot state is achieved. Thus, in this example the Hugoniot state depends only on whether the dissipative pressure resides solely in the solid or in the fluid. In general, the Hugoniot state depends upon the explicit constitutive relations which are used for Q_ξ. A more specific example of this point will be presented in the Example Calculation Section.

NUMERICAL PROCEDURE

The architecture of many numerical wave propagation codes incorporates the constitutive and conservation of energy equations within a subroutine which is relatively easily accessed and modified. Alterations to the conservation of mass and momentum equations are far more difficult. The theory presented within the last section is constructed with this fact in mind. Both the mass and momentum equations in the theory are identical to their counterparts in conventional wave problems. Certain modifications to this theory can vastly complicate its introduction into an existing code. For example, if relative motion between the mixture constituents is included, multiple mass and momentum equations will have to be solved. Instead of modifying an existing code, construction of an entirely new code would be more appropriate.

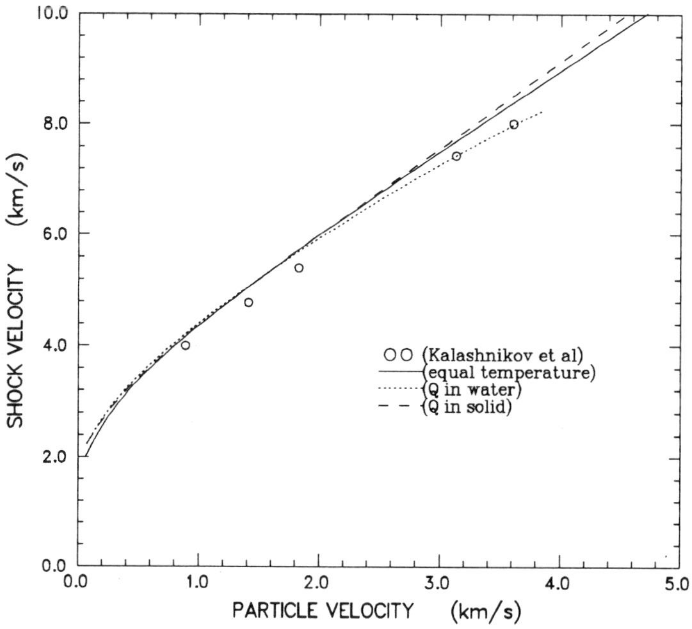

Fig. 1. Computed Hugoniot States and Comparison to Data of Kalashnikov and others.

There are two general classes of numerical wavecodes which are based on either a Lagrangian description X_i or an Eulerian description x_i. Eulerian codes are far harder to modify. Two problems must be addressed before models of the type described in the previous section can be introduced into this class of code. First, new variables are introduced into the theory; for example, the volume fraction ϕ_ξ and the constituent internal energy E_ξ. Accurate algorithms must be developed to convect these new variables through the computational mesh. Second, mixing schemes must be developed for numerical meshes which contain both single-constituent and multi-constituent materials. Addressing these problems is beyond the scope of this work.

Within the Lagrangian framework these problems do not exist; however, introduction of the model into this class of code does require a subcycling scheme. Consider a particular mesh within the computational grid. In an explicit finite-difference algorithm, values are known at the last time step n for all of the variables. The numerical algorithm must then provide values of these variables at the new time step $n+1$. The code first computes new values of position and density. Next, the subroutine containing the mixture description is called. This subroutine must provide new values for ϕ_ξ, P_ξ, Q_ξ, and E_ξ. To do this, estimates for ρ and $\dot{\rho}$ between the old time t^n and the new time t^{n+1} are made by using a linear interpolation between the old and new values of ρ. The Equations (15), (16), (17), (21), and

$$\dot{\phi}_\xi = \alpha(P + Q - P_\xi - Q_\xi) \tag{23}$$

are then used to compute the variables at the new time. This is accomplished by numerically integrating these ordinary differential equations between t^n and t^{n+1} with an auxillary subroutine. An Adams method (Shampine and Gordon, 1974) has proven useful for this purpose.

Equation (23) provides an estimate of ϕ_ξ which approximates the solution to (18). The parameter α is chosen so that the error in this estimate is small. Values of α may be chosen which are too large. This will greatly increase computational time requirements and even result in numerically stiff problems. At present, selection of values for α are based solely on experience and experiment with each individual problem.

Finally, wavecodes usually incorporate an artificial viscosity term which is designed primarily to spread shock waves over several computational meshes and stabilize the calculation. In general, the structure of the artificial viscosity term is in the form of a constitutive equation for Q, and from the previous arguments, it is clear that the artificial term can produce the correct Hugoniot state. In the absence of alternative physical representations for the Q_ξ, these quantities can be represented by artificial viscosity terms. Alternatively, if physical representations are available for the Q_ξ, they should replace the artificial expressions. The mesh size in the computation must then be selected so as to allow resolution of the physical thickness of the shock wave.

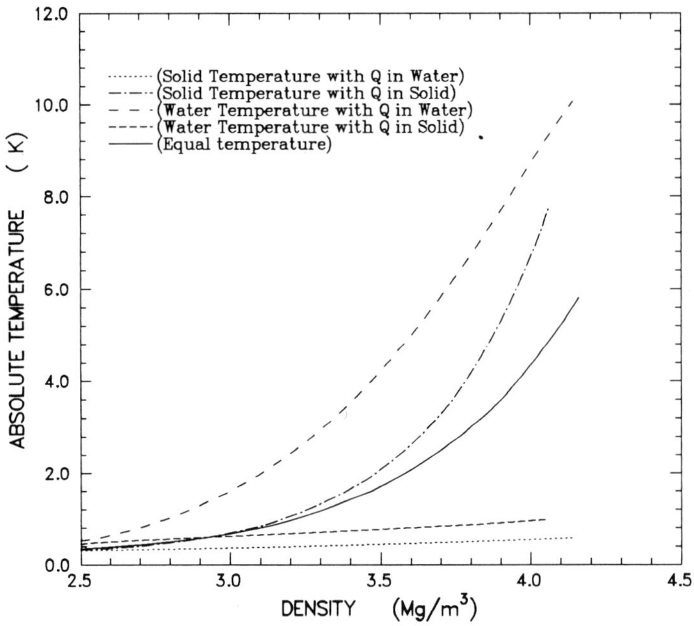

Fig. 2. Computed Hugoniot Temperatures.

EXAMPLE CALCULATIONS

In this section, two types of results will be presented. They are calculations of Hugoniot states from a steady wave analysis, and stress wave calculations from a Lagangian wave propagation code, WONDY (Kipp and Lawrence, 1981). In both cases the material under consideration is a saturated mixture of water and powdered calcite. Since the only purpose of this section is to provide illustrations of mixture calculations, a detailed discussion of material parameter selection and experimental technique will not be included. This information is contained in Drumheller and Grady (1985); Grady, Moody, and Drumheller (1986); and Kalashnikov, Pvlovskiy, Simakov, and Trunin (1973).

The Hugoniot states are determined from Equations (9), (10), and (11) coupled either with the isothermal or adiabatic mixing rule. When the adiabatic mixing rule is used, two conditions are considered, either the dissipative pressure in the water is zero or the dissipative pressure in the calcite is zero. Under these circumstances, the energy equation from (21) which does not contain the dissipative pressure term is integrated numerically, and an explicit definition of the constitutive relationship for the nonzero Q_ξ component is not required.

The constitutive model for the equilibrium pressure in calcite is a Mie-Grüneisen formulation with a linear relationship between U and v_1. The parameters for solid calcite are determined by a fit to the data of Kalashnikov, Pvlovskiy, Simakov, and Trunin (1973). Determination of this fit is reported in Drumheller and Grady (1985). The water is modeled with a tabular equation of state developed by Kerley (1985). No parameters within the mixture model were adjusted in these calculations with exception of the constitutive relations for the dissipative pressures.

The results for a mixture with a mass density of 2200 kg/m^3 are shown in Fig. 1. This corresponds to a solid volume fraction of approximately 30 percent. The data of Kalashnikov, Pvlovskiy, Simakov, and Trunin (1973) are also shown in this figure. While the mass density used in the calculation agrees with the value reported in Kalashnikov, Pvlovskiy, Simakov, and Trunin (1973), the solid volume fraction differs significantly from the value reported in that work. Their values are not self consistent. This contradiction has not been resolved.

The sensitivity of the calculation to the mixing rule is evident at pressures above 40 Gpa. The corresponding Hugoniot temperatures which are plotted in Fig. 2 help to explain the differences in the calculated shock wave velocities. For example when the dissipative heating is introduced into the water alone, temperatures within the water rise to more than 10,000 K for shock pressures of about 70 GPa. These temperatures are high enough to result in dissociation of the water. In contrast, if the dissipative heating is introduced into the calcite alone, the water remains near room temperature and the ice phase boundary.

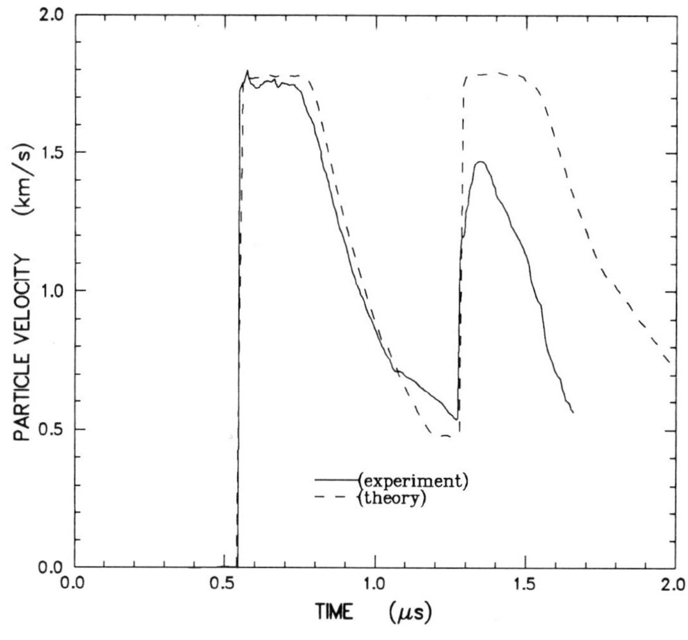

Fig. 3. Computed Stress Wave Profiles and Comparison to Data of Grady and others.

The results of a stress wave calculation are shown in Fig. 3. They are compared to experimental data which, due to timing errors in the experiment, have been time shifted so that time of arrival of the first signal is in good agreement with the calculation. The data are from a flyer-plate impact experiment reported by Grady, Moody, and Drumheller (1986). The mass density of the water-calcite sample is about 1750 kg/m^3. This corresponds to a dry porosity of about 56 percent. A uniaxial state of strain is introduced into the sample by planar impact of two aluminum flyer plates which are separated by a thin void space. This produces the two wave structure that is observed in both the experiment and the calculation. The back side of the first flyer plate was subjected to stresses which may have led to extensive spall damage through about half of its thickness. This spall may have contributed to the errors observed in the comparison of the second pulse. The calculation uses an adiabatic mixing rule with all of the dissipation introduced into the calcite. The dissipative constitutive equation was the artifical viscosity representation reported in Kipp and Lawrence (1981).

SUMMARY

Large amplitude shock waves generate large amounts of heat within many media. When these media are mixtures, the distribution of this heat among the constituents is the central issue in modeling the response of the material. A model has been proposed which addresses this issue. By ignoring relative motion and pressure differences between the constituents, the resulting theory is tractable enough to be introduced into many numerical wave propagation codes. Extension of this theory to materials with strength and unequal pressures between the constituents is possible (Drumheller, 1986), but has not been discussed in this work.

Example calculations of Hugoniot states from a steady-wave analysis and of stress-wave profiles from a numerical wave propagation code are given. They are also compared to experimental data.

REFERENCES

This work performed at Sandia National Laboratories supported by the U.S. Department of Energy, under contract DE-AC04-76DP00789.

Drumheller, D. S., and Bedford, A. (1980). A Thermomechanical Theory for Reacting Immiscible Mixtures, Archive for Rational Mechanics and Analysis, **73**, 257–284.

Shampine, L. F., and Gordon, M. K. (1974). Computer Solutions of Ordinary Differential Equations: The Initial Value Problem, Freeman, San Francisco.

Kipp, M. E., and Lawrence, R. J. (1981). WONDY V-A One-Dimensional Finite-Difference Wave Propagation Code, Sandia National Labs., SAND81-0930, unpublished.

Drumheller, D. S., and Grady, D. E. (1985). The Dynamic Response of Porous Calcium Carbonate Minerals, Proceedings of the Fourth APS Topical Conference on Shock Waves in Condensed Matter, July 22-25, 1985, Spokane, WA, in press.

Grady, D. E., Moody, R. L., and Drumheller, D. S. Release Equation of State of Dry and Water-Saturated Porous Calcite, in preparation.

Kalashnikov, N. G., Pvlovskiy, G. V., Simakov, G. V., and Trunin, R. F. (1973). Dynamic Compressibility of Calcite-Group Minerals, Izv., Earth Physics, **2**, 23-29.

Kerley, G. I. (1985). Theoretical Equations of State for the Detonation Products of Explosives, Proceedings of the Eighth Symposium on Detonation, July 15-19, 1985, Albuquerque, NM.

Drumheller, D. S. A Theory for Dynamics Compaction of Wet Porous Solids, Int. J. Solids Struct., in press.

Int. J. Impact Engng Vol. 5, pp. 269–275, 1987
Printed in Great Britain

0734-743X/87 $3.00 + 0.00
Pergamon Journals Ltd.

ADVANCED ALGORITHMS FOR COMPUTER SIMULATION OF HYPERVELOCITY IMPACT

W. J. Flis

Dyna East Corporation
3132 Market Street, Philadelphia, PA 19104

ABSTRACT

Presented are several algorithms that allow a Lagrangian hydrodynamic computer code to simulate hypervelocity impact. Such impacts typically involve very large material deformations and ordinarily can be handled only by Eulerian codes, in which the materials flow through a stationary grid. In Lagrangian codes, the mesh is attached to the material and deforms with it; in problems involving severe distortions, the mesh can become tangled, resulting in loss of accuracy or even a breakdown of the computational scheme. However, Lagrangian codes present several advantages in economy, ease of programming and use, and interpretation of results.

This paper describes several modifications of the DEFEL code (Flis, Miller, and Clark, 1984). This code, a descendant of EPIC-2 (Johnson, 1976), was originally developed to simulate low- to medium-velocity impact and explosive-metal interaction. Modifications include automatic rezoning, a surface-erosion model, and treatment of fracture (cracking) by automatic introduction of sliding lines between elements. These features allow DEFEL to handle a variety of problems in the hypervelocity regime. Included are discussions of several example problems.

INTRODUCTION

Lagrangian hydrocodes offer several advantages, including:

- Material boundaries are described explicitly, and results are easy to interpret.
- Material history, including work-hardening and failure, may be recorded.
- They generally require less computer time and memory than Eulerian codes.

The major drawback of these codes is that, in problems involving large material distortions, the grid also becomes distorted, resulting in decreased accuracy, increased computer time, and possibly failure of the computational scheme. These difficulties may be prevented by periodic rezoning of the grid. Rezoning serves to re-organize the distorted grid, thereby restoring elements to a more efficient shape and improving the computational efficiency and accuracy.

Several schemes for rezoning have been developed in the past and the theoretical aspects are well established. Presented here are several strategies for rezoning that alleviate the drawbacks of previous schemes, namely:

1. Rezoning requires a lot of computer time.
2. Repeated rezoning smears details of properties, losing accuracy.
3. For rezoning to be effective, the user must intervene.

The scheme presented here overcomes the first objection and, to an extent, the second by "surgical" rezoning. The code, in each time step, examines the condition of the grid and decides whether to rezone and then also what <u>parts</u> of the grid to rezone. The rezoning operation is performed only on the most distorted parts of the grid and other parts are not affected. The amount of rezoning is thereby limited in frequency and area.

The third objection is answered by a completely automatic algorithm. Even the construction of the new grid has been automated, a task made feasible by DEFEL's use of triangular elements exclusively. Most problems can thus be simulated in a single run, with no restarts or other user intervention needed.

Some problems, however, cannot be economically handled even with rezoning. For example, in deep penetration by shaped-charge jets, some jet material remains as a very thin layer on the wall of the crater; even if this layer is gridded by only a single layer of elements, the resulting time step would be too small for the simulation to be affordable.

Often, such fine details are not important to the phenomenon of interest and can safely be deleted from the simulation. Portions of the computational grid can be deleted either manually by the user or automatically by a surface-erosion routine, such as developed by Snow (1982) and Kimsey (1983). In such a routine, finite elements are dynamically deleted from the simulation as they fail, according to some criterion based on, say, total strain (to simulate fracture) or temperature (to simulate melting or vaporization). DEFEL contains an independently developed routine, largely similar to that of Stecher and Johnson (1984).

Another weakness of existing hydrocodes in general is the treatment of material failure and, especially, post-failure behavior. Common practice is to treat a failed material as a continuum with modified properties, as, for example, no strength in tension or shear. This does not always provide a good simulation.

Included here is a technique for simulating cracking or brittle fracture. When a specified criterion is met, a crack is automatically introduced between the finite elements. Specifically, when the value of maximum normal stress at some point exceeds the material tensile strength, a crack is assumed to initiate. The cracking routine then parts the finite-element grid along the intended crack direction and defines a sliding interface to prevent subsequent cross-over of material. Provisions for crack propagation, multiple cracks, and crack interaction have been included. This scheme may be considered an extension of the method of Ringers (1983), which propagates a single fracture to simulate adiabatic shear failure in the plugging mode of plate perforation.

Details of the rezoning and cracking algorithms are presented below.

REZONING

Methodology

DEFEL's rezoning routine has three steps:

1. A criteria step, deciding when and what portions of the grid to rezone.
2. A mesh-generation algorithm, geometrically constructing the elements of the new mesh.
3. Determination of the properties of the new mesh.

Each of these steps is performed automatically (with no user intervention) by the rezoning routine. Each is described briefly here; details are given by Chou, Liang, and Flis (1985).

In the criteria step, performed each computational cycle, the distortion of each element is evaluated to decide whether a rezoning is needed. Specifically, the element's aspect ratio (altitude to base) is compared to a specified minimum value, usually taken as 0.05. An element is flagged for rezoning if it:

1. meets the aspect-ratio criterion (a "bad" element)
2. has a node in common with a bad element (a "neighbor" element)
3. has three nodes in common with the above elements.

This flagging process builds up rezoning "regions." Elements are also flagged for rezoning if they touch more than one region, possibly resulting in the joining of regions. An example of this process is shown in Fig. 1. Optionally, rezoning regions may be specified by the user.

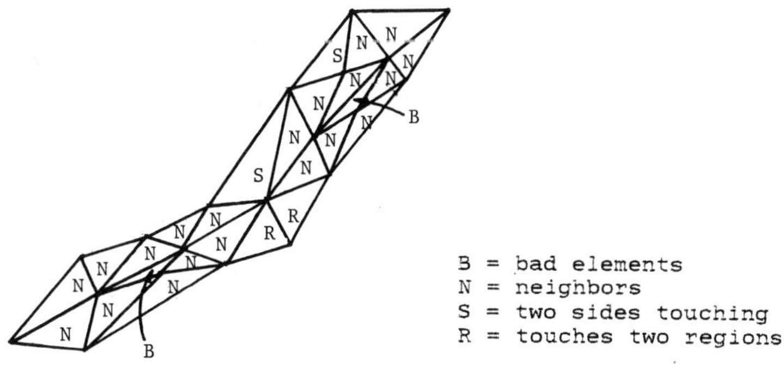

Fig. 1. Defining rezoning regions by flagging elements.

In the mesh-generation step, the arrangement of finite elements within each rezoning region is modified on either of two levels, at the user's option. The old mesh may be merely redistributed (or "smoothed") in space to restore the existing elements to better shape. A number of such smoothing techniques have been developed, mainly for application to the generation of initial grids, as reviewed by Thompson (1984). Typically, these involve an iterative solution of an elliptic differential equation, of the Laplace or Poisson type.

DEFEL uses a different, simpler scheme. Each interior node is moved to a new position such that all elements attached to it will have equal area. Mathematically, this is done by adjusting each node's position to minimize the sum of the squares of the differences between the areas of the elements attached to the node and the average of these areas. This procedure is applied to all nodes several times until the grid has been sufficiently smoothed.

Alternatively, a completely new mesh may be constructed within the polygonal boundary of each rezoning region. DEFEL uses an automatic triangulating process, as described by Bykat (1976) and Tracy (1977), comprising the following steps:

1. Conditioning the rezoning boundary by dividing any boundary segment that is much longer (by a factor of 1.4) than the average boundary segment length and by redistributing nodes uniformly along the axis of symmetry.

2. Constructing elements from the boundary inward, as shown in Fig. 2. Acute boundary angles are closed, as in (a), with a node added if the new side is too long, as in (b); obtuse angles are bisected with a node added, as in (c). As each element is constructed, the "active" boundary is moved inward, and the process continues.

3. Dividing the region into subregions. If the node added as in Figure 2(c) is too close to the opposite side of the boundary, it is moved away and the existing boundary is divided into two new boundaries, or subregions, as shown in Fig. 3. Each subregion is then treated in succession until done.

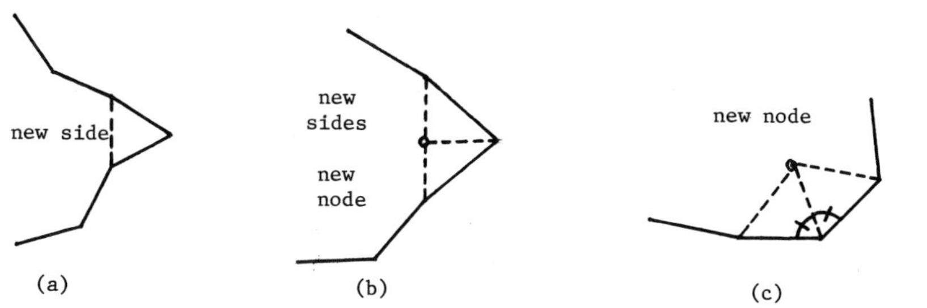

Fig. 2. Construction of new elements within rezoning boundary.

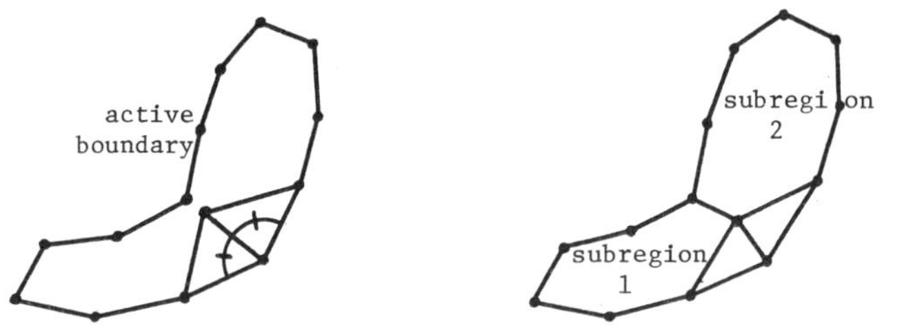

a) Newly constructed node is too close to opposite side of boundary.

b) New node's position is adjusted and two subregions are defined.

Fig. 3. Dividing a rezoning region into subregions.

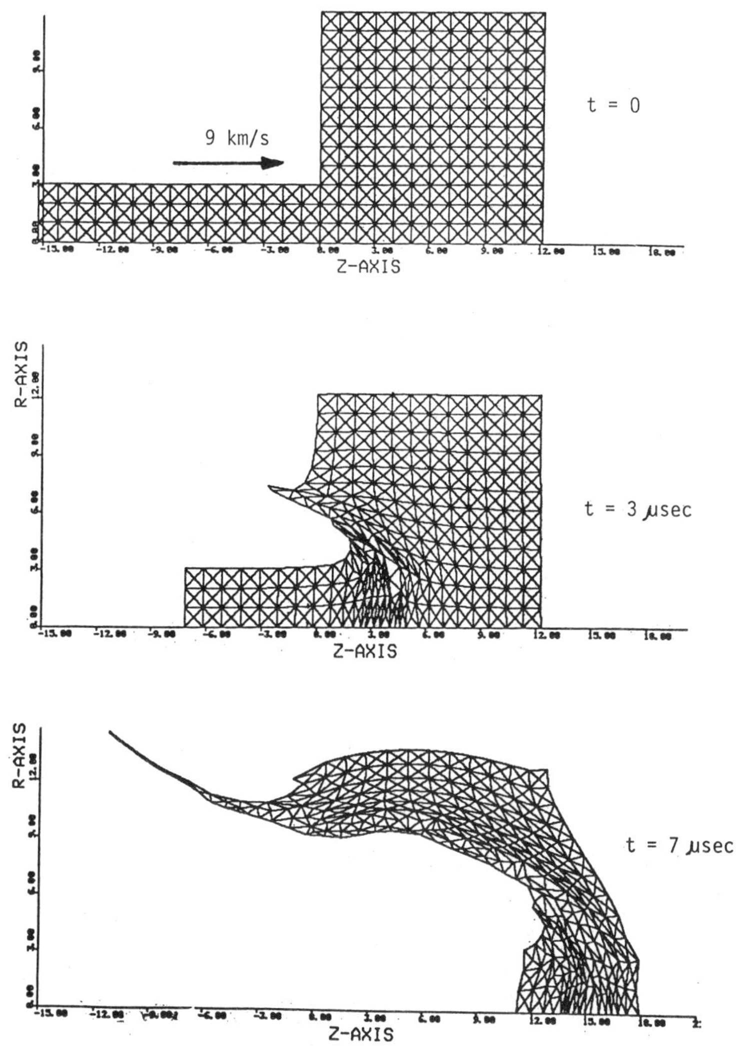

Fig. 4. Penetration of a thick steel target by a copper rod at 9 km/s simulated by the DEFEL code with rezoning.

4. Closing each sub-region when its active boundary has been reduced to three or four segments. If the boundary has only three segments left, a single element is defined. If four segments, then either two elements, or four elements with a node at the center, are defined, depending on the lengths and arrangement of the sides.

Construction of the new mesh may be followed by the smoothing process described above.

In the final step of rezoning, the state variables of the new mesh (properties of the nodes and elements) are determined from those of the old. DEFEL uses interpolation for node properties (consistent with the element shape function) and rule-of-mixture for element properties.

Example Simulation with Rezoning

Figure 4 illustrates the use of rezoning in the simulation of deep penetration of a thick steel target by a copper rod at 9 km/s. The erosion model was also used here, with erosion allowed only in the penetrator material that remains along the crater wall, since this material forms a layer too thin for rezoning. Eight partial rezonings were performed in the course of this simulation. Simulation of such deep penetration would not be possible without rezoning or erosion.

CRACKING MODEL

Methodology

The cracking-model routine begins, in each computational cycle, by evaluating the maximum normal stress (principal stress) at each node. Each tensorial component of stress, an element property, is found at a node by averaging the stresses in the attached elements. If the magnitude of the principal stress (tensile only) exceeds a critical value, taken as the material tensile strength (or spall strength), a crack is assumed to initiate along a direction normal to the stress vector, as in Fig. 5(a).

The routine then searches for the two element boundaries at the node that run nearest to the intended crack direction, as in Fig. 5(b). The node is split in two, and elements are re-attached appropriately across the crack. The mass of the original node is divided between the two new nodes in proportion to the number of elements attached to each; positions and velocities of the new nodes are taken as those of the old node. Finally, a sliding interface is defined across the crack; material on opposite sides of the crack may thus separate or slide (with or without friction) relative to each other, but may not cross over.

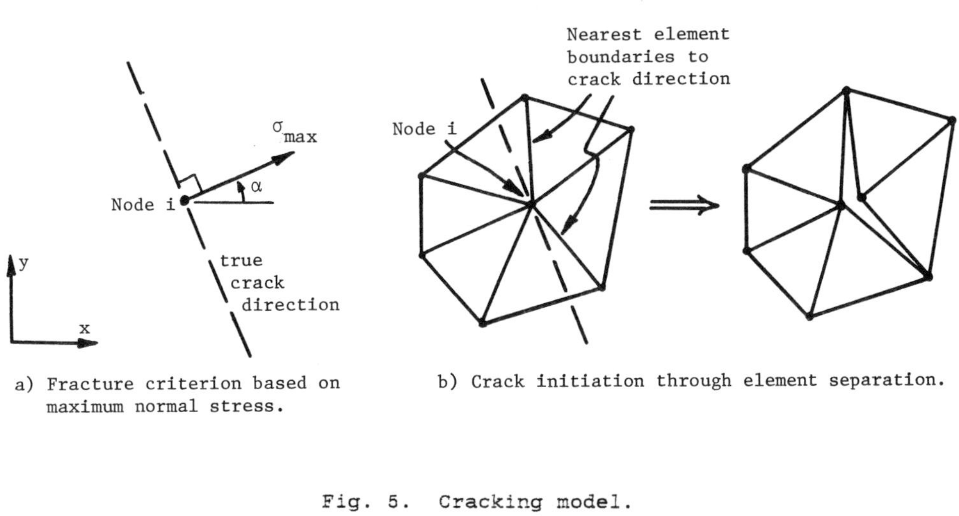

a) Fracture criterion based on maximum normal stress.

b) Crack initiation through element separation.

Fig. 5. Cracking model.

Special routines treat propagation and interaction of cracks, as shown in Fig. 6. Extension of a crack involves splitting the end node and extending both sides of the existing slide line to the new node. On branching of a crack, a previously split node is split again, one side of the existing slide line is extended into the branch, and a new slide line is defined in the branch. Joining of cracks end-to-end simply involves splitting the common node, with no adjustment of slide lines necessary. When cracks intersect, the slide line of the intersecting crack (running horizontally in the figure) is extended to the split node, and the appropriate side of the slide line of the intersected (vertical) crack is extended one segment into the intersecting crack. Each of these mechanisms is assumed governed by the same stress criterion as crack initiation.

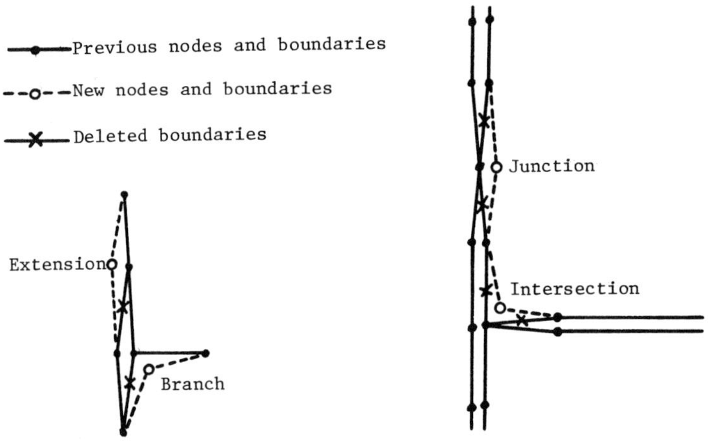

Fig. 6. Treatment of crack propagation and interaction in cracking model.

Example Simulation with Cracking

An example DEFEL simulation using the cracking model is presented in Fig. 7. A thin steel disk impacts a larger steel plate at 3 km/s. In the resulting finite-element grid, a number of fragments have been spalled from the rear of the target, in which about 150 cracks have been created. The spalling phenomenon would not be apparent without the generation of the cracks.

CONCLUSIONS

The applicability of a Lagrangian hydrocode has been extended to hypervelocity impact by several modifications, including a rezoning routine, a surface-erosion model, and a novel fracture algorithm. The rezoning is fully automatic and "surgical," in the sense that only parts of the grid are rezoned, as selected by the algorithm. The erosion model eliminates pathological features that rezoning cannot remedy. The fracture model provides both a failure criterion and a realistic description of post-fracture behavior by parting the grid and inserting sliding interfaces to simulate cracks; multiple cracks can be so generated.

ACKNOWLEDGEMENTS

The author thanks Mr. J. Hashemi for some improvements to the fracture routine. Development of the fracture and erosion models was sponsored partially by the Air Force Armament Laboratory under Contract No. F08635-82-C-0215.

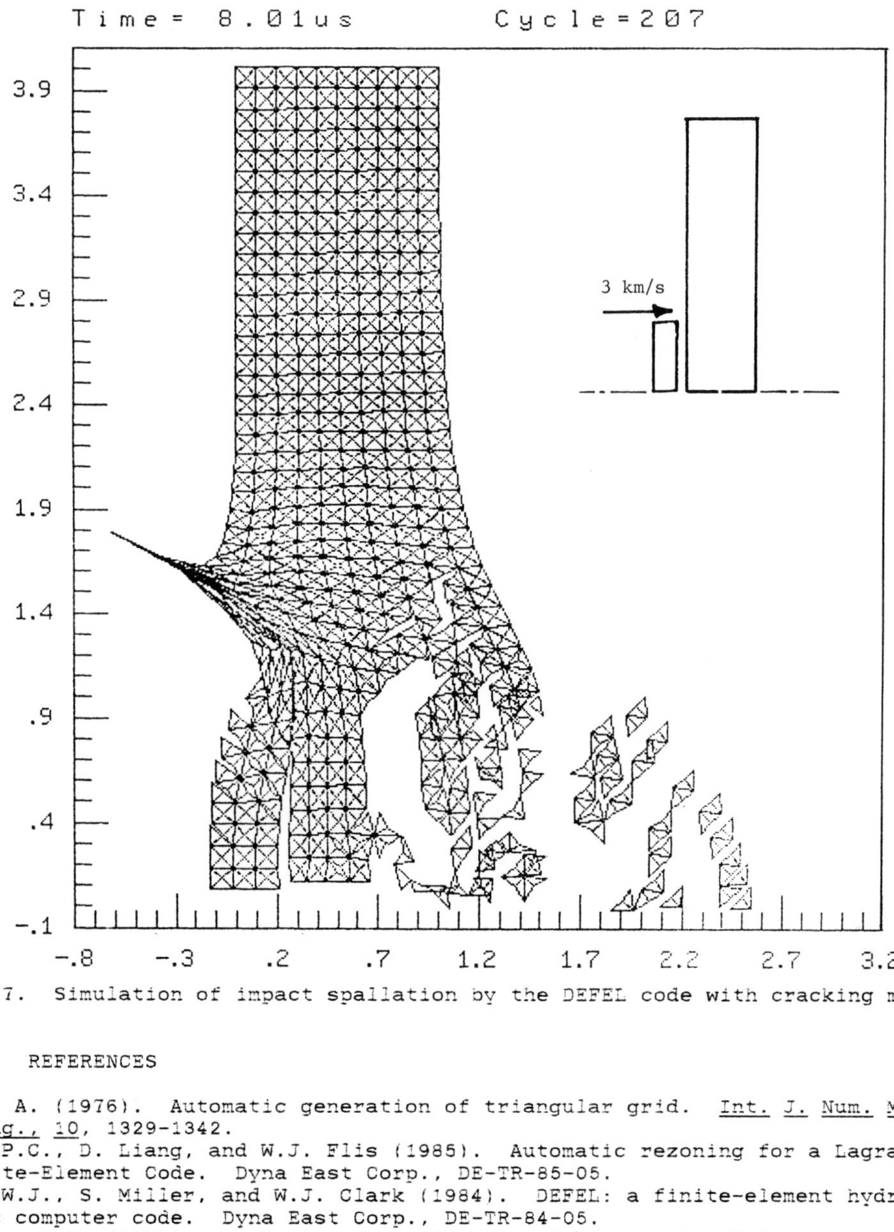

Fig. 7. Simulation of impact spallation by the DEFEL code with cracking model.

REFERENCES

Bykat, A. (1976). Automatic generation of triangular grid. Int. J. Num. Meth. Engng., 10, 1329-1342.

Chou, P.C., D. Liang, and W.J. Flis (1985). Automatic rezoning for a Lagrangian Finite-Element Code. Dyna East Corp., DE-TR-85-05.

Flis, W.J., S. Miller, and W.J. Clark (1984). DEFEL: a finite-element hydrodynamic computer code. Dyna East Corp., DE-TR-84-05.

Johnson, G.R. (1976). Analysis of elastic-plastic impact involving severe distortions. J. Appl. Mech., 98(3), 439-444.

Kimsey, K.D. (1983). Calculation of Penetration. Proc. Workshop on the Theoretical Foundation for Large-Scale Computations of Nonlinear Material Behavior, Evanston, Illinois.

Ringers, B.E. (1983). In J. Chandra and J.E. Flaherty (Eds.), Computational Aspects of Penetration Mechanics, Springer-Verlag, Berlin. Chap. 3, pp. 36-46.

Snow, P. (1982). KEPIC-2. Kaman Sciences Corp., K82-46U(R).

Stecher, F., and G.R. Johnson (1984). Lagrangian computations for projectile penetration into thick plates. Proc. ASME International Computers in Engineering Conf. and Exh., Las Vegas, Nevada.

Thompson, J.F. (1984). Grid generation techniques for computational fluid dynamics. AIAA J., 22(11).

Tracy, F.T. (1977). Graphical pre- and post-processor for 2-dimensional finite element method programs. Siggraph 77, 11(2), 8-9.

Int. J. Impact Engng Vol. 5, pp. 277–284, 1987
Printed in Great Britain

0734-743X/87 $3.00 + 0.00
Pergamon Journals Ltd.

ENERGY-EFFICIENT PENETRATION AND PERFORATION OF TARGETS IN THE HYPERVELOCITY REGIME

K. Frank and J. Zook

U.S. Army Ballistic Research Laboratory
Aberdeen Proving Ground, Maryland, USA

ABSTRACT

In general, the performance of a KE penetrator against most targets increases with velocity regardless of the particular penetrator-target interaction mode. It is possible to show that there exists an optimum velocity which maximizes the performance of an impacting penetrator for a given expenditure of kinetic energy. Simple graphical methods are described that determine the optimum velocity from general performance-velocity plots. These graphical methods may also be applied to experimental data alone. In addition, simple analytic models which describe the velocity dependent penetration/perforation performance of KE penetrators are examined and extended. These models may be used to explicitly assess the influence of parameters such as target strength and density and penetrator mass, strength and density. For some of the models in this analysis, the explicit relations between the optimum striking velocity for a specific kinetic energy value and the penetrator-target parameters are described.

INTRODUCTION

When a kinetic energy (KE) projectile is shot against a homogeneous barrier (target), its performance is, in general, a function that increases monotonically with impact velocity. The performance may be taken as penetration depth, a limit thickness just perforated with negligible residual mass/velocity (energy) behind the barrier, or some other appropriate measure of effectiveness. Nearly all physical penetrator-target interactions admit a performance description in terms of a relation like:

$$P = L \; S(V, \rho_t, \rho_p, H, Y, \ldots) \, , \tag{1}$$

where P is the performance, L is a characteristic dimension of the penetrator, and S is a characteristic function that depends on velocity, V, the density of the target material, ρ_t, and of the penetrator, ρ_p, the strength of the target material, H, and of the penetrator, Y, and possibly other characteristics describing the interaction. But, S is not dependent on the dimensions of the penetrator. This will be illustrated later.

The function S increases with velocity, generally in a monotonic fashion, except in cases where the character of the penetrator-target interaction changes in some velocity regime. For example, at some velocity, the penetrator may change its form rather suddenly, say by breaking up, and the S function may dip in this case and then continue again to increase. The formalism introduced by Eq. 1 implies geometric scaling to be applicable. Inter alia, this means that the same performance, P, can be achieved either with a larger penetrator at a lower velocity (smaller S function value), or a smaller penetrator at a higher velocity (larger S function value). By applying additional constraints, like keeping the impact energy constant, it is possible to show that for most S functions, there exists an optimum velocity where the performance of the penetrator, subject to that constraint, is maximized.

This remarkable fact was discovered in our laboratory (BRL) about 10 years ago and has been used effectively since that time in many studies. We have presented it in various "closed" meetings, reported briefly on some of its main features (Wright, 1983), and used its main conclusions in commenting on the significance of some recent hypervelocity results (Silsby, 1984). We take this opportunity to present these findings to a wider audience.

KE PROJECTILE PERFORMANCE SUBJECT TO CONSTRAINTS

One constraint of interest is that of constant penetrator energy. With some launcher systems, like electromagnetic guns (EMG), it is possible, in principle, to launch projectiles over a very wide range of mass and velocity holding the KE constant, without a significant impact on the system size and cost. The KE of a projectile is simply:

$$E = 0.5\ \pi\ (D/2)^2\ L\ \rho_p\ V^2 \quad = \quad D^3\ f\ V^2\ /\ K^3\ , \tag{2}$$

for a cylindrical projectile of diameter D, length L and material density ρ_p moving with the velocity V. The L/D ratio, often referred to as the fineness ratio, is represented by f. K is a combination of penetrator density and numerical constants. From Eq. 2, we may calculate the diameter of a constant energy projectile as:

$$D = K\ E^{1/3}\ /\ (f^{1/3}\ V^{2/3})\ . \tag{3}$$

With L = f D and Eq. 1, we obtain:

$$P = K\ E^{1/3}\ f^{2/3}\ S(V,\ \ldots)\ /\ V^{2/3}\ . \tag{4}$$

The extrema of Eq. 4 are found by solving $\partial P/\partial V = 0$. The equation for the extrema of P, subject to the constraint E = constant and f = constant is:

$$V\ S' = (2/3)\ S \qquad \text{for} \quad V = V_{opt.}\ , \tag{5}$$

with $S' = \partial S/\partial V$. The solution of Eq. 5 yields the velocity (or velocities) where P' = 0, and therefore, P assumes an extremum. For the kind of S functions described earlier, this extremum is a maximum of P subject to the constraints noted above.

The form of Eq. 5 suggests a simple graphical means to find the optimum when the S function is not given as an explicit analytic function of V, but exists only as a graph of experimental data. A line drawn from the origin of the S-V curve to the 2/3 value of $S(V_{opt.})$ is parallel to the tangent of the S-V curve at $V_{opt.}$. This is illustrated in Fig. 1a. By simple graphical means alone, the optimal velocity may be found. Applying this technique to the data of Silsby (1984), we find an optimum velocity of about 2.2 km/s for tungsten rods of f=23 which were shot for penetration into rolled homogeneous armor (RHA).

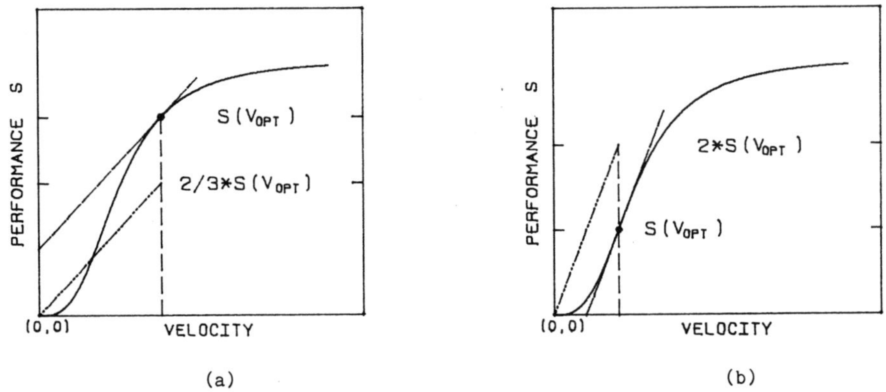

Fig. 1 Method for determining the optimum velocity for a given S function when kinetic energy and (a) L/D or (b) D are held constant.

Another set of constraints that is useful minimizes the energy required to achieve equal performance. By taking P=constant and f=constant and proceeding in the same manner as before, we again obtain Eq. 5. The optimum velocity (representing the minimum energy to achieve the same performance P) is found by solving Eq. 5 as before.

The next case of interest is that of a projectile with the diameter and KE held constant. Calculating f from the projectile energy:

$$f = K^3 E / (D^3 V^2) \tag{6}$$

and using:

$$P = f D S = K^3 E D^{-2} S(V, \dots) / V^2 \tag{7}$$

and setting P' = 0, we obtain the extrema condition to be:

$$V S' = 2 S \quad \text{for} \quad V = V_{opt.} \tag{8}$$

A graphical solution analogous to that outlined for Eq. 5 can be developed (as illustrated in Fig. 1b) but not all S curves yield a solution. A unique solution requires an S-V curve with a relatively steep slope.

The methodology outlined above is most general in nature. The S curves may be analytic forms, may be given in graphical form based on experimental data or may be generated by numerical methods. An even simpler geometric construction is possible if the data are plotted using logarithmic scales, i.e, ln P vs. ln V. Since the log transformation is monotonic, the extrema of P vs. V occur at the same values as those of ln P vs. ln V. Following the same procedures as used above, we find that maximum performance at constant energy and constant L/D occurs when $\partial(\ln P)/\partial(\ln V) = 2/3$ and the maximum for constant energy and constant diameter occurs when $\partial(\ln P)/\partial(\ln V) = 2$. In all cases, the underlying assumption is that geometric scaling applies.

PENETRATION OF HOMOGENEOUS TARGETS

Next, we shall provide some examples using explicit analytic formulations that are suitable throughout most of the hyper-velocity regime. For simplicity, we will restrict the discussion to the most simple case - penetration of homogeneous targets. First, we deal with the short penetrator of constant mass. Vitman and Zlatin (1962) and Belyakov, Vitman and Zlatin (1962, 1963) have shown that the resistance to penetration is given by:

$$H + k_0 \rho_t V^2 , \tag{9}$$

where H is a dynamic material property which is only weakly dependent on velocity. They showed that the resulting simple relation:

$$P = L\ (1/2\ k_0)\ (\rho_p/\rho_t)\ \ln(1 + 2\ k_0\ Z_0^2) \tag{10}$$

where $Z_0^2 = \rho_t V_0^2 / (2 H)$ and V_0 is the striking velocity, describes the penetration behavior of different penetrator and target materials rather well to very high velocities. In the following, we take $2 k_0 = 1$ and add to Eq. 9 a term that accounts for the effect of the target material moving in front of the penetrator in a non-steady flow (Goodier, 1964; Hanagud and Ross, 1971; Byrnside, Torvik and Swift, 1972; Johnson, 1972; Tate, 1979). In passing, we note that the target resistance term is directly related to the work required to make a unit volume cavity in the target (Johnson, 1972; Hill, 1980; Hopkins, 1960). For plastically deforming materials, it can be determined from bi-linear approximations to quasi-static stress-strain data (Hill, 1980; Hopkins, 1960).

The resulting relation for a non-deforming constant mass penetrator is given by (Goodier, 1964):

$$P(Z_0,Z) = L\ (\rho_p/\rho_t)\ [1 + \rho_t D/(3\ \rho_p L)]\ \ln[(1 + Z_0^2)/(1 + Z^2)] \tag{11}$$

where the quantity $(\rho_t D/(3 \rho_p L)$ accounts for the additional accelerated target mass in front of the penetrator (the factor 1/3 implies a hemi-spherical penetration interface). If we subject Eq. 11 to analysis, we find that an inflection point of $S(Z_0,0)$ occurs at $Z_0 = 1$ and that, according to Eq. 5, the optimum Z at constant KE is found by solving:

$$(1 + Z^2)\ \ln(1 + Z^2) - 3\ Z^2 = 0\ .$$

The approximate solution is Z = 3.975 . For a target material of armor steel, the value of $2H/\rho_t$ is about $(1000\ m/s)^2$, so that optimum penetration performance occurs at an impact velocity near 4000 m/s. The curve in the vicinity of the optimum velocity is nearly flat. Therefore, near optimum penetration for constant impact energy can be achieved over a wide range of velocities. This is illustrated in Fig. 2 for a 1 mega-Joule iron penetrator impacting armor steel.

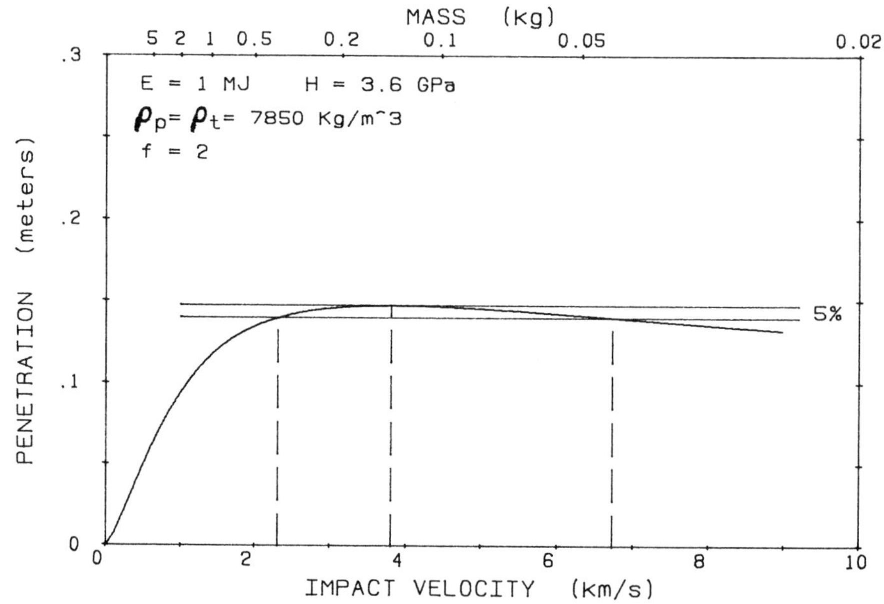

Fig. 2 Penetration performance of an iron projectile striking a steel target with one mega-joule of kinetic energy showing optimum performance and 5% below optimum.

Of much greater interest in practice, are deforming or eroding penetrators. There have been many papers published on this subject beginning with the classical works of Alekseevskii and Tate in the 1960's (Alekseevskii, 1966; Tate, 1967,1969) with more refinements which remove some of the earlier limitations being added even today (Tate, 1986). In the following, we shall present formulations (using a consistent notation) which we have found useful.

Following the basic ideas of Christman and Gehring (1966), the penetration of a long deforming/eroding rod penetrator may be separated according to the important identifiable phases or stages of the penetration process. For example,

$$P(Z_0,Z) = m_1\ D\ S_{D1}(Z_0,Z_1) + (L - m_2\ D)\ S_{L2}(Z_1,Z_2) + m_3\ D\ S_{D3}(Z_2,Z) \tag{12}$$

where $Z_0 > Z_1 > Z_2 > Z \geq 0$. The first term describes the transient where m_1 diameters of the rod are "used" up. The second term is the quasi-steady state phase which erodes the length $L - m_2\ D$ of the rod. The final stage is the penetration by a $m_3\ D$ constant mass penetrator. It is not obvious that the equation described above can be constructed in a consistent manner. However, we have found that the following simpler form gives satisfactory results:

$$P(Z_0,Z) = (L - m\ D)\ S_i(Z_0,Z) + m\ D\ S_D(Z_0,Z)\ . \tag{13}$$

This may be considered an approximation, in the spirit of Christman and Gehring, which holds over the whole velocity range, including low velocities where their original formulation fails. The diameter function S_D, is based on Eq. 11:

$$S_D(Z_0,Z) = (\rho_p/\rho_t + 1/(3\ m))\ \ln[(1 + Z_0^2)/(1 + Z^2)]\ . \tag{14}$$

Here, m is considered an adjustable parameter with typical values between 1 and 2. In this formulation, S_D attempts to describe both stage 1 and stage 3 of the penetration process. When used by itself (non-eroding long rod penetrator), m is set equal to L/D. For the S function denoted as $S_i(Z_0,Z)$, we will present several functions appropriate to special cases of interest.

The first function, S_1, is that for a penetrator with no strength ($Y \rightarrow 0$ using the Tate/Alekseevskii notation) representing, for example, a non-stretching jet (Rostoker, 1953). Since such a penetrator does not slow down during penetration, S is a function of Z_0 only, and the process terminates when the length L - m D is used up.

$$S_1(Z_0) = (Z_0\ (Z_0^2 + 1 - \mu^2)^{1/2} - \mu)\ /\mu\ (1 + Z_0^2)\ , \quad \text{for} \quad Z_C < Z_0 \leq \infty \tag{15}$$

$$\text{and} \quad S_1(Z_0) = 0 \qquad \text{for} \quad 0 < Z_0 < Z_C = \mu \quad ,$$

where $\mu^2 = \rho_t/\rho_p$ (Tate/Alekseevskii notation) and Z_C is the normalized cut-off velocity below which no penetration occurs.

The second S function for the length term is Tate's special case for Y = H, applicable to the entire velocity range from 0 to infinity. (EXP is the exponential function.)

$$S_2(Z_0,Z) = (1 - \text{EXP}[(Z^2 - Z_0^2)/\mu(1 + \mu)])/\mu \quad . \tag{16}$$

If we subject $S_2(Z_0,0)$ to the optimization procedure for constant energy and L/D , i.e., solve Eq. 5, using S_2, we obtain the transcendental equation:

$$\ln(1 + 3\ y) - y = 0\ ,$$

where $y = Z^2_{opt.}\ /\mu\ (1 + \mu)$ which has the solution y = 1.9038 . With $\mu < 1$ (high density penetrator impacting lower density targets), the optimum velocity $V_{opt.} = Z_{opt.}\ (2\ H/\ \rho_t)^{1/2}$ can be considerably lower than that for the case of the constant mass penetrator discussed earlier.

Tate also gave a solution for Y equal to fractional values of H, but only for the restricted case of $\mu = 1$ (target and penetrator densities equal). Using an approximate solution of the Tate/Alekseevskii "Bernoulli equation", but with a better approximation than that of Alekseevskii's, that preserves the cut-off velocity, a closed form solution can be developed for the case $Y = H\ /\ (2 + \mu)$, which is:

$$S_3(Z_0,0) = (1 - (1 + \mu)/a + 1/a\ \text{EXP}[\mu - a]\ )/\mu \qquad \text{for} \quad a_c \leq a \leq \infty, \tag{17}$$

$$\text{and} \quad S_3(Z_0,0) = 0 \qquad \text{for} \quad 0 \leq a \leq a_c$$

where $a = Z^2\ (2 + \mu)/\ \mu(1 + \mu)$ and $a_c = \mu$.

More functions of this kind can be given, but only for Y values that are smaller than that for S_3. Since the S_1 and S_3 curves are not greatly separated, these new functions are not of great value. What is needed, however, is a family of functions that fill the large gap between S_2 and S_D. Functions of this kind can be generated by using what must be considered a radical, if not crude, approximation to the variable mass penetrator problem. In experiments with high density long rod penetrators against finite thickness plates, it is noted that the length of rod lost and eroded is proportional to the target thickness perforated. If systematic variations with velocity are ignored, we can start with:

$$L(x) = L_0 - r\ x \tag{18}$$

where L is the instantaneous rod length, L_0 is the initial length, x is the penetration depth and r is a dimensionless constant. Differentiating Eq. 18 and noting that $dL/dt = -(v - u)$ where (using Tate's notation) v is the instantaneous velocity of the rear of the penetrator and u is the penetrator-target interface velocity, dx/dt, we find that $r = (v - u)/u$. (This will be referred to as the constant v/u approximation.)

The kinematics of the eroding rod are approximated by using Tate's "Bernoulli equation" and noting that $Y = -\rho_p L\, dv/dt$:

$$-\rho_p L\, dv/dt + 0.5\rho_p (v - u)^2 = H + 0.5\, \rho_t u^2 .$$

This is now interpreted as an equation of motion and, as with the S_D function, supplemented on the right-hand side with the added target mass term:

$$-\rho_p L\, dv/dt + 0.5\, \rho_p (v - u)^2 = H + 0.5\, \rho_t u^2 + (1/3)D\, \rho_t\, du/dt . \qquad (19)$$

With the earlier constant v/u approximation, we have $v = (1 + r)\, u$ and $(v - u)^2 = (r\, u)^2$.

Therefore, Eq. 19 can be integrated easily after multiplying both sides by $u\, dt$ and separating the variables. (Note that $u = dx/dt$.) By defining:

$$c^2 = (\mu^2 - r^2) / \mu^2 (1 + r^2) \geq 0 ,$$

$$b = r (1 + r) / (\mu^2 - r^2) \geq 0 ,$$

$$e = \rho_t D / 3\rho_p (L - m D) = \mu^2 / 3(f - m)$$

and

$$d = e / (1 + r) ,$$

the result is:

$$S_4(Z_0,Z) = (1 + d)/r \left[1 - \left\{(1 + c^2Z^2)/(1 + c^2Z_0^2)\right\}^b\right] . \qquad (20)$$

The primary parameter is r which is the proportionality factor between rod mass loss and penetration. It can be shown that as $r \rightarrow \mu$, S_4 approaches S_2. Also, as $r \rightarrow 0$, the differential equation (Eq. 19) approaches that for S_D. Therefore, the range $0 \leq r \leq \mu$ generates curves between S_D and S_2. The physical relevance of Eq. 20 can not be assessed at this time, so it should not be taken too seriously. The constant v/u approximation which is the basis for deriving S_4 is certainly open to severe criticisms. Never-the-less, the collection of S functions given above, together with Eq. 13, allows for some interesting parametric studies. The character of these functions is shown in Fig. 3. For the case e=0, S_1, S_2 and S_3 tend to the high velocity limit of $1/\mu$, S_4 approaches the value 1/r at very high velocities while S_D tends toward infinity in a logarithmic fashion.

An example of the application of these analytical forms is shown in Fig. 4. This is a plot of the maximum penetration P (Eq. 13) as a function of the impact velocity for the case where E = 1 MJ, f=20, μ =0.7, r= $\mu/2$, m=1 and ρ_t = 7850 Kg/m^3 (armor steel). The performance maxima occur at P = 0.30, 0.41, 0.34, and 0.64 meters with the optimum velocities of 2.0, 1.5, 1.72 and 1.52 km/s for the functions S_1, S_2, S_3 and S_4 respectively. The non-dimensional Z values are 2.09, 1.57, 1.79 and 1.59 respectively. The penetration at the optimum velocity increases with penetrator strength Y since the functions $S_1 < S_3 < S_2 < S_4$ are associated with the penetrator strengths $0 < Y_3 < Y_2 < Y_4$.

CONCLUDING REMARKS

These examples illustrate some of the capabilities of the methodology presented in this paper. Another obvious application is the first order analysis of multiple penetrators, that is, the performance of one long penetrator may be compared to N shorter penetrators of equal material length and mass (segmented penetrators). This last case can not consider the effects of separation between penetrator segments, nor can it account for penetrator mass remaining in the crater from

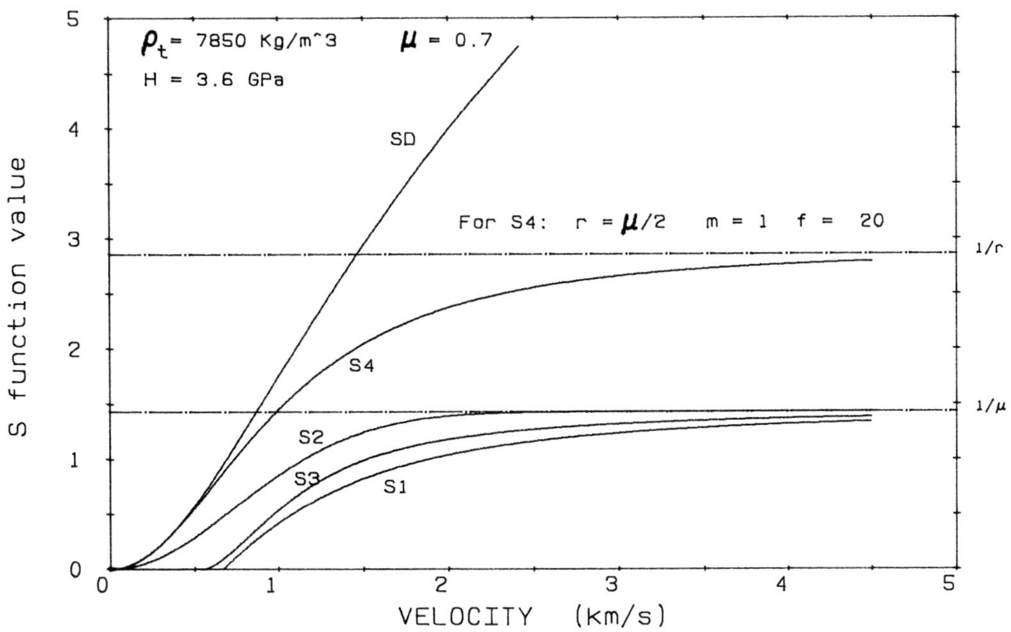

Fig. 3 Characteristic S functions related to penetrator length and penetrator material strength (S1 - S4) and related to penetrator diameter (SD).

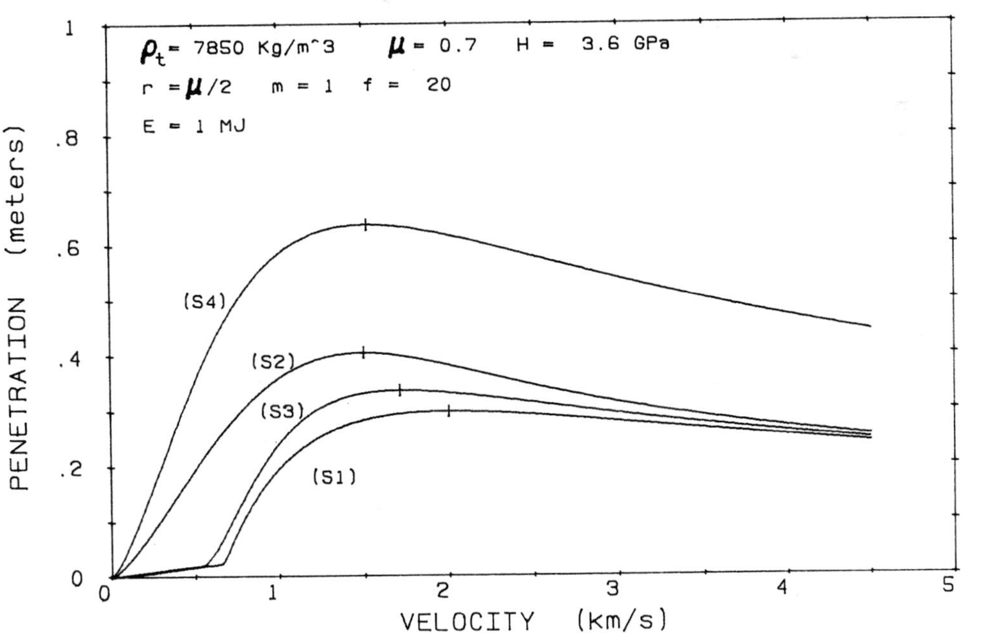

Fig. 4 Two phase penetration performance $(L - m\,D)\ S_i(Z_0,0) + m\,D\ S_D(Z_0,0)$. (Note: f = L/D). The mark at the peak of each curve shows the velocity for optimum penetration.

previous impacts in a successive series of impacts. Never-the-less, it provides for the initial assessment of the efficacy of segmented penetrators, especially in the higher velocity regime.

Finally, we will attempt to answer one obvious question. In this day and age of "super computers" and refined computational techniques, why bother with approximate closed form analytic expressions? In the exploratory stages of research and development, there are many occasions where a large number of parametric investigations need to be made. The requirement is not necessarily the non-plus-ultra in accuracy, but a consistent set of equations that reflect the major trends reasonably well and does not misrepresent the important physical interdependencies. For this kind of use, the formulations presented in this paper prove to be superior. At their best, they provide reasonable first estimates requiring a minimal amount of material parameters. At their worst, one can delegate them into the realm of semi-empirical fitting formulas, but with parameters that have some direct physical relation to the penetrator-target interaction process.

REFERENCES

Alekseevskii, V. P. (1966). Penetration of a Rod into a Target at High Velocity. Fizika Goreniya i Vzryva, Vol. 2, No. 2, pp 99-106. (Combustion, Explosion and Shock Waves, pp 63-66.)

Belyakov,L. V., Vitman, F.F. and Zlatin, N.A. (1962). Collision of Deformable Bodies and its Modelling: II. The Modelling of the Impact of a Sphere and a Half-Space. Zhurnal Tekhnicheskoi Fiziki, Vol. 33, No. 8, pp 990-995.

Belyakov,L. V., Vitman, F.F. and Zlatin, N.A. (1963). Collision of Deformable Bodies and its Simulation : III. Similitude of the Instantaneous Parameter Values for the Original and Modelling Processes. Zhurnal Tekhnicheskoi Fiziki, Vol. 34, No. 3, pp 519-522.

Byrnside, N. C., Torvik, P. J. and Swift, H. F. (1972). Impact Crater Formation at Intermediate Velocities. Journal of Basic Engineering, pp 398-399.

Christman, D. R. and Gehring, J. W. (1966). Analysis of High-Velocity Projectile Penetration Mechanics. Journal of Applied Physics, Vol. 37, pp 1579-1587.

Goodier, J. N. (1964). On the Mechanics of Indentation and Cratering in Solid Targets of Strain-Hardening Metal by Impact of Hard and Soft Spheres. Proceedings of the Seventh Hyper-Velocity Impact Symposium, Volume III - Theory, Orlando FL.

Hanagud, S. and Ross, B. (1971). Large Deformation, Deep Penetration Theory for a Compressible Strain-Hardening Target Material, AIAA Journal, Vol. 9, No. 5, pp 905ff.

Hill, R. (1980). Cavitation and the Influence of Headshape in Attack of Thick Targets by Non-Deforming Projectiles. J. Mech. Phys. Solids, Vol. 28, pp 249-263.

Hopkins, H. G. (1960). Dynamic Expansion of Spherical Cavities in Metals. I.N. Sneddon and R. Hill (Ed.), Progress in Solid Mechanics, Vol. I, North-Holland Publishing Company, Amsterdam; Interscience Publishers Inc. New York , chapter III.

Johnson, W. (1972). Impact Strength of Materials. Edward Arnold, London; Crane, Russak, New York, Chapter 9.

Rostoker, N. (1953). The Formation of Craters by High-Speed Particles. Meteoritics, Vol. 1, No. 1, pp 11-27.

Silsby, G. F. (1984). Penetration of Semi-Infinite Steel Targets by Tungsten Long Rods at 1.3 to 4.5 km/s. Proceedings of the Eighth International Symposium on Ballistics, Orlando FL.

Tate, A. (1967). A Theory for the Deceleration of Long Rods After Impact. J. Mech. Phys. Solids, Vol. 15, pp 387-399.

Tate, A. (1969). Further Results in the Theory of Long Rod Penetration. J. Mech. Phys. Solids, Vol. 17, pp 141-150.

Tate, A. (1979). A Comment on a Paper by Awerbuch and Bodner Concerning the Mechanics of Plate Perforation by a Projectile. International Journal of Engineering Science, Vol. 17, pp 341-344.

Tate, A. (1986). A Theoretical Estimate of Temperature Effects During Rod Penetration. The Proceedings of the 9th International Symposium on Ballistics, Part 2.

Vitman, F. F. and Zlatin, N.A. (1962). Collision of Deformable Bodies and its Modelling: I. Status and Theory of the Problem. Zhurnal Tekhnicheskoi Fiziki, Vol. 33, No. 8, pp 982-989.

Wright, T. (1982). A Survey of Penetration Mechanics for Long Rods. J. Chandra and J.E. Flaherty (Ed.), Proceedings of the Army Research Office Workshop on Computational Aspects of Penetration Mechanics, Ballistic Research Laboratory, Aberdeen Proving Ground, MD, 27-29 April 1982, C.A. Brebbia and S. A. Orszag (Ed.), Lecture Notes in Engineering, Vol. 3 , Springer-Verlag, 1983.

Int. J. Impact Engng Vol. 5, pp. 285–292, 1987
Printed in Great Britain

0734-743X/87 $3.00 + 0.00
Pergamon Journals Ltd.

FRAGMENTATION OF RAPIDLY EXPANDING JETS AND SHEETS

D. E. Grady

Sandia National Laboratories
Albuquerque, New Mexico

ABSTRACT

A common consequence of hypervelocity impact phenomena, or other intense loading process, is the creation of rapidly expanding hydrodynamic jets or sheets. Theoretical results are derived for the time to failure and fragment size associated with the catastrophic breakup of jets and sheets. Physical effects leading to the enhanced ductility of materials during dynamic expansion are discussed. Theoretical results are compared with impact plumes and shaped-charge jets.

INTRODUCTION

Deformation of ductile metals under very high loading rates such as occurs during the formation of shaped-charge jets or explosively-formed projectiles can lead to extensive stable straining before failure through instability and fracture occurs. Even semibrittle metals can exhibit increased ductility and flow during similar high loading rates. The conditions which lead to such enhanced ductile deformation and the subsequent failure and fragmentation process are of interest in numerous hypervelocity impact and explosive loading applications.

Theoretical literature related to a clearer physical understanding of deformation and failure under high rate loading is fairly extensive, particularly on the subject of shaped-charge jets. Analytic studies on the stability and breakup of shaped-charge jets have been provided by Chou and Carleone (1977), Carleone, Chou, and Ciccarelli (1977), Miller (1982), and Walsh (1984). Similar investigations, more numerical in nature, have been put forth by Karpp and Simon (1976), Pfeffer (1980), Osborn and others (1981). The deformation and breakup of rapidly straining ductile rings and shells have been considered by Mott (1943), Ivanov (1981), Hirsch (1981), Kipp and Grady (1984), and others.

Although progress has been considerable, some of the features associated with the deformation of rapidly expanding jets and sheets are still not well understood. What are the physical and/or material conditions which lead to the enhancement of stable ductile flow under dynamic deformation? What are the properties which determine the transition from stable to unstable expansion and ultimately the time and deformation to failure? And finally, what determines the size and shape of fragments produced during the catastrophic failure process?

The present paper is intended to shed some light on these issues. We have in mind the dynamic deformation of rapidly stretching filaments (jets) and films (sheets) of ductile metals with no specific application under consideration. We describe an approach based on a balance of plastic work and kinetic energy which leads to a method for predicting fragment size in the failure process. This result combined with other considerations leads to a prediction of the time and deformation to failure. Possible explanations for the enhanced stability and ductility observed in the dynamic expansion process are discussed. Finally, several applications of the results are considered.

DIMENSIONAL CONSIDERATIONS

In considering the breakup of a rapidly stretching film or filament, some brief preliminary dimensional considerations similar to Walsh (1984) will be useful in scaling the phenomenon and in establishing the class of problems relevant to the analysis. Referring to Fig. 1, consider a stretching film or filament which, at the moment of examination, has a thickness or diameter of dimension a. A measure of the kinematic state of the body is provided by the relative velocity V across the thickness or diameter – a measure of the rate of thinning of the film or filament.

In addition to the geometric parameter a and the kinematic parameter V, assume that the material is incompressible with a constant density ρ, and the material response is completely described by constant flow stress Y. Other material or thermodynamic complications such as temperature or strain-rate sensitivity will be ignored or else assumed to enter only implicitly through the magnitude of Y.

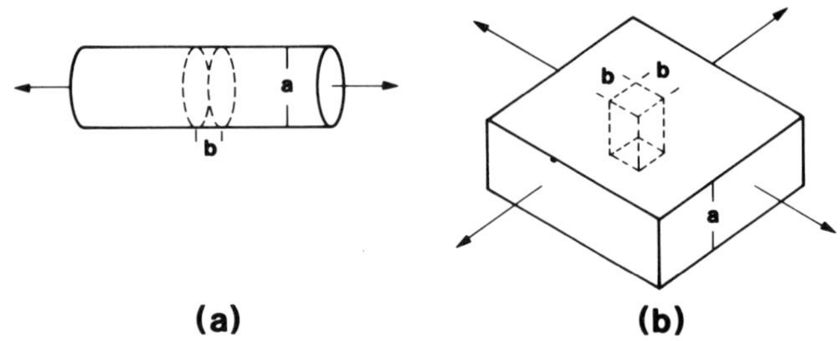

Fig. 1. A rapidly stretching filament (a) or film (b). The diameter or thickness a at time t is assumed to be small compared to the length or extent of the filament or film and consequently boundaries can be considered at infinity. The elements of dimension b indicated in the figure represent the volume of an average fragment.

With the four dimensional parameters a, V, ρ, Y, only one dimensionless parameter can be formed which we will write as

$$J = \frac{\rho V^2}{Y} \quad . \tag{1}$$

The number J provides a measure of the kinetic energy content of the body relative to its ability to dissipate energy in plastic work. In the present study, interest will focus on values of J well in excess of unity. This is characteristic of shaped-charge jets, explosively-formed projectiles, or ductile debris plumes created under hypervelocity impact. In contrast, most explosive or magnetic loading geometries of metal rings and shells lead to values of J less than unity and other analytic methods are required (Kipp and Grady, 1984).

In this work, interest will be in the size of fragments created in the breakup process b, the time to breakup t_b, and perhaps the length or extent that the body stretches before breakup l_b. Accordingly, these properties are expected to depend on J like

$$b = a f_1(J) \tag{2}$$

$$t_b = \frac{a}{\sqrt{Y/\rho}} f_2(J) \tag{3}$$

$$l_b = a f_3(J) \tag{4}$$

Although attention in this study will focus primarily on the breakup of plastic bodies characterized by a flow stress Y, some generalizations are worth noting. If the bodies in Fig. 1 are fluids, characterized primarily by a Newtonian viscosity η, then another dimensionless number $R = \rho V a/\eta$ is obtained which is equivalent to the Reynolds number in the theory of fluids. Alternatively, if breakup is primarily governed by capillary forces through a surface tension γ, then a number $W = \rho v^2 a/\gamma$ is obtained, similar to the Weber number commonly used to characterize the breakup of high velocity fluid particles in an atmosphere. Although some of the physical issues differ, most of the observations to be described in this work carry over to these descriptions.

FRAGMENT SIZE PREDICTION

Consider first the stretching filament illustrated in Fig. 1a. In the subsequent analysis, subscripted variables $a_o, \dot{\epsilon}_o$, *etc.*, are used to represent initial $t = 0$ values. Choice of an initial time is, of course, somewhat arbitrary. The same variables without subscripts are used to represent the variables later in time. The initial diameter of the element is a_o. The axial strain rate $\dot{\epsilon}_o$ will be used to determine the velocity measure $V_o = \dot{\epsilon}_o a_o$ defined within the dimensionless parameter J_o in the previous section. It is assumed that the axial Lagrangian strain rate is time independent.

An element of diameter a_o and length b_o, as indicated in Fig. 1a, is chosen to represent the average volume of a fragment. Since incompressibility is assumed, $a^2 b = a_o^2 b_o$.

An estimate of the average fragment size is based on a balance of the kinetic energy available for fragmentation and the plastic work dissipated in the fragmentation process (Grady, 1982). The local kinetic energy of the element in Fig. 1a relative to its center of mass is given by

$$T = k\rho\dot{\epsilon}_o^2 a_o^2 b_o^3 \ , \tag{5}$$

where $k \approx 0.04$. This kinetic energy is assumed to be dissipated in plastic work when breakup occurs and the element becomes an isolated fragment.

This plastic work occurs when the stretching filament becomes unstable and the initially uniform filament deforms into nominally ellipsoidal-shaped fragments. The plastic work is

$$W = gY a_o^2 b_o \ , \tag{6}$$

where g is approximately the average plastic strain in the fragment incurred during the breakup process. The value of g is expected to be between about 0.01 and 0.1.

If we assume that the local kinetic energy in the fragment is totally consumed as plastic work, then

$$k\rho\dot{\epsilon}_o^2 a_o^2 b_o^3 = gY a_o^2 b_o \ . \tag{7}$$

Equation 7 can be solved for the average initial fragment length of the filament,

$$b_o = \sqrt{\frac{gY}{k\rho\dot{\epsilon}_o^2}} \ , \tag{8}$$

or, in the form of Equation 2,

$$b_o = a_o\sqrt{\frac{g}{k}}J_o^{-\frac{1}{2}} \ . \tag{9}$$

The fragment mass is

$$m \approx \rho a_o^2 b_o = \rho a_o^2\sqrt{\frac{g}{k}}J_o^{-\frac{1}{2}} \ . \tag{10}$$

For the case of a biaxially stretching film characterized by a strain rate $\dot{\epsilon}_o$, as illustrated in Fig. 1b, a similar analysis applies. The element of thickness a_o and sides b_o representing the volume of a fragment prior to breakup has a local kinetic energy

$$T = k\rho\dot{\epsilon}_o^2 a_o b_o^4 \ , \tag{11}$$

where $k \approx 0.08$. The plastic work incurred during the failure process is given by

$$W = gY a_o b_o^2 \ , \tag{12}$$

where again g is expected to be between about 0.01-0.1. Equating energies provides the same expressions for fragment size as Equations 8-10.

THE TIME AND EXTENT OF STRETCHING TO FAILURE

In the analysis of the previous section it was assumed that a rapidly stretching film or filament was unstable with respect to variations in the thickness or diameter and that fragmentation would eventually occur. Although a nominal fragment size was calculated, the time to the onset of instability, growth, and ultimate breakup was not considered. It is well known that the time (extent of deformation) from the formation of a shaped-charge jet to breakup can be considerable. Similarly, in hypervelocity impact of a projectile on a thin plate of ductile metal, the time from impact to breakup of the membrane-like impact plume can be substantial.

Focusing on an initially uniform stretching filament of infinite extent and constant cross section, it is reasonable to question whether the motion is intrinsically unstable for all values of the set of parameters (a, b, ρ, Y). The stability of uniformly stretching plastic jets has been considered by Chou and Carleone (1977). They have shown in the long wavelength limit that all perturbation wavelengths in a linear stability analysis are unstable and exhibit unbounded growth.

There are two issues to consider in this type of analysis: First, since the governing equations of motion have been limited to a description of deformation where variations in cross section vary slowly (in length) in comparison with the jet diameter, the analysis makes no conclusion concerning the stability of perturbation wavelengths short with respect to the jet diameter. Second, the stability analysis is complicated by time dependence of the coefficients in the linearized equations.

The first issue was treated in further work by Carleone, Chou, and Ciccarelli (1977) and later by Walsh (1984) where an effect of increased stress in the cross section of a neck due to non-axial straining was included in the stability analysis. Both Carleone (1977) and Walsh (1984) showed that when this feature, typically quantified through the Bridgman formula, was included in the long wavelength stability analysis of uniformly stretching jets, a critical wavelength of approximately twice the jet diameter is calculated for which unstable growth is a maximum. Some alternative thoughts on this issue will be considered in the next section.

The second issue of time dependence of the coefficients in the linearized governing equations results in an added complexity in the analysis of dynamic systems. Assuming that the coefficients vary slowly in time, and can be considered constant, is frequently invalid. Occasionally a tractable ordinary differential equation in time is obtained and can be analyzed (Chou and Carleone, 1977). An estimate of the onset of catastrophic growth of instabilities leading to failure can be made for stretching jets (and sheets), however, without considering the time dependence in detail. Assuming constant coefficients in the linearized equation of motion reveals that if the subsequent motion of sinusoidal perturbations in the cross-sectional area of wavelength λ is exponential with a growth factor α, then

$$\alpha = \dot{\epsilon} / \left(\sqrt{1 + \frac{Y\beta^2}{\rho\dot{\epsilon}^2}} - 1 \right) \quad , \tag{13}$$

where $\beta = 2\pi/\lambda$.

The validity of the analysis leading to Equation 13 requires that the characteristic time of instability growth be small compared to the characteristic time of uniform deformation. This translates to a requirement that the ratio $\alpha/\dot{\epsilon}$ be substantially larger than unity in which case Equation 13 can be written

$$\frac{\alpha}{\dot{\epsilon}} \simeq \sqrt{\frac{4\pi^2 Y}{\rho\lambda^2\dot{\epsilon}^2}} \quad . \tag{14}$$

Carleone and coworkers (1977) and Walsh (1984) showed that the fastest growing wavelength occurs at the critical value of $\lambda_c \simeq 2a$, where a is the current filament diameter. This is also observed experimentally. Accepting this, Equation 14 can be written as

$$\frac{\alpha}{\dot{\epsilon}} = \sqrt{\frac{\pi^2}{J}} \tag{15}$$

where $J = \rho\dot{\epsilon}^2 a^2/Y = J_o/(1 + \dot{\epsilon}_o t)^3$ is the stretching intensity defined earlier.

Note that as stretching proceeds, J decreases with time from its reference J_o value. We suggest that catastrophic growth initiates when $\alpha/\dot{\epsilon}$ in Equation 15 is of order unity. Although the analysis leading to Equation 15 is certainly suspect for the reasons stated above, the trend toward reduced growth rate relative to stretching rate when extrapolating away from values of J for which the analysis is valid, suggest that this should provide a reasonable criterion for the onset of catastrophic growth and failure.

Thus, we will assume that a value of $J = \pi^2$ (see Equation 13) determines the time at which catastrophic growth and failure initiates and through $J \simeq J_o/(\dot{\epsilon}_o t)^3$ by letting $J = \pi^2$ at $t = t_b$ yields,

$$\bar{t}_b \simeq \frac{1}{2} J_o^{-\frac{1}{6}} \quad , \tag{16}$$

where $\bar{t}_b = t_b\sqrt{Y/\rho a_o^2}$.

An alternative estimate of the time to breakup follows by assuming that the critical wavelength and, consequently, the fragment dimension is given by $\lambda_c = b = 2a$. This is observed experimentally for stretching jets and also seems reasonable for the breakup of expanding films. Using the expression for the initial dimension of a fragment from Equation 8 and the relation $b \simeq b_o \dot{\epsilon}_o t_b$, a time to breakup is obtained,

$$\bar{t}_b = \sqrt[3]{\frac{4k}{g}} J_o^{-\frac{1}{6}} \tag{17}$$

The coefficient in Equation 17 is approximately a factor of 2 to 5 larger than that of Equation 16, depending on the assumed value of the plastic strain g.

THE DUCTILITY OF RAPIDLY STRETCHING JETS AND SHEETS

The enhanced ductility observed in the dynamic expansion or elongation of metal jets or sheets is of fundamental importance. This feature is the single important property which renders the techniques of impulsive metal forming so useful in several applications. Generally, improvements in dynamic ductility leads to improvements in the application. Consequently, a fundamental understanding of the material and kinematic properties affecting the dynamic ductility is desired.

Two issues emerge in consideration of the stability of ductile stretching films or elongating rods. The first is the onset of geometric instability controlled by the mechanical behavior of the material. Strain hardening and strain-rate sensitivity in metals are known to delay the onset of instability. Regazzoni and Montheillet (1984) have shown for materials like copper and tantalum that the exponents governing strain hardening and strain-rate sensitivity increase with strain rate, leading to an increasingly stable ductile flow with increasing strain rate. The second issue, commonly called the intrinsic ductility, relates to the degree of plastic deformation a material can accommodate before failure through intergranular fracture initiates. Intrinsic ductility can also be sensitive to the strain rate and is further dependent on the degree of confining pressure under which flow occurs (Hancock and Mackenzie, 1976). Although some rate sensitivity of instability and intrinsic ductility is suggested by intermediate rate studies on metals, the magnitude is insufficient to explain the extreme ductility observed in expanding jets and sheets. Another mechanism is necessary and the following ideas are suggested.

Consider again an extended jet or sheet of diameter or thickness a stretching with a constant Lagrangian velocity gradient, $\dot{\epsilon}_o$. It is straightforward to show, through integration of the momentum equation, that an inertial pressure exists within the body of the form,

$$P(r,t) = \frac{3}{32}\rho\dot{\epsilon}^2 a^2 \left(1 - 4\frac{r^2}{a^2}\right) , \tag{18}$$

where time dependence enters through $\dot{\epsilon} = \dot{\epsilon}_o/(1 + \dot{\epsilon}_o t)$ and $a = a_o/\sqrt{1 + \dot{\epsilon}_o t}$ and r is the radial coordinate. This observation has been made by Walsh (1984) and Frankel and Weihs (1985). It is informative to write the pressure on axis ($r = 0$) in terms of the current stretching intensity

$$P = \frac{3}{32} Y J \ . \tag{19}$$

Recall that J is a decreasing function of time. It is interesting to note for one criterion for the onset of failure ($J \simeq \pi^2$; see previous Section) that the inertial pressure is approximately equal to the flow stress at the time of breakup. As an example, a copper jet 10 mm in diameter stretching at $\dot{\epsilon} \sim 10^5/s$ yields a stretching intensity of $J = 100$ and implies an inertial pressure of $P \sim 10Y$. A similar inertial pressure is achieved in a rapidly expanding sheet of material.

The inertial confining pressure to which the material in a rapidly stretching jet or sheet is subjected has two stabilizing effects: First, the intrinsic ductility of the material is enhanced because of the increased state of compressive stress under which flow proceeds. Failure through internal void nucleation and fracture is suppressed by the increased confining pressure and for many metals is not expected to occur until the inertial pressure reduces to values on the order of the flow stress or less.

Secondly, because of the parabolic form of the pressure distribution in the jet or sheet, a positive pressure gradient directed from the surface into the material exists. Thus, small surface variations leading to perturbations in the diameter or thickness will be stable in the sense of Rayleigh-Taylor stability. The surface represents a low-to-high density decelerating interface. The stability of perturbations with wavelength small compared to the diameter is clear. Studies of stretching liquid jets where capillarity provides the destabilizing forces (Frankel and Weihs, 1985; Romero, 1986) indicate inertial pressure stabilization for surface perturbations of all wavelengths provided the stretching intensity J (its equivalent for capillary forces) is sufficiently high, with a transition to unstable growth when a critical J is achieved. It is assumed that similar initial stability and transition behavior would occur for the incompressible plastic jet.

In summary, it appears that the inertial pressure persisting within a rapidly stretching jet or sheet is critical to the stability of the deformation. The Bridgman effect may certainly play a role in stabilizing the growth of short wavelength surface perturbations as suggested by Carleone and coworkers (1977) and Walsh (1984); however, inertial pressure through Rayleigh-Taylor stabilization of the surface may provide a more profound effect. Furthermore, inertial pressure may provide a stabilizing mechanism for surface perturbations of all wavelengths, accounting for the enhanced stability of stretching jets and sheets before failure through necking occurs. Finally, the inertial confinement pressure would tend to enhance the intrinsic ductility because of its ability to suppress the formation of internal fracture leading to failure.

APPLICATIONS

There are numerous impulsive loading applications, leading to multiple ductile fragmentation events that can be approximately modeled by the fragmentation geometry described in the present work. Several of these applications will be compared with the results of the present study.

Fragmentation in High-Velocity Impact

Impact experiments with lead projectiles on lead plates have been performed with normal impact velocities in the range of 800-1600 m/s (Grady, Bergstresser, and Taylor, 1985). Following impact, because of the low flow stress and high ductility of the lead, a portion of the plate and projectile merge into a rapidly extruding and expanding bubble of material. This bubble of lead expands until it is quite thin ($\approx$ 0.2 mm) before surface instabilities lead to breakup into thin plate-like fragments of metal which constitute the impact-induced debris plume.

In a series of such experiments, Grady and coworkers, (1985) determined, through photographic imaging and image process methods, an average fragment diameter of the plate-like fragments in the debris cloud. The average fragment size was correlated with the expansion velocity (not impact velocity) of the debris cloud. The experimental results from this study are shown in Fig. 2.

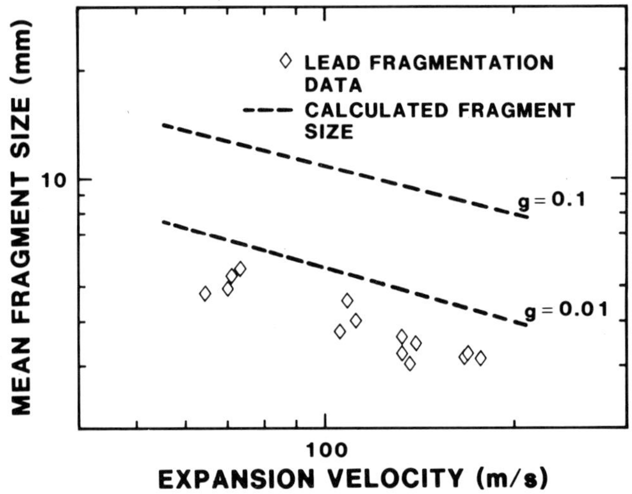

Fig. 2. Fragment size data for normal impact of lead projectile on lead plates (Grady, Bergstresser, and Taylor, 1985).

A comparison of the current theory with this data can be made by starting with Equation 10. Dividing through by ρa_o^3 and inverting the result provides an estimate of the number of fragments. $J_o = \rho V^2/Y$ is estimated from the density and flow stress for lead and the expansion velocity. The flow stress Y is the most poorly constrained. Duffy, Hardy, and Frantz (1972), obtain a value of about $1.5 - 3.0 \times 10^7 Pa$ in dynamic torsion experiments. This would not account for reduction in flow stress due to the excessively higher temperature occurring in the impact experiments. We will assume $Y \approx 10^7 Pa$ which is probably on the high side. These values, with an expansion velocity of 100 km/s, for example, lead to $J_o \simeq 45$ and a fragment number between 6 to 19 for g between 0.01-0.1. To compare with the data in Fig. 2, we have made the experimental observation that breakup occurs when the expanding bubble achieves a diameter of approximately 20 mm.

Dividing the surface area at this time by the number of fragments calculated leads to a nominal fragment size. These results are compared with the data in Fig. 2 and are reasonably encouraging.

The time to failure calculated from Equation 16 is ~ 45 μs and from Equation 17 is ~ 130-270 μs. The observed breakup time is about 50-75 μs.

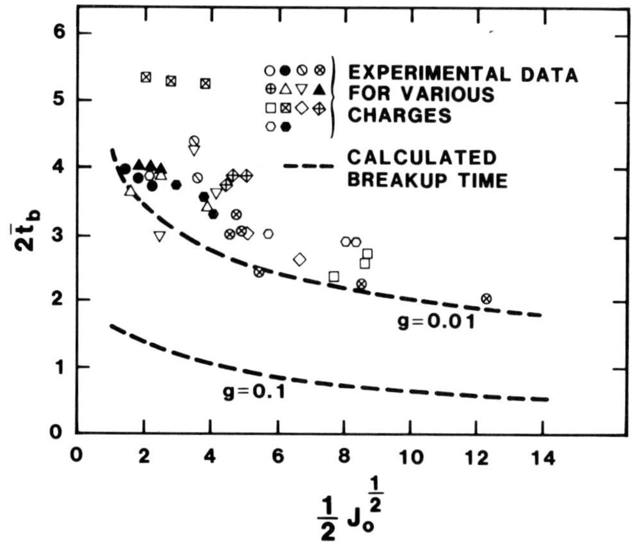

Fig. 3. Time-to-failure data for shaped charge jet breakup (Chou, 1985).

Fragmentation of Shaped-Charge Jets

Chou (1985) has collected time-to-breakup data from several sources in which the explosive charges were varied such that a relatively broad range of jet stretching intensities were achieved. The data, as presented by Chou, are shown in Fig. 3, but with the ordinate and absicca expressed in variables developed in the present text.

In terms of the calculations for stretching jets presented earlier, the nondimensional breakup time is dependent on the initial stretching intensity according to

$$\bar{t}_b = \sqrt[3]{\frac{4k}{g}}J_o^{-\frac{1}{6}} \quad . \tag{20}$$

The coefficient k is approximately 0.04; however, g, which is a measure of the inhomogeneous strain, is not well constrained. As stated earlier, values of g between 0.01 and 0.1 seem reasonable.

In Fig. 3 calculated curves for the breakup time from Equation 20 are shown based on values of $g = 0.01$ and $g = 0.1$. The functional dependence appears to be in accord with the data and reasonable qualitative agreement is achieved for a value of $g = 0.01$. This would imply an average nonhomogeneous strain in a fragment of about 1%. Although somewhat lower than expected, the value is perhaps not unreasonable.

SUMMARY

In the present study, some tentative thoughts concerning the time to breakup and the intensity of fragmentation in rapidly stretching ductile jets and sheets have been broached. In addition, some ideas relating to observed stability and enhanced ductility are discussed.

The prediction of fragment size is based on a local balance of kinetic energy and inhomogeneous plastic work during the breakup process. This energy approach has been used previously for other fragmentation geometries with some success (Grady, 1982; Kipp and Grady, 1984). A difficulty in the present effort is an estimate of the magnitude of the plastic work associated with the fragmentation event. This has not been looked at in detail and further study is needed.

An estimate of the time of failure is provided in the present study by examining linear stability analysis for the point at which the rate of instability growth diverges from the rate of uniform contraction or thinning. This is probably only slightly better than an order-of-magnitude estimate. Studies alluded to earlier in this report, however, (Frankel and Weihs, 1985;

Romero, 1986) suggest an actual transition from stable to unstable deformation. Further analysis is needed here in terms of such a transition in stretching plastic bodies and how this transition relates to a time to failure.

Finally, it is suggested that the enhance ductility observed in rapidly stretching jets and sheets is due to an inertial confining pressure which both suppresses failure through internal fracture, and stabilizes breakup by surface perturbations through the particular parabolic form of the pressure.

REFERENCES

This work performed at Sandia National Laboratories supported by the U.S. Department of Energy, under contract DE-ACO4-76DP00789.

Carleone, J., P. C. Chou, and R. D. Ciccarelli (1977). U.S. Army Ballistic Research Laboratory Contract Report No. 351.

Chou, P. C. and J. Carleone (1977). J. Appl. Phys., **48**, 4181-4195.

Chou, P. C. (1985). Impact Dynamics of Metals and Composites (lecture notes), UCLA Extension, Los Angeles, California.

Duffy, J. R., R. H. Hardy, and R. R. Frantz (1972). J. Appl. Mech., 651-656.

Frankel, I. and D. Weihs (1985). J. Fluid Mech., **155**, 289-307.

Grady, D. E. (1982). J. Appl. Phys., **53**, 322-325.

Grady, D. E., T. K. Bergstresser, and J. M. Taylor (1985). Sandia National Laboratories Report SAND84-1545.

Hancock, J. W. and A. C. Mackenzie (1984). J. Mech. Phys. Chem., **24**, 147-169.

Hirsh, E. (1979). Propellants and Explosives, **4**, 89-94.

Ivanov, A. G. (1976). Strength of Materials, **8**, 1303-1306.

Karpp, R. R. and J. Simon (1976). U.S. Army Ballistic Research Laboratory, Report No. 1893, AD # B012 141.

Kipp, M. E. and D. E. Grady (1984). J. Mech. Phys. Sol., **33**. 399-415.

Miller, C. W. (1982). Ballistic Research Laboratory Symposium on Ballistics, November.

Mott, N. F. (1948). Proc. Roj. Soc., A189, 300-308.

Osborn, J. J. (1981). 7th Annual Technical Meeting, DEA-AF-71-FIG-7304, Lawrence Livermore National Laboratory.

Pfeffer, G. (1980). Proc. 5th Int. Symp. on Ballistics, Tonbuse, France.

Regazzoni, G. and F. Montheillet (1984). Proc. 3rd Conf. on the Mechanical Properties of Materials at High Rates of Strain, Oxford, England.

Romero, L. (1986). In preparation.

Walsh, J. M. (1984). J. Appl. Phys., **56**, 1997-2006.

Int. J. Impact Engng Vol. 5, pp. 293–305, 1987
Printed in Great Britain

0734-743X/87 $3.00 + 0.00
Pergamon Journals Ltd.

HOIL: A TWO STAGE, HYDRODYNAMIC, NUMERICAL SCHEME LINKING HEMP AND SOIL

John T. Harrison
U.S. Army Ballistic Research Laboratory
Aberdeen Proving Ground, Maryland 21005-5066

and

Wallace E. Johnson
Computer Code Consultants, Inc.

ABSTRACT

A two stage system of two-dimensional, hydrodynamic computer codes, HEMP and SOIL (from the OIL family of codes) are combined to form the HOIL method which is designed to predict shaped charge performance. The numerical procedure and its use are discussed. This technique proves to be very cost effective when compared to similar techniques, and will be extremely useful in providing insight into the entire analytical solution of the shaped charge problem. As a result, the HOIL method will speedily provide the designer with sufficient information to optimize warhead performance in a parametric study.

INTRODUCTION

A two-stage Lagrangian and Eulerian numerical technique called HOIL (Harrison, 1984) is named after the two hydrodynamic computer codes used in this approach, HEMP (Wilkins, 1964; Giroux, 1973) and SOIL (Johnson, 1982a). HOIL has been developed to study shaped charge performance. A primary consideration was to decrease the total calculational time thereby improving cost effectiveness, while at the same time providing a solution of the entire collapse process.

The successful treatment of shaped charge phenomena has been a long standing and challenging problem in the field of numerical hydrodynamics. Many codes have been developed over the past ten to fifteen years in an effort to solve this problem. However, they have either fallen short of the complete problem solution, or are very costly to use in terms of computational system time required to calculate a complete solution of the problem.

This paper presents the numerical linking procedure designed to read an output file from the Lagrangian HEMP code, and overlay its nodal positions onto the Eulerian coordinate system, the SOIL code. A complete solution to the shaped charge problem was obtained with a significant decrease in the computational system time.

BACKGROUND

The ability of explosives with metal-lined cavities to penetrate hard, thick materials, such as steel plates, rock or concrete has intrigued military and civilian scientists for a long time. In recent years simple analytical models (Harrison, 1981) have been developed to explain the prominent features of shaped charge performance. Their effectiveness is predicated upon empirical results. Much still remains to be done from an analytical point of view in understanding the mechanisms at work in the shaped charge collapse. Such as the effects of compressibility, shock waves, material strength, and finite acceleration time for liner elements. Numerical hydrodynamic codes provide the source from which significant advances have been made in these areas of study. It is to such numerical modeling that the present effort is aimed.

It should be noted that many investigators have considered the shaped charge performance over a long period of time. A part of this problem are impact and target penetration results. Several investigators (Johnson and Johnson, 1981; Matuska and Osborn, 1981) have linked the same two finite-difference, numerical schemes except in the reverse order, i.e. Eulerian to Lagrangian for those solutions. These two methods involved collecting a large number of material velocity station points, recorded as a function of time, from the Eulerian code; then, the aforementioned points were applied to the surface nodes in the Lagrangian code. The HOIL method, linking Lagrangian to Eulerian, will transfer all material properties from one coordinate system to the other.

The first stage is a Lagrangian numerical method, the HEMP code. It has been applied to conventional warheads such as fragmenting munitions, Miszney-Shardin devices, and shaped charges (Harrison and Karpp, 1977). For all but the shaped charge problem, it has successfully made accurate predictions. In the shaped charge problem, where large deformations occur at very high strain gradients, the Lagrangian codes are required to use extremely small time-steps, and the accuracy is suspect because of the distortions. Therefore, solutions in jet formation are often unsuccessful, unless some sort of rezoning routine is employed. This is found to be an unattractive process because of the time consuming rezoning procedure. The Lagrangian codes, such as HEMP, have not been effective in the entire shaped charge predictive process. However, Lagrangian codes do provide very good approximations for the detonating explosive, the impulsive loading on the liner, liner accelerations, and collapse velocities (Karpp, 1973; Harrison and Karpp, 1977). This information has been used as inputs into several one-dimensional, analytical, shaped-charge models (Harrison and Karpp, 1977; Harrison, 1981) which then use the classic theory for jet formation after Pugh, Eichelberger and Rostoker (1952). However, in this process jet material effects such as density, temperature and strength are lost. Thus, this too proves to be unsatisfactory in many cases.

The second stage is an Eulerian numerical technique, the SOIL code. The Eulerian grid is fixed for all times and is not susceptible to distortions. Material interfaces are not well defined, but given sufficient initial conditions such as density, velocity, and position, an Eulerian code should be able to predict the entire shaped charge performance from explosive initiation through the detonation process to jet formation. However, predictions have fallen short of expectations. In the past, Eulerian codes have artificially overestimated the specific internal energy, which then produces errors in the approximations for density (Harrison, 1978). Even when better approximations for specific internal energy were implemented (Schmitt, 1979), the shaped-charge jet tip velocity was approximately 15% lower than observed. But the most unattractive aspect of these codes is the long computer system calculational time required for a complete solution. A typical hemispherical shaped charge calculation (Kiwan and Arbuckle, 1977), will consume about 5 hours of a CDC 7600 computing system running time, to obtain nearly total collapse of the liner.

The combination of these two well known numerical techniques, in such a way as to exploit the best attribute of each is the main thesis of this paper. No changes were necessary in the two hydrodynamic codes, only an interface routine was developed to transform from the Lagrangian coordinate system to the Eulerian coordinate system. It should be noted that there seems to be no limitation to the use of these transformation processes. An entire solution from shaped charge liner collapse through jet/target impact and penetration could be given by the linking of these two hydrodynamic computer codes. One could use the HOIL method (Lagrangian to Eulerian) for solutions to the shaped charge liner collapse, jet formation, and penetration. Then the reverse procedure, (Eulerian to Lagrangian), could be used to reconstruct the jet after penetration for residual performance effects.

HOIL NUMERICAL TECHNIQUE

At the beginning of the present investigation, the two hydrodynamic computer programs were well established, as has been stated earlier. The main problem for the authors was to have a universal interfacing procedure linking the two numerical techniques. The algorithm developed uses the Lagrangian node point parameters of position and velocity, and zone center parameters of pressure, specific internal energy, and density at any particular time. It then maps these parameters into an Eulerian grid, (rectangular coordinates). This procedure also obeys the conservation laws. The general outline of the linking algorithm, a subroutine in the Eulerian grid generator program (CLAM4), developed by the second author (1982b), will be presented here.

Figure 1a shows a typical Lagrangian quadrilateral (zone A) bound by its node points 1, 2, 3, and 4 which are overlaid upon a fixed, Eulerian grid. The coordinates Ri and Zi are the i'th Lagrangian node point positions from zone A. All of the Eulerian physical properties are to be located at the center of each of the square Eulerian cells, designated as k. In addition, the Eulerian grid generator code, CLAM4, produces (n) material particles to partition the various materials within the Eulerian cell k.

It must be determined whether the Lagrangian zone, A, is a rectangle (case 1) or a general quadrilateral (case 2). The density and specific internal energy transfer over to CLAM4 directly in both cases, since these are cell centered quantities. The velocities are calculated differently depending on whether it is case 1 or case 2; velocities must be made cell-centered quantities in the SOIL code. From Fig. 1a, a test procedure is introduced to ascertain if zone A is a rectangle:

$$\begin{aligned} &\text{If } R1 = R4 \text{ and } R2 = R3 \\ \text{and } &\text{If } Z1 = Z2 \text{ and } Z4 = Z3 \end{aligned} \tag{1}$$

If this test is met, the Lagrangian zone A become a rectangular package (case 1) for the Eulerian grid generator code, CLAM4. If it is not met, then zone A is a general quadrilateral, and two triangles (case 2) are generated as inputs into CLAM4. We will examine these two cases in detail. The velocity packages, U and V, for each material particle n, contained within the Eulerian cell k, will be calculated after determining the average direction of the two components of velocity, SU and SV, as;

Case 1 - Rectangle

$$SU = \frac{\dot{R}1 + \dot{R}2 + \dot{R}3 + \dot{R}4}{|\dot{R}1 + \dot{R}2 + \dot{R}3 + \dot{R}4|} \qquad SV = \frac{\dot{Z}1 + \dot{Z}2 + \dot{Z}3 + \dot{Z}4}{|\dot{Z}1 + \dot{Z}2 + \dot{Z}3 + \dot{Z}4|} \tag{2}$$

where $\dot{R}i$ and $\dot{Z}i$ are the associated i'th Lagrangian node point velocities from zone A (nodes 1-4).

Case 2 - Triangles

From Fig. 1a, triangle A is constructed from node points 1, 2, and 4; and triangle B is from node points 2, 3, and 4.

Triangle A | Triangle B

$$SU = \frac{\dot{R}1 + \dot{R}2 + \dot{R}4}{|\dot{R}1 + \dot{R}2 + \dot{R}4|} \qquad SU = \frac{\dot{R}2 + \dot{R}3 + \dot{R}4}{|\dot{R}2 + \dot{R}3 + \dot{R}4|} \tag{3}$$

$$SV = \frac{\dot{Z}1 + \dot{Z}2 + \dot{Z}4}{|\dot{Z}1 + \dot{Z}2 + \dot{Z}4|} \qquad SV = \frac{\dot{Z}2 + \dot{Z}3 + \dot{Z}4}{|\dot{Z}2 + \dot{Z}3 + \dot{Z}4|} \tag{4}$$

The Eulerian velocity components U and V for each material particle n are:

$$Un = .5*SU\,(\dot{R}1^2 + \dot{R}2^2 + \dot{R}3^2 + \dot{R}4^2)^{.5} \tag{5}$$

$$Vn = .5*SV\,(\dot{Z}1^2 + \dot{Z}2^2 + \dot{Z}3^2 + \dot{Z}4^2)^{.5} \tag{6}$$

The mass, M, assigned to the Eulerian cell, k, is the sum of the material particles, (assume n = 4) as,

$$Mk = Ma + Mb + Mc + Md \tag{7}$$

where a, b, c, and d are the n material particles in the Eulerian cell, k. The velocity components for each Eulerian cell k used in the SOIL code are now computed from conservation of momentum:

$$Uk = \frac{Ua*Ma + Ub*Mb + Uc*Mc + Ud*Md}{Mk} \tag{8}$$

$$Vk = \frac{Va*Ma + Vb*Mb + Vc*Mc + Vd*Md}{Mk} \tag{9}$$

Finally, the Lagrangian node point positions (R,Z) then are assigned to the SOIL code's passive particle positions for plotting purposes.

NEW TRIANGULAR TRANSFER FOR HOIL

A new scheme has been devised which is more accurate in mapping zones where the velocity gradient is large. Referring to Fig. 1b, the Lagrangian zone A (nodes 1,2,3 and 4) make up two triangles A (nodes 1,2 and 4) and B (nodes 2,3 and 4). Define an Eulerian material particle n (by the CLAM4 code) somewhere inside the triangle A. Triangle A is divided into three possible sub-areas as area 1 (AR1), area 2 (AR2) and area 4 (AR4). As an example area 2 is

$$AR2 = .5 \, [\, Zn(R1-R4) + Z1(R4-Rn) + Z4(Rn - R1) \,] \, , \tag{10}$$

where Zn and Rn are the coordinates of the Eulerian material particle n. Similar terms are calculated for area 1 and area 4.

The velocity packages for the Eulerian cell k containing the material particle n are calculated as

$$Un = \frac{\dot{R}1*AR1 + \dot{R}2*AR2 + \dot{R}4*AR4}{AR1 + AR2 + AR4} \tag{11}$$

$$Vn = \frac{\dot{Z}1*AR1 + \dot{Z}2*AR2 + \dot{Z}4*AR4}{AR1 + AR2 + AR4} \tag{12}$$

The density, specific internal energy, and mass are zone centered quantities, thus they transfer into the Eulerian cell containing the particle n. The mass and velocity components of the Eulerian cell are found by solving equations 7 - 9. The advantage of this new triangular transfer is that each of the n particles will have a different velocity (functions of the 4 nodes of the Lagrangian cell.) Our old scheme gives the same velocity Un and Vn to all of the n particles within the triangle. This new technique has improved the accurate transfer of data from other Lagrangian codes to the SOIL code, in particular, cases where large pressure, density and velocity gradients appear, usually in long and narrow zones.

NUMERICAL SOLUTION OF THE HEMISPHERICAL SHAPED CHARGE

The problem chosen to demonstrate the advantage of the two stage HOIL technique is the collapse of a hemispherical shaped charge. This is particularly useful in that a complete set of experimental data has been obtained (Grace, Golaski and Scott, 1984), also, the collapse time is such that the complete formation process can be observed. The dimensions, shown in Fig. 2a, of the hollow hemispherical, copper liner are 63.5 mm (2.5 inch) in diameter, and a uniform 2.06 mm (.081 inch) thick wall. The liner is bounded by a cylinder of bare Octol (75/25) explosive with 3.175 mm (.125 inch) subcalibration. The initiation of the explosive is at a point on the axis of symmetry. This experiment is designated as HEMI 100.

The two stages of HOIL use a Mie-Gruneisen approximation for the equation of state of the liner material (Bothell, 1963). The initial stage, HEMP, uses the JWL equation of state approximation for the explosive fill (Lee, Finger and Collens, 1973). The Lagrangian initial grid is shown in Fig. 2d. It contains 42 k-lines (3 in the liner) in the Z direction, and 41 J-lines in the R direction. The K-max line, line K=42, is the free surface of the liner, and the K-min, K=1, is the back surface of the explosive. The J-min line, J=1, is on the axis of symmetry, and the J-max, J=41, is the top explosive surface. Figures 2e and 2f show the HEMP calculations at 15.1, and 22.2 microseconds respectively, compared with experimental flash x-ray photographs taken at the aforementioned times, shown on Fig. 2b and 2c. Figure 3a shows calculations of liner collapse velocities at various J-line positions as functions of time after initiation of the explosive. The velocity of the front face of the liner (the free surface of the liner) is plotted at various J-line positions on the liner. The J-line positions monitored are 1, 20, and 40 measured from the axis of symmetry of the liner to its explosive interface, respectively. Figure 3b shows the allocation of the energies, kinetic and specific internal, as functions of time. These indicate that the entire liner surface has been accelerated to nearly it terminal collapse velocity at approximately 20 microseconds after initiation. Also, approximately 83% of the system's total energy is kinetic, with 90% of the kinetic energy in the liner material at 20 microseconds. This indicates that the explosive loading may be neglected after this time.

However, it is utilized in the second stage, SOIL, where it is modeled as an expanding gas using a gamma-law equation of state to track the kinematic motion of the explosive. Therefore, 20 microseconds is chosen as the time when the transformation from the HEMP code to the SOIL code occurs. Figure 4a shows a suitable fixed Eulerian grid, in the upper half of the axis of symmetry, overlaid upon the Lagrangian node points of the hemispherical liner and explosive, in the lower half of the axis of symmetry. The SOIL code's, Fig. 4b, passive tracers of the hemispherical liner only is shown in the lower half of the axis of symmetry with the Eulerian grid in the upper half. SOIL calculations are then continued from this time. Figures 5-6 show the passive tracer positions of the liner surface and plotted with a solid interior from the SOIL code at 20, 30, 40, and 50 microseconds, respectively. Each plot also shows the initial HEMP grid for comparison.

As another correlation, a second experiment was performed using the same hemispherical body, and these are compared against HOIL. This second experiment (HEMI 150) had one change in its design, which was an increased liner wall thickness by 50% (i.e. 3.12 mm wall thickness). The HOIL calculations of liner motion are shown in the next series of figures, i.e. Figs. 7-10. In particular, the HEMP code calculations are shown in Figs. 7 and 8 at times of 0, 18, 20, and 24 microseconds, respectively. Using 24 microseconds as the change over time form HEMP to the SOIL code, Figs. 9 and 10 show SOIL code calculations continuing with times of 26, 30, 40, and 50 microseconds, respectively. Again, the SOIL code's plots give the initial HEMP grid along with it for comparative purposes.

HOIL COMPARED AGAINST EXPERIMENTS

Figure 2 shows the excellent, early time, comparisons between the HEMP code and experiment, i.e. the first experiment HEMI 100. This substantiates what Harrison and Karpp (1977) found earlier, that the HEMP code can predict early liner motions accurately.

Figure 11 shows a comparison between experimental data (Fig. 11a, HEMI 100 and Fig. 11b, HEMI 150) and SOIL code results taken from output at 50 microseconds. It is a graph of cumulative mass of the liner in units of grams verses jet velocity distribution in millimeters per microsecond. The experimental data where reduced from flash x-rays which were measured under the assumption that the jet particles are symmetrical in shape and possess the initial liner density. Due to reading errors in the particle measurements, a 10% error bar is expected on each data point. The calculated result is the solid line; excellent agreement is found between the numerical results and both experimental tests.

Another comparison which can be made between numerical and experimental results is the jet tip velocity. These are shown in Figs. 11a and 11b as the lead particle velocities and are summerized as follows:

(a) HEMI 100: an observed jet tip velocity of 4.22 km/s compared to the HOIL calculation of 4.18 km/s.

(b) Hemi 150: an observed jet tip velocity of 3.50 km/s compared to the HOIL calculation of 3.45 km/s.

Finally, Fig. 12 shows the SOIL calculations compared against the flash x-ray picture of HEMI 100 taken at 60.5 microseconds after the initiation of the explosive. The calculation is shown in the upper half of the axis-of-symmetry and the outline of the experimental, flash x-ray is in the lower half of the axis-of-symmetry.

These comparisons show excellent agreement between experiment and calculations. They also demonstrate that the combination of HEMP and SOIL codes, the HOIL method, can produce results that are well within a 10% error bound, thus providing the warhead designer with a tool to be utilize as an experimental supplement.

HOIL COMPARED WITH ANOTHER NUMERICAL TECHNIQUE

This last comparison to be made is the cost savings using the HOIL approach as opposed to using either the Lagrangian method or the Eulerian method as a stand-alone calculation. The Lagrangian method cannot accurately calculate the total collapse of the liner and the Eulerian method can be a very costly approach. Table 1 is a tabulation of the CDC 7600 Computing System's running times (at a cost of approximately $400 per system hour) for the Eulerian numerical method, the HELP code (Hageman and others, 1975) and the HOIL code. These two codes calculated the entire collapse of the hemispherical liner.

Table 1

Complete Hemispherical, Shaped Charge

Simulated Time Compared Against CDC 7600

Computing System Hours and Cost

	Code	Simulated Time (microseconds)	System Computing Time (hours)	Cost at $400/hr.
1.	Eulerian (HELP) [*]	30	5	$2000
2.	HOIL	50	.5	$ 200

* (Kiwan and Arbuckle, 1984)

This table clearly shows that an order or magnitude difference in cost and computing time is the payoff by using the HOIL code, and the results agree very well with experiments. In addition, the HOIL computed results show a 50% improvement over the HELP code in the error bound between experiments and calculations with respect to jet tip velocity.

CONCLUSION AND SUMMARY

"According to HOIL," a new, two stage procedure is shown to be accurate and cost effective in its calculation of the collapse of a hemispherical shaped charge and can be utilized in the solution of shaped charge problems. The methodology also has applications to many other types of hydrodynamic problems. With the extension to three-dimensions, the HOIL technique will be able to increase its effectiveness to real problems facing the warhead designer today.

REFERENCES

Bothell (ed) (1973). AFSWL-TDR-62-147.

Giroux, E. D. (1973). Lawrence Livermore Laboratory, UCRI-51079 (Rev 1).

Grace, F. I., S. K. Golaski, and B. R. Scott (1984) in Proc. 8th International Symposium on Ballistics, Orlando, Florida.

Hageman, L. J., D. E. Wilkins, R. T. Sedgwick, and J. L. Waddell (1975). Systems, Science and Software, SSS-R-75-2654.

Harrison, J. T. (1978). Ballistic Research Laboratory, ARBRL-MR-02841, (ADA059711).

Harrison, J. T. (1981). Ballistic Research Laboratory, ARBRL-TR-02300.

Harrison, J. T. (1984). in Proc. of the 8th International Symposium on Ballistics, Orlando, Florida.

Harrison, J. T. and R. R. Karpp (1977). Ballistic Research Laboratory BRL-R-1984.

Johnson, W. E. (1982a). Ballistic Research Laboratory Contract Report ARBRL-CR-0476.

Johnson, W. E. (1982b). Computer Code Consultants, Inc., CCC-BRL, BRL Contract No. DAAK11-82-0001.

Johnson, G. and W. E. Johnson (1981). Private Communication.

Karpp, R. R. (1973). Ballistic Research Laboratory, BRL-MR-2268.

Kiwan, A. A. and A. L. Arbuckle (1977). Ballistic Research Laboratory, BRL-R-2028.

Lee, E., M. Finger, and W. Collins (1973). Lawrence Livermore Laboratory, Report No. UCID-16189.

Matuska, D. A. and J. J. Osborn (1981). Ballistics Research Laboratory, Contract Report ARBRL-CR-00467, AFATL-TR-71-81.

Pugh, E. M., R. J. Eichelberger and N. Rostoker, (1952). in J. of Appl. Physics, Vol. 23, No. 5.

Schmitt, J. A. (1979). Ballistic Research laboratory, ARBRL-TR-02168.

Wilkins, M. L. (1964). in Alder, B., Fernback, S., and Rotenberg, M. (ed), Methods in Computational Physics, Vol. 3. Academic Press, NY.

LAGRANGIAN-EULERIAN SKETCH

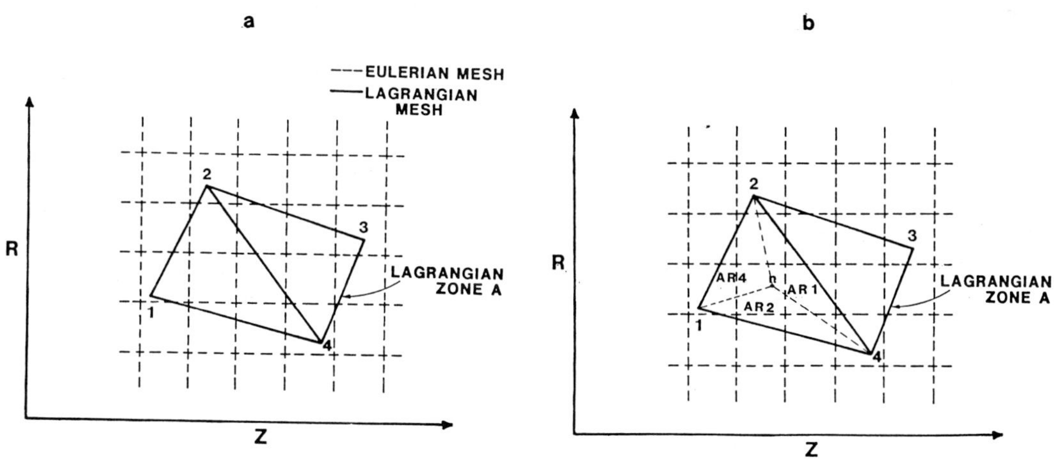

Figure 1. A Typical Lagrangian Zone A Overlayed onto an Eulerian Mesh.

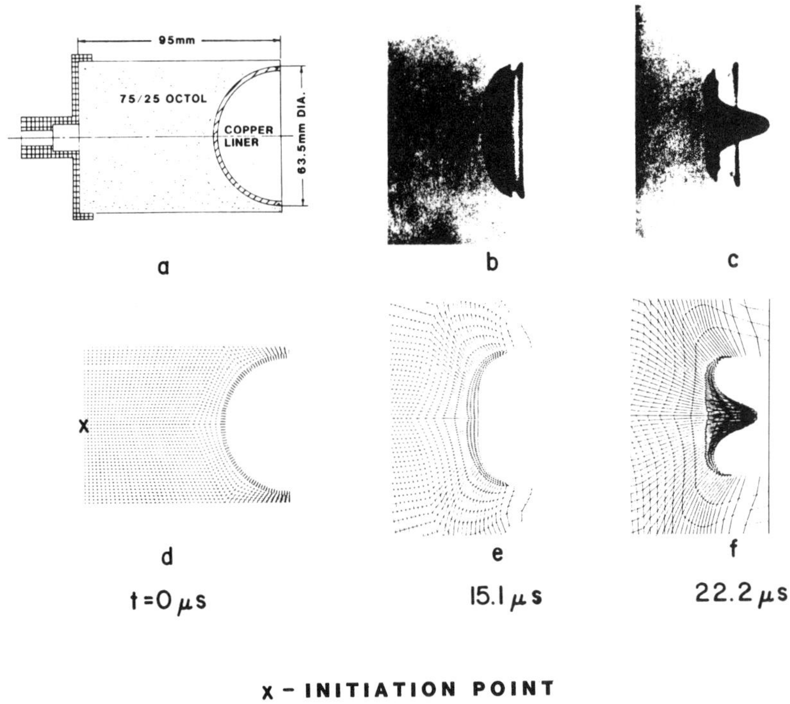

Figure 2. HEMP Calculations of Liner Motion and Experimental Comparisons

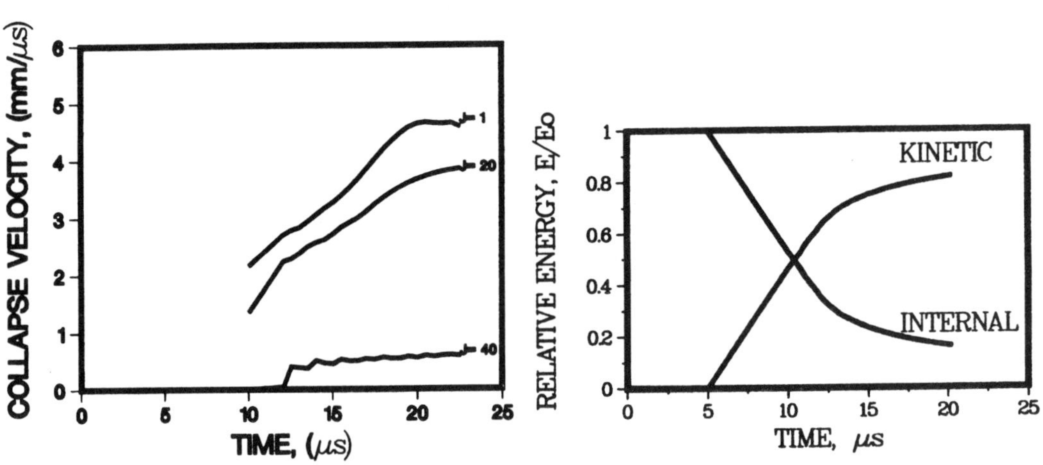

Figure 3. HEMP Code Calculations of Velocity and Energy vs Time

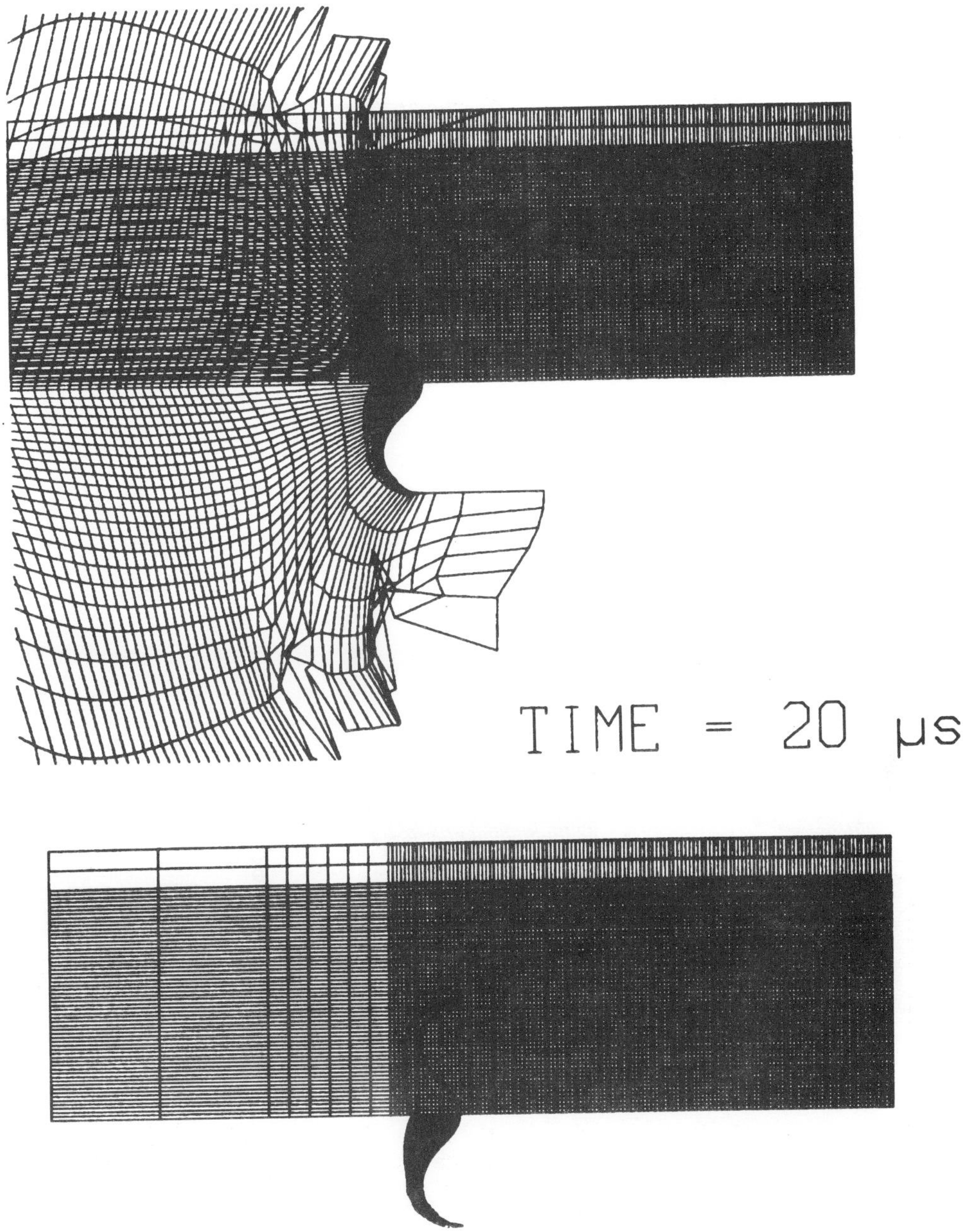

Figure 4. HOIL Transformation at t = 20 Microseconds

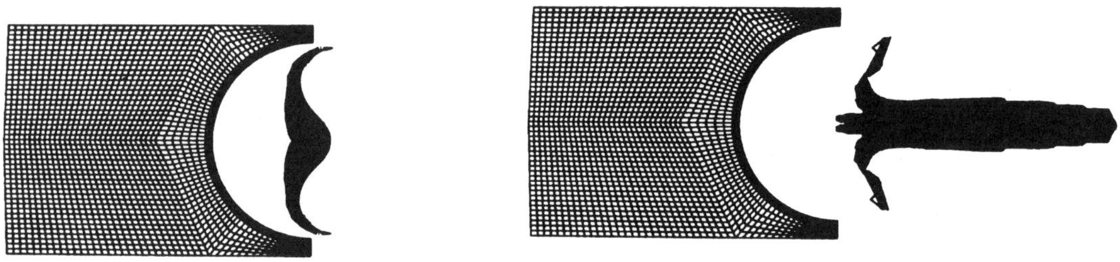

Figure 5. SOIL Code Liner Motion, t = 20 and 40 Microseconds

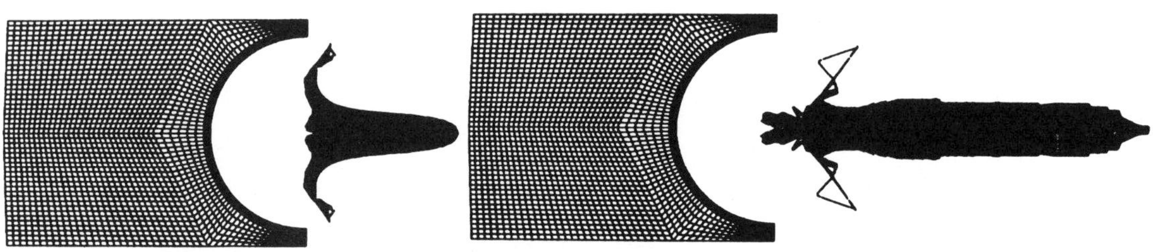

Figure 6. SOIL Code Liner Motion, t = 30 and 50 Microseconds

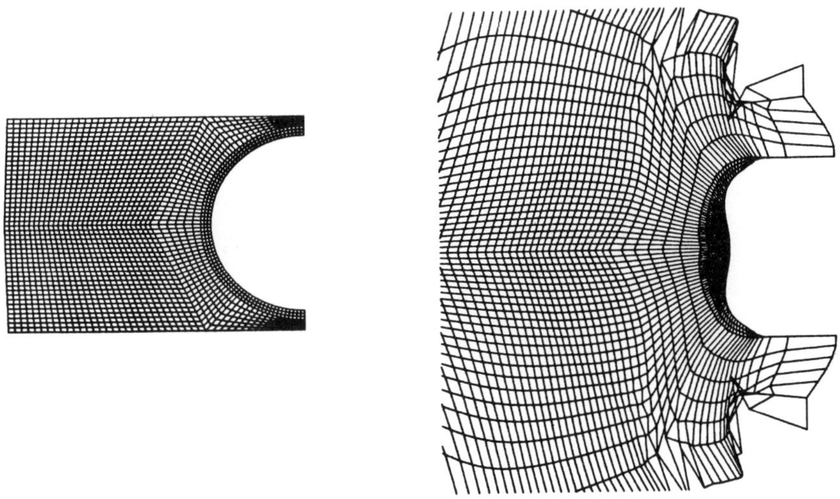

Figure 7. HEMP Code Liner Motion, (HEMI 150), t = 0 and 18 Microseconds

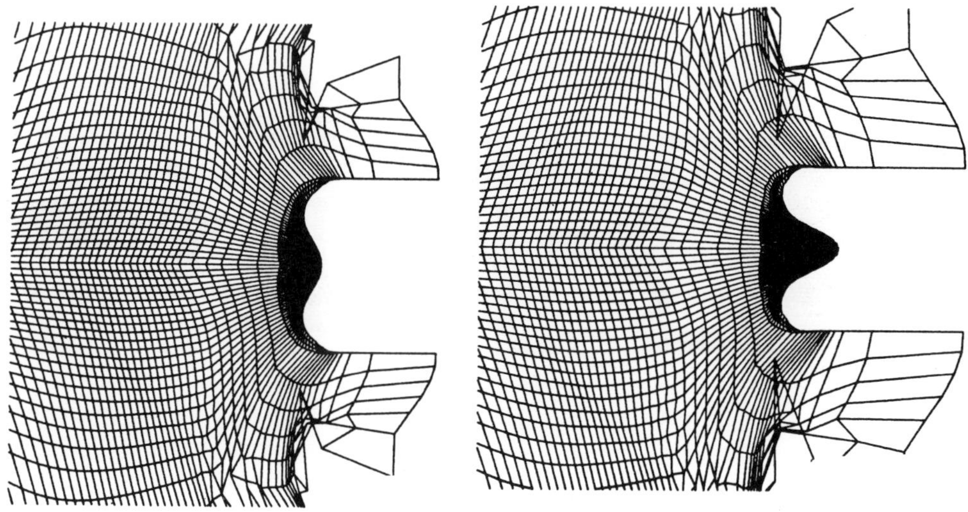

Figure 8. HEMP Code Liner Motion, (HEMI 150), t = 22 and 24 Microseconds

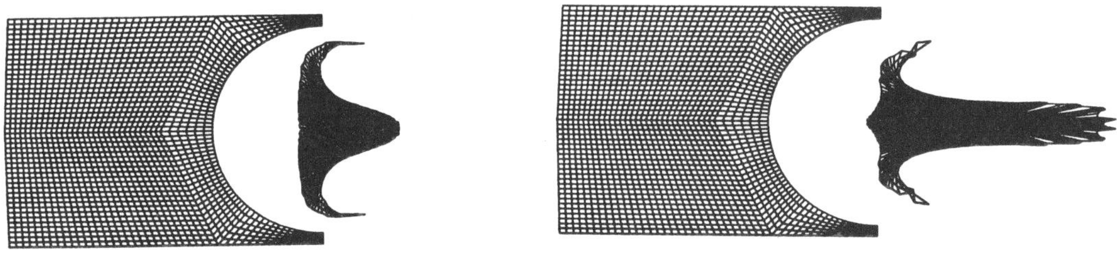

Figure 9. SOIL Code Liner Motion, (HEMI 150), t = 26 and 40 Microseconds

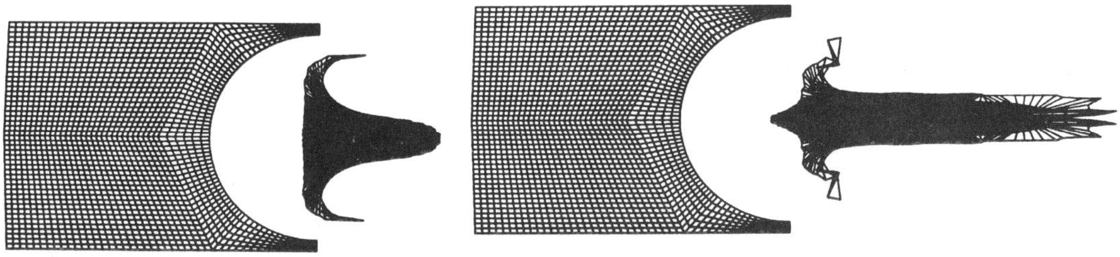

Figure 10. SOIL Code Liner Motion, (HEMI 150), t = 30 and 50 Microseconds

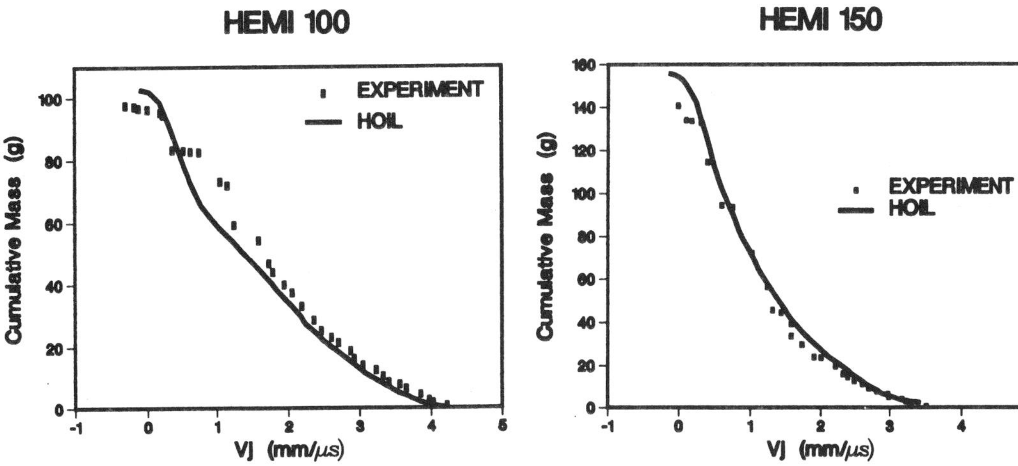

Figure 11. Comparison Between HOIL Calculations and Experiment of Cumulative Mass vs Jet Velocity

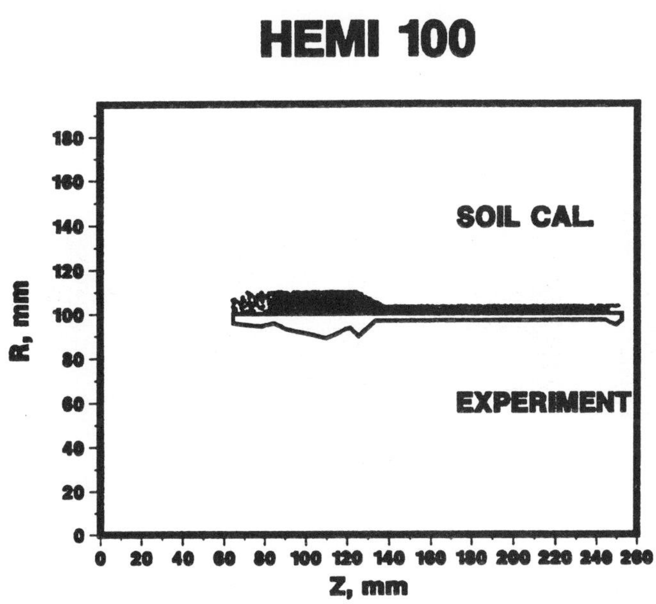

Figure 12. Comparison Between HOIL Calculation and Experimental Flash X-Ray at 60.5μs

Int. J. Impact Engng Vol. 5, pp. 307–322, 1987
Printed in Great Britain

0734-743X/87 \$3.00 + 0.00
Pergamon Journals Ltd.

Review of Hypervelocity Penetration Theories

Walter Herrmann† and James S. Wilbeck‡
†*Engineering Sciences Directorate, Sandia National Laboratories*
Albuquerque, New Mexico 87185
‡*Winzen International, San Antonio, Texas 78213*

Abstract

A review is given of the empirical and approximate theoretical expressions which have been developed to describe various aspects of impact at hypervelocities where the projectile and some of the target materials undergo massive plastic deformation, breakup, melting or vaporization. Various stages of the penetration process are identified on the basis of experimental evidence. Empirical fits to experimental data, or, at velocities above the experimental range, to results of finite difference calculations of the penetration event, are reviewed. In some cases simple theories, usually requiring the evaluation of some undetermined parameters from experimental or numerical data, have been developed. These are described, with emphasis on those which have found use in the design of offensive or defensive systems.

Introduction

An excellent understanding of the phenomenology of hypervelocity impact processes, based on experimental studies usually up to about 7.5 km/s, was obtained in the period spanning the previous hypervelocity impact symposia, from 1956 to 1969. This period also saw the development of detailed numerical solution methods for impact problems, which are reviewed elsewhere [*Anderson and Johnson (1986)*]. Constitutive equations governing the behavior of materials under the extreme conditions pertaining to hypervelocity impact have been developed intensively since that time, and are also reviewed elsewhere [*Asay and Kerley (1986)*], so that detailed numerical solutions may now be generated which are in reasonably good agreement with experiments. At least for some simple materials like metals, the numerical methods should provide reliable results at velocities above those obtainable in the laboratory. While many open questions remain, particularly for complex materials like organics and composites, it might seem that the field is relatively well in hand. However, numerical solutions are extremely expensive in terms of supercomputer resources, and it is impractical to generate the numerous solutions needed for design and optimization of either defensive or offensive systems.

It is very desirable, then, to have simple empirical or theoretical expressions to describe the impact process, for use in systems design and tradeoff studies. Unfortunately, the impact process inherently involves complex large multidimensional deformations, and the material responses are strongly nonlinear. For this reason, theoretical treatments short of the integration of the complete partial differential equations governing the process are very restricted to a few special cases, and usually involve a number of undetermined parameters which must be determined empirically. Such theories either appeal to gross momentum and energy arguments, or make sweeping simplifying assumptions regarding the form of the motion resulting from impact to reduce the dimensionality of the problem. However, when properly normalized, either to experimental data or, beyond the experimental range, to detailed numerical solutions, they and purely empirical fits can provide simple and useful expressions for use in systems studies within their ranges of validity.

It is the purpose of this paper to review the simple empirical and semi-empirical theories which have found use in this connection. We restrict attention to hypervelocity impacts. Without entering into controversial definitions of this regime, we imply that we consider only conditions under which the projectile material and some of the target material undergo disruption, that is, extensive plastic deformation, breakup, melting or vaporization. We specifically exclude conditions under which the projectile may be considered to be rigid, or related theories in which the projectile undergoes only slight deformation or erosion, applicable at lower velocities pertaining to **ballistic armor penetration.**

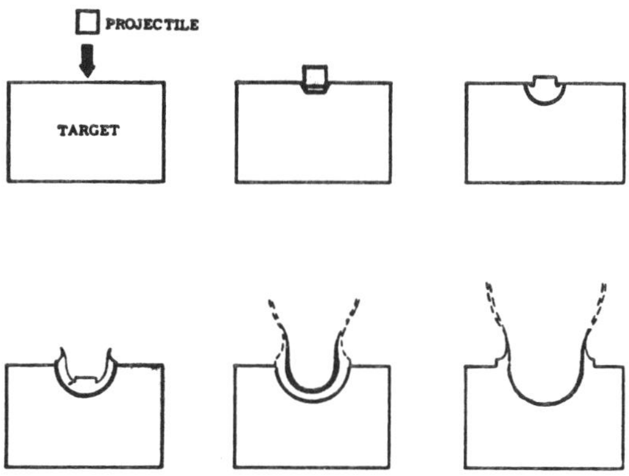

Fig. 1 Schematic of Cratering Process [*Eichelberger and Gehring (1962)*].

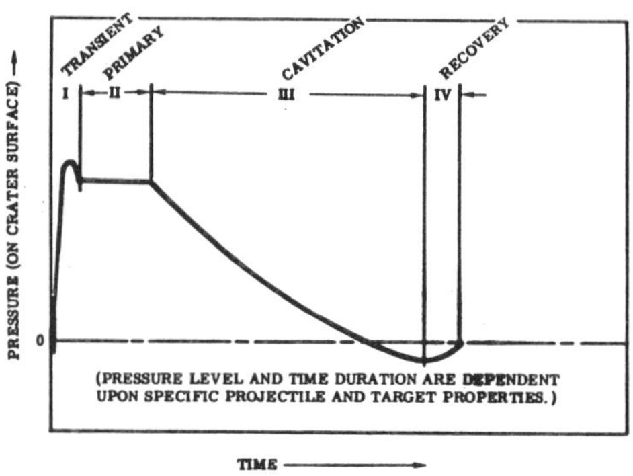

Fig. 2 Impact Pressures during Cratering [*Gehring, Meyers and Charest (1965)*].

The material may be conveniently divided into three categories, i) the target may be considered to be effectively semi-infinite, so that the cratering process is not affected by lateral or rear surfaces, ii) the target may be considered to be a plate of infinite lateral extent of a thickness of order of the ballistic limit, ie. the thickness at which the plate is barely penetrated, iii) the target is a plate of infinite lateral extent with thickness much less than the ballistic limit. While in this case the plate is easily penetrated, at hypervelocities it serves to disrupt the projectile, the resulting debris being potentially less lethal to underlying structure. For this reason the plate is generally referred to as a shield in this case. The projectile may be compact, with a length to diameter ratio near unity, it may be a long rod, with a length to diameter ratio much larger than unity, or it may be a flat disk, with a length to diameter ratio much less than unity. The first two projectile configurations are of greater interest than the third, which will not be discussed. In each case, the phenomenology will be reviewed briefly, followed by a discussion of the empirical and semi-empirical approximate theories that have been found to be useful. Synthesis of these separate expressions into systems level analyses of complex targets will not be considered.

Semi-Infinite Targets

During the early 1960s, an extensive analytical and experimental program was conducted to develop an understanding of hypervelocity cratering. During that period, a phenomenological model of this event was formulated by the hypervelocity community. The following is a brief summary [*Gehring, Meyers and Charest (1965)*]. A simple schematic of the cratering event is given in Fig. 1. The resulting pressure history generated at the interface between the projectile and target is shown schematically in Fig. 2. In this last figure, the pressure profile is seen to divide naturally into four separate regimes as follows.

Transient Shock Regime At the instant of impact, the particles on the impacting surface of the projectile are brought to rest relative to the particles on the impact surface of the target. Shock waves are formed at the interface which propagate into both the projectile and target. Initially, pressures in the shocked regions are given by the Hugoniot shock relations, and depend only upon the densities and compressibilities of the materials as well as the impact velocity. For hypervelocity impacts, the shock pressures are great enough to cause extensive plastic deformation, melting or vaporization. The duration of this transient shock regime is dependent upon the time required for pressure release (rarefaction) waves generated at the free surfaces of both the projectile and target to relieve the high shock pressures. This first regime is usually accompanied by a brief impact flash arising from the initial high pressures and temperatures.

Steady State Regime After several reflections of rarefaction waves in the projectile which release the very high initial pressures, a condition of steady flow is established, during which the projectile, which is being steadily eroded, is the causative force driving the crater formation. The duration of this regime is based upon the relation of the length of the projectile to its lateral dimensions, and for a short projectile may be entirely absent. Pressures in the projectile and target are approximated by Bernoulli's relation for fluid jets, and **depend primarily upon the material densities and the impact velocity.**

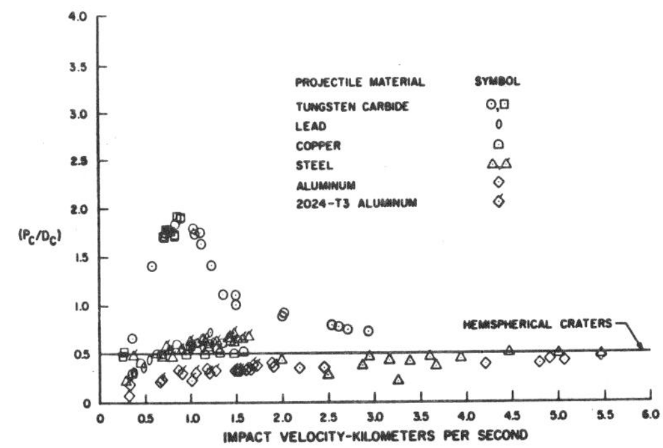

Fig. 3 Crater Depth to Diameter Ratio versus Velocity for Impacts into Copper Targets [*Bruce (1961)*].

Cavitation Regime After the projectile has been completely eroded the crater continues to expand under its own inertia. This third regime continues until the energy density of the material surrounding the crater becomes too small to overcome the intrinsic resistance to deformation of the material.

Recovery Regime After the crater has reached its maximum dimensions, the final crater dimensions may diminish slightly due to elastic or anelastic rebound. The rebound may produce tensile stresses which may cause spallation fracturing. Finally, the high temperatures to which the crater material has been subjected may cause recrystallization of material in an area beneath the crater.

Compact Projectiles

For compact projectiles with a length to diameter ratio near unity, the projectile is entirely eroded during the first phase, which establishes the momentum field in the target for the subsequent cavitation phase which dominates the cavity growth process. The motion in both the transient and cavitation regimes is inherently complex. For this reason, simplified theoretical treatments have not been found to be useful for this problem.

During the early 1960s, extensive experimental studies established most of the features associated with cratering by compact projectiles in thick effectively semi-infinite targets. It was generally found, at sufficiently high velocities, that the crater shape was unaffected by the projectile shape as long as the length to diameter ratio was near unity, and that the crater depth was the same if normalized by the diameter of a sphere of the same volume.

At relatively low impact velocities, it was observed that, if the projectile has a higher density or strength than the target, then the crater depth tends to be greater than its radius at the original surface. Conversely, if the projectile has a lower density or strength than the target, then the crater tends to be shallow. A number of experimental studies have found that the crater approaches a hemispherical shape as the projectile velocity increases, and the attainment of a hemispherical crater has sometimes been used as a criterion for the definition of the hypervelocity regime. A typical example is shown in Fig. 3. If the crater is indeed hemispherical, then $V_c/V_p = 4(p/d)^3$, where V_c and V_p are the volumes of the crater and projectile, and p and d are the depth of the crater and the equivalent diameter of the projectile. However, some numerical solutions of cratering have indicated that high density projectiles continue to produce deep craters even at very high velocities.

A number of studies have suggested that the volume of the crater becomes proportional to the kinetic energy of the projectile at sufficiently high velocities. A typical example is shown in Fig. 4. Several studies have further suggested that the proportionality constant is inversely proportional to a measure of the strength of the target material. An example is shown in Fig. 5, where the Brinell hardness is used as a convenient characterization of strength. Denoting the target strength by Y_t, all of the above observations together would imply $V_p/V_c = 4(p/d)^3 \propto \rho_p v^2/Y_t$, where ρ_p is the density and v the velocity of the projectile. It is convenient to nondimensionalize these relationships by introducing the ratios $\rho^* = \rho_p/\rho_t$ and $v_y^* = \sqrt{\rho_t v^2/Y_t}$ where subscript p refers to the projectile, and t refers to the target material.* The crater volume and depth can then be written in the general forms

* Note that ρ_t has been used in v_y^*, in contrast to usual practice, to separate the dependencies on ρ_p and v. The nondimensional ratio v_y^* then can be interpreted as the ratio of energy per unit mass of projectile to energy required to deform unit mass of target material.

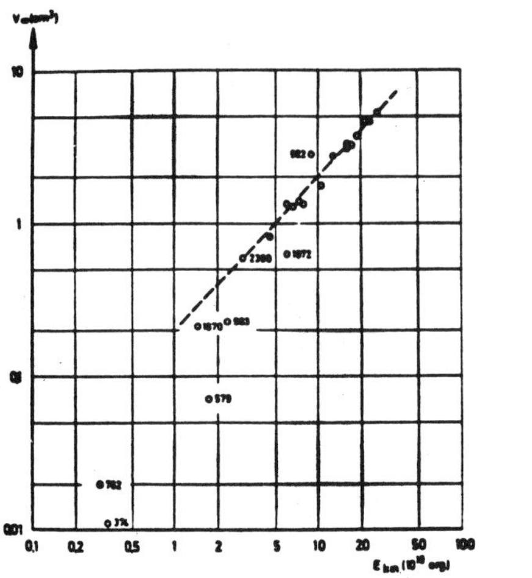

Fig. 4 Crater Volume versus Projectile Kinetic Energy for the Impact of Steel Spheres into Steel Targets [*Schneider (1979)*].

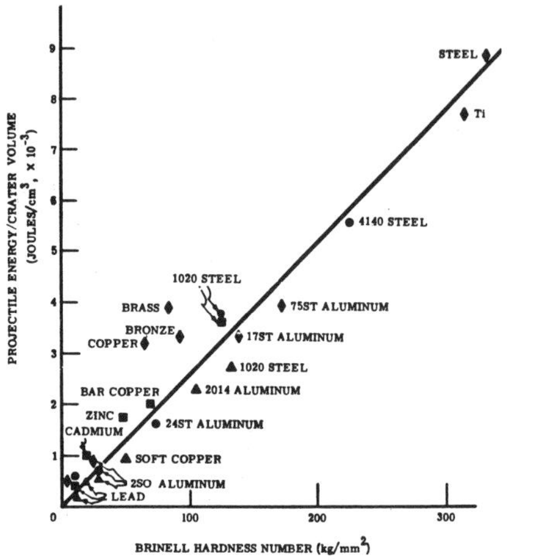

Fig. 5 Projectile Energy per Unit Crater Volume versus Target Strength [*Feldman (1960)*].

$$\frac{p}{d} = K_1 (\rho^*)^m (v^*)^n \tag{1}$$

$$\frac{V_c}{V_p} = K_2 (\rho^*)^r (v^*)^s \tag{2}$$

where $m = 1/3$, $n = 2/3$, $r = 1$, $s = 2$ if the previous observations are valid. While many researchers have used Brinell hardness as a characterization of the target properties, a few have preferred to use the sound speed c_t. In this case, the nondimensional ratio $v_c^* = v/c_t$ may be used in (1) and (2).

Extensive studies have been undertaken to develop empirical expressions for predicting crater depth and crater volume for the hypervelocity impact of compact projectiles, either by fitting experimentally measured or numerically predicted crater dimensions. It is convenient to recast the resulting empirical expressions in the forms (1) and (2) above for comparison. The values of m, n, r and s found by a number of researchers are given in Table 1 and 2, together with an identification of the form used for the nondimensional velocity v^* and whether the data which were fitted were derived from experiments or from detailed numerical solutions of the penetration event.

It may be seen that most of the empirical fits for projectile velocity tend to support dependence of crater size on kinetic energy and the attainment of a hemispherical crater, that is $n = 2/3$, $s = 2$. However, some of the fits to numerical solutions for the crater depth provide velocity exponents between 2/3 and 1/3, the last of which which would be attained if the craters size were assumed to be proportional to the momentum of the projectile. An explanation of why a range of exponents may be obtained for different cases has been proposed on the basis of scaling arguments [*Holsapple (1986)*]. There is greater variety in the fits for ρ_p, probably because of the more restricted variation of projectile density in many of the data set which were used by different researchers. Most of the fits are not seriously in disagreement with the values $m = 1/3$, $r = 2/3$ suggested above.

There is also more disagreement regarding the dependence of crater size on target properties, some researchers preferring the use of v_y^* while others use v_c^*. Experiments using different strength alloys or different temperatures of the same target material, all other factors being the same, clearly indicate a dependence of crater size on target strength up to the highest experimental velocities [*Halperson (1964), Piacesi, Waser and Dawson (1964)*]. It may be expected that the crater size will be influenced by target strength at any velocity for compact projectiles striking semi-infinite targets, since the cratering event is dominated by the cavitation regime where the inertial crater growth process is resisted by the strength of the material. Evidence of the dependence of crater sizes on the sound speed of the target material is much less convinving, empirical fits using v_c^* generally having been made to restricted sets of data.

Table 1

Empirical Penetration Relations

Authors	m	n	v^*	Source
Summers and Charters (1958)	2/3	2/3	v_c^*	exp
Charters and Summers (1959)	2/3	2/3	v_y^*	exp
Herrmann and Jones (1961)	2/3	2/3	v_y^*	exp
Bruce (1961)	1/2	2/3	v_c^*	exp
Eichelberger and Gehring (1962)	1/3	2/3	v_y^*	exp
Loeffler et al (1963)	1/2	2/3	v_c^*	exp
Christman and Gehring (1966)	2/3	2/3	v_y^*	exp
Walsh and Johnson (1964)	1/3	0.58	v_c^*	num
Riney and Heyda(1964)	1/3	2/3	(1)	num
Sedgwick (1980) (2)	0.537	0.576	v_c^*	num

(1) Not nondimensionalized

(2) Sedgewick introduces another term $(\rho_t c_t^2/Y_t)^{0.235}$

Table 2

Empirical Crater Volume Relations

Authors	r	n	v^*	Source
Herrmann and Jones (1961)	3/2	2	v_y^*	exp
Bruce (1961)	2/3	2	v_c^*	exp
Eichelberger and Gehring (1962)	1	2	v_y^*	exp

A number of other empirical fits have been obtained. For example, a logarithmic expression has been found to fit data for both low and high velocity impacts by ductile projectiles [*Herrmann and Jones (1961)*]. Bjork has found that results of finite difference calculations of impacts in aluminum to very high velocities could be fitted by an exponential expression in the velocity [*Bjork (1963)*]. When taken together with a scaling law for dissimilar material impacts [*Bjork and Olshaker (1961)*], this result provides another method of estimating crater sizes in semi-infinite targets. Both of these fits predict lower penetration at very high velocities than the power laws discussed above. As has frequently been pointed out, extrapolation of empirical fits is uncertain, and resolution of such uncertainties must await experimental data in the velocity range of interest.

Impact Obliquity

In real environments, normal impacts are the exception. A number of experimental studies have been conducted to evaluate oblique impacts in effectively semi-infinite targets [*Summers and Charters (1958), Summers (1959), Kineke (1960), Bruce (1961), Gehring and Lieblein (1963), and Nysmith and Summers (1962)*]. Typical results are shown in Fig. 6, where the penetration depth normalized by the projectile diameter is plotted versus the component of the projectile velocity normal to the surface for various angles of obliquity, together with results for normal impacts. The crater which is formed continues to be roughly hemispherical in the hypervelocity range, and the data correspond well with those for normal impacts. As the normal component of the velocity decreases with increasing obliquity, the data depart from those for normal impacts, and the crater becomes elongated. However, it appears that an increase in projectile velocity at constant obliquity tends to restore the hemispherical crater shape. Insufficient data exist at sufficiently high impact velocities to determine the limits of this behavior at shallow angles.

Long Rod Projectiles

For a long rod projectile, the penetration process is dominated by the steady state regime. At hypervelocities, the strength of the materials has been neglected, and the process has been assumed to be similar to that of a fluid jet striking a fluid half space. The theory developed for shaped charge jets has been applied with some success

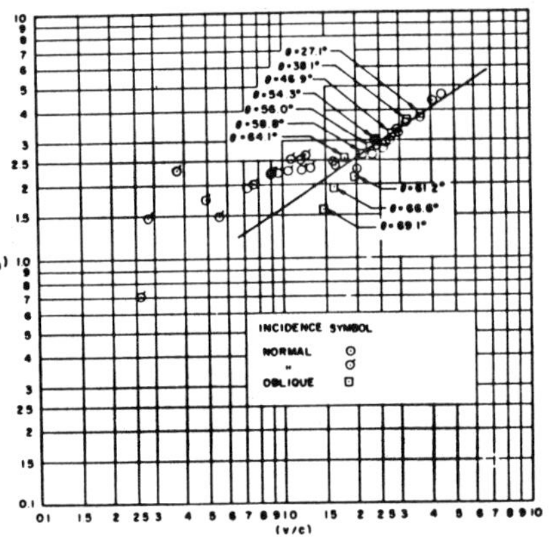

Fig. 6 Penetration Versus Normal Velocity Component for Steel Impacts into Lead Targets [*Bruce (1961)*].

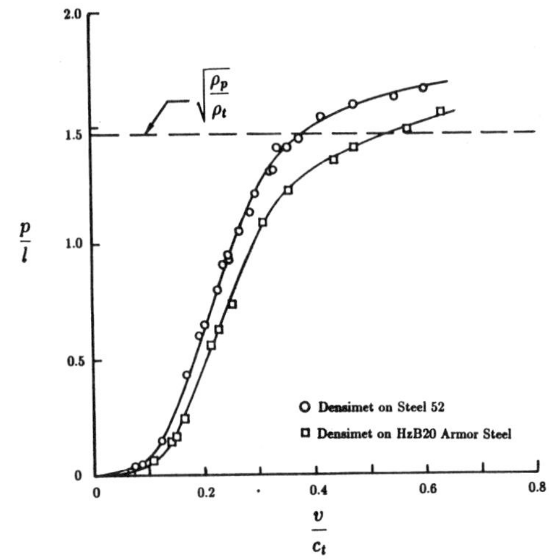

Fig. 7 Depth of Penetration of a Long Rod, Steel into Steel [*Hohler and Stilp (1977)*].

[*Birkhoff, McDougal, Pugh and Taylor (1948)*]. It is assumed that the jet, with initial velocity v_0, impinges on the fluid half space. The stagnation point between jet and target recedes with velocity v_i. The pressure P at the stagnation point in the projectile is given by Bernoulli's relation

$$\int_0^P (1/\rho_p)\, dP' + \int_{v_0 - v_i}^{0} v\, dv = 0 \tag{3}$$

If we assume incompressible flow, the integral is

$$P = \frac{1}{2}\rho_p (v_0 - v_i)^2 = \frac{1}{2}\rho_t v_i^2 \tag{4}$$

where the second equation follows by considering the pressure at the stagnation point in the target. If it is assumed both that steady state is reached almost immediately, and that penetration begins when the first particle of the rod impacts the target and stops when the last particle strikes the target, the duration of impact is given by $t = l/(v_0 - v_i)$ where l is the undeformed rod length. The depth of penetration p is the distance that the interface has moved into the target $p = v_i t$. Solving these equations

$$\frac{p}{l} = \sqrt{\frac{\rho_p}{\rho_t}} \tag{5}$$

Equation (5) is the classic jet penetration equation which relates depth of penetration to the projectile length and the projectile and target densities, but not the impact velocity. Experimental data show that the penetration normalized by the rod length increases with projectile velocity, tending to level off at higher velocities in qualitative agreement with the jet penetration equation, which is expected to hold only at very high velocities since it neglects material strength, as seen for example in Fig. 7 [*Hohler and Stilp (1977)*]. However, the penetration tends to exceed predictions at high velocities.

At lower velocities the depths of penetration predicted by (5) are incorrect because the model fails to account for target strength. Attempts to include target strength have been based on a generalization of Bernoulli's equation [*Eichelberger (1956), Allen and Rogers (1961)*] of the form

$$\frac{1}{2}\rho_p (v_0 - v_i)^2 = \frac{1}{2}\rho_t v_i^2 + Y \tag{6}$$

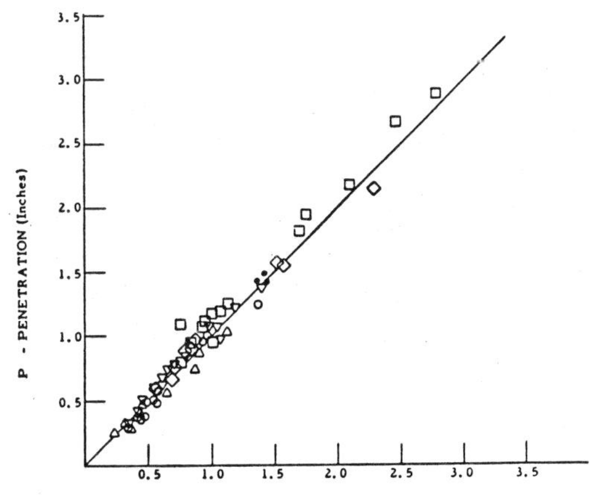

Fig. 8 Long Rod Penetration Comparison with Theory [*Christman and Gehring (1965)*].

where Y is considered to be a dynamic yield strength of the target material. Solution gives an expression for the predicted depth of penetration of the form

$$\frac{p}{l}=\frac{-\rho_p v_0+\left[\rho_p^2 v_0^2-2\left(\rho_p-\rho_t\right)\left(\frac{1}{2}\rho_p v_0^2-Y\right)\right]^{1/2}}{\rho_t v_0-\left[\rho_p^2 v_0^2-2\left(\rho_p-\rho_t\right)\left(\frac{1}{2}\rho_p v_0^2-Y\right)\right]^{1/2}} \tag{7}$$

At high velocities, the effect of target strength becomes negligible and (7) reduces to (5).

The theory leading to (7) takes no account of the strength of the rod. In fact, stresses which are supported by the rod material account for the observed deceration of the rod during the impact process. The theory has been further modified to include a strength term for the rod [*Alekseevski (1966), Tate (1967, 1969)*]

$$\frac{1}{2}\rho_p\left(v_0-v_i\right)^2+Y_p=\frac{1}{2}\rho_t\left(v_i\right)^2+Y_t \tag{8}$$

In the actual impact event, compressive stress waves propagate down the rod, thereby decelerating the rod throughout the process. If this deceleration is ignored, then (8) reduces to (6) with $Y=Y_t-Y_p$, and the solution is given by (7). However, both Tate and Alekseevski took rod deceleration into account by solving the equation of motion of the rod

$$\frac{dv}{dt}=-\frac{Y_p}{\rho_p x} \tag{9}$$

where the resisting stress is taken equal to the projectile strength and the uneroded portion of the rod of length x is treated as a rigid body. The depth of penetration is obtained by solving the relation

$$p=\int_0^t v_i\,dt=\frac{\rho_p}{Y_p}\int_v^{v_0} v_i x\,dv \tag{10}$$

which is integrated numerically, once the expression for the remaining projectile length, x, has been obtained by solving (9) in terms of v_i. Unfortunately, it is not clear what values of yield strengths should be used in these equations, since yield strength is generally expected to be a function of the local strain rate. Various expressions have been suggested.

At high velocities, the underprediction of penetration by the jet equation has been attributed to the neglect of the cavitation regime [*Christman and Gehring (1965)*]. They assumed that the crater continued to be enlarged during

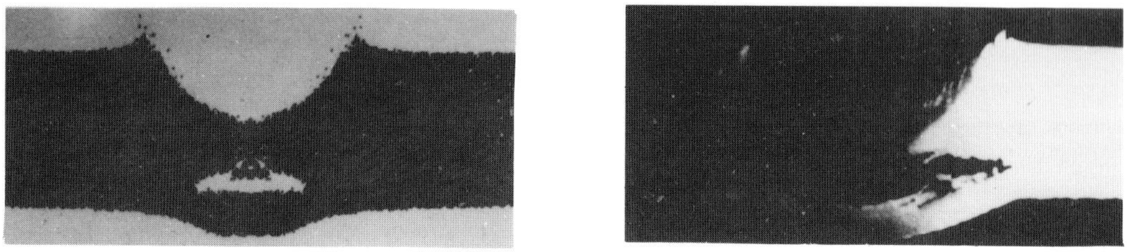

Fig. 9 Impact of a Nylon Ball on Steel Plate at 5 km/sec, left: Numerical Solution, right: Post-impact Photograph of Etched Target [*Bertholf et al (1975)*]

the cavitation phase and moreover that the cavitation phase was identical with that due to a cylindrical projectile of the same diameter d as that of the rod. Returning to (5), they wrote

$$p = (l - d)\sqrt{\frac{\rho_p}{\rho_t}} + p_c \tag{11}$$

where the first term on the right represents the penetration of a rod of length $(l - d)$, and p_c is the crater depth due to a cylindrical projectile of $l/d = 1$. Using their relation from Table 1, they obtain good correlation with data over a wide range of impact velocities, as seen in Fig. 8. One advantage is that the resultant penetration relation reduces to that for a compact projectile as the rod length approaches its diameter. The same modification can be made to expressions derived from (6) or (8) to provide expressions which should apply over a wide range of projectile velocities.

A number of other researchers have developed analytical models of rod penetration to include compressibility [*Haugstad (1981)*] which has been found to alter predictions by only a small amount, or to incorporate material strength by other means [*Perez (1978), Walters and Majerus (1978)*].

Ballistic Limit Penetration

The phenomena associated with the penetration of a target plate that is of finite thickness, but still thicker than the length of the projectile, are initially the same as those associated with impact on a semi-infinite target, described in the last section. The spherical shock wave propagating into the target from the impact point will reach the rear surface first. The reflection of the shock from the free surface as a release wave, and the development of tensile stresses, possibly leading to spall fracturing, are well known [*Chou, Sidhu and Zajac (1965), Kinslow (1969)*]. The initial reflection on axis is often approximated as the reflection of a plane wave, although this is not strictly correct. Since the shock is followed by a rarefaction, its reflection produces tensile stresses which increase with depth below the surface. If the surface is sufficiently close to the impact point, then the tensile stress may reach the dynamic fracture strength of the material, and a spall fracture is initiated. As the spherical wave progressively reflects from the surface at points away from the axis, its strength, and hence the tensile stresses it produces, decay to the point where the material no longer fractures, and the spall fracture will be arrested. The circular disk of material fractured from the target is still attached around its periphery, but carries entrapped momentum, which may be sufficient to tear the disk free. If the spherical wave is of sufficient amplitude, further reflection may take place at the newly formed spall plane, and a second spall fracture may occur.

The proximity of the rear free surface of the target may also affect the growth of the crater, in that the arrival of the release wave reflected from the rear surface assists the crater growth process. For progressively thinner target plates, both the spall process and the crater growth process are intensified, until the material between them is ruptured, and a through hole is formed. Since fracture is a random process depending on the presence of random flaws in the target material, some variation from one shot to the next is usually experienced in the details of the process. The ballistic limit of a particular projectile and target combination is therefore often defined as the projectile velocity at which 50% of the targets are penetrated.

The penetration process is very complex, necessarily so, since it depends critically on the relatively low stress plastic deformation and fracture phenomena near the rear surface of the target. In fact the description given above is highly

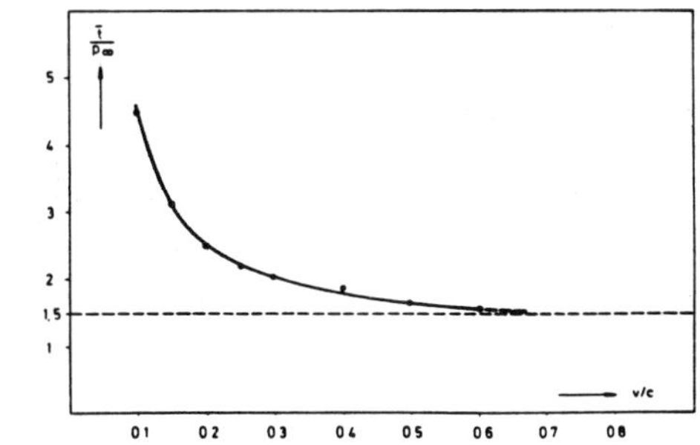

Fig. 10 Ratio of Ballistic Limit Thickness to Semi-infinite Target Penetration Depth versus Velocity for Steel on Steel Impact [*Hohler, Schneider and Stilp (1975)*].

simplified. Shown in Fig. 9 is a photograph of a sectioned crater in a plate of rolled homogeneous armor which was impacted by a nylon ball at 15,000 ft/sec, together with a mass-density plot of an Eulerian finite difference calculation [*Bertholf et al (1975)*]. The formation of a spall fracture is clearly shown. The spall layer has not torn loose from the target, but carried enough momentum to have been plastically deformed. The target exhibits a variety of other fracture phenomena. The photograph shows that the crater is surrounded by a multitude of shear bands following maximum shear trajectories which have initiated shear fractures which undoubtedly weakened the material during the crater formation. These shear fractures have intersected near the bottom of the crater to detach a conical plug of material, so that the crater bottom is conical, not spherical. Moreover, there is a curious cylindrical fracture between the crater and the spall forming a hole between the two. This fracture was also found to be predicted by the finite difference calculation only when the pressure induced polymorphic phase transformation of the iron target was included. In fact, the photograph shows material which has undergone the phase transition as darkly stained material surrounding the crater. The phase transition qualitatively alters the wave motion to produce the cylindrical fracture.

It is not surprising that no reasonable simplified analytical descriptions of plate penetration near the ballistic limit have emerged. In fact, this problem remains as a very challenging one for numerical solution techniques, demanding detailed descriptions of plastic flow and fracture with accompanying accurate material property determinations at the appropriate strain rates. For this reason, it is necessary to appeal to purely empirical relations for simple descriptions.

Many observations of ballistic limit tests have been reported, some of these at sufficiently high velocities to be considered to be in the hypervelocity regime. Typical results suggest that the ballistic limit thickness approaches 1.5 times the crater depth in a semi-infinite target for the same projectile-target combination as the velocity increases Fig. 10.

Thin Target Perforation

Interest in thin target penetration stems from the suggestion that a hypervelocity projectile could be disrupted (fractured, melted or vaporized) by a thin plate of material [*Whipple (1947)*]. The debris emanating from the rear of the plate would be finely divided and diverge radially, producing a distributed load on underlying structure that might be much less lethal than the original projectile. Such "Whipple Meteor Bumpers" have been used extensively as shields to protect vehicles in space, and have been considered for protection from other hypervelocity projectile threats.

The qualitative phenomena related to the hypervelocity penetration of target plates that are thin relative to the dimensions of the projectile are relatively well understood, at least in the range of velocities amenable to laboratory experimentation [*Maiden and McMillan (1964), Gehring (1970), Swift (1982)*].

Early Time Phenomena

The initial phenomena have been well described, see Fig. 11, which shows a schematic of a cylindrical projectile striking a flat shield. At the instant of contact, shock waves S_1 and S_2 are driven into the projectile and shield, respectively. The material between the shocks moves at half the projectile velocity if the projectile and shield

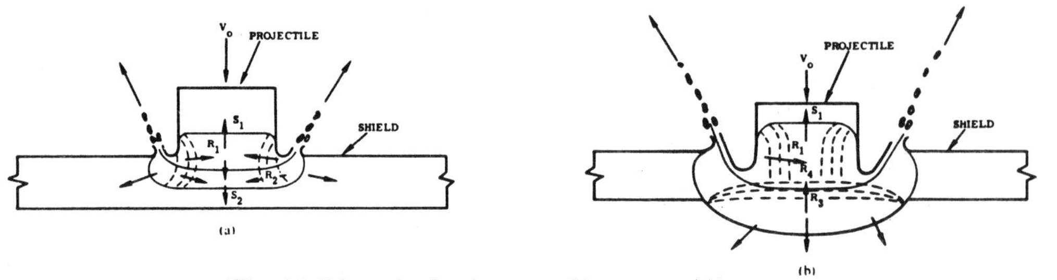

Fig. 11 Schematic of early stages of impact on a thin target a) shortly after impact; b) after reflection of the shock in the target [*Maiden and McMillan (1964)*].

are made of the same material. Initial shock wave speeds and the conditions behind the shocks may be readily calculated by one dimensional shock wave theory [*Zwarts (1965), Bjork and Olshaker (1965)*] for like or unlike material impacts, given the equations of state of the materials.

Lateral rarefactions R_1 and R_2 emanate from the unconstrained free surfaces of the projectile and shield, reducing the stresses and inducing lateral velocities in the materials. If the proportions are such that the shock S_2 reaches the rear surface of the shield before these lateral rarefactions have converged on the centerline, then it reflects as a rarefaction R_3, as shown. Properties of isentropic rarefactions of this type may also be calculated from the equation of state of the material. Since the shock compression process is dissipative, the temperature on rarefaction to zero pressure is higher than the initial temperature. For strong shocks involved in hypervelocity impact, the residual temperatures may be sufficiently high to weaken, melt or even vaporize the released material.*

The initial pressure and temperature, and the residual temperature, depend on the impedances (product of density and sound speed) of the projectile and shield materials. As the impedance of the shield material is reduced, the pressures and residual temperatures may no longer be sufficient to ensure disruption of the projectile, thereby increasing the damage potential of the resulting fragments [*Swift and Hopkins (1968)*]. At sufficiently high velocities, of course, pressures and residual temperatures are high enough to melt or vaporize all realistic shield materials.

Rarefactions are dispersive, with the head travelling into the compressed material at the local sound speed, which is greater than the shock speed to the same compressed state. Thus, R_3 will tend to overtake shock S_1 in the projectile. If this occurs before S_1 has reached the end of the projectile, R_3 will progressively weaken S_1, eventually to the point where the projectile material is no longer disrupted. If the shield is too thin, some of the projectile material may emerge as relatively large solid fragments from the encounter with the shield, with potential to penetrate subsequent structure. A criterion for an effective shield thickness, therefore, may be based on a one dimensional calculation of the overtaking of shock S_1 by rarefaction R_3, conservatively taking the point at which R_3 just overtakes S_1, or more realistically, estimating the attenuation of S_1 by R_3 to a stress where the material is no longer disrupted on release to zero pressure [*Maiden and McMillan (1964)*].

The lateral rarefactions R_1 and R_2 complicate the issue. Since the shock and sound speeds can be calculated, the point at which the heads of the rarefactions R_1 and R_2 meet on the axis at the shock S_1 can be calculated, in the absence of overtaking by rarefaction R_3. These lateral rarefactions will progressively weaken the shock also, and contribute to conditions under which solid fragments of projectile material may survive the penetration of the shield.

On the other hand, if the shield is too thick, then shock S_1 may reflect from the end of the projectile, and overtake the shock in the shield S_2. Alternately, the side rarefactions R_1 and R_2 may attenuate the shock in the shield before it reaches the rear surface. Both factors may reduce the pressures in the shield material near the impact so that disruption may be incomplete, and solid fragments of shield material may be ejected at high forward velocities which may be lethal to underlying structure. For a given threat, it is clear that there exists an optimum

* The calculation of temperature is particularly uncertain, since most equations of state are based on mechanical measurements along Hugoniots, and the temperature is estimated from theoretical arguments [*McQueen, Marsh et al (1970)*]. It is more direct to calculate internal energies, and to compare residual energies with melting and sublimation energies, but the same uncertainties arise in integrating along the release isentrope. Most estimates are based on very simple equations of state. However, care must be taken to use a realistic equation of state when phase changes are present [*Kerley (1986)*].

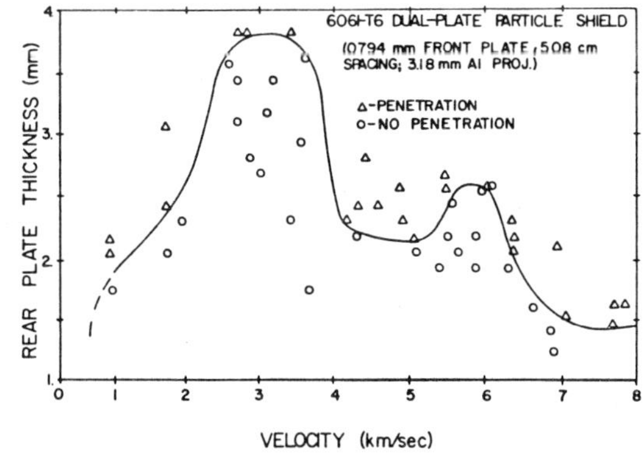

Fig. 12 Damage Potential of the Debris Cloud Behind Thin Targets as a Function of Velocity, Measured by the Ballistic Limit of a Witness Plate [*Swift, Preonas, Dueweke and Bertke (1970)*].

shield thickness, lesser or greater thicknesses offering less protection. Of course, as the shield thickness is increased further, the ballistic limit of the shield is approached, and the phenomena discussed in the previous section pertain.

The above description has been given for a cylindrical projectile. The situation will be more complex for other shapes. However, both experiments and finite difference code calculations suggest that the projectile shape has little effect on the gross phenomena, as long as the aspect ratio remains near unity.

Debris Cloud Formation

Radiographs of the debris issuing from the rear of a thin plate target subjected to hypervelocity impact show that projectile and target material is distributed mainly in a thin shell on the surface of a growing elongated bubble attached to the lip of the hole forming in the target plate [*Swift, Preonas et al (1970)*]. Radiographic evidence, as well as observations of the damage produced in a witness plate placed some distance behind the thin plate target, show that the material in the debris cloud is fractured into small solid fragments, or is melted or vaporized in qualitative agreement with expectations based on residual temperatures calculated from one dimensional shock wave theory described above [*McMillan (1968), Swift and Hopkins (1968), Gehring (1970)*]. At relatively low velocities, the debris is initially in the form of solid fragments. As the velocity is increased and hence as the peak impact pressure and the residual temperature after release of pressure increase, the fragment size decreases, although the damage potential of the debris appears to increase. With further increase of velocity, conditions are approached where the residual temperature is sufficiently high to induce melting, at which point the damage potential appears to decrease abruptly. The melted material is broken up into droplets, the droplet size decreasing as the impact velocity increases, although the damage potential again appears to increase. Finally, the residual temperature becomes sufficient to vaporize the material in the debris cloud, with another drop in damage potential. The damage potential of the debris cloud, as measured, for example, by the ballistic limit of a witness plate placed behind the shield, tends to show several peaks related to the fragmentation and changes of state of the materials, Fig. 12.

The damage potential of the debris cloud depends on the stand off distance between the shield plate and underlying substructure as well as the state of the materials in the cloud. Solid particles will produce small impact craters on the underlying structure, and may penetrate if the substructure is thin. The intensity of damage will reduce with distance of stand off only until the craters are sufficiently separated so that synergistic effects do not occur between adjacent craters. The same will be true of liquid droplets, except that the droplets may subdivide further with distance of flight if the region between the shield and substructure is filled with air. Theories of fragmentation [*Grady (1986)*] hold promise of predicting the fragment or droplet size distribution in an expanding debris cloud of this type, given the velocity distribution in the debris material at the time that it reaches zero pressure as it issues from the rear of the shield. Of course, vaporized material will produce no cratering in the substructure. At very high velocities, it is expected that the vaporized material may be dissociated and ionized, and that radiation from the very hot plasma may begin to play a role in the damage it may inflict on underlying structure [***Swift, Bamford***

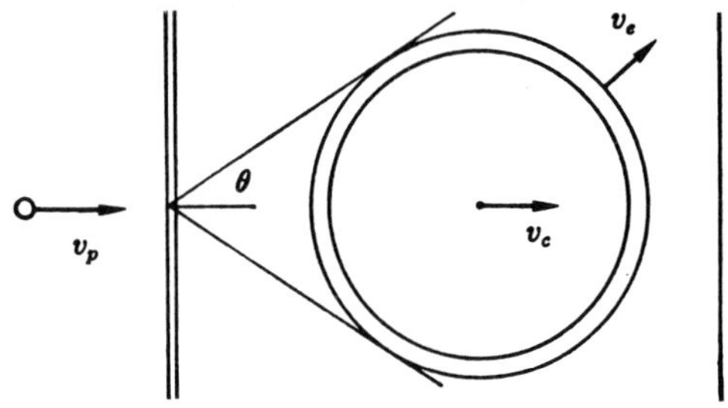

Fig. 13 Schematic of Debris Cloud Behind a Thin Target [*Swift, Bamford and Chen (1982)*].

and Chen (1982), Spillman (1986)]. In all cases, increasing the standoff will distribute the momentum imparted to the substructure over a wider area.

An elegant approximate theory [*Richardson (1969), Swift, Bamford and Chen (1982)*] has been obtained from simple momentum and energy arguments to estimate the characteristics of the debris cloud, which will be summarized here. They assumed that the projectile and a portion of the shield subjected to high pressures is disrupted, the finely divided material being uniformly distributed in a thin spherical shell, see Fig. 13. The forward momentum of the projectile is distributed to the debris material behind the shield after impact, neglecting any momentum transfer to the remainder of the shield, and any momentum carried by debris emanating from the front of the shield. The forward velocity of the center of mass of the debris v_c can be calculated from momentum conservation

$$v_c = \frac{v_p}{1 + KG^2} \tag{12}$$

where v_p is the original velocity of the projectile, K is ratio of mass per unit frontal area of the shield to that of the projectile, and G is the ratio of the radius of the shield material contributing to the debris to the radius of the projectile. If the kinetic energy of the projectile is denoted E_p, and the kinetic energy associated with the forward momentum of the center of mass of the debris cloud is denoted E_c, then there is an excess energy $E_e = E_p - E_c$ given by

$$E_e = \frac{KG^2}{1 + KG^2} E_p \tag{13}$$

This energy goes into heating, fracturing, melting or vaporization of the debris material, and into expansion of the debris cloud about its center of mass. Assume that a fraction Q of the energy E_e appears as kinetic energy of expansion, and that the debris material is uniformly distributed on the surface of a hollow sphere expanding about its center of mass. Then the expansion velocity v_e is readily calculated from the available excess energy as

$$v_e = \frac{G\sqrt{QK}}{1 + KG^2} v_p \tag{14}$$

The half-angle θ subtended by the debris cloud at the center of the shield is simply $\sin\theta = v_e/v_c$ or

$$\theta = \arcsin\left(G\sqrt{QK}\right) \tag{15}$$

The forward momentum of the debris cloud may load subsequent structure. It varies from a maximum on axis to a minimum at the edges defined by the half angle θ. The peak forward velocity of the debris material at the leading edge of the debris cloud is simply $V_{max} = V_c + V_e$. The momentum per unit area associated with this material is

$$P_m = \frac{m_p v_p \left(1 + G\sqrt{QK}\right)^3}{4\pi h^2 QKG^2} \tag{16}$$

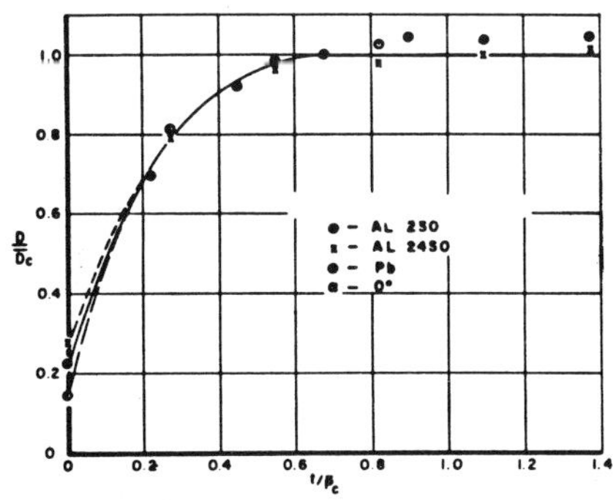

Fig. 14 Entrance Diameter Normalized to the Diameter of a Crater in a Semi-Infinite Target versus Target Thickness Normalized to Penetration in a Semi-Infinite Target [*Vitali, Becker and Watson (1961)*].

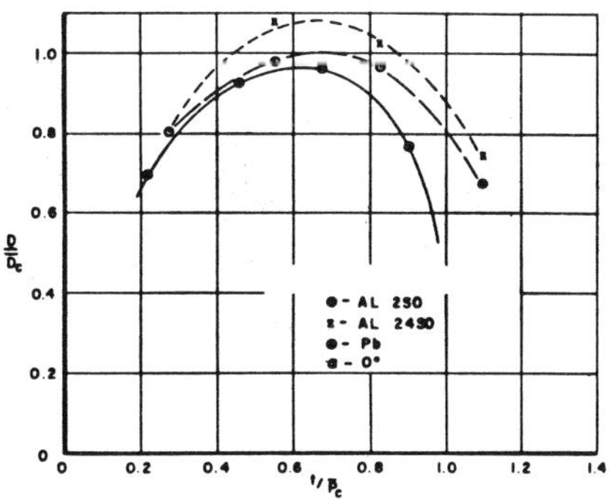

Fig. 15 Exit Diameter Normalized to the Diameter of a Crater in a Semi-Infinite Target versus Target Thickness Normalized to Penetration in a Semi-Infinite Target [*Vitali, Becker and Watson (1961)*].

where m_p is the mass of the projectile and h is the spacing between the shield and substructure. This equation shows that the damage potential of the debris depends on the areal density of the shield through K, not on its mass density, as observed experimentally [*Swift and Hopkins (1968)*].

The theory has two undetermined parameters, G, the ratio of radius of shield material contributing to the debris to the radius of the projectile, and Q, the ratio of energy available for debris expansion to the available energy. Both can be expected to be functions of all the parameters of the problem, namely, the projectile velocity and the materials and relative sizes of the projectile and shield, as indeed will the range of applicability, if any, of the theory. The factor G is assumed to be near unity on the basis of experimental evidence to be cited in the next subsection. Lawrence assumes that, at very high velocities, the only significant energy loss is due to vaporization of the materials, and proceeds to investigate the consequences of the theory in some detail [*Lawrence (1986)*].

Perhaps the most suspect assumption of the theory is that the material is distributed uniformly over the surface of the spherical debris cloud. At least in the range of velocities reachable by experiments, the experimental evidence suggests that more material is concentrated near the axis at the front of the cloud, while the cloud elongates and remains attached to the lip forming in the shield. Finite difference calculations at the same velocities show the same qualitative phenomena, although the codes predict that the debris near the axis of the leading edge of the cloud is too thick, but carries too little momentum.* [*Rosenblatt, Kreyenhagen and Romine (1968), Swift, Preonas, Turpin and Carson (1970), Wagner and Kreyenhagen (1979)*]. Whether or not the debris cloud will become more uniformly distributed at higher velocities is not known. Indications from code calculations are that they continue to predict a concentration of material near the axis.

Hole Formation

Experiments show that the high velocity material in the debris cloud emanates from an area of the shield not much larger than the projectile [*Swift, Preonas, Turpin and Carson (1970)*]. However, as time proceeds, the disturbance propagating into the shield continues to cause the shield material to flow radially, with concomitant plastic thickening. Material continues to be ejected from both the front and back lips of the hole in the shield at continuously decreasing velocities, until the disturbance decays below the elastic limit. Some elastic rebound may occur, although experiments suggest that this is small [*Turpin (1969)*]. Empirical fits of the final hole size D have been reported, fitted to restricted sets of experimental data, [*Maiden and McMillan (1964)*]

$$\frac{D}{d} = 0.45\, v \left(\frac{t_s}{d}\right)^{2/3} + 0.9 \tag{17}$$

where t_s is the shield thickness, and [*McHugh (1966)*]

*Larger discrepancies are obtained when the materials vaporize, probably because the calculations did not include a detailed treatment of the multiphase equation of state.

$$\frac{D}{d} = 0.658\, v \left(\frac{t_s}{d}\right)^{0.14} \exp\left\{0.63 \left(\frac{t_s}{d}\right)^{0.43}\right\} \tag{18}$$

However, it has been found that an extensive collection of hole size data could not be fitted successfully by these expressions [*Swift, Preonas, Dueweke and Bertke (1970)*]. In fact, these expressions predict a very large hole size when extrapolated to very high velocities. Experiments suggest that the entrance hole diameter increases from that of the projectile for very thin shields to that of the crater in a semi-infinite target as the shield thickness increases, Fig. 14, while the exit hole diameter at first increases to a diameter close to that of the semi-infinite crater and then decreases again to zero at the ballistic limit, Fig. 15, [*Vitali, Becker and Watson (1961)*]. It is reasonable to expect qualitatively the same behavior at higher velocities, and this qualitative idea has been used [*Swift, Bamford and Chen (1982)*].

Acknowledgement

This work performed in part at Sandia National Laboratories supported by the U.S. Department of Energy under contract number DE-AC04-76DP00789.

References

Alekseevski, V. P., *Penetration of a Rod Into a Target at High Velocity,* Combustion, Explosion and Shock Waves, 2; 63–66 (1966).

Allen, W. A. and Rogers, J. W., *Penetration of a Rod Into a Semi-Infinite Target,* J. Franklin Inst., 275–284, (1961).

Anderson, C. and Johnson, W., *Overview of Hydrocode Development and Hypervelocity Impact,* 1986 Hypervelocity Impact Symp. San Antonio TX, Oct. (1986).

Asay, J. R. and G. I. Kerley, *The Response of Materials in Hypervelocity Impact,* 1986 Hypervelocity Impact Symp. San Antonio TX, Oct. (1986).

Bertholf, L. D., Buxton, L. D., Thorne, B. J., Byers, R. K., Stevens, A. L. and Thompson, S. L., *Damage in Steel Plates from Hypervelocity Impact. II. Numerical Results and Spall Measurement,* J. Appl. Phys. 46: 3776–3783, (1975).

Birkhoff, G., MacDougalll, D. P., Pugh, E. M., and Taylor, G., *Explosives With Lined Cavities,* J. of Appl. Phys., 19; 563–582, (1948).

Bjork, R. L., *Review of Physical Processes in Hypervelocity Impact and Penetration,* Rand Corp. RM–3529–PR, July (1963).

Bjork, R. L. and Olshaker, A. E., *A Proposed Scaling Law for Hypervelocity Impacts between a Projectile and a Target of Dissimilar Material,* Rand Corp. RM–2926–PR, May (1965).

Bjork, R. L. and Olshaker, A. E., *The Role of Melting and Vaporization in Hypervelocity Impact,* Rand Corp., RM–3490–PR, (1965).

Bruce, E. P., *Review and Analysis of High Velocity Impact Data,* Fifth Symp. on Hypervelocity Impact, Denver CO, Oct. (1961).

Charters, A. C. and Summers, J. L., *Some Comments on the Phenomena of High Speed Impact,* Dicennial Symp., U. S. Naval Ordnance Laboratory, White Oak, Maryland (1959).

Chou, P. C., Sidhu, H. S. and Zajac, L. J., *The Strong Plane Shock Produced by Hypervelocity Impact and Late Stage Equivalence,* Seventh Symp. on Hypervelocity Impact, Tampa FL, Nov. (1965).

Christman, D. R. and Gehring, J. W., *Penetration Mechanisms of High Velocity Projectiles,* General Motors Res. Lab. TR-65-50 (1965).

Christman, D. R. and Gehring, J. W., *Analysis of High-Velocity Projectile Penetration Mechanics,* J. Appl. Phys., 37; 1579–1587, (1966).

Eichelberger, R. J. and Gehring, J. W., *Effects of Meteoroid Impacts on Space Vehicles,* ARS Journal, 1583–1591, Oct. (1962).

Eichelberger, R. J., *Experimental Test of the Theory of Penetration by Metallic Jets,* J. Appl. Phys., 27; 63–68, (1956).

Feldman, J. G., *Volume—Energy Relations from Shape Charge Jet Penetrations,* Fourth Symp. on Hypervelocity Impact, Eglin AFB FL, Sept. (1960).

Gehring, J. W. and Lieblein, S., *Preliminary Results on Effects of Hypervelocity Impact on Space Radiator Tubes,* Progress in Astronautics and Aeronautics Vol 11, Power Systems for Space Flight, ed. M. A. Zipkin and R. N. Edwards, Academic Press, (1963).

Gehring, J. W., Meyers, C. L., and Charest, J. A., *Experimental Studies of Impact Phenomena and Correlation with Theoretical Models,* Seventh Symp. on Hypervelocity Impact, Tampa FL, Nov. (1965).

Gehring, J. W., *Theory of Impact on Thin Targets and Shields and Correlation with Experiment in High Velocity Impact Phenomena,* in High Velocity Impact Phenomena, ed. R. Kinslow, Academic Press (1970).

Grady, D. E., *Fragmentation of Rapidly Expanding Jets and Sheets*, 1986 Hypervelocity Impact Symposium, San Antonio TX, Oct. (1986).

Halperson, S. M., *Comparisons between Hydrodynamic Theory and Impact Experiments*, Seventh Symp. on Hypervelocity Impact, Tampa FL, (1964).

Haugstad, B. S., *Compressibility Effects in Shaped Charge Jet Penetration*, J. Appl. Phys., 52; 1243–1246, (1981).

Herrmann, W. and Jones, A. H., *Correlation of Hypervelocity Impact Data*, Fifth Symp. on Hypervelocity Impact, Denver CO, Oct. (1961).

Hohler, V., Schneider, E. and Stilp, A., *Endballistik von Splittern V*, Ernst Mach Institut, EMI Bericht E 7/75, (1975).

Hohler, V. and Stilp, A. J., *Penetration of Steel and High Density Rods in Semi-Infinite Steel Targets*, Proc. of the Third Intl. Symp. on Ballistics, Karlsruhe, FRG, March (1977).

Holsapple, K., *Analytical Techniques and Theory*, 1986 Hypervelocity Impact Symp., San Antonio TX, Oct. (1986).

Hopkins, A. K., Lee, T. W., and Swift, H. F., *Material Phase Transformation Effects Upon Performance of Spaced Bumper System*, J. of Spacecraft and Rockets, 9; 342–345, (1972).

Kerley, G. I., *Theoretical Equation of State for Aluminum*, 1986 Hypervelocity Impact Symposium, San Antonio TX, Oct. (1986).

Kineke, J. H., *An Experimental Study of Crater Formation in Metallic Targets*, Fourth Symp. on Hypervelocity Impact, Eglin AFB, Florida, Sept. (1960).

Kinslow, R., *Properties of Reflected Stress Waves*, Proc. AIAA Hypervelocity Impact Conference, Cincinnati OH, May (1969).

Lawrence, R. J., *A Simple Model for the Optimization of Stand-off Hypervelocity Particle Shields*, 1986 Hypervelocity Impact Symposium, San Antonio TX, Oct (1986).

Loeffler, I. J., Lieblein, S., and Clough, N., *Meteoroid Protection for Space Radiators*, Progress in Astronautics and Aeronautics Vol 11, Power Systems for Space Flight, ed. M. A. Zipkin and R. N. Edwards, Academic Press, (1963).

Maiden, C. J. and McMillan, A. R., *An Investigation of the Protection Afforded a Spacecraft by a Thin Shield*, AIAA Journal 2; 1992–1998, (1964).

McHugh, A. H., *Evaluation of Hypervelocity Impact Damage to a Thin Sheet by Multiple Regression Analysis*, NAA/S&DI Report STR 152, June (1966).

McMillan, A. R., *Experimental Investigations of Simulated Meteoroid Damage to Various Space Structures*, NASA–CR–915, Jan. (1968).

Nysmith, C. R. and Summers, J. L., *Experimental Investigation of the Impact Resistance of Double-Sheet Structures at Velocities to 24,000 ft/sec*, NASA TN–D–1431, (1962).

Perez, E., *A Theory for Armor Penetration by Hypervelocity Long Rods*, Proc. of the Fourth Intl. Symp. on Ballistics, Monterey CA, Oct. (1978).

Piacesi, R., Waser, R. H. and Dawson, V. C. P., *Determination of the Yield Strength as an Effective Mechanical Strength Property in the Cratering Process of Hypervelocity Impact*, Seventh Symp. on Hypervelocity Impact, Tampa FL, (1964).

Richardson, A. J., *Theoretical Penetration Mechanisms of Multisheet Structures Based on Discrete Debris Particle Modeling*, AIAA Hypervelocity Impact Symposium, Cincinatti OH, April (1969).

Riney, T. D. and Heyda, J. F., *Hypervelocity Impact Calculations*, Seventh Symp. on Hypervelocity Impact, Tampa FL, Nov. (1964).

Rosenblatt, M., Kreyenhagen, K. N. and Romine, W. D., *Analytical Study of Debris Clouds Formed by Hypervelocity Impacts on Thin Plates*, AFML–TR–68–266, Dec. (1968).

Schneider, E., *Velocity Dependencies of Some Impact Phenomena*, Comet Halley Micrometeoroid Hazard Workshop, European Space Agency ESA SP–153, Oct. (1979).

Sedgwick, R. T., *Numerical Techniques for Modeling High Velocity Penetration and Perforation Processes*, in Ballistic Materials and Penetration Mech., ed. R. C. Laible, Elsevier, (1980).

Spillman, G. R., *Hypervelocity Impact Bumpers as Countermeasures*, Los Alamos Natl Lab LA–10629–MS, Jan (1986).

Summers, J. L. and Charters, A. C., *High Speed Impact of Metal Projectiles in Targets of Various Materials*, Third Symp. on Hypervelocity Impact, Chicago, Ill., Oct. (1958).

Summers, J. L., *Investigation of High Speed Impact: Regions of Impact and Impact at Oblique Angles*, NASA TN–D–94 (1959).

Swift, H. F. and Hopkins, A. K., *The Effects of Bumper Material Properties on the Operation of Spaced Hypervelocity Particle Shields*, AIAA Hypervelocity Impact Conference, Cincinatti OH, April (1968).

Swift, H. F., Preonas, D. D., Dueweke, P. W. and Bertke, R. S., *Response of Materials to Impulsive Loading*, AFML–TR–70–135, May (1970).

Swift, H. F., Preonas, D. D., Turpin, W. C., and Carson, J. M., *Debris Clouds Behind Plates Impacted by Hypervelocity Pellets*, J. Spacecraft Rockets, 7; 313–18, (1970).

Swift, H. F., *Hypervelocity Impact Mechanics*, in Impact Dynamics ed. J. A. Zukas, et al, John Wiley & Sons, (1982).

Swift, H. L., Bamford, R. and Chen, R., *Designing Dual Plate Meteoroid Shields—A New Analysis*, JPL Publication 82–39, March (1982).

Tate, A., *A Theory for the Deceleration of Long Rods After Impact*, J. Mech. and Phys. of Solids, 15; 387–399, (1967).

Tate, A., *Further Results in the Theory of long Rod Penetration*, J. Mech. and Phys. of Solids, 17; 141–150, (1969).

Turpin, W. C., *Hypervelocity Perforation Mechanics of Thin Metal Plates*, AFML–TR–69–203, July (1969).

Vitali, R., Becker, K. R. and Watson, R. W., *Perforation of Finite Targets by High Velocity Projectiles*, Fifth Symp. on Hypervelocity Impact, Denver CO, Oct. (1960).

Wagner, M. H. and Kreyenhagen, K. N., *Review of Hydro-Elastic-Plastic Code Analysis as Related to the Hypervelocity Particle Impact Hazard*, Proc. Comet Halley Micrometeoroid Hazard Workshop, European Space Agency ESA SP–153 April (1979).

Walsh, J. M. and Johnson, W. E., *On the Theory of Hypervelocity Impact*, Seventh Symp. on Hypervelocity Impact, Tampa FL, Nov. (1964).

Walters, W. P. and Majerus, J. N., *Impact Models for Penetration and Hole Growth*, BRL–TR–02069, (1978).

Whipple, F. L., *Meteorites and Space Travel*, Astronomical J. No. 1161:131 (1947).

Zwarts, F. A., *The Initial One-Dimensional Expansion of the Shocked States Generated by the Impact of Cylindrical Pellets with Thin Plates*, NASA–CR–54209, Jan (1965).

Int. J. Impact Engng Vol. 5, pp. 323–331, 1987
Printed in Great Britain

0734-743X/87 $3.00 + 0.00
Pergamon Journals Ltd.

HYPERVELOCITY IMPACT OF ROD PROJECTILES WITH L/D FROM 1 TO 32

V. Hohler and A. J. Stilp

Fraunhofer-Institut für Kurzzeitdynamik - Ernst-Mach-Institut -
Terminal Ballistics and Impact Physics Division
Eckerstr. 4, D-7800 Freiburg, FRG

ABSTRACT

Beside a short remark on the "hydrodynamic theory of rod projectiles", the paper deals with the terminal ballistic behaviour of cylindrical projectiles against semi-infinite targets. Experimental data of EMI, completed by results of some other authors, are presented. Crater parameters like depth, diameter and volume and their dependence on projectile velocity (up to 5000 m/s), projectile and target material properties, as well as L/D-ratios (1 - 32), will be discussed. Mainly the projectile materials steel and tungsten sinter-alloys are considered. Target materials are mild steel and high strength steel, an Al-alloy and a tungsten sinter-alloy. The results show that the influence of material density on the crater dimensions is considerably greater than the influence of strength. The L/D ratio determines the velocity dependence of crater depth, diameter and volume. At high velocities in the hydrodynamic regime, the crater depth of short cylinders (L/D ~ 1) is approximately proportional to $v_p^{2/3}$ (v_p = projectile velocity). With increasing L/D-ratio, the slope of the penetration curves decreases and converges for rods (L/D >> 1) versus a saturation, i. e. becomes nearly independent on v_p. A consequence of this saturation is the existence of a so-called "tangent velocity", above which an optimal increase of efficiency is only realized by increasing the projectile mass and not the velocity. Furthermore, ballistic limits of real targets like single plates and symmetric double plates meteorite bumper shield) are taken into account. The expected better performance of "segmented rods" is also discussed.

INTRODUCTION

Based on their high terminal ballistic efficiency, rod projectiles are becoming increasingly important in all caliber ranges. The reason for that is the use of projectile materials of high density and high strength at sufficient toughness. Furthermore, due to improved launching techniques, increased velocities and length-to-diameter ratios have become feasible. This paper gives an overview on the penetration of cylindrical projectiles in semi-infinite targets as related to velocity v_p, L/D-ratio and material properties. Crater parameters like depth p as well as diameter d and volume V_T will be discussed. These considerations are completed by results dealing with real targets like single and double plates.

SEMI-INFINITE TARGET

Theoretical Background

The process of high velocity impact can be represented by a model developed by Eichelberger and Gehring (1962). It consists of four phases, which are illustrated qualitatively in Fig. 1. The "transient phase" describes the early projectile-target interaction. During this phase (generally less than several microseconds) the impact flash and shock waves are generated and a high stress peak is produced. The following "primary phase" is marked by a quasi-stationary flow with a nearly constant stress level behind the shock front. The duration of this process increases with increasing L/D-ratio. The "secondary or caviation phase" begins after the projectile has been completely deformed. The last "recovery phase" is defined by the contraction of the crater at the end of the penetration. Within the last two phases, the penetration process is instationary.

The so-called "hydrodynamic penetration theory of rods" has been derived from the penetration of Munroe-jets (Birkhoff and others (1948); Pack and Evans (1951); Rostocker (1953); Eichelberger (1956)) and is applied to the quasi-stationary incompressible flow phase, shown schematically in Fig. 2. The one-dimensional theory is based on the Bernoulli-equation for pressure equilibrium at the stagnation point

$$\frac{1}{2}\rho_T\, u^2 + R = \frac{1}{2}\rho_p\, (v - u)^2 + Y \quad (1)$$

Y and R represent the dynamic strengths of projectile and target materials, ρ_p and ρ_T the material densities, u the penetration velocity of the stagnation point and v the residual velocity. From equation (1) Allen and Rogers (1961) derived the penetration formula

$$p = L \cdot f_1\, (v_p, \rho_p, \rho_T, \underline{R - Y}) \quad (2)$$

Alekseevskii (1966) and Tate (1967, 1969) have taken into account the deceleration of the residual rod during penetration. The penetration depth p is given by

$$p = L \cdot f_2\, (v_p, \rho_p, \rho_T, \underline{R, Y}) \quad (3)$$

Fig. 3 shows some calculated penetration curves for equal projectile and target material densities (steel) and different Y- and R-values (Williams (1976)). In the case R > Y, the typical S-shaped curves occur. For R < Y, a maximum penetration exists, which is very interesting. At high velocities, the curves converge versus a velocity independent value, postulated by the penetration of shaped charge jets

$$p = L\, (\rho_p/\rho_T)^{1/2} \quad (4)$$

Fig. 4 shows representatively a flash X-ray picture of a tungsten sinter-alloy rod, penetrating a steel target. The analogy to Fig. 2 is obvious.

An improved description (Christman and Gehring (1966)) can be obtained if one takes into consideration both instationary end phases (s. Fig. 1), approximated by the penetration of a short cylinder of length D. Now the quasi-stationary phase is restricted to a rod length (L - D) and the penetration formulas are

$$p = (L - D)\, f_1 + p_{cyl} \quad (2)'$$
$$p = (L - D)\, f_2 + p_{cyl} \quad (3)'$$
$$p = (L - D)\, (\rho_p/\rho_T)^{1/2} + p_{cyl} \quad (4)'$$

For p_{cyl}, experimental data or formula (5) can be used (Charters and Summers (1958); Christman and Gehring (1966)).

$$p_{cyl}\, [cm] = 0.33\, (\rho_p/\rho_T)^{1/3}\, (E_{kin}/B_{max})^{1/3} \quad (5)$$

where E_{kin} is the kinetic energy [J] and B_{max} the maximum target material hardness [kp/mm^2]. Fig. 5 shows a comparison of equations (2)'...(4)' with experimental data. Experimental data for p_{cyl} are used. The best fit is obtained by equation (3)'. Equation (4)' can only be applied at high velocities.

Crater Depth, Diameter and Volume

The typical influence of the L/D-ratio of cylindrical projectiles on the crater shape can be seen in Fig. 6. The crater depth-to-diameter ratio p/d increases with increasing L/D. The craters become more and more tube-shaped like shaped charge jet craters.

Figs. 7 and 8 present penetration data of steel- (115 CrV3, 35 CrNiMo) and tungsten sinter-alloy-projectiles (D17, D17.6) fired into high strength steel targets dependent on velocity, L/D-ratio and material properties (Hohler and Stilp (1977, 1980, 1984)). Penetration data published by Silsby (1984) with L/D = 23 rods of tungsten sinter-alloy are also included. The L/D-influence is clearly demonstrated. p/L decreases with increasing L/D and the p/L-curves are converging versus the S-shaped curve of the ideal rod with a saturation at high velocities. The ideal rod is defined by L/D >> 1 with a penetration behaviour according to the hydrodynamic jet theory. The S-shape of the curves for these material combinations means R > Y. The material strength influence is small compared with that of density.

Fig. 9 presents penetration data of tungsten sinter-alloy and steel rods for the following target materials: Al-alloy, mild steel and high strength steel and tungsten sinter-alloy. In the case of high density metal rods fired against Al-alloy targets, R < Y is observed. The maximum of p/L, postulated theoretically (s. Fig. 3) seems to be indicated, but is not proved because of lacking data at velocities $v_p > 4000$ m/s. The existence of this maximum has been proved experimentally by Tate (1969) for Al-rods penetrating into lead targets. Referring to target material properties, Fig. 9 shows that their influence is comparable to that of projectile material properties (s. Fig. 7 and 8).

In Fig. 10 crater diameters d, normalized by the projectile caliber D, are plotted versus v_p. A nearly linear increase of d/D with v_p is observed for cylinders having L/D = 1. On the other hand, the curves of the rod projectiles are characterized by the typical saddle at velocities, where p/L has its strongest increase with v_p. In the high velocity regime the curves of rod projectiles become also approximately linear. The influence of density is comparable to that of strength.

In Fig. 11 the normalized crater volumes $\rho_T V_T/\rho_p V_p$ (V_T, V_p = crater- and projectile volumes) are plotted as a function of v_p on a double logarithmic scale. In this normalization the densities are nearly eliminated. At lower velocities the crater volumes are dependent on the L/D-ratio. This dependence disappears at higher velocities and the curves in Fig. 11 converge versus a slope of approximately 2. This means that the crater volumes become proportional to the kinetic energy of the projectiles. It is interesting that in the saturation regime crater volumes increase only by expanding crater diameters.

Additional aspects are obtained by plotting p/d versus v_p (s. Fig. 12). The L/D = 1-curve indicates a hemispherical crater at high velocities (for $\rho_p = \rho_T$). On the other hand, the curves of rod projectiles show a maximum. The maximum value increases with increasing L/D-ratio and with increasing projectile material density and strength. The target material properties have an opposite effect.

Energy Diagram

In the p/L diagrams of Figs. 7, 8 and 9, the dependencies of p on all projectile parameters are implied. More explicit information about the influences of v_p, L/D, m_p and of the material properties on p is given in the so-called "energy diagram" (Fig. 13). It shows p as a function of the kinetic projectile energy at constant projectile mass m_p = 100 g on a double logarithmic scale. Since m_p = const., the v_p-scale is also plotted. Based on the scaling law $p \sim m_p^{1/3}$, p-curves for other m_p-values can be obtained by shifting the m_p-curves along straight lines with slope 1/3 to the right (m_p > 100 g) or to the left (m_p < 100 g). These lines are lines of constant velocities. Each curve in Fig. 13 has a tangent with a slope of 1/3. Only p-values on and below the tangent can be reached. There is only one pair of m_p- and v_p-values, called m_t and v_t, that delivers a maximum penetration for a given kinetic energy. These m_t- and v_t-values are determined by the intersection point between the tangent and the vertical line E_{kin} = const. This point of intersection is identical with the tangent point of the p-curve for $m_p = m_t$. The tangent represents the line of optimal velocity v_t. The v_t-values decrease with increasing L/D-ratio and with increasing projectile material strength and density. v_t-values of 2000 m/s and 2800 m/s are observed for high density metal and steel rods of L/D = 10, respectively. Above the tangent velocity an optimal increase of efficiency at constant projectile energy is only realized by increasing the rod mass and not the velocity. The tangent velocity is identical with the maximum of p/d in Fig. 12.

It seems to be significant that the curves in Fig. 13 have an intersection for different L/D-ratios at low velocities. If $(L/D)_1 > (L/D)_2$, then $p(L/D)_1 > p(L/D)_2$ is only observed above the intersection velocity. The projectile material with the better penetration performance has the lower intersection velocity. For example, the comparison of L/D = 1 and 10 leads for D17.6-projectiles to an intersection at about 650 m/s and for 35 CrNiMo-projectiles at about 900 m/s, respectively.

For rod projectiles the tangent velocity is a consequence of the saturation of p in the high velocity range. Since the slope of the curves in Fig. 13 increases with decreasing L/D-ratio, the tangent velocity is shifted more and more to higher velocities. For short projectiles (L/D = 1-cylinders or spheres) Öpik's theory (1969) is postulating $V_T \sim m_p v_p$. From this, $p \sim (m_p v_p)^{1/3}$ should be observed. However, many experiments (e. g. Christmann and Gehring (1966); Gault (1973)) confirm a correlation like equation (5), which is based on $V_T \sim \frac{1}{2} m_p v_p^2$ (s. also Fig. 11). Therefore, at high velocities the penetration of cylinder having a L/D = 1 should increase with energy along a slope of 1/3, no matter whether m_p or v_p or both are increasing.

Segmented Rod

Concluding this chapter, some remarks may be made about the so-called "segmented rod". The idea is based on the fact given in Figs. 7 and 8, where p/L increases with decreasing L/D-ratio. Therefore, e. g. 10 cylinders with L'/D = 1, shot one after the other into the same crater would yield a larger depth than the corresponding homogeneous rod of L/D = 10 with L = 10 L'. The penetration curve of such a segmented rod (material D17.6) is drawn in Fig. 13, assuming simple addition of the penetration of the 10 single segments. There are three papers about segmented rods. Perez and Giraud (1980) did not observe an improved efficiency comparing L/D = 12 data with two segments of L'/D = 6 and three segments of L'/D = 4 at 2000 m/s. Charters (1985) has shown at 3500 m/s that six segments reach about a 50 % better performance than the corresponding homogeneous rod. De Rosset (1981) predicted an even higher efficiency of the segmented rod at 5000 m/s, based on computer code calculations. It seems that for velocities v_p > 2000 m/s the performance of the segmented rod increases with velocity more than the homogeneous rod, probably because at higher velocities the segments are consumed and no residual mass interferes with penetration by subsequent segments. More experiments need to be done in this field.

REAL TARGETS

The semi-infinite target is very suitable for basic studies on projectile target interaction. For application, however, real targets are of more interest. Therefore, the principal terminal ballistic behaviour of single and double plate targets will be discussed, too. Fig. 14 shows the ballistic limit thickness of a single plate target $\bar{t}$ and of a symmetric double plate target $\bar{t}_{12}$ of short steel cylinder projectiles (115 Cr V3, L/D = 1) versus velocity v_p at normal impact. Symmetric means $t_1 = t_2$ with $\bar{t}_{12} = t_1 + t_2$. The plate distance is about 20 caliber. For comparison the penetration into the semi-infinite target is also plotted. In all cases, mild steel targets have been used. The interaction of the projectile with the rear side of the single plate target causes the effect $\bar{t} > p$. At low velocities, $\bar{t} \sim \bar{t}_{12}$ is observed. Here the projectile is only deformed and not destroyed at the first plate of the double plate target. In the high velocity regime, an increasing destruction of the projectile at the first plate causes more and more $\bar{t}_{12} < \bar{t}$. The very good protection efficiency of double plate targets against short projectiles (L/D ~ 1) is applied in meteorite bumper shields.

Fig. 15 presents the corresponding curves for steel rods of L/D = 10. S-shaped curves are also observed for $\bar{t}$ and $\bar{t}_{12}$. That means basic principles of the penetration mechanism of the semi-infinite target (s. Fig. 2) can also be applied in this case. This is obvious from Fig. 16 a, where a flash X-ray picture of a steel rod behind a steel plate in the high velocity regime is shown. The rod is not destroyed but only reduced in length and velocity. A destruction of the residual rod only occurs for brittle projectile materials (elongation smaller than a few percent). In contrast to that behaviour, Fig. 16 b gives an impression about the fragmentation of a steel cylinder of L/D = 1 at a steel plate. Therefore, rods appear to be more efficient against multiple plate targets than short cylinders.

Further information about the content of Fig. 14 and 15 is given in Fig. 17 with the plot $\bar{t}/p$ versus v_p. In addition, a corresponding curve of tungsten sinter-alloy rods is shown. All curves decrease with increasing velocity and converge versus a constant value in the high velocity regime. However, in case of rod projectiles, the $\bar{t}/p$-curves range below that of the L/D = 1-cylinders, and show a more distinct convergence.

CONCLUSIONS

Up to now, a series of results exist about the terminal ballistic behaviour of cylindrical projectiles against semi-infinite targets in the high velocity regime. Only little is known about real targets which are of more interest for applications. Therefore, further investigations have to be done mainly in the field of real targets, using, beside armor steel and Aluminum, modern materials like ceramics and fibre reinforced materials. Especially homogeneous rods and segmented rods should be taken into account.

REFERENCES

Alekseevskii, V. P. (1966). Fizika Goreniya Vzryra, 2, 99
Allen, W. A., and J. W. Rogers (1961). J. Franklin Inst., 272, 275
Birkhoff, G., D. P. Mac Dougall, E. M. Pugh, and G. I. Taylor (1948). J. Appl. Phys., 19, 563
Charters, A. C., and J. L. Summers (1958). Proc. 3rd Symp. Hypervelocity Impact, Armor Research Foundation, Chicago, Illinois, USA
Charters, A. C. (1985). Private communication.
Christman, D. R., and J. W. Gehring (1966). J. Appl. Phys., 37, 1579
De Rosset, W. S. (1981). Proc. 6th Int. Symp. Ballistics, Orlando, Florida, USA
Eichelberger, R. J. (1956). J. Appl. Phys., 27, 63
Eichelberger, R. J., and J. W. Gehring (1962). Am. Rocket Soc. J., 32, 1583
Gault, D. E. (1973). The Moon, 6, 32
Hohler, V., and A. J. Stilp (1977). Proc. 3rd Int. Symp. Ballistics, Karlsruhe, FRG

Hohler, V., and A. J. Stilp (1980). Proc. 5th Int. Symp. Ballistics, Toulouse, France
Hohler, V., and A. J. Stilp (1984). Proc. 8th Int. Symp. Ballistics, Orlando, Florida, USA
Öpik, E. J. (1969). Ann. Rev. Astron. Astrophys., 7, 473
Pack, D. C., and W. M. Evans (1951). Proc. Phys. Soc. (London), B 64, 298
Perez, E., and M. Giraud (1980). Proc. 5th Int. Symp. Ballistics, Toulouse, France
Rostocker, N. (1953). Meteoritics, 1, 11
Silsby, G. F. (1984). Proc. 8th Int. Symp. Ballistics, Orlando, Florida, USA
Tate, A. (1967). J. Mech. Phys. Solids, 15, 387
Tate, A. (1969). J. Mech. Phys. Solids, 17, 141
Williams, A. E. (1976). Proc. 27th Meet. Aeroballistic Range Association, Centre d'Etudes de Vaujours, Sevran, France

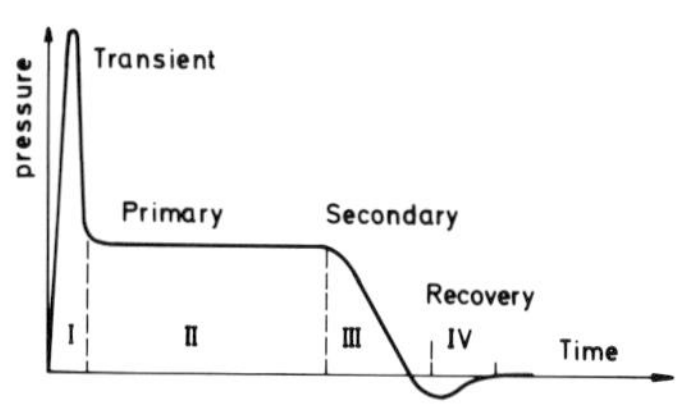

Fig. 1. Schematic pressure-time presentation of the four phases of high velocity penetration.

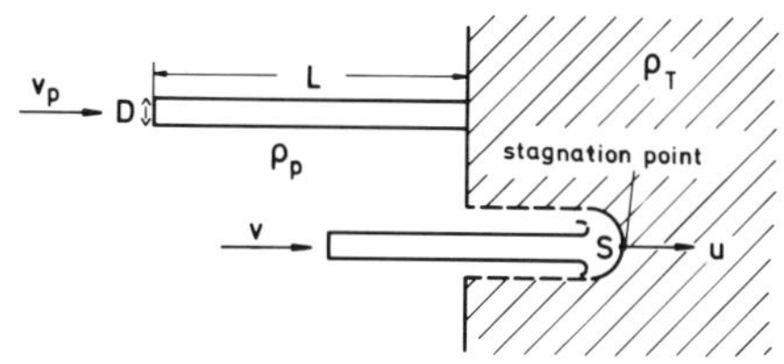

Fig. 2. Quasi-stationary flow of rod penetration.

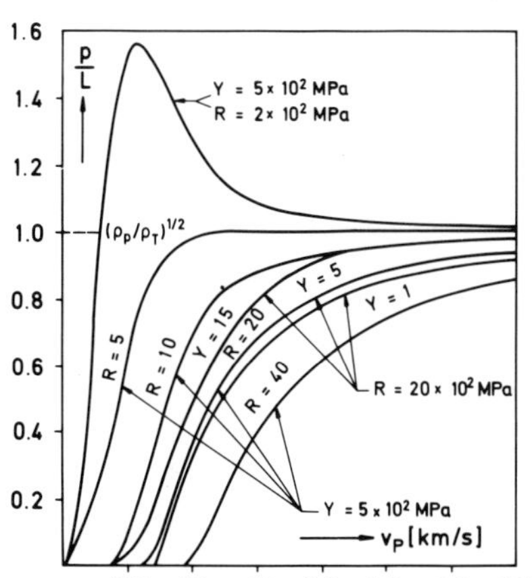

Fig. 3. Variation of crater depth p, normalized by L, with impact velocity v_P for steel rods against semi-infinite steel targets. The curves were calculated by equation (3).

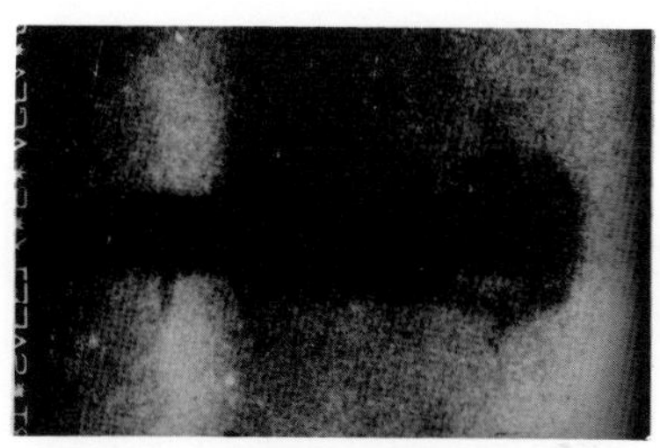

Fig. 4. Flash X-ray picture of a tungsten sinter-alloy rod, penetrating into a steel target (v_P = 1700 m/s).

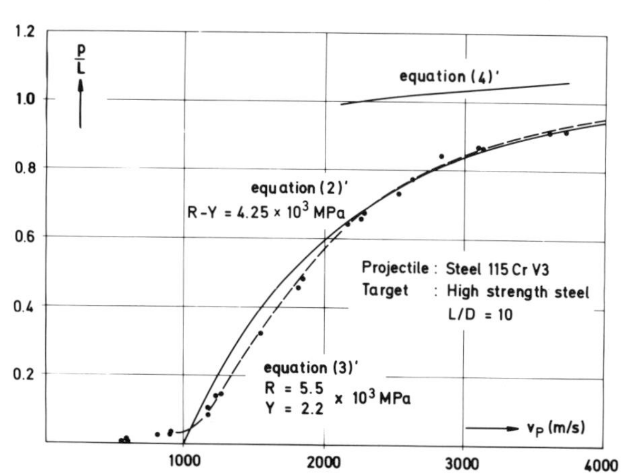

Fig. 5. Penetration depth p of steel rods (115 CrV3, L/D = 10, BHN = 230 kp/mm²) into semi-infinite high strength steel targets (BHN = 360 kp/mm²) versus velocity v_P. The experimental data (circles) are compared with the formulas (2)', (3)' and (4)'.

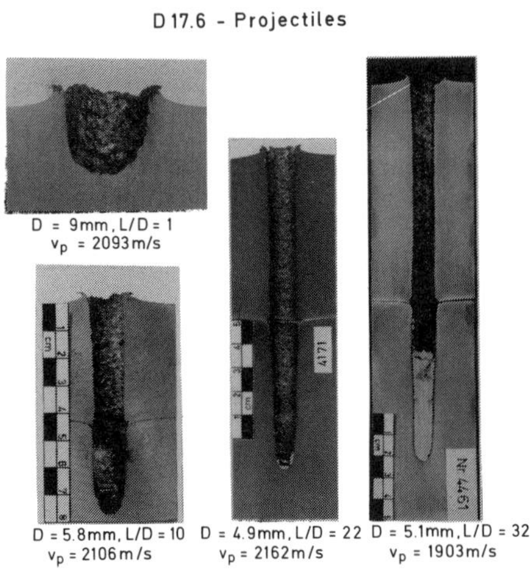

Fig. 6. Crater shapes of tungsten sinter-alloy projectiles (D17.6) in semi-infinite high strength steel targets at different L/D-ratios. The craters are scaled up to equal projectile mass.

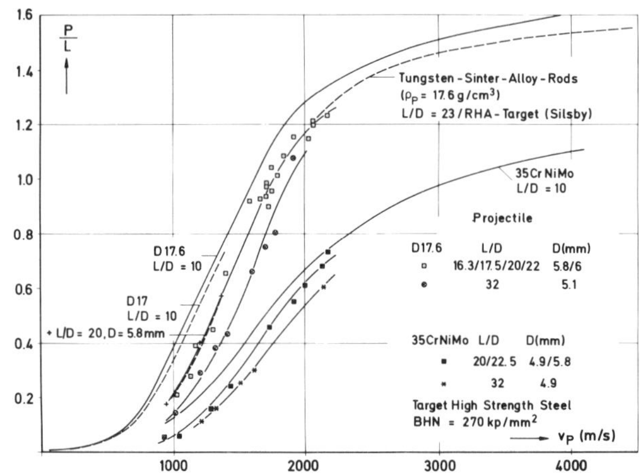

Fig. 8. p/L versus v_p in dependence on L/D and on material properties.

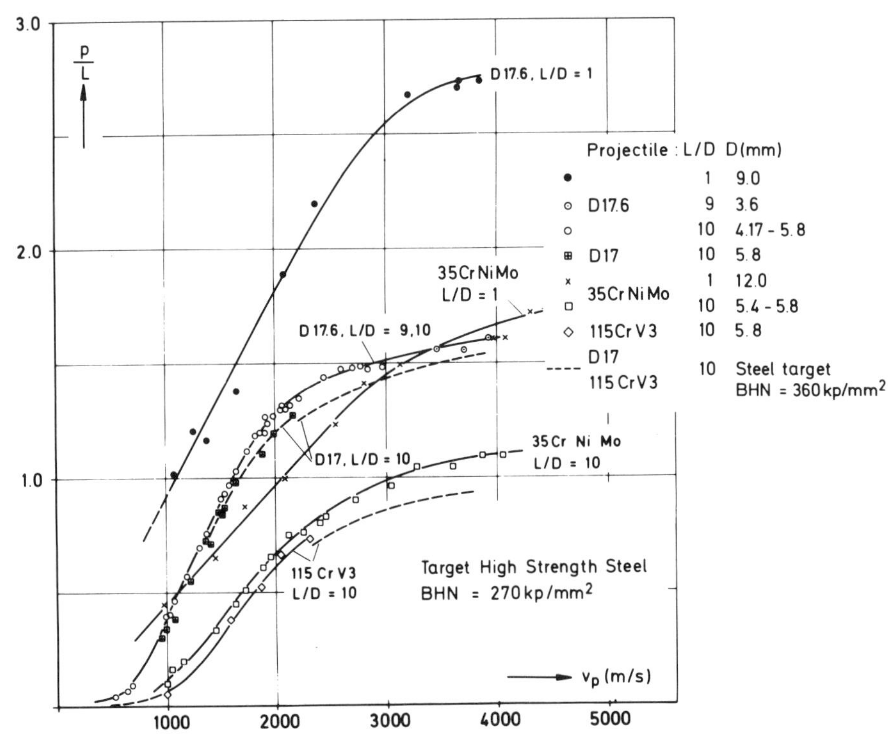

Fig. 7. p/L versus v_p in dependence on L/D = 1 and 10 and on material properties (35 CrNiMo, BHN = 590 kp/mm^2; tungsten sinter-alloy D17, ρ_p = 17 g/cm^3, BHN = 290 kp/mm^2; D17.6, ρ_p = 17.6 g/cm^3, BHN = 420 kp/mm^2).

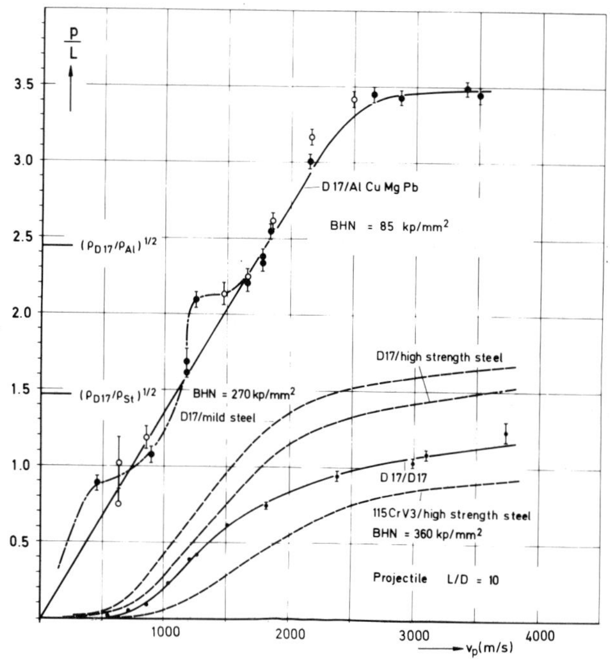

Fig. 9. p/L of tungsten sinter-alloy and steel rods against targets of different materials versus v_p. D17/mild steel means: Projectile/target material.

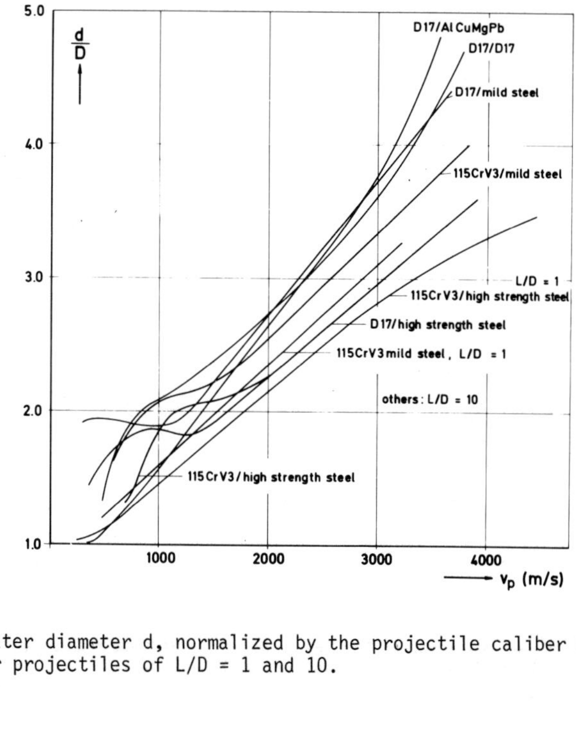

Fig. 10. Crater diameter d, normalized by the projectile caliber D, versus v_p for projectiles of L/D = 1 and 10.

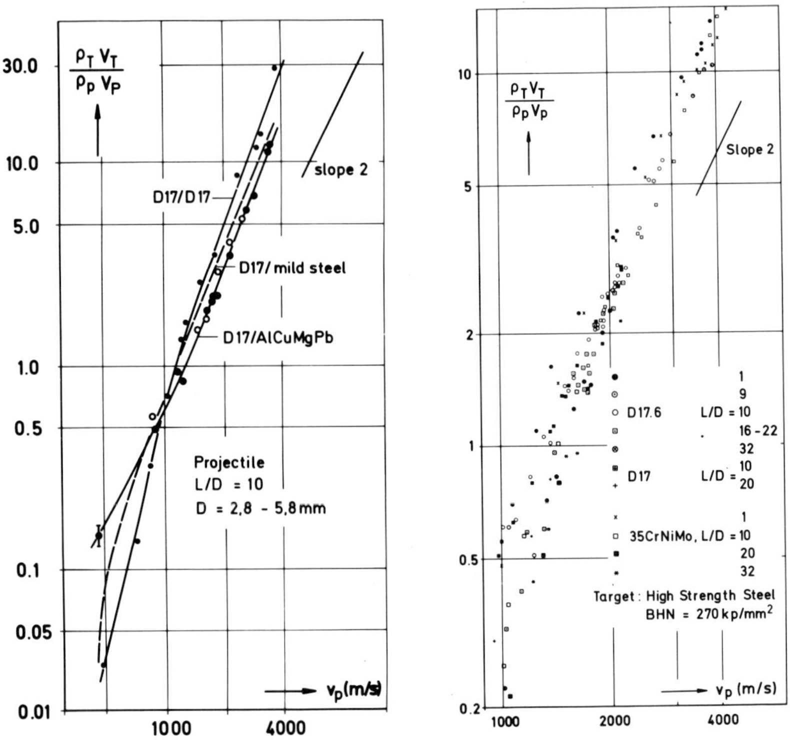

Fig. 11. $\rho_T V_T / \rho_P V_P$ versus v_p in dependence on L/D and material properties (V_T, V_p = crater and projectile volume).

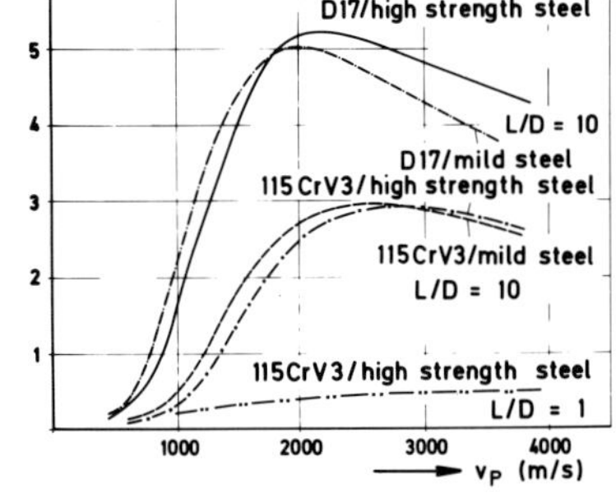

Fig. 12. p/d versus v_p in dependence on L/D and material properties.

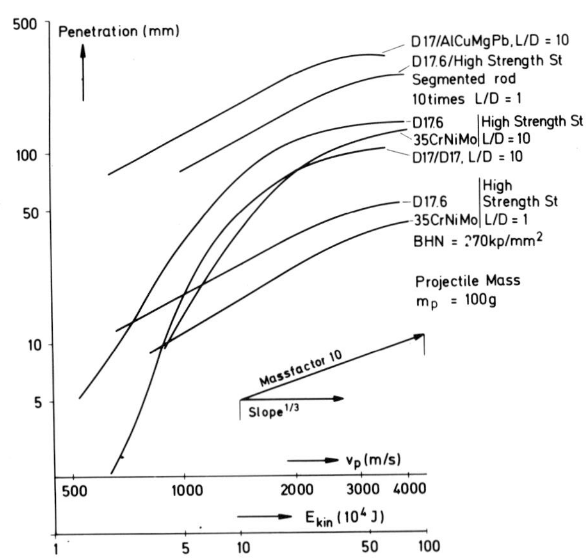

Fig. 13. The "energy diagram". p versus kinetic projectile energy E_{kin} at constant projectile mass m_p = 100 g.

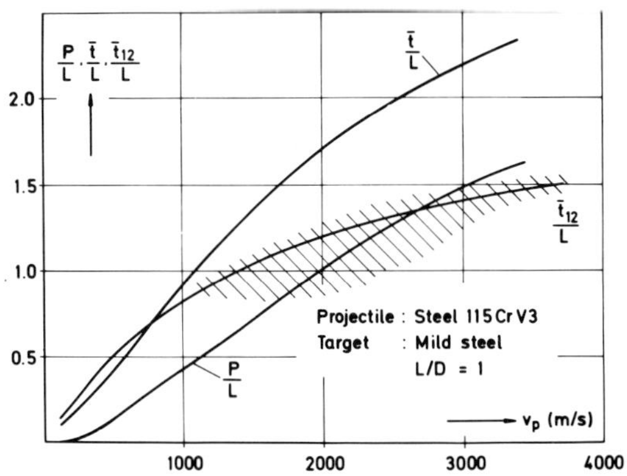

Fig. 14. Projectile: L/D = 1

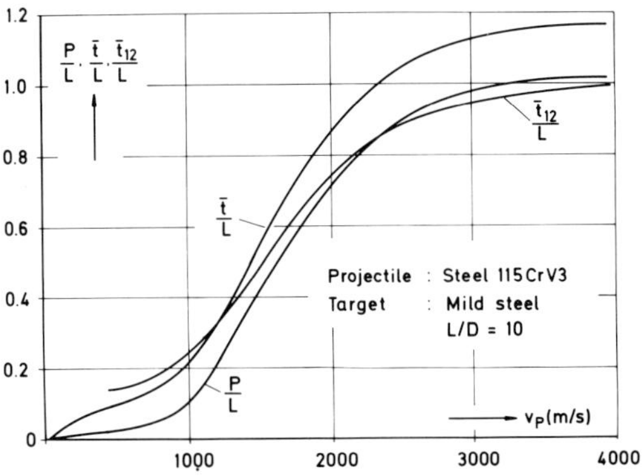

Fig. 15. Projectile: L/D = 10

Ballistic limit thickness of the single plate target $\bar{t}$ and of the symmetric double plate target $\bar{t}_{12} = t_1 + t_2$ (plate distance 20 calibers) at normal impact versus v_p. For comparison the penetration p in the semi-infinite target is also shown.

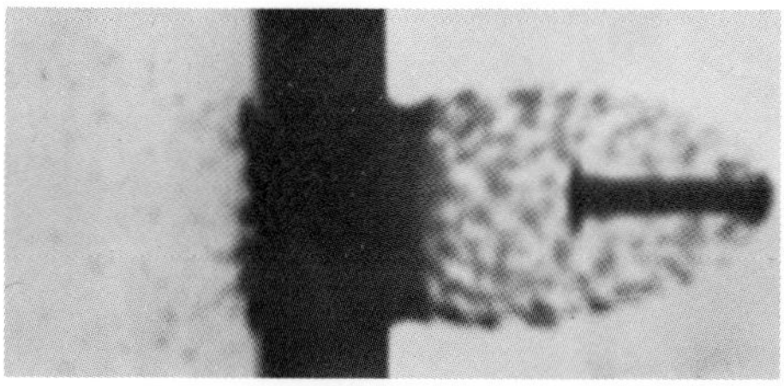

Fig. 16 a. Flash X-ray picture of a residual steel rod behind a steel plate at 3400 m/s.

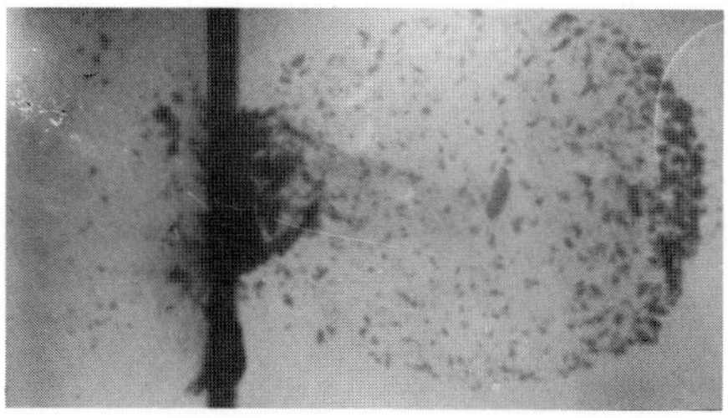

Fig. 16 b. Flash X-ray picture of a fragmented steel cylinder (L/D = 1) behind a steel plate at 3600 m/s.

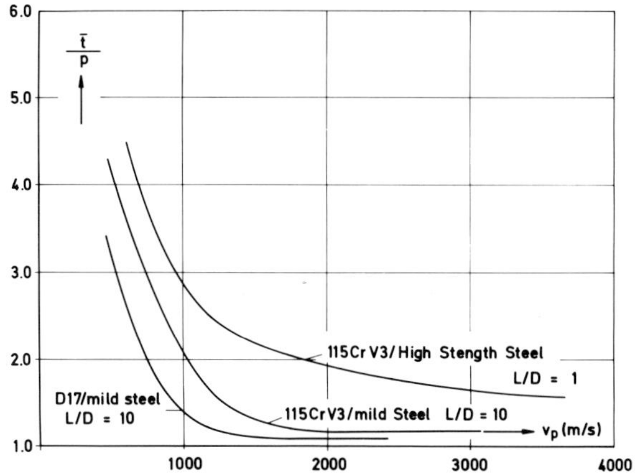

Fig. 17. $\bar{t}/p$ versus v_p.

Int. J. Impact Engng Vol. 5, pp. 333–341, 1987
Printed in Great Britain

0734-743X/87 $3.00 + 0.00
Pergamon Journals Ltd.

SENSITIVITY OF HYPERVELOCITY IMPACT SIMULATIONS TO EQUATION OF STATE

Kathleen S. Holian and Michael W. Burkett

University of California, Los Alamos National Laboratory,
Los Alamos, New Mexico 87545

ABSTRACT

We investigate the sensitivity of hypervelocity impact simulations to differences in equations of state (EOS's), comparing our hydrocode calculations to recent experiments on lead at G. M. Delco (Pomykal, 1986). The results suggest that the hydrocode does not correctly predict all the details of debris cloud structure, even when these are significant EOS differences. We used three different equations of state, one of which is new. It was generated just for sensitivity studies using the most recent experimental EOS data.

INTRODUCTION

In recent years, Los Alamos National Laboratory programs have begun to focus on the equation of state of metals in expansion at high temperature. In particular, there has been interest in simulations of a solid projectile hitting a target at high velocities. The question has been asked, can we simulate this impact with hydrocodes, and, if so, how sensitive are the calculations to the equations of state of the target and impactor materials?

The metals of greatest interest are steel and aluminum, and the impact velocities currently being considered by theorists are in the range of 10-15 km/s or higher. Under this scenario, steel and aluminum will vaporize. Since these velocities are not yet achievable in the laboratory, we are interested in finding out whether we can simulate experiments involving impacts of metals such as lead which will vaporize at lower impact velocities.

If a material is shocked to a sufficiently high pressure, it will release to a state in which there is partial or complete vaporization. The liquid-vapor two-phase region of an equation of state is particularly sensitive to the model used to calculate it. That is, in fact, the region of an EOS in which there is the most uncertainty. Therefore, we would like to determine if a hydrocode simulation in which there is partial vaporization of the material is sensitive to the accuracy of the two-phase equation of state.

In this paper we will compare three different equations of state (EOS) for lead: a Tillotsen analytic EOS, an older Thomas-Fermi tabular EOS from the computer-based SESAME library (Holian, 1984), and a new high-quality tabular EOS produced just for EOS sensitivity studies. The new lead EOS has a carefully constructed mixed-phase region. The critical point is as accurate as possible based on current experimental estimates (it has not been directly measured yet). Also, isobaric expansion data for the liquid have been reproduced, giving us confidence in the boundary of the coexistence region on the liquid side. This new EOS will help determine whether an accurate representation of the two-phase region is at all important in hydrocode simulations.

We will compare calculations made by the two-dimensional Eulerian hydrocode LASOIL with an experiment recently performed by GM Delco with a two-stage gas gun. The experimental configuration (shown in Fig. 1) is a 20-g "spherical" lead projectile hitting a 0.64-cm lead target plate at 6.6 km/sec. X-ray photographs of the debris cloud were taken at 30 and 40 μs after impact. Photodensitometer measurements were made to determine mass per unit area as a function of z and r. (The z-axis is along the direction of the moving projectile.) We will compare these measurements with the results of three hydrocode calculations -- each using a different equation of state for lead.

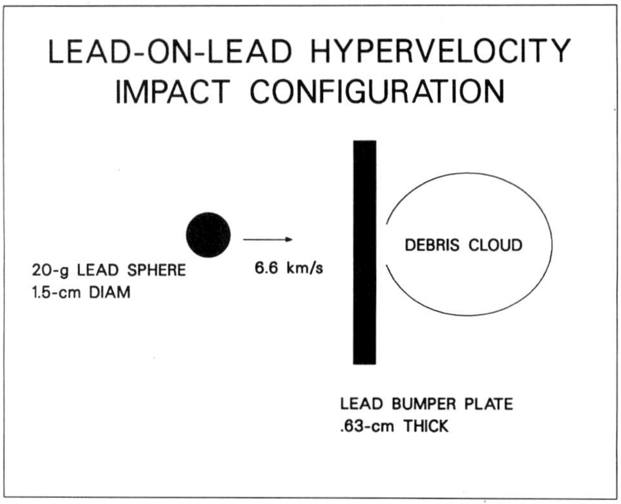

Fig. 1 Schematic of the experiment.

TILLOTSEN EOS

Historically, the Tillotsen equation of state (Tillotsen, 1962) has been used most commonly in hydrocode calculations of hypervelocity impacts involving metals. Since it is an analytic form, it is fast and easy to use. It has been thoroughly documented, so only a brief description will be given here.

The Tillotsen equations give pressure as a function of density and energy. In compression ($\rho > \rho_o$), the equation of state has the form

$$P_c = \left[a + \frac{b}{\frac{E}{E_o \eta^2} + 1} \right] E\rho + A\mu + B\mu^2,$$

where P is the pressure, E is the specific internal energy, ρ is the density, $\eta = \rho/\rho_o$, and $\mu = \eta - 1$. The constants a, b, and A are derived from experimental equation-of-state data for each material (e.g., Hugoniot experiments). E_o and B are then adjusted to provide the best fit for the EOS surface. This formulation has the advantage of being able to both reproduce Hugoniot experiments at low pressures, yet also asymptotes to a Thomas-Fermi EOS at high pressures.

The above formula is used for both compression and for a small region of expansion; in which $\rho_s < \rho < \rho_o$ and $E < E_s$ (E_s is the sublimation energy). This models the situation in which a material is shocked to a low enough energy so that upon release, it returns isentropically to zero pressure and remains solid.

At higher energies in expansion ($E > E_s'$), the gas phase of the material is represented by the following formula:

$$P_e = aE\rho + \left(\frac{bE\rho}{\frac{E}{E_o \eta^2} + 1} + A\mu \, e^{-\beta[(\rho_o/\rho)-1]} \right) e^{-\alpha[(\rho_o/\rho)-1]^2}$$

where α and β are constants controlling the rate of convergence of this equation to an ideal gas. Notice that at $\rho = \rho_o$, the two equations are identical. Therefore, at energies higher than E_s', the equation of state is continuous. However, at lower energies ($E < E_s$), there is an abrupt, discontinuous "phase transition" at $\rho = \rho_s$.

In between E_s and E_s', there is a "mixed phase" (gas-solid) region, in which the pressure is given by

$$P = \frac{P_e (E - E_s) + P_c (E_s' - E)}{E_s' - E_s}$$

While the Tillotsen equation of state is easy to use and asymptotically correct, it does not accurately describe the gas-liquid and liquid-solid phase transition regions. It also does not accurately reproduce isobaric expansion experiments, which measure enthalpy and density at constant pressure. In Fig. 2, the abrupt transition from the solid to the gas form of the EOS produces unphysical behavior.

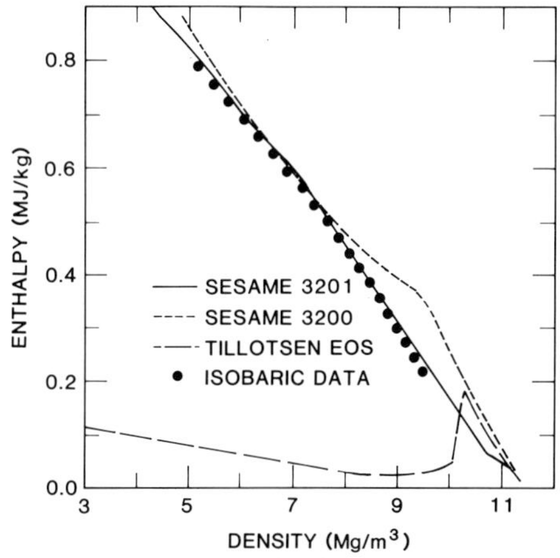

Fig. 2 Theoretical 0.1 GPa isobars for the three lead equations of state are compared to isobaric expansion experiments (Gathers, 1982). The abrupt discontinuity in the Tillotsen isobar is the transition from the solid to the gas formula. SESAME 3201 has a melting transition in the lower right hand corner of the plot.

SESAME 3200

The second equation of state used in this comparison study resides in the SESAME computer-based EOS library at Los Alamos National Laboratory. The EOS actually consists of two tables: one for pressure and one for energy as functions of temperature and density. It was constructed in 1975 by J. Barnes and J. Rood, and the procedure they used to generate it is documented (Holian, 1984).

Briefly, this EOS is the sum of three parts: a cold curve (zero Kelvin isotherm), an electronic thermal EOS, and a nuclear thermal EOS. The electronic part of the EOS was calculated entirely with Thomas-Fermi-Dirac (TFD) theory (Barnes, 1967). The nuclear contributions are from a global model developed by R. D. Cowan above 1 eV and from modified Grueneisen theory below 1 eV. The cold curve contribution was calculated with a modified Morse model (Barnes, 1984) at low density and TFD theory at high density.

In expansion below 1 eV, Barnes and Rood used a virial series EOS. The coefficients of the virial expansion were determined by matching the thermal pressure and the slope of the thermal pressure with respect to density at the reference density ρ_o. The van der Waal's loops in the lead EOS were replaced by vapor pressures from a Maxwell construction.

This EOS reproduces Hugoniot experiments well. However, the character of the EOS in expansion in the gas-liquid two-phase region varies considerably from that of the new tabular EOS for lead which is described in the next section.

SESAME 3201

SESAME 3201 is a tabular EOS for lead which was constructed especially for this sensitivity study. It has been added to the SESAME library as material #3201. Since it has not been documented, we will describe it in some detail.

We chose three models to describe the physics in three different regions. (Fig. 3 shows the temperature-density range covered by each of the theories described below.) We generated smaller EOS subsurfaces with each of the three EOS computer codes, and then joined these together. Sometimes numerical interpolation was required to join two subsurfaces smoothly (indicated by the shaded region in Fig. 3).

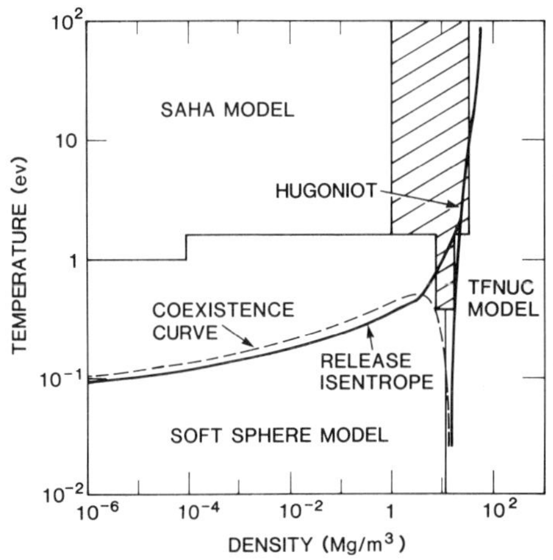

Fig. 3 A temperature vs. density plot of the EOS subregions calculated with the different models. The coexistence curve, Hugoniot and release isentrope from 250 GPa are superimposed. The shading indicates the region in which interpolation was required to smoothly join the models.

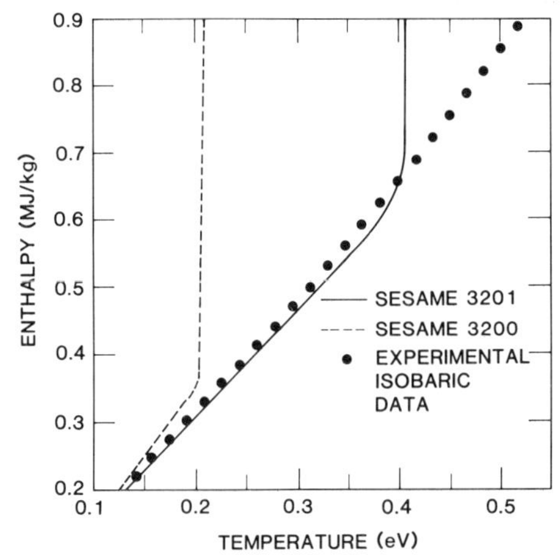

Fig 4 Enthalpy vs temperature comparison of the theoretical 0.1 GPa isobars of SESAME 3200 and SESAME 3201 with experimental data (Gathers, 1982). The discontinuity occurs as the isobar goes into the liquid-vapor two-phase region.

TFNUC MODEL

We generated the compression half of the EOS with the TFNUC code, itself a composite of several different models. TFNUC first calculates the contribution to the EOS from the electrons with the Thomas-Fermi statistical atom theory, and then corrects for quantum and exchange effects with the Kirzhnits formulation (McCarthy, 1965). TFNUC then adds ionic motion contributions (Ree, Daniel, Haggin, 1978) to the electronic EOS. These are based on Grueneisen-like theory at low temperature and one-component plasma theory at high temperature, with an interpolation scheme in between.

In the low temperature region, the Grueneisen parameter γ is represented by three different formulas. At low density,

$$\gamma_G = 2.629\ (V/V_o)^{1.2} \quad .$$

The constants for this equation were determined from fluid cell experiments (Ramakrishnan, Boehler, Higgins, Kennedy, 1978) in which small changes in temperature are measured as a function of small adiabatic pressure changes under hydrostatic conditions. The thermodynammic Grueneisen parameter can then be determined from the relation

$$\gamma_G = \frac{B_s}{T} \left(\frac{\partial T}{\partial P} \right)_s ,$$

where B_s is the adiabatic bulk modulus.

At high density, γ_G is a polynomial function of volume determined by a least squares fit to Kopyshev's numerical data (Kopyshev, 1965). Kopyshev calculated γ_G by displacing the ions by small amounts, then calculating the change in the resulting lattice potential under the Thomas-Fermi approximation.

For the intermediate densities, γ_G is described by the expression

$$\gamma_G = C_0 + C_1\rho^{1/4} + C_2\rho^{1/2} + C_3\rho^{3/4} + C_4\rho$$

The constants C_i are determined from the conditions that γ_G and the derivative of γ_G with respect to density be continuous across the boundaries between the three formulas.

TFNUC also requires that the Lindemann melting rule be obeyed:

$$\frac{d \ln T_m}{d \ln \rho} = 2 \, (\gamma_G - 1/3) \; .$$

SOFT SPHERE MODEL

The gas and liquid phases of the lead EOS were calculated with the soft-sphere model for liquid metals developed by D. Young (1977) at Lawrence Livermore National Laboratory. The soft-sphere model is based on Monte Carlo computer calculations (Hoover, Ross, Johnson, Henderson, Barker, and Brown, 1970) of fluids with interparticle potentials of the form $\phi(r) = \epsilon \, (r_o/r)^n$, where $4 \le n \le 12$, ϵ and r_o are constants, and r is the interparticle distance. The computer calculations were fit with a simple analytic function that consisted of ideal gas, static lattice, and configurational heat capacity terms. To make the model more applicable to liquid metals, two modifications (Hoover, Stell, Goldmark, and Degnani, 1975) were made: a negative cohesive mean-field term was added to the equation, and the soft-sphere heat-capacity term was multiplied by the factor Q. In the original soft sphere theory, Q was a constant. In order to improve the fit with experimental EOS data, we made Q a function of density and temperature.

In short, the following formula describes the free energy for the soft sphere model:

$$A = NkT \, [-3/2 \ln (kT/\epsilon) - \ln V + C_n\rho^{n/3} \, (\epsilon/kT)$$
$$+ (n/2 + 2) \, Q \, \rho^{n/9} \, (\epsilon/kT)^{1/3} - \rho^m(\epsilon/kT)] + E_{coh},$$

where

$$Q = Q_0 + Q_1 \, \rho^{n/3} \, (\epsilon/kT) + Q_2\rho^{2n/3} \, (\epsilon/kT)^2 \; .$$

The first two terms are the ideal gas; the third is the Madelung energy; the fourth is the soft-sphere heat-capacity term; the fifth is the van der Waals attractive term; and the sixth is the mean-field cohesive energy. C_n is the fcc Madelung constant, and $\rho = Nr_o^3/\sqrt{2}\,V$. The pressure and internal energy can easily be computed from the derivatives of the free energy with respect to volume and temperature.

We determined the best values for n, m, Q_0, Q_1, and Q_2 by comparing the theoretical EOS to isobaric expansion experiments. We used temperature versus enthalpy measurements from Hixson (1986) and Gathers (1982); sound speed versus density measurements from Hixson, Winkler, and Shaner (1986); sound speed versus temperature measurements from Gitis and Mikhailov (1965) and Markov (1975); and enthalpy versus density measurements from Gathers (1982). The fitting procedure in the soft-sphere code was automated by including a nonlinear least-squares package.

Given values for n, m, Q_0, Q_1, and Q_2, the parameters ϵ and r_o may be determined iteratively from the conditions that

$$P(V_M, T_M) = 0, \qquad E(V_M, T_M) = H_M,$$

where V_M, T_M, and H_M are the volume, temperature, and enthalpy at melt.

After the fitting procedure, a tabular EOS was calculated using the analytic equations for pressure and energy, and then, in the liquid-vapor region, the van der Waals loops were replaced with a Maxwell construction. Figs 2 and 4 show very good agreement between the theoretical 0.1 GPa isobar and the isobaric experiments (Gathers, 1982). The abrupt discontinuity is where the theoretical EOS enters the two-phase region, and is not reflected by the experiments. However, the higher experimental temperatures are probably not accurate; only the temperature measurements below 4000 K (0.34 eV) are to be trusted (Hixson, 1986).

SAHA MODEL

As temperatures increase, materials will ionize, and an EOS will begin to exhibit shell structure. So, for the lead EOS at temperatures above about 1 eV, we used a Saha model. This model calculates the concentration of all the possible ions and electrons at a given density and temperature. Then, using an ideal gas EOS for each species, it adds the EOS contributions together to arrive at a total equation of state. The Saha model used for this lead EOS was modified with the Debye-Hueckel correction for ionization potentials and Planck's theoretical partition function (Rouse, 1961, 1962). The EOS generated by this model is exact in the ideal gas limit, accurate in the weakly coupled Coulomb regime, but becomes increasingly inaccurate with the onset of strong Coulomb interactions and electron degeneracy.

COMPARISON OF LEAD EQUATIONS OF STATE

In this section we will compare the three equations of state described previously, and point out some differences which may affect hydrocode simulations of hypervelocity impacts. The Hugoniot and various release isentropes can give one a feel for the situations in which one might see differences. But the problem of a spherical projectile crashing through a thin metal plate is very complicated due to geometric considerations. A hydrocode user cannot expect to know how a simulation will differ with a different EOS just from simply inspecting the Hugoniot and release isentropes. To do a straight-forward EOS sensitivity study, it would be better to look at a one-dimensional problem in which one infinite metal plate impacts another, causing two oppositely-moving shock waves to propagate from the impact interface which then release from the back surfaces into the void. However, since experimental data exist for the sphere-plate impact, we decided to examine that first.

We chose for this comparison to look at the Hugoniot, the 0.1 GPa isobar and a release isentrope starting from 250 GPa on the Hugoniot (lead reaches a pressure of about 200 GPa when an impact occurs at 6.6 k/sec).

The Hugoniot describes the final values of the pressure, energy, and density of a material for a given shock compression. Fig. 5 shows the comparison between the theoretical Hugoniots and experiments. All three equations of state are similar in this region, since the models used to calculate the compression half of the equations of state are partly based upon Hugoniot experiments. So, one would expect that the impact simulation itself should be similar for all three equations of state.

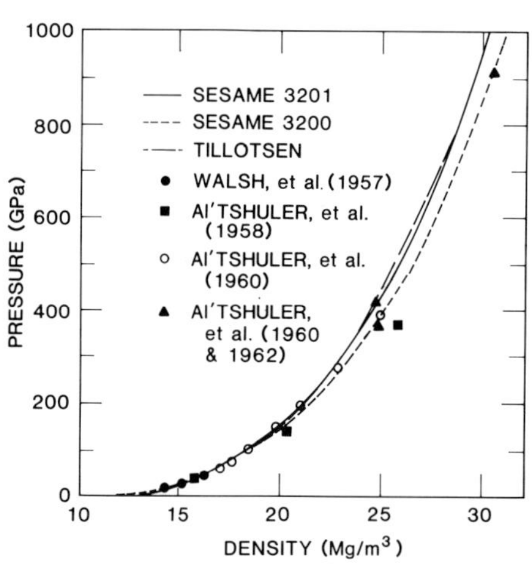

Fig. 5 The theoretical Hugoniots for the three lead equations of state are compared with experiments: the solid circles are from Walsh, et. al. (1957), the solid squares are from Al'tshuler, et. al. (1958), the open circles are from Al'tshuler, et. al. (1960), and the solid triangles are from Al'tshuler, et. al. (1960 and 1962).

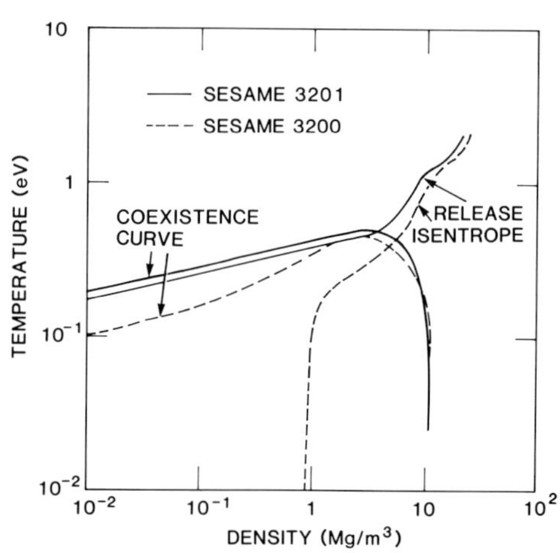

Fig. 6 Temperature vs density comparison of the coexistence boundaries for SESAME 3200 (solid line) and SESAME 3201 (dashed line). Also shown are release isentropes for the two EOS's starting at 250 GPa on the Hugoniot. Notice that the isentrope for SESAME 3201 barely penetrates the two-phase region.

We might expect more noticeable differences though, when the expanded region ($\rho < \rho_o$) of the EOS is accessed by the hydrocode. The main experimental tests we have for the EOS in expansion are isobaric experiments which measure density, temperature, enthalpy and sound speed at constant pressure. Fig. 2 is a comparison of the 0.1 GPa theoretical isobars (entropy vs density) of SESAME 3200, SESAME 3201 and the Tillotsen EOS with experiments. SESAME 3201 reproduces the data perfectly since the data were actually used to define some of the parameters in the EOS model. SESAME 3200 also fits the experiments fairly well, but the Tillotsen EOS shows unphysical behavior. The sharp discontinuity indicates the "phase transition" from the solid to the gas equation in the Tillostsen model.

Fig. 4 is a plot of the theoretical isobars (enthalpy vs temperature), which points out a major difference in the expansion equation of state for SESAME 3200 and SESAME 3201. The abrupt change in the two isobars occurs at the point of entry into the two-phase region. Notice that for SESAME 3200, the entry point is at approximately 2300 K, while for SESAME 3201 it occurs at about 4640 K. This is different by a factor of two, and shows that the boundary of the liquid vapor two-phase region is significantly different. (The fact that neither match the higher temperature data is not necessarily worrisome because of the experimental uncertainty in the temperature measurement.)

Fig. 6 shows the coexistence boundary and a release isentrope for SESAME 3200 and SESAME 3201. The release isentropes are from 250 GPa on the respective Hugoniots of the two EOS's. The isentrope for SESAME 3201 just barely enters the coexistence region, and then skims along the left-hand side, while that of SESAME 3200 enters more into the middle. To the right of the two-phase boundary, the material is in the liquid state; to the left it is in the gaseous state; inside, both phases coexist. So, upon release from a pressure of 250 GPa on the Hugoniot, the SESAME 3200 EOS will show a much higher proportion of liquid-to-gas than SESAME 3201. One can also see that for release from a slightly higher pressure, the isentrope for SESAME 3201 might not enter the two-phase region at all (material would be totally gaseous), while that of SESAME 3200 would still enter it (some material would be liquid).

Ideally, a hydrocode should treat the solid, liquid, gas, and two-phase states differently with regard to movement of material across a void. For example, if the energy deposited into a material is sufficient to cause partial or complete vaporization, a significant increase will occur in the material velocity. For hypervelocity impacts then, the thermodynamic state of the material will establish the radial expansion of the debris cloud. So the placement of the coexistence boundary in an EOS could have important implications in hydrocode calculations depending on the materials involved and the impact velocity.

HYDROCODE CALCULATIONS

The two-dimensional version of LASOIL was used to simulate the debris cloud resulting from the impact of a 20-gm lead sphere with a thin (0.63-cm-thick) lead bumper plate. The impact velocity was approximately 6.6 km/s. The debris cloud is the fragment pattern of projectile and bumper plate material that emerges from the rear surface of the plate following the impact. The LASOIL axisymmetric mesh for the simulation contained 37,500 computational cells--300 cells in the axial direction and 125 cells in the radial direction. A uniform cell size of 0.1 cm was used in both the r and z directions.

Three debris cloud characteristics are compared with experimental data for each calculation: mass per unit area ($\int \rho dx$) as a function of axial position (all positions measured from the rear of the bumper plate), and the debris cloud front centerline axial velocity and position. The experimental mass per unit area, as a function of z and r, was determined by a photodensitometer analysis of the debris cloud radiographs.

Figs. 7 and 8 are comparisons of the mass per unit area as a function of axial position z along the centerline (r=0) at 30 and 40 μs after impact. Notice how similar the hydrocode calculations are to each other. However, there are some striking differences between the calculations and experiment. There are two peaks in the experimental debris cloud mass along the centerline -- one essentially at the leading edge of the cloud and the other at about 7.5 cm in front of the target plate. We interpret this as a hollow, balloon-like cloud with a disk or band of material nearer the target. The peak density of the calculations, on the other hand, occurs only in one place and is substantially behind the cloud front (10-11 cm in front of the target). We interpret this as a ball of gas, most dense at the center of the cloud. Part of the discrepancy between experiment and calculation is due to the non-spherical projectile -- it partially melted on its way down the gun barrel, and was deformed into a bolt-like shape (head first). However, the problem cannot be totally explained by considerations of projectile geometry. (The small blip in the calculation near the front of the cloud is a numerical artifact, and is therefore not physically significant.)

The position and velocity (~ 6 km/s) of the debris cloud front are well reproduced by the hydrocode calculations. Actually, in a problem such as this, the cloud front velocity and, thus, position are largely determined by the ballistic velocity of the projectile, and so we would expect these quantities to be well reproduced. A more interesting comparison is that of the radial expansion velocity. We do not have the experimental values for this velocity, but we can compare the diameter of the cloud at its widest point at 30 and 40 μs. The cloud edge of the SESAME 3201 calculation does seem to be expanding slightly faster -- it is about 8 % wider at 30 μs. This is even more significant when one considers that the hole size in the target is at least one cm smaller for the calculations than the experiment.

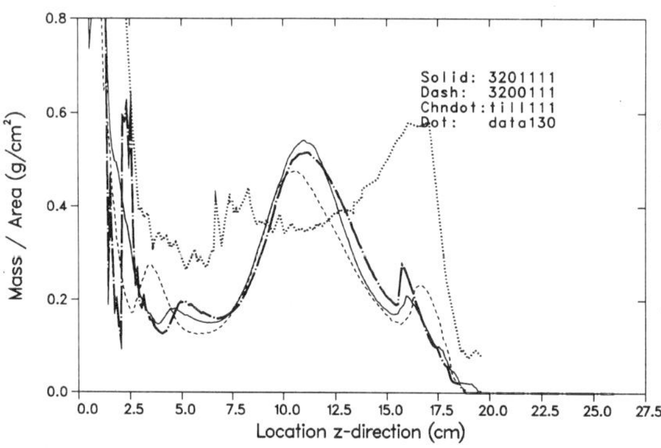

Fig. 7 Comparison of mass per unit area (r=0 cm) values as a function of axial position at 30 μs. The solid line (SESAME 3201), dashed line (SESAME 3200), and the chain-dot line (Tillotsen) are from the hydrocode calculations. Photodensitometer analysis of the experimental x-ray photographs is the dotted line.

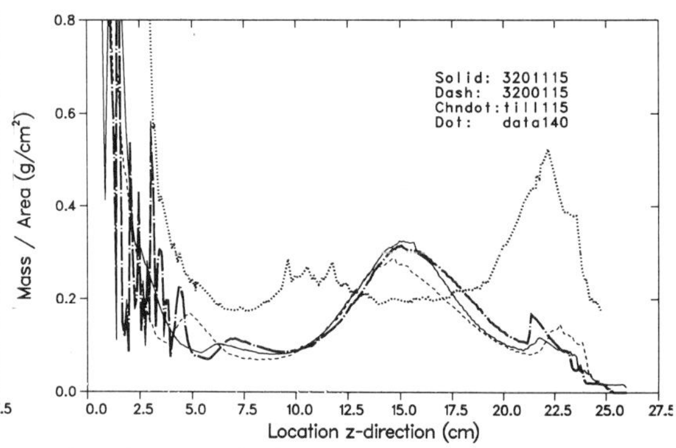

Fig. 8 Comparison of mass per unit area (r=0 cm) values as a function of axial position at 40 μs. The solid line (SESAME 3201), dashed line (SESAME 3200). and the chain-dot line (Tillotsen) are from the hydrocode calculations. Photodensitometer analysis of the experimental x-ray photographs is the dotted line.

CONCLUSION

For a 250 GPa shock, lead enters the two-phase region and appears to nearly completely vaporize upon release. (See Fig. 6.) Since all three equations of state have quite different two-phase regions, one might expect significant differences in the calculations. However, there are only 10-15% differences in the mass distribution predicted by the three hydrocode calculations along the center line of the cloud. Qualitatively, they are the same. In fact, the calculations are more similar to each other than they are to the experiment. The principle difference is that the experimental debris cloud appears to be hollow, while the calculated cloud is not.

There are a number of possible reasons for the qualitative discrepancies noted between theory and experiment. For example, since the projectile is believed to have partially melted on its way down the barrel, its actual shape is different from the spherical geometry used in the calculation. Also, since the lead debris cloud seems to be in the liquid-gas coexistence region, there are, in reality, small liquid droplets suspended in the expanding gas. The hydrocode, on the contrary, models the cloud as a uniform ball of gas. This might account for the faster radial expansion in the calculation. If the hydrocode could transport liquid and gas separately (two-phase flow), using the equation of state to determine the fractions of gas and liquid in the cloud, there might be more significant sensitivity to the EOS used in this problem.

Future study should include the effect of projectile geometry on the shape of the debris cloud. Also, further calculations and experiments should be performed at a variety of projectile velocities to investigate the EOS sensitivity over a wider range of conditions. In particular, we would like to investigate the sensitivity for impacts producing only partial or incipient vaporization. A more difficult physics problem that also needs attention is the treatment of two-phase flow in a hydrocode.

REFERENCE

Al'tshuler, L. V., S. B. Kormer, A. A. Bakanova, and R. F. Trunin (1960). Sov. Phys. JETP 11, 573.
Al'tshuler, L. V., S. B. Kormer, A. A. Bakanova, and R. F. Trunin (1962). Sov. Phys. JETP 15, 65.
Al'tshuler, L. V., S. B. Kormer, M. I. Brazhnik, L. A. Vladimirov, and A. I. Funtikov (1960). Sov. Phys. JETP 11, 766.
Al'tshuler, L. V., K. K. Krupnikov, and M. I. Brazhnik (1958). Sov. Phys. JETP 7, 614.
Barnes, J. F. (1967). Phys. Rev. 153, 269.
Barnes, J. F. (1984). Los Alamos National Laboratory, Los Alamos, NM (private communication), unpublished.
Gathers, G. R. (1982). Lawrence Livermore National Laboratory, Livermore, CA (private communication), unpublished.
Gitis, M. B. and I. G. Mikhailov (1965). Akustichaskii Zh. 11, 434 [Sov. Phys. - Acoustics 11, 372 (1966)].
Hixson, R. S. (1986). Los Alamos National Laboratory, Los Alamos, NM (private communication), unpublished.
Hixson, R. S., M. A. Winkler, and J. W. Shaner (1986). International J. of Thermophys. 7, 161.
Holian, K. S. (1984). T-4 Handbook of Material Properties Data Bases, Vol. 1C: Equation of State. Los Alamos National Laboratory Report LA-10160-MS, unpublished.
Hoover, W. G., M. Ross, K. W. Johnson, D. Henderson, J. A. Barker, B. C. Brown (1970). J. Chem. Phys. 52, 4931.
Hoover, W. G., G. Stell, E. Goldmark, and G. D. Degnani (1975). J. Chem. Phys. 63, 5434.
Kopyshev, V. P. (1965). Sov. Phys Doklady 10, 338.
LASOIL (2D version) is an Eulerian hypervelocity impact computer code that we extensively modified and is maintained by Ken Duerre of the Los Alamos National Laboratory. The original code, SOIL2D, was developed by W. Johnson of Computer Code Consultants.
Markov, B. G. (1975). Teplofizika Vysokikh Temperatur 13, 1108.
McCarthy, S. L. (1965). The Kirzhnits Correction to the Thomas-Fermi Equation of State. Lawrence Livemore National Laboratory Report UCRL-14362, unpublished.
Pomykal, G. W. (1986). Digitized Data for Delco Test 4007, Lead-on-Lead. Lawrence Livermore Laboratory Report DDV-86-0010, unpublished.
Ramakrishnan, J., R. Boehler, G. H. Higgins, and G. C. Kennedy (1978). J. of Geophys. Res. 83, 3535.
Ree, F., D. Daniel, and G. Haggin (1978). NUC - a Code to Calculate Ionic Contributions to P and E. Lawrence Livermore National Laboratory Internal Document HTN-291, unpublished.
Rouse, C. A. (1961). Astrophys. J. 134, 435.
Rouse, C. A. (1962). Astrophys. J. 136, 636.
Tillotsen, J. H. (1962). Metallic Equations of State for Hypervelocity Impact. General Atomic Division of General Dynamics Report GA-3216, unpublished.
Walsh, J. M., M. H. Rice, R. G. McQueen, and F. L. Yarger (1957). Phys. Rev. 108, 169.
Young, D. A. (1977). A Soft-Sphere Model for Liquid Metals. Lawrence Livermore National Laboratory Report UCRL-52352, unpublished.

Int. J. Impact Engng Vol. 5, pp. 343–355, 1987
Printed in Great Britain

0734-743X/87 $3.00 + 0.00
Pergamon Journals Ltd.

THE SCALING OF IMPACT PHENOMENA

Keith A. Holsapple

Aeronautics and Astronautics , University of Washington FS-10, Seattle, WA 98195;
and
Consultant, Boeing Aerospace Company , MS 3H-47, PO Box 3999, Seattle, WA 98124.

ABSTRACT

A basis for scaling laws for impact phenomena is the approximation of the actual problem with a point-source problem. Point-source solutions are characterized by a single scalar *coupling-parameter* measure $aU^{\mu}\delta^{\upsilon}$ of the radius a, velocity U and mass density δ of the impactor. Point-source solutions that are known to exist are discussed. The implications to scaling are developed. Recent numerical code calculations that were designed to investigate the scaling and the application of the coupling parameter are presented. Finally, a summary of some recent applications of the theory is given.

INTRODUCTION

In hypervelocity-impact problems, it is often necessary to extrapolate from results for one size or velocity to others significantly larger. This extrapolation uses *scaling theories*, which specify a particular functional dependence, invariably power-law, of a given phenomena on the impactor velocity and size. The proper exponents are necessary to obtain the right results.

In the twenty years since the last hypervelocity symposium, there have been significant advances in our understanding of the origin and applicability of scaling theories, following from the recognition of *point-source* solutions as their basis. It is now recognized that the simple assumption of *energy scaling* (based on the *a-priori* assumption that the sole measure of the impactor is its kinetic energy) is generally unsuitable. Indeed, it can be stated that there is not one single example in the impact literature where such an assumption can be demonstrated to be strictly valid. The same can be said of the alternate (and different) assumption of *momentum scaling*: that the momentum of the impactor is a suitable single measure. Instead, more general scaling theories based on more general point-source solutions have been developed *(Holsapple and Schmidt, 1986)* and are presented here.

For perfect gases, point-source solutions have been known and used for some time. The classical solution for the spherical expansion of a perfect gas due to a point deposition of energy has been known since the 1940's *(Taylor,1941,1950; von Neumann,1944; Sedov,1946)* and is commonly used for scaling the airblast from a nuclear weapon. In that case, the energy of the source does determine the point-source solution, so that energy scaling does apply.

Rae (1970) gave an extensive review of point-source and self-similar solutions for infinite half-spaces, with the emphasis on perfect gases and fluids. Dienes and Walsh developed their "late-stage equivalence" concept in a number of papers. In their approach, a single combined scalar measure of the impactor, of the power-law form aU^{μ}, is assumed to govern impacts at different impactor radii a and velocities U. *Dienes and Walsh (1970)* reported a number of numerical calculations for impacts into metals, and reported that the value $\mu=0.58$ successfully correlated the results. Rae showed that the Dienes and Walsh approach was the same as would follow from self-similar solutions.

These contributions focused primarily on the intermediate flow fields of impact problems: the velocity and pressure fields and the shock position and decay. It was recognized that the approach had important implications regarding the scaling of parameters such as the penetration depth and crater size, but there was not general agreement as to whether these results were valid.

Holsapple and Schmidt (1986) proved that a single scalar measure of the form $aU^{\mu}\delta^{\upsilon}$ (δ denotes the impactor mass density) is the general measure of any point-source solution. *Holsapple (1981,1982,1983a)* coined the terminology "coupling parameter" for this single scalar measure, since it characterizes the coupling of the impactor energy and momentum into the target material. It is the exponents μ and υ that determine the resulting scaling laws.

Holsapple and Schmidt (1986) extend the approach and results of the 1970's papers in two ways. In the first place, a much wider variety of point-source solutions exists than was presented earlier. There are solutions for idealized porous materials, parts of which have been reported by *Kompaneets (1956)* and *Zeldovich and Raizer (1967)*. These solutions are particularly important because they show that the scaling exponent μ can (for three-dimensional problems) range from a lower limit of *1/3* (equivalent to momentum scaling) to an upper limit of *2/3* (energy scaling) and further, that the value of μ that governs any particular problem *depends specifically on the material properties of the target material.* Highly porous (and very dissipative) materials can approach momentum scaling, while non-porous materials (non-dissipative for the simple model of the theory) are governed by energy scaling. Thus, the exponent $\mu=0.58$ as reported by *Dienes and Walsh* should not be expected to hold for all materials, and in particular would not be expected to hold for highly dissipative materials including geological materials such as ordinary dry soils. This observation is corroborated by the experimentally-obtained scaling of dry soil cratering problems, where the value of μ is approximately *0.4* (*Schmidt,1980*). Those results will be reviewed elsewhere in this conference *(Schmidt and Housen,1986)*.

The second extension to the earlier papers is in the application of the point-source solutions and the coupling parameter to the scaling of a large number of later-stage and final phenomena; in metals, water, soil, ice and other materials. Final crater sizes in dry sand, crater dynamic growth histories in water craters, melt volumes in impact cratering, ejecta blankets in soil cratering, craters in ice, craters in viscous materials, perforation of plates, formation of underground cavities by nuclear explosions and catastrophic collisions of asteroids are all examples where the scaling has been applied and found to work well.

In this paper, the point-source solutions that are known to exist will be discussed. The implications to scaling will be developed. Some recent numerical code calculations that were designed to investigate details of the scaling and the application of the coupling parameter will be presented. Finally, a summary of some recent applications of the theory will be given.

GENERAL IMPACT PROBLEMS

In an impact problem, an impactor ("projectile") with radius a, velocity U, and mass density δ impacts a material ("target") with mass density ρ and strength Y (assumed to have stress units). The projectile parameters also determine the projectile kinetic energy K, its mass m, and its momentum M (the non-zero component perpendicular to the target). The target is assumed to be of infinite thickness, the projectile to be spherical, and the impact to be normal to the target surface.

There are a variety of phenomena of interest. For definiteness, consider the maximum penetration d of the target. The value of d depends in some way on the parameters of the problem listed above:

$$d = f(a,U,\delta,Y,\rho) \tag{1}$$

It is easy to think of other parameters of the problem that might enter into (1) as independent variables: other material properties of the projectile or the target, the gravity field strength, an angle of oblique impacts, the properties of an overlying atmosphere, and so on. The effects of gravity only become important for extremely large impacts, such as in planetary-scale events (see, for example, *Holsapple and Schmidt, 1982)*. Those are not of primary interest here, so that gravity is excluded. Then, with only one important exception to be discussed below, it can be shown (*Holsapple and Schmidt,1986*) that <u>any additional parameters will not affect the form of the velocity or the size dependence in the final scaling relations</u>. Instead, additional parameters will introduce into the final functional forms corresponding nondimensional groups, and <u>the scaling results will continue to hold when these groups are held constant</u>; *e.g.*, for all cases of oblique impacts at a fixed angle, or for all cases of fixed materials of the impactor and the target. The important point is that the inclusion of these additional parameters will not affect the scaling laws relating to either the <u>size</u> or the <u>velocity</u> of the problem. The results derived from (1) are completely general in that regard.

The one exception to these statements is for cases of "rate-dependent" materials. In that case, an additional material property parameter with the dimensions of viscosity (or of time) must be included in (1). For simplicity, this case is specifically excluded here.[1]

A dimensional analysis of (1) can be used to restrict its form. For example, one such choice of dimensionless groups gives

$$\frac{d}{a} = f\left(\frac{Y}{\delta U^2}, \frac{\rho}{\delta}\right) \tag{2}$$

As discussed above, the inclusion of additional material properties in the beginning relation (1) would restrict the results to scaling only between different problems with fixed materials. In that case ρ/δ is also fixed, so there is little point to showing it explicitly in (2). For the remainder of the paper, the

[1]An example where rate effects may be significant is for hypervelocity cratering in ice or rock materials. *Holsapple and Housen (1986a,1986b)* consider the effects of the inclusion of a rate parameter in an approach similar to the one here.

dependences on this dimensionless group will not be shown.

While the reduction to (2) is significant; even for the case of fixed materials there remains an unknown functional dependence on the strength parameter $Y/(\delta U^2)$. If experiments are not possible at the velocity range of interest, then this unknown function cannot be determined experimentally. In this case it is important to have a viable way to further restrict the form of the function f of (2). These restrictions have as their basis additional assumptions about the nature of the coupling of the energy and the momentum of the impactor into the target.

POINT SOURCE APPROXIMATIONS

The phenomena of interest in impact mechanics often occur on a length scale large compared to that of the projectile. The penetration depth is often many projectile radii. A final crater may have one-hundred times the volume of the impactor. In these cases, a further approximation is warranted, based upon an equivalent *point-source* disturbance. The key to this approach is, of course, the determination of this "equivalent" point-source disturbance. Can the effects of the actual impactor be approximated by a point-source with the same *energy* (as for the perfect-gas case mentioned above), with a point-source of the same *momentum*, or is there *some other* appropriate point-source? The answer to this question is crucial, since it will fix the scaling rules that are subsequently obtained.

In order to illustrate this point, the special cases of energy scaling and of momentum scaling are now considered, even though it has been claimed that these are not generally valid for impacts into real solid materials. In the case of energy scaling it is assumed that, in the point-source limit, the single measure of the projectile is its kinetic energy K. Since $K \sim \delta\, a^3 U^2$ then the values of δ, a and U cannot separately occur in (1), but must occur in only this single combination. If those three variables are replaced by K, and the dimensional analysis redone, then there is only one dimensionless group which must be constant among all impact problems with fixed materials of projectile and target. That group is given by

$$\left(\frac{d}{a}\right)\left(\frac{Y}{\delta U^2}\right)^{1/3} = constant \tag{3}$$

(where a power of ρ/δ has been dropped) so that

$$\frac{d}{a} \propto \left(\frac{\delta U^2}{Y}\right)^{1/3} \tag{4}$$

Therefore, the assumption of energy scaling results in the penetration depth scaling with the velocity to the $2/3$ power.

In the case of momentum scaling, the three projectile parameters are replaced by the one combination giving its momentum M. Then it is easily found that

$$\frac{d}{a} \propto \left(\frac{\delta U^2}{Y}\right)^{1/6} \tag{5}$$

Therefore, if the momentum scaling assumption is valid, then the depth scales with the velocity to the $1/3$ power. Both of these mutually exclusive results are well known, and both follow from specific assumptions about the appropriate measure of the impactor.

Any real problem of interest has, of course, a finite source with some characteristic dimension a. The solution at range r will depend on the dimensionless ratio r/a. The limit as r becomes large can also be interpreted as a limit as a goes to zero. The latter interpretation leads to a point-source limit. Point-source solutions of impact problems are limits of sequences of problems where the projectile has a radius a approaching zero, and a velocity approaching infinity. These limits are considered in *Holsapple and Schmidt (1986)*, where it is shown that there will exist such a limit only if a particular measure C of the projectile is held constant in the limit process. This measure C, called the *coupling parameter*, must depend on the projectile radius, velocity and mass density specifically as

$$C \propto a U^{\mu} \delta^{\nu} \tag{6}$$

Special cases include the one where C is the cube-root of the kinetic energy, with $\mu=2/3$, $\nu=1/3$, and the case when it is the cube-root of the momentum, with $\mu=1/3$, $\nu=1/3$. Intermediate values of μ will occur in the more usual cases.

If some parameter of a problem such as the projectile penetration depth d is determined from the point-source solution, then in (1) there will not be separate dependences on a, U, and δ; instead those three are replaced by the single scalar coupling parameter C. Since the dimensions of C depend on the values of the exponents μ and υ, the results of a dimensional analysis will now depend on those values. In this case, the resulting dimensionless form of (1) is obtained as

$$\left(\frac{d}{a}\right)\left(\frac{Y}{\delta U^2}\right)^{\mu/2} = constant \tag{7}$$

where again a power of ρ/δ has been dropped, and the relation (6) has been used. Thus, in this general case the depth of penetration scales with the velocity to the power of μ. The previous (4) becomes a special case of (7) with μ=2/3, and (5) is a special case with μ=1/3. The general case is given by (7), and the remaining fundamental question becomes that of the appropriate value of the scaling exponent μ (or, what is the same thing, of determining the measure C of the governing point-source measure).

Thus the result (7) shows the position of energy scaling and of momentum scaling theories as two special and bounding limits to general scaling rules. Since the projectile material is held fixed in the scaling exercise, the value of υ is of little consequence: it will only occur in the powers of ρ/δ that have been dropped. What remains to be determined is the appropriate value for the exponent μ, which is possible in a number of ways. One can study specific, special theories where analytical results are available. Some of these are discussed in the next section. One can work out the scaling that follows from the assumption of the existence of a governing coupling parameter and compare to experimental results to determine the value of μ. This approach is also to be presented below. Its success will depend on the existence of a sufficient body of results with sufficient range of the parameters of interest. Finally, one can perform numerical experimentation using "hydrocodes" to search for the cases where the results are well-modelled by the theory of the coupling parameter. In this latter approach, one does not need to rely on the codes to give absolute results, but only results that are qualitatively correct.

EXAMPLES OF POINT-SOURCE SOLUTIONS

Analytical point-source solutions for general materials and for impacts into half-spaces and plates are not available. However, there are analytical solutions for simple geometrical configurations with simple material models, including perfect gases and idealized perfectly-porous materials. In addition, there have been numerical studies using wave-propagation "hydro-codes" to attempt to approximately determine the governing point-source solution for "real" materials. Solutions of these types are discussed here.

Perfect-Gas Solutions

Probably the most well-known point-source solution is for the instantaneous deposition of energy at a point in a perfect gas, with a subsequent outward-growing spherical shock (*Taylor,1941,1950; von Neumann,1944; Sedov,1946*). *Sedov(1959)* also presents solutions for the cases of cylindrical and planar symmetries. The case of planar symmetry corresponds to a problem where a thin sheet explosive is detonated against a half-space target, and the material is confined at the original half-space surface. A different case, of more revelance for impact problems, is when the material is allowed to expand across the initial surface. This case will be also be discussed here.

For the present purpose, the details of these solutions are not of interest. What is important is the observation that (in contrast to examples to be given shortly) they are all "energy" solutions: the coupling parameter measure C of the source is its energy (or some power of that energy) and is independent of all other characteristics of that source. For the spherical case, the measure is the cube-root of the energy, so that from (6) the scaling exponent μ=2/3. In the cylindrical case, the measure C is the cube-root of the energy per unit length, which gives that μ=1. Finally, for the planar case, C is the cube-root of the energy per unit area, which gives that μ=2. Once these scaling exponents are known, dimensional analyses can be used to generate the applicable scaling laws of those solutions. Those include, for example, the well-known results in the spherical case that the shock pressure decays with the range to the - 3 power, and the shock expands with time to the *0.4* power.

The other type of solution in the planar case for a perfect gas governs a problem where the free surface of a half-space of a perfect gas is subjected to a pressure pulse of magnitude P_0 for a short length of time t_0, which generates a shock that propagates down into the half-space away from the free surface. After the time t_0 the surface has zero pressure, and, in contrast to the solution discussed above, the gas is free to expand upwards across the plane that was the original free-surface. The point-source solution arises as the limit with $P_0 \to \infty$ and $t_0 \to 0$. The problem is meant to model, among others, the hypervelocity impact of a thin plate on the half-space.

The coupling parameter in the form (6) is for projectile impacts. For a pressure-pulse "source" there is an equivalent coupling parameter with a power-law form $C \sim P_0^a t_0^b$, with the exponents a and b chosen to give C the same units as the relation (6). It is easily shown that the consistent choice is to take

$$C \propto P_0^{(1+\mu)/2} t_0 \tag{8}$$

and $\upsilon=(1+\mu)/2$.

This problem is discussed by *Zeldovich and Raizer (1967)* (where the original references are cited) and in *Barenblatt (1979)*. If $\gamma=7/5$, then an analytical solution can be obtained *(von Hoerner, 1955)*. That solution shows that the pressure P_0 and the time t_0 of a finite source must satisfy $P_0 t_0^{\beta}= constant$, with $\beta=4/5$, in order that the point-source limit exists. When an appropriate root is taken to put this result into the form of C in (8), then one finds that $\mu=(2-\beta)/\beta$, which, with $\beta=4/5$ gives $\mu=3/2$, precisely midway between the case when C is equal to the cube-root of the energy per unit area (with $\mu=2$), and the case when it is the cube-root of the momentum per unit area (with $\mu=1$).

For other values of γ, the point-source (and self-similar) solution has been obtained by a sequence of transformations and a numerical solution of a resulting non-linear eigenvalue problem. This procedure is not of interest here, what are significant are the final results for the scaling exponent μ. Those results are as shown in Fig. 1. Note that for the case of $\gamma=1$, the solution is the "momentum" solution. As the value of γ increases, the corresponding value of μ increases, but it never reaches the "energy" limit of $\mu=2$. Indeed, as $\gamma \rightarrow \infty$, $\mu \rightarrow 1.79$. In this problem, energy scaling is never appropriate.

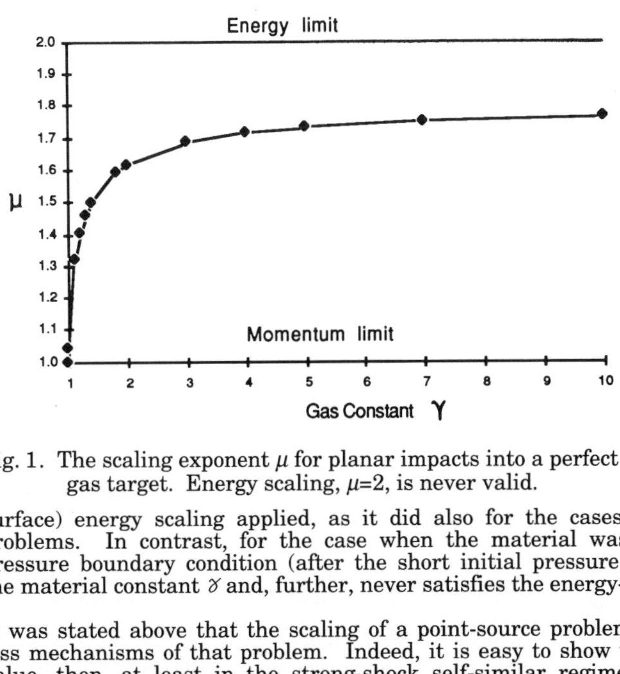

Fig. 1. The scaling exponent μ for planar impacts into a perfect gas target. Energy scaling, μ=2, is never valid.

There have also been numerical code solutions of this one-dimensional problem. These studies allow one to follow the evolution of a finite-source problem to the point-source approximation. References to these studies can be found in *Rae (1970)*.

The differences between the two planar problems that have been presented above should be discussed. In the case of a confined problem (when the material could not pass across the original half-space surface) energy scaling applied, as it did also for the cases of symmetrical cylindrical and spherical problems. In contrast, for the case when the material was allowed to expand freely under a zero-pressure boundary condition (after the short initial pressure pulse), the solution depends explicitly on the material constant γ and, further, never satisfies the energy-scaling conditions.

It was stated above that the scaling of a point-source problem depends in a crucial way on the energy-loss mechanisms of that problem. Indeed, it is easy to show that when μ is equal to the energy scaling value, then, at least in the strong-shock self-similar regime, any energy measure of the problem is constant in time. Conversely, if μ is less than that value, then any energy quantity described by the point source problem must decrease in time. As a consequence, energy-loss mechanisms are necessary if the scaling is not to be energy scaling.

Recall that the point-source problem is the limit of a sequence of problems with a finite source as the size of that finite source goes to zero. Consider the solution for a perfect gas that is constrained. In that case the limit solution is a *uniform* limit for the entire mass of target material, and all of the material is described by that solution. Since the actual finite-source problems have finite and constant total energy, then so must the point-source solution.

In the case of the impulsive load, there is an expansion of the material into the free space above the original half-space. As Zeldovich and Raizer show, for a finite-source problem with a finite pressure and time of application, there is a mass of material that was originally near the free surface that expands upward, carrying with it internal and kinetic energy. In the limit of a sequence of problems with the source size approaching zero the limit is not uniform , and the point-source limit never applies to some small amount of this mass. In the actual problem with finite pressure and finite source size, the total energy is constant in time after the pressure is applied; but in the point-source limit solution, the energy decreases continually in time, corresponding to the fact that a small mass is carrying a non-zero and increasing energy "away from the problem". Correspondingly, energy scaling does not hold,

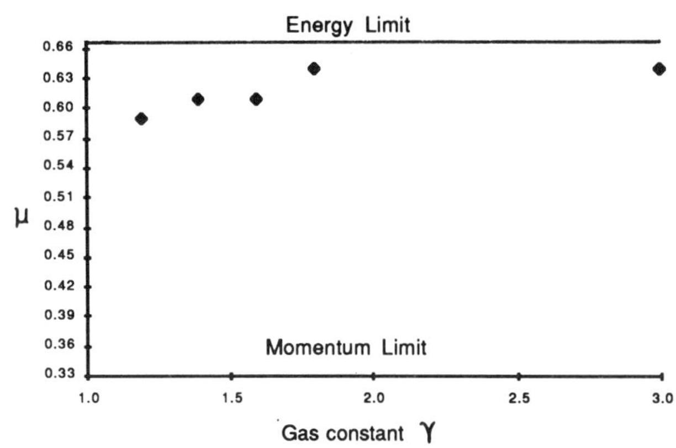

Fig. 2. Numerical calculations for the scaling exponent μ for explosive sources on a perfect gas half-space. Again, energy scaling, μ=2/3, is not observed.

and any energy measure of the point-source solution decreases in time.

For other problems involving perfect gases, such as the impact of a compact projectile into a half-space target, analytical techniques are not sufficient for determining the point-source solution, and one must resort to additional approximations or to direct numerical calculation using hydro-codes. Approximate techniques have been discussed by *Rae and Kirchner (1963)* and by *Raizer (1963)*. Hydro-code numerical solutions have been determined for impacts by *Walsh and Johnson(1965) for γ=1.5,* by *Chou and Burns (1967)* with γ=*1.1, 1.4*, and *2.0*, and, more recently, for an explosion source and for various γ by *Nutt et al. (1981).*

The results of *Nutt et al. (1981)* are reproduced in Fig. 2, and indicate a slight dependence of the scaling exponent on the ratio of specific heats γ. More importantly, it is seen that energy scaling is not obtained, but for the usual values of γ the scaling is close to energy scaling. It is most probably the transport of energy by the upward moving ejecta that is the mechanism for the energy losses in this problem.

Perfectly-Porous Materials

A highly instructive example of point-source solutions arises in the analysis of a so-called "perfectly-porous" material. The virgin material has an initial density ρ_0 but crushes at zero pressure to a final density ρ_1 and subsequently remains incompressible at that density. The porosity is defined as $k=1-(\rho_0/\rho_1)$.

Cases of planar, cylindrical, or spherical symmetry have been analyzed by the present author (unpublished). This analysis is an extension of those presented by *Kompaneets (1956)* and by *Zeldovich and Raizer (1967)*, who considered only some aspects of the spherical problem.

The source is assumed to be an initial high-pressure source region of pressure P_0 and of finite dimension a_0. As the source cavity expands, the pressure is allowed to drop following a γ-law pressure boundary condition.

In contrast to the general case of a perfect gas discussed above, an analytical solution can be obtained for a finite source, and the approach of that solution to a point-source approximate solution (corresponding to the case when $P_0 \to \infty$ and $a_0 \to 0$) can be studied. Here it is the scaling exponents of the point-source solutions that are of interest. The "source" of this problem is measured by the cavity radius a_0 and the pressure P_0 The power-law combination that has the same units as (6) defines the coupling parameter for this problem. It is given by

$$C \propto P_0^{\mu/2} a_0 \tag{9}$$

In the case of a planar impact, it is found that the exponent μ of the coupling parameter is given by[2] $\mu=1$ for all values of $k>0$, but is equal to *2* when $k=0$. For cylindrical impacts μ is given by the formula

$$\mu = \frac{\ln k}{(\ln k + k - 1)} \tag{10}$$

and, finally, for spherical symmetry by

$$\mu = \frac{2}{\left(k + k^{2/3} + k^{1/3} + 3\right)} \tag{11}$$

---

[2]There are actually two possible solutions, depending on the source γ. The solution given here is for $\gamma \geq 2/n\mu$ for which the pressure in the source region drops "fast-enough" to be the same as for an impulsive loading.

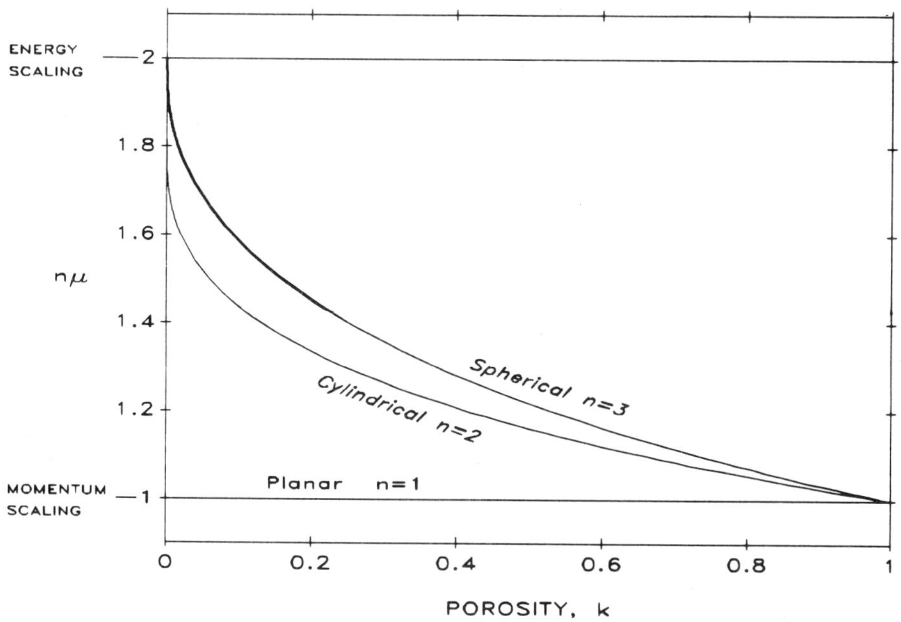

Fig. 3. The dependence of the scaling exponent μ on the porosity, k, of a perfectly-porous material, for cases of spherical *(n=3)*, cylindrical *(n=2)* and planer *(n=1)* symmetry.

This last result has been given by *Zeldovich and Raizer (1967)*.

All three results are conveniently written in terms of a geometry parameter n, which is equal, respectively, to *1, 2, or 3* for problems of planar, cylindrical, or spherical symmetry. Then the energy limit is given by $2/n$ for all three cases, and the momentum limit is $1/n$. Figure 3 shows the results of the above three equations as a plot of $n\mu$ versus the porosity k. The energy limit is obtained when $k = 0$, corresponding to no porosity. At the other extreme, momentum scaling is obtained when $k = 1$, which corresponds to complete porosity. For planar impacts, momentum scaling always governs, for any value of k except at the limit of $k=0$ for which the material is incompressible.

In this important example, the energy-loss mechanism is the porosity of the material. The total energy (internal plus kinetic) in the actual finite-source problem is constant in time. However, as the shock moves out, it encompases new material and imparts to that new material an internal energy which is subsequently fixed in time. This energy must come from the kinetic energy of the previously engulfed material. Thus, this kinetic energy must decrease in time. The energy loss mechanism is the crushing of the material.

General Materials

For material models other than perfect gases or perfectly-porous materials, numerical experimentation using hydro-codes must be used to study the problem. The simplest case is for a planar impact. *Chou and Burns (1967)* performed calculations using a characteristics code for planar impacts into both aluminum and copper materials modelled by a Tillotson equation of state. For the copper target, they obtained a value of $\mu=1.5$. Since their material model approaches a perfect gas with $\gamma=1.5$ at high pressures, and the impact velocities were several tens of *km/sec* for which the pressures are high, their exponent would be expected to match that for a perfect gas with that γ, which from Fig. 1 given earlier, is $\mu=1.53$. Therefore, their calculation gives good agreement with the expected value. In the case of the aluminum, however, they obtained $\mu=1.27$, considerably lower than the expected value of *1.53*.

This discrepancy was the object of a recent investigation by *Bjorkman and Holsapple (1983)*, who redid calculations matching those of *Chou and Burns (1967)*, but using the finite-difference hydro-code *WONDY*. Again, a Tillotson equation of state was used, and the calculation performed for a variety of impact velocities from *20 km/sec* and higher. Contrary to the results of Chou and Burns, the value of the scaling exponent obtained was $\mu=1.53$, as would be expected. It is unknown why the earlier

VARIABLE	POWER-LAW FORM	EXPONENT a
Shock pressure	$p \propto x^a$	$-2/\mu$
	$p \propto t^a$	$-2/(1+\mu)$
Shock velocity or Particle velocity	$v \propto x^a$	$-1/\mu$
	$v \propto t^a$	$-1/(1+\mu)$
Kinetic Energy	$K \propto x^a$	$(3\mu-2)/(1+\mu)$
	$K \propto t^a$	$(3\mu-2)/\mu$
Momentum	$M \propto x^a$	$(3\mu-1)/(1+\mu)$
	$M \propto t^a$	$(3\mu-1)/\mu$
Shock position or Crater Size	$x \propto t^a$	$\mu/(1+\mu)$

Table 1. Power law results for the strong shock regime in terms of the scaling exponent μ.

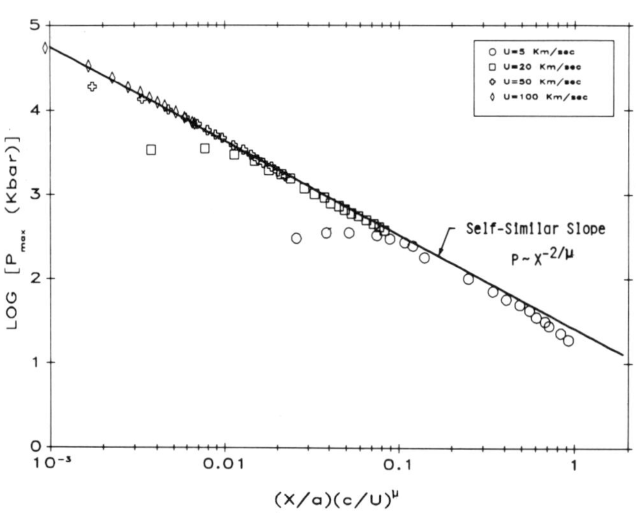

Fig. 4. The shock pressure decay versus range for planar impacts of a plate into a porous half-space with 30% porosity. The various cases superimpose to the same point-source solution with μ=1.7

calculations did not agree.

Therefore, it can now be said that the results of calculations for planar impacts using code models of real materials substantially agree with the analytical results for the perfect gas case. Even when the actual material models have detail and features not present in the simpler perfect gas case, if those parts of the model are not exercised for the earlier parts of a hypervelocity impact problem then the codes simply trace the evolution of a finite-source problem into the same point-source solution as governs a perfect gas. At the same time, these calculations also illustrate the use of codes to study the evolution of a real problem into the point-source approximation, and show that scaling exponents can be determined in that manner.

For the case of the impact of a compact projectile into an infinite half-space, a large number of individual calculations have been performed. However, the intent of most of those problems was not to study the scaling issues, but to model a particular problem. If the impact velocity is sufficiently large, then the solution is in the so-called "strong-shock" regime, and a number of problem variables (such as the shock pressure) have a power-law decay in time, with the value of that power directly related to the scaling exponent μ of interest here (see *Holsapple and Schmidt,1986)*. Table 1 lists some results of that type. As a consequence, if a calculation is done with a very large impact velocity, then the scaling exponent can be obtained from that calculation by a comparison of the numerical results with the results of Table 1.

On the other hand, for typical impact velocities into solid materials, the pressures are usually not large enough to be in this strong shock regime. For typical solids, that would require velocities of many *tens of km/sec* or so. In those cases, a comparison of at least two calculations, at different impact velocities, is necessary in order to determine the scaling coefficient μ.

Dienes and Walsh (1970) used this latter approach for the purpose of determining the scaling exponent μ. They performed a series of calculations for a number of metals, and for several impact velocities. Their primary conclusions included that *1)* the scaling exponent was μ*=0.58* for all the materials; *2)* the projectile density was of no consequence at fixed projectile mass and velocity; and, *3)* different compact-shaped projectiles give the same results at the same mass and velocity. They also made the important observation that the scaling exponent would imply a penetration dependence on the impact velocity to the *0.58* power (as per the relation (7)), between that of energy scaling and momentum scaling.

More recent calculations designed to study the scaling question have been performed by the present author. One goal was to investigate the scaling for porous materials, since the idealized solutions showed a strong dependence on that particular property. Figure 4 shows the results of a number of one-dimensional planar calculations of an impact into a porous material. The material had an air-filled porosity of *30%*. The fully-compacted material was modelled by a standard Mie-Gruniesen form. The various impact conditions were scaled in the way corresponding to the assumption of a coupling parameter, as are to be presented in the next section. Surprisingly, in contrast to the ideal solution for a perfectly-porous material (which is momentum-scaled for planar impacts) the solutions gave the value μ*=1.7*. (Recall that in one-dimension, the energy and momentum limits are $\mu=2$ *and* $\mu=1$, respectively). Either the real aspects of the material model, or the code approach modified the scaling considerably.

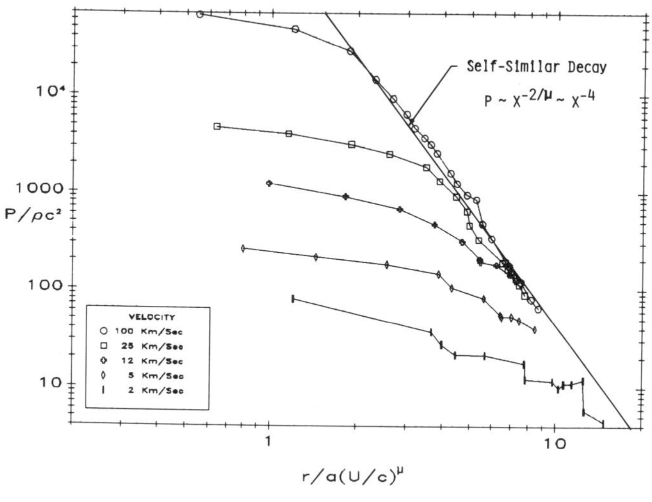

Fig. 5. Code calculations of a spherical aluminum impactor into a half-space of a 30% porous target. The various velocity cases superimpose and match a point-source solution with μ=0.5

Figure 5 shows numerical results for impacts of a spherical projectile into an infinite half-space. Again the porosity was *30%*, but in this case a physically realistic material model was used. In this case, the perfectly-porous model gave, for the case of spherical symmetry, the scaling exponent μ*=0.45*, while, as indicated on the figure, the code calculations were correlated by the value μ*=0.5*. While these values are not substantially different, it was expected that, if anything, the presence of the free surface and the resultant loss of material into upward moving ejecta would give a scaling below that of the idealized spherical solution. Instead, the exponent was above. As for the planar impacts, the code calculation was much closer to the energy scaling than was expected at the outset of the study. The use of codes and realistic material models for the determination of the scaling laws will require more study.

SCALING RESULTS FROM POINT-SOURCE SOLUTIONS

If one simply assumes that a point-source solution is a valid approximation for some aspect of an impact problem, then all scaling results can be worked out in terms of the exponent μ that governs that point-source solution. In the equations (1) through (7) above, the basic method was illustrated. One replaces a dependence on the three impactor measures *a, U* and δ by a dependence on the single coupling parameter *C* of the point-source solution, and performs a dimensional analysis. The results are obtained in terms of the value of the exponent μ .

Consider as an example the pressure *p* versus range *r* at time *t*. At all times prior to the final stages of the impact problem, *p* will then be a function of the coupling parameter *C* (itself a power-law function of *a,U*and δ) of the range *r* and time *t*, and on material properties of the target. Assuming that the target has no rate effects, then there are at most two material properties with independent dimensions[3], which are taken to be the mass density ρ and a material sound speed *c*. These two properties can be used to form dimensionless ratios with any further material properties, and the occurence of those further ratios will not affect any results in a given target material. As a consequence, one can write

$$p = f(C,\rho,c,r,t) \tag{12}$$

[3] Except for the final stages of an impact problem, the role of material strength is negligible. For early or intermediate-time phenomena, the results depend on the wave speed *c* .

This general expression for the pressure has six variables, which can be rewritten in terms of three dimensionless groups. Any three of the independent variables can be chosen to form power-law combinations to be used to nondimensionalize the remaining *3*. Generally one will want to use the coupling parameter C and the mass density ρ as two of the three. Then there are only three remaining useful choices, depending on whether one picks c, r or t as the final variable. For example, choose C, ρ and c. Using these, the combinations with the same units as p,r and t are given by, respectively, ρc^2, $a(U/c)^{\mu}$, and $(a/c)(U/c)^{\mu}$. Using those to nondimensionalize (12) gives

$$\frac{p}{(\rho c^2)}=f\left[\left(\frac{r}{a}\right)\left(\frac{c}{U}\right)^{\mu},\left(\frac{ct}{a}\right)\left(\frac{c}{U}\right)^{\mu}\right] \tag{13}$$

as the general form for the point-source solution.

Special cases are easily obtained; e.g. for the shock pressure as a function of range, one simple drops the dependence on the time t in (12) and therefore in (13). In the strong-shock regime, where the sound speed c can be ignored and must therefore drop out, then, in addition, the dependence on the remaining first variable must be power-law to make the c parameter cancel out. A little algebra shows that the previous result given in Table 1 is found.[4]

A complete description of these methods is given in *Holsapple and Schmidt(1986)*. Here, the attached Table 2 (a simplified version of one in that reference) is offered for the convenience of the reader. It shows various useful ways to construct the correct dimensionless functional forms. The case *1* of the first row gives the general case, without any assumption about the source. The following cases *2* through *4* give the three possibilities mentioned above assuming that the coupling parameter measure is suitable. Finally, the case *5* is to be used for any final variable depending on a material strength Y of the target.

As a final application, this case *5* is illustrated. Consider the final crater volume V for the impact into a half-space target. Then for the two material properties it is convenient to choose ρ and Y, so that

$$V=f(C,\rho,Y) \tag{14}$$

In this case, there are only three independent variables, so that they can be used to nondimensionalize the single dependent variable V. Since V has the units of the cube of a length , one simply takes the third power of the entry in the column labeled as "length", to get a single group which must be constant:

$$\left(\frac{V}{a^3}\right)\left(\frac{Y}{\rho U^2}\right)^{3\mu/2} = constant \tag{15}$$

so that the crater volume must be proportional to the impact velocity to the power of 3μ, to the projectile radius to the power of *3*, and to the material strength to the inverse power of $3\mu/2$. Similar approaches can be used for any other variables of interest.

EXPERIMENTAL RESULTS

There are a number of experimental results that indicate the scaling of impact problems. It was the results from a number of impact (up to *6 km/sec*) and explosive cratering events in a porous dry sand that furnished the motivation for the study of general scaling relations by the present author. In a number of papers, (see *Holsapple and Schmidt, 1986* for references) the results for each of final crater volume, crater ejecta scaling, and crater dynamic growth histories were amazingly consistent with the coupling parameter theory with $\mu=0.4$, and with each other.[5]

Impact experiments were also performed into water, primarily at elevated gravity (*Holsapple and Schmidt, 1982*). In that case the results for maximum transient cavities were completely correlated by $\mu=0.55$. More recent studies of the cavity growth histories (*Schmidt,1983*) for water craters also gave that same value of μ. This value for water was close to the values of *0.58* given by *Dienes and Walsh* for impact problems , but the sand results were substantially lower. Consequently, there arose an important question as to the source of the value of μ. The illustration by the porous model of the strong dependence of the value of μ on the porosity seemed to answer that question.

---

4 The same result could have been obtained more directly by choosing C, ρ, and r to construct the scale factors, and then dropping the parameters containing t and c.

5 In contrast to strength-dominated materials considered here, the crater volumes in dry sand are determined by the gravitational field, not the material strength.

Case	Constructed from:	For Velocity v_o, c_o (a)	For Pressure P_o, K_o (b)	For Length $r_o, V_o^{1/3}$ (c)	For Time t_o (c)/(a)	For Strength Y_o same as (b)	For Mass m_o bc^3/a^2
1.	ρ, U, a *(General Case)*	U	ρU^2	a	$\frac{a}{U}$	ρU^2	ρa^3
2.	ρ, aU^μ, c	c	ρc^2	$a\left(\frac{U}{c}\right)^\mu$	$\left(\frac{a}{c}\right)\left(\frac{U}{c}\right)^\mu$	ρc^2	$\left(\rho a^3\right)\left(\frac{U}{c}\right)^{3\mu}$
3.	ρ, aU^μ, r	$U\left(\frac{a}{r}\right)^{\frac{1}{\mu}}$	$\left(\rho U^2\right)\left(\frac{a}{r}\right)^{\frac{2}{\mu}}$	r	$\left(\frac{a}{U}\right)\left(\frac{r}{a}\right)^{\frac{1+\mu}{\mu}}$	$\left(\rho U^2\right)\left(\frac{a}{r}\right)^{\frac{2}{\mu}}$	ρr^3
4.	ρ, aU^μ, t	$U\left(\frac{a}{Ut}\right)^{\frac{1}{1+\mu}}$	$\left(\rho U^2\right)\left(\frac{a}{Ut}\right)^{\frac{2}{1+\mu}}$	$a\left(\frac{Ut}{a}\right)^{\frac{\mu}{1+\mu}}$	t	$\left(\rho U^2\right)\left(\frac{a}{Ut}\right)^{\frac{2}{1+\mu}}$	$\left(\rho a^3\right)\left(\frac{Ut}{a}\right)^{\frac{3\mu}{1+\mu}}$
5.	ρ, aU^μ, Y	$\left(\frac{Y}{\rho}\right)^{\frac{1}{2}}$	Y	$a\left(\frac{\rho U^2}{Y}\right)^{\frac{\mu}{2}}$	$\left(\frac{a}{U}\right)\left(\frac{\rho U^2}{Y}\right)^{\frac{1+\mu}{2}}$	Y	$\left(\rho a^3\right)\left(\frac{\rho U^2}{Y}\right)^{\frac{3\mu}{2}}$

Table 2. Scale factors for constructing nondimensional groups and scaling laws.

Space does not permit a presentation of all of the experimental results and applications that have been compared to the analysis of the coupling paramter and the point-source solutions. Many are presented in *Holsapple and Schmidt (1986)* . Those applications include explosive cratering in sand *(Schmidt and Holsapple,1980)*, impact cratering in each of sand *(Schmidt, 1980; Holsapple and Schmidt, 1982)*, water *(Holsapple and Schmidt,1982)*, metals *(Holsapple and Schmidt,1982)*, ice *(Gaffney,1984; Holsapple and Housen,1985,1986b)* and in a rate-dependent silicon oil *(Fink et al.,1984)*; transient crater growth histories in general *(Holsapple, 1983b)*, in water *(Schmidt,1983)*, in wet sand *(Schmidt,1984)*, in dry sand *(Schmidt, et al.1985)*, and in metals *(Kineke,1961)*; crater time of formation *(Schmidt,1981; Holsapple,1983b)*; ejecta scaling for explosive and impact craters *(Housen et al.,1982,1983)*; melt volume scaling for impact craters *(Bjorkman,1983,1984; Bjorkman and Holsapple,1986)*; nuclear cavity formation *(Holsapple,unpublished)*, and explosively-generated water waves *(Schmidt and Holsapple, 1984)*. A recent application is to the catastrophic collision and breakup of asteroids and other large planetary bodies *(Holsapple and Housen, 1986a)*. An application of the cylindrical point-source solution to plate-perforation experiments is considered in these proceedings by *Arione and Bjorkman(1986)*.

SUMMARY

The general theory of impact scaling laws and their origins from point-source approximations and the resultant coupling parameter measure can be said to be well understood at the present time. The positions of energy and momentum scaling as ideal limits to those scaling laws is clear. It is seen that neither is likely to be valid for an impact into a real material half-space or plate. It is also clear that the appropriate scaling law for a particular application is likely to depend strongly on the material of that problem, and one should not expect a "universal" law for all materials. Increased material dissipation probably causes the scaling law to deviate increasingly from energy scaling. It is still not clear whether one can predict with a code calculation the exact value of the scaling exponent that will govern particular applications. Indications are numerous that the value of μ for problems with non-porous materials is in the range of *0.55 to 0.6,* and for porous materials may be as low as *0.4*.

ACKNOWLEDGEMENTS

This research was supported by NASA grant NAGW-328 and contract NAGW-3291.

REFERENCES

Arione, S.E. and M.D. Bjorkman (1986). Scaling flow fields from the impact of thin plates. *This conference.*

Barenblatt, G.I. (1979). *Similarity, Self-similarity, and Intermediate Asymptotics.* Consultants Bureau, New York.

Bjorkman, M.D. (1983). Dependence of crater melt on impact velocity. *EOS Trans. AGU 64*, p747.

Bjorkman, M.D. (1984). Feasibility of determining impact conditions fron total melt volume. *Lunar Planet. Sci. XV*. pp64-65.

Bjorkman, M.D. and K.A. Holsapple (1983). Plane hypervelocity impact and source similitude dependence on a Tillotson EOS. *Proc. 3rd APS Topical Conf. on Shock Waves in Condensed Materials.* North Holland, Amsterdam.

Bjorkman, M.D. and K.A. Holsapple (1986). Velocity scaling impact melt volume. *(These proceedings).*

Chou, P.C. and B.P. Burns (1967). Late-stage equivalence in one-dimensional impacts. *J.Applied Phys. 38.* No.2. pp553-560.

Dienes, J.K. and J.M. Walsh (1970). In High-Velocity Impact Phenomena, R.Kinslow, Ed., *Theory of impact: Some general principles and the method of Eulerian codes.*Academic Press, New York. pp 46-104 .

Fink, J.D., D. Gault, and R. Greely (1984). The effect of viscosity on impact cratering and possible application to the icy satellites of Saturn and Jupiter. *J.Geophys Res. 89*,(B1). pp417-423.

Gaffney, E.S. (1984). Effect of strain-rate dependent yield strength on crater scaling relations. *Geophys. Res. Lett. 11:2.* pp121-123.

Holsapple, K.A. (1981). Coupling parameters in cratering mechanics. *EOS Trans. AGU 62*, p949.

Holsapple, K.A. (1982). A comparison of scaling laws for planetary impact cratering: Experiments,calculations and theory. *Lunar Planet. Sci. XIII.* p. 331.

Holsapple, K.A. (1983a). On the existence and implications of coupling parameters in cratering mechanics. *Lunar Planet. Sci. XIV.* p. 319.

Holsapple, K.A. (1983b). The scaling of impact crater growth. *EOS Trans. AGU 64.* p747.

Holsapple, K.A. and K.R. Housen (1985). Scaling considerations for cratering in ice. *Lunar Planet. Sci. XV.*

Holsapple, K.A. and K.R. Housen (1986a). Scaling laws for the catastrophic collisions of asteroids. In *Memorie della Societa Astronomica Italiana,57*, No.1. pp. 65-86.

Holsapple, K.A. and K.R. Housen (1986b). Cratering estimates for ice. *In preparation.*

Holsapple, K.A. and R.M. Schmidt(1982). On the scaling of crater dimensions 2: Impact processes. *J. Geophys. Res. 87* (B3), pp 1849-1870, 1982.

Holsapple, K.A. and R.M. Schmidt (1982) . On the scaling of crater dimensions-2: Impact processes. *J.Geophys.Res.87*(B3). pp 1849-1870.
Holsapple, K.A. and R.M.Schmidt (1986) . Point-source solutions and coupling parameters in cratering mechanics. In press, *J. Geophys. Res.*.
Housen, K.R., R.M. Schmidt, K.A. Holsapple and A.J. Piekutowski (1982). Scaling of explosion crater ejecta and velocities. *EOS (Trans. AGU) 63*, p1020.
Housen, K.R., R.M. Schmidt and K.A. Holsapple (1983). Crater ejecta scaling laws 1: Fundamental forms based on dimensional analysis. *J. Geophys. Res. 88*, (B3). pp2485-2499.
Kineke, J.H. (1961). Observations of crater formation in ductile materials. *Proc. 5th Symp. on Hypervelocity Impacts,* v1, pt2. pp339-370.
Kompaneets, A.S. (1956). Shock waves in a plastic compacting medium. *Dokl. Akad. Nauk. SSSR 109,* pp49-52.
Nutt, G., L. Klein and A.E.Ratcliffe (1981). Explosions on a gas-vacuum interface. *Phys. Fluids 24* (12). pp2150-2153.
Rae, W.J. (1970). Analytic studies of impact-generated shock propagation: Survey and new results. In *High-Velocity Impact Phenomena*, edited by R.Kinslow, Academic Press, New York. pp. 213-291.
Rae, W.J. and H.P.Kirchner (1963). *Cornell Aeron. Lab. Rept. No. RM-1655--M-4*, N 63-16887.
Raizer, Y.P. (1963). In *General Dynamics Rpt. No. GA-tr-5081*, AD 440513, 1964.
Schmidt. R.M. (1980). Meteor Crater: energy of formation. *Proc. Lunar Planet. Sci. Conf. XI*. pp. 2099-2128.
Schmidt. R.M. (1981). Scaling crater time of formation (abstract). *EOS (Trans. AGU) 62*, p944.
Schmidt. R.M. (1983). *EOS (Trans. AGU)64*. p747.
Schmidt, R.M. (1984). Transient crater motions: saturated sand centrifuge experiments. *Lunar Planet. Sci. XV*. pp722-723.
Schmidt, R.M. and K.A. Holsapple (1980). Theory and experiments on centrifuge cratering. *J.Geophys. Res.85*.(B1). pp235-252.
Schmidt, R.M. and K.A. Holsapple (1984). Explosively generated water waves. *EOS (Trans. AGU)65*. p221.
Schmidt, R.M. and K.R. Housen (1986). Some recent advances in the scaling of impact and explosion craters. *(this conference).*
Schmidt, R.M., K.R. Housen and A.J. Piekutowski (1985). Experimental observation on crater growth regimes. *EOS (Trans. AGU) 66*, p296.
Sedov,L.I. (1946). *Prikladnaya Matematika i Mekhanika*, Vol 10, No.2,241-250.
Sedov,L.I. (1959). *Similarity and dimensional methods in mechanics.* Academic, New York.
Taylor, G.I. (1941). The formation of a blast wave by a very intense explosion. *British Report RC-210*.
Taylor, G.I. (1950). The formation of a blast wave by a very intense explosion. *Proc. Royal Soc., Series A,* Vol. 201, part 1.
von Hoerner, S. (1955). "Losung der hydrodynamischen gleichungen mit linearem verlauf der geschwindigkeit" *Z.Naturforsch 10a*. pp687-692.
von Neumann, J. (1944). The point source solution. In *Shock Hydrodynamics and Blast Waves,* Bethe,H.A., Ed., AEDC-2860.
Walsh,J.M. and W.E.Johnson (1965). *Proc.Hypervelocity Impact Symp. 7th. 2.* pp1-75.
Zeldovich, Y.B. and Y.P.Raizer (1967). *Physics of Shock Waves and High-Temperature Hydrodynamic Phenomena,Vol.II*, Academic Press, New York.

Int. J. Impact Engng Vol. 5, pp. 357–362, 1987
Printed in Great Britain

0734-743X/87 $3.00 + 0.00
Pergamon Journals Ltd.

EXPLOSIVE GUN LAUNCHER FOR IMPACT STUDIES

W. H. Holt*, W. G. Soper*, and W. Mock, Jr.*

*Naval Surface Weapons Center, Dahlgren, Virginia, U.S.A.

ABSTRACT

We have modified the design of an explosive gun launcher described at the Third Symposium on Hypervelocity Impact for use with pre-cast explosive charges and to provide performance information for steel projectiles. Other modifications include additional confinement of the detonation products, a longer barrel and longer conical transition from breech to barrel, provision for dynamic sealing of the access hole for the detonator wires, and additional buffering to prevent spall fracture of the projectile in the barrel. The modified gun launched a 12.1 gram steel projectile to a measured velocity of 3.2 km/s. Computational simulations have been performed to determine the effects of changes in the projectile density, confinement material density, and explosive type. The use of recyclable confinement material is discussed.

INTRODUCTION

At the Third Symposium on Hypervelocity Impact in 1958, several explosive actuated guns were described. In the typical propellant gun, one of the limitations on the design is the pressure at which the gun is blown apart. Since an explosive actuated gun is intentionally designed for one-shot use, much higher gas pressure can be applied to the projectile, with consequent higher muzzle velocity. A gun described by Willig and Semon (1959) (Fig. 1) had an overall length of 260 mm, a barrel length of 89 mm, and used about 0.7 kg of Composition C-4 explosive to accelerate 6.35 mm diameter, 6.35 mm long aluminum cylinders to velocities in the range of 4.6 to 5.5 km/s. They reported that similar guns had accelerated 12.7 mm diameter x 12.7 mm long aluminum cylinders to velocities in the range of 3.7 to 4.3 km/s.

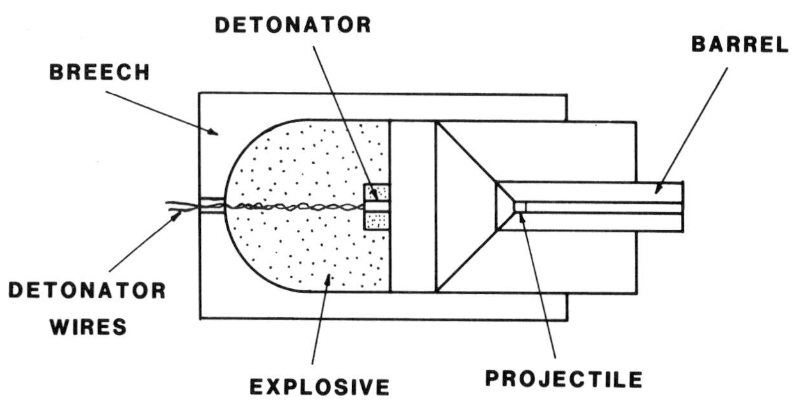

Fig. 1. Explosive gun launcher of Willig and Semon (1959).

We have modified the gun of Fig. 1 for use with pre-cast cylindrical explosive charges and to obtain performance information for steel projectiles. Willig and Semon reported results for aluminum projectiles. The modifications include additional confinement of the detonation products, a longer barrel and longer conical transition from breech to barrel, provision for dynamic sealing of the access hole for the detonator wires, and additional buffering to prevent spall fracture of the projectile in the barrel.

MODIFIED LAUNCHER

Figure 2 is a schematic of the modified explosive gun launcher. The overall length of the assembled launcher is 648 mm, with a barrel length of 374 mm and a 12.7 mm bore. The breech, barrel, and collar were machined from annealed 4340 steel (7.83 Mg/m^3) and assembled with six 7/8 - 9 UNC bolts. The cavity in the breech and the conical transition from the breech to the barrel were designed to have the same volume as in the launcher of Fig. 1. The half-angle of the conical section was decreased from 45 degrees to 30 degrees to provide a more gradual transition from the breech to the barrel. The 12.1 gram projectile was a 12.63 mm diameter, 12.47 mm long cylinder of 4340 steel (Rockwell C 42 hardness) with a 15 degree chamfer extending 0.25 mm back from the leading edge of the projectile. A 5 mm thick polycarbonate disk backed by a 0.8 mm thick mild steel disk formed a 1.5 gram buffer/obturator for the projectile. The cylindrical surfaces of the projectile and buffer parts were heavily coated with fast-setting epoxy and pressed into the bore until the buffer was about 13 mm from the conical transition. The epoxy was allowed to cure completely prior to firing the gun. The explosive charge was 0.82 kg of Composition B, cast into a 89 mm diameter x 100 mm long mild steel cylinder. The hole in the breech for the detonator wires was dynamically sealed via a hollow copper cylinder with tapered ends; the explosive gases collapsed the inner end of the plug, sealing the hole. Figure 3 shows the assembled gun. A plastic fixture on the muzzle, covered by a 0.35 mm thick polyester film, permited evacuation of the barrel prior to firing; this was to minimize the mass of gas ahead of the projectile in the barrel.

The projectile velocity was determined by observing the penetration of a spaced array of 1.6 mm thick aluminum plates via a high-speed camera operating at 40,600 frames/s. The projectile was observed to remain intact while passing through five of the seven plates, breaking apart on the sixth plate. The measured velocity was 3.2 km/s, after penetration of the first plate, suggesting that the muzzle velocity was even higher.

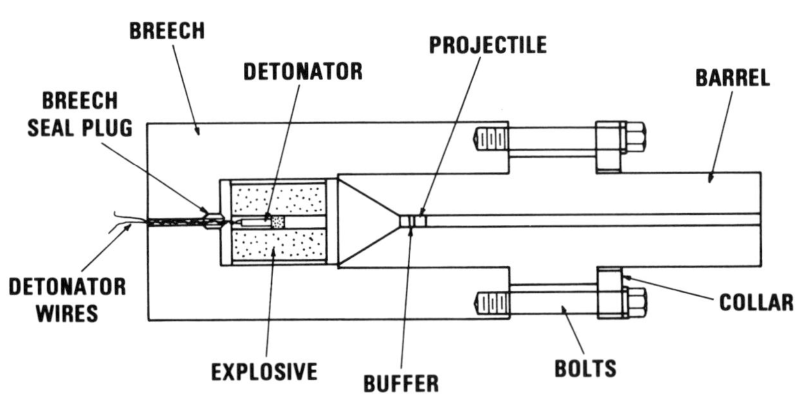

Fig. 2. Modified explosive gun launcher.

Fig. 3. Assembled modified launcher, ready for firing.

COMPUTATIONAL SIMULATIONS

Computational simulations of the modified launcher have been performed using the HEMP two-dimensional Lagrangian finite difference code (Wilkins, 1964). In these simulations, it was assumed that at time zero, before projectile motion, the explosive was completely reacted and the quiescent product gases filled the region behind the buffered projectile. The gases in this region were assumed to have the same total energy as the detonation energy of the explosive and a reduced density. Gas properties were calculated from the Jones-Wilkins-Lee equation of state (Giroux, 1973). It was also assumed that there was no gas leakage at joints in the launcher, around the projectile, or at the detonator wire access hole, while the projectile was being accelerated in the barrel. It was further assumed that the launcher parts did not fracture, the projectile moved without friction in an evacuated barrel, and the 4340 steel was replaced by iron.

The effect of changes in projectile density for iron launcher parts (breech and barrel) was studied by calculating the projectile velocity as a function of distance traveled in the barrel for projectiles made of aluminum, iron, and tungsten, having densities of 2.79, 7.85, and 19.22 Mg/m^3, respectively. The results are shown in Fig. 4. The differences in velocity are believed to reflect primarily the differences in areal density of the projectiles. Wave transit time along the length of the projectiles was in the range of 2 to 3 microsec, while a typical run time for the calculations was 200 microsec. The result for the longer times was a projectile state in which wave activity had been largely damped out, the projectile moving essentially as a rigid body. Also shown in Fig. 4 is the experimental result obtained with the buffered 4340 steel projectile and 4340 steel launcher parts. The higher experimental projectile velocity compared to the calculated velocity may be due to the actual high velocity flow of the detonation product gases into the conical transition and barrel, compared to the initially quiescent gases in the simulation.

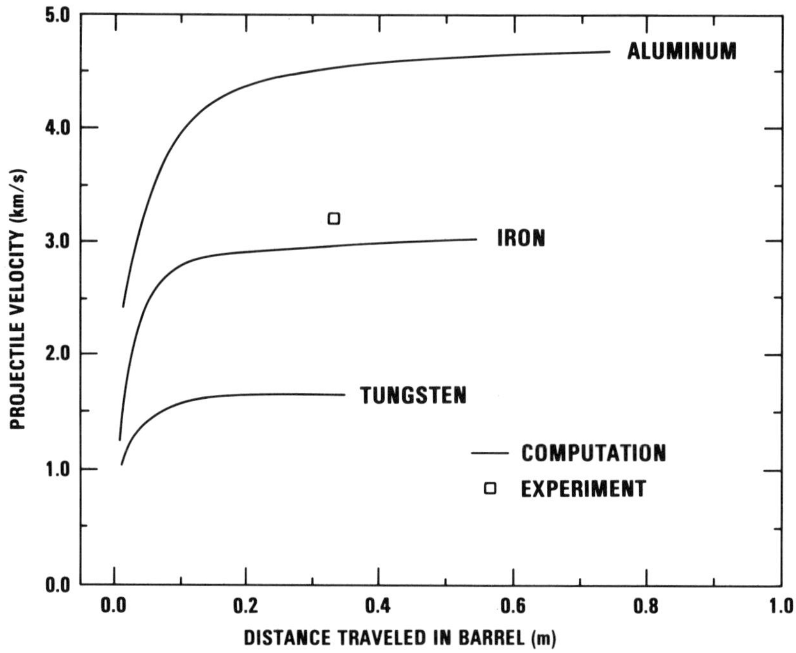

Fig. 4. Effect of projectile density on projectile velocity. The experimental point is for a 0.34 m barrel length with a 4340 steel projectile and 4340 steel launcher parts.

The effect of changes in the density of the confinement material was studied by calculating the velocity of an iron projectile as a function of distance traveled in the barrel for three different launcher materials, as shown in Fig. 5. The differences in velocity clearly indicate that confinement plays a major role in gun performance. The plausibility of these results was investigated by seeking upper and lower bounds on performance. One upper bound was established by taking the gun chamber in the HEMP calculation to be rigid, ensuring no change in chamber volume during projectile acceleration. This curve labeled "rigid confinement" is seen to lie well above that for tungsten confinement. Another upper bound, independent of the HEMP code, was computed from the equations of Seigel (1965) for a rigid gun with infinite chamber volume. The Comp B detonation products were assumed to be initially in the expanded state, and a specific heat ratio of 3 was used throughout. A lower bound on performance was computed by simulating zero confinement of the explosive products. Here, Seigel's relationships between pressure and projectile velocity were applied until a relief wave in the products could traverse the radial dimension of the chamber. At this time, pressure on the projectile was assumed to vanish, and velocity was taken to be constant thereafter. The upper and lower bounds based on Seigel's gun theory for an ideal gas are shown in Fig. 5 as dashed curves. Even though they have been obtained from very approximate models, these bounds are reasonable when compared with the HEMP results.

The use of recyclable cast low melting temperature alloys for inertial confinement around thinner steel breech and barrel parts is being considered, to provide low cost and simple launchers. Confinement material would be collected after a test, melted, and recast around breech and barrel parts. We have formed composites of No. 8 lead shot with Cerrolow 147 alloy[1] that melts near 64 degrees Celsius and has a density of 9.45 Mg/m^3. The theoretical density for the composite, assuming hexagonal close packing of the lead spheres, is 10.8 Mg/m^3. The performance curve for a launcher using the castable composite for confinement would be above that for iron in Fig. 5. Castable materials with higher densities and higher melt temperatures are also being considered.

---

[1]Cerro Metal Products, Bellefonte, PA 16823

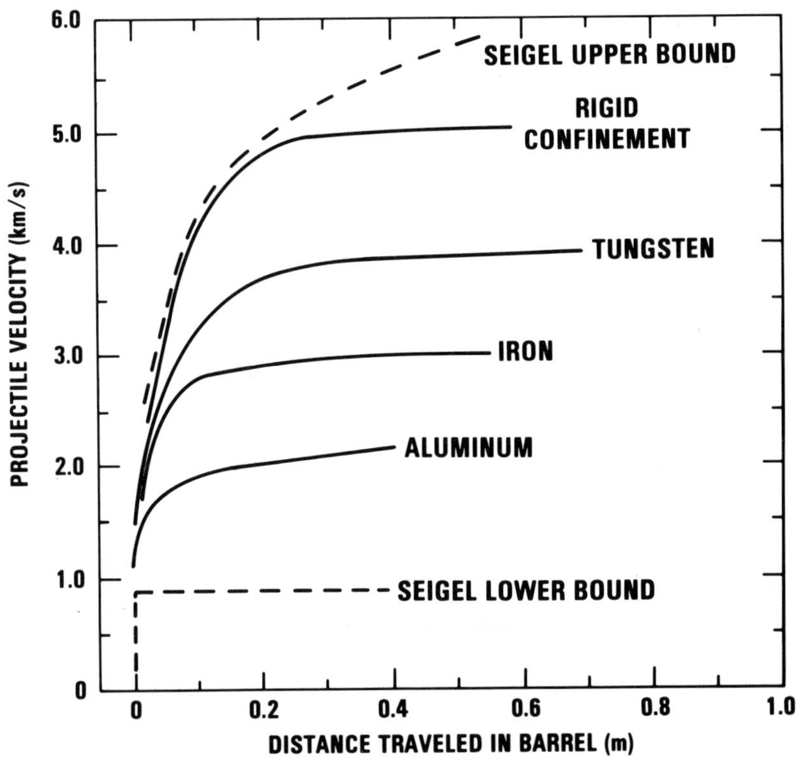

Fig. 5. Effect of confinement material density on projectile velocity for an iron projectile and three launcher materials.

The effect of changes in explosive type was studied by calculating the velocity of an iron projectile as a function of distance traveled in the barrel for iron launcher parts and equal volumes of three explosives having different densities and energies: TNT (1.63 Mg/m^3, 4.30 kJ/gm), Composition B (1.72 Mg/m^3, 4.95 kJ/gm), and Cyclotol (1.75 Mg/m^3, 5.25 kJ/gm) (Giroux, 1973). The results are shown in Fig. 6, where it is noted that the less energetic TNT gives a final projectile velocity that is about 20% less than the other two explosives considered.

In Figs. 4 through 6, only minor increases in projectile velocity are obtained for projectile travel distances greater than 0.3 m. Thus there is little to be gained in performance by using barrel lengths in excess of this value.

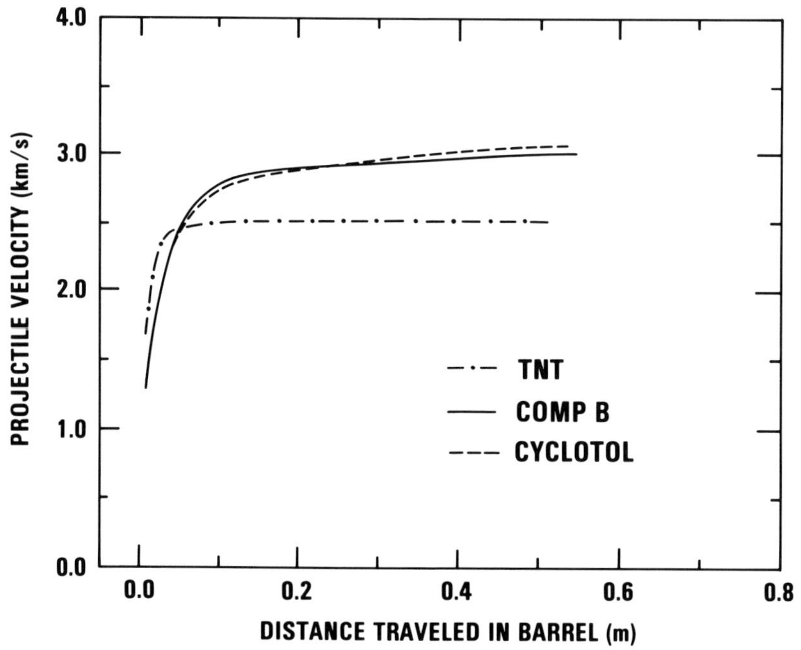

Fig. 6. Effect of explosive type on projectile velocity for equal explosive volumes.

SUMMARY

Modifications to a previously reported explosive gun launcher have been described. These include the use of pre-cast explosive charges, a longer barrel, more breech to-barrel tapering, dynamic sealing of the access hole for the detonator wires, more projectile buffering, and additional confinement of the detonation products. The measured velocity for a steel projectile was 3.2 km/s. HEMP code simulations have also been performed. These include projectile velocity versus distance traveled in the barrel for selected projectile and confinement materials (aluminum, iron, and tungsten). The effect of selected explosive materials (TNT, Compositon B, and Cyclotol) on projectile velocity was also computed.

REFERENCES

Giroux, E. D. (1973). HEMP Users Manual. Lawrence Livermore Laboratory, Rept. UCRL-51079, Rev. 1. (Available from National Technical Information Service, U.S. Dept. of Commerce, 5285 Port Royal Road, Springfield, VA 22161).

Seigel, A. E. (1965). The Theory of High Speed Guns, NATO Advisory Group for Aerospace Research and Development (AGARD), AD 475660 p.45. (Available from National Technical Information Service, U.S. Dept. of Commerce, 5285 Port Royal Road, Springfield, VA 22161).

Wilkins, M. L. (1964). In B. Alder, S. Fernbach, and M. Rotenberg (Eds). Methods of Computational Physics, Vol. III. Academic Press, New York. pp 211-263.

Willig, F. J. and Semon, H. W. (1959). A multi-stage H. E. actuated hypervelocity gun. Proceedings Third Symposium Hypervelocity Impact, Vol. I, Chicago, Illinois.

Int. J. Impact Engng Vol. 5, pp. 363–369, 1987
Printed in Great Britain

0734-743X/87 \$3.00 + 0.00
Pergamon Journals Ltd.

HYPERVELOCITY IMPACT VISUALIZATION WITH A NEW CRANZ-SCHARDIN CAMERA

Gary R. Hough

LTV Aerospace and Defense Company
Vought Missiles and Advanced Programs Division
P. O. Box 650003, Mail Stop TH-85
Dallas, Texas 75265-0003

ABSTRACT

A new Cranz-Schardin high-speed camera system is being used to record a variety of hypervelocity impact phenomena in the LTV Aerospace and Defense Company ballistic research range. Unique features of the camera are described, as are appropriate triggering schemes. Examples of several kinds of hypervelocity visualizations are shown to illustrate the remarkable flow field details which can be observed. These include cylindrical rod and sphere impacts as well as the resulting behind-target debris patterns.

INTRODUCTION

Much of the impetus for development of high-speed cameras has come from the field of ballistics. While the frame rate capabilities of such cameras have increased dramatically, so has their cost. Flash X-ray systems are also often used to visualize ballistic impacts, but are equally expensive. A relatively simple, mobile, and inexpensive means of taking photographs of pre- and post-impact events is desired in many laboratories.

The Cranz-Schardin spark camera is a method for achieving this goal. It is named after the great German ballistician, Carl Cranz, and his equally famous student, Hubert Schardin, who developed the camera in the early 1930's. Essentially a cine version of a focused shadowgraph camera, it provides high-quality photographs in a unit much less expensive than conventional framing cameras. Particular advantages include a framing rate which is independent of exposure time, the ability to use full resolution of photographic film, and an absence of moving mechanical parts.

A new, improved Cranz-Schardin unit is being used at the Vought Missiles and Advanced Programs Division of the LTV Aerospace and Defense Company to record terminal ballistics phenomena. The system consists basically of source and recording units, each of which will be described, with concentration on the unique features. Framing rates as high as 3.5×10^6 pictures per second are attainable. The camera is being used in conjunction with an enclosed solid propellant ballistic gun range to photograph the flight and impact of projectiles launched at velocities up to 2.7 km/sec. It is railmounted in a horizontal plane perpendicular to the range axis, and is positioned to view the range target tank through optically transparent armored windows.

Examples of several kinds of hypervelocity visualization are shown in this paper. They include spherical and cylindrical rod impact on aluminum and steel target plates. Resulting projectile and target deformations as well as behind-target debris can be observed in detail and quantified. Such knowledge leads to improved understanding of hypersonic impact phenomena.

CRANZ-SCHARDIN CAMERA

Basic Principle

The Cranz-Schardin camera is one of the most effective designs for the recording of a sequence of back-lit photographs. It consists of a light source unit and a recording unit. An external trigger must be supplied to initiate the photo sequence. A sketch of the basic Cranz-Schardin system (taken from Swift, 1982) is shown in Fig. 1. A number of light sources are mounted in a plane. As each source fires, it is focused by a condensing lens through the subject volume to a single objective

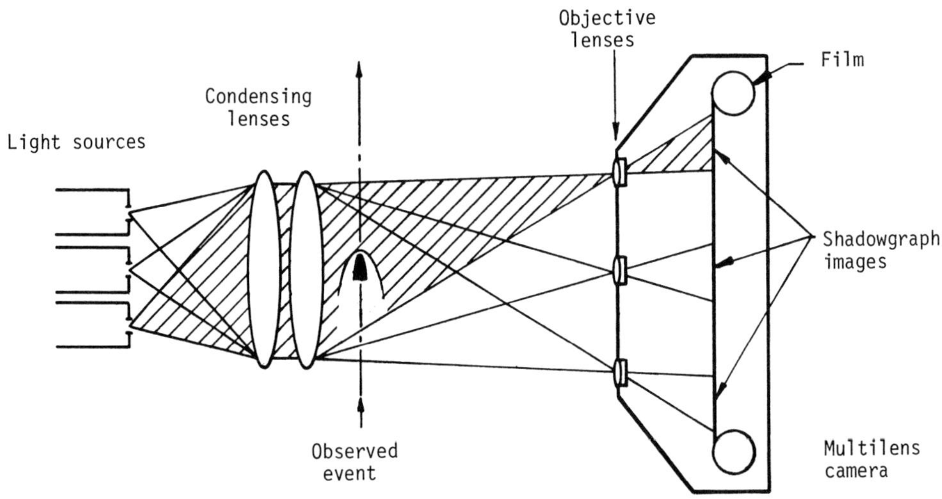

Fig. 1. Cranz-Schardin camera schematic.

lens in the recording unit. Each objective lens is located at or near the image point of the corresponding light source. With an objective lens of appropriate focal length, the shadow of the object in the sample volume is imaged onto the photographic film. This produces a sharp, back-lit photographic record.

By firing the light sources in rapid sequence, a series of frames can be recorded at different times. In principle, the framing rates (or time duration Δt between firings) can be varied independently. In practice, it is easier and so more common to make each Δt the same. In the early Cranz-Schardin cameras, spark light sources were used. These have since been replaced by sources having much shorter exposure times.

The light sources are usually arranged in a rectangular grid pattern. Swift (1982) has noted that Cranz-Schardin cameras containing as many as 121 (11 x 11) channels have been constructed and used successfully. Typically, 10 to 30 channels are employed.

Light Source Unit

The basic Cranz-Schardin camera described here was designed and built by Special Illumination Systems of Dayton, Ohio. The source unit is the key element of the system and contains some unique features. The light sources are pulsed semiconductor laser diodes having a peak output power of 6 watts each. There are 10 of them, arranged in 2 rows of 5 units. The diodes operate at a nominal wavelength of 850 nm, and have an effective photographic exposure time of about 40 nanoseconds. This permits blur-free images of objects with lengths greater than about 1 cm and having velocities as high as 10 km/sec to be recorded at full film resolution.

A custom timing circuit is used to set the framing rate, which can be varied from about 100 up to 3.5×10^6 frames/second. Such high frame-rate capability places this Cranz-Schardin unit in the class of the better rotating mirror cine cameras. Frame rate adjustment is user controlled by a rotary decade switch plus an analog vernier, is monitored by a frequency counter, and recorded on a digital oscilloscope.

The main condensing lens consists of two 15-cm diameter plano-convex lenses spaced 3 cm apart. The camera system has been designed to have an effective sample area approximately 9 cm in diameter at a distance of 20 cm from the condensing lenses. For alignment purposes, visible light-emitting diodes (LEDs) which are fitted with fiber optics are provided. The ends of these fiber optics plug into the laser diode holder in such a way that the light from the LEDs fills the main condensing lens in the same way that the light from the lasers does. Therefore, the visible light from the LEDs can be used in place of the infrared laser light to define the object volume. Finally, a separate power supply is used as a voltage source for the integrated circuit chips of the clock and driver boards and for the rectifiers which drive the laser diodes.

Recording Unit

The recording unit is built upon an Omega 4" x 5" View Camera body. The film holder fits in the rear section of the camera body and is loaded with Kodak High Speed Infrared film 4143 (nominally ASA 400). The normal Omega front lens plate has been replaced by a complex assembly consisting of a plate which holds the multiple camera lenses, a narrow band-pass optical interference filter which is tuned to the laser emission wavelength, and a shutter mechanism. The ten camera lenses are each 1.9 cm in diameter, with a focal length of 19 cm. The magnification of the lenses is such that a 4 cm diameter object space is imaged on a 1 cm diameter circle on the film. The laser diodes are fired so that images are recorded first across the top row from left to right (as viewed from the source unit) and then across the bottom row in the same direction.

Camera/Range Mount

Projectiles are accelerated to hypersonic velocities in the LTV Aerospace and Defense Company enclosed, solid propellant ballistic gun range. The range has interchangeable 30 cal., 20-mm, and 30-mm barrels which permit packages to be launched from 0.3 to 2.7 km/sec. Typical projectiles include spheres and cylindrical rods mounted in sabots (which subsequently are removed from the flight path by a sabot separator). For the studies reported herein, total launch masses ranged from about 20 to 100 grams, with resulting velocities from 1 to 2 km/sec.

The camera units are bolted to each end of an aluminum channel which is guided on shockless bearings along horizontal rails mounted at the top of the range target tank, see Fig. 2. The entire camera unit can thus be repositioned to view a particular object sample area or to allow the target tank to be opened for cleaning, etc. Optical viewing of the selected sample area is through transparent armored Lexan windows in the range tank. These also serve to protect the camera lenses from stray projectile/target debris pieces which could cause damage.

It has been our experience that any vibrations which might be transmitted to the camera due to projectile launch or target impact either are negligible, or occur after all photographs have been taken. No pictures have been spoilt by vibrations.

Camera Triggering

A common procedure is to use a breakscreen, optical path break, or similar arrangement, located close to the phenomena to be viewed. For example, a thin metal foil (usually aluminum) is attached to the front face of the target surface for investigation of ballistic spall formation. Prior to

Fig. 2. Camera mounted on enclosed ballistic range.

impact, a current is passed through the foil. When the projectile strikes, the foil is shattered, the current circuit is broken, and the camera is triggered. This has the advantage that delay times before camera firing can be minimized and so one is virtually assured of capturing the phenomena of interest during the 10 frame Cranz-Schardin sequence (providing of course that the interframe rate has been reasonably chosen). It is relatively inexpensive and easy to implement and has been used for many of the photographs taken in our facility.

This approach, while generally useful, does have some disadvantages in certain instances. For example, consider the problem of photographing behind-target debris many projectile diameters downstream of the rear target surface. If a foil screen is used, pieces of the broken screen may appear in the shadowgraphs and become indistinguishable from the real target debris. If an optical beam (e.g., a laser) is used, it may not be broken by any debris particles and so the camera will not be triggered. To alleviate the possibilities, a new trigger delay scheme was worked out. Usually, such delays are not very often used because of uncertainties in launch velocity which can cause the pictures to be taken before or after the desired time.

The system described here is based on the results of the projectile velocity which has already been measured upstream of the target tank. A simple circuit was designed and constructed to take as input the time difference between the two velocity stations used. Then, having determined in advance the distance from the second, or downstream, station to the object plane, the circuit scales this time difference to select the time delay which triggers the Cranz-Schardin camera at the desired time. The velocity station-object distance is of course variable and is "dialed" into the circuit in advance of each range shot. Since this approach is based on instantaneous knowledge of the actual projectile velocity, it provides precise control of the camera triggering. It has been checked out and tested in numerous range shots and has worked well to velocities over 2 km/sec. In principle this approach is applicable over the entire hypervelocity range. An important advantage is that it is non-intrusive.

Camera Shutter

As received, the camera was equipped with a 13-cm capping shutter which had to be opened and closed before and after photographs were taken. This was done electrically with an input signal to the system power unit which came from the range operator. That is, the shutter was opened just prior to range firing and was closed as soon as possible after the firing occurred. This left the shutter open for several seconds, during which time the film was exposed to a considerable amount of muzzle flash and impact-generated light. In order to keep effects of this extraneous light to a minimum, a faster closing shutter is required.

Such a shutter was designed and works as follows. Two thin steel plates coated with teflon are mounted just in front of the recording unit lenses so that they overlap each other. In the "open" position, holes drilled in each plate are aligned to permit light to pass through the lens openings. To "close" the lenses, a trigger signal is sent to two electromagnets mounted near the plates. When activated, the magnets draw the plates across each other to cause the holes to become misaligned.

Just before the range is fired, all room lights are turned off, and the plates put in the open position so that the film is exposed. The range is fired and, at the instant the projectile passes one of the velocity measuring stations, a closing signal is sent to the electromagnets. Timing experiments show that this magnetically-driven shutter system closes within a few hundredths of a second, while the projectile reaches the target area from the velocity station in a few milliseconds. Hence the lens system is completely closed off very shortly after the photographs are taken.

This has proven to be a dependable arrangement for minimizing muzzle flash and impact-generated light. Furthermore, it has the advantage of not reducing at all the light output from the source unit as the photographs are taken.

HYPERVELOCITY IMPACT VISUALIZATION

The Cranz-Schardin camera has been used to photograph a number of high velocity impact situations. Some examples are shown here and represent spherical and cylindrical rod projectiles launched at velocities of about 1.25 km/sec against relatively thin target plates. The projectile material was steel, while both steel and aluminum target plates were used. Typical camera frame rates were set between 40,000 and 100,000 sec^{-1}.

In Fig. 3 is shown the impact at a Mach number of 3.4 of a steel sphere on an aluminum plate. The sphere is just emerging from the target rear surface after perforating it rather easily. The bow shock is clearly seen as are acoustic waves propagating outwards from both the plate front and rear surfaces. An impact "splash" of fine material particles off the front target surface is also evident; this disappears progressively in subsequent pictures. Portions of the projectile initial

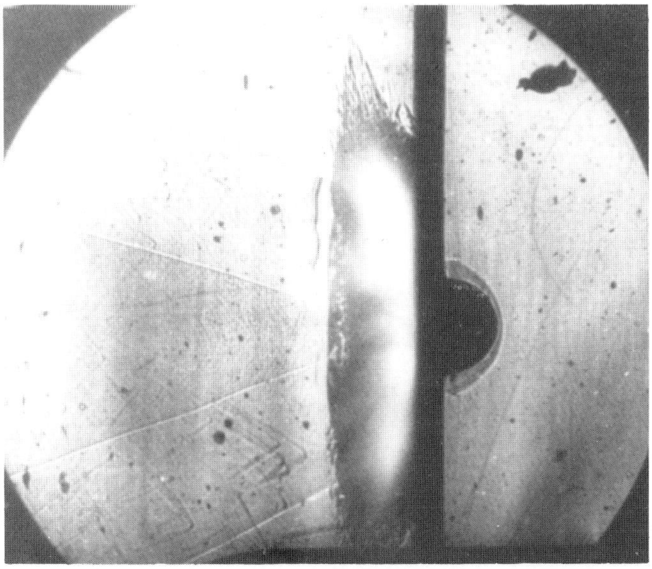

Fig. 3. Spherical projectile impact on thin target plate.

shock wave can also be seen as can some of the impact-generated light flash. Scratches and other defects in the Lexan window material show up as dark background spots.

Following in Fig. 4 is a photograph which catches a cylindrical rod having a length/diameter of 4 passing through a thin steel plate. Much of the same phenomena appear as for the spherical case. Note as before the weak waves propagating outward from the target surfaces. Here, however, the front surface "splash" is not nearly as evident, consisting of a few larger pieces of target debris. Again, the remarkable flow field details are captured quite clearly.

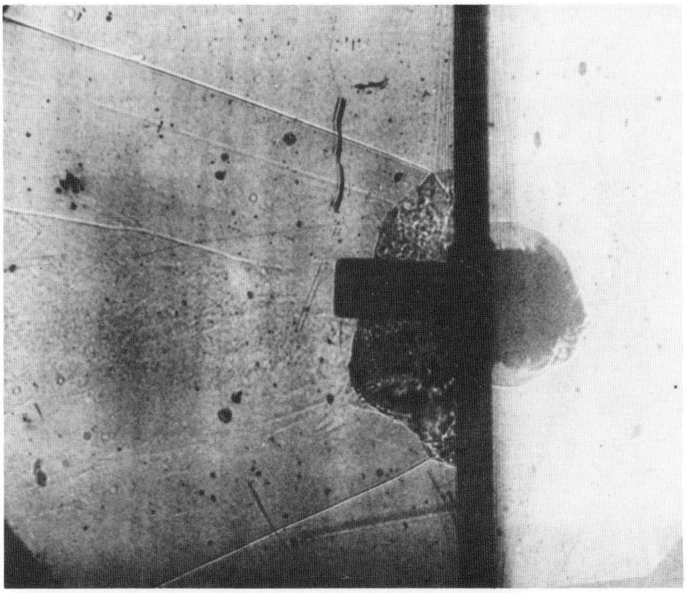

Fig. 4. Cylindrical rod projectile impact on thin target plate.

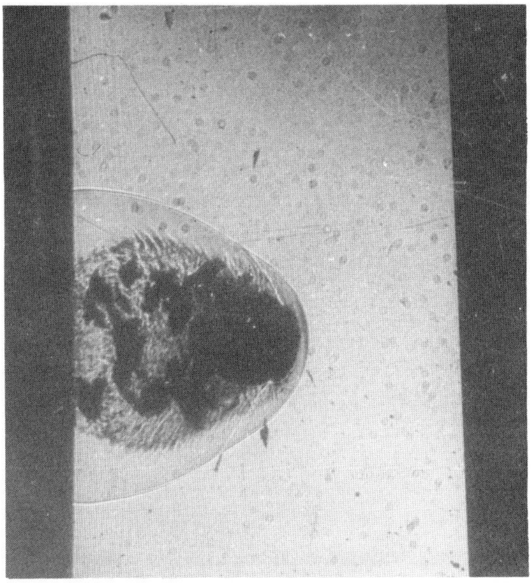

Fig. 5. Initial state of behind-target debris formation.

Moving on to look at some examples of behind-target debris, Fig. 5 illustrates the initial stage of debris formation arising from target penetration by a steel sphere. Part of the target material is being pushed along in front of the sphere; as time progresses, this slides off and is eventually left behind. Some of the target debris pieces are travelling supersonically, others at subsonic velocities. Debris pieces range from several millimeters in diameter down to "dust" size (the sphere was 1.27 cm in diameter). The striking sphere was recovered intact.

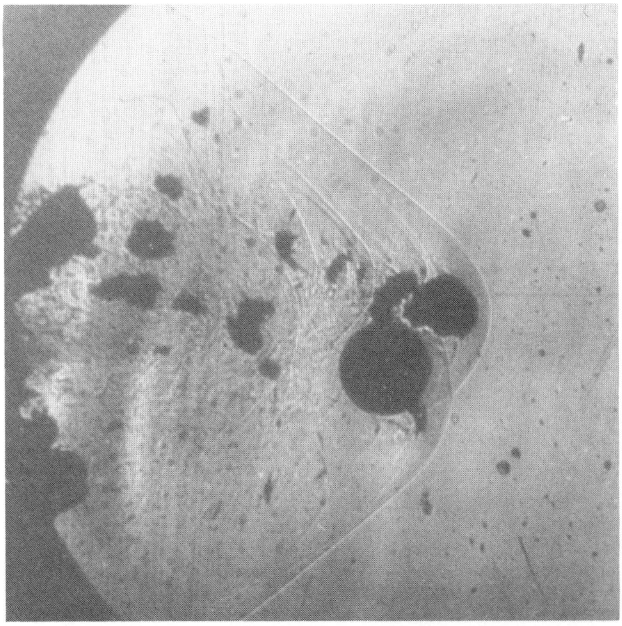

Fig. 6. Behind-target debris pattern downstream of target plate.

Figure 6 shows the debris pattern far downstream (>10 sphere diameters) of another target plate. The portion of the target "pushed" ahead of the sphere during perforation has separated from the sphere surface and is on its way to being left behind. There are perhaps a dozen relatively large fragments following the projectile down range, and they are moving at various velocities whose components along the projectile line of flight can be measured, typically within less than a 10% error, (Fortner and Hough, 1986). Considerable turbulence exists of course behind the projectile.

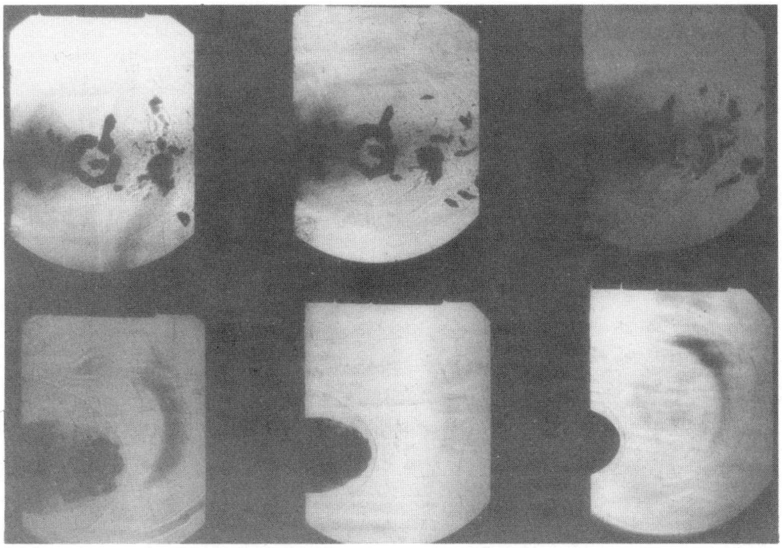

Fig. 7. Cranz-Schardin record of spall-ring formation.

The final illustration, Fig. 7, shows a time-sequenced history of a spall-ring formation behind a thin target plate struck by a steel sphere travelling at 1.2 km/sec. The time duration between photographs is approximately 20 μsec. Both the projectile and spall-ring were recovered intact. The bright portion in the photographs is impact-generated light which has managed to pass through the lens filter. The dark portions in the photographs to the right of the projectile are a result of slight misalignments in the lens system. Variations in overall brightness from frame to frame are the result of differences in output power of the laser diodes.

SUMMARY

The Cranz-Schardin camera system permits the visualization of hypersonic flow and impact phenomena in detail, in an easily-variable time-sequenced history, and at a cost which is much less than alternate approaches. Its use, especially in ballistics studies, will lead to improved fundamental understanding of impact phenomena.

ACKNOWLEDGEMENTS

The author wishes to thank M. L. Fortner and D. M. Gustafson for their considerable assistance with the work reported here.

REFERENCES

Fortner, M. L. and G. R. Hough (1986). Application of pulsed laser shadow photography to terminal ballistic investigations. High Speed Photography, Videography, and Photonics IV, SPIE Volume 693.

Swift, H. A. (1982). Image forming instruments. In J. A. Zukas and others, Impact Dynamics, John Wiley and Sons, New York, pp. 241-276.

Int. J. Impact Engng Vol. 5, pp. 371–380, 1987
Printed in Great Britain

0734-743X/87 $3.00 + 0.00
Pergamon Journals Ltd.

LAUNCHER TECHNOLOGY, IN-FLIGHT VELOCITY MEASUREMENT AND IMPACT DIAGNOSTICS AT THE TUM/LRT

E. Igenbergs, S. Aigner, A. Hüdepohl, H. Iglseder,
H. Kuczera, M. Rott, U. Weishaupt

Lehrstuhl für Raumfahrttechnik (LRT),
Technische Universität München,
D-8000 München 2, Richard-Wagner-Str. 18, FRG.

ABSTRACT

Four different types of electromagnetic accelerators are described, giving their performance data and limitations: electromagnetic, electrothermal, rail gun and plasma gun. Different techniques are used to measure in-flight velocity: Light barrier with pulsed laser, in-flight photography with N_2 laser, and in-flight radar at 35 and 90 GHz. These devices also are described. For some specific investigations, e.g., charge generation by impact of small particles with velocities above 10 km/s, the impact location of every impact (for multiple impact tests) is required along with the projectile impulse and velocity. Techniques for accurate measurement of these parameters are discussed.

INTRODUCTION

In 1971, a hypervelocity accelerator was installed in the laboratory of the Lehrstuhl für Raumfahrttechnik of Technische Universität München for the simulation of micrometeoroid impacts. Since then several launcher systems have been developed, all using stored electrical energy as their energy source. The launchers are described in the next section.
The in-flight measurement techniques, and the impact diagnostics, which are described in the paper were developed for small projectiles (diameter less than 1 mm) and high velocities (10 - 20 km/s).

DESCRIPTION OF ACCELERATORS

Electromagnetic Accelerator

The equivalent circuit for the electromagnetic accelerator is shown in Fig. 1. The following values are representative of the operational system:

Capacitor:	42 μF
Charging voltage:	8 kV
Frequency:	30 kHz.
Masses:	10^{-9} ... 1.0 g
Velocities:	10 ... 400 m/s

The transient current in the accelerator coil (primary coil) generates a current in the conductive driver (secondary short), and the driver is accelerated through the interaction between the magnetic fields. A coil had to be developed which was not destroyed by the current pulses so that it could be reused for a number of operational cycles. An example of the coil developed is shown in Fig. 2. A retainer is placed to stop the accelerated driver. Any material that is placed on the driver can fly through a hole in the retainer and impact a target downrange.
Velocities which can be obtained repetitively, are shown in Fig. 3 as a function of the charging voltage. A specific advantage of this accelerator is that it operates under nearly any environmental condition - underwater or in vacuum. The launcher also can accelerate liquids as well as metals. The current development work is aimed at increasing the velocity of the driver. The facility,

as well as theoretical and experimental investigations, are described by Igenbergs (1979) and Lell (1983).
More recently, experiments with a large capacitor bank have shown that it is possible to accelerate flat rings to velocities on the order of 1.5 km/s, before either the coil or the driver were destroyed.
It is possible to operate this accelerator under very clean conditions at very low temperatures. If the velocity can be increased to 3 km/s, then this system would be a viable injector of frozen hydrogen pellets into fusion reactors.

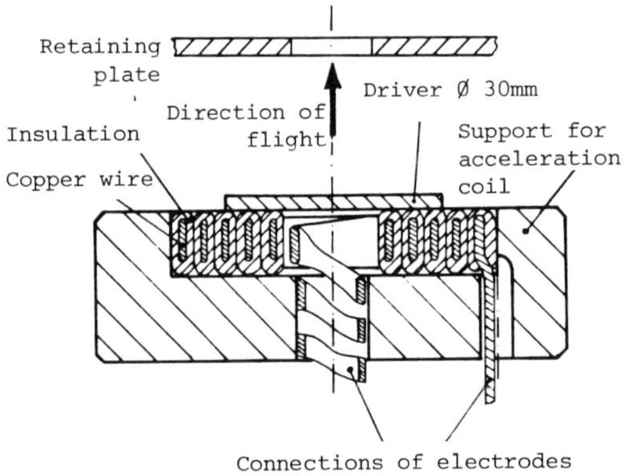

Fig. 1. Eqivalent electrical circuit of the electromagnetic accelerator.

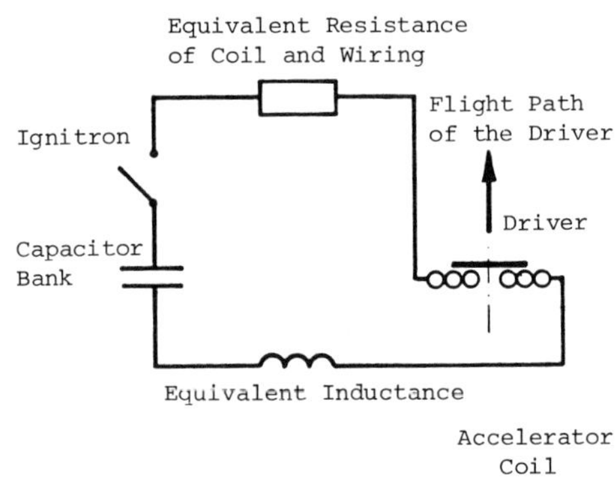

Fig. 2. Accelerator coil.

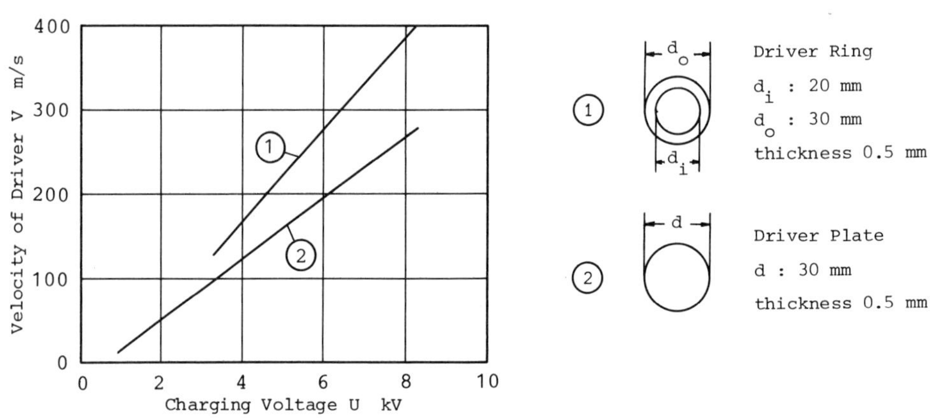

Fig. 3. Performance of the electromagnetic accelerator.

Exploding Wire Gun

The electrothermal accelerator (using the principle developed by ETI) was developed as a first stage for a two-stage plasma gun. The accelerator is able to accelerate particles in the mass range of 10^{-4} g < m < 10^{-2} g up to velocities of 3 km/s.
The principle of the electrothermal accelerator is shown in Fig. 4. An aluminum wire with a diameter of about 0.2 mm is enclosed in a cylindrical chamber, 20 mm long and 6 mm in diameter. The wall of the cylinder is lined with an insulator. The axis of the cylinder coincides with the axis of the barrel to minimize the loss due to friction in the plasma flow. The steel barrel is one electrical contact and the other is a solid copper electrode. These two contacts press slightly against the wall and are in contact with the aluminum wire. The particle to be launched is placed in the barrel, adjacent to the wire. If particles smaller than the diameter of the barrel bore are to be accelerated, they are placed in a sabot.
The electrodes of the exploding wire gun are connected with low inductive cables to a capacitor bank (56 μF, 12 kV max.); the bank is discharged through an ignitron switch. As in the electromagnetic accelerator, a GE 7703 ignitron is used. Following the discharge, a high electrical current

vaporizes the aluminum wire, generating a hot plasma by Joule-heating. The expansion of the plasma accelerates the projectile through the barrel.
The exploding wire gun operates in either air or vacuum. The photograph of a unit is shown in Fig. 5, and the performance of this gun is shown in a mass/velocity diagram in Fig. 6. The present development work is aimed at launching a projectile mass around 0.2 g between 4 and 6 km/s. These are the requirements for the first stage of the intended two-stage plasma gun.

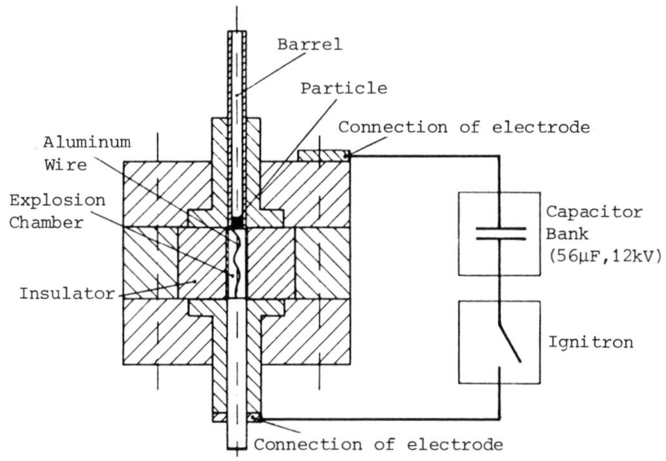

Fig. 4. Cross section of the electrothermal accelerator.

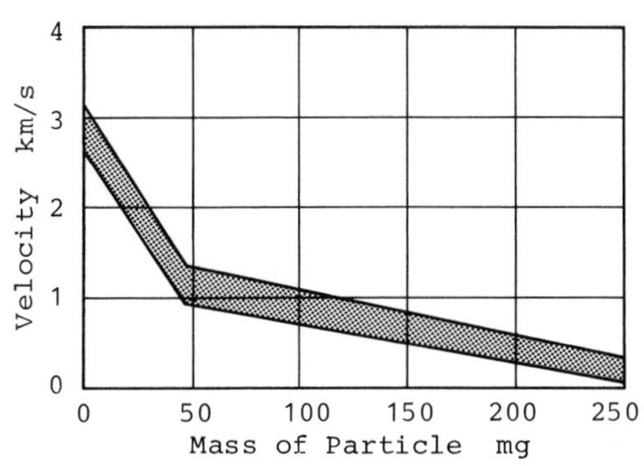

Fig. 5. Performance of the electrothermal accelerator.

Fig. 6. Photograph of the electrothermal accelerator.

Rail Gun

A short version of the rail gun was built and tested in the last two years. The equivalent circuit for this launcher is shown in Fig. 7. The index (0) denotes constant parameters, R_{El} and L_{EL} are the resistance and the inductance of the electrodes, and R_{Pro} is the resistance of the projectile. The installation of the rail gun in the vacuum test tank is shown in Fig. 8 together with the coaxial feed-through. A schematic of an end-on view of the accelerator is shown in Fig. 9. This rail gun is presently operated with non-conducting projectiles; the accelerating plasma is an ionized aluminum foil. Characteristic values of the components are:

Bank capacitance:	C_0	352 μF
Inductance:	L_0	330 nH
Resistance:	R_0	8 mΩ
Length of the accelerating section		240 mm

The facility as well as the theoretical and experimental investigations are described by Aigner (1984) and Aigner/ Kuczera (1984). The performance of this accelerator is shown in Fig. 10 as a function of the discharge current.

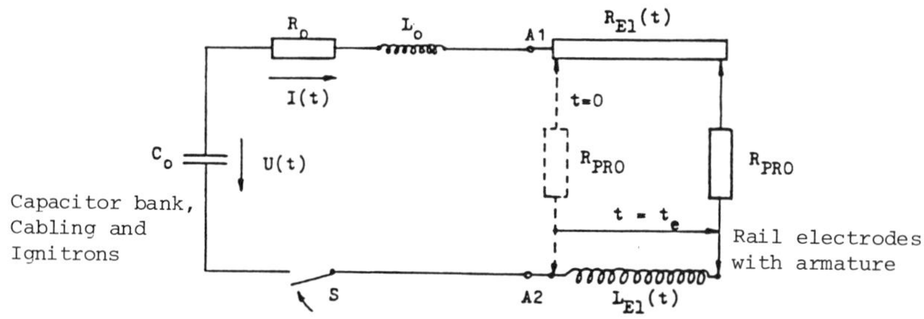

Fig. 7. Equivalent electrical circuit of the rail gun.

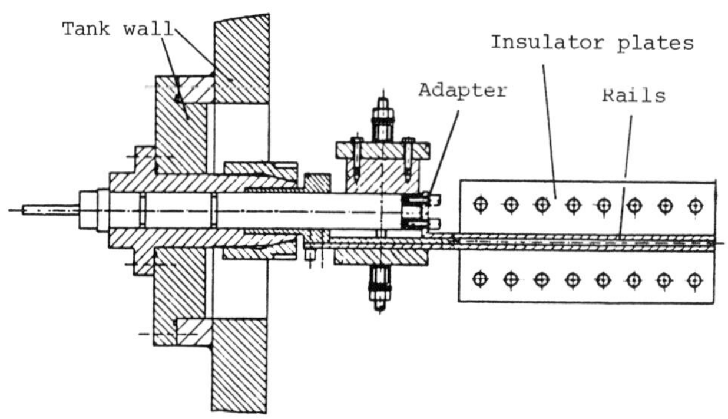

Fig. 8. Installation of the rail gun.

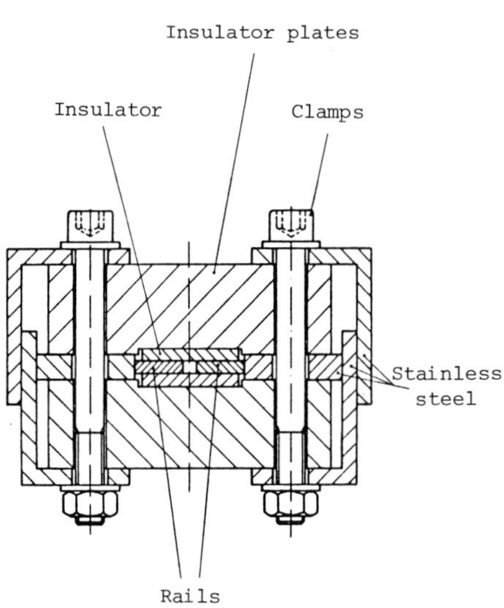

Fig. 9. Barrel cross section of the rail gun.

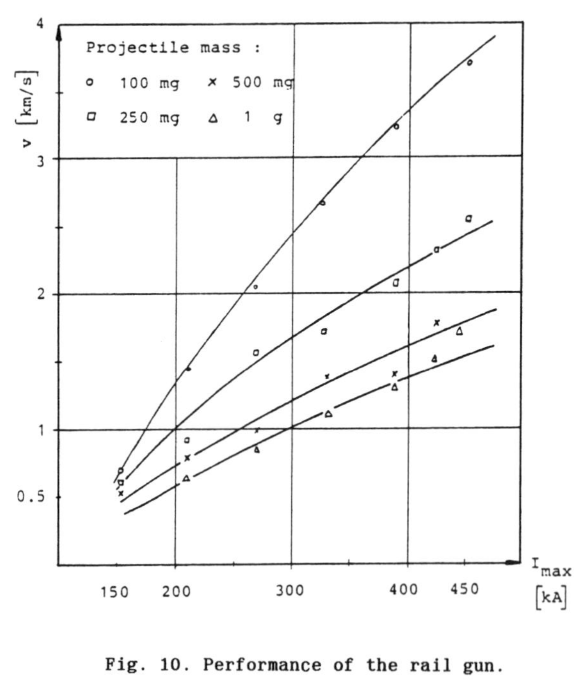

Fig. 10. Performance of the rail gun.

Plasma Accelerator

The theoretical and experimental investigations using the plasma accelerator are described by Kuczera (1980) and Hüdepohl (1985). A schematic of the gun and test range is shown in Fig. 11. Approximately the same constant parameters apply as for the rail gun, since both are operated in the same test range.

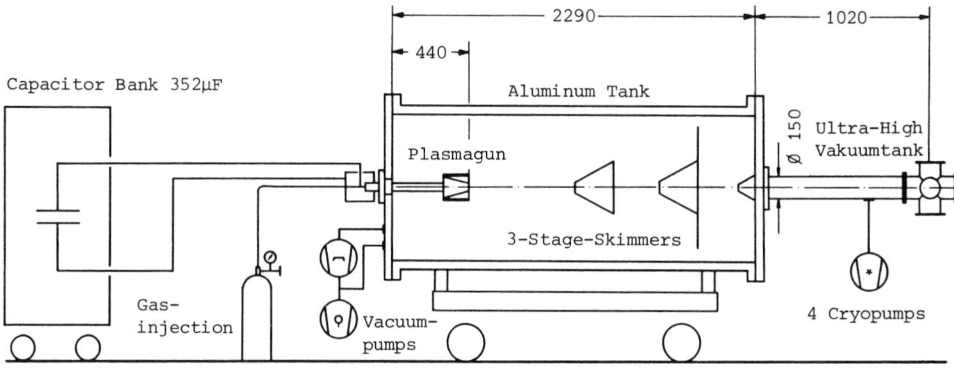

Fig. 11. Plasma accelerator facility with test range.

The plasma accelerator consists of a coaxial accelerator, to which a compressor coil is attached via a ground loop, as shown in Fig. 12. The plasma is generated in the coaxial part of the accelerator, either by evaporation and ionization of a metallic foil or by the injection through the center electrode and subsequent ionization of a gas; see Fig. 12.
The plasma is accelerated in the coaxial accelerator by the interaction of the radial current from the center electrode to the annular outer electrode and the azimuthal magnetic field of the center electrode. The plasma then enters the compressor coil. This is simulated by the equivalent circuit, which is shown in Fig. 13. A current then starts to flow between the center electrode and the windings of the coil. This generates a magnetic field in the compressor coil, producing a high density plasma. At the end of the compressor coil the velocity of the plasma is around 50 km/s.

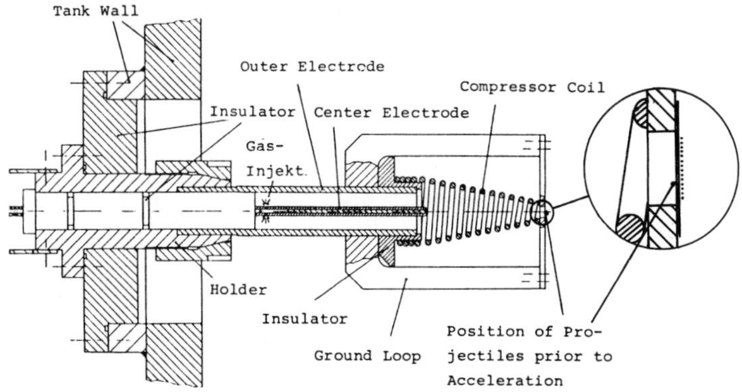

Fig. 12. Plasma accelerator with gas injection.

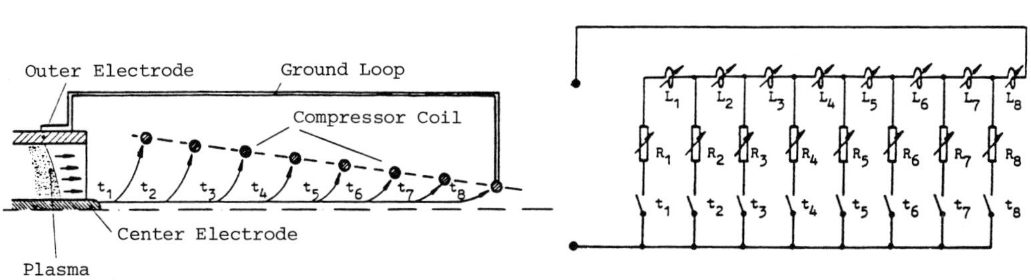

Fig. 13. Equivalent electrical circuit of the compressor coil.

This plasma flow is then used to accelerate glass beads, which are placed across the narrow end of the compressor coil, as shown in Fig. 13 and 14. The glass beads are drag accelerated by the high velocity plasma flow. The performance envelope of this accelerator is shown in Fig. 15.

Fig. 14. Photograph of the plasma accelerator.

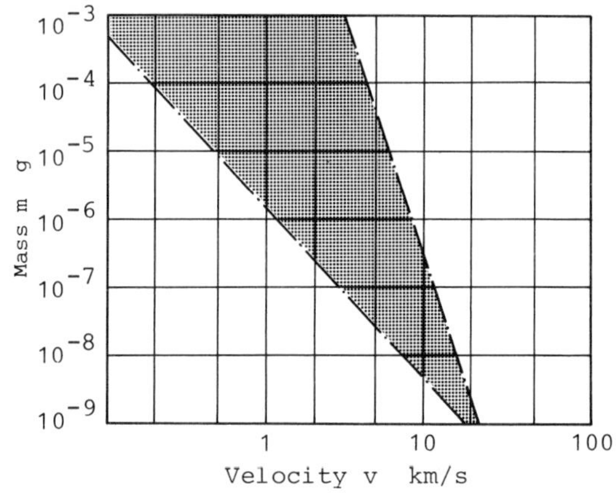

Fig. 15. Performance of the plasma accelerator (15 kV charging voltage, 15 kHz discharge frequency).

IN-FLIGHT VELOCITY MEASUREMENTS

For the plasma accelerator and the rail gun, the in-flight velocity measurements (as well as the impact diagnostics, which are described in the next section) require shielding from the extremely high electromagnetic radiation and interference. The experimental techniques, on the other hand, need appropriate diagnostics of the plasma. The work done in these fields is described by Kuczera (1980) and Hüdepohl (1985).

Light Barrier and In-flight Photography

In order to determine the velocity and the shape of the particle prior to the impact on the target, a system has been developed and put into operation (Schröder 1982) consisting of:
A multiple light barrier, generated by a Nd-YAG-laser, which reflects several times between two mirrors; a particle crossing these beams reflects light into a photomultiplier.
A delay-time computer computes the velocity of the particle from the subsequent passage through the light beams, and then computes the time at which the particle will reach the downrange photographic station.

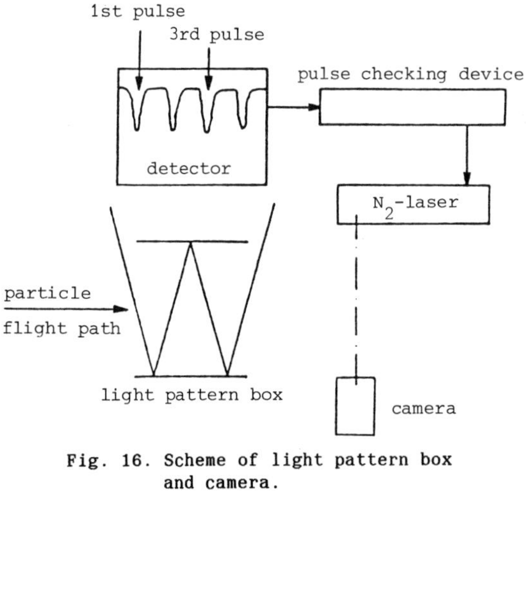

Fig. 16. Scheme of light pattern box and camera.

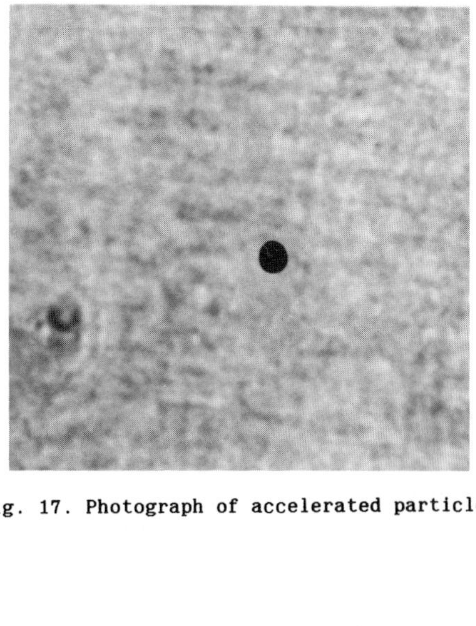

Fig. 17. Photograph of accelerated particle.

A photographic setup uses a pulsed N_2-laser as the light source, and a large-format camera with high-resolution film.

Figure 16 depicts the general arrangement. The light pattern box has an entrance window to protect the light pattern volume from plasma radiation and the plasma flow. This restricts the flight path to less than 2.5° from the center line of the test range. The flight path from the first to the third beam of the zigzag pattern is 55.5 mm long. The 2.5° deviation from the center line leads to path changes of ± 0.5 mm or ± 1%. The maximum possible error of velocity measurement between first and third signal pulse is of this order of magnitude.

The time interval counter (counting frequency 40 MHz) is started by the first signal pulse and stopped with the third signal pulse. The counter then immediately counts back with half the frequency. When the counter reaches zero again, the particle has passed a length of 111 mm since the third signal pulse. At this position the photography is performed.The system has succesfully performed in the size/velocity range between 710 μm/2.7 km/s and 180 μm/16 km/s. As an example of system capabilities, a photograph of a 400 μm glass bead at a velocity of 3 km/s is shown in Fig. 17.

Doppler-Radar Velocity Measurements

The setup for in-flight velocity measurements with Doppler radar is shown in Fig. 18 for the 35 GHz radar, and in Fig. 19 for the 90 GHz radar system. The output of the oscillator is fed via a circulator to the antenna. The reflected signal from the fast particle has a Doppler shift which is superimposed upon the unshifted frequency in a mixer. The resulting Doppler frequency is then available at the output of the mixer. The Doppler signal is recorded with a transient recorder at a

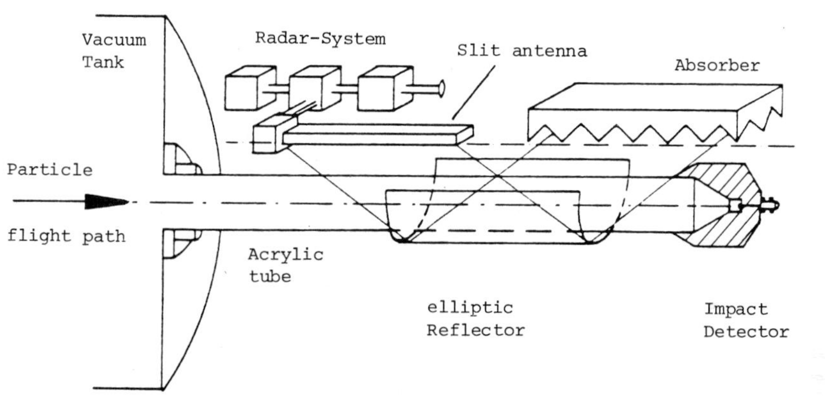

Fig. 18. 35 GHz Doppler-Radar.

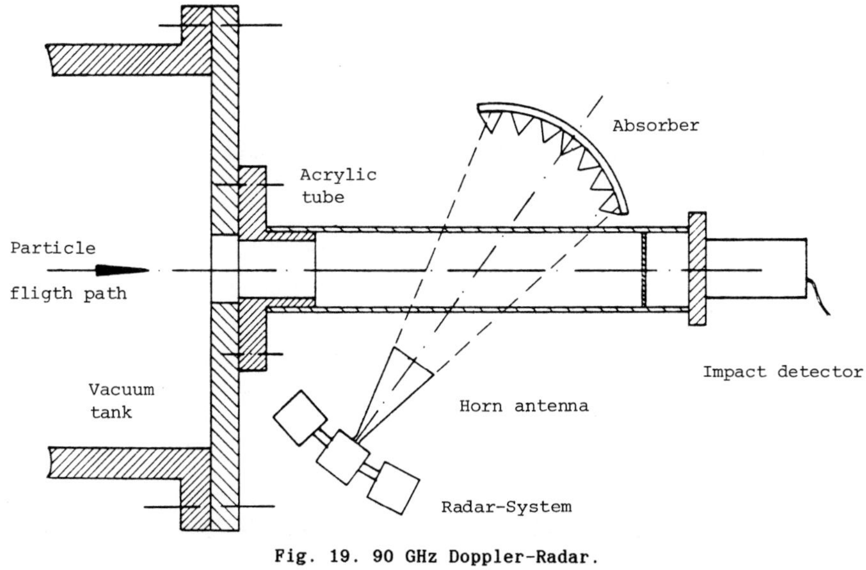

Fig. 19. 90 GHz Doppler-Radar.

maximum sample rate of 100 MHz and processed by a computer. Triggering is provided by an impact detector, which triggers the transient recorder in the pretrigger mode. Figure 20 shows a Doppler signal measured with the 35 GHz system.
The effective cross section of a target drops sharply, when the diameter of the particle is smaller than the wavelength of the radar signal. Thus, the 90 GHz Radar is better suited for the smaller and faster particles accelerated by the plasma accelerator than the 35 GHz system, but it is limited by the lower output of the oscillator.
Both systems have been tested with particle size/velocity ranges between 0.5 mm/3 km/s (plasma gun); and 2.5 mm/6 km/s (light gas gun). A detailed description of the velocity measurements with Doppler radar is given by Kaufmann (1985).
The difference between the Doppler radar measurement and a light barrier measurement in a light gas gun experiment was in the order of 1%.

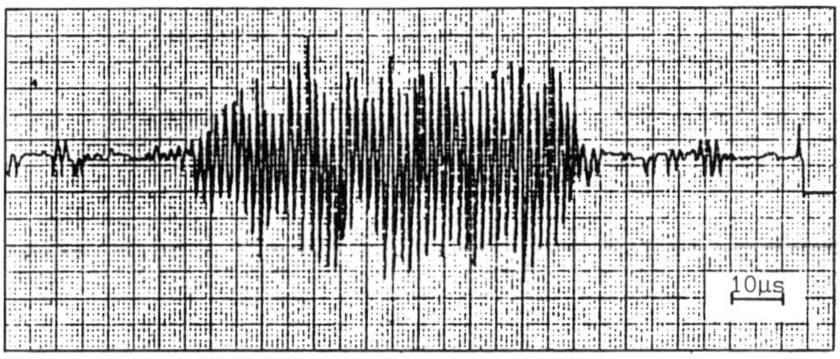

Fig. 20. Doppler signal, aluminum bead, d = 2 mm, v = 3.48 km/s.

IMPACT DIAGNOSTICS

Mass Determination

In the case of the plasma accelerator, the mass of the projectile is not known exactly, because of erosion during the acceleration phase. Thus, it has to be determined before the impact on the target. From a knowledge of impact velocity and a measurement of the diameter of the penetration hole in a thin foil with a thickness on the order of 1/500 of the diameter of the projectile, and a density that is comparable to the density of the projectile, the projectile diameter can be determined to an accuracy of about 10%. The mass of the projectile is then calculated with the known density of the projectile.
Figure 21 shows a photograph of the penetration hole of a glass bead in a foil of nitro-cellulose. The foil thickness is less than 0.1 μm; the initial diameter of the projectile prior to acceleration was about 70 μm at velocities on the order of 10 km/s.

Location of the Impact

The present single-stage operation of the plasma accelerator requires a large number of projectiles be launched in order to obtain some impacts on a 50 - 100 cm^2 target located 2 to 3 m downrannge. The plasma accelerator flow expands at the end of the compressor coil where the magnetic field diverges. The projectiles are therefore accelerated both axially and radially, and the number of particles that impact the target is uncertain. A single impact on the target can be obtained by loading a very small number of particles; many experiments will show no impact at all, and occasionally a single impact will occur. This operation is very time-consuming; thus, it generally is better to have multiple impacts from one acceleration experiment.
The signal from an impact will arrive at different times at four piezoelectric detectors that are mounted at the corners of the target plate. This is used to locate the impact on the target plate. A system, as it is shown in Fig. 22, is mainly limited by the sampling rate of the recorder. Using Nicolet 2090-203A Digital Recorders with a maximum sampling rate of 20 MHz, an impact can be easily located within 1 mm.

Crater Volume and Projectile Energy

The peak amplitude of a signal from an impact, detected by a piezo-electric sensor, can be correlated with the volume of the crater that has been generated by this impact. The following exponential expression is used to correlate the maximum amplitude A_{max} (mV) and the crater volume V (μm^3):

$$A_{max} = c \cdot V^n$$

where c and n are constants. For example, the experimental results shown in Fig. 23 indicate for glass projectiles impacting an aluminum target:

$$c = 3 \cdot 10^{-3}$$
$$n = 0.78$$

A similiar correlation can be found relating the kinetic energy, E, of the projectile (J) to the maximum amplitude A_{max} (mV):

$$E = c \cdot A_{max}^n$$

The corresponding values are:

$$c = 10^{-5}$$
$$n = 1.14$$

as derived from Fig. 24.

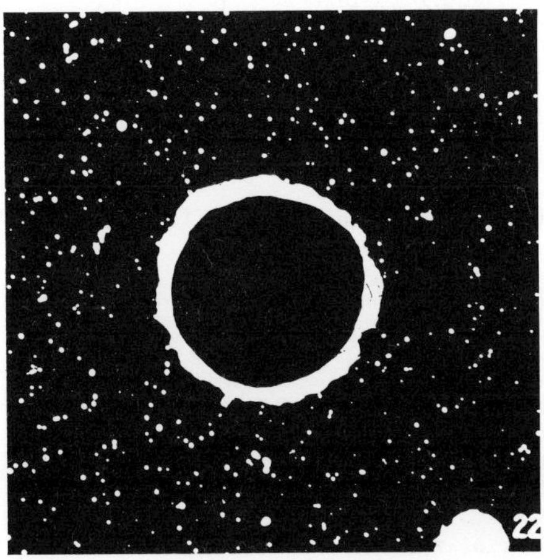

Fig. 21. Penetration hole in a 0.08 μm NC-Foil, d = 63 μm, v = 10.3 km/s.

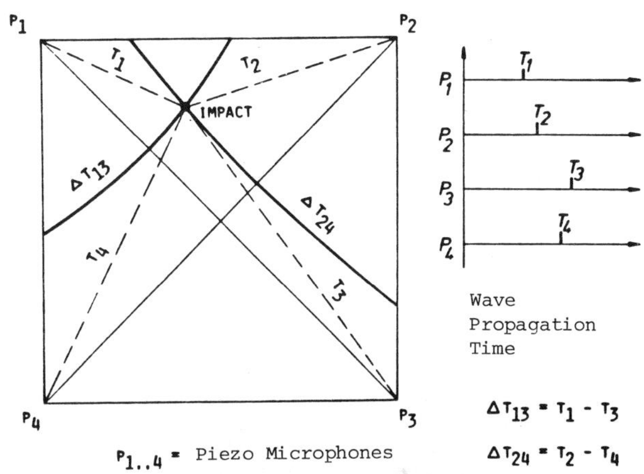

Fig. 22. Evaluation of impact location

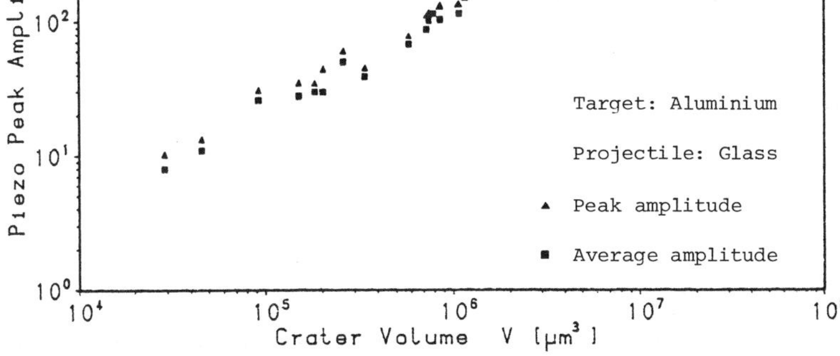

Fig. 23. Calibration of the acoustic impact detector, crater volume.

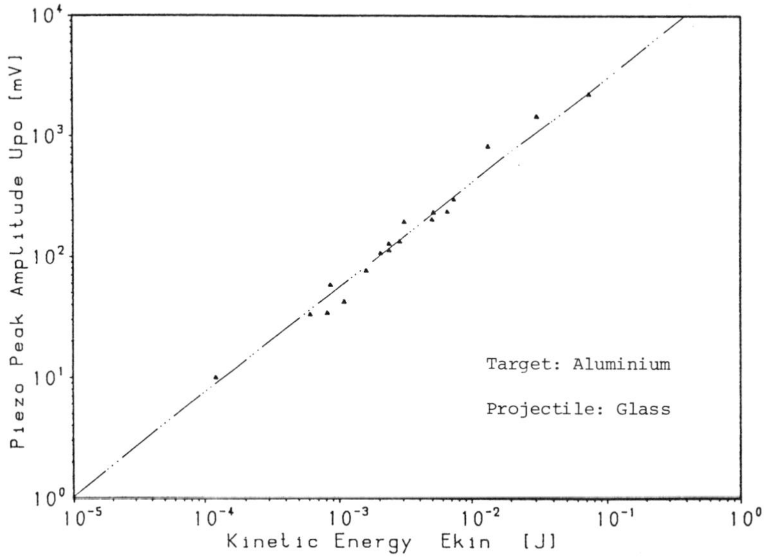

Fig. 24. Calibration of the acoustic impact detector, energy.

CONCLUDING REMARKS

The facilities described in this paper have been used in basic research to simulate micrometeoroid/space debris impacts on surfaces exposed in space. Major projects are the Capture Cell Experiments on board of the Long Duration Exposure Facility for the National Aeronautics and Space Administration (NASA), and dust experiments on board of the European Space Agency (ESA) Spacecraft GIOTTO for its flyby mission to the comet Halley.
Other experiments have been conducted for technological tests of application satellites including the future space station. Basic research includes both the development of new concepts for electromagnetic hypervelocity launchers and the simulation of impacts on simulated celestial bodies.

REFERENCES

Aigner, S. (1984). Entwurf und Auslegung eines elektromagnetischen Leitschienenbeschleunigers. Lehrstuhl für Raumfahrttechnik, Bericht RT-DA 84/3.

Aigner, S., and H. Kuczera. (1984). Erste Railgun-Experimente an der TU München - Ergebnisse und Perspektiven. Nationales Symposium über elektromagnetische Beschleunigung, Euskirchen, 14./15. November 1984. Lehrstuhl für Raumfahrttechnik, Bericht RT-KB 84/8.

Hüdepohl, A. (1985). Der Stand des koaxialen Plasmabeschleunigers mit Kompressionsspule im Jahre 1984. Lehrstuhl für Raumfahrttechnik, Bericht RT-TB 85/1.

Igenbergs, E. et al (1979). Development of an Electromagnetic Pellet-Accelerator. Lehrstuhl für Raumfahrttechnik, Bericht RT-TB 79/1.

Igenbergs, E. et al (1984). Experimental Results Obtained During the Calibration of Dust Experiments on Board of the Halley Cometary Mission Spacecraft GIOTTO. 14th ISTS Conference, Tokyo.

Igenbergs, E. et al (1983). Results of Impact Simulation for the GIOTTO Mission. Acta Astronautica, Vol. 10, No. 12.

Iglseder, H. (1983). Entwicklung und Erprobung eines akustischen Mikrometeoriten-Detektors. Lehrstuhl für Raumfahrttechnik, Bericht RT-DA 83/8.

Kaufmann, K. (1985). Untersuchungen zur Messung kleiner Teilchen im Hochgeschwindigkeitsbereich mit Millimeter-Dopplerradar-Systemen. Dissertation, TU München.

Kuczera, H. (1980). Theoretische und experimentelle Untersuchungen an einem Plasma- und Teilchenbeschleuniger. Dissertation, TU München.

Kuczera, H. (1985). Simulated Micrometeorite Exposure of Solar Arrays. Estec Study Report, Contract No. 6238/85/NL/AN.

Lell, P. et al (1983). An Electromagnetic Accelerator. J. Phys. E.: Sci. Instrum., Vol. 16, GB.

Schröder, B. (1982). Optische Messung von Größe und Geschwindigkeit kleiner Glaskügelchen. Dissertation, TU München.

Int. J. Impact Engng Vol. 5, pp. 381–388, 1987
Printed in Great Britain

0734-743X/87 $3.00 + 0.00
Pergamon Journals Ltd.

MEASURED CHARGE GENERATION BY SMALL MASS IMPACT AT VELOCITIES BETWEEN 1 AND 45 KM/S

H. Iglseder, E. Igenbergs

Lehrstuhl für Raumfahrttechnik, TU München
D-8000 München 2, Richard-Wagner-Str. 18, FRG.

ABSTRACT

The electrical charge that is generated by the impact of a small mass at velocities between 1 and 45 km/s was investigated using the Electrostatic Dust Accelerator of the Max-Planck-Institut in Heidelberg (MPI) and the Plasma Accelerator of the Lehrstuhl für Raumfahrttechnik (LRT) of the Technische Universität München (TUM). Glass beads were accelerated, and the targets were of different materials i.e., (Au, W, Fe, Al). The mass/velocity range of the accelerated small masses was:

MPI : 10^{-15}g < m < 10^{-10}g / 1 km/s < v < 45 km/s
TUM/LRT : 10^{-10}g < m < 10^{-5}g / 2 km/s < v < 20 km/s.

The total yield of the impact charge $Q^{\pm}_{total}$ of both polarities can be described by the empirical formula:

$$Q^{\pm}_{total} = C_{\rho}\, m^{\alpha}\, v^{\beta} \quad \text{[Coulomb]},$$

C_ρ being a function of the density ratio of target/projectile, α is approximately unity and β between 2.92 and 3.77. The charge detector is described and the results that were obtained in test series at both facilities are discussed in relation to the empirical formula.

1. INTRODUCTION

The generation of electrical charges by a hypervelocity impact (impact ionisation) was experimentally observed by Friichtennicht and Slattery, 1963. Subsequently this effect was successfully used in detectors for micrometeoroid detection on board of spacecraft (Auer and Sitte, 1968; Dietzel et al., 1973; McDonnell, 1978; Grün 1976, 1981).

A schematic drawing of the detector that has been used for these experiments, is shown in Figure 1.

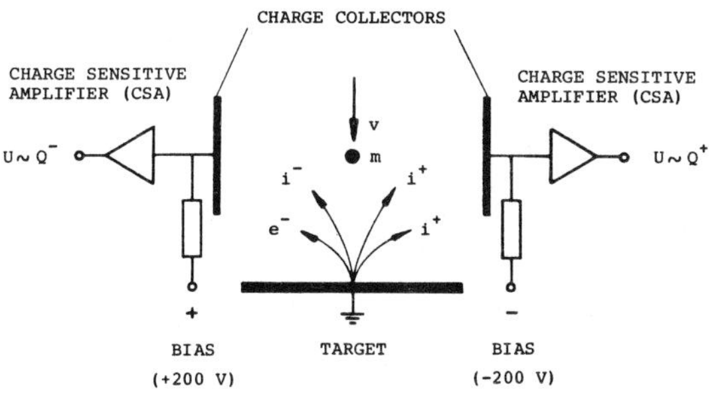

Fig.1. Schematics of an impact-ionisation detector.

An impact charge detector is a combination of a target (semiinfinite or foil) with a set of charge collectors. The hypervelocity impact on the target generates a plasma and the negatively charged components of the plasma (electrons, negative ions) are then accelerated towards the positively biased collector, the positive ions are accelerated towards the negatively biased collector.

Prior to the experiments that are described here, the simulation of the hypervelocity impact was performed in electrostatic accelerators (Fechtig, 1978). The mass/velocity range, that can be simulated in this type of accelerator is $m = 10^{-15}$ g to 10^{-10} g and v = 1 ... 70 km/s.

It was found (Friichtenicht, 1963), that the total charge $Q^{\pm}_{total}$ of both the negative and the positive charges that are generated by the impact of a mass m at the velocity v can be described by the empirical formula:

$$Q^{\pm}_{total} = C_{\rho} \bullet m^{\alpha} \bullet v^{\beta}$$

The parameter C_{ρ} is a function of the density ratio of the materials of the impacting particle and the target.

The exponent α was found to be very close to unity and the exponent ß is a function of the material combination of the particle and the target.

In Table 1, values of the exponent ß are listed together with the impact velocity range from different publications. The impact velocities as well as the mass range from 10^{-15} g to 10^{-10} g indicates that those values have been obtained by experiments at electrostatic accelerators.

The investigations that are described in this paper were performed to extend the mass/velocity range beyond the regime that can be covered by electrostatic accelerators.

Authors	Exponent β	Velocity range v [km/s]
(Friichtenicht 1963)	3.0	2 < v < 15
(Auer, Sitte 1968)	2.6 ± 0.6	2 < v < 10
(Grün et al. 1976)	2.6 - 3.6	1 < v < 70
(Dietzel et al.1973)	3.5 ± 0.2	2 < v < 40
(McDonnell 1978)	2.3 - 3.5	1 < v < 10

Table 1. Exponents ß, velocity range and combination of materials. The exponent α = 1.0, particle mass measurement between 10^{-15} and 10^{-10} g. Iron particles on different target materials.

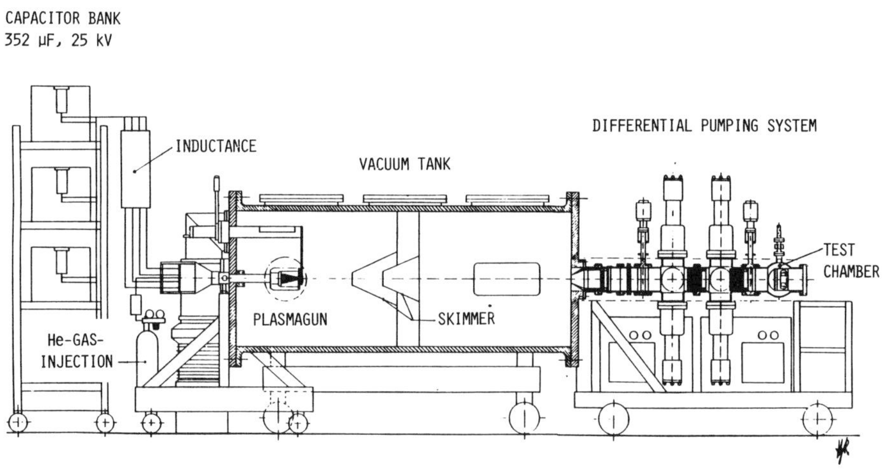

Fig.2. The plasmadynamic accelerator with compressor coil.

2. DESCRIPTION OF THE EXPERIMENTS

It was the intention to establish a clear correlation between (1) the experiments in the range between 10^{-10} and 10^{-5} g, and at velocities between 2 and 20 km/s that could be performed at the plasma accelerator of the Lehrstuhl für Raumfahrttechnik of the Technische Universität München, and (2) the previously mentioned results, that were obtained in electrostatic accelerators.

The same impact charge detector was therefore used in both ranges. The calibration experiments were performed with masses between 10^{-10} and 10^{-15} g at the electrostatic dust accelerator of the Max-Planck-Institut für Kernphysik in Heidelberg (Fechtig et al., 1978).

The plasma acclerator in München is shown schematically in Fig. 2 (Igenbergs et al., 1973, 1986). In this accelerator, the particles are exposed to a plasma flow and drag-accelerated. This plasma flow had to be eliminated, before it could reach the impact charge detector, in order to obtain clean quantitative results.

Skimmers and baffles were installed and a submicron foil in front of the test section was used. The differential pumping system and the test section of the plasma accelerator are shown in Fig. 3. A vacuum of 2.10^{-6} mbar and a residual charge in the order of 10^{-8} C could be maintained during the experiments.

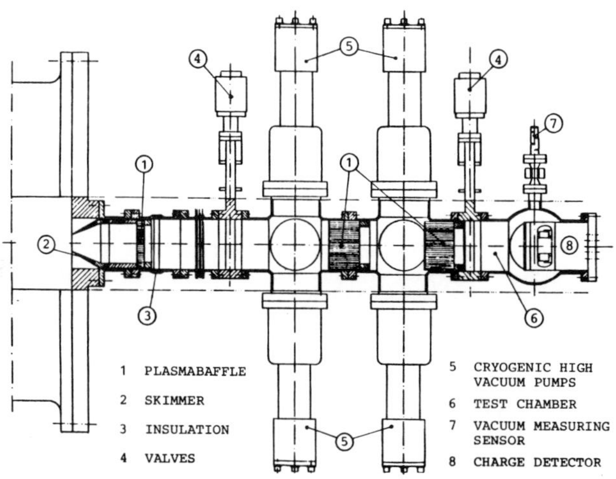

Fig.3. Differential pumping system with test section.

An operational characteristic of the plasma accelerator is that the hot plasma flow ($T \sim 10^4$ K, $v \sim 50 - 100$ km/s) does ablate the accelerated particles. The mass and the velocity of the particles must therefore be determined prior to the experiment in the impact charge detector.

DETERMINATION OF THE PARTICLE PARAMETERS
AT THE PLASMADYNAMIC-PARTICLE-ACCELERATOR

Mass

The submicron penetration foil in front of the detector, with a thickness of less than 0.1 μm will show a penetration hole. If the ratio of the projectile diameter to the foil thickness lies within the range of 100 to 1000, then it can be assumed that the penetration hole diameter equals the diameter of the particle within 10%, and that the particle is not destroyed by the interaction with the foil. The mass of the particle is then determined from this experimentally measured diameter together with the known density of the particle.

A penetration hole of a particle with a diameter of 36 μm and a velocity of 13.0 km/s is shown in Fig. 4a, and Fig. 4b shows the impact of the same particle in a gold target. The diameter of the crater is 130 μm. This type of experiment proved that the submicron foil can be used.

a.

500x

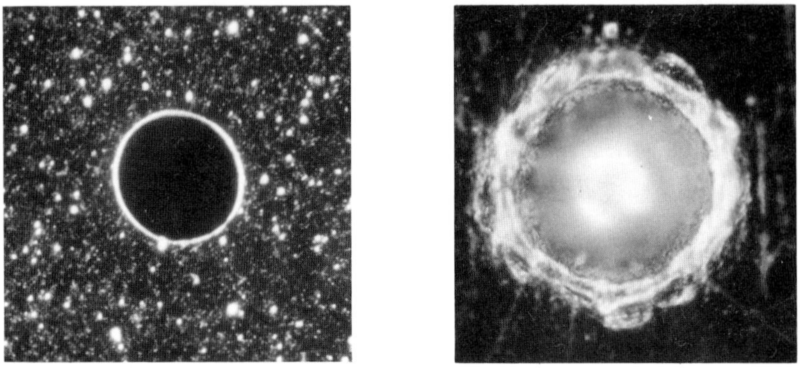

b.

250x

Fig.4a. Penetration hole in a 0.08 μm submicron film; $d_{projectile}$ = 36 μm; $v_{projectile}$ = 13.0 km/s.
4b. Impact crater on gold target; D_{crater} = 130 μm.

Velocity

The velocity of the particle is obtained through time-of-flight measurement. The launch time is known to an accuracy of ± 20 μs. The distance to the target is 3350 mm. At a velocity of 10 km/s this gives a time of flight of 335 μs. The accuracy of the velocity data is therefore better than 5% at velocities in the order of 10 km/s.

Charge

Charge sensitive amplifiers were especially developed and calibrated for the particle mass charge between 10^{-10} g up to 10^{-5} g and a particle speed range from 2 km/s up to 20 km/s. Impact charges from 10^{-10} C up to 10^{-5} C were registered. The impact detector is shown in Fig. 5. Typically registered signals of both polarities are shown in Fig. 6.

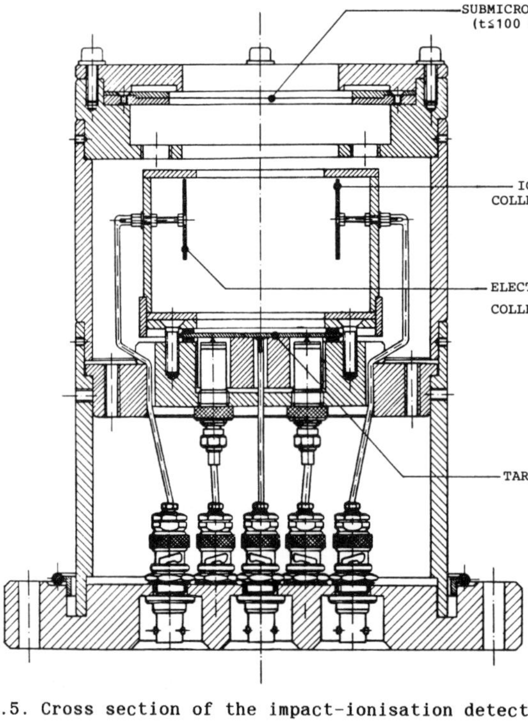

Fig.5. Cross section of the impact-ionisation detector.

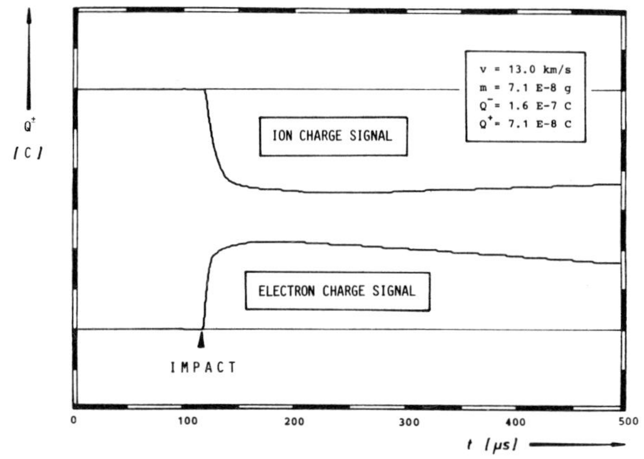

Fig.6. Impact charge signals measured at the plasmadynamic particle-accelerator.

3. DISCUSSION OF THE RESULTS

Fig. 7 shows the mass/velocity range covered by the electrostatic accelerator in Heidelberg and the plasma accelerator in München. Only glass beads were used for the experiments; 280 impacts with a mass between 10^{-15} g and 10^{-10} g and 140 impacts with a mass between 10^{-10} g and 10^{-5} g were analyzed. The charge yield varied from 10^{-15} C to 10^{-5} C. Generally the charge yield decreased with the target density.

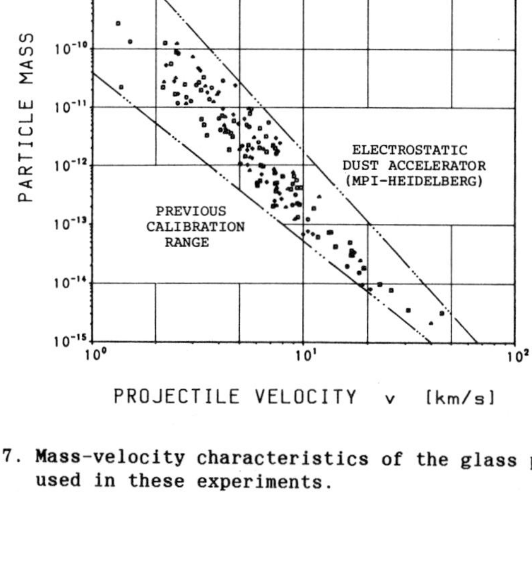

Fig.7. Mass-velocity characteristics of the glass particles used in these experiments.

One of the 'expected' results of the experiments in the higher mass range was that the mass exponent was very close to unity as in the previous experiments in the lower mass range.

The results in the higher mass range vary from $\alpha = 0.96$ to $\alpha = 1.09$: the assumption of a value of $\alpha = 1.0$ seems to be as good as in the lower mass range.

The measured values for the charge in units of Coulomb are therefore presented here divided by the mass as function of the velocity of the impacting particle and the parameter $C\rho$.

The electron yield is shown in Fig. 8a, and the ion yield in Fig. 8b, both for glass beads impacting on a gold target. The corresponding values for ß are 3.54 and 3.47 respectively.

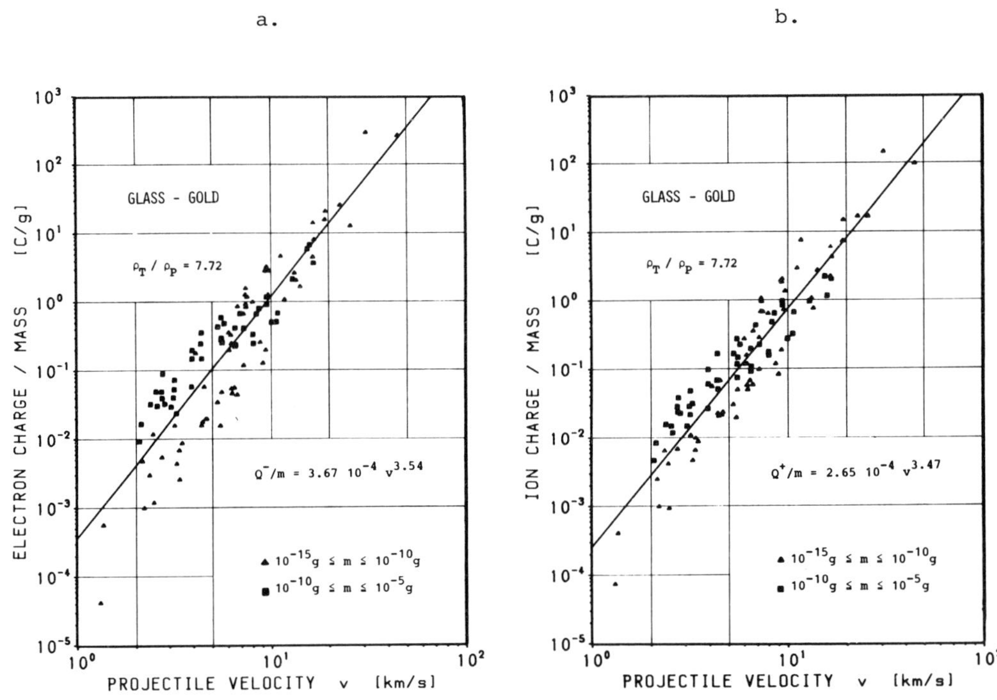

Fig.8. Charge to mass ratio Q/m plotted against particle velocity for glass particles impacting a gold target.
a. (left) Negative charge yield.
b. (right) Positive charge yield.

The experimental data in Fig. 8b deviate from the linear fit as shown in the histogram of Fig. 9. The distribution is approximately gaussian on a logarithmic scale with a standard deviation Δ. For glass → gold experiments the Δ_{ION} is 2.3 and $\Delta_{el} = 2.2$.

It should be noted that no significant contribution of the negative ions to the total yield of the positive collector could be found. As the ions need more time to reach the collector than the electrons, negative ions would have produced a second increase of the negative charge signal. This was not observed.

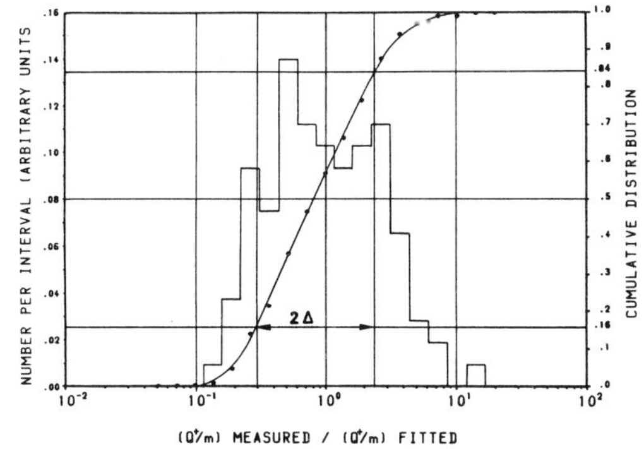

Fig.9. Histogram and cumulative distribution of the ratios (Q^+/m) measured to (Q^+/m) fitted. The fitted charge yields correspond to the linear fit shown in Fig.8b with a standard deviation Δ_{ION} of a factor 2.3.

The results for the impacts of glass beads on tungsten, iron and aluminium are shown in Fig. 10a and 10b.

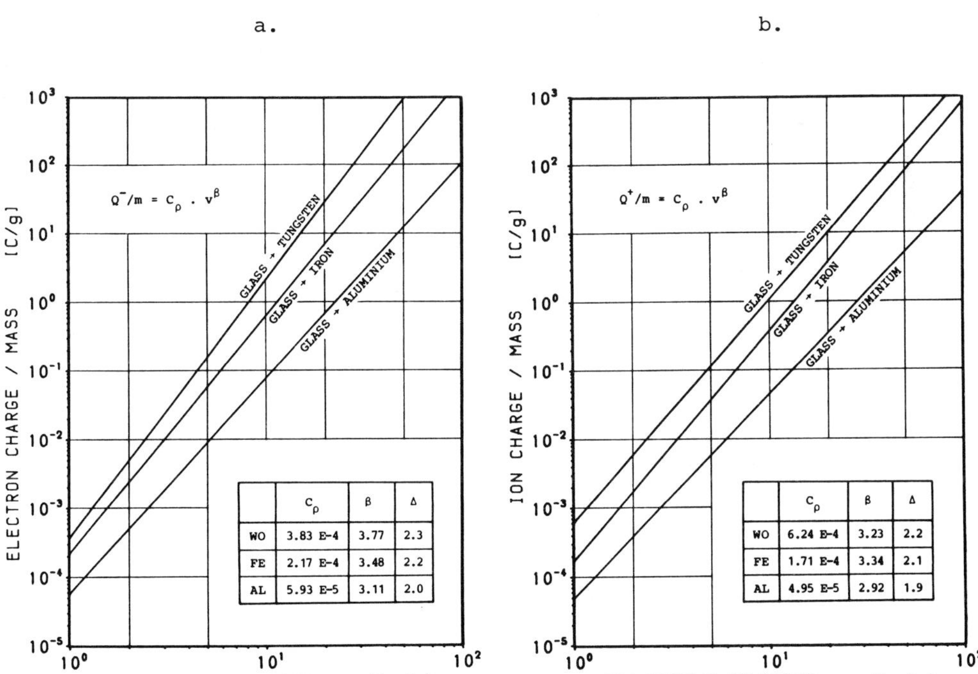

Fig.10. Dependence of the charge yield for the target/projectile density ratio ρ_T/ρ_P.

a. (left) Negative charge yield
b. (right) Positive charge yield

The ratio of the collected negative charges and the collected positive charges is close to unity, as it would be expected from the impact model, which implies the conservation of the charges. It was observed however, that this ratio tends to exceed the value of 1.0 at high velocities.

This probably indicates, that not all positive ions can be collected at the negatively biased collector plate at high impact velocities.

It should finally be indicated that the transport of charges by the ejecta that are generated by a hypervelocity impact must be taken into account.

The ejecta does also generate secondary charges, when they impact on the collector surface. These must be measured separately from the charges that were generated by the primary impact on the target.

4. CONCLUDING REMARKS

The results of the experiments that are described in this paper do indicate that the empirical formula that had been suggested by Friichtenicht, 1963 can be used to approximate the results beyond the initial mass range (10^{-15} to 10^{-10} g), up to 10^{-5} g.

The exponent α of the mass seems to remain very close to unity and the charge yield increases with the density of the target material.

A further extension of the mass/velocity range in which the impact charge is measured depends upon the performance of the hypervelocity accelerators.

Recent theoretical analysis (Krueger, 1984) shows tendencies towards a rather good understanding of the measured phenomena. There exists, however, at this time probably no theory that would provide a general derivation of the empirical formula for the charge generation.

The experiments that are presented in this paper were part of the investigations which have been conducted within the project GIOTTO, that is the flyby of a spacecraft of the European Space Agency ESA at the Comet Halley in March 1986. The authors would like to express their thanks to their colleagues, who were involved in this project, especially E. Grün, J. Kissel, J. McDonnell and H. Kuczera.

This work was performed at the Lehrstuhl für Raumfahrttechnik of the Technische Universität München (Director H. O. Ruppe) and the Kosmophysic Group of the Max-Planck-Institut in Heidelberg (Director H. Fechtig). It was supported by the Bundesministerium für Forschung und Technologie (Förderungsnummer WF/A 01 OF 1026).

5. REFERENCES

[1] Auer, S. and Sitte, K. (1968). Detection technique for micrometeroids using impact ionisation. Earth Planet. Sci. Letters, 4, p. 178.

[2] Dietzel, H., Eichorn, G., Fechtig, H., Grün, E., Hoffman, H.J., and Kissel, J. (1973). The HEOS 2 and HELIOS micrometeoroid experiments. J. Phys. E. Sci. Instrum. 6, p. 209.

[3] Fechtig, E., Grün, E. Kissel, J. (1978). Laboratory simulation, in 'Cosmic Dust', ed. J.A.M. McDonnell, John Wiley & Sons, Chichester, p. 607.

[4] Friichtenicht, J.F. and Slattery, J.C. (1963). Ionisaion Associated with Hypervelocity Impact. NASA TND 2091.

[5] Grün, E., Kisssel, J., Fechtig , H., Gammelin, P., Hoffmann, H.J. (1976). Preliminery results of the micrometeoroid experiment on Helios A, Lecture Notes in Physics, 48, p. 159.

[6] Grün, E. (1981). Experimental studies of impact ionisation. ESA-SP-155, p. 81.

[7] Igenbergs, E., Shiver, E. L. (1973). Magnetgasdynamic compression of a coaxial plasma flow for micrometeoroid simulation. J. Appl. Phys., 44, p. 5.

[8] Igenbergs, E., et al. (1986). Launcher technology, in-flight velocity measurement and impact diagnostics at the TUM/LRT, 1986 Hypervelocity Impact Symposium.

[9] Krueger, F. (1984). Ion emission from solid surfaces: comparison of dust impact with other excitations, ESA-SP-224, p. 49.

[10] McDonnell, J.A.M. (1978). Microparticle studies by space instrumentation. Cosmic Dust, Ed. J.A.M. McDonnell, Wiley, Chichester,GB, p. 337.

Int. J. Impact Engng Vol. 5, pp. 389–410, 1987
Printed in Great Britain

0734-743X/87 $3.00 + 0.00
Pergamon Journals Ltd.

HISTORICAL OVERVIEW OF HYPERVELOCITY IMPACT DIAGNOSTIC TECHNOLOGY

W.M. Isbell
General Research Corporation
Santa Barbara, California

ABSTRACT

The ability to understand hypervelocity impact phenomena and to validate predictive models of material behavior is largely dependent on the diagnostic tools available to the experimenter. These tools range from simple, post-impact examination of targets to extremely sophisticated and complex techniques, which simultaneously measure a myriad of impact parameters on a time scale of nanoseconds to milliseconds. This wide range of available techniques represents the challenge and opportunity of hypervelocity impact experimentation.

There has been a continual challenge to develop diagnostic techniques with ever-increasing resolution, as higher velocities and pressures are achieved. Today's techniques provide the experimenter the means of measuring most hypervelocity impact parameters, including velocity, displacement, temperature, radiation, volumetric change, impulse, stress, and strain of the materials involved. However, the prospective quantum leap in impact velocities to be produced by electromagnetic and electrothermal launchers will require corresponding advances in diagnostic systems.

This paper examines the capabilities and limitations of the major diagnostic techniques for hypervelocity impact experimentation and traces their evolvement as useful laboratory tools.

INTRODUCTION

Impact diagnostics have a history as old as that of mankind, when ancient man observed the deleterious effects to his enemies and prospective food sources caused by the impact of a rock, club, or spear. For millennia, terminal observations of penetration depth, hole diameter, damage to surrounding material, etc. remained the major diagnostic tool available. Because impact processes at all but the lowest velocities are far too rapid for the human eye to resolve and, because the human brain tends to become overwhelmed by the magnitude of the energy released, it was difficult or impossible to time-resolve the impact event until mechanical, optical, and electrical systems were developed to augment human senses.

With the evolvement of photographic and electronic techniques during the 19th and 20th centuries it became possible to time-resolve impact parameters on the necessary scale. During the last fifty years, impact measurement techniques have proliferated, with a corresponding increase in understanding of the impact processes.

Before continuing with a discussion of impact diagnostics, we will examine some of the parameters measurable on hypervelocity impact events (see Fig. 1).

Initial conditions before impact include the physical and geometrical properties of the projectile and target, as well as the velocity, yaw, and integrity of the projectile. Upon impact, shockwaves are created in the target and projectile and propagate with velocities characteristic of the materials involved and the impact pressures created. Characterizing these velocities and the profiles of the shockwaves has proven extremely helpful in understanding shock interactions and the penetration process itself. Numerous studies have been devoted to determining the dynamic strengths of materials of both fundamental and practical interest.

During the initial stages of impact, radiation is emitted from the shocked surfaces. This impact flash is today a relatively poorly understood phenomenon which is currently receiving increased attention as an indicator of target kill for kinetic energy weapons.

As the projectile penetrates the outer target, a cloud of fragments is formed, which impacts inner surfaces. Characterization of this cloud allows calculation of the loading of interior

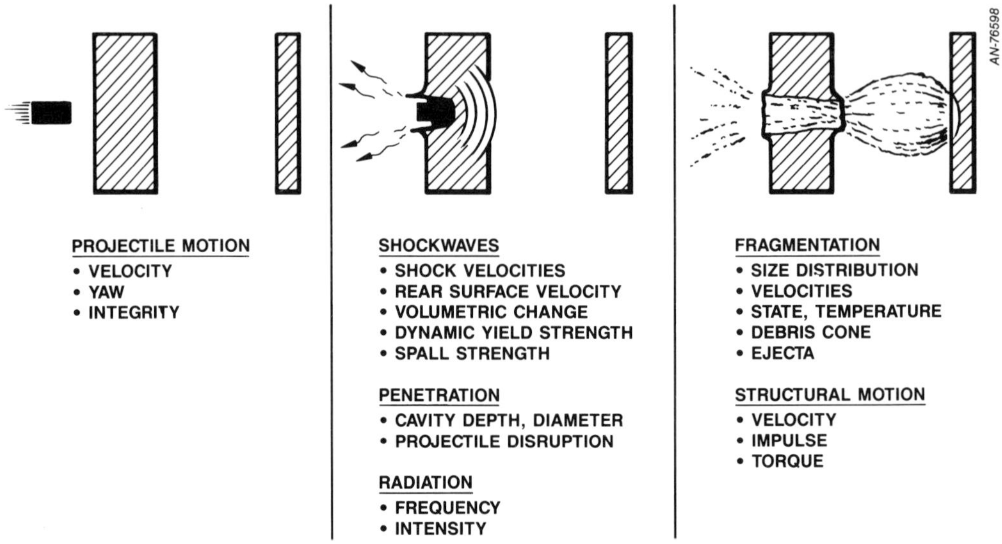

Figure 1. Measureable Parameters in the Impact and Penetration Processes

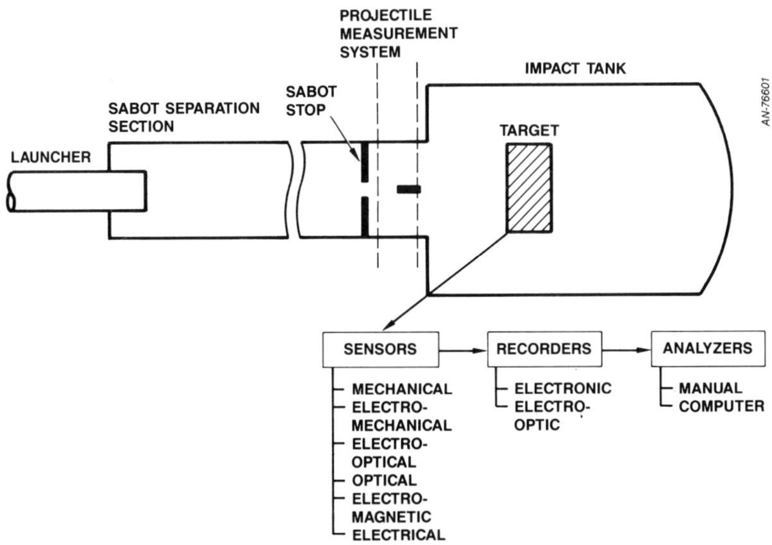

Figure 2. Impact Range Components

components and their possible penetration or fracture. This loading, as well as the initial impact, sets the structure into motion.

Measurements of these parameters are made in dozens of laboratories around the world. While the impact testers differ in their capabilities, their basic components remain the same, as shown in Fig. 2. A projectile, usually placed in a lightweight, discardable sabot for protection during launch, is propelled at a target with pre-determined velocity and orientation.

The launcher is usually (although not always) followed by a sabot separation section. An impact tank holds the target and allows access by a variety of sensors, which may be divided into the general categories shown in Table I. A system of recorders, data analyzers, and computers complete the sensor ensemble.

Table I indicates the wide variety of diagnostic techniques available to the experimenter in various areas of impact research. It is evident that, with the exception of radiation measurements, there are numerous techniques applicable to each study field. In this paper we shall discuss only a few of the more important techniques. The reference list will direct the reader to numerous additional studies.

TECHNIQUE	PROJECTILE MOTION	STRUCTURAL MOTION	SHOCKWAVES, EOS	FRAGMENTATION	PENETRATION	RADIATION
	STUDY AREA					
MECHANICAL						
BALLISTIC PENDULUM		●		●		
WITNESS PLATE			●	●	●	
WITNESS CARDS				●		
ELECTRICAL						
STRESS GAGES			●	●		
CAPACITANCE GAGE			●			
VELOCITY SCREENS	●			●		
STRAIN GAGES		●				
ARRIVAL PINS		●	●			
THERMOCOUPLES			●			
TRANSIENT RECORDERS	●	●	●	●		●
ELECTROMECHANICAL						
DISPLACEMENT TRANSDUCERS		●				
OPTICAL						
SHADOWGRAPH	●	●		●	●	
FRAMING CAMERAS	●	●	●	●	●	
STREAK CAMERAS	●	●	●		●	
MOVING FILM CAMERAS	●	●		●		
MOIRE FRINGE		●	●			
LIGHT SOURCE	●	●		●	●	
ELECTRO-OPTICAL						
INTERFEROMETERS	●	●	●			
HOLOGRAPHY	●	●	●	●		
PYROMETER						●
TELEVISION		●		●		
ELECTROMAGNETIC						
ELECTROMAGNETIC GAGE			●			
MAGNETIC PICKUPS	●	●				
X-RADIOGRAPHY	●	●	●	●	●	
MICROWAVE	●	●		●		

Table 1. Partial Listing of Techniques Applicable to Various HVI Studies

NON TIME-RESOLVED TECHNIQUES

Until the late 1930s, terminal observations were the major diagnostic of impact phenomena available to the experimenter. This technique has remained relatively unchanged over the centuries although, as shown in Fig. 3, profilometer techniques now allow us to accurately map and plot the crater surface and techniques are available to accurately measure the material mass loss.

On the right of the figure is a sectioned crater made in a layered particulate material bonded with epoxy. The deformation of the material around the crater is evident in the photograph.

The ballistic pendulum technique, shown in Fig. 4, was applied to determine the amount of momentum transferred to the target by the projectile. The multiple pendulum technique shown in the figure was developed with the advent of hypervelocities to measure the momentum distribution in the ejecta, the target, and interior components of the target. (Summers and Denardo, 1970)

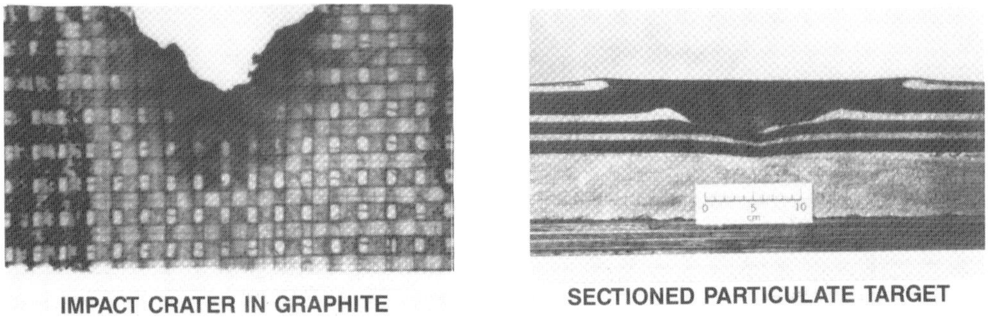

Figure 3. Crater Mapping Techniques Remain the Mainstay of HVI Investigations

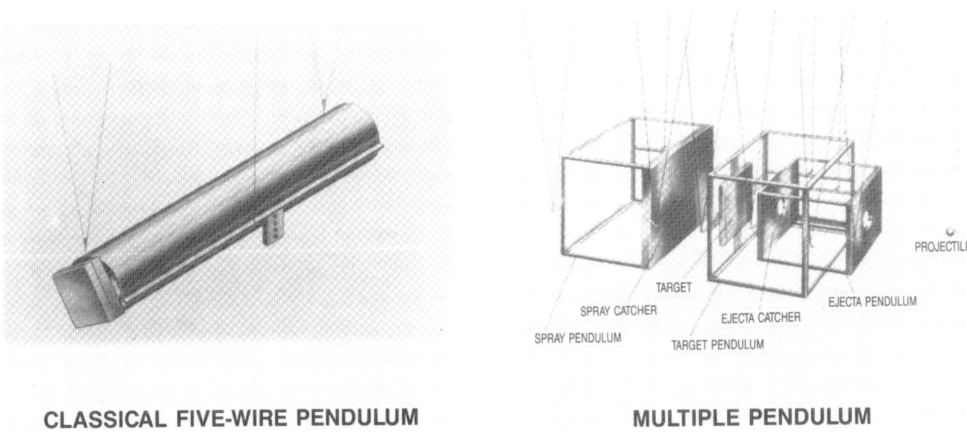

Figure 4. Non-Time-Resolved Impact Test Techniques

SINGLE FRAME IMAGING TECHNIQUES

It undoubtedly did not escape the early researchers that time-resolving the impact process would greatly benefit their understanding. The impact event was much too fast to be resolved by the human eye, however, which is limited to perhaps 50 ms time resolution.

Also, since the human brain tends to become overwhelmed by the magnitude of the energy released, it is a curious fact of our profession that no one has ever actually seen with the naked eye the processes we study. Mechanical, optical, and electronic aids are necessary for all but the most rudimentary evaluations.

Thus, the development of short exposure imaging echniques gave experimenters their first view of the impact process. The first high speed photograph was taken in 1852 by Fox Talbot of the U.K. Talbot mounted a copy of the Times newspaper on a rapidly spinning wheel and "froze" the motion by exposing the page to a high intensity spark and an open shutter camera.

The flash photography techniques were pursued by a number of researchers over the years, but gained their greatest popularity with Harold Edgerton's flash photography, examples of which are shown in Figure 5. (Gehring, 1986).

Single and multiple flash systems allowed researchers to study processes never before visualized. Photography of this type has matured for hypervelocity impact studies with an adaptation of the Cranz-Schardin camera system, in which multiple pictures of an event are recorded on nearly the same optical axis by a series of pulsed light sources and an equal number of individual lenses.

Camera arrays of up to 11 x 11 have been constructed, taking 121 separate exposures (see Fig. 6). (Swift, 1982; Wiley & Sons, 1982).

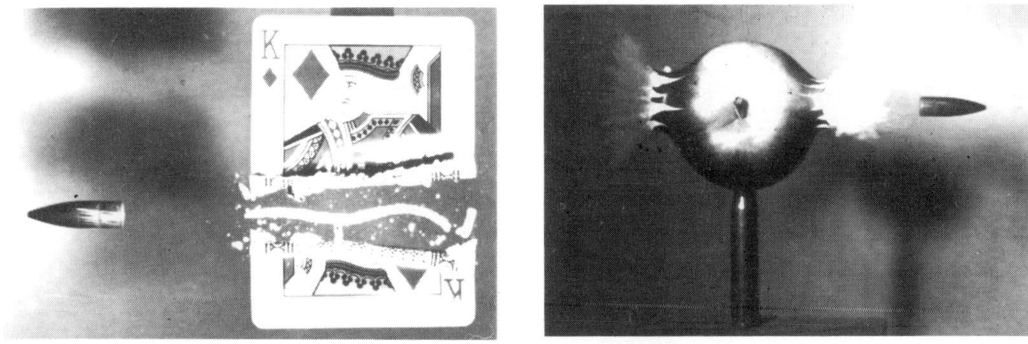

Figure 5. Flash Photograph by Edgerton

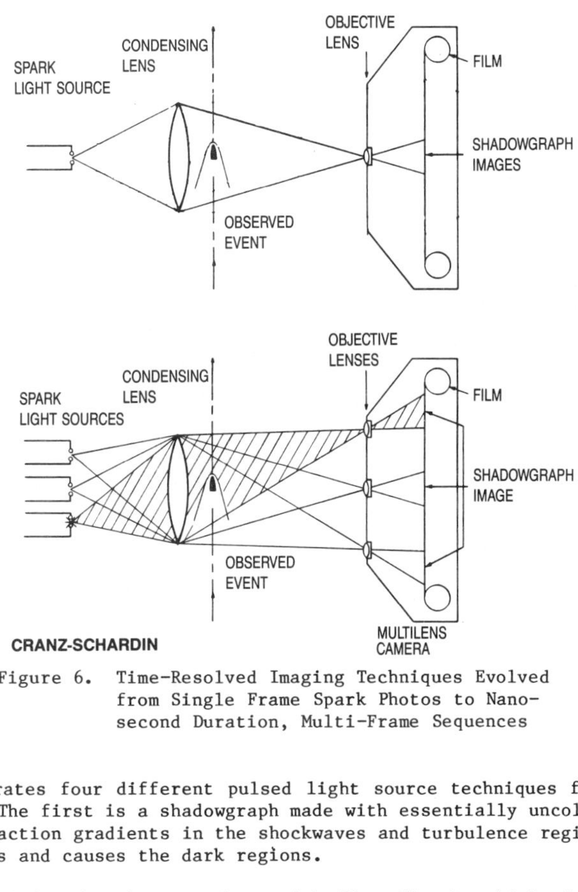

Figure 6. Time-Resolved Imaging Techniques Evolved from Single Frame Spark Photos to Nanosecond Duration, Multi-Frame Sequences

Figure 7 demonstrates four different pulsed light source techniques for obtaining inflight photos of projectiles. The first is a shadowgraph made with essentially uncollimated backlighting. The strong index of refraction gradients in the shockwaves and turbulence regions refract the light away from the camera lens and causes the dark regions.

Compare the definition in these regions with Fig. 7b, in which the light is carefully collimated. The collimation considerably enhances the observable detail.

A certain amount of quantitative information is available from these techniques on the pressure gradients in the atmosphere; however, for detailed information, a Mach-Zender interferometer system (Fig. 7c) is used, in which a laser beam is split into two paths, one of them passing through the test chamber, the other passing around the chamber. Upon recombination, the beams interfere to form a series of parallel fringes. The change in optical path length caused by the disturbed atmosphere around the hypervelocity projectile causes the fringes to shift in the manner shown in this beautiful photograph. Such interferograms furnish additional quantitative information on projectile flight. (Seiff, 1970).

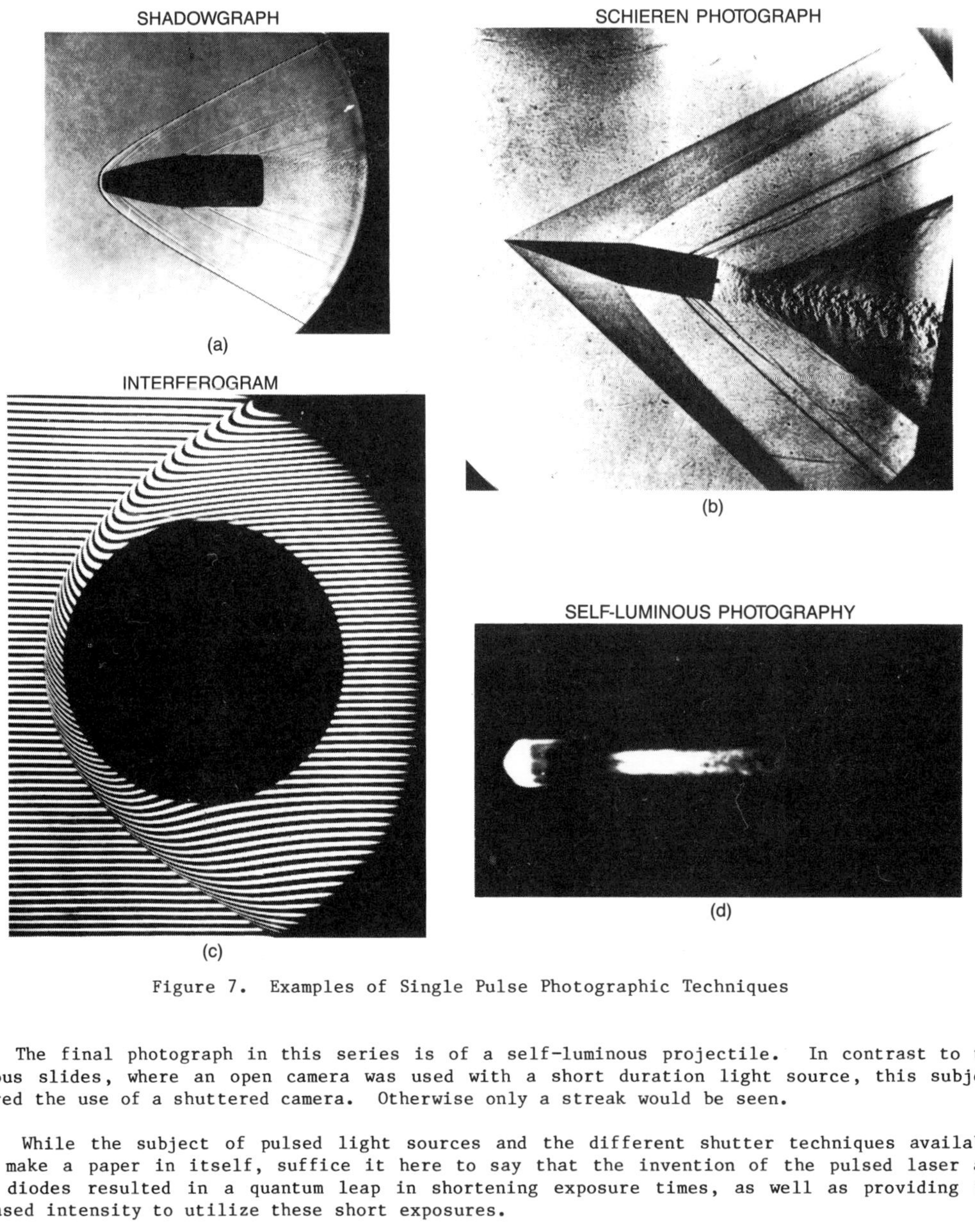

Figure 7. Examples of Single Pulse Photographic Techniques

The final photograph in this series is of a self-luminous projectile. In contrast to the previous slides, where an open camera was used with a short duration light source, this subject required the use of a shuttered camera. Otherwise only a streak would be seen.

While the subject of pulsed light sources and the different shutter techniques available would make a paper in itself, suffice it here to say that the invention of the pulsed laser and laser diodes resulted in a quantum leap in shortening exposure times, as well as providing the increased intensity to utilize these short exposures.

When we speak of time-resolution, we generally mean obtaining the value of a parameter on a time scale short compared to the duration of the event. As shown in Fig. 8, the exposure time necessary to reduce motion blur of a moving object is dependent on the object's velocity and size (as well as other parameters dealing with the camera optics and recording medium). The value of one percent blur shown here is chosen somewhat arbitrarily. A blur of ten percent would generally be judged unacceptable for identifying position or obtaining useful information on surface features.

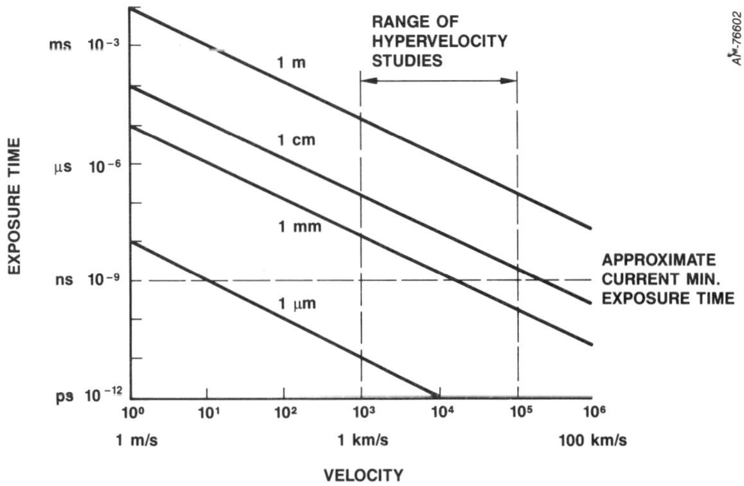

Figure 8. Exposure Times Required for One Percent Motion Blur of Moving Objects

For the range of hypervelocity studies, and with a reasonable range of projectile sizes, 1 mm to 1 cm, the range of required time resolution is 100 ns to 1 ns for velocities to 10 km. This value reduces to 100 ps for higher velocity systems.

If we wish to resolve ejecta spray in the micron range, exposure times below 100 ps are required. Current technology provides near nanosecond exposures, with some instruments reaching into the range of tens to hundreds of picoseconds, although usually at a loss of image quality.

MULTIPLE FRAME IMAGING TECHNIQUES

While single frame photos provide a snapshot of the impact process at one point in its evolution, the development of multiple frame techniques was necessary to further understanding. Until about 1930, the intermittent action camera movement was the only source of cinematography. It was found difficult to obtain framing rates greater than about 100 frames per second without damaging the camera mechanisms. Even today, framing rates do not exceed approximately 500 frames per second. The corresponding exposure times of 400 s are too long for most hypervelocity studies.

Tuttle provided one answer to this problem in 1932 with his invention of the rotating prism, continuous film camera system. In this camera, the film was pulled through continuously, rather than in steps (see Fig. 9). Precise registration of the image on the moving film was enabled by reflecting the image from a multi-faced prism, which was synchronized with the film, and moved the image along the film surface at the same speed as the film.

With this technique, framing rates were increased to 2000 frames per second and then, as new cameras were introduced, to 11,000 full frames per second, or 44,000 quarter-frames per second. Exposure times as low as 10 microseconds can currently be obtained with these simple to operate cameras.

Industry and the military soon found numerous uses for the cameras, although the exposure times were suited primarily for relatively low velocity impact processes and the measurement of structural motion after impact.

It was not until the invention of the rotating mirror camera in 1938 by C. David Miller that framing rates of over one million per second were achieved and hypervelocity events could be photographed. Of course, in 1938 hypervelocity impacts had yet to be invented, so the cameras initially were used for studies of combustion.

During the 2nd World War, rotating mirror camera technology improved rapidly as the cameras were used in the development of both conventional and nuclear weapons. Today, there are many versions of these cameras available, with framing rates to ten million frames per second and exposure times down to 50 nanoseconds. (ICIASF 1969, 1971, 1973, 1975, 1977, 1979) (ICHSP 1972, 1974, 1976, 1978).

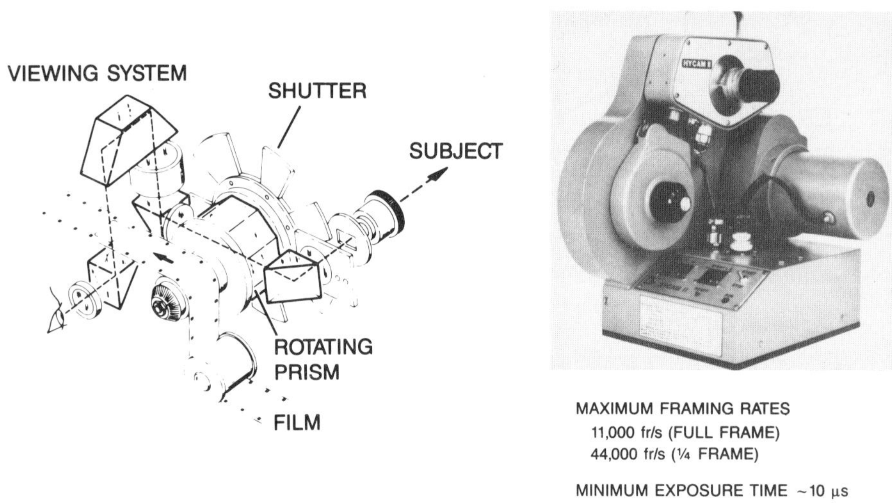

Figure 9. Rotating Prism Cameras Allow Moderate Framing Rates and Long Coverage

As shown in Fig. 10a, rotating mirror camera is actually a series of individual cameras, mounted in an arc about a rotating mirror, with a common objective lens and optical axis. The mirror is driven by a gas turbine and acts as a high speed switch, sequentially directing the image from camera to camera and exposing a strip of film.

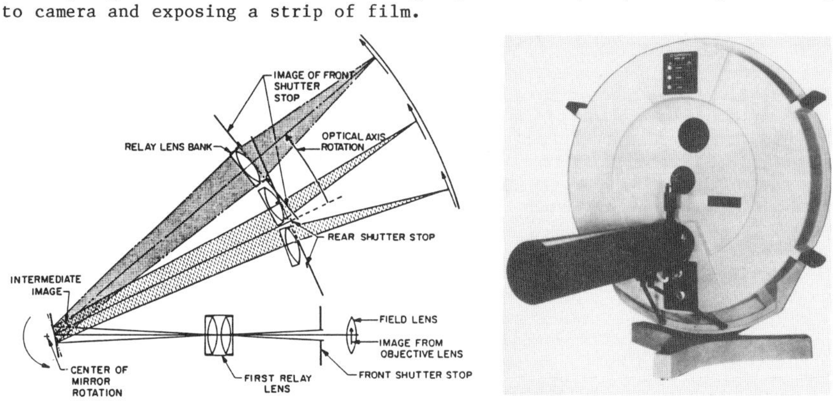

Figure 10. Rotating Mirror Cameras Represented a Breakthrough in High-Speed Imaging Techniques

Although this type of camera became a workhouse instrument for hypervelocity impact and shockwave research, it too suffered from limitations.

- Practical considerations limit the number of frames to around one hundred, which limits the total recording time.
- The illumination system for the camera must be extremely intense, due to the optical system's relatively high f-number.
- The cameras must be synchronized with the event, making their use with guns difficult.

The last restriction on these cameras, which would have made them essentially useable for most impact tests, was removed when asynchronous cameras were developed. An example of an asynchronous camera is shown in the figure. Note that the film plane extends 360 around the optical axis. In the original (synchronous) design, the film was exposed over only 60 to 90 of the circle surrounding the rotating mirror. The event had to be synchronized with the position of the camera mirror so that the film could be properly exposed. Since guns are difficult to synchronize precisely, the development of the asynchronous camera was helpful.

The literature is filled with examples of high speed photographs of hypervelocity phenomena. Figure 11 is a typical example. A 1 mm diameter glass bead enters from the left at 1.8 km/s and impacts a carbon nosetip sample. The impact flash is seen on frame 2. A very high speed spray of ejecta can be seen leaving the impact point on frame 3, followed by much more substantial ejecta leaving at later times and at lower velocities. The study was of the erosion of reentry vehicle nosetip materials as they reentered through a storm cloud. The intent of the tests was to determine whether the ejecta from an impact would act as a screen against rain drops arriving later. These photographs allowed quantification of the screening process.

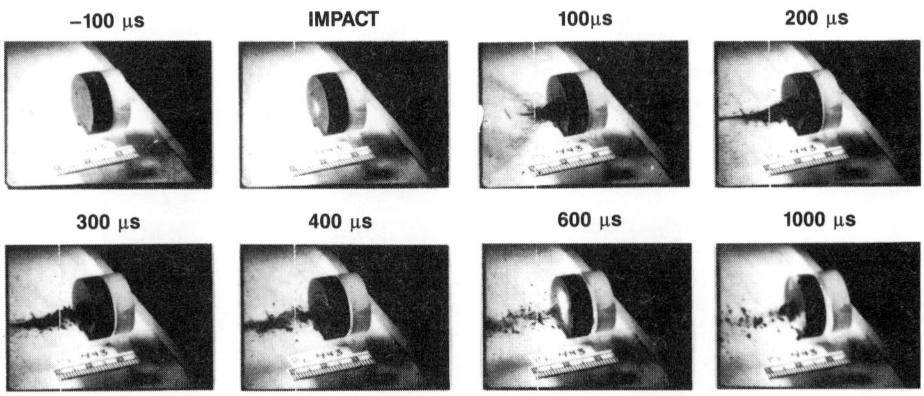

Figure 11. Glass Sphere Impacting Carbon Nosetip Material at 1.8 km/s

A clever means of shortening the exposure time has been used with a variety of cameras. A pulsed laser is synchronized with the camera shuttering mechanism. The camera shown in Fig. 12 has a framing rate of "only" one million frames per second and an exposure time of 300 ns. The exposure time is reduced to 25 ns by the laser illuminator. This technique is being applied to a variety of cameras, including very low framing rate systems, where long coverage is desired, but short exposure times are needed.

ELECTRONIC CAMERAS

A new type of camera was introduced during the 1960's, when image intensifier tubes, similar to those used for night viewing, were electronically shuttered to give short exposure times. Figure 13 demonstrates the major components of the camera.

The advantages to this technique are significant. Exposures down to 1 ns are possible and, with the addition of an additional amplifying stage, light amplification to perhaps 100,000 X is available, making possible the use of these cameras with only modest illumination. (Kotecki & Goranson, 1983)

A penalty is paid for these advantages, however, in the substantially decreased spatial resolution of the systems. The electro-optical systems for image converter cameras is typically a factor of 5 to 10 less than that of rotating mirror cameras and is 10 to 100 times less than still photography. However, the state of the art continues to improve.

As it has with our private lives, television has also entered our professional lives with the fairly recent development of high speed video equipment. The graph in Fig. 14 shows how the framing rates of these systems has increased during the past 10 years to 2000 frames per second for full frames and 12,000 frames per second for a one-sixth frame format. Recording times have been extended to 45 seconds even at the highest framing rates. This technique allows instant recording and immediate playback of high speed events. (Sanders, 1983; Foley, 1986).

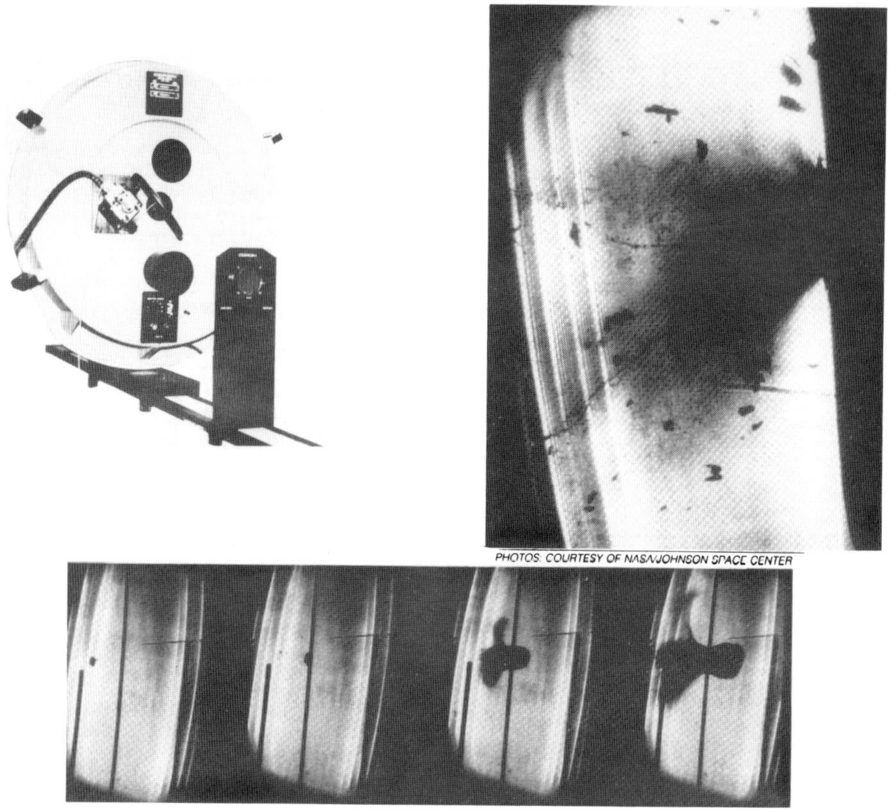

Projectile:	.07" nylon pellet
Velocity:	7.1 Kilometers/sec.
Framing Rate:	1 million frames per second
Target:	.032" aluminum

Figure 12. Pulsed Laser Illumination for 25 ns Exposures

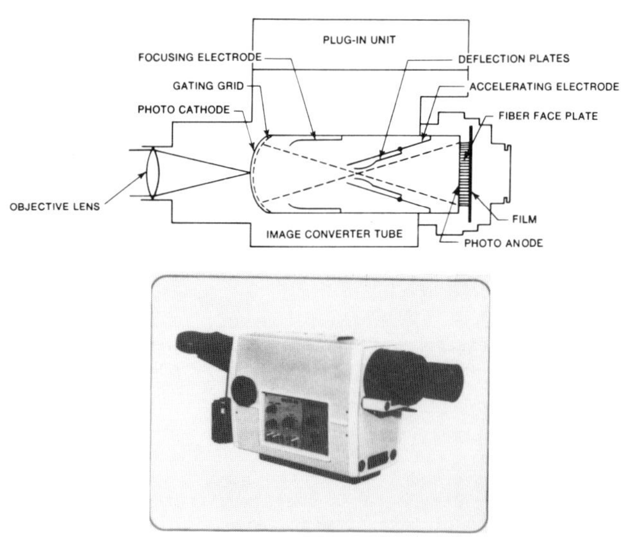

Figure 13. Electronic Framing Cameras Increased Framing Rates to 3×10^9 Frames/s

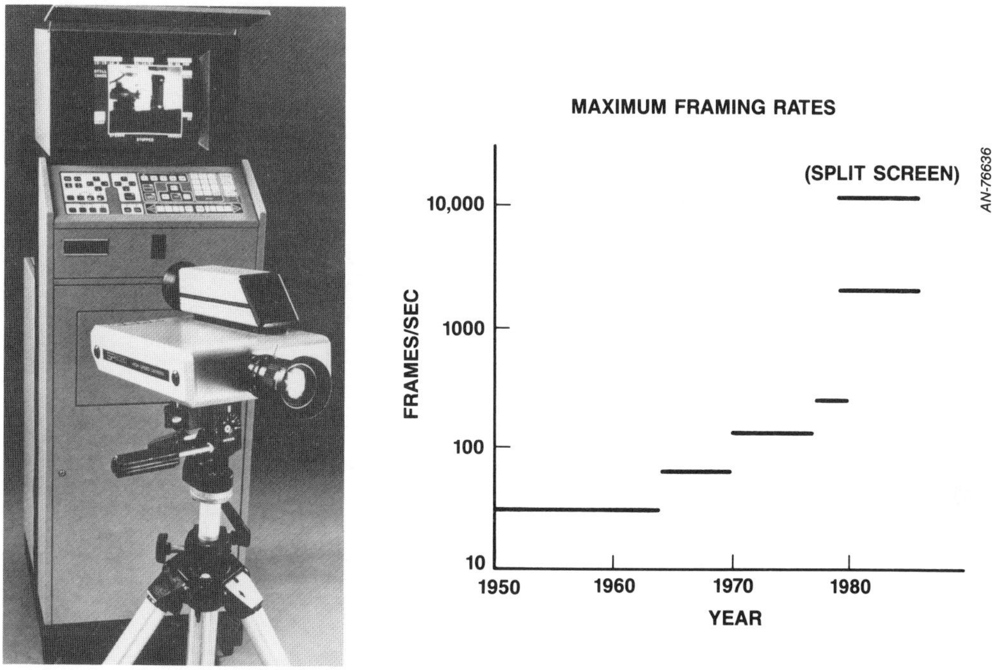

Figure 14. High-Speed Video systems are a Promising New Development

The new video cameras are relatively simple to operate and have a low operating cost compared to film cameras, due to the reusability of the recording tape and the fact that film needs to be developed, adding to its cost. This is counterbalanced to some degree by the higher initial cost of the system, compared to film cameras with comparable framing rates.

Perhaps the most intriguing characteristic of high speed video systems is the ability for their images to be easily digitized, opening up a world of possibilities for computer manipulation of the data. On-screen measurements of position and motion are immediately available to the operator, when the video is used in the "freeze-frame" mode.

The relatively low framing rates of video cameras make motion blur excessive for hypervelocity events. While the relatively long time interval between frames cannot be overcome, the blur can be essentially eliminated by using a pulsed laser illuminator similar to that discussed earlier. Figure 15 shows a bullet front-illuminated with a 30 ns laser pulse. Notice the bullet rotation, as evidenced by the movement of the band on the bullet´s side.

FLASH X-RAY TECHNIQUES

Visible wavelength photography has the obvious limitation that it cannot penetrate to the interior of targets and is blocked by the debris arising from the impact. The invention of the flash x-ray provided a fundamental measurement tool which solved these problems. While continuous wave, long duration x-radiography had been in use for decades, it was the invention of cold-cathode x-ray tubes, capable of operating at thousands of amperes, that provided the short duration, high intensity x-ray fluxes required for hypervelocity impact measurements.

Two types of tubes were developed, the field emission tube and the vacuum arc tube. Primarily because of its superior stability, the field emission tube is the most commonly used tube in

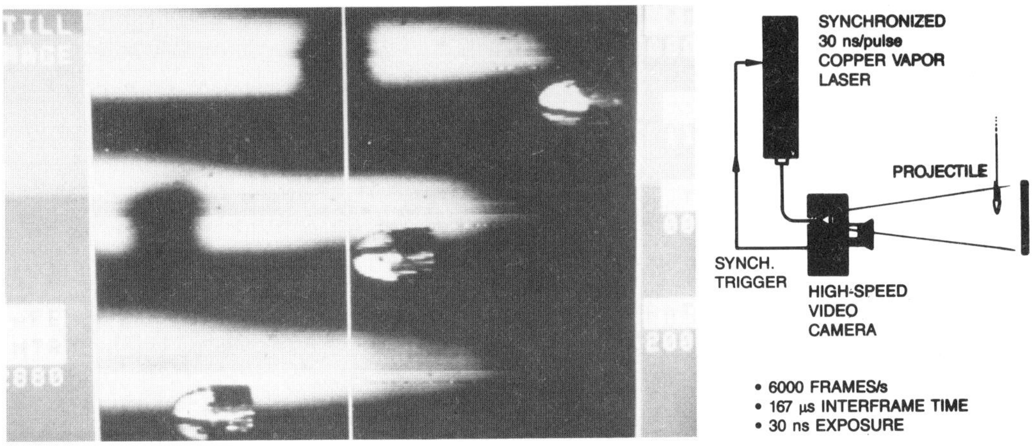

Figure 15. Motion Blur Stopped by Pulsed Laser Illumination Source

modern instruments. As shown in Fig 16 this tube uses a conical anode, surrounded by a cathode consisting of many wires with fine points. The intense fields generated by the application of high voltage causes the emission of electrons from the cathode wires. Collision of the electrons with the anode causes the emission of an intense burst of X-rays. A conducting plasma is formed between the anode and cathode, reducing the voltage and ending the X-ray pulse. (Swift, 1982).

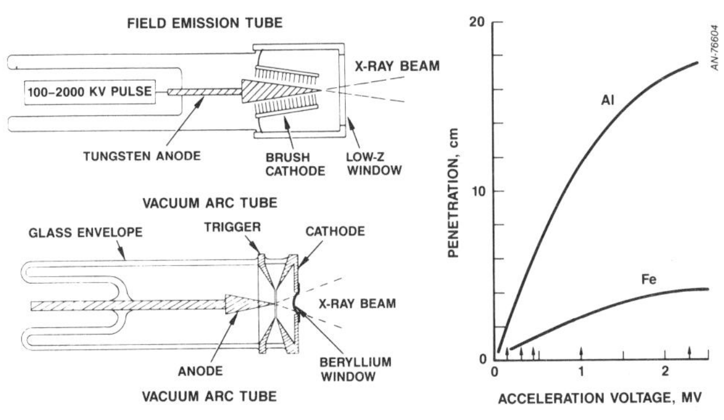

Figure 16. Flash X-Radiography . . . A Logical Next Development

Shaped charge studies in 1943 at the Ballistic Research Laboratories used these new devices and hypervelocity impact ranges were instrumented in the late 50's and early 60's. Projectile and sabot separation radiographs soon became common, as did the radiographs of impacts against simulated space structures and reentry vehicles (see Fig. 17). (Bryant & Lucero, 1983).

In the typical setup shown in Fig. 18, a shadowgraph is taken of the object. In order to stop the X-rays so that the film may be exposed, an image intensifier screen is usually used, made of rare-earth phosphors and high density salts. The geometry and the materials used for high resolution radiographs is generally regarded as much an art as a science.

The degree of detail and the amount of understanding obtained from the radiographs can be exceptional, as exemplified by the tests made by the Delco Systems Operations group shown in Fig. 19. On the left is a debris bubble composed of vapor and liquid particles created by the impact of a lead sphere onto a thin lead plate. On the right are solid particles from the impact of a molybdenum sphere on a molybdenum plate. A second set of X-rays were also taken, at a later time and orthogonal to the ones shown. This allowed measurements to be made of the velocity of the debris bubble elements. Densitometer measurements of the radiographs provided the density of the particles comprising the cloud. Such measurements provide a direct check on the equations of state and transport capabilities of hydrocodes used to predict the event.

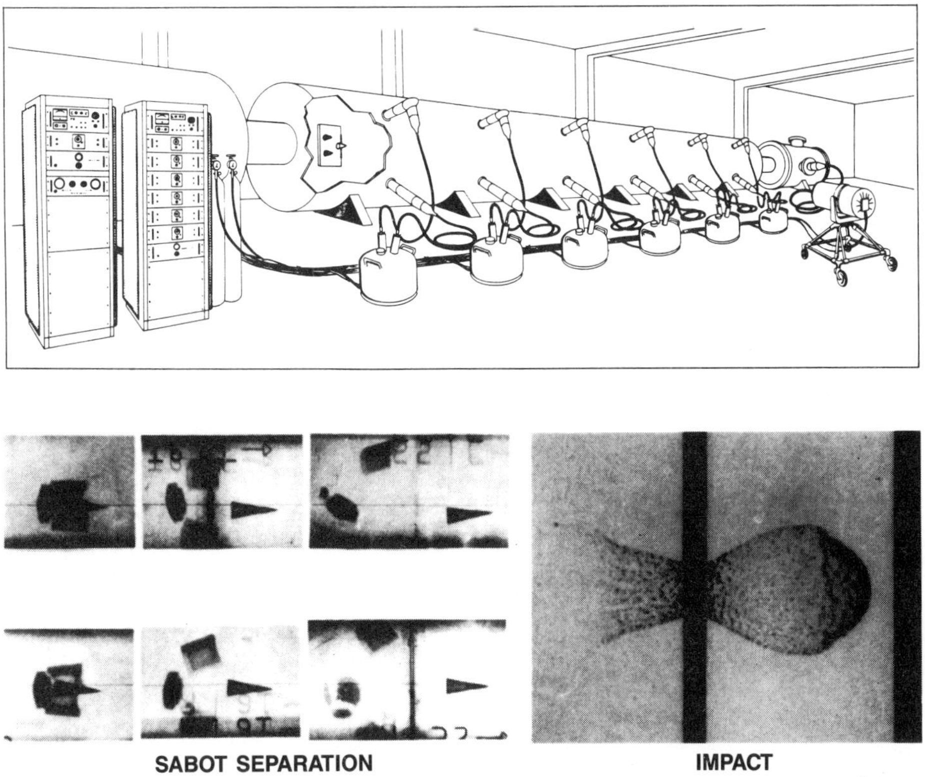

Figure 17. Flash X-Ray Systems Quickly Become a Fundamental Tool of HVI Research

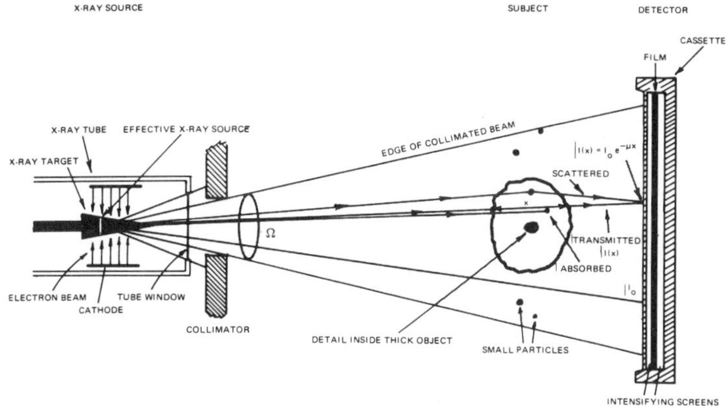

Figure 18. Typical Flash X-Ray Setup

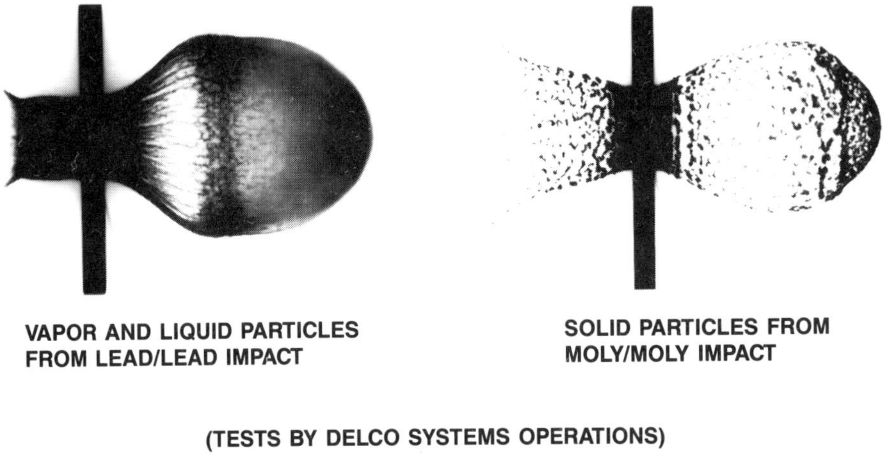

Figure 19. Vapor Phase and Solid Phase Debris Bubbles

Flash X-rays have also proven invaluable in allowing us to visualize the penetration mechanism deep within a solid target, such as the radiograph of a long rod penetrator entering a block of aluminum, shown in Fig. 20.

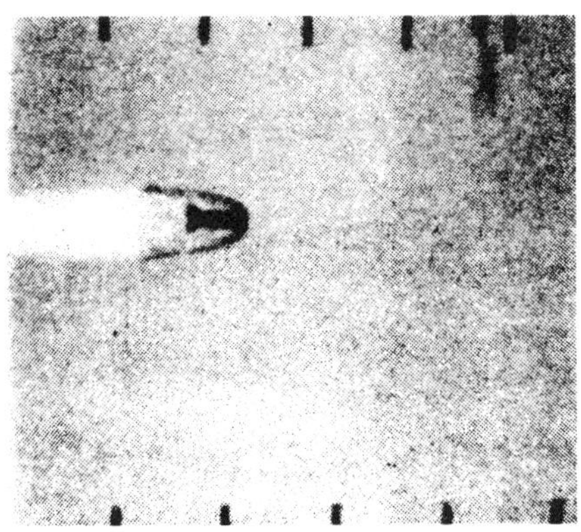

Figure 20. X-Ray Showing Penetration Interior to the Target

During the sixties, flash X-ray technology was exploited in a variety of ways, two of which are shown here. The first technique used the high electron current densities developed for flash X-rays to produce an intense burst of light. A CdS target was irradiated with a pulse of electrons. Approximately 1% of the electron energy was converted to visible light. The duration of the emission followed that of the input pulse. A relatively uniform spectrum was emitted, peaked at 532 nm, with a bandwidth of 6 nm (Fig. 21). The intensity of the light was sufficient to front-light objects, with exposures as short as 2 ns. (Photographs courtesy Hewlett-Packard Co.)

The second technique utilized x-rays and electrons, both emitted from the same source. The technique used electrons to provide high resolution images of small particles (normally lost in the more penetrating x-radiographs) and used x-rays to penetrate the debris cloud and image the denser components interior to the test. Two films were used; the first (used without a covering) was exposed by the electrons (and visible light, if any) X-rays passed through the film without causing exposure. The second film had the intensifier screen normally used for x-radiographs and recorded the X-ray image.

The advantages of short exposure, flash X-ray movies was not lost on the researchers, who developed several methods of providing sequential radiographs of hypervelocity processes. In one technique, a series of tubes is arranged in a row and pulsed sequentially. In another, multiple flashes are produced from one tube. While the latter technique has the advantage of providing a common source for all radiographs, heating of the tube generally limits either the number of pulses or their charging voltages. (Bryant & Medecki, 1983).

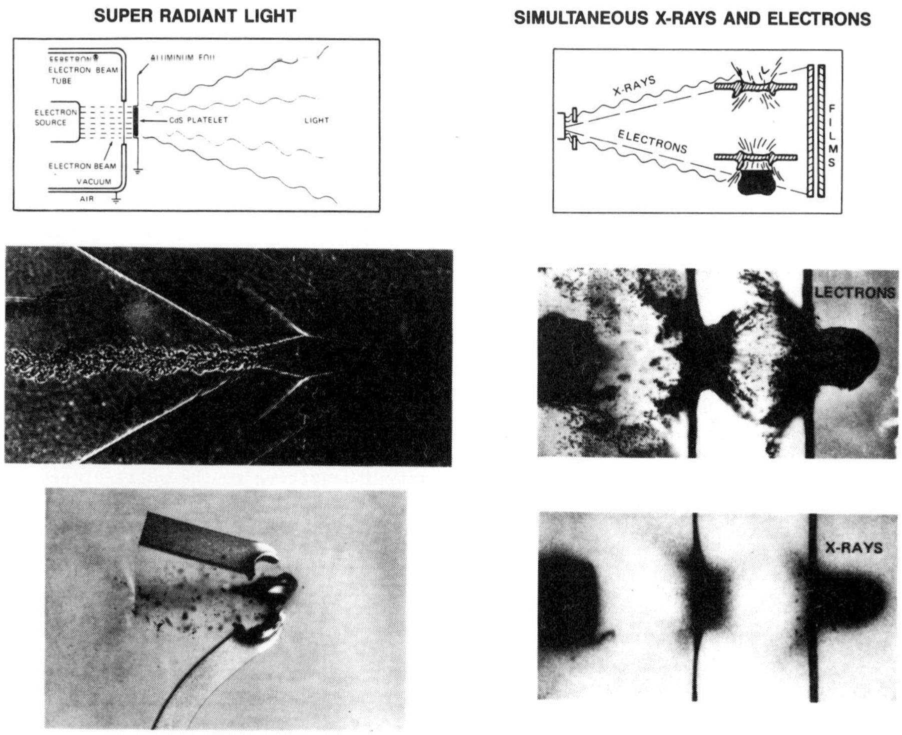

Figure 21. Flash X-Ray Technology Gave Rise to a Variety of Techniques

A second method of obtaining x-ray movies is currently being exploited, using a continuous x-ray source to directly illuminate a large x-ray intensifier tube. This technique and others are discussed more fully in the paper by H.F Swift in these Proceedings.

WAVE PROFILE MEASUREMENTS

Because the parameters measured by imaging techniques are insufficient for describing the shockwave processes involved in hypervelocity impact, a different class of instruments and techniques was developed. Highly accurate, highly detailed measurements of shockwave profiles are performed using a number of techniques, three of which are shown schematically in Fig. 22.

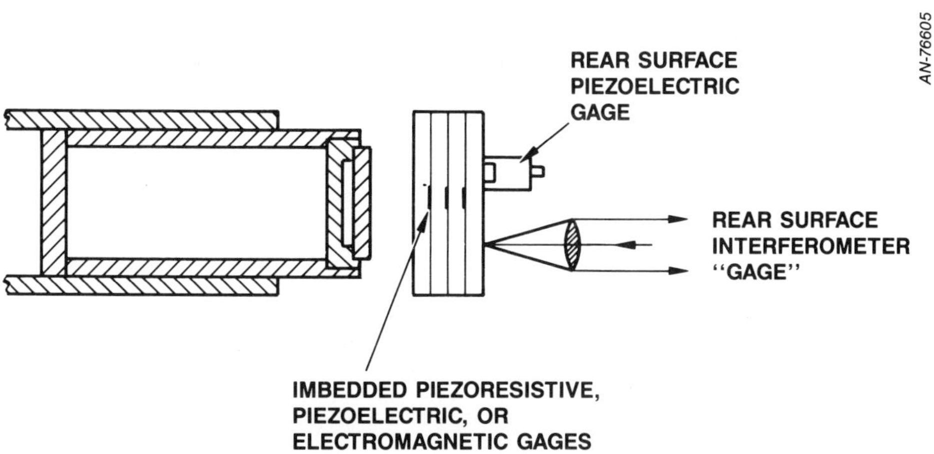

Figure 22. Wave Profile Measurement Techniques

Rear surface gages, consisting of discs of piezoelectric materials were developed (primarily by Robert Graham and coworkers of Sandia National Laboratories), which produced an electric current proportional to the stress transmitted from the target into the gage surface. Although the technique is limited to maximum stresses of 30 to 100 kb due to the properties of the materials involved, it is appealing in its simplicity and its potential for nanosecond resolution of wave profiles.

Imbedded piezoresistive gages are also used extensively in impact work. As an in-situ measurement, they have the advantage of only minimally distorting the wave as it passes through the plane of measurement.

A third "gage", developed by Lynn Barker of Sandia,Corp., Albuquerque, New Mexico, uses a laser beam reflected from the target rear surface to measure the surface motion. As the surface moves, the frequency of the light is Doppler-shifted in proportion to the velocity of the surface. The reflected beam is returned to an interferometer, which creates optical fringes in response to the Doppler shift. (Barker and Hollanback 1974)

This technique has provided the most accurate measurements of wave profiles to date. For hypervelocity impact use, it has the added advantage that it measures motion at a point on the surface, rather than over an area, and thus is nearly insensitive to the multi-dimensional motions frequently accompanying impact processes.

These gage techniques are described in more detail below:

PIEZORESISITIVE GAGES

The use of gages to measure dynamic strain dates back to the mid-30's and the development of a relatively simple device, called a "strain gage" by its inventor, E.T. Simmons, then a student at California Institute of Technology. These gages are used extensively in measuring strains in structures, but they have found only limited use in hypervelocity impact studies due to their limited bandwidth and hence the difficulty in interpreting their records.

The use of gages to measure dynamic stress is more recent. Douglas Keough at Stanford Research Institute and researchers in the U.K. began development of in-situ piezoresistive gages during the early 1960's to measure shockwaves in granite, generated by underground nuclear tests. The early gages were constructed with a small diameter manganin wire attached to conductive leads and sandwiched between two thin insulating layers, frequently made of mica. (Keough. De Carli, Rosenberg, 1981); (Lee, Jenrette, 1981); (Keough, Wong, 1970).

A constant current passed through the gage element gives rise to a constant voltage, until the resistance of the gage is changed by the passage of the shockwave. Static calibrations of the piezoresistive coefficient for manganin, done by Bridgeman in the 1930's, were found to be inadequate for dynamic measurements and calibrations under shockwave conditions were necessary. Careful calibration experiments were performed to relate the observed change in voltage to the amplitude of the stress wave.

Developments over the past 25 years have taken several forms. The original low impedance manganin gages were taken to higher pressures, near 1 Mbar. A 50-ohm manganin gage was developed to give higher sensitivity for lower pressure measurements. Other, more sensitive materials were utilized for low pressure tests, utilizing carbon and ytterbium. The carbon gage has since been extensively developed, especially for underground nuclear tests, where its low atomic number makes it possible to expose the gage to radiation. (Fig. 23). (Engberg, Steele, Charest, 1981; Williams, 1968).

Thinner gage packages were developed to decrease the risetimes of the gages to the 50-100 ns range. Techniques and materials were developed to minimize the hysterisis effects on the onloading portion of the wave. (Erickson, Weingart, Vantine, Johnson, Lee, 1981; Grady, Ginsburg, 1977; Lilley, Stephens 1971; Ginsburg, 1973; Bauer, 1981).

Shown in Fig. 24 are three typical gage constructions - a grid type of folded wire and two examples of strip type gage elements; a 50 ohm carbon gage and a very low impedance manganin gage, which employs four leads.

Note that these gage techniques were developed primarily for one-dimensional measurements. For many studies, conditions of one-dimensional strain are carefully preserved, and the gage calibrations and analysis assumptions are correct. In impact physics, however, we are usually dealing with two or three-dimensional stress and strain fields and these techniques must be used with great caution.

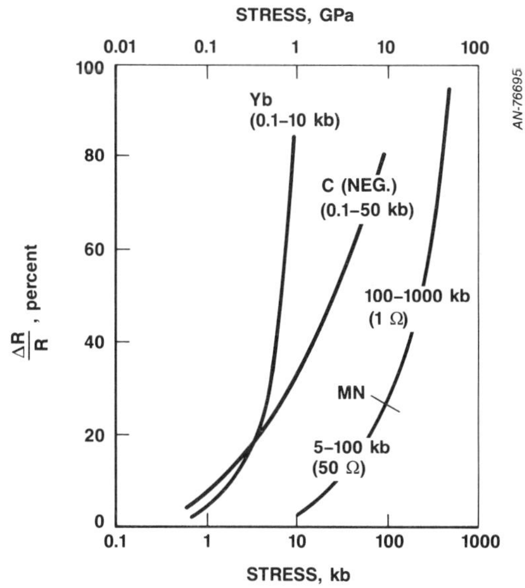

Figure 23. Imbedded Piezoresistive Gates Provide In-Situ Measurement of wave Profiles

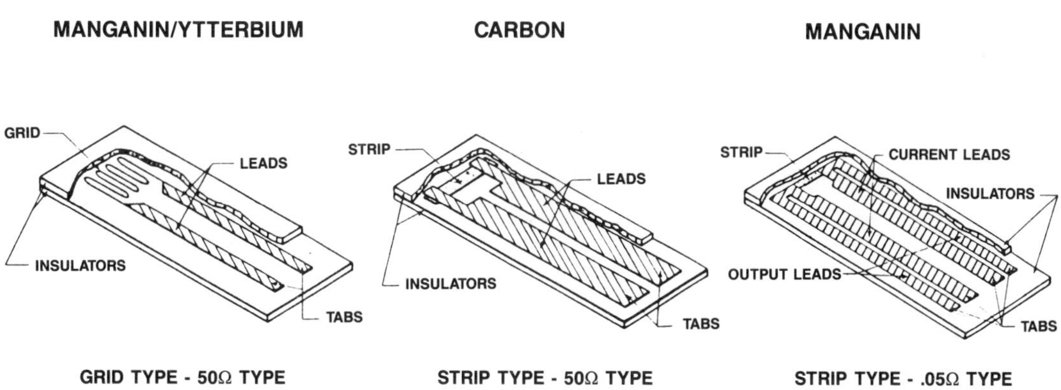

Figure 24. Typical Gage Constructions

Over the past 5 years, a concentrated effort has been made develop to the imbedded gage technique for use in multi-dimensional strain fields. One promising technique developed by Charest and others is shown in Fig. 25. (Charest, 1979; Trimble, 1981; Lockheed Missiles & Space Co., 1976; Gupta, 1980).

A piezoresistive gage of manganin or ytterbium is interwoven with a strain-sensitive material such as constantan. Both gages see the same stress and strain fields.The manganin gage produces a record that is proportional to both stress and strain. The constantan gage records only the strain. These records, combined with a knowledge of the strain factors for both gages, allow the manganin record to be modified to remove the contribution due to strain. Several clever adaptations of this technique have been developed, including a 50 ohm gage which is comprised of two 25 ohm elements in series. The piezoresistive element, ytterbium, is balanced by a strain-sensitive material, Lohm, which has an equal but opposite strain factor. The gage output is thus automatically compensated for strain. (Gupta, Keough, Henley, Walker, 1980; Rosenberg, Yaziv. Partom, 1980).

INTERFEROMETRY

Laser interferometers have been made in literally dozens of configurations to meet varying experimental needs. The accuracy of these systems is frequently an order of magnitude or more

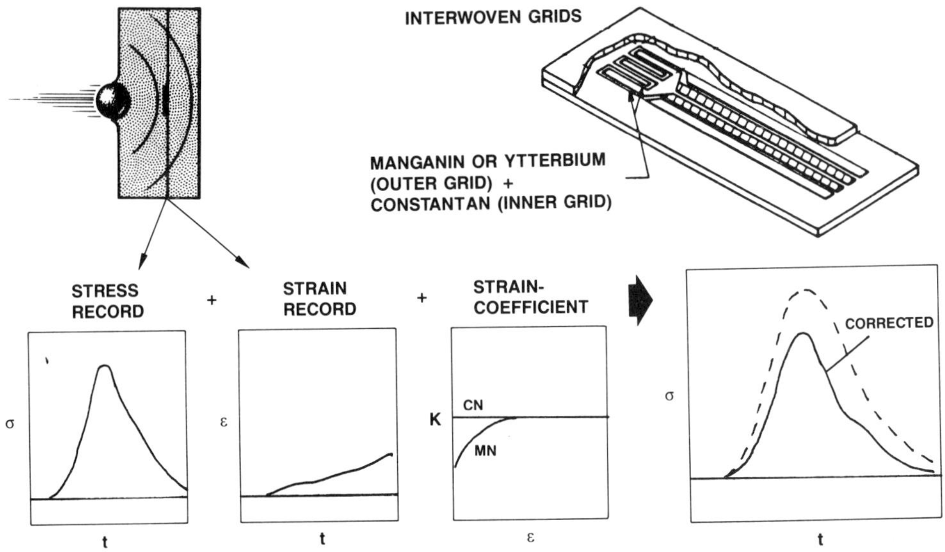

Figure 25. Strain-Compensated Gages Allow Measurement in 2-D Fields

greater than other techniques for measuring the dynamic response of materials to impulsive loading. Impact testing has ranged from measurement of structural motion at a few meters per second to measurement of shockwave profiles and of projectile acceleration at kilometers per second. Time resolution for wave profile measurements is typically a few nanoseconds.

Three basic types of interferometers have been fielded:

- Displacement interferometers, used with specular (mirror-like) surfaces
- Velocity interferometers, used with specular surfaces
- Velocity interferometers, used with diffuse surfaces

Of these three types, the last has proven the most versatile and useful. Dubbed the "VISAR" by its developers (for "Velocity Interferometer System for Any Reflector") (Barker and Hollanbach, 1974), this interferometer collects laser light reflected from the surface of a specimen to be impulsively loaded.

As the surface moves, the frequency of the reflected light is Doppler-shifted in proportion to the surface velocity. The VISAR creates interference fringes at a rate proportional to the Doppler-shift. The fringes are detected photoelectrically in the form of sinusoidally varying signals, the number of which is proportional to the surface velocity and the frequency of which is proportional to the surface acceleration. Changes in sensitivity of the instrument, covering nearly three orders of magnitude, are possible by adjustment of dimensions between optical components.

In contrast to other wave profile measurement techniques, the VISAR is relectively unaffected by multi-dimensional motion of the surface. In its "normal" configuration, the instrument measures only the component of motion along the incident beam. Other orientations of the viewing direction, however, have been used to measure not only motion normal to the surface, but also the lateral motion induced by shear waves.

Improvements to the original system allow measurements to be made in the presence of extraneous light (explosives, exploding foils, etc.) (Hemsing 1979), at greater sensitivities (Amery 1976), for the measurement of the acceleration of projectiles (Munson and May 1976; Isbell and Fuller 1983), and for semi-automated operation in production test use (Isbell 1984).

Figure 26 shows a VISAR arrangement at the Centre d´ Etudes de Gramat, France (Chartagnac 1983). An argon ion laser is used to illuminate a gun-mounted target in an adjacent room. The VISAR is to the left of the photo, with the VISAR controls and transient digitizers mounted in the electronics cabinet.

Figure 26. A VISAR Interferometer System at the Centre d´ Etudes de Gramat (France) Impact Laboratory

ELECTRONIC RECORDING INSTRUMENTATION

In general, the ability to record electronic transients has kept up with, or only slightly lagged, the needs of the sensors used. Wide bandwidth, single sweep oscilloscopes were developed during WWII and resulted in the first commercial product in 1946, a 10 MHz triggered oscilloscope by Tektronix, Inc. By 1952, specialized oscilloscopes had attained bandwidths to 500 MHz by bypassing the then low bandwidth amplifier sections and placing the signal directly on the plates of the cathode ray tube. The lack of vertical amplification severely limited the utilization of these instruments for observing small signals (< 5v).

Wider bandwidths became available during the 1960´s and 1970´s, now featuring adequate vertical amplification. GHz bandwidth oscilloscopes are currently the state of the commercial art and technology is no longer progressing rapidly, due primarily to the advent of an alternative means of recording analog electrical signals, the transient digitizer.

Analog to digital conversion of electrical signals has numerous advantages (and a few disadvantages) over methods using oscilloscopes. They include:

- Conversion of the analog data into formats suitable for computer entry.
- Long, expandable memories which increase recording time without sacrificing "sweep speed".
- Compact size and relatively low cost per data channel.
- Instrument settings controllable by computer.
- Continuously active recording, allowing the capture of data by triggering after the event.

Until fairly recently, wide bandwidth was not a characteristic of transient digitizers. Bandwidth has been steadily increasing, however, and has now surpassed that of oscilloscopes with the announcement of a 6 GHz bandwidth, 17 GHz sampling rate digitizer.

The cost of digitizers in the highest bandwidth range (>200 MHz sampling rate) has also impeded their introduction into the laboratory. Costs will decrease, however, and the majority of facilities are now equipped with multi-channel digitizing systems for data capture. Rapid advances in the capabilities of this type of instrument are expected.

REFERENCES

Amery, B.T. (1976) Wide Range Velocity Interferometer, Sixth Symposium on Detonation; Office of Naval Research, Dept. of the Navy, Arlington VA, 1976

Barker, L.H., and R. Hollenbach (1972) Laser Interferometer for Measuring High Velocities of Any Reflecting Surface; J. Appl. Phys. 43, 11, 4669 (1972)

Barker, L. M. (1983) Velocity Interferometry for Time-Resolved High-Velocity Measurements; Proceedings of SPIE - The International Society for Optical Engineering, Vol. 427, August 23-25, 1983

Bauer, F. (1981) Behavior of Piezoelectric PVF_2 Polymers Under Shock Wave Action; Textes des Communications Proceedings, A.D.E.R.A., Bordeaux, France, Sept. & Oct. 1981

Bryant, L.E. Jr., (1983) High-Speed Videography Combined with an X-Ray Image Intensifier for Dynamic Radiography; Ziff-Davis Publishing Co., Nov. 1983

Bryant, L.E. Jr., J.P. Lucero (1983) A Survey of Flash X-Ray Techniques & Applications; Proceedings of SPIE - The International Society for Optical Engineering, Vol. 427, Aug. 23-25, 1983

Charest, J.A. (1979) Development of a Strain-Compensated Shock Pressure Gauge, Final Report; Dynasen, Inc., Goleta, CA Feb. 1979

Charters, A.C. (1986) Private Communication

Engberg, K.M., E.J. Steele; J.A. Charest (1981) Development & Calibration of a Carbon Piezoresistive Gauge with Berylium Leads for use at Elevated Temperatures; Effects Technology, Inc, 5383 Hollister Ave., Santa Barbara, CA 93111 USA and Dynasen Inc., 20 Dean Arnold Place, Goleta, CA 93117 USA

Erickson, L.M., R.C. Weingart; H.C. Vantine; D.B. Johnson; R.S. Lee (1981) A Summary of Low Resistance Manganin Gauge Development at Lawrence Livermore National Laboratory; Textes des Communications Proceedings, A.D.E.R.A, Bordeaux, France, Sept. and Oct. 1981

Foley, J. (1986) Private Communication; Spin Physics Co., Sorrento Valley Rd., La Jolla, CA, Oct. 1986

Gehring, W. (1986) Pvt. Communication, E.G. & G. Inc., Santa Barbara, CA.

Ginsburg, M.J. et al (1973) Effects of Stress on the Electrical Resistance of Ytterbrium & Calibration of Ytterbrium Stress Transducers; Final Report Contract DNA001-72-C-0146, SRI International, Menlo Park, CA, 1973

Goranson, R.W. (1983) Ten Million Frame per Second IR Videography; Proceedings of SPIE - The International Society for Optical Engineering, Vol. 427, Aug. 23-25, 1983

Grady, D.E., M.J. Ginsburg (1977) Piezoresistive Effects in Ytterbrium Stress Transducers; Journal Appl. Phys., 48(6), 2179, (1977)

Gupta, Y.M. (1980) Determination of the Impact Response of PMMA Using Combined Compression and Shear Loading; Journal Appl. Phys. 51(10), 5352, 1980

Gupta, Y.M., D.D. Keough; D. Henley; D.F. Walker (1980) Measurement of Lateral Compressive Stresses Under Shock Loading; Appl. Phys. Lett., 37(4), 395 (1980)

Hemsing, Willard F. (1979) Velocity Sensing Interferometer (VISAR) Modification; Rev. Sci. Instrum, Vol. 50, No 1, January 1979

International Congress on Instrumentation in Aerospace Simulation Facilities (ICIASF) (1969) Proceedings of the 1969 ICIASF; IEEE Transactions on Aerospace & Electronic Systems, Farmingdale, NY (USA), May 1969

Isbell, W. M. (1981) Advances in Laser Interferometer Techniques for Measurement of Dynamic Material Properties; Proceedings of the International Congress on Instrumentation in Aerospace Simulation Facilities, Dayton, Ohio, Sept. 30, 1981

Isbell, W. M., P.W.W. Fuller (1983) Wide-Range, High-Resolution Measurements of Projectile Motion Using Laser Interferometry: Proceedings of SPIE - The International Society for Optical Engineering, Vol. 427, August 23-25, 1983

Isbell, W M (1982) Laser Interferometric Techniques for Precision Measurement of Shock waves in Material, and Structures; Photomethods Magazine, May 1982, and High Speed Photography and Photonics Newsletter, Vol. 2, No. 1, Winter 1982

Isbell, W.M., The VISAR as a Production Test Instrument, Final Report on Sandia Contract 35-8752; ATA Associates, Santa Barbara CA, 1984

Keough, D.D., P. De Carli; J.T. Rosenberg (1981) Development of a High Modulus Piezoresistance Gage for Dynamic In-Situ Soil Stress Measurements; Textes des Communications Proceedings, A.D.E.R.A., Bordeaux, France, Sept. & Oct. 1981

Keough, D.D., J.Y. Wong (1970) Variation of the Shock Piezoresistance Coefficient of Manganin as a Function of Deformation; Journal Appl. Phys. 41(8), 3508, 1970

Kotecki, D.E., R. Lear (1983) Optical Shutters Using Microchannel Plate (MCP) Intensifier Tubes; Proceedings of SPIE - The International Society for Optical Engineering, Vol. 427, Aug. 23-25, 1983

Lee, L.M., B.D. Jenrette (1981) Carbon Gauge Shock Piezoresistance Behavior Below 0.5 Gpa; Textes des Communications Proceedings, A.D.E.R.A., Bordeaux, France, Sept. & Oct. 1981

Lilley, E.M., D.R. Stephens (1971) Electrical Resistance of Ytterbrium as a Function of Temperature & Pressure; UCRL-51006 LLNL, Livermore, CA 1971

Lockheed Missiles & Space Co. (1976) Strain-Compensated Stress Gage Development Final Report; LMSC-D506298, July 1976

Medecki, H., G.E. Phillips; J.F. Bushman (1983) Development of Soft X-Ray Streak Cameras at Lawrence Livermore National Laboratory; Proceedings of SPIE - The International Society for Optical Engineering, Vol. 427, Aug. 23-25, 1983

Menna, T.L. (1986) Pvt. Comm., General Research Corp., Santa Barbara, CA.

Munson, D.E. and R.P. May, Interior Ballistics of a Two-Stage Light Gas Gun Using Velocity Interferometers, AIAA Journal; Vol. 14, No. 2, Feb. 1976

(1971) Proceedings of the 1971 ICIASF; IEEE/AES Rhodes Saint-Genese, Belgique, June 1971

(1972) Actes du 10´e Congres International Congress on High Rapide; Nice, France, Sept. 1972

(1973) Proceedings of the 1973 ICIASF; IEEE/AES Pasadena, CA (USA), Sept. 1977

(1974) Proceedings of the 11th International Congress on High Speed Photography; London, England, Sept. 1974

(1975) Proceedings of the 1975 ICIASF; IEEE/AES Ottawa, Canada, Sept. 1975

(1976) Proceedings of the 12th International Congress on High Speed Photography; Toronto, Canada, Aug. 1976

(1977) Proceedings of the 1977 ICIASF; IEEE/AES Shrinveham, England, Sept. 1977

(1978) Proceedings of the 13th International Congress on High Speed Photography and Photonics; Tokyo, Japan, Aug. 1978

(1979) Proceedings of the 1979 ICIASF; IEEE/AES Monterey, CA (USA)

Rosenberg, Z., D. Yaziv; Y. Partom (1980) Direct Measurement of Strain in Plate Impact Experiments by a Longitudinal Resistance Gauge; Journal Appl. Phys., 51(9), 4790, 1980

Saunders, J., J. Eldridge (1983) Recent Advances in Utility & Application of a High Speed Video Motion Analysis System; Proceedings of PSIE - The International Society for Optical Engineering, Vol. 427, Aug. 23-25, 1983

Seiff, A., Shadow, Schlieren, And Interferometer Photographs And Their Use In Ballistic Ranges; Agardograph No. 138, AGARD-AG-138-70 (1970)

Setchell, R. E. (1983) Velocity Interferometer System for Any Reflector (VISAR) Studies of Wave Growth in Granular Explosives; Proceedings of SPIE, Vol. 427, August 23-25, 1983

Summers, J.C. and B.P. Denardo, Terminal Ballistics; AGARDograph No. 138, AGARD-AG-138-70, 1970

Swift, H.F. (1982) Image Forming Instruments; Impact Dynamics, John Wiley & Sons, New York, NY 1982

Trimble, J.J. (1981) The Effects of Gage Deformation on Manganin Gage Pressure Measurements; Textes des Communications Proceedings, A.D.E.R.A., Bordeaux, France, Sept. & Oct. 1981

Williams, E.O. (1968) An Etched Manganin Gauge System for Shock-Pressure Measurement in a High-Noise environment; ISA Transaction, Vol. 7, No. 3, pp. 223-30

Int. J. Impact Engng Vol. 5, pp. 411–421, 1987
Printed in Great Britain

0734-743X/87 $3.00 + 0.00
Pergamon Journals Ltd.

ERODING INTERFACE AND IMPROVED TETRAHEDRAL ELEMENT ALGORITHMS FOR HIGH-VELOCITY IMPACT COMPUTATIONS IN THREE DIMENSIONS

Gordon R. Johnson and Robert A. Stryk

Honeywell Incorporated, Defense Systems Division
Edina, Minnesota 55436

ABSTRACT

This paper presents a three-dimensional, eroding interface algorithm which allows Lagrangian computations to be performed for projectile penetration/perforation of thick plates. It also presents an improved tetrahedral element algorithm which provides for very accurate results under mild distortions and allows for continued computations under severe distortions. These algorithms have been incorporated into the EPIC-3 code and examples are presented to illustrate their usefulness.

INTRODUCTION

Many finite difference and finite element codes are currently being used for computations of projectiles impacting targets. Lagrangian computations (in which the grid moves with the material) are often preferred if they have the capability to adequately account for the various phenomena which occur. Generally, they require less CPU time, provide a more accurate description of the interfaces, and predict the time-history of material responses better.

Until recently, however, Eulerian computations (in which the material moves through a fixed grid) have been the only option available for high-velocity impact computations involving erosion of material from both the projectile and the target. This is because Lagrangian codes previously have not had the capability to adequately represent the erosion process.

In spite of the fact that Eulerian codes are well suited to handle these high-velocity impact problems, much effort has been expended during the past several years to extend the capabilities of Lagrangian codes to include these problems. In addition to the Lagrangian advantages mentioned previously, some other desirable features are associated with extending the capability of Lagrangian codes. It is more convenient and efficient for the user if a wide range of problems can be analyzed with the same code, the use of identical material models and data gives computational consistency for the range of problems, and it eliminates the need to "couple" or "link" an Eulerian code to a Lagrangian code.

The primary purpose of this paper is to present a three-dimensional eroding interface algorithm which will give a Lagrangian code the capability to simulate severe distortions and erosion in a routine manner. The other portion of the paper presents an improved algorithm for tetrahedral elements subjected to severe distortions.

IMPROVED ELEMENT ALGORITHM

The Lagrangian EPIC-3 code has been used extensively for three-dimensional problems involving projectile-target impact (Johnson; 1977, 1981a). One distinguishing feature of this code is that it uses constant-strain tetrahedral elements. These elements do not allow the Lagrangian grid to become tangled or overlapped under severe distortions, as often happens with other types of elements. This is due to the favorable geometric characteristics of a tetrahedral element. It is apparent that the lines connecting the nodes cannot become overlapped unless a node crosses the plane containing the other three nodes of that element. This would require the element to go through a zero volume configuration. Since the hydrostatic pressure in an element is inversely proportional to the volume of the element, as the volume approaches zero the pressure approaches infinity. As a result, there is a very large resistance to this mode of distortion and the tetrahedral element will not go through a zero volume configuration provided the integration time increment is properly controlled. Unfortunately, however, it has been shown that these elements can introduce significant inaccuracies (Nagtegaal, Parks and Rice, 1974; Johnson, 1981a).

Before examining three-dimensional element types, the effects of various two-dimensional element types and arrangements are shown in Fig. 1. Cases A and B are reported by Hallquist (1979) to show the difference between results obtained with triangular elements and the more accurate constant-stress quadrilateral elements. It can be seen that the deformed shapes and strain distributions are significantly different for these two cases. A two-dimensional EPIC-2 solution, using triangular elements in the same arrangement, is essentially the same as that of Case A.

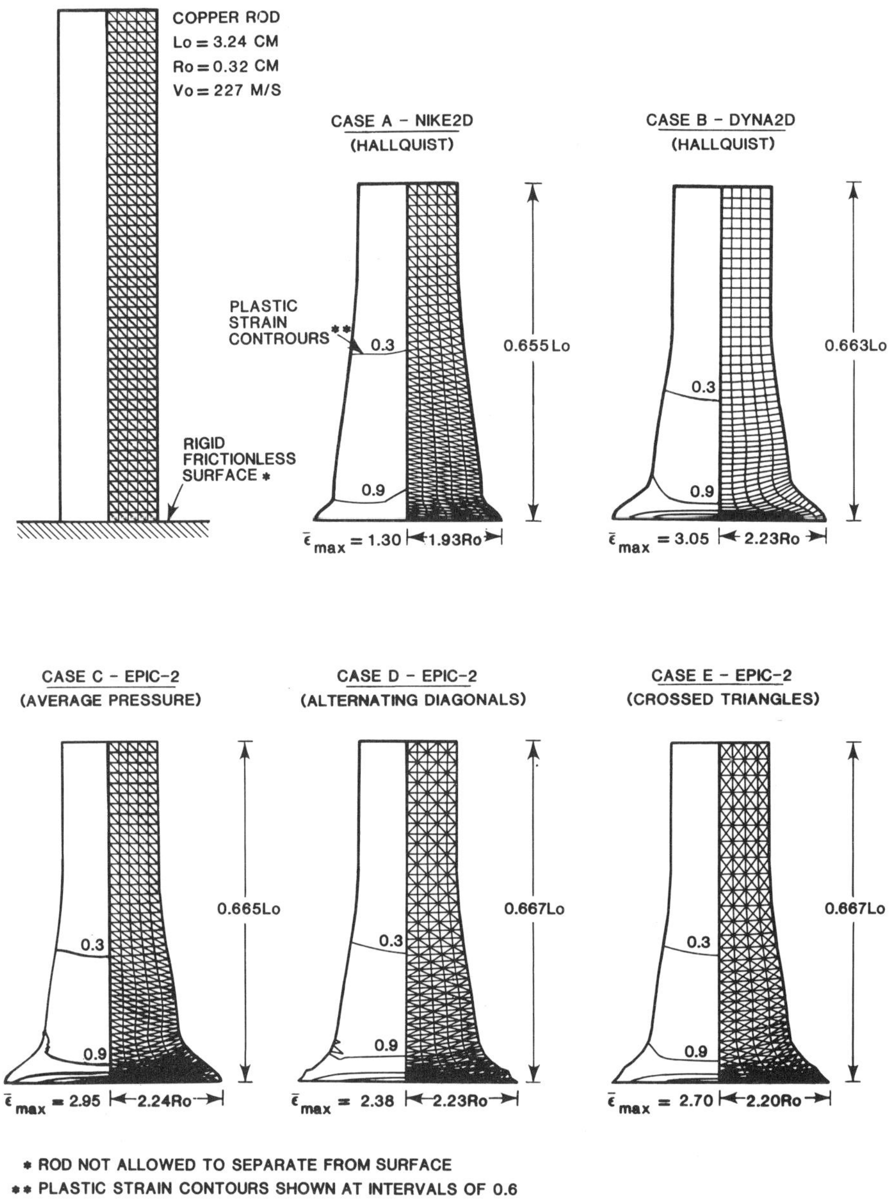

Fig. 1. Effects of various two-dimensional element algorithms.

Cases C, D, and E show other two-dimensional EPIC-2 solutions; these cases use triangular elements, which are in much better agreement with the more accurate solution of Case B. The average pressure solution of Case C computes an average pressure for each group of two adjacent triangular elements, instead of an individual pressure for each element. This reduces the number of incompressibility constraints (Nagtegaal, Parks and Rice, 1974) and provides a solution almost identical to that of Case B. The disadvantage of this approach is that it does not necessarily keep the grid from becoming tangled or overlapped, and therefore cannot always be used when severe distortions are experienced. This basic approach, modified to account for severe distortions, will later be applied to the three-dimensional tetrahedral element.

The preferred approach for two dimensions is to use the crossed-triangle arrangement of elements, as shown in Case E. This also tends to reduce the number of incompressibility constraints and provides a solution showing good general agreement with that of Case B. Some inaccuracies do, however, remain near the centerline axis. The alternating diagonal solution of Case D also has the effect of arranging the elements in a crossed-triangle configuration.

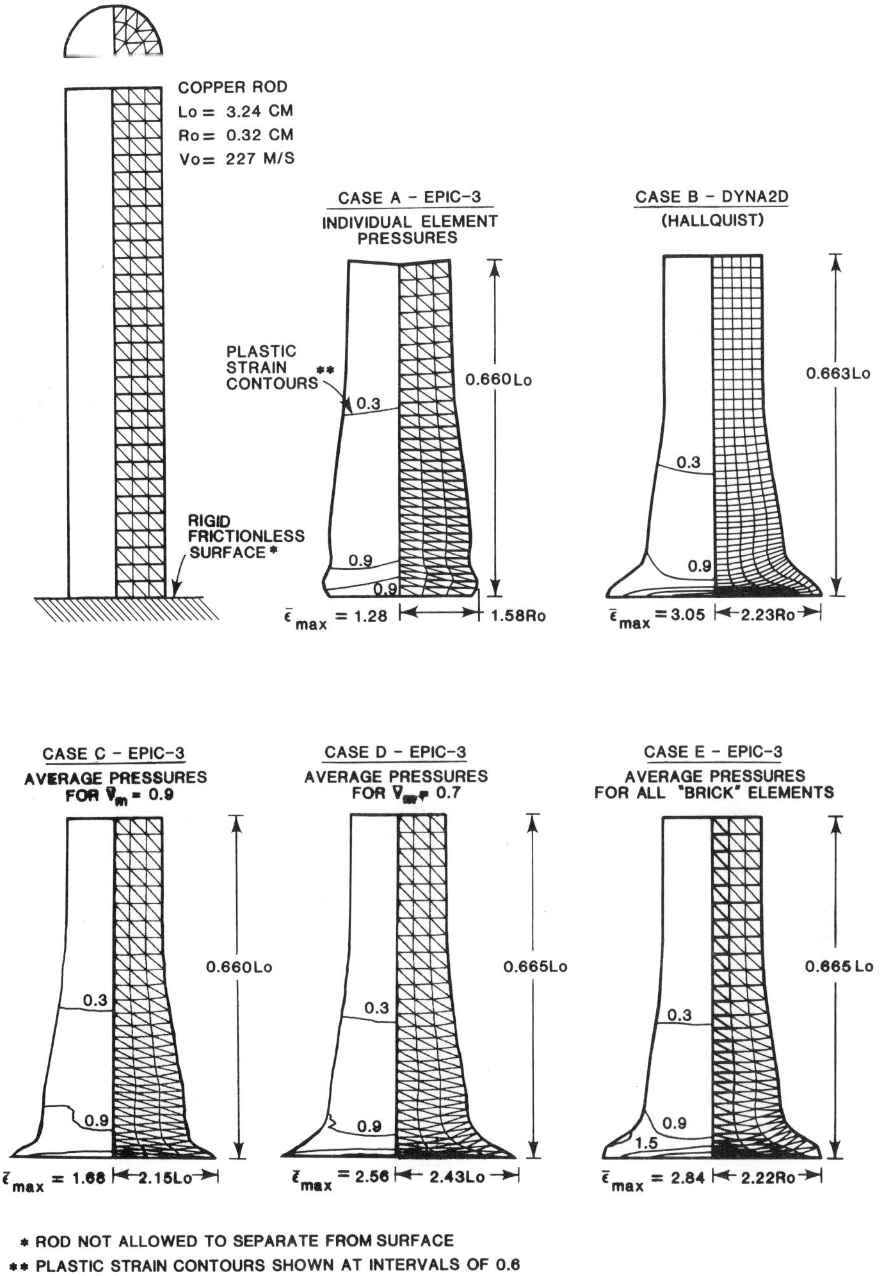

Fig. 2. Effects of various three-dimensional element algorithms.

Figure 2 shows the effect of various three-dimensional tetrahedral element algorithms. Case A shows a solution in which each brick is composed of six tetrahedral elements, and each element has an individual pressure computation. This algorithm also suffers from too many incompressibility constraints and produces an inaccurate solution, as can be seen in a comparison with Case B. Unfortunately, there is no known three-dimensional arrangement of tetrahedral elements which is analogous to the two-dimensional, crossed-triangle arrangement.

The improved tetrahedral element algorithm is based on a modified average pressure approach. The solution for Case E, in Fig. 2, is obtained by using an average pressure for each group of six tetrahedral elements which compose a brick element. This solution is analogous to the two-dimensional average pressure solution shown in Case C of Fig. 1, and it also provides good correlation with the accurate solution of Case B.

It was previously noted that this average pressure approach may not be adequate for severe distortions because it would not necessarily keep the grid from becoming tangled or overlapped. It would be possible, for instance, for one element in the brick to go to a zero or negative volume while the other five expanded. Although this may provide an accurate average pressure, the integration time increment would be driven to zero and the individual element computations for the shear and deviator stresses could not be performed.

The improved algorithm allows the average pressure approach to be used until the relative volume, $\overline{V} = V/V_0 - 1$ (where V is the current volume and V_0 is the initial volume), of an individual element falls below a user-specified input variable, $\overline{V}_m$. When this occurs, the element is "marked" and an individual pressure computation is performed for that element for the remainder of the computations. This ensures that the grid will not become tangled or overlapped, because the resisting pressure will rise as the relative volume decreases.

The pressure, P, in an individual element is generally determined as a function of the relative volume, $\overline{V}$, and the internal energy, E, such that

$$P = f(\overline{V}, E) \tag{1}$$

This same form is used for computing average pressures for a group of individual elements.

After the relative volume, $\overline{V}$, of a specific element falls below $\overline{V}_m$, the element is marked and an individual element pressure is determined from

$$P = f(\overline{V}', E) \tag{2}$$

where the modified relative volume, $\overline{V}'$, is defined as $\overline{V}' = \overline{V}/\overline{V}_m$. The remaining elements within the brick, whose relative volumes have not fallen below $\overline{V}_m$, continue to receive an average pressure. Cases C and D in Fig. 2 show results when the individual pressure computations begin at $\overline{V}_m = 0.70$ and 0.90. It can be seen that these results provide relatively good agreement with the accurate solutions of Cases B and E. The net result is that this algorithm provides the severe distortion advantages of a tetrahedral element, combined with the improved accuracy associated with an average pressure element formulation.

It should be noted that the average pressure algorithm could allow "hourglassing" of the grid (zero-energy modes of deformation) under some circumstances. Generally, this will not occur because the shear and deviator stresses are computed for each individual tetrahedral element, eliminating any zero-energy hourglass modes of deformation.

For computations of liquids, or other instances when the shear and deviator stresses are much smaller than the hydrostatic pressure, hourglassing could occur. This can be resisted by incremental stresses (normal and shear) with the form

$$\Delta\sigma_x = 2\mu\,(\dot{e}_x - \overline{\dot{e}_x}) \tag{3}$$

$$\Delta\tau_{xy} = \mu\,(\dot{\gamma}_{xy} - \overline{\dot{\gamma}_{xy}}) \tag{4}$$

The viscosity coefficient is $\mu = C_H \rho c_s h_{min}$, where C_H is a dimensionless constant ($C_H \approx 0.02$), ρ is the density, c_s is the sound velocity, and h_{min} is the minimum altitude of the tetrahedral element. In Equation (3), $\Delta\sigma_x$ is the incremental normal stress in the x direction, $\dot{e}_x$ is the deviator strain rate, and $\overline{\dot{e}_x}$ is the average deviator strain rate of all the elements in the brick which use the average pressure. In Equation (4), $\Delta\tau_{xy}$ is the incremental shear stress, $\dot{\gamma}_{xy}$ is the individual strain rate, and $\overline{\dot{\gamma}_{xy}}$ is the average strain rate. The other stress components ($\Delta\sigma_y$, $\Delta\sigma_z$, $\Delta\tau_{yz}$, $\Delta\tau_{zx}$) have the same form. These stresses are then converted to nodal forces in the standard manner (Johnson, 1977). An extended discussion of hourglassing for various types of elements is provided by Belytschko and others (1984).

ERODING INTERFACE ALGORITHM

In recent years several organizations have been involved with developing eroding interface algorithms for the EPIC-2 and EPIC-3 codes. The initial work was with the EPIC-3 code (Johnson, 1982) but its usefulness was limited in that the eroded area had to be predetermined. This was followed by general eroding algorithms for the EPIC-2 code (Snow, 1982; Stecher and Johnson, 1984; Kimsey and Zukas, 1986). Recently, a three-dimensional erosion algorithm has been incorporated into a modified version of EPIC-3, which uses hexahedral elements (Belytschko and Lin, 1985).

Examples of two-dimensional EPIC-2 erosion computations, performed by Stecher and Johnson (1984), are shown in Fig. 3. It can be seen that the projectiles are completely eroded and that the targets are significantly eroded. When an individual element is eroded (or totally failed), it cannot develop stresses or pressures. The element essentially "disappears" except that mass is retained at the nodes. The criterion for erosion is generally taken to be an equivalent plastic strain in the range of 1.0 to 2.0. For a limited number of problems performed to date, it appears that the results are not highly dependent on the strain required for erosion.

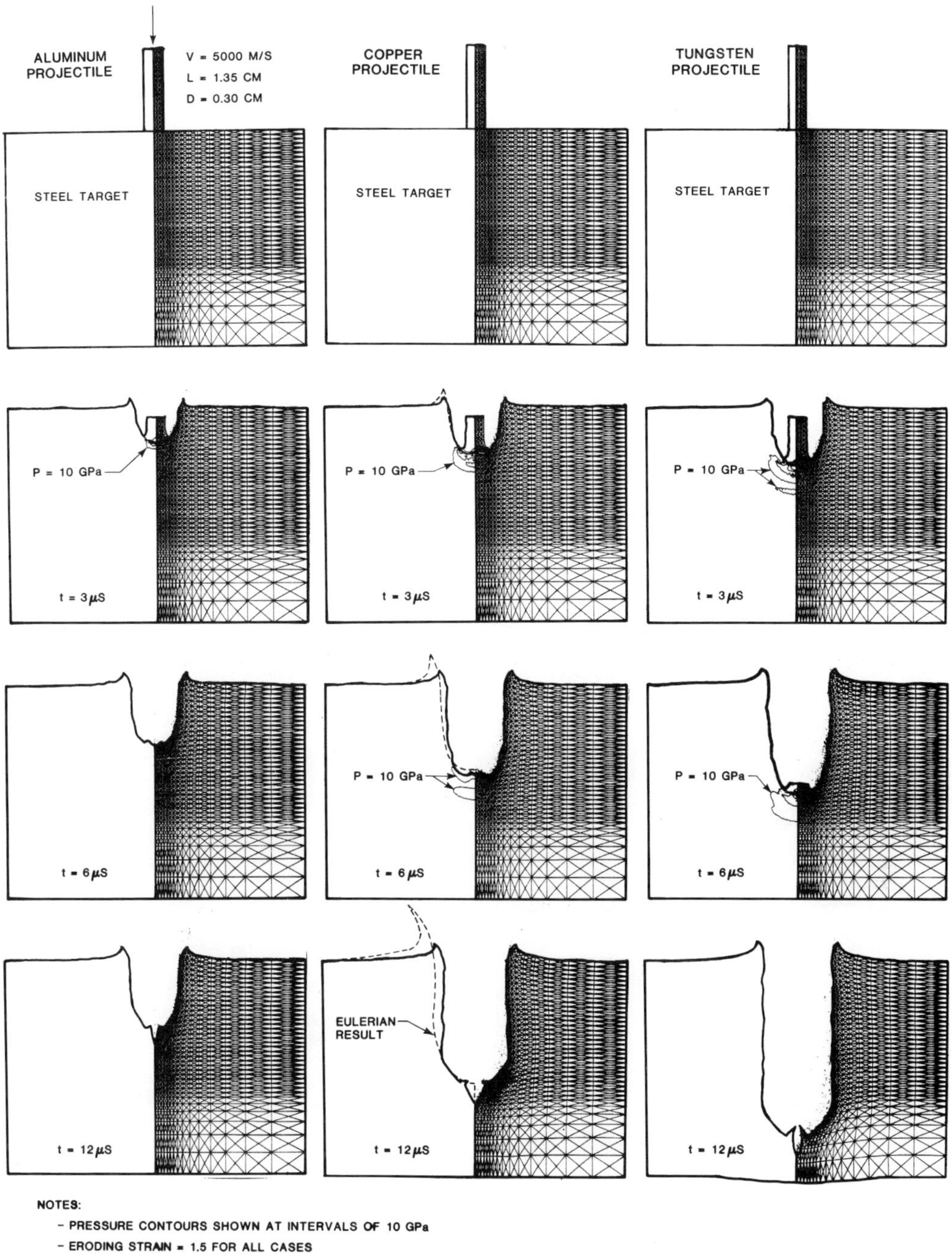

Fig. 3. Examples of two-dimensional computations using the eroding interface algorithm in the EPIC-2 code.

Because this computational erosion process requires some significant approximations (such as disappearing elements), much uncertainty exists about the range of problems for which it is applicable, and the accuracy of the results it produces. One indication is provided in Fig. 3, in which the results of an Eulerian computation by Kutcher (1982) are superimposed on the EPIC-2 computation. Also, in Fig. 4, the penetration depth is shown as a function of time for all three cases. It can be seen that the rate of penetration agrees very well with the one-dimensional approximation given by Birkhoff and others (1948). The specific form of the approximation is $U = V/(1 + \sqrt{\rho_t/\rho_p})$, where U is the penetration velocity, V is the impact velocity, ρ_t is the density of the target, and ρ_p is the density of the projectile.

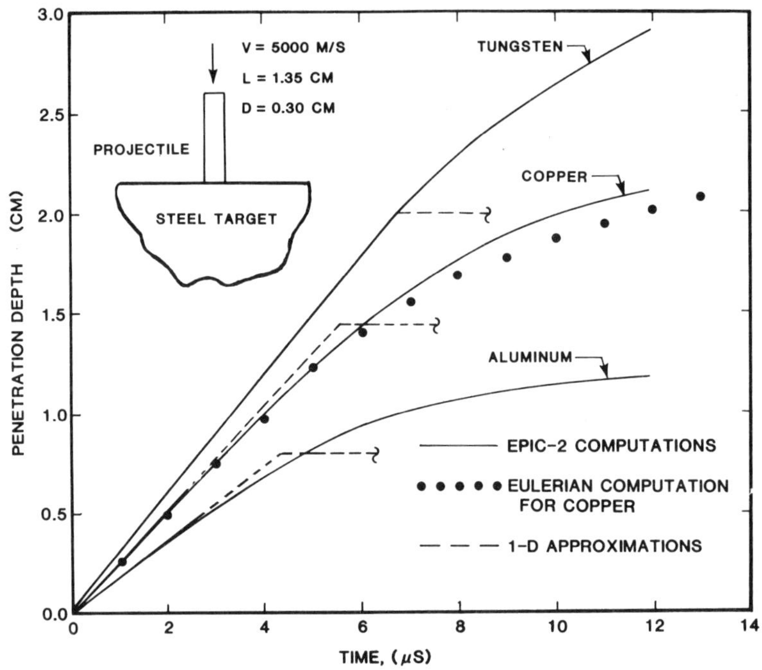

Fig. 4. Penetration versus time for various computations.

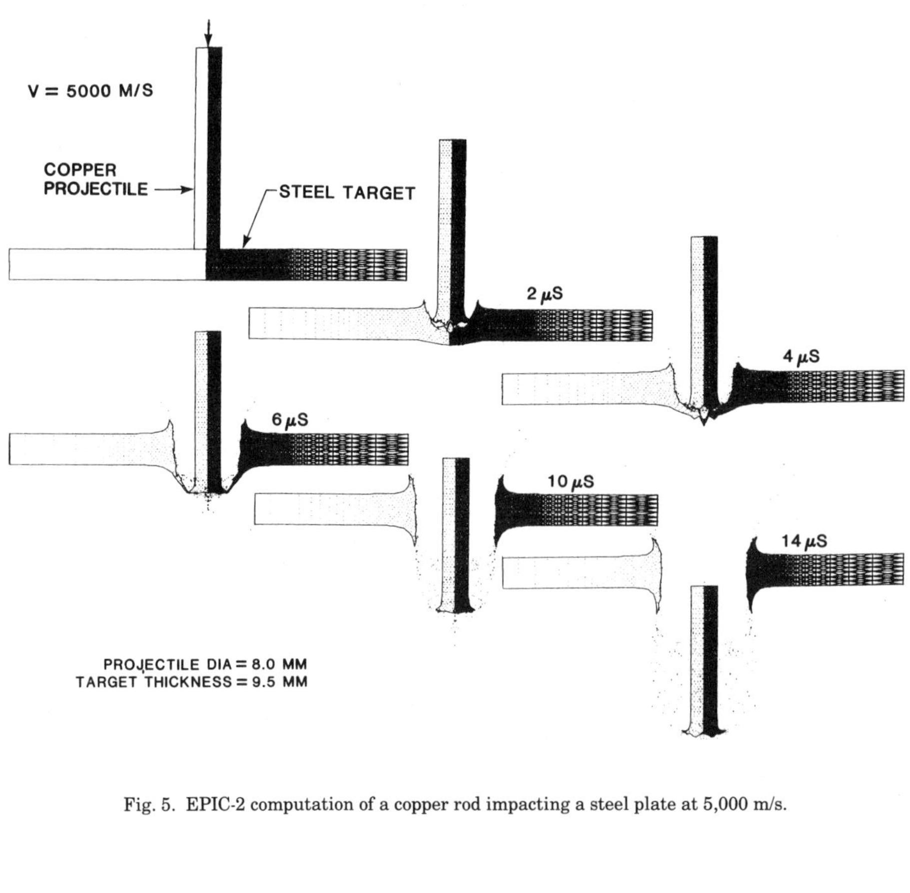

Fig. 5. EPIC-2 computation of a copper rod impacting a steel plate at 5,000 m/s.

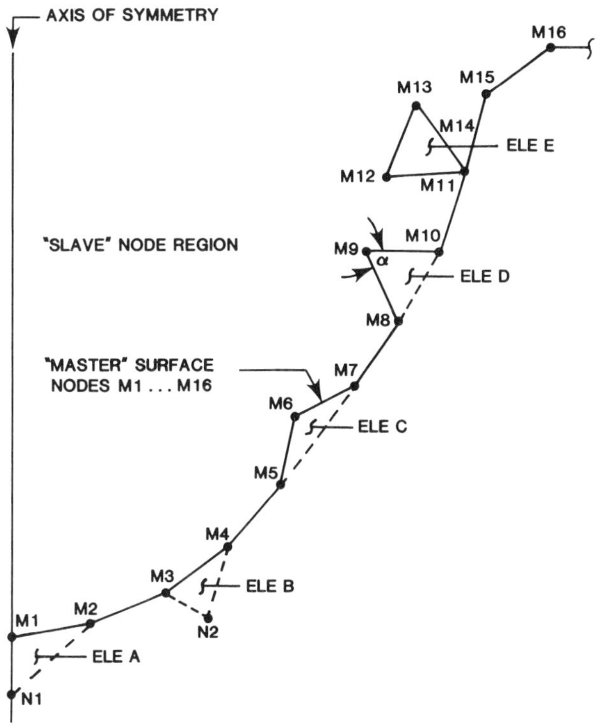

Fig. 6. Schematic representation of the two-dimensional eroding interface algorithm in the EPIC-2 code.

Based on a limited number of computations of projectiles impacting semi-infinite targets, it appears that the rate of penetration is accurately computed throughout the erosion process. The computed hole diameter is often too small, however, and the computed depth is often too deep. A problem also exists at the centerline, as shown in Fig. 3. Other work by Kimsey and Zukas (1986) has shown their erosion algorithm to give good agreement with experimental results. More work remains to be accomplished to adequately assess the accuracy of these erosion algorithms.

Another example is shown in Fig. 5, where the projectile perforates the target. The "dots" appearing on the right side of the deformed geometries are nodes whose associated elements have been eroded. These nodes retain their mass, however, so it is possible to approximate some of the debris characteristics on the back side of the target. The postprocessor for the latest version of the EPIC-2 code has the capability to plot mass/area, momentum/area, and kinetic energy/area (Johnson and Stryk, 1986).

Before presenting the three-dimensional eroding interface algorithm, it would perhaps be helpful to briefly describe the EPIC-2 algorithm as shown in Fig. 6. The solid line (M1 . . . M16) represents the "master" surface through which no "slave" nodes are allowed to pass. The dashed lines represent the revised master surface after elements A . . . E have been eroded. An expanded discussion of eroding interfaces, and the complete EPIC-2 algorithm, is given by Stecher and Johnson (1984).

The three-dimensional algorithm that follows should be applied only to problems in which erosion is the primary mode of penetration. For many problems, the master surface is initially on the top (or any other side) of a plate and it cannot involve other sides (except as special cases) during erosion. The precise restriction for the general algorithm to follow is that no element with a free surface can be eroded unless that free surface was initially designated as a master surface.

Figure 7 shows a partially eroded master surface defined as a set of adjacent triangles. The initial master surface included the entire top surface of the plate. The slave nodes are essentially above the master surface and must always view the associated master triangles so that the nodes of the triangles (M1, M2, M3) are in a counterclockwise order. The slave nodes can be randomly ordered and are not allowed to penetrate through the master surface. The master surface is allowed to erode by totally failing tetrahedral elements which have one, two, three, or four sides on the master surface. It should be recalled that totally failed elements do not develop stresses or pressures; they essentially disappear except that mass is retained at the nodes.

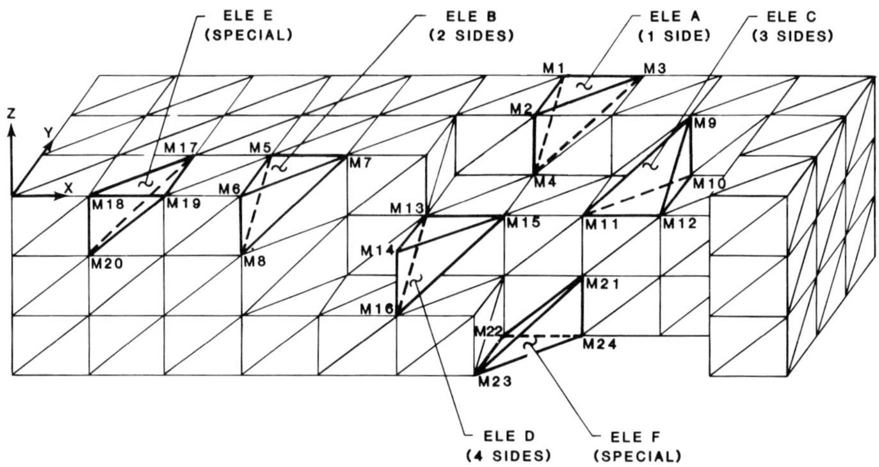

Fig. 7. Schematic representation of the three-dimensional eroding interface algorithm in the EPIC-3 code.

Before the computations begin, the preprocessor must store the groups of three nodes defining the individual triangular master planes. It must also permanently "mark" each node contained on the master surface, and it must count and store the number of elements attached to each node. If three or four nodes of an element are marked, then the element is on the master surface and is eligible for failure if it exceeds the specified erosion strain. When an element is totally failed, the element count for each of the four nodes is decremented, so that the count reflects the number of unfailed elements attached to the node. When the count reaches zero, the node is made a slave node because it is no longer associated with the master surface. This new slave node (which continues to have mass, momentum, and kinetic energy) can then interact with the master surface but cannot pass through it.

The six elements in Fig. 7 (elements A, B, C, D, E, F) have three or four nodes on the master surface and have one, two, three, or four sides on the master surface. The following describes the changes necessary to redefine the master surface and the slave nodes:

- Element A has one side (M1-M2-M3) on the master surface. All four nodes—M1, M2, M3, and M4—are marked. Because only one side (M1-M2-M3) is contained in the list of master triangles, it is known that the other three sides are not on the master surface. If the equivalent plastic strain of element A exceeds the erosion strain, the element is totally failed. The updated master surface has one triangular side (M1-M2-M3) removed and three triangular sides of adjacent elements (M1-M2-M4, M1-M4-M3, M2-M3-M4) added. The nodes of the new sides must be given in the proper order so that the slave nodes see a counterclockwise ordering. The element counts of nodes M1, M2, M3, M4 are decremented, but all four nodes remain on the master surface because none of the element counts goes to zero.

- Element B has two sides (M5-M6-M7, M6-M8-M7) on the master surface. The four nodes M5, M6, M7, and M8 are marked. If the equivalent plastic strain of element B exceeds the erosion strain, the element is totally failed. The previous master surface has two sides (M5-M6-M7, M6-M8-M7) removed and two other surfaces (M5-M6-M8, M5-M8-M7) added. The element counts of nodes M5, M6, M7, and M8 are decremented, but none of the counts goes to zero.

- Element C has three sides (M9-M10-M11, M9-M11-M12, M9-M12-M10) on the master surface. The four nodes M9, M10, M11, and M12 are marked. If the equivalent plastic strain of element C exceeds the erosion strain, the element is totally failed. The previous master surface has three sides (M9-M10-M11, M9-M11-M12, M9-M12-M10) removed and one other surface (M10-M11-M12) added. The element counts of nodes M9, M10, M11, and M12 are decremented, but none of the counts goes to zero.

- Element D has four sides (M13-M14-M15, M13-M16-M14, M14-M16-M15, M13-M15-M16) on the master surface. The four nodes M13, M14, M15, and M16 are marked. If the equivalent plastic strain of element D exceeds the erosion strain, the element is totally failed. The previous master surface has all four sides removed and none added. The element counts of nodes M13, M14, M15, and M16 are decremented. The element counts of nodes M13, M15, and M16 do not go to zero and therefore remain on the master surface. The element count on node M14 does go to zero, however, so it is designated to be a slave node.

- Element E is a special case because it has one side (M18, M19, M20) on a plane of symmetry at Y = 0, which was not initially designated as a master surface. The general algorithm would indicate one side (M17-M18-M19) to be deleted and three sides to be added (M17-M18-M20, M17-M20-M19, M18-M19-M20). The side on the plane of symmetry at Y = 0 (M18-M19-M20) should not be added, however. This side is readily identified and deleted by noting that Y = 0 for nodes M18, M19, and M20. For this case the element counts of nodes M17, M18, M19, and M20 are again decremented. Node M20 was not previously marked and must therefore be marked to identify it as a part of the updated master surface. It should be noted that the general algorithm would handle this special case if the triangular element sides on the plane of symmetry would initially be designated as part of the master surface. Because this would significantly increase the size of the master surface (and the associated "searching" time), it is instead treated as a special case.

- Element F is another special case because it is eroded through a free surface (bottom of the plate) which was not initially designated as a master surface. Here the general algorithm would remove two sides (M21-M22-M23, M21-M23-M24) and add two new sides (M21-M22-M24, M22-M23-M24). The side on the bottom of the plate (M22-M23-M24) should not be added, however. This side can be readily identified if the nodal numbering system proceeds downward by layers, thus placing the highest node numbers on the bottom surface of the plate. Under these conditions, node numbers M22, M23, and M24 would all be greater than (or equal to) the lowest node number on the bottom of the plate, and that triangle would therefore not be added to the list of master triangles. Other approaches could be used for different nodal numbering systems. Again it should be noted that the general algorithm would handle this special case if the element sides on the bottom of the plate would initially be designated as part of the master surface.

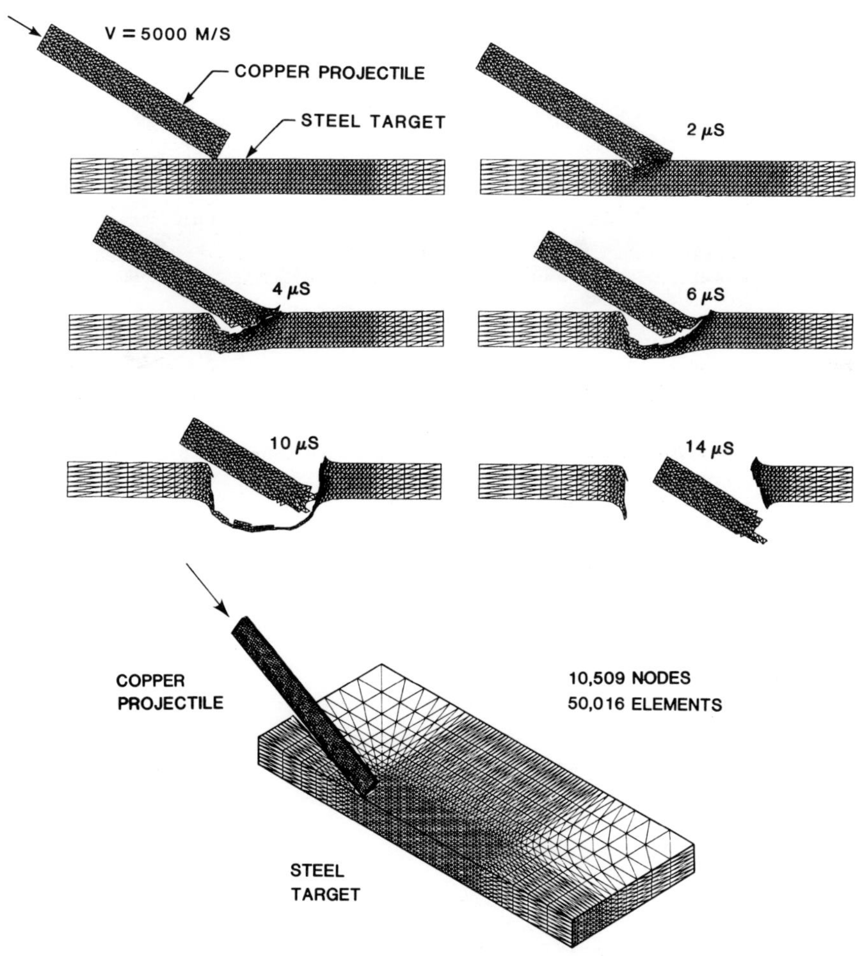

Fig. 8. EPIC-3 computation of a copper rod impacting a steel plate at 5,000 m/s.

After the master surface and slave nodes have been updated, it is necessary for each slave node to find the appropriate master surface with which it must interact. These search routines are described by Johnson (1981b).

When a slave node is found to interact with a master surface, the velocities of the slave node and the three master nodes are adjusted to conserve momenta and to equate the normal velocity of the slave node to the normal velocity of the master surface, at the slave node location. The position of the slave node is also adjusted to place it on the master surface. The specific algorithm for adjusting the velocities is given by Johnson, Colby, and Vavrick (1979).

Figure 8 shows an EPIC-3 example of a copper rod impacting a steel target plate at 5,000 m/s. It can be seen that material is eroded from both the projectile and the target, and that the projectile perforates the target. Another example is shown in Fig. 9. There the target is identical to that of Fig. 8, but the projectile has a smaller diameter and a higher impact velocity of 9,000 m/s. Both of these examples used an erosion strain of 1.5. It should be noted that only the intact elements are shown in Figs. 8 and 9. The "free" nodes, whose associated elements have eroded, are not shown as they are in Fig. 5.

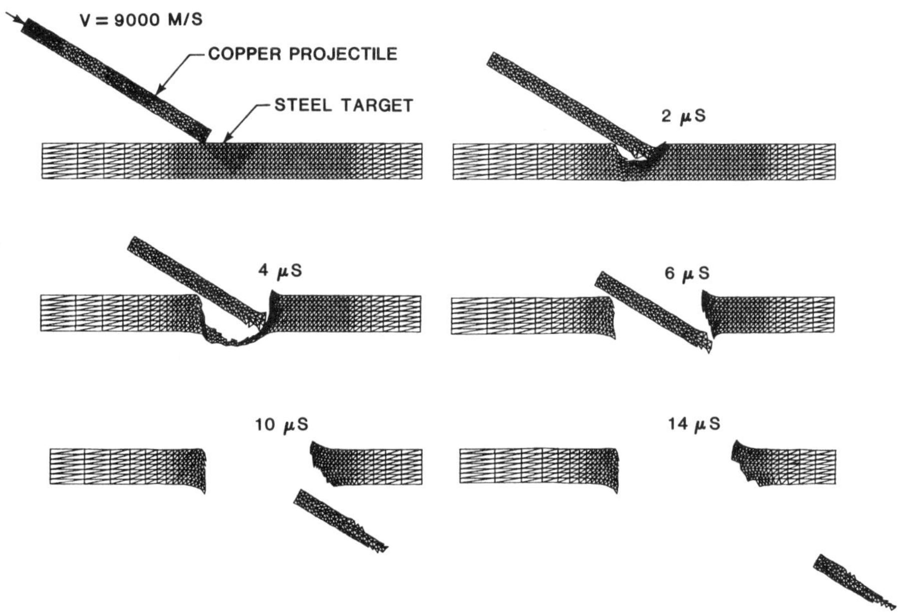

Fig. 9. EPIC-3 computation of a copper rod impacting a steel plate at 9,000 m/s.

SUMMARY AND CONCLUSIONS

An eroding interface algorithm and an improved tetrahedral element algorithm have been developed for high-velocity impact problems in three dimensions. The eroding interface algorithm significantly extends three-dimensional Lagrangian computing capabilities, but the accuracy has not yet been fully assessed. The improved tetrahedral element algorithm provides increased accuracy for tetrahedral elements.

ACKNOWLEDGEMENTS

This work was primarily funded by a Honeywell Independent Development program. The example problems in Figs. 5, 8, and 9 were funded by Dyna East Corporation and the U. S. Air Force.

REFERENCES

Belytschko, T., J. S. Ong, W. K. Liu, and J. M. Kennedy (1984). Hourglass control in linear and the nonlinear problems: *Computer Methods in Solid Mechanics and Engineering, 43,* 251-276.

Belytschko, T., and J. I. Lin (1985). A new interaction algorithm with erosion for EPIC-3, BRL-CR-540.

Birkhoff, G., D. P. MacDougall, E. M. Pugh, and G. Taylor (1948). Explosives with lined cavities. *Journal of Applied Physics,* Vol. 19, 563-582.

Hallquist, J. D. (1979). NIKE2D; An implicit, finite deformation, finite element code for analyzing static and dynamic responses of two dimensional solids, UCRL-52678.

Johnson, G. R. (1977). High velocity impact calculation in three dimensions, *Journal of Applied Mechanics,* ASME, Vol. 99, No. 1, 95-100.

Johnson, G. R. (1981a). Recent developments and analyses associated with the EPIC-2 and EPIC-3 codes. In S. S. Wang and W. J. Renton (Ed.) *1981 Advances in Aerospace Structures and Materials—AD-01,* 141-147.

Johnson, G. R. (1981b). Dynamic analysis of explosive-metal interaction in three dimensions. *Journal of Applied Mechanics,* ASME, Vol. 103, No. 1, 30-34.

Johnson, G. R. (1982). Status of EPIC codes, material characterization and new computing concepts at Honeywell. In J. Chandra and J. E. Flaherty (Ed.), *Computational Aspects of Penetration Mechanics (Lecture Notes in Engineering 3),* Springer-Verlag, Berlin, Heidelberg, New York, Tokyo. pp. 24-35.

Johnson, G. R., D. D. Colby, and D. J. Vavrick (1979). Three dimensional computer code for dynamic response of solids to intense impulsive loads, *International Journal for Numerical Methods in Engineering,* Vol. 14, 1865-1871.

Johnson, G. R., and R. A. Stryk (1986). User instructions for the EPIC-2 code. AFATL-TR-86-51.

Kimsey, K. D., and J. A. Zukas (1986). Contact surface erosion for hypervelocity problems. BRL-MR-3495.

Kutcher, V. (1982). Multiple impacts on monolithic steel. ARBRL-TR-02406.

Nagtegaal, J. G., D. M. Parks, and J. R. Rice (1974). On numerically accurate finite element solutions in the fully plastic range. *Computer Methods in Applied Mechanics and Engineering,* Vol. 4, 153-177.

Snow, P. (1982). KEPIC-2. Kaman Sciences Corporation Report K82-46U (R).

Stecher, F. P., and G. R. Johnson (1984). Lagrangian computations for projectile penetration into thick plates. In W. A. Grover, (Ed.) *Computers in Engineering,* Vol. 2, ASME, 292-299.

Int. J. Impact Engng Vol. 5, pp. 423–439, 1987
Printed in Great Britain

0734-743X/87 $3.00 + 0.00
Pergamon Journals Ltd.

HISTORY AND APPLICATION OF HYDROCODES IN HYPERVELOCITY IMPACT

Wallace E. Johnson*
Charles E. Anderson, Jr.**

*Computer Code Consultants
Chisholm, MN
**Southwest Research Institute
San Antonio, TX

ABSTRACT

Hydrocode enhancements have evolved as a result of the desire of researchers to be predictive, generally for some particular application. A review of the development of hydrocodes is given from a historical viewpoint. The paper then ends with a discussion of possible future enhancements in the codes.

INTRODUCTION

Hypervelocity impact has been investigated almost solely with Eulerian codes because of the problems associated with grid deformation with Lagrangian descriptions (Anderson, 1986). Early application of the codes investigated phenomena which are still being investigated today. In particular, there is interest in phase transitions at hypervelocity impacts (Holian and Burkett, 1986; Bjork 1986; Trucano and Asay, 1986), the dynamics and final state of impact cratering (Holsapple, 1986; Bjorkman and Holsapple, 1986; Schmidt and Housen, 1986), and the dynamics of debris clouds (Spillman, 1986; Arione and Bjorkman, 1986; Piekutowski, 1986; Trucano and Asay, 1986; Schultz, et al., 1986). These three topics were investigated with two dimensional hydrocodes in the very late 1950's and 1960's.

This paper summarize the historical development of multi-dimension hydrocodes as they pertain to hypervelocity impact. Deficiencies of the analysis methods, and what innovations and modifications are required in order to advance the state of the art are noted. In any assessment of history, the judgement and knowledge of the authors invariably come into play, and undoubtedly a few "milestones" have been missed. However, we deliberately have not tried to discuss every hydrocode--many hydrocodes have been developed to various stages of completion over the years, and many have died a death through neglection, for one reason or another. Hydrocodes evolve--their complexity increases as features are incorporated in order to be more versatile and predictive. What is evident historically, for a hydrocode ultimately to be useful in studying the physics of events, there has been a "champion" of that particular hydrocode (e.g., Wilkins, W. Johnson, Thompson, G. Johnson, Matuska and Osborn, Herrmann/Bertholf/Swegle/Lawrence/Kipp, Hallquist, etc.), and the names of these champions appear over and over again in the literature.

At the end of the historical overview, techniques are presented that perhaps have some promise for improving our ability to analyze hypervelocity impact. Many of these techniques grew out of older techniques. Some of these techniques are relatively new and are still developing. And, as in the past, some of the techniques will not be viable except in very special circumstances--but perhaps they will pave the way for other ideas in the future.

PARTICLE-IN-CELL

A methodology called the particle-in-cell (PIC) technique was developed by Evans and Harlow (1957) at Los Alamos. There was a desire to solve problems in two or more space dimensions which had large material distortion, problems which could not be solved with existing techniques (Harlow, 1964). Lagrangian descriptions were not acceptable due to the large material distortion. Eulerian descriptions had the disadvantage that the treatment of free surfaces and projectile-target interfaces had not been resolved. This led to a mixed Euler-Lagrange system which attempted to take advantage of the better features of both fixed and moveable coordinates.

The hydrodynamic continuum equations in Eulerian coordinates are derived from application of the conservation laws of mass, momentum, and energy:

$$\frac{\partial \rho}{\partial t} + v_i \frac{\partial \rho}{\partial x_i} + \rho \frac{\partial v_i}{\partial x_i} = 0 \tag{1}$$

$$\frac{\partial v_i}{\partial t} + v_j \frac{\partial v_i}{\partial x_j} + \frac{1}{\rho}\frac{\partial P}{\partial x_i} = 0 \tag{2}$$

$$\frac{\partial E}{\partial t} + v_i \frac{\partial E}{\partial x_i} + \frac{P}{\rho}\frac{\partial v_i}{\partial x_i} = 0 \tag{3}$$

where ρ is the density, v_i is the particle velocity, P is the pressure, and E is the specific internal energy. Standard tensorial notation has been used in Eqs. (1-3); the summation convention is implied by the repeated indices. An equation of state provides the relationship between pressure, density, and internal energy: $P = P(\rho,E)$.

Though the motivation may have been primarily for fluid flow, the PIC technique was quickly adopted for hydrodynamic impact. The computational framework for the PIC methodology is an Eulerian mesh of cells through which many discrete particles move. Each particle is assigned a distinct mass; the coordinates of each particle are stored in computer memory. These coordinates change in time in accordance with the subsequent motion of the particles through the fixed grid. Density is simply the total mass in a cell divided by the cell volume. Mass is conserved automatically so long as particles do not leave the computational grid. Continuum properties are computed by averaging the particle properties over the finite cell volumes.

The calculations typically were done in three phases. In the first phase, the difference forms of the momentum and energy equations, exclusive of the transport terms [the second term in Eqs. (2) and (3)], are solved for each grid cell. This gives tentative values for the velocity and internal energy of the cells. Phase II is then concerned with the movement of particles, usually referred to as the advection phase. The particles are moved according to the velocity of the cell in which they are located and the velocities of neighboring cells. In moving, the particles carry their share of the cell energy and momentum with them. Phase III then repartitions the momentum and energy carried by the particles to update cell velocity and internal energy.

The first impact problem investigated numerically used the PIC method. Bjork (1959) examined the impact of steel into steel, and aluminum into aluminum, to investigate meteoroid impact at 5.5, 20 and 72 km/s. One important point from Bjork's paper, though refined by subsequent work (e.g., Walsh and Tillotson, 1963), showed that crater volume transitioned from being a function of momentum at low velocities to being proportional to kinetic energy at high velocities. Bjork next analyzed a 11,000 metric ton cylindrical iron projectile striking a semi-infinite target of tuff at 30 km/s to simulate the impact which resulted in the giant meteor crater in Arizona (Bjork, 1961). This hydrodynamic calculation predicted a crater approximately 20 percent less than the dimensions of the actual crater, so Bjork estimated that the impact of a 64,000 metric ton iron meteorite was required to match the crater dimensions. This paper has withstood the test of time. The findings of Bjork's paper are consistent with those of the most recent numerical work investigating the meteor crater (Roddy, et al., 1980).

Bjork (1963), Walsh and Tillotson (1963), Riney (1963), Riney and Heyda (1965), and Riney (1970) used the PIC method for further investigations of hypervelocity impact in both semi-infinite targets and thin targets. In particular, a concept called "late-stage equivalence" was examined using hydrocodes; Riney (1970) also examined the dynamics of the debris cloud as a function of the geometric and material construction of a bumper shield.

LATE-STAGE EQUIVALENCE

In the early to mid sixties, considerable work was performed (Bjork, 1963; Walsh and Tillotson, 1963; Walsh and Johnson, 1965; Dienes, 1965; Chou, et al., 1965) to investigate "late-stage equivalence," i.e., do impacts at two different velocities have late-stage asymptotic solutions which are essentially the same except for a simple scale factor? The impetus for this came from the theory of strong explosions where the details of burning, asymmetries, and other details of the energy release do not significantly affect the process at late times. It was postulated that such an effect might also hold for hypervelocity impact (Whipple, 1958). That is, if two flows, established during the hydrodynamic phase of impact, become indistinguished except for a scale factor (late-stage equivalent), then the terminal effects should also be similar. If late-stage equivalence held, then simple scaling could extrapolate experimental data to the very high velocities of interest.

The initial theory, in analogy with explosions, assumed that the scaling parameter was impact energy. But as Dienes and Walsh (1970) report, the energy hypothesis is somewhat unsatisfactory because of the intuitive feeling that as the momentum of a projectile goes down while its energy remains constant, the overall effect should somehow also decrease. Similitude theory suggested that late-stage equivalence might hold if Lv_0^{α} remained constant between two hypervelocity impacts, where L is the projectile size and the v_0 the impact velocity (Walsh and Tillotson, 1963). Considerable debate took place over the value of α, assuming late-stage equivalence held. If the flow depended solely on energy, α would be 2/3; and if the flow depended solely on momentum, α would be 1/3.

The hydrodynamic codes were used to perform parametric studies, changing the velocities, masses, materials, shapes, and densities. The value of α was determined such that at late times, the velocity, pressure, and density profiles in both the axial and radial directions were asymptotically the same for pairs of projectiles characterized by the same value of Lv_0^{α}. For impact velocities at least two to five times the target sound speed, the parameter α was found to be 0.58 for a wide range of velocities and materials (Dienes and Walsh, 1970). (If the aspect ratio of the projectile became relatively large, e.g., a flat plate or a long rod, slight deviations from the 0.58 power dependence were observed.) The results of these studies led to the simple penetration formula for semi-infinite targets (Dienes and Walsh, 1970):

$$P/D = K(\rho_p/\rho_t)^{1/3}\ (v/c_o)^{5/8} \tag{4}$$

where P is the crater depth, D the diameter of the projectile assuming a sphere of the same mass, ρ_p and ρ_t the densities of the projectile and target, respectively, and c_0 the sound speed of the target. The parameter K is determined from a single impact experiment, and is a constant of the target material, depending on its strength. It also was found that the ratio of total target impulse to initial projectile momentum increases as the 0.74 power of the impact velocity. Subsequent work by Sedgwick and co-workers (1978) modified Eq. (4) by explicitly incorporating the target strength, Y_t:

$$P/D = 0.482\ (\rho_p/\rho_t)^{.537}\ (v/c_o)^{0.576}\ (Y_t/\rho_t c_o)^{-0.235} \tag{5}$$

The reader is also referred to the most recent work by Holsapple (1986), Bjorkman and Holsapple (1986), and Schmidt and Housen (1986).

Continuous Eulerian Computer Calculations

The 1960's and early 1970's saw a very rapid development in hydrocodes. In the late 1950's and early 1960's, one dimensional Lagrangian codes were formulated and developed. In particular, Mark Wilkins at Lawrence Livermore and Walter Herrmann, then at MIT, were very active in the formulation of finite difference and method of characteristic wave codes. Concurrent with this work, considerable effort was being devoted to understanding hypervelocity impact, which required Eulerian calculations. Figure 1 depicts a "road map" of Eulerian code development during the 1960's and '70's, and Table 1 gives a brief description of the codes and references.

In the 1963-1964 time frame, W. Johnson and coworkers developed a continuous Eulerian scheme (Walsh et al., 1964; Walsh and Johnson, 1965). The impetus for a true Eulerian code derived from the requirement of computing small variations in cell density and rarefaction. In the PIC scheme, each particle represents a finite mass, so that if a one-percent variation in density is to be resolved, 100 particles per cell are required. Of course, computer time increases dramatically as the number of particles per cell increases. If fewer particles are used, the solutions obtained with a PIC calculation tend to exhibit unreal cell-to-cell fluctuations in density, velocity, and pressure, Fig. 2, though integrated or suitably averaged quantities appear reliable.

The continuous Eulerian hydrocodes have their roots in the PIC methodology; thus, all the early Eulerian codes performed the computations in phases to advance the solution in time. In the first phase, velocities and energies are updated from the pressure terms. Just like the PIC methodology, the next phase involves mass transport, except that a cell density is used (as opposed to discrete particles), to compute the term $\rho \bar{u} A \Delta t$, where ρ is the density, $\bar{u}$ is the transport velocity, A the surface area through which the mass is advecting, and Δt is the time increment.

The difficulty in developing a continuous Eulerian code involved the exact details of the mass transport, and in particular, what to use as the transport velocity, $\bar{u}$. A number of different approaches were tried before a suitable expression for $\bar{u}$ was found. A representation had to be found which could describe the shock front as well as the details behind the front (rarefaction). For example, if the average velocity of the donor cell was used as $\bar{u}$, then instabilities were found behind the shock front, (Johnson, 1965). The expression finally used was found by performing a Taylor series expansion and averaging the velocity between two cells:

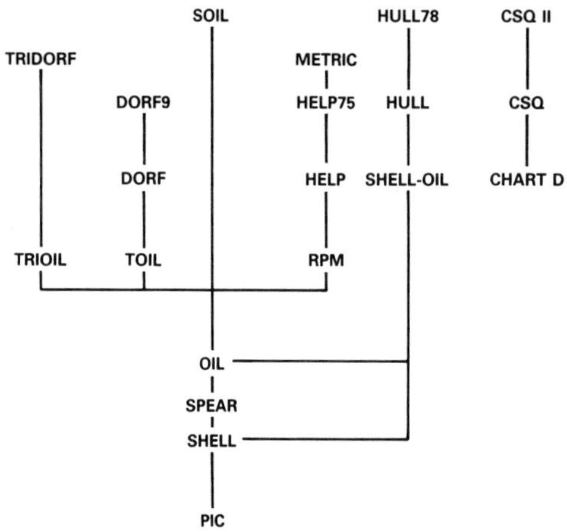

Fig. 1. Evolution of Eulerian hydrocodes.

Table 1. The Development of Eulerian Hydrocodes.

Name	Authors	Description
PIC	Evans and Harlow, 1957	PIC, 2-D hydrodynamic
SHELL	Johnson, 1959	Version of PIC
SPEAR	Johnson, 1963	Improved version of SHELL
OIL	Walsh and Johnson, 1965, Johnson, 1965	Continuous mass transport, 2-D hydrodynamic, 1 material
TOIL	Johnson, 1967b	OIL with 2 materials
TRIOIL	Johnson, 1967	3-D OIL, 1 material
RPM	Dienes, et al., 1968; 1980	OIL method with rigid, perfectly plastic material model
DORF	Johnson, 1971	TOIL code with rigid, perfectly plastic material model
HELP	Hageman and Walsh, 1971	RPM with elastic-plastic material model and Lagrangian interfaces
DORF9	Johnson, 1971b	9 material DORF
CHART D	Thompson, 1969; 1973	1-D Lagrangian with sophisticated thermodynamic equation of state package
CSQ	Thompson, 1975	2-D Eulerian code with material strength and thermodynamic consistent EOS package
TRIDORF	Johnson, 1976	TRIOIL with rigid, perfectly plastic material, 2 materials
HELP75	Hageman, et al., 1975	Improved HELP
METRIC	Hageman, et al., 1976a; 1976b	3-D HELP
HULL	Durrett and Matuska, 1971	Atmospheric response to nuclear weapons, 1 material
SOIL	W. Johnson, 1977b	OIL with operator splitting
HULL78	Durrett and Osborn, 1978	Multi-material HULL with strength, 2-D
CSQII	Thompson, 1979	Improved version of CSQ, 2-D

$$\bar{u} = \frac{(u_{k+1} + u_k)/2}{1 - \dfrac{(u_{k+1} - u_k)\,\Delta t}{\Delta x}} \tag{6}$$

where the subscripts refer to the k-th and (k+1) grid cells, and Δx is the distance between the cells. Another gratifying trait of the scheme in Eq. (6) was that it correctly computed the two components of velocity (along a ray 45° in a rectangular grid) for a spherically symmetric problem, i.e., the transport scheme did not introduce significant preferential treatment in the radial or axial direction.

A comparison between the continuous Eulerian code, OIL,[1] and a PIC calculation is shown in Fig. 2. The grid size and starting conditions were the same for the two calculations; cell dimensions were 0.25 cm and the PIC method used 16 particles per cell. The OIL solution is relatively smooth and in general agreement with the PIC solution, with the cell-to-cell fluctuations in the PIC solution a consequence of the mass resolution of 16 particles. These results are typical of those obtained from a number of comparisons (Walsh and Johnson, 1965). The oscillations in the PIC results could be eliminated by going to more particles; but, even in this comparison between the continuous Eulerian and the PIC method, the PIC calculations took 15 times longer because of the time associated with processing each individual particle. Thus, the economics of continuous mass transport essentially dealt the PIC method a death blow for impact calculations.

Hypervelocity Impact Into Wax

The first comparison between an impact experiment and numerical hydrodynamic calculations (besides the celestial impact event which resulted in the meteor crater in Arizona, and where the initial impact conditions are not known) used OIL. A series of hypervelocity impacts into semi-infinite wax targets was conducted by Frazier and coworkers at the Ballistic Research Laboratories (Frazier, 1961; Frazier and Karpov, 1961; Karpov, 1963; Frazier, Karpov, and Holloway, 1965). Comparisons were made between the experimental data and the computed results in Frazier, et al. (1965), but subsequent computations by W. Johnson (unpublished) showed even better agreement.

In the experiments, a plastic projectile impacted wax at 4 km/s. Since OIL was formulated only for one material, the numerical code simulated the impact of wax into wax. Two calculations were performed. The first was a simple hydrodynamic treatment. In the second calculation, tensor

[1] A secretary decided that "continuous Eulerian computer code" was too long to type each time, so she shortened the title to just the first syllable of Eulerian, except that she thought everyone was saying "Oil-erian" thus, the name of the OIL code was born.

forces due to Newtonian viscosity were included. The pressure and particle velocities from the code calculations are compared against experimental data as a function of depth into the target, Figs. 3 and 4. There is good agreement for pressures between 100 kilobars and 10 kilobars, with better agreement for the calculation with viscosity effects included. Of course, the calculations did not include any strength effects (though the addition of viscosity permits a strain rate effect), thus, it is not surprising that there exists some discrepancy between experiments and calculations at the low pressures. Figure 5 compares crater size from flash radiographs with the code calculations at three times using the viscous-hydrodynamic model.

The importance of this calculation and comparison in the mid 60's should not be minimized. There had been no direct confirmation of the validity of the numerical work, and with wax, a hydrodynamic approximation certainly seemed appropriate. The reasonably good agreement was extremely encouraging for further code calculations and code development. Also, the introduction of a new generation of computers (e.g., CDC 6600) in the late 60's and early 70's provided the computer hardware necessary for more extended hydrocode work.

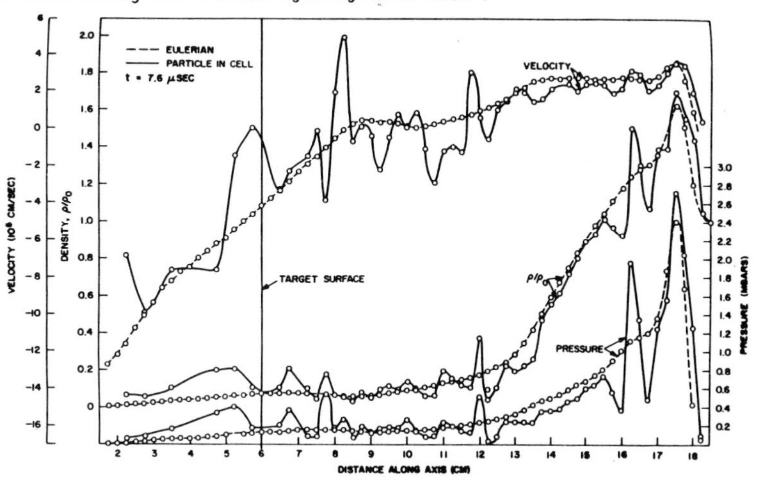

Fig. 2. Comparison of axial velocity, compression, and pressure profiles from particle-in-cell and continuous Eulerian codes (from Walsh, et al., 1964).

Multiple Materials

As already mentioned, the OIL code was a one material code. The two-material version of OIL, TOIL (Johnson, 1967b) required modifications in the advection phase. In an Eulerian formulation, the elements of mass are no longer individually identified from cycle to cycle; thus, the location of material interfaces is a problem. Because of numerical diffusion, a special

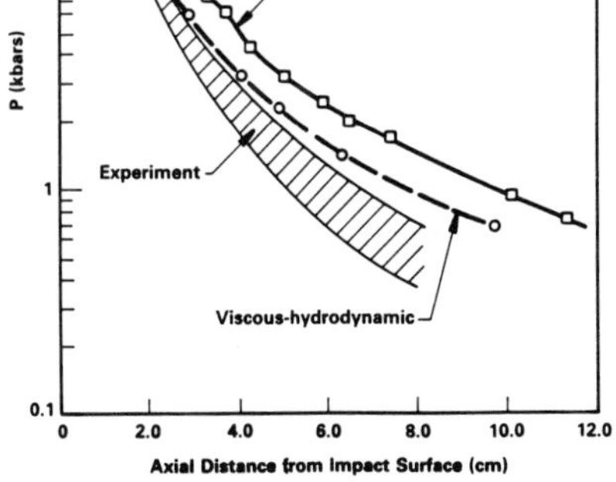

Fig. 3. Pressure attenuation of hypervelocity impact into wax: comparison of experiment and hydrocode analysis.

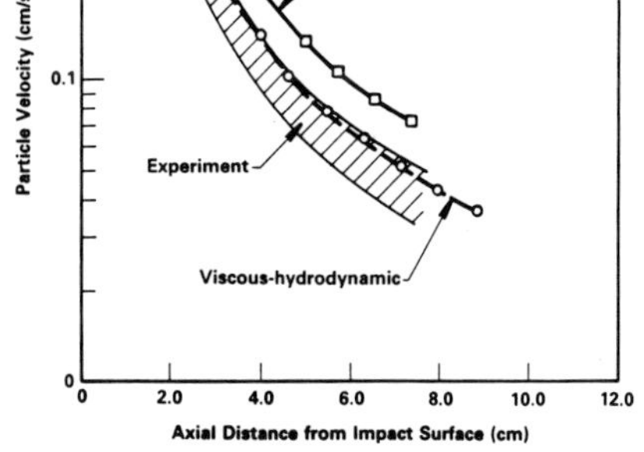

Fig. 4. Particle velocity attenuation of hypervelocity impact into was: comparison of experiment and hydrocode analysis.

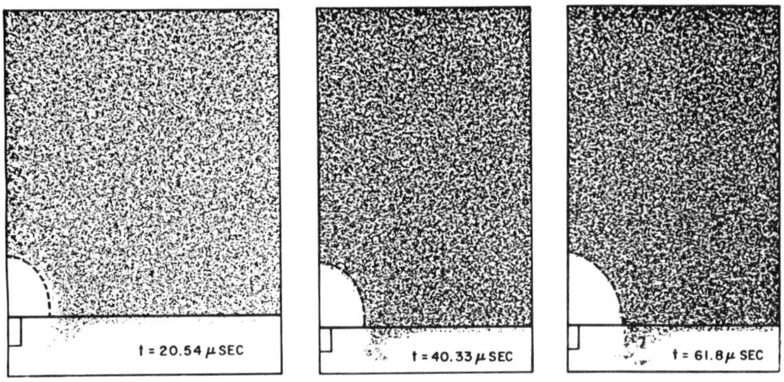

Fig. 5. Hypervelocity impact into wax: comparison of calculated crater size with flash radiographs.

diffusion limiter algorithm had to be devised to prevent the artificial smearing of material interfaces (see Anderson, 1986). Material moving from a pure cell to a mixed cell required no modification; however, material moving from a mixed cell to a pure cell required special treatment. One does not know, a priori, the positions of the materials in a mixed cell, as they are considered to be "mixed" uniformly within the cell. However, their respective masses and volumes are known. For material moving from a mixed cell to a pure cell (e.g., advection from a cell containing materials 1 and 2, into a pure cell containing only material 2), it is assumed that material 2 in the mixed cell is adjacent to material 2 in the pure cell. As the fractional volumes and masses are known, the fraction of each material advected can thus be computed. Material 2 is preferentially transported first, and if any remaining mass needs to be transported, then material 1 is advected. For mass motion between mixed cells, the various mass fluxes are proportional to the masses within the mixed cells.

All Eulerian codes use some sort of "look ahead" for the preferential advection of materials in mixed cells, though calculational details may differ. For example, HULL (Durrett and Matuska, 1978) explicitly computes a material gradient as part of its diffusion limiter scheme (though in essence, all the schemes, in one manner or another, compute a material gradient). Material diffusion can still result in numerical errors, but the diffusion limiter schemes retards the diffusion process considerably.

The other complication in the treatment of mixed cells is that some equilibrium state is assumed; it is generally assumed that the materials are in pressure equilibrium (sometimes, temperature equilibrium is also invoked). The masses and specific internal energies of the constituents are known, along with the total cell volume and energy. An iterative scheme is then used to find the fractional volumes of each material, and thus densities, subject to the constraint that $P_1 = P_2$.

Strength Effects

One of the justifications of using purely hydrodynamic theories was that at sufficiently high impact velocities, the impact generated pressures were considerably above the material yield strength (Anderson, 1986). Thus, the hydrodynamic assumption was a good first approximation. While the high impact pressures were a justification for the hydrodynamic approximation, the real reason for using the hydrodynamic approximation was that the details of how to treat the stress and strain tensors in a numerical scheme had not been formulated.

In a cratering event, as the impact disturbance propagates into the target material, the hydrodynamic pressures will eventually become comparable to the yield stresses. An analysis of experimental data at hypervelocities show that both crater depth and crater volume can be scaled to a single curve if strength effects are incorporated (Sedgwick, et al., 1978; Westine and Mullin, 1986). At late times, the hydrodynamic approximation no longer provides a valid solution as it is inappropriate to neglect material strength. Riney (1963) showed how rapidly the shock strength decayed in the target due to geometric divergence. Rand, et al. (1983), in hypervelocity impact experiments in polycarbonate, in conjunction with hydrocode analyses, demonstrated the effects of dynamic plasticity. It has also been demonstrated that strength effects must be included to predict hole diameters from hypervelocity impact of long rod penetrators (Misey, 1977).

At about the same time W. Johnson and coworkers were developing the Eulerian numerical formulation of the continuum equations, and Riney was incorporating a visco-plastic model into a PIC code (Riney, 1963), Mark Wilkins at Lawrence Livermore worked out the equations for elastic-plastic constitutive modeling to be incorporated into a Lagrangian computer code (the one-dimensional KO code, and the two-dimensional HEMP code). This work was reported at the Sixth Symposium of Hypervelocity Impact (Wilkins and Giroux, 1963), and then was reported more fully, with numerous computational examples, in "Calculations of Elastic-Plastic Flow" by Wilkins (1964, 1969). This work resulted from Wilkins' investigations into the discrepancies between experimental data and the hydrodynamic computer code predictions of initiation and expansion of the detonation products of high explosives. Realizing that strength effects could be important at late times in the expansion process, Wilkins developed the procedures for incorporating strength of materials into the computer calculations. [The major discrepancy, however, was not the exclusion of strength effects, but in the equation of state of the detonation products--Wilkins work, with a refinement by Lee, resulted in the Jones-Wilkins-Lee (JWL) equation of state for the detonation products of high explosives (Lee, et al., 1968).]

EULERIAN CODE ENHANCEMENTS

It was recognized almost from the beginning the desirability of tracking material history, at least in some sort of gross sense. Again, the idea of Lagrangian particles moving through an Eulerian grid proved attractive; however, now the particles are assigned zero mass. These massless "tracer particles" move with an area-weighted velocity, i.e., the tracer particle velocity is interpolated between the average cell velocities as a function of its coordinate position. As opposed to forces being computed, tracer particles are moved according to the local velocity. The tracer particles usually are used entirely for edit purposes, and thus, do not enter the physics of the calculations. However, they have been used to obtain estimates of strains, and to define the location of material interfaces and free surfaces. By connecting tracer particles, the movement of free surfaces and material interfaces can be approximated. Note, however, that accuracy is proportional to the number of tracer particles used in the computation, and the more tracer particles used, the more the computational burden. [Creating particles also assists in post processing. If 100 particles are placed randomly in the cell with the largest density (or any other variable of interest), then the number of particles in any one cell depicts the relative density. The number of particles can be used as a filtering technique. For example, all cells with densities less than $\rho max/100$ will appear as a void. This permits density plots of the grid, and assists in the identification of free surfaces (where the plotted density rapidly goes to zero over one to two cells). If materials have substantially different densities across an interface, then the interface also can be pictured.]

During the late 1960's and early 1970's, considerable work went into Eulerian code enhancements. Two paths were taken, both were derivatives of OIL. Dienes and coworkers (1968, 1970) first added strength effects by incorporating a rigid plastic model to OIL, called RPM. Then, Hageman and Walsh (1971) developed HELP, a multi-material Eulerian code. In HELP, material interfaces and free surfaces propagate through the calculational mesh (using massless Lagrangian tracer particles) as discrete interfaces across which the materials are not allowed to diffuse. HELP was used rather extensively during the 1970's to compute a number of impact and shape charge problems (see Sedgwick, et. al., 1978, for a partial list of references).

Concurrent with the development of RPM and HELP, Johnson developed TOIL (Johnson, 1967b). Next, following the work of Dienes and Walsh, Johnson included a rigid, perfectly plastic model, which later was extended to an elastic, perfectly plastic model. These were the DORF codes (Johnson, 1971, 1971b). A nine material version of DORF became known as DORF9 (Johnson, 1971b).

In the same time period, Johnson developed a three-dimensional version of OIL (single material), called TRIOIL (Johnson, 1967) to investigate oblique impacts. An early calculation of the oblique impact of aluminum into aluminum was reported at the AIAA Hypervelocity Impact Conference (Johnson, 1969) for two velocities and three angles. Typical grid sizes were 27x30x16 for a total of approximately 13,000 grid cells. Running times on a CDC 6600 were between 30 and 60 minutes per problem. Johnson then extended TRIOIL to two materials with a rigid, perfectly plastic model, TRIDORF (Johnson, 1976). An elaborate look ahead was required in the advection phase to insure that there would be no preferential mass transport due to the choice of the direction to calculate the mass flux. This usually involved the two cells above the cell in question, and adjacent cells in the column to the right. This formulation did not allow for corner coupling (mass moving at 45° to the cell above and to the right). Some computations were performed for oblique impact (W. Johnson, 1977; Johnson and Kucher, 1977a; 1977b; 1977c; 1977d; Kucher, 1978), but it is not unfair to say that the major effort was to get the problems to run to completion, and plot the results.

Two other Eulerian codes had their genesis in the 1970's, HULL (Matuska and Durrett, 1971; Durrett and Matuska, 1978) and CSQ (Thompson, 1975; 1979). The original HULL code was a hydrodynamic code used to investigate atmospheric response to nuclear weapons. In 1978, Durrett and Matuska reformulated HULL into a multi-material, elastic-plastic material code; this reformulation is the basis for the HULL code used today.

Thompson (1969; 1973) developed the one-dimensional Lagrangian hydrocode CHART D, which has a thermodynamically consistent equation-of-state package. In order to compute problems with large distortion, Thompson developed CSQ, incorporating many of the features of CHART D, in particular, the equation-of-state package. CSQ probably has the most sophisticated equation-of-state package incorporated in any hydrocode today. In one sense, the roots of CSQ lie in CHART D, and as a result, the numerical treatment of the difference equations are treated philosophically somewhat differently. First, the mesh is treated as material fixed, and a complete Lagrangian calculation is performed. After this phase, a rezoning phase is entered in which the Lagrangian mesh is mapped back to the original mesh, employing a geometrical interpretation subject to the conservation laws. Another feature of CSQ, since incorporated into some other Eulerian codes, is the retention of cell void volumes. Edges of solids, and interior cracks and spall, can be handled in this manner.

One of the first problems examined with CSQ was the hypervelocity impact of a nylon sphere into a steel plate. This problem was also investigated using the Lagrangian code TOODY (Swegle, 1978). Both codes, using fundamentally different mathematical and numerical descriptions, produced results in excellent agreement with experiment, including the extent of spallation, and the distance of the spall plane from the rear surface of the steel plate (Bertholf, et al., 1975). Also, both codes showed the agreement between the numerical results and experiment required modeling the polymorphic phase change in iron, the correct spall strength, and vaporization of the nylon (Bertholf, et al., 1975). This was an important comparison between the predictive capabilities of Lagrangian and Eulerian hydrocodes.

Other improvements in Eulerian codes have been the incorporation of second-order-accurate advection algorithms (e.g., McGlaun, 1982), and transmitting boundaries (e.g., McGlaun, 1982). The next major advance in Eulerian calculations was with SOIL (W. Johnson, 1977b). The elaborate look ahead required by TRIOIL and TRIDORF was inefficient and cumbersome. Johnson devised an operator splitter, i.e., a decoupling of the spatial coordinated system, which permits a series of one-dimensional passes through the grid. Any one-dimensional algorithm can be incorporated into two and three dimensions simply by writing a driver that will call the one-dimensional passes, one at a time, in the various coordinate directions. The elaborate look ahead to remove preferential mass treatment is no longer required. The one-dimensional pass is alternated (the direction changed from cycle to cycle) to remove any preferential mass treatment associated with systematically cycling through the grid in a preferred direction. The one-dimensional advection phase automatically introduces corner coupling to the Eulerian technique, thereby enabling mass transport at 45° in one time step, as opposed to multiple time steps. The simplicity of the operator splitter concept results in savings in programming complexity and CPU time. SOIL has become the basis for LASOIL, used by researchers at Los Alamos to investigate hypervelocity impact (e.g., Holian and Burkett, 1986; Meier, 1986).

LAGRANGIAN CODE ENHANCEMENTS

Lagrangian codes have undergone little algorithm enhancement until recently, though certain features have been extended considerably. The concept of slidelines was introduced in the early 1960's (Wilkins, 1964; 1969). Wilkins was investigating the dynamic loading from high explosives. Physically, it is expected that large transverse motion will occur along the explosive/metal interface. If the "sliding" is not permitted between the adjacent material, then severe mesh distortion will result. Normal components of the acceleration and velocity are required to be continuous, but the tangential components may be discontinuous. Opposite sides of the sliding interface are referred to as the "master" and "slave" surfaces. The master surface is advanced using its own force field and the force field of the slave surface. The slave surface is advanced along the master surface by treating it as a fixed surface, but lateral (tangential) motion is permitted according to a friction law. The friction coefficient can be zero, have intermediate values, or infinite. If the friction coefficient is infinite, the grid is tied together, i.e., no sliding is allowed--this would be used if the slideline was being used to change the size of the mesh.

Slideline treatments have matured over the years (Giroux, 1973; Belytschko, 1976; Swegle, 1978; Hallquist, 1978; Warren, 1983). Sliding routines now permit two materials to initially be in contact and separate, or if initially separated, the void between them can close. If the void option exist in a particular code, it usually is treated point by point so that part of the two materials may meet while part still may be in the void condition. The void may be generated in the open or closed condition, and if closed, allowed to open. It may be allowed to reopen after closing, or it may be forced to remain closed. Intersecting slidelines have also been introduced (Warren, 1983).

Almost from the beginning of two-dimensional Lagrangian codes, there has been a desire to rezone the tangled grid, and rezoners were devised (Leibee, 1974; Thorne and Holdridge, 1975; Warren, 1981; Swegle, 1981; Swegle, 1982). The essential feature of most rezoning procedures is that the new mesh forms a geometric overlay of the old mesh. The rezoning program then proceeds through the new grid structure one zone at a time, determining those zones of the old grid which intersect the particular zone of the new grid. The properties of the zone in the new grid are then computed from the "partial" zones of the old grid from which it is constructed. The three conservation equations are used to conserve mass, momentum, and energy from the old grid to the new grid. It should be mentioned that in conserving mass and momentum, the kinetic energy has already been determined, and is less than the kinetic energy before the rezoning. The kinetic energy loss during the rezone process is determined and then added to the internal energy in order to conserve total energy. (A "switch" is usually provided where internal energy can be explicitly conserved, as opposed to total energy.) Besides straightening tangles meshes, both Eulerian and Lagrangian programs use rezoning to regrid a problem for optimizing stress resolution versus computational run times.

Notwithstanding these improvements, the gains in Lagrangian computations have been largely in the incorporation of improved material constitutive models, including failure. A review of material modeling in hydrocodes is given by Anderson (1986). Recently, however, considerable promise has been demonstrated in modeling deep penetration by finite element Lagrangian codes (see Anderson, 1986, for a summary). For hypervelocity impact, material response is characterized by total material failure (erosion). Figure 6 depicts a copper projectile impacting a steel target at 5 km/s. This calculation was done, without rezoning, with the finite element code EPIC-3. (Johnson and Stryk, 1986). Calculations like this could never be performed with traditional Lagrangian techniques unless many rezones were performed along the way (Bertholf, et. al., 1977; 1979). To date, this capability exists only with the finite element hydrocodes, though there appears to be no reason a similar algorithm could not be developed for the finite difference codes.

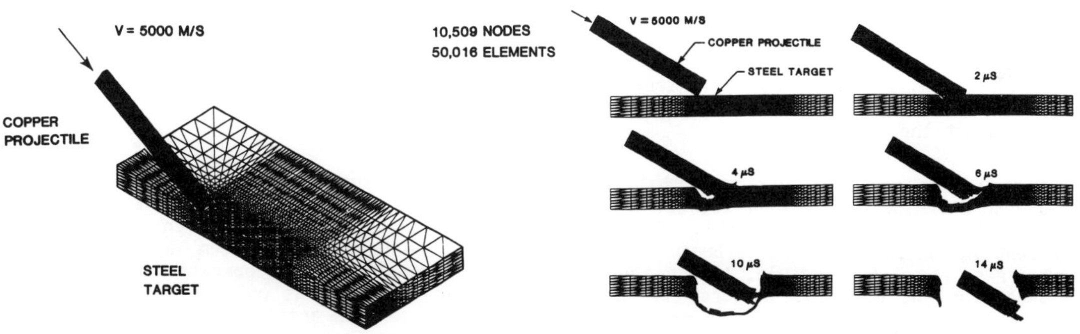

Fig. 6. Lagrangian deep penetration (courtesy of G. Johnson).

THE FUTURE

As the number of physical dimensions increase, there is an attendant increase in computer resources, both in CPU time and memory requirements. Herrmann (1977) summarized code requirements very nicely by comparing one, two, and three-dimensional Lagrangian codes, Tables 2 and 3. Table 2 gives the computer time and storage requirements for each grid cell. This information is used to estimate total computer requirements for the same calculation in one, two, and three dimensions.

Increasing spatial resolution is difficult. For example, a 50,000 cell grid requires approximately one million words of memory for a two-dimensional code. This would require approximately 19.2 hours of Cray 1 time for 3000 cycles, not an uncommon number of cycles. Increasing the grid to 250,000 cells quadruples the memory requirement, and increases the run time to 154 hours (to run to the same physical time). Yet this is only a factor of two increase in resolution. The question then becomes, "What does the future hold?" Obviously, this question cannot be answered definitively, but some insight into trends can be given.

Hardware

Use and evolution of hydrocodes have paralleled computer growth. Computer speeds increased by 10^3 in the 1940's as the technology shifted from relays to vacuum tubes (Kuck, et al., 1986). ENIAC, the first electronic computer, built by the US Army Ballistic Research Laboratories for performing ballistic calculations, had a peak rate of about 10^3 floating-point operations per second (flops).

With changes from vacuum tubes to transistors, and then to integrated circuit technology, computer speeds increased. The CDC 6600, the "supercomputer" of the mid to late 60's, had a maximum rate of 10 Mflops. With the CDC 6600, two-dimensional codes could be executed, but run times were generally measured in hours. In the late 60's through the mid 70's, the CDC 7600 had reached speeds of the order of 100 Mflops. With the 7600, three-dimensional codes could be run, but the grid size was generally limited. Two-dimensional codes could be run with good grid resolution, and it was not unusual to have 15 hours of CPU time to solve a finely zoned two-dimensional problem (Herrmann, 1977).

Table 2. Code Requirements per Mesh Point Cray 1

Code	CPU msec/mesh calc.	Words/mesh point
1D Lagrangian	0.22	10
2D Lagrangian	0.46	19
3D Lagrangian	0.70	27

Table 3. Code Requirements Cray 1

Code	No. of Meshes	No. of Time Cycles	No. of Point Calculations	Memory	CPU Time
1D Lagrangian	10^2	200	20K	1K	4.4 sec
2D Lagrangian	$10^2 x 10^2$	200	2M	190K	15.3 min
3D Lagrangian	$10^2 x 10^2 x 10^2$	200	200M	27M	40 hours

By the early 1980's, with architectural enhancements such as increased clock speeds, the Cray 1, Cray-X-MP, Cyber 205, etc., peak rates of 10^3 Mflops have been achieved (Kuck, et al., 1986). This is an improvement factor of 10^6 in 40 years, or an average factor of 10 every 7 years. But large improvements over these speeds are not seen. It is estimated that in the next 5 to 10 years, clock speeds (which is the rate at which basic computer operations are performed) are not expected to increase by a factor of 5, even with new materials such as gallium arsenide (Kuck, et al., 1986). It took 9 years to achieve a clock speed of 4.1 nsec/period in 1985 with the Cray 2, which is only 3 times that of the Cray 1 (Kuck, et al., 1986). But clock speed is not the whole story.

Though the Cray 2 has a clock speed that is more than twice the speed of the Cray-X-MP, most initial users of the Cray 2 will not see the performance of the Cray-X-MP because of the Cray 2's architecture (Kuck, et al., 1986). Typically, the delivered performance is often only 5 to 15 percent of the peak performance because of "bottlenecks" somewhere in the system, e.g., I/O, data transfers, etc. For example, though there was a factor of 100 in peak operation performances between the CDC 6600 and the Cray 1, actual improvement in speed was only on the order of a factor of 10 (Herrmann, 1977). Because of optimization of certain architectures (both hardware and firmware), computers with slower clock speeds can outperform faster computers.

For a period, when memory was expensive, virtual memory systems permitted the user to use disc storage as a substitute for active core memory. As disc storage becomes large, however, there is a tendency for the machine to spend most of its time in data transfers and very little in computation. In recent years, memory has become very cheap; thus, memory is not the limiting factor it was 5 to 10 years ago.

The biggest decrease in running times in recent years has been from parallel processing, coupled with faster clock speeds and larger direct access memories. For example, Bjork is reported to have used approximately 45 hours of IBM 704 CPU time to perform the meteor impact study (Bjork, 1961). W. Johnson, using SOIL on a Cray 1, ran the same problem with finer zoning in 12 minutes of CPU time.

The vectorization process of the Cray and other supercomputers allows rapid computation of DO loops and other operations suited to parallel processing. Ganging many microprocessors together would permit simultaneous computations similar to vectorization, but with the ability to make decisions in addition to register arithmetic. A computer with 65,536 individual processors has been used to compute a fluid mechanics problem with a two-dimensional grid 4000 by 4000 cells containing 32 million particles. A simulation of 100,000 time steps allegedly took less than 30 minutes (Waldrop, 1986). The computational methodology, called "cellular-automata" (Margolus, et al., 1986), appears to be similar to Harlow's PIC method. Thus, it would appear that the largest gains in computational speeds in the near future will occur from architecture changes.

Firmware Enhancements

The debate on higher order languages has continued in the last fifteen to twenty years, and will certainly continue. Heated debates are waged over whether FORTRAN, the language of hydrocodes, should be retired and replaced (McGraw, 1984; Kuck and Wolfe, 1984). The debate often centers on structural programming capabilities, but has intensified because of the requirements for higher-level languages to adapt to hardware data-flow design philosophies. But, the requirement of a compiler is to optimize a code to the capabilities of a machine; thus, compilers need to evolve with the architecture. FORTRAN certainly has evolved considerably since FORTRAN II, to the current version, FORTRAN '77; many machines now have enhanced versions of FORTRAN '77. It is the authors' opinion that FORTRAN will continue to be the main programming language of hydrocodes; but, initial structuring of the programs and compiler evolution will enable programmers to take advantage of machine architectures.

Code Enhancements

Higher Order Accuracy: Various methods and procedures have been proposed to improve hydrocodes. One possible approach to improvement is the development of better discretization methods. As Herrmann (1977) points out, higher order (greater than second-order) methods in Lagrangian formulations have been disappointing in their performance when artificial viscosity or shock capturing techniques are used. On shock waves, first derivatives do not exist. Thus, near discontinuities, resolution is more important than accuracy. Once the mesh size is small enough to give the required resolution, lower-order accurate schemes suffice (also, see Anderson, 1986). Thus, advances in this area are not expected, as the requirement of resolution is expressed in terms of memory and CPU time.

However, Eulerian codes historically have use first order accurate advection schemes. In recent years, second order accurate advection schemes have been incorporated into Eulerian formulations (e.g., McGlaun, 1982). In a first order accurate scheme, cell quantities such as mass, internal energy, momentum, and kinetic energy have a uniform distribution across the cell. In second order accurate schemes, these quantities are assumed to have a linear distribution across the cell. It has been shown that the second order accurate scheme can dramatically improve the accuracy of a calculation, particularly where large gradients are present (see McGlaun, 1982). Any information that is transported with the material velocity is better resolved with the second order accurate scheme as there is substantially less dissipation, particularly near large gradients. For example, there are indications that the disagreement between hydrocode computations and experients of the melting and vaporization of material in a debris cloud (Holian and Burkett, 1986) may be due, in large part, to the use of a first order accurate advection scheme in the Eulerian calculations. Thus, it would appear that second order accurate procedures are necessary in both Eulerian and Lagrangian formulations, but little is to be gained by going to higher order accurate methods because of the aforementioned reasons.

Shock-Fitting: Artificial viscosity, used by traditional hydrocodes, is a shock-smearing process. Another choice is to explicitly treat the singularities. The smooth solutions are matched across the discontinuities using the Rankine-Hugoniot jump conditions. The limiting case of such a code is a method of characteristics code. This method is excellent for understanding the detailed wave motion; however, if the ultimate state of the material is desired, a finite difference (or finite element) code is generally superior (Karpp, 1972). People have combined the two methodologies (Herrmann, et al., 1971; Hicks and Holdridge, 1972). Internal floating boundaries are introduced in the mesh, and shock fitting, i.e., the shock jump conditions, is used to join the solution on the two sides of the discontinuity. While accuracy is excellent, the difficulty is in providing the logic for tracking all the discontinuities, especially in two or three dimensions (Belytschko, 1976; Herrmann, 1977). It has been over fifteen years since this concept has been introduced, and while it has utility for certain specific applications, it does not seem to be a viable alternative to the general utility of multi-dimensional hydrocodes.

Code Coupling: Three different approaches have been used to couple Eulerian and Lagrangian formulations using an Eulerian calculation during the severe distortion and perforation process, and the Lagrangian code elsewhere. The Eulerian or Lagrangian grids are used in the region, or time regime, where their capabilities are best suited for the problem.

The first approach is not considered as code coupling, but when a Lagrangian code is constantly rezoned, the rezoned area, in effect, can be considered as a quasi-Eulerian grid. The mass and other quantities originally associated with a specific zone can become averaged over a number of cells during the multiple rezone process.

A second method takes existing Eulerian and Lagrangian computer codes and links the output of one hydrocode to provide the input, or "loading conditions" to the other hydrocode. EPIC has been linked to SOIL (Johnson, 1981), and HULL (Matuska and Osborn, 1981). In both cases, the Eulerian code is used for the initial impact calculations, and time-history information is recorded at "stations," which is then used to drive the boundary of the Lagrangian calculation. Similar calculations were performed at Lawrence Livermore to link the Eulerian code JOY (Couch, et al., 1983) to DYNA3D (Hallquist, 1981). Harrison and Johnson (1986) have used the Lagrangian code HEMP to compute the initial collapse of a shaped charge liner, and then used the output from HEMP to define the starting conditions of the Eulerian code SOIL. In this case, the Lagrangian grid is used as part of the mesh generator for the Eulerian code in order to have initial velocities, internal energies, etc., from which the Eulerian calculations begin.

While the code coupling technique has specific applications, it tends to be computer intensive, and does not appear to be as advantageous as thought in the early 1980's. In fact, the technique seems to have fallen into general disuse for hypervelocity impact, though it could be resurrected in the future.

The third approach to code coupling is a true, coupled Euler-Lagrange method (Noh, 1964) which uses pure Lagrangian and pure Eulerian grids which are dynamically coupled together. The coupled code allows for Eulerian flow where material distortion is large, and a full Lagrangian treatment away from the distorted region, e. g., the computer code CHAMP (Edwards, et al., 1978). However, this method generally has been applied to explosive loadings and not to impact problems, though van Thiel and Edwards (1981) used CHAMP to investigate long rod penetration into a steel target.

While not explicitly stated, the Euler-Lagrange methodology can have trouble with the coupling at the Eulerian-Lagrangian interface. A special column of cells is required between the Eulerian grid and Lagrangian grid. These Lagrangian-like cells are used as boundary conditions on the Lagrangian grid, but must receive their information from the Eulerian grid. Herein lies the difficulty of writing a general purpose algorithm. For impact problems, this interface must have many of the attributes of a slideline with friction, but of course, it is not a slideline.

The methodology has considerable pedagogical features, however. The arbitrary Lagrangian-Eulerian (ALE) codes have a computational mesh with vertices that move with the material (Lagrangian), remained fixed (Eulerian), or move in any prescribed manner. That is, the formulation of the codes is similar to Eulerian formulations in that the advection terms are included, but the computational grid may move in any manner prescribed by the user. Trulio (1966) was the first to formulate the technique in a hydrocode called AFTON, though for a number of years, the code was used solely as either Lagrangian or Eulerian due to the lack of algorithms for arbitrary movement of the mesh (Herrmann, 1977). Subsequent work, particularly in fluid mechanics by Hirt and coworkers at Los Alamos (Amsden and Hirt, 1973; Hirt, et al., 1974; Pracht, 1975) has improved the methodology. In 1976, Trulio and coworkers (Trulio, et al., 1976) developed an improved version of AFTON; other ALE codes now exist such as the PISCES code.

With the ALE technique, the difficulties with the Eulerian-Lagrangian boundary, discussed above, can be overcome by having a number of "coupling" cells. These cells transition from Eulerian-like to Lagrangian-like by permitting the grid velocities to transition smoothly from zero to the local particle velocity. More elaborate velocity fields for the grid vertices also can be prescribed. For example, in a impact cratering problem, the general features of the resultant flow field are known. By initially prescribing the overall flow motion to the grid vertices, a Lagrangian-like calculation can be performed without severe distortion of the grid inherent in a full Lagrangian description (Orphal, 1976). Obviously, success with the ALE methodology in this manner requires experience, and often, an iterative procedure with the computer.

Free Lagrangian (Nearest Neighbor): Deep penetration with Lagrangian codes has already been discussed. Certainly this advance has improved the utility of Lagrangian hydrocodes in simulating impact problems. Another scheme, developed by P. Crowley at Lawrence Livermore and R. Clark at Los Alamos, is referred to as the free lagrangian, or nearest neighbor technique, and appears to have considerable promise. In essence, the calculation is Lagrangian. Normally, grid points have to interact with the same nodes every time step. In the nearest neighbor approach, nodes are not forced to interact with the same points all the time; instead, they interact only with their nearest neighbors. Thus, Eulerian-like calculations can be performed. Accelerations or forces are coupled automatically to traditional "fixed" grid points, if these are the nearest neighbors. Also, the nearest neighbor approach will reduce to the usual Lagrangian calculation if the grid does not become highly distorted, i.e., the nearest neighbors stay the nearest neighbors. A visualization of the potential of the free lagrangian technique is shown in Fig. 7, where a copper projectile strikes a steel target at 1 km/s at an oblique angle (Johnson et al., 1986). Benz and coworkers (1985) also have used the technique to investigate impact.

Subcycling and Subpackaging: Subcycling and subpackaging both refer to procedures which can be implemented to improve the efficiency of a hydrocode. Subcycling refers to a procedure where some variable is particularly sensitive to the size of the time step. A typical example where subcycling is required is the computation of strain rate dependent stress relaxation (Herrmann, Lawrence and Mason, 1970; Swegle and Grady, 1985). In stress relaxation, the effective time for relaxation may be smaller than the time step chosen from the stability criterion. The problem could be solved by running the entire calculation at a smaller time step, but it is more efficient to simple subcycle on the equation(s) which are sensitive to the time step. Subcycling is accomplished by dividing the stability time step into k increments, and applying the equation(s) to each of the increments.

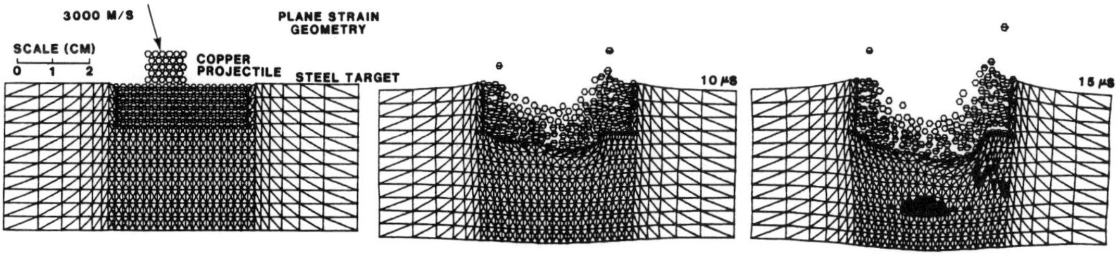

Fig. 7. Nearest neighbor approach to penetration (courtesy of G. Johnson).

In other types of problems, there may be a spatial region where it is extremely important that there be fine zoning for adequate stress resolution. Typically, the time step for integrating the equations is computed by finding the largest time increment which still satisfies the requirement of numerical stability on the entire grid (Anderson, 1986). Fine zoning has the penalty of a small computational time step, and this small time step applies to the entire grid. However, to improve efficiency, it can be worthwhile to subcycle on that portion of the grid where the time step is small, until the summation of the subcycling time steps is equal to the time step for the rest of the grid. Subpackaging can be applied to both Eulerian and Lagrangian codes, though there is probably a tendency to apply the methodology to Eulerian grids with their fixed coordinate system. A difficulty of subpackaging can be encountered with the coupling of the subpackaged region with the rest of the grid. Typically, a sudden change in zone size behaves as an impedance mismatch, introducing reflected waves at the "interface" of the size change. (A "rule of thumb" normally applied in zoning a problem is that the change in zone size be less than ten percent.)

Summary: Considerable enhancement of Lagrangian codes has increased their utility for penetration mechanics and hypervelocity impact work. This enhancement was not predicted, as most "breakthroughs" tend to occur unexpectedly. Eulerian codes now do a better job at limiting diffusion, and treating free surfaces. Operator splitting has improved efficiency, and second order accurate advection has improved accuracy. The new codes which explicitly incorporate both Eulerian-like and Lagrangian-like features in the same code show considerable promise. But there are still problems with obtaining a solution to certain types of problems, for example, the dynamics of debris clouds. Yet, with the difficulties, and sometimes impossibility of experiments, hydrocodes are needed for extending our knowledge of the physics of hypervelocity impact.

CLOSURE

The first multi-dimensional hydrocode combined features of both Lagrangian and Eulerian descriptions. As the codes have developed, there has been a continuing desire to combine the features of a Lagrangian description with the large distortion capability of an Eulerian description. This has been accomplished to some extent with formulations after Noh, and more recently with the direct coupling of two hydrocodes. The direction in this area seems to be concentrating on a more general derivative of the technique, call the arbitrary Lagrangian-Eulerian technique.

Independent of the above work is the extension of existing Lagrangian codes to handle problems which use to be solely in the realm of Eulerian codes. One of the special algorithms developed enables a Lagrangian code to compute deep penetration. A second development, called the free Lagrangian, or nearest neighbor technique, prevents severe distortion of a grid by letting nodes interact only with their nearest neighbors. While showing great promise, it is perhaps too early to tell the general utility of this methodology, and what may be its shortcomings or pitfalls.

A number of other techniques, such as shock fitting, etc., have been proposed over the years; these have been shown to have specific applications, but general utility has not been demonstrated. New techniques in Eulerian codes, such as second order advection and operator splitting, have enhanced these codes. Other concepts such as subpackaging and subcycling have been found to be useful in order to increase the overall efficiency of a computation without the loss of accuracy.

From these observations, the conclusion can be drawn that hydrocodes will remain the primary computational tool of researchers investigating the dynamics of very high rate phenomena, largely because of their general utility to be applied to many types of dynamic problems.

REFERENCES

Amsden, A. A. and C. W. Hirt (1973). "YAQUI: An Arbitrary Lagrangian-Eulerian Computer Program for Fluid Flows at all Speeds," LA-5100, Los Alamos National Laboratory, Los Alamos, NM.

Anderson, C. (1986). "An Overview of the Theory of Hydrocodes," International Journal of Impact Engineering, Vol. 5.

Arione, S. E. and M. D. Bjorkman (1986). "Scaling Flow Fields from the Impact of Thin Plates," International Journal of Impact Engineering, Vol. 5.

Belytschko, T. (1976). "Computer Methods in Shock and Wave Propagation Analysis," in Computing in Applied Mechanics, AMD-Vol. 18, American Society of Mechanical Engineers, pp. 139-161.

Benz, W., W. L. Slattery, and A. G. W. Cameron (1985). "The Origin of the Moon and the Single Impact Hypothesis I," LAUR-85-2922, Los Alamos National Laboratory, Los Alamos, NM.

Bertholf, L. D., L. D. Buxton, B. J. Thorne, R. K. Byers, A. L. Stevens, and S. L. Thompson (1975). "Damage in Steel Plates from Hypervelocity Impact II. Numerical Results and Spall Measurement," Journal of Applied Physics, Vol. 46, No. 9, pp. 3776-3783.

Bertholf, L. D., M. E. Kipp, and W. T. Brown (1977). "Two-Dimensional Calculations for the Oblique Impact of Kinetic Energy Projectiles with a Multi-Layered Target," BRL-CR-333, USA Ballistic Research Laboratory, Aberdeen Proving Ground, Maryland.

Bertholf, L. D., M. E. Kipp, W. T. Brown, and A. L. Stevens (1979). "Kinetic Energy Projectile Impact on Multi-Layered Targets: Two-Dimensional Stress Wave Calculations," ARBRL-CR-00391, USA Ballistic Research Laboratory, Aberdeen Proving Ground, Maryland.

Bjork, R. L. (1959). "Effects of a Meteoroid Impact on Steel and Aluminum in Space," Report P-1662, The Rand Corporation, Santa Monica, CA,.

Bjork, R. L. (1961). "Analysis of the Formation of Meteor Crater, Arizona: A Preliminary Report," Journal of Geophysical Research, Vol. 66, No. 10, pp. 3379-3387.

Bjork, R. L. (1963). "Review of Physical Processes in Hypervelocity Impact and Penetration," Proceedings of the Sixth Symposium on Hypervelocity Impact, Vol. II, Part 1.

Bjork, R. L. (1986). "The Physics of Hypervelocity Lethality," International Journal of Impact Engineering, Vol. 5.

Bjorkman, M. D., and K. A. Holsapple (1986). "Velocity Scaling Volume of Melted Target Material," International Journal of Impact Engineering, Vol. 5.

Chou, P. C., H. S. Sidhu, and L. J. Zajac (1965). "The Strong Plane Shock Produced by Hypervelocity Impact and Late-Stage Equivalences," Proceedings of the Seventh Hypervelocity Impact Symposium, Vol. II, Theory.

Couch, R., E. Albright, and N. Alexander (1983). "The JOY Computer Code," UCID-19688, Lawrence Livermore National Laboratory, Livermore, CA.

Dienes, J. K. (1965). "Late-Stage Equivalence and Similarity Theory for One-Dimensional Impacts," Proceedings of the Seventh Hypervelocity Impact Symposium, Vol. II, Theory.

Dienes, J. K., M. W. Evans, L. J. Hageman, W. E. Johnson and J. M. Walsh (1968). "An Eulerian Method for Calculating Strength Dependent Deformation," GAMD-8497, General Atomic, San Diego, CA.

Dienes, J. K., and J. M. Walsh (1970). "Impact Theory: General Principles and Eulerian Codes," in High-Velocity Impact Phenomena, edited by R. Kinslow, Academic Press.

Durrett, R. E., and D. A. Matuska (1978). "The HULL Code: A Finite Difference Solution to the Equations of Continuum Mechanics," AFATL-TR-78-125, Eglin AFB, FL.

Edwards, L. L., R. B. Hickman, J. K. Hobson, and T. C. Michels (1978). "CHAMP: A Coupled HEMP and Multifluid Eulerian Program for Fluid Flow Simulations," UCRL-52444, Lawrence Livermore Laboratory.

Evans, M. E., and F. H. Harlow (1957). "The Particle-in-Cell Method for Hydrodynamic Calculations," Los Alamos Scientific Lab., Report No. LA-2139, Los Alamos, New Mexico.

Frazier, J. T. (1961). "Hypervelocity Impact Studies in Wax," BRL Report 1124, USA Ballistic Research Laboratories, Aberdeen Proving Ground, MD.

Frazier, J. T. and B. G. Karpov (1961). "Hypervelocity Impact Studies in Wax, Proceedings of the Fifth Symposium on Hypervelocity Impact, Vol. I.

Frazier, J. T., B. G. Karpov, and L. S. Holloway (1965). "The Behavior of Wax Targets Subjected to Hypervelocity Impacts," Proceedings of the Seventh Hypervelocity Impact Symposium, Vol. V.

Giroux, E. D. (1973). "HEMP User's Manual," UCRL-51079, Rev. 1, Lawrence Livermore National Laboratory, Livermore, CA.

Goudreau, G. L. and J. O. Halquist (1982). "Recent Developments in Large-Scale Finite Element Lagrangian Hydrocode Technology," Computer Methods in Applied Mechanics and Engineering, Vol. 33, pp. 725-757.

Hageman, L. J. and E. P. Lee (1976). "Development of Ordnance Velocity Multimaterial, Three Dimensional Perforation Code for Finite Plates," BRL-CR-305, USA Ballistic Research Laboratory, Aberdeen Proving Ground, MD.

Hageman, L. J., J. L. Waddell, and R. G. Herrmann (1976). "Application of the Multimaterial Eulerian Three-Dimensional Metric Code to Two Ballistic Impact Situations," SSS-R-76-2973, Systems, Science and Software, La Jolla, CA.

Hageman, L. J. and J. M. Walsh (1971). "HELP, A Multimaterial Eulerian Program for Compressible Fluid and Elastic-Plastic Flows in Two Space Dimensions and Time," BRL Contract Report No. 39, USA Ballistic Research Laboratory, Aberdeen Proving Ground, MD.

Hageman, L. J., D. E. Wilkins, R. T. Sedgwick and J. L. Waddell (1975). "HELP - A Multi-Material Eulerian Program for Compressible Fluid and Elastic-Plastic Flows in Two-Space Dimensions and Time--Revised Edition," SSS-R-75-2654, System, Science and Software, La Jolla, CA.
Hallquist, J. O. (1978). "A Numerical Treatment of Sliding Interfaces and Impact," UCRL-81305, Lawrence Livermore National Laboratory, Livermore, CA.
Hallquist, J. O. (1981). "User's Manuals for DYNA3D and DYNAP--Nonlinear Dynamic Analysis of Solids in Three Dimensions," UCID-19156, Lawrence Livermore National Laboratory, Livermore, CA.
Harlow, F. H. (1964). "The Particle-in-Cell Computing Method for Fluid Dynamics," in Methods in Computational Physics, Vol. 3, B. Adler, S. Fernbach and M. Rotenberg, eds., Academic Press.
Harrison, J. T., and W. E. Johnson, (1986). "HOIL: A Two-Stage Hydrodynamic Numerical Scheme Linking HEMP and SOIL," International Journal of Impact Engineering, Vol. 5.
Herrmann, W. (1977). "Current Problems in the Finite Difference Solution of Stress Waves," in Proceedings of the Workshop on Nonlinear Waves in Solids, edited by T.C.T. Ting, R. J. Clifton and T. Belytschko, sponsored by US Army Research Office and National Science Foundation.
Herrmann, W., D. L. Hicks, and E. G. Young (1971). "Attenuation of Elastic-Plastic Waves," in Shock Waves and the Mechanical Properties of Solids, edited by J. J. Burke and V. Meiss, Syracuse University Press.
Herrmann, W., R. J. Lawrence, and D. S. Mason (1970). "Strain Hardening and Strain Rate in One-Dimensional Wave Propagation Calculations," SC-RR-70-471, Sandia National Laboratories, Albuquerque, NM.
Hicks, D. L. and D. B. Holdridge (1972). "The CONCHAS Wavecode," SC-RR-72-0451, Sandia National Laboratory, Albuquerque, NM.
Hirt, C. W., A. A. Amsden, and J. L. Cook (1974). "An Arbitrary Lagrangian-Eulerian Computing Method for All Flow Speeds," Journal of Computational Physics, Vol. 14, No. 3, pp. 227-253.
Holian, K. S., and M. W. Burkett (1986). "Sensitivity of Hypervelocity Impact Simulations to Equation of State," International Journal of Impact Engineering, Vol. 5.
Holsapple, K. A. (1986). "On Scaling of Impact Phenomena," International Journal of Impact Engineering, Vol. 5.
Johnson, G. R. (1977c). "EPIC-3, A Computer Program for Elastic-Plastic Impact Calculations in 3 Dimensions," Contract Report No. 343, USA Ballistic Research Laboratory, Aberdeen Proving Ground, MD.
Johnson, G. R. (1981). "Recent Development and Analyses Associated with the EPIC-2 and EPIC-3 Codes," in 1981 Advances in Aerospace Structures and Materials, edited by S. S. Wang and W. J. Renton, American Society of Mechanical Engineers, NY.
Johnson, G. R. and R. A. Stryk (1986). "Eroding Interface and Improved Tetrahedral Element Algorithms for High-Velocity Impact Computations in Three Dimensions," International Journal of Impact Engineering, Vol. 5.
Johnson, G. R., R. A. Stryk and J. G. Dodd (1986). "Dynamic Lagrangian Computations for Solids, with Variable Nodal Connectivity for Severe Distortions," International Journal of Numerical Methods in Engineering, Vol. 23, pp. 509-522.
Johnson, W. E. (1959). Unpublished
Johnson, W. E. (1963). "Computer Development to Improve SHELL Code," GA-4673, General Atomic, San Diego, CA.
Johnson, W. E. (1965). "OIL, A Continuous Two-Dimensional Eulerian Hydrodynamic Code," GAMD-5580, General Atomic, San Diego, CA.
Johnson, W. E. (1967). "TRIOIL, A Three-Dimensional Version of the OIL Code," GAMD-7310, General Atomic, San Diego, CA.
Johnson, W. E. (1967b). "TOIL, A Two-Material Version of the OIL Code," GAMD-8073, General Atomic, San Diego, CA.
Johnson, W. E. (1969). "Oblique Impact Calculations Using a 3D Eulerian Code," AIAA Paper 69-353, AIAA Hypervelocity Impact Conference, Cincinnati, OH.
Johnson, W. E. (1971). "Development and Application of Computer Programs Related to Hypervelocity Impact," 3SR-749, Systems, Science and Software, La Jolla, CA.
Johnson, W. E. (1971b). Unpublished.
Johnson, W. E. (1976). "TRIDORF, A Two-Material Version of the TRIOIL Code with Strength," CCC-976, Computer Code Consultants, Los Alamos, NM.
Johnson, W. E. (1977). "Three-Dimensional Computations on Penetrator-Target Interactions," BRL CR 338, USA Ballistic Research Laboratory, Aberdeen Proving Ground, MD.
Johnson, W. E. (1977b). Unpublished.
Johnson, W. E. and V. Kucher (1977a). "Three-Dimensional Computations. Volume I: 30 Degree Oblique Impact," BRL CR 344, USA Ballistic Research Laboratory, Aberdeen Proving Ground, MD.
Johnson, W. E. and V. Kucher, (1977b). "Three-Dimensional Computations. Volume II: 45 Degree Oblique Impact," BRL CR 354, USA Ballistic Research Laboratory, Aberdeen Proving Ground, MD.
Johnson, W. E. and V. Kucher (1977c). "Three-Dimensional Computations. Volume III: 60 Degree Oblique Impact," BRL CR 355, USA Ballistic Research Laboratory, Aberdeen Proving Ground, MD.
Johnson, W. E. and V. Kucher (1977d). "Three-Dimensional Computations. Volume IV: 77.5 Degree Obliuqe Impact," BRL CR 356, USA Ballistic Research Laboratory, Aberdeen Proving Ground, MD.

Karpov, B. G. (1963). "Transient Response of Wave Targets to Pellet Impact at 4.0 km/sec," BRL Report 1226, USA Ballistic Research Laboratories, Aberdeen Proving Ground, MD.

Karpp, R. R. (1972). "The Method of Characteristics," in Dynamic Response of Material to Intense Impulsive Loading, P. C. Chou and A. K. Hopkins, eds., Air Force Materials Laboratory, Ohio.

Kucher, V. (1978). "One, Two, and Three-Dimensional Impact Computations," ARBRL-TR-02099, USA Ballistic Research Laboratory, Aberdeen Proving Ground, MD.

Kuck, D. J., E. S. Davidson, D. H. Lawrie, and A. H. Sameh (1986). "Parallel Supercomputing Today and the Cedar Approach," Science, Vol. 231, pp. 967-974.

Kuck, D. J. and M. Wolfe, (1984). "Retire FORTRAN? A Debate," Physics Today, Vol. 37, No. 5, pp. 66-75.

Lee, E. L., H. C. Hornig, and J. W. Kury (1968), "Adiabatic Expansion of High Explosive Detonation Products," UCRL-50422, Lawrence Radiation Laboratory, University of California, Livermore, California.

Leibee, J. A. (1974). "SHEMP--A Service Program for the HEMP Program," UCID-30089, Lawrence Livermore National Laboratory, Livermore, CA.

Margolus, N., T. Toffoli and G. Vichniac (1986). "Cellular-Automate Supercomputers for Fluid-Dynamics Modeling," Physical Review Letters, Vol. 56, No. 16, pp. 1694-1696.

Matuska, D. A., and R. E. Durrett (1971). "Version 1, The HULL Code," Air Force Weapons Laboratory, Kirtland AFB, NM.

Matuska, D. A., and J. J. Osborn (1981). "HULL/EPIC3 Linked Eulerian/Lagrangian Calculation in Three Dimension," ARBRL-CR-00467, USA Ballistic Research Laboratory, Aberdeen Proving Ground, MD.

McGraw, J. R. (1984). "Retire FORTRAN? A Debate," Physics Today, Vol. 37, No. 5, pp. 66-75.

McGlaun, J. M. (1982a), "Improvements in CSQII: An Improved Numerical Convection Algorithm," SAND 82-0051, Sandia National Laboratories, Albuquerque, New Mexico.

McGlaun, J. M. (1982b), "Improvements in CSQII: A Transmitting Boundary Condition," SAND 82-1248, Sandia National Laboratories, Albuquerque, New Mexico.

Meier, R. W. (1986). "Numerical Simulations of Small Projectiles Impacting a Multilayered Target at Hypervelocities," International Journal of Impact Engineering, Vol. 5.

Misey, J. J. (1977). "Analysis of Long Rod Penetration at Hypervelocity Impact," AIAA Journal, Vol. 15, No. 12, pp. 1696-1698.

Noh, W. F. (1964). "CEL: A Time-Dependent, Two-Space-Dimensional, Coupled Eulerian-Lagrange Code," Methods in Computational Physics, Vol. 3, B. Alder, S. Fernbach and M. Rotenberg, eds., Academic Press.

Orphal, D. L. (1976). Personal communication.

Piekutowski, A. J. (1986). "Debris Clouds Generated by Hypervelocity Impact of Cylindrical Projectiles With Thin Aluminum Plates," International Journal of Impact Engineering, Vol. 5.

Pracht, W. E. (1975). "Calculating Three-Dimensional Fluid Flows at All Speeds with an Eulerian-Lagrangian Computing Mesh," Journal of Computational Physics, Vol. 17, No. 2, pp. 132-159.

Rand, J. L., C. E. Anderson, and J. S. Wilbeck (1983). "A Parametric Analysis of the Penetration Process," Proceedings of the Seventh International Symposium on Ballistics, The Hague, The Netherlands, pp. 609-616.

Riney, T. D. (1963). "Visco-Plastic Solution of Hyprevelocity Impact Cratering Phenomenon," Proceedings of the Sixth Symposium on Hypervelocity Impact, Vol. II, Part 1.

Riney, T. D. and J. F. Heyda (1965). "Hypervelocity Impact Calculations," Proceedings of the Seventh Hypervelocity Impact Symposium, Vol. II - Theory.

Riney, T. D. (1970). "Numerical Evaluation of Hypervelocity Impact Phenomena," in High-Velocity Impact Phenomena, edited by R. Kinslow, Academic Press.

Roody, D. J., S. H. Schuster, K.N. Kreyenhagen, and D.L. Orphal (1980). "Computer Code Simulations of the Formation of Meteor Crater, Airzona: Calculations MC-1 and MC-2," Proceedings of the Eleventh Lunar and Planetary Science Conference, Pergamon Press, pp. 2275-2308.

Schmidt, R. M., and K. R. Housen (1986). "Some Recent Advances in the Scaling of Impact and ion Cratering," International Journal of Impact Engineering, Vol. 5.

Schultz, J. C., O. E. R. Heimdahl and S. A. Finnegan (1986). "Computer Characterization of Debris Clouds Resulting From Hypervelocity Impact," International Journal of Impact Engineering, Vol. 5.

Sedgwick, R. T., L. J. Hageman, R. G. Herrmann and J. L. Waddell (1978). "Numerical Investigations in Penetration Mechanics," International Journal of Engineering Science, Vol. 16, No. 11, pp. 859-870.

Spillman, G. R. (1986). "A Simple Defense Conservative Model for Mass Requirements of Hypervelocity Projectile Impact Shields for Reentry Vehicles," International Journal of Impact Engineering, Vol. 5.

Swegle, J. W. (1978). "TOODY IV--A Computer Program for Two-Dimensional Wave Propagation," SAND-78-0552, Sandia National Laboratories, Albuquerque, New Mexico.

Swegle, J. W. (1981). "ARTOO--An Automatically Rezoning Two-Dimensional Lagrangian Finite-Difference Wavecode," SAND 81-2235, Sandia National Laboratories, Albuquerque, New Mexico.

Swegle, J. W. (1982), "Verification Problems for ARTOO, an Automatically Rezoning Two-Dimensional Finite-Difference Wavecode," SAND 81-2236, Sandia National Laboratories, Albuquerque, New Mexico.

Swegle, J. W., and D. E. Grady (1985). Shock Viscosity and the Prediction of Shock Wave Rise Times," Journal of Applied Physics, Vol. 58, No. 2.

Thompson, S. L. (1969). "CHART D: A Computer Program for Calculating Problems of Coupled Hydrodynamic Motion and Radiation Flow in One Dimension," SC-RR-69-613, Sandia National Laboratories, Albuquerque, NM.

Thompson, S. L. (1973). Improvements in the CHART D Energy Flow-Hydrodynamic Code V: 1972/1973 Modifications," SLA-73-0477, Sandia National Laboratories, Albuquerque, NM.

Thompson, S. L. (1975). "CSQ--A Two Dimensional Hydrodynamic Program with Energy Flow and Material Strength, "SAND74-0122, Sandia National Laboratories, Albuquerque, NM.

Thompson, S. L. (1979). "CSQII--An Eulerian Finite Difference Program for Two-Dimensional Material Response - Part 1. Material Sections," SAND77-1339, Sandia National Laboratories, Albuquerque, New Mexico.

Thorne, B. J., and D. B. Holdridge (1974). "The TOOREZ Lagrangian Rezoning Code," SLA-73-1057, Sandia National Laboratories, Albuquerque, New Mexico.

Trucano, T. G., and J. R. Asay (1986). "Effects of Vaporization Debris Cloud Dynamics," International Journal of Impact Engineering, Vol. 5.

Trulio, J. G. (1966). "Theory and Structure of the AFTON Codes," AFWL-TR-66-19, Air Force Weapons Laboratory, Albuquerque, NM.

Trulio, J. G., N. K. Perl, and G. N. Balanis (1976). "The AFTON 2A Computer Code Revised User's Manual," AFWL-TR-75-111, Air Force Weapons Laboratory, Kirtland AFB, NM.

Van Thiel, M., and L. Edwards (1981). "Target Response to High Velocity Penetration," UCRL-85904, Lawrence Livermore National Laboratory, Livermore, CA.

Waldrup, M. M. (1986). "The Connection Machine Goes Commercial," Research News, Science, Vol. 232, May 1986, pp. 1090-1091.

Walsh, J. M., W. E. Johnson, J. K. Dienes, J. H. Tillotson, and D. R. Yates, (1964). "Summary Report on the Theory of Hypervelocity Impact," Report GA-5119, General Atomic, San Diego, CA.

Walsh, J. M., and W. E. Johnson (1965). "On the Theory of Hypervelocity Impact," Proceedings of the Seventh Hypervelocity Impact Symposium, Vol. II, Theory.

Walsh, J. M., and J. H. Tillotson (1963). "Hydrodynamics of Hypervelocity Impact," Proceedings of the Sixth Symposium on Hypervelocity Impact, Vol. II, Part 1.

Warren, K. H. (1981), "Users Guide to REHEMP, the HEMP Double Slide Rezoner," C Group Technical Note, CTN-3, Lawrence Livermore National Laboratory, Livermore, CA.

Warren, K. H. (1983). "HEMPDS User's Manual," UCID-18075, Rev. 1, Lawrence Livermore National Laboratory, Livermore, CA.

Westine, P. S. and S. A. Mullin (1986). "Scale Model of Hypervelocity Impact," International Journal of Impact Engineering, Vol. 5.

Whippel, F. L. (1958). "The Meteoric Risk to Space Vehicles," Vistas in Astronautics 1958, pp. 115-125, Pergamon Press.

Wilkins, M.L. (1964). "Calculations of Elastic-Plastic Flow," in Methods of Computational Physics, Vol. 3, edited by B. Adler, S. Fernback, and M. Rotenberg, Academic Press.

Wilkins, M. L. (1969). "Calculation of Elastic-Plastic Flow," UCRL-7322, Rev. 1, Lawrence Livermore National Laboratory, Livermore, CA.

Wilkins, M. L., and R. Giroux (1963). "The Calculation of Stress Waves in Solids," Proceedings of the Sixth Symposium on Hypervelocity Impact, Vol. II, Part 1.

Int. J. Impact Engng Vol. 5, pp. 441–449, 1987
Printed in Great Britain

0734-743X/87 $3.00 + 0.00
Pergamon Journals Ltd.

Theoretical Equation of State for Aluminum

G. I. Kerley

Sandia National Laboratories
Albuquerque, NM 87185

ABSTRACT

Theoretical methods for calculating equations of state are discussed and applied to aluminum. The model includes a treatment of melting, vaporization, and thermal electronic excitation. The thermal electronic terms are shown to cause a plasma phase transition in the hot expanded fluid, affecting the unloading behavior from shocked states. The calculations give good agreement with experimental Hugoniot data, the melting curve, and isobaric expansion measurements.

INTRODUCTION

The theoretical description of hypervelocity impacts requires equations of state (EOS) that cover a large range of temperatures and pressures. Because direct measurements of the EOS are difficult or impossible to obtain over much of this range, it is desirable to develop theoretical models that do not require extensive experimental data for their calibration. Such models should include a realistic treatment of several physical phenomena that can occur during an impact, including not only the initial shock compression and heating, but also the subsequent release wave and expansion. The important processes include melting, vaporization, and electronic excitation and ionization.

In this paper, we discuss a theoretical model that includes a description of these phenomena and present calculations for the EOS of aluminum. Aluminum has already been studied extensively, and many models for it are available. However, our approach contains several new features, particularly with regard to the expansion behavior. The method is similar to that previously used in our calculations for iron (Kerley, 1983). Separate EOS are constructed for the solid and the fluid, and the melting curve is determined by matching the free energies of the two phases. The thermal electronic contributions to the EOS are computed by using the INFERNO model (Liberman, 1979) at high densities and a new version of ionization equilibrium theory at low densities. These models both predict a transformation from a metallic to a nonmetallic electronic structure at fourfold expansion, resulting in a first-order plasma phase transition in the temperature range from 4000 to 7000 K. This transition has not yet been observed, but the calculations are in good agreement with experimental data up to 4000 K.

SOLID MODEL

The solid EOS was constructed using standard options available in the PANDA code (Kerley, 1981). We assume that the thermodynamic functions can be written as the sum of three terms. For example, the Helmholtz free energy is given by

$$F_{solid}(\rho,T) = E_c(\rho) + F_l(\rho,T) + F_e(\rho,T), \tag{1}$$

where E_c is the internal energy on the zero Kelvin isotherm, F_l is the contribution from lattice vibrations, and F_e is the contribution from thermal electronic excitation and ionization. The pressure P and internal energy E can be written in a similar fashion.

The zero-Kelvin curve was represented by the following analytic formula (Rose, *et al.*, 1984):

$$E_c(\rho) = -E_0(1 + \xi + 0.05\xi^3)\exp(-\xi), \tag{2}$$

where

$$\xi = (9\beta_0/\rho_0 E_0)^{1/2}[(\rho_0/\rho)^{1/3} - 1]. \tag{3}$$

E_0 is the cohesive energy, ρ_0 is the density, and β_0 is the bulk modulus of the solid at zero pressure.

Our values for these parameters, obtained from experimental data with corrections for zero-point and thermal lattice contributions, are $E_0 = 12.25$ MJ/kg, $\rho_0 = 2.76$ g/cc, and $\beta_0 = 81.2$ GPa. This expression is in good agreement with band-theoretical predictions, as shown in Fig. 1. In order to obtain the correct asymptotic behavior at high densities, this formula was matched to the Thomas-Fermi interpolation formula in the PANDA code for $\rho > 15$ g/cc.

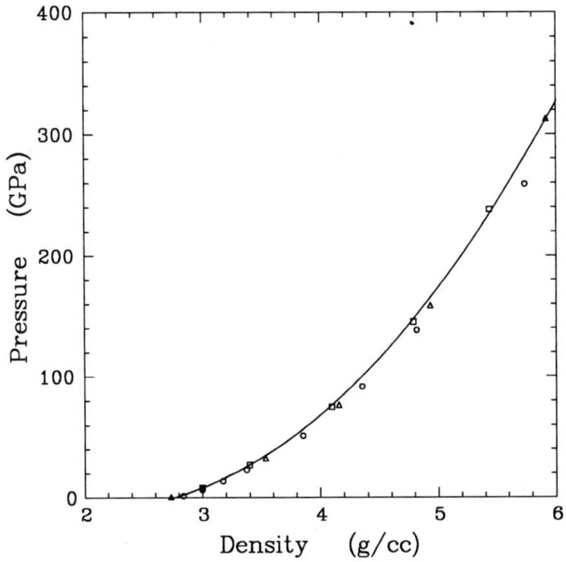

Fig. 1: Zero Kelvin isotherm for aluminum. Band-theoretical calculations are shown as follows: triangles are from Friedli and Ashcroft (1975), circles are from McMahan (1976), and squares are from Pelissier (1981). The solid line was computed from Eq. (2).

Contributions from the lattice vibrations were computed using the Debye model. The pressure expression has the form

$$P_l(\rho, T) = Nk\rho\gamma[9\Theta/8 + 3TD(\Theta/T)], \quad (4)$$

where D is the Debye function, Θ is the Debye temperature, and γ is the Grüneisen function,

$$\gamma = d\ln\Theta/d\ln\rho. \quad (5)$$

Because the calculated melting curve was found to be somewhat sensitive to the lattice term, special attention was paid to the density dependence of γ and Θ. Experimental data show that the Grüneisen parameter drops more rapidly with increasing density than predicted by the simple approximations commonly used. The data of Neal (1976) are shown in Fig. 2, along with results computed from the difference between static isotherm measurements at room temperature (Tups and Syassen, 1984) and at 673 K (Lityagina *et al.*, 1982). Also shown are results calculated from pseudopotential theory and lattice dynamics (Moriarity *et al.*, 1984). It was found that the general features of this data could be represented by the expression

$$\gamma = (\gamma_i - 1.0)(\rho_i/\rho)^{3.5} + 1.0, \quad (6)$$

where γ_i is the Grüneisen parameter at ambient solid density ρ_i. Values of $\rho_i = 2.7$ g/cc, $\gamma_i = 2.16$, and $\Theta_i = 383K$ were obtained from static data (Hultgren *et al.*, 1972; Touloukian, *et al.*, 1975).

FLUID MODEL

Recent developments in the perturbation theory of liquids have led to new models that can be used to calculate the thermodynamic properties in the fluid region. The success of these theories is due to the fact that the structure of a dense fluid is determined principally by repulsive forces, so that an idealized liquid, such as the hard-sphere system, is a useful starting point for expanding the free energy in powers of the intermolecular potential. The terms above first-order can be made quite small when the hard-sphere diameter is chosen in an optimum way. In the variational method, for example, the hard-sphere diameter is determined by minimizing the free energy. These theories have been found to give accurate *a*

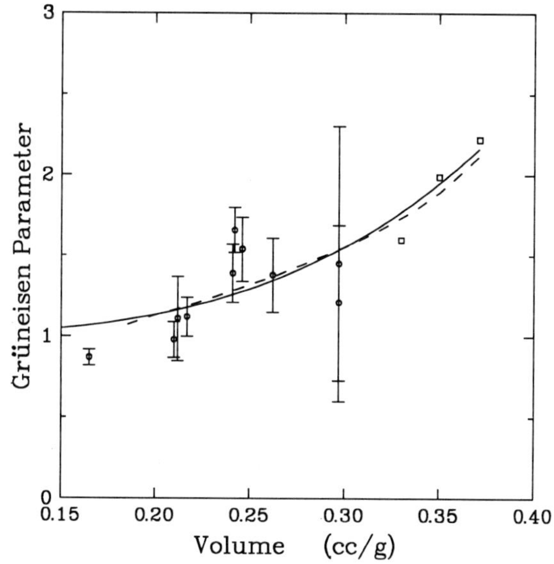

Fig. 2: Grüneisen parameter for aluminum. Circles are data from Neal (1976), squares are computed from static isotherm data (Tups and Syassen, 1984; Lityagina *et al.*, 1982), and dashed curve is from pseudopotential/lattice dynamical theory (Moriarity *et al.*, 1984). The solid curve was computed from Eq. (6).

priori predictions for the properties of liquids and gases from known intermolecular forces, but they have not been widely applied to describe melting and vaporization in metals, for which the intermolecular forces are not well known. Aluminum is normally regarded as one of the simple metals, for which the pseudopotential theory of electronic structure can be used to compute an effective interatomic potential. This approach has been applied successfully to calculate the properties of compressed aluminum (Moriarity, *et al.*, 1984). However, this approximation breaks down for expanded states, especially in the region of the metal-insulator transition.

To circumvent these difficulties, we have developed an approximate form of liquid perturbation theory, called the CRIS model (Kerley, 1983), that does not require explicit knowledge of the intermolecular potentials. In order to calculate the terms in the expansion, the potential energy of a fluid molecule in the cage formed by a particular configuration of its neighbors is related to the the zero-Kelvin isotherm of the solid. The model retains the essential features of a fluid model and is found to give good predictions for the properties of both the vapor and liquid phases, for the coexistence curve, and for the supercritical region of a fluid. Therefore, it is an attractive option for modeling the expanded region of liquid metals.

In this work, the thermodynamic functions for the fluid phase were written as the sum of two terms,

$$F_{fluid}(\rho,T) = F_n(\rho,T) + F_e(\rho,T), \tag{7}$$

where F_n, calculated using the CRIS model, includes the contribution from both the nuclear degrees of freedom and the ground state electronic degrees of freedom (through the intermolecular forces). F_e is the thermal electronic contribution. The CRIS model was employed using the PANDA code (Kerley, 1981). A correction was made to the binding energy, in order to match the experimental melting point, and the zero Kelvin isotherm was modified below solid density, using the Lennard-Jones formula with constants chosen to improve agreement with the liquid density. Values of the parameters used in the calculation, as defined in the PANDA manual (Kerley, 1981), were EFAC = 0.559, RLJ = 2.75, FACLJ = 0.75, and BEXP = 1.3.

THERMAL ELECTRONIC EXCITATION AND IONIZATION

F_e, the contribution from electronic excitation and ionization, becomes more and more important with increasing temperature, especially in the expanded fluid region where our calculations predict that it gives rise to a plasma phase transition. Because none of the existing theories of electronic structure are satisfactory over the entire range of interest, we have used two models to compute this term. The two sets of calculations were joined by constructing a table of the entropy, shown in Fig. 3, and calculating the other EOS variables from thermodynamic relations, thereby ensuring thermodynamic consistency (Kerley, 1981).

The INFERNO model (Liberman, 1979) was used for densities $\rho > 0.64$ g/cc. In this theory, wave functions and energies are computed for both the bound and continuum electronic levels of an atom inside its ion sphere; the ion sphere is embedded in an electron gas that simulates the effects of neighboring atoms. For simplicity, the electronic structure is determined only for an average electronic population, and the entropy is computed from the formula for a noninteracting Fermi gas. The model predicts a decrease in the effective ionization energy as the atom is compressed, so that the bound levels are absorbed into the continuum. Some confidence in the ability of the INFERNO model to describe these effects correctly can be obtained from calculations for xenon (Kerley, 1983). Continuum lowering was found to enhance the importance of the thermal electronic term at high densities and to cause marked softening of the Hugoniot, in good agreement with the experimental data. Band theoretical calculations (Ross, 1968) were found to give much the same result.

In aluminum, a transition to the metallic state is predicted to occur when the atom is compressed to about 0.6 g/cc, where the $3s$ and $3p$ valence shells pass into the continuum. At higher densities, these levels become bands that rapidly broaden with increasing compression, so that the density of states in the continuum takes on more of a free electron character. This broadening of the valence bands causes a rapid decrease in entropy with increasing density at low temperatures that can be seen in Fig. 3.

The average configuration approximation that is used in INFERNO does not give satisfactory results for localized states; for example, it is found to overestimate the entropy in aluminum by a factor of about 2 at low densities and temperatures. Therefore, we have switched to an ionization equilibrium theory (IET) that explicitly sums over all electronic configurations to compute the thermodynamic properties for densities below the transition. The version of IET that is employed here is based upon one-electron orbital energies and average radii for the ground state of the neutral atom that were calculated using the HEX code (Liberman and Cromer, 1971). These data are used to compute ionization potentials and energies for excited states. Corrections for change in ionization state, continuum lowering, and pressure broadening are treated in an approximate way. A detailed description of this model will be given elsewhere.

The IET calculations were found to agree very well with INFERNO results in the vicinity of the metal-nonmetal transition and even to predict the general features of the transition itself. Therefore, it was possible to join the two sets of data together without introducing fictitious behavior. For the sake of smoothness, the IET calculations were used for densities $\rho < 0.25$ g/cc, and the region $0.25 < \rho < 0.64$ was defined by interpolation between the two theories. (Some details of our theoretical predictions for aluminum are dependent upon the interpolation scheme in this region, but the principal features are not.) For densities below the transition, narrowing of the gap for ionization of the valence electrons results in an increase of the entropy with increasing density at low temperatures, as seen in Fig. 3. This effect also results in a negative thermal electronic pressure in this region; the behavior is similar to that in xenon at high pressures.

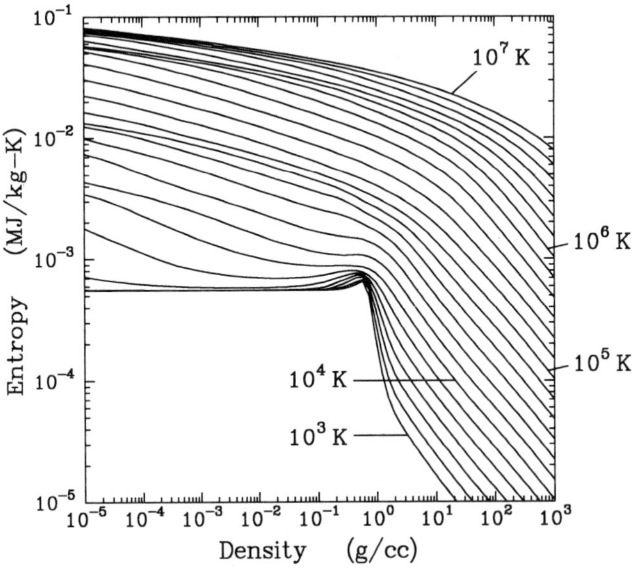

Fig. 3: Calculated thermal electronic contribution to the entropy on isotherms for aluminum. Temperatures range from 10^3 to 10^7 K and are equally spaced in the logarithm.

RESULTS - COMPRESSED REGION

The calculated melting curve and pressure-temperature locus of the Hugoniot curve are displayed in Fig. 4. The onset of melting is predicted to occur at about 125 GPa, where the solid Hugoniot crosses the melting line. This result agrees with the value that has been obtained from measurements of release wave velocities (McQueen *et al.*, 1984). The calculated U_s, U_p curve, shown in Fig. 5, agrees very well with the experimental data. The thermal electronic contribution to the EOS has very little effect on the results in the shock compressed region until very high pressures are reached.

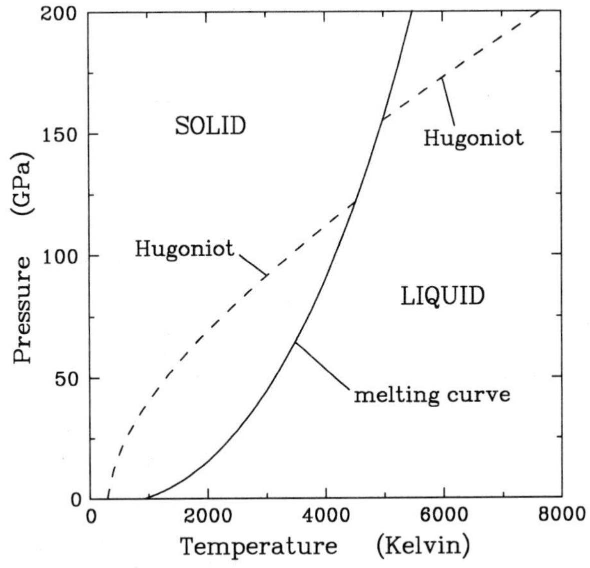

Fig. 4: Shock melting in aluminum. The solid curve is the calculated melting curve, and the dashed curves are the Hugoniot curves for the solid and liquid phases.

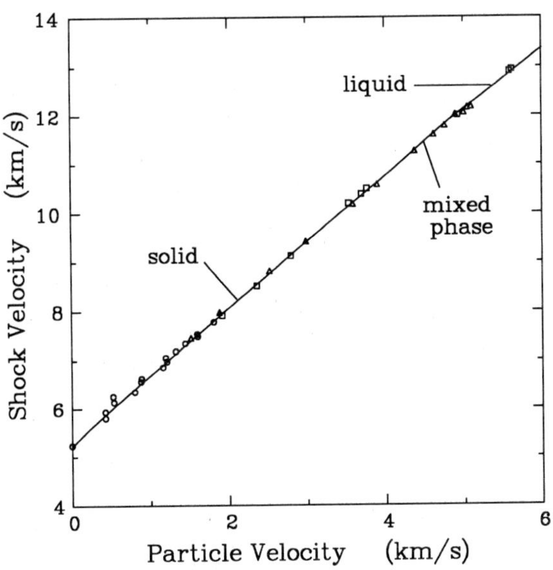

Fig. 5: Hugoniot for aluminum. Circles are from the Los Alamos shock compendium (Marsh, 1980), triangles are data of Mitchell and Nellis (1981), and squares are data of Al'tshuler, *et al.* (1960). The calculated Hugoniot is shown by a solid line.

RESULTS - EXPANDED REGION

The behavior of pressure isotherms for the hot expanded fluid is illustrated in Fig. 6. The dashed lines were computed without including the thermal electronic contribution to the EOS; they exhibit van der Waals loops that are characteristic of simple fluids and lead to the usual description of the vapor-liquid coexistence curve and the critical point. When the thermal electronic term is included, the behavior is markedly different, as shown by the solid lines. In that case, the theory predicts the existence of a nonmetallic liquid phase with a density of about 0.6 g/cc. Three phase transitions are now possible: from the vapor to the nonmetallic liquid, from the nonmetallic to the metallic liquid, and from the vapor directly to the metallic liquid.

For temperatures below 3900 K, we find that a direct transition between the vapor and the liquid metal leads to the lowest free energy, so that the nonmetallic liquid phase is not stable. Therefore, liquid aluminum exhibits the usual type of behavior at low temperatures. The theoretical liquid density and enthalpy as functions of temperature are compared with data from isobaric experiments (Gathers, 1983) in Figs. 7 and 8, respectively. The results that include the thermal electronic term in the calculation (solid lines) are in better agreement with experiment than those that do not (dashed lines).

For temperatures above 3900 K, the nonmetallic liquid phase is found to be stable, and there is a first-order phase transformation between it and the metallic liquid, while vaporization occurs at a lower pressure. This "plasma phase transition" persists to a critical temperature of about 7000 K; the thermal electronic term continues to be very important at higher temperatures, but there is no first-order transformation. The coexistence region between the nonmetallic and metallic liquids is shown by a dotted line in Fig. 6.

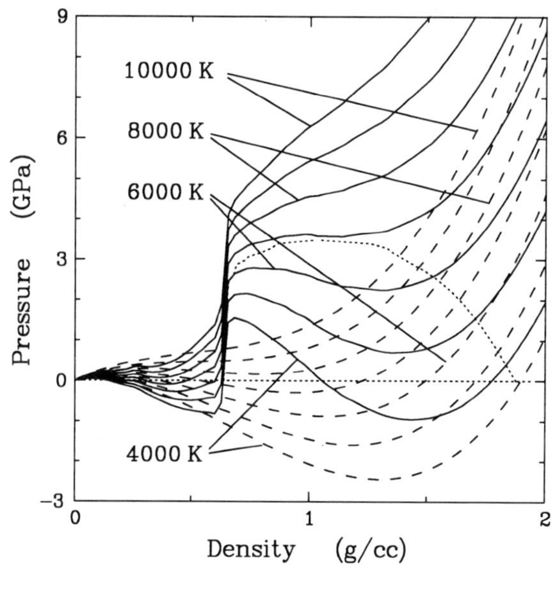

Fig. 6: Isotherms for expanded fluid aluminum. Solid curves show calculations including the thermal electronic term, dashed curves without this term. Temperatures range from 4000 to 10000 K and are equally spaced.

No conclusive experimental evidence for the existence of a plasma phase transition has been reported for aluminum or any other metal. However, the isobaric expansion measurements showed a sudden rapid increase in the sample expansion and resistivity at about 4000 K in aluminum and 4500 K in copper (Gathers, 1983). Similar observations have been made in exploding wire experiments for a number of metals (Lebedev, 1980). The reason for this phenomenon has not been established. We now propose that a plasma phase transition could be the explanation for this behavior. We further note that reflected shock wave measurements on liquid nitrogen are anomalous and suggest the existence of a phase transition in the hot dissociated fluid region (Nellis *et al.*, 1984). Our calculations of the thermal electronic term in monatomic nitrogen exhibit features that are similar to those for aluminum, so a plasma phase transition could be the cause for that behavior as well.

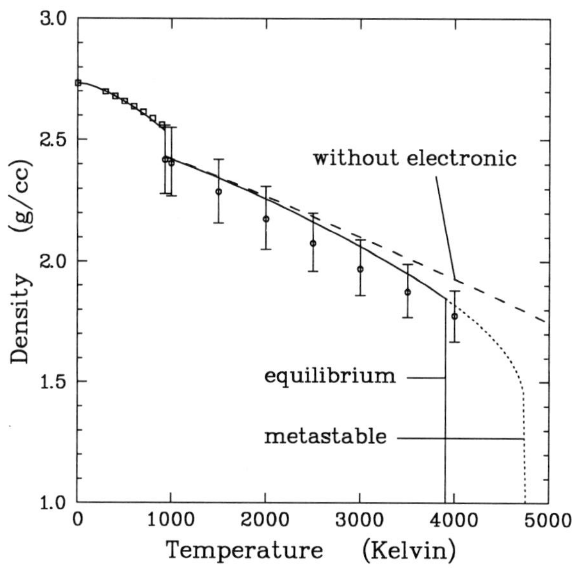

Fig. 7: Thermal expansion of aluminum. Circles are isobaric expansion measurements for the liquid at .3 GPa (Gathers, 1983), and squares are for the solid at zero pressure (Touloukian *et al.*, 1975). The solid curve includes the thermal electronic contribution to the EOS and the dashed curve does not. The dotted line shows the expansion of the metastable metallic fluid.

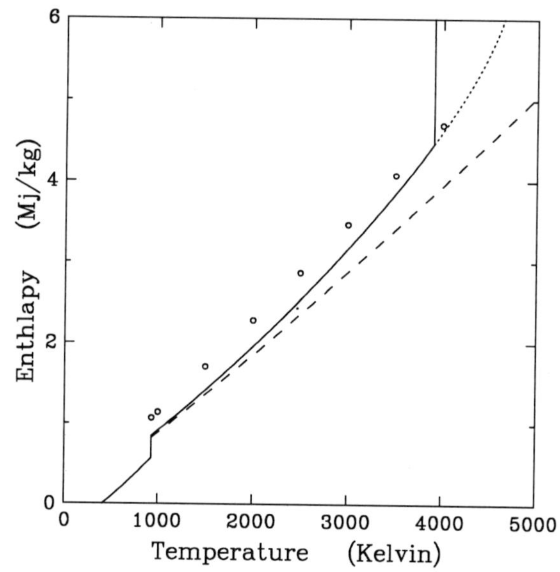

Fig. 8: Enthalpy of expanded liquid aluminum. The curves have the same meaning as in Fig. 7.

The unloading behavior of shocked porous lead and copper has been studied using impedance matching techniques to make simultaneous measurements of pressure and particle velocity on release isentropes (Al'tshuler, *et al.*, 1980). The absence of sharp discontinuities in the experimental curves has been cited as negative evidence for the existence of plasma phase transitions. Although we have not yet made detailed calculations for either of these materials, we believe that the metal-insulator transition should have similar effects on the EOS in both copper and aluminum. Therefore, we have computed release isentropes for shocked porous aluminum with an initial density of 1.6 g/cc. The results are shown in Fig. 9. The unloading curves are found to pass through the metal-nonmetal mixed phase region for shock pressures in the range 50 to 200 GPa. The curves of pressure vs. particle velocity exhibit changes in slope due to the transition but do not have any sharp features that would be easy to identify from impedance matching data. More precise experiments and careful exploration of the EOS surface would be required to prove or disprove the existence of the transition.

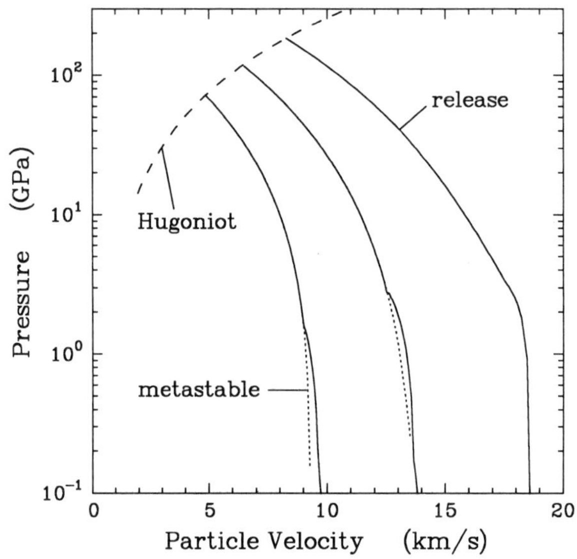

Fig. 9: Release isentropes for shocked porous aluminum. The dashed line is the Hugoniot for aluminum having an initial density of 1.6 g/cc, and the solid lines are release curves from three shocked states. The dotted lines show the expansion of the metastable metallic fluid.

DISCUSSION

Because aluminum is important as a structural material, a good model of the aluminum EOS is needed for calculation of hypervelocity impacts in a variety of applications. The theory described in this paper includes a realistic treatment of all the physical phenomena that can affect the EOS under the conditions of an impact, including melting, vaporization, and thermal electronic excitation and ionization. The most unique feature of our model is the prediction of a metal-nonmetal transition in the hot expanded fluid. The general effects of this transition on impact phenomena should be similar to those for vaporization - absorption of kinetic energy and expansion of the debris cloud. However, the transition pressures are about one order of magnitude higher than those typical for vaporization, and there is less expansion. It is difficult to assess the importance of this transition without performing numerical calculations to study EOS sensitivity.

It should be possible to study the effects of this transition experimentally, provided that the impact velocity is high enough so that the residual temperature upon unloading from the shocked state is above 4000 K. For porous aluminum with a density of 1.6 g/cc, impact velocities of 9 km/s or greater will result in material being shocked and unloaded into the transition region; for normal density aluminum, impact velocities of 13 km/s are required. Hence predictions of hypervelocity impact effects that are based upon extrapolations of experimental results for lower impact velocities may not be reliable.

REFERENCES

This work performed at Sandia National Laboratories supported by the U. S. Department of Energy under contract number DE-AC04-76DP00789.

Al'tshuler, L. V., S. B. Kormer, A. A. Bakanova, and R. F. Trunin (1960). *Sov. Phys. JETP*, **11**, 573-579.

Al'tshuler, L. V., A. V. Bushman, M. V. Zhernokletov, V. N. Zubarev, A. A. Leont'ev, and V. E. Fortov (1980). *Sov. Phys. JETP*, **51**, 373-383.

Brewer, L. (1975). *The Cohesive Energies of the Elements.* Lawrence Berkeley Lab report LBL-3720, unpublished.

Friedli, C., and N. W. Ashcroft (1975). *Phys. Rev. B*, **12**, 5552-5559.

Gathers, G. R. (1983). *Int. J. Thermophys.*, **4**, 209-226.

Hultgren, R., P. D. Desai, D. T. Hawkins, M. Gleiser, K. K. Kelley, and D. D. Wagman (1972). *Selected Values of the Thermodynamic Properties of the Elements.* American Society for Metals, Metals Park, Ohio.

Kerley, G. I. (1981). *User's Manual for PANDA: A Computer Code for Calculating Equations of State.* Los Alamos National Laboratory report LA-8833-M, unpublished.

Kerley, G. I. (1983). In J. M. Hale and G. A. Mansoori, (Eds.), *Molecular Based Study of Fluids*, Am. Chem. Soc., Washington, DC. pp 107-138.

Lebedev, S. V. (1980). *High Temp. (USSR)*, **17**, 222-228.

Liberman, D. A., and D. T. Cromer (1971). *Computer Physics Communications*, **2**, 107-113..

Liberman, D. A. (1979). *Phys. Rev. B*, **20**, 4981-4989.

Lityagina, L. M., Z. V. Malyushitskaya, T. A. Pashkina, and S. S. Kabalkina (1982). *Phys. Status Solidi A*, **69**, K147-150.

McMahan, A. K. (1976). *Bull. APS*, **21**, 1030.

McQueen, R. G., J. N. Fritz, and C. E. Morris (1984). In J. R. Asay, R. A. Graham, and G. K. Straub (Eds.), *Shock Waves in Condensed Matter - 1983*, North-Holland, Amsterdam. p. 95-98.

Moriarty, J. A., D. A. Young, and M. Ross (1984). *Phys Rev. B*, **30**, 578-588.

Neal, T. (1976). *Phys. Rev. B*, **14**, 5172-5181.

Nellis, W. J., N. C. Holmes, A. C. Mitchell, and M. van Thiel (1984). *Phys. Rev. Lett.* **53**, 1661-1664.

Marsh, S. P. (1980). *LASL Shock Hugoniot Data.* University of California, Berkeley. p. 165.

Mitchell, A. C., and W. J. Nellis (1981). *J. Appl. Phys.*, **52**, 3363-3374.

Pelisser, J. L. (1981). *Physica*, **105A**, 607-619.

Rose, J. H., J. R. Smith, F. Guinea, and J. Ferrante (1984). *Phys. Rev. B*, **29**, 2963-2969.

Ross, M. (1968). Phys. Rev., **171**, 777-784.

Touloukian, Y. S., R. K. Kirby, R. E. Taylor, and P. D. Desai (1975). *Thermophysical Properties of Matter.* IFI/Plenum, New York. Vol. 12, pg. 2.

Tups, H., and K. Syassen (1984). *J. Phys. F:Met. Phys.*, **14**, 2753-2767.

Int. J. Impact Engng Vol. 5, pp. 451–461, 1987
Printed in Great Britain

0734-743X/87 $3.00 + 0.00
Pergamon Journals Ltd.

A SIMPLE MODEL FOR THE OPTIMIZATION OF STAND-OFF HYPERVELOCITY PARTICLE SHIELDS

R. Jeffery Lawrence

Strategic Defense Division
Sandia National Laboratories
Albuquerque, New Mexico, USA 87185

ABSTRACT

Impacts at extreme velocities of many tens or even hundreds of kilometers per second are essentially inaccessible to experiment; hence, theory and the resulting scaling laws are the only means to describe the relevant phenomena. One area amenable to this approach involves analyzing the efficacy of stand-off shields for protection against hypervelocity particles. A simple model that allows for optimal design of these shields is developed. Examples and scaling laws that provide estimates for the minimum requirements for such shields are derived from the model.

INTRODUCTION

Currently, there are a number of applications associated with hypervelocity impacts occurring at many tens and even hundreds of kilometers per second. Here we look at a simple model for analyzing one such aspect of the field: the design of stand-off shields for protection against hypervelocity projectiles. The effectiveness of these shields depends on the breakup or vaporization that occurs when hypervelocity particles are intentionally allowed to penetrate thin-walled structures, a concept that has been widely proposed for space-based, dual-walled meteoroid shields. Using both momentum and energy conservation, we describe the relevant phenomenology, and then minimize the mass of the shield configuration required to provide this type of protection. Many of the results are simple enough to allow representation by power-law scaling relations that are applicable over reasonably broad parameter ranges.

PHENOMENOLOGY OF STAND-OFF SHIELDS

Stand-off hypervelocity particle shields function by converting a zero divergence beam of energy (the particle) into a highly divergent fluence (the expanding debris and/or vapor cloud). Thus with the proper separation from the main structure, the loading per unit area can be reduced to a level that can be easily withstood by the underlying layers. This idea has been studied mainly for the protection of space-based assets from meteoroid impact, an application for which Swift, Bamford, and Chen (1982) have provided a detailed analysis that is, in fact, the basis for the present work. We first develop velocity and shield-thickness criteria for debris or vapor generation on penetration, and then examine the subsequent motion of the resulting cloud. The details of this motion then lead to expressions that can be used to minimize the mass needed for protection against given hypervelocity particle threats.

Criteria for Debris Formation

When hypervelocity particles impact and penetrate relatively thin material layers, most observers, e.g. Gehring (1970), have noted that the resulting cloud of debris takes the form of an expanding shell, as suggested in Fig.1. Although it is not necessary to assume this shape, the analysis is more convenient if we follow this lead. Adopting the geometry and notation shown in the figure, we first observe that the amount of mass removed from the shield, m_s, in the penetration process is given by

$$m_s = \pi \left(\frac{D_s}{2}\right)^2 t_s \rho_s ,$$

where ρ_s is the density of the shield material. For an incident particle density, ρ_p, we have, similarly,

$$m_p = \frac{\pi}{6} D_p^3 \rho_p . \qquad (1)$$

We also assume that $D_s = \beta D_p$, where β is the ratio of the diameter of the hole created in the shield to the incident particle diameter and is an undetermined constant of order unity. We can now define a nondimensional shield thickness or mass ratio, M, as

$$M = \frac{m_s}{m_p} = \frac{3\beta^2}{2} \frac{\rho_s t_s}{\rho_p D_p} . \tag{2}$$

The other principal independent variable of the problem is the velocity of the incident particle, u_p. It will be convenient to nondimensionalize this with respect to the average energy per unit mass, $\mathcal{E}_b$, needed to break up or vaporize the materials involved in the penetration process. Hence,

$$V_p = \frac{u_p^2}{2\mathcal{E}_b} , \tag{3}$$

where V_p is actually a nondimensional kinetic energy, although we will continue to refer to it as a velocity. We should emphasize that although we implicitly assume that $\mathcal{E}_b$ is a constant, it will vary from a fragmentation energy for impacts at low velocities, through a melt energy, to a true vaporization energy only for impacts at extremely high velocities. It will likely also vary for the different mass elements involved in a single interaction. Thus, in the present context $\mathcal{E}_b$ should be considered not only an average quantity, but one that is integrated over the space and time involved in the penetration and debris generation process.

The velocity, u_c, of the center of mass of the debris or vapor cloud then follows from conservation of momentum during the penetration process,

$$u_c = \frac{u_p}{1 + M} . \tag{4}$$

By subtracting the kinetic energy associated with the motion of this center of mass, $(m_s + m_p) u_c^2/2$, from the kinetic energy of the incident particle, $m_p u_p^2/2$, we arrive at the excess energy, E_e, that is available both for breaking up or vaporizing m_p and m_s, and for expansion of the debris cloud,

$$E_e = \frac{m_p u_p^2}{2} \frac{M}{1 + M} . \tag{5}$$

Now, the total amount of energy necessary to break up or vaporize m_p and m_s is simply $\mathcal{E}_b (m_p + m_s)$, and E_e must be at least this large. Using Eqs. (3) and (5), we define f as the ratio between E_e and $\mathcal{E}_b (m_p + m_s)$,

$$f = V_p \frac{M}{(1 + M)^2} . \tag{6}$$

The necessary condition for breakup or vaporization to occur can then be expressed as $f \geq 1$. From Eq. (6), the bounding condition, $f = 1$, leads to

$$M = \frac{1}{2} \{V_p - 2 \pm [V_p(V_p - 4)]^{1/2}\} . \tag{7}$$

Since physically meaningful situations require both M and V_p to be positive, the restriction $V_p \geq 4$ is needed to ensure real values for M. Of the total energy associated with this lower limit, which occurs at $M = 1$ or $m_s = m_p$, only one-half is excess and used to break up both m_p and m_s. The remaining half appears as the kinetic energy associated with the debris center of mass, moving at a velocity of $u_c = u_p/2$, as predicted by Eq. (4). In this limiting case the expansion velocity, u_e, of the debris about its center of mass will be zero, as will be seen in the next section.

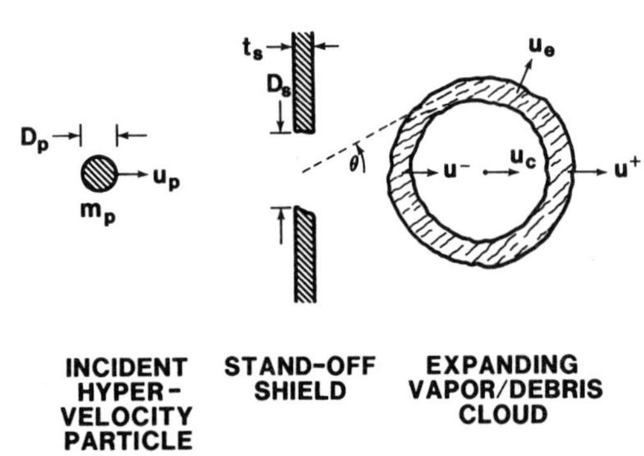

Fig. 1. Schematic for stand-off shield penetration and subsequent debris cloud motion.

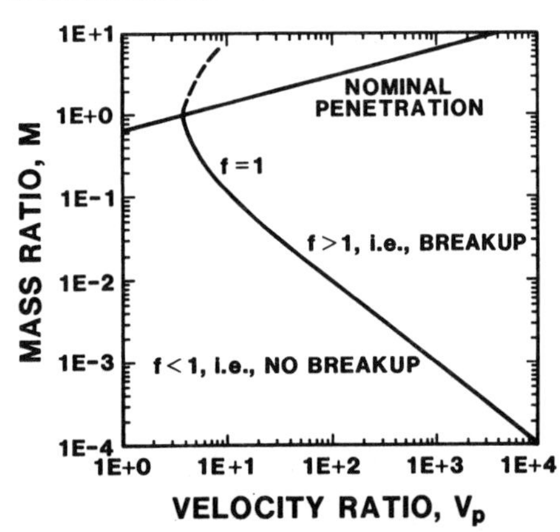

Fig. 2. Criteria for breakup or vaporization on penetration of stand-off shield by hypervelocity particle.

When $V_p > 4$, M is double valued, with the negative sign in Eq. (7) leading to a reasonable lower limit for M. However, the positive sign in Eq. (7) represents shield thicknesses such that actual penetration is not a realistic assumption. To establish a better upper limit on M, we use the hypervelocity penetration model suggested by Swift (1982), in which he notes that the relevant craters are hemispherical in shape and have volumes that correlate linearly with the kinetic energy of the impacting particle. Using the previously defined $\mathcal{E}_b$ as the constant of proportionality for crater formation, the penetration depth, x_{pen}, in a semi-infinite medium becomes

$$x_{pen} = (8\mathcal{E}_b)^{-1/3} D_p u_p^{2/3} , \tag{8}$$

where D_p may have an exponent slightly greater than one, and actual penetration through a material layer probably occurs at a thickness greater than x_{pen} by a factor of ~1.5. Including this additional multiplier and identifying the penetration thickness with the shield thickness, t_s, we can express an appropriate penetration limit in terms of M and V_p as

$$M_{pen} = \frac{9\beta^2}{4} \frac{\rho_s}{\rho_p} \left(\frac{V_p}{4}\right)^{1/3} . \tag{9}$$

For the approximate penetration threshold we require here, the diameter ratio β can be chosen as one, and for convenience we can pick the density ratio $\rho_s/\rho_p = 4/9$, a not unreasonable value. The nominal curve represented by Eq. (9) will then pass through the point $V_p = 4$, $M = 1$, at the knee of the other limit given by Eq. (7).

Combining the lower limit represented by Eq. (7) with the approximate penetration limit obtained from Eq. (9), we can plot the region in M–V_p space over which breakup or vaporization is possible, as shown in Fig. 2. These curves do not suggest precise limits, but rather should be considered relatively broad bands that describe criteria for the process to occur. Further, although we have termed these conditions necessary, for some situations they may not be sufficient. In any event, since normal operation will occur well inside the boundaries of the applicable region, this last observation is somewhat academic.

Debris-Cloud Motion

After penetration of the stand-off shield, we assume that the cloud of debris and/or vapor expands as a hollow shell. The kinetic energy associated with this expansion is obtained by subtracting the breakup or vaporization energy, $\mathcal{E}_b (m_p + m_s)$, from the excess energy, E_e, given by Eq. (5). The expansion velocity, u_e, is then

$$u_e = \left(u_p^2 \frac{M}{(1 + M)^2} - 2\mathcal{E}_b\right)^{1/2} , \tag{10}$$

where we can nondimensionalize with respect to the incident particle velocity, u_p,

$$\frac{u_e}{u_p} = \left(\frac{M}{(1 + M)^2} - \frac{1}{V_p}\right)^{1/2} . \tag{11}$$

As suggested earlier, these expressions lead to $u_e = 0$ at the limiting values of $V_p = 4$ and $M = 1$. When $V_p > 4$ we can use Eq. (11) to plot the debris expansion velocity as a function of the mass ratio, M, with the velocity ratio, V_p, as a parameter, as has been done in Fig. 3. It is evident that u_e has a broad maximum at $M = 1$, where the shield thickness is of the same order as the impacting particle diameter. Here also, from Eq. (4), the center-of-mass velocity is $u_c = u_p/2$, and in the limit of $V_p = \infty$, $u_e = u_c = u_p/2$. We will show later, however, that optimum performance of the shield occurs for mass ratios much smaller than this, and hence in practice the actual debris velocities will be quite different from these values. Note also from this figure that for each of the curves, the threshold for M, where $u_e = 0$, is exactly that value given by Eq. (7) and plotted in Fig. 2.

The divergence associated with the debris cloud can be determined readily from the geometry shown in Fig. 1. The half-angle for debris expansion, θ, is simply

$$\sin\theta = \frac{u_e}{u_c} , \tag{12}$$

or with the aid of Eqs. (3), (4), and (10),

$$\theta = \sin^{-1}\left(M - \frac{(1 + M)^2}{V_p}\right)^{1/2} , \tag{13}$$

where the argument in the latter must be between zero and one. The lower limit is equivalent to the debris formation criterion, $f \geq 1$, whereas the upper limit is $u_e \leq u_c$ and can be expressed in terms of a limiting mass ratio, $M \leq M_\ell$, where

$$M_\ell = \frac{1}{2}\{V_p - 2 \pm [V_p(V_p - 8)]^{1/2}\} . \tag{14}$$

This equation is similar in form to Eq. (7), but in this case with a restriction on the velocity ratio, $V_p \geq 8$, which occurs at $M_\ell = 3$. The positive sign in Eq. (14) leads to large values of M that are inconsistent with penetration and thus need not be considered further. For the negative sign, M_ℓ approaches one as V_p becomes large, and thus yields the maximum value of u_e as shown in Fig. 3. At $M = M_\ell$ we have $u_e = u_c$, and the rear portion of the debris cloud remains essentially stationary and in contact with the shield. When $M > M_\ell$, $u_e > u_c$ and Eq. (12) no longer applies; in this case the net expansion velocity of the debris is outward in all directions, even back onto the shield and through the penetration hole. In Fig. 4 we use Eq. (13) to plot the divergence half-angle as a function of M with V_p as a parameter, as before. The anomalous region that is evident for $M > 1$ and $V_p \leq 8$ occurs under conditions for which penetration is unlikely and thus is of little concern. In any case, stand-off shield operation for mass ratios near one will be very inefficient at best, and further, impact velocities represented by $V_p < 8$ border the lower limit of applicability for this simple approach to hypervelocity interactions.

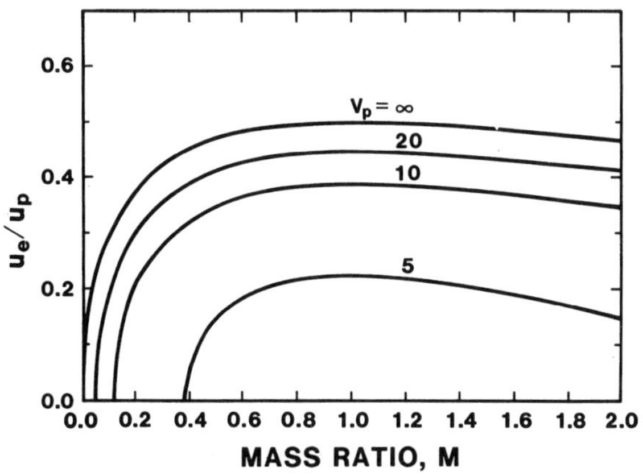

Fig. 3. Debris expansion velocity as a function of M with V_p as a parameter.

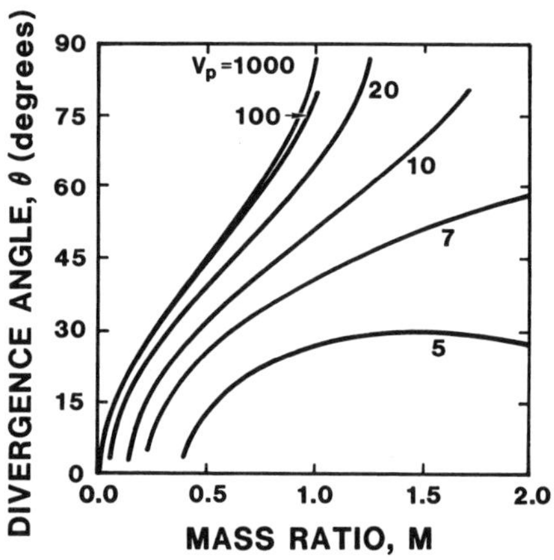

Fig. 4. Debris expansion angle as a function of M with V_p as a parameter.

In general, the peak loads generated on the underlying structure by the debris will occur on the axis of symmetry. With this in mind, we calculate the on-axis debris velocities for both downstream and upstream portions of the expanding cloud. We have $u_{axial} = u_c \pm u_e$, where the plus sign refers to the downstream value, u^+, and the minus sign denotes the upstream component, u^-, as indicated in Fig. 1. Using Eqs. (4) and (10), and nondimensionalizing with respect to u_p, this becomes

$$\frac{u_{axial}}{u_p} = \frac{1}{1 + M} \pm \left(\frac{M}{(1 + M)^2} - \frac{1}{V_p}\right)^{1/2}, \tag{15}$$

an expression that we have plotted in Fig. 5, again with V_p as a parameter. The upper branch is in each case u^+, while the lower branch is u^-, and the dashed line separating the two represents the threshold locus for debris or vapor formation. Although u^- has no extreme value, there is a maximum for u^+, which is evident in the figure. It is interesting to note that in the limiting case of $V_p = \infty$, it is straightforward to show that $u^+_{max} = 1.21\, u_p$ at a mass ratio of $M = 0.172$, a result that was first reported by Swift, Bamford, and Chen (1982). In general, we can show that this maximum occurs at

$$M = \frac{3 - 4/V_p - 2[2(1 - 4/V_p)]^{1/2}}{(1+4/V_p)},$$

and yields a value for u^+_{max} that can be found by substituting into Eq. (15). As suggested earlier, however, normal operating regimes will generally occur for values of M much lower than at these maxima.

It remains to combine these velocities with the divergence of the debris cloud to calculate the loading on the underlying structure. These loads will be expressed in terms of the specific impulse, I, or momentum per unit area. The average mass per unit area in the cloud is simply $(m_s + m_p)/4\pi r^2$, where the cloud radius is $r = u_e \tau$, and τ is the time after shield penetration. At impact on the structure the time is given by $\tau^\pm = x/u^\pm$ for the downstream and upstream portions respectively. Here x is the distance of stand-off between the shield and structure. Combining, the specific impulse becomes

$$I^\pm = u^\pm \frac{m_s + m_p}{4\pi(u_e\, \tau^\pm)^2},$$

which, with the use of Eqs. (11) and (15) and some manipulation, becomes

$$I^\pm = \frac{m_p\, u_p}{4\pi x^2} \frac{(1 \pm U^{1/2})^3}{U}. \tag{16}$$

To simplify this result we have made the substitution

$$U = M - \frac{(1+M)^2}{V_p}, \tag{17}$$

where U is the same as the argument in the expression for the debris cloud divergence angle, Eq. (13), and as before must generally be in the range $0 < U \leq 1$. This last is strictly true only for I^-, and in this case leads to the same limiting value for M as given in Eq. (14). Noting that the specific impulse delivered to the structure scales with $1/x^2$, we can rewrite Eq. (16) in the nondimensional form

$$\frac{(I^\pm)x^2}{m_p\, u_p} = \frac{(1 \pm U^{1/2})^3}{4\pi U},$$

which we then plot in Fig. 6. The upper set of curves describes the variation of I^+, while the lower set refers to I^-. Note that the minimum delivered loads, as well as the increases shown on the right, occur for $M \geq 1$. Again, this is a region that, when the design is optimized, will be inappropriate for stand-off shield operation.

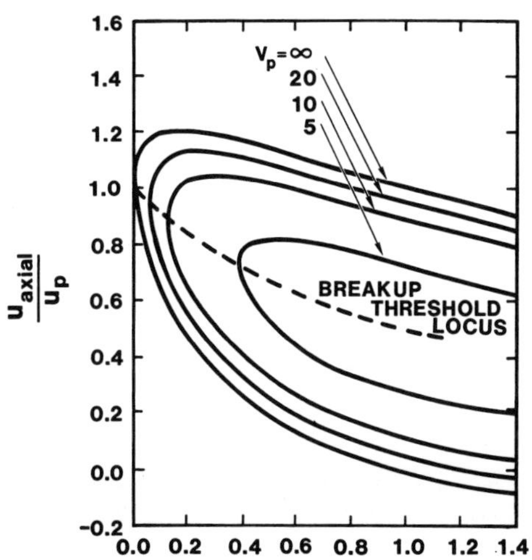

Fig. 5. Axial debris velocity as a function of M with V_p as a parameter.

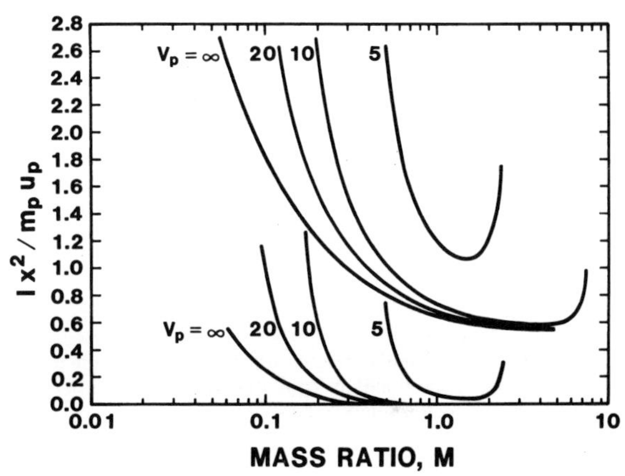

Fig. 6. Downstream and upstream specific impulse as a function of M with V_p as a parameter.

When U is in the range $0 < U \leq 1$, we can be conservative in calculating the impulsive loading on the structure by employing both the downstream and upstream portions of the specific impulse, $I = (I^+) + (I^-)$. Using Eq. (16), we have the simple result

$$I = \frac{m_p\, u_p}{2\pi x^2}\left(\frac{1}{U} + 3\right). \tag{18}$$

From Fig. 6 we see that the upstream component, I^-, represents a relatively small portion of the total load, and hence this expression for I will be both conservative and, in comparison with Eq. (16), somewhat easier to use.

Optimization for Minimum Mass

The principal figure of merit for a stand-off hypervelocity particle shield is undoubtedly its total mass. Based on the expressions derived in the last section we can calculate the loads generated by the shield interaction and subsequently delivered to the underlying structure. The thicknesses and resulting masses that will be needed to withstand a given hypervelocity particle threat then follow. Minimization with respect to the total mass then yields an optimum design for the configuration.

We first determine the optimum parameters for planar or slab geometry, in which case we can work directly in terms of areal density. Recalling from Eq. (2) that the mass ratio was defined as $M = m_s/m_p$, the mass per unit area of the shield, M_S, is given by

$$M_S = \rho_s\, t_s = \frac{2}{3}\rho_p\, D_p\,\frac{M}{\beta^2}, \tag{19}$$

where we also note from Eq. (1) that $D_p = (6m_p/\pi\rho_p)^{1/3}$. The areal density of the main structural layer can be represented similarly as $M_L = \rho_L t_L$. Since at lower levels of loading penetration can usually be correlated with momentum delivery, the thickness of this layer, t_L, is I/i_p, where i_p is the specific impulse per unit thickness necessary to penetrate a given material layer. The mass per unit area of this layer is thus

$$M_L = \frac{\rho_L}{i_p} I . \tag{20}$$

In addition to resisting penetration, this layer must be capable of absorbing the kinetic energy deposited by the debris cloud. As suggested by Swift (1982), this can be accomplished with a crushable foam or honeycomb layer mounted directly beneath the structural layer. This additional energy-absorbing component can be characterized in terms of an energy absorption coefficient, $\mathcal{E}_A$, which is defined as the energy per unit mass that can be dissipated through the crushing of such a material. Since the kinetic energy per unit area of the structural layer is $E_L = I^2/2M_L$, the areal density of the additional energy-absorbing component can be written

$$M_A = \frac{E_L}{\mathcal{E}_A} = \frac{i_p}{2\mathcal{E}_A \rho_L} I . \tag{21}$$

The total mass per unit area that we wish to minimize is then the sum of these three terms,

$$M_{tot} = M_S + M_L + M_A = \frac{2}{3}\frac{\rho_p D_p M}{\beta^2} + K I ,$$

where

$$K = \frac{\rho_L}{i_p} + \frac{i_p}{2\mathcal{E}_A \rho_L} \tag{22}$$

is a reciprocal velocity that is a constant characteristic of the materials used in the underlying structural configuration. Substituting the specific impulse, I, from Eq. (18), the quantity to be minimized becomes

$$M_{tot} = \left(\frac{2}{3}\right)\frac{\rho_p D_p M}{\beta^2} + K\left(\frac{m_p u_p}{2\pi x^2}\right)\left(\frac{1}{U} + 3\right) . \tag{23}$$

In this one-dimensional planar configuration, the single parameter that does not yield a monotonic variation of the mass is the thickness of the shield, or equivalently, M. Hence to optimize the design, we take the derivative of M_{tot} with respect to M and set the result equal to zero. This leads to

$$\frac{4\pi}{3\beta^2}\frac{\rho_p D_p}{m_p u_p}\frac{x^2}{K} = \frac{1}{U^2}\left(1 - \frac{2(1+M)}{V_p}\right) , \tag{24}$$

where we recall from before,

$$0 < U = M - \frac{(1+M)^2}{V_p} \leq 1 . \tag{25}$$

If we define

$$G = \frac{4\pi}{3\beta^2}\frac{\rho_p D_p}{K m_p u_p} , \qquad g = \frac{1}{U^2}\left(1 - \frac{2(1+M)}{V_p}\right) , \tag{26}$$

Eq. (24) becomes simply $Gx^2 = g$, where g is a function of M and V_p only, through Eq. (25). The function G, a reciprocal area, depends on all the other parameters of the problem.

Since we will generally be using u_p or $V_p = u_p^2/2\mathcal{E}_b$ as a parameter, the easiest approach to solving Eq. (24) for the optimum value of M is to first plot $g(M, V_p)$ as a function of M, as we have done in Fig. 7 for several selected values of V_p. Using the first of Eq. (26) we can then calculate the function $g = Gx^2$ for the specific parameters of interest, which with Fig. 7 and the proper value of V_p, yields the desired result. Both the optimized shield mass, M_S, and the minimum total mass of the structure and shield, M_{tot}, can then be obtained directly from Eqs. (19) and (23), respectively. We can repeat this process and plot the results as functions of any of the independent variables that we wish. For a given hypervelocity particle threat and a generic structural design, the stand-off distance, x, will probably be the most logical choice. In the next section we will look at two specific examples that will illustrate this procedure.

In most real applications the stand-off shield will be used to protect a three-dimensional object, a situation in which the mass depends on the geometry as well as on the one-dimensional interaction phenomena. In this case we have the more complicated task of minimizing the actual total mass rather than just the areal density. In addressing this problem we first assume that the geometry is spherical, that the device to be shielded has a radius, R_o, and that shielding will be required

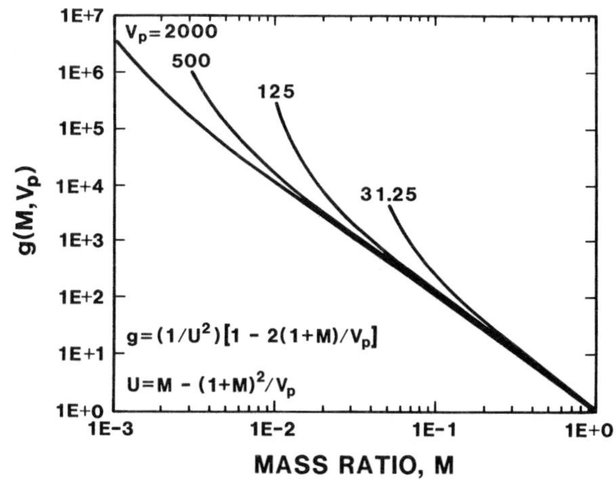

Fig. 7. The function $g(M, V_p)$ from Eq. (26).

over the full 4π-sr solid angle surrounding the target. Using an asterisk (*) to indicate total mass as opposed to mass per unit area, the three mass components become, for the shield,

$$M_S^* = 4\pi(R_o + x)^2 M_S , \tag{27}$$

for the main structural layer,

$$M_L^* = 4\pi R_o^2 M_L , \tag{28}$$

and for the required energy-absorbing component,

$$M_A^* = 4\pi R_o^2 M_A . \tag{29}$$

With Eqs. (19), (20), (21), and Eq. (18) for the specific impulse, we have, analogous to Eq. (23),

$$M^* = \frac{8\pi}{3} \frac{\rho_p D_p M}{\beta^2} (R_o + x)^2 + 2R_o^2 K \frac{m_p u_p}{x^2} \left(\frac{1}{U} + 3\right) , \tag{30}$$

where, as above, both Eqs. (22) and (25) for K and U apply.

Because in this situation M_S^* varies with the stand-off, x, we must minimize M^* with respect to both x and M. Taking the derivative of Eq. (30) with respect to M and setting the result equal to zero yields,

$$\frac{4\pi}{3\beta^2} \left(\frac{\rho_p D_p}{m_p u_p}\right) \frac{(R_o + x)^2 x^2}{K R_o^2} = \left(\frac{1}{U^2}\right)\left(1 - \frac{2(1 + M)}{V_p}\right) . \tag{31}$$

Similarly, setting the derivative of M^* with respect to x equal to zero gives

$$\frac{4\pi}{3\beta^2} \left(\frac{\rho_p D_p}{m_p u_p}\right) \frac{(R_o + x)\, x^3}{K R_o^2} = \frac{1}{M}\left(\frac{1}{U} + 3\right) , \tag{32}$$

where Eqs. (31) and (32) together yield the optimum values for both M and x. Using the substitutions given in Eq. (26) along with

$$h = \frac{1}{M}\left(\frac{1}{U} + 3\right) , \qquad X = \frac{x}{R_o} , \tag{33}$$

where X is the stand-off distance nondimensionalized by the target radius, R_o , we can simplify these two simultaneous equations for M and x, yielding

$$X = \frac{1}{g/h - 1} , \tag{34}$$

$$R_o^2\,G = \frac{(g - h)^4}{g\,h^2}\;. \tag{35}$$

Although these equations appear relatively simple, their direct algebraic solution is complex. However, in the form of Eqs. (34) and (35) there is a graphical procedure that allows us to determine the desired answers in a straightforward manner. We first use Eq. (34) to plot the nondimensional stand-off, X, as a function of the mass ratio, M, as in Fig. 8. In Fig. 9 we use Eq. (35) to plot the dimensionless parameter $R_o^2\,G$, also as a function of M. In both cases we use a reasonable range of values for V_p as a parameter. From Eq. (34) we observe that for the physically meaningful condition of positive stand-off, we require that $g/h > 1$. As a consequence, there are only certain ranges of M and V_p that yield useful results, and it is these ranges that we have shown in these two figures.

In practice we obtain the desired solution by first using Eqs. (22) and the first of (26) to calculate the parameter $R_o^2\,G$ for the conditions of interest. With the appropriate value for V_p, Fig. 9 yields the optimum mass ratio, M. Fig. 8 then gives the associated optimum stand-off, X. The dimensional values for the various masses then follow from Eqs. (27) through (30), with Eqs. (17) and the second of (33) providing the needed subsidiary relations. Repeating the process varying one or more of the independent parameters allows different types of solutions to be generated. Plotting optimum masses and stand-off distances as functions of target radius, R_o, seems to be one of the most useful types of output. Two explicit examples will be given in the next section.

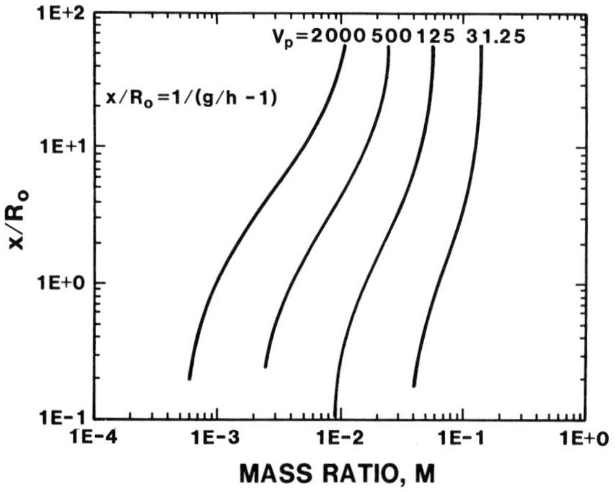

Fig. 8. The dimensionless stand-off distance as represented by Eq. (34) for the parameter range of practical interest.

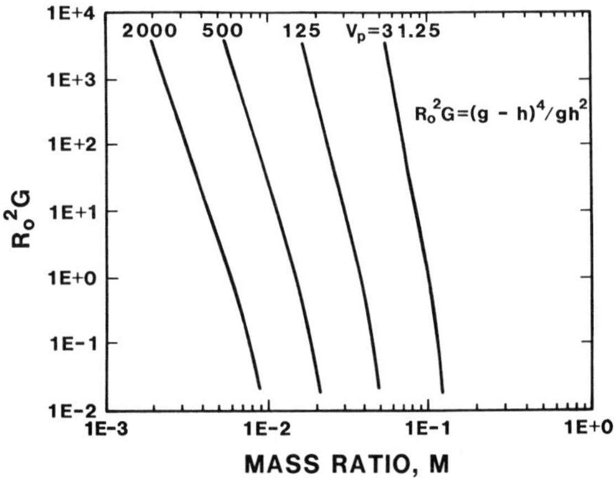

Fig. 9. The function represented by Eq. (35) for the parameter range of practical interest.

SPECIFIC EXAMPLES

In using these models to actually calculate minimum-mass stand-off shield configurations, a number of material properties and other parameters need to be determined. Typical values for the required material properties are listed in Table 1. We should emphasize that each of these values is generic, and in at least one case (i.e. $\mathcal{E}_b$), is a weighted average over different materials. However, they are felt to reasonably represent the phenomena that we are describing. Typical operational parameters that are well into the extremely high velocity regime, and that represent situations that we may wish to investigate, are shown in Table 2.

We first look at one-dimensional planar geometry and determine the minimum areal density of the shield configuration as a function of the stand-off distance. Here we will examine 0.2-g particles impacting at velocities of 100 and 200 km/s. Using the first of Eq. (26) and assuming that the ratio between the diameter of the hole produced in the shield and the diameter of the incident particle is one (i.e. $\beta = 1$), the problem-dependent parameter, G, becomes 444 m^{-2} and 222 m^{-2} for these two cases respectively. We can now follow the procedure outlined in the previous section directly. The major results are plotted in Figs. 10 and 11, which incorporate the masses per unit area of both the total configuration, M_{tot}, and the stand-off shield, M_S, for the two cases considered.

Since the optimum shield design was based on a minimum mass for the entire configuration, it is of interest to compare the resulting values of M_{tot} with the areal densities that would be required to prevent penetration with single layers. An approximate areal density, M_{sp}, for a single-layer shield can be determined by using the multiplier of ~1.5 suggested earlier in connection with Eq. (8), leading to a value of

$$M_{sp} = \frac{3}{4}\,D_p\,\rho_s\left(\frac{u_p^2}{\mathcal{E}}\right)^{1/3}.$$

TABLE 1 Typical Material Properties

Average Vaporization Energy, $\mathcal{E}_b$	10 MJ/kg
Impulse Penetration Coefficient [for metal], i_p	0.4 MPa-s/m
Energy Absorption Coefficient [for metallic honeycomb], $\mathcal{E}_A$	5 kJ/kg
Projectile Density, ρ_p	15 Mg/m^3
Shield and Structural Layer Densities, ρ_s and ρ_L	3 Mg/m^3
Structural Material Constant [from Eq. (22)], K	0.021 s/m

TABLE 2 Operational Parameters Used in Examples

Projectile Mass, m_p	0.2, 1.0 g
Projectile Velocity, u_p	50, 100, 200 km/s
Dimensionless Projectile Velocity [from Eq. (3)], V_p	125, 500, 2000

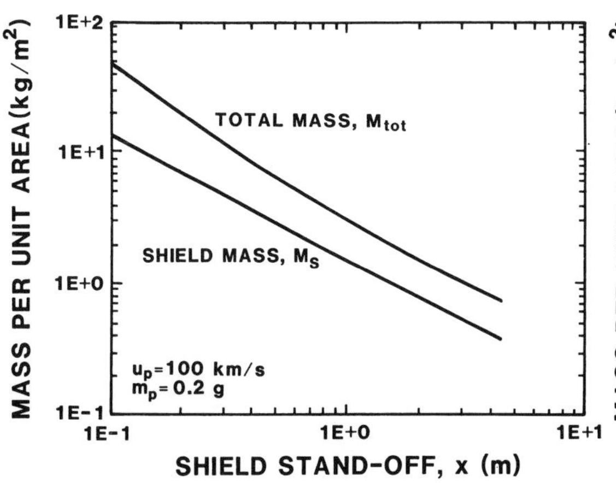

Fig. 10. Minimum areal density stand-off shield configuration for a 0.2-g, 100-km/s incident particle.

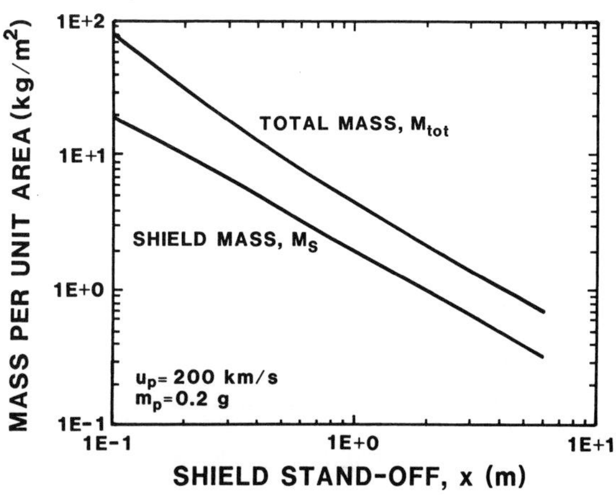

Fig. 11. Minimum areal density stand-off shield configuration for a 0.2-g, 200-km/s incident particle.

A reasonable range of uncertainty follows from letting $\mathcal{E}$ vary from a melt energy to a vaporization energy. For the present example a 100-km/s impact velocity gives a single-shield areal density of from 70 to 140 kg/m^2. From Fig. 10 we can see that this is significantly greater than can be achieved with a stand-off design, where even for the smallest separation shown, 0.1 m, only ~50 kg/m^2 are required. Similarly, 200 km/s leads to a single-shield mass per unit area of 100—200 kg/m^2, which again is considerably larger than the ~80-kg/m^2 M_{tot} given in Fig. 11 for a 0.1-m stand-off. As is evident from both figures, larger stand-off distances lead to progressively more favorable comparisons with single-layer shields.

A detailed examination of the results shows that the shield masses fall off almost linearly with the stand-off distance, while the total configuration masses decrease somewhat more rapidly. Over the range of parameters examined here, very rough averages suggest that

$$M_S \sim x^{-1}, \qquad M_{tot} \sim x^{-1.2},$$

where the latter exponent has a slightly larger variance associated with it. These power-law scaling relations should also apply reasonably well in other parameter regimes as long as they are properly normalized and the range of extrapolation is not too great. We also see that for both velocities the shield mass varies from one-fourth to one-half of the total configuration mass as the stand-off increases over the range shown. Finally, we note that for these conditions, stand-off distances of several meters lead to optimum mass ratios in the range $0.01 < M < 0.03$, which are equivalent to shield thicknesses of from 1/10 to 1/30 of the incident particle diameter. This is considerably thinner than previous estimates have suggested (e.g. Swift, Bamford, and Chen, 1982; Gehring, 1970).

For the more realistic application associated with spherical geometry, we choose examples involving, first, a 0.2-g particle impacting at 100 km/s, and second, a 1-g particle with a velocity of 50 km/s. For these two cases (again assuming $\beta = 1$) the values for the problem-dependent parameter, G, become 444 m^{-2} as before and 355 m^{-2}, respectively. In these situations, where to optimize the configuration we must determine both the mass ratio, M, and the stand-off distance, x, we use the unshielded target radius, R_o, as the independent variable. For the first example (0.2 g at 100 km/s) we plot the masses, M* and M_S^*, as functions of R_o in Fig. 12, and show the associated optimum stand-off, x, in Fig. 13. Similarly, Figs. 14 and 15 show the results from the second example (1 g at 50 km/s).

As with the previous examples, there are several significant observations that we can make from these results. First, for both of the present cases the ratio of the shield mass to the total mass remains essentially constant at $M_S^*/M^* \approx 2/3$ for virtually all of the calculations performed. Second, the thickness of the shield relative to the incident particle diameter, t_s/D_p, ranges from 1/20 to 1/75 for the 0.2-g, 100-km/s case, and from 1/10 to 1/20 for the 1-g, 50-km/s encounter. These ratios follow from optimum values of M that vary from 0.01 to 0.004, and thus substantiate the statement made earlier that the most efficient stand-off shield operation occurs for $M \ll 1$.

Most of the data plotted in these figures appear to obey power-law scaling relations similar to those given with the previous examples. That is,

$$M^* \sim R_o^i \ , \qquad M_S^* \sim R_o^j \ , \qquad x \sim R_o^k \ .$$

For both cases i ranges between 1.2 and 1.3, j is just a little greater than 1.3, and k departs very little from 0.6. Again, with proper normalization and reasonable ranges of extrapolation, these scaling laws should prove useful in many other situations of interest.

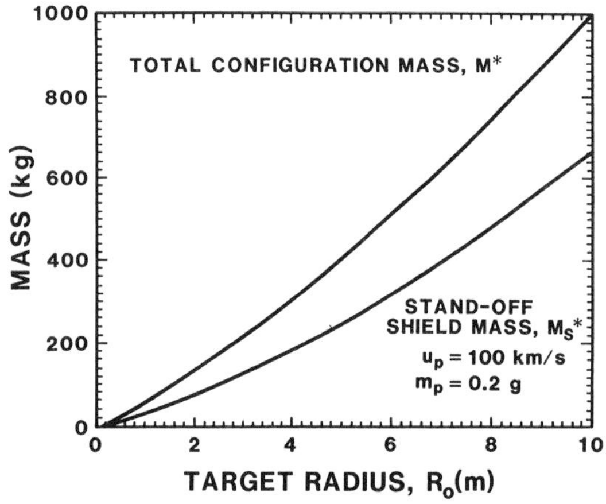

Fig. 12. Minimum mass configuration as a function of target radius for a 0.2-g, 100-km/s incident particle.

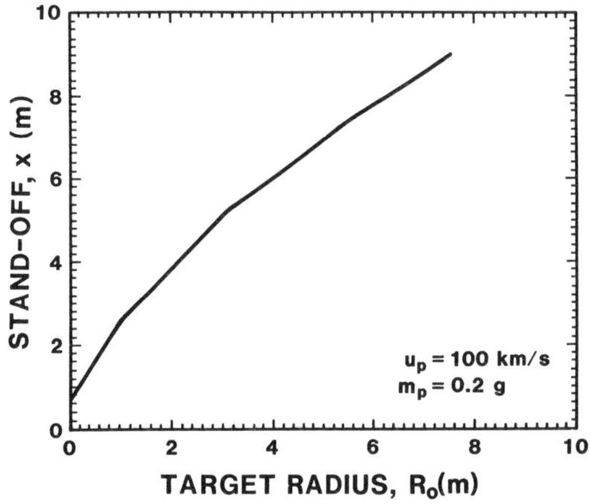

Fig. 13. Optimum stand-off as a function of target radius for a 0.2-g, 100-km/s incident particle.

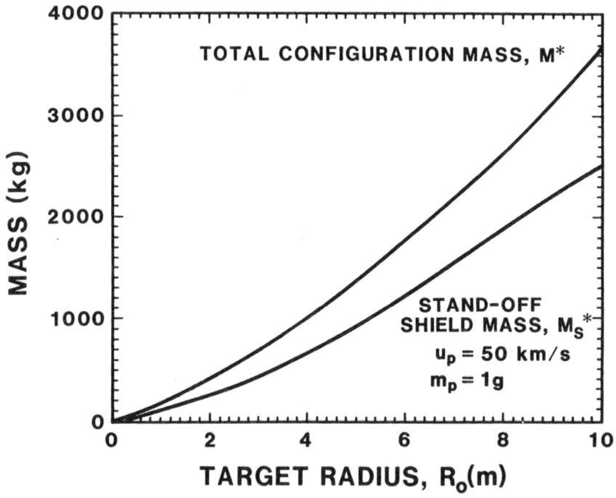

Fig. 14. Minimum mass configuration as a function of target radius for a 1-g, 50-km/s incident particle.

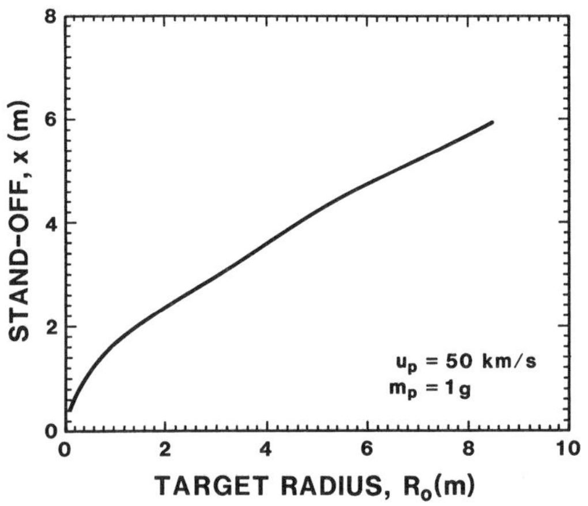

Fig. 15. Optimum stand-off as a function of target radius for a 1-g, 50-km/s incident particle.

A more detailed and extensive analysis along the lines of that presented in this second example has been given by Spillman (1986). Although direct comparisons are difficult because of differences in geometry and many of the other basic assumptions, it appears that the present shield masses are smaller than his by factors of 3 to 4. This is probably due to the fact that Spillman minimized the shield mass only, leaving the vulnerability of the underlying structure as a parameter. When the mass of the entire configuration is minimized, as is the case here, the acceptable loading on the internal structure is constrained by the model. In fact, the specific impulses determined here are substantially lower than those assumed by Spillman, and hence the observed discrepancies are not surprising.

CLOSURE

In the present work we have extended an existing model for the analysis of stand-off hypervelocity particle shields to the point that optimization can be achieved in a semi-analytic fashion. Applying the model, we have also shown that the optimum configuration can involve stand-off components considerably thinner than previously assumed. Scaling laws derived from the examples show how the various areal densities vary with stand-off for planar geometries, and how the optimum masses and stand-off distances vary with target size for spherical geometries.

There are, however, several additional aspects of the model that would prove useful to pursue. First, since many structures are cylindrical in shape rather than spherical, a straightforward modification to Eqs. (27), (28), and (29) would take this into account. Although major differences are unlikely, the results could be important in some situations. Second, for a more realistic comparison with the calculations of Spillman (1986), it would be interesting to alter the analysis by providing for a fixed mass for the underlying structure. This would be accomplished by holding I constant in Eq. (18), substituting the resulting relation between M and x into Eqs. (19) and (27), and then appropriately minimizing M_S and M_S^* (actually, only the latter has a true minimum). Third, sensitivity of the results to some of the other independent variables such as impact velocity and particle mass should be more fully examined. Simple scaling laws may not be adequate to describe the variation over more than a very limited range of values. Finally, comparison with detailed calculations using two-dimensional hydrocodes could help to verify, both qualitatively and hopefully quantitatively, the major features of the phenomena described by the model. In any case, the development and use of simple models such as this will continue to be important until these extremely high velocities become readily accessible to experiment.

ACKNOWLEDGMENT

This work was supported by the U. S. Department of Energy at Sandia National Laboratories under contract DE-AC04-76DP00789. We appreciate helpful discussions with H. F. Swift of Physics Applications, Inc., and S. L. Passman, M. M. Widner, and R. M. Zazworsky of Sandia National Laboratories. Thanks are also due to T. M. Calloway and W. R. Cordwell, also from Sandia, who provided computer verification of both the algebra and some of the graphical solutions.

REFERENCES

Gehring, J. W. (1970). In R. Kinslow (Ed.), *High-Velocity Impact Phenomena*. Academic Press, New York. Chap. IV, pp. 113-117, pp. 133-147, and Chap. IX, p. 493.

Spillman, G. R. (1986). *Hypervelocity Impact Bumpers as Countermeasures*, LA-10629-MS. Los Alamos National Laboratory, Los Alamos.

Swift, H. F., R. Bamford, and R. Chen (1982). *Designing Dual-Plate Meteoroid Shields—A New Analysis*, JPL Publication 82-39. Jet Propulsion Laboratory, Pasadena.

Swift, H. F. (1982). In J. A. Zukas, T. Nicholas, H. F. Swift, L. B. Greszczuk, and D. R. Curran, *Impact Dynamics*. John Wiley & Sons, New York. Chap. 6, pp. 215-239.

Int. J. Impact Engng Vol. 5, pp. 463–470, 1987
Printed in Great Britain

0734-743X/87 $3.00 + 0.00
Pergamon Journals Ltd.

STRESS MEASUREMENTS IN GLASS USING SHAPED-CHARGE JETS

William Lawrence and Robert E. Franz

Ballistic Research Laboratory
Aberdeen Proving Ground, MD 21005-5066

ABSTRACT

Stresses were measured in glass targets in the vicinity of a penetrating shaped-charge jet. Stress levels of approximately 0.3 GPa were measured 12-20mm away from a jet formed by a 35mm copper liner. High speed framing camera photographs showed that the penetration velocity in the glass was 2.57 km/s and the glass fracture velocity was 2.10 km/s.

INTRODUCTION

It has been known for over 40 years that glass targets are unusually effective in stopping shaped-charge jets. Studies of this effectiveness were made during and after World War II[1]. These studies established the principles of defense against shaped-charge jets using glass. Nevertheless, little is known about the physics and chemistry associated with the penetration process in glass.

This report describes some experiments made to measure dynamic stresses in glass in the vicinity of a penetrating jet. Strain-compensated stress gages and simultaneous high-speed framing camera photographs were used for this purpose. These experiments are part of a larger effort to understand the fundamentals of the jet penetration process in glass and glass-like materials. The overall work also includes the use of flash radiography to observe the on-going penetration, and target recovery to observe the material after penetration.

EXPERIMENTAL PROCEDURE

The shaped-charge jets used in these experiments were obtained using a 35 mm diameter conical copper liner with a wall thickness of 1.57 mm and a cone angle of 45 degrees. The explosive used was Composition B. When fired at a standoff of two cone diameters the jet penetrated four cone diameters in stacked 25 mm thick rolled steel plates (Rockwell hardness B81). The jet tip velocity was 6.8 km/s.

The targets consisted of stacked layers of float glass (Density = 2500 kg/m^3). The glass layers were either 25.4 mm or 19.1 mm thick and were approximately 150 mm square. They were stacked either perpendicular to the jet path or parallel to the jet path. The layers were bonded with epoxy cement while clamped in a press. Various cover plates were used. The cover plates were either rolled homogeneous armor (RHA) or polymethyl-methacralate (PMMA), or a combination of the two of different thicknesses. Table 1 gives details of the various thicknesses, orientations and materials used in each experiment. Figure 1 shows drawing of typical target configuration. The shaped charges were fired at a standoff of two cone diameters above the cover plates.

Strain-compensated stress gages were used to measure stresses during jet penetration. These gages consist of two separate interlaced foil grids encased in a polyimide plastic film. One of the grids is made of manganin and is used to measure stress. The other is made of constantan and is used to measure strain. The measured strain is used to correct the stress measurements.

The gages were bonded between the layers of glass in shallow slots approximately 0.38 mm deep. When the glass layers were parallel to the jet path the gage grids were also parallel. Likewise, when the layers were perpendicular the grids were perpendicular to the jet path. Table 1 also

TEST NO.	GLASS PLATES THICK (MM)	GLASS PLATES NO.	ORIENTATION OF TARGET AND GAGE TO THE INCOMING JET	GAGE IN THE SLOT YES	GAGE IN THE SLOT NO	GAGE LOCATION FROM FRONT PLATE(MM)	FRONT PLATE MATERIAL	FRONT PLATE THICK (MM)
3	25.40	5	PERPENDICULAR	YES		76.20	RHA	25.40
4	25.40	5	PARALLEL	YES		76.20	RHA	25.40
5	25.40	5	PARALLEL	YES		82.55	RHA	25.40
8	25.40	5	PARALLEL	YES		107.95	RHA/PMMA	50.80
9	25.40	4	PERPENDICULAR	YES		101.60	PMMA	50.80
				YES		127.00		
10	25.40	4	PERPENDICULAR	YES		95.25	PMMA	50.80
				YES		120.65		
14	19.05	3	PERPENDICULAR	YES		63.50	PMMA	25.40
					NO	44.45		
15	19.05	3	PERPENDICULAR	YES		44.45	PMMA	25.40
					NO	63.50		
16	25.40	4	PERPENDICULAR		NO	101.60	PMMA	50.80

STANDOFF DISTANCE BETWEEN SHAPED CHARGE AND THE FRONT PLATE = 70.10 MM

TABLE 1. TEST CONDITIONS

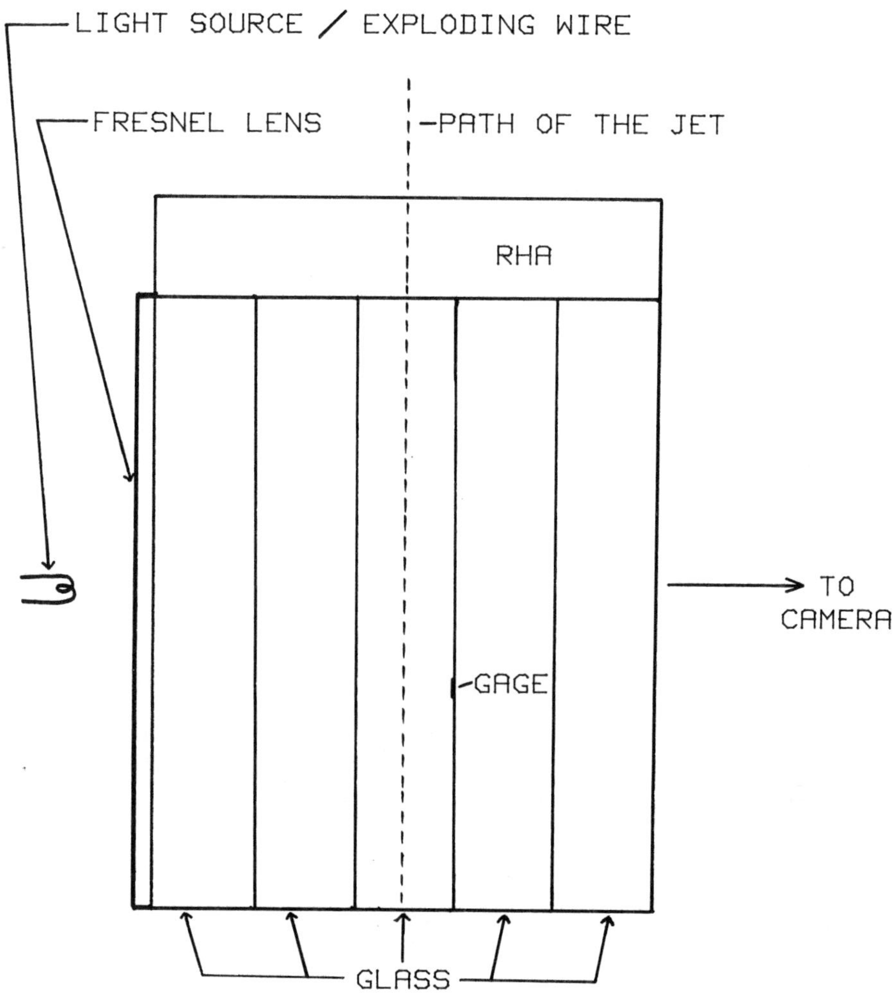

	LENGTH (MM)	WIDTH (MM)	THICKNESS (MM)
GLASS	150	127	25.4
RHA	150	127	25.4

Fig. 1. Parallel-Layer Glass Target. Gage is also Parallel to the Incoming Jet.

includes particulars on gage placement in each experiment. Test No. 15 and Test No. 16 were run without slots in the glass and the offset spaces caused by the thickness of the gage were filled with plastic film and epoxy cement. This was done to see if the slots affected the gage signals. Gupta and Gupta[2] have reported such disturbance in very strong materials. The gages were attached to 93 ohm RG-62u coaxial cables which were connected to bridge circuits similar to those used by Rapacki[3]. The circuits were modified to be used with the 50 ohm Dynasen gages instead of the 120 ohm gages used by Rapacki. The signals were recorded on digital oscilloscopes.

For some of the experiments simultaneous framing camera photographs were made during jet penetration. The glass targets were back-lit with an exploding tungsten wire placed near the focus of a plastic Fresnel lens.

RESULTS

The data from the two signals obtained from each gage were reduced in a manner similar to that used by Rapacki[3]. That is, calibrations were made of each circuit with known dummy gages inserted in the lines instead of the actual gage. Signals from known changes in resistance were then used to determine the circuit constants. These constants along with the known bridge and cable resistances were used to calculate the resistance of each grid at each simultaneous time step of the signals.

To calculate the stress in the glass for each time step it was assumed that the resistance of the constantan grid was not affected by the stress field that surrounded both it and the manganin grid. It was further assumed that the strain of the manganin grid was the same as in the constantan grid. Finally, it was assumed that the stress in the glass was the same as that in the manganin.

The strain for each time step was calculated using the equation[4]

$$ST = -1 + (RC/ROC)^{0.5}.$$

Here

ST = strain in the constantan grid
RC = resistance of the constantan grid
ROC = the initial resistance of the constantan grid

To calculate the stress, the resistance change associated with the strain in the manganin must first be subtracted from the total resistance change. In order to do this the strain gage factor for the manganin grid was determined from a separate static experiment. A strain-compensated stress gage was mounted on an aluminum alloy plate along with conventional strain gages. This plate was pulled in an universal testing machine. The data showed that the manganin grid had a gage factor of 0.7 to a strain of 0.5%. Above 0.5% strain the manganin deformed plastically and the gage factor became the same as for the constantan grid, that is, 2+ST. The apparent gage factor G.F., is defined as

$$G.F. = DRG/(ROG \times ST).$$

Where

DRG = the change in gage resistance,

and

ROG = the initial gage resistance.

The resistance change in the manganin due to the strain was, therefore, calculated by using the following equations.

For $-0.005 < ST < 0.005$

$$DRMS = 0.7 \times ST \times ROM.$$

For $ST > 0.005$

$$DRMS = (2+ST-0.005) \times (ST-0.005) \times (1+0.7\times0.005) \times ROM + (0.7\times0.005) \times ROM$$

$$= (1.995+ST) \times (ST-0.005) \times 1.0035 \times ROM + 0.0035 \times ROM$$

For $ST < -.005$

DRMS = (2.005+ST) x (ST+0.005) x 0.9965 x ROM - 0.0035 x ROM

Here

DRMS = the change in resistance of the manganin grid due to strain

ROM = the initial resistance of the manganin grid.

The stress in the manganin was then calculated using a cubic fit to the data of Charest[5]. This equation is

$$S = 53.19 \times (DRM/RO) - 159.91 \times (DRM/RO)^2 + 493.16 \times (DRM/RO)^3$$

Where

S = stress in the manganin (GPa)

DRM = DRMT - DRMS

RO = ROM + DRMS

and

DRMT = the total resistance change in the manganin.

Figure 2 shows results obtained from one of the tests.

DISCUSSION

The stress measurements reported here were obtained under particularly adverse physical conditions. It was learned in preliminary experiments that the foil gages and their leads had to be protected from a hostile environment which included fracturing glass. The layers of polytetraflouroethylene, used for this protection, restricted the mechanical response of the gages especially to the initial sharply rising shock waves. The response was further restricted by the electronic bandwidth of the recording equipment so that submicrosecond changes in either stress or strain could not be tracked.

The gages were approximately 19 mm to the side of the jet path for perpendicular shots and 12.7 mm from the path for parallel shots. The stresses measured did not exceed approximately 0.3 GPa until the gages were on the verge of destruction from the fracturing glass. The strain in most of the shots did not contribute significantly to the stress compensation until 3 to 4 microseconds after the start of the stress signals. The usual signal showed two compressive peaks followed by either tension or compression. These signals are shown in Figs. 2 and 3.

Although unslotted gages did not show these distinct peaks the measured stress levels were approximately the same as for slotted gages. In most cases the gages started to fail before or near jet arrival, that is, 15 to 20 microseconds after the shock arrived. For two shots the gages survived for longer times. No extremely large stresses were measured on these shots. The parallel gages showed lower initial stress levels than the perpendicular gages.

The framing camera photographs clearly show that fracture occurred around the gages before the arrival of the penetrating jet. This was caused by interaction with the leading shock wave which was formed when the jet impacted the cover plate. There was a delay of about 3 microseconds after shock arrival before cracking was seen at the gage. Data from the film was used to calculate the shock velocity, the initial penetration velocity, and the glass fracture velocity. The measured values are:

Penetration Velocity = 2.57 km/s

Fracture Velocity = 2.10 km/s

Shock Velocity = 5.90 km/s.

The wave profiles determined with the stress gages were difficult to interpret when compared to the high speed framing camera photographs. For instance, at least two shock waves are evident in the photographs of the glass in Test No. 8 where the cover plate consisted of 25.4 mm of RHA and 25.4 mm of PMMA. The first wave was caused by the original jet impact at the RHA surface and the second (probably) by the impact of the jet reaching the PMMA-glass interface. In the steel, the shock velocity was higher than the penetration velocity, whereas, in the PMMA the opposite was true. The jet reached the glass interface approximately 2.2 microseconds after the shock and accounts for the photographic results. The stress profile also showed a two wave structure but

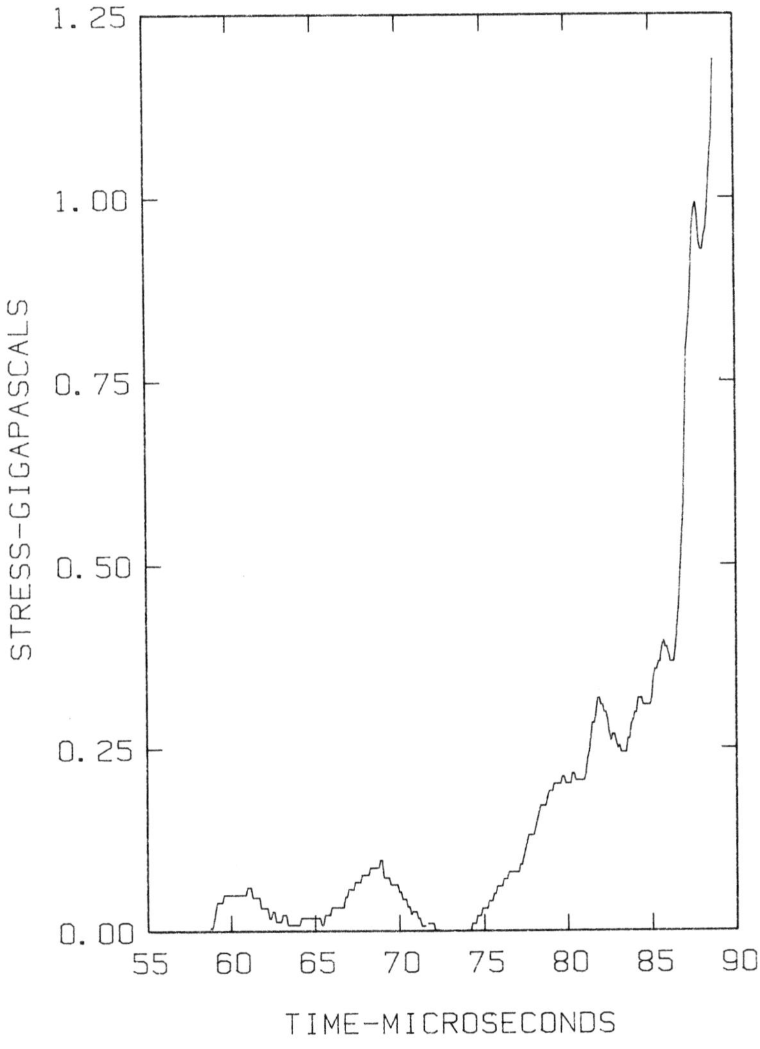

Fig. 2. Stress Measurement from Gage in Parallel Configuration in Test No. 8. Gage was in the Slot.

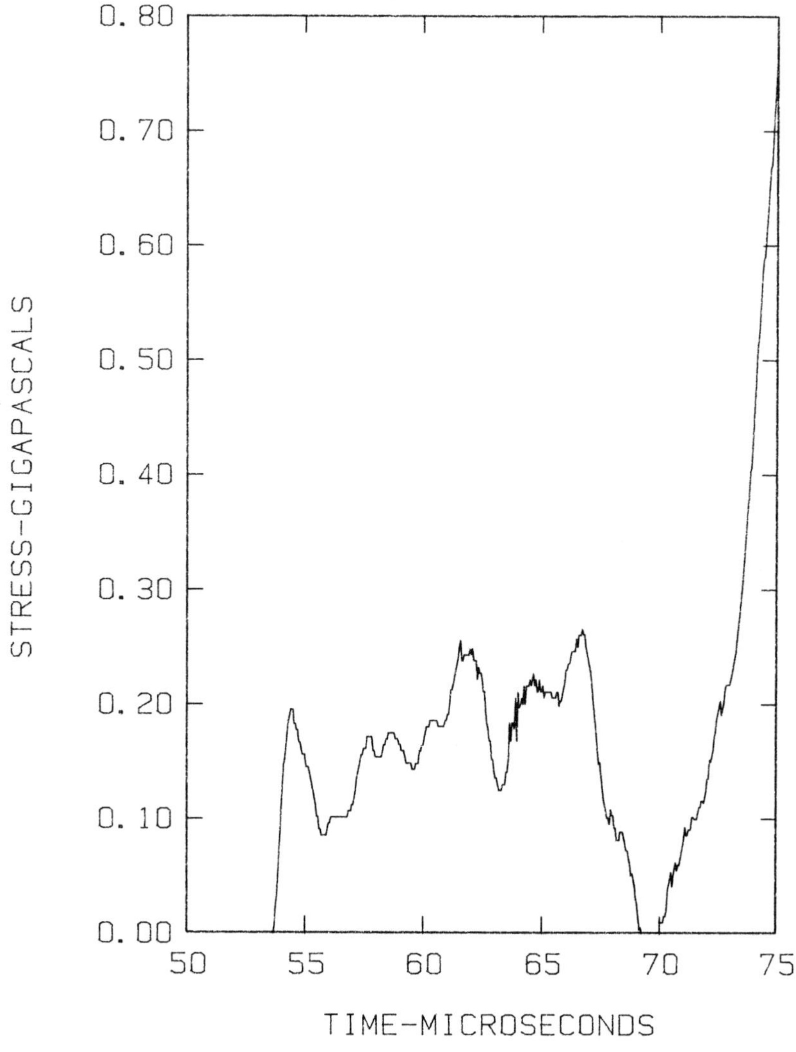

Fig. 3. Stress Measurement from Gage in Perpendicular Configuration in Test No. 16. Gage was in the Slot.

the spacing was 7 microseconds. A cover plate of PMMA 50.8 mm thick was used in Test No. 10. The photographs show a double wave structure but in this case the second shock is much weaker than for Test No. 8. In the PMMA, the penetration velocity was higher than the shock velocity, and the jet tip arrived at the PMMA-glass interface before the shock. The jet impact produced a shock in the glass after which the following shock in the PMMA entered the glass and caused the double wave structure of the photographs. The photographs of Test No. 8 also show a weak third shock which can be ascribed to the same causes.

CONCLUSIONS

The following conclusions were drawn from the experience gained in applying stress gage technology to the measurement of stress during shaped-charge jet penetration in glass, and from the combined photographic and stress results of this study.

First, the stress profiles measured were limited by both the mechanical and electrical response of the instrumentation.

Second, the stress measurements, with their shortcomings, were of sufficient quality to indicate stress levels in the immediate vicinity of a penetrating jet.

Third, although higher stress levels than those measured in these experiments must be present at the jet-glass penetration interface, they will be difficult, if not impossible to measure with foil stress gages.

Fourth, the information presented here can be used in analysis or design considerations for armor or anti-armor applications. The stress measurements can be used to verify analysis that can infer stress levels in inaccessible locations in a target.

REFERENCES

Allison, F. E., Defeat of Shaped Charge Weapons, Final Report, Contract No. DA-36-061-ORD-507, Carnegie Institute of Technology, April 1960.

Gupta, S. C. and Gupta, Y. M. Effect of Matrix Properties on the Piezoresistance Response of Ytterbium and Manganin Foils, Bulletin of the American Physical Society, Vol. 29, No. 4, 741, April 1984.

Rapacki, Jr., E. J., Instrumentation Techniques for Measuring Large, High Rate Strains with Foil Resistance Strain Gages, Ballistic Research Laboratory Technical Report ARBRL-TR-02573, August 1984.

Franz, R. E., The Measurement of Large Strains with Foil Resistance Strain Gages, Ballistic Research Laboratory Technical Report ARBRL-TR-02519, August 1983. ADA 133686.

Charest, J. A., Characterization of an Encapsulated 50-Ohm Manganin Foil Piezoresistive Gauge, Air Force Weapons Laboratory Report No. AFWL-TR-71-81, August 1972.

Int. J. Impact Engng Vol. 5, pp. 471–481, 1987
Printed in Great Britain

0734-743X/87 $3.00 + 0.00
Pergamon Journals Ltd.

NUMERICAL SIMULATIONS OF SMALL PROJECTILES IMPACTING A MULTILAYERED TARGET AT HYPERVELOCITIES

R. W. Meier

Los Alamos National Laboratory, Los Alamos, New Mexico 87545

ABSTRACT

The effects of a hypervelocity (10-80 km/s) projectile of low mass (1 and 5 g) striking a multilayered target have been investigated. The two-dimensional numerical simulation was done with the Eulerian hydrodynamics computer program LASOIL. Target responses that were calculated include the time variant material deformation, pressures, shock-wave movement, and hole sizes. Parametric studies were done to determine the effects of projectile mass, velocity, and shape on target response as measured by these parameters.

INTRODUCTION

The response of a complex target when struck by a projectile is of interest in many different applications. It is useful to know how projectile characteristics interact to determine the damage that is inflicted on a target. Three prime determinants of projectile effectiveness are projectile mass, velocity, and shape. It was also of interest to determine the response of a target when struck by small projectiles traveling at extremely high velocities.

The purpose of this study is to investigate the damage that is inflicted on a target by small projectiles striking at extremely high velocities and to delineate the relative importance of projectile mass, velocity, and shape in causing damage to a complex target.

Numerical hydrodynamics simulations were carried out for 1- and 5-g projectile masses, striking at 10, 40, and 80 km/s, to determine the relative importance of projectile mass and velocity on target damage. The 10 km/s simulation was chosen to approximate the upper limit of velocities at which experimental data might be obtained, and the 40- and 80-km/s velocities were chosen to form a parametric study at higher velocities. The case of a 1-g projectile striking a multi-layered target at 80 km/s was considered for two different aspect ratios to determine the effect of projectile shape on target damage.

METHODOLOGY

The target consisted of a series of cylindrical disks and one void. Each disk had a radius of 128 mm. This radius was considered adequate for preventing edge effects from affecting results. The thicknesses of the target layers are indicated in Table 1, where each layer represents a composite of materials. The central portion of the resultant target and the 1-g projectile are shown in Fig. 1. In subsequent figures, the projectile is shown moving from bottom to top, with time zero being the instant of initial impact.

For the first six cases, the pure tungsten projectile was modeled as a right circular cylinder with equal length and diameter. For the cases with the 1-g projectile, the diameter was 4.04 mm; for the cases with the 5-g projectile, the diameter was 6.91 mm. One impact was simulated for a 1-g cylindrical projectile with a length-to-diameter ratio of eight. This projectile had a diameter of 2.02 mm and a length of 16.16 mm. A finite-difference mesh with uniform 1-mm cells in the radial and axial directions was used in all of the computations. The impact mechanics were simulated with the LASOIL computer program. LASOIL is an Eulerian hydrodynamics computer code based on the SOIL family of codes (Johnson, 1965). The code exists in both two- and three-dimensional versions, but for

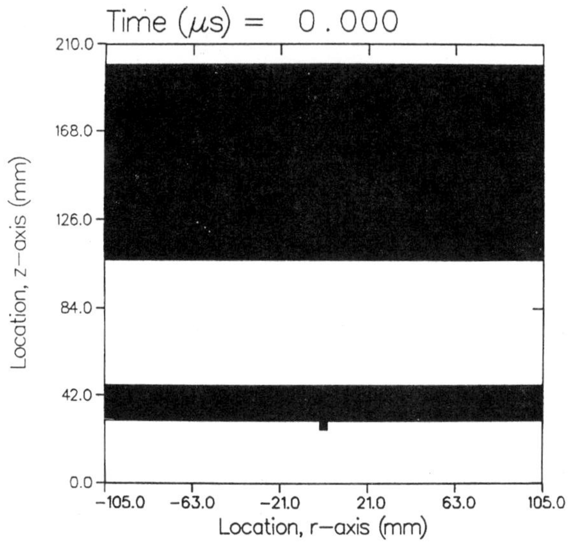

Fig. 1. Cross-sectional view of the central region of the target, with the 1-g projectile at the bottom center.

TABLE 1 Target Components

Layer Number	Thickness (mm)
1	17.0
void	60.0
2	92.5

this analysis the two-dimensional version was used because of axial symmetry. It is capable of including the effects of material strength; at these velocities however, material strength was expected to be negligible compared with hydrodynamic pressures. Thus, material strength was not included in the computations.

The Tillotson equation of state (Tillotson, 1962) was used for all materials. The equation of state takes the form

$$P = \left[a + \frac{b}{\frac{E}{E_o \eta^2} + 1} \right] E\rho + A\mu + B\mu^2$$

for condensed states of the materials and

$$P = aE\rho + \left[\frac{bE\rho}{\frac{E}{E_o \eta^2} + 1} + A\mu e^{-\beta(V/V_o - 1)} \right] e^{-\alpha(V/V_o - 1)^2}$$

for expanded states, where P is pressure, E is specific internal energy, ρ is density, V is specific volume, $\eta = \rho/\rho_o = V_o/V$, $\mu = \eta - 1$, and the subscript "o" refers to the normal state of the material. The parameters α and β are constants that control the rate of convergence to ideal gas behavior. The remaining parameters (a, b, A, and B) are chosen to provide good agreement at low pressures with experimental shock-wave data and at high pressures with theoretical results obtained from the atomic Thomas-Fermi-Dirac model.

Equation-of-state data are not available for the conditions produced by the extremely high-velocity impacts simulated in this study. Thus, the computations required extrapolation beyond the region of verification of the constants, which introduced uncertainties into the equation-of-state calculations.

To provide a measure of projectile effectiveness, a pressure gauge was simulated inside the second target layer. This gauge consisted of three particles for which the pressure was monitored. These particles moved with the material. The pressure for the gauge was the average of the pressures at each of the three particle locations. A 3.175-mm long pressure gauge was placed 46-mm deep in Layer 3 at an axial location of 153 mm in Fig. 1 and offset 2 mm from the centerline. Pressures calculated at this location were used to assess the damage produced by the impacts and to calculate $P^2\tau$ values, where P is the maximum pressure from the pressure- vs-time curve and τ is the pressure pulse width at one-half the maximum pressure. The $P^2\tau$ value is one criterion that is used to determine the likelihood of shock initiation of explosives. A sufficiently high pressure is required for a sufficiently long period of time to achieve detonation. The $P^2\tau$ value combines both of these characteristics into a quantifiable measure. Damage was also assessed by measuring the size of the hole produced in Layer 1 and the depth of penetration into Layer 2 by each of the impacts.

RESULTS OF THE PARAMETRIC STUDY OF PROJECTILE MASS AND VELOCITY

One-gram Projectile Impacting at 10 km/s

Figure 2 shows a graphic of the target deformation 112 μs after the 1-g projectile strikes the target at 10 km/s. After 112 ms the deformation rates were small, and the results shown are a good approximation of the final damage. This figure was used to determine target deformation, projectile penetration, and Layer 1 hole size. However, projectile penetration and hole size were difficult to determine precisely because it was not apparent which material was close to its original density and would be expected to have material strength because of the hydrodynamic approximation. The target was penetrated to a depth of 50 mm from the front of Layer 2; however, we observed no deformation of the back of the target. The impact produced a hole in Layer 1 that was 94 mm in diameter.

The pressure gauge showed a maximum pressure of 100 MPa 72 μs after the projectile struck Layer 1. The $P^2\tau$ value calculated for this case was 0.1 MPa^2•s.

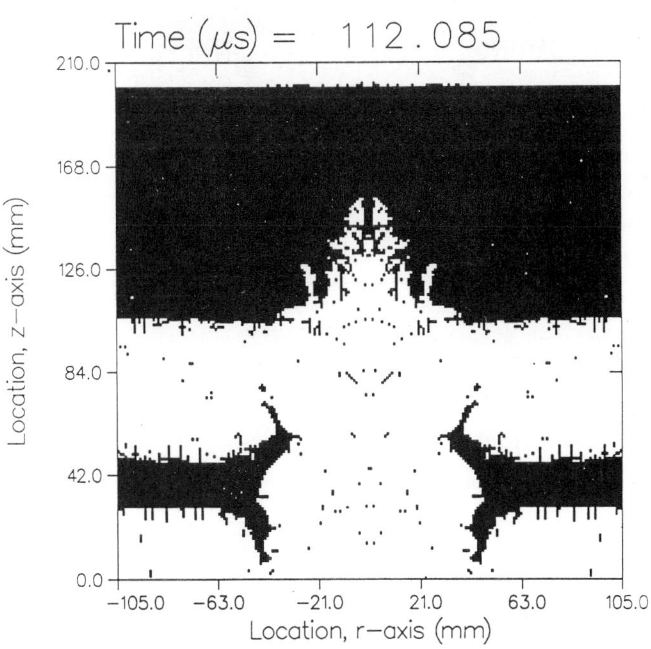

Fig. 2. Target deformation 112 μs after being impacted by the 1-g projectile at 10 km/s.

Five-gram Projectile Impacting at 10 km/s

The results of the computation for the 5-g projectile striking the target at 10 km/s are shown in Figs. 3 and 4. Figure 3 is a material deformation graphic of the central portion of the target after 120 μs, the time at which deformation rates were small. The target was penetrated to a depth of 78 mm from the front of Layer 2. The back of the target was slightly deformed from the impact. The hole left in Layer 1 was 118 mm in diameter. The penetration depth was 50% greater and the Layer 1 hole size was 25% greater than those for the 1-g projectile impacting at 10 km/s.

A plot of the computed pressure at the pressure gauge location is shown in Fig. 4. The pressure reached a maximum of 700 MPa after 28 μs. The maximum pressure is seven times greater than that for the 1-g projectile impacting at 10 km/s and it occurs in about one-third the time. The resultant $P^2\tau$ value for this impact was 5 $MPa^2 \bullet s$.

One-gram Projectile Impacting at 40 km/s

The final simulation times for the higher velocity impact were obtained by taking the ratio of the velocities to 120 μs for the 10 km/s impacts. Figure 5 shows the target deformation predicted by the numerical simulation for the 1-g projectile striking the target at 40 km/s 30 μs after impact. At this time, the target deformation rates were

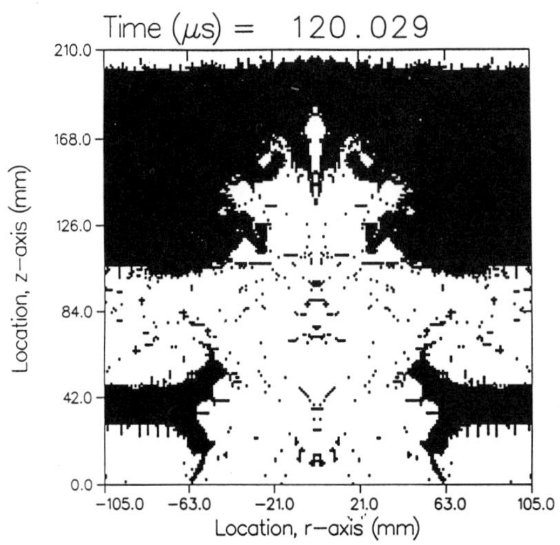

Fig. 3. Target deformation 120 μs after being impacted by the 5-g projectile at 10 km/s.

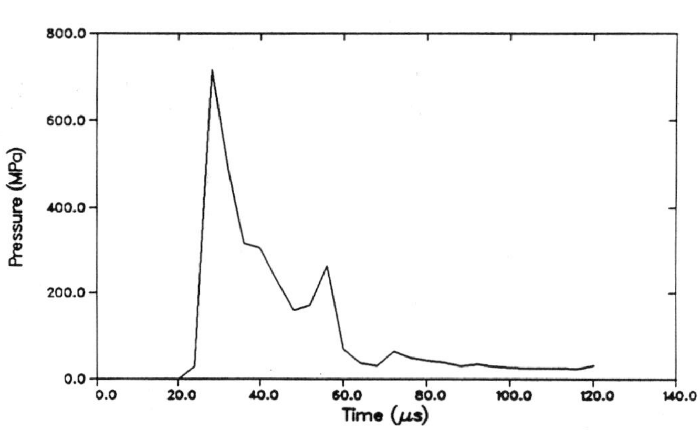

Fig. 4. Gauge pressure for a 5-g projectile impacting at 10 km/s.

small, thus, the results are a good approximation of the final damage. More debris can be seen in the void than is observed from impact of the 1-g projectile at 10 km/s shown in Fig. 2.

The projectile formed a hole in Layer 1 that was 86 mm in diameter and penetrated the second plate to a depth of 29 mm. The Layer 1 hole size is approximately the same as that for the 1-g projectile impacting at 10 km/s, but smaller than that for the 5-g projectile impacting at 10 km/s.

The simulated pressure gauge reached a maximum pressure of 700 MPa 29 μs after initial impact. The resultant $P^2\tau$ value for this impact was 5 $MPa^2 \bullet s$.

The maximum pressure and $P^2\tau$ values are greater than those for the 1-g projectile impacting at 10 km/s, but are the same as those for the 5-g projectile impacting at 10 km/s. The maximum pressure occurs at the same time as that for the 5-g projectile impacting at 10 km/s.

Five-gram Projectile Impacting at 40 km/s

Figure 6 shows the computed deformation for the 5-g projectile striking the target at 40 km/s at 30 μs after initial impact. The Layer-1 hole diameter after 30 μs is 100 mm and the subsequent layer is penetrated to a depth of 53 mm from the front of Layer 2. The back of the target is deformed from the impact. The hole diameter was smaller than that for the 5-g projectile impacting at 10 km/s, but larger than that for the 1-g projectile impacting at 10 km/s.

Figure 7 is a plot of the pressure at the pressure-gauge location. Nine microseconds after initial impact, the pressure reached a maximum of 7500 MPa. The resultant $P^2\tau$ value for this impact was 250 $MPa^2 \bullet s$. The maximum pressure, approximately ten times greater than that for the 5-g projectile impacting at 10 km/s and the 1-g projectile impacting at 40 km/s, occurs in about one-third the time.

One-gram Projectile Impacting at 80 km/s

Figure 8 shows the target deformation 15 μs after that target was struck by the 1-g projectile at 80 km/s. After 15 μs, the hole in Layer 1 is 80 mm in diameter. The hole size is smaller than that for the 1-g projectile impacting at 10 or 40 km/s or for the 5-g projectile impacting at 40 km/s. The target is penetrated to a depth of 29 mm from the front of Layer 2; the back of the target is not deformed. The penetration depth is smaller than that for the 1-g projectile impacting at 10 km/s or the 5-g projectile impacting at 40 km/s and the same as that for the 1-g projectile impacting at 40 km/s.

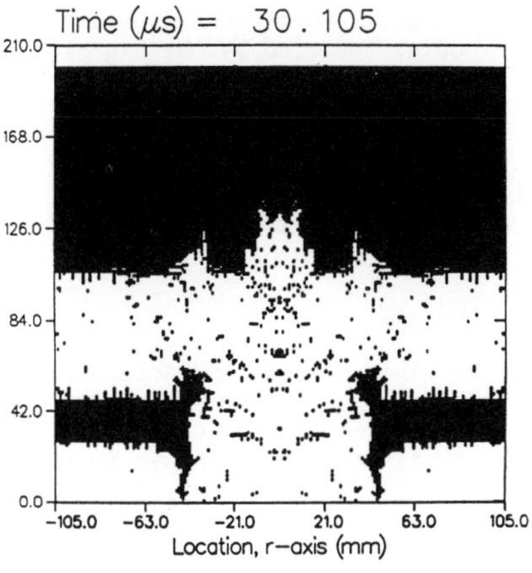

Fig. 5. Target deformation 30 μs after being impacted by the 1-g projectile at 40 km/s.

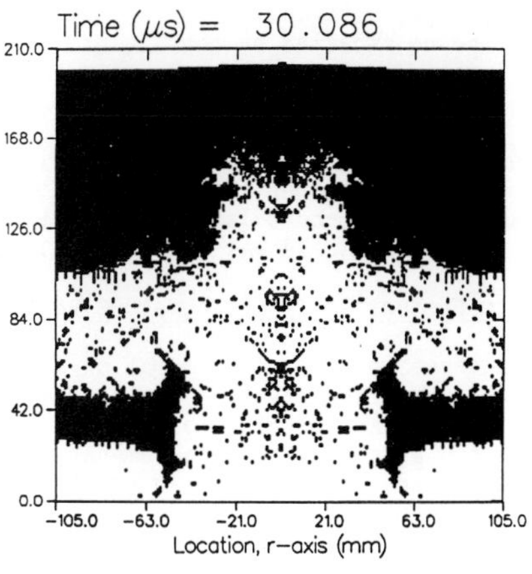

Fig. 6. Target deformation 30 μs after being impacted by the 5-g projectile at 40 km/s.

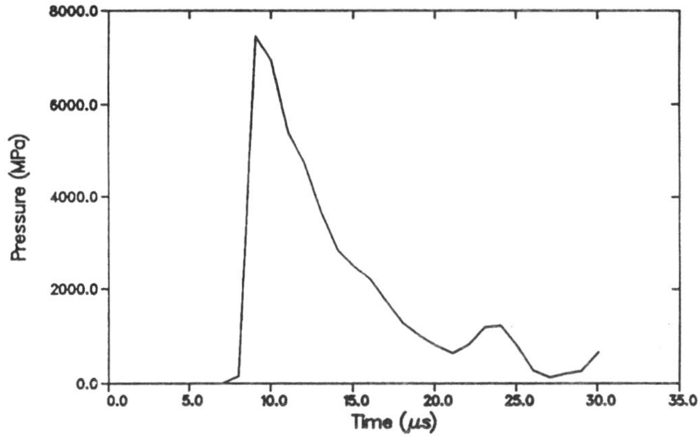

Fig. 7. Gauge pressure for a 5-g projectile impacting at 40 km/s.

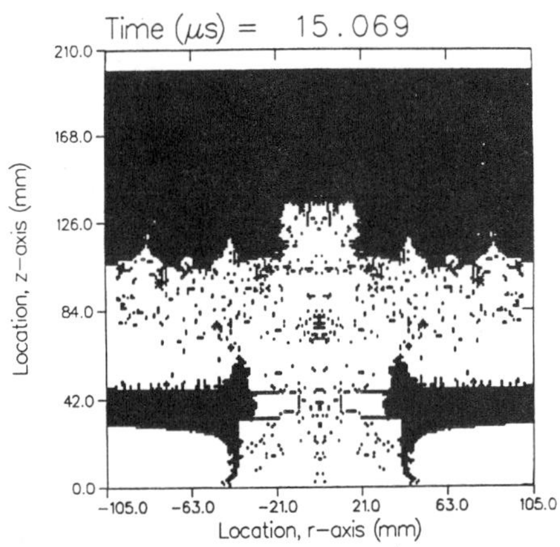

Fig. 8. Target deformation 15 μs after being impacted by the 1-g projectile at 80 km/s.

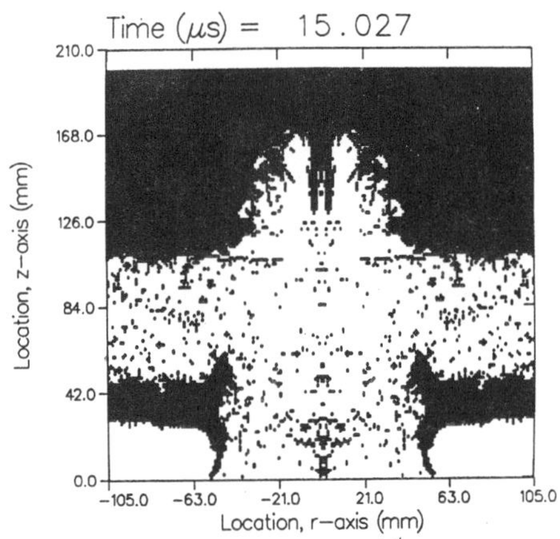

Fig. 9. Target deformation 15 μs after being impacted by the 5-g projectile at 80 km/s.

The pressure gauge showed a maximum of 1800 MPa after 14 μs. The maximum pressure is higher and occurs sooner than that for the 1-g projectile impacting at 10 or 40 km/s, but it is smaller and occurs later than that for the 5-g projectile impacting at 40 km/s. The $P^2\tau$ value for this case was approximately 50 $MPa^2 \bullet s$.

Five-gram Projectile Impacting at 80 km/s

Figure 9 shows the target deformation 15 μs after impact. This deformation occurred when the target was struck by the 5-g projectile traveling at 80 km/s. The hole in Layer 1 was 102 mm in diameter. This hole size was smaller than that for the 5-g projectile impacting at 10 km/s, approximately the same as that for the 5-g projectile impacting at 40 km/s, and larger than that for the 1-g projectile impacting at 80 km/s. The target is penetrated to 60 mm from the front of Layer 2. The back of the target is not deformed after 15 μs, but deformation would occur and penetration depth would increase if the computation were continued to a later time because significant changes were still occurring in the target after 15 μs.

The pressure at the gauge location is shown in Fig. 10. The pressure peaks at 31,000 MPa after 4.5 μs. The maximum pressure is much higher and occurs much sooner than that for any of the other impacts considered. The $P^2\tau$ value for the pressure pulse is 1500 $MPa^2 \bullet s$.

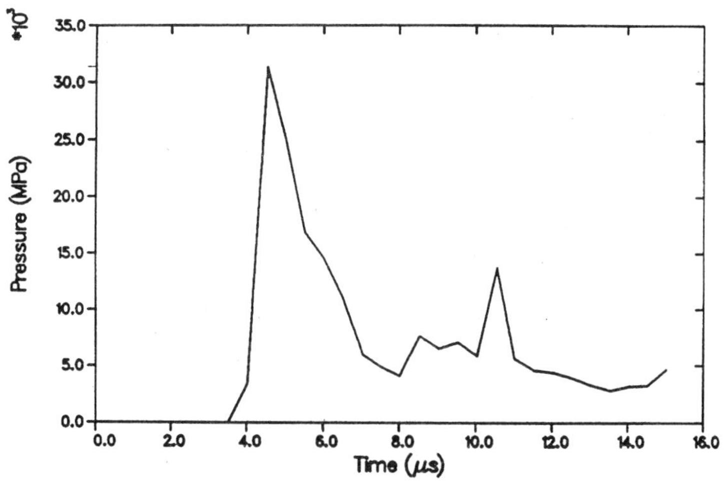

Fig. 10. Gauge pressure for a 5-g projectile impacting at 800 km/s.

Synopsis of the Parametric Study of Projectile Mass and Velocity

We investigated the target response to impact under six different conditions. Two projectile masses (1 g and 5 g) and three projectile velocities (10, 40, and 80 km/s) were considered. Table 2 shows the diameter of the holes formed in Layer 1 under each of these impact conditions. The hole size increases as projectile mass increases, but is not proportional to projectile size. As projectile velocity increases, the hole size decreases. It is interesting to observe, also, that the hole sizes are all in a relatively small range, varying by less than ±20%, which is a range much smaller than the variation in projectile size, momentum, or kinetic energy.

Table 3 shows the effect of projectile mass and velocity on the arrival time of the maximum pressure at the pressure gauge location. This matrix of impact conditions shows that in Layer 2, at each velocity, the effects of the 5-g projectile occur sooner than the effects of the 1-g projectile. For instance, the pressure gauge shows the maximum pressure at about the same time for the 5-g projectile impacting at 10 km/s as it does for the 1-g projectile impacting at 40 km/s. Similarly, the maximum pressure at the gauge occurs sooner for the 5-g projectile impacting at 40 km/s than for the 1-g projectile at 80 km/s.

In addition, the time to maximum gauge pressure varies linearly as the reciprocal of velocity for a given projectile mass between 40 and 80 km/s. However, this linearity does not occur between 10 and 40 km/s.

Table 4 is a summary of the $P^2\tau$ values from the pressure waves at the gauge location produced by each of the six impact conditions. At each velocity, the $P^2\tau$ value produced by the 5-g projectile is significantly higher than that produced by the 1-g projectile. The $P^2\tau$ value produced by the 5-g projectile impacting at 10 km/s is approximately equal to that produced by the 1-g projectile impacting at 40 km/s. In addition, the $P^2\tau$ value produced by the 5-g projectile at 40 km/s is significantly higher than that produced by the 1-g projectile at 80 km/s. The $P^2\tau$ values increase with increasing projectile velocity in all cases.

RESULTS OF THE STUDY OF PROJECTILE SHAPE VS TARGET DAMAGE

To investigate the effect of projectile shape on target damage, a numerical simulation was done of the same multilayered target as shown in Fig. 1. The target was struck at 80 km/s by a 1-g cylindrical projectile with a length-to-diameter ratio of eight. The results of these simulations were compared with the damage produced by a projectile of the same mass with a length-to-diameter ratio of one, impacting at the same velocity. The projectile and target are shown in Fig. 11 at the instant of initial impact.

TABLE 2 Effect of Projectile Mass and Velocity on Layer 1 Hole Diameter

Projectile Velocity (km/s)	Projectile Mass (g) 1	5
	Hole Diameter (mm)	
10	94	118
40	86	100
80	80	102

TABLE 3 Effect of Projectile Mass and Velocity on Arrival Time of Maximum Gauge Pressure

Projectile Velocity (km/s)	Projectile Mass (g) 1	5
	Maximum Gauge Pressure (μs)	
10	72	28
40	29	9
80	14	4.5

TABLE 4 Effect of Projectile Mass and Velocity on P^2t Values ($MPa^2 \bullet s$)

Projectile Velocity (km/s)	Projectile Mass (g) 1	5
	P^2t Values ($MPa^2 \bullet s$)	
10	0.1	5
40	3	250
80	80	1500

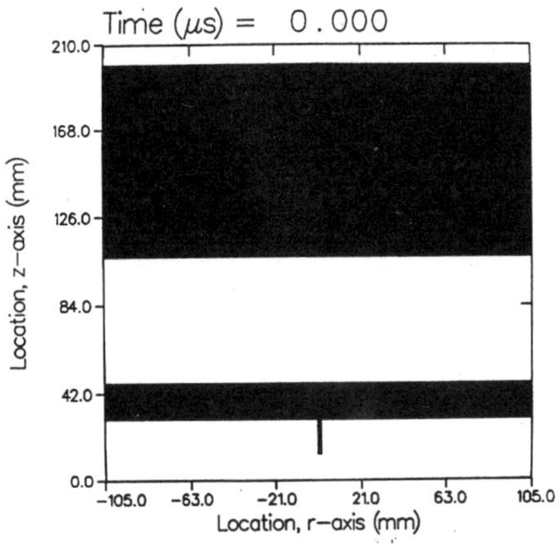

Fig. 11. Cross-sectional view of the central region of the target. The 1-g projectile has an L/D=8 at the bottom center.

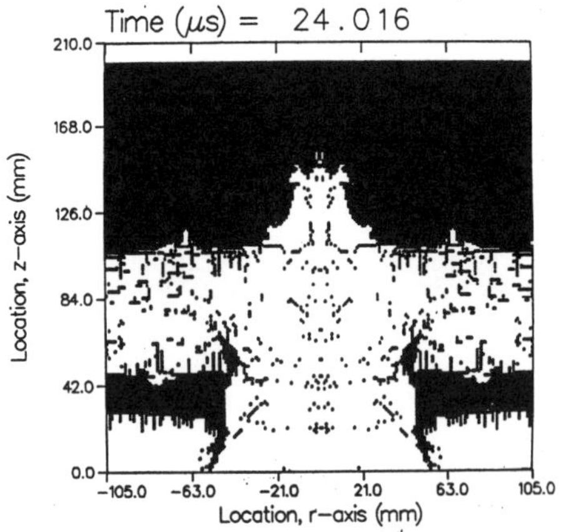

Fig. 12. Target deformation 24 μs after being impacted by the 1-g projectile with L/D=1 at 80 km/s.

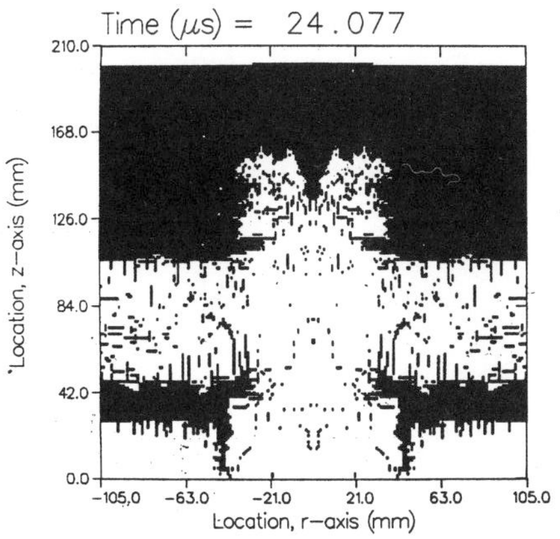

Fig. 13. Target deformation 24 μs after being impacted by the 1-g projectile with L/D=8 at 80 km/s.

The final simulation time for the study of projectile shape effects was increased to 24 μs to be certain that the final deformation rate was small. The material deformation after 24 μs for L/D=8 can be compared with that for unity aspect ratio in Figs. 12 and 13. The L/D=1 case is shown in Fig. 12 and the L/D=8 case is shown in Fig. 13.

Both projectiles penetrate Layer 1 with similar damage; however, the L/D=8 projectile stays much more intact than the L/D=1 projectile. The damage to Layer 2 is dramatically different after the debris impacts this layer because of the relative intactness of the long thin projectile as compared with that of the chunky projectile. After 24 μs, the L/D=1 projectile has penetrated Layer 2 to a depth of only 43 mm and has not deformed the back of the target. The L/D=8 projectile, however, has penetrated to a depth of 54 mm from the front of Layer 2, has slightly deformed the back of the target, and has produced a much larger crater in Layer 2 than the L/D=1 projectile.

The relative strengths of the shock waves produced by the two projectiles are shown graphically in Figs. 14 and 15. These figures show the pressure at all radial and axial locations in the target after 14 μs. The L/D=8 projectile produces a much more intense shock wave in Layer 2 than does the L/D=1 projectile. The pressures produced by the two projectiles in Layer 1 are similar to each other.

The qualitative effects shown in Figs. 12-15 are shown quantitatively in Fig. 16, which shows the pressures produced by the two projectiles at the pressure-gauge location. The L/D=8 projectile produces a maximum pressure of 6000 MPa after only 6 μs, while the L/D=1 projectile produces a peak pressure of 1800 MPa after 14 μs.

The $P^2\tau$ values for the two curves in Fig. 16 are shown in Table 5. The value for the L/D=8 projectile is four times higher than that for the L/D=1 projectile.

Fig. 14. Target pressure contour 14 μs after being impacted by the 1-g projectile with L/D=1 at 80 km/s.

Fig. 15. Target pressure contour 14 μs after being impacted by the 1-g projectile with L/D=8 at 80 km/s.

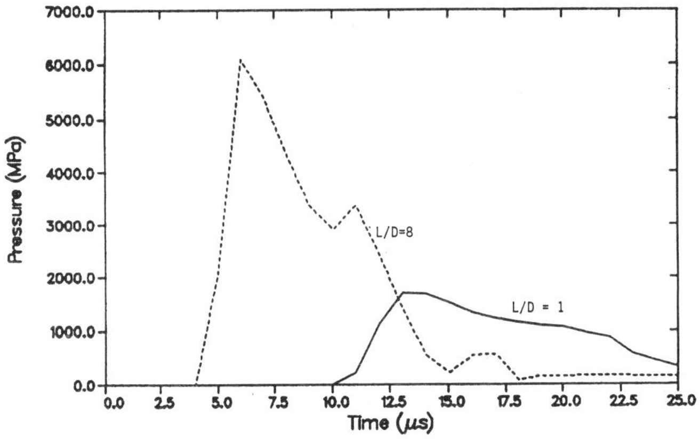

Fig. 16. Gauge pressures after being impacted by 1-g projectiles at 80 km/s. The projectiles have an L/D=1 and L/D=8.

TABLE 5 Effect of Projectile L/D on $P^2\tau$

1-g W Cylinder at 80 km/s

L/D	$P^2\tau$ ($MPa^2 \bullet s$)
1	50
8	200

CONCLUSIONS

This study has used numerical hydrodynamics simulation methods to investigate the relative damage inflicted on a multilayered target by cylindrical projectiles of various masses, velocities, and shapes. The projectile masses considered were 1 and 5 g, the projectile velocities were 10, 40, and 80 km/s, and the projectile aspect ratios were one and eight.

These results must be viewed as preliminary because the conditions being considered are far beyond the range of available equation-of-state data. Also, no experimental data are available from projectile/target interactions at the projectile velocities and masses considered here. Therefore, we are unable to benchmark the numerical simulations against experimental data. While future computations will take advantage of improvements in equation-of-state data, material handling in voids, and other hydrodynamic code developments, we anticipate that the results presented here are useful for making relative comparisons of the effects of projectiles of these masses, velocities, and shapes on multilayered targets.

The damage to Layer 1 that was produced by the projectile impacts does not differ greatly among the cases considered here. However, the heavier projectile penetrates much better than the lighter projectile after penetrating Layer 1 and traversing the void. Therefore, the damage to Layer 2 varies greatly among the cases studied.

The Layer 1 hole size increases as the projectile mass increases but is not proportional to the projectile size. The results also show that the hole size decreases as projectile velocity increases.

The arrival time of the maximum pressure at the gauge location in Layer 2 decreases with increasing projectile mass and with increasing projectile velocity. For both the 1-g and 5-g projectiles, the arrival time is proportional to the reciprical of projectile velocity from 40 to 80 km/s, but not from 10 to 40 km/s.

The gauge pressures and $P^2\tau$ values increase with increasing projectile mass and velocity. However, at the higher velocities, the $P^2\tau$ values for the 5-g projectile are more than an order of magnitude greater than those for the 1-g projectile at the same velocity. Thus the $P^2\tau$ values do not scale linearly with projectile mass. When the projectile velocities are considered, the gauge pressures and $P^2\tau$ values do not scale with projectile momentum or kinetic energy. Computations of more impact conditions are required before projectile scaling parameters can be determined for $P^2\tau$ values in multilayered targets.

The investigation of the effects of projectile shape on target damage show that the long thin projectile does significantly more damage to the target than does the chunky projectile of the same mass and at the same velocity, and it does it significantly quicker. The long, thin projectile penetrates much more than the chunky one, causing more deformation of the deeper layers, resulting in higher pressures, and higher $P^2\tau$ values in much less time. Thus, target damage can be increased with no increase in projectile mass or velocity, by striking the target with a high-aspect-ratio projectile.

REFERENCES

Johnson, W. E. (1965). OIL: A continuous two-dimensional Eularian hydrodynamic code. General Atomic Division of General Dynamics Report GAMD-5580.

Tillotson, J. H. (1962). Metallic equations of state for hypervelocity impact. General Atomic Division of General Dynamics Report GA-3216.

Int. J. Impact Engng Vol. 5, pp. 483–492, 1987
Printed in Great Britain

0734-743X/87 $3.00 + 0.00
Pergamon Journals Ltd.

HIGH-VELOCITY SOLID EJECTA FRAGMENTS FROM HYPERVELOCITY IMPACTS

H.J. Melosh
Lunar and Planetary Laboratory
University of Arizona
Tucson, AZ 85721

ABSTRACT

The recent discovery of meteorites from the moon and the strong probability that the 8 SNC (Shergottite, Nakhlite and Chassignite) meteorites originated on Mars indicate that large hypervelocity impacts eject some solid debris at very high speed (more than 2.5 and 5 km/sec in the above cases). The standard Hugoniot relation between particle velocity and shock pressure predicts that lunar ejecta should be very heavily shocked (40-50 GPa) and Martian ejecta should be vaporized (100-200 GPa). However, the lunar meteorite ALHA 81005 was in fact subjected to less than 15 GPa, while the most highly shocked SNC meteorite was exposed to ca. 50 GPa, while others showing no detectable shock damage at all.

Theoretical work shows that the normal Hugoniot relation doesn't apply in the vicinity of a free surface. The free surface is, by definition, a pressure-free boundary, so shock pressures on it must be identically zero. On the other hand, the acceleration of debris is proportional to the pressure *gradient*, so that near-surface material may be accelerated to high speed and still escape compression to correspondingly high pressure. This process occurs only in a restricted zone near the free surface. The thickness of this zone is proportional to the rise time of the stress-wave pulse generated by the impact.

The rise time of the stress wave generated by a large impact is typically a/v_i, where a is the projectile radius and v_i its impact velocity. The near-surface zone in this case is comparable in thickness to a fraction of the projectile radius. Since the cratering event itself displaces many thousands of times the projectile mass, the quantity of lightly-shocked, high speed ejecta is small, amounting to only a few percent of the projectile's mass (for ejecta speed > few km/sec). The fastest solid ejecta leave at about 1/2 the impact velocity.

Although the total quantity of high speed solid ejecta is thus small in comparison to the total crater ejecta, it is significant because no other process yields such high velocity fragments. Many meteorites appear to be near-surface samples of their parent bodies (many are regolith samples and one is a vesicular lava) and so may have been ejected by this process.

INTRODUCTION

One of the principal features of a hypervelocity impact is the spray of debris excavated from the final crater. This spray is highly organized in both space and time. In general, the highest velocity debris originates from close to the impact site and is launched first. Ejection velocity declines rapidly with increasing distance from the impact site, dropping off as the inverse third or larger power of distance. Ejection time, on the other hand, increases with increasing distance.

It is generally assumed that the fastest ejecta is also the most highly shocked. This assumption is based on the Hugoniot equation that relates shock pressure P to particle velocity u_p and initial density ρ_o:

$$P = \rho_o U u_p \tag{1}$$

where U is the shock velocity. For many materials shock velocity is a linear function of particle velocity:

$$U = c_B + s u_p \tag{2}$$

where c_B is the bulk sound speed and s is a dimensionless material constant. The shock pressure is then a quadratic function of the particle velocity,

$$P = \rho_o (c_B + s u_p)\, u_p \tag{3}$$

In a basalt target, for example, ρ_o=2860 kg/m^3, c_B=2.31 km/sec and s=1.62 (Gault and Heitowit, 1963). If the ejection velocity is equated to the particle velocity, as is commonly assumed (erroneously as I shall soon show), then ejecta launched at 2.4 km/sec (lunar escape velocity), should have been subjected to a minimum shock pressure of P=42.5 GPa -- enough to partially melt the rock.

There is a serious flaw in this argument, however. This flaw was dramatically emphasized by the 1983 discovery of Antartic meteorite ALHA81005, which originally came from the moon. Although this lunar soil fragment must have been ejected faster than 2.4 km/sec to escape the moon's gravity, it shows no detectable shock damage -- there is no petrographic evidence of shock pressures exceeding 20 GPa (Warren and others, 1983; Verkouteren and others, 1983), or perhaps even 15 GPa (Ryder and Ostertag, 1983) in the meteorite's matrix. Furthermore, rapidly accumulating evidence strongly suggests that the eight SNC (Shergottite, Nakhalite, and Chassigny) meteorites came from Mars (see the recent review by McSween, 1985). Although one of these meteorites may have been shocked to 50 GPa, others show no sign of shock damage. Since Mars' escape velocity is approximately 5 km/sec, the problem of how high speed ejecta are launched from an impact crater without severe shock damage becomes even more serious.

Part of the solution of this problem comes from the process of spallation. When the stress wave generated by an impact encounters the free surface of the target, the compressive stress wave is reflected as a tensile rarefaction wave. The interference of these two waves near the surface leads to *velocity doubling* (Walsh and Christian, 1955) in which the ejecta is thrown out at twice the particle velocity of the incident wave. Velocity doubling only works for dense rocks: In highly porous rocks such as lunar regolith the ejection velocity is only slightly greater than the particle velocity due to the irreversible compaction of the rock by the shock wave.

On the basis of velocity doubling and using equation (3), the minimum shock pressure experienced by lunar ejecta must be 14.6 GPa -- uncomfortably close to the lower limit of 15 GPa for AHLA81005. Martian ejecta are subjected to at least 45 GPa, which is plausible for the most highly shocked specimen, but which is much too high for the other seven SNC meteorites.

In this paper I show that not only are ejection velocities increased by stress wave interference in the near-surface spall zone, but maximum pressures are reduced by the same phenonemon. The essential physical process that reduces the pressure is the finite rise time of the stress pulse. The free surface, by definition, is a surface where the pressure vanishes at all times. However, if the stress pulse is viewed conventionally as a sharp-fronted wave with negligible rise time, the pressure jumps discontinuously from zero at the free surface to its maximum value just beneath. This sharp-fronted wave, however, is merely a convenient approximation to a wave with a small, but finite, rise time. Although this approximation is adequate for many purposes, it is highly misleading when near-surface processes are considered.

The interference of waves with finite rise times in the near-surface spall zone produces a thin layer in which the largest pressure ever attained rises gradually from zero at the free surface to the stress wave's full value at some depth. This depth is determined by the rise time, and is roughly equal to half the rise time multiplied by the propagation speed.

Material in this interference zone is both protected from the peak pressures in the stress wave and accelerated to high speed, since its acceleration g,

$$g = \frac{1}{\rho} \frac{dP}{dz} \tag{4}$$

is proportional to the pressure *gradient*, not the pressure itself. The possibility thus exists that some very high speed ejecta is thrown out in a solid condition, never having experienced high shock pressures. Such ejecta were actually noted in a theoretical computation by Ahrens and O'Keefe (1978), but no cause was assigned to the phenonemon. Although the quantity of such ejecta is not large, this material is of great significance for the study of meteorites and planetary geology in general. It may also be of importance in the more technological aspects of impact cratering.

IMPACT-GENERATED STRESS WAVES

Although the rise time of the stress wave generated by an impact plays a major role in governing the near-surface ejection process, very little is known about it. Only a few experimental measurements have been reported in the literature, and the rise times of numerical computations are frequently distorted by artificial viscosity.

A rough estimate of the rise time is obtained by equating it to the time required for the projectile to bury itself in the target. For a spherical projectile of radius a, this time τ is

$$\tau \simeq a/v_i \tag{5}$$

where v_i is the impact vleocity. This estimate is in reasonable agreement with two numerical computations in which artificial viscosity was *not* used (D. Orphal and T.J. Ahrens, 1983 personal communication), as well as VISAR interferometer measurements of stress wave structure in plates struck by small spheres (2.5 - 10 mm diameter) at more than 5 km/sec (Lipkin and Kipp, 1976).

The decay time of the stress pulse is comparable to, but probably longer than τ. It is roughly the time for the release wave from the rear of the projectile to propagate into the region of maximum pressure. Supposing that this distance is of order 2a (the projectile diameter), and that it propagates at speed c_L, then the decay time is $\beta\tau$, where β is a dimensionless number and

$$\beta\tau \simeq \frac{2a}{c_L} \tag{6}$$

The overall duration of the pulse is then $(1+\beta)\tau$.

The maximum velocity in the pulse is initially half the impact velocity, $v_i/2$, if the projectile and target have identical compositions. Many measurements of particle velocity decay curves near underground explosions (Perret and Bass, 1975) as well as impact computations (Austin and others, 1981) demonstrate that peak particle velocity falls as nearly the inverse square of distance away from the impact site. This is a much simpler dependence on distance than that shown by the pressure, and is a direct consequence of momentum conservation in the pulse when stresses greatly exceed material strength.

The stress wave expanding away from the site of the impact must be clearly distinguished from the excavation flow that eventually opens the crater. The stress wave moves quickly, never traveling slower than the bulk sound speed c_B of the target material. Peak particle velocities in this wave decline from a maximum of $v_i/2$, attained near the projectile itself, and are usually 3 to 5 times larger than the excavation flow velocity at the same radius. Pressures are large only in this stress wave, and pressure contours become isolated or "detached" on the surface of an expanding hemispherical shell early in the cratering event (Bjork and others, 1967).

The excavation flow, on the other hand, develops behind the stress wave. Pressures are nearly zero in this region and velocities slow enough that the flow can be treated as subsonic and incompresssible. The radial flow velocity can be represented as a simple power of the distance from the impact site. These assumptions are, in fact, the starting point of Maxwell's Z-model (Maxwell, 1977) which often provides an excellent representation of the excavation flow field. The relation between the stress wave and excavation flow is treated in more detail by Melosh (1985a).

Nearly all of the debris excavated from the crater is thrown out by the excavation flow. The maximum ejection velocity of this material is 1/3 to 1/5 of the peak particle velocity, or between $v_i/6$ and $v_i/10$. Since this material is excavated from well below the free surface, it exhibits the usual Hugoniot relation between peak particle velocity and *minimum* shock pressure. Because flow lines cross contours of the highest pressure attained, ejecta at all speeds contains fragments shocked to higher pressures than predicted by the Hugoniot relation, equation (3). The excavation flow patterns cannot, however, account for the inclusion of debris fragments *less* shocked than the Hugoniot relation predicts.

In addition to the debris thrown out by the excavation flow, a small amount of material is ejected directly by stress wave interference near the free surface. This material is ejected early, at the time the stress wave first reaches a given point on the surface. For this reason I will refer to this material as the *prompt ejecta.* This ejecta is also fast, since the particle velocity driving it is that of the stress wave that leads the excavation flow. This velocity is then doubled by near surface interactions. The shock level in this ejecta varies from zero for debris originating from the surface itself, to values which may approach the Hugoniot prediction if the rise time of the wave is extremely short.

The prompt ejecta is thus the fastest and most lightly shocked debris thrown out from a given distance from the impact site. Its time of ejection is also quite different from that of the excavation flow. Prompt ejecta is launched when the stress wave arrives at a given point, and its ejection time thus depends approximately linearly on distance from the impact site (assuming the stress wave travels at a nearly constant velocity), whereas the ejection time of debris launched by the excavation flow depends upon the square root of the distance for craters in the gravity regime (Housen and others, 1983). Since the velocity of the prompt ejecta declines with depth below the surface (see Fig. 1), the two ejection styles grade into one another with increasing depth below the free surface.

Figure 1 illustrates the relations between the shock pressure and ejection velocity near the site of an impact. The stress wave generated by the impact is modeled as a spherically expanding wave that emanates from a point one projectile radius below the surface. This is a good approximation except for points on the surface close to the impact site. The stress wave profile is assumed to be triangular with rise time τ given by equation (5) and with decay time $\beta\tau = 4\tau$. The effect of the free surface is modeled by a tensile wave emanating from an "image source" located one projectile radius *above* the surface (for details, see Melosh, 1984).

Figure 1a displays contours of the maximum pressure ever attained at a given point at depth z and epicentral distance s. The effect of the free surface in reducing the maximum pressures is evident. In general, pressures are reduced from the maximum occuring at a given radius from the effective center of burst for all depths less than z_p, where

$$z_p = \frac{c_L\tau}{2}\left\{1+\frac{s^2}{d^2-(c_L\tau/2)^2}\right\}^{1/2} \tag{7}$$

where c_L is the mean longitudial wave speed and d is the depth of the effective center of burst.

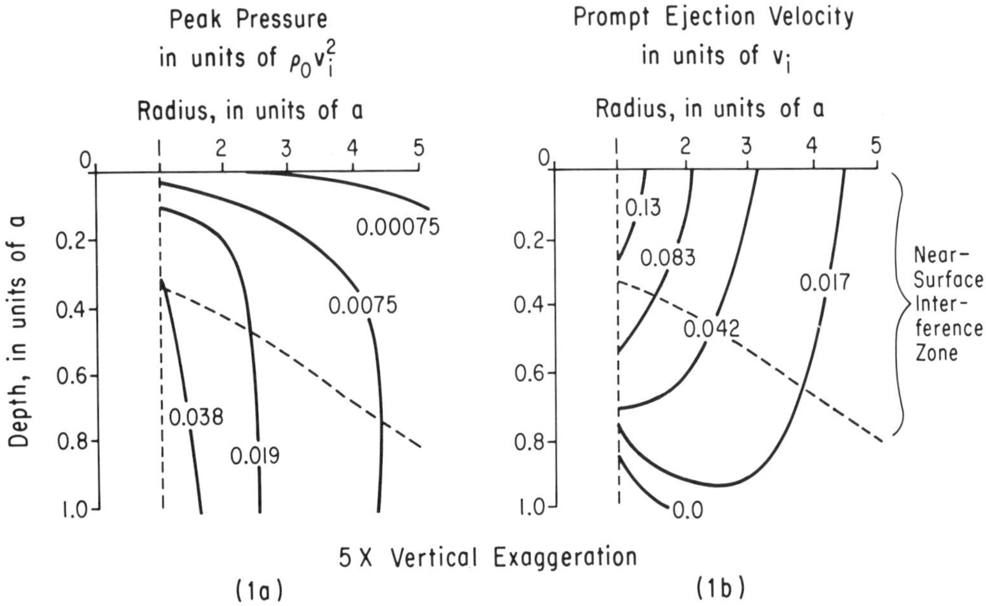

Fig. 1. Contours of peak pressure (in units of $\rho_0 v_i^2$) and ejection velocity (in units of v_i) near the surface surrounding the site of a vertical impact. All distances are scaled by the projectile radius a, and the unphysical region within one projectile radius of the origin is left blank (the projectile passes through this region, so the model described in the text does not work here). The dashed line defines the bottom of the near-surface interference zone, whose depth z_p is given by equation (7).

Figure 1b illustrates the prompt ejection velocities derived from this model. A material point in the interference region is considered to be "ejected" when the pressure first drops to zero after attaining its maximum value. This prescription neglects the material's tensile strength, but this is generally small compared to the maximum pressure excursion and little error seems to result from this simplification. In all cases material closest to the surface is ejected first, and progressively deeper-lying material is ejected progressively later.

The prompt ejection velocity falls to zero at some depth below the surface. This does *not* mean that the ejection velocity is zero here, only that stress wave interference does not add any extra velocity component to the normal excavation flow.

It is clear from this figure that small amounts of material near the surface receive large velocity increments but never experience high shock pressures, as asserted earlier. The volume of this material may be roughly estimated from an equation (Melosh, 1985b)

$$\frac{V(\text{velocity} \geq v_e)}{V(\text{projectile})} \simeq 4.8 \; \frac{P_{max}}{\rho_o \beta c_L v_i} \left\{ 1 - \left(\frac{2v_e}{v_i} \right)^{1/3} \right\} \tag{8}$$

where $V(\text{velocity} \geq v_e)$ is the volume of debris ejected faster than velocity v_e and P_{max} is the maximum pressure experienced by the debris. For example, a 22 km/sec impact from which the stress wave propagates at 6 km/sec ejects about 1.6% of a projectile volume of debris shocked to less than 15 GPa at speeds greater than 3 km/sec.

Although this simple model provides some quantitative insight into the effects of near-surface stress wave interactions, many features have been idealized and it is unclear how realistic some of the results are, especially because the model relies upon linear superposition of inherently non-linear stress waves. The next obvious step in the investigation is to construct a numerical model in which the full shock dynamics and material strength properties are taken into account. The next section describes a one-dimensional model which fulfills some of these requirements. Although a two- or three-dimensional model would be more satisfying, full treatment of stress wave interference requires at least 300 computational cells to adequately resolve the stress pulse, making even 2-D computations prohibitively expensive at the present time. Nevertheless, much can be learned from the 1-D results described below.

A SIMPLE 1-D NUMERICAL SPALLATION MODEL

A one dimensional numerical wave propagation model was constructed to study some aspects of near surface spall, ejection, and stress wave interference near impacts. The model assumes a vertically-propagating compressive stress wave of initially triangular profile. Although many of the results scale with pulse width, I report here computations performed for a wave with a 50 m wide rising portion and a 200 m long decaying tail.

The numerical code is a fully explicit Lagrangian hydrocode which uses an algorithm patterned after Los Alamos's 2-D SALE code (Amsden and others, 1980). It incorporates material strength, plastic yielding at the Hugoniot elastic limit, and physical viscosity. The equation of state uses a Murgnahan-type elastic constitutive relation. When stresses become tensional it employs Grady and Kipp's (1980) fragmentation model. This model assumes an original Weibull distribution of crack nucleation centers

$$n(\varepsilon) = k\varepsilon^m$$

where n is the number of flaws that are activated at or below a tensile strain ε. Constants k and m are material properties. As flaws are activated the elastic moduli are decreased by a factor (1-D), where D(t) represents the damage due to crack growth. When D=1 the material is completely fractured. Grady and Kipp's original work required the solution of an integral equation that presents considerable numerical difficulty. In this work I used an approximation to the original equation more suitable for machine integration (Grady and Kipp, 1980).

$$\frac{dD^{1/3}}{dt} = \frac{(m+3)}{3}\,\alpha^{1/3}\,\varepsilon^{m/3} \tag{9}$$

where α is defined in terms of the Weibull parameters k and m and the crack velocity c_g:

$$\alpha = \frac{8\pi\, c_g^3\, k}{(m+1)(m+2)(m+3)} \tag{10}$$

In the original work for this paper I used an algorithm devised by Chen and others (1983) which provides a close approximation to equation (9). However, Chen and others' (1983), equation for fragment size gives results that differ greatly from Grady and Kipp's (1980) analytic results, even for a pulse with constant strain rate. It can be shown that the fracture area A(t), from which fragment size is derived, is related to the rate of increase of the damage D(t) by

$$A(t) = \frac{\dot{D}(t)}{2c_g} \tag{11}$$

This equation is perfectly general and is derived from Grady and Kipp's (1980) full set of integral equations without any approximation whatever. However, equation (11) is difficult to implement numerically with high accuracy. A better algorithm is similar in form (but with different constants) to Chen and others (1983):

$$\frac{dA}{dt} = \frac{(m+2)(m+3)}{2c_g}\,\alpha^{2/3}\,\varepsilon^{2m/3}\,D^{1/3} \tag{12}$$

which is exact for constant strain rate. The fragment diameter at the maximum of the fragment size distribution L_{GK} can be shown to be

$$L_{GK} = \frac{3(m+3)}{(m+2)}\,\frac{1}{A(t_f)} \tag{13}$$

where $A(t_f)$ is the fracture area integrated from equation (12) when the damage D(t) reaches 1.

In the computations reported here a triangular stress pulse was imposed at the bottom of a 100 m deep layer which was divided into 300 cells. The upper surface is stress free. During the computation the stress pulse rises to the surface at 6 km/sec, reflects as a tensile wave which interferes with the remainder of the oncoming compressive wave, and eventually brings the surface material into tension. The fragments that finally result are ejected at speeds that decrease with increasing depth of origin.

A noteworthy aspect of these computations is that a single spall plate is ejected from the surface itself, followed by larger volumes of deeper seated debris broken into fragments whose size is given by equation (13). This spall plate remains intact throughout the course of the calculation for weak incident stress waves, but since it contains considerable amounts of elastic energy when the incident stress wave is stronger, it is observed to break up further following ejection by strong waves.

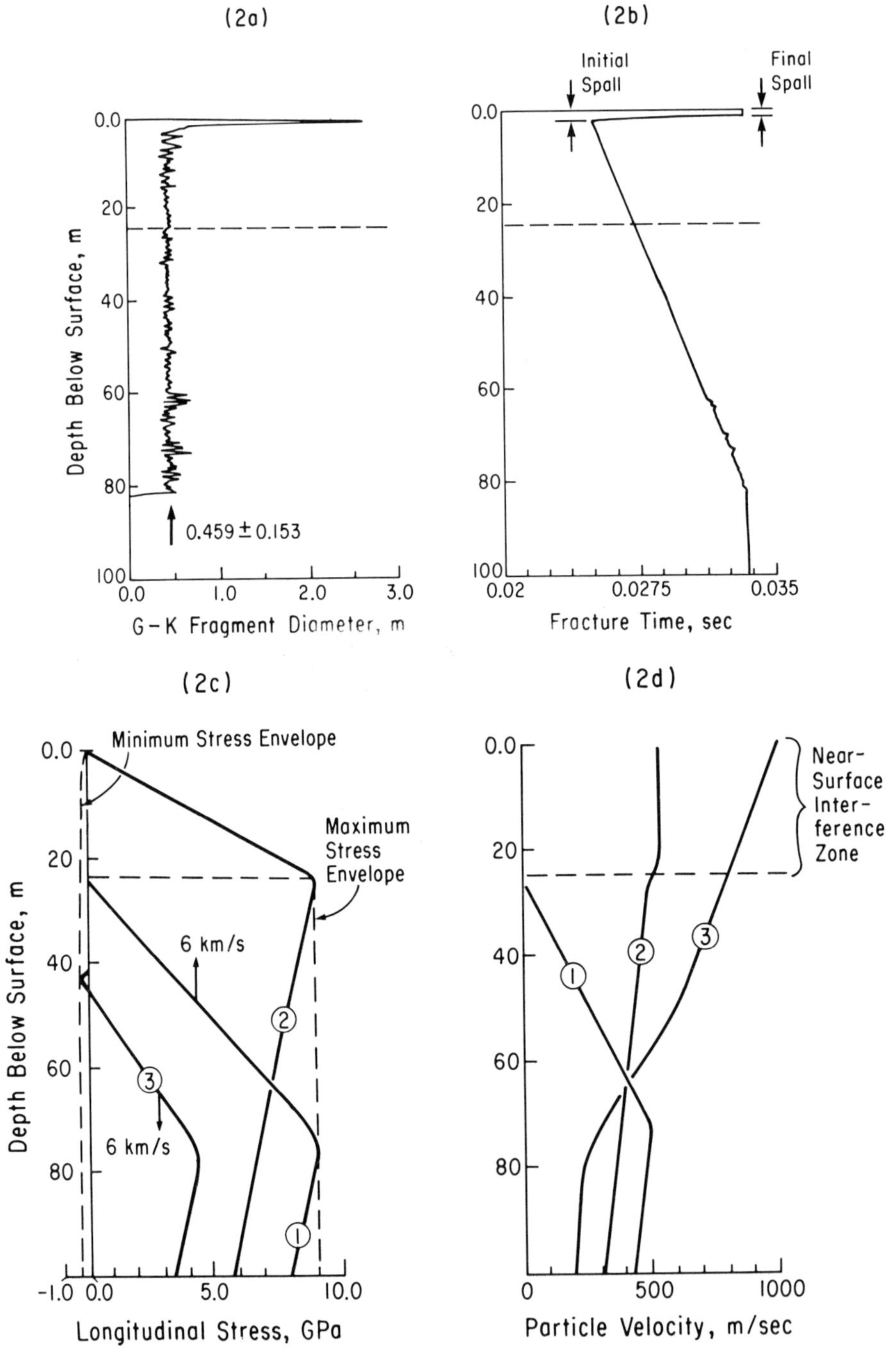

Fig. 2. The four parts of this figure illustrate (a) Grady-Kipp fragment size as a function of depth below a free surface on which a vertically-rising triangular stress pulse impinges, (b) time at which fragmentation occurs, (c) longitudinal stress (compressive stress is positive), and (d) particle velocity. The long dashed lines in (c) are envelopes of maximum and minimum stress and the horizontal short dashed lines mark the bottom of the near surface interference zone. The numbers (1), (2), and (3) in (c) and (d) refer to times 0.0125, 0.0208, and 0.0292 seconds, respectively, after the start of the computation.

Figure 2 illustrates the fragment size, ejection time, and the evolution of longitudal stress and particle velocity for a pulse in which the original peak particle velocity is 500 m/sec. Near-surface material is ejected at 1000 m/sec, exhibiting velocity doubling. Figure 2a shows the Grady-Kipp fragment size, which averages 0.459± 0.153 m. This model uses a relatively strong material compared to Grady and Kipp's (1980) oil shale measurements, for which the dynamic tensile strength is 0.194 GPa at the 10 sec^{-1} strain rate computed in this case. The fragment size is independent of depth, except for one cell just beneath the spall plate in which the computed fragment "size" is 2.6 m, about 6 times larger than the mean size. Since the spall plate itself is only about 2 m thick, this computed size is unphysically large: The algorithm has trouble computing an accurate size in the cell in which the spall fracture forms. Elsewhere, the algorithm works fine.

Figure 2 b illustrates the time at which fracture (D=1) occurs for each cell, if it does at all. In general, fracture begins near the surface and propagates downward, following the progress of the reflected tensile wave. However, very close to the surface the first fracture takes place at a depth of 1.84 m below the free surface. This fracture defines the spall plate, which is ejected vertically at 1000 m/sec. The time-of-fracture plot shows that the lower part of this plate (from which the unphysically large Grady-Kipp fragment described above is derived) breaks up somewhat later, leaving a final 1-m thick plate that survives until the end of the computation.

The plots of longitudinal stress and particle velocity, Figs. 2c and 2d show the propagation of the stress pulse to the surface and its reflection there. Tensile stresses are limited by the dynamic tensile strength of 0.194 GPa. This strength is computed during the calculation from the Grady-Kipp model. The tensile stress excursion is truncated when fracture occurs and the strength drops to zero. This can be seen on the curve labeled (3) on Figure 2c, where the stress dips just below the axis toward tensional stresses, becomes tangent to the minimum stress envelope, then rises to zero in the space of a few meters. Crack growth and fragmentation are occuring in this region, and the final complete failure is marked by the return of the tensile stress to zero.

Figure 2d is a plot of the particle velocities, which vividly demonstrates velocity doubling near the surface and the decline of ejection velocity with depth below the surface. The slope of this decay is 0.8 times the slope of the rising portion of the stress pulse. In general, for a triangular pulse, this slope should be $4/(\beta+1)$ times the slope of the rising portion, so this agrees exactly with the numerical computation for the value $\beta=4$ actually used.

The effect of stress wave interference in limiting the maximum pressure attained is clear from the maximum pressure envelope in Fig. 2c. The maximum pressure in the pulse, 8.99 GPa, is attained everywhere except for a zone near the surface, where it starts to fall at a depth of 25 m (exactly one-half the width of the rising portion of the pulse) and drops linearly until it reaches zero at the free surface. Debris in this zone, which includes the surface spall plate, is ejected at very high speed but suffers only slight shock damage.

Computations of this type were performed at a variety of impact velocities. The results of each computation are broadly similar to the 500 m/sec computation just described, but fragment sizes, of course, decrease with increasing velocity. As a check on the accuracy of the integration, equation (9), the computed dynamic tensile stresses are plotted in Fig. 3 as a

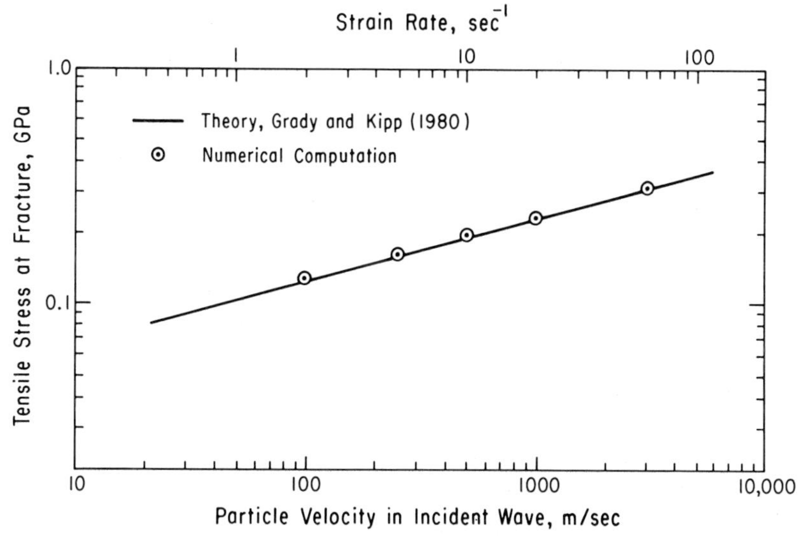

Fig. 3. Comparison of the dynamic tensile strength computed from equation (24) of Grady and Kipp (1980) with the minimum (tensile) stress computed by the hydrocode. Although the hydrocode results (points on the plot) are slightly high, agreement is better than 1% in all cases.

function of incident particle velocity (or, equivalently, strain rate). These stresses computed by the hydrocode are compared to predictions based on Grady and Kipp's (1980) analytic solution for a linearly increasing tensile stress. The numerical results typically fall less than a percent too high: The agreement is impressively good over the entire range of velocities considered, 30 to 3000 m/sec.

Figure 4 is a compilation of all fragment size data over the full range of velocities computed. Although the plots are presented specifically for the pulse considered here, with a rising portion 50 m wide (suitable for describing, for example, the impact of a 100 m radius projectile at 12 km/sec), theoretical considerations indicate that all the fragment sizes scale as $\sigma_t\dot{\varepsilon}$, where σ_t is dynamic tensile strength and $\dot{\varepsilon}$ is strain rate. Note that σ_t itself is a function of $\dot{\varepsilon}$ (Fig. 3) raised to the power 3/(m+3).

The results of these computations may be compared to the analytical results of Melosh (1984). In that paper the spall thickness z_s was predicted to be

$$z_s = \beta \, \frac{w}{2} \, \frac{\sigma_t}{P_{max}} \tag{14}$$

where w is the width of the rising portion of the stress pulse ($w = c_L\tau$), and P_{max} is the maximum stress in the wave. The spall thicknesses predicted by this equation are in excellent agreement with the computed initial spall plate thicknesses (see Fig. 4).

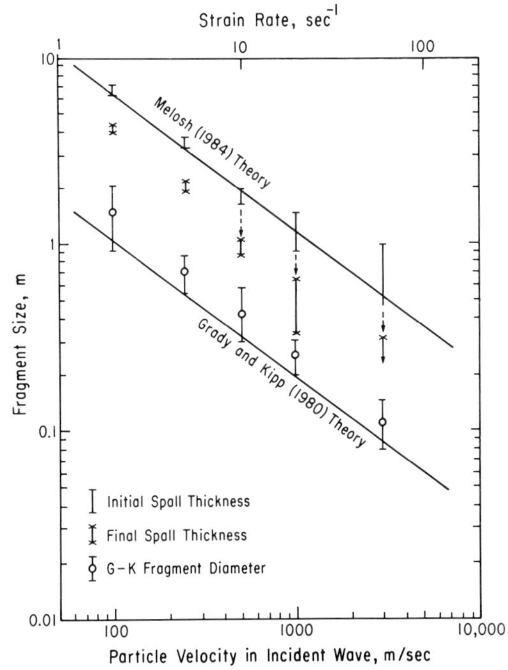

Fig. 4. Initial spall thicknesses, final spall thicknesses, and Grady-Kipp fragment sizes from a number of hydrocode runs at various velocities. The uncertainty in the spall thicknesses is the size of a single computational cell (0.33 m), whereas the Grady-Kipp fragment size uncertainties are computed statistically from the large number of cells fractured. The initial spall size agrees well with the theory of Melosh (1984). The reason for the slight discrepancy between theory and code results for the Grady-Kipp fragment sizes is discussed in the text.

Similarly, the predicted Grady-Kipp fragment size from Melosh (1984) is

$$L_{GK} = A(m)\, w \, \frac{\sigma_t}{P_{max}} \tag{15}$$

where

$$A(m) = \frac{6(m+4)^{[(m+4)/(m+3)]}}{(m+2)(m+3)} \frac{c_g}{c_L} \quad (16)$$

The predicted fragment size is about a factor 1.5 smaller than the median fragment size resulting from the computations, although the computed size lies within one standard deviation of the prediciton. This discrepancy may be an artifact of the one-dimensional computation. In one dimension the failure of one computational cell partially relieves the stresses acting on adjacent cells (leaving extra-large fragments within those cells) because there is no lateral support from surrounding material as there would be in two or more dimensions. The 1-D Grady-Kipp fragmentation model thus has an extreme amount of strain softening near failure and may consequently be somewhat unstable. This instability is responsible for the spiky cell-to-cell variations in G-K fragment size in Figure 2a. In a 2-D model lateral support would smooth out this instability and eliminate the extra-large fragment sizes computed. Since the smallest computed fragment sizes (Fig. 4) agree well with the prediction (Equation 16), there may in fact be no discrepancy.

CONCLUSION

The principal conclusion of this work is that impact events eject a small quantity of solid, lightly shocked, material at very high speed. This material comes exclusively from near the target's free surface and is due to the phenomenon of stress wave interference, or spallation, near that surface. The time at which this material is ejected is also much earlier than that of the main excavation flow, hence the name "prompt ejecta" for this material.

Although the amount of material thus ejected is small, this process has great significance for the ejection of meteorites from their parent bodies. Indeed, it is becoming clear that the preponderance of meteorites were in a near-surface environment immediately prior to their ejection.

The largest fragments that are thus ejected are spall plates from just under the surface. These spall plates are followed by a mass of uniform-size fragments whose diameters are accurately given by the theory of Grady and Kipp (1980). The spall plates themselves break down into smaller fragments, whose size approaches the Grady-Kipp fragment size in the highest speed ejection events.

This sequence of events is in good agreement with observations of secondary craters around large lunar impact craters (Vickery, 1986). These secondaries, whose size implies that they were made by enormous fragments (up to 1 km in diameter for the 93 km diameter crater Copernicus), are probably created by spall plates thrown off the surface near the site of the impact. The breakup of spall plates in flight accounts for the clustering so common among secondary craters. The paucity of secondary craters made by fragments traveling more than about 1 km/sec is explained by the breakdown of spall plates into Grady-Kipp size fragments at high ejection velocities, as seen in Fig. 4. The total volume of material that formed the observed secondaries also agrees with the estimated volume of spalled rock (Vickery, 1986).

The investigation reported here can be extended in a number of ways. The most obvious extension is to perform a full 2-D simulation of a vertical impact and follow the evolution of the stress wave in detail. However, the results of the 1-D computations show that even 300 cells covering the stress pulse is barely adequate to resolve the near-surface stress wave interactions. A 2-D simulation of a complete impact at this resolution would be prohibitively expensive (at a minimum, the projectile radius should be 300 cells, so a grid 5 projectile radii wide and 5 deep must involve at least 2.25×10^6 cells!).

Other extensions that would make the models more applicable to real geologic situations are the addition of layers with varying strength and elastic properties in the target, consideration of the effects of an elastic precursor and the addition of possible air (or vapor)-blast loading on the surface as the stress wave begins to interact with it.

The simple model described here, however, makes it clear that the interaction of the stress wave from an impact with the free surface leads to a variety of hitherto overlooked phenonema that may be of great geologic importance.

REFERENCES

Ahrens, T.J. and J.D. O'Keefe (1978). *Proc. Lunar Planet. Sci. Conf. 9th*, 3787-3802.
Amsden, A.A., H.M. Ruppel, and C.W. Hirt (1980). *Los Alamos Report LA-8095*.
Austin, M.G., J.M. Thomsen, S.F. Ruhl, D.L. Orphal, W.F. Borden, S.A. Larson, and P.H. Schultz (1981). *Multi-Ring Basins* (eds. P.H. Schultz and R.B. Merrill). *Proc. Lunar Planet. Sci. 12A*, Pergamon Press.
Bjork, R.L., K.N. Kreyenhagen, and M.H. Wagner (1967). *NASA Report CR-757*.
Chen, E.P., M.E. Kipp, and D.E. Grady (1983). *Mechanics of Oil Shale* (eds., K.T. Chang and J.W. Smith).
Gault, D.E., and E.D. Heitowit (1963). *Proc. Sixth Hypervelocity Impact Symp.* **vol. 2**, pp. 419-456.
Grady, D.E., and M.E. Kipp (1980). *Int. J. Rock Mech. Min. Sci. and Geomech., Abstr.* **17**, 147-157.
Housen, K.R., R.M. Schmidt, and K.A. Holsapple (1983). *J. Geophys. Res.* **88 B3**, 2485-2499.
Lipkin, J., and M.E. Kipp (1976). *J. Appl. Phys.* **47**, 1979-1986.
McSween, H.Y., Jr. (1985). *Rev. Geophys.* **23**, 391-416.
Maxwell, D.E. (1977). *Impact and Explosion Cratering* (eds. D.J. Roddy, R.O. Pepin, and R.B. Merrill), pp. 1003-1008. Pergamon Press.
Melosh, H.J. (1984). *Icarus* **59**, 234-260.
Melosh, H.J. (1985a). *Icarus* **62**, 339-343.
Melosh, H.J. (1985b). *Geology* **13**, 144-148.
Perret, W.R., and R.C. Bass (1975). *Sandia Report SAND74-0252*.
Ryder, G. and R. Ostertag (1983). *Geophys. Res. Lett.* **10**, 791-794.
Verkouteren, R.M., J.E. Dennison, and M.E. Lipschutz (1983). *Geophys. Res. Lett.* **10**, 821-824.
Vickery, A.M. (1986). *Icarus* **67**, 224-236.
Walsh, J.M. and R.H. Christian (1955). *Phys. Rev.* **97**, 1544-1556.
Warren, P.H., G.J. Taylor, and K. Keil (1983). *Geophys. Res. Lett.* **10**, 779-782.

Int. J. Impact Engng Vol. 5, pp. 493–499, 1987
Printed in Great Britain

0734-743X/87 $3.00 + 0.00
Pergamon Journals Ltd.

THE SIZE DISTRIBUTIONS OF FRAGMENTS EJECTED AT A GIVEN VELOCITY FROM IMPACT CRATERS

John D. O'Keefe and Thomas J. Ahrens

Seismological Laboratory 252-21
California Institute of Technology
Pasadena, CA 91125

ABSTRACT

The mass distribution of fragments that are ejected at a given velocity for impact craters is modeled to allow extrapolation of laboratory, field, and numerical results to large-scale planetary events. The model is semi-empirical in nature and is derived from (1) numerical calculations of cratering and the resultant mass versus ejection velocity, (2) observed ejecta blanket particle size distributions, (3) an empirical relationship between maximum ejecta fragment size and crater diameter, (4) measurements and theory of maximum ejecta size versus ejecta velocity, and (5) an assumption on the functional form for the distribution of fragments ejected at a given velocity. This model implies that for planetary impacts into competent rock, the distribution of fragments ejected at a given velocity is broad, e.g. 68% of the mass of the ejecta at a given velocity contains fragments having a mass less than 0.1 times a mass of the largest fragment moving at that velocity. The broad distribution suggests that in the impact process, additional comminution occurs after the initial shock has passed. This additional comminution produces the broader size distribution in impact ejecta as compared to that obtained in simple brittle failure experiments.

INTRODUCTION

Meteorites come from collisions on objects varying in size from asteroidal to terrestrial planets (Ostertag and Ryder, 1983; Bischoff and Stöffler, 1985; Bogard and Johnson, 1983; Wasson and Wetherill, 1979; Melosh, 1984; McSween, 1985; Ott and Begemann, 1985; Nyquist, 1982). Massive extinctions on the earth may have resulted from large impacts (Alvarez, et al., 1980). The origin of meteorites is dependent on the maximum-size fragment that can be ejected at velocities exceeding the planetary escape velocity, and on the dynamics of atmospheric entry, whereas the biological consequence of large impacts is dependent on the number and mass of fine particles (diameters $\lesssim 10^{-6}$ m) that are ejected high into the atmosphere (O'Keefe and Ahrens, 1982).

A model was developed for the mass distributions of fragments that are ejected at a given velocity for impact events (O'Keefe and Ahrens, 1982). The model is semi-empirical in approach and is consistent with empirical data from small scale impact experiments, nuclear, and chemical explosion data, ejecta fragment size vs. velocity data, largest ejecta fragment size vs. crater size data and total mass ejected vs. impact energy. In addition, the model is based upon calculations of mass ejected as a function of ejection velocity and maximum fragment size as a function of strain rate models. Using the above data base, the distribution function for the fragments was derived. In this paper we have extended our previous approach and have found additional data that strengthen those conclusions.

CUMULATIVE MASS DISTRIBUTION OF FRAGMENTS

The objective of this paper is to determine the distribution of fragments that are ejected at a given velocity during an impact event. Shown in Fig. 1 is a typical impact-induced flow field at a given time. The figure demonstrates the decrease in ejecta velocity, V, with increasing radius, r, described by $V \propto r^{-3}$ (Housen et al., 1983). The crater plume consists of particles ejected at positions, r, at times, t, and the velocity of ejection varies as a function of position along the surface. In cases where the angle of ejection does not vary with r, the plane elongates in time, whereas in the case where the angle varies, the plume changes shape. For most cases the angle is nearly constant (Gault et al., 1963, Melosh 1984).

By definition, the cumulative amount of mass M_c of fragments having mass greater than m ejected at all velocities is given by

$$M_c \equiv \int_{V_{min}}^{V} \frac{\partial M_{cv}}{\partial V} f(m, m_{bv}(V)) dV \tag{1}$$

where $\frac{\partial M_{cv}}{\partial V}$ is the amount of mass ejected at velocities between V and V+dV, and $f(m, m_{bv}(V))$ is the unknown distribution of fragments ejected at V. The expression for the amount of mass ejected at a given velocity, M_{cv}, can be determined from the fits to the cumulative amount of mass ejected at velocities greater than V obtained either from theoretical calculations (O'Keefe and Ahrens, 1982; O'Keefe and Ahrens, 1977a, 1977b, 1986) (see Fig. 2) or experimental data (Stöffler, et al., 1975; Andrews, 1975).

$$\frac{\partial M_{cv}}{\partial V} = \frac{\xi M_T}{V_{min}} \left[\frac{V}{V_{min}} \right]^{-(\xi+1)} \qquad V_{min} \leq V \leq V_{max} \tag{2}$$

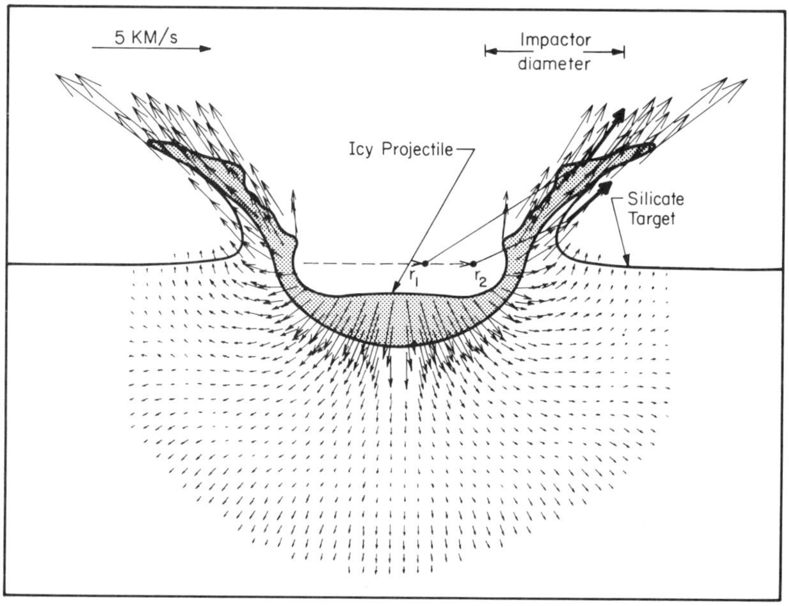

Fig. 1. Cratering flow field at a time, t, given by d/Vt=2.0, where V is impact velocity and d is projectile diameter for water projectile striking silicate half space at 5 km/sec. The relationship between the ejecta plume and surface velocity field is indicated. The velocity, V, of each fragment of ejecta is approximately related to its take-off radius, r, by $V \propto (R/r)^3$, where R is the projectile radius.

where M_T is the total amount of mass ejected, and V_{min} and V_{max} are the minimum and maximum velocities of ejection of solid particles. A discussion of the thermodynamic constraints on the maximum velocity of ejection is given by O'Keefe and Ahrens (1985). The exponent ξ is related to the mass scaling parameter (Housen, et al., 1983), α, by

$$\xi = 6\alpha/(\alpha - 3) \tag{3}$$

The mass scaling parameter specifies the relation between M_c and m_b, the maximum fragment mass.

$$\frac{M_c}{M_T} = [1-(m/m_b)^{\alpha/3}] \tag{4}$$

The minimum velocity of ejection varies with the planetary strength at small scales and Rg at large scales, where R is the radius of the impactor and g is the planetary gravitational acceleration. Shown in Fig. 3 is that data which is described by

$$m_b = 0.2 M_T^{0.8} \tag{5}$$

The key assumption in the theory is that the functional form of the distribution of fragments ejected at a given velocity is similar to the distribution function of the fragments in the ejecta blanket. For explosive events Jaeger et al. (1986) have found that this is the case. With this assumption the cumulative fraction of mass fragments of mass greater than m is given by

$$f(m, m_{bv}(V)) = \left[1 - (m/m_{bv})^{\frac{\beta}{3}} \right] \tag{6}$$

where m_{bv} is the mass of the largest fragment ejected at a given velocity.

In addition, we assume that the mass of the largest fragment ejected at a given velocity is a function of the velocity of ejection and is given by

$$\frac{m_{bv}(V)}{m_b} = (V/V_{min})^{-\delta} \tag{7}$$

where δ is an unknown parameter to be determined.

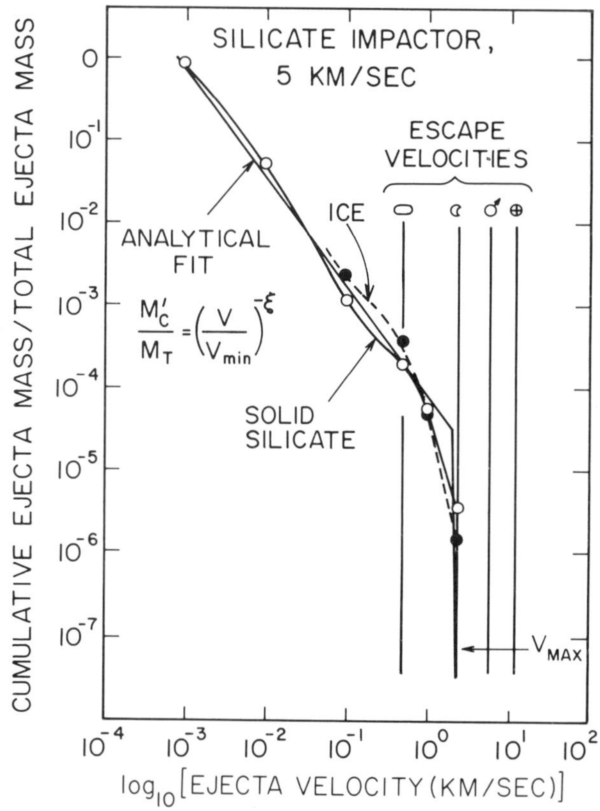

Fig. 2. Cumulative ejecta mass at velocity less than V versus ejecta velocity. Calculations for impacts of solid silicate and ice at 5 km/sec are indicated. Here $M_c/M_T = M'_c/M_T - 1$.

The above functional form can be justified by comparison to explosion cratering data (Schoutens, 1979), planetary cratering observational analysis (Vickery, 1986), and theoretical considerations. Ejecta velocities were measured on a 100 ton TNT above ground explosion called Middle Gust III, and are summarized by Schoutens (1979). The size-dependent ejection velocity limit inferred from film recording and analysis by Wisotski (1975) and corrections for deceleration in air (Seebaugh, 1975) is shown in Fig. 4. The upper envelope on limit of launch velocities correlates with $\delta \cong 3$. In addition, the maximum fragment size ejected at a given velocity can be estimated from analysis of planetary secondary craters distributions (Vickery, 1986). The upper bound on fragment size versus velocity is indicated in Fig. 5 for $\delta = 3$.

The theory developed by Grady and Kipp (1980) can be used to determine an expression for δ. They found that the mass of the largest fragment produced is a function of the strain-rate, $\dot{\epsilon}$,

$$m_{bv} \propto \dot{\epsilon}^{\,3l/(l+3)} \tag{8}$$

where $l \cong 8$ for rocks.

The strain-rate is related to the velocity field by

$$\dot{\epsilon} \propto \frac{\partial V}{\partial r} \tag{9}$$

where r is, again, the surface radial coordinate. The surface velocity field has the following form for impacts (Housen, et al., 1983).

$$V \propto \left(\frac{r}{R}\right)^{-3/\xi} \tag{10}$$

where R is the radius of the impactor. Evaluating the surface strain-rate, the mass of the largest fragment created at a given velocity is

$$m_b \propto \left(\frac{V}{V_{min}}\right)^{\frac{(\xi+3)l}{l+3}} \tag{11}$$

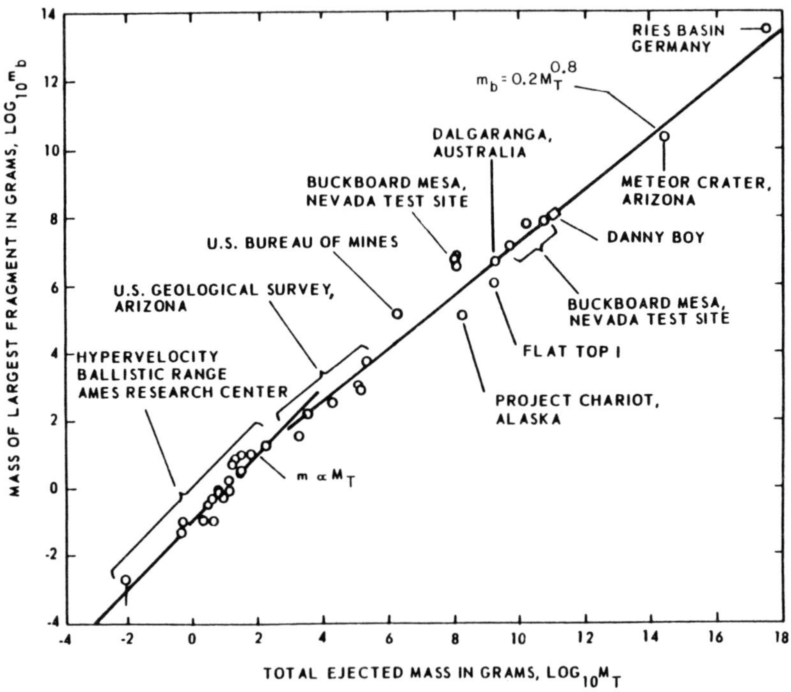

Fig. 3. Mass of the largest fragment versus total ejected mass for explosive and impact cratering events (Gault, et al., 1963).

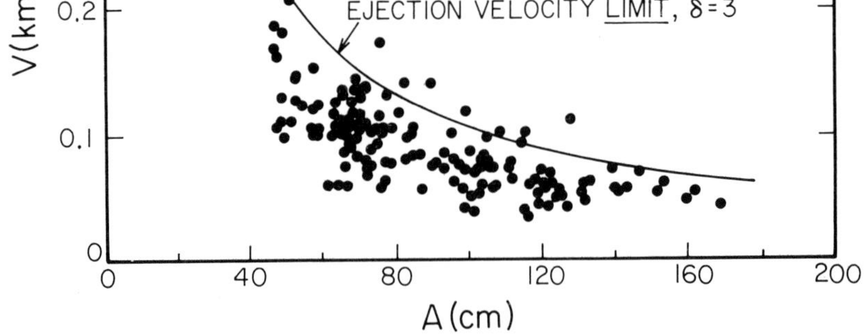

Fig. 4. Observed ejecta velocity, versus, ejecta diameter, A, 5 to 10 sec after explosion of MIDDLE GUST III, a 100 ton TNT surface explosion conducted over partially saturated Pierre shale (Schoutens, 1979). Size dependent ejecta velocity limit is bounded by $\delta \cong 3.0$.

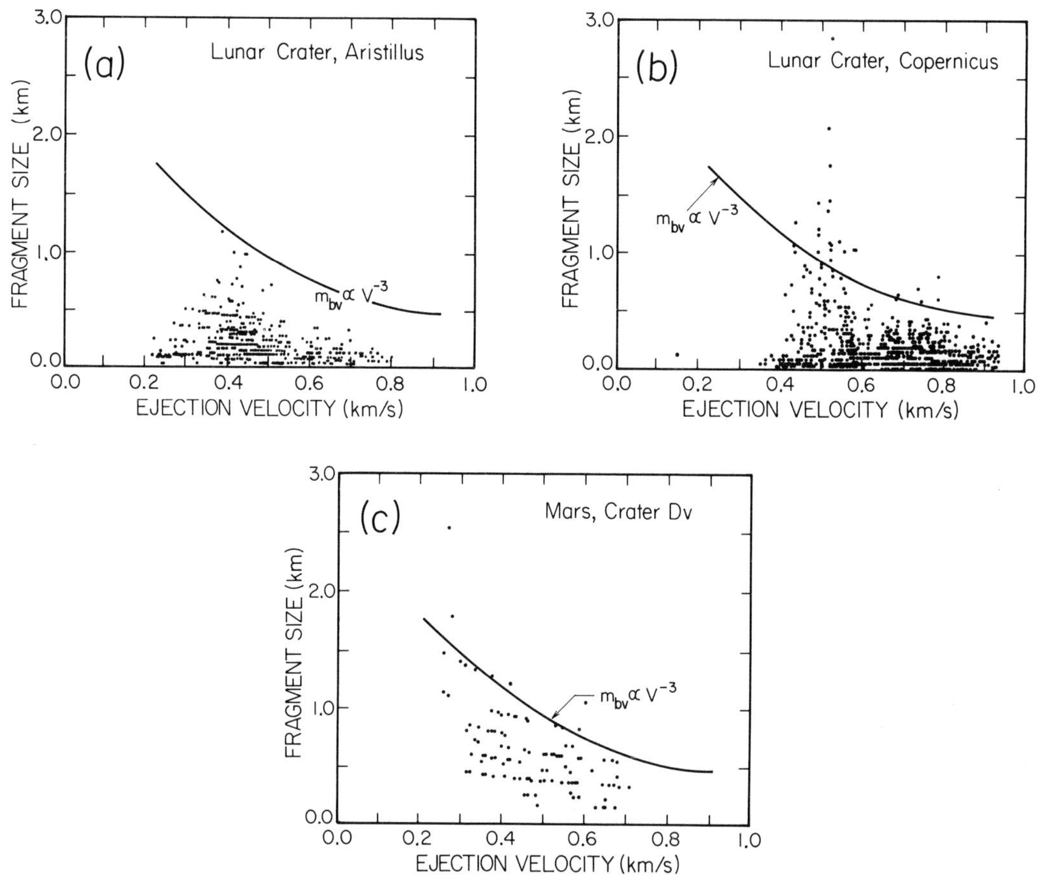

Fig. 5 Ejecta fragment size versus ejection velocity inferred from gravity scaling of secondary impact craters from Lunar (a) and (b), and Martian (c) craters (Vickery, 1986). Mass of fragments for relation $m_b \propto V^{-3}$ shown for $\delta = 3$.

and thus

$$\delta = (\xi+3)\, l\,/(l+3). \tag{12}$$

For a variety of materials, δ ranges from 3 to 3.6, using the l values from Grady and Kipp (1980) and ranges from Housen et al. (1983).

The integral for the cumulative mass distribution in the ejecta blanket can be evaluated by substituting Eqs. 2, 6, and 7 in equation 1, which yields the following

$$\frac{M_c}{M_T} = \left[1 - \left(\frac{m}{m_b}\right)^{+\frac{\xi}{\delta}}\right] + \left(1 - \frac{\delta\beta}{3\xi}\right)^{-1} \left[\left(\frac{m}{m_b}\right)^{+\frac{\xi}{\delta}} - \left(\frac{m}{m_b}\right)^{\frac{\beta}{3}}\right] \quad \text{for } m \geq m_s \tag{13}$$

$$= 0 \qquad \text{for } m < m_s$$

where m_s is the mass of the smallest fragments. Note that in the case of both impact craters (Gault, et al., 1963) and explosion events (Jaeger, et al., 1986) that there is a characteristic scale size, or mass, in which the number of smaller fragments rapidly decreases.

The above expression (equation 12) can be compared to experimental data on the cumulative mass distributions fragments ejected from impact craters (Gault, et al., 1963; Lange and Ahrens, 1981; Fujiwara, et al., 1977). Both impact and explosion fragments distributions can be described by expressions of the following form (O'Keefe and Ahrens, 1977a; Hartmann, 1969; O'Keefe and Ahrens, 1985).

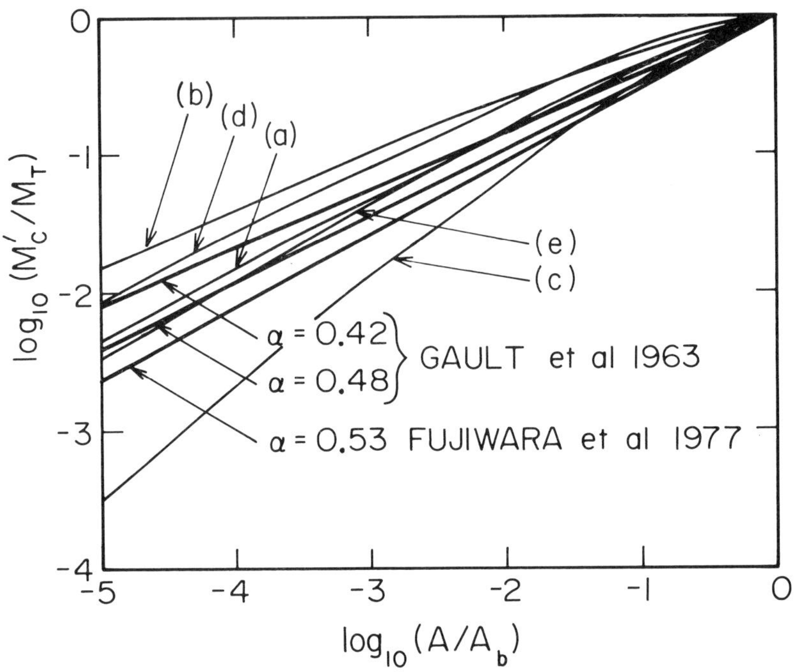

Fig. 6. Cumulative mass M_c (ejected at velocity V) normalized by the total mass M_T (ejected at velocity V) versus fragment size normalized to the largest fragment ejected at velocity V. Data is for a value of $\xi = 1.9$ as indicated and theoretical curves are present for $\beta = 0.4$, 0.5, and 0.6. Here $M_c/M_T = M_c'/M_T$-1. All curves correspond to $\xi = 1.36$ and curve (a) to $\beta = 0.5$ and $\delta = 3$; curve (b) to $\beta = 0.4$ and $\delta = 3$; curve (c) to $\beta = 0.6$ and $\delta = 3$; curve (d) to $\beta = 0.5$ and $\delta = 5$; and curve (e) to $\beta = 5.0$ and $\delta = 8.5$.

$$M_c/M_T = [1-(m/m_b)^{\alpha/3}] \tag{14}$$

where m_b was found by Gault et al. (1963) to correlate with the total mass ejected from the crater for large range of crater sizes.

The expression given by Eq. 13 for cumulative mass distribution of fragments in the ejecta blankets is compared to experimental data in Fig. 6. In the case shown, $\delta = 3$ and $\xi = 1.4$. The only unknown parameter in Eq. 13 is β, which is in the exponent in mass distribution of fragments ejected at a given velocity. Referring to Fig. 6, we get good agreement with the fragment size data when $\beta \cong 0.5$. Note that this has the same value as the cumulative mass distribution exponent α (Eq. 14). A low value of β (0.5) implies that the distribution of fragments ejected at a given velocity is polydispersed and that in the flow and ejection process, grinding and crushing has occurred since fracture processes are more monodispersed (Grady and Kipp, 1980; Hartmann, 1969).

The mass fraction of fragments that are ejected with velocities in fraction range ΔV at V can be obtained from Eqs. 3, 7, and 13, and is given by

$$\frac{\Delta M_{cv}}{M_T} = \xi\left(\frac{\Delta V}{V_{min}}\right)\left(\frac{V}{V_{min}}\right)^{-(\xi+1)}\left[1-\left(\frac{m}{m_b}\right)^{\frac{\alpha}{3}}\left(\frac{V}{V_{min}}\right)^{\frac{\delta\alpha}{3}}\right] \tag{15}$$

CONCLUSIONS

1. The ejecta velocity, V, of surface material achieved upon impact of a projectile of radius, R, varies with radial distance from the impact point as r^{-3}.

2. The mass of the largest fragment ejected at a given velocity is described by a power law function of ejecta velocity where the scaling exponent, δ, is given by $\delta = (\xi+3)l/(l+3)$ where l is the parameter in the expression for the largest fragment m_b produced upon brittle failure at varying strain rates as described by Grady and Kipp (1980). The most massive fragment launched at a given velocity, m_{bv}, is found to depend on velocity as, $m_{bv} \propto V^{-3}$, in agreement with results of Vickery (1986) and Wisotski (1975).

3. Both the cumulative mass of ejecta and the cumulative mass of ejecta launched at a given velocity, V, have similar power law distributions.

4. Brittle fragment distributions obtained in simple failure experiments are less dispersed, as to their size, in comparison to impact ejecta distributions which are very broadly dispersed. We infer that the additional dispersion is a result of comminution that occurs in the velocity shear field which is present in the region that becomes the ejecta material.

ACKNOWLEDGMENTS:

Research supported under NASA Grant NSG7129. We appreciate the computational assistance of Michael Lainhart and receiving preprints from Ann Vickery. We are grateful to Ann Vickery in allowing us to use Figs. 5a-c and the helpful comments of two reviewers. Contribution number 4325, Division of Geological and Planetary Sciences, California Institute of Technology, Pasadena, California 91125.

REFERENCES

Alvarez, L. E., W. Alvarez, F. Asaro, and H. V. Michel (1980). Extraterrestrial cause for the Cretaceous-Tertiary extinction. *Science, 208,* 1095-1108.

Andrews, R. J. (1975). Origin and distribution of ejecta from near-surface laboratory-scale cratering experiments, Report AFWL-TR-74-314, 207 pp.

Bischoff, A. and D. Stöffler (1985). Clast population statistics of the lunar meteorite Yamato 791197 - Sample from a new source region of the lunar highlands [abs.], *Lunar and Planetary Science XVI,* 63-64.

Bogard, D. D., and P. Johnson (1983). Martian gases in an Antarctic meteorite. *Science, 221,* 651-653.

Fujiwara, A., G. Kamimoto, and A. Tsukamoto (1977). Destruction of basaltic bodies by high-velocity impact. *Icarus, 31,* 277-288.

Gault, D. E., E. M. Shoemaker, and H. J. Moore (1963). Spray ejected from the lunar surface by meteoroid impact: *NASA Publication TND-1767,* p. 39.

Grady, D. E. and E. Kipp (1980). Continuum modeling of explosive fracture in oil shale. *Inst. J. Rock Mech. Min. Sci. & Geomech. Abstr., 17,* 147-157.

Hartmann, W. K. (1969). Terrestrial, lunar, and interplanetary rock fragmentation. *Icarus, 10,* 201-203.

Housen, K. R., R. M. Schmidt, and K. A. Holsapple (1983). Crater ejecta scaling laws: Fundamental forms based on dimensional analysis, *J. Geophys. Res., 88,* 2485-2499.

Jaeger, Z., R. Englman, and D. Slotky (1986). Fractal behaviour and velocity-dependence related to spatial variation of fragment sizes, *Proc. Int. Symp. on Fracture and Fragmentation,* Jerusalem, Jan. 5-8, 1986.

Lange, M. A. and T. J. Ahrens (1981). Fragmentation of ice by low velocity impacts, in *Proc. Lunar & Planet. Sci. Conf. 12th,* Pergamon Press, New York, 1667-1687.

McSween, H. Y., Jr. (1985). SNC meteorites: Clues to Martian petrologic evolution?, *Rev. Geophysics, 23,* 391-415.

Melosh, H. J. (1984). Impact ejection, spallation, and the origin of the meteorites, *Icarus, 59,* 234-260.

Nyquist, L. E. (1982). Do oblique impacts produce martian meteorites? *Proc. Lunar Planet. Sci. Conf. 13th,* Suppl. *J. Geophys. Res., 88,* A779-784.

O'Keefe, J. D. and T. J. Ahrens (1977a). Impact induced energy partitioning, melting, and vaporization on terrestrial planets. *Proc. Lunar Sci. Conf., 8th,* 3357-3374.

O'Keefe, J. D. and T. J. Ahrens (1977b). Meteorite impact ejecta: Dependence of mass and energy lost on planetary escape velocity. *Science, 198,* 1249-1251.

O'Keefe, J. D., and T. J. Ahrens (1982). The interaction of the Cretaceous-Tertiary extinction bolide with the atmosphere, ocean, and solid earth. *Proc. Conf. on Large Body Impacts and Terrestrial Evolution, Sp. Paper #190 Geol. Soc. Am.,* 103-120.

O'Keefe, J. D. and T. J. Ahrens (1985). Impact and explosion crater ejecta, fragment size, and velocity, *Icarus, 62,* 328-338.

O'Keefe, J. D. and T. J. Ahrens (1986). Oblique impact: A process for providing meteorite samples of other planets, submitted for publication.

Ostertag, R., and C. Ryder, (1983). ALHA 81005: Petrography, shock, Moon, Mars, Giordano Bruno, and composition (abstr.) in *Lunar and Planetary Science XIV,* Lunar and Planetary Institute, 23-24.

Ott, U., and F. Begemann (1985). Are all the 'martian' meteorites from Mars? *Nature, 317,* 509-512.

Schoutens, J. E. (1979). Empirical analysis of nuclear and high-explosive cratering and ejecta, *Nuclear Geophysics Sourcebook, IV-Part II,* Rep. DNA 65 01H-4-2, Defense Nuclear Agency, Washington, D.C.

Seebaugh, W. R. (1975). *Science Applications, Inc.:* DNA Report 3640 I.

Stöffler, D., D. E. Gault, J. Wedekind, and G. Polkowski (1975). Experimental hypervelocity impact into quartz sand: Distribution and shock metamorphism of ejecta. *J. Geophys. Res., 80,* 4062-4077.

Vickery, A. M. (1986). Size-velocity distribution of large ejecta fragments, *Icarus, 67,* 224-236.

Wasson, J. T., and G. W. Wetherill (1979). Dynamical, chemical, and isotopic evidence regarding the formation location of asteroids and meteorites, In T. Gehrels, (Ed.), *Asteroids,* U. Ariz. Press, pp. 926-974,

Wisotski, J. (1975). Middle Gust dynamic ejecta measurements, Denver Research Institute, Denver, CO. AFWL-TR-75-104. Air Force Weapons Laboratory, Kirtland Air Force Base, N. M.

Int. J. Impact Engng Vol. 5, pp. 501–507, 1987
Printed in Great Britain

0734-743X/87 $3.00 + 0.00
Pergamon Journals Ltd.

APPLICATION OF A 100-kV ELECTRIC GUN FOR HYPERVELOCITY IMPACT STUDIES*

J. E. Osher, H. H. Chau, G. R. Gathers, R. S. Lee†, and R. C. Weingart

Lawrence Livermore National Laboratory
University of California, Livermore, CA 94550

ABSTRACT

The Lawrence Livermore National Laboratory 100-kV electric gun has been used to launch flat-plate projectiles for use in studies of spall and hypervelocity impact penetration of thin plates. Impactors were 0.3-mm thick Kapton with dimensions and velocities ranging from 100 mm^2 at 4 km/s to 10 mm^2 at 18 km/s. A Fabry-Perot laser velocimeter, an electronic streak camera, and a flash x ray were used as diagnostics of the flyer-plate impact on the selected specimen. Experiments generally included the recovery of the remnant specimen and fragments for detailed examination, permitting a study of incipient spall, onset of melting, and fraction fragmented. Experiments to be described include spall measurements on simple and composite target walls at normal and oblique incidence and "reverse ballistics" impacts of the thin-plate impactor on a stationary penetrator (e.g., Kapton impactors at 15 km/s incident on rods of steel, aluminum, and lead) for calibration of hypervelocity impact codes.

INTRODUCTION

The device that we call an electric gun operates by discharging a capacitor bank through a thin metallic foil. Ohmic heating deposits a significant portion of the stored capacitor bank energy into the foil, causing the foil to explode. The explosion of the foil and accompanying magnetic forces drives a thin plate of material placed on top of the foil down a barrel to impact a target. We have described the operation of electric gun systems in some detail in an earlier publication (Chau and co-workers, 1980). Electric guns have been used at Lawrence Livermore National Laboratory (LLNL) since 1975 for high explosive studies (Weingart and co-workers, 1976) and more recently for hypervelocity impact studies of inert materials (Dein and co-workers, 1984; Froeschner and co-workers, 1984). We will describe two types of experiments: studies of spall from both simple and composite structure materials, and "reverse ballistics" impact studies of thin-plate projectiles on rods of steel, aluminum, and lead.

EXPERIMENTAL ARRANGEMENT

The standard experimental arrangement for the electric gun is shown schematically in Fig. 1(a). The exploding foil and flat-plate impactor are part of an expendable transmission line laminate that is clamped into a permanent flat-plate transmission line leading from a capacitor bank. The capacitor bank has been previously described (Chau and co-workers, 1981). The bank has a peak charging voltage of 100 kV, a capacitance of 17.3 μfd, and an inductance that ranges from 24 nH at 100 kV to about 34 nH at 35 kV. Maximum stored energy is about 87 kJ, and the bank is capable of exploding conductors with dimensions as large as 100 × 100 × 0.051 mm. The flat-plate impactor (flyer plate) was 0.3-mm Kapton for the work reported here, but may be of thicker or thinner Kapton[1] or of a Kapton-metal laminate if higher impact pressures are desired. The target assembly is placed at or near the end of the barrel, and the diagnostics employed will depend on the type of experiment conducted. For many of the experiments, the barrel is evacuated to eliminate the air shock ahead of the flyer plate, and if recovery is desired, an appropriate fixture may be employed to catch the target and fragments. Because the momentum and kinetic energy per unit area of the flyer plate are relatively low, compared to projectiles accelerated by a gas gun or a high explosive, recovery of impacted specimens is relatively simple.

Figure 1(b) gives a perspective of the 100-kV electric gun shot-confinement chamber. The 91-cm-i.d. door to the end of the chamber is shown open for a shot setup, and part of the closely coupled capactitor bank is shown in the background. A standard gun-laminate configuration for firing a thin flyer vertically upward is clamped to the input power transmission line (the clamp is hidden from view by the heavy insulating phenolic clamping blocks). Figure 1(c) is a close up of this partially completed shot assembly [similar to the cutaway shown in Fig. 1(a)]. A disposable laminate with a 10-cm^2 (outside dimension), 6.35-mm-long Lucite barrel is positioned over the laminate bridge foil (to-be-exploded) and epoxy bonded to the Kapton-covered laminate. This is covered by a target specimen over the gun muzzle, and finally topped by a diagnostic assembly consisting of a standard aluminum witness plate and a 5-mm-thick LiF crystal combination [for measuring the Al-LiF interface velocity with a Fabry-Perot (F-P) velocimeter for impulse attenuation studies]. For vacuum shots,

* Work supported by DNA/DOE. Work performed under the auspices of the U.S. Department of Energy by the Lawrence Livermore National Laboratory under Contract W-7405-Eng-48.

† Department of Physics, Kansas State University, Manhattan, KS 66506.

[1] Kapton is the tradename for Dupont's high dielectric strength, high temperature rated polyimide film. We use lay-ups of multiple layers of their type H film — each layer is 0.13-mm thick and backed with 0.025 mm of adhesive.

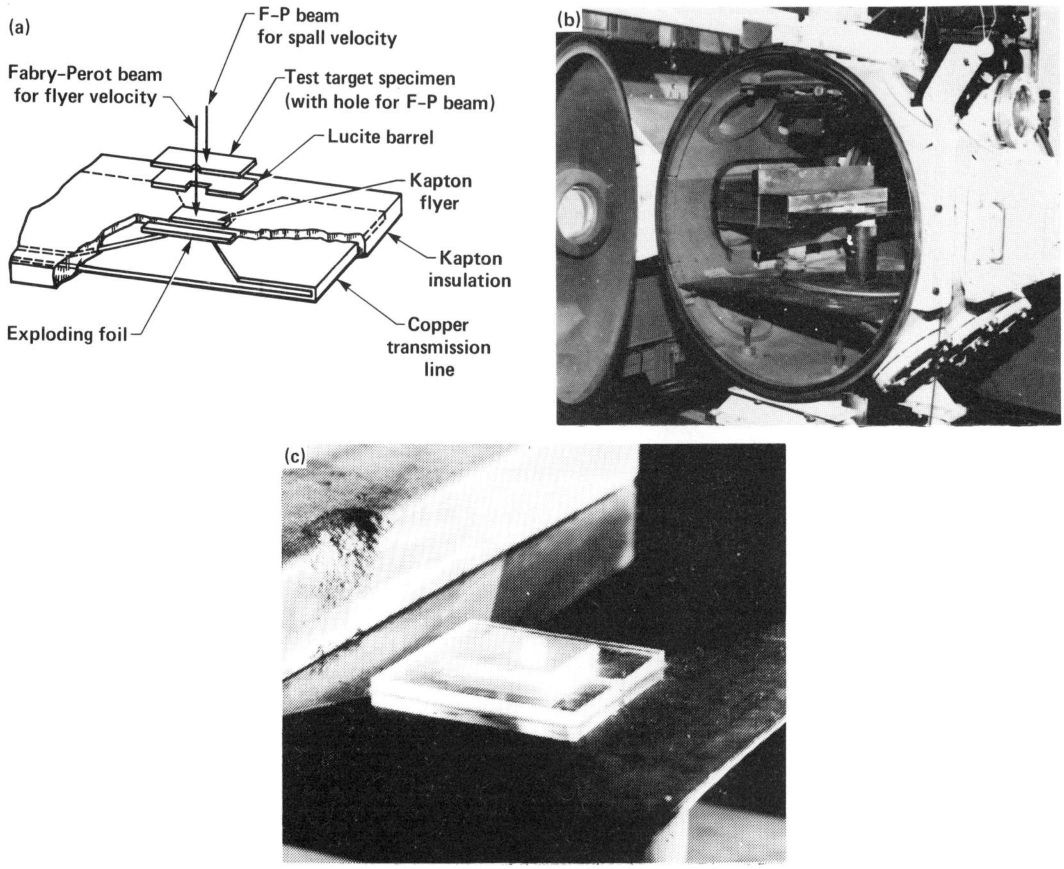

Fig. 1. (a) Exploded and cutaway view of a typical electric gun spall experiment, (b) shot confinement chamber, and (c) close up of partially completed shot assembly.

the setup is enclosed in an evacuated vacuum chuck also bonded to the laminate. The chamber door is closed for firing but the confinement chamber is not under vacuum. The laminate assembly and optical mirrors and windows are destroyed by the shot (a foam catch tank is incorporated over the shot when the specimen is to be recovered).

An F-P laser velocimeter (Goosman, 1979) is used to measure projectile velocities, velocity-time records of spalled surfaces, and jump-off velocities of shocked surfaces. This technique uses an F-P interferometer to measure the Doppler shift of light from an argon ion laser, which has been reflected from a silvered spot on the surface of interest. The collimated reflected light is focused by a cylindrical lens and passes through an F-P interferometer and a spherical lens to form a set of dots that go directly into the slit of an electronic streaking camera. The streak camera record consists of parallel lines that diverge as the surface accelerates, changing the wavelength of the reflected light. The velocity of the intact reflecting surface can be measured to about 1% by measuring the change in the spacing between pairs of lines.

A Thompson CSF Model TSN 506 electronic streaking camera is used to record a time-resolved slit image of the light produced at the collision interface by projectile impacts on flat or stepped blocks of glass or Plexiglas to measure impact simultaneity or as a backup velocity measurement. Figure 2 shows a streaking camera record of the impact of a 50-mm^2, 0.25-mm-thick Mylar flyer-plate at 3.7 km/s on a flat glass surface at a streaking rate of 50 ns/mm. The bow in the flyer front at impact is approximately 0.2 mm. The initial light in the figure is because of light from the exploding aluminum foil plasma passing through the Mylar flyer and glass witness plate. The dark central band is from an opaque silver spot painted on the Mylar that serves as a reflecting surface for the F-P diagnostic. At impact there is a bright surface flash, and then the pulverized glass becomes relatively opaque. The surface error is well within the limits for code calculation input needed for the applications reported here.

For flash x rays, we use either one or two x-ray tubes mounted outside thin Kevlar windows in the confinement chamber door and operated with a pulse width of 50 ns and a peak tube voltage of 100 kV. Because much of our work has been concerned with relatively low atomic-number materials, most of our experiments have used tube voltages in the 15 to 25 kV range. The film cassette is armored with

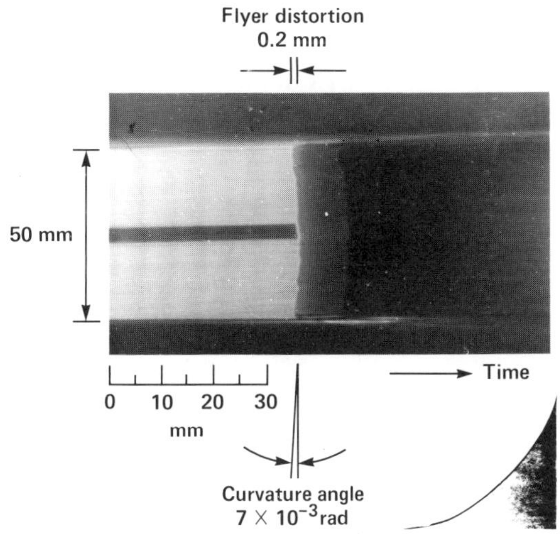

Fig. 2. Impact record of a streaking camera showing a 50-mm², 0.25-mm-thick Mylar flyer hitting a flat glass surface at 3.7 km/s. The sweep speed of the camera was 50 ns/mm.

Lucite and placed near the far laminate edge so that the image is only slightly enlarged. The x-ray image can be used to deduce relative density profiles (line-of-sight integrated mass density) and an average velocity from known timing of any distinguishable features.

HYPERVELOCITY IMPACT APPLICATIONS

Spall Studies

When a strong shock wave that decays rapidly reaches a free surface, the reflected tensile wave required to satisfy the zero pressure boundary condition results in extreme tensile states in the interior of the material. If the resulting tensile stresses exceed the strength of the material, the material ruptures or spalls, producing a new free surface. The process can then repeat itself until the remaining stress wave is no longer strong enough to rupture the material further. The amplitude of the original wave thus determines how many spall layers may occur, and the slope of the tail on the wave will determine the spacing between the ruptured regions.

Figure 1 shows the diagnostic arrangement used for the normal incidence spall experiments. The barrel used was made of Lucite to allow transmission of flash x rays, and the bore was square. A vacuum chamber with a window was added to eliminate air shock.

The shock wave entering a target specimen because of the impact of an electric gun flyer plate can be roughly characterized by an amplitude (stress or particle velocity) and a duration. The duration, τ, is adjustable with flyer thickness from approximately 10 ns to several hundred nanoseconds because it is determined by the travel time for release waves from the back of the flyer to overtake the leading edge of the shock. The amplitude of the pulse depends on the flyer velocity and the choice of density and material for both flyer and target. For most experiments, we can use flyer plates with lateral dimensions much greater than the thickness of the sample, hence the experiments are one dimensional and can be analyzed using a 1-D computer code.

A typical example of such a pulse as it enters a target is shown in Fig. 3, where an F-P velocimeter was used to measure the velocity of the interface between a 5-mm-thick LiF window and a 0.075-mm-thick aluminum buffer target that was shocked by a 0.3-mm-thick Kapton flyer plate with an impact velocity of 2.4 km/s. The purpose of these measurements was to characterize the amplitude and profile of the pulse produced in aluminum by the flyers used in this work. The lower curve shows the observed material velocity at the aluminum-LiF interface. The dip in the top of the curve is caused by the construction of the flyer, which was composed of two layers of 0.13-mm-thick Kapton and two layers of adhesive material that had nearly the same density. Experiments with single-piece flyers showed no dip. The LiF was chosen as a transparent material whose Hugoniot is known and that lies quite close to that of aluminum. We assume that the release isentrope followed in the material from the overtaking wave is approximated by the principal Hugoniot. (This is obviously not correct for releases involving large pressure changes because of the thermal pressures created in the material, but for steeply dropping pressures the difference is not important for the major part of the pulse.) We used the Hugoniot data tabulated by Marsh (1980). The upper curve in the figure showing pressure was calculated in this way. It corresponds to the pulse in the LiF. Calculated values of the expected release pressure and material velocity (U_p) using an approximate Hugoniot for Kapton are in agreement with the peak pressure deduced from the F-P record.

For the Kapton, the Hugoniot for the shock velocity U_s used was

$$U_s = 0.292 + 1.101\, U_p \quad , \tag{1}$$

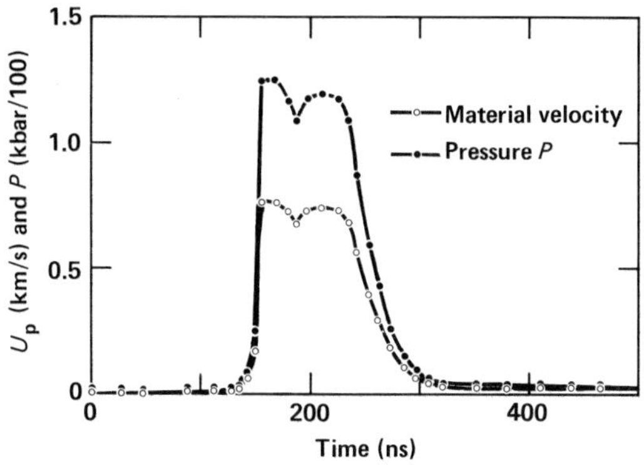

Fig. 3. Pulse shape and amplitude calibration.

TABLE 1 Spall Data for 6.35-mm-thick, 7.6-cm^2, 6061-T6 Al Test Specimens. Variables Tabulated, from Left to Right Are: Bank Charge Voltage, Flyer Velocity, Flyer Momentum, Fractional Mass Spalled, Peak Pressure in the Aluminum, Spall Fragment Velocity, and Angle of Flyer Incidence.

V_o (kV)	V_f (km/s)	I_t (ktap)	dm/M	P_{Al} (kB)	V_s (km/s)	Angle (deg)
			7.3-cm^2 Flyer			
45	0.45	1.94	No spall	16.1	—	90
50	1.05	4.52	Incipient	42.0	—	90
60	1.16	4.99	Incipient	47.3	—	90
65	1.78	7.65	6.2/99.4	80.0	0.37	90
70	2.28	9.80	9.7/97.7	109.7	0.58	90
80	2.89	12.43	12.4/97.7	150.2	0.77	90
90	3.31	14.23	14.1/97.2	180.6	0.93	90
99	3.54	15.22	14.6/97.7	198.2	0.88	90
			3.5-cm^2 Flyer[a]			
80	6.94	29.8	Not measured	—	Not measured	60
90	6.90	29.4	7.9/101.2	—	Not measured	45
80	6.77	24.8	Incipient	—	—	30

[a] Smaller flyer size and angle of incidence to determine spall footprint size.

where the velocities are in cm/μs. (The density of Kapton is rather variable, hence it needs to be determined for the particular batch used. We used a density of 1.43 g/cm^3.)

For the studies of shock loading of structures, a commonly used unit is the kilotap, where 1 tap is a dyn-s/cm^2. The upper impulse range that we have studied varies from 17 ktap for 10-cm^2 flyer plates to 35 ktap for 3.8-cm^2 flyers. With smaller composite flyer plates using tantalum as the impactor, we could generate impulses up to 100 ktap.

We have studied the spall threshold for a wide variety of materials and reports are currently in preparation by Froeschner and co-workers (1986) and Maiden and Froeschner (1986). The work reported here will therefore be limited to recent experiments performed on 6061-T6 aluminum to illustrate the general character of aluminum spall behavior in terms of spall threshold, partition of momentum between the spall scab(s) and parent body, and to describe preliminary measurements of spall produced by a flyer plate incident on a specimen at an oblique angle. The data for a variety of impulse levels near threshold are given in Table 1. The flyer momenta shown were deduced from the density and initial dimensions of the flyer and the observed velocity recorded by the F-P velocimeter. The fractional mass spalled is the mass of the recovered scab divided by the original target mass, both expressed in grams. The edge-corrected ratio of the spalled thickness normalized by the initial thickness can be obtained from Table 1 by multiplying dm/M by 1.08, because the specimen was 76 mm^2 and the flyer was only 73 mm^2. The peak pressures shown in the aluminum were deduced from the observed flyer velocity and the Hugoniots for Kapton and aluminum. The experiments with the LiF showed that for all entries in the table, the initial pressure pulse width at half maximum was approximately 100 ns. While spall is governed by the strength and profile of the wave when it reaches the free surface, for a strongly attenuated wave produced by a pulse of given duration at the impacted surface, the spall threshold can be related to the impulse received. The angle of incidence shown is measured from the plane of the flyer to the normal axis of the target plate (i.e., for 90°, the target and flyer plates are parallel).

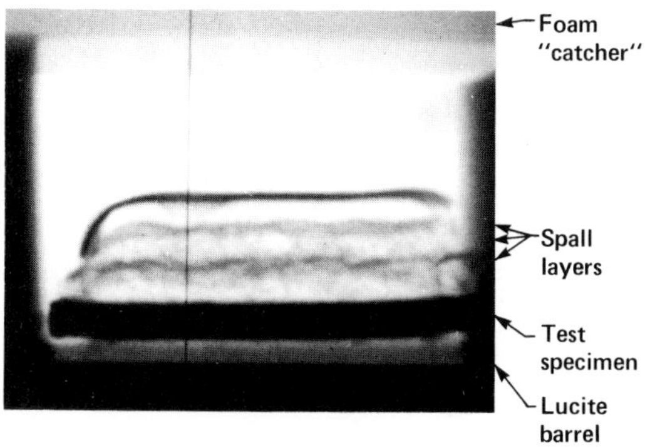

Fig. 4. Fast x-ray picture of spall scab structure produced by a 0.3-mm-thick, 73-mm² Kapton flyer at 2.9 km/s impacting a test specimen of 6.35-mm-thick 6061-T6 Al (76.2 mm²). The x-ray timing was 30 μs after impact at normal incidence.

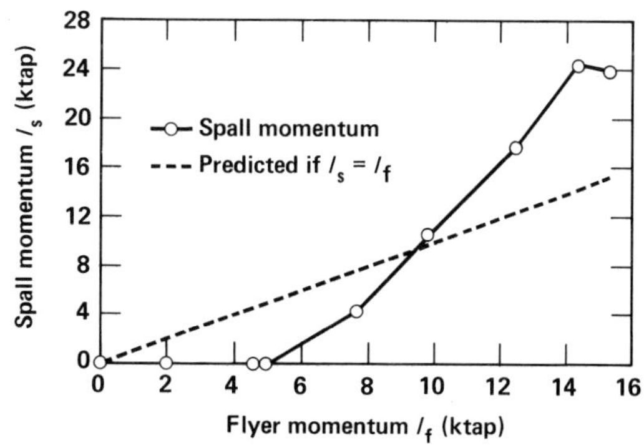

Fig. 5. Spall momentum vs incident flyer momentum.

Spall scabs are recovered by stopping them with a plastic foam catcher block. We can estimate the partition of momentum between the parent specimen and the spall scabs by measuring flyer-plate and scab velocities with a dual-beam F-P velocimeter. A source of error in this method is that only the velocity of the first scab is measured with the F-P. If later scabs have a somewhat lower velocity, transfer of momentum to the target will be underestimated. Flash x-ray photographs show that spall may occur in multiple layers (Fig. 4). Experimentally, these tend to be rejoined when they impact the foam catcher block. A more detailed analysis is then required using the mass distribution and velocities of intermediate spall layers as deduced from the known timing of the fast x-ray picture.

Figure 5 shows a plot of the spall fragment momentum calculated as a single scab versus the incident flyer plate momentum. At threshold, where a single spall scab was produced, the data show that virtually all of the incident momentum is carried away by the spall scab. At higher flyer momenta, the estimated momentum carried by the spall scab (or scabs) is actually higher than the incident momentum (indicated by the line $I_s = I_f$.) This may be partially because Kapton has a lower shock impedance than aluminum and recoils after impact on the aluminum, transferring an additional increment of momentum. Alternatively, surface ablation may occur, also adding to the net impulse. In any event, it appears that the first spall scab carries away most of the incident momentum. For a 6061-T6 Al target of the given thickness, the threshold for spall is encountered when the incident 100-ns duration pulse is between 50 and 80 kbar, or between 5 and 8 ktap.

In the case of flyer impact on an oblique surface, we observe, in the fast x-ray picture, appreciable momentum transfer orthogonal to the direction of the incident momentum of the flyer (near 33% for 45° incidence on 6061-T6 Al in a preliminary test). Another effect of the obliquity is the $1/\sin\theta$ increase in effective specimen thickness (here, θ is the angle measured with respect to the target surface; i.e., 90° at normal incidence). Experimentally, a flyer impacting aluminum at 90° incidence above the spall threshold results in a relatively large sheet spall scab (often the full area of the flyer) and leaves reasonably fine-grain, new free surface on the test specimen body, while the spall scab produced for a flyer at oblique incidence is torn into many small segments, and leaves a very jagged new free-surface on the recovered test-specimen body. The difference in spall scab characteristics apparently is because of the velocity shear in the spall scab as it is torn free.

Reverse-Ballistics Study

To study penetration phenomena when a projectile is difficult to launch at the desired velocity without damage by the acceleration process, one can perform the hydrodynamically equivalent experiment of launching the target instead, at the desired velocity. If the target is a thin flat wall, we can simulate it with the flat flyer plate launched in the electric gun. Such experiments allow us to address penetration effects at velocities that are unattainable by other means. The target object (projectile) can be a simple metal rod or sphere or a more complex assembly, depending on the effects of interest. The orientation of the target assembly with respect to the flyer plate can be precisely chosen. The primary diagnostic is flash x-ray pictures, showing the target object at various stages of the flyer-plate penetration. Further information is obtained from study of the debris and what remains of the target object. The study of the debris now includes the capability of recovery of any sizable fragments, and a new catcher cylinder is under test to collect the vaporized portion of the test specimen. Measurements of the fraction fragmented and the fraction vaporized are crucial for benchmarking computer penetration codes that as yet do not satisfactorily predict this detail. This reverse-ballistics study used a geometry similar to that shown in Fig. 1, except that the target rod in each case was suspended 1 cm above the end of the barrel. A Lucite barrel, 6.35-mm long with a 7.6-mm² bore was used. The F-P was generally used to measure the flyer velocity until impact. In one case, the F-P trace faded just before impact, so a corrected velocity is given from the known x-ray timing. The measurements are made in vacuum to eliminate air shock. Table 2 summarizes the test-rod data for this geometry.

For each shot we obtained flash x-ray pictures at a selected stage in the impact process. In addition, the recovered rod specimens are under detailed metallurgical study by Don Shockey and co-workers (1986). In the case of the lead rod, the entire rod was vaporized or melted into droplets too small to be recovered. The observations have been compared with predicted damage loss from a theory by R. J.

TABLE 2 Reverse Ballistics Tests of 0.3-mm-thick Kapton Flyers Incident on Various 1.59-cm-long by 0.36-cm-diam Rods. The V_f is the Flyer Velocity. The Axis of Each of the Rods was Normal to the Flyer Except for the Aluminum Rod, which was Tilted 18° from Normal. The *M/L* refers to the fractional mass loss.

Material	V_f (km/s)	Fractional *M/L* (%)	Tip Damage
Steel	15	16	Melted, fragmented
Steel	15	18	Melted, fragmented
Lead	15+	100	Nothing recovered
Aluminum	11	28	Melted, fragmented

Lawrence (1986) on the relative energy required for complete disruption of the test specimen. The observations and theory agree that only the lead rod should have been completely melted or vaporized.

Present limitations on the wall mass/area and velocity are primarily because of limitations in the energy available and the current risetime capabilities of the 87-kJ, 100-kV capacitor bank. Nominal performance at 99 kV with a 0.3-mm-thick Kapton flyer (0.043 g/cm^2), 7.6 mm in diam, is 18 km/s as measured independently with the F-P velocimeter and impact on a flasher block. Larger or more massive flyers give correspondingly reduced velocities.

Table 2 is the start of a data base needed for a study of scaling of penetration with target dimensions and flyer thickness, correlating the importance of material properties with penetration, and the importance of object shape and orientation over a range of relative velocities up to at least 18 km/s. This type of laboratory measurement is needed, for example, to guide in selecting the minimum weight micrometeorite shield needed to protect a spacecraft against the bulk of small predicted space debris.

FUTURE DIRECTIONS

A new study on measuring the release isentropes of aluminum through the two-phase region is under way. We propose to make release isentrope measurements by allowing the target to unload into selected materials with known Hugoniots. For the very low pressures we will use argon gas with Hugoniots predicted by the ACTEX code (Rogers, 1981). Alt'shuler and co-workers (1980) have previously used this technique to study copper and lead, but because they used high-explosive drivers, they used porous materials to achieve the high-entropy states of interest. Bushman and co-workers (1984) used focused explosive methods to measure shock adiabats and release isentropes for aluminum. We propose to use normal density materials with fast projectiles. The measurements assume the Hugoniots of all materials used are known. The diagnostics measure shock velocity in the aluminum to determine the initial state of the release, then the shock velocity in a variety of softer materials to determine the release states.

ACKNOWLEDGMENTS

We gratefully acknowledge the support and encouragement of S. Cochran, W. Bookless, G. Polmykal, W. Von Holle, D. Steinberg, and others of LLNL in making these measurements. We also acknowledge the contributions of K. Froeschner of Martin, Froeschner and Associates; D. Maiden of LLNL on spall modeling; and many helpful discussions with D. Curran, D. Shockey, and S. Holmes of SRI, International, and with R. Bjork of Pacifica Technology. Finally, we acknowledge the work and dedication of the laboratory group of C. Lee, R. Diaz, J. Clarr, and C. Trapp in carrying out the operation and electric gun measurements.

REFERENCES

Alt'shuler, L. V., A. V. Bushman, M. V. Zhernokletov, V. N. Zubarev, A. A. Leont'ev, and V. E. Fortov (1980). Unloading isentropes and the equation of state of metals at high energy densities. *Sov. Phys. JETP, 51*, 373–383.

Bushman, A. V., I. K. Krasyuk, P. P. Pashinin, A. M. Prokhorov, V. Ya. Ternovoi, and V. E. Fortov (1984). Dynamic compressibility and thermodynamics of a dense aluminum plasma at megabar pressures. *Sov. Phys. JETP Lett., 39*, 411–413.

Chau, H. H., G. Dittbenner, W. W. Hofer, C. A. Honodel, D. J. Steinberg, J. R. Stroud, and R. C. Weingart (1980). Electric gun: a versatile tool for high pressure shock-wave research. *Rev. Sci. Instrum., 51*, 1676–1681.

Chau, H., G. Dittbenner, K. Mikkelsen, R. Weingart, K. Froeschner, and R. Lee (1982). Performance of a 100 kV, 78 kJ electric gun system. In W. J. Nellis, L. Seaman, and R. A. Graham (Eds.), *Shock Waves in Condensed Matter—1981*. North-Holland, Amsterdam. pp. 691–695.

Dein, J., R. Tokheim, D. Curran, H. Chau, R. Weingart, and R. Lee (1984). Aluminum damage simulation in high velocity impact. In J. R. Asay, R. A. Graham, and G. K. Straub (Eds.), *Shock Waves in Condensed Matter—1983*. North-Holland, Amsterdam. pp. 171–174.

Froeschner, K. E., H. H. Chau, D. E. Maiden, and C. Lee, (1986). Effects of short duration stress pulses on structural materials. Lawrence Livermore National Laboratory, Livermore, CA.

Froeschner, K. E., R. S. Lee, H. H. Chau, and R. C. Weingart (1984). Shock Hugoniot measurements on Ta to 0.78 TPA. In J. R. Assay, R. A. Graham, and G. K. Straub (Eds.), *Shock Waves in Condensed Matter—1983*. North-Holland, Amsterdam. pp. 85–88.

Goosman, D. R. (1979). Measuring velocities by laser Doppler interferometry. In R. C. Haight, P. Adye, R. B. Crawford, J.K. Prono, and J. T. Staehle (Eds.), *LLNL Energy and Technology Review*, UCRL-52000-79-3. Lawrence Livermore National Laboratory, Livermore, CA. pp. 17–24.

Lawrence, R. J. (1986). Sandia National Laboratories, Albuquerque, NM. Private communication.
Maiden, D. E., and K. E. Froeschner (1986). Analysis of spall in metals due to short high intensity pulses. Lawrence Livermore National Laboratory, Livermore, CA.
Marsh, S. P. (Ed.), (1980). *LASL Shock Hugoniot Data.* University of California Press, Berkeley, CA.
Rogers, F. J. (1981). Equation of state of dense, partially degenerate, reacting plasmas. *Phys. Rev. A, 24,* 1531–1543.
Shockey, D. E., D. R. Curran, J. E. Osher, and H. H. Chau (1986). Disintegration behavior of metal rods subjected to hypervelocity impact. In C. E. Anderson, Jr. (Ed.). *Proc. 1986 Hypervelocity Impact Symposium, October 21-24, San Antonio, TX.* Pergamon Press, Oxford, England.
Weingart, R. C., R. S. Lee, R. K. Jackson, and N. L. Parker (1976). Acceleration of thin flyers by exploding metal foils: application to initiation studies. In D. J. Edwards (Ed.), *Proc. 6th Symposium (International) on Detonation,* ACR-221. Office of Naval Research, Washington, DC. pp. 653–663.

DISCLAIMER

This document was prepared as an account of work sponsored by an agency of the United States Government. Neither the United States Government nor the University of California nor any of their employees, makes any warranty, express or implied, or assumes any legal liability or responsibility for the accuracy, completeness, or usefulness of any information, apparatus, product, or process disclosed, or represents that its use would not infringe privately owned rights. Reference herein to any specific commercial products, process, or service by trade name, trademark, manufacturer, or otherwise, does not necessarily constitute or imply its endorsement, recommendation, or favoring by the United States Government or the University of California. The views and opinions of authors expressed herein do not necessarily state or reflect those of the United States Government thereof, and shall not be used for advertising or product endorsement purposes.

Int. J. Impact Engng Vol. 5, pp. 509–518, 1987
Printed in Great Britain

0734-743X/87 $3.00 + 0.00
Pergamon Journals Ltd.

DEBRIS CLOUDS GENERATED BY HYPERVELOCITY IMPACT OF CYLINDRICAL PROJECTILES WITH THIN ALUMINUM PLATES

Andrew J. Piekutowski

University of Dayton Research Institute
300 College Park Avenue
Dayton, Ohio 45469

ABSTRACT

Orthogonal, flash x rays were used to observe the debris clouds produced by the hypervelocity impact of cylindrical aluminum projectiles with thin aluminum sheets or bumpers. Three major structural features were observed in the debris clouds--a front cone, a bulbous main debris cloud, and an inner cone. Inclination of the projectile at impact changed the orientation of these features and the severity of damage to the rear wall of a double-sheet structure; projectiles with the greatest inclination produced the most damage. Two experiments, using aluminum and copper as projectile and target or target and projectile, respectively, were performed to determine the source of material in each of the three structural features of the debris clouds. The front cone and main cloud were shown to consist of bumper debris while the inner cone was composed of projectile fragments.

INTRODUCTION

Early studies directed to protection of spacecraft concentrated on developing systems and structures which would resist penetration by high-velocity, low-mass micrometeoroids. Today, this same micrometeoroid environment has become increasingly burdened with significant amounts of more massive orbital debris. The assessment of the threat posed by this debris to the long-term habitation of space aboard the Manned Space Station is a continuing effort that becomes more complex with the passage of time (Kessler, 1982).

Martin Marietta Michoud Aerospace is currently investigating the penetration resistance of a variety of materials and structural configurations which could be used in construction of an integrated wall for the Manned Space Station. In support of this investigation, hypervelocity impact tests have been performed at the University of Dayton Research Institute (UDRI) Impact Physics Laboratory using the UDRI 50/20 mm, two stage, light-gas gun (Piekutowski, 1985).

Because a "design threat" had not been specified, the tests performed at UDRI have simulated orbital debris with cylindrical and spherical aluminum projectiles. Projectile weights varied from 0.37 g to 2 g and impact velocities ranged from 3 to 7 km/sec. Cylindrical, 1 g projectiles (7.72 mm diameter, length to diameter ratio = 1) of 2024-T4 aluminum were used for most of the tests.

Targets from four early tests using cylindrical, 1 g aluminum projectiles are shown in Fig. 1. A 30.5-cm-square, 2.03-mm-thick, 6061-T6 aluminum bumper or shield was used in each of the targets shown in this figure. Rear walls were 30.5-cm-square, 2219-T851 aluminum plates ranging from 3.81 to 6.35 mm thick. Sheet thicknesses and spacings are noted in Fig. 1. Perforation of the rear wall or spall-induced damage to witness plates behind the rear wall occurred in each of the targets shown.

The damage patterns formed on the front of the rear walls shown in Fig. 1b, 1c, and 1d are different than the symmetric patterns typically shown in the literature, e.g., Maiden, McMillan, and Sennett (1965). Although the desired projectile orientation at impact was with the cylinder axis normal to the target, varying angles of inclination between the projectile axis and the trajectory were observed in high-speed films of the projectiles in flight. In some cases, the projectiles exhibited significant inclination angles. The severity of damage and pattern asymmetry appeared to be related to the angle of inclination of the projectile.

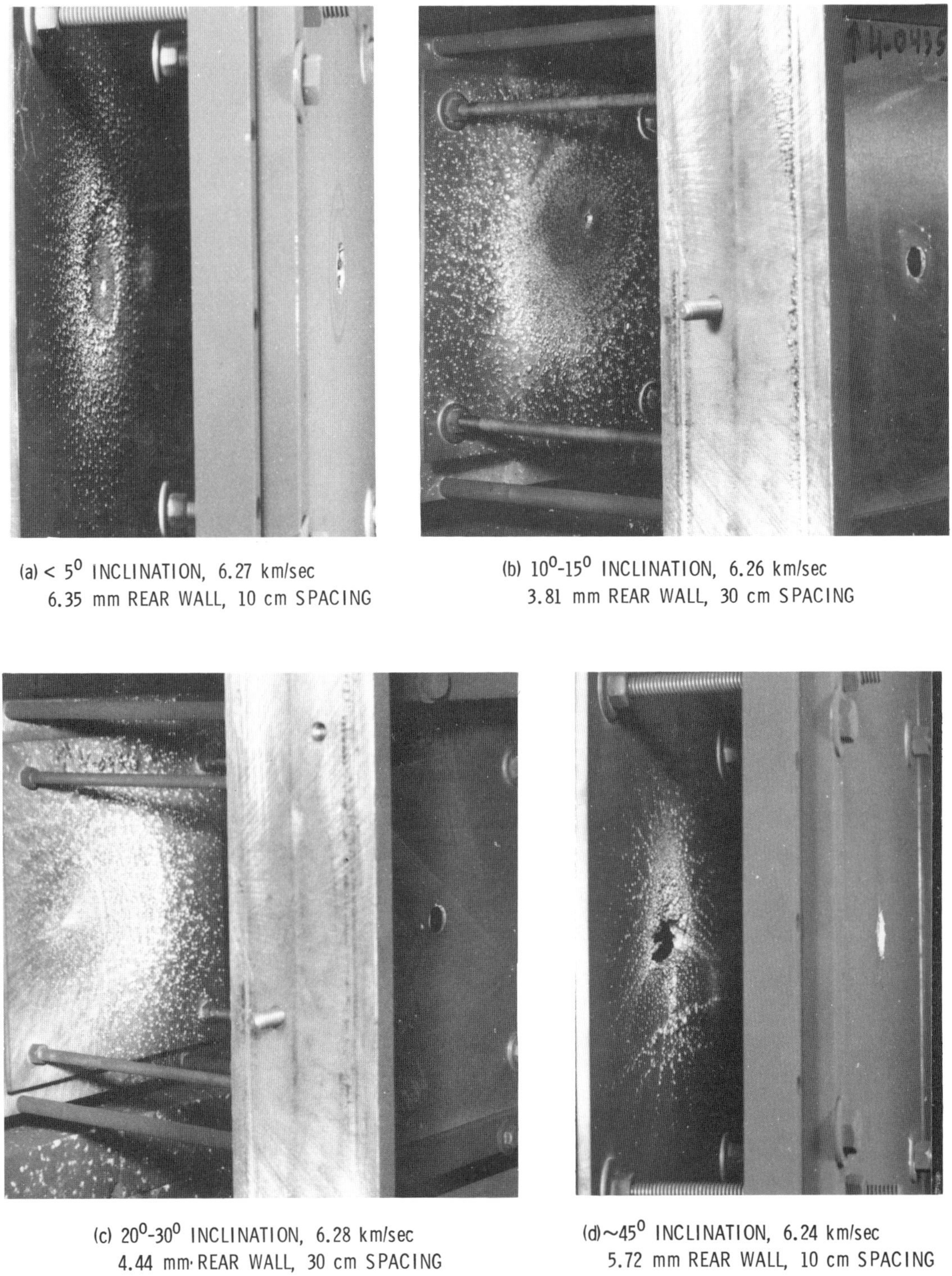

(a) < 5° INCLINATION, 6.27 km/sec
6.35 mm REAR WALL, 10 cm SPACING

(b) 10°-15° INCLINATION, 6.26 km/sec
3.81 mm REAR WALL, 30 cm SPACING

(c) 20°-30° INCLINATION, 6.28 km/sec
4.44 mm REAR WALL, 30 cm SPACING

(d) ~45° INCLINATION, 6.24 km/sec
5.72 mm REAR WALL, 10 cm SPACING

Fig. 1. Various rear wall damage patterns produced by debris from impact of 1 g, cylindrical, 2024-T4 aluminum projectiles with 2.03-mm-thick, 6061-T6 aluminum bumpers.

Most previously reported experimental studies of meteoroid protection systems have employed small spherical projectiles, made from a variety of metallic and nonmetallic materials, to simulate micrometeoroids. However, various other projectile shapes have also been used by other investigators. Rolsten, Hunt, and Wellnitz (1964) used explosively-launched discs in their studies of principles of meteoroid protection. In an earlier work, (Rolsten, Hunt, and Wellnitz, 1962) they reported significant differences in the impact patterns formed by discs that impacted at different orientations. Friend, Murphy, and Gough (1969) used polycarbonate cylinders and various conical- and cup-shaped projectiles in their studies of debris cloud pressure distributions on a secondary surface. Kassel and DiBattista (1971) used cylindrical polycarbonate projectiles and double-sheet targets to demonstrate the usefulness of an ultra-high-speed camera system for impact studies. In one of their test firings, the projectile was inclined about 30 degrees from the normal. The rear plate of the target for that test exhibited many of the features shown in the target in Fig. 1d. Morrison (1972) performed a rather thorough study of the effects of projectile shape on double-sheet structure resistance to penetration. Both he and Maiden, McMillan, and Sennett (1965) reported that greater shield thicknesses were required to defeat cylindrical projectiles (with very low inclination angles) than were required for spherical projectiles having the same mass. High-speed photographs of debris clouds shown by Morrison (1972) appeared to have certain structural features that could produce the damage patterns observed in many of the Space Station integrated wall targets. Flash x rays were used to examine debris clouds produced during subsequent tests of integrated wall targets.

EXPERIMENTAL PROCEDURE

Targets were installed in holders located 6.1 m from the muzzle of the launch tube. Targets used for seven tests, whose results are presented later in this paper, were simple double-sheet structures. Each target consisted of a 15.2-cm-square bumper and a 20.3-cm-square, 6.35-mm-thick, 2024-T351 aluminum rear wall. Bumpers were fabricated from 2.03-mm-thick, 6061-T6 aluminum sheet for six of the tests; the seventh bumper was made from 0.61-mm-thick OFHC copper sheet. Spacing between the bumper and rear wall was maintained at 10.59 cm for these seven tests. A reduced range air pressure of 8 mm Hg was used to aerodynamically separate the sabot from the projectile during its flight down the range. The sabots were stripped from the projectile about 1 m uprange of the target.

Three orthogonal pairs of 180 kV flash x rays were used to view the projectile, target, and debris cloud. As shown in Fig. 2, a single exposure image of the projectile and double-exposed

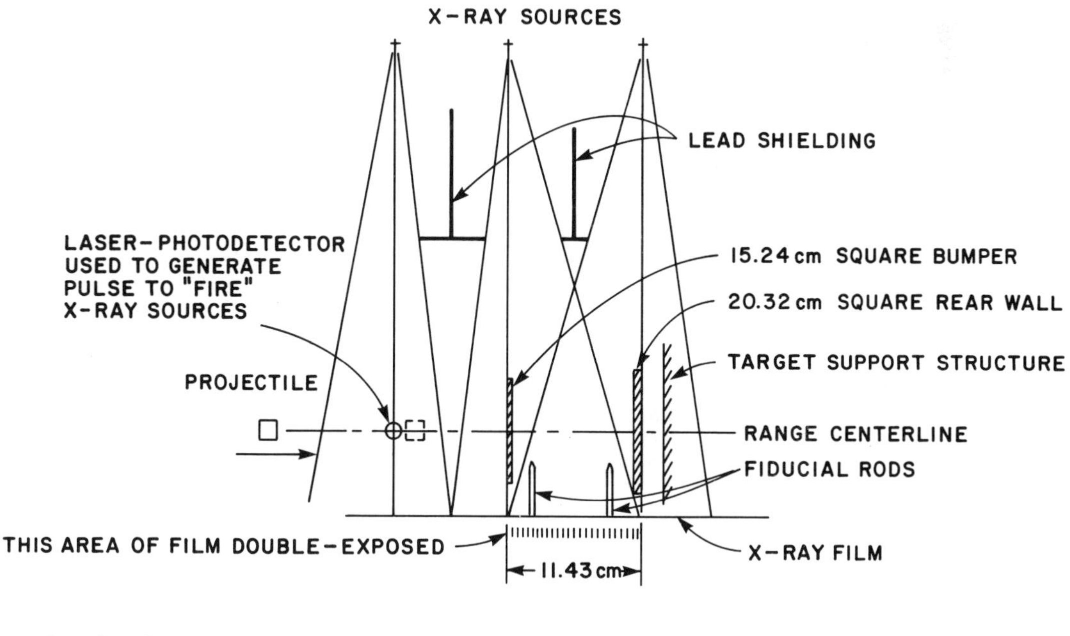

Fig. 2. Experimental setup used to provide radiographs of debris clouds. For clarity of illustration, only half of setup is shown.

images of the debris cloud were recorded on a single piece of film for each view. Use of the fiducial pointers and a small computer program made it possible to obtain quantitative trajectory information for features and/or particles that could be positively identified in the clouds for both views. Careful identification and measurement of the position of selected debris cloud features/particles permitted determination of the velocity of the selected "point" with an accuracy of about ± 0.1 km/sec.

TEST RESULTS AND DISCUSSION

Orthogonal views of the debris clouds produced by the impact of a 1 g, 2024-T4 aluminum sphere and a 1 g, 2024-T4 aluminum cylinder are presented in Figs. 3 and 4, respectively. The external

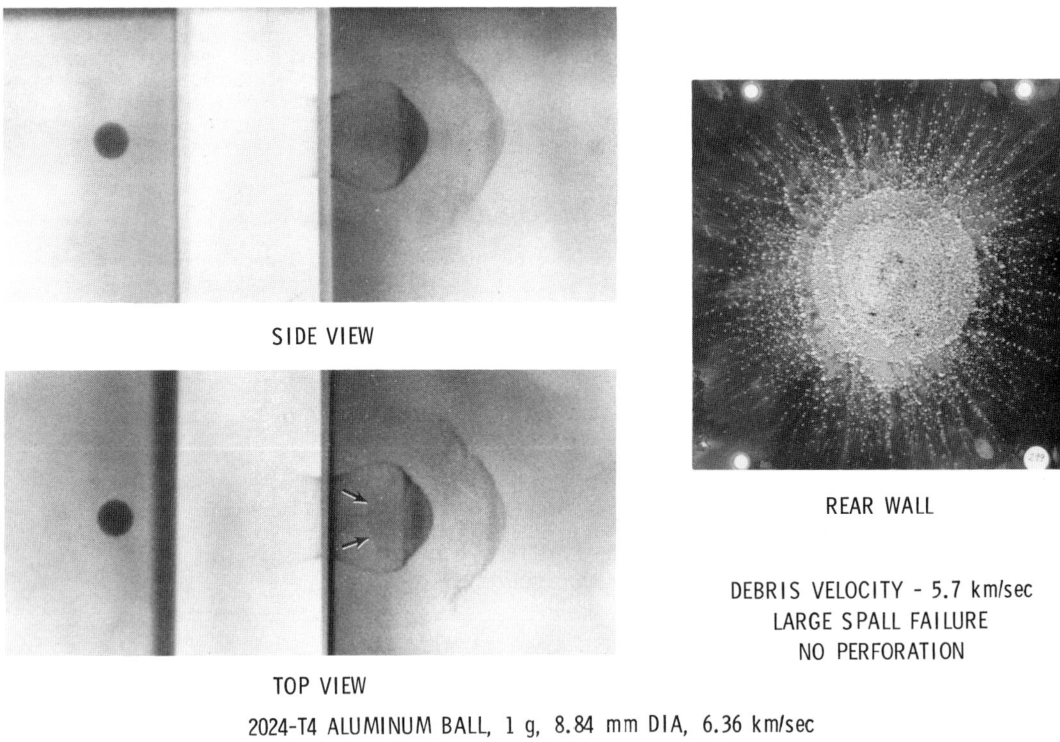

Fig. 3. Radiographs of debris cloud produced by impact of sphere with 2.03-mm-thick, 6061-T6 aluminum bumper. Damage pattern formed on 6.35-mm-thick, 2024-T351 aluminum rear wall also shown. Debris velocity is for leading edge of cloud.

characteristics of the debris cloud produced by the impact of the sphere are typical of those shown in high-speed photographs and radiographs from similar tests performed by other investigators. The faint but significant hemispherical structure inside the debris cloud in Fig. 3 (top view) is not seen in their photographs and radiographs, however. The external features of the cloud shown in Fig. 4 are similar to those in the debris cloud photographs presented by Morrison (1972). The dark, conelike region inside the cloud is evident in some of Morrison's photographs.

Several differences in the damage patterns shown on the rear walls in Figs. 3 and 4 are noteworthy. The pattern produced by the sphere debris cloud exhibits a well defined region with evenly distributed small craters at the center of the rear wall. A small, deep crater was formed near the center of the pattern produced by cylinder debris (Fig. 4). In addition, there is a significant variation in the density of small craters in the region surrounding this central crater. Finally, the rear wall impacted by cylinder debris does not exhibit the pronounced ray pattern evident in Fig. 3.

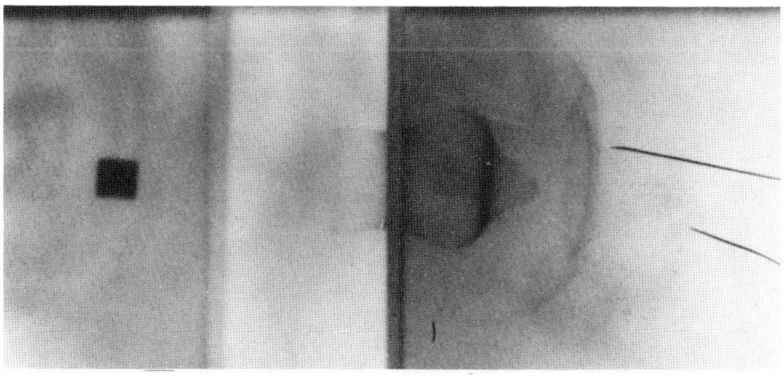

SIDE VIEW - 3^0

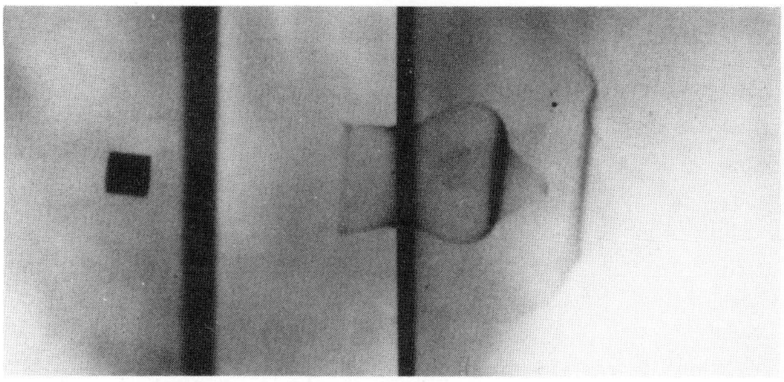

TOP VIEW - 6^0

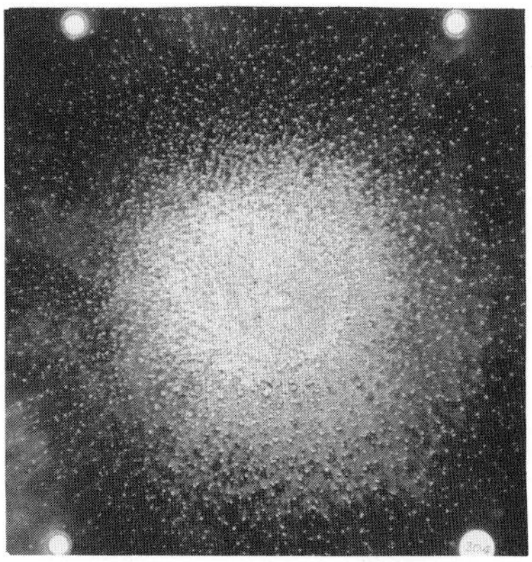

REAR WALL

DEBRIS VELOCITY - 6.7 km/sec
LARGE SPALL FAILURE
NO PERFORATION

2024-T4 ALUMINUM CYLINDER, 1 g, 7.72 mm DIA (L/D=1), 7^0 INCLINATION, 6.48 km/sec

Fig. 4. Radiographs of debris cloud produced by impact of cylinder with 2.03-mm-thick, 6061-T6 aluminum bumper. Damage pattern formed on 6.35-mm-thick, 2024-T351 aluminum rear walls also shown. Debris velocity is for tip of front cone.

An enlarged view of a debris cloud produced by the impact of a cylinder is shown in Fig. 5a. Characteristic structural features of the debris cloud are shown in Fig. 5b. These structural features--a front cone, a main debris cloud, and an inner cone--were observed in all radiographs of debris clouds produced by the impact of cylindrical projectiles.

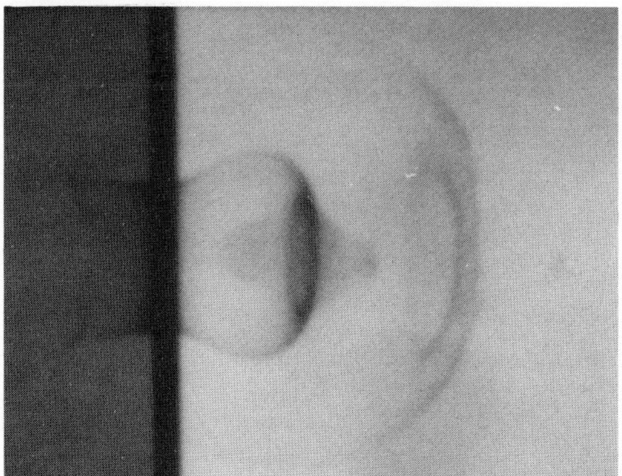

(a) VIEW OF DEBRIS CLOUD

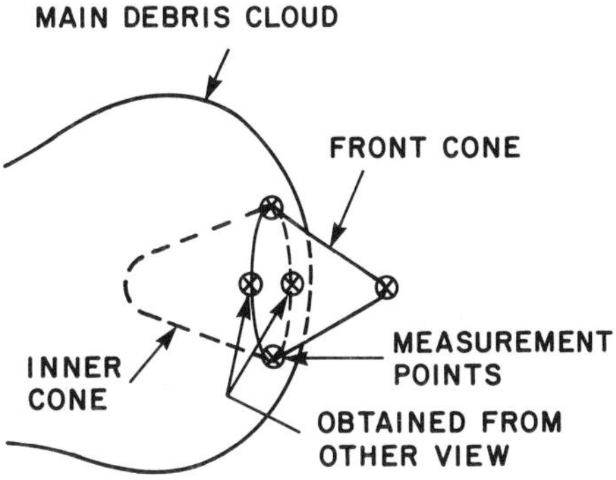

(b) CHARACTERISTIC FEATURES OF DEBRIS CLOUD

Fig. 5. Features of a debris cloud produced by impact of aluminum cylinder with aluminum bumper. Points shown with "x" were used to obtain quantitative debris trajectory data.

As the inclination angle of the projectile increased, changes in the shape and location of the front and inner cone and the main debris cloud were observed. A vew of a debris cloud produced by an inclined cylinder is presented in Fig. 6a. Significant changes in the structural features shown in Fig. 5b are identified for the distorted debris cloud in Fig. 6b.

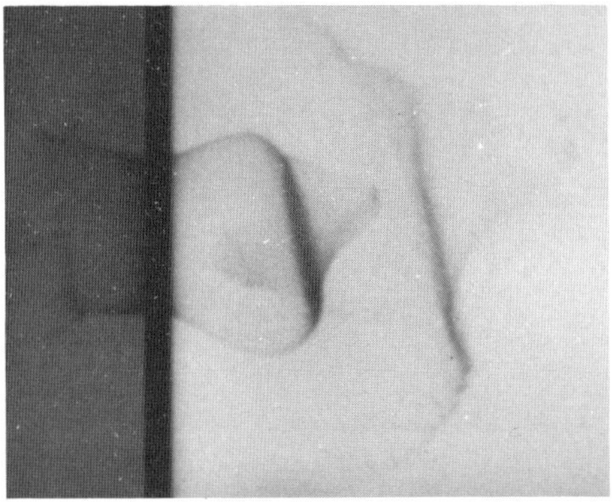

(a) VIEW OF DEBRIS CLOUD

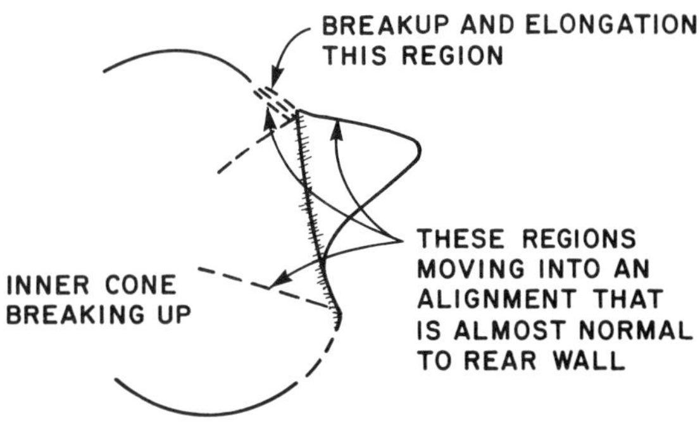

(b) FEATURES OF DEBRIS CLOUD

Fig. 6. Debris cloud produced by impact of inclined cylindrical projectile.

The effects of a change in the inclination angle of the projectile axis on the development of the debris cloud and the damage pattern on the rear wall are shown for three tests in Fig. 7.

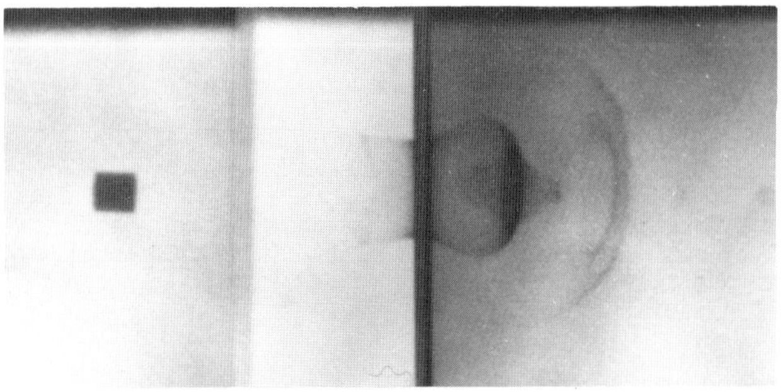

SIDE VIEW - 2^0

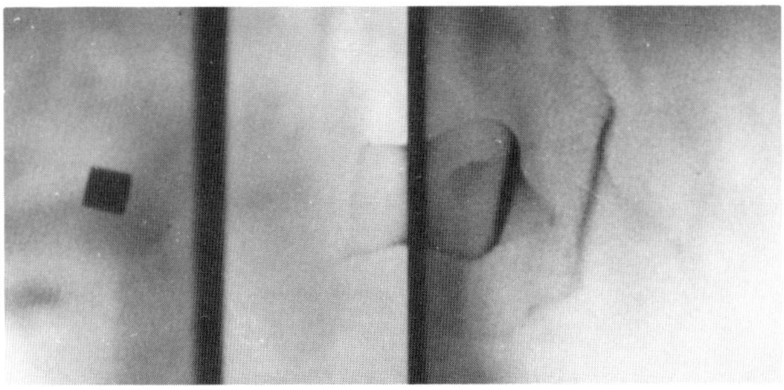

TOP VIEW - 12^0

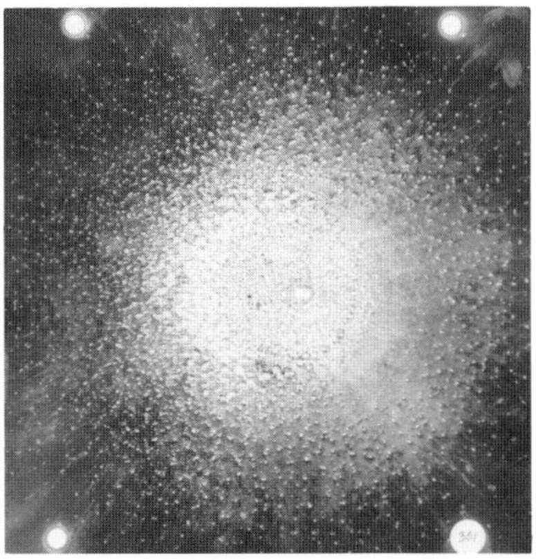

REAR WALL

DEBRIS VELOCITY - 6.4 km/sec
LARGE SPALL FAILURE
NO PERFORATION

(a) 12^0 INCLINATION, 6.30 km/sec

Fig. 7. Radiographs of debris clouds produced by impact of inclined, 1 g, 2024-T4 aluminum cylinders with 2.03-mm-thick, 6061-T6 aluminum bumpers and damage patterns formed on 6.35-mm-thick, 2024-T351 aluminum rear walls. Debris velocity is velocity of leading edge of cloud, normal to the rear wall.

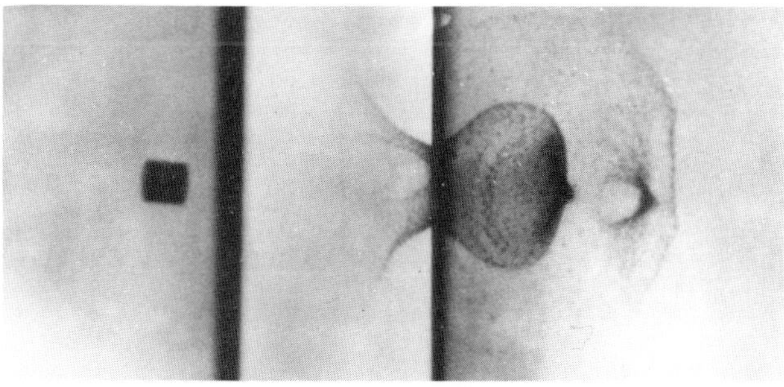

SIDE VIEW - 13^0

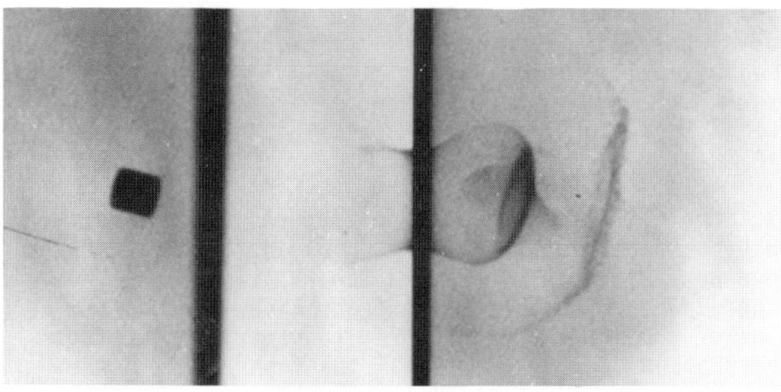

TOP VIEW - 13^0

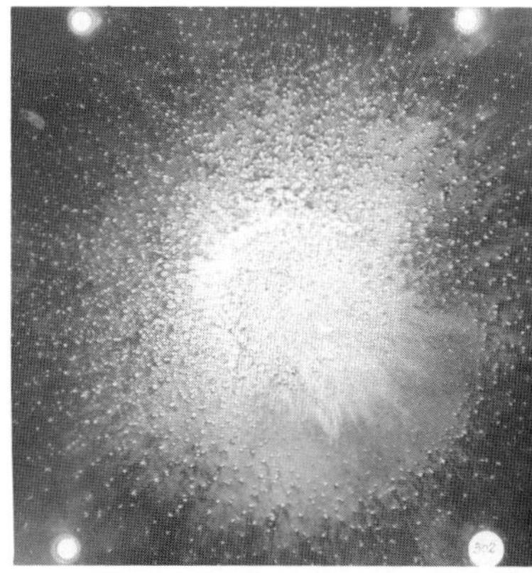

REAR WALL

DEBRIS VELOCITY - 6.4 km/sec
LARGE SPALL FAILURE
CRACK

(b) 18^0 INCLINATION, 6.31 km/sec

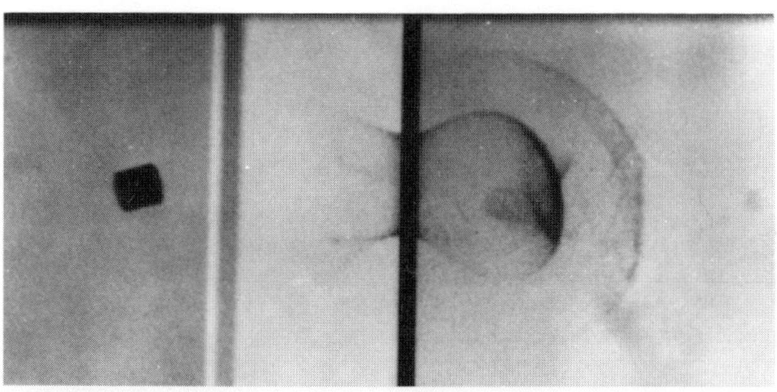

SIDE VIEW - 15^0

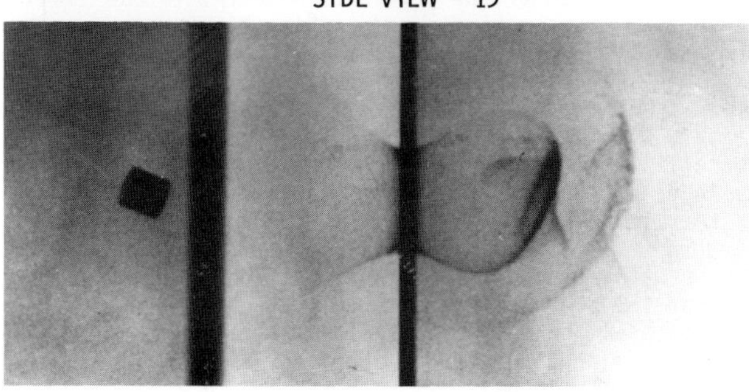

TOP VIEW - 20^0

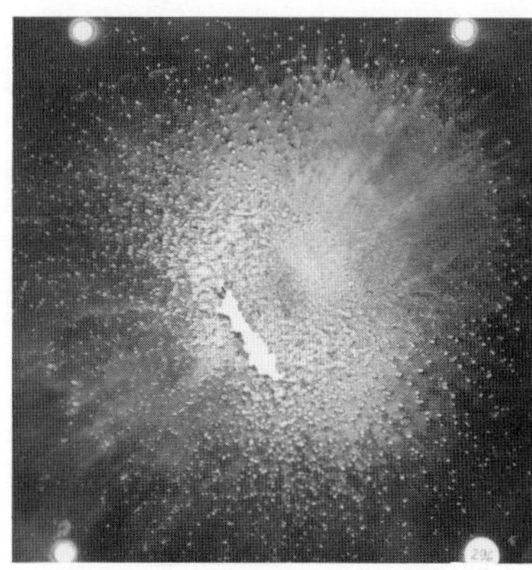

REAR WALL

DEBRIS VELOCITY - 5.6 km/sec
LARGE SPALL FAILURE
LARGE PERFORATION

(c) 25^0 INCLINATION, 6.39 km/sec

Fig. 7 (Concluded). Radiographs of debris clouds produced by impact of inclined, 1 g, 2024-T4 aluminum cylinders with 2.03-mm-thick, 6061-T6 aluminum bumpers and damage patterns formed on 6.35-mm-thick, 2024-T351 aluminum rear walls. Debris velocity is velocity of leading edge of cloud, normal to the rear wall.

Distortion of the front and inner cone, the main debris cloud, and severity of damage to the rear wall increased as the inclination angle increased. The small central crater evident in the rear wall in Fig. 4 is still evident in the rear wall in Fig. 7a, but the number and concentration of craters in the region to the left of the small crater is greater in Fig. 7a than in Fig. 4. In addition, a small region showing little evidence of cratering can be seen to the right of the central crater. Except for significant spall from the rear surface (behind the heavily cratered region), the rear wall shown in Fig. 7a retained its structural integrity.

In Fig. 7b, the "central" crater is smaller and a significant distance from the center of the rear wall. A well defined, heavily cratered, crescent-shaped region is shown to the left of the central crater. The region to the right of the small crater exhibits little evidence of cratering and has increased significantly in size from the area shown in Fig. 7a. Spall of the rear surface is comparable to that for the rear wall shown in Fig. 7a; however, a crescent-shaped through crack has developed along most of the length of the heavily cratered region.

As shown in Fig. 7c, a further increase in the inclination angle of the projectile results in a more dramatic separation of the regions noted in the description of the damage patterns displayed by the rear walls shown in Figs. 7a and 7b. The small central crater is not evident in Fig. 7c and the concentration of craters in the crescent-shaped region has intensified and is dominated by the large hole. The extent of the area which shows little evidence of impact has increased significantly.

Comparison of the debris cloud structural features with rear wall damage patterns identifies the probable source of damage. It is evident that the small central craters, observed for tests with low projectile inclination angles, were formed by the impact of material in the tip of the front cone. In a number of tests, perforation occurred where the tip of the front cone impacted the rear wall. The radiographically dense regions of the inner cone produced the heavily cratered, crescent-shaped regions where cracking or perforation usually occurred. The inner cone is the most significant structure in the debris cloud in terms of potential to cause damage to the rear wall of a double-sheet structure. The randomly cratered region outside the area affected by the front and inner cones was impacted by relatively harmless material from the main debris cloud.

Analysis of the velocity and trajectory data obtained with use of the five "points" shown in Fig. 5b, indicate that the front cone travels as a coherent unit, even for the case of moderately inclined cylinders. For near-normal impacts, the velocity of the tip of the cone is about 5 per cent greater than the projectile impact velocity. As the inclination angle of the projectile increased, the cone tip velocity decreased as shown by the values given in Figs. 4 and 7. The path followed by the cone was along a line that was slightly less inclined than the axis of the projectile.

Two experiments were performed to determine the origin of material in the three regions making up the debris cloud. In the first experiment, a 1 g, 7.72 mm diameter, 7.72 mm long, 2024-T4 aluminum cylinder impacted a 0.61-mm-thick, OFHC copper bumper at 6.39 km/sec. This bumper thickness had the same areal density as the 2.03-mm-thick, 6061-T6 aluminum bumpers used in the other experiments. Copper was selected as the bumper material for this experiment because it is considerably more opaque to x rays than aluminum. The contrast of density of debris fragments in the radiographs would, therefore, facilitate determination of the origin of material in the cloud. The radiographs and rear wall for this experiment are presented in Fig. 8. The radiographs clearly indicate that the front cone and the main debris cloud are composed of copper fragments. The interior cone of debris is visible in the side view radiograph.

In the second experiment, a 1 g, 7.72 mm diamter, 2.26 mm long, OFHC copper disc impacted a 2.03-mm-thick, 6061-T6 aluminum bumper at 6.23 km/sec. Results of this experiment are presented in Fig. 9. Use of a disc instead of a right cylinder of copper for this test permitted the projectile to have the same areal density as the aluminum cylinders used in the other tests. The debris cloud produced by this impact exhibited the three features observed in the aluminum-on-aluminum debris clouds. A very large front cone was produced. The small fragment at the tip of this cone has a velocity 44 per cent greater than the projectile impact velocity. The front cone appears to consist exclusively of aluminum fragments. The main debris cloud also appears to consist mainly of aluminum although some copper fragments are evident where it joins the front and inner cones. The projectile has disintegrated into a cloud that is contained in a well defined region inside the outer cloud of aluminum debris. The velocity of the tip of the forward facing cone is roughly the impact velocity of the projectile. The trailing edge of the rearward facing cone is travelling at about 35 per cent of the projectile velocity.

The tip of the large front cone of aluminum fragments is travelling along an axis that is normal to the surface of the slightly inclined disc. The small hole just off the center of the rear wall in Fig. 9 was produced by a fragment(s) travelling at a velocity of 9 km/sec.

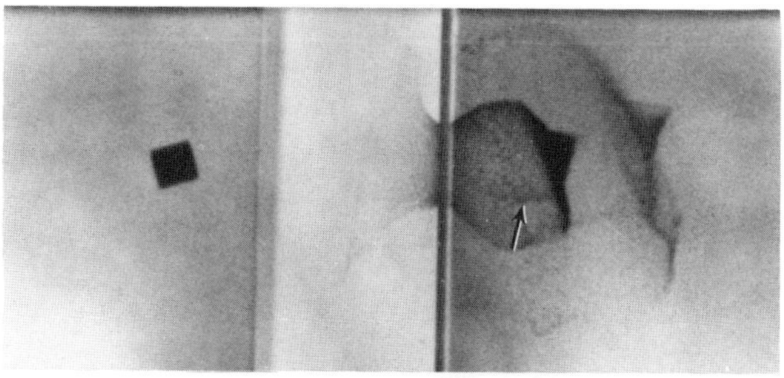

SIDE VIEW - 17^{0}

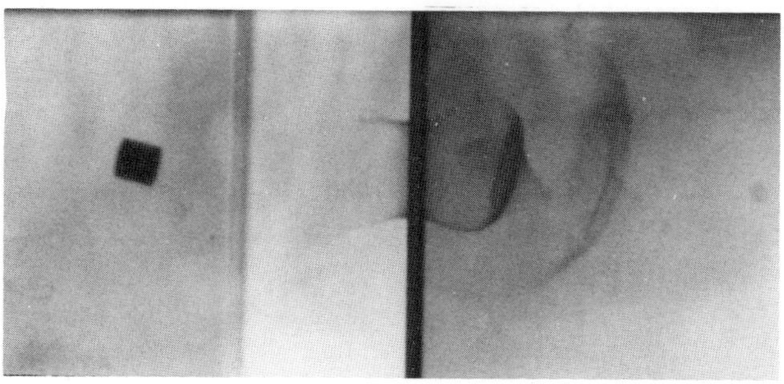

TOP VIEW - 5^{0}

REAR WALL

DEBRIS VELOCITY -
SINGLE POINT - 5.4 km/sec
DOUBLE POINTS - 6.1 km/sec
LARGE SPALL FAILURE
2 SMALL PERFORATIONS

Fig. 8. Radiographs of debris cloud produced by impact of 1 g, cylindrical, 2024-T4 aluminum projectile with 0.61-mm-thick, OFHC bumper at 6.39 km/sec. Damage pattern formed on 6.35-mm-thick, 2024-T351 aluminum rear wall also shown.

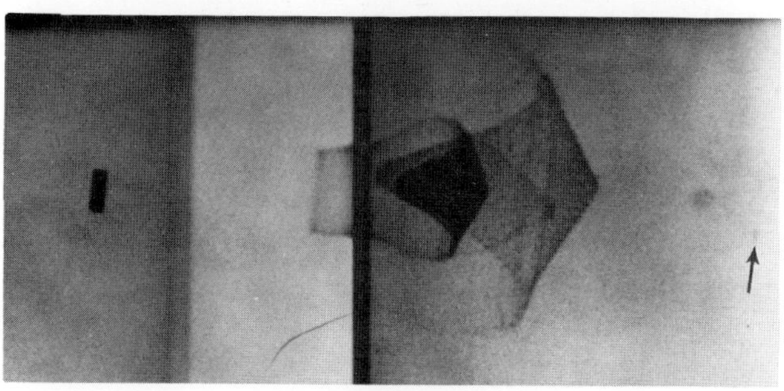

SIDE VIEW - 8^{0}

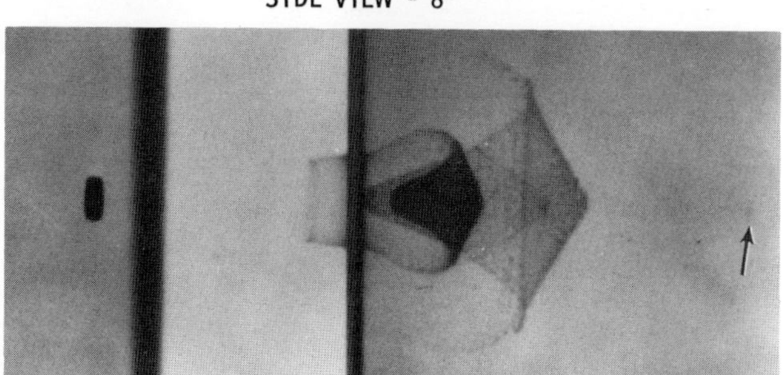

TOP VIEW - 1^{0}

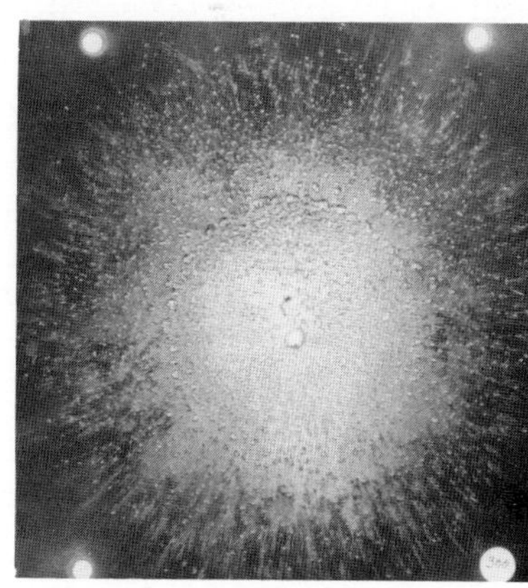

REAR WALL

DEBRIS VELOCITY - 9.0 km/sec
MODERATE SPALL FAILURE
1 SMALL PERFORATION

Fig. 9. Radiographs of debris cloud produced by impact of 1 g, 7.72 mm diameter, 2.26 mm long, OFHC copper disc with 2.03-mm-thick, 6061-T6 aluminum bumper at 6.23 km/sec. Damage pattern formed on 6.35-mm-thick, 2024-T351 aluminum rear wall also shown.

CONCLUSIONS

As concluded by Morrison (1972), cylindrical projectiles were more effective penetrators of double-sheet structures than were spherical projectiles of the same mass. Inclination of the axis of the cylinder to the line of flight resulted in more severe damage to the structure. Use of flash x rays has revealed three significant structural features in debris clouds produced by the impact of a cylinder with a thin bumper or shield. Changes in the orientation of these structural features, resulting from inclination of the projectile, were related to observed damage patterns. Multiple impacts inflicted by the lower velocity fragments in the inner cone were most destructive to the rear wall, probably because these fragments impact an area which has been damaged and thinned by spallation induced by the impact of higher velocity material in the leading edge of the debris cloud.

Although the shape of individual fragments of orbital debris is not known, they most likely are nonspherical and will have their major axis inclined to the spacecraft surface at the time of impact. Further investigation of the effect of projectile shape and orientation during hypervelocity impact of a spacecraft exterior wall should be done to insure that adequate protection is provided to inhabitants of those vehicles.

ACKNOWLEDGEMENT

This work was supported by Martin Marietta Michoud Aerospace under Contract No. A71050. The author wishes to sincerely express his appreciation to Dr. Norman Elfer of Martin Marietta for his continued interest in and support of this effort. In addition, the author would like to express his appreciation to the following: T. Williams and C. Blair, for performing the experiments; T. Klopfenstein, for his care in fabrication of the many components used in the experiments; E. Strader, for his assistance with instrumentation; and R. Tocci, J. Moreau, and R. Strong for their photographic work.

REFERENCES

Friend, W.H., C.L. Murphy, and P.S. Gough (1969). Review of meteoroid-bumper interaction studies at McGill University. NASA CR-54858.

Kassel, P.C. and J.D. DiBattista (1971). An ultra-high-speed photographic system for investigating hypervelocity impact phenomena. NASA TN D-6128.

Kessler, D.J. (1982). Orbital Debris. NASA Conference Publication 2360, pp. 78-83.

Maiden, C.L., A.R. McMillan, and R.E. Sennett (1965). Thin sheet impact. NASA CR-295.

Morrison, R.H. (1972). A preliminary investigation of projectile shape effects in hypervelocity impact of double-sheet structures. NASA TN D-6944.

Piekutowski, A.J. (1985). The University of Dayton Research Institute 50/20 mm, two stage, light-gas gun. Proceedings of the Aeroballistic Range Association, Southwest Research Institute, San Antonio.

Rolsten, R.F., H.H. Hunt, and J.N. Wellnitz (1962). Study of principles of meteoroid protection. Final Report, Contract NAS-8-875, George C. Marshall Space Flight Center, AL.

Rolsten, R.F., H.H. Hunt, and J.N. Wellnitz (1964). Meteoroid bumpers: aluminum. Meteoritics, 2, No. 2, 149-159.

Int. J. Impact Engng Vol. 5, pp. 519–523, 1987
Printed in Great Britain

0734-743X/87 $3.00 + 0.00
Pergamon Journals Ltd.

THE DEVELOPMENT OF A PROJECTILE VELOCITY MEASURING SYSTEM FOR THE RARDE NO. 2 HYPERVELOCITY LAUNCHER

A.J.RICHARDS, A.V.B.BOWDEN

ROYAL ARMAMENT RESEARCH AND DEVELOPMENT ESTABLISHMENT
FORT HALSTEAD, SEVENOAKS, KENT. TN14 7BP. U.K.

ABSTRACT

This paper outlines the princinle of a projectile velocity measuring system fitted to the RARDE No. 2 Hypervelocity Launcher. The system is capable of measuring the velocity of small particles, (1g), travelling at velocities in excess of 6000m/s.

INTRODUCTION

The RARDE No 2 Hypervelocity Launcher is a unique UK facility which was designed in the late 1950's as a hypersonic ballistic range. Previous work has been reported to the Proceedings of the Third Symposium on Hypervelocity Impact. It is capable of firing a 15g projectile in excess of 6000m/s. Over the last three years the launcher has been refurbished and adapted for the study of fragment/target impact.

A general view of the launcher facility is shown in Fig 1 and 2.

Fig 1. General view of launcher comporessor barrel and driving chamber

Fig. 2. General view of launcher barrel and range

With the rebuild nearing completion, careful consideration was given to the method of measuring projectile velocity. The system had to be capable of measuring the velocity of very small fragments travelling at speeds in excess of 5000m/s. It was also indicated that the ability to measure the velocity of several fragments temporally spaced would be desirable.

The system developed has met all of these requirements and during trials has proved to give more information than expected. Having developed the velocity measuring equipment a single channel flash x-ray unit has been installed and linked into the system. It has given valuable data on projectile flight and target penetration characteristics.

VELOCITY MEASURING SYSTEMS CONSIDERED

Several techniques were considered including laser detectors, foil screens, optical and doppler radar. Each of these systems have both advantages and disadvantages when considering their application to the launcher facility. Ease of use was one of the most important considerations. This ruled out foil screen, as they need to be placed in the projectile flight path, necessitating the breakdown of vacuum seals in the range section after each firing, they were also rejected as the projectile would be influenced by penetrating the screen. Laser systems were rejected because of the delicateoptical equipment which may be disturbed by the vibration during firing. Doppler radar seemed a promising method, but no in-house expertise was available for this particular application (high projectile velocity/enclosed range) thus, the idea was not pursued. An optical system was chosen in the form of skyscreens. These are devices used by many UK authorities for the detection of projectiles in flight, be it bullets or large calibre shells. The manufacturers supplied one unit for trials. On the basis of the results it was decided to purchase three units with an agreement to follow a development programme to meet all the requirements.

THE PRINCIPLES OF THE SKYSCREEN SYSTEM FITTED TO THE LAUNCHER

The basis of the system is an array of photo diodes that respond to light near the infra-red band of the spectrum. The detectors are set behind a narrow slit and view a fan of space via a lens system. Any projectile traversing this fan casts a shadow on the detectors. The signal caused by the shadow is amplified and received by an interface unit. The detectors were primarily designed for use outdoors and to receive light from the sky, hence their name. As the light in outdoor conditions fluctuates, the detectors incorporate compensation circuitry and are set to respond to rapid rates of light change rather than to overall changes.

For adaptation to the launcher, light is supplied by d.c. strip lights, but the light compensation circuitry has proved useful as the system is not sensitive to shadows cast when setting up instrumentation or switching off the room lights etc. Output from the detectors is processed by interface units which convert the analogue signal from the detectors to digital pulses that start and stop the velocity computers. They also incorporate a system test facility to ensure correct operation of the detectors prior to a firing. This has also proved useful for setting up other instrumentation running from the skyscreen system.

The interface units have an additional capability to set the computer to start and stop off either the base or nose of the projectile. The velocity computer converts the start and stop pulses to either velocity or time of flight between detectors and then displays the result. The computer incorporates many features including storage for multishots, printout facilities, mean and standard deviation of shot velocities, and a false reading reset function to prevent spurious triggering by dust particles following the projectile.

THE DEVELOPMENT OF THE SKYSCREEN SYSTEM FOR APPLICATION TO THE LAUNCHER

At first only one of the three systems ordered was installed, the detectors were mounted beneath the range and the strip lights above as shown in Fig 3.

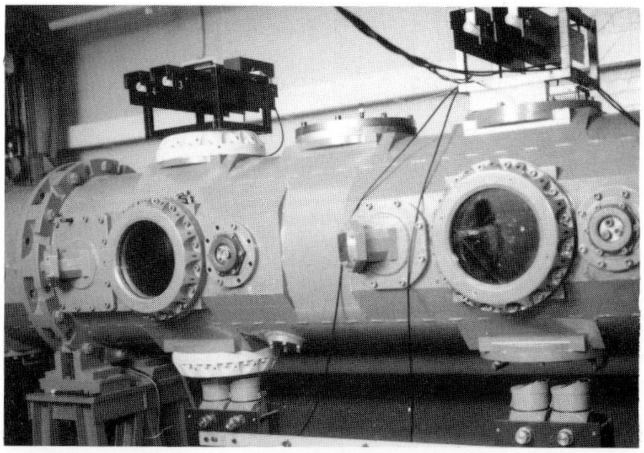

Fig 3. Final instalation of skyscreen/velocity measuring system

The interface units and velocity computers were installed in the control room. An immediate problem became apparent, involving the glass range ports through which the detectors view the projectile. These appeared opaque to the photodiodes as the optical glass used was of sufficient thickness and quality to filter out the infra-red light. This problem was overcome by fitting Perspex ports to the range.

Initial Firings

To test the system a series of low velocity firings (400-500m/s) were carried out using the launcher as a single stage gas gun. The results were promising.

High Velocity Firings

Using the launcher in its full capacity, a series of high velocity firings (2500-3500m/s) were made. The results showed that the detectors were not working correctly. Muzzle flash was thought to be the problem. A series of modifications were made to the detectors but the problem persisted. Baffles and muzzle breaks were fitted but to no avail. To aid identification of the problem, a four channel oscilloscope was coupled to the detector monitoring the analogue signal from the detectors and the digital output to the computer. The trace was very erratic, as shown in Fig 4.

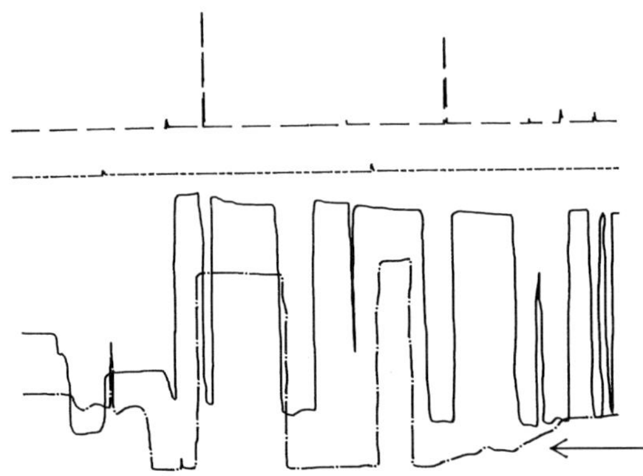

Fig 4. Erratic oscilloscope trace caused by perspex port.

By a process of elimination the problem was traced to the distortion of the Perspex and consequent change of optical properties during firing. The distortion was caused when firing, by a shock wave travelling through the launchers structure, which preceded the projectile down the range. The problem was solved by reverting to glass ports. By utilising a redesigned mounting it was possible to reduce the thickness of the glass to allow sufficient infra-red transmission to operate the detectors. The improvement in the oscilloscope traces is seen by comparing Fig 4 and 5.

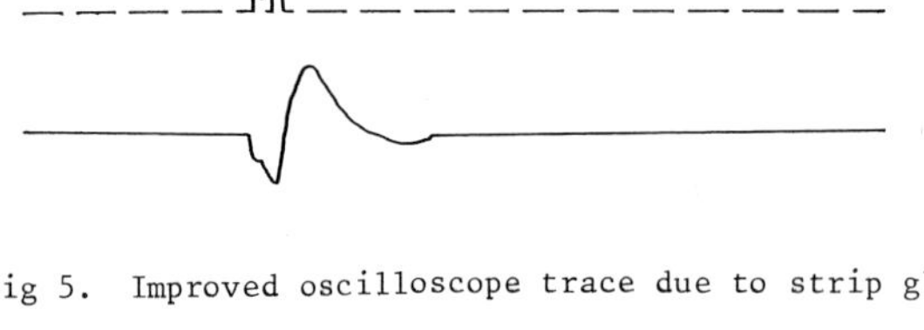

Fig 5. Improved oscilloscope trace due to strip glass windows

The analogue start and stop pulses are a measure of the light level seen by the detectors, a decrease in light level is indicated by a positive pulse on the trace. If more than one temporally spaced projectiles are seen by the detectors they will appear on the trace as a series of discrete pulses. By comparing both the start and stop traces the velocity of each projectile can be assessed. If the projectiles vary in shape and size this is indicated by the amplitude and profile of the pulse. Thus far, no work on temporally spaced projectiles has been carried out however, during sabot stripping trials debris from the sabot sometimes followed the projectile down range thus proving the capacity of the system to detect temporally spaced projectiles.

THE INSTALLATION OF THE FLASH X-RAY SYSTEM

The results of the completed skyscreen velocity measuring system showed that the launcher was producing very consistent velocities for given gas pressures. A 600Kv single channel Field Emision X-ray unit was linked to the skyscreens to provide shadowgraphs of projectiles in flight and impacting the target. Preliminary shadowgraphs were taken of low velocity projectiles to set up the x-ray collimation. The problem of the thick glass of the range ports attenuating the x-rays to an unacceptable degree was overcome by replacing the glass with Perspex ports removed during the skyscreen development programme. As can be seen from Fig 6, excellent pictures of target penetration can be taken.

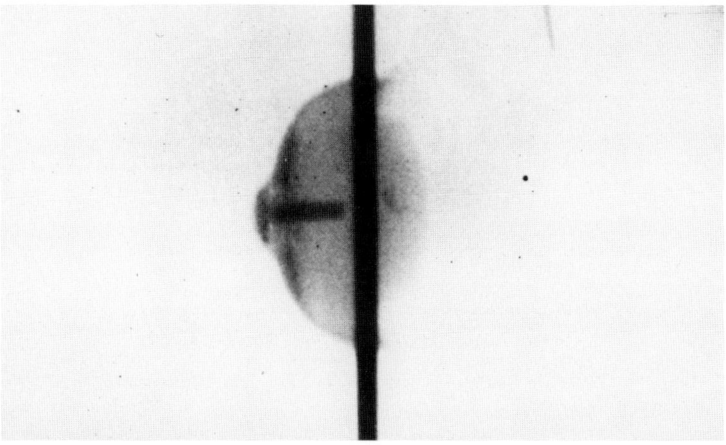

Fig 6. Radiograph of the 15g projectile penetrating a 3mm aluminium alloy target. (taken 12.5 microseconds after impact)

The picture shows a 15g 25mm diameter polythene cylinder penetrating a 3mm aluminium alloy target. The projectile has a small self tapping screw in its nose to aid x-ray definition. The x-ray unit is triggered by using a skyscreen detector mounted 40mm uprange from the target. The digital output from the interface unit is fed into an oscillating crystal delay unit to trigger the x-ray at the required time. The skyscreen is positioned to compensate for a 60 μs delay inherent in the x-ray system. Having gained considerable experience with the single channel flash x-ray system a four channel 150Kv unit has been ordered. This is fitted with soft x-ray tubes that will show much more detail than the 600 Kv unit. The four x-ray heads will be mounted orthogonally in the range either in two pairs displaced laterally or in an octagon formation around the target. This will give projectile orientation.

WORK PROGRAMME

From the results obtained to date computer modelling for the penetration of the 3mm aluminium alloy targets by the polythene projectiles has started. It is hoped that these results will contribute to validation of our theories. Future work will include characterisation of cubic fragments against various targets. In particular, 3.5 gram mild steel and 1.5 gram tungsten alloy cubes will be fired for interaction studies against targets representing missile and aircraft structures and components.

Int. J. Impact Engng Vol. 5, pp. 525–541, 1987
Printed in Great Britain

0734-743X/87 $3.00 + 0.00
Pergamon Journals Ltd.

COMPUTER SIMULATIONS OF LARGE ASTEROID IMPACTS INTO OCEANIC AND CONTINENTAL SITES--PRELIMINARY RESULTS ON ATMOSPHERIC, CRATERING AND EJECTA DYNAMICS

David J. Roddy*, Sheldon H. Schuster**, Martin Rosenblatt**, Lisa B. Grant**, Paul J. Hassig**, Kenneth N. Kreyenhagen**

*U.S. Geological Survey, Flagstaff, Arizona
**California Research and Technology, Chatsworth, Calif.

ABSTRACT

Computer simulations have been completed that describe passage of a 10-km-diameter asteroid through the Earth's atmosphere and the subsequent cratering and ejecta dynamics caused by impact of the asteroid into both oceanic and continental sites. The asteroid was modeled as a spherical body moving vertically at 20 km/s with a kinetic energy of 2.6×10^{30} ergs (6.2×10^7 Mt). Detailed material modeling of the asteroid, ocean, crustal units, sedimentary unit, and mantle included effects of strength and fracturing, generic asteroid and rock properties, porosity, saturation, lithostatic stresses, and geothermal contributions, each selected to simulate impact and geologic conditions that were as realistic as possible. Calculation of the passage of the asteroid through a U.S. Standard Atmosphere showed development of a strong bow shock wave followed by a highly shock compressed and heated air mass. Rapid expansion of this shocked air created a large low-density region that also expanded away from the impact area. Shock temperatures in air reached ~20,000 K near the surface of the uplifting crater rim and were as high as ~2000 K at more than 30 km range and 10 km altitude. Calculations to 30 s showed that the shock fronts in the air and in most of the expanding shocked air mass preceded the formation of the crater, ejecta, and rim uplift and did not interact with them. As cratering developed, uplifted rim and target material were ejected into the very low density, shock-heated air immediately above the forming crater, and complex interactions could be expected. Calculations of the impact events showed equally dramatic effects on the oceanic and continental targets through an interval of 120 s. Despite geologic differences in the targets, both cratering events developed comparable dynamic flow fields and by ~29 s had formed similar-sized transient craters ~39 km deep and ~62 km across. Transient-rim uplift of ocean and crust reached a maximum altitude of nearly 40 km at ~30 s and began to decay at velocities of 500 m/s to develop large-tsunami conditions. After ~30 s, strong gravitational rebound drove both craters toward broad flat-floored shapes. At 120 s, transient crater diameters were ~80 km (continental) and ~105 km (oceanic) and transient depths were ~27 km; crater floors consisting of melted and fragmented hot rock were rebounding rapidly upward. By 60 s, the continental crater had ejected ~2×10^{14} t, about twice the mass ejected from the oceanic crater. By 120 s, ~70,000 km^3 (continental) and ~90,000 km^3 (oceanic) target material were excavated (no mantle) and massive ejecta blankets were formed around the craters. We estimate that in excess of ~70% of the ejecta would finally lie within ~3 crater diameters of the impact, and the remaining ejecta (~10^{13} t), including the vaporized asteroid, would be ejected into the atmosphere to altitudes as high as the ionosphere. Effects of secondary volcanism and return of the ocean over hot oceanic crater floor could also be expected to contribute substantial material to the atmosphere.

INTRODUCTION

Hypervelocity impact has been widely accepted as playing a major role in the early evolution of the terrestrial planets and satellites. However, a number of studies now show that large planetary and terrestrial impact events have continued to occur, although at a greatly reduced rate, throughout geologic time (Hörz, 1971; Roddy et al., 1977; Shoemaker, 1977, 1983; Soderblom, 1977; Silver and Schultz, 1982; Grieve, 1982; Grieve et al., 1985; Shoemaker and Wolfe, in press; and others). Recently, an important set of consequences was proposed for a terrestrial impact by Alvarez et al. (1980), who suggested that a large cratering event occurred 65 million years ago at the end of Cretaceous time and ejected sufficient material into the atmosphere to induce global climatic and biologic changes. Their proposal of such a large event and its effects on the atmospheric, oceanic, geologic, and biologic environments has drawn considerable attention from a number of scientific disciplines (Alvarez, 1986). Unfortunately, the full range of impact processes and the widespread effects of such a large-scale event remain minimally understood.

We have started to examine the first part of the problem, the impact event itself, by completing computer simulations of the passage of a 10-km-diameter asteroid through the Earth's atmosphere and by following the dynamics of the cratering events at both oceanic and continental sites. Our objective is (1) to provide a set of simulations that are as realistic as possible for large-scale terrestrial impacts in two different types of geologic environments, and (2) to determine the initial cratering, ejecta, and geologic effects. Our longer range goal is to provide quantitative data that will help improve understanding of the broader role of large-body impacts in the evolution of the different planetary bodies and, more specifically, improve understanding of the consequences of such events on the atmospheric, oceanic, geologic, and biologic environments on Earth. This paper describes the major results of our preliminary analysis of these computer simulations.

PREVIOUS WORK

Computer-code simulations designed to examine generalized conditions associated with large terrestrial and planetary impacts have been limited to a relatively small number of studies. The most comprehensive simulations of this kind have been completed by Ahrens and O'Keefe (1977, 1983, 1985, this volume) and O'Keefe and Ahrens (1975, 1977a,b, 1978, 1982a,b, 1983, 1985, 1986a,b). Their studies of both terrestrial and planetary impacts concentrated on quantitative aspects of impact processes, early-time cratering, and ejecta dynamics. In one set of studies, they modeled asteroids, comets, and meteoroid swarms involving bodies up to 10 km in diameter impacting into oceanic and continental environments (O'Keefe and Ahrens, 1982a,b) to examine conditions associated with the Cretaceous/Tertiary large impact event proposed by Alvarez et al. (1980). The only computer simulations of a specific terrestrial (and planetary) cratering event are the calculations that describe the impact that formed Meteor Crater (Barringer Crater), Arizona (Bjork, 1961; Bryan et al., 1978; and Roddy et al., 1980). In one study, computer simulations were carried out long enough to determine the final crater profile scaled to actual field data and to establish a new estimate of the energy of formation (Roddy et al., 1980); their calculations also provided graphic displays of major geologic features developed in late time, such as the breccia lens, permanent rim uplift, and ejecta distribution.

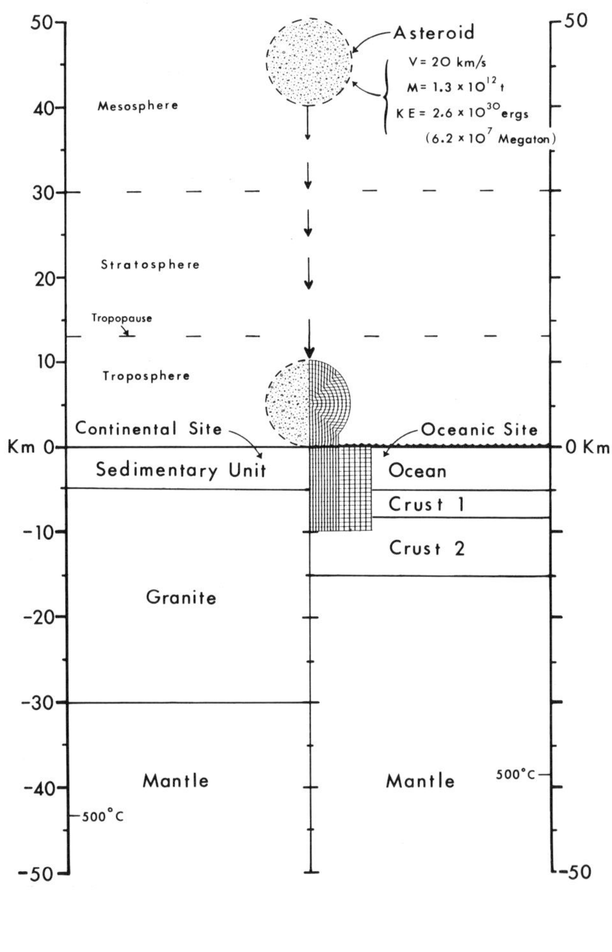

Figure 1. Summary of pre-impact conditions for the atmosphere, asteroid, continental site, and oceanic site. The lower regions of the U.S. Standard Atmosphere, 1962/1976, are shown to an altitude of 50 km. Arrows show descent path of the asteroid used in the atmospheric calculations. The computational grid used in the two-dimensional, axisymmetric cratering simulations is shown for the asteroid and oceanic target. The grid is enlarged as the shock effects expanded throughout the targets. The flat-lying geologic units are shown here only to a depth of 50 km, but calculations followed shock effects to a depth of 200 km and to a radius of 150 km. The depths of the 500° C isotherms are shown for both of the impact sites.

PRE-IMPACT CONDITIONS FOR THE ATMOSPHERE, ASTEROID, AND TERRESTRIAL SITES

Pre-impact conditions for the calculations were chosen to represent a real terrestrial atmosphere, an inferred dynamic state for the asteroid, and realistic terrestrial sites for the two cratering events. The atmosphere was modeled as having normal distributions of air pressure, temperature, and density as a function of altitude by using the values of the U.S. Standard Atmosphere, 1962/1976. In the atmospheric calculation, the asteroid was modeled as a 10-km-diameter, rigid, nondeforming spheroid moving downward at 20 km/sec in a vertical trajectory; the surface of the Earth was assumed as rigid and smooth to air flow.

In the cratering calculations, the asteroid was also modeled as a spherical body 10 km in diameter moving at 20 km/s vertically through the atmosphere. However, detailed material modeling was constructed for the body in this calculation, and it was given the composition of a generic silicate with the properties of quartz, a bulk density of 2.5 gm/cm^3, and a porosity of 8%. The mass of the body was 1.3 x 10^{12} t and the kinetic energy associated with the impact was 2.6 x 10^{30} ergs (6.2 x 10^7 Mt TNT equivalent).

The choices of these conditions for the asteroid involved a number of considerations designed to simulate realistic impact events. For example, a spherical shape for the body was adopted to approximate the oblate shapes observed for many of the asteroids (Delsemme, 1977; Gehrels, 1979). Its size was chosen to be equivalent to that of the impacting body estimated by Alvarez et al. (1980), which permits extrapolation of our results to a number of studies of the Cretaceous/Tertiary impact event proposed by Alvarez and his co-workers. The velocity of the body was chosen to match the root mean square value of the calculated velocities of 25 Earth-crossing asteroids (Shoemaker et al., 1979). We used a vertical impact trajectory, which in turn yielded a lower bound for the kinetic energy required to form the craters; as noted in laboratory experiments by Gault and Wedekind (1978), vertical impact angles require slightly lower kinetic energies than do lower impact angles to form the same size crater. Interpretations of the compositions of a number of Earth-crossing asteroids suggest a silicate mineralogy (Delsemme, 1977; Gehrels, 1979). In our calculations, simplification of the composition to a generic silicate, quartz, is reasonable because experiments have shown that when shock pressures exceed a megabar, both quartz and complex silicates behave virtually the same (van Thiel, 1977). The bulk density and porosity of the quartz asteroid are our estimates based on discussions with E.M. Shoemaker (U.S. Geological Survey, personal comm., 1985) and C. Chapman (Univ. Arizona, personal comm., 1985).

Pre-impact geologic conditions for the impact sites were selected to be broadly representative of oceanic and continental environments. The geologic conditions are shown in Fig. 1, and the calculational parameters are summarized in Table 1. Both sites had flat-lying geologic units that were not tectonically stressed. We modeled the oceanic site to consist of an ocean 5 km deep overlying two crustal units: crust 1 (3 km thick) overlying crust 2 (7 km thick). The Mohorovicic (Moho) discontinuity that separates the crust and mantle was placed at a depth of 15 km, overlying mantle rock that extended semi-infinitely downward. Using appropriate equation-of-state data, we modeled the crust 1 and 2 units as basaltic rocks having slightly different properties and the mantle as ultramafic rock (Table 1). The thicknesses and depths of the lithologic units for this site were generalized from several sources, including D. Anderson (California Institute of Technology, written comm., 1983), S. Biehler (Univ. California, Riverside, written comm., 1983; personal comm., 1985), and T. Wallace (California Institute Technology, written comm., 1983). These workers defined the generalized rock units mainly on the basis of seismic data and density assumptions. From their data, we chose values for lithologic types and thicknesses to simulate a generic site for oceanic crust older than about 20 million yr. The thermal gradient was constructed from data of Pollack and Sass (1985) to fit their estimate for an oceanic slab about 100 million yr old.

We modeled the continental site to consist of a 5-km-thick layer of sedimentary rock overlying granitic rock 25 km thick. The sedimentary unit was modeled with 17% porosity and was water saturated. The Moho discontinuity, at a depth of 30 km, overlay mantle rock that extended semi-infinitely downward. The thicknesses and depths of the lithologic units for this site were taken from several sources, including D. Anderson (written comm., 1983), S. Biehler (written comm., 1983; personal comm., 1985), T. Wallace (written comm., 1983), Pollack and Murthy (1984), and Taylor and McLennan (1985). The units were defined on the basis of seismic data and density assumptions, and we chose values to simulate a generic continental site on a stable platform/shield. The geothermal gradient was estimated from generalized Precambrian shield data of Ringwood (1969). The Moho discontinuity was placed at a depth about 5 km less than normally found in older platform regions in order to model a region of younger age comparable to the oceanic site.

RESULTS OF CALCULATIONS OF THE PASSAGE OF AN ASTEROID THROUGH THE ATMOSPHERE

a) Introduction

Calculations described in this section simulated the atmospheric shock dynamics created by a large asteroid passing through the Earth's atmosphere. Computations were carried to 30 s using the finite-difference, continuum-mechanics, DICE computer code developed and operated by California Research and Technology. DICE is a two-dimensional, axisymmetric, Eulerian code used for advanced analysis of single and multiphase flow dynamics (Rosenblatt et al., 1982). We chose DICE because of its reliability in calculating complex shock-wave conditions in real atmospheres to late times, its multiphase capability to define atmospheric and ejecta/debris interactions, and its extensive testing against nuclear-explosion and high-explosion data.

In our current study, DICE was used to simulate the shock effects generated in the air by the asteroid's passage and to follow the expanding shock waves and related pressure, temperature, and density effects in the atmosphere. After the asteroid reached the Earth's surface, it no longer contributed to the expanding air-shock fronts and shocked air mass and it was removed from the atmospheric simulations. The impact process was then initiated with the cratering-code calculations. This was possible because our results showed that the major features of interest in this preliminary analysis, the shock fronts and most of the shocked airmass, that were generated by the asteroid, expanded consistently well beyond the region being cratered and were not affected. A study of the complex interactions that would occur later between the uplifting rim and ejecta and the low-density, shock-heated region in the air will require linking of the atmospheric and cratering codes, an effort that we hope can be undertaken in the future.

b) Atmospheric Calculations

The DICE atmospheric calculation incorporated a two-dimensional axisymmetric geometry with a computational grid that extended upward to an altitude of 51 km, outward to a range of 71 km, and had grid zones 1 km on a side. The calculation started with the leading edge of the asteroid at 40 km altitude and the trailing edge at 50 km. Computations tracked the 2.0-s descent of the body to the Earth's surface and followed the expanding shocked atmosphere for another 28 s (Fig. 2a,b,c,d). During descent, the asteroid generated a bow shock wave that strongly compressed and heated the atmosphere and set it in motion with velocities as high as 18 km/s (Fig. 2a). The outward radial flow of air, together with the atmospheric path cleared by the passage of the asteroid, created a highly evacuated region along the asteroid trajectory in which the atmospheric densities were reduced to a fraction of their original values; this low-density region expanded behind the shocked air mass and remained well defined at ranges and altitudes of over 50 km throughout the calculational interval of 30 s.

After the bow shock wave reached the Earth (~1.8 s asteroid time; ~-0.2 s impact time), it began to move radially outward in excess of 15 km/s, and its interaction with the Earth's surface caused an additional strong air shock front to develop (Fig. 2b,c). These combined shock fronts were followed by outward radial flow of intensely heated, shocked air that, in turn, was followed by the expanding low-density region. At 2.5 s (0.5 s impact time), cratering calculations showed that the transient uplifting rim and ejecta (mostly vapor) were forming in the part of the low-density region of shocked air that was well behind the passage of the air shock fronts in time and location (Figs. 2b,c).

Representative air-particle velocity, temperature, and overpressure fields at 6.5 s (4.5 s impact time) are shown in Fig. 2d. Peak temperature was calculated to be ~20,000 K over the uplifting rim at a range of 15 km, and temperatures in excess of 10,000 K were reached in a large low-density region expanding radially above the Earth's surface. Winds are calculated to have been as high as 100 m/s at a range of 70 km. Shock fronts and outer shocked regions in the atmosphere shown in Fig. 2d were well beyond the uplifting rim and ejecta and remained unaffected by cratering processes. The rim uplift and ejecta continued to expand into the low-density region in the atmosphere, and they could be expected to modify velocity, temperature, and overpressure fields in the atmosphere in their vicinity. The basic effect of the uplift and ejecta would be to continue to compress and displace the air and its associated temperature and overpressure fields out and away from the impact area; the basic effect of the low-density, shock-heated region in the air would be to vaporize an additional amount of ejecta and uplifting rim material. The immediate region where rim uplift and ejecta field are shown overlain on the atmospheric data should be ignored in Figs. 2b,c,d; the rest of the shocked regions in the atmosphere were unaffected by the cratering and are valid as shown in Fig. 2.

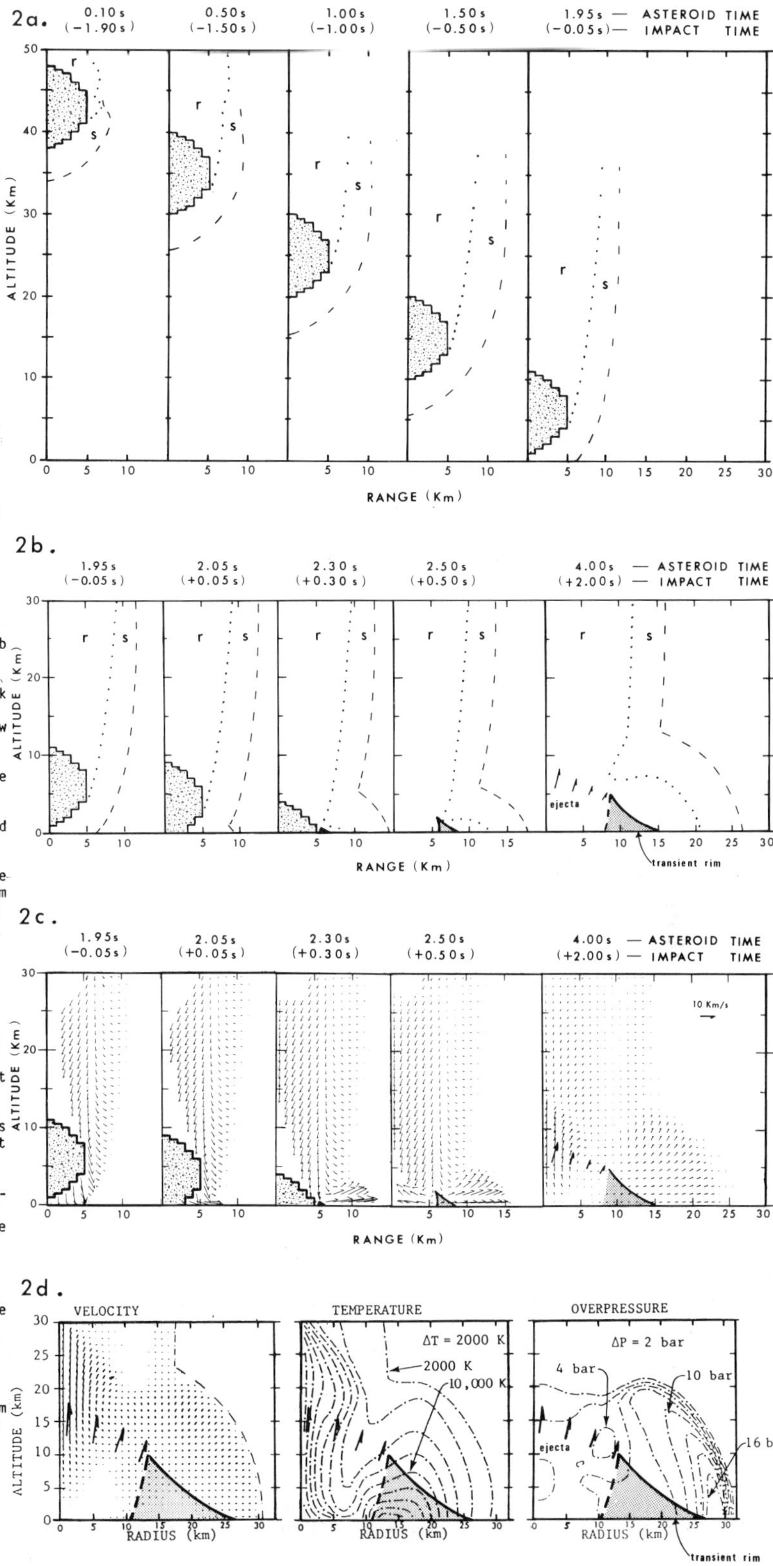

Fig. 2. Sequence of atmospheric shock effects developed by descent of a 10-km-diameter asteoid through a U.S. Standard Atmosphere, 1962/1976. Times for atmospheric and impact calculations are given top of figures. Atmospheric (asteroid) time starts at initial descent from 40 km altitude and is always 2.0 s ahead of impact-cratering time. Calculations are two-dimensional axisymmetric and only half of each dynamic field is shown. Atmospheric calculations defined the bow shock wave (s) and low-density region (r) forming behind the bow-shocked air mass. Dashed line shows location of first arrival calculated for bow shock front in air. Dotted line shows location of first negative value calculated for air overdensity change that represents an expansion of the shocked air, i.e., the edge of the low-density region (r) forming behind the bow shock. In Fig. 2a, downward-directed air velocities as large as ~18 km/s occur near the front of the asteroid; air shock pressures in front of asteroid range from ~0.5 kb at an altitude of 10 km to ~200 kb at an altitude of ~1 km. Fig. 2b shows outward expansion of bow shock wave, low-density region, and air shocked region that developed as bow shock interacted with Earth's surface. Fig. 2c shows velocity-vector flow fields in the atmosphere for same times shown in Fig. 2b. Asteroid was removed from atmospheric simulation as it reached the Earth's surface, and cratering calculations were then used to follow the impact. Positions of the uplifting crater rim were taken from cratering calculations and overlaid on these figures to determine relative positions as a function of time. Transient rim developed beneath the expanding low-density region in the atmosphere but well after passage of air shock fronts. An ejecta field (heavy arrows; not drawn to velocity scale) consisting of vapor and minor amounts of melt is shown at 2.0 s after impact as it expands into low-density region. Fig. 2d shows air velocity-vector, temperature, and overpressure fields at 6.5 s asteroid time (4.5 s impact time). Shock fronts in air, the bulk of the expanding shocked air mass, and the upper part of the low-density region were not affected by the uplifting rim and ejecta because they were farther away and still expanding beyond the cratering sequence. However, the rim and ejecta field will modify the atmospheric conditions in their immediate vicinity, and the atmospheric data in this region should be ignored on Fig. 2c,d; one effect will be to reposition the high temperature (20,000 K) region near the Earth's surface to an altitude of 5 to 10 km on the outer slope of the uplifting rim.

Table 1. Summary of Physical and Numerical Constants Used in Cratering Calculations of Asteroid Impacts at Oceanic and Continental Environments.

Equation of State Input			Materials						
Parameter	Symbol	Units	Asteroid (Quartz)	Ocean (Water)	Crust 1 (Basalt)	Crust 2 (Basalt)	Mantle (Ultramafic)	Sedimentary Unit	Granite
Thickness	-	km	10.0	5.0	3.0	7.0	infinite	5.0	25.0
P wave velocity	Vp	km/sec	5.2	1.5	5.5	7.0	8.0	4.2	4.5
S wave velocity	Vs	km/sec	2.6	0.0	3.2	4.0	4.5	2.4	2.6
Initial *in situ* density	ρ_o	gm/cm^3	2.5	1.0	2.8	3.0	3.3	2.4	2.7
Air voids	-	%	8.0	0.0	0.0	0.0	0.0	0.0	0.0
Initial bulk modulus	K_o	Mb	0.45	0.022	0.46	0.83	1.2	0.24	0.3
Maximum bulk modulus	K_m	Mb	0.75	0.3	0.75	1.0	1.2	0.75	0.7
Exponential factor	μ^*	-	0.03	1.0	0.03	0.03	0.2	0.3	0.02
Coef. of thermal expansion	β	gm/Te	3.0	10.0	3.0	3.0	3.0	3.0	3.0
Minimum energy for P_v	E_m	gm/Te	0.017	0.06	0.014	0.014	0.015	0.022	0.02
Initial shear modulus	G_o	Mb	0.173	0.0	0.276	0.498	0.652	0.137	0.18
von Mises limit (compression)	Y_{vm}	Mb	0.017	0.0	0.017	0.017	0.003	0.003	0.001
Cohesion	C_o	Mb	5.0×10^{-5}	0.0	5.0×10^{-5}	5.0×10^{-5}	3.0×10^{-3}	5.0×10^{-5}	3.7×10^{-4}
Slope of Mohr-Coulomb surface	α	-	0.75	0.0	0.75	0.75	0.0	0.75	0.6
Tensile limit	σ_t	Mb	-6.67×10^{-5}	0.0	-6.67×10^{-5}	-6.67×10^{-5}	-6.67×10^{-5}	-5.0×10^{-5}	-6.67×10^{-5}
Poisson's ratio	ν	-	0.33	0.5	0.25	0.25	0.27	0.26	0.25
Phase change parameter	δ	-	0.304	0.0	0.297	0.234	0.0	0.293	0.328
Phase change parameter	μ_p	-	0.111	0.0	0.071	0.0	0.0	0.250	0.111

Notes: Te = Terraerg, i.e., 10^{12} ergs; the asteroid had V=20 km/s, m=1.3×10^{12} t, and KE=2.6×10^{30} ergs (6.2×10^7 Mt TNT equivalent).

RESULTS OF CALCULATIONS OF THE OCEANIC AND CONTINENTAL IMPACT CRATERING EVENTS

a) Introduction

Calculations described in this section simulated the cratering dynamics of two large asteroid impacts, one at an oceanic site and the other at a continental site. Cratering and ejecta computations were carried to 120 s by using the finite-difference, continuum-mechanics, CRALE2 computer code developed and operated by California Research and Technology. CRALE2 is a two-dimensional, axisymmetric, arbitrary, Lagrangean-Eulerian code used for advanced analytical studies of cratering, ground motions, and ejecta (Schuster, 1984). We chose CRALE2 because of its proven reliability in accurately simulating large-scale cratering and ejecta events to late times, its capability of accepting complex material modeling, and its extensive testing with a wide range of high-explosion and nuclear-explosion data. An earlier form of CRALE was used to numerically simulate the impact event that formed Meteor Crater (Barringer Crater), Arizona (Roddy et al., 1980).

b) Initial and Boundary Conditions

The initial conditions for the cratering calculations are shown in Fig. 1. The CRALE2 computer code, in the form used in these studies, was axisymmetric and the impacts were modeled with a vertical angle of incidence. A normal 1-g gravitational field was used for the impact sites, and appropriate lithostatic stress fields, assumed to be nondeviatoric, were modeled for both oceanic and continental sites. A constant atmospheric pressure of 1 bar was used for the boundary condition at the Earth's surface; as discussed earlier, no atmosphere was included. In the cratering calculations, the asteroid was initially represented in the computational grid by 210 grid zones averaging about 0.5 km on a side (Fig. 1). Target materials were modeled as flat-lying lithologic units, and in the initial target calculations, grid zones were 0.5 km long and 0.35 km wide. Size of target grid zones were increased geometrically with increasing distance from the impact point: grid dimensions were doubled at 4 s (continental) and 7 s (oceanic) and again at 1 min for each calculation, and at the end of calculations, a region 200 km in depth and 150 km in radius was included. Shock conditions impinging on these boundaries were assumed to be elastic and allowed to pass smoothly out of the calculations.

c) Equation of State and Material Modeling

The input parameters for material properties used in this study were chosen in conjunction with averaged seismic and density data and generic Hugoniot curves derived from experimental shock data for representative materials, i.e., quartz, water, sedimentary rocks (shale), granite, basalt, and ultramafic rocks (Table 1). The asteroid and all target materials were modeled with updated versions of the Schuster and Isenberg (1972) and Isenberg (1972) equations of state used extensively in nuclear- and chemical-explosion cratering studies. In these equations of state, the stress-strain-energy behavior is decomposed into a mean stress or pressure relation plus the deviatoric stress tensor. The mean stress is further decomposed into two terms, i.e.,

$$P = P_s + P_v \tag{1}$$

where P_s represents the contribution to the mean stress of the solid and/or liquid phases and P_v represents the contribution of the vapor. Hysteresis, low-energy thermal effects, and reversible solid-solid phase changes are incorporated into the calculation of P_s. For nonhysteretic materials, such as mantle,

$$P_s = K_m\mu - (K_m - K_o)\mu^*(1 - e^{-\mu/\mu^*}) \tag{2}$$

where μ is the excess compression $([\rho - \rho_o]/\rho_o)$. The bulk moduli at zero and infinite compression, K_o and K_m, and μ^*, a free parameter, were selected to fit the Hugoniot data (Table 1). The thermal-energy dependence of the solid is incorporated by using an effective μ_{eff} in Eq 2 where μ_{eff} is formed by adding the effect of thermal expansion to μ.

$$\mu + \beta E \rightarrow \mu_{eff} \tag{3}$$

where E = energy/unit mass and β = coefficent of thermal expansion/unit of energy. This is equivalent to the Grüneisen correction used in other models, with a variable Grüneisen gamma. At a solid-solid phase change, the effective μ is again altered to reflect the decrease in $dP/d\mu$. Hence, μ_{eff} is replaced by μ_Δ such that,

$$\mu_\Delta = \delta(\mu_{eff} - \mu_p) \tag{4}$$

where μ_p and δ are phase-change parameters. Hysteresis is treated here in a similar fashion (see Schuster and Isenberg, 1972) by a shift of the effective μ during loading. The vapor term, P_v, is computed by using a variable gamma-law gas,

$$P_v = (\gamma - 1)\rho E^* \tag{5}$$

where ρ is density, γ is the ratio of specific heats, and

$$\gamma - 1 = 0.4 + 0.23 \log(\rho) + [0.35 \log(E^*/\rho) - 0.464]^2 \tag{6}$$

and E* is a nonnegative effective energy density,

$$E^* = \max \begin{cases} (E - E_m)\left(1 - e^{\left(\frac{E_m - E}{E_m}\right)}\right), & \text{where } E > E_m \\ 0, & \text{where } E \le E_m \end{cases} \tag{7}$$

where E is energy/gram and E_m is minimum energy for P_v. Thus, P_v is zero for materials with energy densities less than E_m.

Incremental deviatoric stresses, $d\sigma'_{ij}$, are computed from changes in the deviatoric strain tensor, $d\varepsilon'_{ij}$, by using the elastic equation,

$$d\sigma'_{ij} = -2G d\varepsilon'_{ij}. \tag{8}$$

The shear moduli are derived from the dynamic bulk moduli and constant Poisson ratios as determined from in situ seismic velocities of each geologic unit. The second invariant of the deviatoric stress tensor, $\sqrt{J'_2}$, is then compared to a plastic yield surface, Y, where

$$Y = \min \begin{cases} C_o + \alpha p \\ Y_{vm} \end{cases} \tag{9}$$

where C_0 is the cohesion, α is the slope of the Mohr-Coulomb surface, and Y_{vm} is the limiting von Mises yield surface. If $\sqrt{J'_2}$ exceeds Y, the material has yielded plastically and the deviatoric stresses are reduced by the standard Drucker-Prager flow rule, i.e., without volumetric plastic strain. Finally, the three individual principal stresses are tested against the tensile limit, T. If a stress exceeds this limit, the material is assumed to crack. The offending stress is set to zero and a crack volume computed. Subsequently, cracks can open perpendicular to this first crack and each can close upon the appropriate reloading of the material. Cracks do not heal, so that if one closes it will reopen as soon as the stress perpendicular to it becomes tensile, i.e., it does not have to reach the tensile limit again.

d) Cratering Calculations

The results of our calculations of asteroid impacts at oceanic and continental sites are shown to 60 s in the velocity-vector and pseudomomentum-vector field plots in Fig. 3, and profiles of the growths of the two craters are shown to 120 s in Fig. 4. The plots in Fig. 3 show the different geologic units separated by solid lines and, in some areas, dashed lines. The solid lines indicate that the materials on both sides of the line have internal energy values such that the materials are still in solid or melt form and have not mixed. Dashed lines show generalized material boundaries at higher levels of internal energy where the materials have become vapor and are permitted to mix across boundaries. The near-horizontal dashed line at the top of the ejecta fields above the craters in Fig. 3c,d,e,f represents the calculational boundary where ejecta (mainly vapor) were allowed to leave the computational grid and motions were no longer tracked; however, total mass is still retained in the calculation. Deleting the high-speed ejecta at altitude is necessary to avoid unreasonably large distortions and mass flow of vapor in the computational grid, as well as to permit calculations of the ejecta processes only where the physics remains well defined.

The profiles of crater growth shown in Fig. 4 are based on ratios of densities and represent another way to characterize the boundaries of the transient craters and rims. Each profile represents a plot of the ratio of $\rho/\rho_0 \leq 0.5$, where ρ_0 is the initial unshocked density and ρ is the shocked density, for a given time and material. The calculations determined field plots of density ratios for the target materials, and we chose a ratio of 0.5 because the calculations showed a rapid change in density ratios over short distances across the approximate boundaries of the crater walls, floors, and rims. The ratio chosen represents a density contrast in which densities of shocked materials decreased by 50% or more. This means that everywhere along a given profile, the density of the material that is shock unloading is on the order of half of its pre-impact density.

The 10-km-diameter asteroid initially produced peak shock pressures on-axis on the order of 4 Mbar at the surfaces of the targets, and impacted with an initial velocity greater than the characteristic shock velocities of either the targets or impactor. Consequently, at 0.5 s, when the trailing edge of the asteroid reached the pre-impact level of the Earth's surface, it was still traveling at 20 km/s (Fig. 3a). At this time, the deformed leading edge of the asteroid had penetrated ~5 km into both the oceanic and continental sites, and it was strongly compressed along the axis of penetration by ~50%. The ocean was compressed to 3.5 times its normal density, and pressures in the water exceeded 3 Mbar along the axis of penetration. The rims at both sites were transiently uplifted to a height of ~2 km and the transient apparent diameters (measured at the original level of the surface of the Earth) were ~12 km at both sites.

At 1.0 s, the asteroid had penetrated to ~11 km in both targets and had greatly compressed crust 1 and crust 2 and the sedimentary unit (Fig. 3b). At the oceanic site, internal energy of the asteroid now exceeded the melt energy, and about half of the asteroid was decompressing and expanding as vapor. Vaporizing ocean water and some vaporized asteroid had also begun to expand into the crater void above the asteroid, whereas in the continental crater the region evacuated by the passage of the asteroid still existed. Peak axial pressures had decayed to ~1.8 Mbar and the crustal and granitic units were compressed to 1.5 to 2 times normal density.

By 2.3 s, the asteroid had penetrated more than 16 km in both the oceanic and continental targets, and in the oceanic impact it was expanding rapidly as vapor to several times its original size (Fig. 3c). At the oceanic site, the asteroid and ocean vapor had not mixed to any great degree. At the continental site, the asteroid still remained a coherent mass lining the expanding crater walls and floor. The differences in asteroid expansions at this time are largely the result of different target interactions with the projectile. Crust 1, crust 2, the sedimentary unit, and the upper part of the granitic unit were severely compressed and deformed. At the oceanic site, both the ocean and crust 1 were involved in the transient uplift of the rim, whereas at the continental site, only the sedimentary unit was uplifted. The continental transient rim was uplifted to an altitude of about 8 km, and the inner part of the rim began to expand as vapor and melt into the void above the crater. Vaporized material was now ejecting at several kilometers per second or more. At the oceanic site, a very small amount of water vapor near the axis of penetration had velocities of about 12 km/s, but the vertical components were less than about 11 km/s and no measurable amount of material from the target or asteroid was identified to have reached escape velocity.

By 5.0 s, the asteroid had penetrated to a depth of ~20 km and the craters contained turbulent mixtures of asteroid and target that were being ejected at velocities as high as ~8 km/s, primarily as vapor (Fig. 3d). The transient rims were uplifted to altitudes of ~15 km and contained both solid, melt, and vapor. Granite had not yet been ejected from the continental site. Near the oceanic axis of penetration, the mantle rock was melted to a depth in excess of 27 km, and the shock wave was ~40 km deep.

At about 28 s, a maximum crater depth of ~39 km was reached in the oceanic site, where mantle rock had been initially at a lithostatic pressure of ~11 Kbar and a temperature of ~450 °C. The peak axial shock pressure at a depth of 40 km was calculated to have been 250 Kbar, and the shock wave had expanded to a range in excess of 200 km. Mantle was melted but none was ejected at the oceanic site. At 29 s, a maximum crater depth of ~37 km was reached at the continental site, where the lithostatic pressure in the mantle was ~10 Kbar and temperature was ~425 °C. Calculations indicate that no mantle rock was melted or ejected at the continental site.

Velocity-vector fields calculated at 30 s after impact (Fig. 3e) and at 60 s (Fig. 3f) have been transformed into pseudomomentum-vector fields by multiplying the velocity vectors by the ratio of shock density to initial density. This greatly reduces the size of vectors in the lower density (higher velocity) vapor regions so that they do not graphically overwhelm displays of the rest of the vector field for the solid-phase (lower velocity) regions. By 30 s, vaporized asteroid, ocean, crust, and sedimentary units expanded upward, and were allowed minor mixing in their respective craters, as shown by the near-vertical dashed-line boundaries in Fig. 3e. At the oceanic site, water and crustal units in the rim were transiently uplifted as solid material with minor amounts of melt to a maximum altitude of ~40 km and the apparent transient-crater diameter was about 64 km. At the continental site, sedimentary and granitic units were transiently uplifted as solid material with minor amounts of melt to nearly 25 km in altitude and the apparent transient-crater diameter was ~60 km. Vapor and minor amounts of melt continued to be ejected to altitudes above the tropopause. Rock as deep as the mantle was severely deformed at the oceanic site, and a large volume was compressed, melted, and moved radially outward. Less severe deformation occurred in the granite at the continental site. Pseudomomentum-vector fields at both sites have similar flow patterns above a depth of about 10 km; below ~10 km, the patterns are substantially different, presumably due to different material responses. This difference requires further analysis, especially in terms of late-stage rebound, central-uplift mechanisms, and final morphologic features in the crater.

After about 28 s at the continental crater and 29 s at the oceanic crater, lithostatic pressures in the crust and upper mantle began to induce strong rebound of subcrater units and collapse of crater walls. The rebound was well defined by 34 s and continued to the end of calculations. By 60 s, rebound of the subcrater rocks had uplifted the crater floors to depths of ~37 km at the oceanic site and ~32 km at the continental site (Fig. 3f). Massive collapse of the crust in the rim coupled to inward and upward flow of the mantle as it converged on the axis of symmetry guaranteed continued uplift of the oceanic floor. A more conservative flow of granite and mantle uplifted the continental floor. Differences in flow paths and flow rates between the two sites lay mainly in different material responses. At 60 s (Fig. 3f), the transient rim uplifts in solid and melt form were now moving radially outward and decaying downward from their former maximum heights at about 30 s. Extensive vapor was still being ejected from the craters but the pseudomomentum vectors display only a small amount of the vapor (Fig. 3e,f). By 60 s, over half of the asteroid had been ejected as vapor, and crust 2 and granite were both contributing vapor and minor amounts of melt to the ejecta.

The growth of the oceanic and continental craters is summarized to 120 s in Fig. 4. By 120 s, the floors of both the oceanic and continental craters had rebounded to a depth of only ~27 km. The transient apparent diameters continued to expand at a decreasing rate to ~105 km for the oceanic crater and ~80 km for the continental crater. At 120 s, the heights of the transient-rim uplifts are estimated to have decayed to altitudes of 25 to 30 km at the oceanic site and 15 to 20 km at the continental site. At the oceanic crater, the large transient uplifted rim of water and crust was still moving radially outward at velocities of over 200 m/s; its decay would eventually develop large tsunami conditions. At the continental crater, the rim consisted of the sedimentary and granitic units moving at similar velocities. At the end of calculations, both transient craters had similar shallow-bowl shapes.

Strong gravity-driven motions and other material responses deep in mantle rocks would continue well beyond 120 s, creating much of the flow inward toward centers of craters and upward as materials converged at the axis of symmetry. The effect of these motions is expected to change the forms of both craters from transient bowl shapes to broader, shallower, flat-floored shapes.

Fig. 3. Velocity-vector fields (Fig. 3a,b,c,d) and pseudomomentum-vector fields (Figs. 3e,f) showing cratering and ejecta sequence of events caused by the asteroid impacting in the continental site (left) and the oceanic site (right). Times after impact are at top right of figures. Velocity and pseudomomentum vectors are determined from cratering calculations and show the direction of motion of asteroid and target material. Lengths of the vectors are scaled to give velocities which, in turn, can be determined from the velocity scale at the top of the figures. Calculations are two-dimensional axisymmetric and only half of each field is shown. In Fig. 3a at 0.5 s, the asteroid and upper target units are greatly shock compressed and target materials are flowing at kilometers per second radially out from the impact. The only materials being ejected are transient uplifting rims that consist of the sedimentary unit with minor amounts of melt at the continental site, and water and minor amounts of water vapor at the oceanic site. In Fig. 3b at 1.0 s, the asteroid at the continental site is compressed well over 50% and beginning to spread out along the uplifting rim; at the oceanic site the upper half of the asteroid is beginning to expand as vapor. The uplifting rim at the continental site consists mainly of solid and minor amounts of melt from the sedimentary unit and the uplifting rim at the oceanic site consists mainly of water and water vapor. At the oceanic crater, ejecta consisting of high-speed water vapor are forming and also were beginning to expand rapidly over the top of the vaporizing asteroid. At both sites, the sedimentary and crustal units at depth are compressed to nearly twice their normal density and are moving radially outward at kilometers per second. In Fig. 3c at 2.3 s, the asteroid at the continental site has smeared out along the transient crater walls and floor. The uplifting continental rim consists mainly of the sedimentary unit, some melt, and a minor percentage of the asteroid (melt and vapor). At the oceanic site, the asteroid has expanded as vapor to fill the crater and is capped by a turbulent flow of oceanic water vapor ejecting upward; mixing of vapors has just begun. The uplifting oceanic rim consists of water, water vapor, and crust 1. The dashed line extending from the oceanic rim crest to the axis of penetration shows the boundary beyond which ejecta is not tracked, but mass is retained in the calculation (see text). Major compression and deformation is occurring in the lower part of the granitic unit (continental) and in the upper part of the mantle (oceanic). In Fig. 3d at 5.0 s, craters now contain a mixture of violently expanding asteroid and target vapor escaping upward as ejecta. The inner part of the continental rim contains sufficient vapor and melt to expand up and over the asteroid and to be ejected upward at several kilometers per second. At the continental site, only the sedimentary unit is being ejected; at the oceanic site, ejecta consist of mixed vapor from asteroid, ocean, and minor amounts of crust 1. In Fig. 3e,f, velocity-vector fields have been transformed into pseudomomentum-vector fields by multiplying velocity vectors by the ratio of shock density to initial density. In Fig. 3e at 30 s, the asteroid is vaporized and is escaping upward as ejecta in both craters, and some mixing of asteroid and target vapor is occurring. Both craters have reached their maximum transient depths of nearly 40 km; at the continental site, deeper granite and upper mantle material have just started return flow that will lead to rebound toward the axis of symmetry. Also, both rims are near their maximum heights of ~25 km for the continental crater and ~40 km for the oceanic crater. The continental rim consists of the sedimentary and granitic units and minor amounts of melt and vapor, and the oceanic rim consists of water, water vapor, crust 1, crust 2, and minor amounts of melt and vapor. Vapor, melt, and solid are all being ejected ballistically from both sites; oceanic ejecta consist of ocean, crust 1, and crust 2; continental ejecta consist of the sedimentary and granitic units. After 30 s, the vector-flow fields below the two craters become increasingly different, due largely to different material responses. In Fig. 3f at 60 s, the completely vaporized asteroid continues to escape upward as ejecta in both craters. Large amounts of mixed asteroid and target vapor are escaping as ejecta from both craters, but their vectors are not shown because of the use of the pseudomomentum vectors. Both craters are experiencing strong gravitational rebound driven by motions in target units whose locations range from the collapsing rim to regions rebounding deep in the mantle. Uplifted rims still over 25 km high, are moving outward at greater than 200 m/s and starting to decay in height. In the continental and oceanic rims, target materials consisting of solid, melt, and vapor from the sedimentary and granitic units, water, crust 1, and crust 2 are still being uplifted and ejected but at a substantially decreasing rate. The amount of vapor, melt, and solid ejected ballistically from both craters is also rapidly decreasing. The vector-flow fields below the two craters are now different in several regions, due largely to different material responses.

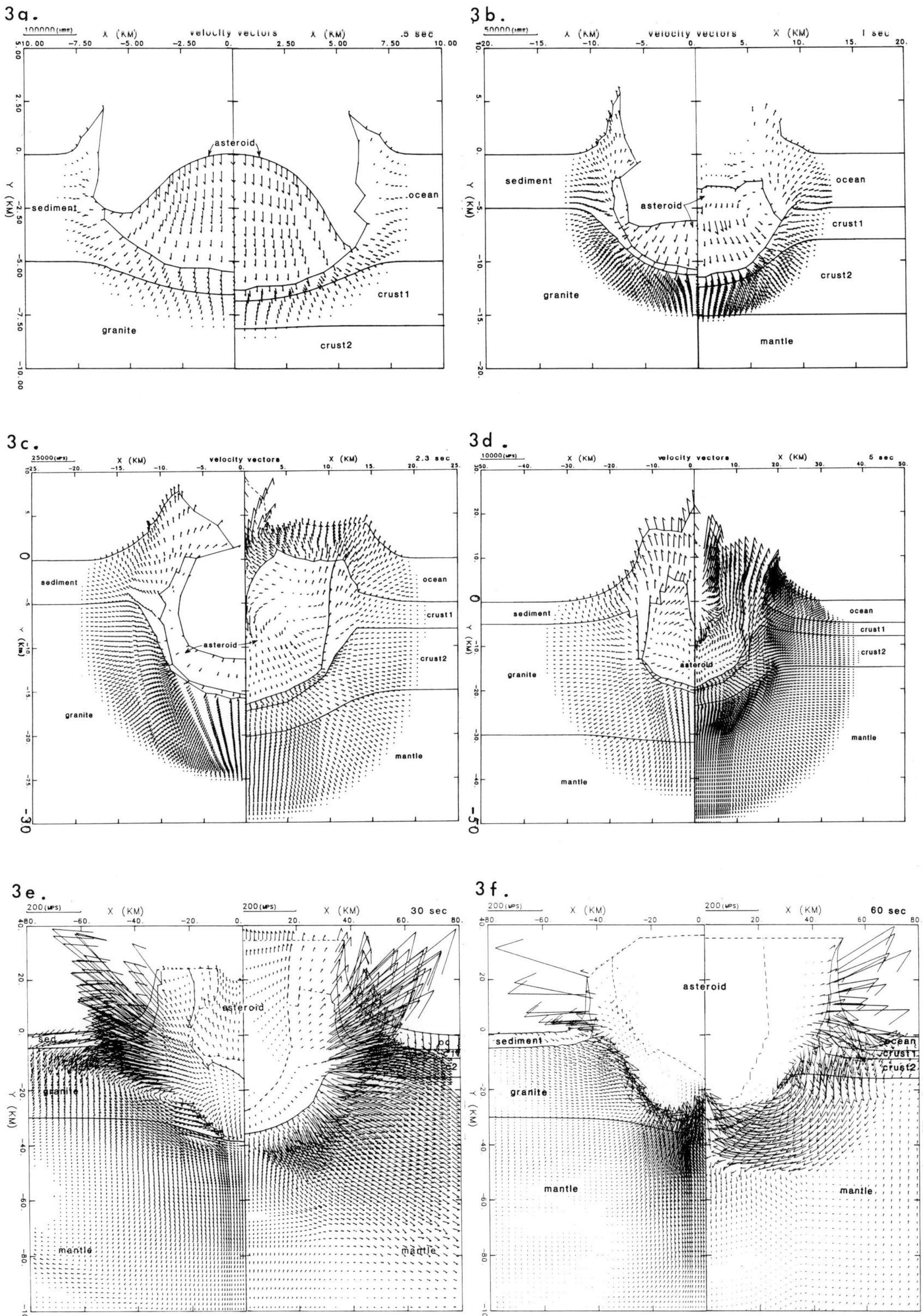
3a.
velocity vectors
.5 sec
X (KM)
Y (KM)
asteroid
sediment
ocean
crust1
granite
crust2
3b.
velocity vectors
1 sec
sediment
ocean
asteroid
crust1
crust2
granite
mantle
3c.
velocity vectors
2.3 sec
sediment
ocean
crust1
asteroid
crust2
granite
mantle
3d.
velocity vectors
5 sec
sediment
ocean
crust1
crust2
granite
asteroid
mantle
3e.
30 sec
sed
asteroid
granite
mantle
3f.
60 sec
asteroid
sediment
ocean
crust1
crust2
granite
mantle

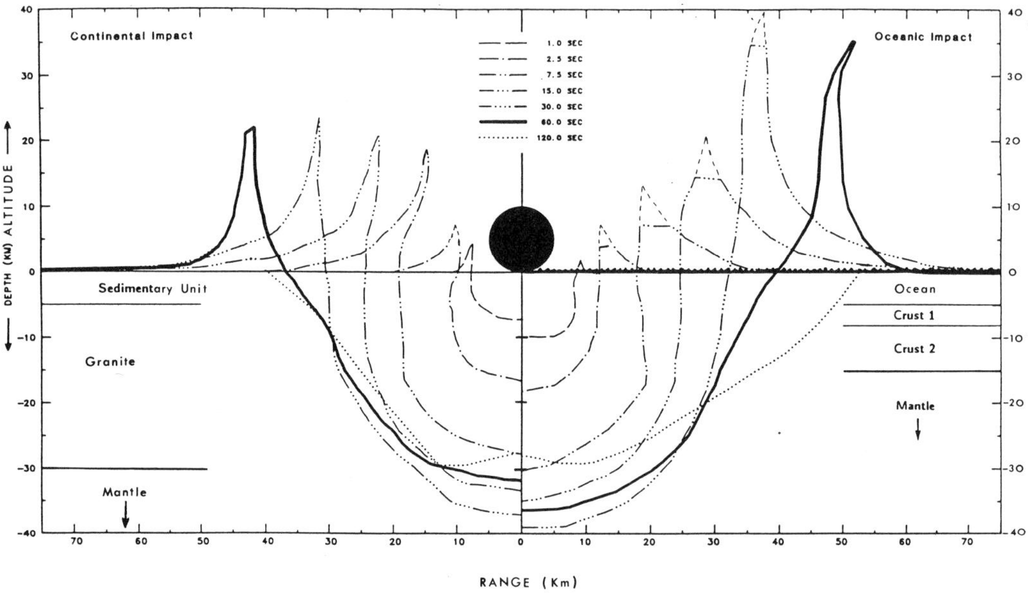

Fig. 4. Profiles showing sequence of formation of the transient craters, uplifting rims, and rebounding floors after asteroid impact at continental and oceanic sites. Profiles were determined from cratering calculations by using a density criterion of $\rho/\rho_0 \leq 0.5$ to plot each line as a function of a specific time; ρ_0 is unshocked density and ρ is shocked density. The early transient uplifted rim of the continental crater consisted mainly of solid and minor amounts of melt from the sedimentary unit, whereas the early transient uplifted rim of the oceanic crater consisted mainly of water and water vapor. By ~30 s, transient uplifted rim of the continental crater also included the granitic unit with minor amounts of melt, and the oceanic rim now included crust 1 and crust 2 with minor amounts of their melt. By ~5.0 s, interiors of both craters had filled with expanding ejecta consisting mainly of asteroid and target vapor and minor amounts of melt. At ~30 s at both sites, a maximum crater depth of ~40 km was reached and massive rebound of the crater floors and walls began. Rebound continued to the end of calculational time at 120 s, modifying the bowl-shaped transient craters to broader, more shallow craters. Maximum rim heights of ~25 km for the continental impact and ~40 km for the oceanic impact were reached at ~30 s. After this time, transient uplifted rims continued to move outward and to decay in height. Large transient uplifted rims existed after 60 s but were not plotted. Horizontal lines in the upper parts of the oceanic rim profiles indicate boundaries above which ejecta were not tracked; dashed lines above the boundaries are extrapolations for rim crests.

e) Ejecta Results

Ejecta dynamics were monitored throughout the calculations, i.e., materials that passed through the pre-impact surface of the Earth were analyzed according to mass, material type, velocity, ejecta angle, internal (thermal) energy, and kinetic energy. These results also include all mass (solid and melt) in the transient uplifted rims of the craters. In this preliminary study, we have analyzed the ejecta to 60 s at both craters, when the velocity and amount of material (including rim) being ejected had decreased substantially and was approaching being exceeded by ejecta moving downward. A reduced amount of target material and the remaining vaporized asteroid were still being ejected after 60 s, so our analysis gives a conservative value for total ejecta.

At 60 s, $\sim 1.92 \times 10^{14}$ t had been ejected by the continental impact and $\sim 8.94 \times 10^{13}$ t by the oceanic impact, i.e., about half as much. The continental crater had nearly reached its maximum volume ($\sim 7 \times 10^4$ km^3) by ~60 s and the oceanic crater grew only 20% more ($\sim 9 \times 10^4$ km^3) by 120 s.

The mass distribution of the ejecta as a function of material type for each impact site is shown in Table 2. As seen in this table, the bulk of the ejecta comes from the upper target units, i.e., the ocean, crust 1, crust 2, and the sedimentary unit. This distribution is consistent with data from other impact and explosion studies, including those of Roddy (1977, 1978), Henny (1981), and Hörz (1982).

The sedimentary (continental) and water (oceanic) units had the same initial thickness, but by 60 s, the mass of sedimentary rock ejected was 3.5 times greater than the mass of water ejected. The difference was largely due to the fact that the target rock was ~2.4 times that of water. Also, the total volume from the continental impact crater was smaller than that from the oceanic impact crater. Yet, the mass ejected from the sedimentary unit alone was more than the total mass ejected from the oceanic crater, largely because, again, the oceanic crater ejected a large volume of low-density target, i.e., water. Cratering and ejection processes influenced by different material responses must account for the remaining differences in ejecta masses between the two impacts.

No mantle material was ejected in either impact calculation. The oceanic impact penetrated deep enough to melt and expose the upper part of the mantle, but it did not have enough upward velocity to be ejected. The continental impact also did not eject any mantle material.

Substantial kinetic and internal energies were carried by the ejecta to different levels of the atmosphere. Total, kinetic, and internal energies were calculated for ejecta from each impact. The total energy calculated for the ejected mass was $\sim 8.7 \times 10^{29}$ ergs for the oceanic impact and $\sim 6.9 \times 10^{29}$ ergs for the continental impact, approximately 33% and 26% of the initial kinetic energy of the asteroid, respectively. The amounts of internal energy of ejecta from both impacts were approximately the same, but the oceanic ejecta consisted mostly of high-velocity water vapor that had a higher kinetic energy than the continental ejecta vapor.

Ballistic calculations were made to determine the distribution of the ejecta (vapor, melt, solid) in the atmosphere without regard to particle size. For both impacts, most of the ejecta reached no higher than the tropopause (~13 km altitude). Calculations show that $\sim 5 \times 10^{13}$ t were ejected to the level of the troposphere by the oceanic impact and $\sim 1.5 \times 10^{14}$ t by the continental impact. Much of these ejecta was made up of the transient-rim uplift of solid and minor amounts of melt, and it would be expected to fall back within ~3 crater diameters to form a continuous ejecta blanket surrounding the crater (Roddy, 1977).

Ejection velocities of the oceanic ejecta vapor were sufficient to lift $\sim 9 \times 10^{12}$ t into the ionosphere (~80 km altitude and higher), $\sim 1 \times 10^{13}$ t into the mesosphere (~30 km to 80 km altitude), and $\sim 1.5 \times 10^{13}$ t into the stratosphere (~13 km to ~30 km altitude). Most of these ejecta consisted of ocean water combined with a smaller amount of crust 1 and crust 2. The distribution at altitude of ejecta from the continental impact was similar to that from the oceanic impact: $\sim 7 \times 10^{12}$ t into the ionosphere, $\sim 9.0 \times 10^{12}$ t into the mesophere, and $\sim 1.6 \times 10^{13}$ t into the stratosphere; most of these ejecta came from the sedimentary unit.

All of the asteroid vaporized, but in both impact events its mass was less than ~1% of the total mass of the target materials ejected. By 60 s, over half of the asteroid vapor had been ejected (Table 2), and in both cratering calculations velocity-vector fields for times from 60 to 120 s show that the vapor continued to move upward at high velocities. This indicated that all, or nearly all, of the asteroid would eventually be ejected as vapor. By 60 s, $\sim 5 \times 10^{11}$ t of vaporized asteroid had been ejected above the tropopause by the oceanic impact and $\sim 6 \times 10^{11}$ t by the continental impact. By 60 s, the total mass of asteroid vapor ejected was $\sim 6.8 \times 10^{11}$ t for the oceanic impact and $\sim 9 \times 10^{11}$ t for the continental impact, i.e., ~52% and 70%, respectively (Table 2).

The amount of vapor phase versus solid/melt phases of continental ejecta was determined using an energy-density criterion. The total ejecta at 60 s was $\sim 1.92 \times 10^{14}$ t (Table 2). We assumed that an energy density of 0.1 terraergs/gm would give a reasonable estimate to define the partition of vapor versus solid/melt. Ejecta processed with this criterion gave $\sim 1.8 \times 10^{12}$ t of ejecta as vapor. Subtracting the asteroid vapor at 60 s from this value gives $\sim 0.9 \times 10^{12}$ t of target ejecta that was vapor which could reach well above the tropopause if initial velocities were sufficiently high. Using another energy density of 0.02 terraergs/gm, we calculate that $\sim 8.5 \times 10^{12}$ t of the total ejecta, or $\sim 7.6 \times 10^{12}$ t of target ejecta, was vapor at 60 s in the continental impact. At this energy density, we may be including some melt.

Calculation of particle sizes of the ejecta was not part of our computer-code simulations in this study. However, most of the material that reached the upper layers of the atmosphere was vapor, and we assume that it would have very small particle sizes. Most of the ejecta that reached only into the troposphere was part of the transient uplifted rim of solid and melt, and would be relatively large fragments of solid and melt. Estimates of particle sizes and atmospheric interactions in similar cratering events have been discussed by O'Keefe and Ahrens (1982a,b).

TABLE 2. Total Mass (t) and Percentages of Asteroid and Target Ejected at the End of 60 Seconds After Impact at Oceanic and Continental Sites.

OCEANIC CRATER					
Asteroid	Water	Crust 1	Crust 2	Mantle	Total
0.00684×10^{14}	0.464×10^{14}	0.224×10^{14}	0.199×10^{14}	0	0.894×10^{14}
0.8%	51.9%	25.1%	22.2%	0%	100%

CONTINENTAL CRATER				
Asteroid	Sedimentary rock	Granitic rock	Mantle	Total
0.00905×10^{14}	1.65×10^{14}	0.264×10^{14}	0	1.92×10^{14}
0.5%	85.8%	13.7%	0%	100%

Note: Pre-impact mass of the asteroid was 1.3×10^{12} t. By 60 s, all of the asteroid vaporized, and ~52% of the asteroid vapor at the oceanic site and ~70% at the continental site had been ejected. Most of the remainder of the asteroid vapor continued to be ejected to the end of the calculation at 120 s.

DISCUSSION

Computer simulations of a 10-km-diameter asteroid descending through the atmosphere at 20 km/s and impacting with the Earth showed equally dramatic effects on the surrounding air mass and on the oceanic and continental sites that were cratered. The atmospheric calculations with the DICE code show that the passage of the asteroid generated a sequence of intense shock effects that moved rapidly out through the atmosphere and over the surrounding area. Effects included an expanding shocked air mass having high velocities, pressures, and temperatures which was followed by a low-density region. By 30 s, the shocked air mass had swept strongly across the Earth's surface to ranges in excess of 70 km, covering an area of over 15,000 km^2. The shock effects continued to expand at several kilometers per second for considerable time, and we estimate that shock velocities in the air were still as high as ~50 m/s at a range of 200 km. Early-time temperatures in this air mass were ~5000 K at 15-km altitude and as high as 20,000 K at the Earth's surface at ranges of 15 to 20 km. As the shocked air mass expanded, temperatures dropped but still remained in excess of several thousand degrees kelvin. Such high temperatures could be expected to produce effects ranging from shock (thermal) metamorphism in the surface materials to surface fires; the latter is consistent with arguments by Anders et al. (1986) regarding the origin of soot at the Cretaceous/Tertiary boundary. Much of the area exposed to the highest temperatures would be destroyed by cratering, but a thermally affected layer might be formed at greater distances (at lower temperatures) and be preserved beneath the ejecta blanket. Although secondary cratering would disrupt the pre-impact surface (Hörz, 1982), additional examination of the bases of ejecta blankets at certain large terrestrial craters, such as the Ries Crater (Germany), could prove useful in determining if such a thermally metamorphosed layer was formed. Evidence of such an effect would at least support the possibility of high temperatures and initiation of widespread fires at continental sites.

Our analysis of the atmospheric and the cratering calculations shows that the bow shock wave, air shock front, and much of the following shocked air mass expanded in range and time well ahead of the uplifting rim and ejecta fields. However, the low-density region developed over the same area where the uplifting rim and ejecta field developed. Consequently, the rim and ejecta were thrown into a region of the atmosphere that had very low densities and high temperatures, overpressures, and particle velocities. Complex interactions could be expected in this region, such as additional heating of the ejecta and rim surface; such heating could substantially increase melting and vaporization. Evaluation of these effects is especially important in estimating the total amount of material vaporized, melted, and lofted to various altitudes. The low-density air would also experience both heating and cooling by contact with the ejecta, and complex flow patterns could be expected. We conclude that ejection into the large, hot, low-density region above the craters can increase the amount of material carried to high altitudes and long ranges, but the actual quantitative results of such effects have yet to be determined. Full analysis of the complex atmospheric/cratering interactions will require linking the DICE and CRALE2 codes so that they include multiphase flow and radiation transport.

The cratering and ejecta calculations made with the CRALE2 code show that, during the first 30 s, oceanic and continental impact events developed with remarkably similar sequences despite the geologic differences at the two sites. Both impacts initially formed large, transient, bowl-shaped craters of comparable depths (~37 vs. ~39 km). Gravitationally driven rebound of the crater floors began after the maximum depths had been reached at ~30 s. The craters continued to expand to the end of calculational time, and rebound continued to raise the floors. At 120 s, velocities in the mantle were still as high as ~5 m/s (continental) and ~50 m/s (oceanic), indicating that substantial deformation could still occur. Aspect ratios of the transient diameter/depth were ~1.6 for both craters at 30 s and ~3.0 (continental) and ~3.9 (oceanic) at 2 min. We conclude that long-term relaxation at these types of sites will lead to broad, flat-floored craters, possibly containing central-uplift or multiring structures. Final dimensions as large as 120 to 150 km in diameter and a few kilometers in depth appear to be reasonable extrapolations to very long relaxation time. It also seems reasonable to conclude that erosional degradation will proceed more rapidly on shallow craters than deep ones, depending on their erosional environment. This may explain in part, why many large, ancient, impact sites are relatively short lived as topographic features, i.e., craters. Oceanic craters are further complicated by the inevitable failure of that part of the crater developed in water and subsequent return of the ocean over the rest of the crater formed on the seafloor. Additional studies are necessary to examine the many different geologic conditions that influence the major processes governing structural deformation and final crater shapes.

Another difference between the two impacts that affected crater dimensions was one of material strength: at the end of calculations, the apparent diameter of the oceanic crater was ~105 km, while that of the continental crater was only ~80 km. This difference is due simply to the fact that the sedimentary unit was stronger than the ocean. This strength difference also caused the transient uplift of the oceanic rim to rise to a height nearly twice that of the continental rim. The ~40-km uplift of the ocean and crustal units could be expected to produce spectacular tsunami effects during final relaxation; the height of the water in the rims will decay to a kilometer or more in tsunami height and these large waves should leave stratigraphic evidence of their passage if they reach coastal environments.

Another difference between the two craters can be seen in the pseudeomomentum vectors in Fig. 3e,f. For the same times, motions in the rims and in subcrater mantle and granitic units have become dissimilar. This suggests that somewhere between 5 s and 30 s, the cratering processes did begin to differ in some respects, presumably due to the differences in material responses at the two sites. Comparisons made at different times would be helpful to find out if similar flow sequences exist at each site but occur at different times and, depending on the different geologic material responses, might thereby lead to different structural deformation and morphologic features in the final craters.

The ejecta phase of the cratering calculations indicated that in excess of 70,000 km^3 (over 10^{14} t) of material was ejected. We estimate that more than 70% of this ejecta would form a massive ejecta blanket lying within ~3 crater diameters of the impact. The remaining ejecta, including the asteroid material, would reside in the atmosphere at different altitudes and lie at extended ground ranges. The calculations indicate that substantial amounts of ejecta (mostly vapor) would reach every layer of the atmosphere up through the ionosphere. The amount and composition of the ejecta at different altitudes and ranges would vary substantially, depending on the geologic conditions at the impact site. Virtually all of the asteroid would be ejected, but its mass would account for less than ~1% of the total ejecta, in which terrestrial components would dominate. Results of our calculations remain consistent with conclusions by Alvarez (1986) and others workers; the geology and generalized location of a hypothesized impact may be reasonably inferred by studying the mineralogy, chemistry, and shock metamorphic features at appropriate stratigraphic horizons, especially where an iridium anomaly (or other extraterrestrial material from the impacting body) is present. This has interesting implications for predicting the composition of fallout from the postulated Cretaceous-Tertiary impact. Presumably, the fallout from a continental impact would be dominantly felsic continental material and trace amounts of exotic material (such as iridium). On the other hand, the fallout from an oceanic impact would contain mainly mafic constituents and trace amounts of asteroid material. Other workers have reached similar conclusions, however, the cratering calculations provide quantitative data that should help place limits on interpretations of shock features, ejecta sources and distributions, and total masses deposited.

Hot crustal and mantle rocks, uplifted nearly instantly, would expose an area of ~15 x 10^3 km^3 on the crater floor and could be expected to degas, adding ash to the atmosphere that may rival other cratering contributions. Presumably, volcanism would continue on crater floors and rims for a considerable time. The return of the ocean over the hot crater and ejecta would also be expected to add substantial water vapor to the atmosphere. The combination of these secondary effects could be expected to produce a large heated area at the crater site that could lift large volumes of ash and water vapor (oceanic) into the atmosphere long after the impact. Consequently, at an oceanic site, volcanic ash and hydrous materials seem likely components to be deposited over longer periods of time after a large impact; these materials would overlie initial fallout of ejecta material and may dominate the lithology of later fallout deposits.

ACKNOWLEDGEMENTS

Reviews of this paper by T. Ahrens, P. Davis, and P. Spudis greatly assisted the writers, and we express our thanks for their time and help. Discussions with P. Davis, D. Orphal, and G. Shoemaker on geologic, cratering, and asteroid problems, with D. Anderson, S. Biehler, and T. Wallace on oceanic and continental seismic and density data, and with J. Sass on geothermal data were especially helpful in providing information for our calculations. We remain indebted to Dr. G. Sevin and Major J. Jones of the Defense Nuclear Agency for their continued support of various aspects of these studies. D. Roddy expresses his appreciation again to J. Roddy and N. Roddy for their continued assistance. Parts of this work were supported by the National Aeronautics and Space Administration under Contract W15,814 and by the Defense Nuclear Agency, Department of Defense.

REFERENCES CITED

Ahrens, T. J., and O'Keefe, J. D. (1977) Equation of state and impact-induced shock-wave attenuation on the Moon. In D. J. Roddy, R. O. Pepin, and R. B. Merrill (Eds.), Impact and Explosion Cratering, Pergamon Press, pp. 639-656.

Ahrens, T. J., and O'Keefe, J. D. (1983) Impact of an asteroid or comet in the ocean and the extinction of terrestrial life, Proc. Lunar Planet. Sci. Conf., 13th, part 2, Suppl., J. Geophys. Res., Vol. 88, pp. A799-A806.

Ahrens, T. J., and O'Keefe, J. D. (1985) Shock vaporization and the accretion of the icy satellites of Jupiter and Saturn. In P. Klinger, A. Dollfus, and R. Smoluchowski (Eds.), Ices in the Solar System, Reidel Press, pp. 631-654.

Ahrens, T. J., and O'Keefe, J. D. Impact on the Earth, ocean and atmosphere. In C.E. Anderson (Ed.), Internat. J. Impact Engin., 5, Pergamon Press, this volume.

Alvarez, L. W., Alvarez, W., Asaro, F., and Michel, H. V. (1980) Extraterrestrial cause for the Cretaceous-Tertiary extinction. Sci., 208, pp. 1095-1108.

Alvarez, W. (1986) Toward a theory of impact crises. EOS, 67, 649-658.

Anders, E., Wolbach, W. S., Lewis, R. S. (1986) Cretaceous Extinctions and Wildfires: Response. Sci., pp. 263-264.

Bjork, R. L. (1961) Analysis of the formation of Meteor Crater, Arizona. Proc. Acad. Natl. Sci. of Philadelphia, Vol. 76, pp. 275-278.

Bryan, J. B., Burton, D. E., Cunningham, M. E. and Lettis, L. A., Jr. (1978) A two-dimensional computer simulation of hypervelocity impact cratering: Some preliminary results from Meteor Crater, Arizona. Proc. Lunar Planet. Sci. Conf., 9th, pp. 3931-3964.

Delsemme, A. H. (Ed.) (1977) Comets, asteroids, meteorites, interrelations, evolution, and origin. The Univ. of Toledo Press, 587 p.

Gault, D. E., and Wedekind, J. A. (1978) Experimental studies of oblique impact. Proc. Lunar Planet. Sci. Conf., 9th, pp. 3843-3875.

Gehrels, T. (1979) Asteroids. The University of Arizona Press, 1177 p.

Grieve, R. A. F. (1982) The record of impact on Earth: Implications for a major Cretaceous/Tertiary impact event. In L. E. Silver, and P. H. Schultz (Eds.), Geol. Soc. Am. Spec. Pap., 190, pp. 25-38.

Grieve, R. A. F., Sharpton, V. L., Goodacre, A. K., and Garvin, J. B. (1985) A perspective on the evidence for periodic cometary impacts on Earth. Earth Planet. Sci. Letts., 76, pp. 1-9.

Henny, R. (1981) The effect of the geologic setting on the distribution of ejecta from the buried nuclear detonation Schooner: AFWL-TR-81-36, 346 p.

Hörz, F. (Ed.) (1971) Meteorite impact and volcanism. J. Geophys. Res., 76, pp. 5381-5798.

Hörz, F. (1982) Ejecta of the Ries Crater, Germany. In L. E. Silver, and P. H. Schultz (Eds.), Geol. Soc. Am. Spec. Pap., 190, pp. 39-55.

Isenberg, J. (1972) Part II Mechanical properties of earth materials, Nuclear Geoplosics: DNA 1285H2, Defense Nuclear Agency, Washington, D.C.

O'Keefe, J. D., and Ahrens, T. J. (1975) Shock effects from a large impact on the moon. Proc. Lunar Planet. Sci. Conf., 6th, pp. 2831-2844.

O'Keefe, J. D., and Ahrens, T. J. (1977a) Impact induced energy partitioning, melting, and vaporization on terrestrial planets. Proc. Lunar Planet. Sci. Conf., 8th, pp. 3357-3374.

O'Keefe, J. D., and Ahrens, T. J. (1977b) Meteorite impact ejecta: Dependence of mass and energy lost on planetary escape velocity. Sci., 198, pp. 1249-1251.

O'Keefe, J. D., and Ahrens, T. J. (1978) Impact flows and crater scaling on the moon. Phys. Earth Planet. Interiors, 16, pp. 341-351.

O'Keefe, J. D., and Ahrens, T. J. (1982a) Impact mechanics of the Cretaceous-Tertiary extinction bolide. Nature, 298, pp. 123-127.

O'Keefe, J. D., and Ahrens, T. J. (1982b) The interaction of the Cretaceous/Tertiary extinction bolide with the atmosphere, ocean, and solid Earth. In L. E. Silver, and P. H. Schultz, (Eds.), Geological implications of impacts of large asteroids and comets on the Earth: Geol. Soc. Am. Spec. Pap., 190, pp. 103-120.
O'Keefe, J. D., and Ahrens, T. J. (1983) Constraints on the impact on Mars origin of SNC meteorites [abs.]: Lunar Planet. Sci., 14th, pp. 578-579.
O'Keefe, J. D., and Ahrens, T. J. (1985) Impact and explosion crater ejecta, fragment size and velocity. Seis. Lab., Calif. Inst. of Tech., Pasadena, California, p. 328-338. Reprinted Icarus, 62, May 1985.
O'Keefe, J. D., and Ahrens, T. J. (1986a) Effect of planetary gravity and impact size on crater maximum penetration and excavation depth [abs.]. Lunar Planet. Sci., 17th, pp. 632-633.
O'Keefe, J. D., and Ahrens, T. J. (1986b) Oblique impact: A process for providing meteorite samples of other planets. Sci., 86, p. 346-349.
Pollack, H. N., and Murthy, V. R. (1984) Structure and evolution of the continental lithosphere. Pergamon Press, NY, 203 p.
Pollack, H. N., and Sass, J. H., Thermal regime of the lithosphere: Submitted to Handbook of Terrestrial Heat-flow Density Determination of Guidelines and Recommendations of the Internatl. Heat Flow Comm.
Ringwood, A. E., 1969, Composition and evolution of the upper mantle. In Pembroke, J. (Ed.), The Earth's crust and upper mantle, Am. Geophys. Union Geophys. Monograph No. 13, pp. 1-17.
Roddy, D. J. (1977) Large-scale impact and explosion craters: Comparisons of morphological and structural analogs. In Roddy, D. J., Pepin, R. O., and Merrill, R. B. (Eds.), Impact and explosion cratering--Planetary and terrestrial implications, Pergamon Press, NY, pp. 185-246.
Roddy, D. J. (1978) Pre-impact geologic conditions, physical properties, energy calculations, meteorite and initial crater dimensions and orientations of joints, faults, and walls at Meteor Crater, Arizona. Proc. Lunar Planet. Sci. Conf., 9th, pp. 3891-3930.
Roddy, D. J., Pepin, R. O., and Merrill, R. B. (Eds.) (1977) Impact and Explosion Cratering--Planetary and terrestrial implications, Pergamon Press, 1301 p.
Roddy, D. J., Schuster, S., Kreyenhagen, K. N., and Orphal, D. (1980) Computer code simulation of the formation of Meteor Crater: Calculation MC-1 and MC-2. Proc. Lunar Planet. Sci. Conf., 11th, pp. 2275-2308.
Rosenblatt, M., Hassig, P. J., and Orphal, D. L. (1982) Soil-air interactions during airblast-induced ground motions. DNA-TR-103.
Schuster, S. (1984) CRALE user's manual: Chatsworth, Calif., Calif. Res. and Tech. 5370-R1, 300 p.
Schuster, S. H., and Isenberg, J. (1972) Equations of state for geologic materials. DNA 2925Z Rept., Defense Nuclear Agency, Washington, D.C.
Shoemaker, E. M. (1977) Why study impact craters?. In Roddy, D. J., Pepin, R. O., and Merrill, R. B. (eds.), Impact and explosion cratering--Planetary and terrestrial implications, Pergamon Press, NY, pp. 1-10.
Shoemaker, E. M. (1983) Asteroid and comet bombardment of the Earth: Ann. Rev. Earth Planet. Sci., 11, pp. 15-41.
Shoemaker, E. M., Williams, J. G., Helin, E. F., and Wolfe, R. F. (1979) Earth-crossing asteroids: Orbital classes, collision rates with Earth and origin. In T. Gehrels (Ed.), Asteroids, Univ. of Arizona Press, pp. 253-282.
Shoemaker, E. M., and Wolfe, R. F., (in press) Mass extinctions, crater ages, and comet showers. In Smoluchowski, R., Bahcall, J., and Matthews, M. (Eds.), The Galaxy and the Solar Sys., Univ. of Arizona Press.
Silver, L. E., and Schultz, P. H. (Eds.), Geological implications of impacts of large asteroids and comets on the Earth. Geol. Soc. Am. Spec. Pap., 190, pp. 461-468.
Soderblom, L. A. (1977) Historical variations in the density and distribution of impacting debris in the inner solar system: Evidence from planetary imaging. In Roddy, D. J., Pepin, R. O., and Merrill, R. B. (Eds.), Impact and explosion cratering--Planetary and terrestrial implications, Pergamon Press, pp. 629-634.
Taylor, S. R., and McLennan, S. M. (1985) The continental crust: its composition and evolution--An examination of the geochemical record preserved in sedimentary rocks: Blackwell Scientific Publications, Oxford, London, 312 p.
U.S. Standard Atmosphere (1962/1976) Washington, D.C., U.S. Government Printing Office.
Van Thiel, M. (1977) Compendium of shock wave data. VCR1-50108, Vol. 1-3, L. Livermore Lab., Livermore, California.

Int. J. Impact Engng Vol. 5, pp. 543–560, 1987
Printed in Great Britain

0734-743X/87 $3.00 + 0.00
Pergamon Journals Ltd.

SOME RECENT ADVANCES IN THE SCALING OF IMPACT AND EXPLOSION CRATERING

Robert M. Schmidt and Kevin R. Housen

Boeing Aerospace Company
MS 3H-47, P. O. Box 3999
Seattle, WA 98124

ABSTRACT

The geotechnical centrifuge is a valuable experimental tool for studying impact and explosion craters. Performing experiments at elevated gravity effectively allows one to simulate crater formation at much larger scales than those otherwise attainable in the lab. The utility of the centrifuge has been demonstrated by successful simulations of explosive field tests with charges equivalent to a few kilotons of TNT. Impact experiments conducted on the centrifuge have provided measurements of final crater size, growth of the transient crater, formation time and material motions. These observations have been analyzed within the framework of a recently-developed theory of source coupling, which relates many aspects of crater formation through a single scaling exponent. For example, the scaling of crater growth and formation time are found to be consistent with that one would predict from the observations of final crater size. The results are summarized and applied to develop improved estimates for scaling laws appropriate to the impacts of large bolides on planetary surfaces.

INTRODUCTION

Geotechnical centrifuge techniques have been widely used for the past 50 years, especially in Europe, England, and the Soviet Union. Until recently, however, the applications have been limited to static or quasi-static soil mechanics problems. Dynamic phenomena were not generally studied because of concerns about Coriolis effects and other potential experimental difficulties. The idea of using elevated gravity to investigate cratering was first suggested by Sedov[60] and more recently by Chabai.[4] The first reported high-G cratering experiments, using a rocket sled, were performed in the Soviet Union in 1960 by Viktorov and Stepenov.[62] Other Soviet experiments using centrifuges to investigate blast loadings and cratering are referred to by Pokrovsky and Fyodorov.[31] However, the experiments were limited and the specific results were vague.

A centrifuge is useful in cratering studies for two reasons. First, under conditions discussed below, crater formation by large impactors or explosive sources in a terrestrial, lunar or planetary gravity field can be directly simulated by much smaller projectiles or explosives if the experiments are performed at elevated gravity. Proper simulations require that various "similarity conditions" are satisfied, as discussed below. Although direct simulations of kilometer-scale and larger impact craters exceed the capabilities of existing centrifuges, the size-scale of explosive field tests is within reach. As a result, a geotechnical centrifuge has been used as a cost-effective tool in studies of explosive crater formation, directly at the size scale of interest. A second, equally important, application of the centrifuge is the determination of scaling laws which can be used to extrapolate cratering experiments to very large sizes. For problems where gravity affects crater formation, variations in gravity can be shown to be equivalent to variations in source size. The centrifuge has been used to construct a data base which effectively embodies much larger variations in source size than those attainable otherwise. These data have been useful in the construction and verification of scaling laws for both explosive and impact cratering.

The purpose here is to summarize recent results obtained in the Boeing centrifuge cratering program. Some of the experimental apparatus are described in the next section. The third section gives a brief review of various direct simulations of explosive field tests. These experiments have demonstrated the usefulness of the centrifuge as a cost-effective simulation tool. The fourth section briefly summarizes recent advances in the development of scaling laws, which are based on a point-source approximation to the impactor or explosive source. Measurements of growth and collapse of the transient crater, and final crater size, are presented in the fifth section. Current "best estimates" for scaling laws are given in the final section.

EXPERIMENTAL TECHNIQUE

The centrifuge used at Boeing is a 600-G centrifuge with a dynamic load rating of 60,000 G-kg (66 G-tons) at 620 rpm and incorporates symmetric swing baskets allowing testing of noncohesive soil materials. The arm radius to the fully extended base plate is 1.4 meters with a maximum payload mass of 250 kg on each rotor end. The centrifuge is powered by a 30-hp Eaton Dynamatic Model ACM-326-9108 drive unit incorporating an adjustable speed, constant torque, eddy-current clutch. The unit also has electric dynamic braking, allowing shut down from maximum rpm in less than 30 seconds. The constant output-speed motor and variable drive unit are shock mounted and coupled to the main shaft with a belt to minimize vibration. A 150-hp Eaton Dynamic Model 444 T-9144 drive unit is in house and a mechanical drive system is currently being developed. This new unit should provide centrifugal acceleration in excess of 1250 gravities.

The rotor shaft is equipped with 24 slip rings for instrumentation channels, three 200 V.a.c power slip rings and a hydraulic slip ring which can accommodate either gas or liquid. A pair of motor driven Nikon F2 35 mm still cameras are hub-mounted in a stereo configuration. Two high-speed movie cameras are fully operational on board the centrifuge rotor arm up to maximum G. A Photosonic pin-registration camera provides moderate framing rates up to 400 pictures/sec. A Fastax II rotating prism camera can be used to obtain framing rates up to 13,000 pictures/sec or can operate in a streak mode.

Hypervelocity projectile guns that are capable of launch velocities up to 7 km/sec at maximum G have been mounted on the centrifuge. These include two conventional powder guns and a two-stage light-gas gun, all with active velocity monitoring.

Craters can be formed in a conventional "half-space" sample container or in a so-called "quarter-space" target container,[50,51] where the crater forms against a transparent window so that the developing crater profile is directly observable. In this manner, the rate of crater growth and time of formation can be determined. The quarter-space test apparatus, shown in Fig. 1, allows both impact and explosive crater formation to be recorded at 13,000 pictures per second at gravity levels in excess of 500G. For more precise time resolution, an ultra high speed Barr and Stroud camera was used in some 1-G experiments to attain framing rates of 1.4 million pictures/sec.

Fig. 1. Quarter-space sample container mounted on centrifuge. A rotating prism framing camera is mounted above rotor arm to left of the main shaft. A pair of 35-mm stereo cameras are mounted on the hub. Illumination is provided from two 28-volt aircraft landing lights.

DIRECT SIMULATIONS

The theoretical basis of centrifuge techniques in explosive cratering are the similarity requirements that subscale models correctly reproduce the behavior of large-scale events. When using the same soil and the same explosive source, a small-scale experiment will correctly reproduce an event n-times larger in each linear dimension if the small-scale experiment is conducted at n-times the normal gravity level. Hence, the simulated explosive mass and energy is a factor of n^3 larger than the actual charge used in the centrifuge experiment. For example a 50-gm chemical high-explosive charge at 600G produces a crater corresponding to a $(600)^3$ x 50 gm =11 kiloton TNT field event. Moreover, specially-shaped high-explosive charge configurations can be used to simulate nuclear loading to the soil. This method introduces an additional large equivalence factor that accounts for the higher mechanical effectiveness of distributed conventional explosives compared to nuclear explosives. Thus, these same experimental conditions can simulate cratering behavior for nuclear yields approaching 0.5 megaton.

At the beginning of our centrifuge program, a concern was whether or not similarity could be achieved in the complete thermomechanical sense. Hence, a similarity analysis of the general field equations for a continuum was performed.[42,45] The analysis showed that subscale similarity is possible with arbitrary scaling of size, and that there are no additional constraints for materials that are both rate and size-independent. In those cases, assuming that the prototypic soil material is used in the centrifuge, "dynamic scaling" prevails: the scale factor for size is equal to that for time; and stresses, velocities and internal energies are equal at corresponding positions in the model and prototype. Therefore, dynamic experiments can be scaled and interpreted in the same manner as the well established quasi-static soil mechanics experiments. Since volume and energy scale with the cube of the gravity, very large prototypic explosive energies can be simulated with small charges. The sufficiency of this scaling has been demonstrated by numerous experiments summarized below.

In order to establish feasibility of cratering experiments on the centrifuge, buried charges were detonated at gravity levels to 480G.[32,33] The results of these high-G experiments clearly showed that gravity plays a dominant role in crater formation and that it must be increased if experiments are performed at subscale. In addition, a dimensionless charge-size parameter was shown to correlate the data over a range of twelve orders of magnitude in explosive energy release, demonstrating the relevance of using subscale laboratory tests performed at elevated gravity to predict large-scale cratering events in the field. These initial results were in good agreement with existing field data for buried explosives.[33,52] Subsequent studies investigated near-surface explosions in dry Ottawa sand.[10,41,42] The functional relationship obtained by varying gravity at fixed charge size was shown to be identical to that obtained by varying explosive charge size at fixed gravity, confirming the primary prediction of the scaling laws.

To confirm that specific field prototypes could be simulated at subscale, various simulations of field events were conducted using both wet and dry soils. These included a subscale simulation of the "JOHNIE BOY" 500-ton nuclear crater,[34] simulations of 20-ton TNT events, "JANGLE HE-2" and "STAGECOACH II"[22] and a simulation of the 500-ton TNT event "PRAIRIE FLAT".[46,54] All showed very good agreement with the prototype field events: the craters obtained on the centrifuge matched the corresponding prototype field event in both shape and size. Crater dimensions were matched to better than 10%. The significant shape differences observed between the dry-site craters and the wet PRAIRIE-FLAT crater was also observed on the centrifuge simulations. Furthermore, control tests performed at 1G confirmed that these same results could not have been obtained in subscale without the high-G environment of the centrifuge. These simulations proved that the centrifuge small-scale technique could successfully simulate large-scale HE field events.

Perhaps a more important and significant test of the technique was the pre-test prediction for the recent "NSS" field event. The actual event used a complicated distribution of 40 tons of HE charge that was designed to simulate a hypothetical 2KT radiative NE weapon. The source involved multiple detonation points and times, variable thicknesses and variable stand-off distances for the explosive. An explosive charge similar to the NSS field charge was designed by calculations for use on the centrifuge, in order to predict the results of the event prior to the actual field test. This small-scale source was geometrically similar to the field source, but was simplified in ways that were assumed to not affect its fidelity. The charge was built-up using thin sheet explosive. A depiction of the charge is shown in Fig. 2., which also shows our prediction for the NSS crater compared to the field event. That prediction was within 10% on all crater-size parameters of interest, including volume, depth, central bowl radius, shelf radius and depth. This result proved that a comparable subscale model for a surface-burst nuclear source can be attained in the centrifuge environment, using the same approach for the charge design as used in the field events. In addition, the small-scale and low costs of the centrifuge technique allow parametric and screening studies that are not feasible at field scales.[53]

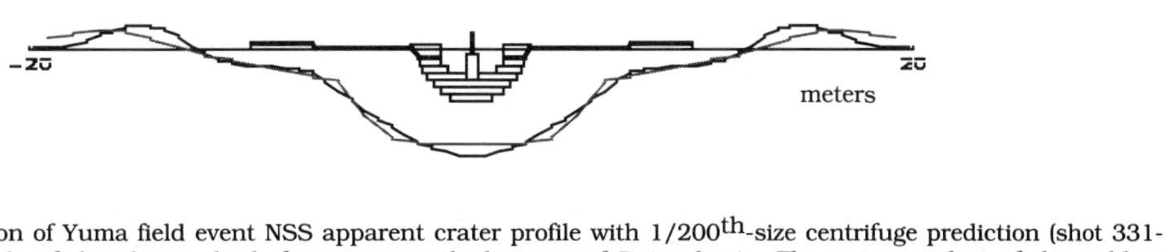

Fig. 2. Comparison of Yuma field event NSS apparent crater profile with $1/200^{th}$-size centrifuge prediction (shot 331-OED). Note details of the charge, built from various thicknesses of Deta sheet. The outer reaches of the airblast simulator rest on foam. A concrete tamping ring rests on the central charge. A microdet was used to fire the assembly.

SCALING STUDIES

The explosion tests mentioned in the previous section, along with impact experiments[11,35,43,56], show a very close power-law relationship between a nondimensional cratering efficiency and a gravity-scaled source size (see Fig. 3a discussed below). The power-law relationship has been shown to imply the existence of a "coupling parameter"[23,24], which depends on the impactor properties. However, the details of the impactor are not very important, except very near the impact point. At ranges beyond a few impactor radii, the coupling parameter alone describes the impactor. More recently, this has been shown to be a equivalent to a "point-source" approximation for the deposition of energy and momentum in the target material by the impactor or explosive.[13,14,15,16] In this approximation, a limiting solution is developed where the impactor is replaced by a body of increasingly high velocity and vanishingly small size. This section gives a brief look at how scaling laws can be developed from this concept. More details can be found in the references noted above.

The coupling parameter, C, is assumed to be a function of the impactor properties. If the impactor radius, velocity and density are a, U and δ, respectively, then

$$C = C\,[a, U, \delta] \tag{1}$$

For the case of explosions, the velocity is replaced by the explosive energy density, and C may additionally depend on the burial depth of the charge. The scaling laws discussed in this paper apply only to the impact case or to explosions where the burial depth of the explosive is nearly zero (discussions of buried charges can be found in ref. 53).

Equation (1) involves four variables with three independent dimensional units. As such, the methods of dimensional analysis can be used to show that C must be a power-law in its arguments. For example, if the units of C are assumed to be [length][velocity]$^{\mu}$[density]$^{\upsilon}$, where μ and υ are constants to be determined, then

$$C \propto a\, U^{\mu}\, \delta^{\upsilon} \tag{2}$$

Note that if μ and υ turn out to be 2/3 and 1/3 respectively then "energy scaling" results, because C is a function only of the impactor energy. On the other hand, if μ=1/3 and υ=1/3 then "momentum scaling" applies, because C depends only on the impactor momentum. For all real materials of interest studied to date, μ has been observed to be between the energy and momentum limits, and appears to be related to the target porosity (which governs energy losses in the target)[24]. As discussed below, μ has been measured to be about 0.4 for porous targets, such as dry soils, and about 0.55 for nonporous materials, such as water or rock.

As an example, a scaling law for crater volume is now derived. For simplicity consider the case of a cohesionless material, or large-scale impacts in a cohesive material, so that the effects of strength can be ignored compared to those of gravity. Under the assumption of a coupling parameter the volume, V, can be written as

$$V = V\,[C, \rho, g] \tag{3}$$

where ρ is the density of the target material and g is the gravity field strength. Equation (3), like eqn. (1), must be a power-law in its arguments. Using eqn. (2), eqn. (3) can be written in nondimensional form as

$$\rho V/m \propto (ga/U^2)^{-3\mu/(2+\mu)} \tag{4}$$

where for simplicity the density ratio, ρ/δ is assumed to be constant. The quantity on the l.h.s of eqn. (4) is a dimensionless cratering efficiency. The term on the r.h.s is a gravity-scaled size parameter. As stated above, this power-law relationship has been observed in experiments, and provides strong justification for the point-source coupling parameter assumption. Notice that by measuring the dependence of V on g the value of μ can be determined for any given material. Centrifuge experiments have made use of this fact in order to determine μ.

The coupling parameter has been used to derive scaling laws for a wide variety of impact-related phenomena such as cratering and fragmentation in the strength regime;[18,19] impact melt generation;[1,2] coupling and cratering flow fields;[3,27] and ejecta velocity and debris problems.[26,28,29]. In many cases, as in the example above, the relations simplify to power-laws. These relations are all determined by the single exponent μ(with the less-interesting exponent υ appearing primarily in terms involving the density ratio). Hence, the coupling parameter provides a common link between many aspects of crater formation. For example, the dependence of crater size on gravity can be used to predict the dependence of crater formation time on impactor size, or the growth of the transient cavity in time. Table 1 lists some scaling results which are pertinent to the discussions presented below. As shown in the next section, the exponents determined from final crater volume agree well with those determined from dynamic growth histories for a wide variety of experimental and numerically-generated craters.[16,17,24]

Table 1

Scaling Relations Assuming a Coupling Parameter of the form:

$$C = C[a, U, \delta] \Rightarrow C \propto aU^{\mu}\delta^{\nu}$$

	Strength Regime	Gravity Regime
Crater Volume	$\frac{\rho V}{m} \propto \left(\frac{\rho}{\delta}\right)^{1-3\nu} \left(\frac{Y}{\rho U^2}\right)^{\frac{-3\mu}{2}}$	$\frac{\rho V}{m} \propto \left(\frac{\rho}{\delta}\right)^{\frac{2+\mu-6\nu}{2+\mu}} \left(\frac{ga}{U^2}\right)^{\frac{-3\mu}{2+\mu}}$
Formation Time	$\frac{UT}{a} \propto \left(\frac{\rho}{\delta}\right)^{-\nu} \left(\frac{Y}{\rho U^2}\right)^{\frac{-(1+\mu)}{2}}$	$\frac{UT}{a} \propto \left(\frac{\rho}{\delta}\right)^{\frac{-\nu}{2+\mu}} \left(\frac{ga}{U^2}\right)^{\frac{-(1+\mu)}{2+\mu}}$
Growth of the Transient Crater	$\frac{\rho V}{m} \left(\frac{\rho}{\delta}\right)^{3\nu-1} \left(\frac{Y}{\rho U^2}\right)^{\frac{-3\mu}{2}} = F\left[\frac{Ut}{a} \left(\frac{\rho}{\delta}\right)^{\nu} \left(\frac{Y}{\rho U^2}\right)^{\frac{1+\mu}{2}}\right]$	$\frac{\rho V}{m} \left(\frac{\rho}{\delta}\right)^{\frac{6\nu-\mu-2}{2+\mu}} \left(\frac{ga}{U^2}\right)^{\frac{3\mu}{2+\mu}} = F\left[\frac{Ut}{a} \left(\frac{\rho}{\delta}\right)^{\frac{\nu}{2+\mu}} \left(\frac{ga}{U^2}\right)^{\frac{1+\mu}{2+\mu}}\right]$
	Prior to the onset of strength or gravity effects	
	$\frac{\rho V}{m} \propto \left(\frac{\rho}{\delta}\right)^{\frac{1+\mu-3\nu}{1+\mu}} \left(\frac{Ut}{a}\right)^{\frac{3\mu}{1+\mu}}$	

Definitions:

V=crater volume	$\Pi_V = \rho V/m$
R=crater radius	$\Pi_R = R\,(\rho/m)^{1/3}$
T=crater formation time	$\Pi_2 = 3.22ga/U^2$
ρ =target density	$\Pi_t = Ut/a$
Y=target strength	$\Pi_T = UT/a$
g=gravity	
a=impactor radius	
U=impactor velocity	
E=impactor energy	
t=time	

EXPERIMENTAL RESULTS

Table 2 summarizes the test conditions and final crater dimensions for a series of 33 experiments which employed the quarter-space technique in order to observe the growth and possible collapse of the transient crater. Both explosive and impact events were conducted in dry sand, saturated sand and water targets.[30,40,50,55,58]

Final Crater Volume

Measurements of final crater size are shown in Fig. 3a for Ottawa Flintshot sand and alluvium, two dry porous materials. The results are plotted as a nondimensional cratering efficiency versus a gravity scaled size (denoted by π_2), in accord with the scaling equation (4) developed in the previous section. The dry sand data follow a power-law trend. The slope of the regression line shown in Fig. 3a is -0.507 which, from eqn. (4), implies that μ=0.41. The trend exhibited by the alluvium results differs from the sand because at small sizes craters formed in alluvium are governed primarily

Table 2. Summary of quarter-space impact and explosive shots in dry sand, saturated sand and water.

		SOURCE						TARGET			CRATER					π-GROUPS					
Shot No. –	Gravity (G)	Mat'l –	a (cm)	DOB (cm)	Mass (gm)	U (km/s)	Density (g/cm³)	Wat.Tab. (cm)	Type –	Density (g/cm³)	V (cm³)	r (cm)	h (cm)	h_m (cm)	T_m (msec)	π_2 –	π_V –	π_r –	π_h –	π_{hm} –	π_{Tm} –
206–QPI	1	Pyrex	0.201	–	0.087	1.88	2.57	–	water	1.00	11190.0	19.70	14.10	14.10	102.0	1.80×10^{-8}	1.29×10^{5}	44.49	31.85	31.85	9.54×10^{4}
207–QPI	1	Pyrex	0.201	–	0.087	1.95	2.56	–	water	1.00	12586.0	20.00	14.90	14.90	74.0	1.67×10^{-8}	1.45×10^{5}	45.14	33.63	33.63	7.18×10^{4}
208–QPI	1	Glass	0.100	–	0.009	4.64	2.05	–	water	1.00	4613.0	14.30	12.10	12.10	83.0	1.47×10^{-9}	5.36×10^{5}	69.80	59.06	59.06	3.85×10^{5}
209–QPI	1	Alum.	0.300	–	0.316	5.13	2.78	–	DFS	1.80	489.0	10.40	2.45	2.85	69.0	3.60×10^{-9}	2.79×10^{3}	18.57	4.38	5.09	1.18×10^{5}
210–QPI	1	Alum.	0.300	–	0.315	5.16	2.78	0.00	WFS	2.09	675.0	10.60	3.37	4.73	35.0	3.56×10^{-9}	4.48×10^{3}	19.92	6.33	8.89	6.02×10^{4}
211–QPI	1	Alum.	0.400	–	0.751	1.99	2.79	–	DFS	1.80	396.0	9.35	3.10	3.16	63.0	3.19×10^{-8}	9.49×10^{2}	12.51	4.15	4.23	3.13×10^{4}
212–QPI	1	Alum.	0.400	–	0.751	2.15	2.79	0.00	WFS	2.09	526.0	8.77	3.80	4.89	54.0	2.73×10^{-8}	1.46×10^{3}	12.34	5.35	6.88	2.90×10^{4}
213–QPI	1	Alum.	0.300	–	0.315	5.07	2.78	–	DFS	1.80	552.0	10.80	2.80	2.96	69.0	3.69×10^{-9}	3.15×10^{3}	19.31	5.01	5.29	1.17×10^{5}
214–QPI	1	Nylon	0.198	–	0.037	2.47	1.13	–	water	1.00	4402.0	13.50	11.70	11.70	56.0	1.03×10^{-8}	1.18×10^{5}	40.44	35.05	35.05	6.97×10^{4}
215–QPI	1	Alum.	0.300	–	0.316	4.07	2.79	0.00	WFS	2.09	361.0	8.46	3.23	4.00	28.0	5.72×10^{-9}	2.39×10^{3}	15.88	6.06	7.51	3.80×10^{4}
216–QPI	1	Glass	0.100	–	0.009	4.54	2.07	–	water	1.00	3654.0	12.80	10.80	10.80	57.0	1.53×10^{-9}	4.25×10^{5}	62.48	52.71	52.71	2.59×10^{5}
218–QXI	50	Poly.	1.164	–	6.326	1.99	0.96	–	DFS	1.78	246.0	8.59	2.05	–	–	4.64×10^{-6}	6.92×10^{1}	5.63	1.34	–	–
219–QXI	50	Poly.	1.164	–	6.340	1.98	0.96	0.00	WFS	2.11	359.0	11.00	2.18			4.69×10^{-6}	1.19×10^{2}	7.62	1.51		
220–QXE	100	Sph. PbN6	0.640	0.000	3.400	1.62	3.10	0.00	WFS	2.11	93.1	6.09	1.74	–	–	7.70×10^{-6}	5.78×10^{1}	5.19	1.48	–	–
222–QXE	100	Sph. PETN	0.712	0.000	2.680	3.27	1.78	0.00	WFS	2.10	187.0	7.84	2.08			2.10×10^{-6}	1.47×10^{2}	7.23	1.92		
231–QPE	1	Hemi. PbN6	0.508	0.000	1.700	1.62	3.10	–	water	1.00	1762.0	9.12	9.84	–	6.3	6.11×10^{-8}	1.04×10^{3}	7.64	8.24	–	2.01×10^{3}
232–QPE	1	Hemi. PETN	0.565	–0.565	1.340	3.27	1.78	–	DFS	1.80	271.4	8.94	2.01	2.38	43.0	1.67×10^{-8}	3.65×10^{2}	9.86	2.22	2.63	2.49×10^{4}
233–QPE	1	Hemi. PETN	0.565	–0.565	1.340	3.27	1.78	0.00	WFS	2.11	103.7	6.80	2.40	2.96	2.7	1.67×10^{-8}	1.63×10^{2}	7.91	2.79	3.44	1.56×10^{3}
234–QPE	1	Hemi. PETN	0.565	0.000	1.340	3.27	1.78	0.00	WFS	2.11	187.6	6.67	3.23	3.95	4.0	1.67×10^{-8}	2.95×10^{2}	7.76	3.76	4.60	2.32×10^{3}
235–QPE	1	Hemi. PbN6	0.508	0.000	1.700	1.62	3.10	0.00	WFS	2.11	105.2	6.62	2.45	3.20	2.5	6.11×10^{-8}	1.31×10^{2}	7.11	2.63	3.44	7.97×10^{2}
289–QOE	10	E–106	0.387	–0.222	0.574	2.41	2.37	–	water	1.00	–	–	–	–	–	2.10×10^{-7}	–	–	–	–	–
294–QOE	50	E–106	0.387	–0.222	0.574	2.41	2.37	–	water	1.00	–	–	–	–	–	1.05×10^{-6}	–	–	–	–	–
295–QOE	125	E–106	0.387	–0.222	0.574	2.41	2.37	–	water	1.00	–	–	–	–	–	2.63×10^{-6}	–	–	–	–	–
296–QOE	250	E–106	0.387	–0.222	0.574	2.41	2.37	–	water	1.00						5.26×10^{-6}					
297–QOE	416	E–106	0.387	–0.222	0.574	2.41	2.37	–	water	1.00	428.0	6.17	5.05	5.50	3.5	8.76×10^{-6}	7.46×10^{2}	7.42	6.08	6.62	2.19×10^{3}
306–QXE	399	Deta Sht	1.270	–1.110	13.130	2.78	1.48	1.00	WBS	2.06	232.0	8.64	2.08			2.07×10^{-5}	3.64×10^{1}	4.66	1.12		
307–QXE	400	Deta Sht	0.890	0.510	4.610	2.78	1.48	1.10	WBS	2.06	289.0	8.86	2.77	4.68		1.46×10^{-5}	1.29×10^{2}	6.77	2.12	3.58	
319–QXE	403	Deta Sht	0.960	0.490	5.650	2.92	1.45	0.70	WBS	2.06	232.0	7.45	2.95			1.43×10^{-5}	8.46×10^{1}	5.32	2.11		
324–QXE	398	Deta Sht	0.880	0.380	4.390	2.92	1.45	1.05	WBS	2.06	202.0	7.21	2.60			1.30×10^{-5}	9.48×10^{1}	5.60	2.02		
326–QXE	398	Deta Sht	0.880	0.380	4.300	2.92	1.45	1.05	WBS	2.06	212.0	7.75	2.40			1.30×10^{-5}	1.02×10^{2}	6.06	1.88		
328–QXE	401	Deta Sht	0.880	0.380	4.410	2.92	1.45	1.05	WFS	2.11	193.0	7.41	2.17	3.64		1.31×10^{-5}	9.23×10^{1}	5.80	1.70	2.85	
329–QXE	100	Deta Sht	0.880	0.380	4.300	2.92	1.45	1.10	WBS	2.06	392.0	9.76	3.25	6.55		3.26×10^{-6}	1.88×10^{2}	7.64	2.54	5.13	
330–QXE	100	Deta Sht	0.880	0.380	4.350	2.92	1.45	1.05	WFF	1.97	320.0	8.04	3.49	5.90		3.26×10^{-6}	1.45×10^{2}	6.17	2.68	4.53	

by target strength, rather than gravity. In this "strength regime", cratering efficiency is independent of π_2, and instead depends on the strength parameter, $Y/\rho U^2$, shown in Table 1. Hence, for constant impactor and target material and impact velocity, impacts of various size in the strength regime should follow a horizontal line on a plot such as Fig. 3a. In the alluvium experiments, the impact velocity (as well as target and impactor materials) were held constant. Hence, the alluvium points are expected to follow the trend shown. As can be seen from Table 1, if the impact velocity is increased then the horizontal line should move upward. Thus, higher impact velocities in alluvium should produce craters which lie above the line shown in Fig. 3a.

Figure 3b shows comparable results for saturated sand and Permoplast modeling clay, two nonporous materials. Results are shown for both half-space and quarter-space experiments. The final crater volumes determined in the quarter-space experiments are not used here because of present uncertainties in an empirical calibration which relates these results to the half-space measurements. The strengths of both the wet sand and the clay are apparently large enough to keep even the highest-G experiments from fully entering the gravity regime. Certainly the experiments at the smaller scaled sizes were dominated by strength effects. Thus, it would appear difficult to determine the exponent μ from a gravity-regime slope. However, μ can be calculated from the measurements in the strength regime. From Table 1, if target strength and density are held constant then cratering efficiency should be proportional to $\delta^{3\upsilon-1}U^{3\mu}$. Comparing the efficiencies for the aluminum (δ=2.7) and polyethylene (δ=0.9) impactors, where the impact velocity was constant at 2 km/s, gives a value of υ=0.46. Using this value of υ, the cratering efficiencies for the lead impactor at 0.5 km/s can then be compared with the 2 km/s results to calculate μ=0.6. Thus the value of μ for wet sand appears to be higher than that measured for dry sand. This is in accord with the expected dependence of μ on target porosity, as noted in the previous section.

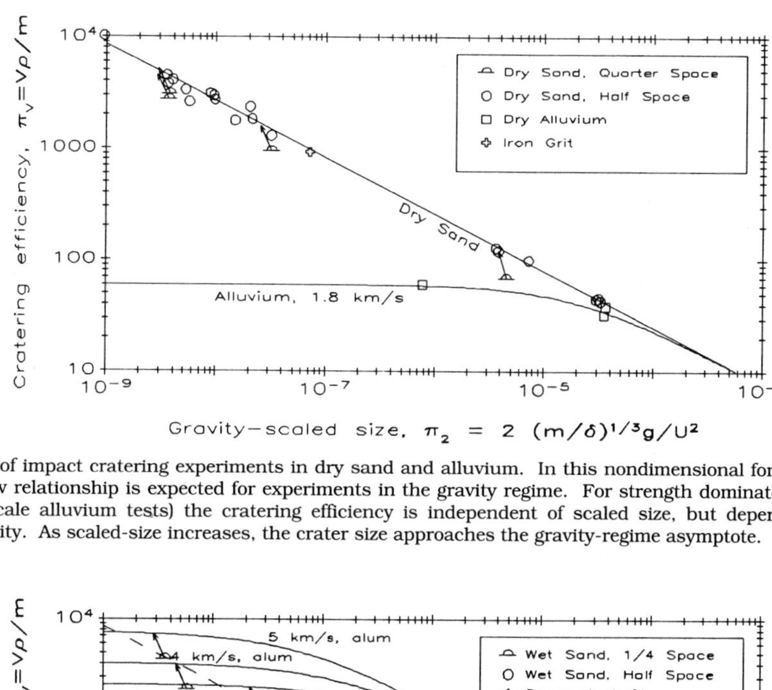

Fig. 3a. Results of impact cratering experiments in dry sand and alluvium. In this nondimensional form, a downward-sloping power-law relationship is expected for experiments in the gravity regime. For strength dominated experiments, (e.g. the small-scale alluvium tests) the cratering efficiency is independent of scaled size, but depends on impactor density and velocity. As scaled-size increases, the crater size approaches the gravity-regime asymptote.

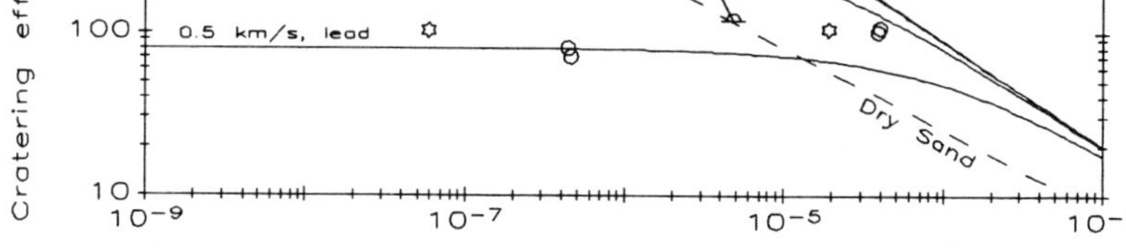

Fig. 3b. Results of impact cratering experiments in wet sand and in clay. For these nonporous material, the gravity-regime slope is expected to be steeper than the slope for the porous materials shown in Fig. 3a.

Transient Crater: Dynamic Experiments

The kinematics of crater formation were investigated using a quarter-space test bed on a centrifuge. This technique allows high-speed movie coverage of target deformations as seen in cross section through a 2-inch Plexiglas window along a diametrical symmetry boundary. Tests have been performed for both explosions and impacts, both at elevated gravity and at 1-G.

In seven of the fourteen explosion tests, the water table was nominally at 1.05 cm below the original ground zero. (This was to correspond to the approximate scale of the 500-ton PRAIRIE FLAT explosive crater in the field, which was modeled with the tangent-above shot 306-QXE.) The objectives of these tests were to dynamically record the growth, the maximum depth of the transient crater and its collapse. Figure 4a shows a preliminary interpretation of the film records for one of these tests: shot 307-QXE, which was conducted in nearly-saturated moderately-dense fine-grained Ottawa "Banding" sand. For this target material, crater floor uplift ("collapse") occurs on the same time scale as the radial motion of the ejecta plume. Because of the impairment of the optical path due to detonation products plating onto the window, the floor uplifting sequence could not be resolved in some of the intermediate frames of the film. It appears to be complete by frame 21 and maybe earlier. At this time the ejecta plume is in the vicinity of the final apparent crater radius and the crater profile is in agreement with the final apparent crater measured after the shot.

Crater size versus time is plotted for shot 307-QXE in Figs. 4b, 4c and 4d. Fom Table 1, crater size is expected to grow as a power of time, at least prior to when gravity or target strength begin to arrest growth. Crater volume displays power-law growth until maximum transient depth occurs (frame 4). More interestingly crater radius shows power-law growth all the way out to final radius (frame 21). Lines corresponding to μ=0.55 were drawn in each figure, and agree well with the data. Note that this is in accord with (although somewhat smaller than) the value of μ determined from final crater volume (see previous subsection).

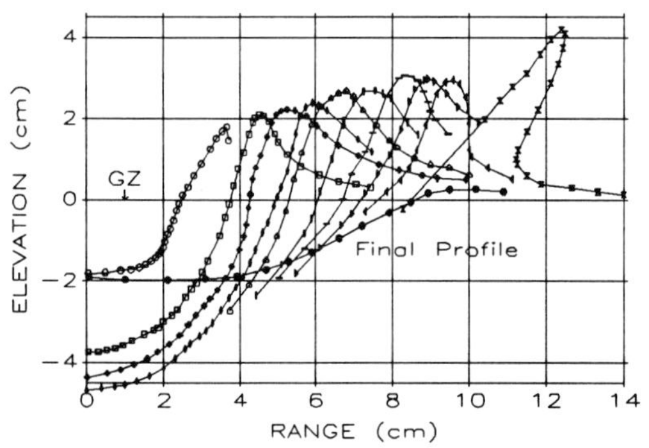

Fig. 4a. Crater growth for shot 307-QXE, a tangent-below explosion in saturated dense sand. Although a complete record was not obtained, the transient cavity apparently collapsed to the final profile shown.

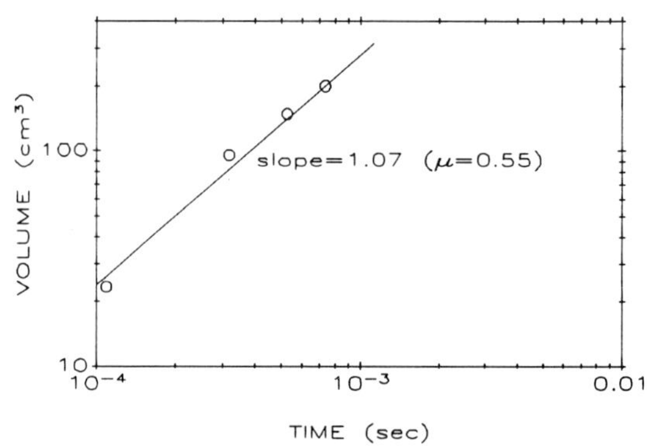

Fig. 4b. Transient crater volume vs time, for 307-QXE.

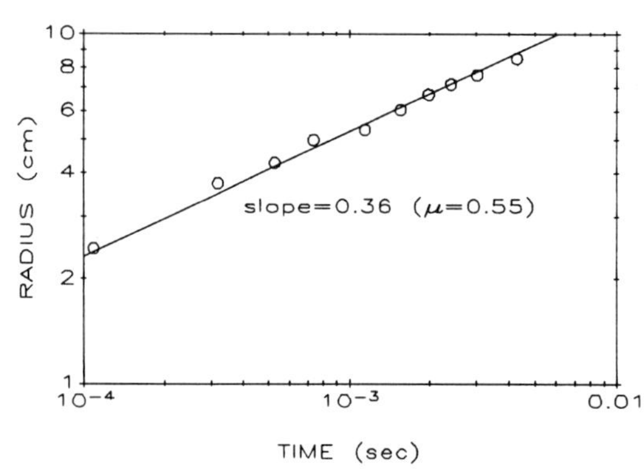

Fig. 4c. Transient crater radius vs time, for 307-QXE.

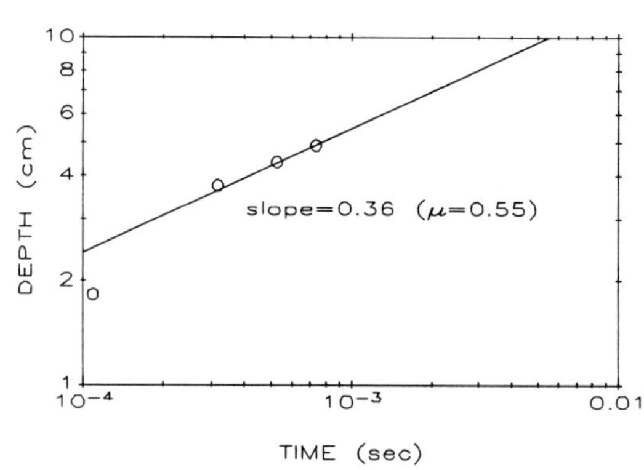

Fig. 4d. Transient crater depth vs time, for 307-QXE.

As an expedient when determining the maximum charge size that would not spall or puncture the window, the explosive charges were built up from Dupont Deta sheet. This is PETN with approximately 25% inert plastic filler, which is presumed to cause much of the smoke and detonation debris products. Pure PETN is now under consideration for future shots to improve resolution by reduction of smoke and opaque detonation products. However, since the Deta-Sheet detonation products deposit on the window as it is exposed to the transient crater, a permanent record of the envelope of maximum crater depth versus range is formed. Preliminary comparisons of the detonation profile with the corresponding final apparent crater profile along with viewing the films indicate that there was no significant difference in the transient crater phenomena over a variation of approximately an order of magnitude in sand-grain size although there was some size variation.

Crater profiles for three shots with each of the different sand sizes is shown in Figs. 5a, 5b and 5c. It is noteworthy that shot 328-QXE was performed at 515G and had the smallest transient crater. However since it was performed with the coarse grained sand, the cause of the smaller transient crater is not clear.

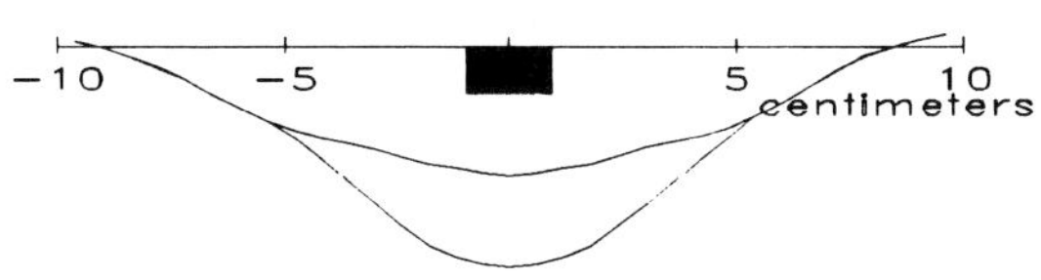

Fig. 5a. Maximum transient crater and final profile for shot 307-QXE.

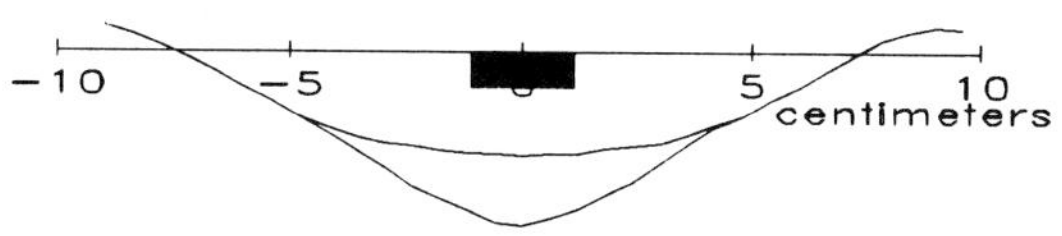

Fig. 5b. Maximum transient crater and final profile for 328-QXE (400G, 1.05 cm water table, wet Flintshot sand, 4.2 m prototype).

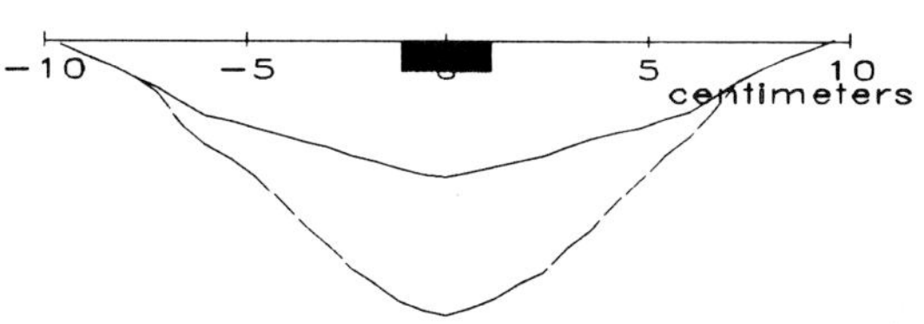

Fig. 5c. Maximum transient crater and final profile for 329-QXE (100G, 1.10 cm water table, wet Banding sand, 212μ grain size).

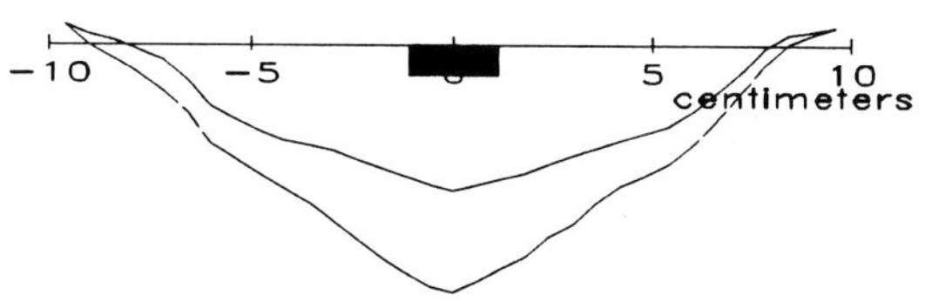

Fig. 5d. Maximum transient crater and final profile for 330-QXE (100G, 1.05 cm water table, wet F-140 sand 106μ grain size, 1.05 m prototype).

Crater growth histories for impacts and explosions in dry sand, wet sand and water are shown in Figs 6a and 6b. Here also, the results follow a power-law trend. Solid lines are shown with a slope calculated from μ=0.55 (see Table 1), corresponding to expectations for nonporous targets (water and wet sand). Dashed lines for μ=0.4 are shown for the dry sand targets. Although some deviations are observed at late times, where strength or gravity are important, the initial growth agrees well with the trend expected from the coupling parameter scaling relations.

In order to extend the observable time regime for crater growth in water, some explosion tests using a common charge mass and configuration were performed. Some of the shots were filmed at 5000 pictures/sec, while others were filmed at 1.4 million pictures/sec. The results, shown in Fig. 7, closely follow the power-law trend over five orders of magnitude in time with a slope corresponding to μ=0.55.

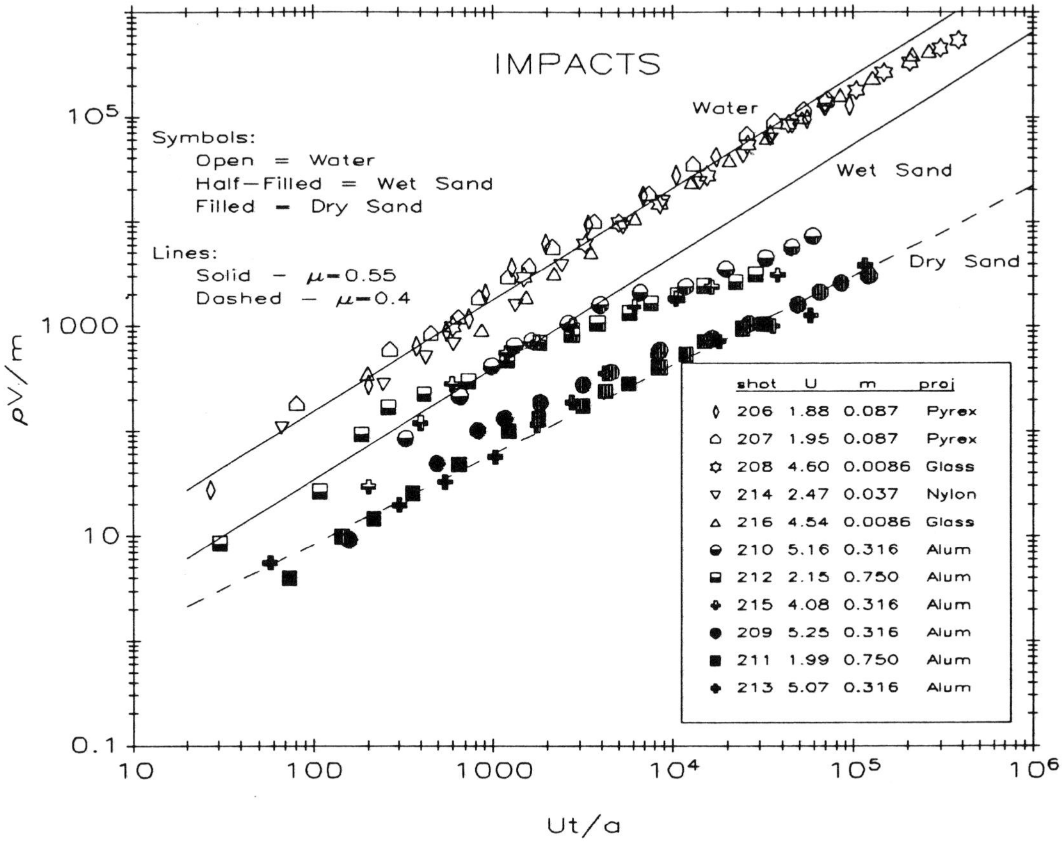

Fig. 6a. Cratering efficiency versus time for impacts in water, wet sand and dry sand. The results are shown in nondimensional form. The lines shown indicate the power-law behavior expected for each material.

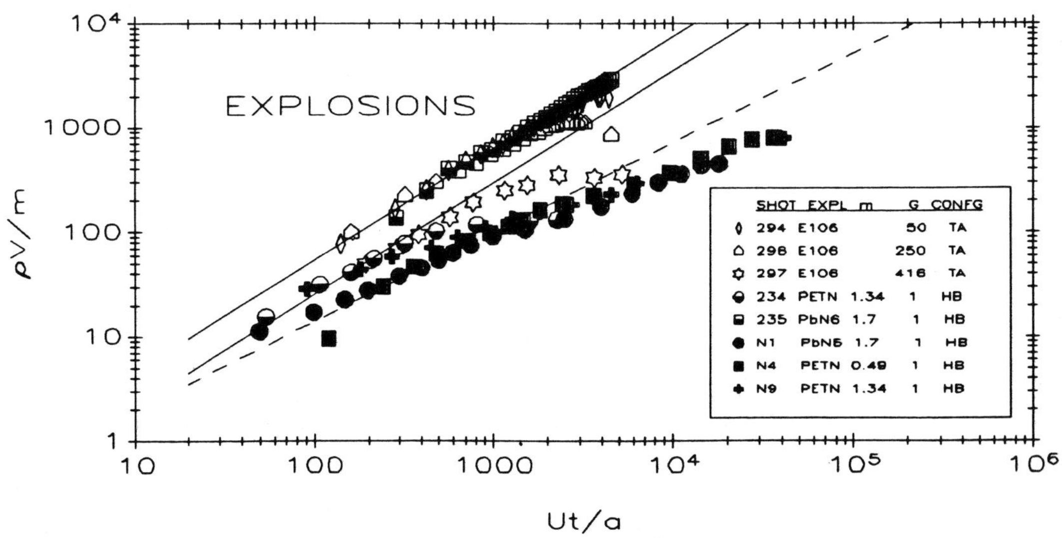

Fig. 6b. Cratering efficiency versus time for explosions in water, wet sand and dry sand. The lines show the expected power-law dependence.

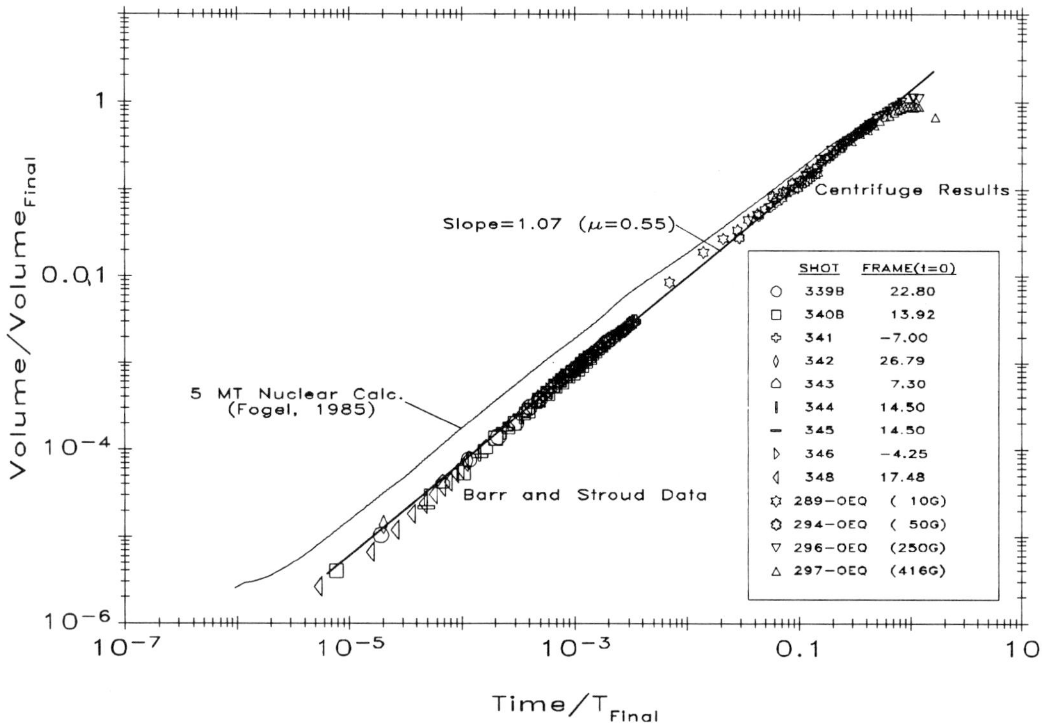

Fig. 7. Comparison of a 5MT code-calculation[5] with subscale laboratory measurements for crater growth in water.

Time of Formation

The time of formation, defined as the time of maximum transient crater volume, is shown in Fig. 8a for impacts and Fig. 8b for explosions. For some experiments, formation time could not be determined because the film record did not carry through to final crater formation. For three of the water experiments in Fig. 8a, a lower bound on the formation time is shown (indicated by vertical arrows). Also, the experiments conducted in the strength regime (e.g. wet sand at 1 G) are not plotted.

For comparison purposes the times of formation for impacts in dry sand (Gault and Wedekind[8]) and in water (Gault and Sonett[7]) are shown in Fig 8a. Both of these data sets pertain to cratering in "half-space" targets. The formation times measured by Gault and Wedekind for dry sand are somewhat larger than our results. This is partly due to the fact that they measured the time at which the ejecta veil passed over the crater rim; an event which occurs after the crater reaches maximum volume. In order to compare the results on an equal basis, their times were reduced by a factor of 1.6. The reduction was based on experiments of Piekutowski[30], who reported values for both definitions of formation time for near-surface explosions in dry sand. Although there are some deviations attributable to the differences between half-space and quarter-space configurations, the results reported here agree reasonably well with those of Gault and Wedekind and Gault and Sonett.

The results shown in Fig. 8 exhibit a power-law trend, as expected based on the coupling parameter scaling relations given in Table 1. Note from Table 1 that the expected slope is rather insensitive to the value of μ: it is about 0.58 for μ=0.4 and about 0.61 for μ=0.55. Hence, precise values of μ are difficult to determine from measurements of time of formation. Note however that the predicted slopes, shown in Fig. 8, agree well with the data.

A general scaling law for time of formation for all materials can be developed as follows. The fact that a power-law relation is observed in Fig. 8 implies that the formation time, T, is governed by the coupling parameter. Hence, for the gravity regime,

$$T = T[C, \rho, g] \tag{5}$$

C can be eliminated from eqn. (5) by using eqn. (3) to obtain

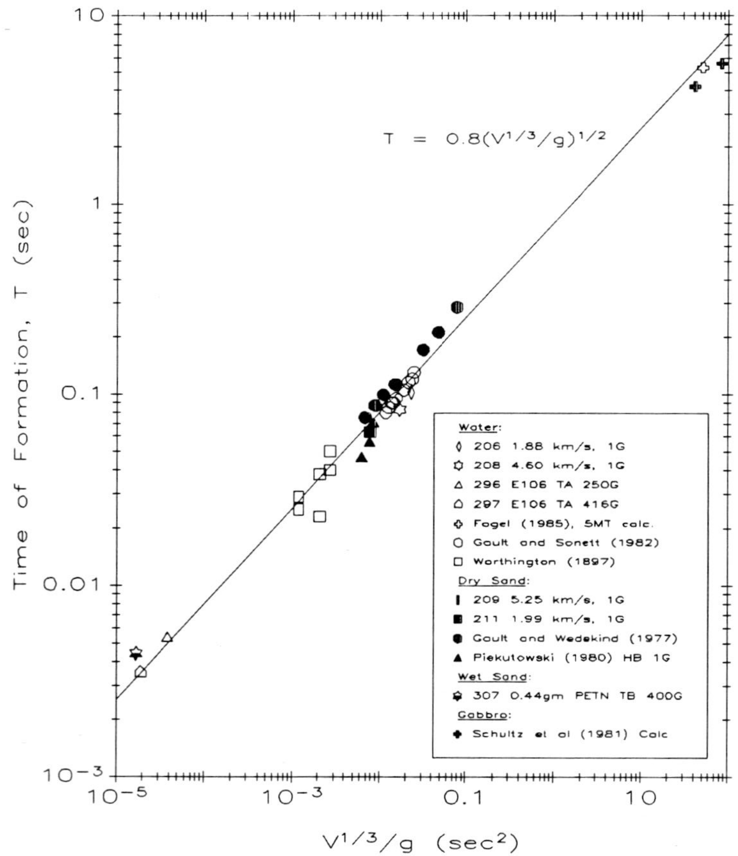

Fig. 9. The time of formation data plotted in a form which correlates the four material types.

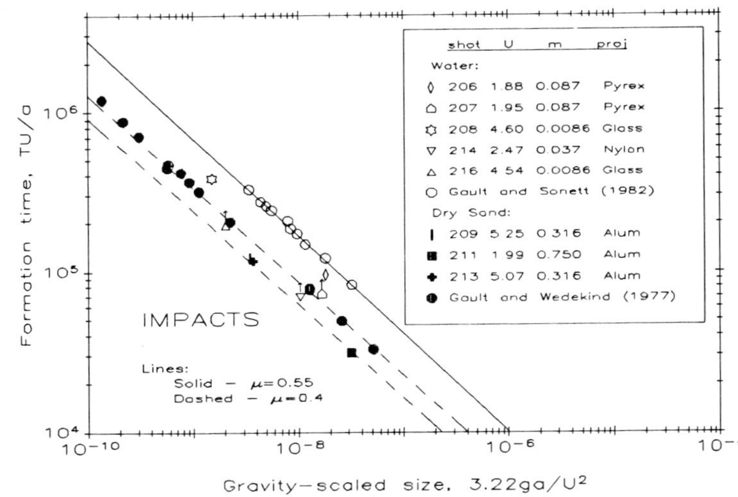

Fig. 8a. Formation time versus gravity scaled size for impacts in water and in dry sand.

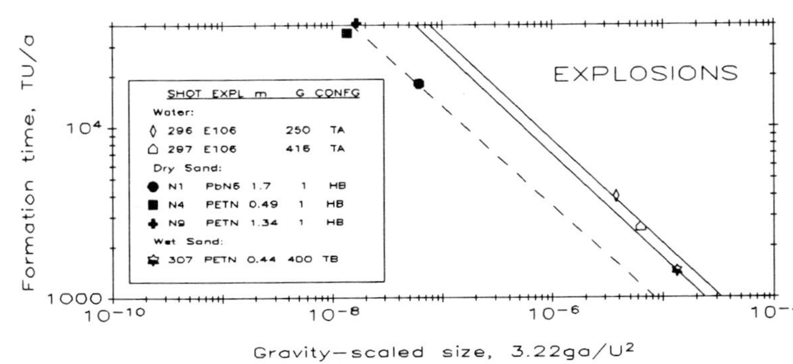

Fig. 8b. Formation time versus gravity scaled size for explosions in water, dry sand and wet sand.

$$T = T[V, \rho, g] \tag{6}$$

From dimensional arguments, ρ can be dropped from eqn.(6) because it is the only variable with units of mass. Thus, eqn. (6) gives

$$T \propto [V^{1/3}/g]^{1/2} \propto V^{1/6} g^{1/2} \tag{7}$$

So, for example, if T is plotted versus $V^{1/3}/g$, then the results for all materials should fall on a common line of slope 1/2. Figure 9 is an example. In addition to the data shown in Fig. 8, Fig. 9 includes the water-drop measurements reported by Worthington[63] and the finite-difference code-calculations of Schultz et al.[59], for impacts in gabbroic anorthosite, and Fogel[5] for a 5 MT explosion in water. The line of slope 1/2 shown in the figure gives a reasonable fit to all of the data shown for water, wet sand, dry sand and gabbro.

Transient crater collapse: marker techniques

Large flat dish-shaped craters have been conjectured to go through a hemispherical bowl-shaped transient stage before subsequently "collapsing". Results and analysis of various code calculations have shown consistency with an uncoupled transient crater formation followed by large scale slumping. This concept is fostered by quasi-static slope stability arguments which allow predictions of the height for which incipient failure occurs. For application to cratering, one must assume that on the time scale of excavation, inertia allows the formation of a statically unstable cavity. For the many centrifuge experiments performed to date, the aspect ratio (radius/depth) of the final crater formed in saturated sand or in oil-base clay has been observed to increase with increasing gravity. This is expected and corresponds to the same trend seen for increasing crater size at fixed gravity. Taking this similarity into account, particular experiments unique to the centrifuge were employed to examine the validity of the commonly accepted sequence of a transient bowl phase followed by collapse and slumping to form the stable large-aspect-ratio crater. Two cratering shots were made with all conditions identical except for gravity. In one case, the crater was formed at low G and then subjected to high G. This result can be compared with the crater formed at high G directly.

Two schemes were devised to passively record the maximum extent of the transient cavity formed prior to the final crater. For the oil-base clay, two identical samples were prepared using "Permoplast" oil-base clay. Along two perpendicular diameters, holes were drilled in the clay, into which melted clay of a different color was poured providing a series of marker columns. A spherical 12.8-gm PETN explosive charge was placed in a tangent-above configuration on both samples which were then mounted on opposing rotor ends of the centrifuge. One was fired at 526 G corresponding to 1.86 kilotons of PETN at 1 G. The other was fired at 10 G, corresponding to 12.8 kg. After measuring crater dimensions for both samples, a thin wire was used to bisect each one along the centerline of one of the sets of precast marker columns, which were then photographed to record the permanent deformations. Both halves of each sample were then rejoined and placed back onto the centrifuge and subjected to 522 G for 104 seconds. The craters were remeasured and then bisected along the centerline of the other set of marker columns.

As expected, the crater formed originally at 10G slumped considerably and the marker columns were known to have gone through the deformation history consistent with a transient crater of roughly twice the size of the final crater. These columns could then be used as a control to compare with the corresponding cross section of the crater formed at high G directly. The latter did not appear to have gone through a transient phase since the columns showed distortion only near the floor of the crater. What was surprising however, was that the low-G crater also did not exhibit any gross distortion of the columns to anywhere near the depths of excavation of the initial low-G crater. The slumping flow was essentially reversible and the columns returned to their original positions and dimensions. Only small second order deviations indicated the prior presence of a transient crater.

Figures 10a and 10b show cross sections of the 10-G and the 526-G craters as observed initially. Figures 10c and 10d show the same two craters respectively after both were subjected to 522-G for 104 seconds. The degree of reversibility of the flow field is remarkable. Close inspection of these markers as well as a second cut through the other cross section does indicate that if a transient crater occurred for the high-G crater its extent was considerably less than the size of the crater formed at low G.

A different scheme was used for saturated sand. Shot, 116-OEA, was designed to address the maximum transient crater and record the degree of slumping that occurs for the saturated sand at high G. In this test, a 17-micron aluminum foil witness plate was placed "vertically" directly beneath the charge. Thereby, explosive excavation of the crater would perforate or cut out the foil. Slumping or flow of the sand would then partially fill in the crater, but would not disturb the foil that was placed along a symmetry axis. The foil definitely indicated that the transient crater extended below the final crater. The extent of the transient crater however could not be quantitatively determined. What is very clear from both of these sets of experiments is that passive markers can only provide final displacements. The clay experiments especially showed that the collapse of the crater can be very systematic and even restore the stratigraphy beneath the uprising floor.

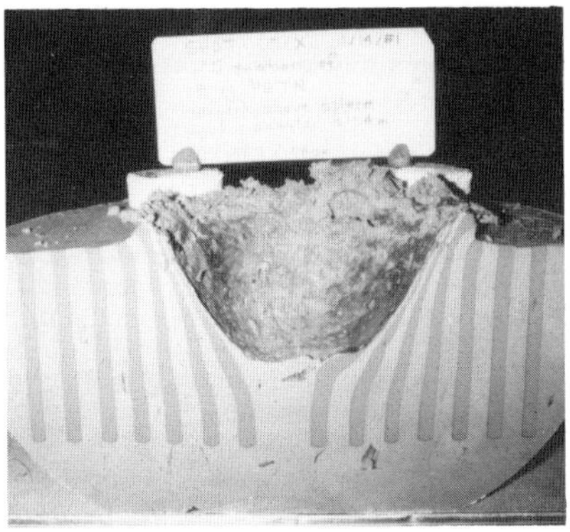

Fig. 10a. Dissection of a crater formed at 10G in Permoplast clay.

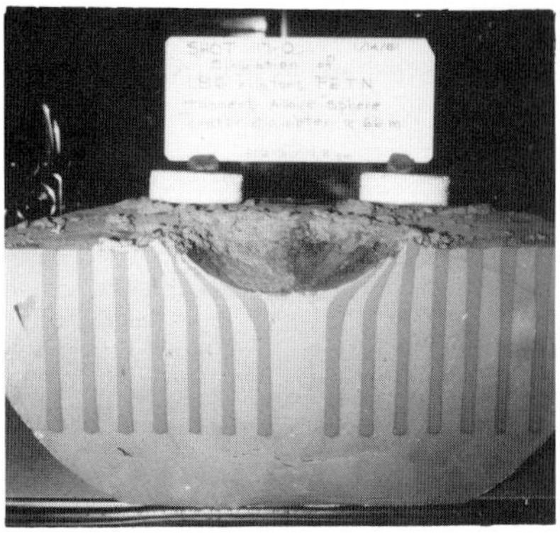

Fig. 10b. Dissection of a crater formed at 526 G in Permoplast clay.

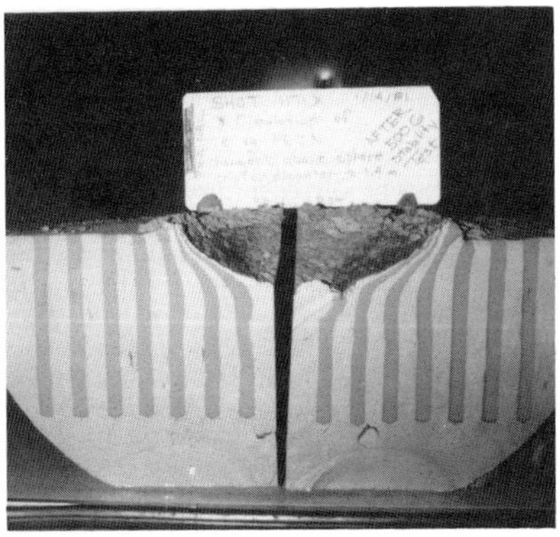

Fig. 10c. The 10G crater after being subjected to 522G for 104 sec. Note the similarity to the 526 G crater and the apparent reversibility of the material motions.

Fig. 10d. The 526G crater after being subjected to 522 G for 104 sec.

SCALING ESTIMATES FOR POROUS AND NON-POROUS TARGETS

Scaling laws for crater size have been proposed by a number of authors[6,8,9,61]. This section summarizes a set of scaling relations based on the experimental results and the point-source coupling parameter concept discussed above. The scaling laws proposed here differ from earlier estimates, as discussed below.

Table 3 lists our scaling estimates in two forms. The upper part of the table gives the results in the nondimensional form shown in most of the figures. The lower part of the table shows the same scaling laws in a dimensional form (with cgs units) in terms of the impactor energy and velocity, and gravity. In constructing these scaling laws, the density ratio was assumed to be constant, because in most cases there were insufficient data to determine the dependence on density. The lower part of Table 3 was calculated with the target density appropriate for each material and a nominal impactor density (δ) of 3 gm/cm^3.

The estimates for crater growth rate and formation time were constructed from the lines shown in Fig. 6a, Fig 8a and Fig. 9. The slopes of the lines in those figures are based on a value of μ=0.4 for dry sand and μ=0.55 for water and wet sand. Note that the equation in the upper part of Table 3 for time of formation, expressed in terms of crater volume and gravity, is independent of material type. This is due to the fact that, through the use of a coupling parameter, the crater volume "accounts" for variations in target and impactor properties.

Table 3
Proposed Scaling Laws for Impact Craters

(Crater dimensions refer to the apparent crater)

	Dry Sand	Wet Sand	Water
Crater Volume	$\Pi_V = 0.24\Pi_2^{-0.51}$	$\Pi_V = 0.2\Pi_2^{-0.65}$	$\Pi_V = 2.1\,\Pi_2^{-0.65}$
Crater Radius	$\Pi_R = 0.84\,\Pi_2^{-0.17}$	$\Pi_R = 0.8\,\Pi_2^{-0.22}$	$\Pi_R = 0.94\Pi_2^{-0.22}$
Volume vs Time	$\Pi_V = 0.16\,\Pi_t^{0.86}$	$\Pi_V = 0.25\,\Pi_t^{1.07}$	$\Pi_V = 1.1\Pi_t^{1.07}$
Time of Formation	$\Pi_T = 1.9\Pi_2^{-0.58}$	$\Pi_T = 1.6\Pi_2^{-0.61}$	$\Pi_T = 2.3\Pi_2^{-0.61}$
	$T = 0.8V^{1/6}\,g^{1/2}$		

	Dry Sand	Wet Sand	Water
Volume	$V=0.20E^{0.83}U^{-0.64}g^{-0.51}$	$V=0.13E^{0.78}U^{-0.27}g^{-0.65}$	$V=2.9E^{0.78}U^{-0.27}g^{-0.65}$
Radius	$R=0.77E^{0.28}U^{-0.21}g^{-0.17}$	$R=0.70E^{0.26}U^{-0.09}g^{-0.22}$	$R=1.1E^{0.26}U^{-0.09}g^{-0.22}$
Vol vs t	$V=0.29E^{0.71}U^{-0.57}t^{0.86}$	$V=0.21E^{0.64}U^{-0.22}t^{1.07}$	$V=1.9E^{0.64}U^{-0.22}t^{1.07}$
Form. Time	$T=0.74E^{0.14}U^{-0.12}g^{-0.58}$	$T=0.62E^{0.13}U^{-0.04}g^{-0.61}$	$T=0.89E^{0.13}U^{-0.04}g^{-0.61}$

The variables are defined in Table 1.

The estimates for crater size for dry sand were taken from Schmidt[35], while those for water were derived from the experiments of Gault and Sonett[7]. The wet sand result is based on the data which were presented in Fig. 3 and are compared in Figs. 11 and 12 to other target materials. As discussed earlier, the effective size scale for the wet sand shots was not large enough to put them fully in the gravity regime. Thus, the proposed scaling estimate was constructed as an asymptote to the data, with a slope calculated from a value of μ=0.55. The estimate is significantly above the dry sand line because of better coupling in the (nonporous) wet sand, but below the line for water targets. The wet sand line is regarded as a current best-estimate for rocky materials, and differs from our earlier estimate[48] where the slope was calculated based on a value of μ more appropriate to porous materials.

Previous scaling estimates for large-scale cratering in rock (see refs. cited above) have assumed that the crater radius is proportional to powers of the impactor energy and gravity, with no additional dependence on impact velocity. The energy exponent is typically reported to be 0.28 to 0.29 while estimates of the gravity exponent range from 0.11 to 0.17. The size scaling laws proposed here differ from earlier ones in three respects. First, a distinct dependence of impact velocity is recognized, even for constant impactor energy. The effects of velocity are smallest for nonporous targets, which approach (but do not reach) the asymptotic case of pure energy scaling. The porosity of materials like dry sand result in velocity-dependent energy losses. Hence, the velocity dependence is greatest for these materials. Second, the present energy exponent for low-porosity targets, like competent rock, is lower than earlier estimates. Although the difference might appear to be small (the present estimate is 0.26 versus the values of 0.28-0.29 mentioned above), the effect of this difference can be significant when extrapolating many orders of magnitude in energy up to kilometer-sized craters. Third, the gravity exponent is recognized here as being related to both the energy and the velocity exponents. As shown in Table 1, these exponents are all determined by the single scaling exponent, μ.

As mentioned above, the wet-sand result is used as a current best-estimate for large-scale cratering in rock. We stress however that this approximation is tenative until a better understanding is achieved of the dominant material properties which govern cratering in the gravity regime (e.g. density and friction angle). Once the role of density and friction angle are understood, reasonable analogues of rock might be constructed for which laboratory experiments can be performed in the gravity regime.

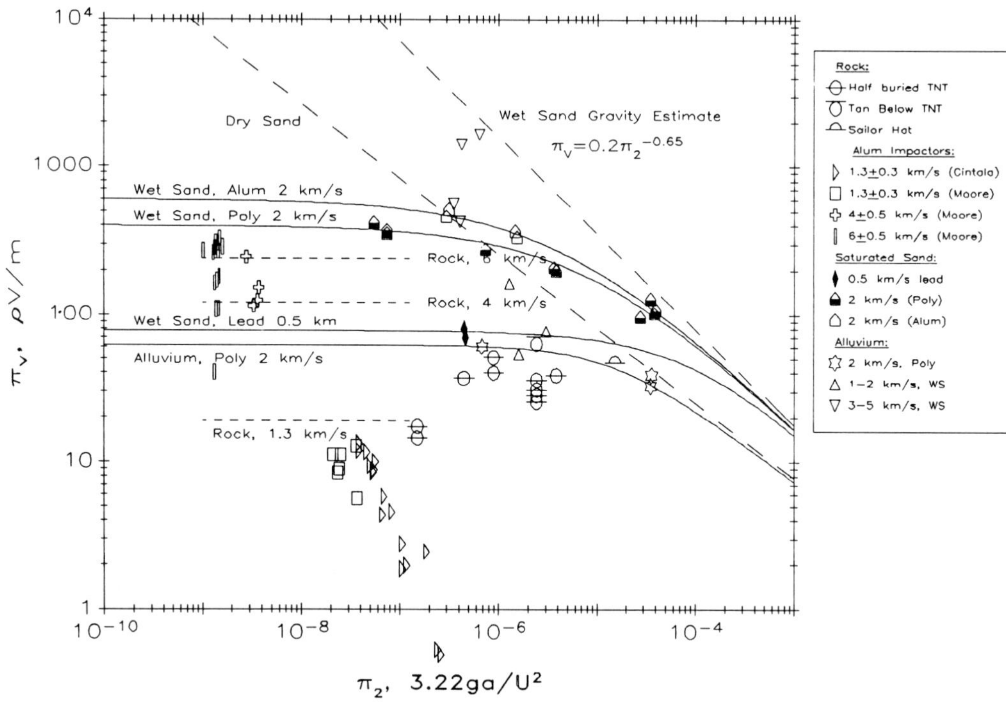

Fig. 11. Cratering efficiency for impacts and explosions in dry sand, alluvium, wet sand, and rock. The dashed line (labeled Wet Sand Gravity Estimate) is proposed as a current best estimate for large-scale cratering in rock.

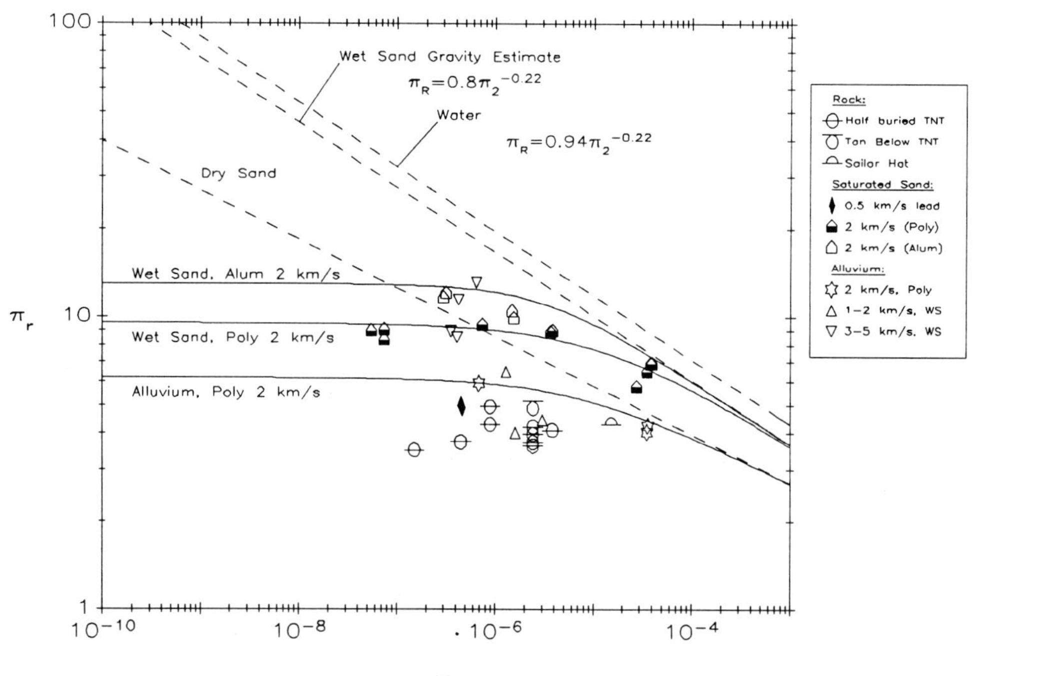

Fig 12. Scaled crater radius results for impacts and explosions in various materials. The dashed line (labeled Wet Sand Gravity Estimate) is proposed as a current best estimate for large-scale cratering in rock.

REFERENCES

1) Bjorkman M. D. (1983) Dependence of crater melt on impact velocity (abstract). *EOS (Trans. Amer. Geophys. Union)* **64**, 747.

2) Bjorkman M. D. (1984) Feasibility of determining impact conditions from total melt volume (abstract). In *Lunar Planetary Science XV*, 64-65.

3) Bjorkman M. D. and Holsapple K. A. (1983) Plane hypervelocity impact and source similitude dependence on a Tillotson E.O.S. *Proc. of the 3rd APS Topical Conference on Shock Waves in Condensed Media*, North Holland, Amsterdam.

4) Chabai A. J. (1965) On scaling dimensions of craters produced by buried explosives. *J. Geophys. Res.* **70**, 5075-5098.

5) Fogel M. (1985) Pacifica Technology, Personal Communication.

6) Gault D.E. (1974) Impact cratering. In *A Primer in Lunar Geology*, (R. Greeley and P. Schultz, eds.). NASA - Ames Research Center, p. 137-175.

7) Gault D.E. and Sonett C.P. (1982) Laboratory simulation of pelagic asteroidal impact: Atmospheric injection, benthic topography, and surface wave radiation field. *Geological Society of America Special Paper 190*, 69-92.

8) Gault D.E. and Wedekind J.A. (1977) Experimental hypervelocity impact into quartz sand-II, Effects of gravitational acceleration. In *Impact and Explosion Cratering*, (D.J. Roddy *et al.*, eds.) p. 1231-1244, Pergamon, New York.

9) Grieve R.A.F. and Dence M.R. (1979) The terrestrial cratering record: II. The crater production rate. *Icarus* **38**, 230-242

10) Henderson B. J., Schmidt R. M., and Konicek D. J. (1978) Statistical and dimensional analysis of crater configuration (abstract). *EOS (Trans. Amer. Geophys. Union)* **59**, 245.

11) Holsapple K. A. (1979) Impact cratering experiments with ambient atmospheric pressure (abstract). *EOS (Trans. Amer. Geophys. Union)* **60**, 871.

12) Holsapple K. A. (1980) The equivalent depth of burst for impact cratering. *Proc. Lunar Planet. Sci. Conf. 11th*, 2379-2401.

13) Holsapple K. A. (1981) Coupling parameters in cratering (abstract). *EOS (Trans. Amer. Geophys. Union)* **62**, 949.

14) Holsapple K. A. (1982) A comparison of scaling laws for planetary impact cratering: Experiments, calculations, and theory (abstract). In *Lunar Planetary Science XIII*, 331-332.

15) Holsapple K. A. (1983) On the existence and implications of coupling parameters in cratering mechanics (abstract). In *Lunar Planetary Science XIV, 319-320.*

16) Holsapple K. A. (1984) On crater dynamics: Comparisons of results for different target and impactor conditions (abstract). In *Lunar Planetary Science XV*, 367-368.

17) Holsapple K. A. and Bjorkman M. D. (1983) The scaling of impact crater growth (abstract). *EOS (Trans. Amer. Geophys. Union)* **64**, 747.

18) Holsapple K. A. and Housen K. R. (1985) Cratering estimates for rock and ice. *J. Geophys. Res.* (in review).

19) Holsapple K. A. and Housen K. R.(1986) Scaling laws for the catastrophic collisions of asteroids. *Memoirie della Societa Astronmica Italiana* (in press).

20) Holsapple K. A. and Schmidt R. M. (1979) A material-strength model for apparent crater volume. *Proc. Lunar Planet. Sci. Conf. 10th*, 2757-2777.

21) Holsapple K. A. and Schmidt R. M. (1980a) On the scaling of crater dimensions--1. Explosive processes. *J. Geophys. Res.* **85**(B12), 7247-7256.

22) Holsapple K. A. and Schmidt R. M. (1980b) A subscale simulation of two 20-ton cratering events (abstract). *EOS (Trans. Amer. Geophys. Union)* **61**, 1021.

23) Holsapple K. A. and Schmidt R. M. (1982) On the scaling of crater dimensions--2. Impact processes. *J. Geophys. Res.* **87**(B3), 1849-1870.

24) Holsapple K. A. and Schmidt R. M.(1985) Point-source solutions and coupling parameters in cratering mechanics. *J. Geophys. Res.* (in press).

25) Holsapple K. A., Schmidt R. M., and Dyrdahl R. L. (1978) Gravity scaling methods applied to crater induced ground motions. *Proc. of the Nuclear Blast and Shock Simulation Symposium*, San Diego. Report DNA 4797P, Defense Nuclear Agency, Washington, D.C.

26) Housen K. R. (1983) Crater ejecta scaling laws (abstract). In *Lunar Planetary Science XIV*, 333-334.

27) Housen K. R. (1984) Cratering flow fields: A general form and the Z-model (abstract). In *Lunar Planetary Science XV*, 377-378.

28) Housen K. R., Schmidt R. M., Holsapple K. A., and Piekutowski A. J. (1982) Scaling of explosion crater ejecta velocities (abstract). *EOS (Trans. Amer. Geophys. Union)* **63**, 1020.

29) Housen K. R., Schmidt R. M., and Holsapple K. A. (1983) Crater ejecta scaling laws: Fundamental forms based upon dimensional analysis. *J. Geophys. Res.* **88**(B3), 2485-2499.

30) Piekutowski A. J. (1980) Formation of bowl-shaped craters. *Proc. Lunar Sci. Conf. 11th*, 2129-2144.

31) Pokrovsky G. I. and Fyodorov I. S. (1969) *Centrifugal Model Testing in the Construction Industry*. Vol. I, II, Niedra Publishing House, Moscow. (English translation by Building Research Establishment Library Translation Service, Great Britain, August 1975.)

32) Schmidt R. M. (1976) A centrifuge cratering experiment (abstract). *Flagstaff Symposium on Planetary Cratering Mechanics*, 126-128. Lunar and Planetary Institute, Houston.

33) Schmidt R. M. (1977) A centrifuge cratering experiment: Development of a gravity-scaled yield parameter. In *Impact and Explosion Cratering* (D.J. Roddy,*et al.*, eds.) p. 1261-1278, Pergamon, New York.

34) Schmidt R. M. (1978) Centrifuge simulation of the JOHNIE BOY 500-ton cratering event. *Proc. Lunar Planet. Sci. Conf. 9th*, 3877-3889.

35) Schmidt R. M. (1980) Meteor Crater: Energy of formation--implications of centrifuge scaling. *Proc. Lunar Planet. Sci. Conf. 11th*, 2099-2128.

36) Schmidt R. M. (1981) Scaling crater time of formation (abstract). *EOS (Trans. Amer. Geophys. Union)* **62**, 944.

37) Schmidt R. M. (1982) Dynamic scaling relationships for impact crater formation (abstract). In *Lunar Planetary Science XIII*, 687-688.
38) Schmidt R. M. (1983a) Strength-gravity transition for impact craters in wet sand(abstract). In *Lunar Planetary Science XIV*, 666-667.
39) Schmidt R. M. (1983b) Centrifuge quarter-space cratering results: A gravity criterion for kinematic similarity (abstract). *EOS (Trans. Amer. Geophys. Union)* **64**, 747.
40) Schmidt R. M. (1984) Transient crater motions: Saturated sand centrifuge experiments (abstract). In *Lunar Planetary Science XV*, 722-723.
41) Schmidt R. M. and Holsapple K. A. (1978a) Ottawa sand cratering experiments in centrifuge with gravity variation to 450 G's (abstract). *EOS (Trans. Amer. Geophys. Union)* **59**, 245.
42) Schmidt R. M. and Holsapple K. A. (1978b) Centrifuge cratering experiments I: Dry granular soils. Report DNA 4568F, Defense Nuclear Agency, Washington D.C.
43) Schmidt R. M. and Holsapple K. A. (1978c) A gravity-scaled energy parameter relating impact and explosive crater size (abstract). *EOS (Trans. Amer. Geophys. Union)* **59**, 1121.
44) Schmidt R. M. and Holsapple K. A. (1979) Centrifuge crater scaling experiment II: Material strength effects. Report DNA 4999Z, Defense Nuclear Agency, Washington, D.C.
45) Schmidt R. M. and Holsapple K. A. (1980) Theory and experiments on centrifuge cratering. *J. Geophys. Res.* **85**(B1), 235-252.
46) Schmidt R. M. and Holsapple K. A. (1981a) An experimental investigation of transient crater size (abstract). In *Lunar Planetary Science XII*, 934-936.
47) Schmidt R. M. and Holsapple K. A. (1981b) Gravity scaling with centrifuge data: Minutes of the sixth meeting of DNA ad hoc cratering working group. RDA-TR-120004-001, Marina Del Ray, CA.
48) Schmidt R. M. and Holsapple K. A. (1982) Estimates of crater size for large-body impact: Gravity scaling results. *Geological Society of America Special Paper 190*, 93-102.
49) Schmidt R. M. and Holsapple K. A. (1984) Explosively generated water waves (abstract). *EOS (Trans. Amer. Geophys. Union)* **65**, 221.
50) Schmidt R. M. and Piekutowski A. J.(1982) Quarter-space impact experiments: Particle-motion and crater-growth measurements (abstract). *EOS (Trans. Amer. Geophys. Union)* **63**, 1020.
51) Schmidt R. M. and Piekutowski A. J. (1983) Development of the quarter-space technique for cratering experiments on a centrifuge (abstract). In *Lunar Planetary Science XIV*, 668-669.
52) Schmidt R. M., Holsapple K. A., and Housen K. R. (1983) Optimum depth of burst cratering: Strength-gravity transition identified (abstract). *EOS (Trans. Amer. Geophys. Union)* **64**, 255.
53) Schmidt R. M., Holsapple K. A., and Housen K. R. (1986) Gravity Effects in Cratering. Draft Report, DNA 001-82-C-0301, Defense Nuclear Agency, Washington, D.C.
54) Schmidt R. M., Holsapple K. A., and Piekutowski A. J. (1980) Centrifuge simulation study of the PRAIRIE FLAT multi-ring crater (abstract). *Conference on Multi-ring Basins: Formation and Evolution*, 71-73. Lunar and Planetary Institute, Houston.
55) Schmidt R. M., Housen K. R., and Piekutowski A. J. (1985) Experimental observations on crater growth regimes (abstract). *EOS (Trans. Amer. Geophys. Union)* **66**, 296.
56) Schmidt R. M., Watson H. E., and Wauchope C. R. (1979) Projectile density/target density correlation for impact cratering (abstract). *EOS (Trans. Amer. Geophys. Union)* **60**, 871.
57) Schmidt R. M., Housen K. R., and Webbeking V. T. (1985) Growth of water cavities over a range of 10^6 in time (abstract). *EOS (Trans. Amer. Geophys. Union)* **66**, 947.
58) Schmidt R. M., Housen K. R., Luke M., and Piekutowski A. J. (1984) Scaling consistency tests for cratering experiments (abstract). *EOS (Trans. Amer. Geophys. Union)* **65**, 983.
59) Schultz P.H., Orphal B., Miller B., Borden W.F. and Larson S.A. (1981) Multi-ring basin formation: Possible clues from impact cratering calculations. In *Multi-ring Basins, Proc. Lunar Planet. Sci.*, 181-195.
60) Sedov L. I. (1959) *Similarity and Dimensional Methods in Mechanics*. 4th ed., p. 256, Academic, New York.
61) Shoemaker E.M (1981). Personal communication.
62) Viktorov V. V. and Stepenov R. D. (1960) Modeling of the action of an explosion with concentrated charge in homogeneous ground. *Inzh. Sb.* **28**, 87-96. (English translation, Sandia Report SCL-T-392, Albq., NM, 1969.)
63) Worthington A.M. and Cole R.S. (1897) Impact with a liquid surface, studied by the aid of instantaneous photography. *Philos. Trans. R. Soc. London Ser. A*, 193.

Int. J. Impact Engng Vol. 5, pp. 561–568, 1987
Printed in Great Britain

0734-743X/87 $3.00 + 0.00
Pergamon Journals Ltd.

MICROMETEORITE IMPACT SIMULATION AT EMI - A REVIEW

E. Schneider, A. Stilp

Fraunhofer-Institut für Kurzzeitdynamik
- Ernst-Mach-Institut - EMI
Eckerstr. 4, D-7800 Freiburg, FRG

ABSTRACT

Activities at EMI in the field of hypervelocity impact techniques are reported. Optimization experiments have been carried out with a light gas gun in order to achieve projectile velocities up to 10 km/s. Different methods for measuring the projectile velocities have been developed and adapted according to respective velocity and mass ranges of projectiles. Experimental efforts have been undertaken to accelerate also microgram particles in light gas guns. Masses as small as 37 μg can be accelerated as individual particles. As examples, several contributions to recent space projects are described.

INTRODUCTION

Experimental meteorite simulation has been studied since the beginning of the space age in order to gain insight into nature, origin and properties of meteorites, as well as into physical processes occuring under meteorite impact on natural and artifical space objects. Meteorite simulation programs at EMI have been performed with respect to two aspects:

1) Protection of space vehicles against micrometeorite and space debris impact

2) Detection and analysis of interplanetary particles.

In both fields of interest impact experiments are needed. High standards of acceleration technology are obligatory in order to simulate space conditions as closely as possible.

Several techniques exist to accelerate particles up to high velocities. Essentially, three of them are applied for impact simulation experiments (Fechtig and co-workers, 1972):

a) Electrostatic dust accelerators
   (particle mass range 10^{-15} g - 10^{-9} g; velocities up to 60 km/s)
b) Electromagnetic plasma guns
   (particle mass range 10^{-8} g - 10^{-4} g; velocities up to 20 km/s)
c) Light gas guns
   (particle mass range 10^{-3} g - 10 g; velocities up to 12 km/s)

At EMI, three two-stage light gas guns are operated to perform hypervelocity experiments. Recent improvements of equipment and technical developments as well as applications in space projects will be described in this paper.

EXPERIMENTAL TECHNIQUES

Upgrading of a Test Light Gas Gun

Optimization experiments have been carried out on a light gas gun in order to obtain projectile velocities sufficiently high for meteorite simulation. System components have been constructed and tested which allow a broad variation of gun operation conditions. Fig. 1 shows a schematic of the gun. The operation principle of light gas guns has been described in several papers (e. g. Crozier and Hume, 1957; Fechtig and co-workers, 1972). The most important geometrical parameters are the pump tube volume, and the ratio of pump tube diameter to launch tube diameter, and the length to caliber ratio of the launch tube. For finding an optimal gun geometry, three types of pump tubes (length 2 m, diameters 20, 30 and 40 mm), and two launch tubes (cal. 8 mm, lengths 1200 and 1000 mm) have been tested. Special emphasis was given to the inner surfaces of the launch tubes. They have been honed until a smoothness of < 1 μm was reached, a procedure which is most essential for obtaining maximum impact velocities. Concerning the operational conditions of the gun, pump tube pressure, mass of the compression piston, and powder mass have been varied for each geometrical test configuration. Helium and, in the later phase of the tests, hydrogen have been used as driver gases.

Fig. 2 shows a diagram of the projectile velocities as function of the pump tube pressure for the two types of launch tubes tested. In this case the projectiles were Lexan-cylinders with a mass of 520 mg. The piston weight was 123 g and 60 g of gun powder was used. Over the range tested projectile velocities increase with decreasing pump tube pressures. However, at pump tube pressures of 2.5 bar and smaller,pressure peaks occur which cause intolerable material flow effects within the high pressure section and the launch tube. Thus, material load limits are one of the most severe constraints in velocity optimization of a light gas gun.

Taking into account the experiences obtained from gun construction, and the results of parametric variations of the gun's operating conditions, the following dimensions and operation parameters have been found optimal for our 8.4 mm bore gun:

Pump tube	Diameter 40 mm Length 2000 mm
Launch tube	Diameter 8.4 mm Length 1200 mm
Propellant gas	Hydrogen (Helium)
Piston	Material: Polyethylene Weight: 130 g (with shear disc)
Diaphragm	High ductile stainless steel, with four-petal opening configuration
High pressure section	Sealing: Steel delta rings Coupling: Hydraulic clamb Cone angle: 5^0
Powder chamber	Volume: 2000 cm^3 Powder mass: max. 150 g

Fig. 3 shows the maximum velocities which are obtained for respective masses accelerated. At the moment, the upper limit of velocities attainable with this gun is 9.5 km/s.

Velocity Measurement Techniques

Usually, projectile velocities are measured by means of laser light barriers. During projectile passage, the laser beam is interrupted and the photodiode signal produced is used to trigger an electronic counter and a very short duration high voltage spark. The spark is used to take an inflight shadowgraph of the projectile by means of a Polaroid camera. Thus, the exact position of the projectile at two light barriers and the time of flight can be used for a velocity determination at an accuracy of better than 1 %.

This method cannot be applied if a) the projectiles are too fast (> 5 km/s) and/or b) the projectiles are too small. Very fast projectiles produce a bright light trace due to excitation of residual gas molecules. In this case light filters adapted to the spark spectrum in connection with image converter cameras are used to obtain inflight shadowgraphs.

In case of very small projectiles (< 1 mm), the portion of the laser beam covered by the projectile may be so small that the light intensity variation ranges below the sensitivity level of the system. To overcome this difficulty, one can use the muzzle flash and the impact flash as trigger signals, and thus get a velocity measurement. Unfortunately, no inflight shadowgraphs of the projectile are obtained. The muzzle flash, occuring when the projectile leaves the launch tube, delivers a well-defined, reproducible signal (Fig. 4). The signal rise time is in the order of 2 μs. At a free distance of about 2 m between launch tube and target, the muzzle flash signal is sufficiently short to be used as a start trigger.

Another method to measure the velocities of small and fast projectiles has been developed on the basis of Radar techniques. A 35 GHz-Microwave Doppler Radar system has been designed and tested in a collaboration with the Technical University Munich (Kaufmann and co-workers, 1983). A microwave slit antenna is placed along the projectile trajectory emitting a microwave signal under a defined angle with respect to the projectile trajectory. A Doppler-shifted signal, reflected at the surface of the projectile is detected, which can be used to determine the projectile velocity with an accuracy of about 1 %. This method can be applied for metallic as well as non-metallic projectiles with diameters as small as 0.4 mm.

Acceleration of Microgram-Particles

Particles in the mass range of micrograms are usually accelerated by means of electromagnetic plasma accelerators. However, because of the high plasma temperatures which are necessary for the acceleration, projectile materials which can be used are restricted to non-metals. Therefore, mainly glass particles are used for acceleration in plasma guns. In order to accelerate metallic particles of this size range, light gas guns must be used.

The following processes can cause problems in launching such small projectiles from light gas guns:

- embedding of particles in sabots
- projectile-sabot separation
- free flight stability
- inflight visualization and velocity measurements.

Considerable effort has been undertaken to overcome these difficulties, and to determine what kinds of particles can be launched, either individually or as particle ensembles.

Acceleration of individual particles. For light gas gun acceleration, the projectile is usually embedded in a plastic sabot. After leaving the launch tube, projectile and sabot parts are separated by aerodynamic forces, with the sabot parts being deflected off of the projectile trajectory. In case of small projectiles (< 1 mm), sabot separation is the most critical process. The projectiles can stick to the sabot due to adhesive forces, or they are deflected off the shot axis if the sabot separation is slightly asymmetrical. Sticking can be avoided by applying a thin layer of µm-sized SiO_2 powder. To avoid asymmetries, the hole for the projectile must be carefully centered within the polycarbonate sabot. Figs. 5 a and b show a 0.2 mm Cu-sphere embedded in a sabot. The sabot is slit twice, i. e. it separates into 4 parts after leaving the launch tube. The photos demonstrate the extreme size of the sabot relative to the projectile.

In the following, minimum diameters (masses) of projectiles which have been launched individually are given (all spheres):

Al: 0.5 mm (177 µg)
Glass: 0.5 mm (157 µg)
Steel: 0.35 mm (126 µg)
Cu: 0.2 mm (37 µg)

Maximum velocities in the order of 7 km/s have been achieved. The velocity measurement methods for small particles have already been described in chapter "Velocity Measurement Techniques".

Acceleration of particle ensembles. Particles smaller than those mentioned in the previous section can be accelerated as particle ensembles. In this case, a large number of particles (50 ... 100) are embedded into a sabot hollow.

After sabot opening, a particle cloud develops, impacting the target randomly. Figs. 6 a and b show such a particle ensemble of glass projectiles with diameters between 80 µm and 120 µm within a polycarbonate sabot. In order to avoid dropping out of particles during insertion into the launch tube, the particles are slightly pressed into the sabot, along with Al_2O_3 powder which acts as a weak bonding agent.

Since a particle cloud is produced, impact velocities show some scattering. However, for equally sized particles it can be estimated from crater diameters that the velocities range within about 25 % of a maximum velocity. The respective maximum velocity can be determined in test shots with a larger "tracer particle" added to the sabot filling. This "tracer particle" delivers a sufficiently bright flash signal. Fig. 7 shows an Al test target impacted by a glass projectile cloud. The impact craters are randomly distributed. Impact velocities range between 3.0 km/s and 3.6 km/s.

In some applications, the acceleration of irregularly shaped microparticles (e. g. crystalline fragments) is required. Since such projectiles are usually brittle and friable, they must be embedded in a coherent medium for acceleration. Wax was found to be a suitable embedding material. After sabot separation, the wax evaporates and is removed from the projectiles under residual gas drag forces. Fig. 8 shows this process in an in-flight shadowgraph taken shortly after sabot separation. The sabot parts are still within the field of view.

EXPERIMENTAL SIMULATION PROGRAMS

Lunar and Planetary Impact Studies

It has been confirmed by numerous space missions that impact cratering is the most prevalent process on many planetary surfaces. During the Apollo program, large quantities of lunar soil and rock samples were brought to earth. Thus, detailed laboratory investigations became possible. Among others, large numbers of micrometeorite impact craters were found on the surface of these moon rocks. In order to interprete these craters and to draw conclusions on the interplanetary particles which had impacted, extensive experimental simulation programs were performed. Many kinds of metal, glass and rock targets have been bombarded by metal and silicate hypervelocity projectiles.

Relations between static and dynamic projectile parameters (e. g. size shape, density, velocity) and crater parameters (depth, diameter, morphological features etc.) have been established. Respective results are published in various publications (e. g. Hörz and co-workers, 1975, Schneider and co-workers, 1973, Fechtig and co-workers, 1972). As an example, Figs. 9 a and b show a laboratory produced crater in an Al target in comparison with a crater in a piece of iron-meteorite from the lunar surface (Apollo 16 mission).

The experimental results and empirical relations which have been derived, form a good data base for the interpretation of lunar microcraters and the calculation of interplanetary dust fluxes (Hörz and co-workers, 1975).

LDEF Simulation

The NASA project LDEF (Clark, 1981) (long duration exposure facility) consists of a drum-like structure (Fig. 10), placed in orbit by the space shuttle. Parts of the surface of this instrument serve as passive impact detectors, where interplanetary particles leave impact craters. After recovery (during a future shuttle mission), these craters will be investigated in the laboratory. From their morphology, conclusions on mass, shape, density and velocity of the interplanetary particles can be drawn. The crater interior will be studied by means of X-ray analytical methods in an effort to detect residues of projectile material. In a collaboration with the NASA-Johnson Space Center, Houston, Texas, calibration experiments have been carried out. Various high purity metal targets have been impacted with particles of a defined mineralogical composition in an impact velocity range between 3 km/s and 8 km/s. The analytical investigation of these laboratory-produced craters is performed at the Johnson Space Center, to answer two questions:

a) What is the quantity of projectile material remaining in a crater as a function of impact velocity?

b) What is the extent of chemical fractionation and the final disposition of volatile projectile constituents?

The answer to these questions is of fundamental importance for the interpretation of LDEF impact craters.

Tests and Optimization of the Giotto Meteoroid Bumper System

In a collaboration with ESTEC (European Space Research and Technology Centre), impact experiments were performed to assist in the design of an effective shield configuration for protecting the Giotto spaceprobe against cometary particles travelling at a relative velocity of 68 km/s. Testing of a variety of shield materials and configurations lead to the selection of a system which consists of a front plate of aluminum, 1 mm thick, and a rear plate of Kevlar in a sandwich arrangement, 13.5 mm thick, with a spacing of 25 cm between the front and rear plates (Lainé and Felici, 1982). Thus, the principle of "meteoroid bumper" has been applied.

A particle perforates the front plate, and is thereby highly fragmented. The momentum of the expanding fragment/vapor is transferred to the rear plate, with the load spread over a large area. Fig. 11 shows this mechanism in a series of flash X-ray photos. In the case of Giotto, complete vaporization was to be expected due to the very high velocity of the cometary particles.

Micrometeorite Impact on Solar Panels

Short currents have been detected in the solar cell panels of several ESA-telecommunication satellites (e. g. Marecs A). Micrometeorite and space debris particle impact have been considered to be one of the possible causes. Therefore, the European Space Research and Technology Centre (ESTEC) initiated an experiment program to simulate micrometeorite impacts to study damage phenomena and failure mechanisms on solar panels.

In a first series of tests, hypervelocity projectiles were shot into solar cell samples at impact velocities of up to 6 km/s. Projectile masses ranged between 0.04 mg and 0.5 mg.

Transient short currents occuring during the solar cell perforation process were expected, since a hot plasma cloud is created by hypervelocity projectile impact. Permanent short circuits can occur if target material fragments form mechanical contacts between the system electrical ground and the cell material, and by vapor deposits of conductive materials evaporated during the impact process.

Both types, transient as well as permanent short circuits, have been observed within the first phase of investigations. Transient short circuits were detected by means of a special detector circuit working on the basis of a triggered capacitor discharge. Under realistic voltages between the cells and system ground level, as applied in orbit, impact induced arc discharges were observed. Maximum arc burning durations were in excess of 1 minute. The damage phenomena which occurred depend on the impact velocity and on the geometrical arrangement of the solar cell panels. Figs. 12 a and b show examples of craters in solar cell samples. From the results obtained to date, it can be concluded that failures caused by impact induced short circuits can be minimized by improving the geometrical arrangement of the panels and by optimizing the electrical circuitry.

DISCUSSION

It has been shown that, by means of high performance light gas guns, realistic meteorite simulation experiments can be performed. Many applications exist in the field of micrometeorite studies which deal with the nature and dynamic behavior of these interplanetary particles. Applications for the protection of space vehicles are becoming increasingly important since the population of space debris is continuously increasing.

It would be desirable to increase light gas gun velocities beyond 10 km/s. However, considerable efforts and expenses are required to achieve velocities beyond this limit. To some extent, the experimental results can be extrapolated for higher impact velocities (Schneider, 1979). In combination with numerical codes a good characterization of micrometeorite impact effects was achieved.

ACKNOWLEDGMENT

The authors would like to thank the Deutsche Forschungsgemeinschaft for financial support during the light gas gun upgrading studies.

REFERENCES

Crozier, W. D., and W. Hume (1957). High velocity light gas gun. J. Appl. Phys., 28,

Fechtig, H., D. E. Gault, G. Neukum, and E. Schneider (1972). Laboratory Simulation of lunar craters. Naturwissenschaften, 59, 151 - 157.

Hörz, F., D. E. Brownlee, H. Fechtig, J. B. Hartung, D. A. Morrison, G. Neukum, E. Schneider, J. F. Vedder, and D. E. Gault (1975). Lunar microcraters: Implications for the micrometeoroid complex. Planet. Space Sci. 23, 151 - 172.

Kaufmann, K., E. Schneider, I. Detlefsen, and E. Igenbergs (1983). Mikrowellen - Dopplerradar zur Messung schneller Teilchen. Nachrichtentechnische Zeitschrift, Band 36, Heft 9, 578.

Lainé, R., and F. Felici (1982). The Giotto Dust Protection System. ESA Bulletin, 32, 36 - 42

Schneider, E., D. Storzer, J. B. Hartung, H. Fechtig, and W. Gentner (1973). Microcraters on Apollo 15 and 16 samples and corresponding cosmic dust fluxes. Proceedings of the Forth Lunar Science Conference, Supplement 4, Geochimica et Cosmochimica Acta, Vol. 3, 3277 - 3290.

Schneider, E. (1979). Velocity Dependencies of some Impact Phenomena. ESA SP-153. 101 - 107

Clark, G. (1981). LDEF Mission I Experiment Description. Compiled and Edited by NASA Langley Res. Center.

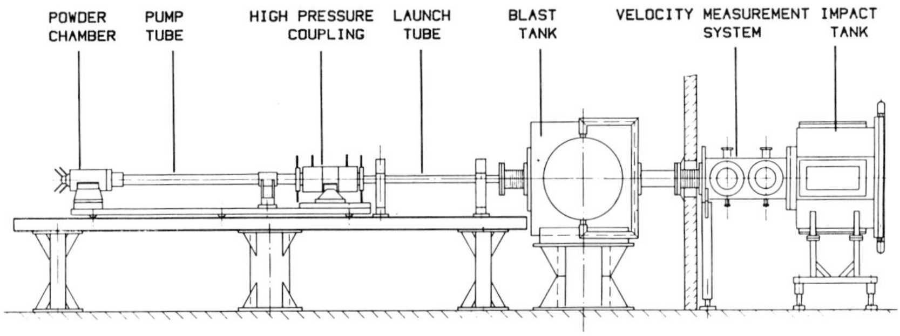

Fig. 1. Schematic of the light gas gun.

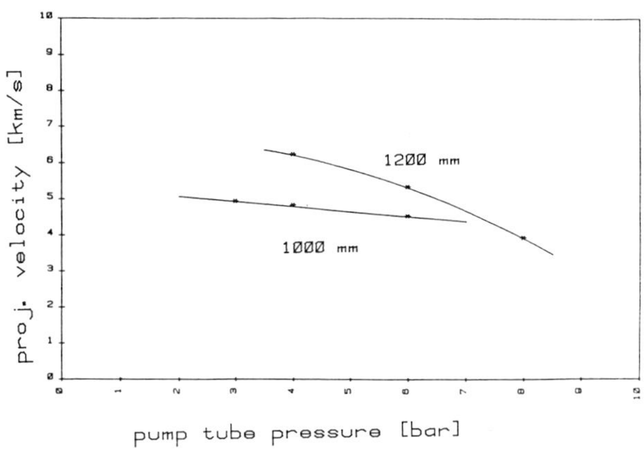

Fig. 2. Projectile velocity as a function of pump tube pressure.

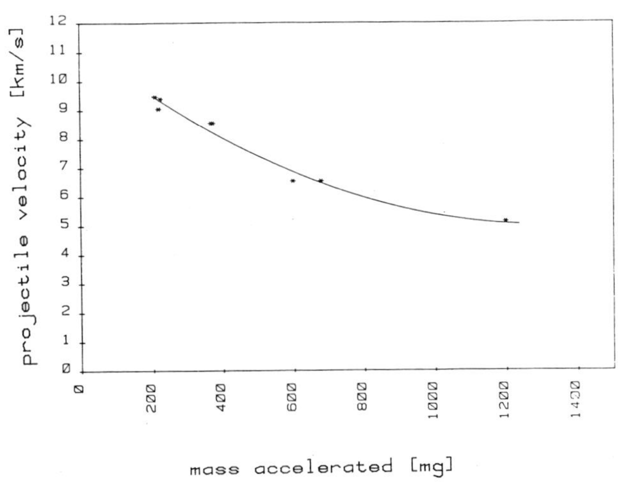

Fig. 3. Maximum velocities achieved for different projectile masses.

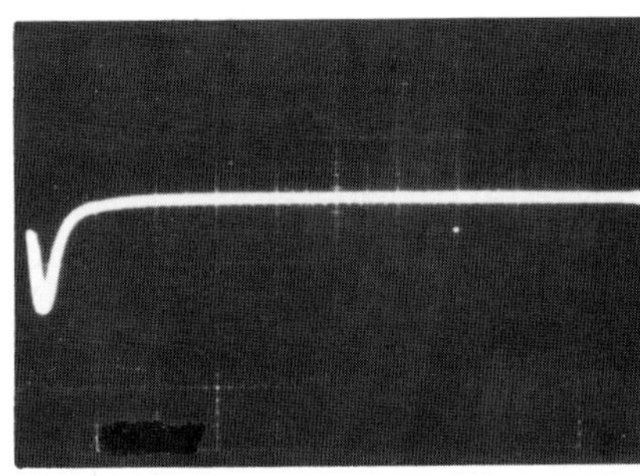

Fig. 4. Intensity signal of the muzzle flash (5 μs/div., 500 mV/div.)

a

b

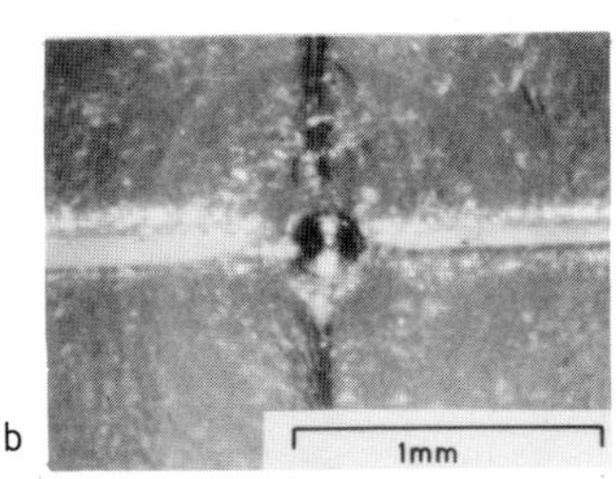

Figs. 5. Copper sphere (diameter 0.2 mm) embedded in Lexan sabot.

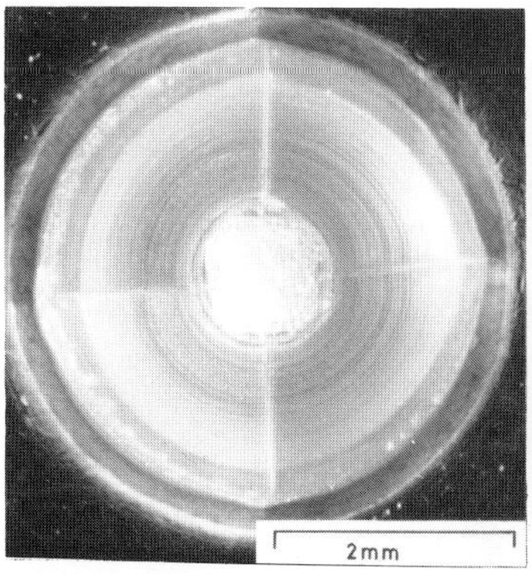

a

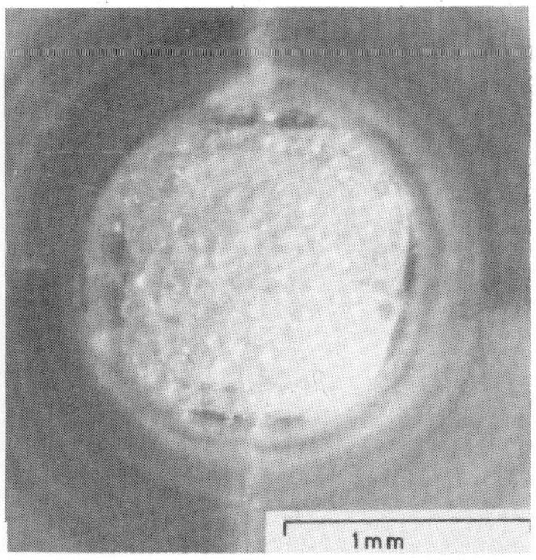

b

Figs. 6. Particle ensemble of glass projectiles (diameter 80 μm ... 120 μm) embedded in Lexan sabot.

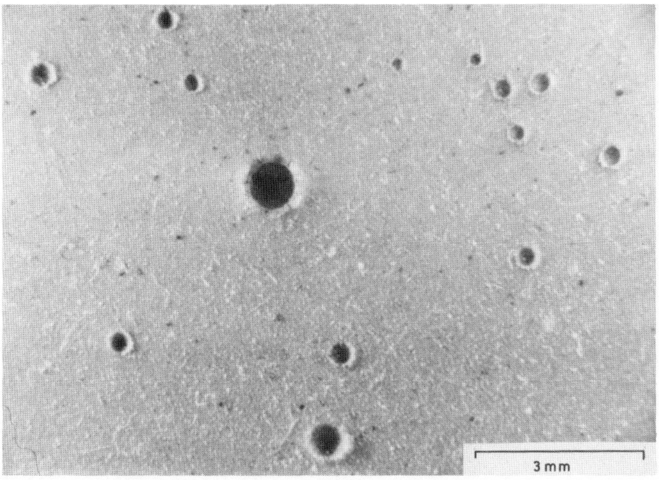

Fig. 7. Al-test target impacted by a cloud of glass particles (particle diameter: 150 μm ... 300 μm). The larger crater in the center is made by a 0.9 mm tracer projectile for velocity determination.

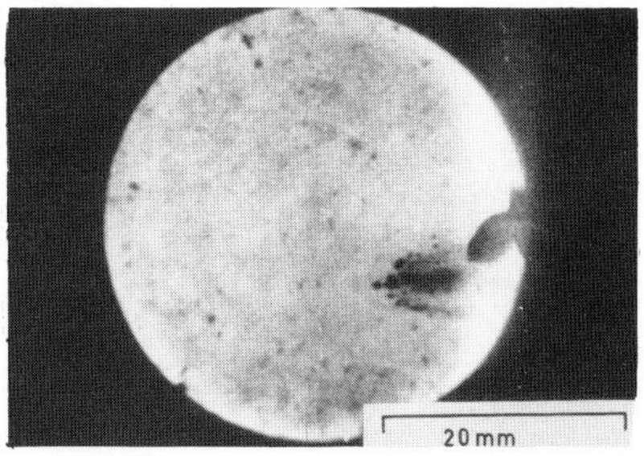

Fig. 8. Inflight shadowgraph of a particle ensemble embedded in wax.

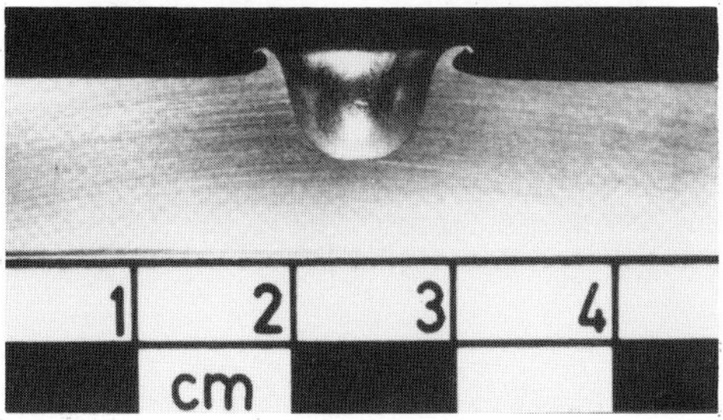

Fig. 9 a. Labbratory produced hypervelocity crater in an Al-target. (Projectile: Al-sphere; diameter 2 mm; velocity 4.85 km/s)

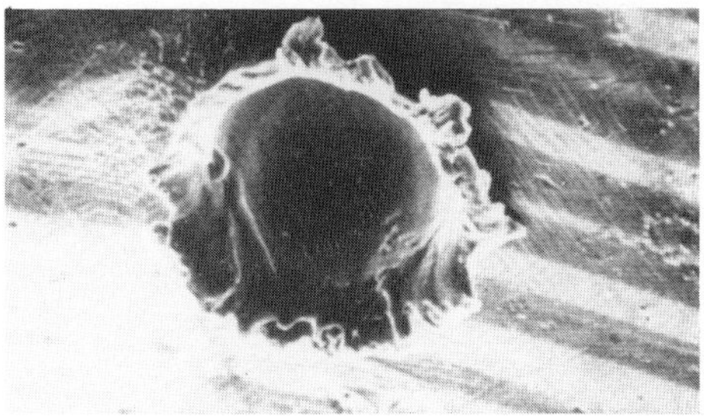

Fig. 9 b. Crater on a piece of iron meteorite (crater diameter $\sim$ 50 μm)

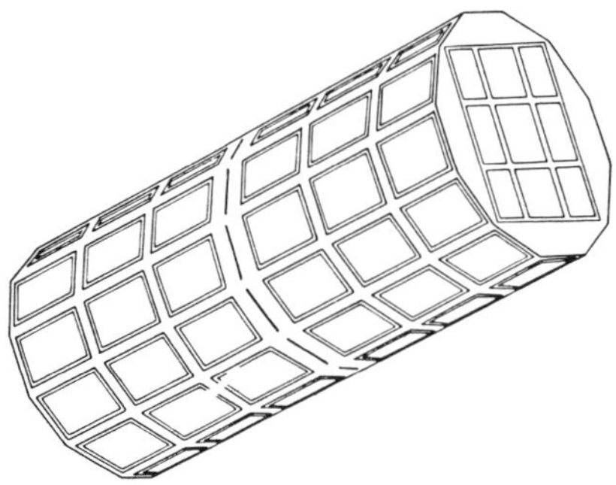

Fig. 10. Schematic of the LDEF probe.

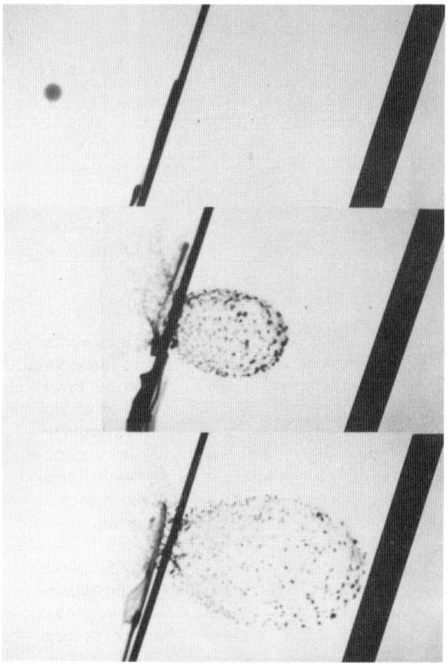

Fig. 11. Flash X-ray series of a particle perforating a brass plate of 1 mm thickness. The particle was a 6 mm diameter Al-sphere with an impact velocity of 5 km/s (exposure time: 30 ns).

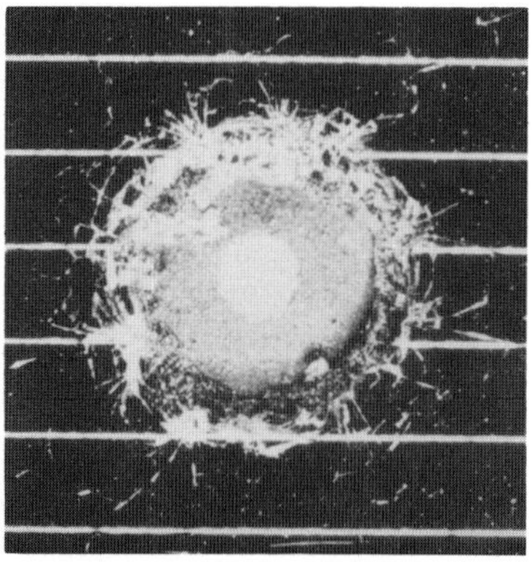

Fig. 12 a. Impact perforation hole of a solar cell made by a 0.5 mm diameter glass sphere at 5.3 km/s impact velocity. (line spacing: 1 mm)

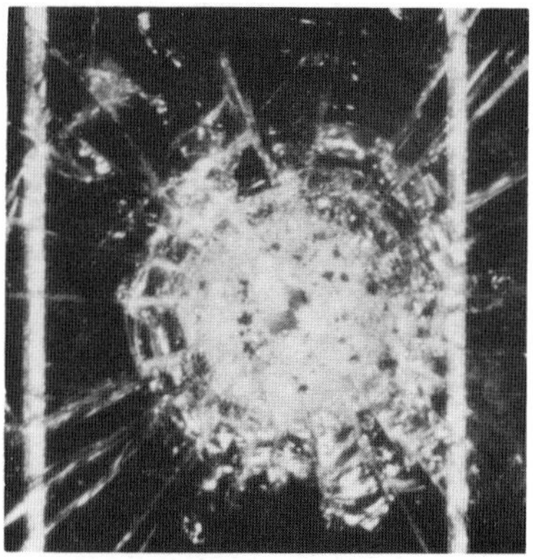

Fig. 12 b. Impact crater on a solar cell panel made by 0.35 mm diameter steel sphere at 400 m/s impact velocity. (line spacing: 1 mm)

Int. J. Impact Engng Vol. 5, pp. 569–576, 1987
Printed in Great Britain

0734-743X/87 $3.00 + 0.00
Pergamon Journals Ltd.

EXPERIMENTAL PLANETARY IMPACT RESEARCH

P. H. Schultz
Dept. of Geological Sciences, Brown Univ.,
Providence, RI

ABSTRACT

Laboratory experiments provided an invaluable tool for simulating cratering processes on planets subject to a wide range of impactor, target, and environmental conditions.

INTRODUCTION

The largest laboratories for hypervelocity impacts are not in California, New Mexico or Texas. Planetary missions have documented a wide range of environments (near-vacuum to 100 atmospheres), targets (lavas to ice), and impactor conditions (5 km/s-85 km/s) where the results of impact cratering have been preserved for billions of years. Planets are not the ideal laboratories: we only see the aftermath and except for the Moon have yet to systematically analyze pre-impact and post-impact conditions. Yet, planetary surfaces provide a unique record of events unthinkable in scale. Although understanding 20 km/s collisions by 100 km-diameter impactors (equivalent to 10^{11} megatons) or impact under a 100 bar atmospheric environment may seem unconstrained, such extremes permit fundamental tests for understanding the impact process at smaller scales. The purpose of this invited review is to illustrate ongoing laboratory experiments that are designed to isolate, if not simulate, aspects of planetary cratering.

Under vacuum environments, the martian satellites (Phobos, Deimos), the Moon, and Mercury provide similar well-preserved target surfaces with contrasting gravitational accelerations (g) of 1 cm/s^2, 162 cm/s^2, and 370 cm/s^2 respectfully. The respective rms impact velocities for the most probable impactor type are 19 km/s (Mars-crossing asteroids), 22 km/s (Earth-crossing asteroids), and 87 km/s (parabolic comets) that reflect the size of planetary body (mass), the most likely type of impact (asteroids or comets), and the heliocentric gravitational setting as summarized in (Hartmann and others, 1981). Laboratory experiments ideally could isolate the effects of impactor size, density, and g and then can be calibrated with observed planetary impact craters. Limitations in the range of laboratory conditions and incomplete or mixed planetary observational data (e.g., variable resolution, incomplete coverage, absence of topographic data), however, leave many definitive tests for the future.

Under non-vacuum conditions, Mars, the Earth, and Venus provide environments with atmospheric conditions of 6 mbars, 1 bar, and 90 bars, respectively and corresponding rms impact velocities of 19 km/s, (Mars crossing asteroids), 25 km/s (Earth-crossing asteroids), and 62 km/s (parabolic comets) (Hartmann and others, 1981). Additional complications include surface temperature (230 K, 290 K, 740 K), and substrate variables (possible volatiles, volatile-free, possible volatiles). Moreover, the crater retention age for craters smaller than 20 km is estimated to be less than about 300 my for the Earth (Grieve, 1984), about one billion years for Venus (Ivanov and others, 1986), and 4 billion years for Mars (Hartmann and others, 1981). Laboratory experiments designed to address first-order effects of impact velocity, volatile-rich targets, and substrate viscosity help to address the significance of these various environmental parameters. Fortunately, the satellites of the outer planets present very different conditions including volatile-rich surfaces over a variety of inferred temperatures, from well below to near the solidus.

Observed trends in planetary impact crater data establish a fundamental data base that may not directly reveal the effects of impactor/target variables but help constrain possibilities. Similarly, laboratory and computer experiments cannot directly simulate all planetary impact phenomena but allow controlled separation of variables and extrapolations by complementary theoretical considerations.

Most hypervelocity experiments for planetary studies are performed at the NASA-Ames Vertical Gun Range (AVGR), a national facility funded by the Planetary Geosciences Programs of NASA and operated through a joint agreement between NASA-Ames Research Center, California, and the Lunar and Planetary Institute, Houston. This unique facility permits impacts at various angles (in 15^0 increments) from vertical to horizontal; consequently, experiments can be designed to simulate crater growth in low-strength targets. Other horizontal facilities capable of hypervelocities are used for research concerned with equations of state (e.g., Ahrens and others, 1977; Hörz, 1970). Because many problems in planetary studies are concerned with late-time phenomena, considerable work also has been done at lower velocities at the AVGR as well as other facilities: NASA-Johnson Space Center with a vertical powder-gun assembly (Cintala and others, 1985) and a low-gravity airborne facility (Cintala and others, 1986); and Boeing-Seattle with a unique high-gravity centrifuge (see Schmidt, 1980). It is not possible to adequately review all on-going experimental programs in the limited space available. Consequently, the following review emphasizes my personal research in two broad aspects: scaling of crater dimensions and impact phenomena relevant to planetary cratering. My apologies to other active planetary researchers and the reader is referred to the Proceedings and abstract volumes of the Lunar and Planetary Science Conference (Houston) as sources for additional information.

CRATER SCALING

On a given planet under near-vacuum conditions, the post-impact morphology dramatically changes with size (Fig. 1). These changes are identified in the shape/morphology of the crater interior and the emplacement of ejecta (e.g., see Schultz, 1976;

Pike 1980). For the Moon, craters smaller than about 0.8 km in diameter contain rubbled interiors and possible expressions of strength discontinuities. Ejecta rubble surrounding the crater is dispersed and transparent to most pre-existing features except very near the rim. Craters between about 1 km and 15 km contain flat floors typically covered with impact melt and sloping walls of talus without terraces. Ejecta deposits have a distinctive dune-like appearance with blocks generally smaller than about 1 m. Between 15 km and about 120 km, the flat, melt-covered floor includes an increasing fraction of the crater interior surrounded by extensive slump terraces. Central peaks protrude above the flat floor with a variety of expressions: large central cones, low-relief mounds, annulus of knobs, and central pits. Ejecta deposits are complex with secondary craters having maximum sizes about 5% of the primary crater and impact melt draped or cascaded down the wall and rim to form ponds and flows. Craters larger than about 120 km commonly have relatively shallow floors with a broken annulus of 0.1-1 km high knobs. The wall is a broad annulus of terraces, and ejecta deposits commonly contain evidence for considerable lateral basal flow. The largest lunar craters have multiple rings of mountains in the interior with a prominent scarp-ring marking the transition to a complex sequence of ejecta deposits.

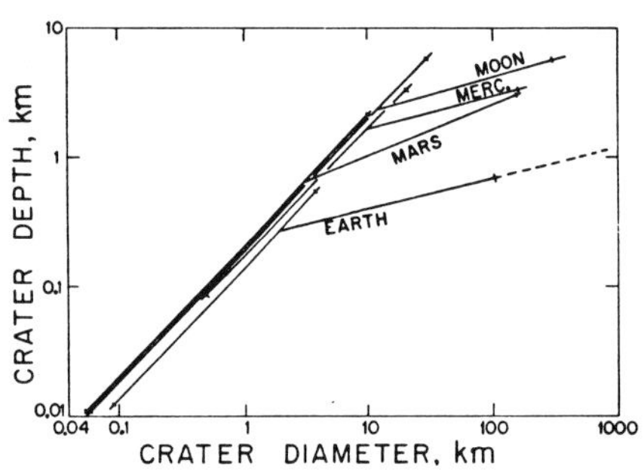

Fig. 1. Change in crater shape as a function of crater size on different planets (after Pike, 1980).

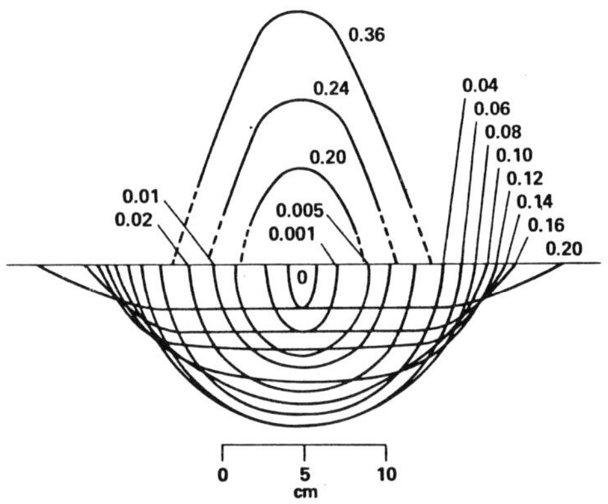

Fig. 2. Evolution of crater shape for impact into water (from Gault and Sonett, 1982). Crater reaches a maximum depth and begins to shallow prior to collapse. Such experiments help to understand and end-member of late-stage growth/collapse in a strengthless medium.

In terms of the cratering process, lunar craters demonstrate that with increasing size crater profiles become shallow, central peaks and rings develop, brittle wall failure increases in importance, impact melt becomes more evident on the floor and rim, and ejecta deposits are composed of blocks that may be larger in absolute dimensions but typically are smaller relative to the size of the crater. Are such changes reflecting fundamental changes in crater excavation processes, are these changes indicating the dynamic response of the crust to the crater cavity, or is it a combination of these processes?

Low-velocity and hypervelocity impacts into water and low-viscosity targets have documented crater growth and collapse (Gault and Sonett, 1982). Low-velocity impacts (< 1 km/s) provide a useful record of the dynamic response of a strengthless medium to a transient cavity but caution should be exercised in all applications because such velocities are below the sound-speed velocity in both target and projectile. Results for impacts into water (Gault and Sonett, 1982) indicate that cavity collapse begins after reaching maximum crater depth but prior to achieving maximum diameter (Fig. 2). Following crater collapse a large central mound develops. Studies of impacts into viscous targets indicate that the height of the mound depends directly on impact energy and inversely on target viscosity and strength (Greeley and others, 1980). Collapse of the mound produces an oscillating structure.

Impacts into water and other viscous target are not necessarily designed to simulate planetary surfaces but the effective response of solid substrates to impact-induced shock waves. Theoretical studies (Melosh, 1977) suggest that even large craters with diameter-depth ratios of 5:1 are stable against collapse under static conditions. The seismic (or acoustic) energy retained in the post-impact environment, however, may serve to fluidize the target, thereby providing an analogy with experiments using low-viscosity targets (Fink and others, 1981). As the acoustic energy dissipates, material strength falls below the yield strength and the central peak or multi-ringed crater morphology is established. The inferred collapse of a bowl-shaped transient cavity is consistent with studies of terrestrial impact structures and possibly the onset of certain crater morphologies (central peaks, rings, etc.) with size on different planets with different gravitational accelerations. There is still debate, however, whether crater collapse triggers the development of central peaks and wall failure or floor rebound (central peak formation) induces crater collapse (Pike, 1983).

An implicit assumption in such models is that all impacts create a deep bowl-shaped transient cavity analogous to impact into low-strength but statically stable targets. This assumption of "proportional crater growth" has two elements: proportional growth throughout time in a given crater and proportional growth in different craters but at the same scaled stage in growth. The first assumption is probably incorrect. Analyses of impacts into water (Gault and Sonett, 1982) as well as theoretical codes (Orphal and others, 1980) demonstrate that strictly proportional growth occurs only during the intermediate stages, i.e., after transfer of momentum/energy from the projectile to the target and before the onset of collapse or material strength effects. After maximum crater depth is achieved, the crater continues to grow laterally, thereby resulting in a gradual increase in the diameter/depth ratio. If gravity arrests the limit of growth before collapse without altering the flow field, then the diameter/depth ratio of the transient cavity will be less. In other words, with increasing crater size or gravity the transient cavity becomes tumbler-shaped (Schultz and others, 1981).

The second implicit assumption requires that impactor/target conditions do not alter the cratering flow field. This assumption has been frequently challenged on the basis of analogies with near-surface explosions (Roddy, 1977), certain terrestrial impact craters (Roddy, 1968), and preliminary experiments under high g (Schmidt and Holsapple, 1981). Additional experiments by Schmidt (1984), however, subsequently revealed little effect of the gravity term on crater shape except during the collapse stages. Recent experiments with low-density impactors have documented changes in the cratering flow field in granular targets at hypervelocities (Schultz and Gault, 1985a) and provide an analogy with static near-surface explosions. However, the extremely low densities required probably have only limited applications to the planets: atmospheric break-up prior to impact (Melosh, 1981) or very small (< 0.1 mm) and fragile cosmic dust. Although comets may have an effective low density, the nucleus has a density near 1 g/cm^3. At hypervelocities factors of three to five in density have little effect.

Fig. 3. Change in crater aspect ratio (diameter/depth) as a function of projectile diameter (2r) and velocity (v) for impacts into No. 24 sand. Below or above a critical value of 2r/v craters become shallower. Trends suggest that large-body (km) impacts may produce transient cavities shallower than expected for small-body (0.3 cm) impacts into strengthless targets.

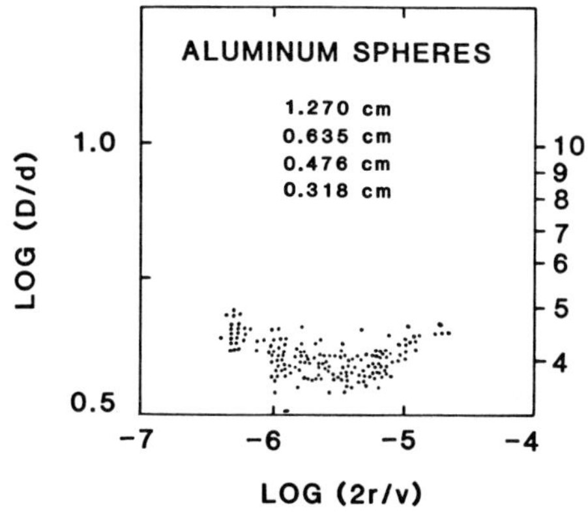

Ongoing experiments at the NASA-Ames Vertical Gun are revealing new evidence for fundamental changes in the cratering flow field as a function of impactor size and velocity. Earlier studies (Gault and Wedekind, 1977) revealed two different scaling relations perhaps depending on projectile deformation. More recent analysis (Schultz and Gault, 1985b; 1986a) reveals different regimes of dependencies that depend on a critical value of 2r/v, where 2r is diameter and v is the velocity of the projectile. For values of 2r/v < 5x10^{-5} sec, crater aspect ratio (diameter/depth, D/d) increases with increasing impactor velocity and decreasing size following the dimensionless π_2 parameter (Holsapple and Schmidt, 1982). This domain can be viewed as equivalent to trends resulting from static point-source explosions because the time for momentum/energy transfer is essentially instantaneous. Above the critical limit, D/d increases directly as impactor size and inversely with velocity (Fig. 3); moreover, there is a change in the cratering efficiency (Fig. 4). Impactor density appears to play a role only if the projectile does not deform and the velocity is not high enough to exceed the crushing strength of the granular target. These

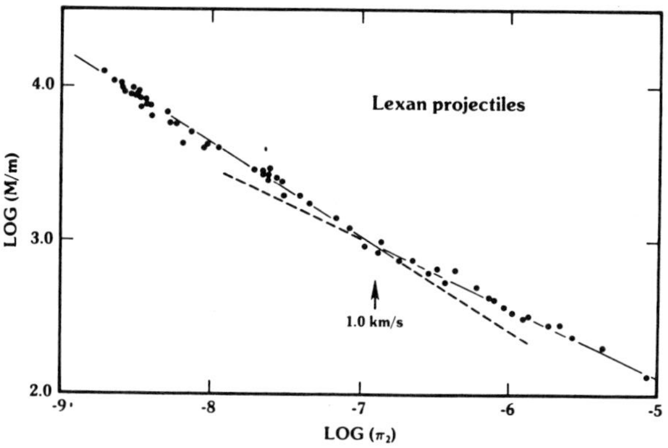

Fig. 4. Change in cratering efficiency as a function of the gravity-scaled parameter, π_2 = 3.22 gr/v^2 where g = gravitational acceleration, r = projectile radius, v = impact velocity. The change in slope occurs near a critical value of 2 r/v which also depends on the strength of the projectile.

conditions are always met for hypervelocity impacts and dictate caution in extrapolating scaling relations from low velocity to hypervelocity impacts if crushing and deformation has not occurred. The dependence of D/d on 2r/v suggests that the time for projectile penetration is important. Physically penetration time may indicate that the flow field changes with shock duration. It is important to note that in this regime, scaled crater depth (d/r) relations closely matches relations predicted by computer codes for iron impacts into anorthosites ($d/r \sim v^{.4}$); however, the diameter, which is affected by an altered lateral flow field, takes on a completely different relation ($D/r \sim v^{.12}$).

Most existing experimental facilities cannot launch large solid objects (> 1 m) at hypervelocities exceeding the critical limit for 2r/v. However, larger low-density debris clouds have been produced at hypervelocities; results for these data are consistent with the observed trends for lower velocity impacts by large solid projectiles undergoing complete deformation at impact (Schultz and Gault, 1985a).

At planetary scales, a 10 km-diameter object impacting at 15 km/s requires about 0.8 seconds for penetration. By contrast, a 0.635 cm object in the laboratory impacting at 6 km/s requires 6 orders of magnitude less time. The new laboratory findings predict that the transient cavity shape will be fundamentally different and shallower than normally envisioned. This change in flow field again may resemble the shallow depth-of-burst explosions but now does not require unusual impactor configurations.

IMPACT SIMULATION AND PROCESSES

Rather than surveying all experimental programs in planetary hypervelocity phenomenology in detail, the following account summarizes three areas in order to illustrate the rationale and approach: impact vaporization, atmospheric effects, and clustered impacts.

Impact Vaporization: Existing launch facilities generally restrict velocities to below 8 km/s for masses between 0.1 and 0.3 g. For vertical impacts into most materials, such velocities result in minimal vaporization. Most impacts on planetary surfaces, however, exceed 7 km/s and typical velocities range between 15 km/s and 50 km/s. Impact-generated vaporization may play an important role in crater scaling relations, emplacement of ejecta, launching material from the surfaces of other planets,

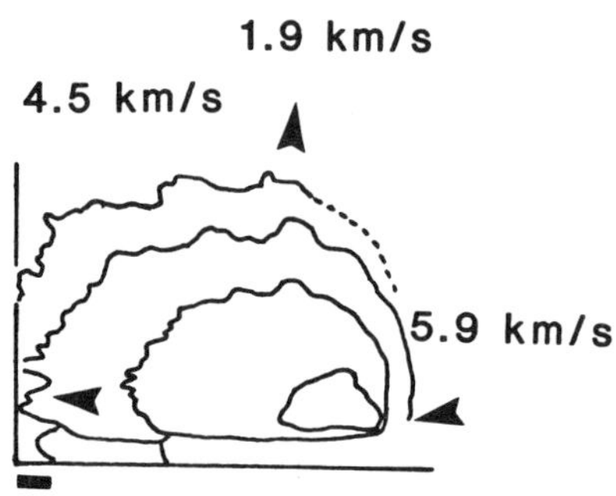

Figure 5. Early-time evolution of a self-luminous ionized cloud produced by a 5.86 km/s impact by a 0.635 cm aluminum sphere into dry-ice powder on top of No. 24 sand at an angle of 15^0. The cloud expands at nearly 2 km/s and travels downrange at 4.5 km/s. The bar scale corresponds to 5 cm.

and perhaps modifying the terrestrial or other planetary climate through time. The role of impact vaporization might be most evident on Mars where craters have been preserved over billions of years and where the existence of volatiles, whether in a free state or chemically bound, seems highly likely. Recent experiments by Schultz and Gault (1985c; 1986b) have used easily volatized dry-ice to explore the conditions controlling vaporization and the resulting phenomenology.

Impact-induced vaporization has been examined in three ways: the evolution of the ionized vapor cloud under self-illumination; the modification of the ejecta curtain under artificial illumination; and the generation of a magnetic field by the ionized cloud. Under self-illumination, vertical impacts (0.376g Al spheres at 5-6 km/s) into No. 24 sand under low atmospheric pressure (< 25mm, Hg) produce virtually no ionized cloud, whereas impacts into dry-ice powder or blocks produce only a faint ionized cloud visible at 35,000 fps (frames per second) and 7,000 fps. Oblique (15^0) impacts with the same impactor conditions, however, generate a brilliant ionized cloud that expands at 2 km/s and travels downrange at velocities comparable to the initial impact velocity. From expressions derived by Taylor (1950) for the expansion rate of the ionized cloud, it is estimated that the cloud contains nearly 60% of the initial projectile kinetic energy, a value providing an upper limit of 40 projectile masses of vaporized target material (Fig. 5). In contrast, the expansion rate of the cloud produced by a vertical impact indicates an energy fraction of about 3% or less than 0.3 projectile masses. Vertical impacts at the same velocity into dry-ice under 0.92 atmospheres of Argon also produce a brilliant ionized ball that moves upward, but the estimated energy fraction is only 0.01%. The ionized cloud in this case may be in part ionized Argon, rather than CO_2, trapped in the aircap and wake of the projectile. Consequently, 15^0 impacts appear to vaporize nearly 100 times the mass of a vertical impact for the same impactor energy.

Measurements were also made to explore the possibility that the downrange ionized cloud would generate a magnetic field. Preliminary analysis indicates that oblique impacts into dry-ice generate a highly localized and transient field associated with the ionized cloud with a magnitude of the field estimated to be 10^{-3} gauss. Although previous studies (Bianchi and others, 1984) identified a possible impact-generated magnetic field for aluminum impacting vertically into hard targets, the impact velocities were significantly higher (10 km/s) and the intensity unspecified. Experiments are underway to characterize further the properties and origin of the field. The results will be particularly significant in planetary studies because certain very young deposits on the Moon are closely associated with strong remanent magnetic fields and because impact-induced vaporization/ionization should characterize planetary-velocity impacts.

Atmospheric Effects: The presence of atmospheres on Mars, Earth, Venus introduces an additional variable to the impact process and ejecta emplacement. Although the present atmospheric pressure on Mars is only 6 mb (or equivalent to the pressure at 30 km altitude on the Earth), atmospheric effects on ejecta emplacement at first may seem negligible. Such a presumption, however, is probably incorrect. Aerodynamic drag increases as v^2 and ejection velocities at the same scaled position in a gravity-controlled crater increases approximately as $R^{1/2}$ where R is the excavation crater radius (Post, 1974; Schultz and Gault, 1979; Housen and others, 1983); consequently, the effects of air drag on given size ejecta increase roughly as the size of the crater. A simplified model of crater growth and ejection applied to Mars (Schultz and Gault, 1979) revealed that for a given size crater, a critical ejecta size exists below which most ejecta will be restricted to ballistic ranges less than a crater diameter. As crater size increases, the critical ejecta size increases. For Mars, the critical ejecta size for a 5 km diameter crater is about 2 cm under current atmospheric conditions. This simplified particle-by-particle ballistic model, however, fails to model the entire ejection process since it predicts that ejecta much smaller than the critical size will never leave the crater. As pointed out in Schultz and Gault (1979), such a result simply demonstrates that ejecta are part of a flow of debris within a curtain and cannot be treated simply as individual particles until the curtain becomes sufficiently tenuous or until the trajectories become perpendicular to the curtain. Such considerations require experiments to isolate controlling variable in a complex system.

Not all facets of cratering in an atmosphere can be modeled in the laboratory simultaneously. Table I permits comparison of parameters in the laboratory and on Mars. The experimental model fails to simulate particle-atmosphere interactions completely because the Froude, Reynolds, and Weber numbers are not matched. For example, a small water droplet ejected in a laboratory crater will not be disrupted during ballistic flight, whereas such a droplet will be atomized on Mars. Moreover, the effects of ejecta impacting at 50 cm/s on the emplacement process in the laboratory may be very different from

TABLE I

LABORATORY SIMULATION

VARIABLE	EXPERIMENT	MARS
impact velocity	1-6 km/s	15-40 km/s
target	pumice	"sand"→ bedrock
ejecta size	100μ	2 cm
ejecta velocity	70 cm/s	140 m/s
curtain velocity	50 cm/s	100 m/s
crater diameter	20 cm	10 km

SCALING PARAMETERS

Froude Number	10	3
Reynolds Number	>300	10^8
Weber Number	10^{-10}	>10

the effects of 100 m/s ejecta on the martian surface. The experimental model, however, does provide useful information about the response of the atmosphere to the moving ejecta curtain where the Froude numbers can be matched and the Reynolds numbers are sufficiently high. It also provides information about the interplay between early-time and late-time processes.

Experiments (Schultz and Gault, 1982; 1984) have been performed for a variety of impact velocities, target grain sizes, atmospheric pressures and atmospheric densities (He, CO_2, N_2, Ar, and air). Three observations are particularly significant: interaction of the ejecta curtain with the atmosphere; changes in ejecta morphology with changes in impactor/target parameters; and the effects of an atmosphere on crater scaling and dimensions. Film (400 fps-8000 fps) and high-speed video (200 fps) records permit comparing the evolution of the ejecta curtain. One particularly useful technique uses digitized film records and image processing to reveal sequential changes in the ejecta curtain at the same stages in growth for different impact conditions (Schultz and others, 1984). With increasing atmospheric pressure, the late-time ejecta curtain becomes increasingly distorted from the classical funnel-shaped profile in a vacuum to a kettle-shaped profile under one atmosphere (Fig. 6). During crater formation, a funnel-shaped plume similar to vacuum conditions is evident although the angle from the horizontal is higher. After crater formation, the bulge at the base of the curtain develops and outpaces the curtain formed in a vacuum for the same impact conditions. Analysis of such records, stereo-views, quarter-space experiments and barometric records indicates that the outward-moving ejecta curtain creates a partial vacuum in the crater. A strong recovery airflow is produced that resembles an outward-moving rolling vortex entraining ejecta and modifying emplacement.

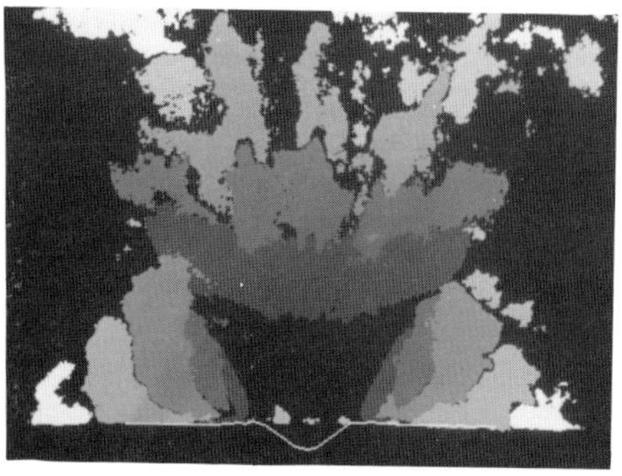

Fig. 6. Evolution of the ejecta curtain in the presence of an atmosphere of one bar. Different shades of grey correspond to different stages of growth: 0.0-7.5 ms (black), 7.5-15 ms (dark grey), 15-30 ms (medium grey), 30-60 ms (lighter grey), 60-120 ms (light grey), and 120-240 ms (white). This time sequence was produced by computer processing the film record and subtracting successive frames. Although the impact velocity was about 1.5 km/s (0.635 cm aluminum into compact pumice), the same late-stage phenomenology is observed for hypervelocity impacts: vacuum-like conical growth during early stages with basal flow developing after crater formation. Bar scale represents 5 cm.

A systematic change in ejecta morphology is observed with increasing atmospheric pressure (Fig. 7), and the various types of ejecta morphologies bear striking resemblence with martian impact ejecta morphologies (Schultz and Gault, 1984). Between 0.1 and 0.3 atmospheres (P_0), a contiguous ridge or rampart encircles the ejecta. As atmospheric pressure increases in this range, the rampart diameter relative to the crater diameter increases. Between 0.3 and 0.7 P_0, individual flow lobes develop and extend many crater radii. Above 0.7 P_0, the ejecta deposits are radially scoured. These systematic changes reflect increased energy transferred to the recovery airflow and entrained ejecta cloud. The overall trends hold important clues for conditions of ejecta emplacement on Mars; however, because the Reynolds number is very small for individual ejecta-air interactions, further experiments are necessary to detail constraints. In general, the experiments suggest that if a certain fraction (<25%) of the ejecta is sufficiently small to be decelerated below the critical size (2 cm for a 5 km-diameter crater) and to be suspended in the vortex, then various ejecta morphologies can result. The existence of liquid water does not appear to be a necessary condition for many of the observed ejecta patterns.

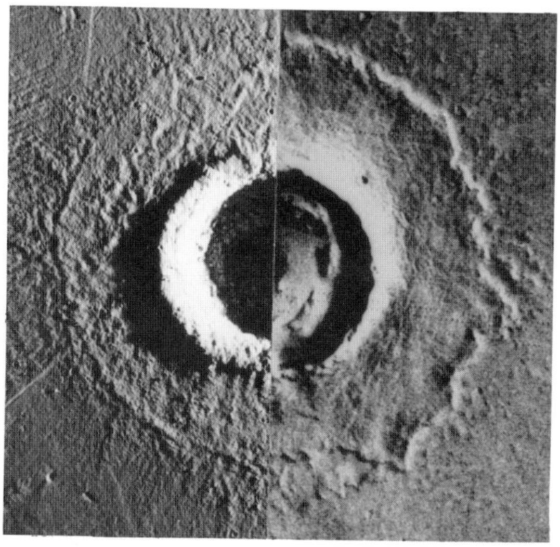

Fig. 7. Comparison of contiguous rampart surrounding craters on Mars (right half) and in the laboratory (left half). Laboratory experiment used 5.3 km/s aluminum sphere (0.635 cm diameter) impacting compacted pumice under 0.25 bar of argon. Martian impact crater is 12 km in diameter (Viking Frame 10A61).

Clustered Impacts: The atmospheres of Mars, Earth and Venus act as barrier for incoming asteroids and comets below a critical size (Baldwin and Schaeffer, 1971; Tauber and Kirk, 1976; Moloch, 1981). Below this size, incoming objects will disrupt and form an impactor with an effectively low density. Consequently, experiments were performed to examine the effects of hypervelocity clustered impactors on cratering efficiency and crater shape (Schultz and Gault, 1983; 1985a).

Pyrex projectiles (0.635 cm diameter) were fragmented during passage through 2.5 mil paper. Because the launch tube was rifled for spin-induced sabot separation from the projectile, centrifugal force dispersed the fragments to different diameters depending on velocity and the distance between the paper and target. Witness plates in separate experiments with identical velocities, and high-frame rate film records provided a record of projectile dispersal and size distribution. The estimated cluster densities ranged from 0.2 to 10^{-4} g/cm^3. Both compacted pumice and No. 140-200 sand were used as targets; the higher shear strength of the former target type perserved subtle slopes resulting from the event.

Early-time crater growth for clustered impacts was markedly different from solid-body impacts. Immediately after a 6 km/s impact (60 μs) a hemispherical filamentary cloud rose above the surface at 1 km/s. A diffuse cloud subsequently developed over the impact site, followed by an ejecta curtain with a very low (5-10^0) angle from the surface. The early-time phenomena bear striking resemblance to results calculated by O'Keefe and Ahrens (1982). At late-times, however, the ejecta curtain angle increased to 45^0. This evolution was unique to very low density clusters (10^{-3} g/cm^3).

Crater-aspect ratio and morphology were found to change systematically with increasing values (δ_t/δ_p) (2r/v): from 8:1 with a mounded floor to 50:1 with a multi-ringed floor pattern for compacted pumice. Subsequent re-analysis reveals that the parameter 2r/v not only provides a self-consistent relation for clustered impactors but also is consistent with solid-body impactors. Cratering efficiency, however, remained remarkably consistent with data for solid impactors if the gravity-scaled dimensionless scaling parameter, π_2 (Holsapple and Schmidt, 1982), is used with the radius of the impactor at impact (Fig. 8).

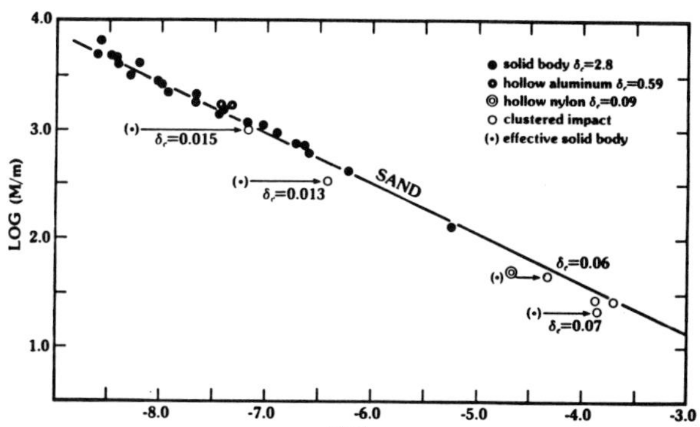

Fig. 8. The crater efficiency of very low density clouds of debris at hypervelocities is consistent with the efficiency for solid-body impacts into No. 140-200 sand provided that the cluster radius (open circles) is used. From Schultz and Gault (1985a).

Clustered impacts may provide an alternative explanation for understanding certain very shallow terrestrial structures where conventional interpretations are strained. As high resolution views of the surface of Venus become available, the reality of this process can be tested further.

CONCLUDING REMARKS

Planetary surfaces permit invaluable tests for concepts in large-scale cratering. Experiments provide only one approach and must be complemented by theoretical codes and geological observations of terrestrial and planetary craters. Eventually synthesis of the results will permit craters to be used as diagnostic indicators of geologic and atmospheric histories as well as a reliable tool for dating surfaces and exploring the flux history in the early solar system.

REFERENCES

Ahrens, T. J., I. Jackson, and R. Jeanloz (1977). *Proc. Lunar Sci. Conf. 8th,* 3437-3455.

Baldwin, B., and Y. Schaeffer (1971). *J. Geophys. Res. 76,* 4635-4668.

Bianchi, R., F. Cappaccioni, P. Cerroni, M. Coradini, E. Flamini, P. Hurren, G. Martelli, and P. N. Smith (1984). *Nature 308,* 830-832.

Cintala, M. J., F. Hörz, S. Smrekar, and F. Cardenas (1985). *Lunar and Planetary Science XVI* (abstract), 129-130. Lunar and Planetary Institute, Houston.

Cintala, M. J., F. Horz, and T. H. See (1986). *Lunar and Planetary Science XVII* (abstract), 131-132. Lunar and Planetary Institute, Houston.
Fink, H. J., R. Greeley, and D. E. Gault (1981). *Proc. Lunar Planet. Sci. 12 B,* 1649-1666.
Gault, D., and Wedekind (1977). In *Impact and Explosion Cratering* (D. Roddy, R. O. Pepin, R. B. Merrill, Eds.), 1231-1244, Pergamon, N.Y.
Gault, D., and C. Sonett (1982). In *Geol. Soc. Amer. Sp. Paper 190* (L. Silver and P. Schultz, Eds.), 69-92.
Greeley, R., J. Fink. D. E. Gault. D. B. Snyder, J. E. Guest, and P. H. Schultz (1980). *Proc. Lunar Planet. Sci. Conf. 11th,* 2075-2097.
Grieve, R. A. F. (1984). *J. Geophys. Res. 89,* B403-B408.
Hartmann, W. K., R. G. Strom, S. J. Weidenschilling, K. R. Blasius, A. Woronow, M. R. Dence, R. A. F. Grieve, J. Diaz, C. R. Chapman, E. M. Shoemaker, and K. L. Jones (1981). In *Basaltic Volcanism Study Project 1976-1979,* 1050-1129, Pergamon, N.Y.
Holsapple, K., and R. M. Schmidt (1982). *J. Geophys. Res. 87,* (B3), 1849-1870.
Hörz, F. (1970). A small ballistic range for impact metamorphism studies. NASA TN D-5787, 16 pp.
Housen, K. R., R. M. Schmidt, and K. A. Holsapple (1983). *J. Geophys. Res. 88* (B3), 2485-2499.
Ivanov, B.A., A. T. Basilevsky, V. P. Kryuchkov, I. M. Chernaya (1986). *Proc. Lunar Planet. Sci. Conf. XVI,* D413-D430.
Melosh, H. J. (1977). In *Impact and Explosion Cratering* (D. Roddy, R. O. Pepin, R. B. Merrill, Eds.), 1245-1260, Pergamon, N.Y.
Melosh, H. J. (1981). In *Multi-Ring Basins* (P. Schultz and R. B. Merrill, Eds.), 29-35, Pergamon, N.Y.
Orphal, D. L., W. F. Borden, S. A. Larson, and P. H. Schultz (1980). *Proc. Lunar Sci. Conf. 11th,* 2309-2323.
O'Keefe, J. D., and T. J. Ahrens (1982). Cometary impact on planetary surfaces. *J. Geophys. Res., 87,* 6668-6680.
Pike, R. J. (1980). *Proc. Lunar Planet. Sci. Conf. 11th,* 2159-2189.
Pike, R. J. (1983). *J. Geophys. Res. 88* (B3), 2500-2504.
Post, R. L. (1974). Rep. AFWL-TR-74-51, Air Force Weapons Lab, Kirtland AFB, N.M.
Roddy, D. J. (1968). In *Shock Metamorphism of Natural Materials* (B. M. French and N. M. Short, Eds.), 291-322, Mono Book Corp., Baltimore.
Roddy, D. J. (1977). In *Impact and Explosion Cratering* (D. J. Roddy, R. O. Pepin, and R. B. Merrill, Eds.), 185-246, Pergamon, N.Y.
Schmidt, R. M. (1980). *Proc. Lunar Planet. Sci. Conf. 11th,* 2099-2128.
Schmidt, R. M., and K. A. Holsapple (1981). *Lunar and Planetary Science XI* (abstract), 934-936. Lunar and Planetary Institute, Houston.
Schmidt, R. M. (1984). *Lunar and Planetary Science XV* (abstract), 722-723. Lunar and Planetary Institute, Houston.
Schultz, P. H. (1976). *Moon Morphology,* U. Texas Press, 626 pp.
Schultz, P. H., and D. E. Gault (1979). *J. Geophys. Res., 84,* 7669-7687.
Schultz, P. H., D. Orphal, B. Miller, W. F. Borden, and S. A. Larson (1981). In *Multi-Ring Basins* (P. Schultz and P. B. Merrill, Eds.), 181-195, Pergamon, N.Y.
Schultz, P. H., and D. E. Gault (1982). In *Geol. Soc. Amer. Sp. Paper 190* (L. Silver and P. H. Schultz, Eds.), 153-174.
Schultz, P. H., and D. E. Gault (1983). *Lunar and Planet. Sci. XIV* (abstract), 674-675. Lunar and Planetary Institute, Houston.
Schultz, P. H., A. Meloy, and D. E. Gault (1984). *EOS,* 35.
Schultz, P. H., and D. E. Gault (1984). *Lunar and Planet. Sci. XV* (abstract), 732-733. Lunar and Planetary Institute, Houston.
Schultz, P. H., and D. E. Gault (1985a). *J. Geophys. Res., 90,* (B5), 3701-3732.
Schultz, P. H., and D. E. Gault (1985b). *Lunar and Planet. Sci. XVI* (abstract), 742-743. Lunar and Planetary Institute, Houston.
Schultz, P. H., and D. E. Gault (1985c). *Lunar and Planet. Sci. XVI* (abstract), 740-741. Lunar and Planetary Institute, Houston.
Schultz, P. H., and D. E. Gault (1986a). *Lunar and Planet. Sci. XVII,* 777-778. Lunar and Planetary Institute, Houston.
Schultz, P. H., and D. E. Gault (1986b). *Lunar and Planet. Sci.* (abstract), 779-780. Lunar and Planetary Institute, Houston.
Tauber, M. E., and D. B. Kirk (1976). *Icarus 28,* 351-357.
Taylor, G. T. (1950). *Proc. Royal Ac. 201A,* 175-186.

Int. J. Impact Engng Vol. 5, pp. 577–584, 1987
Printed in Great Britain

0734-743X/87 $3.00 + 0.00
Pergamon Journals Ltd.

COMPUTER CHARACTERIZATION OF DEBRIS CLOUDS RESULTING FROM HYPERVELOCITY IMPACT

J. C. Schulz, O. E. R. Heimdahl and S. A. Finnegan
Research Department
Naval Weapons Center
China Lake, California 93555

ABSTRACT

A computational study to examine the effects of parameter variation on debris cloud formation and perforation behavior in a simple sphere-plate impact situation is described. The parameters varied were impact velocity, thickness: diameter ratio, projectile-target material combination and number of plates. In particular, effects associated with changes from strength-dominated to hydrodynamic behavior were considered.

INTRODUCTION

Impact at sufficiently high velocity of a compact projectile against a thin plate target results in fragmentation of the projectile, perforation of the plate and formation of a cloud of plate and projectile debris behind the target. This debris cloud, which may consist of solid fragments, liquid drops and/or vapor, spreads transversely as well as longitudinally, significantly reducing the impulse intensity that can be delivered against a subsequent target element. (The "bumper plate" concept for armoring space vehicles against meteoriod or other impact makes use of this fact.) Nevertheless, such a debris cloud may still possess considerable damage potential. Assessment of this potential requires either extensive experimentation or, alternatively, detailed theoretical numerical analysis.

Projectile launch becomes increasingly difficult as velocities increase from the ordnance range into the hypervelocity regime. (Backman, 1978, gives approximate limits for these ranges.) Two-stage light gas gun accelerators are capable of launching laboratory projectiles at speeds up to about 8.5 km/s, and the ultimate limit for projecting macroparticles is about 12.5 km/s (Swift, 1979). Considerably higher velocities appear likely in the near future through the use of electromagnetic rail gun launchers currently under development. Data acquisition at high speeds usually involves cloud dissection coupled with flash radiography.

Continuum mechanics computer codes have been used for some time to model hypervelocity impact events. These codes are commonly referred to as "hydrocodes," although this term is something of a misnomer since current codes include strength as well as hydrodynamic effects. Early computational studies (Rosenblatt, 1969; Swift, 1970), though based on simple models, agreed qualitatively with experiment. Considerable work has been done to develop improved computational capabilities, particularly in the areas of equation of state and failure characterization. This has led to more reliable code results, especially for materials whose behavior is reasonably well understood (Bertholf, 1975; Sedgwick, 1978; Shockey, 1975).

Our primary area of interest lies in the phenomenological characterization of ordnance and ultra-ordnance impact events (Alexander, 1983; Backman, 1986). We are not code developers, but have used codes as a tool for understanding perforation and fragmentation processes. In this paper we briefly describe our code experiences and present the results of a computational study of debris cloud behavior. We considered a simple impact situation, namely, the impact at normal incidence of a sphere against a plate, and we examined the effects of varying velocity, thickness, materials and number of plates on debris cloud formation. In particular, we tried to observe effects associated with changes from strength-dominated to hydrodynamic behavior. To our knowledge, a single study covering this range has not been presented previously. This is a computation-only study, though experimental verification is planned.

CODE EXPERIENCES

We have applied two Eulerian hydrocodes in our impact work, the HELP code, developed by Systems, Science, and Software and documented in Hageman (1975), and CSQII (an updated version of CSQ) developed at Sandia National Laboratories by Thompson (1979). We experienced greater success in using CSQII, and all of the computations described in this paper were performed using this code.

CSQII is an Eulerian finite difference wave propagation code employing a mesh fixed in space through which mass, momentum and energy flow. However, during each cycle of a computation, calculations are first carried out treating the mesh as imbedded in the material (Lagrangian). Rezoning is then used to return the mesh to its original location. CSQII employs a complete, thermodynamically consistent equation of state. A unique feature of the code is the retention of cell void volumes. This helps to maintain material definition along boundaries and free surfaces.

SPHERE-PLATE IMPACT

The basic problem treated in this paper is that of a solid spherical projectile impacting a plate target at normal incidence. Projectiles of steel (St) and aluminum (Al) were considered. A number of code computations were made to look at effects of variation of impact parameters on debris cloud formation and perforation behavior. The parameters varied were impact velocity, thickness :diameter ratio, projectile-target material combination and also the number of plates. The sphere diameter was 0.6 cm (0.25 in.) in all cases. Thickness:diameter ratio was determined by the thickness of the plate. Two materials were considered: low-carbon steel with a yield strength of 0.69 GPa (100,000 psi) and aluminum alloy with a yield strength of 0.48 GPa (70,000 psi). Equation of state properties for these materials came from the CSQII library. The starting geometry for an example problem, a steel sphere impacting a single half-diameter aluminum plate at 5 km/s, is diagrammed in Fig. 1. (Only the portion of the impact space to the right of the axis of revolution was modelled.) The mesh size for all computations was 81 x 121 cells. Runs used approximately 10 to 15 minutes of CPU time on the CRAY-XMP.

VELOCITY EFFECTS

The character of an impact event changes dramatically with impact velocity. Computational runs were made for the same thickness:diameter ratio and projectile-target material combination (1:2, St-Al) as in Fig. 1, but at four different impact velocities (1, 2.5, 5 and 10 km/s). Deformed shapes (density plots) of the impact system versus time at the four velocities are given in Fig. 2. (Only a portion of the modelled space is shown.)

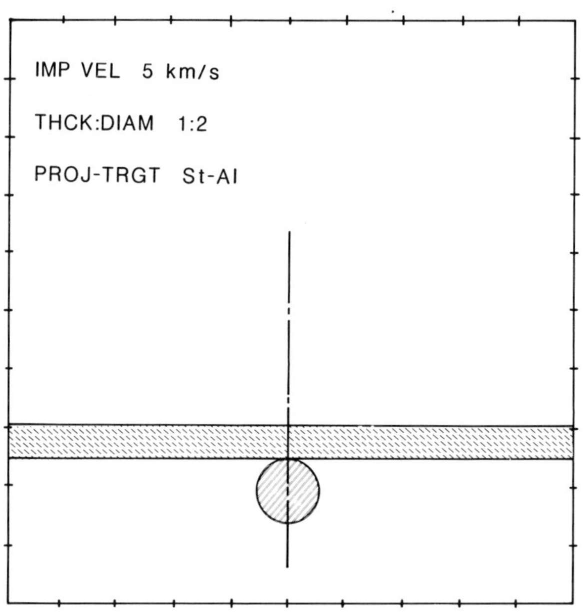

Fig. 1. Steel sphere impacting single half-diameter aluminum plate at 5 km/s.

Fig. 2. Velocity effects for sphere-plate impact. (St-Al, 1:2, time in μs in upper right of each box.)

Fig. 3. Thickness effects for sphere-plate impact. (St-Al, 5 km/s, t:D at upper left of each row.)

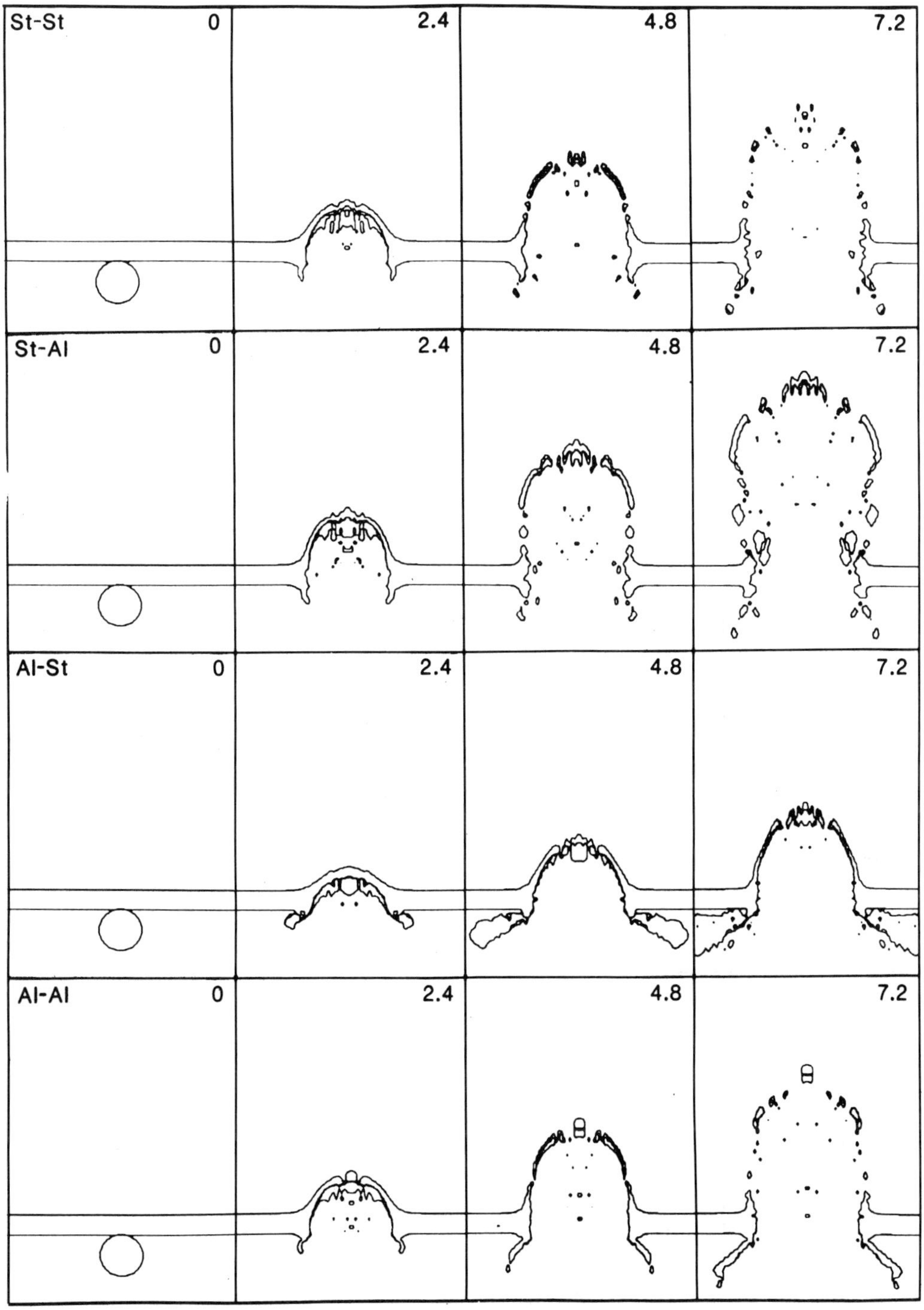

Fig. 4. Material effects for sphere-plate impact. (5 km/s, 1:2, projectile-target material at upper left of each row.)

Fig. 5. Multi-plate effects for sphere-plate impact. (St-Al, 1:2 for both plates.)

At 1 km/s, which is well within the ordnance range, strength effects largely dominate. The central portion of the plate bulges, then fractures into a disk and surrounding ring. The projectile, though severely deformed, remains in one piece (at least until the solution end). Considerable bending of the plate occurs at this relatively low velocity. The computed behavior agrees qualitatively with experimental observations. Fracture at this velocity tends to result from adiabatic shear (Backman, 1986; Shockey, 1975) rather than being tensile in nature as assumed in CSQII. The fracture surface between disk and ring should perhaps be angled inward rather than directly across the surface as the code predicts. At 2.5 km/s, in the ultra-ordnance range, the projectile is starting to break up. A fairly large central fragment persists, as well as surrounding smaller fragments. Oscillations, especially on the rear surface of the projectile, are probably code-induced. At 5 km/s, at the lower end of the hypervelocity range, breakup of projectile and target material into small solid and/or liquid fragments is apparent. A "cloud" or "bubble" of debris characteristic of hypervelocity events has clearly formed. At 10 km/s the cloud of debris consists largely of vapor, as evidenced by its extreme size. The code indicates that at about 1.2 microseconds a glob of high energy material is "spalled" off the rear of the projectile.

THICKNESS EFFECTS

Target plate thickness strongly influences the impact process. A series of runs was made for the same impact velocity and projectile-target material combination (5 km/s, St-Al) as in Fig. 1, but with four different thickness:diameter ratios (1:4, 1:2, 1:1 and 2:1). Deformed shapes of the impact system versus time for the four ratios are given in Fig. 3.

At a 1:4 ratio, the target is obviously overmatched by the projectile. Plate material ahead of the projectile is vaporized. It is not clear whether the bullet-shaped projectile fragment remaining at the solution end is real or a code artifact. At a 1:2 ratio (also shown in Fig. 2), both projectile and target have broken up into solid and/or liquid fragments. The hole in the plate is larger as more energy is transferred laterally. At a 1:1 ratio, considerable energy is expended in the perforation process. The plate material appears to stretch thinner before fragmentation occurs. The debris cloud is travelling at a considerably lower velocity. At a 2:1 ratio, the impact event largely involves cratering and penetration. It is not clear at the solution end whether perforation of the plate will ultimately occur.

MATERIAL EFFECTS

Different material combinations alter the nature of an impact event. Calculations were made for the same impact velocity and thickness:diameter ratio (5 km/s, 1:2) as in Fig. 1, but with four different projectile-target material combinations (St-St, St-Al, Al-St and Al-Al). Deformed shapes of the impact system versus time for the four combinations are given in Fig. 4.

For St-St, the debris cloud consists of relatively small fragments. Moreover, both projectile and target fragments appear of roughly similar size. For St-Al, the aluminum forms larger fragments and these appear to be partly liquid and/or vapor. For Al-St, perforation of the plate is more difficult. A sizable central fragment is ejected. Considerable projectile material is ejected in front of the target. For Al-Al, the shape and size of the debris cloud is quite similar to that for St-St (although the degree of fragmentation differs). This is to be expected because the ratio of projectile to target density is the same in both cases (i.e., unity) and the density ratio is probably the most significant property in the hypervelocity range. The difference in degree of fragmentation is no doubt due to differences in failure criteria for the two materials.

MULTI-PLATE EFFECTS

Debris from a single plate perforation often retains considerable damage potential. Consider the impact system in Fig. 1 (1:2, St-Al) with the addition of a second identical plate spaced five thicknesses from the first plate. Runs were made for this system at the four previous impact velocities (1, 2.5, 5 and 10 km/s). Deformed shapes versus time are given in Fig. 5.

At 1 km/s, perforation of the second plate produces a single projectile fragment capped with a disk of target material. Bending is slightly more pronounced in the second plate. At 2.5 km/s, a central disk and two ring fragments (one incipient) result. At 5 and 10 km/s, perforation also occurs. Impact debris from the first plate can be observed travelling laterally along the front face of the second plate. Both these runs stopped shortly after the last time shown, apparently due to difficulties in the equation of state subroutines.

CONCLUSION

A computational study to examine the effects of impact parameter variation on debris cloud formation and perforation behavior in a simple impact situation has been described. For the most part the results have been discussed qualitatively in an attempt to characterize the phenomenonology involved. Nevertheless, one question that need be raised is that of accuracy.

Sources of error in code calculations are numerous. Adequate mesh refinement is no longer a problem if one is willing to pay the "cost." (The mesh we used, though relatively coarse, was probably not a major source of error.) Reliable equation of state data is now readily available. The CSQII code, in particular, features a sophisticated, thermodynamically consistent equation of state. The major source of error remaining in code computations is associated with failure characterization. Under extremely high-rate loading, failure can occur through a variety of mechanisms. As mentioned, adiabatic shear is one such mechanism in the ordnance range. The tensile failure criterion used in CSQII can not be expected to adequately treat all likely failure modes. It is surprising that the code is capable of handling the impact situations considered here as well as it does.

The authors are planning a series of experiments in the ordnance and ultra-ordnance range to check the validity of some of the simulations presented in this paper. (The particular steel and aluminum used were selected with these tests in mind.) Test results will be available at a later date.

REFERENCES

Alexander, M. D., and S. A. Finnegan (1983). A model for multiple fragment interactions in 3-D targets, Proceedings Seventh International Symposium on Ballistics, 295-299, The Hague, The Netherlands.

Backman, M. E., and W. Goldsmith (1978). The mechanics of penetration of projectiles into targets, Int. J. Engng Sci., 16, 1-99.

Backman, M. E., S. A. Finnegan, J. C. Schulz and J. K. Pringle (1986). Scaling rules for adiabatic shear. In L. E. Murr (Ed.), Metallurgical Applications of Shock-Wave and High-Strain-Rate Phenomena, 675-687, Marcel Dekker, Basel, Switzerland.

Bertholf, L. D., L. D. Buxton, B. J. Thorne, R. K. Byers, A. L. Stevens and S. L. Thompson (1975). Damage in steel plates from hypervelocity impact. II. Numerical Results and Spall Measurement, J. Appl. Phys., 46, 3776-3783.

Hageman, L. J., D. E. Wilkins, R. T. Sedgwick and J. L. Waddell (1975). HELP-A multi-material Eulerian program for compressible fluid and elastic-plastic flows in two space dimensions and time, Systems, Science and Software, SSS-R-75-2654, La Jolla, California.

Rosenblatt, M., N. K. Kreyenhagen and W. D. Romine (1969). Numerical studies of ejecta char acteristics behind thin plates, AIAA Hypervelocity Impact Conference, No. 69-357, Cincinnati, Ohio.

Sedgwick, R. T., L. J. Hageman, R. G. Herrmann and J. L. Waddell (1978). Numerical investiga tions in penetration mechanics, Int'l J. Engr. Sci., 16, 859-869.

Shockey, D. A., D. R. Curran and P. S. De Carli (1975). Damage in steel plates from hyper-velocity impact. I. Physical Changes and Effects of Projectile Material, J. Appl. Phys., 46, 3766-3775.

Swift, H. F., D. D. Preonas, W. C. Turpin and J. M. Carson (1970). Debris clouds behind plates impacted by hypervelocity pellets, J. Spacecraft and Rockets, 7, 313-318.

Swift, H. F. (1979). Experimental technology for investigating meteoroid threats to deep space probes, Comet Halley Micrometeoroid Hazard Workshop, 131-136, Paris, France.

Thompson, S. L. (1979). CSQII-An Eulerian finite difference program for two-dimensional Material Response, Sandia National Laboratories, SAND77-1339, Albuquerque, New Mexico.

Int. J. Impact Engng Vol. 5, pp. 585–593, 1987
Printed in Great Britain

0734-743X/87 $3.00 + 0.00
Pergamon Journals Ltd.

DISINTEGRATION BEHAVIOR OF METAL RODS SUBJECTED TO HYPERVELOCITY IMPACT

D. A. Shockey*, D. R. Curran*, J. E. Osher** and H. H. Chau**

*SRI International
Menlo Park, CA 94025

**Lawrence Livermore National Laboratory
Livermore, CA 94550

ABSTRACT

The disintegration behavior of projectile rods when impacting plates at velocities above 10 km/s is largely unexplored. Possible material responses include vaporization, melting, solid state phase changes, plastic flow, shear localization, and ductile and brittle tensile fracture. Two primary difficulties have prevented experimental studies of projectile response in this velocity regime, namely the lack of a technique for accelerating a projectile to velocities above 10 km/s, and the lack of a procedure for recovering the projectile remains after impact for examination. These difficulties were overcome recently by using an exploding foil technique to drive an impactor plate and by using a reverse ballistics test arrangement to allow projectile recovery. This paper describes this experimental procedure for investigating the disintegration behavior of projectile rods under hypervelocity impact conditions, presents preliminary results for steel, aluminum and lead, and discusses requirements for computational material response models.

INTRODUCTION

The debris emanating from the rear surface of a plate impacted by a projectile is of concern because of its importance in determining the damage to components behind the plate. Practical examples of interest include encounters between meteorites and satellites, kinetic energy penetrators and target vehicles, and long-rod projectiles and armored structures.

Because many experiments have been performed at conventional ordnance velocities (around 1 km/s), debris formation phenomenology at these velocities is at least partially understood. The material failure mechanisms observed to operate include homogeneous plastic flow, shear strain localization (shear bands), and ductile and brittle tensile fracture. Shear localization and tensile fracture have been recognized as kinetic processes involving nucleation, growth, and coalescence of shear bands and cracks. Thresholds and mathematical expressions and parameters describing these kinetics have been measured in characterization experiments (Shockey, Seaman, and Curran, 1979), and the results have been used in computational simulations of various ballistic impact scenarios (Dullum and Seaman, 1985). Thus far, a measure of success has been achieved, in that calculated damage/penetration modes correspond well with those observed in experiments: ballistic limit velocities are accurately represented in some cases, and fragment size distributions appear approximately correct.

At velocities substantially above conventional ordnance velocities disintegration behavior is considerably more complex and computational models are less reliable. Experiments performed in the 5 to 8 km/s range by Swift and Hopkins (1970) and Swift, Preonas, Turpin, and Carson (1970), for example, show that certain structural materials may melt and that materials of low sublimation energy such as polymers, lead, cadmium and tin may vaporize. Thus at these and higher impact velocities, tensile fracture and shear localization may be accompanied or even supplanted by two additional material disintegration mechanisms, melting and vaporization.

Projectile response to velocities greater than 10 km/s has not been studied in experiments, and hence is even less well understood. Definitive experiments, which are necessary both for developing an understanding of the phenomenology through observations and for generating a data base, are difficult for two reasons: first is the problem of accelerating materials to velocities above 10 km/s, and second is the problem of recording material response and recovering the material for post-test analysis. The work reported here aimed at establishing projectile disintegration behavior at impact velocities above 10 km/s. We first present the results of theoretical considerations, then describe and present the results of experiments to check the theoretical considerations, and finally discuss requirements for modeling material disintegration.

THEORETICAL CONSIDERATIONS

It is possible to show from first principles that solid fragments will always be produced downrange from a hypervelocity impact if the target plate is thinner than the penetrator dimensions. Furthermore, estimates of the impact velocities required for significant melting and vaporization can easily be made.

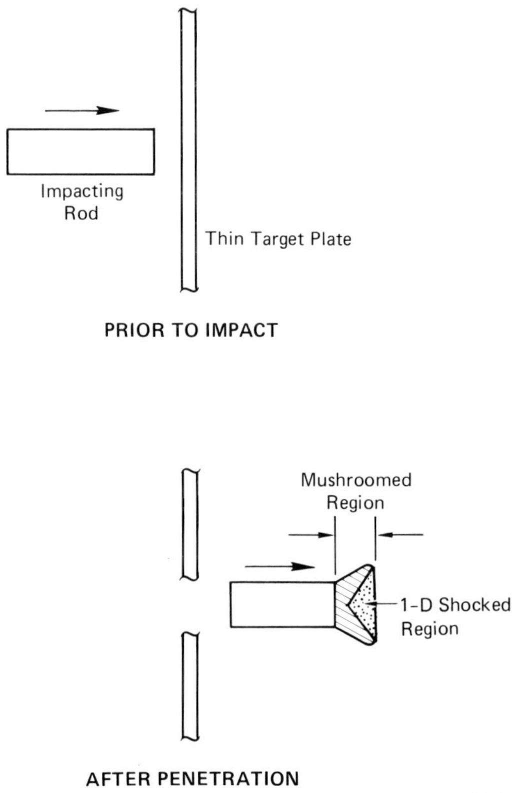

Fig. 1. Impact phenomenology for rod impacting a thin plate of the same material at hypervelocity.

For simplicity, consider a right circular rod impacting end-on at hypervelocity a thin sheet of similar material, Fig. 1. At the instant of impact, the energy density in a thin layer of material on either side of the impact interface is large, and extreme radiation intensities are possible. However, because most of the area of this high energy surface is adjacent to cool unshocked material that is opaque to radiation, very little energy is lost. Only that material at the periphery of the impact site is exposed and therefore able to radiate energy; furthermore, because the energy level in material at the periphery is almost immediately reduced to a low level by release waves at the free surface, the duration of radiation is quite short, and negligible energy is lost. This phenomenon is manifested as the "impact flash" routinely recorded in photographs of high-speed impact events.

The energy in the impact surface now begins to move into the projectile and target plate in the form of intense shock waves moving in opposite directions. The portions of the projectile and target subjected to the impact energy can melt or vaporize if the fusion or sublimation energies of the materials are exceeded (collision velocities required to melt or vaporize the material are calculated later). However, the amount of rod material that experiences the impact pressure is limited by isentropic pressure-release waves that emanate from the free surfaces of the rod (the rod periphery). Simple analyses based on overtaking rarefaction waves suggest that only a conical volume whose base is the impact surface of the rod (as indicated in Fig. 1) can experience one-dimensional impact pressure of significant duration, and only a region about one rod diameter deep can be melted or vaporized by shock heating.

This deduction leads us to conclude that, in general, some portion of the projectile will always remain solid, regardless of the magnitude of the impact velocity, provided the rod length exceeds the rod diameter and the target plate thickness. Moreover, the relative amount of vaporized material can be controlled to some extent by varying the projectile geometry: shorter, larger diameter, "disk-like" projectiles will be more completely vaporized; longer, smaller diameter, "rod-like" projectiles will tend to remain in the solid state.

Rod material outside the conical volume but near the impact surface will be accelerated outward, resulting in large deformations and a "mushroomed" region (Fig. 1) characterized by tensile and shear stresses that will tend to cause fragmentation. Because large deformation waves travel very slowly in rods (essentially at the velocity given by the tangent modulus of the stress-strain curve), we expect the amount of material in this mushroomed region to also be small. The remainder of the rod will experience more moderate deformations and consequently is expected to break up into several large fragments or to remain intact.

The portion of the rod remaining in the solid state can undergo several stress-induced changes, including solid-state phase transitions, plastic flow, fracture and shear localization. For example, at pressures above 13 GPa, steel undergoes a solid-state, polymorphic phase change to a hexagonal-close-packed structure that is 4% denser than the ambient-pressure body-centered-cubic structure. Other metals may exhibit similar phase changes, and most changes are reversible as pressure drops below some critical level. These changes modify the stress history in the shock wave, and may affect the failure behavior [(Bertholf and co-workers (1975)].

Plastic flow results in projectile deformation and may cause temperature rises of several hundred degrees. Tensile fracture, the separation of material by the nucleation, growth, and coalescence of voids and cracks, is expected to occur under this predominantly compressive loading by the passage of bending waves and by the intersection of pressure-release waves. Shear localization, the nucleation, growth, and coalescence of narrow surfaces of intense shear strain observed in materials under high-rate compressive loads, is another response mode expected to be activated in hypervelocity projectiles. All these solid-state changes will affect the fragmentation behavior of the projectile and will need to be more closely examined to obtain such details as the number and sizes of fragments.

We next estimate the impact velocities required to melt or vaporize the penetrator material exposed to the strong shock pulse. The equations for conservation of mass, momentum, and energy through a planar shock front propagating into material at zero pressure can be combined to yield

$$E_s = u^2/2 \ , \qquad (1)$$

where E_s and u are the internal energy and particle velocity increases behind the shock, respectively. When the rod and plate are of the same material, u is one-half the impact velocity, v. Thus,

$$v = [8\ E_s]^{1/2} \ . \qquad (2)$$

When the rarefaction waves reduce the shock pressure to zero along an isentrope, the internal energy is also reduced, but a significant amount is retained as heat. The actual amount retained depends on the high pressure equation of state (HPEOS) of the material. Unfortunately, good HPEOS data in the many-terapascal pressure region are lacking for most materials. An exception is the case of aluminum, for which a reasonably reliable HPEOS has been compiled. In a recent calculation (Tokheim, Curran, and Dein, 1983), a 0.3-mm-thick Kapton flyer was assumed to impact a plate of aluminum at an impact velocity of 10 km/s. A 100-GPa shock pulse of 34 ns full width at half maximum duration propagated into the aluminum plate. The value of E_s was 2×10^7 J/kg, and the residual internal energy behind the shock was 2.3×10^6 J/kg. Thus, about one-tenth the shock energy was retained behind the shock as heat.

Bjork and Olshaker (1965), on the other hand, in calculating threshold impact velocities for incipient and complete melting in many metals found that the fraction of maximum shock internal energy remaining in the metal on release is approximately constant and equal to about 0.2. Thus, as a rule of thumb, let us assume that one- to two-tenths of the shock energy is retained as heat in all cases. Then, we can let E_s in Eq. (2) be equal to 5 to 10 times the incipient melt or vaporization energies to find the impact velocities required to produce the onset of melt or vaporization. We consider three materials, steel, aluminum and lead. The values of the relevant energies assumed are given below (Stull and Sinke, 1956).

TABLE 1 Energies Required to Begin and Complete Melting and Vaporization of Three Metals

	Iron	Aluminum	Lead
Incipient Melting Energy (J/kg x 10^4)	105	66.5	4.07
Complete Melting Energy (J/kg x 10^4)	132	106	6.35
Incipient Vaporization Energy (J/kg x 10^4)	241	300	25.9
Complete Vaporization Energy (J/kg x 10^4)	869	1390	112

We now put 5 to 10 times the above energies into Eq. (2) to get a rough estimate for the impact velocities required to cause the onset of melt and vaporization. The results are:

TABLE 2 Impact Velocities Required to Produce Melting and Vaporization of Three Metals

	Iron	Aluminum	Lead
Incipient Melting (km/s)	6.6 - 9.2	5.2 - 7.3	1.3 - 1.8
Complete Melting (km/s)	7.1 - 10	6.6 - 9.2	1.6 - 2.2
Incipient Vaporization (km/s)	10 - 14	11 - 15	3.3 - 4.6
Complete Vaporization (km/s)	19 - 26	24 - 33	6.8 - 9.5

The impact velocities in the above table are for symmetric impacts, where the impactor and target are of the same material. In the experiments to be reported below, Kapton flyer plates impacted metallic rods. In these cases the significant mismatch of shock impedance causes the Kapton impact velocity to be the equivalent of a metallic flyer impacting at about one-fourth to one-half the flyer velocity. Therefore, significantly higher velocities than those listed in the table are required for a Kapton flyer to cause melting and vaporization of a metal rod.

In summary, for a rod-on-thin-plate impact configuration, geometrical effects are expected to limit the amount of material melted or vaporized by the shock to relatively small values. The amount of material experiencing large plastic deformation and fragmentation is similarly limited. Thus, when like materials collide, the downrange fragment cloud for impact velocities about 10 to 14 km/s for steel, 11 to 15 km/s for aluminum, and 3.3 - 5.6 km/s for lead is expected to consist of a small amount of vapor and liquid droplets, followed first by a spray of solid fragments and later by the intact distal rod end. When the metal is impacted by a polymer, these velocities are substantially higher.

We expect that varying projectile geometries and impact configurations (yaw and obliquity) will not change these basic conclusions, as long as the projectile dimensions are larger than the target plate thickness. In the next section we describe an experimental technique that can be used to check these inferences; preliminary results are also presented.

HYPERVELOCITY IMPACT EXPERIMENTS

The electric gun facility at the Lawrence Livermore National Laboratory (Chau and co-workers 1980; Osher and co-workers, 1986) was used to achieve velocities in the 10 to 20 km/s range. In this technique thin foils are accelerated by hydrodynamic and magnetic forces when a capacitor bank is discharged into a thin metallic foil adjacent to the flyer foil. The discharge time is kept shorter than the thermal diffusion time in the foil, which causes the foil to explode, and the rapidly expanding gas plus the resulting Lorentz force drive the flyer foil to velocities that can approach 30 km/s, depending on the mass of the flyer.

The experimental arrangement used in this work is diagrammed in Fig. 2. The evacuated plastic (Lucite) gun barrel was 5.0 mm long and 8.3 mm in diameter. The exploding laminate consisted of a 0.05-mm-thick aluminum bridge foil next to the flyer and an eleven layer sandwich containing nine 0.125-mm-thick Kapton foils between two 0.25-mm-thick copper foils. The 100-kV, 17.3-μF capacitor bank, which stores about 87 kJ of energy, was connected to the exploding foil by a rail-gap switch and a large, flat-plate transmission line. Flyer plate velocities were measured by a Fabry-Perot laser velocimeter.

Steel Rod

A rod of 4340 HRC 40 steel, 15.9 mm long by 3.56 mm in diameter, suspended about 1 cm above the gun barrel in the LLNL electric gun facility, was impacted end-on at about 15 km/s by a 0.3 mm-thick Kapton plate. The distal portion of the rod was recovered after the test; debris from the proximal end was not recovered.

Figure 3 shows a side view and the impacted end view of the rod. About 15% of the original length (about 70% of a rod diameter or 8 Kapton plate thicknesses) is missing, presumably melted and fragmented. The mushroom shape of the impacted end attests to the occurrence of plastic flow, and the shiny, faceted appearance of the surface suggests shear failure. The distal three-fourths of the rod showed no evidence of damage of any kind. These observations are in accord with our expectations from theoretical considerations, as discussed earlier.

Figure 4(a) is a magnified view of a polished cross section of the rod near the impacted end, revealing a population of internal tensile cracks. Figure 4(b) is the same area after etching in nital, showing the grain structure of the steel and the white etching bands that indicate strain localization. The figure shows that the mushroom shape of the impacted end is caused by homogeneous plastic flow, shear bands, and tensile fracture. Figure 5 provides close-up views of an adiabatic shear band and the grain structure in the vicinity.

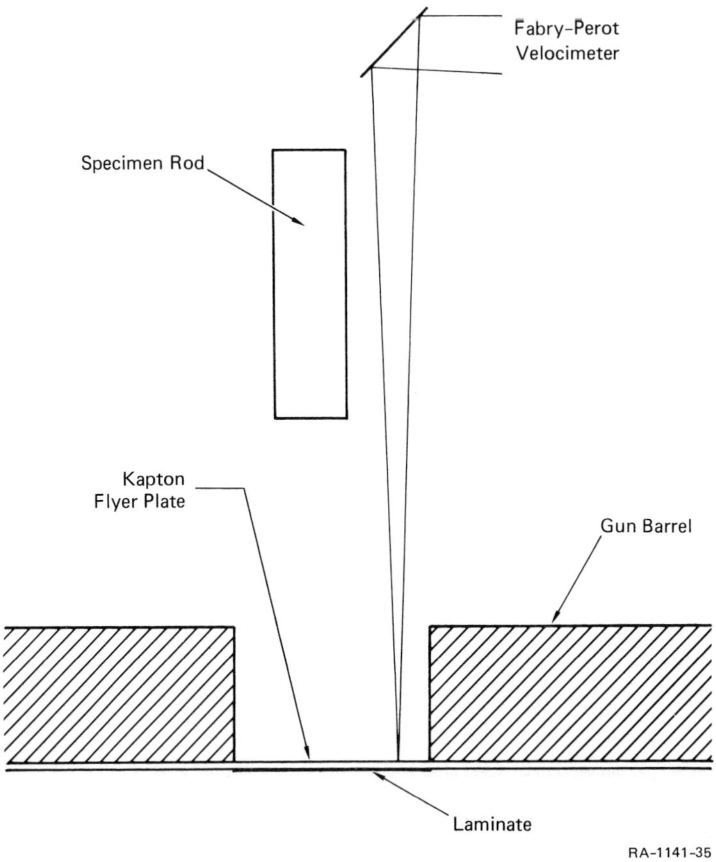

Fig. 2. Experimental arrangement to investigate material response to hypervelocity impact.
(When the laminate explodes, it punches out a section of the Kapton plate and drives it up the barrel to strike the target rod.)

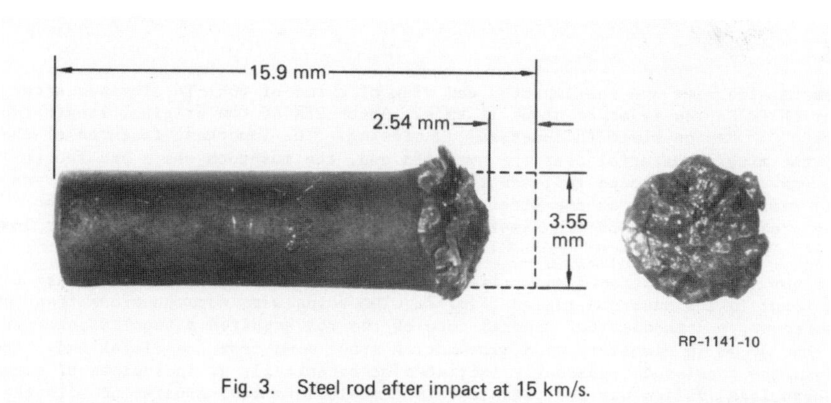

Fig. 3. Steel rod after impact at 15 km/s.

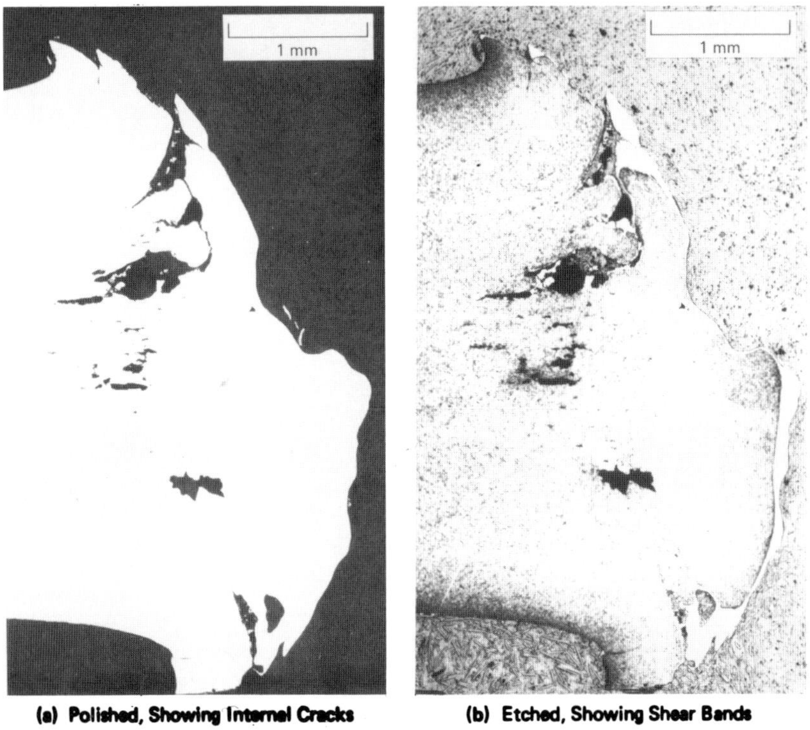

Fig. 4. Cross section of 4340 steel rod impacted by a Kapton plate at 15 km/s.

Since the rod debris was not recovered, we cannot say with certainty how the proximal portion of the rod failed. The white etching layer covering the surface of the rod [and seen in cross section in Fig. 4(b)] indicates high local temperatures and suggests that the debris material adjacent to this surface was in the form of solid fragments formed by intersecting shear bands. The cracks and shear bands below this surface in the mushroom region represent an incipient stage of fragmentation, and thus support this hypothesis.

These observations clearly indicate that rod remnants produced by impact with a thin (with respect to rod dimensions) plate at 15 km/s include a large, intact distal end. Furthermore, although solid fragments were not recovered, the cracks and shear bands observed in the recovered rod suggest that fragments are also produced. The shock energy considerations discussed earlier suggest that some portion of the rod may have been melted, but not vaporized, by the impact.

Aluminum Rod

Figure 6 shows a side view and the impacted end view of a rod of 6061-T6 aluminum after collision with a 0.3-mm-thick Kapton flyer at about 11 km/s. About 28% of the original length (about 1.3 rod diameters or 15 Kapton plate thicknesses) is missing. The important features of the recovered rod, namely, the missing material near the impacted end, the mushroom shape and faceted surface of the proximal end of the recovered rod, and the apparently undamaged distal portion, are qualitatively similar to those of the steel rod. The aluminum rod, however, suffered significantly greater damage than the steel rod, even though the impact velocity was lower.

When the aluminum rod was sectioned and polished to look for internal damage, a dense population of voids was found in the mushroom region, Fig. 7. The voids were more numerous than in the 4340 steel rod; moreover, a 1-mm-diameter central core of the rod acquired a population of smaller voids (less than 30 μm in diameter) to a distance of about 6 mm from the distal end. These voids, not observed in the steel rod, apparently initiated preferentially at inclusions of magnesium silicate. Shear localization was not prevalent in the aluminum rod, consistent with the low propensity of aluminum for this failure mode and in accord with observations at conventional impact velocities. Conclusions from the shock energy considerations discussed earlier suggest that a portion of the rod may have melted, but no vaporization occurred.

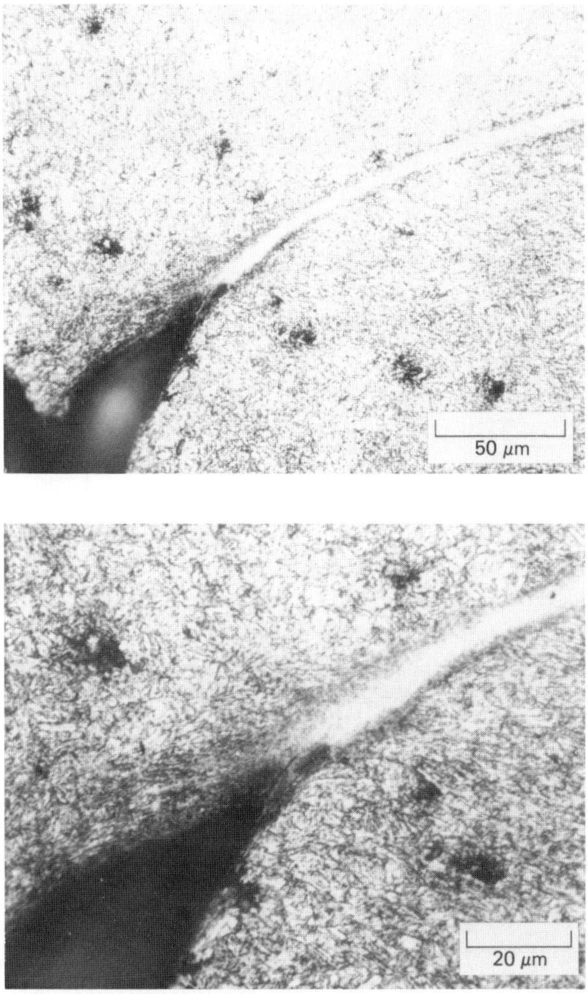

Fig. 5. Close-up views of an adiabatic shear band in a 4340 steel rod after hypervelocity impact.

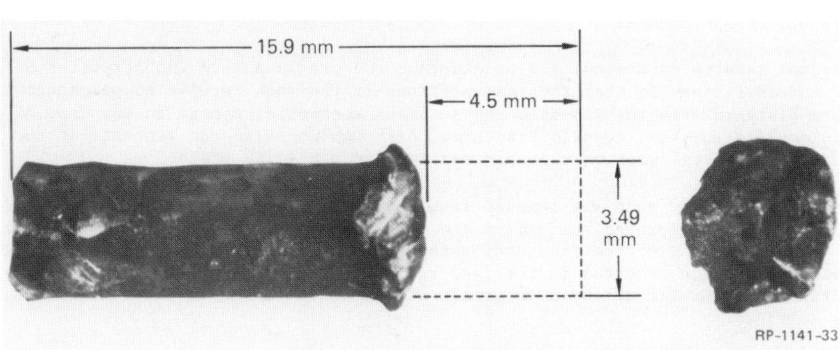

Fig. 6. Aluminum rod after impact at 11 km/s.

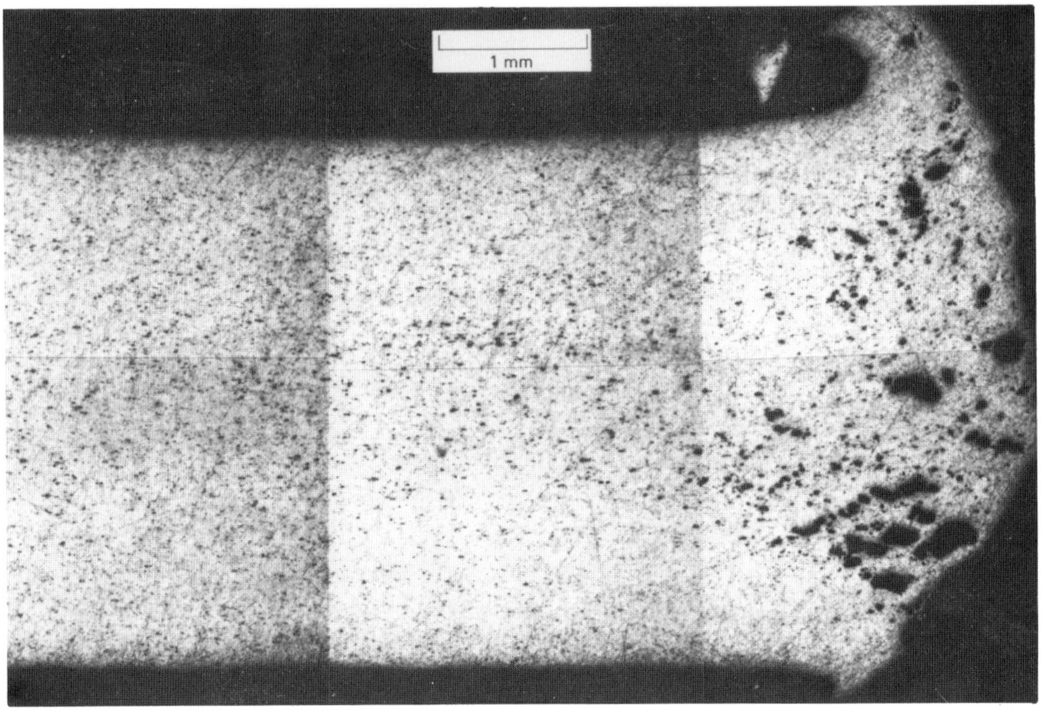

Fig. 7 Cross section of 6061-T6 aluminum rod impacted by a Kapton plate at 11 km/s.

Lead Rod

When a lead rod having dimensions similar to the steel and aluminum rods was impacted at a velocity greater than 10 km/s, no solid distal portion was recovered. Based on the results of the theoretical considerations discussed earlier, we expect that only that portion of the rod within one or two rod diameters of the proximal end was vaporized and melted. Our inability to identify rod remnants in the solid state may be attributed to the very low strength and low fusion energy of lead, causing rod disintegration under subsequent impacts with the chamber walls. More care should be taken in future experiments to recover solid lead debris.

DISCUSSION AND SUMMARY

The experimental results on a steel and an aluminum rod are in accord with expectations from theoretical considerations in that the rear portions of the rods receive no permanent damage, and material some distance from the impacted end sustains mechanical damage in the form of plastic flow, shear localization, and tensile fracture. Although no solid rod remnants of the lead rod were recovered, the distal portion probably remained in the solid state.

Material near the impacted ends was removed from the rods by the impact and was not recovered. Material closest to the impacted surface of the steel and aluminum rod could have melted; material closest to the damaged end of the recovered rods probably fragmented. Theoretical considerations suggest that vaporization occurred in the lead rod, but not in the steel and aluminum rods. The impact vaporization behavior of these materials should be confirmed in experiments designed to contain the debris.

Thus a capability for computing the disintegration behavior of material under hypervelocity impact conditions, in general, must include mathematical models for nucleation, growth, and coalescence of voids and shear bands as well as adequate high-pressure equations of state for melting and vaporization.

Existing dynamic failure models that treat void and shear band evolution in materials impacted at lower velocities appear capable of describing behavior at hypervelocities because the morphology and phenomenology of mechanical damage is similar. Future efforts should be directed to applying these models to simulate well-monitored hypervelocity experiments in which the mechanical damage is quantitatively assessed and the fragment size distribution is determined. Future experimental efforts should focus on improved material recovery techniques in the reverse ballistic electric gun tests.

ACKNOWLEDGMENTS

This work was performed for the Air Force Armament Laboratory, Eglin AFB, under a subcontract to Science Applications International Corporation (SAIC). The interest and encouragement of Major D. Apo of the Defense Nuclear Agency, J. Gagliano of Eglin, AFATL, S. Brockway of the U.S. Army, and N. Banks of SAIC are greatly appreciated.

REFERENCES

Bertholf, L. D., L. D. Buxton, B. J. Thorne, R. K. Byers, A. L. Stevens, and S. L. Thompson (1975). Damage in steel plates from hypervelocity impact. II. Numerical results and spall measurements. J. Appl. Phys., 46, 3776.

Bjork, R. L., and A. E. Olshaker (1965). The role of melting and vaporization in hypervelocity impact. Rand Corporation Memorandum, RM-3490-PR, unpublished.

Chau, H. H., G. Dittbenner, W. W. Hofer, C. A. Honodel, D. J. Steinberg, J. R. Stroud, and R. C. Weingart (1980). Electric gun: a versatile tool for high-pressure shock-wave research. Rev. Sci. Instrum., 51 (2), 1676.

Dullum, O., and L. Seaman (1985). Simulations of projectile penetrations with a shear banding model. SRI International Poulter Laboratory Technical Report No. 004-84, unpublished.

Osher, J. E., H. H. Chau, G. R. Gathers, R. S. Lee, and R. C. Weingart (1986). Application of the 100-kV LLNL electric gun for hypervelocity impact studies. This symposium proceedings.

Shockey, D. A., D. R. Curran, and P. S. DeCarli (1975). Damage in steel plates from hypervelocity impact. I. Physical changes and effects of projectile material. J. Appl. Phys., 46 (9), 3766-3775.

Shockey, D. A., L. Seaman, and D. R. Curran (1979). Microfailure models and their application to nonlinear dynamic fracture problems. In Nonlinear and Dynamic Fracture Mechanics, Vol. 35. The American Society of Mechanical Engineers, New York. pp. 79-104.

Shockey, D. A., L. Seaman, and D. R. Curran (1985). The microstatistical fracture mechanics approach to dynamic fracture problems. Int. J. Fract., 27, 145-147.

Stull, D. R. and G. C. Sinke (1956). Thermodynamic properties of the elements. American Chemical Society, Washington, DC.

Swift, H. F., and A. K. Hopkins (1970). Effects of bumper material properties on the operation of spaced meteoroid shields. J. Spacecraft, 7 (1), 73.

Swift, H. F., D. D. Preonas, W. C. Turpin, and J. M. Carson (1970). Debris clouds behind plates impacted by hypervelocity pellets. J. Spacecraft, 7 (3), 313.

R. E. Tokheim, D. R. Curran, and J. L. Dein. (1983). SRI International, Menlo Park, CA. Unpublished.

Int. J. Impact Engng Vol. 5, pp. 595–602, 1987
Printed in Great Britain

0734-743X/87 $3.00 + 0.00
Pergamon Journals Ltd.

A SIMPLE DEFENSE CONSERVATIVE MODEL FOR MASS REQUIREMENTS OF HYPERVELOCITY PROJECTILE IMPACT SHIELDS FOR REENTRY VEHICLES

George R. Spillman

Los Alamos National Laboratory, Los Alamos, NM 87545

ABSTRACT

Simple analytical modeling of the physics of interaction of hypervelocity (50-100 km/s) projectiles with a bumper shield countermeasure is given. The interaction of projectile and bumper is discussed briefly. Expansion of bumper/projectile debris in the region between bumper and underlying vehicle and interaction of bumper/projectile debris cloud with vehicle are examined. Expansion of debris is treated as an expansion superimposed upon a translation with partition derived from a simple inelastic collision model. The effect of nonunity aspect ratio of compressed debris is included. Debris colliding elastically with the vehicle will impart momentum equal to twice the incident normal component. A steady-state diffusion model is used to estimate the effect of stagnation radiative loss on collision elasticity. Impulse may be reduced up to a factor of 2 by stagnation radiative losses for small projectiles and large bumper/vehicle stand-off. Stagnation radiation loss is small for larger projectiles and smaller stand-off. Impulse can be enhanced by vehicle ablation from radiative coupling, shock heating (inadequate stand-off), or liquid droplet microcratering (inadequate bumper thickness). Estimates of required bumper mass are given for a specific example.

INTRODUCTION

We consider the physics of the problem of bumper shield countermeasuring reentry vehicles against high hypervelocity (>50 km/s) defensive projectiles. The hypervelocity impact bumper is a concept borrowed from the design of meteoroid impact protection for spacecraft [Zukas and colleagues (1982) and Swift and colleagues (1982)]. The concept as applied to a conical reentry vehicle (RV) is illustrated in Fig. 1, in which we indicate a small, high-velocity projectile incident upon a thin balloon (the bumper) surrounding an impulse-hardened conical RV. The strong shock interaction between bumper and projectile causes them to vaporize, which results in an expanding vapor cloud that blast-loads the underlying vehicle. The basic assumption of the projectile interaction with the vehicle is changed from that of a solid, penetrating collision to that of the blast load from a low-density cloud, which makes the interaction problem more analogous to that of the impulse loading because of x rays.

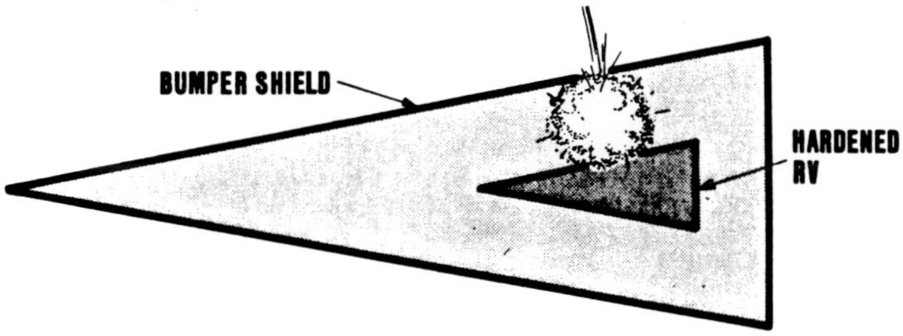

Fig. 1. The bumper shield concept.

In the following, we will consider the example of tungsten for both incident projectile and bumper shield. We will assume a simple right circular cylinder with unit aspect ratio (length = diameter) incident normally with its circular face striking the bumper. Since the specific kinetic energy of the projectile is well above vaporization energy, we will use ideal gas equations for treating the various interactions with nominally $\gamma = 1.5$ and we will use strong shock approximations for all shocks. The geometry variables used in the text are as follows: For the projectile, r_p = radius; ℓ = length; v_p = initial velocity; and m_p = mass. For the target, R = base radius and ϕ = cone half-angle for the underlying vehicle; h = stand-off between bumper and underlying vehicle and the bumper thickness is $\varepsilon\ell$.

INTERACTION OF BUMPER AND PROJECTILE

The impact produces equal strong shocks in the bumper and projectile with equal thermal and translational velocities, and with a compression of about a factor of 5 for Υ = 1.5 (as appropriate for tungsten at these high velocities). Conditions behind the shock will be modified by edge rarefactions and, after the shock emerges from the bumper or back of the projectile, it will be modified by rarefactions from those surfaces.

The resulting shock decay can result in incomplete vaporization of the projectile for sufficiently thin bumpers. We have assumed that this results in an undesirable interaction with the underlying vehicle and that this condition should place a lower limit on the bumper thickness. We find, however, that the limitation has little influence on mass penalties in our range of principal interest and, since the subject of shock decay has been treated elsewhere (Gehring, 1970; Fowles, 1960; Bjork, 1965), we will not present details of that aspect of the modeling here.

We will make a "cookie cutter" approximation to determine the mass of bumper principally involved in the collision; i.e., we assume that the bumper mass involved in the collision is that in direct line of the incoming projectile. A mass, m_p, is incident and a mass, $m_p(1 + \varepsilon)$, emerges containing the incident momentum and energy. Note that this assumption pertains only to the mass sharing the main energy and momentum and not to the size of a hole produced in the bumper. One-dimensional synthesis suggests that at our high velocities, less than 2.5% of the energy is transferred laterally into a bumper of ε = 1/4 and 2D steady-state approximations (Taylor, June 1986) and 2D hydrodynamics calculations (Oyer, 1985) support the cookie-cutter assumption.

VOID REGION EXPANSION

As the projectile and ejected bumper material emerge from the back of the bumper, it is compressed and heated, but still has a net translational velocity. Some of the existing heat energy will be converted to kinetic energy of expansion and the overall expansion may be viewed as a superposition of the net translational velocity and the local expansion velocity. We first consider how the debris would expand if it emerged from the back of the bumper as a uniformly compressed and heated sphere and then discuss the effect of nonspherical nature of the initial debris and non-uniform shocking by thin bumpers. Consistent with our strong shock and ideal gas assumptions, for the case ε = 1 the energy partition will be one-half translational energy and one-half internal energy with all internal energy convertible to expansion kinetic energy. If the sphere expands as a thin shell, then the shell remains tangent to the bumper at θ = 180° (see Fig. 2) and expands at the projectile velocity (v_p) at θ = 0°. In the center-of-mass frame, the expanding balloon has a uniform mass distribution in solid angle.

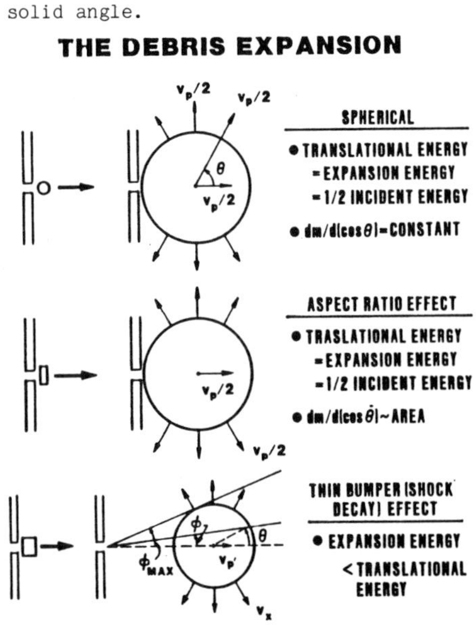

Fig. 2. The debris expansion.

If we could ignore the edge rarefactions until the shocks emerged from front and back, for the case of $\varepsilon = 1$ the emerging debris could be characterized as a disk of diameter equal to the projectile diameter and a length 40% of the projectile length (a linear compression of twice the projectile length by a factor of 5). We could then expect translational and expansion velocities to each be one-half of the projectile velocity, but the uniformity of mass distribution with solid angle would be lost. Since rarefactions in any direction will move the same distance in a given amount of time, the mass moving in a particular direction will be approximately proportional to the area of the compressed debris normal to that direction. For a thin face, however, the effective area will be reduced about a factor of 2 by rarefactions from the other face. From this discussion, we make an assumption that the ratio of dm/dcosθ in the axial and radial directions is proportional to areas normal to $\theta = 0°$ and $\theta = 90°$ gives

$$\left(\frac{dm}{d\cos\theta}\right)_{0°} / \left(\frac{dm}{d\cos\theta}\right)_{90°} = r_p/\tilde{\ell} \quad \text{where } \tilde{\ell} = \text{disk thickness} \quad . \tag{1}$$

We define $\mu = \cos\theta$ and arbitrarily fit the two directions with the form $a + b\mu^2$. Additionally, we assume that the expansion is as from a disk starting at uniform compression η, given from the strong shock equations such that $\tilde{\ell} = \ell(1 + \varepsilon)/\eta$, and using $\eta = 5$ for $\gamma = 1.5$ we obtain

$$\frac{dm}{d\mu} \sim 1 + \mu^2 \left(\frac{1.5 - \varepsilon}{1 + \varepsilon}\right) \quad . \tag{2}$$

Swift, Bamford, and Chen (1982) use a very simple inelastic collision model for the effect of bumper thickness on expansion velocity from which they obtain an expression for the maximum of the angle ϕ. We follow their procedure here and combine with Eq. (2) to obtain an estimate of the angular distribution of momentum. We assume that the expansion will again be uniform in direction in the center-of-mass coordinate system. Defining v_p' as center-of-mass translational velocity and v_T as expansion velocity allows us to obtain from energy and momentum conservation

$$v_p' = v_p/(1 + \varepsilon); \; v_T = v_p \varepsilon^{1/2}/(1 + \varepsilon) \tag{3}$$

and simple geometry gives (see Fig. 2)

$$(\tan\phi) = \varepsilon^{1/2}\sin\theta/(1 + \varepsilon^{1/2}\cos\theta) \tag{4}$$

and the maximum value is

$$(\tan\phi)_{max} = \sqrt{\varepsilon/(1 + \varepsilon)} \text{ at } \theta = \cos^{-1}(-\sqrt{\varepsilon}) \quad . \tag{5}$$

The item of most interest in the expansion is the angular distribution of the axial component of the momentum. This we may write as

$$\frac{d(mv_N)}{d\cos\phi} = v_N \frac{dm}{d\mu} / d(\cos\phi)/d\mu \tag{6}$$

where v_N is the axial velocity given by

$$v_N = v_N' + v_T\cos\theta = v_p\,(1 + \varepsilon^{1/2}\cos\theta)/(1 + \varepsilon) \tag{7}$$

under assumption of expansion as a thin shell at velocity v_T in all directions from the center of mass. Since θ is double valued for any ϕ, there are two values to sum for Eq. (6). From Eq. (4), we obtain

$$\frac{d\cos\phi}{d\mu} = \left\{\frac{\sqrt{\varepsilon}\cos\theta}{1 + \sqrt{\varepsilon}\cos\theta} + \frac{\varepsilon\cos^2\theta}{(1 + \sqrt{\varepsilon}\cos\theta)^2}\right\} \frac{\sin\phi\cos^2\phi}{\sin\theta} \quad , \tag{8}$$

which is most convenient to use in this mixed form. We use Eqs. (2), (4), (5), (6), and (7) to numerically evaluate dm v_N/dcosϕ and normalize to the incident momentum; this is plotted in Fig. 3. The plots show a peak at $\phi = \phi_{max}$ because of the assumption of infinitesimal debris shell thickness. Finite thickness will tend to remove this double-peaked structure. We may note the tendency toward uniform distribution for $\phi < \phi_{max}$ for small ε, even for the infinitesimal shell thickness. Note also that the strict limitation of debris to angles $\phi \leq \phi_{max}$ is an artifact of assigning the rms value of velocity to expansion for all mass points. From Eq. (5), we may obtain for small ε

$$\frac{1}{m_p v_p}\frac{dmv_N}{d\cos\phi} = \frac{2}{\varepsilon} \qquad \phi < \phi_{max} \; (\varepsilon \ll 1) \tag{9}$$

$$= 0 \qquad \phi > \phi_{max} \quad .$$

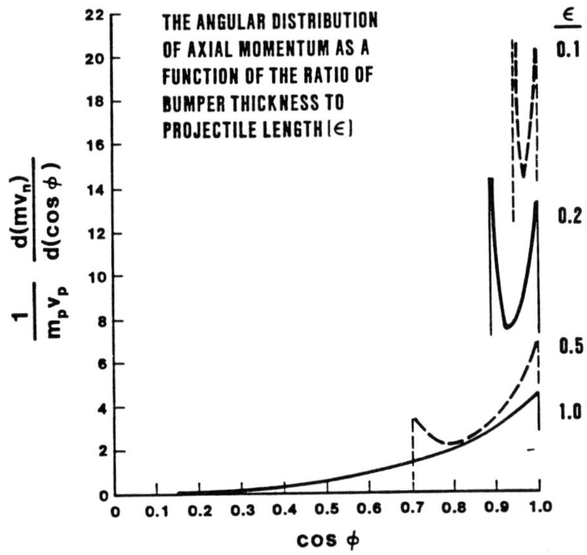

Fig. 3. The angular distribution of axial momentum as a function of the ratio of bumper thickness to projectile length (ε).

INTERACTION OF DEBRIS WITH UNDERLYING STRUCTURE

The debris cloud striking the underlying structure may interact elastically or inelastically. If the collision is elastic, twice the normal component of the incident momentum will be delivered to the structure. The momentum transfer to the structure may be reduced to the normal incident component if the debris can radiate away essentially all internal energy on stagnation (a result given for a specific case by Swift, 1982) without causing any ablation of the underlying structure. The momentum may be enhanced by several possible mechanisms causing ablation, including shock heating, radiative transfer from the stagnating debris, and microcratering by liquid droplets or solid fragments. We assume that avoidance of each of these potential momentum enhancement mechanisms is a countermeasure design goal. We discuss each mechanism to some degree.

To be able to estimate impact and stagnation densities and times, we will assume that the expanding debris shell is 0.2 times a radius thick. Shock ablation may be circumvented by achieving a large density discontinuity between the debris cloud and the material of the underlying structure. This is achieved by debris cloud expansion via an adequate combination of separation between bumper and underlying structure and bumper thickness. Our calculations indicate that reducing the incident vapor impulse/area to a reasonable level seems likely to avoid shock ablation. At least, it is defense conservative to assume so.

At 70 km/s impact velocity, the stagnation temperature of a low-Z debris cloud may be of order 3 X 10^5 K, and that of a high-Z debris cloud may be twice as high. At such high temperatures, there is a possibility of significant radiative energy transfer.

This problem has been considered by Swift, Bamford, and Chen (1982) for small meteoroids. They conclude that most of the kinetic energy of a meteoroid striking a dual-plate shield is expended as radiation at the stagnation zone. In their analysis, they assume that an appropriate model for radiation loss is that of a blackbody flux from the stagnation layer at the stagnation temperature, provided that the stagnation layer is optically thick. They use cold absorption coefficients of aluminum in the vicinity of 430-eV photon energies to argue for optical thickness of the layer. Additionally, they state that energy radiated toward the target is absorbed.

The conclusions reached by Swift, Bamford, and Chen (1982), are probably correct for their application; however, the model they use is not generally applicable to our problems. As the authors realized, their flux is an overestimate if the stagnation layer is optically thin (reduced approximately by the factor 2τ, where τ is the optical thickness). It is also reduced if the layer is optically thick. Although the surface may appear to radiate as a blackbody at some effective temperature, that effective temperature may be significantly less than the interior temperature because interior photons must go through a diffusion process to escape (Zeldovich, 1966). Also, generally the applicable absorption coefficients are not those of cold material, but are changed (usually reduced significantly) because the material becomes highly ionized. Typically, significant ionization occurs at bound electron energy levels many times the average thermal energy of free electrons [Bond, Watson, and Welch (1965)]. Also, although radiation incident upon the target may be absorbed, it may also be reradiated. Net radiation absorption on the target can also be a diffusion problem and that diffusion may be small if the target is denser and/or of higher atomic number than the incident debris.

For our model of radiation loss, we use a steady-state diffusion model. The diffusion equation for radiative flux is

$$F = -\frac{4}{3}\sigma\lambda\nabla T^4 \tag{10}$$

$$= F_+ - F_- \tag{11}$$

where F_+ is a forward flux and F_- is a backward flux, given by

$$F_+ = \sigma T^4 - \frac{2\sigma\lambda}{3}\nabla T^4 \tag{12}$$

and

$$F_- = \sigma T^4 + \frac{2\sigma\lambda}{3}\nabla T^4 \quad . \tag{13}$$

Here, σ is the Stephan Boltzman constant, λ is the Rosseland mean free path, and T is the temperature (Zeldovich, 1966). We now consider a problem of steady-state radiation flow in planar geometry like that of the Milne problem (Zeldovich, 1966), except that we assume a source of energy that is uniform in space, but nonzero in contrast to the Milne problem. The plane is bounded by vacuum at $X = X_0$, and at $X = X_I$, the flux is zero. Then at the boundary $X = X_0$, we set the two expressions for flux equal (as in the Milne problem solution), and obtain

$$F_0 = \frac{4}{3}\sigma\lambda_0 (\nabla T)_0^4 = \sigma T_0^4 - \frac{2\sigma\lambda}{3}(\nabla T)_0^4 \quad , \tag{14}$$

which, as for the Milne problem, gives

$$F_0 = 2\sigma T_0^4 \tag{15}$$

where subscripts zero means evaluation at $X = X_0$. We now define an optical depth τ

$$\tau = \int_0^X \frac{dx}{\lambda} \tag{16}$$

and write for the net diffusion flux in one dimension

$$F(\tau) = -\frac{4\sigma}{3}\frac{dT^4}{d\tau} \quad . \tag{17}$$

Our source assumption gives for the flux divergence in steady state

$$\frac{dF}{dx} = \text{constant} \tag{18}$$

and if we assume for this divergence equation that τ is linear with x, we may integrate and use Eq. (17) and obtain

$$F(\tau) = F_0\tau/\tau_0 = -\frac{4\sigma}{3}\frac{dT^4}{d\tau} \tag{19}$$

and integrate again and use equation

$$F_0\tau_0/2 = \frac{4\sigma}{3}\left[T_I^4 - T_0^4\right] \tag{20}$$

$$F_0 = \frac{T_I^{\ 4}}{1+3\tau_0/4} \tag{21}$$

where an I subscript means evaluated at $X = X_I$ (inside).

If the stagnation layer loses one-half its internal energy by radiation, the rebound momentum may drop a factor of 2 for a total momentum transfer reduction of ~25%. We will somewhat arbitrarily define that condition as the transition between radiative and nonradiative stagnation. Since the radiation rate is a very strong function of temperature, while the specific internal energy is a relatively weak function of energy, radiative rates that are initially high can be quenched rapidly with little internal energy loss. Therefore, to meet our condition above, to be crudely self-consistent, we will estimate the radiation loss rate at the temperature corresponding to a drop of a factor of 2 from the specific internal energy at the temperature that would exist nonradiatively. For our equation of state, this temperature is about 0.6 times the nonradiative stagnation temperature. For our nonradiative stagnation temperature, we assume the collision of a very low density debris on a high-density vehicle and, considering on axis only in a planar approximation we set, the EOS energy equal to the incident specific kinetic energy. The density dependence in the equation of state, we note, is very weak.

For the final assumptions, we have to define an approximately equivalent steady-state problem so that we may apply Eq. (21) to the dynamic stagnation problem. We assume that the appropriate problem is a layer that is one-half the maximum stagnation layer thickness (to represent a very crude time averaging) with a uniform internal stagnation heating source. For the internal temperature (as discussed above) we choose 0.6 times the nonradiative stagnation temperature. Then, we define the radiative/nonradiative transition condition from

$$F_0 = \frac{1}{2}\frac{dE_K}{dAdt} \qquad (22)$$

where $dE_K/dAdt$ is the rate of incidence of debris-kinetic energy per unit area per unit time.

For the equation of state and opacity of the stagnating tungsten debris, we take

$$E\ (J/kg) = 7.3X10^5\ T^{1.6}\ (eV)\ \rho^{-0.16}\ (kg/m^3) \qquad (23)$$

$$K\ (cm^2/g) = 2.4X10^7\ T^{-1.8}\ (eV) \qquad (24)$$

where K is the Rosseland mean opacity defined by

$$K = \frac{1}{\rho\lambda} \qquad (25)$$

where the above expressions are fits to old Los Alamos tables (Huebner, 1969). Note that the equation of state has a very weak density dependence and opacity in the region of interest has a relatively weak density dependence that we have ignored. From the assumption of the debris occupying a shell that is 0.2-radius thick, following the same procedure as for momentum, we may obtain its angular distribution at time of impact. For the separation h between bumper and target and $\varepsilon = 1/4$, the debris density at impact on axis is approximately ρ_I where

$$\rho_I = 12\ m_p/h^3 \qquad (26)$$

and for this purpose only we assume the debris velocity is that of the projectile, which gives for the incident kinetic energy flux on axis

$$\frac{dE_K}{dAdt} = \frac{1}{2}\rho_I v_p^3 \quad . \qquad (27)$$

We assume a strong shock approximation for the stagnation compression so that the stagnation density ρ_S is

$$\rho_S = \rho_I \left(\frac{\gamma+1}{\gamma-1}\right) \qquad (28)$$

and applying the condition of Eq. (25) for the case $\varepsilon = 1/4$, from Eqs. (21), (23), (24), (26), and (27), we obtain for the transition condition

$$h(m) = 2.6X10^3\ m_p^{0.44}\ (kg)/v_p^{1.3}(km/s) \quad . \qquad (29)$$

Separations larger than h will result in significant radiation losses. For $m_p = 1$ g, $\varepsilon = 1/4$, and $v_p = 70$ km/s, h ≈ 50 cm. For tungsten projectiles in the mass range of 1 to 10 g, stagnations on underlying structure such that the momentum/area incident is less than about 4 ktap tend to lose energy radiatively. On structures closer such that incident momentum is much higher, radiative loss is small.

We will not consider the debris/target shock coupling in detail here, but simply state that in the regime of concern here, shock vaporization of the target seems avoidable.

For high-density particles incident (such as liquid droplets) ejecta from craters can give significant momentum enhancement (e.g., about a factor of 5 for projectiles incident on aluminum at 70 km/s) (Dienes and Walsh, 1970). The minimum size of an atomic cluster that acts as a cratering projectile may be estimated by finding the minimum aggregate of atoms that have sufficient energy to melt a hemisphere of depth equal to the atomic range. Such estimates of minimum mass corresponds to large molecules.

A more restrictive condition on the size of liquid droplets that may give momentum enhancement comes from a requirement that they must be able to penetrate the stagnating vapor. One restriction is that the areal density of the drop should be much higher than that of the vapor, to prevent slowing down. This condition we may express in terms of the vapor impulse/area and projectile velocity. The maximum areal density of stagnating vapor is $m/A = L/v_p$, where L is the momentum/area in the stagnating vapor. The droplets will not slow significantly if

$$\rho_D R_D >> \frac{1}{2} m/A = L/2v_p \quad , \tag{30}$$

where ρ_D and R_D are droplet density and radius, i.e., if $R_D >> L/2\rho v_p$ or, equivalently, $m_D >> \pi L^3/6\rho^2 v_p^3$. For v_p = 70 km/s, ρ = 19.3 g/cm^2 and L = 10^5 taps, this requirement is $m_D >> 4 \times 10^{-12}$ kg. Unless the droplets are vaporized in the stagnation layer before reaching the underlying structure, even very small droplets can cause momentum enhancement by microcratering. The effect of possible droplet vaporization by the radiation field or other heat transfer in the stagnation layer has not been adequately considered and may lead to requirements of larger droplet masses for microcratering. For the present, we assume that the bumper is failing to perform adequately if the shock in the projectile drops below that required for full vaporization on release.

APPLICATIONS

We have applied the above physics to estimating minimum bumper mass to reduce blast loading below an assumed lethal impulse/area level for projectiles of various mass and velocities incident on a specific target. The impulse per unit area (L) will be given by

$$L = \frac{1}{\pi h^2} \frac{dmv_N}{d\cos\phi} \tag{31}$$

for nonradiating stagnation or

$$L = \frac{1}{2\pi h^2} \frac{dmv_N}{d\cos\phi} \tag{32}$$

for complete radiation loss on stagnation. Or, if L is a lethal impulse, the required stand-off is

$$h \geq \left[\pi L / \frac{dmv_N}{d\cos\phi}\right]^{1/2} \tag{33}$$

for nonradiating stagnation and a factor of $\sqrt{2}$ higher for complete radiation loss. Here $dmv_N/d\cos\phi$ is obtainable from Fig. 3 or asympotically from Eq. (9).

The target selected is a cone of 2-m length and 10° cone half-angle. The mass of the bumper shield will be its areal density times its area, and is given by

$$M_B = \rho_B \varepsilon \ell \pi \left[R + h\left(\tan\phi + \frac{1}{\cos\phi}\right)\right]\left[R(1+\mathrm{ctn}\phi) + h\left(1+\tan\phi+\frac{1}{\cos\phi}+\frac{1}{\sin\phi}\right)\right] \tag{34}$$

where ρ_B is bumper density. For very small stand-off (h) the area essentially does not vary with increasing h, but the impulse/area falls off as h^{-2}. In such a region, the bumper weight can be decreased by reducing thickness and increasing stand-off (provided that we do not drop below the full vaporization regime). At large stand-off, the area increases as h^2 and, using our asymptotic expression for the angular distribution of momentum [Eq. (9)], we see there is no further gain to be made by increasing the stand-off further. We have used a shock decay model to place a lower limit on bumper thickness that will provide complete shock vaporization of the projectile. We use a rough estimate of 19 Mbar as the minimum shock pressure for which tungsten will completely vaporize on relief (Bennett, May 1984). For projectile masses of 1 g or more or impulse levels below 120 ktap, this limit has a small effect on bumper mass estimates. Those interested in the details may find them in a Los Alamos National Laboratory internal report (Spillman, 1986).

This minimization process has been applied to estimation of required bumper shell mass as optimized against specific projectile mass and velocity for the conical vehicle described above.

Some results for a projectile velocity of 50 km/s and an impulse loading of 120 ktap on the underlying vehicle are shown in Fig. 4. Here, the curve labeled leakage probability of 1 - ε corresponds to normal impact of the projectile as modeled herein. Other values of leakage probability correspond to the probability of the angle of incidence for a random direction of the projectile resulting in less than 120-ktap impulse. In modeling oblique impact, we assume that the effective thickness of the bumper is increased by the reciprocal of the cosine of the angle between the trajectory and the normal to the surface and that the angular distribution about the trajectory line is unchanged. Since the debris cloud (Gehring, 1970) is deflected toward the normal, we believe this results in an overly conservative estimate of leakage.

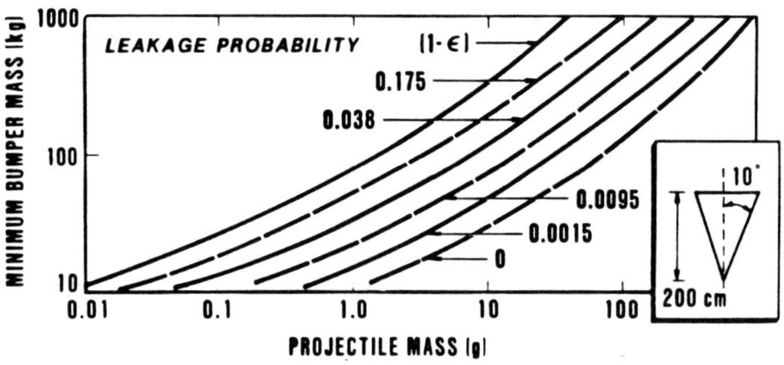

Fig. 4. Minimum required bumper mass to achieve specified survival (leakage) probability against random-direction hit of tungsten projectile with velocity of 50 km/s.

REFERENCES

Zukas, J. A., Nicholas, T., Swift, H. F., Greszczuk, L. B., and Curran, D. R. (1982). Impact Dynamics, John Wiley and Sons, New York.

Swift, H. F., Bamford, R., and Chen, R. (March 1, 1982). Designing dual-plate meteoroid shields--A new analysis, JPL Publication 82-39.

Gehring, J. W. Jr. (1970). In R. Kinslow (Ed.) High-Velocity Impact Phenomena, Academic Press, New York.

Fowles, G. R. (1960). Attenuation of the shock wave produced in a solid by a flying plate, J. App. Phys. 31, 655.

Bjork, R. L., Rosenblatt, M., Brooks, M. B., and Kreyenhagen, K. N. (April 1965). Hypervelocity impact of end-oriented rods, U. S. Naval Research Laboratory, NRL Report 6310.

Taylor, J. W. (June 1986). Personal communication, Los Alamos National Laboratory.

Oyer, A. T. (May 1986). Personal communication, Los Alamos National Laboratory.

Bond, John W., Jr., Watson, Kenneth M., and Welch, Jasper A., Jr. (1965). Atomic theory of gas dynamics, Addison-Wesley Publishing Company, Inc.

Zel'dovich, Y. B. and Raizer, Y. P. (1966). Physics of shock waves and high-temperature hydrodynamic phenomena, Vol. I, Academic Press, New York.

Huebner, W. (1969). Personal communication, Los Alamos National Laboratory.

Dienes, J. K. and Walsh, J. M. (1970). In R. Kinslow (Ed.) High-Velocity Impact Phenomena, Academic Press, New York.

Bennett, B. (May 1984). Personal communication, Los Alamos National Laboratory.

Spillman, G. R. (January 1986). Hypervelocity impact bumpers as countermeasures, Los Alamos National Laboratory report LA-10629-MS.

Int. J. Impact Engng Vol. 5, pp. 603–611, 1987
Printed in Great Britain

0734-743X/87 $3.00 + 0.00
Pergamon Journals Ltd.

CONSTITUTIVE MODEL USED IN COMPUTER SIMULATION OF TIME-RESOLVED, SHOCK-WAVE DATA*

D. J. Steinberg

Lawrence Livermore National Laboratory
University of California
Livermore, California 94550

Abstract

We have designed a constitutive model for use with high-speed, hydrodynamic computer codes. The model, valid at high-deformation rates, accounts for pressure and temperature dependence of the yield strength and shear modulus, work hardening, pressure-dependent melting, Bauschinger and strain-rate effects, and spall. There are a minimum number of parameters needed to implement the model, and most can be determined without recourse to shock-wave data. At Lawrence Livermore National Laboratory, we assembled a library of these material properties for 44 metals, alloys, mixtures, and compounds. Shock and release data from plate-impact experiments for Be, U, Ta, Cu, 1100-0, and 6061-T6 Al, with peak stresses from 6.4 to 230 GPa, are successfully compared against calculations.

Introduction

About 15 years ago, we developed a constitutive model for metals applicable at high-strain rates (Cochran and Banner, 1977; Steinberg, Cochran, and Guinan 1980; Steinberg and Sharp, 1981). The model is designed for use with high-speed, hydrodynamic computer codes, and therefore is based on continuum mechanics. There are a minimum number of material parameters necessary to implement it, the majority of which can be determined using well-known techniques; for example, ultrasonics, compression-tension tests, and the Hopkinson split-bar. However, a few are best evaluated from shock-wave experiments. For these, a single experiment, done at a peak loading stress of a few times the Hugoniot elastic limit (HEL), is usually sufficient. Over the years, we have built a library of material properties for 44 metals, alloys, mixtures, and compounds.

The solid line in Fig. 1(a) schematically illustrates a typical, experimental, velocity-time profile that one might observe in a plate-impact experiment. The dashed line similarly depicts what a calculation of this experiment might look like using an idealized, elastic-plastic model under the condition of uniaxial strain. The two bear faint resemblance to one another.

For the idealized, elastic-plastic behavior, the yield strength Y and shear modulus G are constant. We have made Y and G functions of pressure, temperature, and strain, and also taken into account the fact that the elastic and plastic waves cannot always be treated as separate, independent identities. No part of the model is considered to be a unique representation of the physics; we are looking only for simple ways to reproduce measured results. Using the typical experimental profile in Fig. 1(a) as a guide, we can see how our model will operate.

Upon loading, the time difference between the elastic and plastic waves is decreased because of work-hardening and the rise in pressure, and increased because of the rise in temperature. In addition, the clear distinction between these idealized waves is removed because of rate-dependent effects. The pressure and temperature also affect the arrival time of the elastic release; the pressure decreases the arrival time, while the temperature increases it. The Bauschinger effect removes the clear distinction between the elastic and plastic unloading waves, creating a continuous decrease in the measured velocity. Finally, spall can manifest itself through one or more oscillations in the experimental data.

Figure 1(a) is recast as 1(b) into the more familiar coordinates of stress-total strain, $\sigma - \varepsilon_t$. The stress can be written as $\sigma = -P + S$, where P is the pressure and S is the stress deviator. Using the Von Mises yield criterion, S cannot exceed $2/3Y$. If $S < 2/3Y$, then $\sigma = -P + 2G\varepsilon_D$, where ε_D is the deviator strain. For strains in excess of ε_D at the HEL, $\sigma = -P + 2/3Y$.

Shock-wave data provide excellent proof tests of the model because the shock process simultaneously affects so many thermodynamic and mechanical variables. For example, work-hardening, the Bauschinger effect, and the pressure dependence of Y all affect the shape of the unloading curve. A material such as Be exhibits significant work-hardening and has a large Bauschinger effect, but the pressure dependence of Y is small. On the other hand, Al shows modest work-hardening and Bauschinger effect, but Y has a substantial pressure dependence. Therefore, a good way to validate the model is to successfully calculate time-resolved shock-wave data for a wide variety of materials and initial conditions.

* Work performed under the auspices of the U.S. Department of Energy by the Lawrence Livermore National Laboratory under Contract W-7405-Eng-48.

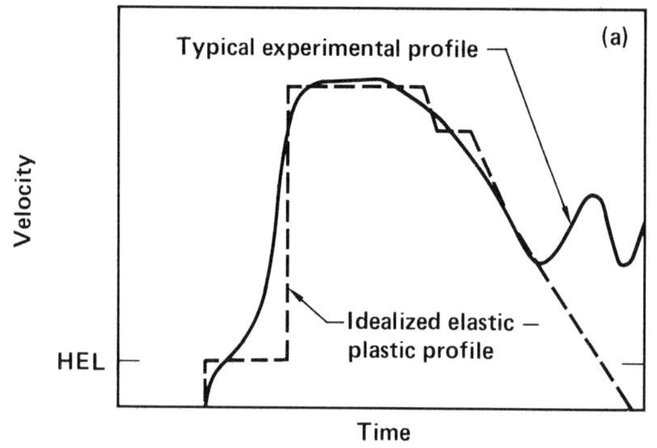

Fig. 1.(a) **Comparison of an idealized, elastic-plastic, velocity-time profile and a schematic representation of a typical experimental profile for a plate-impact experiment.**

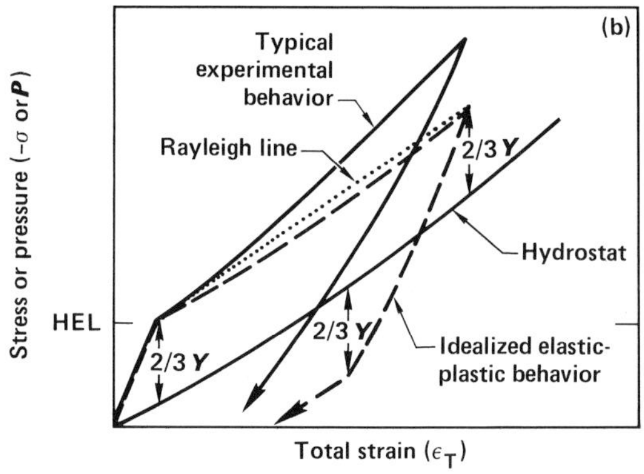

Fig. 1.(b) **The information in Fig. 1.(a) in stress-strain coordinates. The HEL is the Hugoniot elastic limit.**

When this model was first described in the literature, very little time-resolved, shock-wave data were available for peak loading stresses in excess of 10 GPa. This problem has since been rectified, primarily by the Sandia, Albuquerque shock-wave group. Our primary purpose here is to show how well our model can predict these new data. We will discuss the various parts of the model and compare experimental and calculational velocity-time shock loading and release profiles for Be, U, Ta, Cu, 6061-T6, and 1100-0 Al. (There is the advantage of simplicity and accuracy in working directly with the velocity-time data rather than first integrating the measurements to get stress-strain.) In addition, we will point out model deficiencies, and suggest directions for future work to remedy these shortcomings.

Pressure and Temperature Dependence of the Shear Modulus and Yield Strength

The model for the pressure P and temperature T dependence of the shear modulus G and yield strength Y is described fully by Steinberg, Cochran, and Guinan (1980). Briefly, in the typical experimental range ($P < 1$ GPa), G is linear in P. Similarly, for most engineering materials, G is also nearly linear in T all the way to the melting point. To extrapolate the pressure dependence in a reasonable fashion, G is made to vary as $P/\eta^{1/3}$, where η is compression. In this way, even at Thomas-Fermi conditions, G cannot become unreasonably large. We write G as

$$G = G_0\left[1 + \left(\frac{G_P'}{G_0}\right)\left(\frac{P}{\eta^{1/3}}\right) - \frac{G_T'}{G_0}(T - 300)\right] . \tag{1}$$

The subscript 0 refers to the reference state (T = 300K, P = 0), and primed parameters with the subscripts P and T imply derivatives of that parameter with respect to pressure or temperature at the reference state. The parameters G_0, G_P', and G_T' are determined from ultrasonic data.

While there are extensive data showing the pressure and temperature variation of G, there are no such definitive data for Y. What theoretical and experimental guidance that does exist indicates that

$$\left.\frac{1}{Y_0}\frac{dY}{dX}\right)_0 \approx \left.\frac{1}{G_0}\frac{dG}{dX}\right)_0 , \tag{2}$$

where X stands for either P or T. Consequently, we have assumed that the pressure and temperature dependences of Y are the same as that for G. The parameter Y_0 is determined from HEL data whenever such data are available. Otherwise, the highest strain-rate or the lowest temperature data are used that most nearly approximate dynamic conditions.

Because of the very large value of G_P' for Al, the wave profile for 6061-T6 Al shocked to 41.2 GPa (Asay and Chhabildas, 1980) is the clearest illustration of the behavior of G and Y. For G, Fig. 2 compares our best calculation and the experimental data. The agreement for the elastic release is excellent. (The discrepancy for the plastic release will be discussed later.) The first point to be made is that the calculated elastic release arrives at the correct time. This is because the pressure dependence has increased G from 27.6 GPa to 81.0 GPa. If this pressure dependence is reduced by 75%, the elastic release comes much too late, as illustrated by the third curve in Fig. 2.

Figure 3 shows the effect of removing the pressure dependence from Y. There is a much smaller drop in velocity during the elastic release because the calculated strength has dropped from 1.16 to 0.22 GPa. (This latter figure is actually less than Y at the HEL, 0.29 GPa, because the temperature effect is still operating, and it lowers Y.)

Until melting occurs, the effect of temperature on Y and G is typically much less significant than the pressure dependence. Figure 4 illustrates the effect of turning off the temperature dependence in both Y and G. The elastic release arrives about 20 ns early. While observable, we need more wave-profile data to see if this effect is outside experimental uncertainties.

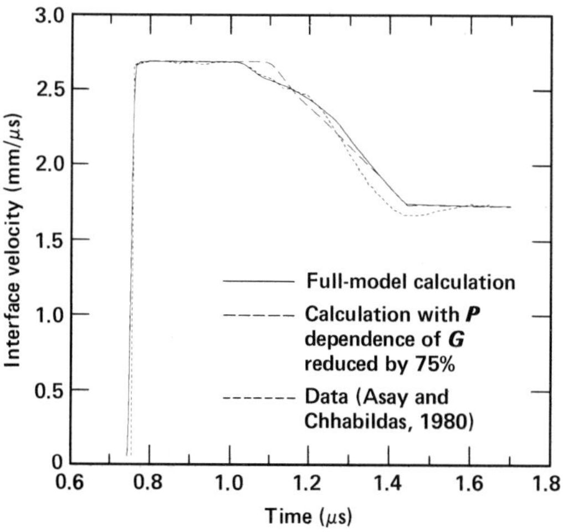

Fig. 2. Comparison of calculations and experiment illustrating the effect of the pressure dependence of the shear modulus for 6061-T6 Al.

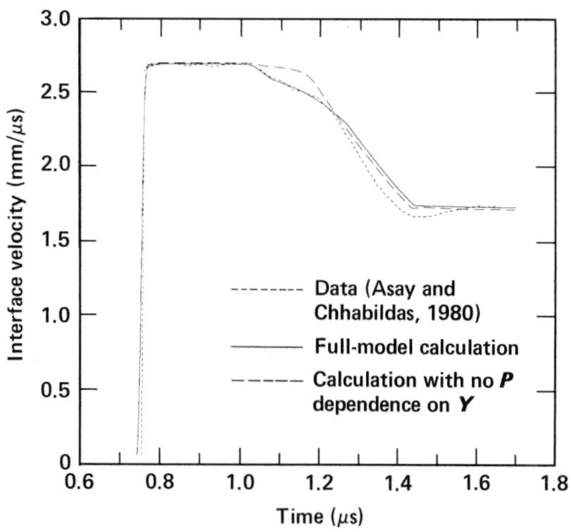

Fig. 3. Comparison of calculations and experiment illustrating the effect of the pressure dependence of the yield strength for 6061-T6 Al.

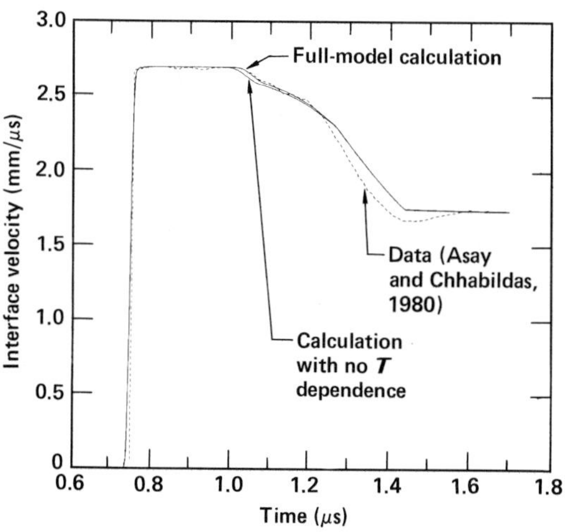

Fig. 4. Comparison of calculations and experiment illustrating the effect of the temperature dependence of the yield strength and shear modulus for 6061-T6 Al.

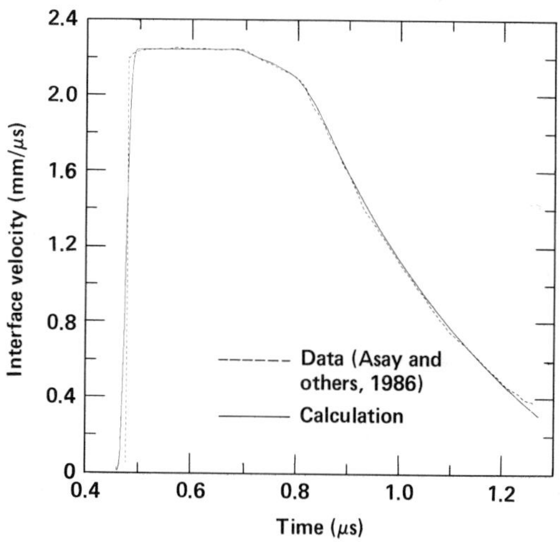

Fig. 5. Comparison of calculations and experiment for 1100-0 Al initially shocked to 50.0 GPa.

We believe our inability to calculate the plastic release profile is an experimental problem. Figure 5 shows a more modern experiment (Asay and others, 1986) for 1100-0 Al initially shocked to 50 GPa. The total release path is very well reproduced by the calculation (Y = 0.65 GPa, G = 89.2 GPa, where Y_0 = 0.06 GPa, as estimated from static data). This experiment used a LiF window, while the earlier one used PMMA. Perhaps this is a significant difference. The only other experiment where we had trouble reproducing the plastic release profile was one with PMMA; an experiment using Cu initially shocked to 93.7 GPa (Chhabildas and Asay, 1982). This comparison is illustrated in Fig. 6 (Y = 0.88 GPa and G = 137.8 GPa). The shape of the calculated elastic release agrees well with the data, but is lower overall. However, when the elastic release arrives at the Cu/PMMA interface, the experimental velocity increases slightly, which does not seem reasonable.

Figure 7 shows a more complicated release profile, where Be has been initially shocked to 34.4 GPa (Wise, Chhabildas, and Asay, 1982). Here Y has increased from 0.33 to 1.49 GPa, and G from 151 to 216.7 GPa. Again, the total experimental release profile, observed using a LiF window, is in excellent agreement with the calculation. Further discussion appears in Steinberg, Cochran, and Guinan (1980).

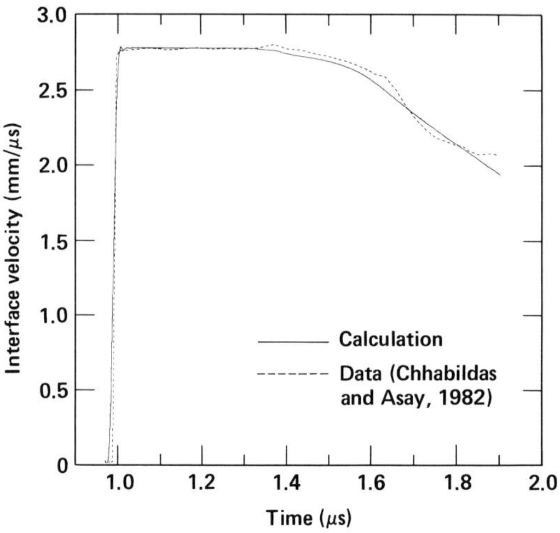

Fig. 6. Comparison of calculations and experiment for Cu initially shocked to 93.7 GPa.

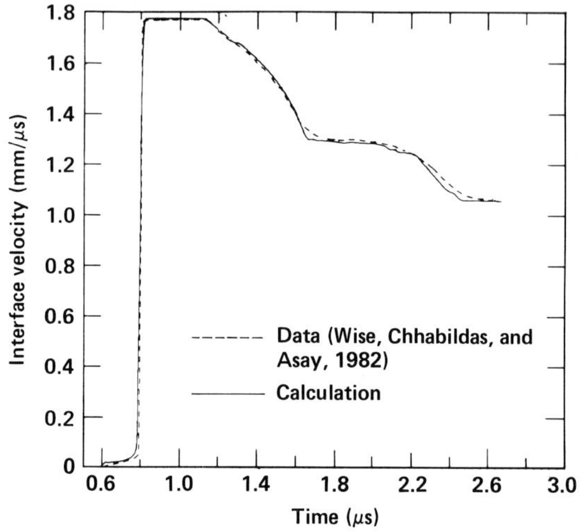

Fig. 7. Comparison of calculations and experiment for Be initially shocked to 34.4 GPa.

Work Hardening

Our two-parameter, work-hardening function $(1 + \beta\varepsilon)^n$ is empirical and appears to do a reasonable job of fitting the available data. The parameter ε is the equivalent plastic strain, and the amount of work hardening is limited so that $Y_0(1 + \beta\varepsilon)^n$ does not exceed some maximum work-hardened strength Y_{max}. Unless there is specific data for Y_{max}, it is simply the largest value for Y in the literature. A detailed discussion on the determination of Y_{max}, β, and n can be found in Steinberg, Cochran, and Guinan (1980). Beryllium exhibits considerable work hardening (Steinberg, Breithaupt, and Honodel, 1985). Figure 8 shows the effect of work hardening on the release profile in a plate-impact experiment where Be was initially shocked to 17 GPa (Wise, Chhabildas, and Asay, 1982). At peak stress, Y is 0.93 GPa and G is 185 GPa. With this model, we have also successfully calculated Taylor anvil tests (Steinberg, Breithaupt, and Honodel, 1985).

Our constitutive model, so far described, has two major deficiencies. The first is that it is unable to predict the elastic precursor seen on re-shock (Asay and Chhabildas, 1981). This is because a Von Mises yield criterion is used in our typical hydrodynamics code, and once a material has yielded, there is no way a second precursor can be calculated. To allow an elastic precursor on re-shock, either the yield surface must increase, or the stress deviator must decrease prior to the arrival of this wave. Swegle and Grady (1985a) have made an interesting start on this problem, but much more remains to be done.

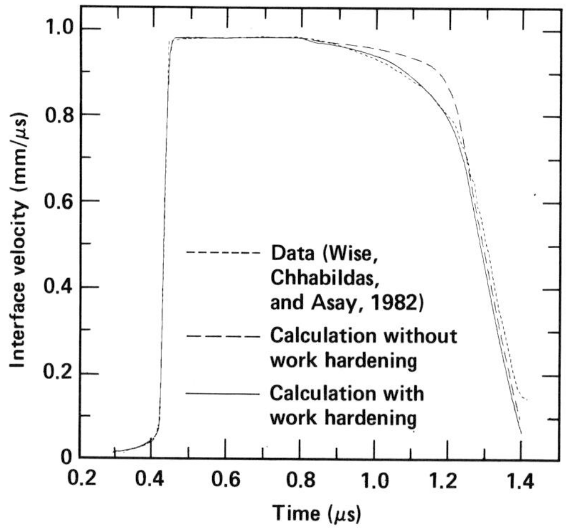

Fig. 8. Comparison of calculations and experiment illustrating the effect of work hardening in Be initially shocked to 17.0 GPa.

The second deficiency is termed thermal softening. Asay and colleagues (1986) note that the strength of 1100-0 Al dropped as the initial loading stresses were raised to levels close to what is required to shock-melt Al. Our model for Al has Y always increasing with increasing loading stress because the pressure dependence and work hardening increase Y more than the temperature dependence decreases it. Successful reproduction of Asay's data could only be accomplished by decreasing Y_0, which makes Y_0 an arbitrary parameter rather than a material constant. The data of Asay have some experimental problems and the elastic release is difficult to see, but the calculations show that $Y = 0.6$ and 0.5 GPa for initial loading stresses of 86 and 104 GPa, respectively.

Obviously, an additional temperature-dependent parameter could be added to the model to allow us to calculate these high-stress wave profiles with one value of Y_0. However, we do not know of any such model that has a physical basis. This is a place where additional work and ideas would indeed be welcome.

Bauschinger Effect

The Bauschinger effect is the phenomenon of elastic-plastic materials behaving differently upon unloading than when they were loaded. Our Bauschinger model is described by Cochran and Guinan (1976), and Steinberg, Cochran, and Guinan (1980). Unlike the usual approach that acts on Y, this model modifies G. True deviator stress-deviator strain data ($S_1 - \varepsilon_1$) taken from quasistatic tension-compression experiments were differentiated to produce an effective G, $G_{eff} = 1/2\ dS_1/d\varepsilon_1$. When G_{eff}/G_0 is plotted against S_1/S_1^{max}, where S_1^{max} is the maximum value of S_1 upon loading, it was observed that G_{eff}/G_0 could be represented quite well by two straight lines, the first for $1 > S_1/S_1^{max} > 0$, and the second for $0 > S_1/S_1^{max} > -1$.

When shock unloading begins, the normal G_0 is replaced by G_{eff}. To completely specify the release path, only three values of G_{eff} are needed. For the values of S_1/S_1^{max} in parentheses, $G_{eff}(+1) = G_0$, $G_{eff}(0) = G_1$, and $G_{eff}(-1) = G_2$, where G_2 is normally equal to zero. Therefore, this is essentially a one-parameter (G_1) model. In addition, we found for most materials, $G_1 = G_0$ gave good agreement with experiment (i.e., no additional parameters).

However, materials such as U, Be, and Ti exhibit large Bauschinger effects. The velocimeter data of Wise, Chhabildas, and Asay (1982) show the motion of a Be/LiF interface that was initially shocked to 17.6 GPa (Fig. 9). In the calculation, $Y = 0.95$ GPa and $G = 186.2$ GPa. These data are compared against calculations done with and without the Bauschinger model. The importance of the model in correctly shaping the release profile is clearly evident.

The principal drawback of this model is that it cannot be generalized to two- and three-dimensional codes. This is because there is only one independent strain in one-dimension, so loading and unloading are well defined. However, in three-dimensions, there are five independent deviator strains, so one must invent some new, unambiguous loading/unloading criterion. Moss and Glenn (1983), have proposed such a model suitable for use in large, multidimensional computer codes. However, their comparison of the model with data was limited to one experiment, so more work is needed to verify its predictive capability.

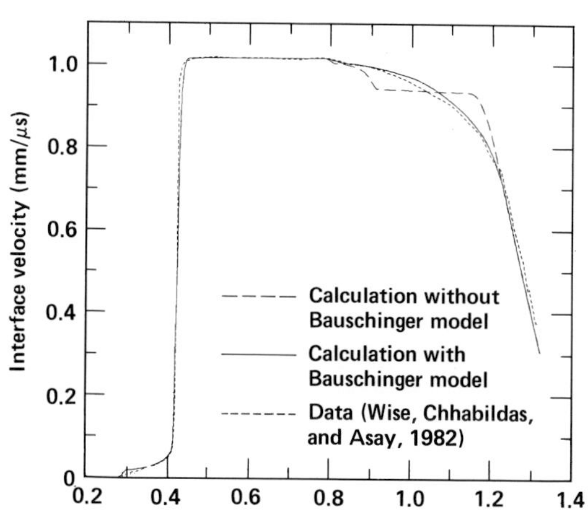

Fig. 9. Comparison of calculations and experiment illustrating the Bauschinger effect in Be initially shocked to 17.6 GPa.

Melting

Our model for melting as a function of compression is described by Steinberg, Cochran, and Guinan (1980). We generally use a Lindemann-type relation, but for special cases, where a Lindemann law is patently wrong, we have developed polynomial representations of melt energy as a function of compression based on the known phase diagrams. Experimental data for melting at high pressure

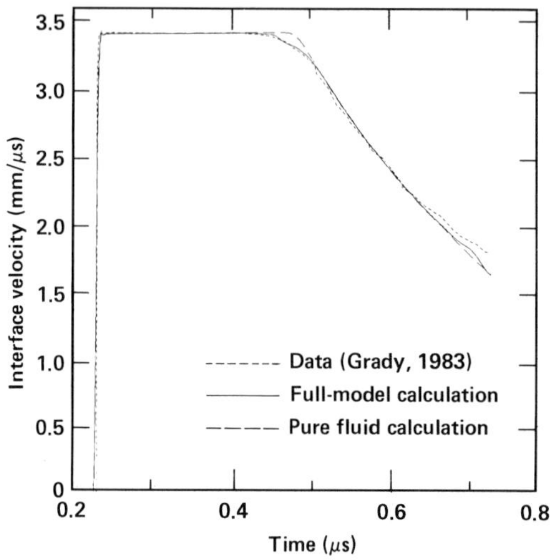

Fig. 10. Comparison of calculations and experiment for Ta initially shocked to 230.2 GPa (Y = 2.54 GPa, G = 198.0 GPa).

are sparse. Shaner, Brown, and McQueen (1983) show data for melting on the Hugoniot for Ta and 2024 Al, and Bernier and Lalle (1982) give similar data for Pb. The Lindemann predictions for these metals are reasonably good. If the Lindemann law is not sufficient, it is easy to replace it with any other melt criterion because the question of melting is not intimately connected to the rest of the constitutive model.

Grady (1983) has performed an experiment on Ta, which is shown in Fig. 10. The short dashed line is the experimental velocimeter record of a Ta/LiF-window interface as a function of time. The 1.388-mm-thick Ta was initially shocked to 230.2 GPa by a 1.013-mm-thick Ta plate driven to 3.4 mm/μs by a two-stage gas gun. The constitutive model, shown by the solid line, correctly calculates the elastic and plastic release. To simply illustrate the effect of melting on calculated wave profiles, we also treat Ta as a liquid. This is shown by the long dash line in Fig. 10, and the elastic release is now no longer evident.

Rate-Dependent Model

A full description of our elastic-viscoplastic model is given by Steinberg and Sharp (1981). In this model, we wish to calculate the following experimental observations.

- Upon shock loading to a few times the HEL, there is no clear distinction between the elastic and plastic waves.
- As the peak stress increases, the plastic wave becomes sharper and more distinct.

The standard, linear solid model is used to represent the stress deviator; that is,

$$\dot{S} = 2G\dot{\varepsilon} - \left(S - S_{eq}\right)/\tau \quad . \tag{3}$$

Dots over S and ε denote Lagrangian derivatives with respect to time, and S_{eq} is the equilibrium or strain-rate-independent stress deviator.

$$\dot{S}_{eq} = 2G\dot{\varepsilon}, \text{ when } \left|S_{eq}\right| < 2/3Y \quad , \tag{4}$$

otherwise $\left|S_{eq}\right| = 2/3Y$. The parameter τ is written as

$$\tau = A_1 \exp\left(Q/E_{in}\right) \quad , \tag{5}$$

where the instantaneous thermal energy is

$$E_{in} = \left(E - E_c\right) + A_2\left|S\dot{\varepsilon}\right| \quad . \tag{6}$$

The parameters E and E_c are the total and zero-Kelvin-compression energies, respectively, and A_1, A_2, and Q are material constants.

The constant Q is an activation energy whose value is chosen to be the same as for other rate-dependent phenomena in solids such as creep and self-diffusion. Our calculations are not very sensitive to the choice of Q. The constant A_2 can be estimated to a factor of two

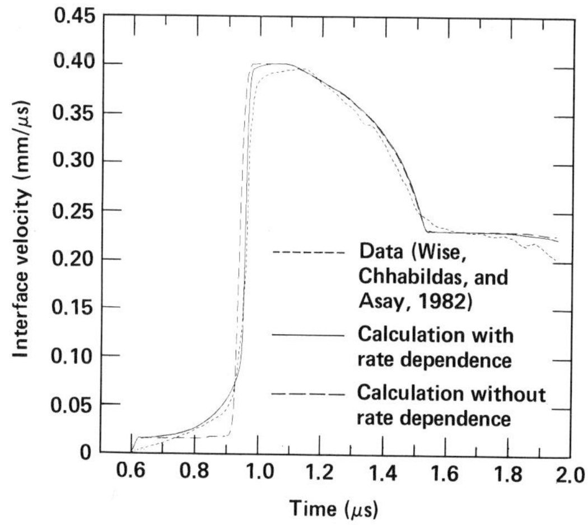

Fig. 11. Comparison of calculations and experiment illustrating rate-dependent effects in Be initially shocked to 6.38 GPa.

from the grain size X and diffusion coefficient D of the sample material; that is, $A_2 \approx X^2/D$. The determination of A_1, and some adjustment of A_2, are done by normalizing the calculation to at least one shock-loading profile.

Steinberg and Sharp (1981) showed a number of shock-loading and release profiles for S-200 Be initially shocked to 1.1 to 2.4 GPa. The calculated profiles compared very well with the experimental data. In Fig. 11, the same values of A_1, A_2, and Q were used to calculate a 6.38-GPa experiment (Wise, Chhabildas, and Asay, 1982) for a Be very similar to S-200. (The calculated Y is 0.56 GPa, and G is 164 GPa.) The calculated shock-loading curve agrees very well with the data, attesting to the predictive abilities of the model. Without the rate-dependent model, neither the shape nor the timing is as good.

The parameter τ has some relationship to solid viscosity (Steinberg, Breithaupt, and Honodel, 1985). However, the major drawback of the elastic-viscoplastic model is that A_1 must be determined through normalization to experiment and it does not have any obvious physical meaning. More recent work, such as that of Swegle and Grady (1985b), may prove to be better models.

Spall

The spall model is described completely by Cochran and Banner (1977). Briefly, the process of spall begins when the tensile stress normal to the spall plane exceeds a given spall strength Σ. The process continues through the formation and growth of microcracks. In our model, after the Σ criterion is met, any increase in volume is assumed to be an increase in the volume of microcracks. In a one-dimensional, finite-difference calculation, we define a damage function at a given location x and time t by

$$D(x,t) = \int_0^t dV/A,\; dV > 0 \quad , \tag{7}$$

where A is the unit area, and V is the volume of microcracks.

The material weakens as D increases until complete spall occurs. This weakening is accomplished by altering material properties (for example, Y, G, and Σ) by multiplying them by a strength function

$$F = 1 - (D/D_0)^{2/3} \quad , \tag{8}$$

where D_0 is a critical value of damage that corresponds to total fracture.

The model requires only two parameters, Σ and D_0. The parameter Σ is related to the pullback velocity, and D_0 to the size distribution of voids when they coalesce.

One change has been made in our constitutive model since our last paper (Steinberg and Sharp, 1981). We found that when both the spall and rate-dependent models were operating simultaneously, the code calculations were too zone sensitive. Spall is by nature an unstable process, difficult to interface with other physical models. We found that this zone sensitivity problem disappeared if the rate-dependent model was turned off in zones that had begun to spall.

The rate-dependent model now affects spall only by its effect on the shape of the tensile wave through the sample at the start of spall. If the rate-dependent model is not used, incipient spall concentrates in just a few zones, creating a steep, tensile-wave profile across the spall plane. With the rate-dependent model operating, this tensile-wave profile is broader, involving more zones. As pointed out by Steinberg and Sharp (1981), the shape of the tensile-wave profile plays an important role in how a material spalls.

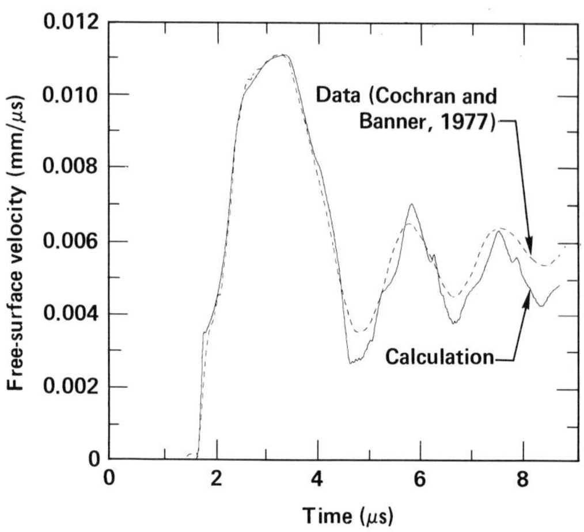

Fig. 12. Comparison of calculations and experiment illustrating spall effects in U initially shocked to 3.2 GPa.

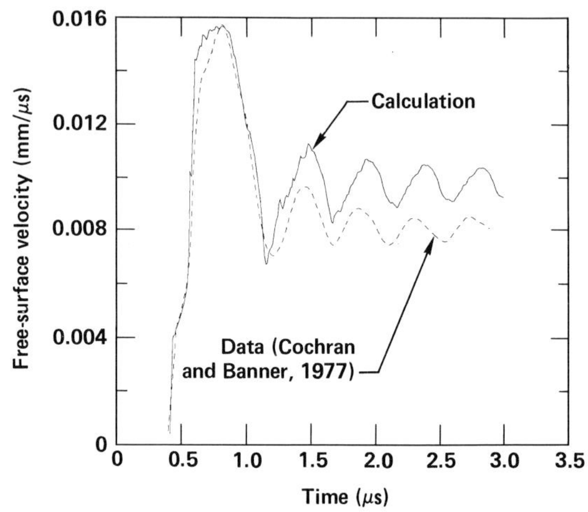

Fig. 13. Comparison of calculations and experiment illustrating spall effects in U initially shocked to 3.9 GPa.

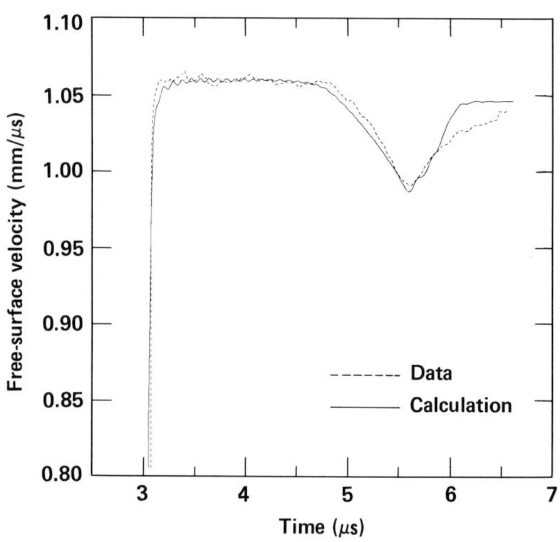

Fig. 14. Comparison of calculations and experiment illustrating spall effects in U initially shocked to 33.8 GPa. Note suppressed zero.

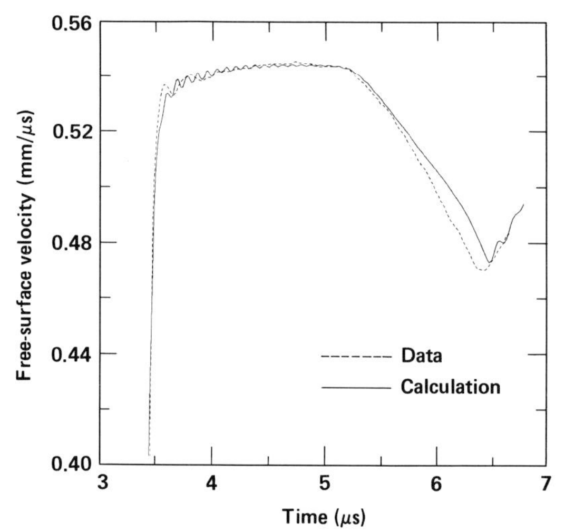

Fig. 15. Comparison of calculations and experiment illustrating spall effects in U initially shocked to 15.7 GPa. Note suppressed zero.

Two experiments described by Steinberg and Sharp (1981), were recalculated with this new version of the constitutive model. The experimental data are compared with the calculations in Figs. 12 and 13. The values for Σ and D_0 are 2.5 GPa and 25 μm, respectively. The calculated spall peaks are higher than the data in Fig. 13, but this is the first time we have been able to calculate even the frequency of oscillation correctly for this experiment. (See Cochran and Banner, 1977; Steinberg and Sharp, 1981.)

The vast majority of spall data is for initial stress-loading below 5 GPa. To test our model at higher stress, we performed two plate-impact experiments on U at peak stresses of 15.7 and 33.8 GPa. In both experiments, a 5.029-mm-thick U plate struck a U target, and a Fabry-Perot velocimeter followed the motion of the target free surface. The low-stress experiment had a target 10.013-mm-thick, and the flyer velocity was 0.55 mm/μs. The corresponding parameters for the high-stress experiment were 10.027 mm and 1.06 mm/μs.

The data are compared against our model calculations in Figs. 14 and 15 and the agreement is very good. At these stresses, the effect of the rate-dependent model is negligible, so these experiments are good tests for our choice of D_0. A D_0 value of 25 μm is especially good in terms of the time when spall occurs and the trough width in the 33.8-GPa experiment. Using 55 μm, as Cochran and Banner (1977) did in their original work at low stresses, makes the bottom of the trough occur too late. At lower stresses, where the rate-model is important, this same value of D_0 also works well, as we have seen. Consequently, the rate-dependent model allows us to cover the stress range up to 33.8 GPa with one set of material properties for the Cochran-Banner spall model.

Considering that this model contains only two parameters, it works remarkably well. However, it is obvious that the onset of spall cannot be completely described by the single parameter Σ. Steinberg and Sharp (1981) show the data for a U sample initially loaded to 2.7 GPa. There is a small, incipient-spall peak that this simple model cannot reproduce unless Σ is lowered to about 1.8 GPa. We need a more sophisticated way to represent the onset of spall if we want to calculate these details in shock-wave profiles.

Conclusions

The constitutive model as presented does an excellent job of reproducing experimental shock and release wave profiles. This was verified by comparing the calculations with data for U, Be, Cu, Ta, 1100-0 Al, and 6061-T6 Al from peak loading stresses of 6.4 to 230 GPa. Though there are deficiencies in this model, an improved version can certainly be developed.

Acknowledgments

We would like to thank Leroy Erickson and Norval Parker for the high-pressure spall data on U.

References

Asay, J., and L. Chhabildas (1980). Some new developments in shock wave research. In B. Vodar and Ph. Marteau (Eds.), *High Pressure Science and Technology*, Vol. 2. Pergamon Press, New York. pp. 958–964.

Asay, J., and L. Chhabildas (1981). Determination of the shear strength of shock compressed 6061-T6 aluminum. In M. Meyers and L. Murr (Eds.), *Shock Waves and High-Strain-Rate Phenomena in Metals*. Plenum, New York. pp. 417–431.

Asay, J., L. Chhabildas, G. Kerley, and T. Trucano. High pressure strength of shocked aluminum. In *Proceedings of the 4th APS Topical Conference on Shock Waves in Condensed Matter, Spokane, WA*.

Bernier, H., and P. Lalle (1982). Experimental Hugoniot of lead in the pressure range 20 GPa – 80 GPa — melting on the Hugoniot. In C.-M. Backman, T. Johannisson, and L. Tegnér (Eds.), *High Pressure in Research and Industry, 8th AIRAPT Conference, 19th EHPRG Conference, 17–22 Aug. 1981, Institute of Physical Chemistry, University of Uppsala, Sweden*, Vol. 1. Arkitektkopia, Uppsala, Sweden. pp. 194–197.

Chhabildas, L., and J. Asay (1982). Time-resolved wave profile measurements in copper to megabar pressures. In C.-M. Backman, T. Johannisson, and L. Tegnér (Eds.), *High Pressure In Research And Industry, 8th AIRAPT Conference, 19th EHPRG Conference, 17–22 Aug. 1981, Institute of Physical Chemistry, University of Uppsala, Sweden*, Vol. 1. Arkitektkopia, Uppsala, Sweden. pp. 183–189.

Cochran, S., and M. Guinan (1976). *Bauschinger Effect in Uranium*. Lawrence Livermore National Laboratory, Livermore, CA. UCID-17105.

Cochran, S., and D. Banner (1977). Spall studies in uranium. *J. Appl. Phys., 48*, 2729–2737.

Grady, G. (1983). Sandia National Laboratory, Albuquerque, NM. Private communication.

Moss, W., and L. Glenn (1983). A Bauschinger effect model suitable for use in large computer codes. In J. Asay, R. Graham, and G. Straub (Eds.), *Shock Waves in Condensed Matter — 1983*. North-Holland, Amsterdam. pp. 133–136.

Shaner, J., J. Brown, and R. McQueen (1983). Melting of metals above 100 GPa. In C. Homan, R. MacCrone, and E. Whalley (Eds.), *High Pressure in Science and Technology*, Part 3. North-Holland, Amsterdam. pp. 137–141.

Steinberg, D., S. Cochran, and M. Guinan (1980). A constitutive model for metals applicable at high-strain rate. *J. Appl. Phys., 51*, 1498–1504.

Steinberg, D., and R. Sharp (1981). Interpretation of shock-wave data for beryllium and uranium with an elastic-viscoplastic constitutive model. *J. Appl. Phys., 52*, 5072–5083.

Steinberg, D., D. Breithaupt, and C. Honodel (1985). Work-hardening and effective viscosity in solid beryllium. In N. J. Trappeniers S. N. Biswas, H. R. van den Berg, C. Prins, K. O. Prins, and J. A. Schanten (Eds.), *Xth AIRAPT International High Pressure Conference*. North-Holland, Amsterdam. pp. 762–765.

Swegle, J., and D. Grady (1985a). Calculation of thermal trapping in shear bands. In *Proceedings of the 4th APS Topical Conference on Shock Waves in Condensed Matter, Spokane, WA*.

Swegle, J., and D. Grady (1985b). Shock viscosity and the prediction of shock wave rise times. *J. Appl. Phys., 58*, 692–701.

Wise, J., L. Chhabildas, and J. Asay (1982). Shock compression of beryllium. In W. Nellis, L. Seaman, and R. Graham (Eds.), *Shock Waves in Condensed Matter — 1981*. American Institute of Physics, New York. pp. 417–421.

DISCLAIMER

This document was prepared as an account of work sponsored by an agency of the United States Government. Neither the United States Government nor the University of California nor any of their employees, makes any warranty, express or implied, or assumes any legal liability or responsibility for the accuracy, completeness, or usefulness of any information, apparatus, product, or process disclosed, or represents that its use would not infringe privately owned rights. Reference herein to any specific commercial products, process, or service by trade name, trademark, manufacturer, or otherwise, does not necessarily constitute or imply its endorsement, recommendation, or favoring by the United States Government or the University of California. The views and opinions of authors expressed herein do not necessarily state or reflect those of the United States Government or the University of California, and shall not be used for advertising or product endorsement purposes.

Int. J. Impact Engng Vol. 5, pp. 613–621, 1987
Printed in Great Britain

0734-743X/87 $3.00 + 0.00
Pergamon Journals Ltd.

REVIEW OF MODERN HYPERVELOCITY IMPACT FACILITIES

A. J. Stilp

Fraunhofer-Institut für Kurzzeitdynamik
- Ernst-Mach-Institut - EMI
Terminal Ballistics and Impact Physics Division
Eckerstr. 4, D-7800 Freiburg, FRG

ABSTRACT

The paper describes three hypervelocity impact facilities, their instrumentation, the launcher systems, the launch techniques for a large variety of projectile geometries, and the launcher performance. A very brief review is given of the research in the areas of military applications, meteorite impact simulation, basic impact research, and efforts to improve the performance of the light-gas guns and measurement techniques. An outline at future research programs is given.

INTRODUCTION AND GENERAL DESCRIPTION

Since the last Hypervelocity Impact Symposium in 1965, many facilities working in hypervelocity impact research have been closed and new facilities have been built. Until 1965, the research in this area was mainly restricted to the US and to the Soviet Union, conditioned by the requirements of their space and weapons programs. In the meantime, other nations have made efforts to step into this area of research, to catch up with the new developments, and to make their own contributions.

In West Germany, the EMI has become active in hypervelocity research in the last 15 years. The EMI operates 7 closed indoor ranges; three of them are equipped with two-stage light-gas guns, with launch tube calibers varying from 4.5 - 40 mm. It is the purpose of this paper to describe these ranges, their instrumentation and subjects of activities in order to give an overview of a modern hypervelocity impact facility. These activities can be grouped into the following main categories:

- Terminal ballistic tests in support of the Ministry of Defence (MOD) research and development activities
- Hypervelocity impact test in support of the European Space Agency and industries
- Basic research in penetration mechanics and in behavior of materials under high dynamic loading.

A characteristic of the EMI facilities is the high flexibility as far as the geometry, the masses and the velocities of the projectiles and their broad diversified instrumentation are concerned.

FACILITY DESCRIPTION

All three ranges are closed indoor ranges with the same basic construction. They consist of three main components: the launcher system, an interface with a blast tank, a sabot stripper and velocity measurement system, and the impact tank. Many parts of the ranges are interchangeable. This concept gives a high measure of flexibility.

Range I

Range I, in operation since 1968, is equipped with the largest light-gas gun. The launcher is a hybrid-type, which means it has different energy storages. For low velocity impact, up to velocities of about 500 m/s, the launch tube is connected by a hydraulic clamp to a high-pressure compressed gas reservoir. This technique has many advantages; for instance acceleration of large masses and low strength projectiles is possible. Accurate velocities are attained, and costs are low. To obtain velocities up to 2.5 km/s, solid-propellant is used; for this application the smooth-bore launch-tube is connected by the hydraulic clamp to a powder chamber. For high impact velocities and low acceleration forces, compressed and heated helium in a two-stage light-gas gun is used. All three techniques are available with simple modifications, and provide a wide range of velocities and applications. For all three techniques, the same hydraulic coupling device is used.

Smooth-bore launch tubes with calibers from 10 to 40 mm are in use. The pump tube of the first stage of the light-gas gun has a caliber of 65 mm and a length of 6.5 m. The high pressure section is a joint-fit of two high quality steel parts with a transition angle of 5 degrees. New techniques for high-pressure sealing (Δ-ring seals) are adapted to the critical connections between launch tube, diaphragm, and high pressure section. This sealing technique shows large dynamic behavior. The principal dimensions of the EMI 65/15, 20 mm light gas gun are shown in Fig. 1.

Powder chamber, pump tube, high pressure section and launch tube are all made of a high-strength CrNi-steel. All parts are mounted on a solid steel foundation. The pump tube and high pressure section are supported on rollers, which allow them to be moved axially to insert the projectile or to connect the pressure reservoir or powder chamber to the launch tube for ordnance applications. The launch tube mount restricts the axial motion of the gun to a very small amount.

Fig. 2 is a photograph of the light-gas gun. A portion of the range is visible in the right-hand corner. A polyethylene piston, with lead inserts to increase its mass to about 5 kg, is used. The launch tube is connected to the blast tank by an axial compensator. This blast tank is 3 m long and 1.5 m in diameter and has a 20 mm thick steel wall. It has four windows 200 mm in diameter for two shadowgraph stations and an access door which permits easy installation of heavy sabot catchers.

A second velocity measurement station is installed at the entrance of the impact tank. Like the first station in the blast tank, it consists of two shadowgraph systems. Each system has its own trigger delay electronic device. A 5 mW He-Ne laser is aligned with the trajectory of the projectile and a photodiode on the opposite position of the station. The projectile interrupts the laser beam and the electrical signal of the diode triggers a spark point light source, which is in the focus of a spherical mirror.

Parallel light of the light source enters the two windows of the station. A lens with the diameter of the window collects the parallel light in a camera. Optical elements of the shadowgraph station and the laser light barrier are sketched in Fig. 3. The two point light sources start and stop an electronic counter. From the distance of the projectiles and the flight time between the two stations, the velocity of the projectile can be calculated with an accuracy of 0.5 %. In addition, the shadowgraph pictures give information about the sabot separation and the orientation of the projectile.

The last element of the range is the impact tank, it is a configuration of three modules. The chamber houses the sabot catcher, target holders, X-ray film holders, fiducial line holders, and catchers to stop the fragments and debris generated by the impacts. The three modules of the impact chamber are about 3 m long and 1 m in diameter. They are fabricated from 15 mm thick steel walls. Different windows on all sides permit the observation of the impact process under different angles and positions for flash X-rays or other high-speed photographic instruments.

The dimensions of the chamber allow the installation of a special ballistic pendulum, which is able to measure the momentum transferred to the target and the momentum of the fragments formed in front of and behind the target. In combination with flash X-ray arrangement, sketched in Fig. 4, the ballistic pendulum permits the measurement of energy and momentum balances for different impact situations. An efficient exhaust and vacuum pumping system is connected to the range to reduce the internal pressure of the range. The lowering of the pressure has two advantages: first, it reduces the effects of muzzle blast, and, second, it influences the aerodynamic forces which act on the discarding sabot.

The main instrumentation used in Range I includes the following:

1. Four stereo 180 kV X-ray channels
2. One stereo 300 kV X-ray channel
3. One 600 kV X-ray channel
4. TRW-image converter camera for streak and framing
5. Special ballistic pendulum
6. Multichannel target temperature and pressure measurement system with transient recorders.

An overall view of Range I can be seen in Fig. 5.

Range 2

In 1982, it became apparent that the development of a new range for hypervelocity impact studies with the highest reachable velocities was necessary. In addition to the high performance, high flexibility for wide spread application and low cost operation was desirable. A schematic drawing of the two-stage light-gas gun and the range with the main dimensions is shown in Fig. 6. Table I summarizes the main dimensions and construction details of the gun.

TABLE I

Pump tube diameter	30, 34, 40 mm
Pump tube length	200 cm
Transient angle of the high pressure section	5^0
Propellant gas	Hydrogen (1 - 15 bars)
Piston weight	100 - 300 g (Polyethylene)
Diaphragm	High ductile stainless steel with four petal opening configuration
High pressure scaling	Δ-steel rings
Powder weight	∿ 150 g (maximum)
Coupling	Hydraulic clamp

An experimental program was established to optimize the loading conditions for high velocity performance. Parallel to this program, the best designs of sabots for different projectile masses, geometries, and densities were selected.

In Table II, typical standard values of this program are summarized.

TABLE II Launch-tube 8.5 mm cal.

Projectile Type	Diameter [mm]	Mass [mg]	Total Mass Sabot + Proj. [mg]	Velocity [km/s]
Al-Sphere	2	12	225	9.35
Al-Sphere	1.5	10	220	9.6
Al-Sphere	4	95	375	8.5
Glass-Sphere	1.5	5	220	9.0
Glass-Sphere	4	90	370	8.5
Ti-Sphere	4	180	680	6.6
Hard-Metal-Sphere	4	500	1200	5.5
Steel-Cylinder	3	130	600	6.6

At the moment, we are optimizing the loading conditions and the sabot constructions for higher performance. Fig. 7 shows a photograph of the gun with its support and the blast tank in the rear.

Like range I, this range consists of a blast tank, a velocity measurement station and an impact tank. Blast tank and impact chamber are connected with separate vacuum pump systems. With this arrangement, different pressure levels in both systems are maintained. The light-gas gun and the blast tank are in a separate room.

The velocity measurement station is an in-flight, interrupted laser beam combined with the projectile photographic system. Beckman and Whitley spark gap light sources are used to back light the projectile, and Beckman and Whitley image converter cameras take 5 - 10 ns exposures. The scheme of the arrangement is shown in Fig. 8.

The target chamber is a steel tank approximately 80 cm in diameter and 75 cm deep. Four 150 kV flash X-ray channels are arranged in a horizontal plane perpendicular to projectile trajectory or, alternatively in a vertical plane. The X-ray film cassettes are inside the tank, protected by Pertinax shields. Fig. 9 is a photograph looking at the target chamber and velocity measurement station.

For firing low density materials such as aluminium or glass, we use two or four-piece polycarbonate sabots cut with a razorblade. The cutting-device has a mechanism to control the depth of the cut. For heavy density projectiles such as steel, copper, tungsten etc. we use a two-piece aluminium cup fitted into the polycarbonate sabot. The aluminium cup has a hole for the projectile. With this construction we prevent punching of the heavy density projectile through the polycarbonate sabot.

Range III

The hypervelocity impact range III is the smallest range, developed for low cost hypervelocity impact experiments, especially for micrometeorite and space debris impact simulations. Only one technician is required to operate the gun and the range. The launcher is a table-size two-stage light-gas gun with an overall length of about 1.5 m. The caliber of the pump tubs is 15 mm and calibers of the launch tubes very between 4.5 and 4.8 mm. For cleaning and renewing the eroded interior surface of the launch tubes of Range II and III, a semi-automatic honing device was developed. This device guarantees high quality bore finish and results in a significant reduction of launcher problems.

The range itself consists of the following parts: blast tank, velocity measurement station, sabot catcher box, and impact chamber. The latter is a stainless steel tank with windows on four sides and a door at the rear side to permit easy installation and alignment of the target. A high vacuum pumping system is capable of reducing the pressure in the impact tank to levels of high altitude (10^{-6} torr).

For some applications different pressure levels are maintained between the blast and impact tanks. By applying this technique, the sabot can easily be eliminated prior to impact. Fig. 10 is a photograph of the range.

To detect small particles, a special trigger and velocity measurement system was developed. The trigger system consists of two bright light sources focused on the trajectory of the projectile. If the projectile passes these light spots, light reflects from the projectile surface and is reflected into photomultipler tubes. The electric signals trigger two sparks with a duration of about 20 ns. The sparks are the light sources for two shadowgraph stations.

RESEARCH PROGRAM

Besides the activities in terminal ballistics (KE-Armor/Antiarmor, fragment ballistics, armor design, material response under high dynamic loading, etc.) the division is active in the following general areas of impact physics:

- Hypervelocity Impact Research
  Impacts into thin plates, thick targets (impact cratering), composite structures, multi-sheet perforation, study of all phases of the impact process, planar impact (EOS, high pressure strength) impact flash studies, effects of cumulative impacts of small hypervelocity particles

- Meteorite bumper design
  (Protection of space vehicle components)

- Studies of the damage potential of simulated micrometeorites and space debris particles

- Improvement of the range capabilities and instrumentation

Future activities:

- Study of target material and sheet spacing on perforation behavior

- Study of optimum energy and momentum transfer to the target by varying projectile mass, density, shape, and velocity

- Study of the effects of cumulative impacts of segmented projectiles in different types of targets.

CONCLUSION

Since the beginning of impact physics research at EMI, there has been a steady growth with permanently increasing requirements. Today, the Terminal Ballistics and Impact Physics Division at EMI operates 7 modern ballistic indoor ranges with efficient launchers and instruments to study impact phenomena. The subjects investigated cover a remarkably wide range of applied and basic research.

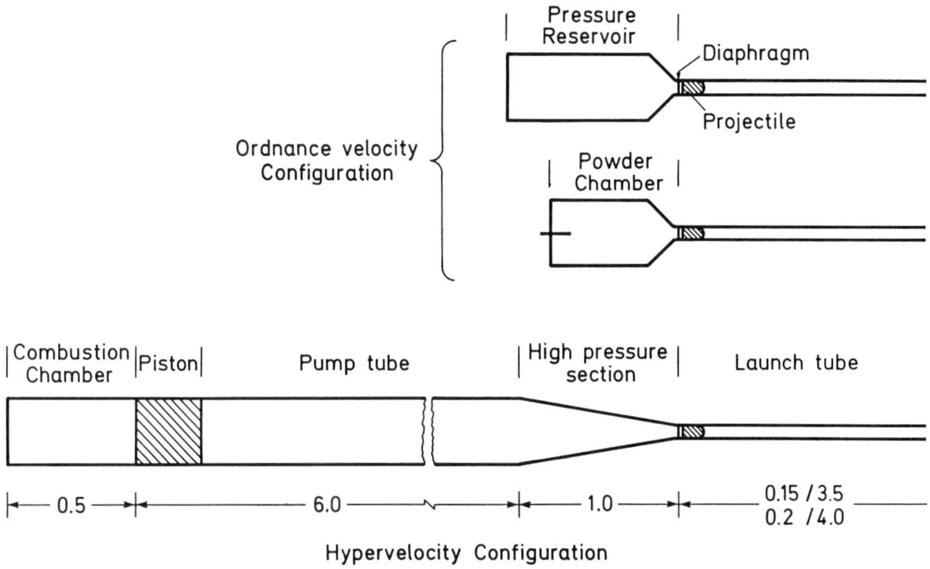

Fig. 1. Launcher System Range 1

Fig. 2. Photograph of the Light-gas gun, Range I

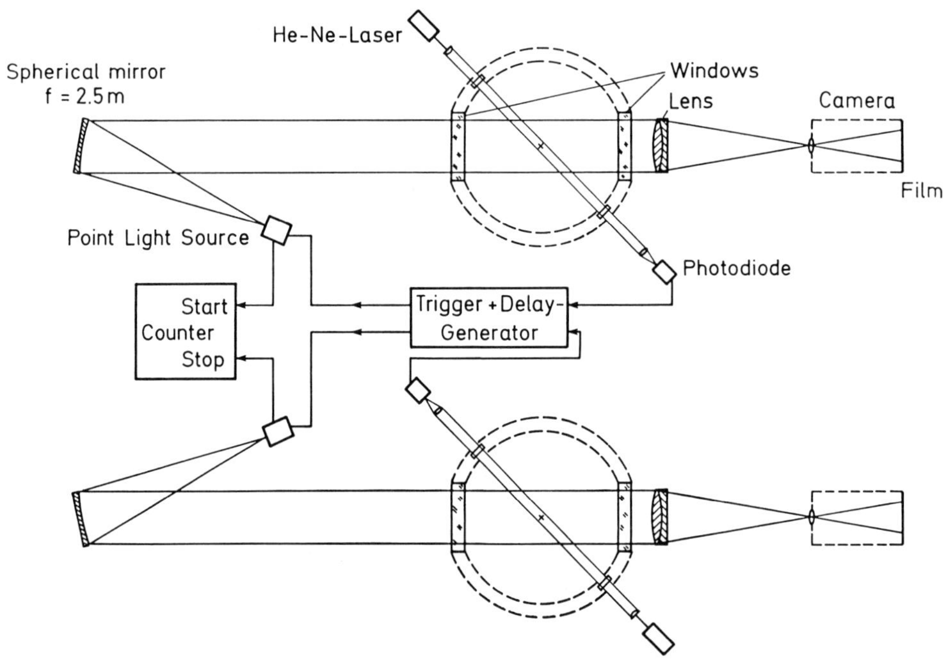

Fig. 3. Laser light barriers and shadowgraph stations of the velocity measurement system, Range I.

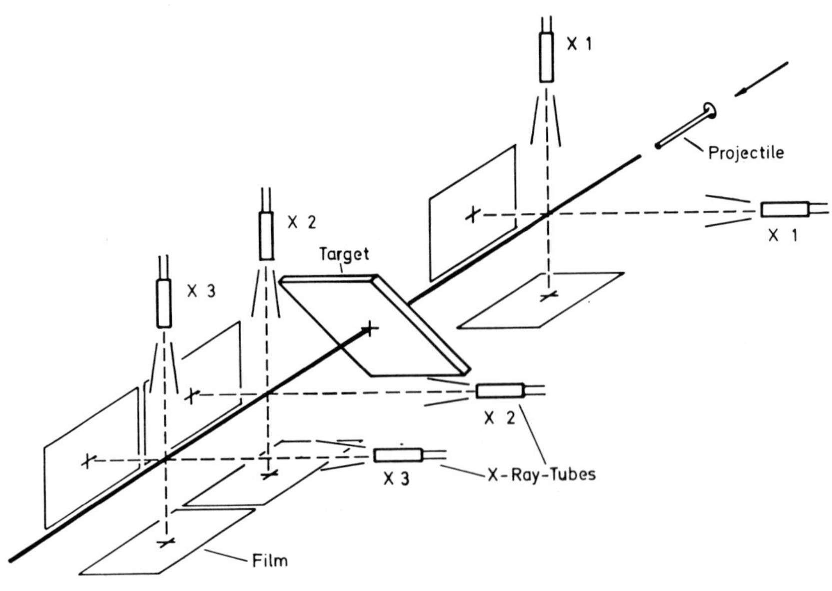

Fig. 4. X-ray channel arrangement, Range I, arrangement for behind target effects.

Fig. 5. Overall view of the Range I with blast tank, velocity measurement stations and the three-modul impact tank in front.

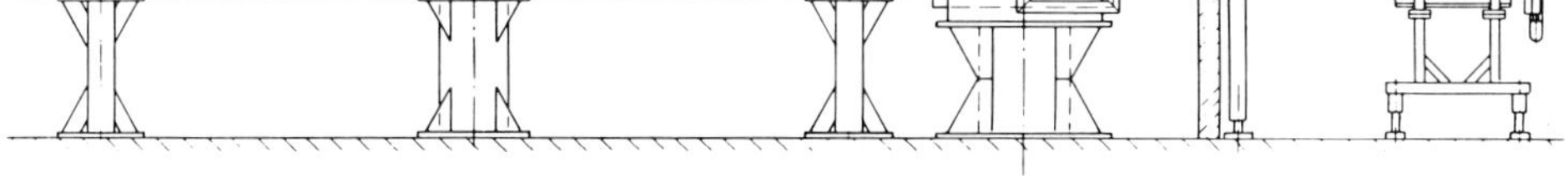

Fig. 6. Sketch of Range II.

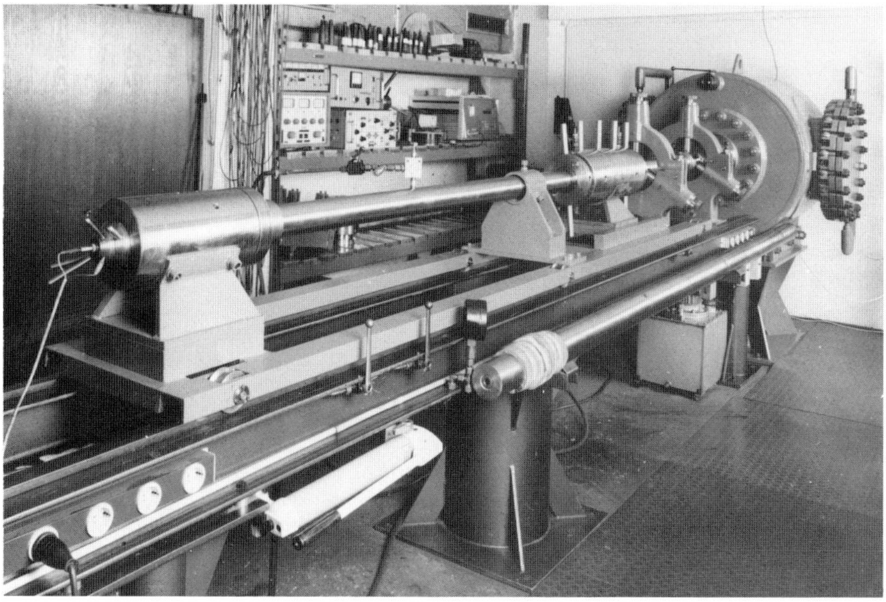

Fig. 7. Photograph of the Light-gas gun, Range II, with blast tank.

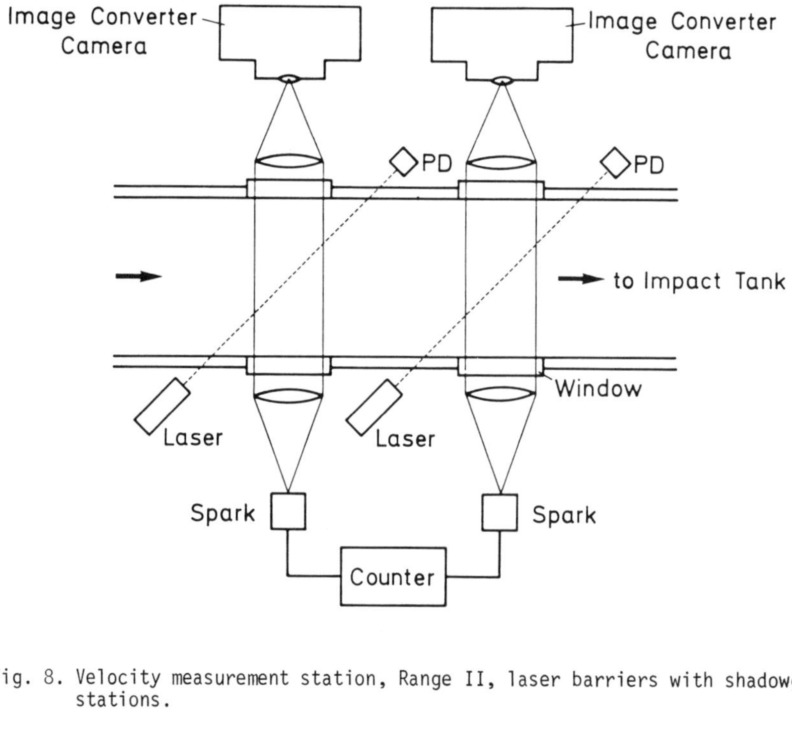

Fig. 8. Velocity measurement station, Range II, laser barriers with shadowgraph stations.

Fig. 9. Photograph Range II, Impact chamber, velocity measurement station, exhauster and flash X-ray channels in the rear.

Fig. 10. Photograph of Range III, Light-gas gun, velocity measurement station and stainless steel impact tank.

Int. J. Impact Engng Vol. 5, pp. 623–634, 1987
Printed in Great Britain

0734-743X/87 $3.00 + 0.00
Pergamon Journals Ltd.

HIGH-SPEED IMAGE-FORMING INSTRUMENTATION FOR HYPERVELOCITY IMPACT STUDIES

H. F. Swift

Physics Applications Inc.
Dayton, Ohio

ABSTRACT

Fast image-forming instrumentation is, by far, the most widely used means for studying hypervelocity impacts when time-resolved information is sought. Single frame (snapshot) cameras are available whose exposure times are short enough to effectively "freeze" material movements in even the fastest impact events. Flash X-radiography equipment is available for taking "snapshot" silhouettes through smoke/dust that defeat visible-light cameras. Similar X-ray equipment operating at higher voltage allows taking flash radiographs through massive materials so that internal features of impacts can be observed. A second series of imaging techniques take cine' sequences of photographs of dynamic events analogous to conventional motion-picture photography. Currently, available equipment operating in the visible is fast enough to time-resolve almost any conceivable macro-impact. X-ray cine' equipment is available, but it is in a considerably earlier stage of development. Streak photography has found wide use in the shock wave analysis community and has also been applied successfully to diagnosing hypervelocity impacts. Streak cameras produce position-vs.-time plots of visualized events that occur along a pre-selected line across the image of an impact process. To the author's knowledge, streak photography has been applied to ballistic research applications exclusively in the visual spectral region ... although appropriate energy sources and electronic streak cameras for operating in the X-ray spectral regime have been available for at least a decade.

INTRODUCTION

High-speed photographic instrumentation must meet two severe criteria if images useful for observing hypervelocity events are to be produced. Equivalent exposure times must be short enough to limit motion blurring to acceptable values, and sufficient light must be available to expose the photosensitive material (film) used to record the images. In the past, either one or both of these criteria served to limit the range of hypervelocity impact events that could be observed photographically. Today, both of these limits can be exceeded for almost any hypervelocity process of interest by a variety of photo-instrumentation (Zucas, et al, 1982).

Let us consider the limitations in some detail before we examine technology that has been developed for overcoming them. The critical parameter determining the motion blur limit for photographing a moving object is the time required for the object to move its own length. This parameter, τ , controls the maximum resolution of an object's image taken by a camera with a finite exposure time rather than the absolute speed of the object. If the effective exposure time of the camera system observing the object is less than approximately .001 τ, image quality is limited to that which can be achieved by a quality 35 mm camera photographing a stationary object. Images as good as can be reproduced with conventional television may be achieved when the exposure time is less than .002 τ . Images capable of distinguishing only gross shapes of moving objects (such as the difference between a circle, a triangle, a square, etc.) require exposure time shorter than 0.2 τ. Finally, images that are capable only of detecting the presence of an object without revealing any information about its shape require exposure times somewhat shorter than, τ , itself. Fig. 1 is an old, but classic, presentation of the above arguments (Zucas et al, 1982). Time required for an object to travel its own length is plotted vs. image exposure time over broad ranges which encompass both conventional and high-speed photographic regimes. Typical objects which may be photographed, are identified on the ordinate, and exposure capabilities of typical photo-instrumentation appear along the abscissa. The fastest-moving object considered is a planet moving along its orbital path. Since the planet is so large, (and takes such a long time to move its own diameter), it poses little photographic difficulty from the standpoint of motion blurring. One of the slowest moving objects considered is an activated relay contact. Its photography poses relatively serious exposure-time problems because of the small size of the relevant area. Of course, extreme requirements for exposure times occur when photography is attempted of very small objects moving at very high velocities (such as simulated micro-meteoroids). The diagonal lines on Fig. 1 represent the various quality images described in the preceding paragraph. They, together with the ordinate and abscissa calibrations, serve to establish exposure time requirements for photographing particular objects at particular resolution levels. A companion plot to Fig. 1 appears in Fig. 2. Here are identified the typical exposure times and framing rates required by instrumentation to resolve effectively phenomena associated with a range of dynamic mechanical events.

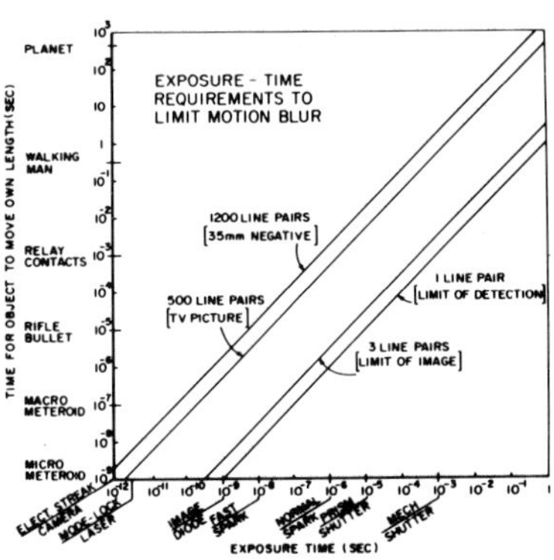

Fig. 1 Exposure Time Requirements for Limiting Image Motion Blurring.

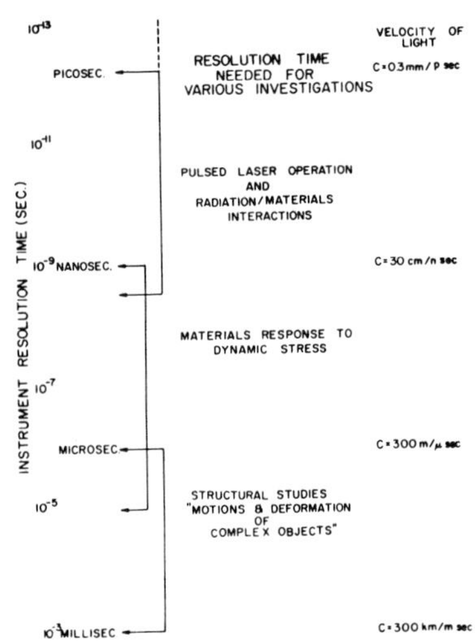

Fig. 2 Exposure Time and Framing Rate Requirements for Conducting Dynamic Mechanical Studies.

Let us consider next lighting requirements for achieving specific levels of photographic exposure. A useful rule-of-thumb for evaluating photographic systems from a light economics viewpoint is that 0.1 ergs/sq.cm. of luminous energy must be deposited on the film to be sure of producing a usable image (Cross, et al, 1982). This factor varies, of course, with the speed of the film and the relationship between the spectral content of the radiation and the spectral sensitivity of the photographic emulsion being used. A phenomenon called "reciprocity failure" also comes into play when exposure times are shorter than 10 u.sec. Film sensitivity drops slowly as exposure times below this value are employed (Brekhov, et al, 1980). The effect is of enormous theoretical importance, but its magnitude is small enough to ignore in all but true photometric applications (where film exposure levels are used to determine incident light intensities). These factors not withstanding, the 0.1 ergs/cm criterion allows systems which easily can be made to function properly, to be separated from those which must either function marginally, or cannot be made to function at all.

Let us close this discussion by considering the basic analysis used for conducting light economics calculations. If applied carefully, these techniques can be used for making at least approximations of the light levels reaching film planes of almost any optical system usable for high-speed photography (Swift, 1979). Eq. 1 states, simply, that energy reaching the film, E_i, is equal to the total image brightness over the spectral region of film sensitivity, B_i, (measured in units of watts per sq/cm times the equivalent exposure time, τ, (measured in seconds). The constant of 10 E 7 is a conversion factor between energy units in joules and in ergs.

$$E_i = 10^7 B_i \tau \tag{1}$$

Eq. 2 represents the heart of the matter. It evaluates the ratio between an object brightness, B_o, and that of its image, B_i, in terms of the taking lens parameters.

$$\frac{B_i}{B_o} = \frac{T}{8F\#(M+1)^2} \tag{2}$$

where:

F# = taking lens F number;
m = optical magnification; and,
T = transmission efficiency of the optics. (T = 1 = perfect optics, T = 0 = opacity.)

Eq. 2 can be applied to virtually any optical system although multi-lens systems require careful consideration of their effective F# and their total transmission efficiencies.

The final question for considering light economics is "how bright is the source"? Sources useful for high-speed photography today may be generally divided between those whose outputs simulate black bodies to one degree or another, and monochromatic ones (usually Lasers). Laser power outputs are specified for continuously operating units in watts and energy per pulse is specified for pulsed Lasers in joules. Both of these specifications can be employed directly for considering object brightness...or amounts of energy available to photographic emulsions.

The situation is somewhat more complex for black body sources such as: heated metal filaments in conventional lamps, burning metal in flashbulbs, intensely radiating plasmas in arc discharge lamps, etc. Here the "black body" equivalent temperature, T_s, is usually specified. Spectral output between wave length limits defining film sensitivity may be calculated using Eq. 3.

$$B_S = 3.741 \times 10^{-12} \epsilon \int_{\lambda_{MAX}}^{\lambda_{MIN}} \lambda^5 \left[e^{(1.4391 \lambda T_s)} - 1 \right]^{-1} d\lambda \tag{3}$$

where:

B_o = source brightness in w/cm
η = the source emissivity coefficient (η = 1 for a black body); and,
λ min.= minimum wave length of interest; and
λ max.= maximum wave length of interest.

Eq. 3 must be integrated numerically for each value of black body equivalent temperature considered. A specialized slide rule is available for making these computations for equivalent temperatures below 10,000 K (G.E., 1970).

PHOTOGRAPHIC TECHNOLOGY

Let us now consider the three main types of high-speed photographic equipment that are currently available, single-frame "snapshot", cine' cameras, and streak cameras.

Single-Frame Photography

Single-frame photography is by far the oldest form of photography applied to high-speed mechanical events. Before the turn of the 20th century, Faraday employed triggered electric sparks to take shadowgraph images of projectiles (Faraday, 1872). The system employed electrical energy generated by a Wimsherst static electric generator, stored in a Leiden Jar (capacitor) to power a spark gap whose light output fell upon a bare glass emulsion plate. The projectile, whose trajectory passed just in front of the film, cast a shadow image when the spark was triggered which was recorded. The system was triggered in an elegantly simple manner. A second spark gap, in series with the first one, was mounted so the oncoming bullet short-circuited it thereby switching the capacitor voltage to the light source gap which broke-down spontaneously. Modern analysis of Faraday's experiment indicates that his effective exposure time was near $\tau = 0.5$ u.sec. which effectively "froze" bullet motion. Systems analogous to Faraday's are used today to very good advantage (see Fig. 3). The Wimsherst Machine is, of course, replaced by a high-voltage electronic power supply, the Leiden Jar with a low-inductance capacitor, and the spark gap trigger, with one of many mechanical electronic/opto-electronic projectile detectors (Swift, 1982).

A major advance was made in shadowgraph photography during the 1920's when Cranz invented the focused shadowgraph where a more-or-less conventional camera was set up to view the object as shown in Fig. 3. The object was "lit" from behind by a spark-condensing lens system which focused an even illumination field into the camera's objective lens. Silhouette images of the projectile are generated as with the original shadowgraph system, but the camera could now be shuttered which allowed the system to operate in subdued lighting, rather than requiring absolute darkness. Light from the object area could be discriminated effectively by stopping-down the objective lens without affecting back-illumination efficiency and chromatic filters could be used to "clean-up" the photographic process since the filter need only be as large enough to cover the objective lens aperture. Image clarity could now achieve the standards of conventional photography (within the range of motion blurring limitations) because the recorded image was a truly focused one, rather than being a shadow illuminated by a light source of finite size. Finally, image magnification relative to the object could be controlled readily by changing the focal length of the objective lens.

Two relatively recent advances have served to revolutionize single-frame photographic capability, Laser light sources, and electronic shuttle tubes. The replacement of spark or xenon flash light sources with pulsed Lasers reduced photographic durations to one n.sec. and below which effectively freezes motions of almost any high-speed objects of interest. Radiative energy available per pulse can be increased almost without bound which opens the possibility of using front lighting techniques for photography even at nanosecond exposure time durations (Cross, 1982). Finally, Laser illumination is virtually monochromatic which allows narrow-band filters to be installed at the taking camera's objective lens which can prevent almost any amount of object illumination from affecting object photography. One problem unique to pulsed Laser illumination, is "speckling" of the resulting photographic images caused by spacial coherence in the Laser illumination (Paisley, 1976). Areas of the image on the photographic emulsion are brightened through constructive interference of illumination arriving from several points on the object while adjacent areas are darkened by similar destructive interference. The result is a severe mottling of the image which can obscure the fine image detail. Several optical techniques are available for breaking up spacial coherence in the laser beam before it is allowed to fall upon the object (Paisley, 1976).

The second major advance in single-frame photography is the advent of high-quality electron image shutter tubes which process data from individual image pixels in parallel. The simplest of these tubes, the biplaner image diode, consisted of a flat photocathode onto which is projected the image to be recorded, mounted 2 or 3 mm in front of a flat phosphor screen similar to ones found on oscilloscope faces (Swift, 1980). When a high positive potential is switched onto the phosphor, an electrostatic field is produced within the tube which draws electrons emitted by the photocathode directly across to conjugate points on the phosphor surface where their energetic impacts produce substantial light output. The electron analogue of the image falling on the photocathode is, thus, reproduced in an intensified form on the phosphor surface. The tube operates only during the time voltage is applied to the phosphor so the tube can be shuttered down to 1. n.sec. by using electronic technology to produce a short-duration voltage pulses. A complete camera using a biplaner image diode is shown schematically in Fig.4. An objective lens images the object to be photographed onto the photocathode. A relay lens transfers the image from the phosphor to a stationary film. The tubes can be made to operate with intensity gains of up to 80X, but this advantage is almost completely negated by the limited efficiency of the relay lens which is at best, just over 1%.

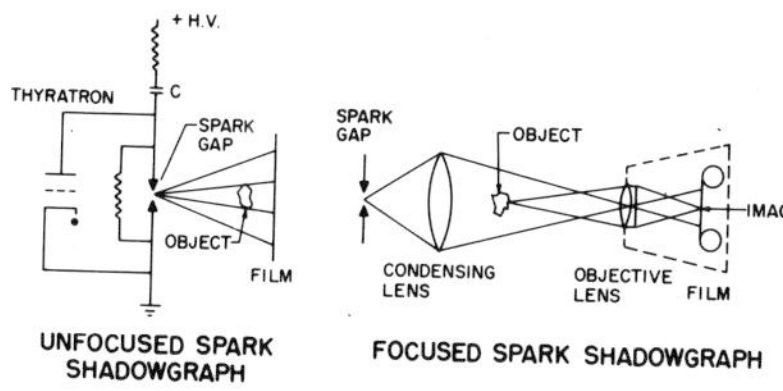

Fig. 3 Modern Spark Shadowgraph Systems.

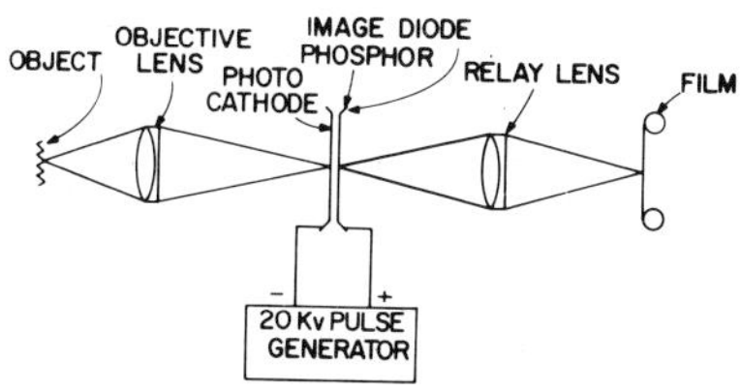

Fig. 4 Biplaner Image Diode Camera.

Recently, enormous improvements have been made in biplaner image tubes. The most important is addition of a micro-channel plate between the photocathode and the phosphor. The micro-channel plate is, in essence, an array of up to 1,000,000 electron multiplier channels which increase the density of electrons in the image analogue cloud. This density increase leads directly to increases in image intensification capabilities of such tubes from 80X above 200,000 X.

Coherent fiberoptic plates are also being used in image diodes to transfer images from the phosphor layer to a film emulsion pressed against the rear surface of the tube. Fiberoptic transmission is over 60% efficient as opposed to relay lens efficiencies of less than 1%. Thus, nearly the entire benefit from micro-channel plate image intensification may be realized photographically. Scientists and engineers now have available the most powerful image intensifiers that will ever be built! The reason is that individual light photons detected by the intensifier tube produce enough light to expose photographic emulsions. Further intensification beyond this point becomes superfluous.

High efficiency image intensifier tubes have found a number of important applications in dynamic mechanics research. First, they allow photography in situations where intense illumination is either difficult or impossible. Second, they allow launched projectiles that emit visible and near-infrared radiation to be photographed "in their own light". Arnold Engineering Development Center has used this capability to evaluate temperature profiles on projectiles protected by heat shields that experience rain/ice/dust impacts during simulated aerodynamic re-entry heating (Hendrics, 1982). Finally, image intensification is being used to good advantage for developing high-speed cine' radiographic capability as described in a subsequent paragraph.

Attempts to photograph high-velocity impact phenomenon are often frustrated by clouds of dust/smoke which envelop the area to be observed. Such clouds effectively prevent photographic analysis. This problem can be alleviated by employing high-speed X-radiography. This technology employs pulsed X-ray generators as sources for taking unfocused radiographs of the object space. The X-ray pulses are unattenuated by dust/smoke and are absorbed only by materials with substantial mass. Thus, the original projectile, its principle fragments and substantial target fragments from a hypervelocity impact may be visualized clearly. Pulse generators that produce more penetrating ("harder") X-ray bursts may be used to produce instantaneous images of interiors of targets subjected to impacts. Results can provide unique insights into processes controlling hypervelocity mechanics.

Two problems must be solved to produce any X-ray image. First, the radiation must be penetrating enough to pass through extraneous material surrounding the object to be recorded and illuminate the recording emulsion. Second, the penetrating power of the X-radiation must be slight enough to allow significant attenuation by objects to be observed so that their images stand out against the background (have sufficient contrast). Thus, we have both upper and lower limits on allowable X-ray penetrating power.

X-ray penetration capability is roughly proportional to the voltage applied to the X-ray tube so an optimum X-ray voltage is generally definable for radiographic situations. Situations occasionally arise during hypervelocity impact experiments where the upper and lower voltage limits approach each other closely so that X-radiation voltage becomes a critical experimental design factor. Unfortunately, these criteria occasionally cross one-another for critical experimental situations which effectively eliminates flash X-radiography as a candidate technology for extracting the desired data.

High-Speed Cinematography

Individual photographs of a dynamic mechanical event can provide a wealth of information about the event's development...especially if they are timed to coincide with critical developments. They lack fundamentally, information about the event before and after their exposure periods, however. This problem has been solved classically by taking series of individual photographs at different times during the event (cine' sequences). Rules-of-thumb exist for estimating the rates at which photographs should be taken to provide smooth motion simulation via moving picture visualization, and for extracting numerical information. In general, at least 25 frames of data should span the portion of a dynamic event of interest if the results are to be observed as a smooth moving-picture sequence. Exposure times of the individual frames may be as long as 1/4 to 1/3 the inter-frame period to keep motion blurring from becoming a serious problem. Between 5 and 10 frames of information provide adequate coverage of an event when numerical data concerning object velocities and trajectories is to be determined. Relative exposure periods should be somewhat shorter than for motion visualization although techniques are available for extracting precise object position data from slightly motion-blurred images (Swift, 1974).

The oldest high-speed cinematography cameras were those developed by Cranz and Schardin (Cranz, et al, 1928). These cameras are, effectively, a series of focused shadowgraph units which share a single condensing optic, but use separate spark gaps and objective lens pairs as shown in Fig. 5. Each camera takes a single shadowgraph image of the object space when its light source is triggered. Firing the light sources in a sequence with constant inter-frame times, provides a classic cine' sequence of the object space. Potential also exists for synchronizing the exposure times with events within the experiment so that the framing rate can be varied to maximize temporal information available from a limited number of frames. Conventional spark-powered Cranz-Schardin camera systems have been built with as many as 121 individual frames although units with 64 frames and 25 frames are more common. Recently, inexpensive solid-state pulsed Lasers have been employed to reduce the cost of Cranz-Schardin cameras substantially while opening the possibility of using narrow-band optical filters to protect the film from spherious radiation originating in the object space (Cross, 1982).

The Cranz-Schardin camera system has two disadvantages which may or may not be important for any particular application. First, it is strictly a shadowgraph recorder so only silhouette images can be produced. Second, each camera observes the object space from a slightly different viewpoint which complicates producing moving-picture sequences for motion visualization and extracting numerical information from frame sequences. These difficulties are eliminated in a unique all mechanical camera technology developed by Miller during the 1930's (Miller, 1934). The Miller rotating mirror camera principle employs individual cameras to take individual frames of information, but uses a turbine-driven rotating mirror as an optical switch to direct light from the event to be photographed into each of the taking cameras sequentially (see Fig. 6). A primary objective lens images the object space onto the face of the turbine-driven rotating mirror. Light from the mirror face is reflected sequentially to a battery of secondary objectives which image the mirror face onto a stationary, photographic emulsion. Miller's original camera took a series of 24 images at rates up to 100,000 frames/sec. Modern cameras are available which take sequences of up to 144 individual images at peak rates as high as 10,000,000 per/sec. (Cordin, 1986). Experiments began during the 1940's to use conventional illumination to provide frontlet photographic sequences using Miller cameras. The relatively high F numbers of the Miller cameras, plus the relatively short image exposure times (typically 1/4 to 1/2 the inter-frame period) led to extreme illumination requirements...which were solved in a variety of interesting manners. During the 1960's attempts were first made to take color photograph sequences. After some initial difficulties, this technology proved highly successful both from an aesthetic viewpoint and for adding a new dimension to extracting quantitative data from image sequences. Technology continues to develop in this area. Recent advances have involved enlarging the mirror formats while maintaining photographic taking rates by employing beryllium as an effective engineering material for mirror bodies...thus allowing larger units to be spun at high angular rates. Cameras are currently available with 70 mm formats that can take image sequences at rates well over 1,000,000 frames/sec. (Nebecker, 1978).

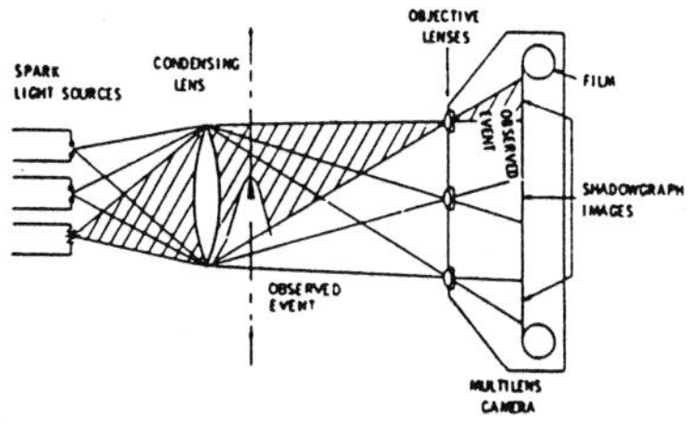

Fig. 5 Cranz-Schardin Cine' Camera System.

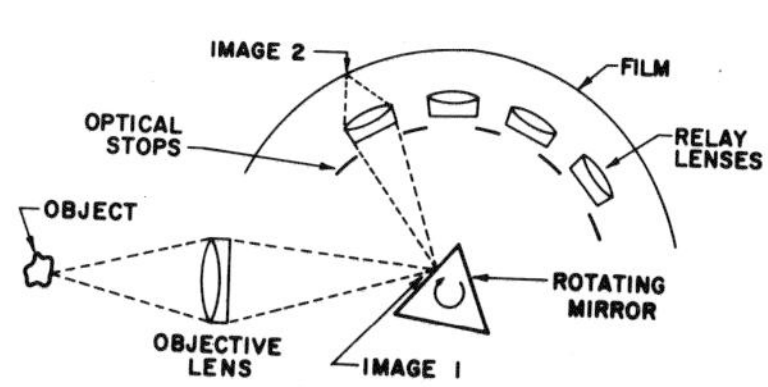

Fig. 6 Miller-principle Rotating Mirror for Cine' Camera.

Miller-principle cameras can be built using two-sided mirrors in the form of flat plates or multiple-faced mirrors in the form of polygons. Cameras with flat-plate mirrors are referred to as "synchronous" because they spend up to five-sixth of their time "blind", i.e., the rotating mirror is out of position for directing light into any of the secondary objective lenses for all but 60° of each rotation. Traditionally, such cameras are equipped with a mirror position detector which emits an electrical pulse a fixed time before the camera becomes sensitive. This pulse is used to trigger the event to be photographed and the illumination system which assures that the available photographic frames are used efficiently. Synchronous cameras have a second advantage in that the blind period is available to allow illumination of the object space to build up to optimal photographic exposure level before the camera becomes sensitive and to die-away over an extended period after the sequence without risk of re-exposing the film during the next sequence. Of course, many dynamic mechanical events cannot be triggered by any piece of diagnostic instrumentation. Typically, events depending upon gun-launched projectiles fit into this category. Synchronous cameras are unusable for observing such events. Fortunately, continuous access Miller-principle cameras are available which use rotating mirrors in the form of polygons to produce continuous image switching. As one mirror face illuminates the last secondary objective lens, the next face illuminates the first one. These cameras are referred to as "continuous access" units. Their turbines are brought to operating speed well before the event occurs. The event itself is used to trigger a light source which comes up to photographing intensity as quickly as possible and illuminates a single series of frames during which the event of interest occurs. Means must usually be provided for blinding the camera immediately after the last frame is taken to prevent the film being double exposed. A wide variety of techniques have been used to either clip the illumination from the taking light source (Preonas, et al, 1970) or to "cap" the main camera objective lens with a fast-acting shutter (2nd HSPC 1952).

Continuous access Miller-principle cameras, are in general, more expensive than synchronous units because of the added complexity produced by precision alignment of the multi-faced mirror. They also produce individual photographs of lower spacial resolution because compromises must be made in design of the optical systems when mirror motion occurs about an axis located some distance from the mirror surface. Care should always be taken in selecting a Miller-principle camera to assure that an appropriate unit is chosen for the task or tasks to which it will be applied.

Very high-speed mechanical events may well require framing rates substantially above the 10,000,000 frames/sec. limitation of Miller-principle cameras to be visualized properly. Often, the extreme illumination levels required for taking ultra high-speed sequences become difficult to achieve...especially in laboratory applications. These two problems led to development of electronic cine' cameras with no mechanical parts. These cameras have achieved framing rates in excess of 600,000,000 frames/sec. Other units have employed micro-channel plate technology to achieve terminal intensification levels (where individual photons are recorded on the film emulsions). Operation of electronic "image triode" cameras is sketched schematically in Fig. 7.

One important application of ultra-sensitive high-speed cine' cameras has been in the area of high-speed X-ray cinematography. The use of X-radiation rather than visible radiation to sense dynamic mechanical events allows the same benefits enjoyed by flash X-ray equipment to accrue to cine' radiographic studies. Two basic approaches have been used successfully to achieve taking rates high enough to support hypervelocity impact investigations. The first is essentially an X-ray version of the Cranz-Schardin cine' camera. The camera consists of a tight grouping of flash X-ray tube heads set to illuminate an object area from nearly a single direction when they are fired sequentially. The X-radiation passing through the object area illuminates a free-standing phosphor screen which produces a visible-light rendition of each X-ray shadowgraph in turn. A high-sensitivity electronic cine' camera with its operation synchronized with the X-ray sources records the image sequence emitted by the phosphor screen. The images produced on the phosphor screen are not inherently bright, so substantial image intensification must be employed to produce usable photographic exposures. The system is limited in most cases by the performance of the phosphor screen. These screens illuminate immediately upon receiving X-radiation but the illumination decays exponentially in time after radiation ceases. Typically, the X-ray sources have durations of 0.25 u.sec. but normal screens have decay time-constants in excess of 1.m.sec.! Such long time-decays render normal phosphor screens useless for high-speed cine' radiographic applications. Special screens are now available with decay time-constants less than 2 u.sec. Such screens have been used successfully for producing cine' sequences at rates up to 400,000 frames/sec. (Trimble, 1980). Such speeds are more-than-adequate for many hypervelocity impact investigations but fall short for several important ones. A second approach capable of operating at inter-frame times as short as X-ray exposure periods involves using several individual pairs of flash X-ray sources and film-packs to observe dynamic mechanical events from a variety of directions (Prater, 1974). Shrewd use of basic symmetries of the process being investigated enables radiographs taken from different directions to be interrelated with one another. Let us consider an example where a small aluminum target in a form of a cylinder is struck axially by an aluminum sphere traveling at hypervelocity. Four X-ray generators are mounted to illuminate the target across approximate cylinder diameters to provide complete profiles of the developing crater at specific times after initial projectile/target contact. Six other flash X-ray generators are aligned to produce a cluster of bursts nearly parallel to the original projectile trajectory. The resultant images are recorded on a film-pack mounted around the projectile's trajectory. The resultant six images show only the instantaneous diameters of the evolving crater. Results from previous studies plus those from the four crater profile images show that the craters expand at constant shape so that instantaneous crater diameters can be used to infer instantaneous crater depths.

The system is, of course, capable of operating at extremely high framing rates (limited by X-ray exposure time only). Effective framing rates in excess of 1,000,000 frames/sec. have been achieved using older generators (Prater, 1974) and nearly 10,000,000 frames/sec. are now possible using modern high-speed X-ray generators. The obvious disadvantage of this system is that it requires the observed event to possess reliable spacial symmetries so that data from views taken from markedly different directions can interrelated. Little or no potential exists for visualizing motion using this approach. One interesting speciality of this technique which alleviates some of the symmetry problems is to group the X-ray generators in pairs with relatively small angles between their viewing directions. The generator pairs are fired simultaneously and the resulting images are analyzed stereoscopically. This technique allows the event to be interpreted in three dimensions which eliminates the need for event symmetries to perform effective data reduction.

Fig. 8 is a plot of cinematographic framing rate vs. spacial information content within the images for a variety of high-speed cinematographic technologies currently available. The areas marked out for each technology represent the performance extremes that have been reported thus far in the technical literature.

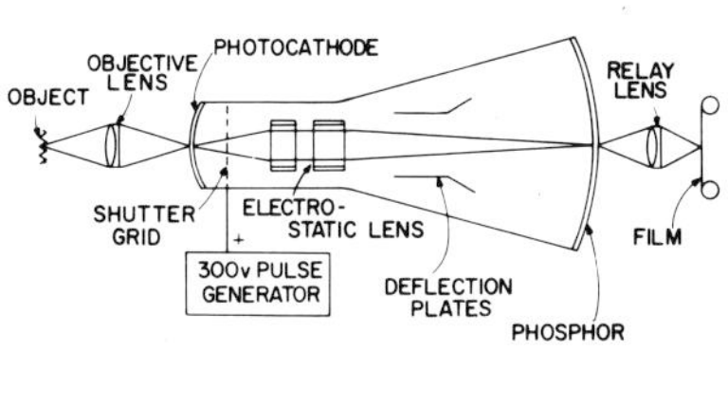

Fig. 7 Electronic Image Triode Cine' Camera.

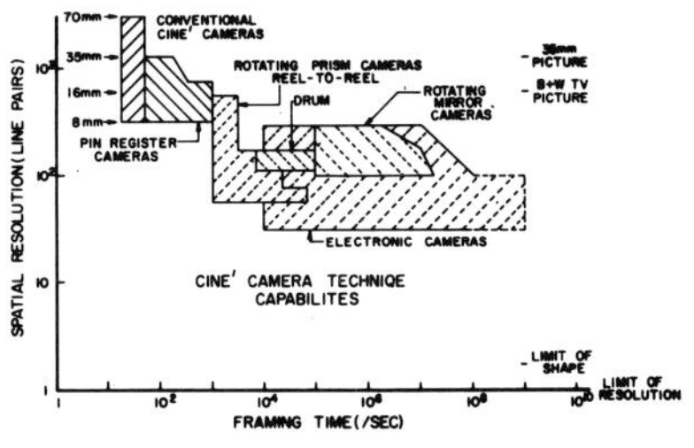

Fig. 8 Capabilities of High-speed Cine' Cameras.

Streak Photography

Streak photography produces plots of position vs. time for critical objects lying along a straight line within a mechanical event. Temporal resolution of cameras using particular image-moving technologies is finer, by orders of magnitude, than similar resolutions produced by cine' cameras. Another interesting feature of streak photography is that it has no analogue which addresses the problems that occur at ordinary speeds. The author has been at a loss for some time to understand why this is so!

Let us consider how streak cameras operate by examining a simple moving film camera sketched in Fig. 9. The objective lens images the object space onto an opaque plate through which is pierced a narrow slit. In this case, the slit is oriented so that movements of interest occur along it. A relay lens images the slit onto a film plane that moves continuously perpendicular to the slit. When the slit is illuminated along its entire length, it exposes a broad stripe extending along the film in the direction of film motion. Images moving along the slit produce slanted lines within this stripe image which are plots of object position vs. time. The angles these line images make with the direction of film motion reveal the instantaneous velocities of the objects making them. Thus, we have a device that produces automatically, position vs. time plots for all objects whose images pass along the slit during the period that the camera is sensitive.

A simple example is presented in Fig. 10. Here an explosive generator is being used to launch one flat plate of material against another. Questions to be answered are: (1) What velocity is achieved by the explosively-driven plate? (2).What is the average shock-wave velocity in the target? and, (3) What is the material velocity of the shocked target material (one-half of the rear surface recoil velocity)? A simplified sketch of the resulting streak photograph is also presented in Fig. 10. The known speed of the slit image across the film, together with the known magnification of the optical system, are used to calculate the velocity of the driven explosive plate from the angle its image makes with the direction of slit image motion after acceleration commences. (Curvature of this line image is an observation of plate acceleration under action of the explosive charge.) Impact time with the target plate is depicted by the driver plate image intersecting the edge-on target plate image. The transit time of the shock wave through the target plate may be derived from the displacement between target impact and initial movement of the rear target surface. Finally, target recoil velocity is measured from the angle of the target image as it and the remainder of the experiment proceed to the right out of the camera view-field.

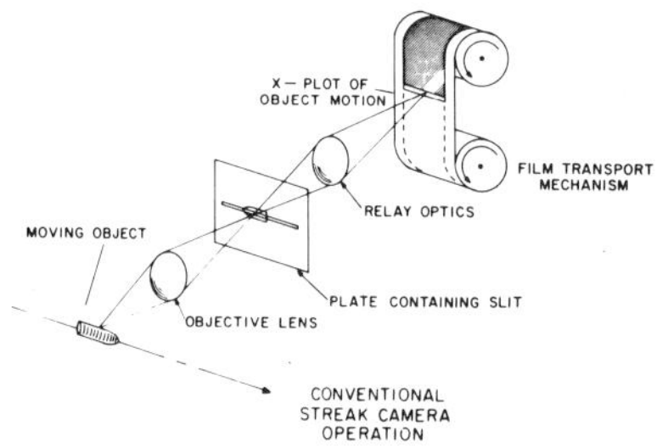

Fig. 9 Sketch of a Simple "Parallel-slit" Streak Camera Setup Observing Bullet in Flight.

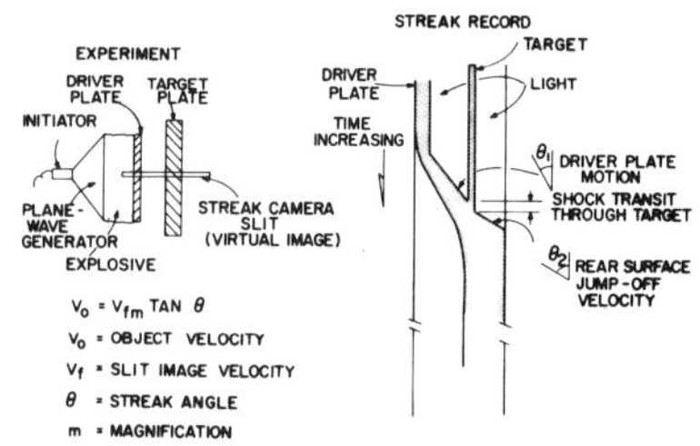

Fig.10 Streak Camera Applied to Observing a Flat Plate Impact Experiment.

Streak cameras can operate in a completely different manner if the slit image is rotated 90° relative to the primary image of the mechanical event so that the motion of interest occurs perpendicular to the slit, rather than along it. Now, the camera produces an image of each object which passes the slit while the camera is sensitive. A sketch demonstrating this process is presented in Fig. 11. The imaging system is unique in that it requires no significant synchronization with the mechanical event (passing of a projectile in this case) to achieve a picture...as is required by any shuttered photographic technique. The camera records everything passing this slit while it is sensitive. This feature has proven extremely useful for "Hall Intervalometers" which uses perpendicular-slit streak photography to take at least two photographs of projectiles in flight as they pass stations spaced along their trajectory. Elapsed time between the photographs (as measured by their separation along the film) and the separations between the stations are used to compute projectile velocities. Extreme reliability in producing projectile images, achieved by using a non-synchronized photographic technique, leads to measurement reliabilities of better than 99%. This reliability is enhanced further by the fact that the system eliminates the possibility of reporting the velocity of an object preceding the projectile along its trajectory rather than the projectile velocity itself, since projectiles are readily identified by their images.

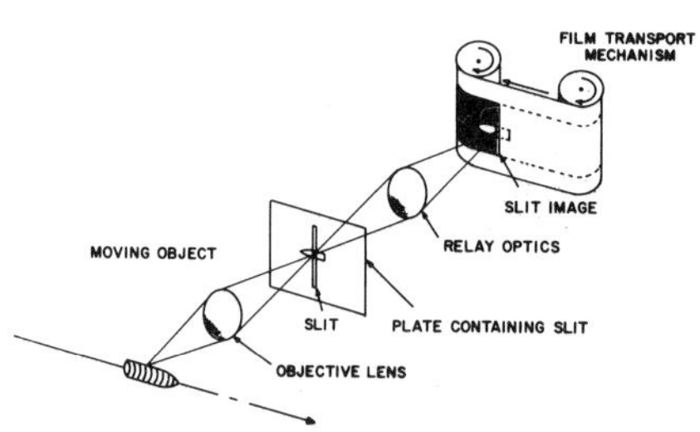

Fig.11 Sketch of a Simple "Perpendicular-slit" Streak Camera Setup for Photographing a Bullet in Flight.

The time resolution capabilities of streak cameras are amazing! The temporal resolution for both parallel-slit and perpendicular-slit configurations equals the time required for the slit image to move its own width across the film plane. Narrowing the slit image until its width just matches the resolution of the optical system maximizes this resolution level. Medium-speed streak cameras that use reel-to-reel film transport to achieve slit image motion along the photographic emulsion, produce temporal resolution limits approaching 100 n.sec with sensitivity times that may exceed one second. Drum cameras with film strips mounted on the inner lip of a high-speed fly wheel, can produce temporal resolutions approaching 30 n.sec. with sensitivity times of several milliseconds. Rotating mirror technology can be used to achieve fundamental time resolution limits near 1.0 n.sec. (Schardin, 1962) and electronic streak cameras have achieved resolutions significantly below 100 f.sec. (4-5 x 10 E-13 sec). The most commonly used streak camera for recording details of hypervelocity impacts uses rotating mirror technology. Synchronous cameras with flat-plate rotating mirrors are generally used for observing explosively powered events; and continuous access cameras are used for gun powered ones. Electronic cameras are finding increased use for hypervelocity impact studies because they are commandable... and require no special treatment in order to achieve synchronization or prevent image rewrite. They are also far more efficient optically than rotating mirror cameras so they can be used with significantly simpler event illumination systems.

CONCLUSIONS

As stated in the introduction, image-forming instrumentation is currently, and has always been, the most widely used technology for extracting information from hypervelocity impacts. Originally, image-forming instrumentation operated exclusively in the visible and near ultra-violet spectral regions. Today, a substantial fraction of high-speed image-forming systems operate in the X-ray spectral region and a small but increasing amount of activity is being conducted in the near infrared.

Non-image-forming instrumentation has much more recent origins and is currently developing rapidly. For this reason at least, the number of measurement efforts employing non-image-forming instrumentation is currently increasing relative to image-forming efforts. It is too early to say which of these two approaches will ultimately become dominate but the importance of both of them for diagnosing future experiments is assured.

REFERENCES

- Brekhov, O., et al, (1980). On energy and time calibration of high-speed streak cameras. Proc. 14th High-Speed Photography Congress (HSP), Moscow, USSR.
- Cranz, E., Schardin, H. (1928). Multiple picture spark photography. Proc. Royal Soc., Vol. 2, pp. 143-145.
- Cross, L., Strader, E. (1980). High rep. rate high res. shutterless photos. for gasdynamic studies. Proc. 15th HSP, SPIE, Vol. 343, San Diego, California.
- Cordin Corp., (1986). Commercial literature, Salt Lake City, Utah.
- Faraday, E., et al, (1872). Electrical sparks used to photograph bullets in flight. Proc. Royal Soc., Vol 1, pp. 142-146.
- General Electric Co. (1972) product literature.
- Hendrics, R., Dugger, P. (1982). Role of H.S. Phot... Proc. 15th HSP, SPIE, Vol. 348, San Diego, California.
- Miller, E. A. (1934). A New Tech. for H.S. Cine' Photos, RSl, Vol. 6 #3.
- Nebecker, S. J. (1978). Tech. & appl. of rotating mirror cameras. Proc. 13th HSP, Tokyo, Japan.
- Paisley, D. L. (1976). Pulsed laser stereophotos of miniature explod. foils. Proc. 12th HSP, Toronto, Canada, 1976.
- Prater, R. E. (1974). Hypervelocity impact processes in thick al. targets... AFML Tech Report 5570.
- Preonas, D., Swift, H. F. (1970). A high intensity point light source. Proc. 9th HSP, Denver, Colorado.

- o Schardin, H. (1962). Res. limit of rotating mirror cameras. Proc. 6th HSP, The Hague, Netherlands.
- o Swift, H. F. (1974). Precision meas. using H.S.P. tech., Proc. 11th HSP, London, UK.
- o Swift, H. F. (1980). Appl of new electro-optical tech. to HSP. J. SMPTE, Vol. 89 #10.
- o Trimble, J., Aseltine, C. (1980). Flash X-ray cine' radiography at 100,000 F.P.S. Proc. Nat. Fall. Conf. ASNT, Houston, Texas.
- o Zucas, J., et al, (1982). Impact dynamics. John Wiley & Sons, New York City, pp. 242-245.

Int. J. Impact Engng Vol. 5, pp. 635–644, 1987
Printed in Great Britain

0734-743X/87 $3.00 + 0.00
Pergamon Journals Ltd.

HYPERVELOCITY IMPACT TESTING USING AN ELECTROMAGNETIC RAILGUN LAUNCHER

M. M. Tower, G. L. Jackson, L. K. Farris, C. H. Haight

LTV Aerospace and Defense Company
Vought Missiles and Advanced Programs Division
P. O. Box 650003
Dallas, Texas 75265-0003

ABSTRACT

An electromagnetic (EM) railgun launcher facility has been developed to routinely conduct hypervelocity impact tests. Two types of completely reusable EM launchers have launched sabot/impactor packages between 2 and 43 grams to velocities between 1.5 and 8.5 km/s. The highly reliable railguns have conducted over 250 projectile launchings and have established a projectile/launcher data base covering interior as well as exterior ballistic considerations. A conventional type instrumented ballistics range is compatible with the EM launcher and can be used to conduct anti-armor and lethality experiments at hypervelocities.

ELECTROMAGNETIC LAUNCHER DEVELOPMENT

With the re-emergence of electromagnetic (EM) railgun development, significant advances have demonstrated the potential for a break-through in the capability to achieve hypervelocity launching of projectiles of a few grams to several kilograms. Conventional chemical propellant guns and two-stage light-gas guns, in general, have been extended to their thermodynamic limits. The launcher muzzle velocity is limited by the expansion rate of the propellant gases. The EM launcher using the Lorentz force extends the muzzle velocity above acoustic or thermodynamic limits to barrel stress or material melting limits or projectile strength limits, which significantly exceeds the velocity capability of conventional and light-gas guns.

The simplest form of an EM launcher is the single-pulse, breech-fed railgun demonstrated at the Australian National University by Barber (1972) and Marshall (1978). They accelerated a 3 gm projectile to 5.9 km/s in the mid-1970's using a launcher powered by a homopolar generator. Single shot explosive magnetic flux compressors and expendable railgun barrels have achieved the highest EM launcher velocities in a cooperative Lawrence Livermore National Laboratory and Los Alamos National Laboratory (LANL) experiment by Fowler (1980) and Brooks (1980) and duplicated recently at LANL by Schnurr (1986) with the acceleration of 3 gm projectiles to 11 km/s. In 1981 at LTV Aerospace and Defense Company (LTV) Tower and Haight (1984) fired a 2.5 gm projectile over 8 Km/s with a breech fed railgun powered by a 660 kJ capacitor bank, the highest velocity yet attained with an EM launcher facility that is completely reusable.

The high velocity was achieved with a single-pulse, breech-fed railgun having a 12.7 mm square bore cross-section and 3.65 meters in length. The railgun barrel has copper conducting rails and epoxy-glass insulators which are rigidly contained by an aluminum structure. A major contributing factor according to Jackson (1986) for extending railgun life is injecting the projectile into the railgun. By injecting the projectile at sufficient velocity (500 to 1000 m/s), rail erosion associated with projectile standing-starts is significantly reduced. The projectile injection is accomplished by rupturing the diaphragm at the orifice of a high pressure helium reservoir. The projectile is accelerated by a one meter gun barrel. The injection gun is mated to the railgun breech as shown in Fig. 1 and 2.

The breech fed railgun, however, has a deficiency for a barrel length greater than a few meters. As the projectile accelerates down the barrel the current path through the rail conductors lengthens, and an ever increasing portion of the input energy is dissipated in Joule heating of the rails, thus decreasing the energy conversion efficiency. A distributed energy store (DES) railgun has energy sources distributed along its length as opposed to feeding all the energy in at the breech. The energy sources in principle can be any device capable of storing electrical energy and subsequently delivering on-demand current in a modular fashion to the launcher.

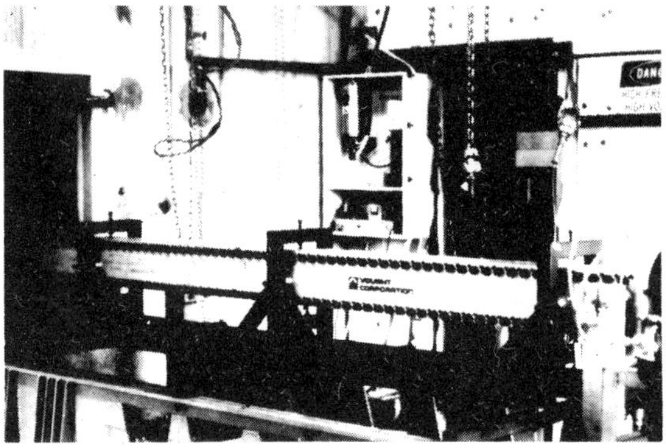

Fig. 1. Single-pulse, breech fed electromagnetic railgun (12.7 mm bore).

Fig. 2. Distributed energy source electromagnetic railgun (20 mm bore).

A typical capacitor-type energy module schematic is shown in Fig. 3. The modular capacitor bank is pre-charged to the desired voltage and discharged through a closing trigger switch by a projectile location sensing trigger circuit logic or by a pre-set time delay trigger pulse. The capacitor bank voltage is sufficiently large as to permit real-time inductor charging with minimal energy transfer delays to the railgun. The inductance value of each module may vary to match back EMF, projectile velocity-travel time, and launcher impedance. A second switch is placed in the circuit to crowbar the capacitor bank near zero voltage in order for the inductor to dissipate its energy into the railgun, rather than recharging the capacitor bank.

Each energy source is switched-on as the projectile passes its energy input connector as shown in Fig. 4. Parker (1982) analytically showed that segmentation of the rails is not necessary because current is prevented from flowing in the reverse rail direction by the high impedance value of previously activated energy sources.

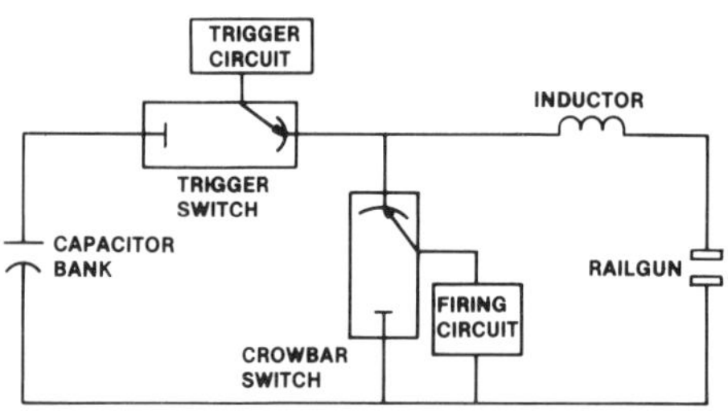

Fig. 3. Single-stage circuit diagram.

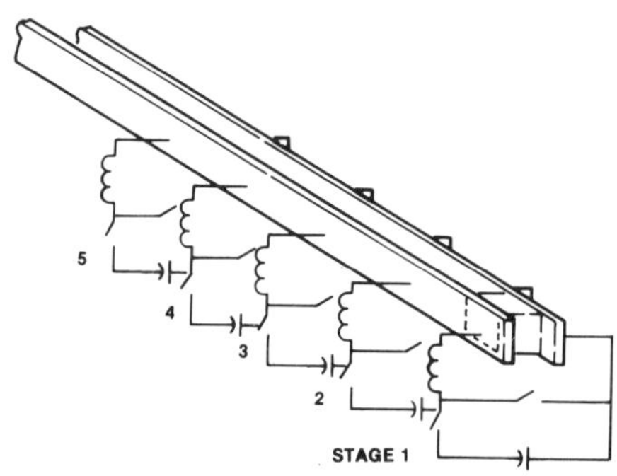

Fig. 4. Distributed energy input.

Several benefits have been described by Tower and Holland (1984) as a result of distributing the energy along the railgun length as opposed to feeding all the energy in at the breech. The DES railgun provides enhanced energy conversion efficiency compared to a breech fed railgun where launcher lengths are longer than a few meters. The DES launcher also offers the capability of shaping the armature current profile as reported by Haight (1986), and thus the acceleration force on the projectile. This feature is critical in accelerating delicate payloads in hypervelocity impact experiments, or in achieving the maximum terminal velocity from very long barrels.

In three years of operation the DES launcher has been fired over 200 times in developing EM launcher technology and gathering hypervelocity impact data.

PERFORMANCE CAPABILITY

Analytical simulation of the DES railgun is performed by simultaneous solution of equations involving the force on the projectile, the motion of the projectile, the electrical input parameters to the gun, and the thermal properties of materials used in the gun. The simulations yield the projectile's acceleration, velocity, and position as a function of length along the gun and/or time. Fig. 5 compares the predicted performance of the gun against actual velocity data measured in the gun using B-dot probes to monitor the armature current and position in time. For this simulation and corresponding test the 6.9 gram projectile was injected into the gun with an initial velocity of 1250 m/s and advanced approximately 0.3 meters into the gun before the firing of the first capacitor bank stage. Subsequent stage inputs increased the initial velocity of the projectile from 1250 m/s to 4020 m/s.

The forces acting on a projectile in a railgun system include many effects. In the simplest approximation, the force acting on the projectile is the Lorentz force as shown in equation 1, which excludes the effects of friction and mass accretion in plasma armatures on the motion of the projectile.

$$F = \frac{1}{2} L' I^2 \tag{1}$$

Where F is the force (N) acting on the projectile; L' is the inductance gradient (H/m) of the railgun; and I is the current (A) in the rails. The integral of the force on the projectile over the distance that this force acts is the kinetic energy imparted to the projectile by the railgun. Assuming a constant current is supplied to the rails over the entire length of the railgun and using the product of the current density in the rails and the bore size of the rails gives equation 2 for the maximum kinetic energy available from a railgun system.

$$E = \frac{1}{2} L' (I')^2 d^2 \ell \tag{2}$$

Where E is the kinetic energy (J) of the projectile; I' is the current density (A/m) in the rails; d is the bore size (m); and ℓ is the active acceleration length (m) of the gun. Using a value of 43 MA/m for the theoretical maximum current density in copper rails operated at room temperature before gross rail melting occurs due to Joule heating as shown by Hawke (1986); a gun length of 3.65 m with an inductance gradient of 0.384 μH/m; and a bore size of 0.0127m implies that the maximum kinetic energy available in such a system is approximately 210 kJ (disregarding any energy delivered to the projectile from any preinjection scheme). For a projectile mass of 2.5 gm this amount of energy corresponds to a velocity of about 13 km/s.

In practice it is usually not feasible to use the maximum rail current density as given above if plasma armatures are used because radiative heating from the hot plasma armature can significantly increase the surface temperature of the rails and cause gross melting, which limits the lifetime of the system. Also, operating a railgun at near maximum current density imposes severe dynamic stress loadings on the containment structure (a current density of 43 MA/m corresponds to a dynamic repulsive magnetic pressure on the rails of approximately 1 GPa).

Experience at LTV with the 12.7 mm breech fed railgun has shown that operating a railgun at a current density of no more than 33 MA/m (with copper rails) results in a highly reliable and reusable system. Using the parameter values given in the example with this current density and including pre-injection kinetic energy of 2.8 kJ yields a maximum kinetic energy of 125 kJ or a velocity of about 10 km/s for a 2.5 gram projectile.

If the effects of ablation and mass accretion in plasma armatures caused by radiative heating of the sidewalls are included in the performance predictions for railguns, then the maximum velocity is even further decreased as shown by Parker (1986). Parker's predictions indicate that the maximum velocity obtainable in a railgun system can be shown to be given by equation 3.

$$V_m = 0.371 \frac{(L')^{0.7} d^{0.3} I_0}{(\alpha V_a)^{0.4} (C_f m_p)^{0.3}} \tag{3}$$

Where α, is the ablation constant of materials (kg/J); V_a is the armature voltage (V); C_f is the friction coefficient of the plasma with the sidewalls; and m_p is the projectile mass (kg). Using the example case with V_a = 260 V, C_f = 0.003, and α = 23.4 x 10^{-9} kg/J the maximum velocity in this instance would be 8.5 km/s compared to 10 km/sec shown above. The significance of ablation problems in railgun systems is shown in Fig. 6 which compares the maximum velocity predictions from equations 2 and 3 with data from the LTV DES railgun at an average current density of 17.5 MA/m.

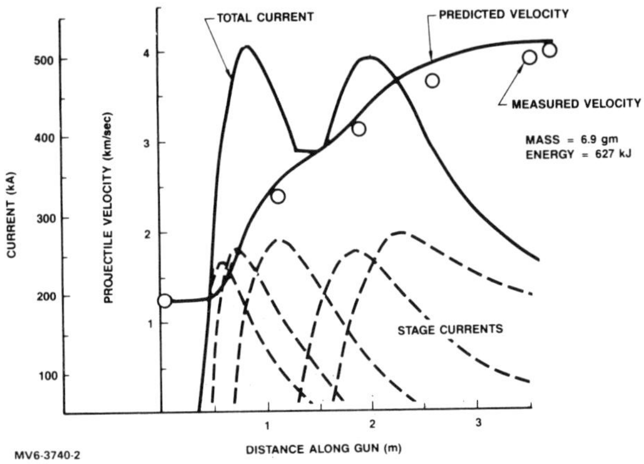

Fig. 5. Analytical simulation of a DES railgun.

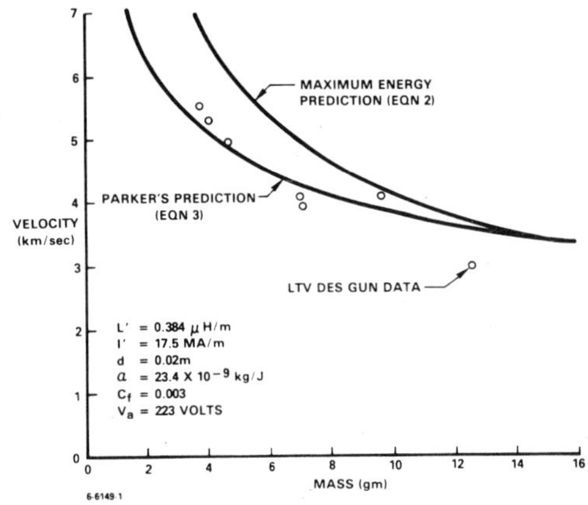

Fig. 6. LTV 20mm DES railgun predicted and maximum velocity comparison.

RANGE DESCRIPTION

The LTV Electromagnetic Launcher Laboratory, established in 1981, was originally designed to conduct penetration mechanics and impact physics experiments in the hypervelocity region. The scope of the facility was broadened to include projectile launch dynamics, which required developing the capability to launch payloads with discarding sabots and to contribute to EM launcher technology development.

The facility, Fig. 7, is comprised of a modular capacitor bank, the EM launcher, the ballistics range, and the instrumentation/control room. The capacitor bank is divided into five modules. Each module stores 132 kJ of energy. The modules may be configured to operate independently or in parallel with one or more other modules, thus serving as the power source for either the breech fed gun or the DES gun. The ballistics range consists of an instrumentation chamber, a foil gate/sabot stripper chamber, a Hall camera section, and the target/flash x-ray chamber. The entire range and gun may be evacuated or backfilled with selected gases. Characteristics of the two launchers are listed in Table 1.

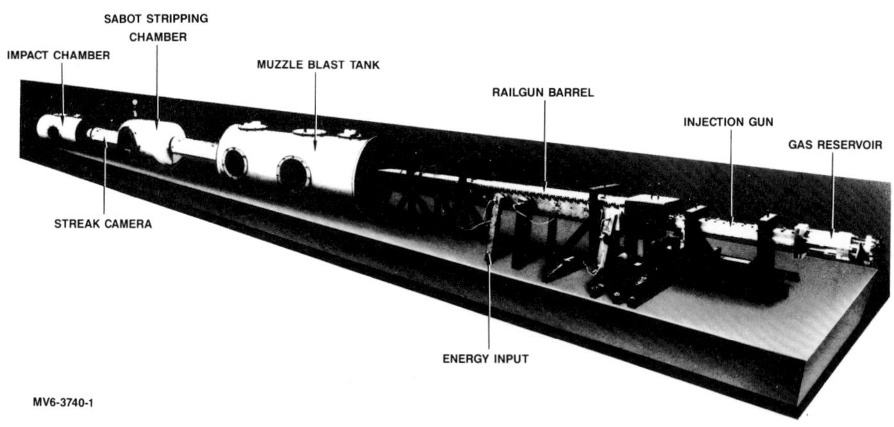

Fig. 7. LTV Hypervelocity EM launcher and ballistics range.

Table 1. Railgun Characteristics

	Breech Fed	DES
Bore	12.7 mm	20 mm
Length	3.7 m	3.7 m
Velocity Range	2-8 km/s	1-6 km/s
Proj. Mass	2-20 g	4-60 g
No. of Stages	1	5
Energy/Stage	660 kJ	132 kJ
Current/Stage	600 kA	200 kA

DIAGNOSTICS

Several diagnostic sensors are used to assess the performance of the launcher and power supply. Data generated can be used to predict the muzzle velocity of the projectile. As the current from a bank flows into the gun, a Rogowski coil, which encompasses the input conductor terminal, produces a voltage which is proportional to the time rate of change of current. This is integrated by a passive, in-line integrator, and produces a signal which is proportional to the current.

Also during the firing process, the voltage difference across the rails is monitored, by voltage dividers attached to the rails at each current input terminal and at the gun muzzle. The output of each voltage divider is fed into a differential voltage amplifier. Data received from the current and voltage probes are used to evaluate the performance of the launcher against the ideal operating conditions and computer simulation models.

While the projectile traverses the gun, its position is detected as a function of time by magnetic field pick-up coils which monitor the passage of current in either the rails or the armature depending upon the orientation of the probe. By utilizing a series of these coils along the length of the gun, average velocity and acceleration may be obtained.

Once the projectile has exited the gun, it enters the muzzle blast expansion chamber, transits a drift tube, and then enters the diagnostic and sabot stripping chamber. Ballistic foil gates are used, if desired, to measure velocity. The foil gates also serve as a sabot separation analysis tool. When a sabot separates from the main projectile, the sabot pieces will penetrate the foil gates, and after the shot one may visually determine how well the sabot separated from the projectile. Down range of the foil gates a sabot stripper prevents the separated sabot from traveling further down the range.

Next the projectile enters the Hall Velocimeter section. Measuring projectile velocity has been a chronic problem with hypervelocity EM launchers due to muzzle flash, electromagnetic interference, and short event times. Traditional ballistic screens suffer from interrupting the projectile's flight and flash x-rays must rely on the dependability of a trigger pulse. A Hall camera system was installed in the LTV EM launcher ballistic range and has proved to be completely reliable. The high speed camera records the passage of the projectile as it passes two slits, thus providing projectile velocity and integrity as well as orientation in one plane as shown in Fig. 8.

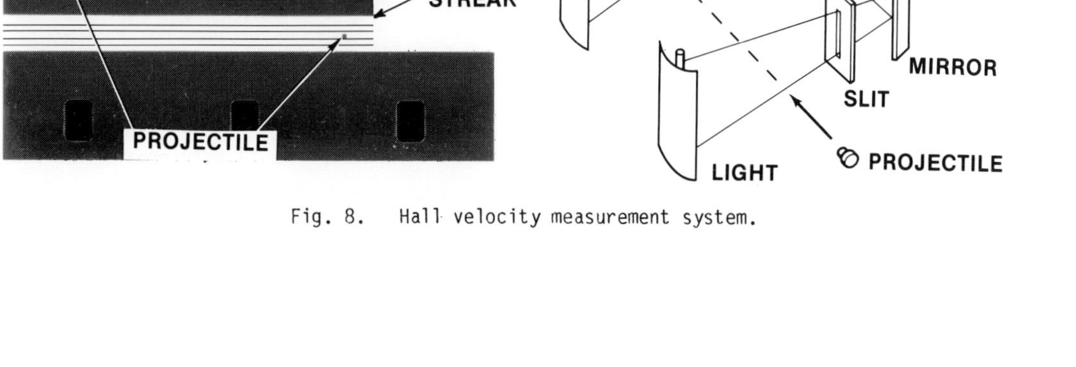

Fig. 8. Hall velocity measurement system.

The projectile exits the Hall station area and enters the target chamber. It immediately interrupts a laser beam, which triggers an orthogonal view flash x-ray system. The flash x-ray system produces photographs which may be used to determine projectile integrity and orientation, or to analyze post-impact target ejecta. The latter is accomplished by varying the time delay between interrupting the laser beam and triggering the x-ray system.

EML PROJECTILE DESIGN

Projectile launch packages for electromagnetic railguns experience a more complex environment than corresponding projectiles for light gas guns by virtue of the higher velocities, finer tolerances required, potentially longer length barrels, high sustained acceleration loads and transient electromagnetic field interferences.

The projectile package is typically composed of three elements: payload, sabot, and armature. Projectile packages of this type have been successfully launched by Tower (1986) with subsequent sabot separation shown by the in-flight shadowgraph of Fig. 9.

The payload is the component of primary interest while the sabot and armature are viewed as parasitic mass required in the launch process. The payload may vary from simple geometrical shapes (e.g., spheres, cubes, cylindrical rods), to aerodynamic configurations or to guided projectiles with guidance and control components packaged internal to a structural frame.

The sabot is used to carry the payload, protect the launcher bore, insulate metallic payloads, seal driving armature plasmas, and distribute the accelerating loads. The payload/sabot package must survive three critical factors associated with the launch process: heat input from the accelerating armature plasma; longitudinal acceleration stresses induced by the launching impulse: and lateral stresses resulting from projectile ballot, side-to-side bumping caused by the irregularities of the launcher barrel. The sabot design is complicated by the requirement for separating the sabot from the payload immediately after launch.

The driving armature function is to supply a current path from one rail conductor to the other. The current in the rails generates a magnetic field which interacts with the current in the armature to produce the accelerating force. The current carrying armature can be either a metallic element, a plasma, or a combination of both called a hybrid or transitioning armature.

The launcher environment imposes severe constraints on the projectile design. The projectile is typically exposed to extremely high acceleration loads (10^5 to 10^6 g's), transient magnetic fields (0.1 to 30 Tesla), and heat inputs from sliding friction and hot plasma armatures (10,000 to 20,000K) over the acceleration period. Fortunately the heat flux is usually on the order of several milliseconds depending upon barrel length and muzzle velocity and heat transfer is minimal. The structural loads are the most critical issue.

Structural loads introduced to the projectile arise from several sources during the launch process. The axial acceleration loads are the most critical and may exceed one million g's in electromagnetic launchers. To the uninitiated this may seem high, however light-gas guns typically reach 1.5 million g's. For short projectiles (length/diameter = 1) with monolithic composition, launch induced loads of a million g's do not present a particular problem, but when multi-component projectiles, long rods or guided projectile electronic packages are involved, the maximum allowable g-loading or setback loads fall below 100,000 g's with current component technologies. The rate of applied acceleration load or jerk rate is also an important dynamic consideration. The rate at which current rises to the maximum and subsequently decreases can introduce shock loadings within the projectile, causing catastrophic failure. During initial movement, a compression wave can cause structural buckling and the subsequent reflected tensile wave can cause spalling at the projectile nose or fracturing of the internal structural.

Another failure associated with jerk occurs at the muzzle. If the acceleration loads remain quite high as the projectile leaves the muzzle, the rapid lateral stress relief can cause the projectile to disintegrate. High energy EM launcher operators have learned that the driving current must be reduced as the projectile nears the muzzle, to provide a coast section at the end of the barrel, thus allowing the projectile stresses the opportunity to relax. Electromagnetic railguns do provide the benefit of shaping the accelerating current as reported by Tower and Trzaska (1985), thus tailoring the projectile axial launch loads.

Lateral acceleration loads are less than longitudinal loads yet are still significant. Transverse loads are attributed to bore curvature, surface irregularities, projectile center of gravity misalignment, and possible gun section misalignment. Typical conventional guns impart side loads of 3-5% of the axial loads. The military specifications allow a 14% maximum side load limit. Ballotting is aggravated as the bore/projectile tolerance increases, due to wear/erosion of the launcher bore surfaces.

The most important interface between the projectile and the launcher is the driving armature. Issues that must be addressed in armature designs are armature mass to projectile mass ratio, armature efficiency and rail damage effects. Three types of armature designs have been tested, the solid metallic armature with brush type, sliding contacts; the plasma armature; and to a lesser extent the hybrid/transitioning armature using a combination of solid and plasma types. Each of these armature types has its good and bad characteristics depending upon the range of velocity desired of the launcher. The choice between solid or plasma armature can be made on the basis of efficiency, as suggested by Barber (1986), by comparing the projectile kinetic energy (KE) to the total energy required to accelerate the projectile package. This total energy includes: projectile KE, armature KE, armature resistive losses, and launch package frictional losses. An efficiency factor can then be determined for each armature type as a function of velocity. On this basis, metallic armatures are much more efficient at the lower velocities, whereas plasma armatures are more efficient above 7 km/s, as shown in Figure 10.

In practical situations metallic armatures have demonstrated a higher efficiency than plasma armatures for velocities less than 1.5 km/s. However, above this velocity solid armatures have yet to maintain good sliding electrical contact and have transitioned to plasma armatures. Plasma armatures provide good electrical contacts to velocities in excess of 8 km/s as demonstrated in railgun firings at LTV and LANL; however, they have high energy losses and cause ablation of the launcher bore materials, which detracts from this performance and limits the barrel lifetime. Theoretically hybrid armatures should provide an intermediate solution between the efficiency of metallic armatures at low velocities and plasma armatures at high velocities.

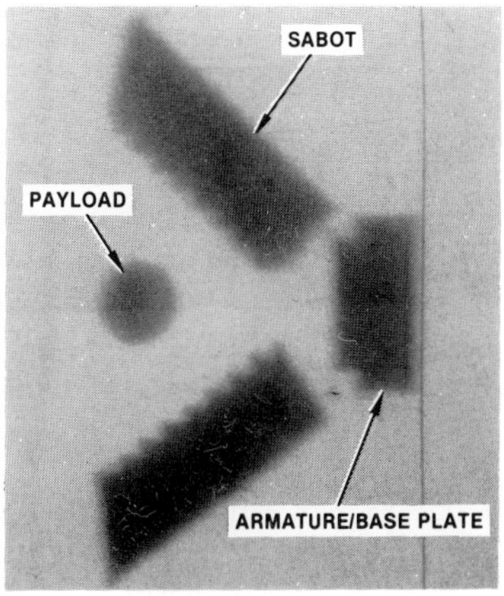

Fig. 9. Shadowgraph of in-flight sabot separation process.

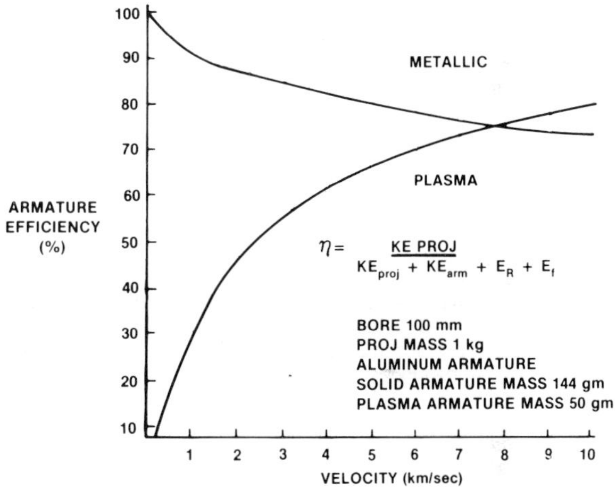

Fig. 10. Solid versus plasma armature efficiency.

The design of the sabot is as much an art as it is a science and frequently reduces itself to trial and error. The design requirements for the sabot of an EM launched projectile are very similar to that of conventional chemical guns and thus the design approach is nearly the same. Requirements include minimum weight, high strength, dielectric material and non-abrasive surface. Other design considerations are a balanced center of gravity location, a length to diameter ratio greater than approximately 0.7, and non-interference separation geometries for discarding designs. The most critical problem with sabot performance is structural failure during the launch. Failure modes include: compressive buckling, shear plane failure due to inertia mis-match between payload and sabot, spall caused by shock loading, and tensile failure resulting from instantaneous stress relief at muzzle exit. Several materials have been found acceptable including: Polycarbonate, glass-filled polycarbonate, Xydar, Zytel, and G-10 glass-epoxy.

Discarding sabots were launched from the EM railgun to support hypervelocity impact testing. The sequence in Fig. 11 shows an in-flight shadowgraph of a 3.5 gram steel ball and polycarbonate sabot package. Impact of this projectile package-type at 3.9 km/s in aluminum is shown in Fig. 12. In actual impact tests the sabot pieces are stopped by a stripper plate, while the penetrator passes through a hole and continues on to impact the target. This discarding sabot package achieved nominally 1 mil radian dispersion in a ballistics flight range evacuated to about 100 torr.

HYPERVELOCITY IMPACT TESTING

During the past five years the LTV EM railguns have launched over 150 projectiles intended to demonstrate the effects of hypervelocity impacts on armor plating and aerospace vehicle components. Projectiles have included Lexan, steel, aluminum, tungsten, ceramic and composite aluminum-epoxy

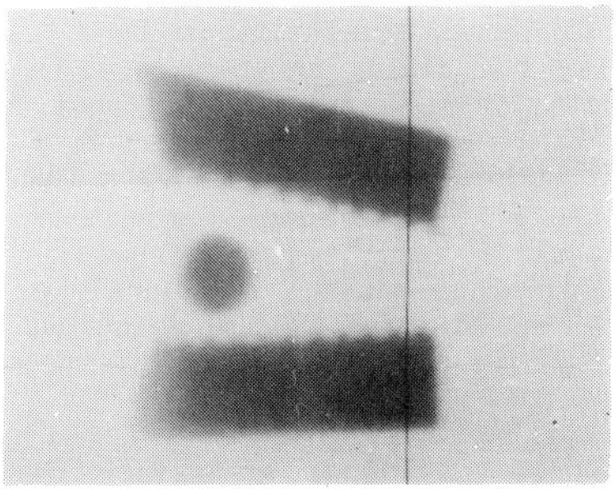

Fig. 11. Free-flight shadowgraph of aerodynamic sabot separation.

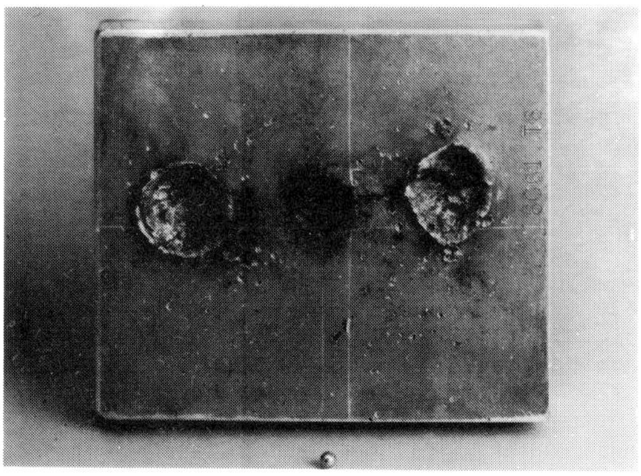

Fig. 12. Impact crater of steel sphere (center) and two Lexan sabot fragments.

materials with various shapes (spheres, cubes, rectangles and hollowed cylinders) and sizes (masses ranging from 2 to 43 grams and length to diameter ratios of 0.6-2.0). Aerospace vehicle and advanced armor targets tested include: aluminum; steel; ceramic and diffusion-bonded metal laminates of various thickness as well as multiple plate targets at various angles of obliquity.

Tests against thin aluminum materials which simulate booster walls are shown in Fig. 13 where a 10 gram projectile impacted the target at 2.5 km/s leaving large holes in both the front and back walls. Impacts of various projectiles on steel plates are shown in Fig. 14 as examples of impact tests performed by EM railguns.

The standard method for quantifying impact penetration performance is to compare the impact crater depth, P, to projectile length, ℓ, and plot with respect to impact velocity. Fig. 15 summarizes the penetration performance of various projectile materials with masses from 2 to 43 gm over a velocity range of 1 to 7 km/s against steel plates. Projectile materials and shapes are indicated on the figure.

The data shown is for "chunky" type projectiles (i.e, spheres, cubes, cylinders). The projectile length to diameter ratio for the data shown in Fig. 15 vary from 0.7 to 2.0. The crater for a hypervelocity impact is not significantly influenced for chunky projectiles of various geometries. Curves were fit through the data to represent trends for various pr ectile materials. Two curves for Lexan projectiles are shown in Fig. 15. When the target plate thickness is on the order of the projectile length, the data follows the curve labeled "finite target". Penetration measurement indicate deeper craters due to back surface bulges which do not occur in semi-infinite targets.

The data indicates that impact velocities are still in the linear penetration regime except for the ceramic projectile which may have reached a transitioning point where the projectile "shatters" and its penetration versus velocity slope decreases.

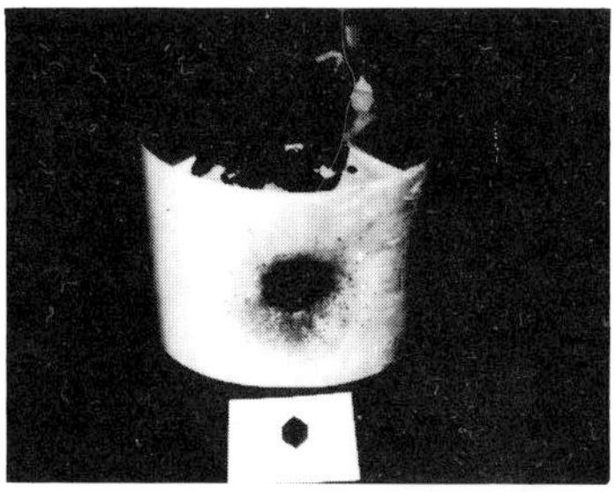

Fig. 13. Booster component impacted by 10 gm at 2.5 km/s.

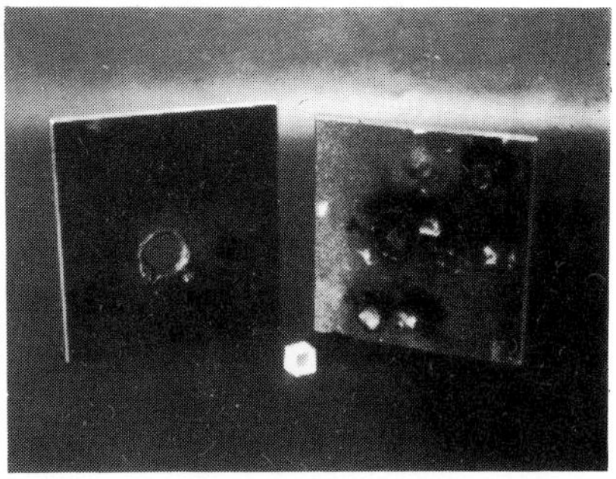

Fig. 14. Steel plate penetration (6.4 mm) by a 2.5 gm Lexan projectile traveling at 8 km/s.

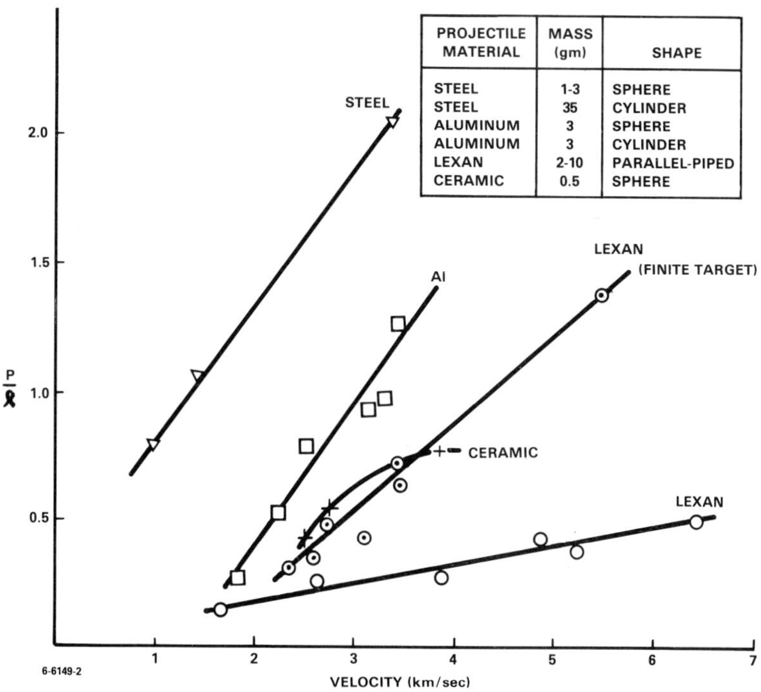

Fig. 15. Steel target penetration/projectile length.

RECOMMENDATIONS

The electromagnetic railgun has demonstrated that hypervelocities can be achieved for impact testing of current interest (3-10 km/s) and extension to yet higher velocities 10-20 km/s are likely in the next several years. The EM railgun can be mated to a ballistic range in a similar manner as conventional powder guns or light-gas guns.

While many of the features of a classical ballistic range are relevant to an EM launcher lethality range, there are some unique differences that must be considered in relation to instrumentation and launcher interface. With the eventual extension of velocity goals to above 10 km/s, shorter measurement times for instrumentation will be encountered. Many photographic techniques presently available will not be applicable at the higher velocities. Cameras and detector diagnostics must continue to evolve to meet the launcher velocity goals. The launch velocity is also in a regime where no aerostability data base exists for projectile and sabot packages. Launcher bore and (possibly) flight ranges must be evacuated which will further influence sabot separation techniques and introduce problems not encountered in partial vacuums.

The EM launcher by its very nature introduces EMI/EMP effects and muzzle arc/flash that will influence diagnostic and photographic instrumentation. The muzzle design of the EM launcher will require that muzzle exit dynamics or tip-off effects be minimized.

REFERENCES

Barber, J. P. (1972). The acceleration of macroparticles and hypervelocity electromagnetic accelerator. PhD Thesis, The Australian National University.

Barber, J. P., and C. L. McDonald (1986). A comparison of armature performance. IEEE Transactions on Magnetics, MAG-22.

Brooks, A. L., R. S. Hawke, J. K. Scudder, and C. D. Wozynski (1982). Design and fabrication of large and small bore railguns. IEEE Transactions on Magnetics, MAG-18, No. 1.

Fowler, C. M., D. R. Peterson, R. S. Caird, D. J. Erickson, B. L. Freeman, and J. C. King (1982). Explosive flux compression generator for railgun power sources. IEEE Transactions on Magnetics, MAG-18, No. 1.

Haight, C. H., and M. M. Tower (1986). Distributed energy store (DES) railgun development. IEEE Transactions on Magnetics, MAG-22.

Hawke, R. S. (1979). Devices for launching 0.1g projectile to 150 km/s or more to initiate fusion. Lawrence Livermore National Laboratory, Report No. UCRL-52779 Part 2.

Jackson, G. L., L. K. Farris, and M. M. Tower (1986). Electromagnetic railgun extended-life bore material tests results. IEEE Transactions on Magnetics, MAG-22.

Parker, J. V. (1982). Electromagnetic projectile acceleration utilizing distributed energy sources. J. Appl. Phys.

Parker, J. V., W. M. Parsons, C. E. Cummings, and W. E. Fox (1985). Performance loss due to wall ablation in plasma armature railguns, Proc. of AIAA 18th Fluid Dynamics and Plasmadynamics and Lasers Conference, AIAA-85-1575.

Marshall, R. A., and S. C. Rashleigh (1978). Electromagnetic acceleration of macroparticles to high velocity. J. Appl. Phys. 49(4).

Schnurr, N. M., J. F. Kerrisk, and J. V. Parker (1986). Numerical predictions of railgun performance including the effects of ablation and arc drag. IEEE Transactions on Magnetics, MAG-22.

Tower M. M., L. D. Holland, and T. J. Trzaska (1984). Distributed energy source electromagnetic railgun. U.S. Army Armament Research and Development Command Report, ARLCD-CR-84025.

Tower, M. M., and C. H. Haight (1984). Development of a high energy distributed energy source electromagnetic railgun with improved energy conversion efficiency. IEEE Transactions on Magnetics, MAG-20 No. 2.

Tower, M. M., and T. J. Trzaska (1985). Accelerator assembly/distributed energy stores. U.S. Army BMDATC Final Report, Contract No. DASG60-83-C-0138.

Tower, M. M., L. K. Farris, and C. H. Haight (1986). Electromagnetic launched projectile compatibility. IEEE Transactions on Magnetics, MAG-22.

Int. J. Impact Engng Vol. 5, pp. 645–653, 1987
Printed in Great Britain

0734-743X/87 $3.00 + 0.00
Pergamon Journals Ltd.

EFFECTS OF VAPORIZATION ON DEBRIS CLOUD DYNAMICS*

T. G. Trucano and J. R. Asay

Sandia National Laboratories
Albuquerque, New Mexico

ABSTRACT

Experimental techniques are discussed for studying the debris products produced by plane shock loading of specimens to final temperatures sufficient to cause vaporization. These techniques have been applied to a study of the shock-induced vaporization of cadmium. The expansion products produced by shock reflection from a free surface were studied by allowing the debris to stagnate against a time-resolved interferometer gauge that allowed measurement of the stagnation pressure history. The experimental results were modeled in one dimension with Eulerian and Lagrangian computer wave codes utilizing detailed equations of state for cadmium. These comparisons illustrate that vaporization of cadmium occurs under essentially equilibrium conditions and that good agreement with experimental results is obtained under the assumption of instantaneous vaporization. Another observation is that the vapor expansion appears to be a centered expansion wave, with a density distribution differing from the computer calculation.

INTRODUCTION

In this paper we present preliminary results of our current experimental and theoretical investigations into the role played by vaporization on the dynamics of debris formed during hypervelocity impact. Our efforts are marked by attention to the coupling of details of the equation of state to details of the resulting hydrodynamic flow.

The canonical hypervelocity impact event that we have in mind is the impact of a metallic projectile upon a thin metallic "bumper" or "meteor" shield. Such an impact produces back surface debris having varying characteristics, depending upon geometry, materials, and impact velocity. Hopkins, *et al.*, (1972) have described these effects and identified the general characteristics of the debris cloud in terms of the penetration of a second metallic plate lying behind the shield in the direction of projectile motion. These authors have found that the effectiveness of penetrating the second plate by the debris (a hydrodynamic effect) decreases as the projectile velocity increases, thus driving the debris through successive transformations from solid, to liquid, to vapor (which is a thermodynamic effect). It is likely that the equation-of-state (EOS) details of the liquid/vapor coexistence region of the materials determine the penetration behavior of the debris. For this reason, close examination of controlled experiments is necessary to improve our ability to predict (through computational or analytical means) the interaction of hypervelocity debris with material behind the shield.

The need to characterize the dynamic properties of debris products in hypervelocity impact experiments has been previously recognized. To address this issue, Swift and Hopkins (1970) developed ballistic impact techniques to measure the momentum distribution in debris clouds. Although these techniques are useful for estimating off-axis distributions of debris momentum, they are inadequate for critically evaluating theoretical models. In this paper, we describe a technique for making time-resolved measurements of debris momentum under planar loading conditions, that provides a much more sensitive measurement of debris dynamics. The technique separates the geometric and thermodynamic influences on debris dynamics and is therefore useful in evaluating theories describing shock-induced vaporization.

Because of its low boiling point, which allows us to produce shock-induced vapor states with current laboratory technology, cadmium is a metal that facilitates such an investigation. Other metals of greater practical interest (tungsten, tantalum, or steel) lie beyond current laboratory capabilities for producing substantial vaporization upon impact. By coupling careful experimentation on a metal such as cadmium with adequate theoretical modeling of the process, we hope to understand the details of how the relevant phenomenology scales to the metals of greater practical interest.

In this paper we describe experiments performed on cadmium that are designed to provide time-resolved information on the dynamics of the debris cloud formed when a sheet of this metal is vaporized under impact loading. We discuss computational simulations of the experiments, performed to analyze the experimental results, and draw conclusions about the hydrodynamic implications of cadmium vaporization. We refer to this work as a study of the *vaporization dynamics* of cadmium. We then provide some general conclusions concerning the study and point out additional investigations that are currently feasible.

EXPERIMENTAL PROCEDURE

The experimental technique used for studying shock-induced vaporization states of cadmium is illustrated in Fig. 1. In this method, a two-stage light gas gun (Asay, 1981) is used to accelerate a projectile with a tantalum facing to velocities near 5 km/s. The tantalum disc impacts a thin cadmium specimen in a chamber evacuated to an initial pressure of about 0.1 Torr. At these impact velocities, a shock pressure of approximately 200 GPa and a shock temperature of nearly 15,000 K (as estimated from our theoretical EOS models) are produced in the cadmium specimen. Upon arrival of the shock wave at the back surface of the specimen, the shock pressure decreases to near zero (limited by the vapor pressure of the specimen) and the temperature decreases to about 1,500 K. A rarefaction wave is formed at this time that propagates back through the specimen, unloading the initial shocked state to the unloaded final pressure. Since the specimen is thin, this unloading occurs in very short times (typically 0.05 μsec), so that the entire specimen is shock-heated to vaporized states very rapidly. The vapor states produced in this way were studied by letting the vapor stagnate against a time-resolved gauge, which provided a measure of vapor velocity and mass distribution. A technique similar to this has been used by Spaulding (1981) to study the expansion of explosive products.

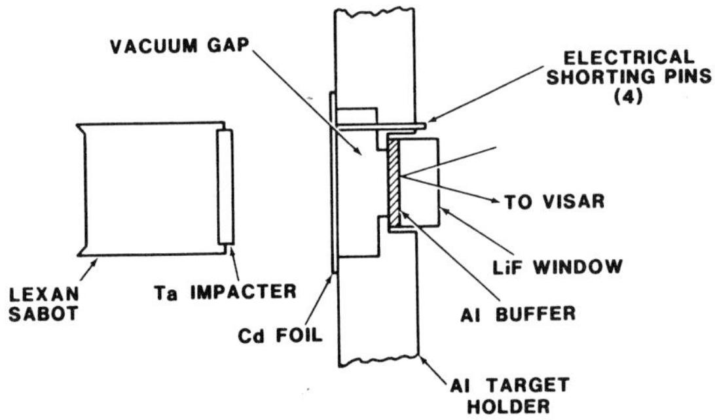

Fig. 1. Cadmium vaporization experimental schematic.

In the present experiments, the diameter of the tantalum impactor was 30.5 mm with a nominal thickness of 2 mm. The impactor was mounted in a Lexan sabot of 32.5-mm diameter and 25-mm length. The cadmium specimens, of 99.99% purity, were received in the form of flat and parallel foils, with a thickness of about 0.25 mm. The samples were mounted in an aluminum target plate of 12.7-mm thickness, over a cavity of 35-mm diameter. This configuration prevented impact and subsequent wave propagation in the aluminum plate and also enhanced one-dimensional conditions for the expanding cadmium vapor. To determine initial shock arrival at the cadmium back surface, four electrical shorting pins with a 90-degree separation were spring-loaded against the back surface of the specimens. These pins also allowed determination of impactor misalignment, resulting in a small correction to the timing of the VISAR, since measurements were performed at the centerline of the impact axis. A vacuum gap, which varied from 3 to 9 mm, separated the cadmium sample from the time-resolved interferometer gauge. The gauge consisted of a 2-mm thick buffer of 6061-T6 aluminum, which was about 19 mm in diameter and 2.000 $\pm$ 0.005 mm thick. The buffer was bonded to a single-crystal LiF laser window (Wise, 1986), 19 mm in diameter by 12.7-mm thick, with the $< 1\,0\,0 >$ axis of the crystal aligned in the direction of motion.

The purpose of the buffer was to smooth out irregularities in the loading pressure produced during stagnation of the cadmium vapor. Since the shock impedance of aluminum is almost identical to that of LiF (Asay and Chhabildas, 1981), wave reverberations in the buffer were minimized, so that this technique provided a nearly unperturbed measurement of the front-surface pressure history produced by stagnation of the vapor. The main limitation is that the nonlinear shock response of aluminum produces a steepening of the input pressure profile that can result in loss of information regarding the initial stagnation pressure history if a shock wave is formed in the aluminum buffer.

Table I. Experimental Impact Conditions

Exp. No.	Impact Vel. (km/s)	Impactor Thick. (mm)	Sample[a] Thick. (mm)	Gap Sep. (mm)	Shock[b] Press. (GPa)	Peak[c] Gauge Press. (GPa)
CD2	4.91	2.000	0.252	8.959	187.6	43.8
CD3	5.23	2.000	0.262	8.979	206.9	44.7
CD4	5.21	2.001	0.255	3.010	205.7	64.2

[a]The aluminum buffer thicknesses were 2.000 ± 0.005 mm in all experiments.
[b]The reported Hugoniots for cadmium and tantalum were used for defining impact conditions. Cd: $P = 21.32u + 14.26u^2$, GPa;
Ta: $P = 56.86u + 19.99u^2$, GPa.
[c]The Hugoniot for aluminum used for these calculations was:
$P = 14.41u + 3.59u^2$, GPa.

The particle velocity at the aluminum–LiF interface was recorded with a VISAR velocity interferometer (Barker and Hollenbach, 1972). The time resolution of this system was about 2 ns in the present configuration and the particle velocity accuracy was 1–2 % of peak velocity. Although the interferometer records particle velocity at the recording surface, pressure histories can be determined from the dynamic equation of state for LiF (Wise and Chhabildas, 1986). The experimental impact conditions for the different experiments studied are given in Table I.

The particle velocities recorded in the different experimental studies are illustrated in Fig. 2a–2c. The particle velocity histories generally consist of a small initial motion (precursor), with an equivalent pressure of about 0.5 GPa, followed by a large discontinuous increase in particle velocity. A gradual decrease in particle velocity is observed after which a second large increase in velocity occurs. The second increase in particle velocity corresponds to arrival of the tantalum impactor at the compressed cadmium vapor–aluminum interface.

The small precursor that arrives at the aluminum-LiF interface prior to the large increase in particle velocity is not a result of mass velocity distributions in the cadmium vapor, but is an artifact of the aluminum buffer response to the initial loading history induced by the vapor. Aluminum exhibits elastic-plastic response under dynamic loading, with a Hugoniot elastic limit of about 0.5 GPa (Lipkin and Asay, 1977). For initial *shock* loading to the peak pressure of about 50 Gpa occurring in these tests, a single plastic shock wave propagating with a velocity of about 8.2 km/s should be produced in the aluminum buffer. For initial *continuous* loading to the same pressure, an elastic wave would be produced, followed by a continuous plastic compression "fan" that steepens with propagation distance because of nonlinear material response (Barker, 1984). The compression fan ultimately forms a plastic shock wave that has a velocity exceeding the elastic wave and eventually overtakes it. This results in a single elastic–plastic shock wave similar to that produced by direct shock loading to the final pressure (Barker, 1984).

The fact that an elastic precursor is observed in the measured particle-velocity histories and the known aluminum equation of state provide important additional information about the mass distribution of the cadmium debris. The precursor can only arise if the initial pressure loading is continuous, rather than abrupt. Otherwise, the elastic precursor would be overdriven and a single plastic shock wave having the final amplitude would be observed, as discussed above. The results therefore strongly suggest that the stagnation pressure produced by the vapor is not an abrupt increase in pressure, but is dispersed over a short interval in time. It is not possible to accurately determine the input pressure history produced by the vapor from the present data because of the resulting shock wave produced in the aluminum buffer, but it is possible to place bounds on the risetime of the initial loading. From the known equation of state for aluminum (McQueen, *et al.*, 1970) and the observed time separation between the elastic and bulk waves, the risetime of the initial pressure loading is estimated to be about 0.13 μs. Although additional experiments with a thinner buffer are required to precisely determine the risetime, this preliminary estimate indicates that the density distribution at the vapor front increases gradually and not discontinuously as the vapor expands freely into a vacuum.

COMPUTATIONS

Full details of the hydrodynamics simulations can be found in Trucano and Asay (1986), so that we will not dwell upon many of these details here. Instead, we will point out some aspects of the EOS of cadmium that we have explored in our computations.

We have performed several simulations, corresponding to major variations in the cadmium EOS. The nature of the cadmium EOS and its liquid/vapor regime can influence shock amplitudes (the initial stagnation pressure jump and the pressure increase produced by the impact of the tantalum on the gauge) and timing in this experiment. It is useful to make significant modifications in the cadmium EOS used in our calculations, in order to study the nature of the changes which occur in the computed hydrodynamics. This provides correlation of experimental data with features of the EOS that might be difficult to isolate. Basically, we have considered three distinct models of cadmium:

Model 1: Cadmium treated as a frozen solid – no melting or vaporization is allowed to occur.

Model 2: Cadmium treated as a solid/vapor – the mixed liquid/vapor regime is ignored, and the melting behavior is neglected.

Model 3: Cadmium treated as a three-phase material in a thermodynamically consistent fashion. This allows us to study the influence of critical and fusion properties of the material upon the computed hydrodynamics.

Each of these models treats the cadmium as a hydrodynamic material, which is certainly reasonable, considering the final states produced. Model 1 uses a Mie-Gruneisen EOS for the solid, referenced to Hugoniot data, which is a standard option in the WONDY wavecode. Model 2 uses the ANEOS EOS, a standard description in the CHART D and CSQ wavecodes for the solid, coupled to an instantaneous transformation to the vapor state when the internal energy of the cadmium exceeds a threshold (typically, the sublimation energy). The absence of melting and mixed liquid/vapor regions does not accurately reflect real material response, and the solid is unstable above the melt energy and below the sublimation energy. Model 3 uses ANEOS with the solid/liquid region included. This provides a thermodynamically consistent three-phase EOS for cadmium. Comparisons of the predictions of these models for the CD3 experiment are given in Figs. 3 and 4.

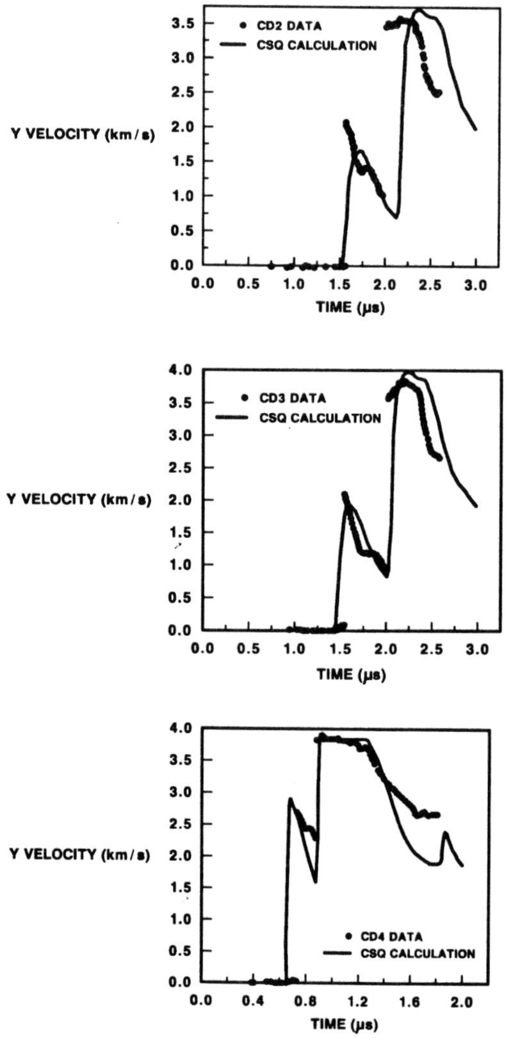

Fig. 2. Experimental particle velocities (dots) and CSQ calculations (solid lines) for experiments (a) CD2, (b) CD3, and (c) CD4.

We have simulated the above experiments using the 2 D Eulerian wavecode CSQ (Thompson, 1977), in a 1 D configuration, and the 1 D Lagrangian wavecodes WONDY (Kipp and Lawrence, 1982) and CHART D (Thompson and Lawson, 1977). We have plotted the CSQ results for the three experiments conducted, along with the experimental data, in Fig. 2. The calculations shown in Fig. 2 were performed with Model 3 for the EOS of cadmium. Other calculations discussed later explored the sensitivity of the computations to equation-of-state variations.

We point out that the CSQ calculations show, in general, earlier arrival of the debris than the experimental records for CD3 and CD4. Although not apparent in the figure, the calculated vapor arrival time agrees well with the first arrival of the elastic precursor, as observed experimentally. On the other hand, the arrival of the flyer plate, indicated by the second sharp rise in the experimental and computed records, is delayed with respect to the experimental data.

It is instructive to compare the calculations performed with an Eulerian code (CSQ) with those resulting from a Lagrangian code, as shown in Fig. 3. There are several curves plotted there, representing various calculations for the CD3 experiment (Fig. 2, (b)). In both CD2 and CD3, the CSQ calculation (Model 3) underestimates the initial vapor stagnation pressure, as compared to the data. On the other hand, we see that a Lagrangian calculation (CHART D, Model 3) results in an increased pressure. Part of this increase is the noise in the Lagrangian calculation which is not present in the Eulerian calculation. The noise is also apparent during the impact of the flyer upon the gauge. The mean of these oscillations lies closer to the data than the Eulerian calculation. We also note that the risetimes are sharper for the Lagrangian calculation. This is due to the manner in which mass is transported in the different codes. Lagrangian codes, which maintain sharp material interfaces, are characterized by sharp risetimes (basically fixed by artificial viscosity). Such codes generally show oscillations of finite amplitude in impact-generated shock waves. These oscillations can be reduced, for example, by adjusting the zoning so that very little mass is present in the zones that directly impact the aluminum gauge for both the tantalum and the cadmium, and by dropping the time-step just before the various impacts. More detailed discussion of this point can be found in Trucano and Asay (1986).

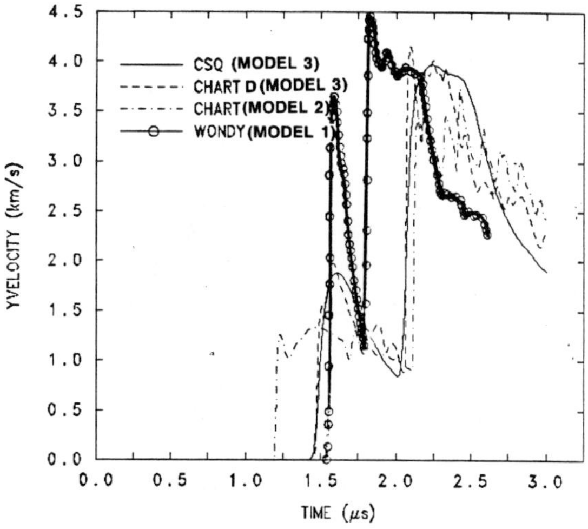

Fig. 3. Computational comparisons for CD3, with different cadmium models. (a) CSQ, Model 3; (b) CHART D, Model 2; (c) CHART D, Model 3; and (d) WONDY, Model 1.

CSQ yields smoother profiles with no visible oscillations at the times of impact. The risetimes are also broader than those associated with the Lagrangian calculation. This is due to the manner in which mass is fluxed through the Eulerian grid, which numerically diffuses or disperses the materials of the calculation. The cadmium vapor travels over a relatively large distance in these experiments, which increases the amount of diffusion that occurs in the Eulerian fluxing calculations. Because of the mass diffusion, we can argue that the Eulerian computations tend to underestimate impact pressures because of a more extended spatial mass gradient than is present in the experiment. While this is true in CD2 and CD3, it is not true in CD4. However, CD4 represents a smaller propagation gap than the other two experiments, so that the numerical dispersion of cadmium debris is smaller. The amplitude of the second shock is subtly coupled with the cadmium debris mass gradients generated by the experiment, because the cushioning of the flyer plate impact on the gauge by the stagnated cadmium vapor is affected by this. It is important to point out that all of the calculations miss the small elastic precursor in the aluminum gauge that is visible in the data. This means that the computed debris that stagnates against the gauge produces pressures that overdrive the elastic wave. Thus, there is too much mass at the leading edge of the debris cloud in the calculations, which could be due to badly chosen zoning, or to inadequacies in the cadmium EOS.

Models 1 and 2 suffer from obvious limitations, and will not describe cadmium in a realistic way. On the other hand, comparisons of the predictions of these models with each other and the data immediately reveal the dominant thermodynamic mechanisms of the expanding debris. The elucidation of such mechanisms is a major objective of this work. Thus, computer simulations using these models can be viewed as thought experiments that attempt to clarify the complex coupling of equation of state with hydrodynamics in the current experiments.

A comparison of the different thermodynamic models is given in Fig. 3 for experiment CD3. There, we see that Model 2 and Model 3 yield similar records, while Model 1 (cadmium treated as a frozen solid) overestimates the amplitude of the shock in both of the impacts in the experiment. Also, note that there is a major discrepancy in the timing of the second shock with Model 1 (but a relatively small discrepancy in the timing of the first shock). From these results, we must conclude that there is little or no metastable solid material present in the vapor expansion front in the experiments. A similar conclusion is expected to hold for metastable liquid, but this issue is still under investigation.

Uncertainties in the numerical predictions for cadmium can also be directly coupled to uncertainties in the EOS of cadmium, one example being uncertainty of critical point data. For the computed hydrodynamics, we have studied the implications of making thermodynamic changes in the Model 3 EOS for cadmium. In Fig. 4, we compare our baseline Model 3 (called Model 3-I) calculation with a modification in which the 0 kelvin contribution is changed, and the melt transition is modified. The effect of this new model, Model 3-II, is actually an improvement in the overall EOS for cadmium: the computed melt boundary is in better agreement with published data, the boiling point is more accurate, agreement with Hugoniot data remains close, and pathologies evident in the vapor dome of Model 3-I (see Fig. 5; specifically, we note the change in convexity of the dome boundary) disappear. The net effect of all the changes is to raise the amplitude of the vapor stagnation pressure by almost 20%, which is in somewhat better agreement with our experimental data.

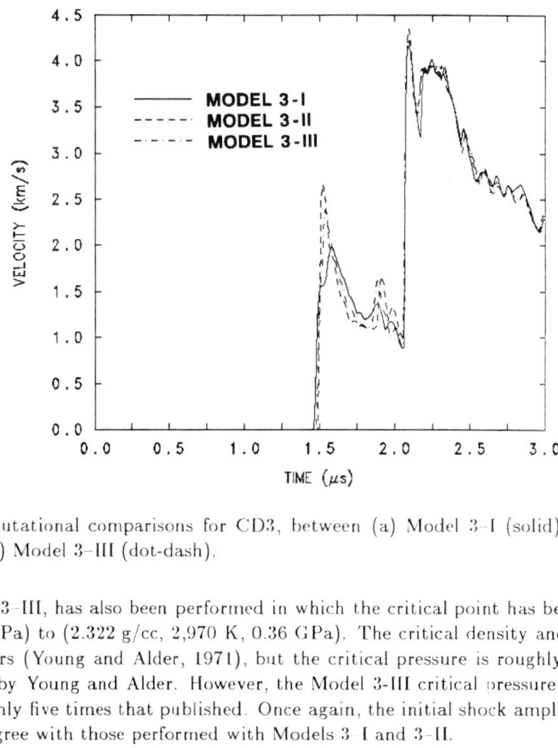

Fig. 4. Computational comparisons for CD3, between (a) Model 3-I (solid), (b) Model 3-II (dashed), and (c) Model 3-III (dot-dash).

Another adjustment, Model 3-III, has also been performed in which the critical point has been shifted from (ρ_c, T_c, P_c) = (1.974 g/cc, 5,857 K, 0.54 GPa) to (2.322 g/cc, 2,970 K, 0.36 GPa). The critical density and temperature of Model 3-III agree with published numbers (Young and Alder, 1971), but the critical pressure is roughly three times higher than the theoretical prediction given by Young and Alder. However, the Model 3-III critical pressure is somewhat below the value of Model 3-II, which is roughly five times that published. Once again, the initial shock amplitude changes, but the overall features of the calculation agree with those performed with Models 3-I and 3-II.

DISCUSSION

The present experimental data and numerical calculations allow conclusions to be made about shock-induced vaporization that heretofore have not been reported. Previous experimental investigations of shock-induced vaporization have been directed, for the most part, to measurements of unloading adiabats from high-pressure shocked states into regions of liquid and liquid-vapor response for the purpose of studying the EOS of hot, expanded states (for example, see Al'tshuler and colleagues, 1977; Fortov and colleagues, 1974). These data did not allow evaluation of kinetic effects associated with rapid unloading into regions of mixed phase response. The present study allows direct evaluation of kinetic effects associated with shock-induced vaporization.

As illustrated in Fig. 3, the expected peak impact pressures of the cadmium debris products with the time-resolved interferometer gauge depend crucially on whether or not vaporization occurs. The calculation performed, assuming metastable solid response, produces a stagnation pressure in the gauge that is about a factor of two higher than that corresponding to equilibrium behavior in which vaporization is assumed to occur instantaneously in the mixed phase region. The experimental data are found to be in significantly better agreement with calculations based on equilibrium behavior, providing strong evidence that kinetic effects of vaporization are not important in this regime of material response. Furthermore, calculations based on solid-vapor behavior utilizing a simple sublimation energy appear to significantly deviate from experimental results. These results have important implications for the numerical simulation of hypervelocity lethality in problems involving shock-induced vaporization of cadmium. Further experimentation is required to determine whether this conclusion applies generally to metals, and perhaps to other materials.

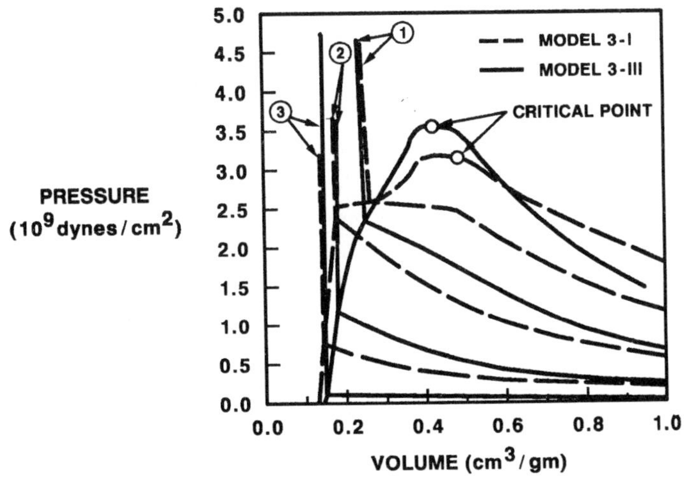

Fig. 5. Vapor dome comparison between Model 3-I (solid) and Model 3-III (dashed) for cadmium, with release isentropes from (1) 4 Mbar, (2) 2 Mbar, and (3) 1 Mbar.

Another observation from the present study is that the late-time response of the tantalum impactor, which undergoes spall after the initial impact, can be reasonably calculated in both amplitude and history during recompression of the cadmium vapor products, although the experimental results appear to occur slightly earlier in time. Again, there is a definite sensitivity of the debris recompression history to the occurrence of vaporization, as indicated in Fig. 3. The experimental results appear to be in much better agreement with calculations assuming equilibrium vaporization, which gives further support to the conclusion that kinetic effects of vaporization are minimal in this problem. The slight discrepancy between experimental results and numerical simulation may be due to small kinetic effects, to slight errors in the equation of state, to possible two-dimensional effects, or to inadequacy of the spall model for tantalum used in our calculations (which was simply a minimum stress criterion).

A small discrepancy between experiment and theory involves the initial arrival time of the debris products upon the gauge, as measured by particle velocity at the aluminum-LiF interface. In the two experiments having a 9-mm separation between the cadmium specimen and the gauge, the timing error is about 0.1 μs for the calculated arrival of the major increase in stagnation pressure. Close examination of the results reveals that the calculated arrival time agrees closely with the experimentally observed arrival of the elastic precursor. As indicated earlier, the precursor separation with the main pulse indicates that the experimental risetime of the stagnation pressure pulse is about 0.13 us, which is not predicted in the calculations. Thus, the apparent timing error may be due to fine details in the expansion characteristics of the expanding vapor cloud. This is specifically seen in the failure of our calculations to predict the observed elastic precursor and appears to be the only major deviation from predicted response. The discrete shoulder to the experimental results observed in all experiments (see Fig. 2), but not predicted theoretically, is also another minor deviation that is not presently understood.

One of the major observations of the present study is that both Lagrangian and Eulerian wavecodes employing the proper thermodynamic properties give reasonable agreement with experimental data. This is an important point, because many practical problems of interest in hypervelocity impact require the use of Eulerian codes which traditionally lose some of the fine details of material response due to material diffusion through the numerical grid.

A final observation is that the experimental technique discussed here appears to be fairly sensitive to fine details in the vapor dome physics, as is apparent from Fig. 3 and Fig. 2c. This is encouraging from a scientific point of view, since vapor-state properties are difficult to measure experimentally. Our work is also a probe of the sensitivity of hydrodynamic impact effects to details of the liquid-vapor EOS region, an issue currently under debate. We claim that the hydrodynamics are sensitive to this EOS region. However, considerably more experimental and theoretical study remains before definite conclusions can be drawn regarding the sensitivity of determining vapor-state properties with this method.

Since the stagnation pressure against subsurface materials behind bumper shields is of direct interest in lethality studies, the present data provide direct evaluation of simulation capabilities, without the complications of divergent flow of the vapor products. Although the present experimental results are considered preliminary and further work is in progress, they are encouraging regarding the calculational ability to correctly describe the major effects of vapor cloud dynamics. However, further experimentation is needed to address questions of reproducibility, response for states deeper into the mixed-phase region, fine details of kinetic response, and the implications of the hypothesized centered compression fan for the debris products. The latter observation may be related to mixed phase flow between liquid and vapor products in the debris cloud. Experiments are currently in progress that are designed to address these questions.

ACKNOWLEDGEMENTS

We would like to thank Robert D. Hardy for assistance in performing the experiments, Samuel L. Thompson for consulting on the use of the CSQIII and CHARTD wavecodes and for helpful discussions concerning the equation-of-state of cadmium, and Thomas K. Bergstresser and Douglas S. Drumheller for critically reviewing our manuscript.

REFERENCES

*This work performed at Sandia National Laboratories supported by the U.S. Department of Energy, under contract DE-AC04-76DP00789.

Al'tshuler, L. V., A. A. Bakanova, A. V. Bushman, I. P. Dudoladov, and V. N. Zubarev (1977). Evaporation of Shock-Compressed Lead in Release Waves, Sov. Phys. JETP, **46**, (5), 980-983.

Asay, J. R. (1978). Thick Plate Technique for Measuring Ejecta from Shocked Surfaces. J. Appl. Phys. **49**, 6173-6175.

Asay, J. R. and L. C. Chhabildas (1980). Determination of the Shear Strength of Shock Compressed 6061-T6 Aluminum. In Marc A. Meyers and Lawrence E. Murr (eds.), Shock Waves and High-Strain-Rate Phenomena in Metals, Plenum Publ. Co., NY.

Asay, J. R. (1981). The Sandia National Laboratories Shock Thermodynamics Applied Research (STAR) Facility. Sandia National Laboratories Report SAND81-1901.

Barker, L. M. and R. E. Hollenbach (1972). Laser Interferometer for Measuring High Velocities of any Reflecting Surface. J. Appl. Phys. **43**, 4649-4675.

Barker, L. M. (1984). High Pressure Quasi-Isentropic Impact Experiments. In J. R. Asay, R. A. Graham, and G. K. Straub (eds.), Shock Waves in Condensed Matter-1983, Elsevier Science Publ., B.V.

Fortov, V. E., A. A. Leonit'en, A. N. Dremin, and S. V. Preshin (1974). Isentropic Expansion of Shock-Compressed Lead. JETP Lett., **20**, (1), 13-14.

Hopkins, A. K., T. W. Lee, and H. F. Swift (1972). Material Phase Transformation Effects Upon Performance of Spaced Bumper Systems, J. Spacecraft **9**, 342-345.

Kipp, M. E. and R. J. Lawrence (1982). WONDY V - A One-Dimensional Finite- Difference Wave Propagation Code. Sandia National Laboratories Report SAND81-0930, unpublished.

Lipkin, J. and J. R. Asay (1977). Reshock and Release of Shock-Compressed 6061-T6 Aluminum. J. Appl. Phys. **48**, 182-189.

McQueen, R. G., S. P. Marsh, J. W. Taylor, J. N. Fritz, and W. J. Carter (1970). The Equation of State of Solids From Shock Wave Studies. In Ray Kinslow (ed.), High-Velocity Impact Phenomena, Academic Press, NY.

Spaulding, Robert L., Jr. (1981). Precursors in Detonations in Porous Explosives. In James M. Short (ed.), Seventh Symposium on Detonation, U.S. Government Printing Office, Washington, DC.

Swift, H. F. and A. K. Hopkins (1970). Effects of Bumper Material Properties on the Operation of Spaced Meteoroid Shields. J. Spacecraft **7**, 73-77.

Thompson, S. L. and H. S. Lawson (1971). Improvements in the Chart D Radiation-Hydrodynamic Code II: A Revised Program. Sandia National Laboratories Report SC-RR-71 0713, unpublished.

Thompson, S. L. (1977). CSQII - An Eulerian Finite Difference Program for Two-Dimensional Material Response - Part I. Material Sections. Sandia National Laboratories Report SAND77-1339, unpublished.

Trucano, T. G. and J. R. Asay (1986). In preparation.

Wise, J. L. and L. C. Chhabildas (1986). Laser Interferometer Measurements of Refractive Index in Shock Compressed Materials. In Proceedings of the 1986 Topical Conference on Shock Waves in Condensed Matter. Spokane, WA, July 23-26, 1985.

Young, D. A. and B. Alder (1971). Critical Points of Metals From the van der Waals Model. Phys. Rev. A, **3**, 364-371.

Int. J. Impact Engng Vol. 5, pp. 655–661, 1987
Printed in Great Britain

0734-743X/87 \$3.00 + 0.00
Pergamon Journals Ltd.

EFFECT OF AN IMPACT-GENERATED GAS CLOUD ON ENTRAINED SURFACE BOULDERS AND SPALL EJECTA

A.M. Vickery
Lunar and Planetary Laboratory
University of Arizona
Tucson, AZ 85721

ABSTRACT

The purpose of this investigation is to study the effect of an impact-generated gas cloud on the size-velocity distribution of large, solid ejecta fragments that become entrained in it, with particular reference to the possible Martian origin of the SNC meteorites. The entrainment both of loose surface boulders and of the early-time, large-size, high-velocity spall component of the crater ejecta is modeled numerically. Surface boulders from as far as 40 km from the center of impact can be accelerated by the high velocity leading edge of the gas cloud to velocities in excess of Martian escape velocity (5 km/s), but are generally crushed by the acceleration. Spall fragments become entrained later and nearer the center of the gas cloud, where gas velocities are much less. High velocity spalls are decelerated by the gas and low velocity spalls are accelerated by the gas, but no spalls ejected at < 5 km/s are accelerated to velocities > 5 km/s. An impact-generated gas cloud is thus expected to scour the pre-existing surface of loose material and to change the size-velocity distribution of spall ejecta, but does not sufficiently enhance the velocities of crater ejecta to explain the Martian origin of SNC meteorites as large rocks.

INTRODUCTION

In recent years, an increasing body of evidence strongly suggests that the shergottites, nakhlites, and Chassigny, or SNC meteorites, are fragments of the planet Mars (see McSween, 1985, for a review). The major problem for the Mars-origin hypothesis has always been the dynamical difficulty of accelerating large, coherent masses of rock, some of which show no petrologically detectable evidence of shock damage, to Martian escape velocity, 5 km/s. How large the original rocks must have been is not well-constrained. Some size estimates are based on interpretations of the ejection history(ies) from the shock age recorded in the shergottites (~200 Ma, Chen and Wasserburg, 1985; Shih and others, 1982; Wooden and others, 1982) and the cosmic ray exposure ages of the various meteorites (10 Ma for the nakhlites and Chassigny and 2.5 Ma and 0.5 Ma for the shergottites, Bogard and others, 1984). For instance, if all the SNC's were ejected in a single event ~200 Ma ago, the proto-shergottites must have been at least 15 m in diameter to have shielded the inner portions from cosmic rays until collisional fragmentation in space 0.5 to 2.5 Ma ago (Wetherill, 1984). On the other end of the spectrum, in terms of the size required, is the hypothesis that the SNC meteorites were ejected in three separate events at times corresponding to their cosmic ray exposure ages; in this case, the minimum required size, constrained by the recovered mass of the meteorites, reduces to a few tens of centimeters.

In general, shock pressures great enough to accelerate material to > 5 km/s are expected to melt or vaporize it (O'Keefe and Ahrens, 1977). Nyquist (1983) speculated that high-velocity, ricocheting projectile material from a very oblique impact could accelerate entrained rocks to very high velocity, but numerical modeling of this process showed that only small (< 10 cm) rocks lying on the surface before the impact could become entrained and survive acceleration to 5 km/s (O'Keefe and Ahrens, 1986). Melosh (1984, 1985) demonstrated that near-surface material, subjected to low shock pressures but high stress gradients, is ejected at high velocity and may be in the form of relatively large fragments. Since the SNC meteorites have young crystallization ages (see references cited in McSween, 1985), they are almost certainly from the young terrane of Mars, on which the largest crater is ~30 km in diameter. The maximum size spall from the near-surface zone ejected at > 5 km/s from a 30 km crater is < 1 m (Melosh, 1986). A study of the drag acceleration of rocks entrained in an impact-generated vapor cloud, with initial ejection velocities of 0 to 5 km/s and sizes of 0.5 to 15 m, similarly showed that only rocks < 1 m in diameter survive acceleration to 5 km/s (Vickery, 1986a).

Thus the results of very different kinds of dynamical studies clearly converge to a < 1 m maximum size rock expected to be ejected from a crater ≤ 30 km in diameter. This apparently favors the hypothesis that the SNC meteorites were ejected as tens-of-centimeters rocks in three separate events at times corresponding to their cosmic ray exposure times. This requires three craters 10 to 30 km in diameter on the young terrane of Mars within 10 Ma. Craters in this size range may be capable of ejecting tens of centimeter size rock at ≥ 5 km/s, but the total mass ejected as such fragments is several orders of magnitude smaller than that required to explain the mass flux of SNC meteorites (Vickery and Melosh, 1986; Wetherill, 1984). Models requiring fewer cratering events but larger fragment sizes are therefore more likely.

The purpose of the present study is to extend previous work on the acceleration of rocks by an impact-generated gas cloud (Vickery, 1986a) in two ways: First, some of the previous calculations were repeated, using ejection velocities and fragment sizes predicted by the Melosh (1984) spallation model. Second, the entrainment of pre-existing surface boulders was studied in more detail.

THE CALCULATIONS

All the calculations were done for a 30 km crater. Previous results (Vickery, 1986a) showed that increasing crater diameter by a factor of four made, at most, a 20% difference in the final velocities of the entrained rocks. The procedure is outlined below; for more details, see Vickery (1986a). The radius of the impactor required to produce a 30 km diameter crater for various combinations of impactor and target material and impact velocity was estimated using the Schmidt-Holsapple gravity-regime scaling law (Schmidt and Holsapple, 1982). This radius was converted to a mass by assuming a spherical impactor and a density (103 kg/m3 for water ice and 3x103 kg/m3 for andesite). Published values of the ratio of mass vaporized to mass of the impactor (Gault and others, 1972; O'Keefe and Ahrens, 1977), were used to estimate the mass vaporized. The initial pressure, density, and internal energy of the vapor were estimated from the Gault and Heitowit (1963) impedance-matching scheme. The impact types used were the 10 and 20 km/s impacts of andesite onto andesite and the 25 and 45 km/s impacts of ice onto andesite. The mass of the vapor cloud and its initial pressure, density, and internal energy are shown in Table 1 for each impact type.

The vapor was assumed to expand into a vacuum as a sphere of initially uniform pressure and density. Its expansion was calculated using a one-dimensional, explicit Lagrangian hydrocode developed for this purpose, which uses the Tillotson equation of state (Tillotson, 1962); the material parameters used in the Tillotson EOS are given in Table 2. Although this problem is not susceptible to analytic solution, the similar problem of a gas cloud of initially uniform density and a parabolic distribution of pressure is (Zel'dovich and Razier, 1966). When the code was tested with this second problem, the numerical results agreed with the analytic ones to within 0.01%. The hydrocode results became input for a second code that placed rocks of various sizes and initial velocities into the expanding gas and calculated the acceleration of the rocks according to:

$$a = \frac{3}{2}\frac{C_D}{\rho_r d}\rho_g (u_g - u_r)^2 \tag{1}$$

where a is the acceleration, C_D is the drag coefficient, d is the diameter of the rock, ρ is density and u is velocity. The subscripts r and g refer to the rock and the gas, respectively. C_D=0.5 was used in all calculations, because this is the value for spheres (which the rocks were assumed to be) over a wide range in Reynolds number.

TABLE 1 Mass and initial pressure, density, and internal energy of the gas cloud

Impactor Material	Impact Velocity (km/s)	Target Material	$\frac{Mv}{Mi}$	Mass of Vapor (10^{14}kg)	Initial Pressure (10^2GPa)	Initial Density (10^3kg/m^3)	Initial Internal Energy (10^8J/kg)
andesite	10	andesite	1	2.80	1.37	5.32	0.125
andesite	20	andesite	1	1.22	4.70	6.35	0.500
water ice	25	andesite	1	1.08	4.50	3.85	1.16
water ice	45	andesite	10	5.30	12.9	7.52	1.53

TABLE 2 Material Parameters for Tillotson Equation of State

Material	a	b	A (10GPa)	B (10GPa)	Eo (10^7J/kg)	ρ_o (10^3kg/m^3)
Andesite[1]	0.5	1.3	1.8	1.8	1.6	2.7
Water Ice [2*]	0.3	0.1	1.07	6.5	1.0	1.293

[1]Hageman and Walsh, 1970
[2]Anderson, 1968; Larsen and others, 1973; Gaffney and Ahrens, 1980
*High pressure phase at -10°C.

The first set of calculations modeled the entrainment of pre-existing surface boulders, from 0.5 to 15 m in size, placed in the gas cloud with zero initial velocity at times corresponding to the arrival of the leading edge of the gas at their initial positions. These initial positions ranged from 3 to 40 km from the center of the impact. The results are presented in Fig. 1-4 as graphs of final (maximum) velocity vs. initial position.

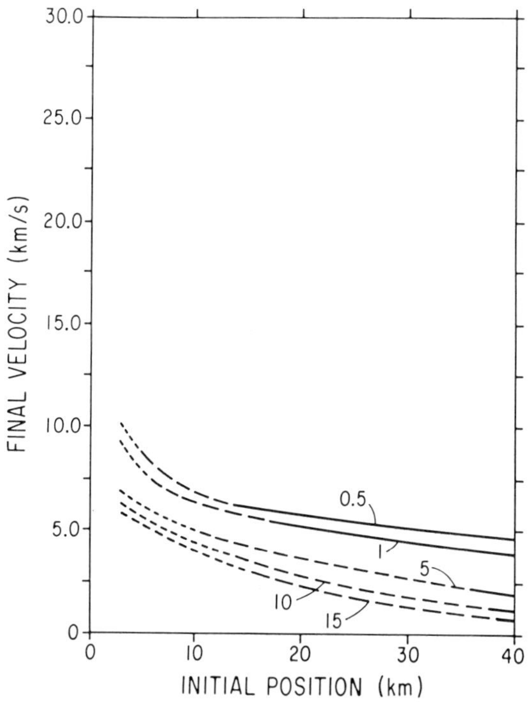

Fig. 1. Maximum (final) velocity vs. initial position (distance from center of impact) for boulders entrained in the gas cloud produced by the 10 km/s impact of andesite onto andesite. Labels on the curves are boulder sizes, in meters. Short-dash portions of the curves indicate the rocks experienced accelerations great enough to fragment a rock with a density of 3×10^3 kg/m^3 and a crushing strength of 100 MPa; long-dash portions of the curves indicate accelerations great enough to crush a rock with a strength of 10 MPa; and solid portions indicate no crushing.

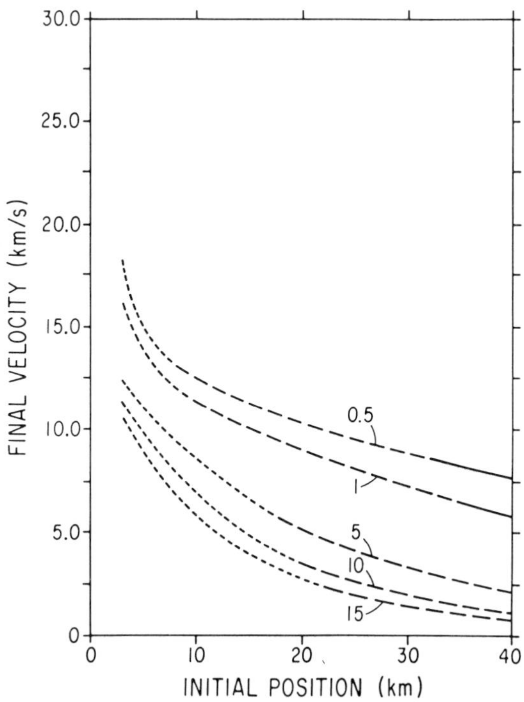

Fig. 2. Same as Figure , for the 20 km/s impact of andesite onto andesite.

The second set of calculations modeled the entrainment of near-surface spalls. The ejection times, velocities, and fragment sizes were calculated from the Melosh (1984) spallation model. Since Melosh's theory predicts that spalls ejected at > 1 km/s will further break-up into "Grady-Kipp" fragments because of stored elastic stresses, the calculations were done for both the original spall size and for the (generally) smaller Grady-Kipp fragment size. The spall sizes and ejection velocities are shown in Fig. 5, and the results are shown in Fig. 6 and 7 as graphs of final (maximum) velocity vs. fragment size.

An important result of the previous study of gas acceleration of solid ejecta (Vickery, 1986a) was that the entrained rocks may be subjected to dynamic pressures that exceed their crushing strength. The manner in which solid bodies break up in a gas stream has not been well-studied. In this paper, an alternate method of determining which rocks are fragmented was used: in studies of the strength of a lithosphere required to support a load which is not isostatically compensated, the strength, σ, is estimated as

$$\sigma = \rho g h, \quad (2)$$

where ρ and h are the density and height of the load, respectively, and g is the gravitational acceleration. This formula can be rearranged to find the maximum size rock of a given density and strength that can survive a given acceleration. In general, this method of estimating which rocks will be crushed agreed well with the results obtained using the simpler assumption that rocks subjected to a given dynamic pressure will be crushed.

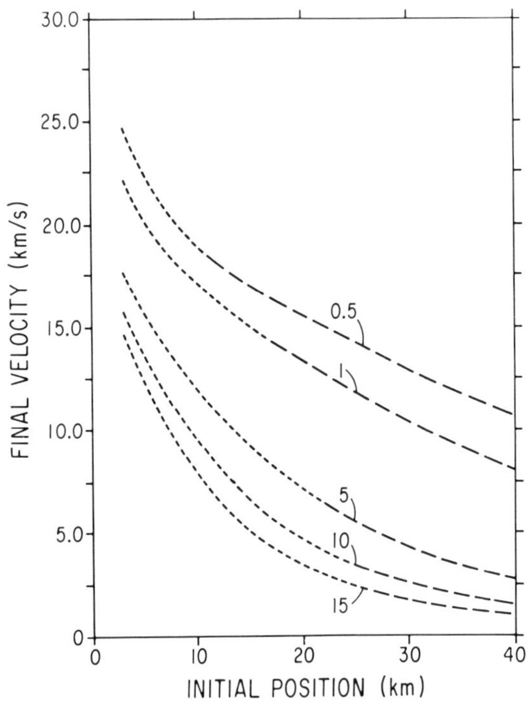

Fig. 3. Same as Figure 1, for the 25 km/s impact of ice onto andesite.

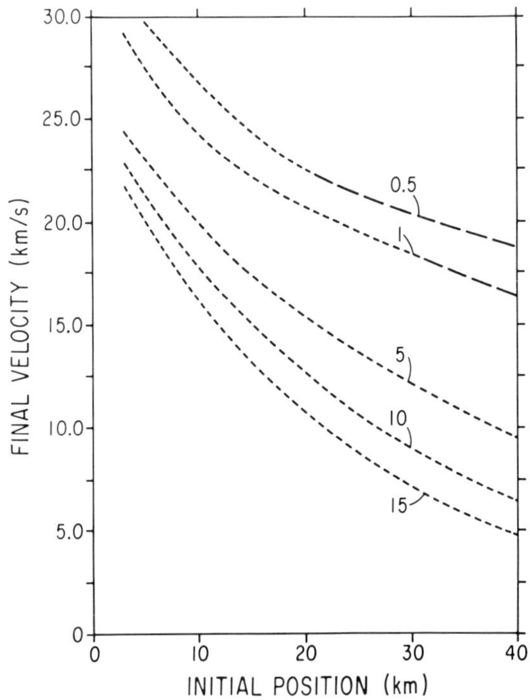

Fig. 4. Same as Figure 1, for the 45 km/s impact of ice onto andeiste.

RESULTS

Surface Boulders

In general, the results for the entrainment of surface boulders (Fig. 1-4) show that such rocks can be accelerated to high velocities by an impact-generated vapor cloud. The maximum velocities achieved vary, as expected, inversely with rock size and directly with impact velocity (for a given size crater). The latter result is reasonable because the gas's initial pressure, energy, etc., and therefore its expansion velocity, increase with impact velocity. Even rocks from as far away as 40 km from the center of impact can be significantly accelerated. Most of the rocks, however, are not expected to survive these accelerations intact. For the 10 km/s impact of andesite onto andesite (Fig. 1), rocks up to 1 m in diameter can be accelerated to velocities ≥ 5 km/s; for the 20 km/s impact of andesite onto andesite and the 25 km/s impact of ice onto andesite (Fig. 2 and 3) even 5 m rocks can be accelerated to ≥ 5 km/s if the crushing strength is at least 100 MPa. For the 45 km/s impact of ice onto andesite (Fig. 4), the accelerations are so great that only rocks ≤ 1 m can survive acceleration to ≥ 5 km/s and only if the crushing strength is ≥ 100 MPa.

The velocities shown in Fig. 1 through 4 were calculated assuming the rocks remained intact and are therefore minimum velocities. If, for instance, the 15-m rocks were reduced to 0.5 m fragments upon crushing, the curve for the 15 m rocks would move up in velocity more nearly to to coincide with the 0.5 m curve. Given the paucity of information on rock fragmentation under such conditions, any attempt to model the ultimate size-velocity distribution of the entrained rocks would be pure speculation. It can be said, however, that entrainment in an impact-induced vapor cloud can be expected to cause the acceleration of a significant mass of loose surface material to planetary escape velocities (cf., Wasson and Wetherill, 1979). Whether any of this material is in the form of large enough fragments to survive collisional fragmentation in space for long enough to evolve into earth-crossing orbits, and thus contribute to the observed meteorite flux on earth, is impossible to say.

This mode of ejection from a planetary body is probably irrelevant to the problem of the Martian origin of SNC meteorites because many of them have cumulate textures, which indicates that they were portions fairly deeply seated intrusions (see

McSween, 1985 and references therein). It could be argued that the SNC meteorites were lying on the Martian surface as ejecta from a previous impact that was large enough to have excavated a large enough volume to include several disparate rock types. A study of various possible ejection/cosmic ray exposure histories of the SNC meteorites (Vickery and Melosh, 1986), however, shows that there is negligible probability of two such overlapping craters in the time available.

Spalls

The ejection time for the spalls was calculated as the arrival time of a constant velocity stress wave at their initial positions (Melosh, 1984). The velocity of the wave varies in a complicated and poorly-constrained manner, especially near the center of impact. Using a constant stress wave propagation velocity of 5.3 km/s should, however, give a reasonable, first-order estimate of ejection times, and the degree of approximation is consistent with that of the many other approximations that went into constructing this model. These ejection times are considerably later than the times of entrainment of the surface boulders, so that the spalls were ejected closer to the center of the expanding gas cloud than to the leading edge. The gas densities were therefore higher, and the gas velocities lower than for the pre-existing boulders. Because the spalls are ejected with some velocity (Fig. 5), the difference between the gas velocity and rock velocity was also less, and the dramatic increase in velocity that occurs for the boulders does not occur for the spall fragments.

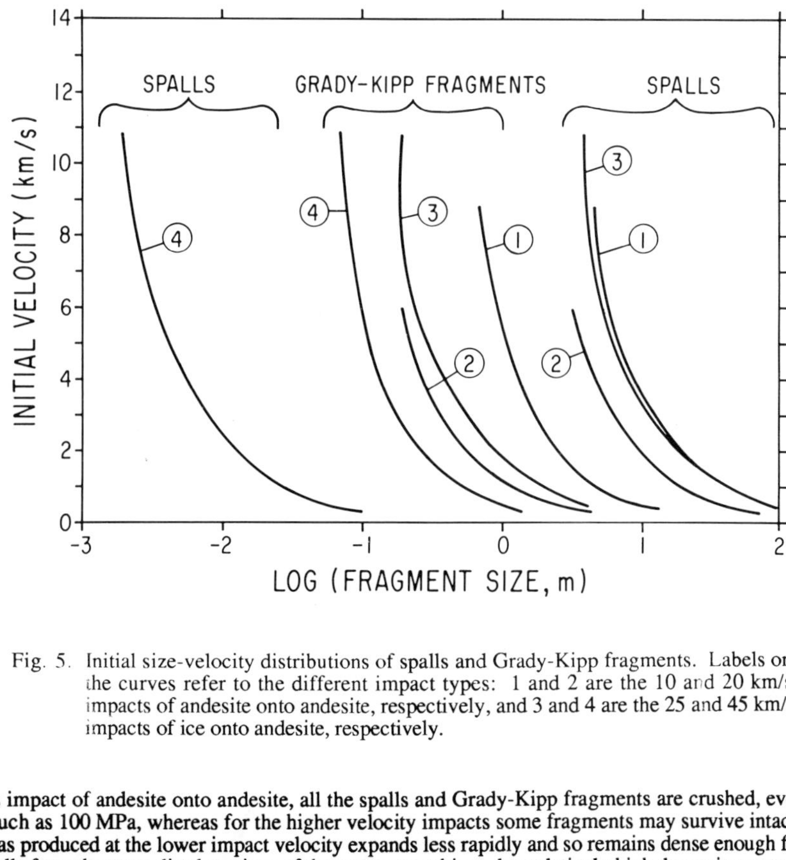

Fig. 5. Initial size-velocity distributions of spalls and Grady-Kipp fragments. Labels on the curves refer to the different impact types: 1 and 2 are the 10 and 20 km/s impacts of andesite onto andesite, respectively, and 3 and 4 are the 25 and 45 km/s impacts of ice onto andesite, respectively.

For the 10 km/s impact of andesite onto andesite, all the spalls and Grady-Kipp fragments are crushed, even if the crushing strength is as much as 100 MPa, whereas for the higher velocity impacts some fragments may survive intact. The reason for this is that the gas produced at the lower impact velocity expands less rapidly and so remains dense enough for long enough to that even the spalls from the more distal protions of the crater are subjected to relatively high dynamic pressures. Furthermore, the lower velocity impact produces larger spalls which are less likely to sustain a given acceleration without fragmentation than smaller rocks.

An impact-generated gas cloud thus tends to homogenize the velocity distribution and decrease the average size of the spalls. None of the spalls (Fig. 6) or the Grady-Kipp fragments (Fig. 7) with ejection velocities < 5 km/s were accelerated to > 5 km/s, and all of those with ejection velocities ≥ 5 km/s were decelerated by the gas (compare Fig. 5 with Fig. 6 and 7). This is because the gas velocity near the center of the cloud is < 5 km/s, even for the 45 km/s impact. The spalls with very high ejection velocities are decelerated by the gas. As before, the final velocities shown in Fig. 6 and 7 were calculated assuming that the rocks did not break up. The smaller fragments produced by crushing would more closely approach dynamic equilibrium with the gas, and the size-velocity curves would shift to smaller sizes and a more nearly uniform velocity of ~4 to

5 km/s. Thus an impact on a volatile-rich target actually reduces the fraction of spall material that achieves escape velocity, and is unlikely to be an explanation for the dynamic problem of ejecting large rocks from Mars as is probably required to explain the Martian origin of SNC meteorites.

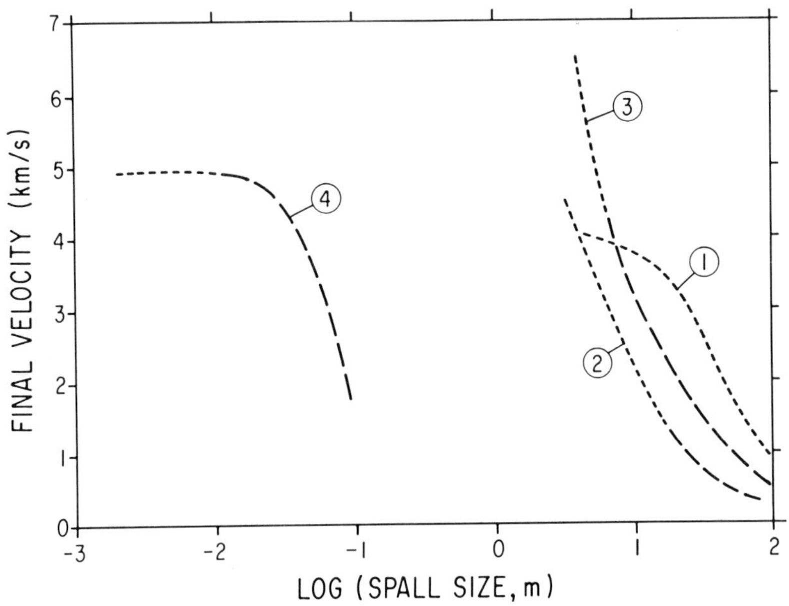

Fig. 6. Final size-velocity distributions of spalls. Labels on curves refer to the different impact types, as in Figure 5.

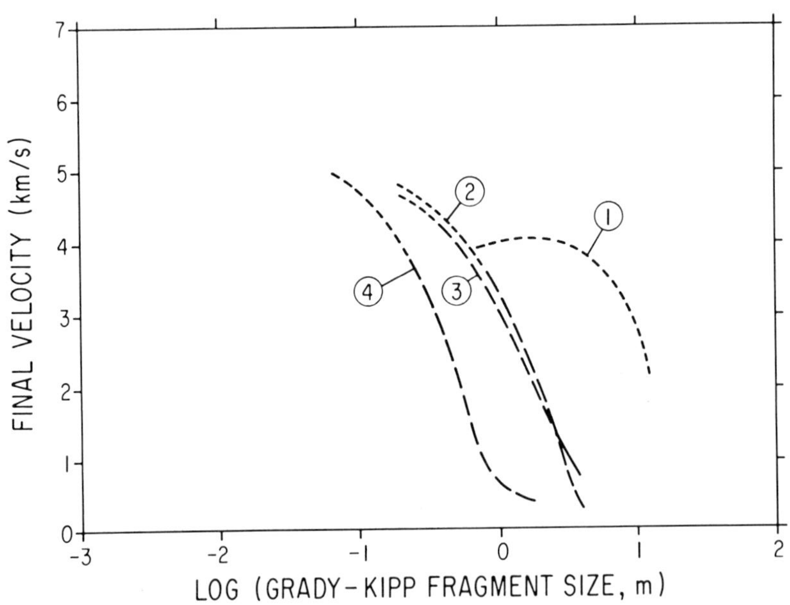

Fig. 7. Final size-velocity distributions of Grady-Kipp fragments. Labels on curves refer to the different impact types, as in Figure 5.

The change in the size-velocity distribution of spall material caused by entrainment in an impact-generated gas cloud may be expected to change the ejecta patterns observed around craters formed in volatile-rich targets. A study of the size range distribution of secondary craters around large primaries on the Moon and Mars (Vickery, 1986b) strongly suggests that the majority of secondary craters are produced by the impact of spall fragments from the primary. Deceleration of the highest velocity spalls will reduce the maximum range of the secondary craters, and the reduction in size of the spalls would tend to reduce the size of secondary craters. The acceleration of lower velocity spalls would tend to move the inner boundary of the secondary crater field to a greater distance from the primary crater. Because the continuous ejecta deposits from the main excavation flow field may bury the most proximal secondaries, this latter effect may not be observable except in extreme cases of homogenization of spall velocities.

CONCLUSION

An impact onto a volatile-rich target will tend to scour the surface of pre-existing loose boulders and to loft a significant portion of this material to planetary escape velocities. These fragments may generally be too small to survive collisional fragmentation in space for long enough to be perturbed into earth-crossing orbits (cf. Wetherill, 1984), and thus may not contribute significantly to the flux of meteorites on earth. Even if this is a potentially viable mechanism for the production of Martian meteorites, it is unlikely to have been the means by which the SNC meteorites were lofted from Mars because it is unlikely that they were ejected by a previous impact to lie on the surface as boulders, which were accelerated to escape velocity by the vapor produced by a second impact a short (by geological standards) time later. The size-velocity distribution of spalls will be changed by an impact-generated gas cloud and this may change the resulting ejecta pattern, especially the distribution of secondary craters, but gas acceleration of spalls cannot explain the Martian origin of SNC meteorites.

ACKNOWLEDGEMENTS

This work was supported by NASA grant NAGW 428.

REFERENCES

Anderson, G.D. (1968). U.S. Army CRREL Research Report , 257.
Bogard, D.D., P.Johnson, and L.E. Nyquist (1984). Lunar Planet. Sci. XV, 68-69.
Chen, J.H., and G.J. Wasserburg (1986). Geochim. Cosmochim. Acta 50, 955-?.
Gaffney, E.S., and T.J. Ahrens (1980). Geophys. Res. Lett. 7, 407-409.
Gault, D.E., and E.D. Heitowit (1963). Proc. Sixth Hypervelocity Impact Symp. 2, 419-456.
Gault, D.E., F. Horz, and J.B. Hartung (1972). Proc. 3rd Lunar Sci. Conf., 2713-2734.
Hageman, L.J., and J.M. Walsh (1970). Internal Report, 3SR-350, vol. 1, Systems, Science, and Software, LaJolla, CA.
Larson, D.B., G.D. Bearson, and J.R. Taylor (1978). North American Contributions, 2nd International Conference on the Permafrost, Yakutsk, p. 318-325.
McSween, H.Y., Jr. (1985). Rev. Geophys. 23, 391-416.
Melosh, H.J. (1984). Icarus 59, 234-260.
Melosh, H.J. (1985). Geology 13, 144-148.
Melosh, H.J. (1986) Numerical study of impact spallation. In preparation.
O'Keefe, J.D., and T.J. Ahrens (1977) Proc. 8th Lunar Planet. Sci. Conf., 3357-3374.
O'Keefe, J.D., and T.J. Ahrens (1986). Submitted.
Schmidt, R.M., and K.A. Holsapple (1982). Geol. Soc. Amer. Spec. Pap. 190, 93-102.
Shih, C.-Y., L.E. Nyqquist, D.D. Bogard, G.A. McKay, J.L. Wooden, B.M. Bansal, and H. Wiesmann (1982). Geochim. Cosmochim. Acta 46, 2323-2344.
Tillotson, J.H. (1962). General Atomic Report, GA3216.
Vickery, A.M. (1986a). Effect of an impact-generated gas cloud on the acceleration of ejecta. Submitted.
Vickery, A.M. (1986b). Icarus, in press.
Vickery, A.M., and H.J. Melosh (1986). The large crater origin of SNC meteorites. Submitted.
Wasson, J.T., and G.W. Wetherill (1979). Asteroids (T. Gehrels, ed.) University of Arizona, Tucson.
Wetherill, G.W. (1984). Meteoritics 19, 1-13.
Wooden, J., C.-Y. Shih, L. Nyquist, B. Bansal, H. Weismann, and G. McKay (1982). Lunar Planet. Sci. XIII, 879-880.
Zel'Dovich Ya.B. and Yu.P. Razier (1966). Physics of Shock Waves and High-Temperature Hydrodynamic Phenomena. Academic Press, N.Y.

Int. J. Impact Engng Vol. 5, pp. 663–670, 1987
Printed in Great Britain

0734-743X/87 $3.00 + 0.00
Pergamon Journals Ltd.

HYPERVELOCITY IMPACT OF SMALL MASSES ON LARGE SURFACES OF PIEZOELECTRIC CERAMICS

U. Weishaupt

Lehrstuhl für Raumfahrttechnik (LRT),
Technische Universität München,
D-8000 München 2, Richard-Wagner-Str. 18, FRG.

ABSTRACT

The hypervelocity impact of small masses on large surface piezoceramics was investigated to study the impact behavior of hypervelocity projectiles. From a linear elastic model obtained at lower velocities, solutions were found for the hypervelocity case which determine both the size and the momentum of impacting projectiles from the rising slope of the charge signal generated by the impact. The results lead to the development of a new generation of impact detectors for small masses at hypervelocities which consist only of a plate of piezoceramic material.

INTRODUCTION

Piezoelectric transducers mounted on targets have been used in many applications to detect projectiles impacting with hypervelocities and to study the behavior of the impacting process. During the impact the kinetic energy and the momentum of the particle are partly transformed into acoustic waves, which can be observed by pickups. Due to this transformation, however, primary characteristics of the projectile, like mass or size, are lost. Therefore, the determination of unknown parameters of the projectiles by means of such acoustic techniques is mostly restricted to the momentum or the energy of the projectile.

The application of targets made out of piezoceramic material makes it possible to study directly the hypervelocity impact process by means of acoustic methods, at least for small masses with diameters $\lesssim$ 2 mm. The momentum transfer is without loss for this device and allows the determination of more parameters of the projectile and higher sensitivities than in former measurement concepts.

BASIC PRINCIPLE

Due to the piezoelectric effect the mechanical displacement in a piezoceramic material caused by an impacting projectile will be directly proportional to internal electrical charges. This charge displacement can be measured on the electrodes of the target disc (fig. 1), and can be used for triggering or for detecting the impact event. Additionally the amplitude and the risetime of the first slope of the charge signal can be used to evaluate the momentum and size of the impacting particle.

THEORY

Evaluation of the parameters of the impacting projectile from the rising slope of the charges generated by the impact plasma is well understood from measurements of hypervelocity impacts of cosmic dust on nonpiezoelectric targets. However, there is no theoretical or experimental base to derive size or momentum from the rising slope of an acoustic signal. Theoretical studies concerning hypervelocity impacts of raindrops on airplanes (Röhrle, 1971) lead to a solution for treating the new principle (Weishaupt, 1985).

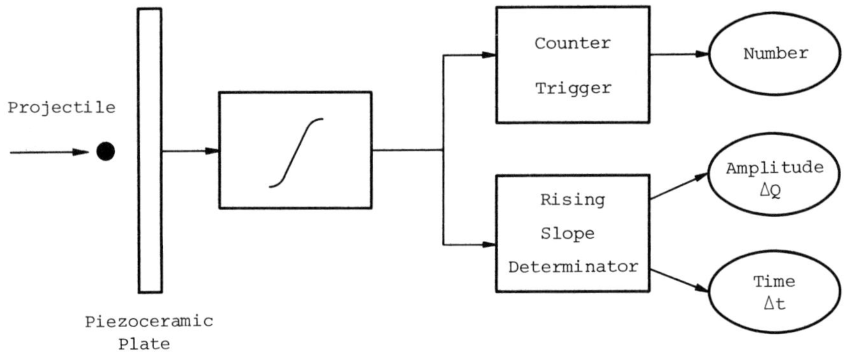

Fig.1: Basic Principle

The impact parts are assumed to remain intact during the impact process. In addition, it is assumed that the dimensions of the target are much larger than the projectile ones. When a small mass impacts (fig. 2 a) a target, pressure waves propagate into both the target and the particle (fig. 2 b). The pressure wave travelling through the particle is reflected as a rarefaction wave at the free end of the rear of the particle and travels back again through the particle. On the other side the pressure wave inside the target is travelling unchanged deeper into the target (fig. 2 c). This represents the transport of momentum from the particle to the target. The amount of transported momentum depends on both the acoustic impedances of impacting parts and the available contact area between them. In the real case of imperfectly matched conditions on both sides of the contacting area, the momentum is transferred to the target as "pressure packets" for times corresponding to wave reverberations in the projectile (fig. 2 d). These "pressure packets" travel from the contacting area as longitudinal compression waves into the target and increase the internal pressure of the target.

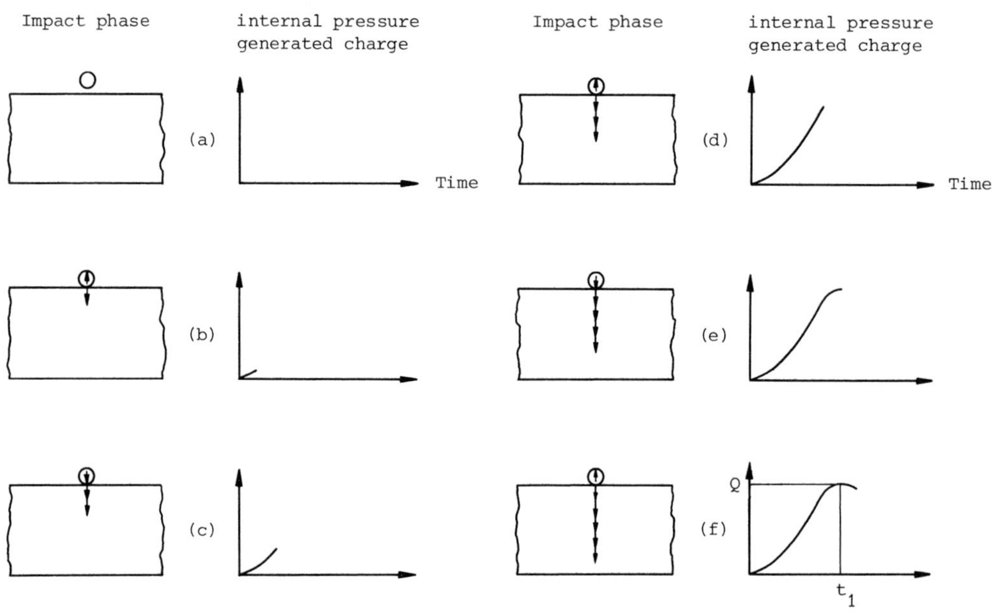

Fig.2: Impact on a piezoelectric target

Due to the piezoelectric effect, a piezoelectrical material, which is polarized in the direction of impact, will respond to the increased internal pressure with a monotonic increasing charge displacement. When the momentum transfer from the particle to the target is finished, there is still the pressure from the elastic reaction of the particle, which causes it to move back from the contacting area. At this time, t_1, the internal forces and displacements, and therefore the generated charge, begin to decrease (fig. 2 e - f). If the first front of compression waves has not yet reached a free end of the target, the amplitude Q of the generated charge signal at the time t_1 is proportional to the force, which is caused by the momentum totally transferred by the projectile with mass m and velocity v prior to the impact, divided by the piezoelectric constant d*. This leads to an easy equation for the momentum of the projectile:

$$m \cdot v \simeq \text{constant} \cdot \frac{Q \cdot t_1}{d^*} \qquad (1)$$

A more exact equation can be obtained if the target is described as a spring-massystem with a time constant t_2 which depends on the dimensions of the target. On the other hand the projectile is also described by a spring-massystem with time constant t_1. The time t_1 depends on the travel time of pressure waves through the projectile, and as explained above, is equal to the risetime of the charge signal. The target-system is forced to an oscillation by the projectile; the oscillation results from the impact of a small mass on a large one. From Jacobsen and Ayre (1958) it is known that the first slope of the oscillation follows the parameters of the forcing system. This leads to the equation:

$$m \cdot v = -\frac{2}{\pi} \cdot \frac{Q}{d^*} \cdot \frac{t_1 \cdot (t_1^2 - t_2^2)}{t_2 \cdot \left(t_2 - t_1 \sin \frac{\pi \cdot t_1}{t_2} \right)} \qquad (2)$$

$$Q \cdot t^*$$

It can be seen easily that for $t_2 >> t_1$, equation 1 is a short form of equation 2.

EXPERIMENTAL

In all experiments a circular disc 50 x 3 mm was used for the target. It was fabricated from Lead-Zirconate-Titanate (PZT) Ceramic, Sonox P4, obtained from Hoechst CeramTech, Germany. The charge signal generated during impact was measured as the voltage drop across a low-Ohm resistor to ensure fast risetimes. The rising slope was determined by means of a 100-MHz transient recorder (fig. 3.). The mass and the velocity of the projectile were measured prior to the impact to obtain its momentum and energy.

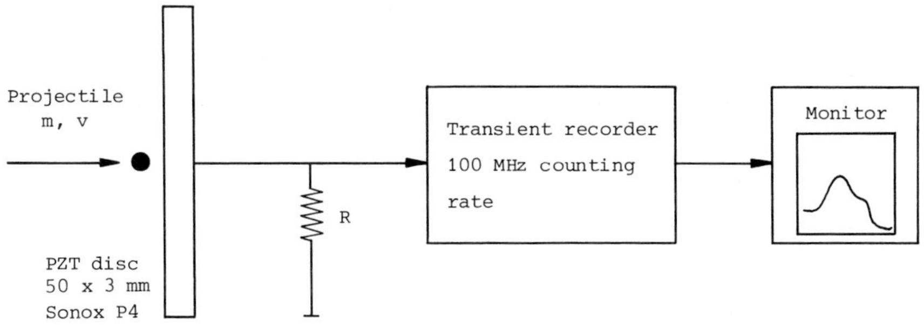

Fig. 3: Measurement Configuration

In the initial series of experiments, the basic physics of the elastic impact of small spheres (diameters of 0.1 to 1.5 mm) were investigated at velocities from 20 to 400 m/s. The projectiles were accelerated with the electromagnetic accelerator of the LRT (Igenbergs et al., 1986).

The experiments of the second phase were carried out in the hypervelocity range from 2 to 10 km/s with glass spheres (diameters of 0.04 to 0.2 mm) to compare the elastic impact results with the hydrodynamic ones. The projectiles were accelerated in the plasma dynamic launcher of the LRT (Igenbergs et al., 1986).

RESULTS

Fig. 4 shows the charge signal obtained from the impact of a glass particle with a diameter of 330 μm and an impact velocity of 60 m/s onto the piezoelectric crystal. The dot-dashed curve shows the result of the computer simulation of the spring-massystems. It is easy to distinguish between the loading phase of the impact until the saddle point of the curve and release after the impact. If the pressure wave in the target reaches the rear side, the charge signal changes sign due to introduction of tensile states in the specimen.

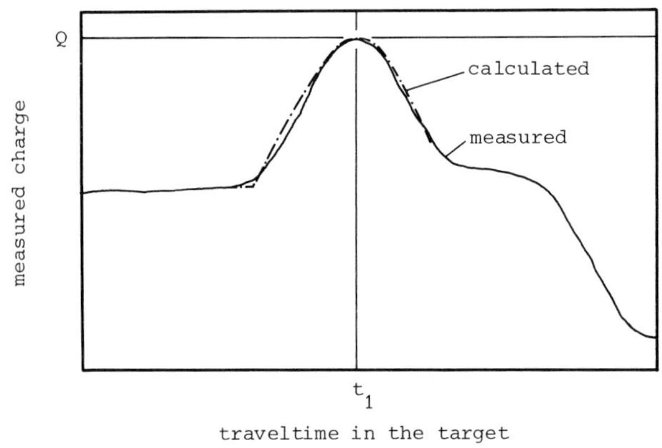

Fig.4: Charge signal, obtained from a glassbead with a diameter of 330 µm and a velocity of 60 m/s

In fig. 5 the dependence of risetime is depicted as a function of the particle velocity and size. Up to 60 m/s there are strong elastic effects, but over an impact velocity of 150 m/s the risetime of the charge signal depends primarily on the projectile size.

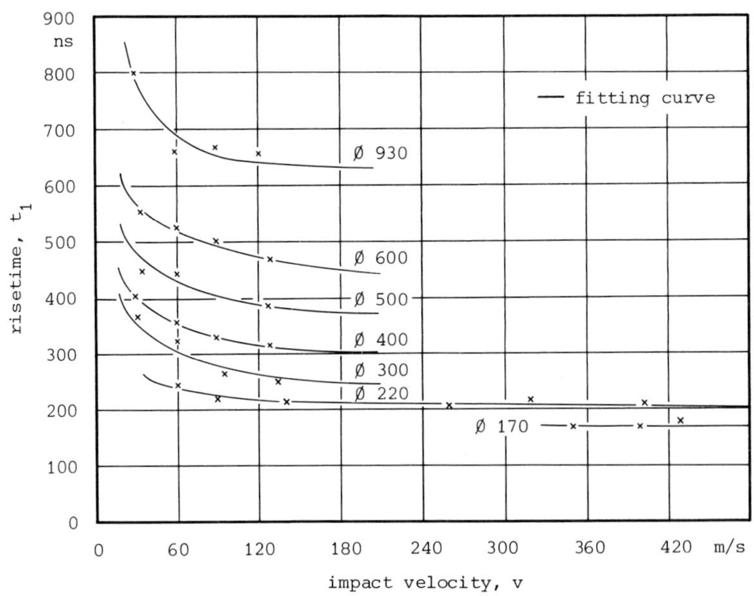

Fig.5: Dependence of the risetime as a function of the projectile velocity and size

Evaluation of the experimental results by means of equation 2 is demonstrated in the fig. 6. Very good linearity is observed between the real momentum (m·v) of the projectile and that calculated by equation 2. The measurement values are plotted as crosses in the diagram, the dashed line represents their fit, obtained by means of the method of least squares. The thin solid line shows the exact proportionality between measurement and theory.

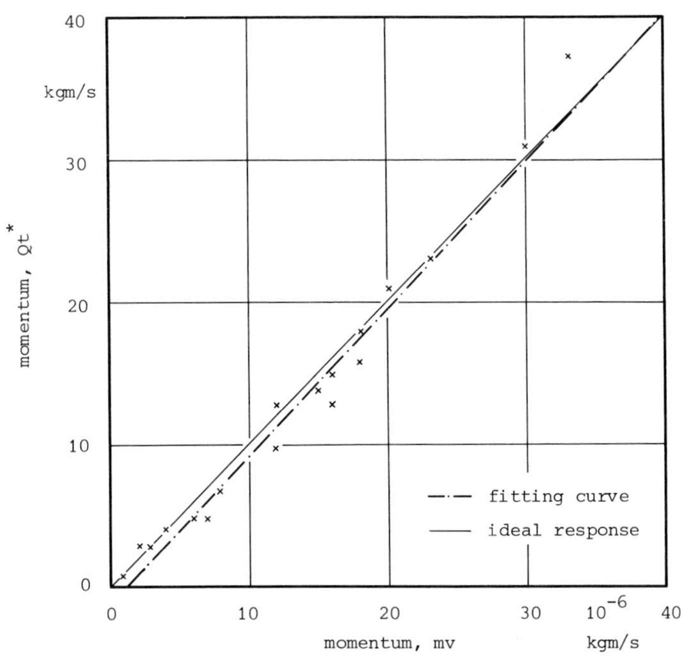

Fig.6: Results of the elastic experiments

The charge signal in fig. 7 was generated by the hypervelocity impact of a projectile with a diameter of 120 μm and a velocity of 6.6 km/s. The shape of the curve looks very similiar to the one in fig. 4 obtained at a velocity of 60 m/s.

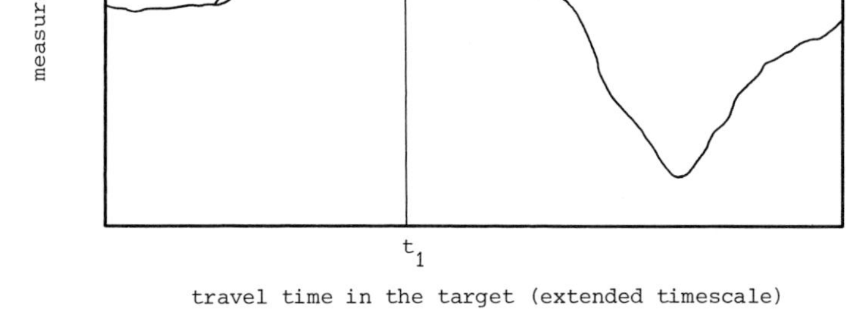

Fig.7: Charge signal of a glassbead with a diameter of 120 µm and a velocity of 6.6 km/s

We applied equation 2 of the elastic model to several hypervelocity impact tests to compare the elastic impact measurements with hydrodynamic ones. The results are shown in fig. 8. Results of these comparisons demonstrate a good linearity between the calculated and measured momentum of the projectiles. The correlation between the size of the projectiles and the risetime of the first slope of the charge signal is shown in fig. 9 for a series of hypervelocity impacts. It also illustrates good linearity, within the sample rate of the transient recorder which is 10 ns. At risetimes below 100 ns, the systematic error is higher for the smaller the projectiles. This observation is also the probable reason for the stronger slope of the fitting curve with respect to the ideal line shown in fig. 8.

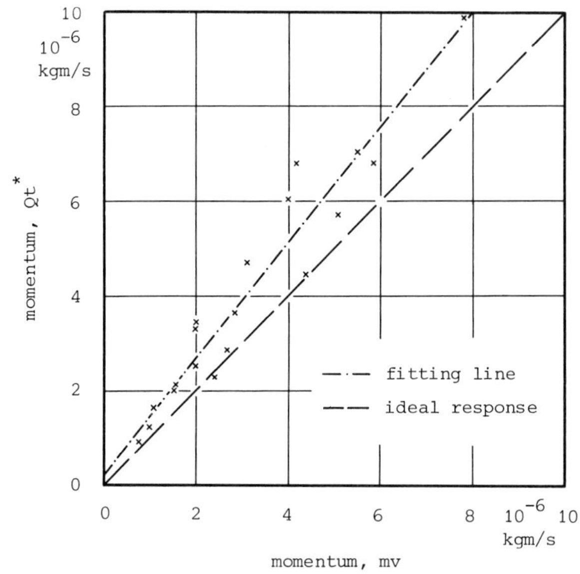

Fig.8: Results of hypervelocity experiments

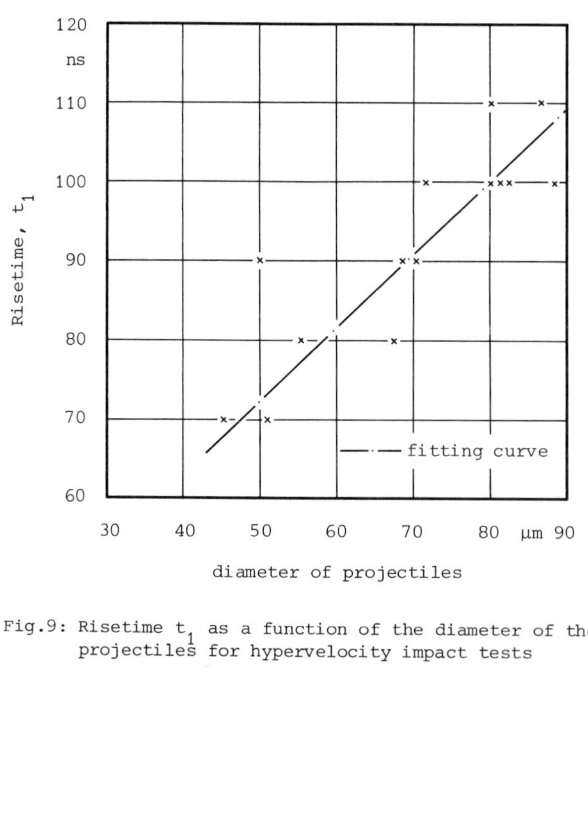

Fig.9: Risetime t_1 as a function of the diameter of the projectiles for hypervelocity impact tests

CONCLUSIONS

The experimental results of the impact investigations at low and at hypervelocities show that, in both cases, it is possibile to derive the momentum and size of a projectile from the first slope of the charge signal generated during impact on a piezoceramic plate.
The shape of low-velocity signals are very similiar to those of hypervelocity signals. It is therefore useful to apply equation 1 and 2 for both the elastic model and hypervelocity cases, although the physical processes are very different. A possible model for the hypervelocity case results from substitution of a hard projectile for the impact-generated debris cloud, which acts as transient load to the piezoelectric target, with similiar charge generating characteristics (fig. 10).

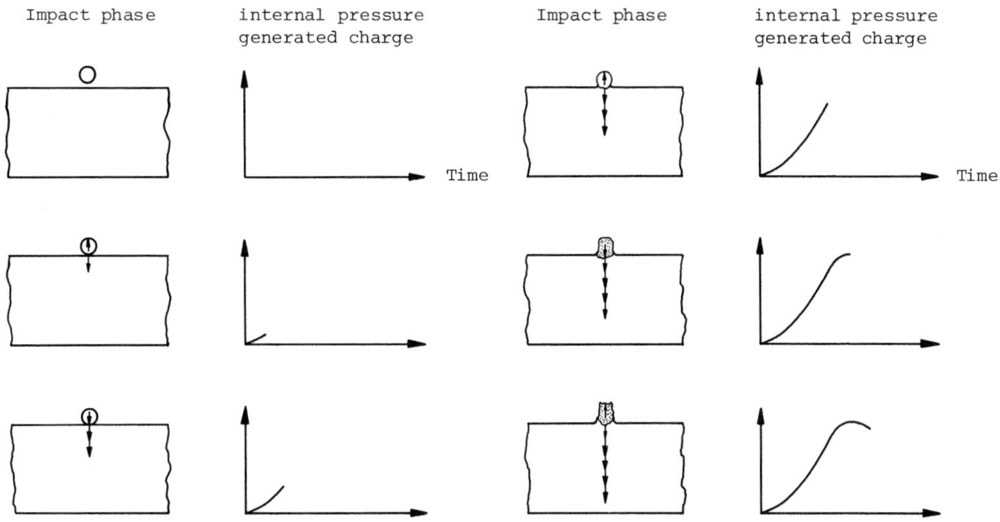

Fig.10: Hypervelocity impact on a piezoelectric target

Besides the basic results of the investigations presented here, there is also application of this technique to a new generation of dustimpact detectors for space research, which consists only of a plate of piezoceramic material: Because of the high-charge rate per unit loading force, and the efficient momentum transfer, maximum sensitivity of an acoustic detector can be achieved. Since the technique is not critically related to mechanical properties of the impacting particle, there is a wide range of space applications. In fig. 11 the schematic diagram of a proposed dust detector system for space research is shown (Weishaupt, 1985).

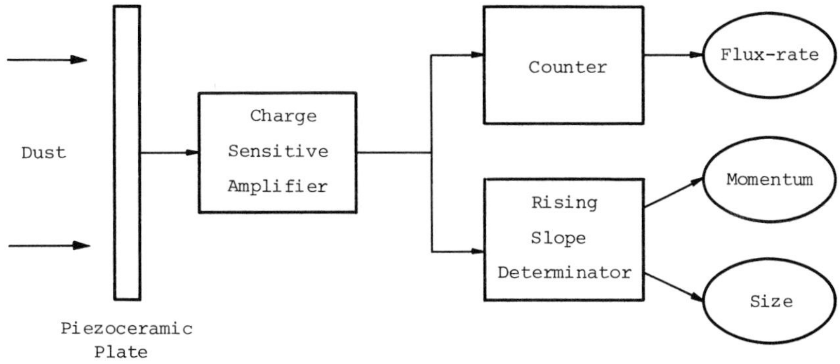

Fig.11: Dust impact detector for space research application

ACKNOWLEDGEMENT

The authors are indebted to Martin Rott for his valuable help as a good colleague.

REFERENCES

Röhrle, H. (1971). A contribution to the finite-element-method and to elastic impact on semi-infinite targets. Dissertation, TU Berlin.

Weishaupt, U. (1985). Hypervelocity impact of small masses on large surfaces of piezoelectric ceramics. Dissertation, TU München.

Jacobsen, L.S. and Ayre R.S. (1958). Engineering Vibrations with Applications to Structures and Machinery. Mc Graw-Hill, New York.

Igenbergs, E. et al. (1986). Launcher technology, in-flight velocity measurement and impact diagnostics at the TUM/LRT. Int. Journ. of Impact Engin. 5.

Weishaupt, U. (1986). Cosmic dust detection with large surface piezoceramics. In F. Hörz (Ed.) Lunar and Planetary Institute Technical Report 86-05, Houston.

Int. J. Impact Engng Vol. 5, pp. 671–679, 1987
Printed in Great Britain

0734-743X/87 $3.00 + 0.00
Pergamon Journals Ltd.

DEVELOPMENT OF HYPERVELOCITY ELECTROMAGNETIC LAUNCHERS

William F. Weldon
Director, Center for Electromechanics
Professor, Electrical and Computer Engineering
Professor, Mechanical Engineering

Center for Electromechanics
The University of Texas at Austin
Balcones Research Center
Bldg. 133, EME 1.100
Austin, TX 78758-4497

ABSTRACT

Interest has increased substantially during recent years in the application of electromagnetic launch (EML) technology for a variety of purposes. In part this increased interest is due to the recent availability of compact pulsed power supplies suitable for driving such launchers. Also, several successful EML experiments have provided encouraging results.

The history of electromagnetic launch is reviewed, the current status of the railgun is presented, and plans for the next generation of electromagnetic launchers are discussed.

INTRODUCTION

The prospect of using electromagnetic rather than thermodynamic forces to accelerate projectiles has excited the imagination of scientists and engineers throughout the twentieth century. Of course the single stage induction accelerator capable of launching a copper or aluminum ring into the air was a popular demonstration in the physics laboratory before the turn of the century (Fig. 1). In 1882 C. A. Cheever was issued a patent for an "electromagnetic dispatch tube"[1] and in 1902 the Norwegian scientist Birkeland was issued several patents on an electromagnetic cannon.[2] Written by Arthur B. Reeve in 1915, the science fiction novel The War Terror gives a technically accurate description of an electromagnetic accelerator.[3]

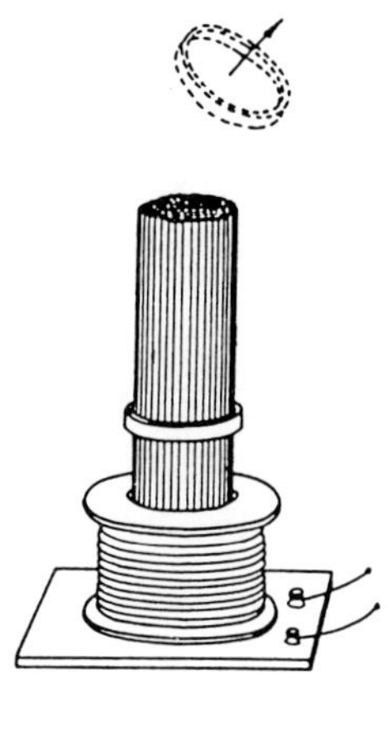

Fig. 1. Repulsion coil induction accelerator.

World War I stimulated a strong interest in the improvement of guns in general and in particular the first serious investigations of electrically-driven guns. In the period between 1917 and 1922 the Frenchman A. L. O. Fauchon-Villeplee was issued several U.S. patents on "electric gun apparatus for propelling projectiles"[4] (Fig. 2) and wrote a book on electromagnetic guns.[5] A magazine article published in 1921 indicates that during the war, Fauchon-Villeplee successfully accelerated a 50-g projectile to 200 m/s with his electric gun.[6] By 1918, Birkeland also had built an electromagnetic cannon (Fig. 3)[7] and in 1921 F. B. MacLaren was issued a quite detailed patent on an electric gun[8] (Fig. 4). Early efforts to demonstrate the electromagnetic acceleration of projectiles were limited by the availability of suitable sources of electric power. As electric power became more commonly available and its application better understood, experiments with electric guns became more common also; but the limited availability of suitable power sources has plagued the development of electromagnetic projectile accelerators to the present day.

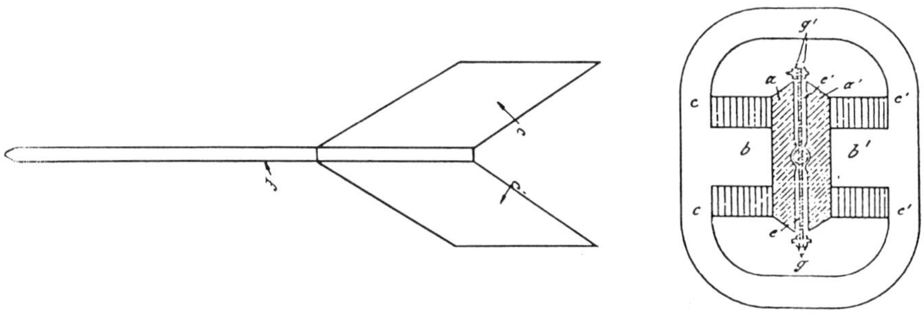

Fig. 2. Fauchon-Villeplee's electric cannon for launching finned projectiles.

Fig. 3. Birkeland's cannon (1918), the earliest example of a tubular motor.

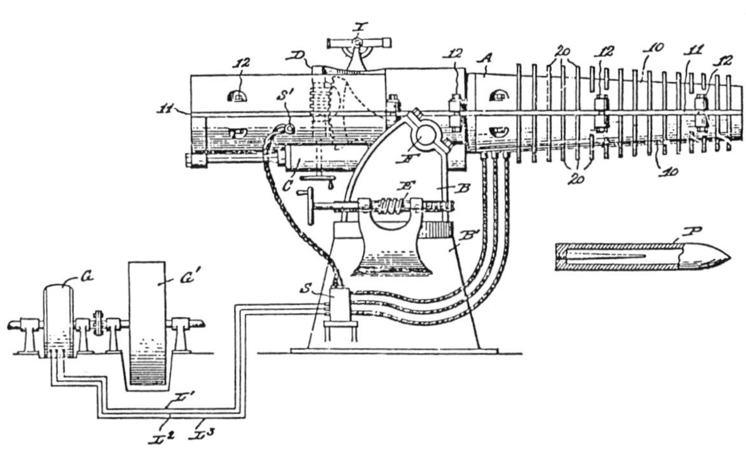

Fig. 4. MacLaren's electric gun.

The Russians reportedly built an electric cannon called "Ivan the Terrible" around 1929 although little is known of the device. In the early 1930's the Yugoslavian War Ministry reportedly repeated Fauchon-Villeplee's experiments. In 1934 engineers at The University of Texas at Austin (UT) demonstrated "a new type of machine gun, operated entirely by electromagnetic power and without the aid of gun powder."[9] In 1937 Professor Edwin Fitch Northrup of Princeton University published Zero to Eighty in which he described a series of electromagnetic gun experiments actually performed in his laboratory and then extrapolated this experience to an earth-to-moon launcher in the form of a fictional autobiography of an individual performing the work.[10] The book is illustrated with photos of electromagnetic guns built and tested by Northrup (Fig's. 5a and b) and includes a technical supplement which presents the scientific principles and calculations upon which both the actual experiments and extrapolations were based.

As did World War I, World War II again served as the catalyst for more ambitious electromagnetic gun experiments. The Japanese considered a linear induction motor for launching finned, 10 kg anti-aircraft projectiles at a velocity of 2 km/s. A model was built to accelerate a 2-kg projectile to a velocity of 500 m/s but only attained 335 m/s (Fig. 6).[11] As a result, the full scale launcher project was abandoned, but the model continued to be used for testing projectiles. The most intense effort during the war was that of Dr. Joachim Hansler in Berlin under contract to the German Air Force.[12,13] Although the project goal of developing an electric gun to accelerate a 40 mm, 6.5-kg projectile to a muzzle velocity of 2,000 m/s was not achieved, Hansler did electromagnetically accelerate 10-g projectiles to velocities up to 1,200 m/s. The work proceeded from 1943 until it was interrupted by the end of the war. Hansler's papers indicate that he anticipated most of the developments responsible for the recently renewed interest in electromagnetic guns.

In 1957 General Electric published the results of their study of electromagnetic accelerators concluding that there was no fundamental limitation preventing the attainment of velocities of 7.6 km/s by either the ac or dc accelerators that they had analyzed.[14] They further observed "however, practical limitations must be solved to increase the conversion efficiencies." The 1960s and early 1970s saw a flurry of interest in electromagnetic accelerators both as potential hypervelocity weapons and as a means of investigating hypervelocity impact of meteorites.[15,16,17] Most of these studies were primarily analytical in nature until the landmark experiments performed under the direction of Dr. R. A. Marshall at the Australian National University (ANU) during the mid-1970s.[18] Marshall and his colleagues succeeded in accelerating a 3 g plastic projectile to a velocity of 5.9 km/s in an electromagnetic railgun 4-m long.

The principle of the railgun accelerator is shown simply in Fig. 7. Basically, the current flowing in the rails produces a magnetic flux density between the rails and this magnetic field interacts with the current $\bar{J}$ flowing in the armature which, for the ANU railgun, was an arc. The resulting J x B or Lorentz force accelerates the armature together with the projectile along the rails. The magnitude of this driving force is given by $\frac{1}{2}L'I^2$, where L' is the inductance per unit length of the parallel rails and I is the current. A typical value for L' is 0.4×10^{-6} H/m, so that the driving force for a typical railgun may be expressed as 2×10^{-7} N/A^2. For currents in the kA range, the forces produced in a railgun are uninteresting, but as current levels approach a megamp, extremely large forces result. Thus, only with the advent of pulsed megamp power supplies has this type of electromagnetic gun become of interest for practical applications.

In the ANU railgun circuit (Fig. 8) the homopolar generator, having been brought up to speed by an external power source, is the source of the total electrical energy required by the circuit as well as the high electrical current required. However, the output voltage of the homopolar generator is not high enough to produce the necessarily rapid rise of current in the railgun armature, nor can it drive the requisite current against the rising back electromotive force (speed voltage) of the armature. This then is the function of the storage inductor in the circuit. Once charged to the required current, it produces the voltage necessary to maintain that current in the dynamically changing railgun discharge circuit. The switching armature is large and heavy by comparison with the railgun armature so that it can carry the inductor charging current without overheating. It also serves to rapidly introduce the railgun into the circuit so there is no heating and/or creeping of the railgun armature until peak current is reached in the circuit, and therefore maximum acceleration can be achieved.

The important technical contributions of the ANU railgun are the use of a homopolar generator to provide the required high energy and current (originally proposed by Hansler to the Germans during WW II), the use of a storage inductor rather than the capacitors used in earlier experiments to provide the voltage necessary to maintain the driving current against the rising back electromotive force or speed voltage of the railgun and, perhaps most important, demonstrating that the sliding metallic armature of earlier railgun experiments could be replaced with a metal-vapor arc allowing much higher velocities to be reached.

Early railgun experiments, including the first experiments at ANU, used a sliding metal armature to carry the current between the rails of the accelerator. The armature was typically designed so

Fig. 5a. Northrup's electric gun.

Fig. 5b. Horizontal gun coils in Northrup's laboratory.

Fig. 6. Japanese linear induction accelerator.

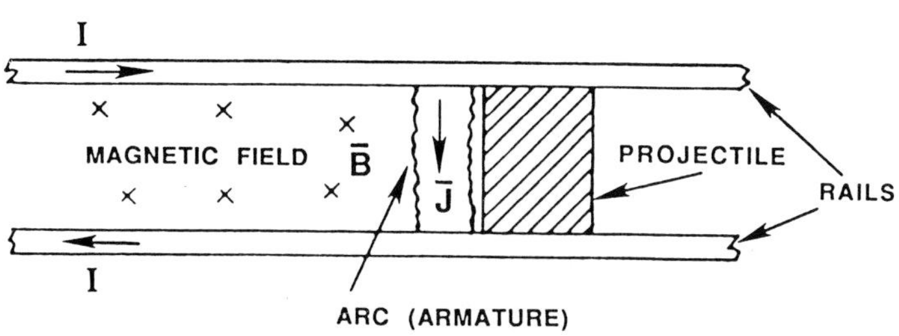

Fig. 7. Principle of the railgun.

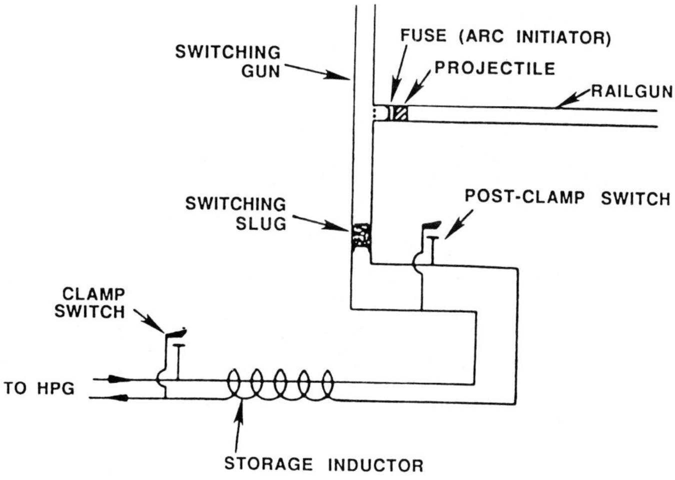

Fig. 8. Schematic circuit of the ANU railgun.

that the magnetic field produced by the current passing through it forced the armature into contact with the rails. Marshall and others found that at sliding velocities around 1 km/s a phenomenon known as "gouging" occurred wherein the rail and armature material apparently welded together -- resulting in substantial amounts of material being pulled off of the rails, degrading the rail surface, and effectively limiting the velocities attainable.

Marshall, et al., began experimenting with thin metal fuses which were vaporized by the rapidly rising current as the railgun was switched into the circuit and formed a metal-vapor arc with relatively low voltage drop (about 160 V in the ANU gun). This technique successfully drove nonmetallic projectiles (typically polycarbonate plastic) to much higher velocities than had previously been achieved.

In the late 1970s and early 1980s, reports of the ANU experiments inspired several small scale railgun experiments in the U.S. With one exception, these experiments were powered by capacitor banks, but all used the storage inductor, parallel rail geometry, polycarbonate projectiles, and fuse initiated arc armatures pioneered so successfully at ANU. Capacitor-driven railguns were built and operated at Westinghouse Electric Corporation, Vought Corporation, The University of Texas at Austin, and Lawrence Livermore National Laboratory (LLNL). The Vought railgun, designed at UT and using an existing capacitor bank originally built for shock-tube research, accelerated a 6 g polycarbonate projectile to 8 km/s in a 4 m long railgun in 1981 (Fig. 9). This represented a mass/speed record for capacitor-driven railguns at that time.

In 1980, a unique team was formed with a cooperative effort between LLNL and Los Alamos National Laboratory (LANL). The LLNL/LANL team combined an LLNL group interested in railgun development with an LANL group having the expertise to design and build explosive flux compression generators or magnetocumulative generators (MCG) capable of delivering megampere currents at reasonably high voltages. Although it is not attractive for many applications because of the explosive driver and because the MCG and often the rear portion of the railgun are destroyed with each shot, the MCG-driven railgun did allow the LLNL/LANL team to explore a very high-performance regime for railguns early in the program before more conventional power supplies were available. In 1980, the LLNL/LANL experiments succeeded in accelerating 3 g polycarbonate projectiles to velocities in excess of 10 km/s.[19]

In 1980-82, a system quite similar in concept to the ANU railgun, but operating at much higher currents and accelerating heavier projectiles, was built by Westinghouse Electric Corporation under joint sponsorship and direction of the Defense Advanced Research Projects Agency (DARPA) and the U.S. Army Armament Research, Development, and Engineering Center (U.S. Army ARDEC). This laboratory demonstration electromagnetic launcher, called EMACK, is now located at the Picatinny Arsenal near Dover, NJ (Fig. 10). The EMACK railgun is reported to have accelerated a 317-g projectile to 4.2 km/s at a peak current of 2.1 MA early in 1982.[20]

Fig. 9. Vought railgun.

The technical achievements of the electromagnetic gun program during the past eight years are documented in the proceedings of the three symposia.[21,22,23] At the present time, several railguns around the country are routinely operating in the 3- to 5-km/s range with projectile masses of a few grams to a few hundred grams. Research on these guns at present generally centers around rail and insulator material investigations, projectile/armature design, and plasma armature studies although some are occasionally used for impact studies. Of more interest for hypervelocity impact studies are the next generation of electromagnetic guns now being designed and built.

The DoE sponsored SUVAC gun at Westinghouse Research and Development Labs has accelerated 1 g to around 8 km/s in a 3.2-m barrel and is being expanded with the goal of reaching 20 to 30 km/s. The SUVAC power supply is a 6-MJ capacitor bank. The GEDI facility under construction at UT, supported by SDIO, DARPA, and U.S. ARDEC, has a goal of accelerating a 1-g projectile to 50 km/s, (Fig. 11). The power supply for GEDI is a system of six, 10-MJ homopolar generators, each charging an individual inductive store to an energy of 4.6 MJ at a current of 1.2 MA. The inductive stores will then be sequentially switched into the railgun, using explosively driven opening switches, to provide an essentially constant driving current during the launch. Average projectile acceleration levels of 2 x 10^6 gees or more are anticipated. The power supply and range should be completed by the end of 1986 with the first experiments in the early part of 1987.

Fig. 10. EMACK railgun.

Fig. 11. GEDI facility at The University of Texas at Austin

This same 60-MJ power supply will be used to power a single shot laboratory gun capable of providing 9 MJ of kinetic energy to projectiles in the 2.5 to 4.0 km/s range, (Fig. 12). This facility which is scheduled to become operational in August, 1987 is being built under the sponsorship of the DARPA/Army/USMC Armor/Antiarmor Program. A capacitor powered facility with similar goals is to be constructed at Maxwell Laboratories. Both systems will be used for development and testing of antiarmor projectiles.

The Lethality Test System under construction at LANL has the goal of accelerating 20-to 30-g projectiles to velocities upto 15 km/s, (Fig. 13). This railgun will be powered through pulse transformers by 28 commutated dc machines with external flywheels. The total energy stored inertially is 95 MJ. A similar facility called Thunderbolt is under construction by Westinghouse. Both facilities are supported by SDIO, DNA, and SDC.

Electromagnetic gun technology has developed to a point that railguns now offer repeatable hypervelocity performance. The next generation of electromagnetic guns, now under construction, promises to extend the capabilities for hypervelocity impact research into regions not previously available.

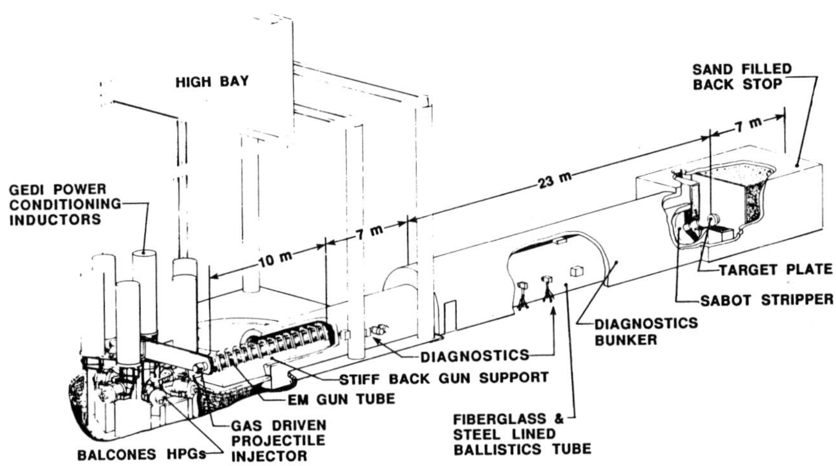

Fig. 12. The University of Texas at Austin single shot laboratory gun.

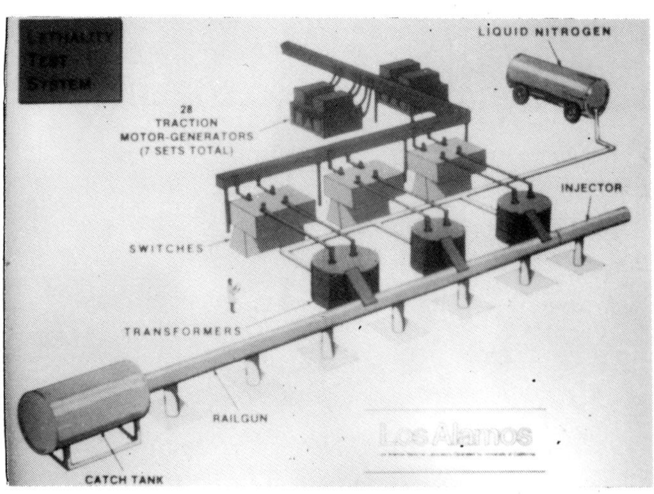

Fig. 13. Los Alamos National Laboratory Lethality Test System.

REFERENCES

1. U. S. Patent #259,817.
2. Norwegian Patent #'s 11,201, 11,228, 11,342, 13,052, and U. S. Patent #754,637.
3. Reeve, A. B., The War Terror, Harper and Brothers Publishers, New York, 1915.
4. U.S. Patents 1,370,200, 1,421,435, and 1,422,427.
5. Fauchon-Villeplee, A.L.O., Canons Electriques, Berger-Levrault, 1920.
6. Electrical World, September 10, 1921.
7. Laithwaite, E. R., Propulsion Without Wheels, The English Universities Press, Ltd., 1970, pp. 147.
8. U.S. Patent #1,384,769.
9. Program from "The University of Texas Engineers' 25th Annual Power Show," May 3-4, 1934.
10. Northrup, E. F., Zero to Eighty, Scientific Publishing Company, Princeton, NJ, 1937.
11. Delaosso, L., "Japanese Experiments with the Electromagnetic Gun," Ordnance Technical Intelligence Report No. 17, Office of the Chief Ordnance Officer, General Headquarters, Army, Pacific, Tokyo, Japan, January 31, 1946.
12. Pollack, M., "Electric Gun and Power Source," Armor Research Foundation, Project No. 15-391-E, Report No. 3.
13. Hansler, J., "A Contribution to the Problem of the Electrical Cannon and Test Reports," September 18, 1944, November 7, 1944, and January 18, 1945.
14. Brate, D. C., et al., "Electromagnetic Accelerator Systems," General Electric Engineering Laboratory, Report No. 57 GL 127, November 1957.
15. Radnik, J. L. and Lathan, B. F., "Electromagnetic Projector Study," Armor Research Foundation, Report No. ASD-TR-61-35, July 1961.
16. Thom, K. and Norwood, J., "Theory of an Electromagnetic Mass Accelerator for Achieving Hypervelocities," NASA Technical Note D-886, June 1961.
17. Kouroukiis, G., et al., "Electromagnetic Propulsion of Ordnance," NSWC/DL Technical Report TR-3497, August 1976.
18. Rashleigh, S. C. and Marshall, R. A., "Electromagnetic Acceleration of Macroparticles to High Velocities," J. Appl. Phys., Vol. 49, page 4, March 1984.
19. Hawke, R. S., et al., "Results of Railgun Experiments Powered by Magnetic Flux Compression Generators," pp. 82-93.
20. Deis, D. W. and Scherbarth, D. W., "EMACK Electromagnetic Launcher Commission," pp. 245-248.
21. 1980 Conference on Electromagnetic Guns and Launchers, IEEE Trans. Mag., Vol. MAG 18, No. 1, January 1982.
22. 2nd Symposium on Electromagnetic Launch Technology, IEEE Trans. Mag., Vol. MAG 20, No. 2, March 1984.
23. 3rd Symposium on Electromagnetic Launch Technology, to be published in IEEE Trans. Mag., November 1986.

Int. J. Impact Engng Vol. 5, pp. 681–692, 1987
Printed in Great Britain

0734-743X/87 $3.00 + 0.00
Pergamon Journals Ltd.

A REVIEW OF EXPLOSIVE ACCELERATORS FOR HYPERVELOCITY IMPACT

Alex B. Wenzel

Southwest Research Institute
6220 Culebra Road
San Antonio, Texas 78284

ABSTRACT

A technical overview of experimental methods using high explosive techniques for conducting hypervelocity impact studies is presented. The explosive techniques use the explosive detonation fronts as means of accelerating the projectile, or as means of compressing a light gas which is then used to launch the projectile.

The explosive launchers are in six subdivisions: high explosive pellet accelerators, flyer plate accelerators, shaped charges, explosive-formed projectiles, fragment and microparticle accelerators, and explosive gas guns. Each one of the subdivisions presents the various techniques, their advantages and disadvantages, the range of mass and velocity capable of being accelerated, and whether the technique can be scaled for larger or smaller masses.

INTRODUCTION

The objective of this paper is to review the various experimental methods which use explosive techniques for conducting hypervelocity impact. These techniques play an important role in advancing our knowledge of penetration mechanics. This field is still largely empirical, and fully comprehensive theories are yet to be developed; the limits of experimental techniques mark the boundaries of our understanding. A brief review of some of the historical developments will help place the explosive launchers in perspective.

At the end of the 19th Century and beginning of the 20th Century, investigation in Europe, primarily Germany, and the U.K. (Munroe, 1888; Patents, 1910), showed that significant penetrations into steel blocks could be achieved by using an explosive block with a hollow cavity. The depth of the crater was increased by displacing the hollow charge some optimal distance from the target, thereby optimizing the focus of the explosive gases (detonation products). By lining the hollow cavity with a metal, it was observed that the penetration increased substantially over the hollow design.

Lined-cavity shaped charge research accelerated tremendously between 1935 and 1950 due primarily to World War II. Franz Rudolf Thomanek and Henry Hans Mohaupt (Thomanek, 1959; Mohaupt, 1947) perfected this concept and developed the first effective lined-cavity shaped-charge. By 1955, these lined-cavity shaped-charges were capable of accelerating long jets with tip velocities up to 8 km/sec. These jets have an inherent velocity gradient from tip to origin, causing the jet to break up into distinct particles as the travel distance increases.

In the 1960's, a surprisingly large gain in projectile velocity was achieved as a result of a vigorous effort to make the study of hypervelocity impact phenomena in the laboratory a reality. This was primarily because of the interest to protect space vehicles against meteoroids and to create weapons capable of defeating thick armors. During this time, variations to the hollow charge designs were developed to launch discrete projectiles up to 16.5 km/sec (Kronman, 1962; Wenzel, 1965a, 1965b). Most recently, the U.S. Strategic Defense Initiative (SDI) and the U.S. space program has focused on hypervelocity impact due to the speeds associated with satellites and missiles as well as space debris and meteoroid hazards. Many of the explosive projection techniques developed in the 60's are being looked at again.

The current interest is to apply the explosive launchers: laboratory tools to evaluate new materials and structures subjected to hypervelocity impacts; to determine equation of state of various materials; to use them as micrometeoroid, meteoroid or space debris simulators; to determine if explosive energies can be used in combination with other launchers to achieve higher velocities; and to develop them as warheads for intercept devices.

This overview examines the explosive launcher under six subdivisions: high explosive accelerators, flyer plate accelerators, shaped charges, explosive-formed projectiles, fragment and microparticle accelerators, and explosive gas guns.

Techniques such as exploding wire guns or exploding foil guns will not be discussed because these methods are considered to depend more on rapid vaporization of the wire or foil as the means for acceleration and not on the reaction of chemical explosives. Also, the use of high explosives to generate plasmas or act as the driver to generate electromagnetic fields or as augmenters to powder and light-gas guns will not be discussed.

Each of the subdivisions will discuss the various techniques developed, their advantages and disadvantages, the range of mass and velocity capable of being accelerated, and whether the technique can be scaled for larger or smaller masses.

High Explosive Pellet Accelerators

Pellet accelerators consist of a projectile (i.e., sphere, cylindrical pellet, or a plate) placed in direct contact with an explosive charge. The pellet is either laid on the surface or embedded in the surface of a high explosive charge (Poulter, 1957 and Willing, 1957), as shown in Fig. 1. The impulse delivered by the blast load imparts a velocity to the projectile.

These accelerators are dependent upon the explosive properties and the ratio of pellet mass to explosive mass per unit area. This latter ratio is known as the mass-to-charge ratio (M/C). If the pellet mass is kept constant, its velocity increases nearly linearly with a decrease in the M/C ratio, and then approaches a constant value as M/C becomes smaller. This asymptotic value corresponds to the detonation front having reached detonation velocity.

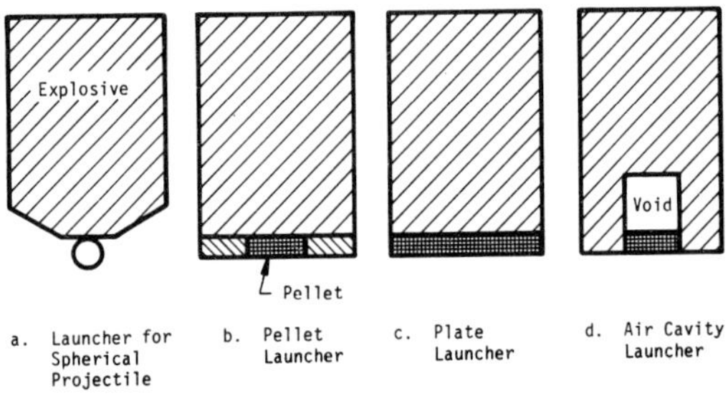

Fig. 1. Pellet and plate accelerators.

Pellet masses ranging from 0.15 to 8.86 gm at velocities of approximately 2 to 5.5 km/sec have been reported with the pellet and air cavity techniques (Clark, 1960 and Kineke, 1960). Projectile damage can occur with these methods because of the shocks propagated through the projectile. Damage can be reduced by embedding the projectile in the explosive or by having inert material between the projectile and the explosive charge. Finally, it should be noted that the above accelerators can be scaled to provide capabilities for launching pellets and plates weighing from 10^{-5} kg up to 34 kg (Garza, 1982).

The main advantage of this technique is that it is relatively inexpensive for launching spherical, cube, and plate type projectiles for a wide range of masses and velocities. It can also be used to project simultaneously more than one projectile. The disadvantage is that it is limited to the three mentioned projectile geometries. Also, the distance between the launcher and the target must be short in order to insure normal impacts of the cube and plate projectiles. This restricts the type of targets one can use to those materials or structures which are not affected by the blast or non-recoverable targets. For testing fragile targets, the use of a blast deflector is a must.

Flyer Plate Accelerators

Flyer plates operate on the same principle as the high explosive pellet accelerators. The plate to be accelerated is placed in contact with the explosive. The explosive is then initiated along one edge with a line wave generator. A line wave generator is an explosive device designed to create a plane line wave through the explosive onto the surface of the plate to be accelerated. As the detonation front progresses in a straight line down the explosive, the plate to be accelerated is folded off, eventually reaching a flat configuration when the entire explosive has detonated. Due to edge effects, the plate will not be perfectly flat, but a large central area of the plate is sufficiently flat to permit the gathering of useful data. This technique is illustrated in Fig. 2.

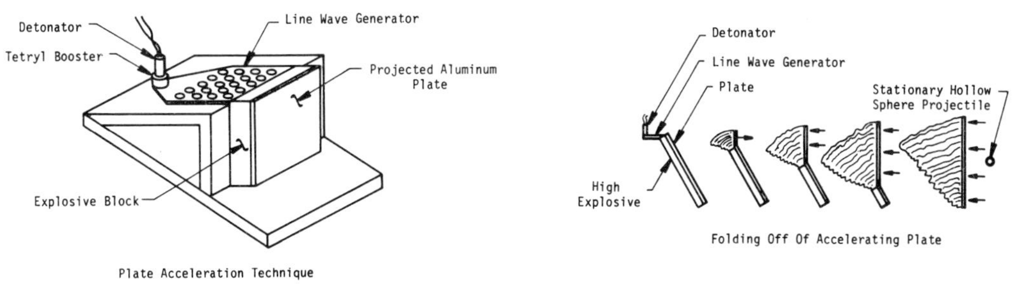

Fig 2. Flyer plate accelerator (Kreyenhagen, 1963).

Flat plates have been accelerated by investigators for the purpose of introducing strong, planar shock waves in various materials to determine dynamic compressibilities (i.e., Hugoniot equations-of-state). The technique has its theoretical background in the Gurney approximation (Walters, 1986). The Gurney approximation establishes the relationship between the mass of the flyer plate and the explosive weight needed to accelerate the plate. Velocities up to 3.5 km/sec have been reported (Kreyenhagen, 1963). Damage to the plate is observed in flight with flash radiographs subsequent to impact.

Reverse ballistic techniques have been adapted for penetration mechanics studies by suspending stationary projectiles in the path of the flying plate. This technique is useful where the projectile material is too fragile, difficult or impractical to launch using guns or other explosive techniques. For example, it has been possible to examine the impact effects on very thin metal sheets or foils, long rods, washers, projectiles consisting of bare high explosives, and large projectiles of any type.

When used as a "reverse ballistics" technique, the limitation of projectile mass is dependent on the purpose of the experiment and the type and mass of flyer plate of interest. The mass of the flyer plate is controlled by the M/C ratio. To achieve velocities ranging from 1 km/sec to 3.5 km/sec mass ratios ranging from 0.1 to 1.0 are needed (Walters, 1986).

Impact velocities up to 8 km/sec have been obtained (Kreyenhagen, 1963) by a modification of the technique in which the flyer plate is projected towards another projectile launched from another accelerator, such as a gas gun or explosive cavity charge.

Shaped Charge Accelerators

One of the most commonly used explosive projectors is the shaped charge. The shaped charge is generally a conical metal liner around which a cylindrical explosive charge is cast or pressed as illustrated in Fig. 3. Detonation is initiated at the end near the core apex. As the detonation wave progresses down the charge, it exerts pressure on the liner, collapsing it to form a high speed jet and a slower slug of the liner material. With the conical liner design, the jet flies out as a plastically deforming stem of metal. Because of the conical geometry of the liner and the path of liner collapse, the jet has an inherent velocity gradient from the tip of the jet to the slug; thus, the jet stretches as it flies and breaks up into individual projectiles as illustrated in Fig. 4. Maximum penetration into targets is obtained by placing the target at an optimum standoff distance from the shaped charge in order to minimize the effect of jet breakup due to the velocity gradient.

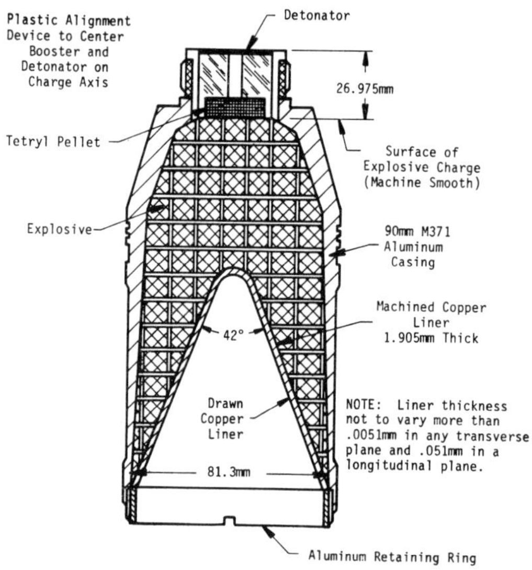

Fig. 3. 81.3mm BRL precision charge (Aseltine, 1978).

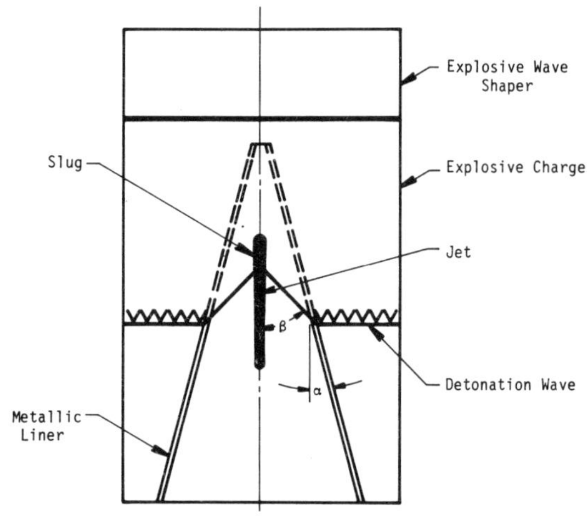

Fig. 5. Jet formation from the collapse of a metallic liner by plane converging detonation waves (Wenzel, 1965).

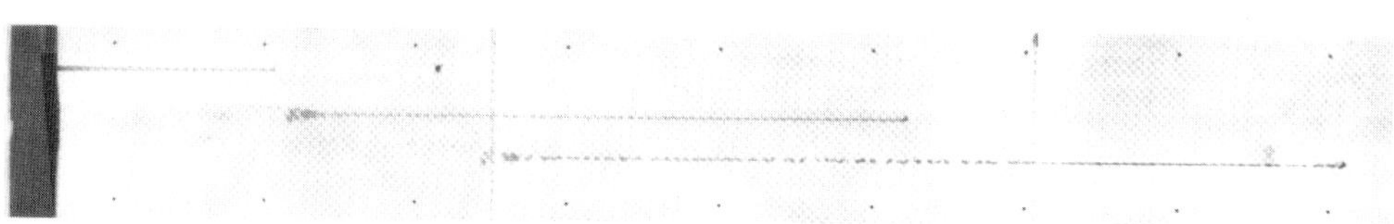

Fig. 4. Radiograph series showing jet breakup from normal shaped charge (courtesy of Los Alamos National Laboratory).

The theory of jet formation by shaped charges has been described extensively (see Birkhoff, 1948). In essence, the theory visualizes the liner being collapsed upon itself by the pressure of the detonation products. Since the pressure is applied progressively, rather than instantaneously, to various parts of the liner, the liner collapses upon the axis at an angle β greater than its initial angle α as illustrated in Fig. 5. Taylor (Birkhoff, 1948) derived, from steady-state hydrodynamics principles, the relation between the velocity with which each element approaches the axis and the direction it travels. If it is assumed that the detonation velocity, V_D, is constant and that the velocity, v_0, is the same for all elements, then the phenomenon will appear to be a simple steady-state case as viewed from an inertial coordinate system whose origin is fixed as the junction of the walls collapsing on opposite sides of the axis of symmetry.

Following the arguments presented by Birkhoff, one obtains from the conservation laws and the principles of streamline flow the equations that represent the steady-state theory:

$$V_j = V_D \frac{\sin(\beta - \alpha)}{\cos\alpha} \{\csc\beta + \cot\beta + \tan\tfrac{1}{2}(\beta - \alpha)\} \tag{1}$$

$$m_j = \frac{m(1-\cos\beta)}{2} \tag{2}$$

where V_j is the velocity of the jet, m_j is the mass of the jet and m is the total mass of the liner. For cylinders ($\alpha = 0$) Eq. (1) reduces to

$$V_j = 2\,V_D/\cos\beta \tag{3}$$

and jet velocity approaches a maximum as $\alpha \to 0$. However, as $\alpha \to 0$, $\beta \to 0$ and $m_j \to 0$ and Eq. (3) becomes

$$V_j = 2\ V_D \tag{4}$$

and Eq. (4) states that the jet velocity cannot exceed twice the detonation velocity. This theoretical description does not consider the details of engineering design such as the use of explosives, wave shaping, and metallurgical considerations regarding the properties of the metal liners. Therefore, in order to design shaped charges that will launch fragments of various masses, densities, and shapes at velocities greater than the standard conical copper liner designs, it is necessary to understand the physical phenomena which take place in the shaped charge, taking into consideration explosive properties such as detonation velocity, pressures, geometrical configurations and the metallurgical properties of the liner material.

Over the years, investigators (Kronman, 1962, 1963; Wenzel, 1965; Berus, 1968; Kreyenhagen, 1963; and others) have systematically explored the effects of each of the shaped charge parameters, i.e., cone angles, liner properties, explosive properties, methods of manufacture, etc. Explosives such as Composition B, Octol, HMX9404, ERS-9, KEL-F, LX-10, and PBX-9010 are used because of their high detonation velocities, ranging between 7 to 9 km/sec. Numerous liner materials have been tested, including the heavy metals, beryllium, nickel, copper, brass, aluminum, steel, lead, iron, calcium, magnesium, beryllium-aluminum alloys, high purity single crystal grown precision liners, bi-metallic, etc. Cone angles from 0° (cylindrical) to 90° have been tested, forming high speed jets.

Various liner geometries have been tested departing from the standard conical designs. Geometries such as bi-conical, hyperbolic fluted, hemispherical, pyramidal, etc., as well as bi-explosive designs, wave-shapers and tandem charges have been used successfully. Velocities up to 16.5 km/sec have been achieved using a beryllium cylindrical liner combined with an HMX9404 explosive. Single particle jets of L/D ranging from 1 to 13 and masses ranging from 0.15 to 0.2 gm were achieved with this design as shown in Fig. 6 (Wenzel and Gehring, 1965).

To obtain single pellets from shaped charge accelerators using liner geometries other than cylindrical, inhibited charge designs are used (Wenzel, 1965 and Woodall, 1970). One of the inhibited shaped charge designs is shown in Fig. 7.

An "inhibitor ring" can be inserted at the base of the liner to control the collapse of the liner such that only the front portion of the jet is formed. The length of the leading pellet is controlled by varying the height of the inhibitor. The semi-circular increment of explosive, located at the base of the main charge, is used to deflect latter portions of the jet or slug material. The slug, whose velocity is 1 to 2.5 km/sec, can also be deflected by manufacturing an explosive deficiency around one-half of the base of the metallic liner (Pack, 1965).

Beryllium Pellet

M = 0.15 gm
V = 16.5 km/sec

Fig. 6. Radiograph of a beryllium cylindrical liner (Wenzel and Gehring, 1965).

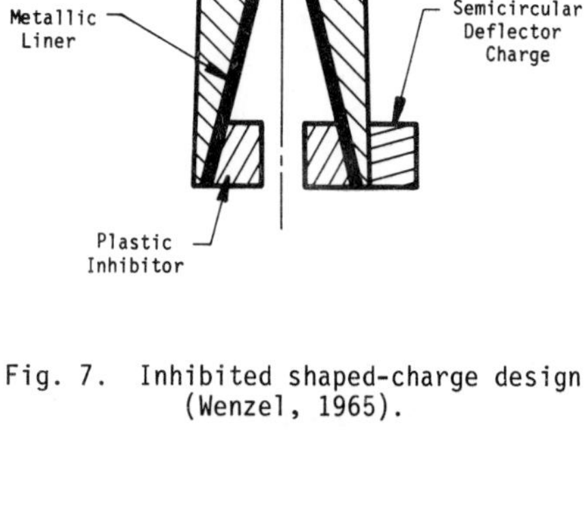

Fig. 7. Inhibited shaped-charge design (Wenzel, 1965).

Another technique used to deflect the slug is to recess the liner into the explosive charge allowing an explosive doughnut to overhang in front of the metallic liner. The jet is allowed to pass through a hole in a blast deflector designed to absorb the impacts of the deflected slug.

The velocity of the inhibited jet projectiles can be controlled over an appreciable range by varying the liner cone angle and explosive material. Pellet velocities measured for different aluminum cone angles using this design are given in Fig. 8. Figure 9 shows typical single pellet projectiles formed by the inhibited shaped charge design.

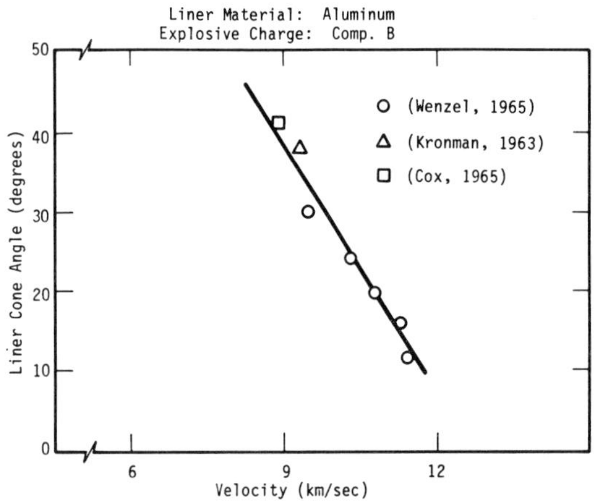

Fig. 8. Pellet velocity versus cone angle for conical liner (Wenzel, 1965).

Fig. 9. Radiographs of inhibited shape-charge jets (Wenzel, 1965 and Merendino, 1963).

Another design to generate single pellets was investigated by Woodall and Berus in 1970 (Woodall, 1970). Their work concentrated primarily in developing a one-gram aluminum pellet at velocities up to 15 km/sec. Velocities up to 10.7 km/sec were achieved using a 40° hyperbolic inhibited shaped charge design, incorporating a slug and debris trap, with an inhibitor for controlling the length-to-diameter ratio (L/D) of the pellet. This design did not achieve the intended goal of 15 km/sec.

The hyperbolic design is illustrated in Fig. 10. Note that the hyperbolic liner has a conical exterior and the inside surface is a hyperboloid. This is done to maintain a constant cross-sectional area relative to the explosive necessary to generate integral pellets. Similarly, the explosive charge loaded around the liner had to be a hyperboloid to maintain a constant cross-sectional area. The slug and debris trap isolates the slug and other low velocity jet material by collapsing an auxiliary shaped charge liner around the slug and debris material, thus stopping and propelling this debris in the direction opposite the pellet.

To control the L/D, a composite liner inhibitor was shown to be an effective method. However, external skirting of the liner was also found to be satisfactory. The composite liner consisted of introducing a discontinuity into the liner material by cementing a material such as Loctite 307 at a desired position in the liner, thereby controlling the length of the pellet formed. The skirted liner basically consists of machining the liner such that additional material is left at the end of the liner as shown in Fig. 11. Also, during their studies, it was found that employing large-angle liners can produce cohesive jets, and, for the first time, cohesive nickel and iron jet material was observed at velocities approximately 12 km/sec. To achieve higher velocities, these investigators combined a 90°, 0.030 in. wall thickness hyperbolic aluminum liner with a bi-explosive to produce a pellet at approximately 15 km/sec.

The bi-explosive design is illustrated in Fig. 12. This design depends on the use of two explosives with different detonation rates and an inert wave shaper (detonation barrier). This combination produces a converging conical detonation wave, causing the effective detonation rate to increase and be constant as it sweeps the liner.

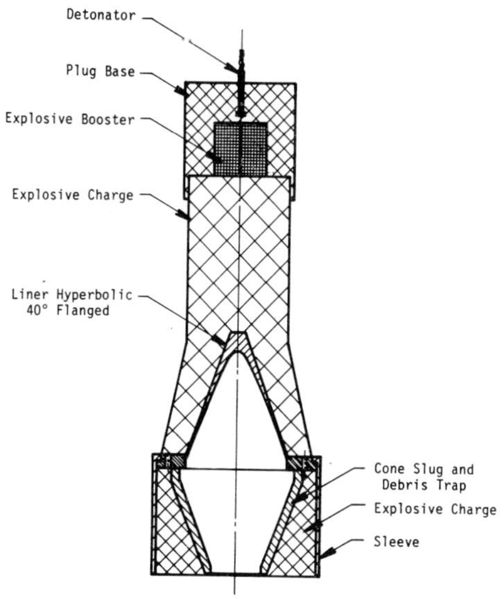

Fig. 10. 40° hyperbolic inhibited shaped charge (Berus, 1968).

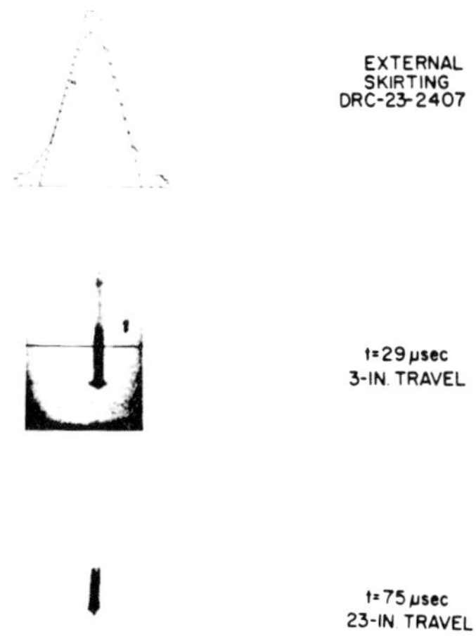

Fig. 11. Radiographs of a 30°, 9.8 mm wall thickness, 1100-0 aluminum, hyperbolic skirted inhibited jet-pellet (Berus, 1968).

In general, for most shaped charge designs, the projectile mass, shape, and velocity can be changed by varying the explosive effective detonation velocity, liner material properties, liner dimensions, and liner angles; by changing inhibitor heights; recessing the liner in the explosive; and using a slug and debris traps. A shaped charge can be scaled up and down, and its limitations are primarily in the minimum and maximum thicknesses of the liner to produce useful jets. The limits of liner thickness will also vary with liner material properties. Optimum standoff distances need to be determined for each design to maximize penetration and minimize the deleterious effects of jet breakup, ablations, and velocity gradients.

The main disadvantage associated with shaped charge accelerators is the uncertainty of determining accurately the mass, shape, and density of the projectile being launched.

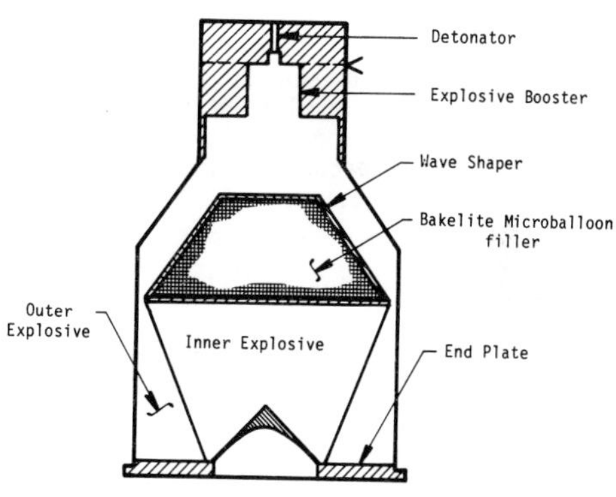

Fig. 12. Sketch of a 90°, hyperbolic liner, bi-explosive shaped charge (Berus, 1968).

Explosive Formed Projectiles

The explosively formed projectile (EFP) evolved from the shaped charge. Where the shaped charge utilizes a metal liner of either cylindrical or conical shape, whose angle can vary from 0° to 90°, the EFP liner is usually of a curved shape or wide angle conical liners usually ranging from 120° to 170°. The EFP uses the mechanism of explosively forming a metal liner into a single compact fragment, and projecting it with sufficient velocity to defeat armor at standoff distances up to 1000 calibers. Velocities approaching 3 km/sec have been achieved with EFP's. EFP includes devices known as the "P Charge," "Misznay-Schardin," "Ballistic Disc," "Mass Focus Warhead," "Implosion Fragment," "Self-Forging Fragments," and others. The formation of the fragment takes place either by inverting the liner, or by a process basically comparable to shaped charge jet formation, but with little or no velocity gradient. A typical EFP liner design is shown in Fig. 13 (Herman, 1977). Liner materials commonly used are copper, steel, tantalum, and depleted uranium. Current work has shown that an EFP liner can be formed into a slug with sufficient mass and velocity to defeat 185 mm of armor. The shape and velocity of EFP's are very sensitive to

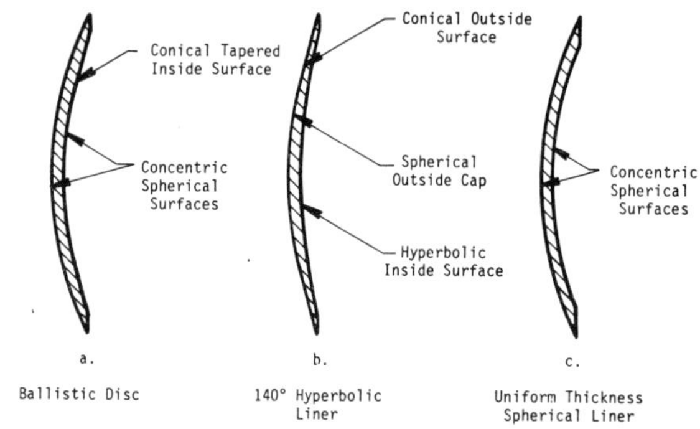

Fig. 13. Sketch of self-forging fragment liner designs (Herman, 1977).

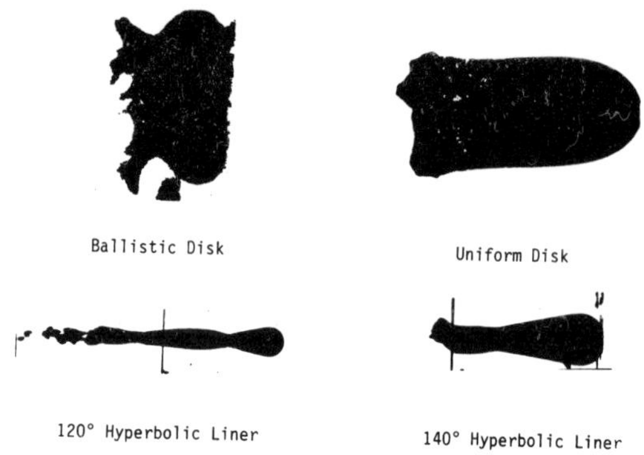

Fig. 14. Radiographs of explosive formed steel projectiles (Herman, 1977).

small changes in liner design and local dimensions. The material properties, strength, and density of the liner strongly affect the final shape of the slug. Materials of high density and high ductility make effective liner materials for EFP application. Softer materials converge easily on the axis, but are more susceptible to longitudinal stretching in response to velocity gradients. Typical projectiles formed from EFP's are shown in Fig. 14. These types of devices are very effective in penetrating relatively thick targets, and also generate large holes in targets.

The main advantage of EFP's is the aerostable shape that can be produced. This allows them to be launched at standoff distances as great as 125 meters; also EFP's can readily be scaled (Herman, 1977). Their disadvantage is that reproducibility of shape, mass, and density are sensitive to liner geometry, and explosive loading techniques. There is an uncertainty in determining accurately the density of the projectile being launched. Note that this is a common problem for explosively formed projectiles and jets. Also, EFP's can produce non-uniform density and asymmetric projectiles.

<u>Fragment and Microparticle Accelerators</u>

This type of accelerator is based on placing a continuous metal casing around an explosion charge. The casing is either internally or externally notched or is preformed to produce one discrete fragment size upon detonation. Fig. 15 illustrates typical designs of a notched and preformed accelerator. These types of accelerators are generally cylindrical in shape, but need not be; spherical, elliptical, or other shapes have been used. Fragment accelerators can be scaled. The performance parameters of fragment distribution masses and velocities are based on the mass-to-charge ratio (M/C), length-to-diameter ratios (L/D), explosive filler, and initiation condition. Fragment velocities up to 2.5 km/sec can be achieved, and as was the case with the flyer plate accelerators, to achieve velocities of 1 to 2.5 km/sec, mass-to-charge ratios ranging from 0.1 to 1.0 are needed (Walters, 1986). Types of initiations which are used are single end, center, dual end, triple (ends and center), and multipoint. In general, test results of cylindrical geometries indicate the following (Waggener, 1978):

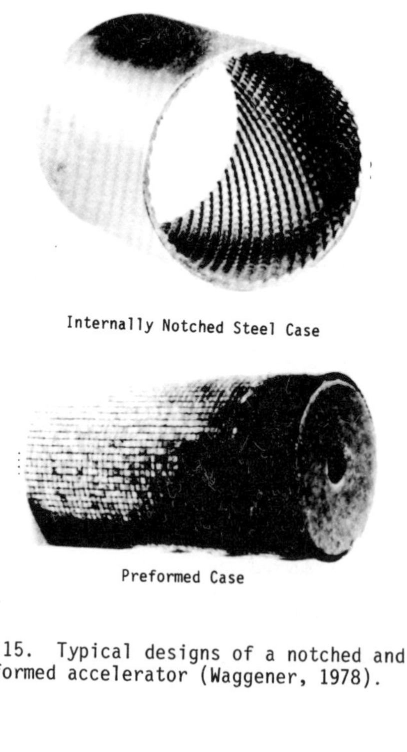

Fig. 15. Typical designs of a notched and preformed accelerator (Waggener, 1978).

- Fragment velocity and ejection angles are smooth functions of positions along the accelerator.
- Varying the initiation technique significantly changes the polar distribution of fragment mass and velocity.
- Fragment velocities are approximately 10% greater for the internally notched cases than for the performed cases, for accelerators of similar M/C, L/D, and initiation techniques.
- Tests of explosive fillers produced the following relative rankings in terms of ability to generate fragments, using Composition B as a standard: PBXN-3 (1.01), C-3 (1.00), Comp B (1.00), A-3 (1.00), PBXW-108 (0.96), AFX-108 (0.94).

Microparticle Accelerator

This type of accelerator was conceived for the production of microparticles for use in micrometeoroid impact studies (Richards, 1960).

The charge designs are essentially the same as those used to produce shaped charges, with the exception that a very thin brittle liner must be used. Typical charge designs are shown in Fig. 16, and are constructed of cast iron with a 250 micron thick wall. The technique has been used to accelerate clusters of 100 to 200 micron-sized particles to 10 km/sec using a 20° conical liner, and to 12 km/sec using a cylindrical liner. Masses from 10^{-9} to 10^{-2} g at velocities up to 15 km/sec have been launched (Gehring, 1960). The microparticles are produced having little or no velocity gradient and the size distribution is reproducible. The size distribution of the particles is determined by recovering in paraffin wax and measuring the diameter under a microscope. A similar charge is then fired against a target, and a crater size distribution is determined by microscopic examination. A comparison of these two distributions defines the size of particle which has produced a given crater. These designs can be scaled to produce somewhat larger sized particles; however, to the author's knowledge, there has never been a study of scaling limitation.

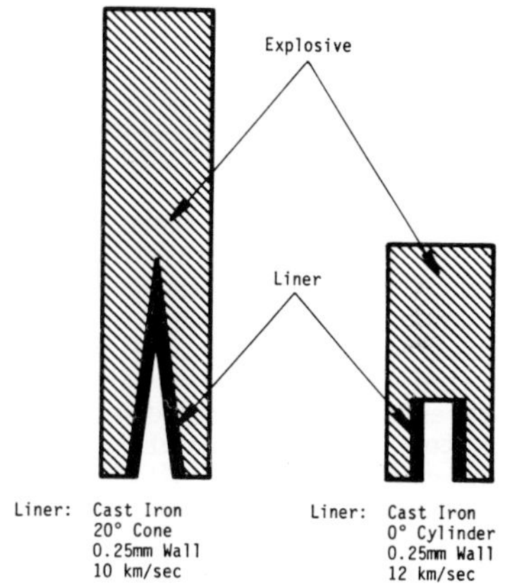

Fig. 16. Charge designs for microparticle acceleration.

Explosively Driven Gun

This technique is based on converting a substantial portion of the chemical energy of an explosive to kinetic and internal energy of a gas. The energy densities in the gas are produced by a strong shock generated by the progressive collapse of a tube. This collapsed tube may be considered as a piston driving a gas.

The concept of a single-stage, explosively driven launcher is illustrated in Fig. 17. A thin-walled, metal tube containing the driver gas is surrounded by a chemical explosive. After initiation of the explosive, the detonation wave accelerates the tube wall towards the axis, sealing the tube and forming a conical-shaped piston. The motion of the piston

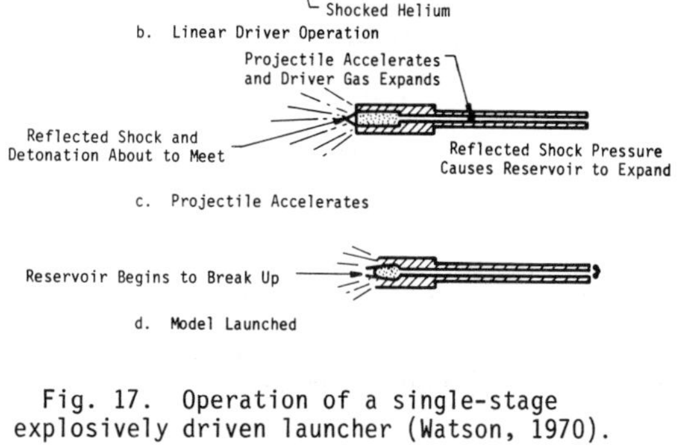

Fig. 17. Operation of a single-stage explosively driven launcher (Watson, 1970).

generates a strong shock wave in the driver gas, creating a high enthalpy gas. A barrel coupled to the explosive driver houses the projectile. The reservoir of high enthalpy gas is expanded to accelerate the projectile.

The length of shocked gas generated by the explosive driver is usually limited to less than 10 tube diameters to preclude effects of boundary-layer growth or pressure tube expansion. With the single-stage gun system, saboted lithium-magnesium models up to 4.5 in. diameter have been launched to 4.8 km/sec (Watson, 1970). Also, McCall reported a method for producing shockless acceleration of a 1 gm disk to predicted velocities of 20 km/sec using a driver containing a 1 kilogram of explosive (McCall, 1984). Basically, the system consists of a two-component explosive lens for producing higher phase velocities than the detonation velocity of a single lens driver. The shock wave produced by the collapse of a nickel cylinder travels through a cavity of foam or air to produce a forcing function that does not fracture the disk. To date, performance data have not been published.

Two-stage explosive gun concepts for accelerating small projectiles to velocities in excess of 10 km/sec were conceived in the late 1960's. The first stage is the standard linear gun described earlier. The second stage is formed by an explosive lens which collapses the barrel behind the accelerating projectile. With this type of system, a 2 gm cylindrical projectile was accelerated to 12.2 km/sec. Currently this type of system is limited to the 12.2 km/sec of unsteady, turbulent boundary layer developed behind the projectile, from the formation of the second-stage piston. Perhaps additional work can overcome this limitation.

The inherent high sound speed in light gases (especially, high temperature light gases) has been used to achieve high launch velocities in a conical gas ejector, as shown in Fig. 18. A high explosive charge surrounding a metal liner is initiated from the end to produce a shaped-charge effect; the cone angle of the collapsing liner is too low for the metal to form a jet, but produces a "piston velocity" in excess of 20 km/sec, accelerating hydrogen or helium gas. This gas is then used to accelerate a projectile. Velocities of 7 km/sec have been obtained for intact projectiles and 9 km/sec for projectiles broken in two or three pieces (Moore, 1965).

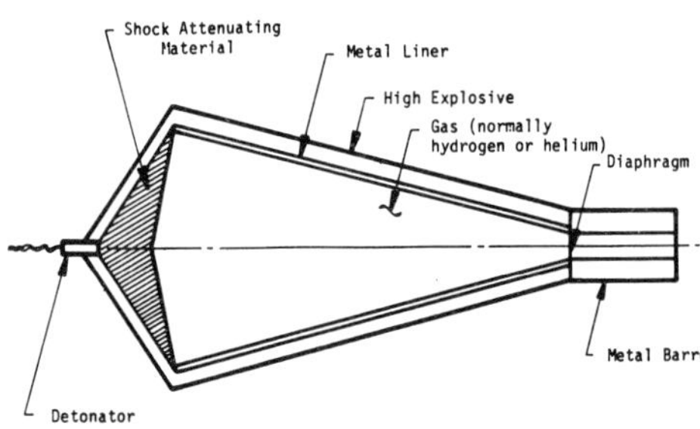

Fig. 18. Conical gas ejector (Moore, 1965).

Hybrid concepts have been advanced that coupled explosive drivers with a two-stage light gas gun. Basically, the light gas gun is augmented by the initiation of a series of explosive drivers which inject a very highly heated light gas gun into the pump tube through which the piston travels. The heated gas collected in the pump tube is compressed by the piston and applied to the base of the projectile. Although these concepts can, in theory, deliver exceptional performance, they are fraught with massive engineering problems and need further development.

CONCLUDING REMARKS

To summarize, the following general observations are made about explosive accelerators:

- Explosive shaped charge accelerators currently are capable of launching discrete projectiles up to 16.5 km/sec. However, the type of projectiles formed is thin long rods of L/D = 3 to 30 whose density mass and shape are not well defined. Useful integral projectiles of one gram or greater can readily be formed at velocities ranging from 3 to 12 km/sec, with the same limitations.

- It is possible to develop accelerators that will launch discrete projectiles in excess of 20 km/sec. To accomplish this, additional research and development is needed. Candidate designs to achieve this objective are inhibited shaped charges using cylindrical or conical bi-metallic liners with well controlled physical and metallurgical properties, coupled with advanced bi-explosive materials, with or without wave shapers.

- Current explosive accelerators have a limited capability for firing projectiles with variable shape and density.

- o Explosive accelerators using the shaped charge principle (i.e., shaped charges and explosively formed projectiles) are very effective in penetrating relatively thick targets and generating large holes in thick or thin targets. A disadvantage of the type of acceleration is the uncertainty of determining accurately the mass, shape, and density of the projectile being launched. However, this can be somewhat overcome through the use of advanced computational methods combined with experiments to arrive at designs that produce projectiles with consistent properties.

- o Most of these accelerators need short standoff distances, thereby requiring the placement of a blast deflector between charge and target, in order to test relatively fragile targets.

- o Even though explosive guns have accelerated projectiles up to 12 km/sec in the past, they cannot do this reliably. Other gun accelerators (i.e., light gas guns and electromagnetic guns) currently being used and under development will approach these velocities more readily.

- o Until other launcher technologies are fully developed, explosive accelerators offer now the capability of testing and generating an impact data base at velocities greater than 10 km/sec.

- o Through scaling, explosive accelerators can generate a wide variety of masses and velocities. The limits of scaling are determined by the minimum dimensions of the projectile that can be integrally accelerated, the mass-to-charge ratio (M/C), and the minimum thickness of the metal liner that will form a jet in a shaped charge design. Geometric scaling applies. The upper limits are defined by explosive weights, fabrication limitations, and availability of materials and cost.

REFERENCES

Berus, E. R. (1968). Shaped charge approach to meteor simulation for impact and luminous efficiency studies. NASA Meteoroid Impact Penetration Workshop, NASA Manned Spacecraft Center, Houston, Texas.

Birkhoff, G., D. MacDougall, E. Pugh, and G. Taylor (1948). Explosives with lined cavities. J. Appl. Phys., 19, No. 6, 563-582.

Clark, E. N., A. Mackenzie, F. H. Schmitt and I. L. Kintish (1960). Studies of hypervelocity impact on lead. Symp. Hypervelocity Impact, 4th, Eglin AFB, Florida.

Garza, L. R. (1982). Feasibility study of defeating flyer plate. Final Report SwRI Project 06-7076-004, Contract No. N62474-82-C-8228, Southwest Research Institute, San Antonio, Texas.

Gehring, J. W. and L. R. Richards (1960). Further studies of microparticle cratering in a variety of target materials. Proceedings of Fourth Hypervelocity Impact Symposium. See also Ballistic Research Laboratories Report Nos. 1134, 1286 and 1380.

Herman, J. W., Jr., G. Randers-Pherson and E. R. Berus (1977). Experimental and analytical investigation of self forging fragments. Proceedings of the Third International Symposium on Ballistics.

Kineke, J. H., Jr. (1960). An experimental study of crater formation in metallic targets. Symp. Hypervelocity Impact, 4th, Eglin AFB, Florida.

Kreyenhagen, K. N., J. E. Ferguson, R. R. Randall and J. P. Joyce (1963). Special explosive projectors, I. Shaped charge accelerator, II. Target plate accelerator. Symp. Hypervelocity Impact, 6th, Cleveland, Ohio.

Kronman, S. and J. H. Kineke (1962). Explosive devices for projecting hypervelocity pellets up to 21.0 km/sec (U). Proceedings of 5th Hypervelocity Symposium, Vol. II, Nonr-(6)-0020-62(x), Colorado School of Mines, Golden, Colorado (Confidential).

Kronman, S. (1963). Jet pellet projection technique (U). Proceedings of 6th Hypervelocity Impact Symposium, Vol. IV, DA-31-124-ARO(D)-16, Firestone and Rubber Co. (Confidential).

Kronman, S. and A. Merendino (1963). Inhibited jet charge. Proceedings of 6th Hypervelocity Impact Symposium, Vol. I, DA-31-124-ARO(D)-16, Firestone Tire and Rubber Co.

McCall, G. H. (1984). A method for producing shockless acceleration of masses to hypervelocities using high explosives. Report LJI-TM-84-106, La Jolla Institute, La Jolla, California.

Merendino, A., J. M. Regan and S. Kronman (1963). A method of obtaining a massive hypervelocity Pellet From a Shaped Charge Jet. BRL Report No. 1508, Ballistic Research Laboratories, Aberdeen Proving Ground, Maryland.

Mohaupt, H. H. (1947). Projectile. U.S. Patent No. 2,419,414.

Moore, E. T., D. Mumma, C. S. Godfrey and D. Bernstein (1965). Explosive Gas-Guns for Hypervelocity Acceleration. Fourth Hypervelocity Techniques Symposium, Arnold Air Force Station.

Munroe, C. E. (1888). Modern explosives. Scribner's Magazine, Vol. III, 563-576.
Patent (1911). Improvement in Explosive Charges or Bodies. Patent No. 28,030 (UK) to Westfolisch-Anhaltische Sprengstoffe A.G. (WASAG), Berlin.
Poulter, T. C. (1957). Acceleration of small particles and high explosives. Symp. Hypervelocity Impact Effects, 2nd, Washington, D.C.
Richards, L. G. and L. S. Holloway (1960). Study Hypervelocity Microparticle Cratering. BRL Memorandum Report No. 1286, Ballistics Research Laboratories, Aberdeen Proving Ground, Maryland.
Thomanek, F. R. (1959). Die Erste Hohlladungswaffe--The First Hollow Charge Weapon. Explosivstoffe, No. 1, 9-11.
Waggener, S. (1978). The performance of axially initiated cylindrical warheads. Fourth International Symposium on Ballistics.
Walters, W. P. (1986). Explosive Loading of Metals and Related Topics. BRL-SP-56, U.S. Army Ballistic Research Laboratory, Aberdeen Proving Ground, Maryland.
Watson, J. D. (170). A Summary of the Development of Large Explosive Guns for Reentry Simulation. PIFR-155, Final Report, Physics International Company, San Leandro, California.
Wenzel, A. B. (1965). Hypervelocity Impact Studies Against Apollo-Type Structures Up to 16.5 km/sec. Final Report P. O. M458XA-406033, NAA-SISD, GM Defense Research Laboratories Report No. TR-65-56.
Wenzel, A. B. and J. W. Gehring (1965). Techniques for launching 0.01 to 25 gm discrete Projectiles at Velocities Up to 54,100 ft/sec. Proceedings of the Fourth Hypervelocity Techniques Symposium, Arnold Air Force Station.
Willing, F. J. (1957). Acceleration of masses to hypervelocities by explosive Means. Symp. Hypervelocity Impact Effects, 2nd, Washington, D.C.
Woodall, R. L. and E. R. Berus (1970). Development of Meteoroid Simulators for Hypervelocity Impact Studies. Final Report Contract No. NAS9-9343, The Firestone Tire & Rubber Co., Defense Research Division, Akron, Ohio.

Int. J. Impact Engng Vol. 5, pp. 693–701, 1987
Printed in Great Britain

0734-743X/87 $3.00 + 0.00
Pergamon Journals Ltd.

SCALE MODELING OF HYPERVELOCITY IMPACT

Peter S. Westine
Scott A. Mullin

Southwest Research Institute
San Antonio, Texas

ABSTRACT

The significance of different Pi terms which result from a dimensional analysis of the parameters involved in hypervelocity impact is discussed. The consequences of distorting some physical phenomena in models are analyzed, and experimental verification is presented for the use of replica and dissimilar material models.

INTRODUCTION

This paper discusses use of scale models to study experimentally kinetic energy fragment or projectile hypervelocity impacts into semi-infinite, shielded, or specific targets, such as reentry vehicles and spacecraft. In hypervelocity impacts, many different mechanical and thermal processes take place, such as plastic flow, fragmentation, spalling, melting and vaporization.

Included are discussions of the pi terms, or dimensionless ratios, which follow from a hypervelocity impact model analysis; three physical phenomena (strain rate, thermal conductivity, and crack propagation effects) which are distorted in most of these models; and the use of replica and dissimilar material models in simulating hypervelocity impact problems.

PI TERMS

Three types of pi terms result from any dimensional analysis using the Pi Theorem--similarity ratios, response pi terms, and energy ratios. The similarity ratios and response pi terms never determine the interrelation and scaling of parameters. Examples of similarity ratios for geometry are length/diameter, ℓ/d; density ratios, $\rho_{target}/\rho_{penetrator}$, ρ_t/ρ_p; heat of fusion ratios, n_t/n_p; and material strength ratios S_t/S_p. These similarity ratios exist for geometry, density, strength, temperatures, specific heats, heats of fusion, heats of vaporization, speeds of sound, and other parameters. These ratios distribute thermal energy and mechanical energy in their various forms between target and penetrator and also within various target or penetrator layers. Effectively, these ratios indicate that all pertinent physical processes must be scaled everywhere.

Response pi terms do not determine a model law either; instead, they tell investigators how to interpret the results. Examples of response pi terms might be a scaled depth of penetration x/d, or a scaled time ratio tv_o/d. Pressures, depth of penetration, temperature increases, strains, or other responses as functions of time are all predicted in the prototype from measurements in a model through response pi terms. Response pi terms do not scale or interrelate physical processes. Instead, they show how to relate measured responses in corresponding model and prototype systems.

Energy ratios, the third type of pi term, determine a modeling law. For a projectile which liquifies or is on the verge of liquifying, four major energy ratios result from a model analysis. Table A, where ρ is mass density, v_o is impact velocity, S is mechanical strength, C_p is specific heat, θ is the temperature change to melt, n is the heat of fusion, and a_o is the speed of sound, summarizes these ratios. These parameters may be in target or penetrator, because both are included when similarity ratios are scaled. Notice that five different energies per unit volume are interrelated by these four ratios. If pi terms 1 and 4 in Table A are squared, the denominator is the mechanical energy per unit volume, becoming clear when S is multiplied by strain ε. The numerator of pi term 1 squared is the kinetic energy per unit volume, the pi term 2 numerator the thermal energy per unit volume associated with bringing a body up to the phase change, the numerator of pi term 3 the energy per unit volume expended in the phase change, and the numerator of pi term 4 squared is associated with the energy per unit volume radiated away in waves.

A model system is equivalent to a prototype system when the pi terms are the same. In this case, the five physical processes are all in the same proportion to one another in both systems. This observation explains why a model works. The processes are proportional because the energy ratios are the same; hence, everything is distributed appropriately between target and penetrator or in various layers because of the similarity ratios.

Table A. Main Energy Ratios for Liquifying Impacts

$\pi_1 = \dfrac{\rho^{1/2} V_o}{S^{1/2}}$	$\dfrac{\text{inertia}}{\text{strain energy}}$
$\pi_2 = \dfrac{\rho C_p \theta}{S}$	$\dfrac{\text{thermal effects to bring to phase change}}{\text{strain energy}}$
$\pi_3 = \dfrac{\rho n}{S}$	$\dfrac{\text{thermal effects in phase change}}{\text{strain energy}}$
$\pi_4 = \dfrac{\rho^{1/2} a_o}{S^{1/2}}$	$\dfrac{\text{energy propagated away in waves}}{\text{strain energy}}$

The pi terms in Table A can be divided or multiplied by one another to give a different reference energy in the denominator. Irrespective of the reference energy, the important physical processes are interrelated when the pi terms are the same in model and prototype. Thus, no one way of expressing a model analysis is unique. What matters is what processes are represented, rather than through which parameter physical processes are interrelated.

If a prototype and model are to vaporize, two more energy ratios arise, one for each of the additional two properties that must be scaled. These additional terms are summarized in Table B where C_p is the specific heat for a liquid, and N the heat of vaporization. Pi terms 5 and 6 are analogous to pi terms 2 and 3 except now vaporization is involved.

Table B. Additional Energy Ratios for Scaling Vaporization

$\pi_5 = \dfrac{\rho C_p \theta_v}{S}$	$\dfrac{\text{thermal energy to bring to vaporization}}{\text{strain energy}}$
$\pi_6 = \dfrac{\rho N}{S}$	$\dfrac{\text{thermal energy in vaporizing}}{\text{strain energy}}$

A model analysis such as this one is valid whether crater descriptions, debris clouds, loads imparted to a target from penetration of a shield, or other responses are of interest. The same physical processes must be interrelated irrespective of whether responses involve ejecta cloud descriptions, loads, or penetration depths; hence, the same dimensionless ratios are involved. Most of the comparisons made subsequently will use crater depth because more data are available to make these comparisons; whereas, very little scaled data exists defining debris clouds and loads.

DISTORTED PHENOMENA

Strain rate, thermal conductivity, and crack propagation effects are not scaled. These effects can be distorted when they cause only insignificant changes to occur in the model relative to responses in a corresponding prototype.

The major effects of strain rate are related to increases in the mechanical strength resisting penetration. Figure 1 from (Lindholm, 1979) is a plot of flow stress versus strain rate for a plastic strain of 10% in 1100 aluminum.

In the region where data exist, the curves are approximately straight lines on a semi-log plot. One example constitutive equation for this region is:

$$\sigma = \sigma_o \ \ell n \ (1 + \beta \dot{\varepsilon}) \tag{1}$$

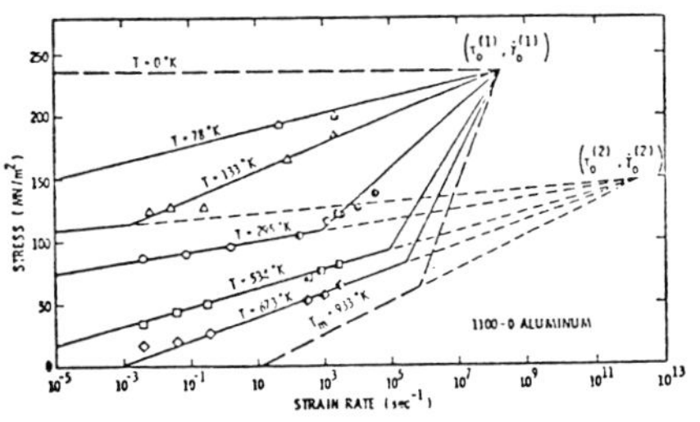

Fig. 1. Flow stress ($\varepsilon p = 0.10$) for 1100 aluminum as a function of temperature and strain rate (Lindholm, 1979).

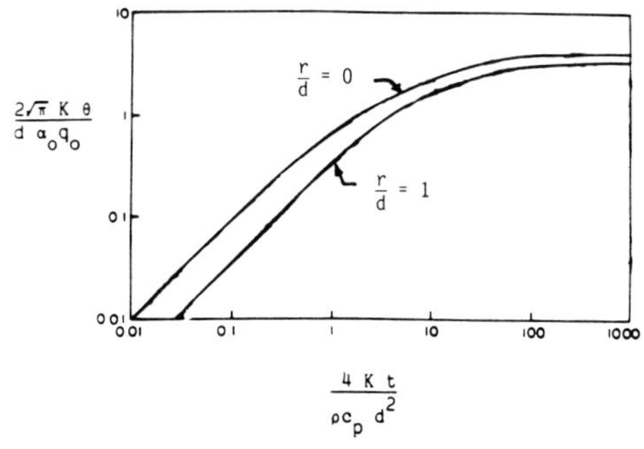

Fig. 2. Ready's solution for heating of thin sheet (Ready, 1971).

At a temperature of 534°K, β would equal 10^{+7} and σ_o would equal 3.26 MN/m^2 based upon results in Figure 1. Other aluminums have a higher σ_o, but similar β. At a rate of strain $\dot{\varepsilon}$ of 10^{+3} per second, the resisting flow stress σ would be 75.1 MN/m^2. An example of a small model would be a 1/10 scale replica model. A replica model, discussed subsequently, would have a strain rate 10 times higher than in a larger prototype. It is ten times larger, because the dimensionless ratio for scaled rate of strain would be $\dot{\varepsilon}d\rho^{1/2}/S^{1/2}$ with $\rho^{1/2}$ and $S^{1/2}$ the same in both systems, the caliber d smaller by 1/10; hence, $\dot{\varepsilon}$ is larger by 10. In this 1/10-scale model, the stress σ would be 85.6 MN/m^2 or 14% higher than in the prototype. Larger models would not have as large an error in σ, as the difference in the strain rates would not be as great. Both model and prototype are at high rates of strain, but the difference in these rates is important, not the absolute values. Strain rate effects largely influence S, and if S is only modified a few percent, adequate answers should result.

The problem of a thermal flux equal to $\alpha_o q_o$ impinging normally on a thin plate is similar to the thermal energy transfer from a projectile entering a target plate. Some energy goes into $\rho C_p \theta$ local heating with its associated temperature increase. If the duration t of heating is long, thermal energy will flow away from the heated spot indicating that thermal conductivity K matters. Assuming that temperatures are essentially equal on rear and front target faces (a thin plate assumption), means that all heat transfer flows laterally. Ready (1971) presents graphically the solution seen in Figure 2 for temperature distributions directly under the spot (r/d = 0) and off to one side (r/d = 1).

The ordinate is a nondimensional temperature and the abscissa is a nondimensional duration. When the scale duration equals 1.0 or less, the Y-axis equals the X-axis in Figure 2, which reduces to:

$$\frac{\rho C_p \theta\ d}{(\alpha_o q_o t)} = \frac{2}{\sqrt{\pi}} \tag{2}$$

For short durations, the temperature is independent of thermal conductivity, a situation arising when scaled durations equal 1 or less. Substituting typical Al values of K = 204 watts/m °K, ρ = 2.70 gm/cc, $C_p = 2.79 \times 10^{-4}$ watts hr/gm °K, and a time for impact t = 10 μs, into the abscissa

$$\frac{4Kt}{\rho C_p d^2} = 1.0 \tag{3}$$

gives a very small spot diameter d of 5.5×10^{-3} cm. Larger diameters in either models or prototype penetrators make conductivity effects effectively insignificant.

Another secondary phenomenon is crack propagation energy. To establish the order of magnitude, assume that the structure fracturing is a hollow sphere or hemisphere. If the energy of fracturing is small relative to the strain energy associated with plastic work, the energy caused by cracking is insignificant. The energy associated with plastic work under a biaxial state of stress is:

$$SE_{ext} = 2(\sigma_y \varepsilon_{max})(\text{Vol.}) = 8\pi\ \sigma_y\ \varepsilon_{max}\ h\ R^2 \quad (4)$$

The energy associated with having F equal size fragments in the sphere is:

$$SE_{crack} = \frac{K_{1c}^2}{E}\ h\ (\Sigma\ \text{length}) = 4\sqrt{\pi}\ \frac{K_{1c}^2}{E}\ \sqrt{F}\ h\ R \quad (5)$$

In equation (5), the total length of cracks is $4\sqrt{\pi}\ R\sqrt{F}$. Assuming that each fragment is a cube of thickness h, the number of fragments equals:

$$F = \frac{4\pi R^2}{h^2} \quad (6)$$

Substituting for N in Equation (5) and finding the ratio of crack energy to extensional strain energy gives:

$$\frac{SE_{crack}}{SE_{ext}} = \frac{K_{1c}^2}{E\ \sigma_y\ \varepsilon_{max}\ h} \quad (7)$$

Typical values such as $K_{1c} = 27.5\ MN/m^{3/2}$, $\sigma_y = 345$ MPa, $\varepsilon_{max} = 0.10$, E = 68,950 Mpa, and h = 6.35 mm indicate that this energy ratio is approximately 0.05, a value too small to be significant. This observation is consistent with fragmenting shell velocity computations such as (Gurney, 1943) that are made without considering crack propagation in energy balances.

MODELING LAWS

Model and prototype systems are equivalent provided the dimensionless ratios in Tables A and B are the same in both systems at homologous locations and at homologous times. Several methods exist for accomplishing this. None of the pi terms in Tables A and B depends upon any geometric parameter. Thus, any size model can be fabricated until secondary physical effects dependent upon size are introduced into the model system.

One method considered in most model evaluation is fabrication of the model from the same materials as in the prototype. In this approach, termed a "replica model," use of the same materials means that the scale factors for density λ_ρ, strength λ_σ, heat of fusion λ_n, speed of sound λ_{ao}, specific heat λ_{Cp}, heat of vaporization λ_N, and temperature λ_θ will be identical in both systems at homologous locations and times. The geometric scale factor λ_L will be less than one in a replica model if the model is to be smaller than the prototype. Provided impact velocities are the same in model and prototype, all the major pi terms will be scaled appropriately. Hence, a replica model works in all hypervelocity impact realms. Phase transitions, crater depth, impact pressures, etc., can all be scaled using a replica model.

A replica model, the easiest model to design and fabricate, has only one serious problem with its use, which arises because impact velocities are the same in both model and prototype systems. This limitation can be serious because of launcher limitations. Light-gas guns currently in use are expensive launchers with an upper velocity range of 8 to 9 km/sec. Aluminum has begun to melt at these velocities; however, no vaporization has begun. Launcher limitations can limit experimental investigations.

Fortunately, use of a replica model is not the only method of satisfying all pi terms in Tables A and B. In a dissimilar material model, one considers replacing prototype material with another of corresponding but not identical mechanical and thermal properties. Some resolution might be lost with this substitution, but models can be tested at a lower impact velocity than the corresponding prototype, which cannot otherwise be simulated because of launcher limitations. For a material which might have a density scale factor $\lambda_\rho = \gamma$ and a speed of sound scale factor $\lambda_{ao} = \beta$, Table C summarizes the material requirements which would result for satisfying all pi terms. This dissimilar material modeling approach is really the general solution. Use of a replica model is a special case of the general solution, with $\lambda_\rho = \gamma = 1$ and $\lambda_{ao} = \beta = 1$. Table C summarizes the modeling requirements imposed by use of either the replica or dissimilar material modeling laws.

Table C. Model Laws for Scaling Hypervelocity Impact

Parameter	Symbol	Scale Factors Dissimilar Materials	Scale Factors Replica Model
Impact Velocity	v_0	β	1.0
Geometry	d	λ	λ
Mass Density	ρ	γ	1.0
Speed of Sound	a_0	β	1.0
Material Strength	S	$\gamma\beta^2$	1.0
Heat of Fusion	n	β^2	1.0
Heat of Vaporization	N	β^2	1.0
Specific Heat Times Temperature	$C_p\theta$	β^2	1.0
Time	t	λ/β	λ

Lead, an example of a dissimilar material model material, can be used in place of either steel or aluminum. Lead, alloyed so that it is the correct strength, would model 8 km/sec prototype aluminum impacts at model impact velocities of 2 km/sec, and 32 km/sec impacts at model velocities of 8 km/sec. More discussion on this topic will be made subsequently when dissimilar material model-prototype comparisons are presented.

REPLICA MODELING COMPARISONS

Because large amounts of data are available in the literature for spheres shot into semi-infinite targets, a few comparisons will be made by plotting scaled depth of penetration P/d versus impact velocity v_0 or scaled impact velocity $\rho^{1/2}v_0/S^{1/2}$. In replica models, velocity, although dimensional, can be plotted on an abscissa because it is the same in model and prototype systems. Figure 3 presents a replica modeling comparison for aluminum spheres of different diameters fired into aluminum targets. Test results from three different investigators (Maiden, 1964; Sorenson, 1965; Halperson, 1965) are presented for only a small variation in penetrator diameters. Two separate families or groupings of data can be seen. High purity 1100 aluminums with a low ultimate strength around 62 MN/m^2 have greater scaled depths of penetration than high strength 2014 and 6061 aluminums with ultimate strengths from 207 to 283 MN/m^2. At velocities of impact from 5.4 to 7 km/sec aluminum liquifies; yet, the mechanical strength matters, at least for impacts into semi-infinite targets.

The data in Figure 3 plus additional data can be telescoped into one curve by plotting $\rho^{1/2}v_0/S^{1/2}$, pi term 1, versus scaled depth of penetration. Material strengths do matter and cannot be treated as invariant, which is done whenever the velocity v_0 is plotted by itself. In Figure 4, a corrected scaled plot is made using scaled velocity for various size aluminum spheres fired into different strength aluminum targets. The range in diameter for the spheres presented in Figure 4 is from 1.59 mm to 9.53 mm, a factor of six in geometric scale factor. Larger variations in scale factor should work, but launcher limitations prevent these comparisons from being made.

Some individuals do not believe in the validity of using a replica model because of some tests conducted at Arnold Engineering and Eglin Air Force Base. Although two sites were used, the material used at both came from the same source. Two different thicknesses of plate (38.1 and 50.8 mm) were used, but stress-strain curves were not reported for either. Figure 5 from (Sorenson, 1965) uses this data to show a systematic error with smaller diameter aluminum projectiles penetrating smaller scaled penetration depths. The thinner plates were used with the smaller diameter projectiles. These thinner plates were apparently stronger, which we believe, but cannot prove, was caused by heat treatment and subsequent quench. The smallest of the projectiles used in Figure 5 was the same size as the smallest of the projectiles used in Figure 4 where everything scales. Unfortunately, the data seen not to scale in Figure 5 was published over and over again in three different hypervelocity impact symposium papers (Engle, 1963; Goodman, 1963; Sorenson, 1965).

Replica models do work, and are not limited to aluminums. Figure 6 from (Sorenson, 1965) demonstrates the validity of replica modeling when copper spheres are fired into copper.

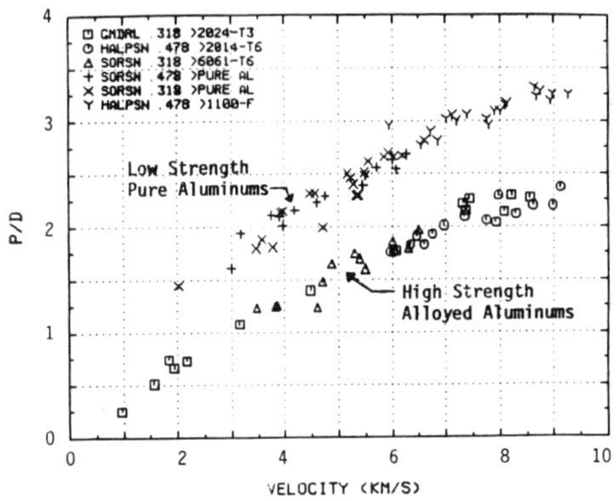

Fig. 3. Aluminum spheres impact into aluminum example of pi terms correlation.

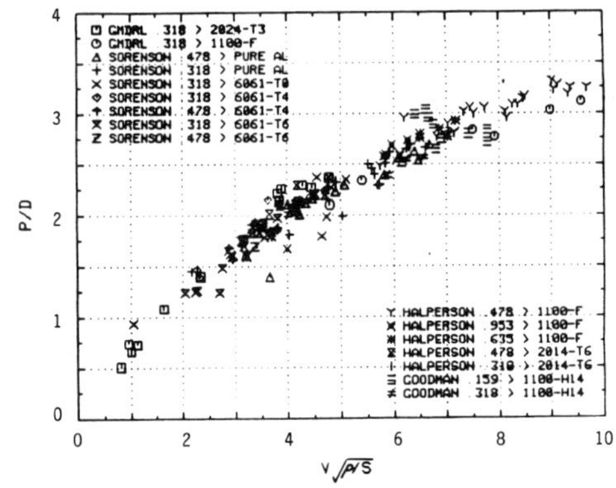

Fig. 4. Aluminum spheres impact into aluminum example of pi terms correlation.

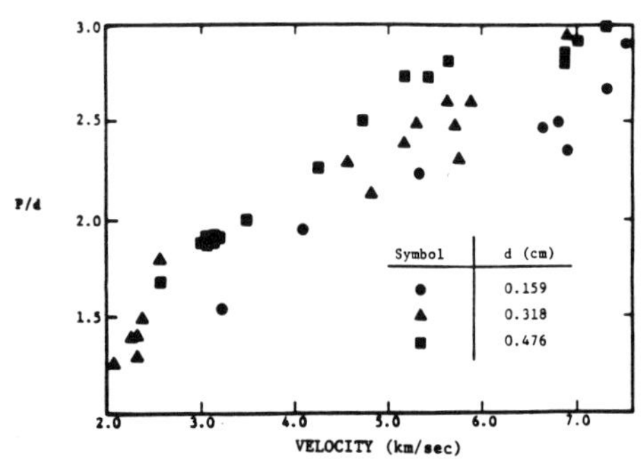

Fig. 5. Depth of penetration for Al spheres into Al targets (Sorenson, 1965).

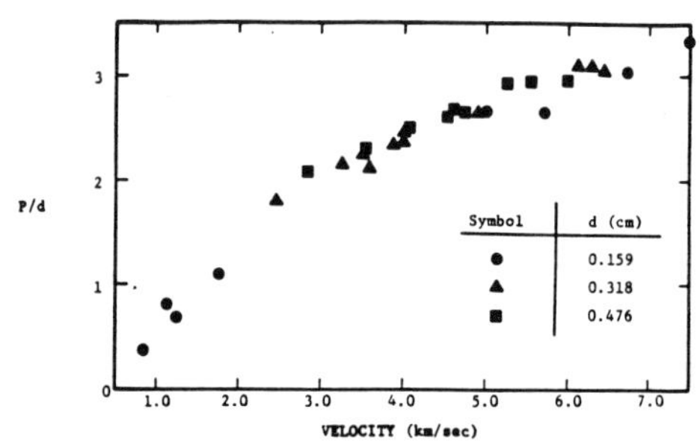

Fig. 6. Depth of penetration for Cu spheres into Cu targets (Halperson, 1965).

DISSIMILAR MATERIAL MODELING COMPARISONS

We have already discussed how five physical effects are interrelated through four pi terms in Table A, and how another physical process can become the reference process in the denominator. An alternate method of presenting Table A would be as in Table D with the heat of fusion n as the reference process. Pi terms 5 and 6 would not be required if the material did not vaporize. Some interesting observations can be made when the pi terms are rearranged in this manner. Table E from (Baker, 1973) shows that for many materials, pi term 2 is approximately constant. The quantities C_p, θ, and n are listed for aluminum, magnesium, copper, lead, tungsten, and steel. When the scaled quantity $C_p\theta/n$ is made up, it equals 2.30 for all, give or take a small variation.

Table D. Alternate List of Energy Ratios

$$\pi_1 = \frac{V_o}{n^{1/2}} \qquad \pi_3 = \frac{S}{\rho n} \qquad \pi_5 = \frac{C_{P\ell}\theta_v}{n}$$

$$\pi_2 = \frac{C_p\theta}{n} \qquad \pi_4 = \frac{a_o^2}{n} \qquad \pi_6 = \frac{N}{n}$$

Heat of fusion has, in turn, been related to the speed of sound by (Palmer, 1960). Figure 7 is a plot which shows that pi term 4 is a constant, equal to 95, for most metals. The only material deviating significantly is paraffin wax, a nonmetal.

Table E. Invariance of $C_p\theta/n$ For Various Materials (Palmer, et al., 1960)

Metal	C_p (cal/gm°C)	θ(°C)	n(cal/gm)	$C_p\theta/n$
Aluminum	0.214	660	76.8	1.84
Magnesium	0.246	651	70.0	2.29
Copper	0.0921	1083	42.0	2.31
Lead	0.0306	327	5.9	1.70
Tungsten	0.034	3370	44.0	2.58
Steel	0.107	1535	65.0	2.48
			Median	2.30

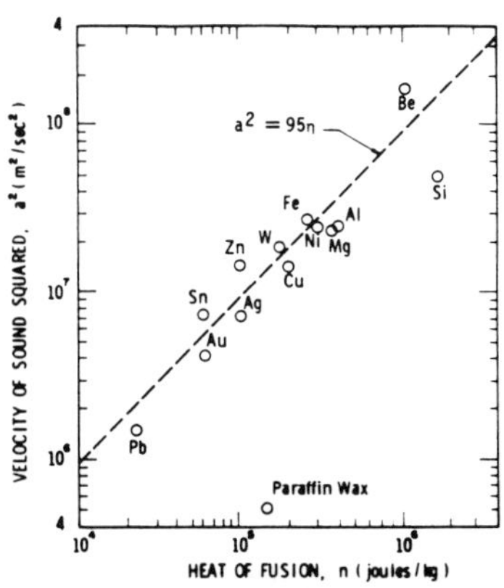

Fig. 7. Velocity of sound squared versus heat of fusion for various material (Palmer et al., 1960).

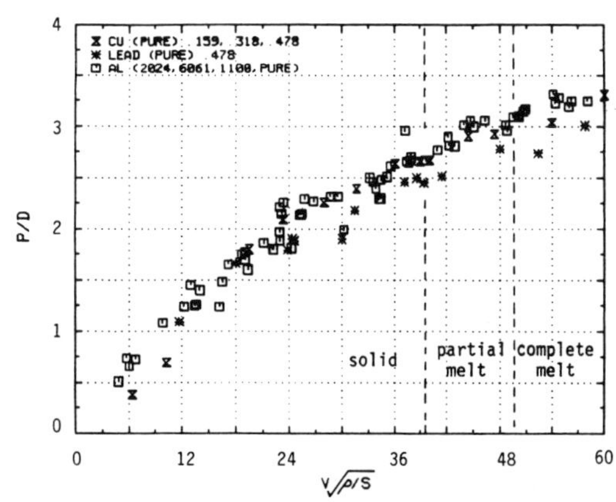

Fig. 8. Comparison of different metal impact phenomena example of pi terms correlation.

Because material strength does matter, but can be varied by alloying without modifying a_0, n, and $C_p\theta$ significantly, the correct selection of materials in a dissimilar material model can be made with pi terms 2 and 4 being held the same in all systems. Pi term 1 $v/n^{1/2}$, or v/a_0, or $\rho^{1/2}v/S^{1/2}$, or $v/C_p^{1/2}\theta^{1/2}$ (because a_0, n, $C_p\theta$, and S/ρ are all interrelated) instructs the modeler how the impact velocity scales.

Two examples demonstrating that the dissimilar material modeling concept works are Figure 8 for scaled crater depth from different material impacts, and Table F for the fraction of momentum transfer from dissimilar material penetrations of shielded targets. These results infer that loads as well as extent of damage can be scaled approximately using dissimilar material models.

In Figure 8 and Table F, large amounts of material melts; similar plots cannot yet be made to demonstrate the use of dissimilar material modeling in scaling vaporization because launchers cannot obtain the required velocities with aluminum and steels. In principle, lead should be a good model material for simulating aluminum or steel vaporization, but this statement can be supported only by theory. Lead at approximately one-fourth the impact velocities should simulate aluminum or steel impacts.

Table F. Momentum Amplification Comparison for Dissimilar Materials

Material	H/d	$V\sqrt{\rho/S}$	MV/mv *
Aℓ	.2	42.5	1.19
	.2	44.5	1.25
	.2	50.0	1.24
Cd	.2	47.5	1.21
Aℓ	.1	49.5	1.31
	.1	50.5	1.32
Cd	.1	52.5	1.26

*This term represents momentum registered at a shielded ballistic pendulum divided by original projectile momentum.

PROBLEMS IN USING MODELS

Some additional practical test problems also are associated with hypervelocity impact model testing. In addition to launcher limitations on the maximum possible velocity, the accelerations associated with launch can be so high that soft materials such as lead and cadmium will be deformed prior to impact. Some soft penetrator materials are actually extruded into joints in sabots at launch. Thus, the projectile at impact is not always the desired shape.

To model hypervelocity impacts, the dimensionless ratios must all be the same at homologous locations and times, which also means they must be constant for the various states (solid, liquid, and vapor) that can exist. This will occur with greater precision in replica models than in dissimilar material models. In dissimilar material models, some accuracy must be sacrificed because so many dimensionless ratios exist that one can only approximately make them equal for all the various physical states. Some resolution is thus sacrificed in exchange for being able to model impacts simulating the otherwise unattainable high velocity domains. Not knowing C_p, η, etc., for all states also means modelers use handbook values established at one atmosphere and room temperature. More appropriate values measured closer to the correct states might indicate that other scale factors or use of other model materials could be more appropriate.

Complex targets such as RV's can have many layers with various materials in them. While any one substrate may be matched, trying to match two, three and more with the same scale factors for density γ and speed of sound β create formidable practical problems.

CONCLUSIONS

Both replica modeling and dissimilar material modeling of hypervelocity impacts are possible. While investigators in the past have had problems, their difficulties appear to have been caused by failure to model material strength. While comparisons have been presented only for penetrations into semi-infinite targets, extent of damage and the penetration process through targets of complex geometry and finite size will also follow these modeling laws. Provided launchers can furnish the appropriate impact velocities, a replica model should be made and tested. When velocity limitations exist, a dissimilar material model properly constructed can give answers with only a slight loss in accuracy.

Data do not exist for demonstrating that vaporizing targets follow these same modeling laws; however, in principle, theory indicates that replica and dissimilar material models are possible. Both mechanical and thermal processes as well as conditions for initiating phase changes must be simulated in hypervelocity impact models.

REFERENCES

Lindholm, U. S. (1979). "Deformation Maps in the Region of High Dislocations Velocity," High Velocity Deformation of Solids, Kawata and Shiori, editors, Springer-Velag, New York.

Ready, J. F. (1971). Effects of High Power Laser Radiation, Academic Press, New York.

Gurney, R. W. (1943). The Initial Velocity of Fragments From Bombs, Shells, and Grenades, BRL-405.

Maiden, C. J., J. W. Gehring, and A. R. McMillan (1964). Damage to Thin Targets by Hypervelocity Projectiles, GM Defense Laboratory Report No. TR63-225.

Sorenson, N. (1965). "Systematic Investigation of Crater Formation in Metals," Proceedings of the Seventh Hypervelocity Impact Symposium, Vol. 6, pg 281.

Halperson, S. M. (1965). "Comparisons Between Hydrodynamic Theory and Impact Experiments," Proceedings of the Seventh Hypervelocity Impact Symposium, Vol. 5, pg 235.

Engle, O. (1963). "Hypervelocity Centering Data and a Crater Depth Model for the Regime of Partial Fluidity," Proceedings of the Sixth Hypervelocity Impact Symposium, Vol. II, pg 337.

Goodman, E. H. and C. D. Liles (1963). "Particle-Solid Impact Phenomena," Proceedings of the Sixth Hypervelocity Impact Symposium, Vol. II, pg 543.

Baker, W. E., P. S. Westine and F. T. Dodge (1973). Similarity Methods In Engineering Dynamics, Spartan Division of Hayden Books, Rochelle Park, NJ.

Palmer, E. P., R. W. Grow, D. K. Johnson, and G. H. Turner (1960). "Cratering: Experiment and Theory," Proceedings of the Fourth Hypervelocity Impact Symposium.

Int. J. Impact Engng Vol. 5, pp. 703–708, 1987
Printed in Great Britain

0734-743X/87 $3.00 + 0.00
Pergamon Journals Ltd.

A NEW HYPERVELOCITY SHOTGUN

A. E. Williams

U.S. Naval Research Laboratory

ABSTRACT

This paper presents a technique for launching multiple, hypervelocity projectiles in a predictable pattern. The technique has been successfully applied to collections of 4-42 projectiles launched at velocities of 2 to 5 km/s. The projectile dispersion is obtained by imparting a pre-determined radial impulse to the collection of projectiles as the sabot exits the gun muzzle.

INTRODUCTION

Some modern weapon platforms and targets attain such high speeds that impact velocities of warhead fragments can be much greater than the velocity generated by the warhead explosion alone. Realistic evaluation of the lethality of a warhead must include tests in which these very high impact velocities are attained. However, the evaluation is most conveniently and economically performed with static tests in which the weapon platform and the target are held steady. Therefore, there is a need for hypervelocity shotguns which can produce both the very high impact velocities and the representative patterns of many fragments or projectiles released by a single warhead. To meet this need, NRL has developed a hypervelocity shot gun which can launch multiple projectiles at a sufficiently high velocity and with a controllable pattern spread. Projectiles are accelerated in a two-stage light-gas (such as hydrogen or helium) gun which supplies very high speeds: 6 km/s and higher. The shotgun effect is achieved with a technique developed at NRL over a two-year period.

CONCEPT

This technique is illustrated in Fig. 1, which shows a cross-section view of a 12 projectile shot just after the sabot has exited from the gun muzzle. In this illustration, each projectile is a quadrant of a hollow cylinder, a design which maximizes the total mass of projectiles that can be fitted into the launch tube. Three cylinders, which separate into a total of 12 projectiles after firing, fit into the sabot. The cavity inside each cylinder is filled with a low-density material (such as polyurethane foam) and the front end of the sabot is split into four sections by longitudinal saw cuts. The energy needed for the radial separation of each set of projectiles is produced by the interaction of the foam with a separation rod prepositioned in a diaphragm near the muzzle of the gun. The radial velocity of each set of projectiles depends upon the velocity, weight, and shape of the projectiles, the density of the cavity material, and the cross section and density of the separation rod. To a good approximation, the ratio of radial to axial velocity of a set of projectiles from a single sabot is proportional to the accumulative sum of debris momentum produced by the passage of the separator rod through the sabot cavities. The separation rod also destroys the base of the sabot, allowing the precut sabot sections to separate. The radiograph in Fig. 1 shows the projectiles, sabot, and separation rod immediately after impact with the separator rod.

The multiple projectile launch technique is adaptable to a variety of projectile materials, shapes, sizes and quantities. Since the early development tests illustrated in Fig. 1, the technique has been successfully applied to launch a pattern of 42 spherical projectiles at 2 km/s from a smoothbore powder gun and a pattern of 24 projectiles at 5 km/s from the two-stage light-gas gun. Figure 2 illustrates an example of one of these sabot designs. The uniform separation that can be produced by this sabot design is indicated by Fig. 3.

ENGINEERING DESIGN

The reaction of the projectile sets to the interaction between the separator rod and the cavity material would appear to be difficult to predict. For example, the projectile sets in Fig. 3 show increasing separation from the front set to the rear set of projectiles, yet the contents of the cavities were generally the same for each set. This suggests that separation is dependent, in some manner, on the accumulative amount of material engaged by the separator rod. Also, a certain amount of separation could be expected because of residual range pressure (typically 20 kPa) and muzzle blast. Test data indicated that radial separation due to these causes was approximately 0.0015 of the axial travel, or about 5 minutes of arc. The radial separation due to the separator rod is defined as

$$X_R = \Delta R - B - 0.0015 X_A$$

where ΔR is the measured separation, B is the initial separation in the sabot, and X_A is the distance from the gun muzzle.

The following model evaluates the momentum of the debris from the impact between the separator rod and the cavity material. The model, illustrated in Fig. 4, is highly simplified and treats the cavity material as incompressible (when it is obviously highly compressible). The separator rod is assumed to have strength (σ) and the length loss to the rod is assumed to be only the loss during penetration (continued flow after perforation is ignored). Based on this model the pressure at the stagnation point is:

$$P = 1/2 \rho_R (V-U)^2 + \sigma_R = 1/2 \rho_c U^2$$

or

$$U = aV - (a^2V^2 - aV^2 - 2a\sigma/\rho_R)^{0.5}$$

where the R and c subscripts refer to the rod and cavity material respectively and

$$a = \rho_R / (\rho_R - \rho_c).$$

The mass of material consumed in each cavity is determined by the rod cross-section, the cavity length, the material densities and the velocity of the rod, V, relative to the penetration rate, U. The total momentum for each material is the product of the mass consumed in a cavity and the vector sum of the debris velocity for that material:

$$\mathcal{M}_R = m_R \,|(U + V) - U|$$

$$\mathcal{M}_c = 2\, m_c \,|U|$$

where $\mathcal{M}_R$ and $\mathcal{M}_c$ are the absolute values of the momentum. The sum of these momentums can best be compared with the radial reaction of the projectile sets, X_R/X_A, by defining a normalized momentum for each set

$$M_1 = (\mathcal{M}_R + \mathcal{M}_c)\, \mathcal{M}_p$$

where $\mathcal{M}_p$ is the axial momentum of a set of projectiles.

As noted earlier, the radial response of the projectiles appears to be dependent on the accumulative interaction between the separator rod and the cavity material. To account for this effect a new term was defined:

$$M_{2n} = \sum_{n=1}^{n} M_{1n}$$

where M_{2n} represents the accumulative momentum in the nth set. A comparison of M_2 with radial response indicated a strong non-linearity due to velocity, resulting in the final term,

$$M_2{}^* = M_2 + (V/3.6)^{0.8}$$

where V is the axial velocity in km/s. This corrected value of accumulative, normalized cavity momentum is compared to normalized radial separation, X_R/X_A, in Figure 5. The results for two sabot designs and five tests suggest a linear relationship between $M_2{}^*$ and displacement:

$$X_R/X_A = k\, M_2{}^*$$

where k is dependent on the shape of the projectiles. For sets of spherical projectiles, k = 0.5.

DISCUSSION

The analysis given here should be sufficient to predict sabot response for a wide range of initial conditions. Perhaps other, even simpler models would be adequate for design purposes. For example, the sum of the accumulative arial density of the cavities and the mass loss to the separator rod compares well with the radial displacement. Once the prototype sabot has been designed and tested the projectile pattern may be varied by simply changing the cross-section of the separator rod or by adding (or subtracting) thin discs of plastic to the contents of a cavity. For the tests presented here the cavity contents have included plastic foam, solid plastic and aluminum discs. The design possibilities are restricted to the "natural" separation rate of the projectiles and the maximum radial impulse the projectiles can endure without damage. The reader is cautioned that small separator rods may not produce adequate destruction and dispersion of the aft portion of the sabot. Sabot separation can be improved by placing higher density material in the aft section. This will increase the violence of the interaction with the separator rod.

Through the use of these sabot separation techniques the experimenter will be able to choose the distribution of damage to a target. The technique can also be used in conjunction with limited-size targets to achieve a multiplicity of data points from a single test. There is also a possibility that the separation technique can be applied directly to actual guns, allowing a gunner, by selecting the mass of the separator rod, to select the spread of the projectile pattern in accordance with the angular size of the target.

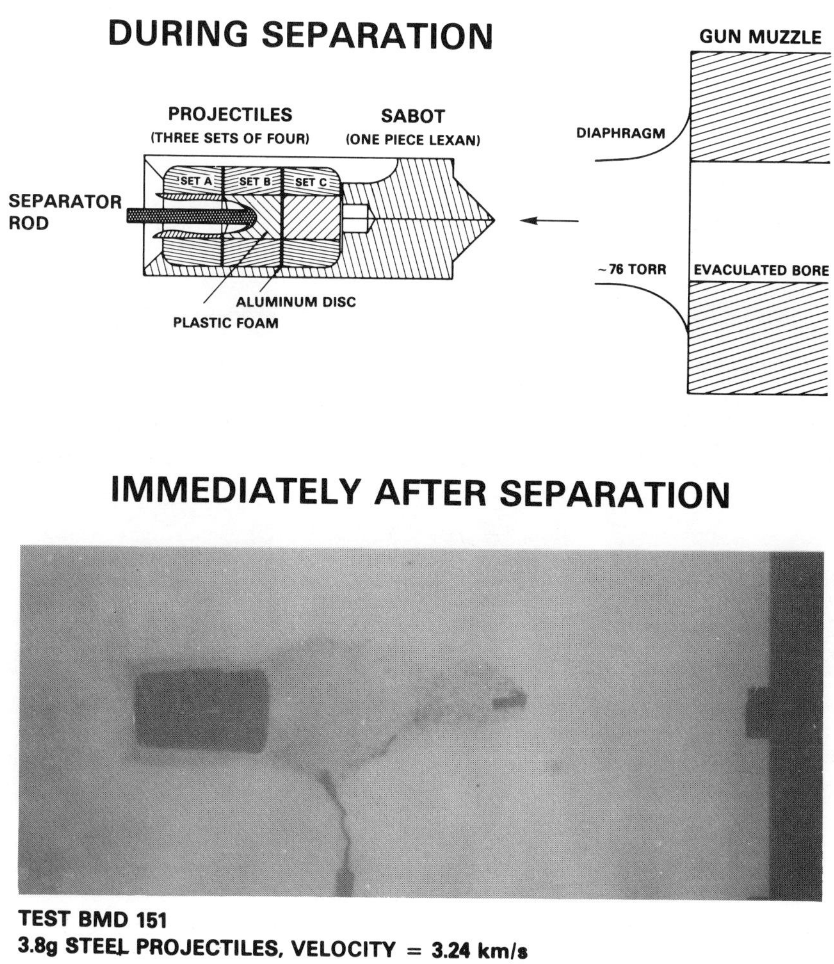

Fig. 1. Radial separation of 12 projectiles

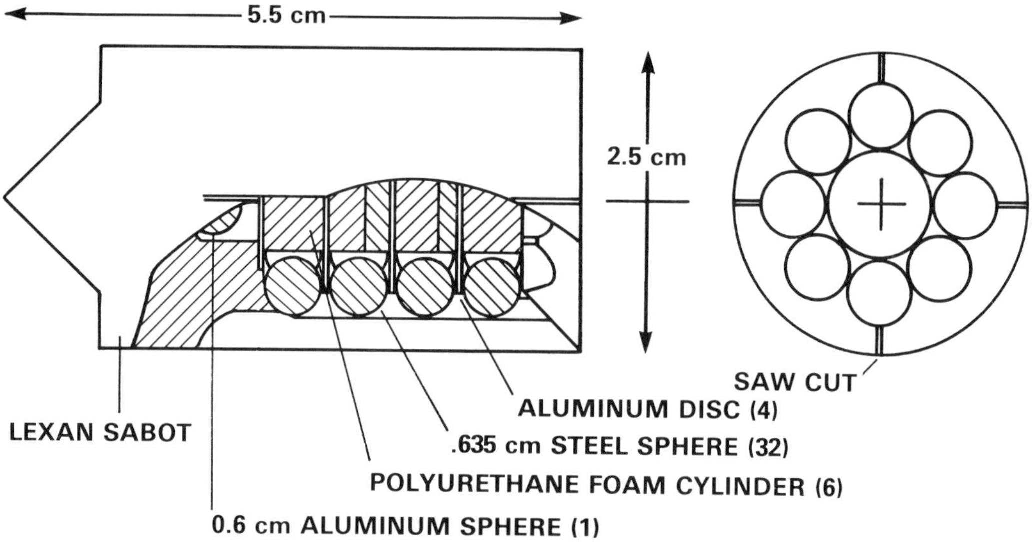

Fig. 2. Sabot design for 32 projectiles.

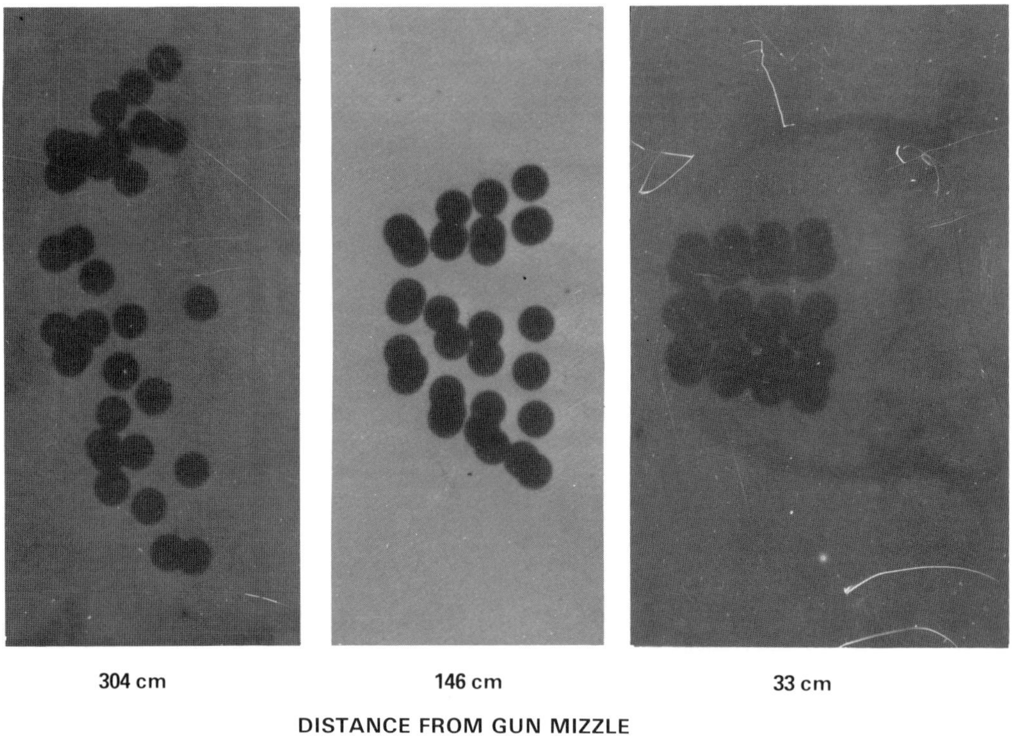

Fig. 3. Radiographs of projectile pattern. Thirty-two 0.635cm steel spheres launched at 3.15km/s from a 2.5cm bore light-gas gun.

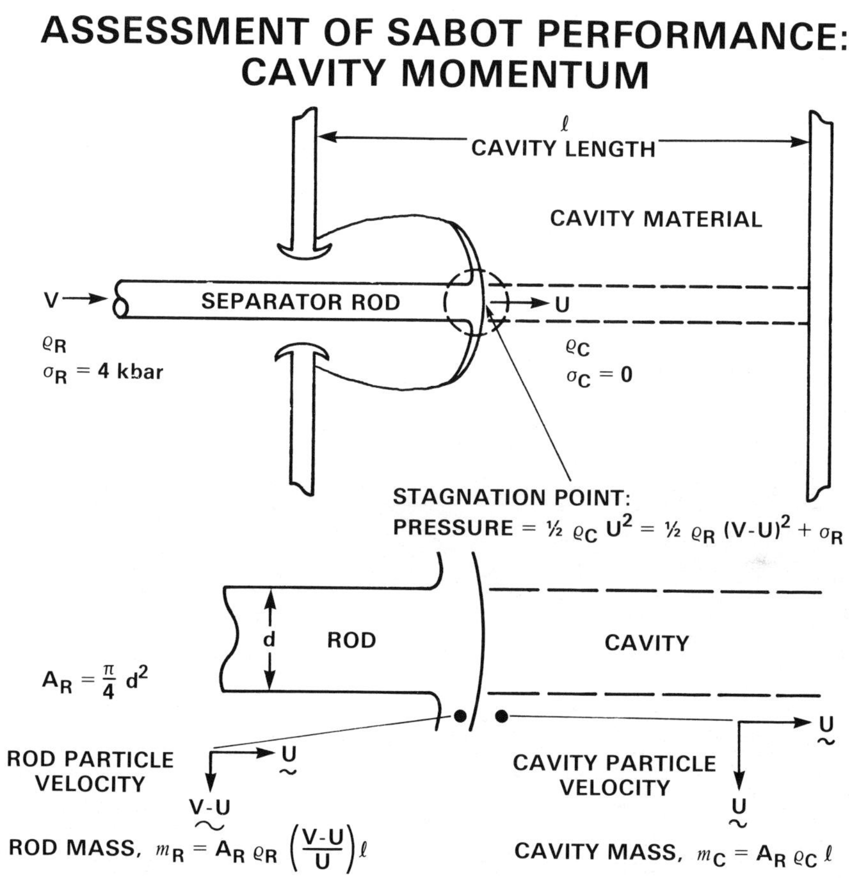

Fig. 4. Model for separator rod interaction with the sabot cavity material.

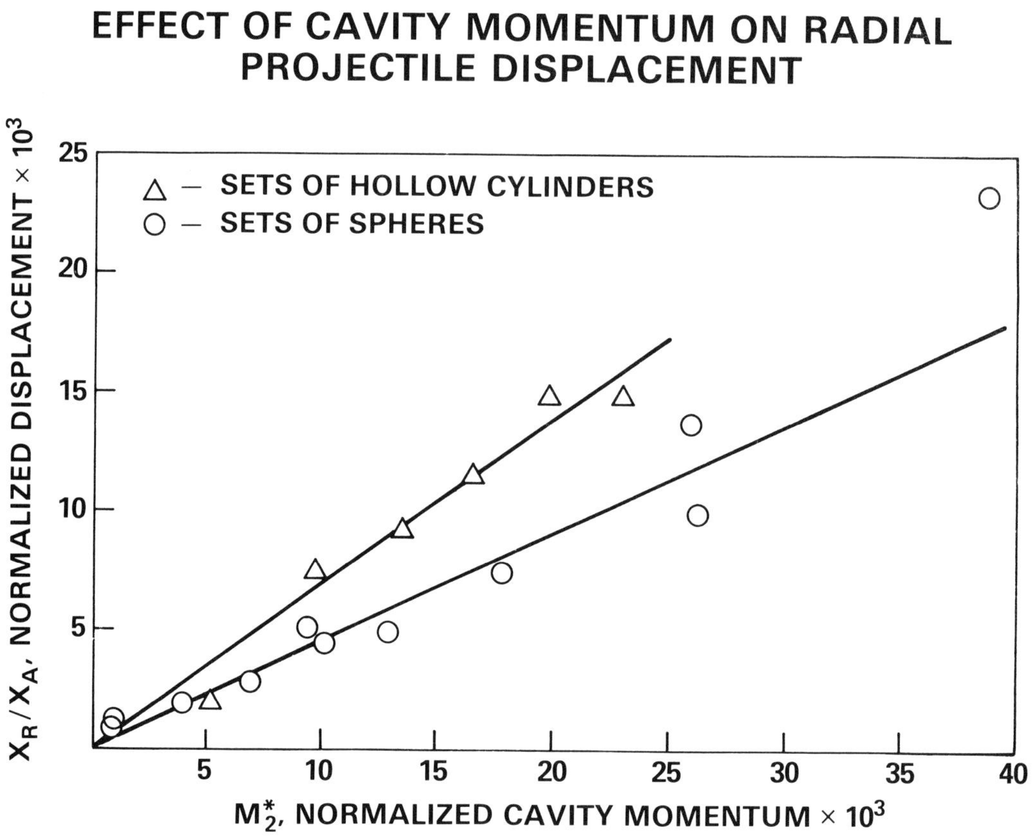

Fig. 5. The effect of cavity momentum on radial projectile displacement. Results from five tests.

Int. J. Impact Engng Vol. 5, pp. 709–728, 1987
Printed in Great Britain

0734-743X/87 $3.00 + 0.00
Pergamon Journals Ltd.

THE EFFECT OF PROJECTILE PROPERTIES ON TARGET CRATERING

A.E. Williams* and M.A. Persechino**

*U.S. Naval Research Laboratory, Washington, DC

**Potomac Research, Inc., Alexandria, VA

ABSTRACT

The effect of projectile properties on target cratering is evaluated for two basic target designs: a simple, half-space target and a half-space target protected by a shield. The comparison of unshielded and shielded target impacts is based on empirical evaluations and emphasizes the velocity range of 4-6 km/sec. A wider velocity range, 0-8 km/sec, is considered for evaluations of projectile property effects. This paper examines the effects of projectile strength, density, shape, size, velocity and particulation. It was observed that the presence of a shield amplifies the influence of projectile density, shape and size but mitigates the influence of projectile velocity. For shielded target impacts the properties of the projectile debris cloud behind the shield were used to predict the crater damage to the target.

INTRODUCTION

This paper examines hypervelocity impact phenomena for two fundamental impact configurations; a simple, half-space target and a half-space target protected by a shield. The impact of such targets generally causes significant craters which are easily measured and, for ductile target materials, can be precisely related to other impact phenomena such as impact velocity or projectile geometry. The commonly accepted definitions of crater parameters are illustrated in Fig. 1. A variety of empirical formulae have been used to show the relationship of these crater dimensions to basic projectile properties. One example is the anlysis by Sorenson (1965) of craters in half-space targets:

$$V/V_o = 0.12\ (\rho_P/\rho_T)^{0.5}\ (\rho_P v^2/S_T)^{0.845} \tag{1}$$

and, assuming hemispherical craters,

$$P/d = 0.311\ (\rho_P/\rho_T)^{0.17}\ (\rho_P v^2/S_T)^{0.285} \tag{2}$$

- V - target crater volume
- V_o - projectile volume
- P - crater depth
- d - projectile diameter
- ρ - density
- v - impact velocity
- S - target static shear strength
- P,T - projectile, target properties.

Equations 1 and 2 are typical of many empirical formulae which show crater parameters dependent upon target density and strength and upon projectile mass, velocity and density. These two equations present crater parameters in non-dimensional form, suggesting that replica scaling will apply to hypervelocity impact.

Other projectile properties can affect target crater dimensions. Obvious examples include projectile strength and shape. A complete generalization of projectile properties would have to include projectile particulation (a cluster of projectile fragments) and projectile material phase (solid, liquid, gas). These last two projectile properties, particulation and mixed phases, are readily observed to be the usual state of a projectile after a hypervelocity impact with a thin target plate.

This paper combines the results of earlier studies of projectile properties and half-space target craters with additional data from NRL research on projectile shape, density and particulation. The empirical relationships between projectile properties and half-space target craters are then compared with test results for the second target configuration: a half-space target protected by a thin plate target or shield placed some distance in front of the main target. A comparison of the effects of projectile properties on damage to the two target configurations demonstrate that shielding grossly modifies and generally amplifies the influence of projectile properties.

IMPACT OF HALF-SPACE TARGET

Projectile Strength

Projectile strength is usually regarded as an insignificant parameter for hypervelocity impact. Indeed, the definition of hypervelocity impact presumes fluidic behavior by the projectile. Under certain circumstances, very strong projectiles can survive the transient shock loads of moderate impact speeds and produce very deep craters. The strength of tungsten carbide projectiles is demonstrated in impacts of weak target materials, Fig. 2. The projectile shatter velocity is arbitrarily taken to be a 5% offset from a rising penetration-velocity graph. These projectiles survive transient shock loads up to 7 times the projectile strength. The impact results with lead targets are an interesting contrast to the results with the other target materials. For lead targets the peak shock pressure is about the same as the projectile strength. Further, an estimation of the hydrodynamic loads on the projectiles, $P=0.5\rho_T v^2$, suggests that projectile failure in aluminum and copper targets might not be due to the initial shock. These hydrodynamic pressures are very similiar to the static strength of the projectile, about 4 GP_a. The impact conditions and estimates of failure pressure are summerized in Table 1.

TABLE 1 Failure Pressures for Transient and Steady Loads on a Tungsten Carbide Spherical Projectile

Target Material	Failure Velocity* (km/s)	Impact Pressure Shock (GPa)	 Steady State (GPa)
1100 Alum.	1.7	27.0	3.9
2014 Alum.	1.7	27.0	3.9
Copper	1.0	28.0	4.5
Lead	0.35	6.8	0.7

* 5% off-set from penetration-velocity graphs.

The low velocity failure in lead targets may be due to low shock speed in the lead which reduces the movement of the target surface ahead of the contact boundary between the spherical projectile surface and the plane surface of the target.

Projectile Density

Equations 1 and 2 are based on several combinations of projectile and target material. Crater volume is shown to be proportional to the square root of projectile density. This conclusion is also supported by data for lead targets impacted at speeds up to 2.5 km/s (Palmer, 1960). Figure 3 compares this relationship with data from Gehring (1963) and an NRL analysis of approximately 200 tests for nylon, aluminum, steel and copper projectile impacts into aluminum targets at velocities of 2.5 - 6.5 km/sec. The analysis showed a decreasing dependance of crater volume on projectile density with increasing impact velocity:

$$V \propto \rho_p^{\,0.46-0.041v} \quad (v \text{ in km/sec}) \tag{3}$$

The equation suggests that projectile density effects will disappear at velocities over 11 km/sec.

Projectile Shape

The effect of projectile shape on crater parameters has been evaluated by Denardo (1968), Williams (1974) and Gehring (1963). These results and some NRL test data are presented in Fig. 4. Two types of data collections are shown here. The first type represents the changing aspect ratio (length to diameter) of a cylinder projectile oriented with the axis of symmetry aligned with the trajectory. The second type is a skewed rod. The data in Fig. 4 is normalized in terms of the crater volume, V_1, produced by a spherical projectile of cross-section A_S which has a volume and density equal to the test projectile. For a skewed rod the projectile cross-section is given by:

$$A = \pi r^2 \cos a + 2\, rL \sin a \tag{4}$$

where r, L and a are the projectile radius, length and skew angle. Figure 4 shows excellent agreement for the three data sets with aligned cylinders. The straight line, representing a large number of tests by Denardo (1968), has the following relationship between the shape of the plastic projectile (length to diameter ratio) and the crater volume in the 2024-T351 aluminum target:

$$V/V_p = 0.36(L/d)^{0.25}\, v^{1.67} \tag{5}$$

or

$$V/V_p \propto (L/d)^{0.25}$$

In contrast to the aligned cylinder results the skewed rod impacts have a considerably different effect. A simple plot of the volume - area parameters for skewed rod impacts shows an approximately linear relationship between crater volume and projectile cross-section, Fig. 5. In comparing the results of these impact studies the most interesting, but tentative, conclusions are: 1) crater volume in aluminum targets impacted by aligned cylinders varies with the fourth root of the projectile aspect ratio; 2) crater volume in aluminum targets is insensitive to the density of aligned cylinders when the crater volume is normalized to impact by a sphere of the same density; 3) crater volume values produced by skewed rods have a different, possibly a linear, relationship to impact cross-section; and 4) the side-on impact of a rod (L/d = 5) produces only 1/2 the crater volume of a disc of equal area. This last conclusion suggests that the smaller crater volume may be due to the larger surface area of an elongated crater. From Figs. 4 and 5 the crater volume - projectile cross-section relationships, normalized to the impact characteristics of a spherical projectile, are:

$$V/V_1 = A/A_S^{\,0.37}, \quad \text{aligned cylinder} \tag{6}$$

$$V/V_1 = 1.64 - 0.7A/A_s, \text{ skewed rod } (L/d = 5) \quad (7)$$

Projectile Particulation

Investigations of metal jets formed by shaped charges have repeatedly shown that target penetration is a function of the distance between the shaped charge and the target. Pictures taken during shaped charge experiments reveal that target penetration improves when the metal jet has broken into distinct fragments. As the distance between the shaped charge and target increases, fragment separation and target penetration also increases. There is an optimum distance beyond which fragment separation has no beneficial effect and other conditions cause a decrease in target penetration.

The effect of projectile particulation was evaluated by impacting arrays of two or three projectiles into HY-100 steel targets. The spacing between the projectiles was controlled by the initial separation within the sabot. The spacing just prior to impact was determined from flash x-rays. The average radial despersion of the projectiles was approximately 0.3 projectile diameters. The shape of the projectiles is shown is Fig. 6. The largest value of spacing, infinite spacing, was obtained by repeated impacts into the same crater.

Large values of projectile spacing produced deep, narrow craters. The crater shape is illustrated in Fig. 7. The effect of axial spacing or particulation on crater depth is shown in Fig. 8. Although the increase in crater depth with increasing spacing does not appear to be large, an interpretation of these results in terms of crater depth increase beyond the crater depth produced by the leading projectile may be more instructive. This increase in depth, shown as the difference between values B and A or D and C in Fig. 8, suggests that penetration by a large array of projectiles, axially aligned, would increase the crater depth by 60%.

Increased projectile spacing decreased the crater volume. This decrease in crater volume, as much as 30% for large spacing, may be due to discontinuities in the flow of target material during crater formation. Some additional impact tests were performed with targets containing machined craters (flat-bottomed holes) to see if work hardening and boundary conditions would influence impact damage. In general, these modified targets had little influence on the subsequent crater damage.

The cratering effect of a radially dispersed array of hypervelocity projectiles has not been examined. One could present arguments for either increased or decreased target damage. Since the debris cloud produced by hypervelocity impact with a plate target is less effective in damaging a second, half space, target as the spacing between targets is increased, it is reasonable to conclude that radial dispersion of a projectile array decreases target cratering.

Projectile Velocity

Projectile velocity is the most important projectile property. Projectile velocity effects have received innumerable evaluations in hypervelocity literature. The shape of the penetration - velocity graphs in Fig. 2 illustrate a particularly complex relationship of target cratering to impact velocity. Even projectile - target combinations which demonstrate hydrodynamic behavior at low impact velocities show significant changes in the relationship of crater damage to impact velocity for different velocity regimes. For example, Halperson's (1965) data for aluminum projectiles and soft aluminum targets disagrees with Denardo's (1967) results at higher impact speeds.

Halperson (1.4-6 km/s)

$$V \propto v^2$$

$$P \propto v^{1/3}$$

Denardo (4 - 8 km/s)

$$V \propto v^{1.545}$$

$$P \propto v^{2/3}$$

Data sets from both sources are extensive and well documented. The differences between the data sets are not as great as the velocity exponents would suggest since the coefficients for these equations are also different. However, the differences do suggest that a simple power law velocity relationship to target cratering may be unrealistic for wide velocity ranges and that velocity effects are more complex than generally presumed.

IMPACT OF SHIELDED TARGETS

If a plate is placed in front of a half-space target the plate will shield the target by shattering the hypervelocity projectile and dispersing the projectile debris over a wide area so that the impact with the target is no longer concentrated. The degree of dispersion will depend upon the projectile and shield materials, the thickness of the shield, the space between the shield and the target and the impact velocity. Most of the prior work on target shielding was motivated by the need for protecting satellites and deep space probes against impact by meteoroids. The impact experiments which evaluated this hazard were characterized by high impact velocities and large spacings between the shield and target. The damage criteria for the target was typically penetration or perforation. Cratering damage to these targets was more a collection of individual impacts by debris cloud particles rather than a single, major crater. In the analysis here the effects of smaller shield - target spacing and lower impact velocities will be evaluated. For these impacts the debris cloud does form a single, major crater and the principal measures of target damage will be the same as impact into an unshielded half-space target. The work shown here was taken from an earlier study by Williams (1983).

A shielded target impact is illustrated in Fig. 9. For most of the tests discussed here the shield is a fiberglas plate with a thin sheet of aluminum bonded to the back of the plate. The shield assembly is defined as a plate of thickness T and an average density of 1.55 g/cc. The shield has an orientation of α with respect to the trajectory and the plate is placed a distance x in front of the target. The target is 2024-T351 aluminum. The impact with the shield reduces the projectile velocity from v_o to v_r. The cloud of projectile debris has a different shape, direction and orientation from the original projectile. The effect of shielding on crater damage is illustrated in Fig. 10. To differentiate the damage to unshielded and shielded targets the subscripts c and cs have been added to the crater parameters of volume, penetration, and diameter. Another useful impact parameter is the mass of the shield intercepted by the initial projectile cross-section:

$$M_S = ((\pi/4)\, d^2)(1.55\ T/\sin\alpha)$$

For a non-spherical projectile the value of d will depend upon the shape and orientation of the projectile. For most applications it is useful to normalize the shield mass M_s, by dividing by the projectile mass, M_p. For the graphs involving projectiles of different density the term V_p, projectile volume, is defined as the volume of a steel sphere, equal to the mass of the test projectile.

The effects of shield - target spacing and shield thickness are illustrated by the condition of the projectile debris cloud after shield perforation, Fig. 11 and 12. Figure 11 shows that the projectile continuously changes shape and size as it travels the space between the shield and target. The impact conditions at the target therefore depend upon the shield - target spacing. Shield thickness affects the shape of the debris cloud. As the thickness of the shield increases the debris cloud transforms from a spherical shape to a hemispherical shell.

The new parameters introduced here are shield thickness and shield - target spacing. Two approaches can be taken to evaluate the effect of these parameters on target cratering: 1) simple comparisions of the initial impact parameters to target damage: 2) an application of the debris cloud characteristics to calculate target damage. Both approaches will be presented here. The evaluation of shield thickness and spacing effects is made more complex by the presence of other impact parameters such as projectile strength, density, shape, particulation, and velocity. To some degree these parameters can be used to evaluate debris cloud characteristics.

The effect of shield thickness on target crater volume is shown in Fig. 13 for an impact velocity of 5.9 km/sec. The data for these graphs was limited to three values of shield thickness, the unshielded condition, and an estimate of shield thickness for threshold perforation. Therefore, the graphs should be viewed as approximations. The effect of shield - target spacing on crater size is shown in Fig. 14 for thin shield impact. Both crater volume and penetration decrease with increased spacing but crater diameter increases by a moderate amount. In the following sections, shield thickness and spacing will be related to other projectile parameters.

Projectile Strength, Particulation and Size

Figure 15 shows the results of 136 tests of tungsten carbide spheres impacting aluminum plates of varying thickness. The curve defines the boundary between no visible damage and the fracture of the projectile. The roller coaster character of the shatter curve demonstrates the interaction of shock waves and projectile/-target boundaries to produce the tensile loading required for spall failure. The transient shock loads at incipient failure vary from 14 to 28 GPa. This is 3-7 times the strength of the very strong tungsten carbide projectile (about 4 GPa). At low impact speeds the strength of the projectile will affect the shatter velocity and, to an unknown extent, the degree of shatter.

The appearance of the debris cloud in Fig. 11 indicates that high speed impact produces a large number of very small particles. It is unclear whether the size of the particles (the degree of particulation) affects the size of the crater but it seems a reasonable premise that larger particle sizes will result in larger craters. Certainly in the extreme case, impact by a vaporized projectile, the target cratering is much less than observed for impact by a cloud of solid particles. Particle size has been observed to affect the size of scaled target craters. Denardo (1967) reports the following size effect for two grades of aluminum targets:

1100-0 aluminum

$$P/d \propto d^{0.056}$$

$$V/V_p \propto d^{0.1375}$$

2024-T4 Aluminum

$$P/d \propto d^{0.056}$$

$$V/V_p \propto d^{0.2065}$$

For the two large scale tests performed for this study the 286g projectile was 5.2 times the size of the 2g projectile. Applying the size effects equation for 2024 aluminum crater depth and volume results in a prediction of:

$$P_{cs}/d(286g) = 1.1\ P_{cs}/d\ (2g)$$

$$V_{cs}/V_p\ (286g) = 1.4\ V_{cs}/V_p\ (2g)$$

These large scale impact tests were performed with thin and thick fiberglas shields with the shields placed at normal incidence, 3 projectile diameters in front of the target. The ratios of penetration and volume values for the two sizes of projectiles are:

Thin shield (T/d = 0.64)

$$P_{cs}/d\ (286g) = 1.5\ P_{cs}/d\ (2g)$$

$$V_{cs}/V_p\ (286g) = 2.8\ V_{cs}/V_p\ (2g)$$

Thick shield (T/d = 2.15)

$$P_{cs}/d\ (286g) = 1.9\ P_{cs}/d\ (2g)$$

$$V_{cs}/V_p\ (286g) = 1.5\ V_{cs}/V_p\ (2g)$$

Based on these few tests the effects of projectile size on shielded target damage would appear to be far greater than has been observed for unshielded target impacts. The reasons for this size dependance are unclear. A possible explanation may be projectile strength and debris cloud particulation. For these tests both sizes of projectiles were 4340 steel, hardened to Rockwell C 35. Gault (1965) has postulated that size effects may be related to flaw size distribution in target materials. A similiar argument could be applied to a shattered projectile, leading to the conclusion that size effects are the consequence of non-scalable particulation in the debris cloud.

Projectile Density

Three values of projectile density were examined in this shielded target study. The results for normal incidence are presented in Figs. 16 and 17. Also, the results of an earlier NRL study of target penetration, shown in Fig. 18, have been included in the penetration - density graph. The graphs indicate that projectile density effects are much more significant for shielded target impact. This is partially due to the smaller size of higher density projectiles which reduces the amount of shield material encountered by the projectile. For thin shields placed 3 projectile diameters in front of the target the crater volume is proportional to $\rho_p^{0.7}$.

Projectile Shape

The initial projectile shapes were not varied by a large degree. The end-on, side-on impact by an L/d = 2 rod represented the greatest range of cross-section, a ratio of 2.5. Figure 19 summerizes the results of several tests with projectiles of different shape and density. The effect of shield thickness is represented by M_s/M_p to normalize the effects of different projectile shapes, orientations and densities as well as shield orientation. The data presented here has a range of impact velocity of 4-6 km/sec. The data points have been normalized to 5 km/sec to facilitate the comparison of projectile shape effects. The most dramatic characteristic of the resulting graph is the separation of spherical projectile data from all other projectile shapes. This is a surprising result that would appear to be attribuatable to a shape effect that has not been accounted for.

A comparison of the end-on, side-on rod impact data shows that a change by a factor of 2.5 in projectile cross-section causes a change by a factor of 2-4 in crater volume.

Projectile Velocity

The effects of initial velocity on target damage are summerized in Figs. 20 and 21. In general, the velocity dependance of the shielded targets is less than for unshielded targets. The data of Fig. 18 indicates that for large spacing the target penetration can decrease at higher velocities (the anomalous behavior for nylon is the result of impacts by shield material).

The velocity loss, Δv , due to shield impact was observed to have nearly linear dependance on shield thickness, Fig. 22:

$$\Delta v/v_o = 0.61 \; (M_s/M_p)^{1.046} \tag{8}$$

For the range of velocities evaluated in this study the ratio of residual velocity to impact velocity was invarient with impact velocity. The differences between the test results and a simple momentum model are greatest for large values of shield thickness. This difference is likely due to the growth in projectile cross-section during thick shield penetration.

Evaluation of the Projectile Debris Cloud

If the cloud of projectile debris particles produced by impact with the shield is treated as an entity then the generalized properties of the debris cloud may be used to predict crater damage to the target. For example, the debris clouds shown in Fig. 12 have new values of velocity, shape, and density in comparison to the steel spheres which impacted the shields. Although the residual projectile velocity does not change as the debris cloud travels from the shield to the target, the shape and density of the cloud changes continuously. Some examples of debris cloud change with position are presented in Figs. 23 and 24. Debris cloud shape is defined in Fig. 12 by the axial dimension d_A and the radial dimension d_R. The

TABLE 2 Comparison of Crater Volume Measurements with Calculated Values of Crater Volume Determined from Residual Projectile Characteristics

IMPACT CONDITIONS						RESIDUAL PROJECTILE						CRATER VOLUMES	
						Velocity		Shape		Density			
Projectile	Test	T/d	$\frac{M_s}{M_p}$	x/d	Veloc. km/s	$\frac{v_R}{v_o}$	$\left(\frac{v_R}{v_o}\right)^a$	$\frac{d_A}{d_R}$	$\left(\frac{d_A}{d_R}\right)^b$	$\frac{\rho}{\rho_o}$	$\left(\frac{\rho}{\rho_o}\right)^c$	$\frac{V_c}{V_{co}}$ Calc.	$\frac{V_c}{V_{co}}$ Meas.
Steel Sphere	29	0.64	0.19	0	5.88	0.893	0.796	0.70	0.93	0.74	0.93	0.69	0.80
	28	0.64	0.19	1	5.86	0.893	0.796	0.54	0.87	0.55	0.86	0.60	0.73
	5	0.64	0.19	3	5.77	0.893	0.796	0.42	0.83	0.32	0.75	0.50	0.51
	27	0.64	0.19	12	5.87	0.893	0.796	0.22	0.72	0.08	0.54	0.31	0.35
	73	0.65	0.19	3	4.33	0.893	0.796	0.44	0.83	0.58	0.85	0.56	0.65
	74	0.65	0.19	12	4.51	0.893	0.796	0.26	0.74	0.29	0.69	0.41	0.54
	76	1.18	0.35	6	4.60	0.797	0.634	0.19	0.69	0.32	0.70	0.31	0.41
	80	2.10	0.63	6	3.65	0.624	0.388	0.38	0.81	0.31	0.65	0.21	0.28
Steel Stepped Cone	21	0.64	0.28	3	5.38	0.84	0.704	0.2	0.70	0.26	0.69	0.34	0.25
	120	1.98	0.87	3.5	5.42	0.646	0.415	0.05	0.52	0.18	0.58	0.13	0.15
Uranium Sphere	19	0.60	0.11	3	5.73	0.939	0.881	0.55	0.877	0.44	0.82	0.64	0.65
	84	0.66	0.12	2.6	4.23	0.934	0.872	0.55	0.877	1.0	1.00	0.77	0.83

Calculations:

$$\frac{V_c}{V_{co}}\,(\text{calc}) = \left(\frac{v_R}{v_o}\right)^a \left(\frac{d_A}{d_R}\right)^b \left(\frac{\rho}{\rho_o}\right)^c$$

where:

$a = 2.01$

$b = 0.22$

$c = 0.46 - 0.041\, v_R$

NOTE:

(1) $v_R/v_o = 1 - 0.61\,(M_s/M_p)^{1.046}$

(2) $\rho_o = 7.8$ for all cases

data in Fig. 23 indicate that projectile debris cloud shape d_A/d_R, has a strong dependance on shield thickness but only a weak dependance on impact velocity. The debris cloud density data in Fig. 24 indicate that cloud density is dependent upon both impact velocity and shield thickness.

Figures 22, 23, and 24 may be used to determine the residual projectile velocity, shape and density. When this information is combined with the impact velocity and Equations 3, 5, and 8 the crater volume may be calculated. That is:

$$V = 2.34\ v^{2.01}\ (d_A/d_R)^{0.22}(\rho/\rho_o)^{0.46-041v}$$

where ρ_o is the density of steel. Applying this equation to the projectile debris cloud is simplified if the relationships are expressed in terms of the ratio of shielded target crater volume to unshielded target crater volume:

$$V_{cs}/V_{co} = (V_R/V_o)^{2.01}\ (d_A/d_R)^{0.22}\ (\rho/\rho_o)^{0.46-041v}$$

The results of these calculations are compared with test measurements in Table 2 and Fig. 25.

SUMMARY AND CONCLUSIONS

In this study the effects of projectile properties on target cratering were evaluated. The results for half-space targets were successfully applied to shielded target experiments to obtain accurate predictions of crater volume.

It was observed that projectile strength is sufficient to withstand transient shock loads as great as seven times the static strength of the projectile material. The role of projectile strength in the degree of particulation during shatter was not determined but it was tentatively submitted that the degree of particulation might not be scalable and that this could account for the large size effects observed for the shielded target tests.

The evaluation of projectile density showed agreement between an NRL analysis and a small set of separate test data. Crater volume was observed to vary with c = 0.46 - 0.41v power of projectile density. This result is in significant disagreement with higher, more commonly accepted values such as c = 0.5. The effect of initial projectile density on shielded target damage was much stronger than was observed for unshielded targets. This was to be expected since dense projectiles have a smaller cross-section and engage less shield material, M_s, than low density projectiles of equal mass.

Comparisons of projectile shape effects were excellant for aligned cylinder projectiles. Three combinations of projectile and target materials gave the same result: crater volume varies with the fourth root of the projectile aspect ratio. An evaluation of skewed rod data yielded considerably different results. There was a factor of two difference between crater volumes produced by discs and side-on rods. The smaller crater volumes produced by side-on rods are probably due to the larger surface areas of the elongated craters. The results with shielded targets were surprising. The spherical projectiles produced twice as much crater volume as other projectile shape for equal values of shield encountered, M_s/M_p.

A study of impact by aligned, spaced projectiles suggested that target penetration could be increased by 60% for large spacing and large numbers of particles. On the other hand, particle spacing was observed to reduce crater volume by as much as 30%.

Although the effect of impact velocity on target cratering is fairly consistent for impact velocities well above the projectile shatter velocity, the presence of a shield produces a wide range of velocity - crater relationships. For very high impact speeds the crater damage may decrease.

The influence of projectile parameters on crater volume in unshielded and shielded targets is summerized in Table 3. The presence of a shield amplifies and distorts the effects of projectile properties such as strength, density, shape and velocity. This study demonstrated that measurements of projectile debris cloud properties would allow the use of simple projectile - target relationships to calculate target cratering.

TABLE 3 Influence of Projectile Properties on Crater Volume

Projectile Property	Unshielded Target	Shielded Target
Density (ρ) (5.85 km/s)	$\rho^{0.2}$	$\rho^{0.7}$
Shape (A/A_s)	$(A/A_s)^{-0.4}$	$A/A_s)^{-0.6}$
Velocity (v)	v^2	$v^{0.4} - v^{1.4}$
Scale (d)	$d^{0.2}$	$d^{0.5}$

REFERENCES

Atkins, W.W. (1960), Hypervelocity Penetration Studies, Proceedings of the Fourth Hypervelocity Impact Symposium, Volume I.

Denardo, B.P., Summers, J.L. and Nysmith, C.R. (1967), Projectile Size Effects on Hypervelocity Impact Craters in Aluminum, NASA TN D-4067, July 1967.

Denardo, B.P. (1968) Projectile Shape Effects on Hypervelocity Impact Craters in Aluminum, NASA TN D-4953, December 1968.

Gault, D.E. and Moore, H.J. (1965), Scaling Relationship for Microscale to Megascale Impact Craters, Proceedings of the Seventh Hypervelocity Impact Symposium, Volume VI (341-351).

Gehring, J.W., Christman, D.R., and McMillan, A.R. (1963), Hypervelocity Impact Studies Concerning the Meteoroid Hazard to Aerospace Materials and Structures, GM Defense Research Laboratories TR63-243, Santa Barbara, CA November 1963.

Goodman, E.H. and Liles, C.D. (1963), Particle - Solid Impact Phenomena, Proceedings of the Sixth Symposium on Hypervelocity Impact, Volume II - Part 2 (543-576).

Halperson, S.M. and Atkins, W.W. (1962), Observations of Hypervelocity Impact, Proceedings of the Fifth Symposium on Hypervelocity Impact, Volume I - Part 2 (497-509).

Halperson, S.M. (1965), Comparisons Between Hydrodynamic Theory and Impact Experiments, Proceedings of the Seventh Hypervelocity Impact Symposium, Volume V (235-258).

Palmer, E.P., Grow, R.W., Johnson, D.K., and Turner, G.H. (1960), Cratering: Experiment and Theory, Proceedings of the Fourth Hypervelocity Impact Symposium, Volume I.

Sorenson, N.R. (1965), Systematic Investigation of Crater Formation in Metals, Proceedings of the Seventh Hypervelocity Impact Symposium, Volume VI (281-325).

Williams, A.E. (1974), Hypersonic Ice Erosion Simulation Study, NRL Memorandum Report 2812, June 1974.

Williams, A.E., Persechino, M.A. and Worman, R.W. (1983), Effects of Hypervelocity Impact on Shielded Targets, NRL Memorandum Report 5114, September 1983.

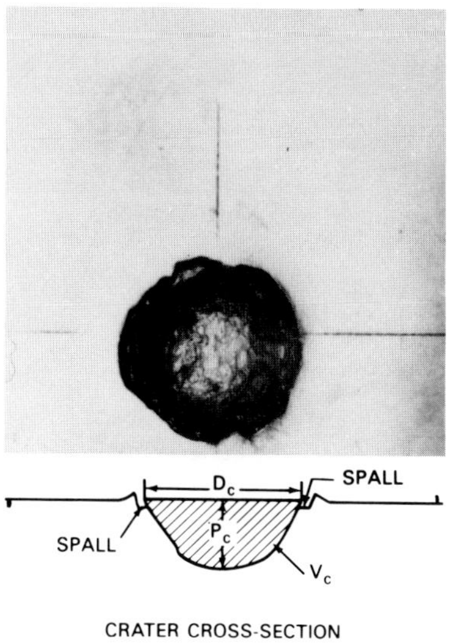

Fig. 1. Crater produced by the hypervelocity impact of a steel sphere and an aluminum target.

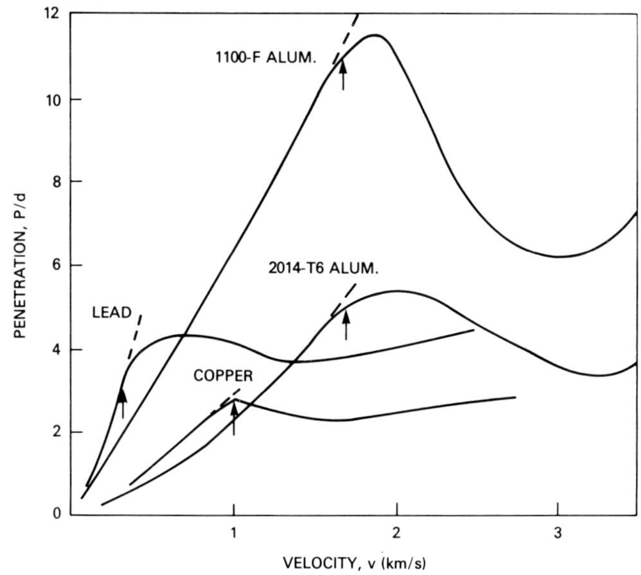

Fig. 2. Penetration of tungsten carbide spheres into half-space targets. Data from Atkins (1960).

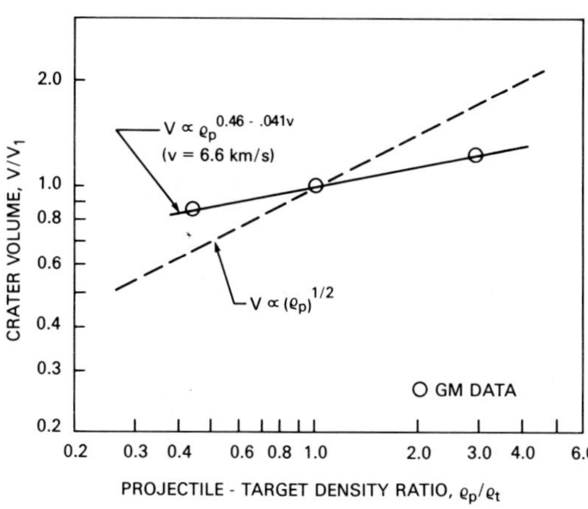

Fig. 3. The effect of projectile density on crater volume in half-space targets. Data from Gehring (1963), Goodman (1963), Halperson (1962, 1965), and NRL data. V_1 is the crater volume produced by like-like impact.

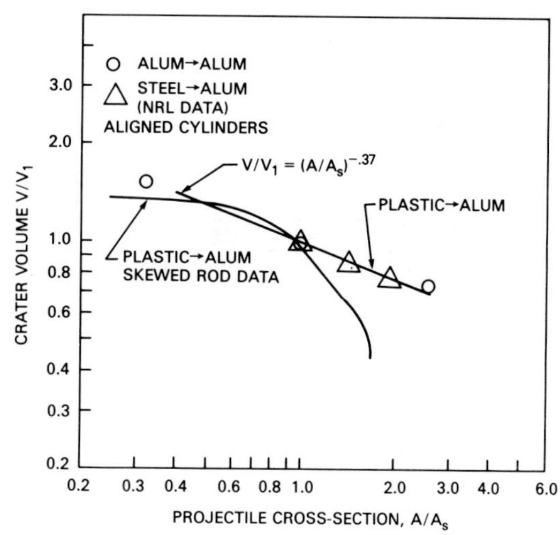

Fig. 4. The effect of projectile shape on crater volume. The graphs represent skewed rods and projectiles of different aspect ratio for which the axis of symmetry is aligned with the trajectory. A_s is the cross-section of a sphere of equal volume to the projectile.

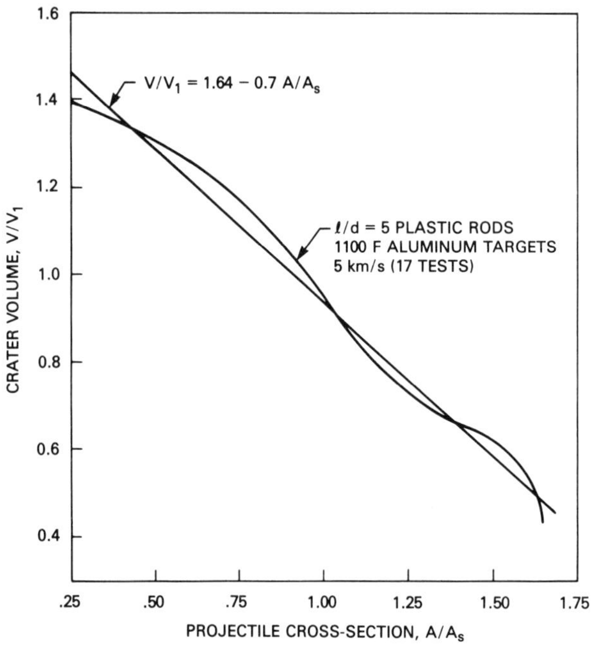

Fig. 5. The effect of rod orientation on crater volume. A_s is the cross-section of a sphere of equal vol. to the projectile which produces a crater of volume V_1 in the target. Projectile cross-section values, A, are calculated from measured values of skew angle and Equation 4.

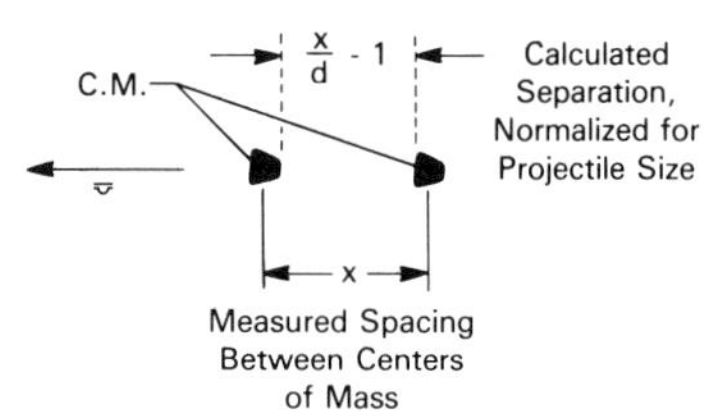

Fig. 6. Radiograph of a two projectile array.

Fig. 7. Photo of section through target crater.

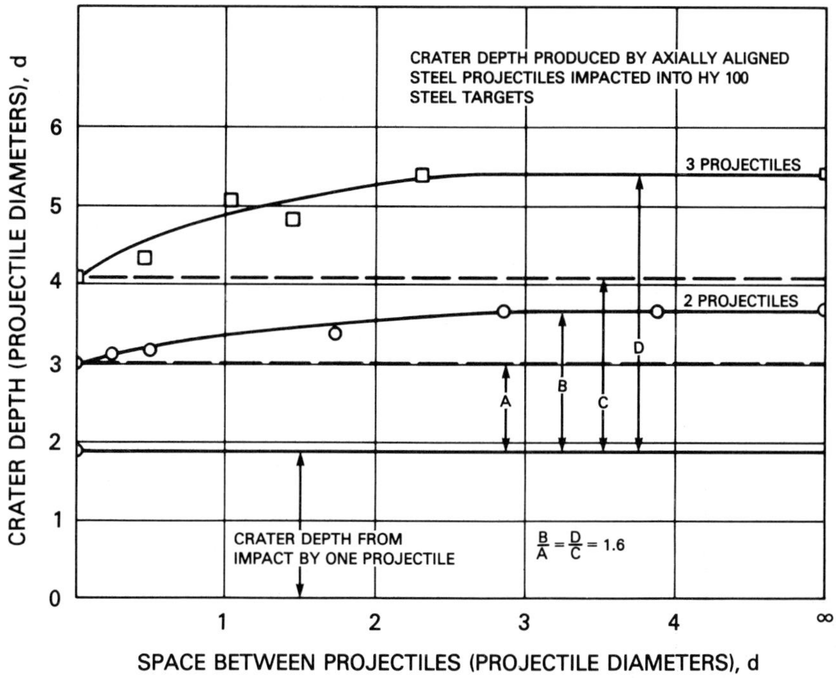

Fig. 8. Crater depth for multiple impacts as a function of axial projectile spacing.

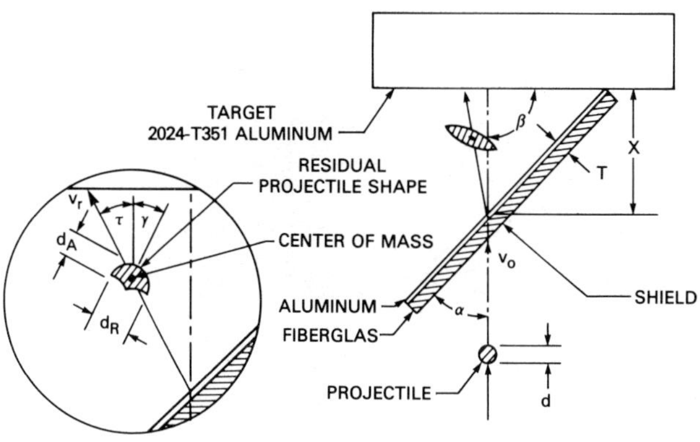

Fig. 9. Shield, target arrangement for shielded target tests.

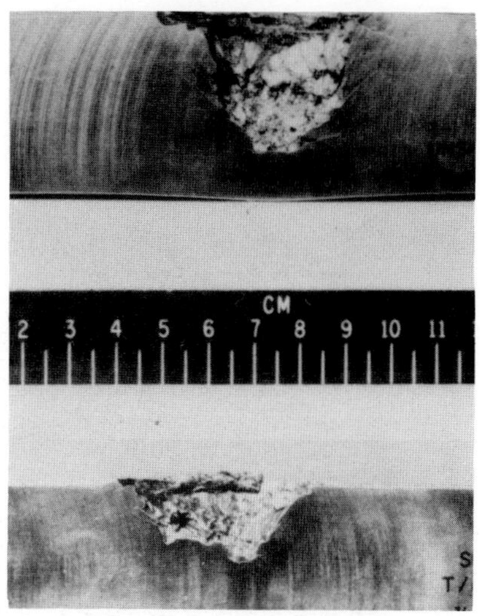

Fig. 10. Comparisons of crater cross-sections from shielded and unshielded 2024-T351 aluminum targets. Top: unshielded target impacted by a 2g steel sphere at 5.63 km/s. Bottom shielded target ($t/d = 0.64$, $\alpha = 45^o$, $x/d = 6$) impacted by a 2g steel sphere at 5.73 km/s.

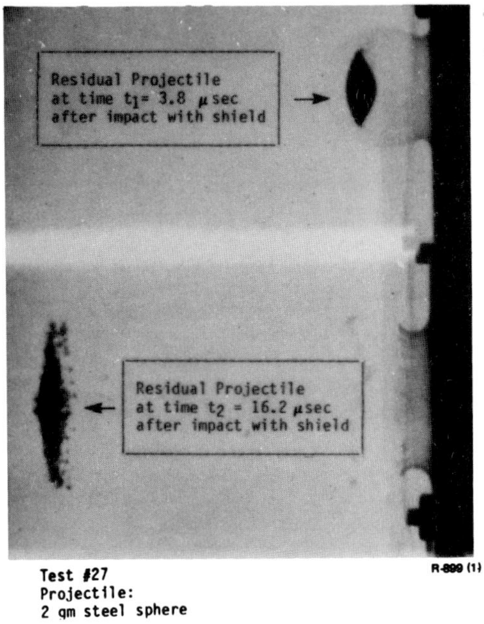

Fig. 11. Radiographs illustrating shape and size effects at spacings of 1.7 and 3.4 cm behind a thin ablative shield v = 5.8 km/sec).

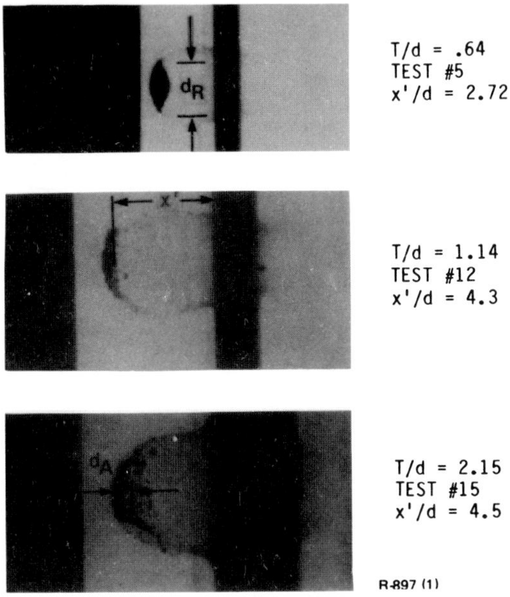

Fig. 12. Radiographs of residual projectiles illustrating the size and shape of the shattered projectile debris for three thicknesses of shield. (2g steel sphere, v = 5.8 km/s).

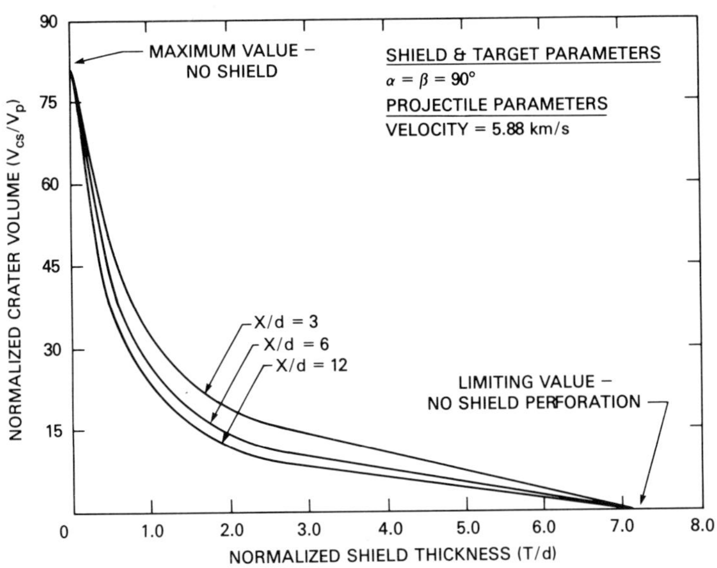

Fig. 13. Crater volume vs shield thickness.

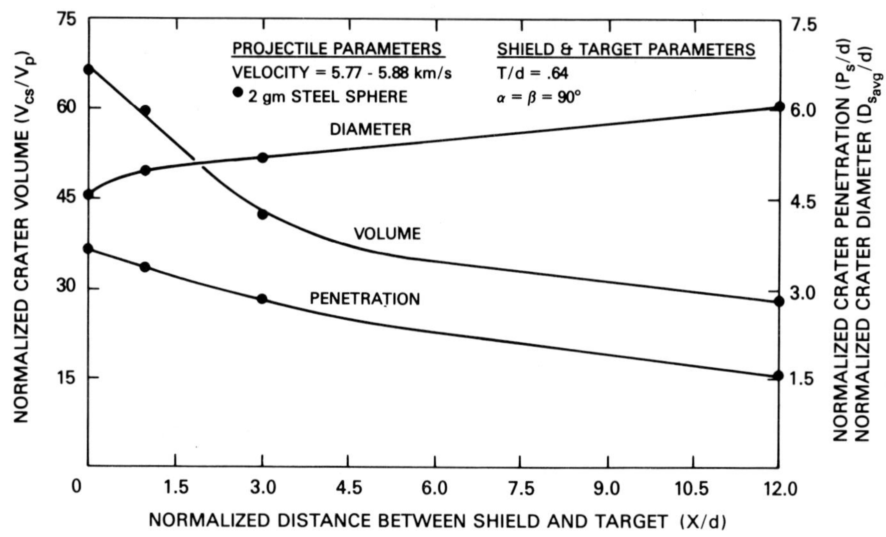

Fig. 14. Target spacing vs crater volume, penetration, and diameter.

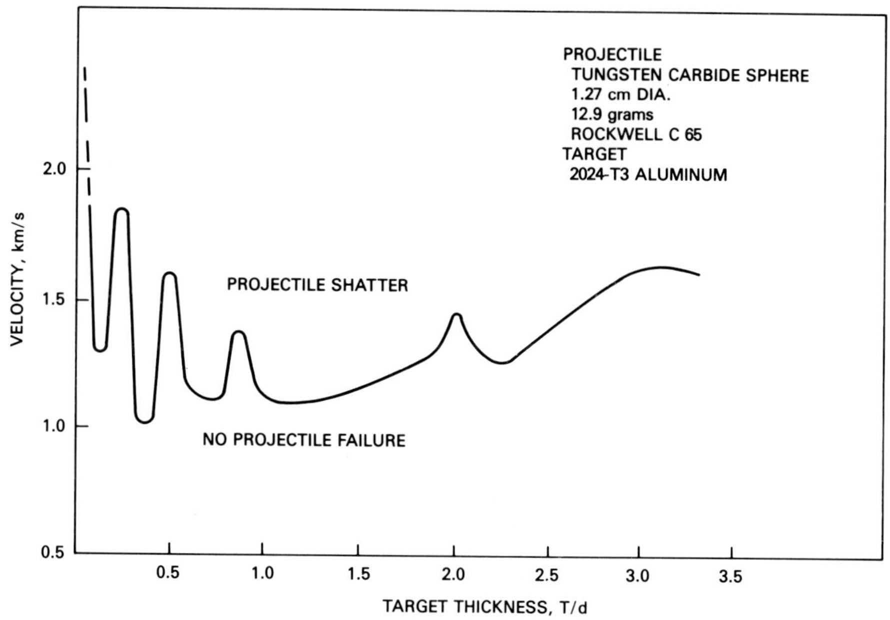

Fig. 15. Shatter velocity for tungsten carbide spheres impacting aluminum plate at normal incidence.

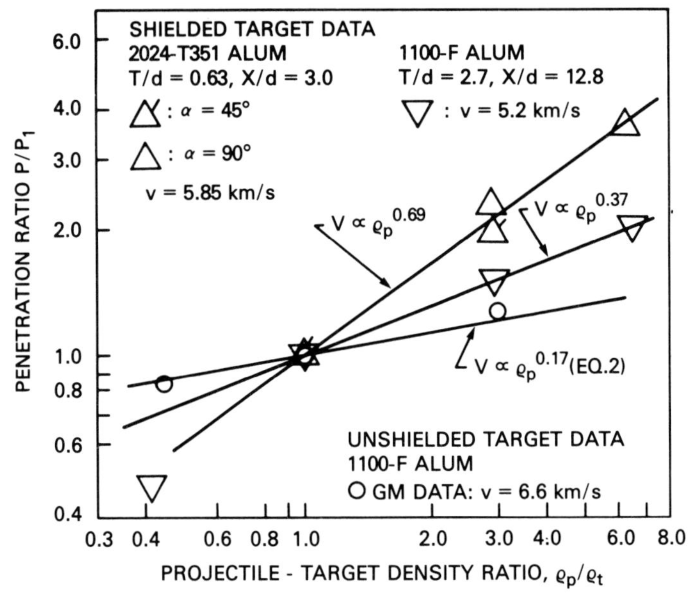

Fig. 16. Effect of projectile density on penetration for shielded aluminum targets (equivalent steel spheres).

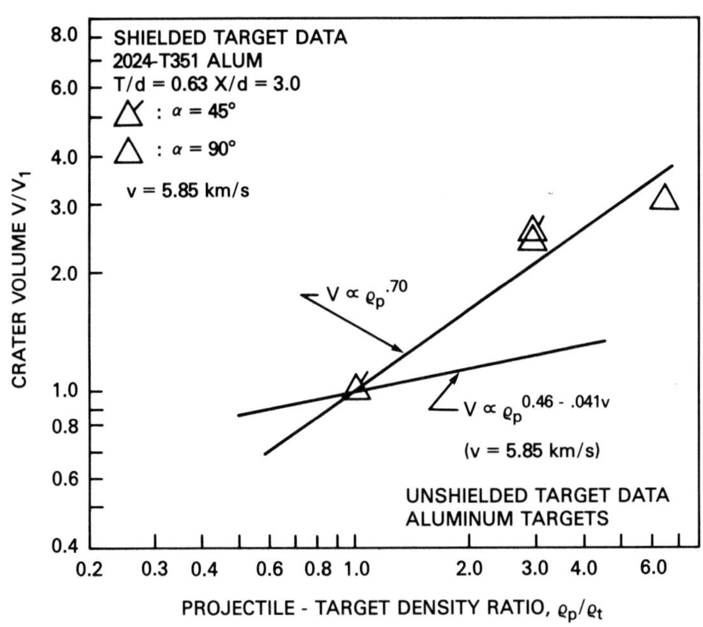

Fig. 17. Effect of projectile density on crater volume for shielded aluminum targets (equivalent steel spheres).

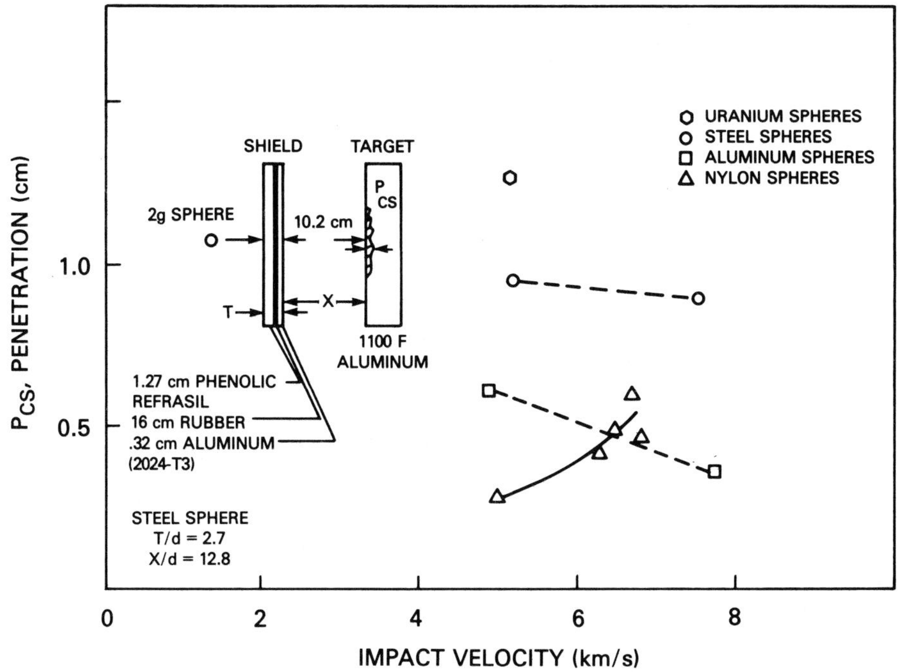

Fig. 18. The effects of projectile density and velocity on penetration into soft aluminum targets protected by a thick shield at a large stand-off distance.

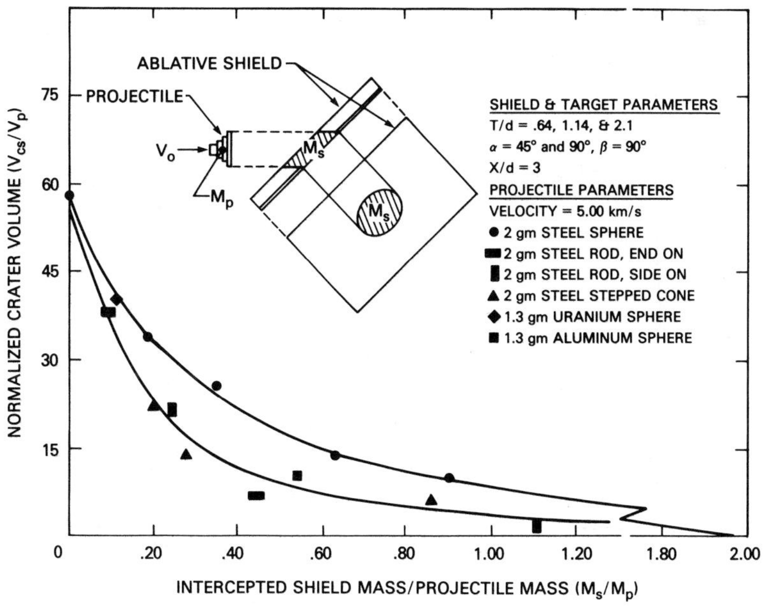

Fig. 19. The effect of initial projectile shape and orientation on crater volume (normalized to v_o = 5 km/sec, x/d = 3).

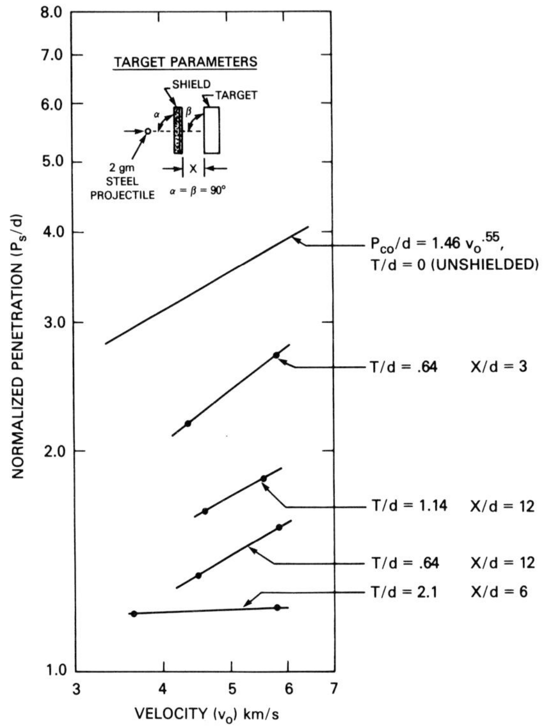

Fig. 20. Effect of velocity on normalized crater penetration for shielded and unshielded targets.

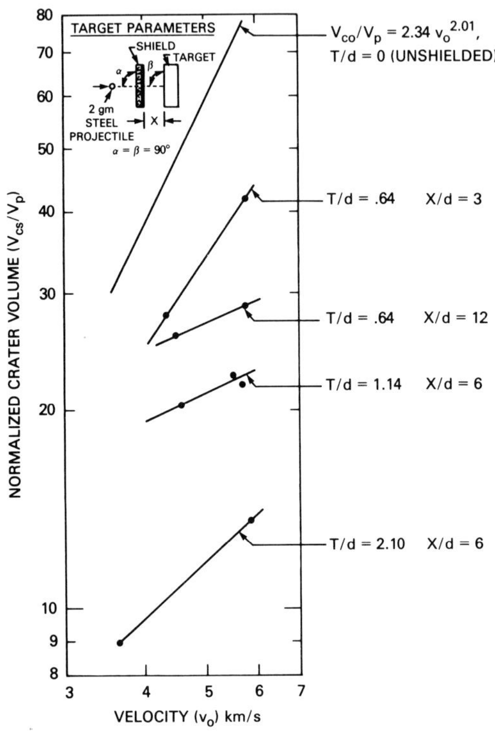

Fig. 21. Effect of velocity on normalized crater volume for shielded and unshielded targets.

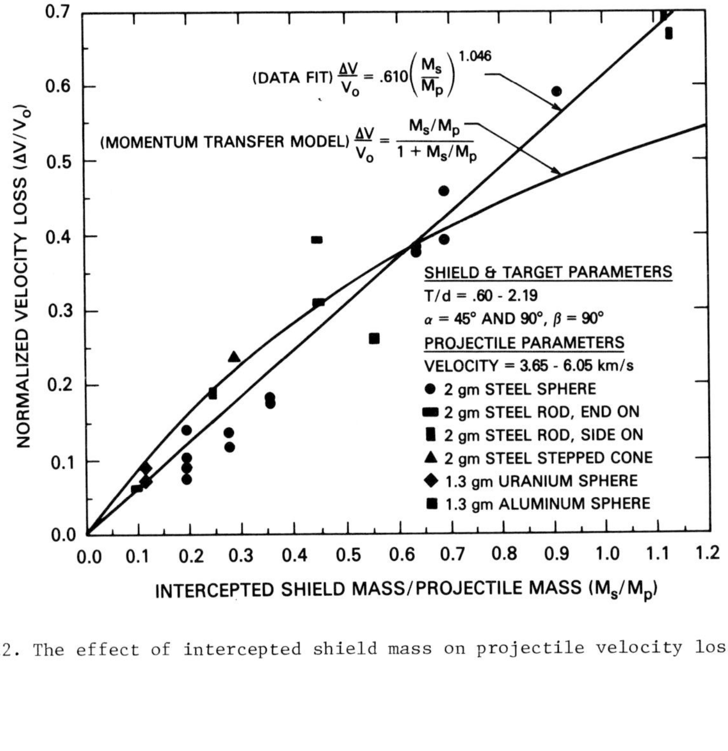

Fig. 22. The effect of intercepted shield mass on projectile velocity loss.

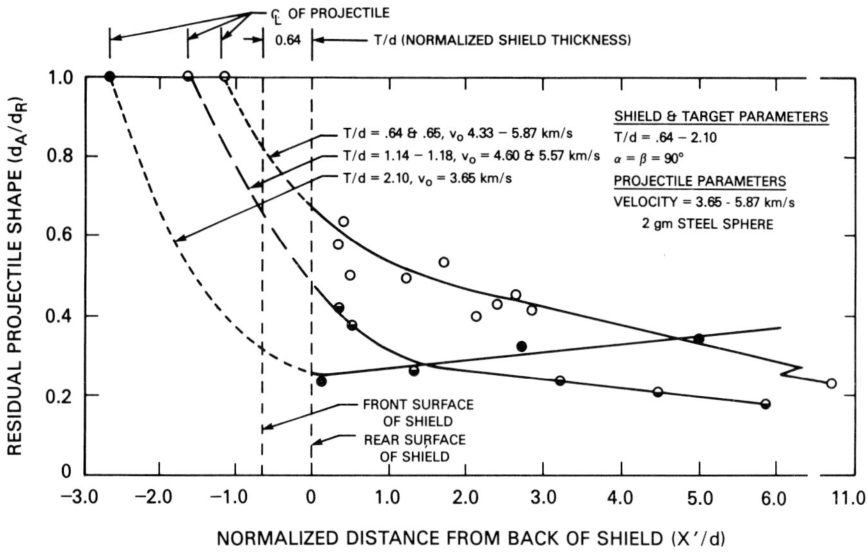

Fig. 23. Effect of distance on residual projectile shape.

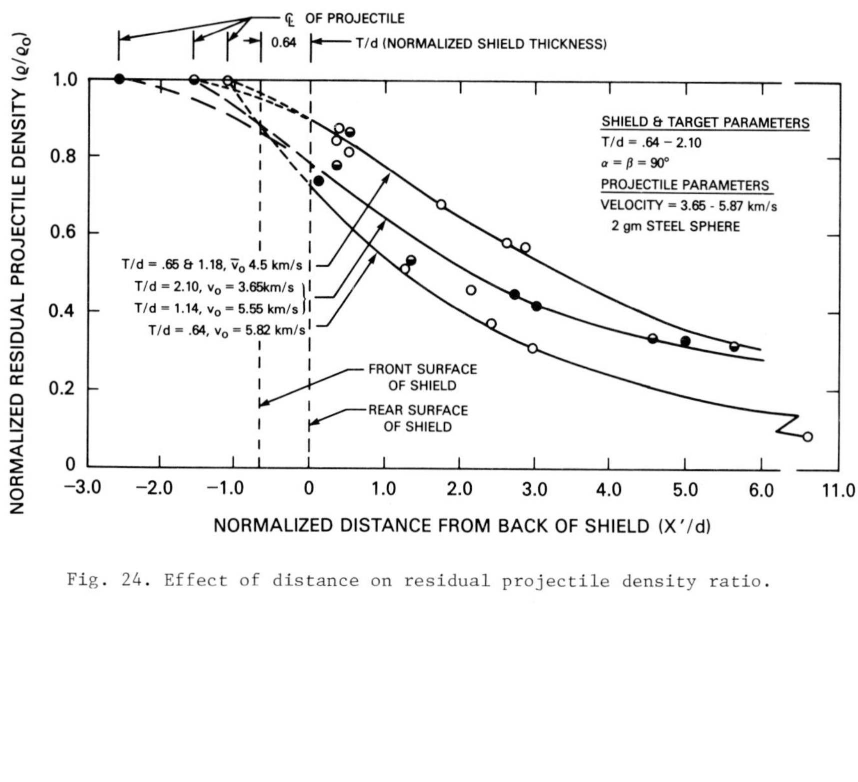

Fig. 24. Effect of distance on residual projectile density ratio.

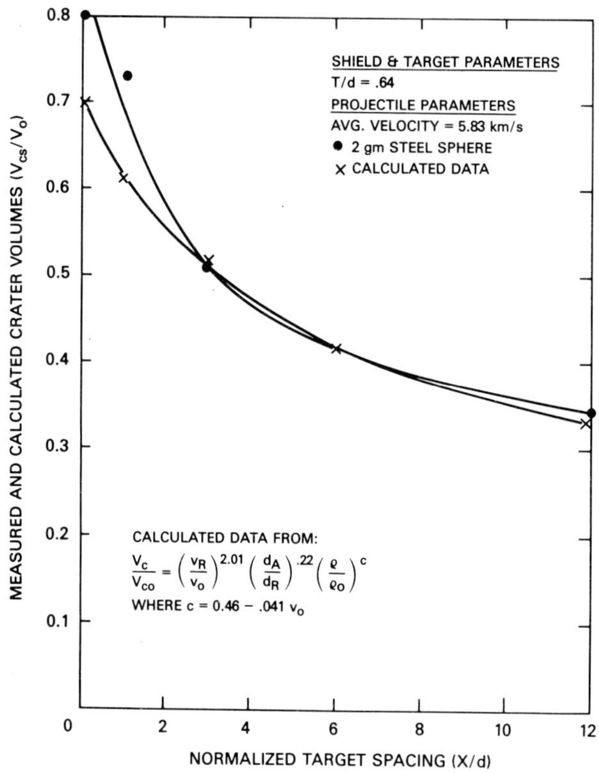

Fig. 25. Comparison of measured and calculated ratios of crater volumes.

Int. J. Impact Engng Vol. 5, pp. 729–738, 1987
Printed in Great Britain

0734-743X/87 $3.00 + 0.00
Pergamon Journals Ltd.

A STUDY OF DAMAGE IN COMPOSITE PANELS PRODUCED BY HYPERVELOCITY IMPACT

Ching H. Yew* and Rodney B. Kendrick**

*The University of Texas at Austin
Austin, TX 78712

**Boeing Company
Seattle, Washington

SUMMARY

A phenomenological observation of the damage in graphite fiber (AS4/3501-6) composite panels caused by hypervelocity impact was made in this study. The panels have a nominal thickness of 2.54, 4.83, 6.6 and 17 mm. The impacts were made with nylon and aluminum projectiles of dimension 1.75 mm (dia) x 1.88 mm (length) with velocity from 3-7.5 km/sec. It was observed that the damage in the plate was caused by multiple breakage and delamination of the laminae and matrix material. The crater or hole area in all panels are approximately 7 to 9 times the area of projectiles for the velocity range used in the testing. The area of multiple breakage and delamination of layers in the panels are much larger than the corresponding crater or hole area, and they increase with the panel thickness and impact velocity.

INTRODUCTION

Composite materials are being used for space applications on an increasing scale. In orbit, these materials are exposed to potentially damaging hypervelocity impacts from meteoroids and orbit debris. The relative impact velocities can range from 3 km/sec to 16 km/sec for the orbit debris, and from 11 km/sec to 72 km/sec for the meteoroids.

Extensive researches into the hypervelocity impact resistance of metals have been conducted (Eichelberger and Gehring, 1962; Gehring, 1970a,b; and Bjork, 1963). However, similar research on fiber reinforced composite materials has been lacking. The preliminary results of such a study on graphite fiber composite material is presented in this paper.

The damage in composite panels produced by low velocity impact has been studied by many authors (Chai and others, 1983; Elber, 1983; Williams and others, 1979; Takeda and others, 1982; Ross and others, 1983; Winkel and Adams, 1985; Davies and others, 1985). These authors found that the failure modes in the composite plate were quite different from that in a metallic panel under the same impact conditions. Instead of producing a dent or a localized plastic deformation at the impact area, the damage in the composite panel was fiber and matrix breakages and layer delaminations. Reed and Schuster (1969, 1970) studied the damage in boron-aluminum composite material caused by hypervelocity impact with a thin mylar flyer plate. They were able to demonstrate that the filament damage near the impact face was caused by the initial passage of the loading pulse, while the matrix spallation was caused by the reflection of the loading wave from the free boundary of the specimen.

This paper presents a phenomenological observation of the damage in composite panels caused by the hypervelocity impact of a simulated space debris.

SPECIMEN, PROJECTILE, AND TEST APPARATUS

Specimen. The composite panels used in this study were the graphite fiber epoxy panels made of AS4/3501-6 (at 62% fiber volume) prepregs. The mechanical properties of a unidirectional prepreg and the laminae lay-up are listed in Table 1. The graphite fibers in the panel have an approximate diameter 0.02 mm. The test plates were cut from a large panel measuring 91.44 x 91.44 cm. The dimension of the plate was so designed that the process of perforation was completed before the arrival of the waves reflected from the boundaries (side) of the plate.

Projectiles. The projectiles used in the test were made of nylon (L = 1.828 mm, R = 0.87 mm, ρ = 1140 kg/m^3) and aluminum (L = R = 0.87 mm, ρ = 2712 kg/m^3). The cylindrical projectiles were machined from a rod stock to fit properly into the launch tube of the hypervelocity gun.

Test Apparatus. The light-gas gun (Curtis, 1962) facility at the Johnson Space Center was used to conduct the hypervelocity impact testing for this study. The impact velocities in this test sequence ranged from 3 to 7.5 km/sec. The projectiles were launched without the protection of a sabot.

TABLE 1

Unidirectional Ply Properties of AS4/3501-6 at 62% Fiber Volume

(Courtesy Hercules Inc.)

Panel	Lay-Ups	Mechanical Properties	Failure Stresses
16-Lamina Plate	[±45, 0_2, ±45, 0, 90], thickness = 2.54 m	E_1 = 1.455 x 10^{11} Pa	X_T = 1.920 x 10^9 Pa
		E_2 = 9.308 x 10^9 Pa	X_C = -2.270 x 10^9 Pa
32-Lamina Plate	[±45, 0, 90, ±45, 0_2, ±45, 0_2, ±45, 0, 90]$_s$, thickness = 4.83 mm	ν_{12} = 0.291	Y_T = 4.216 x 10^7 Pa
		G_{12} = 5.764 x 10^9 Pa	Y_C = -9.308 x 10^7 Pa
48-Lamina Plate	[(±45, 0_2, ±45, 0_2, ±45, 0, 90)$_2$]$_s$ thickness = 6.6 mm	ν_{23} = 0.34	S = ±1.211 x 10^8 Pa
		G_{23} = 3.46 x 10^9 Pa	
96-Lamina Plate	[(±45, 0_2, ±45, 0_2, ±45, 0, 90)$_4$]$_s$ thickness = 17mm	δ(density) = 1577.5 kg/m^3	

PHENOMENOLOGICAL OBSERVATION OF THE DAMAGED SPECIMEN

The damaged plate was first given a visual inspection for surface damage. The front (impacted) and back surface of the plate were photographed. The crater produced by the impact was also photographed using a Zeiss microscope. The image of the crater was magnified eight (8) times for area measurement. The damaged plate was then examined ultrasonically (C-scanned) to determine the extent of the internal damage.

After completing the above gross damage observation, the damaged specimen was sectioned through its thickness across the center of the perforated zone. Cutting was made with a high-speed diamond cutter of 0.508 mm (0.02 in.) in width. The surface of the sectioned plate was then cleaned with a soft cloth to remove any fibers or particles which might have adhered to the surface. The naked section was then treated with a dye-penetrant (Magniflux Zyglo penetrant, Type ZL-22A). Sufficient time (approximately 30 min.) was allowed for the penetrant to seep into the damaged area. The excess dye on the surface was removed with a polishing wheel using 4 micro aluminum powders. After the above described surface preparations, the specimen was placed under the microscope, illuminated with ultra-violet light for visual observation, and photographed.

RESULTS AND OBSERVATIONS

Ejecta. The ejecta were manually collected and examined after testing. The debris appeared totally from the damaged composite plate. They, in general, can be grouped into the following two categories:

1) powdery particles of approximate diameter - 0.02 mm.

2) flakes of length from 1 mm to 50 mm.

The mass spectrum analysis of the powdery particles showed a small amount (cannot be quantified) of aluminum. This finding seems to suggest that the aluminum projectile might have been liquidized and adhered to the shattered composite material particles upon impact in our testing velocity range. The longer flakes were the fall-out from the damaged plate. We noticed that the particles and flakes would continue to fall during the handling of the specimen. The crater does not have a smooth surface as that appeared on the surface of a metallic target. Due to its rough surface and the loose particles and flakes at the bottom of the crater, we are unable to measure the crater depth with a reasonable accuracy. For this reason, only the crater area was measured in our damage analysis. In the testing velocity range, the projectile did not perforate the 96-lamina panels. The crater depth was measured. We found that the crater depth was approximately equal to the radius of the crater.

Surface Damage. Figures 1 and 2 show the photographs of the front (impacted) and back surface of the damaged 32-lamina (thickness = 4.83 mm) plate, respectively. The appearance of the damage on the 16-lamina (2.54 mm) and on the 48-lamina (6.6 mm) plate are identical to those on this plate. The projectile did not perforate the thick 96-lamina (17 mm) plate. The impact produced a crater on the impact surface of the plate. There was no visual damage on the back surface of the plate.

Figure 1

Front Surface of a Perforated
32-Lamina Plate (Nylon Projectile:
V = 6.81 km/sec)

Figure 2

Back Surface of a Perforated
32-Lamina Plate (Nylon Projectile:
V = 6.81 km/sec)

In view of Figures 1 and 2, the damage to the surface area is a crater and peelings of the superficial laminae. Severe damage appeared to be confined to the vicinity of the crater. Multiple delaminations and breakage of the lamina are clearly visible in these photographs. The area (21 sq. mm) of the crater is nearly nine times larger than the impact area (2.4 sq. mm) of the projectile. By comparing Figures 1 and 2, one notices that the area of damage (approximately 85 sq. mm) on the back surface of the plate is substantially larger than that inflicted on the frontal surface. This difference in the damaged areas strongly suggests that spallation may be the major cause of damage to the back side of the plate.

The direction of peelings is consistent with the direction of the fibers in the surface laminate. The peelings were the near-surface damage which occurred only to the first layer of the laminates. Other near-surface damage was the multiple breakages and delaminations of the laminates. The extent of the surface peeling was not measured due to the fact that their length might have been affected by the post-impact handling. A series of tests was also performed on the composite panel with cloth cover on both surfaces. We found that, under the same impact condition, the area of the crater on the cloth-covered panels was approximately the same as those on the regular panels, but the surface peelings were totally suppressed by the cloth cover of the panel.

It is interesting to observe that the shape of the crater in the composite plate is not circular symmetric as that produced in the metallic specimen by high velocity impact. The directional fracture plane of the laminates is clearly revealed in Figure 1 and 2. The area of the crater was measured from the enlarged photograph with a planimeter. The measured areas are plotted versus the kinetic energy of the projectiles in Figures 3, 4, 5 and 6 for the 16-, 32-, 48- and 96-lamina plates, respectively.

Internal Damage. The internal damage of the plate was assessed by the following two methods:

A) the damaged area was determined by scanning the specimen ultrasonically (C-scan); and

B) the modes of the internal damage were studied by observing the dye-treated specimen section under a microscope.

Figure 7 shows a typical C-scan record of a regular panel and a panel with cloth covers impacted with the nylon projectiles of approximately the same velocities. The scan of the regular panel shows a large area of damage. However, much of the area is caused by the peeling of the surface layers. The area enclosed by the elliptical boundary (with the major axis in the peeling direction) around the impact point is the region where the severe internal damage had occurred. This conclusion was later confirmed by examining the sectioned specimen. The damaged zone in the cloth-covered panel has a circular shape, and at a same impact condition, it has approximately the same area as that in the regular panel. The area was measured with a planimeter. The results are shown in Figures 8, 9, 10 and 11, respectively, for the four different thickness panels used in this study. It is interesting to observe from these figures that (1) the C-scanned damage area in a thin panel is smaller than that in a thick panel for a same projectile energy, and (2) the slope of the damaged area versus projectile energy increases with the panel thickness. The C-scan measures an accumulated damage throughout the panel thickness. The above observations seem to imply that a less amount of projectile energy is transferred to the panel when the panel thickness is thin. One thus expects a small zone of internal damage and a high ejecta velocity from a thin panel. In a thick plate, however, most of the projectile energy is imparted to the plate and it, consequently, causes a larger region of internal damage.

Details of the internal damage in the plate were observed by placing the dye-treated specimen section under a microscope and illuminating the section with ultra-violet light. The ultra-violet light causes the dyes to phosphoresce, and it thus reveals the details of the internal damage in the specimen. Figures 12, 13, 14, and 15 show the photographs of the section of a 16, 32, 48 and 96-lamina plate, respectively. The light region surrounding the crater and the tunnel is the region penetrated by the dyes. They thus represent a cross-sectional profile of the damaged region in the plate. The damaged area detected by the C-scan is, therefore, an overall top view of this damaged region. Fig. 16 shows an enlarged view of damaged zone in a 48-lamina plate. Delamination and multiple breakage of laminates are clearly shown in this figure. By examining these photographs, the following observations are made:

(A) Figure 12 shows a sectional view of a thin (16-lamina) plate. The plate was clearly perforated. There is no crater on the impact side of the plate, and the effect of spallation on the back surface of the plate is also small. Furthermore, the zone of lamina separation and breakage around the hole (revealed by the light region) is small and localized in comparison with those found in the thicker plates. This phenomenon clearly suggests that the plate material was locally shattered by the impact, and the damage propagated radially outward from the axis of impact. The plate was not thick enough to reveal the damage caused by the spallation.

(B) The damage caused by the crater forming and spallation can be seen in the sectional view of a thick plate. Figure 14 shows the damage profile in a 48-lamina plate. There is a clearly defined crater at the impact side of the plate. The small amount of surface peelings near the crater are believed to be caused by the "front" ejection of the plate fragments. The existence of a small tunnel at the apex of the crater strongly suggests that the "shear plugging action" (Awerbuch and Bodner, 1974) might have occurred during the process of perforation. The "plugged-out" materials were "rear" ejected with the spalled fragments from the back of the plate. Spallation caused most of the damage to the back region of the plate. Large and multiple delamination of the layers and multiple breakages of the laminae near the back surface reveal clearly the damage mode of spallation. Since the spallation is caused by the reflected waves at the back surface, the damaged region is expected to be larger than that on the impact side of the plate. The light area surrounding the perforation shows clearly the extent of the internal damage in the plate.

(C) As shown in Fig. 15, the projectile did not perforate the 96-lamina panel. The damage inflicted to the panel are a crater at the impact face and the surrounding large zone of broken and delaminated layers. Although there are no visible surface damage on the back surface of the panel, Figure 15 clearly revealed a large delaminated zone near the back surface - an indication of damage caused by the reflected tensile waves from the back surface.

DISCUSSION AND ANALYSIS OF RESULTS

Since the composite material used in this study is a layered brittle material, the damage inflicted by hypervelocity impact is multiple delamination and breakage of the laminae and matrix material. The process of perforation and propagation of damage in the plate can be understood by following the shock wave motions produced by the impact. (1) For a thin plate, according to the observation made by Reed and Schuster (1969, 1970), one can assume that the hole in the plate is generated by the impact and the initial passage of the shock waves. Since the time for the waves reverberating in the thickness of the plate is short, the shock waves quickly convert themselves into the bending and compressive waves radiating outward from the impact point; these waves caused a uniformly distributed damaged region throughout the plate thickness. (2) For a relatively thicker plate (i.e., 32- and 48-lamina plate), a crater is formed upon impact. There is sufficient time, however, for the impact-generated shock waves to reflect from the free back surface. These reflected tensile waves cause a spalled zone in the back region of the plate propagating towards the front surface. When the thickness of the material, which separates the merging crater and spalled region, becomes thin, the material is "shear plugged" and "rear" ejected from the back surface of the plate. The perforation process is thus completed. (3) For a very thick plate, a crater is formed on the impact surface. However, due to a longer travelling distance, the strength of the impact-generated compressive shock is reduced when reaching the free surface of the plate. If the magnitude of the wave is lower than the flow resistance strength of the target plate, there will be no spallation in the back region of the plate; the damage in the plate will thus be limited to delaminations.

REFERENCES

Awerbuch, J. and S. R. Bodner (1974). Analysis of the mechanics of perforation of projectiles in metallic plates. Int. J. of Solids and Str., 10, 671-683.

Baumeister, T. (1978). Mark's Standard Handbook for Mech., 8th ed. McGraw-Hill, New York.

Bjork, R. J. (1963). Review of physical processes in hypervelocity impact and penetration. Proc. Sixth Sym. of Hypervelocity Impact, 2-58.

Chai, H., W. G. Knauss and C. D. Babcock (1983). Observation of damage growth in compressively loaded lamiates. Exp. Mechanics, 23, 3, 329-337.

Curtis, J. S. (1962). An accelerated reservoir light-gas gun. NASA, TN D-144.

Davies, C. K. L., S. Turner and K. H. Williamson (1985). Flexture plate impact testing of carbon fiber-reinforced polymer composites. J. Comp. Mat'l., 16, 4, 279-285.

Eichelberger, R. J. and J. W. Gehring (1962). Effect of meteoroid impacts on space vehicles. Am. Rocket Soc. J., Oct., 1583-1591.

Gehring, J. W. (1970a,b). Theory of impact on the target and shields and correlation with experiment. Chap. 4, 105-156; Engineering considerations in hypervelocity impact. Chap. 9, 463-514, in R. Kinslow (Ed.), High-Velocity Impact Phenomena, Academic Press, New York.

Reed, R. P. and D. M. Schuster (1969). Filament fracture and post-impact strength of boron-aluminum composite. J. Comp. Mat'l., 4, 514-525.

Reed, R. P. and D. M. Schuster (1970). Fracture behavior of shock-loaded boron-aluminum composite materials. J. Comp. Mat'l., 3, 562-572.

Ross, C. A., L. E. Malvern, R. L. Sierakowski and N. Takeda (1983). Finite-element analysis of interlaminar shear stress due to local impact. Am. Soc. Testing and Mat'l., Publication 864, 335-365.

Takeda, N., R. L. Sierakowski, C. A. Ross and L. E. Malvern (1982). Delamination-crack propagation in ballistically impacted glass/epoxy composite laminates. Exp. Mechanics, 22, 1, 19-25.

Williams, J. G., M. S. Anderson, M. D. Rhodes, J. H. Starnes, Jr. and W. L. Stroud (1979) Recent developments in the design, testing, and impact damage tolerance of stiffened composite panels, NASA TM-80077.

Winkel, J. D. and D. F. Adams (1985). Instrumented drop weight impact testing of cross-ply and fabric composites. Composite, 16, 4, 279-285.

ACKNOWLEDGEMENTS

The study was conducted pursuant to Contract NGT 44-021-801 between The University of Texas at Austin and NASA, Johnson Space Center (JSC) at Houston. The study was carried out under the supervision of Ms. Jeanne Lee Crews. Mr. Kendrick was supported by a NASA trainee fellowship during the course of the study.

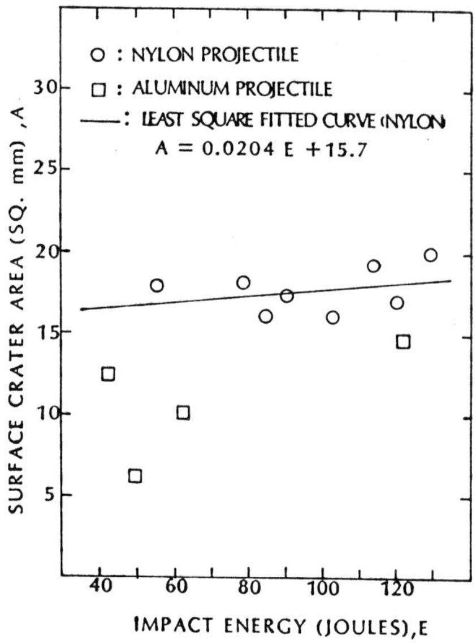

Figure 3

Crater Area on the Surface of a 16-Lamina Plate versus the Kinetic Energy of the Projectile.

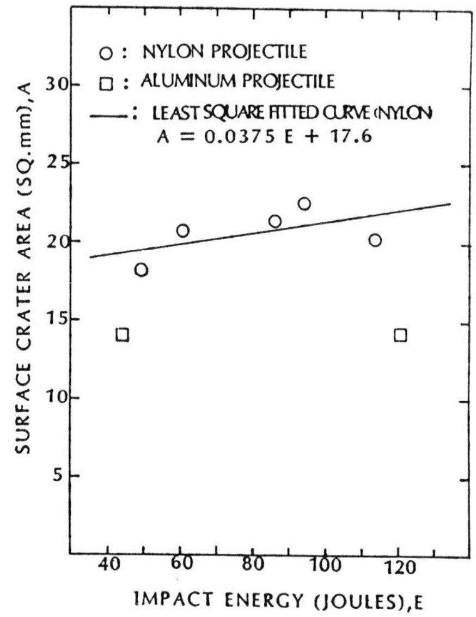

Figure 4

Crater Area on the Surface of a 32-Lamina Plate versus the Kinetic Energy of the Projectile.

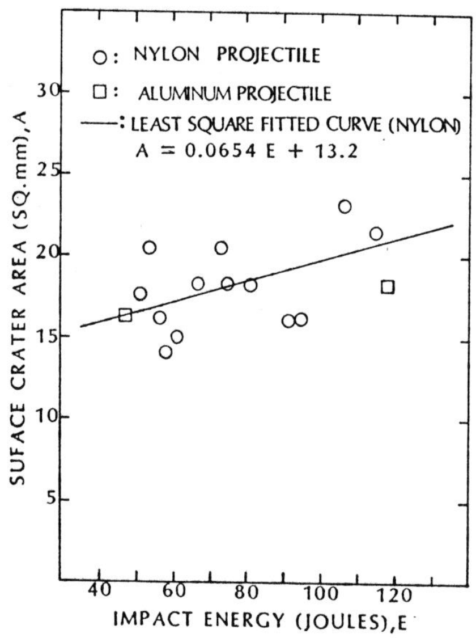

Figure 5

Crater Area on the Surface of a 48-Lamina Plate versus the Kinetic Energy of the Projectile.

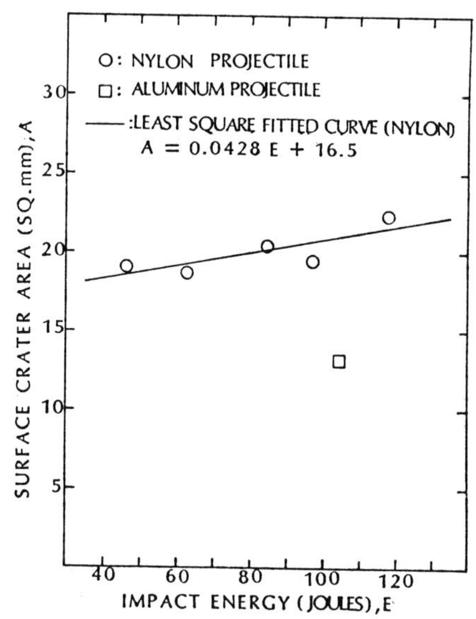

Figure 6

Crater Area on the Surface of a 96-Lamina Plate versus the Kinetic Energy of the Projectile.

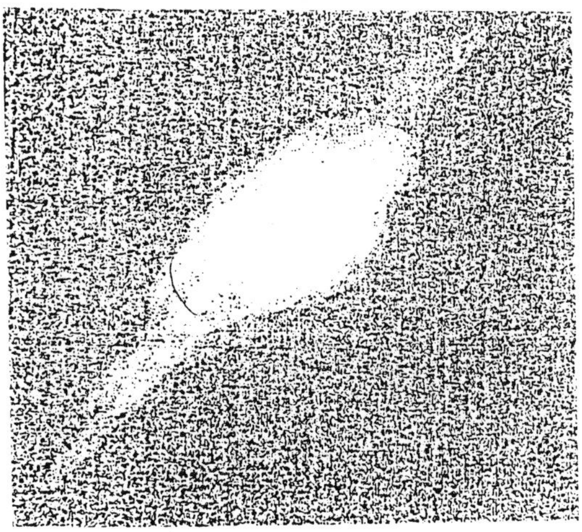

96 Lamina Regular Plate
Nylon Projectile
V_o = 6.95 km/sec

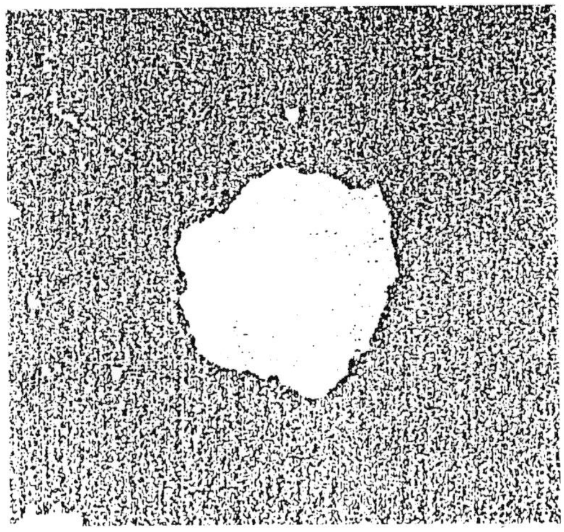

96 Lamina Cloth Covered Plate
Nylon Projectile
V_o = 6.5 km/sec

Figure 7. Typical C-scan records

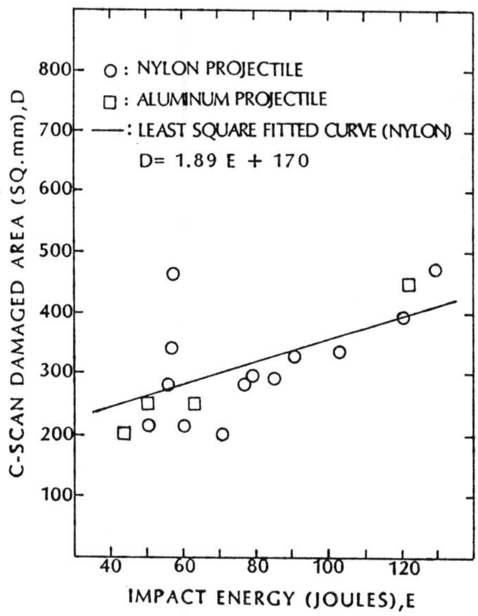

Figure 8

Damaged Area in a 16-Lamina Plate Versus the Kinetic Energy of the Projectile.

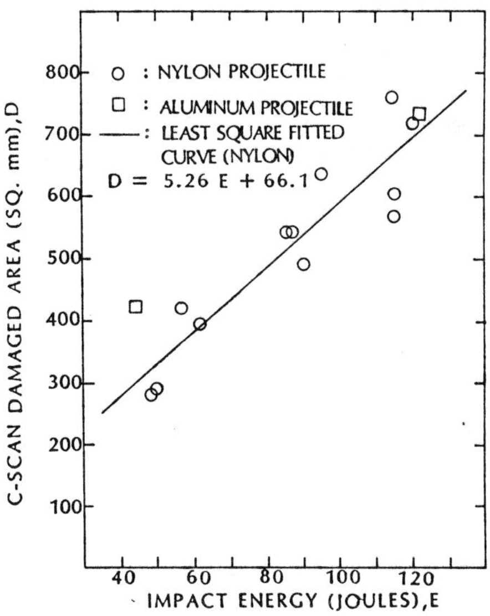

Figure 9

Damaged Area in a 32-Lamina Plate Versus the Kinetic Energy of the Projectile.

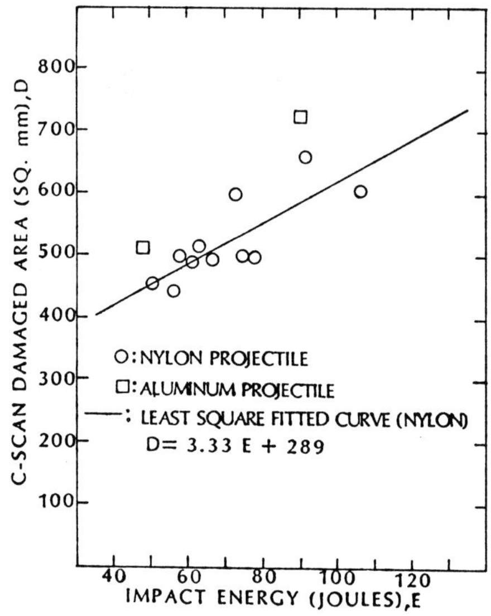

Figure 10

Damaged Area in a 48-Lamina Plate versus the Kinetic Energy of the Projectile.

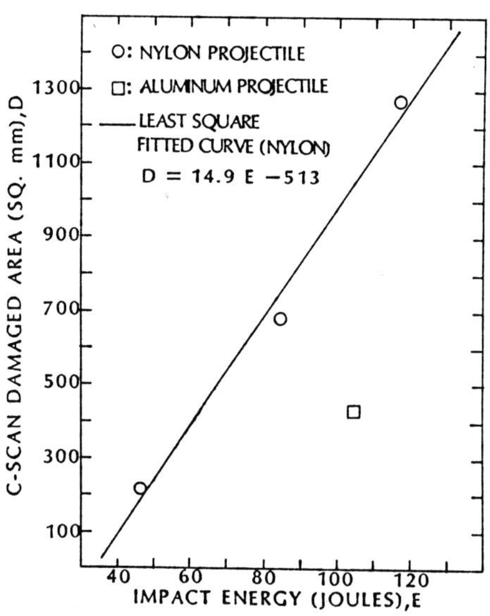

Figure 11

Damaged Area in a 96-Lamina Plate versus the Kinetic Energy of the Projectile.

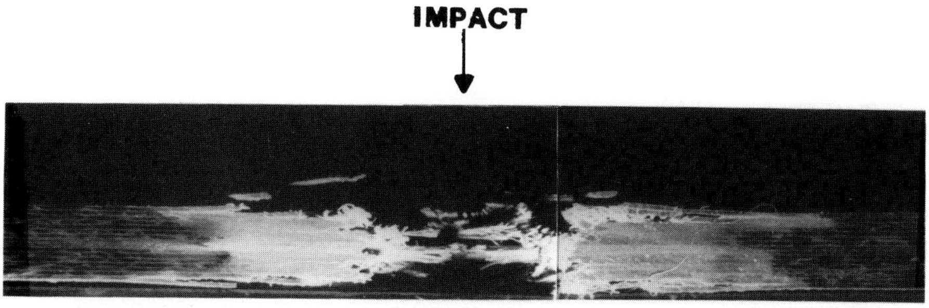

Figure 12. A Sectional View of a Perforated 16-Lamina Plate. (Nylon Projectile: V = 6.7 km/sec) 6.75X Magnification

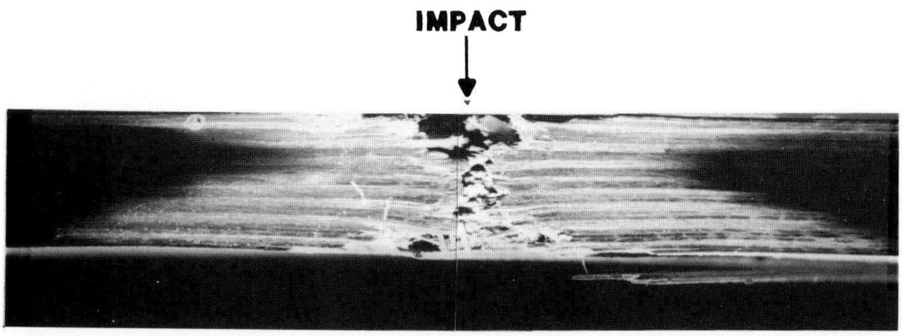

Figure 13. A Sectional View of a Perforated 32-Lamina Plate. (Nylon Projectile: V = 4.4 km/sec) 6X Magnification

Figure 14. A Sectional View of a Perforated 48-Lamina Plate. (Nylon Projectile: V = 4.67 km/sec) 6.5X Magnification

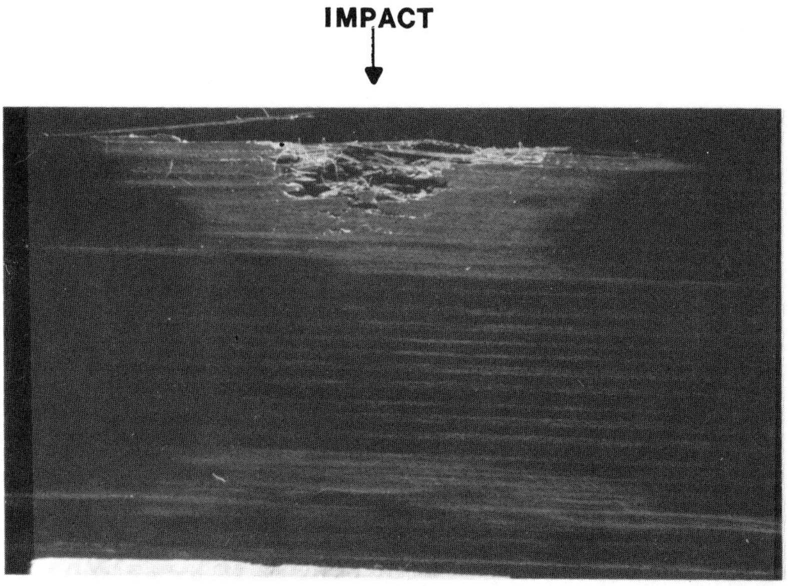

Figure 15. A Sectional View of a Damaged 96-Lamina Plate. (Nylon Projectile: V = 6.78 km/sec) 4.8X Magnification

Figure 16. A Sectional View of the Damaged Zone in a 48- Lamina Plate. (Nylon Projectile: V = 4.67) 30X Magnification

ATTENDEES -- HVIS SYMPOSIUM

Abbatini, Major Nazzareno	Ministry of Defence-Italian Army
Abrams, Martin	Sandia National Laboratories
Adams, Marc A.	Jet Propulsion Laboratory
Adelman, Stephen	System Planning Corporaton
Ahrens, Thomas J.	California Institute of Technology
Aigner, Sebastian	Technische Universitaet Muenchen
Alme, Marvin	R & D Associates
Anderson, Jr., Charles E.	Southwest Research Institute
Anderson, Gordon D.	Lawrence Livermore National Laboratory
Antoniak, Zenen I.	Battelle Pacific Northwest Laboratory
Apo, Douglas K.	Defense Nuclear Agency
Arnaudeau, Francis	MECALOG
Asay, James R.	Sandia National Laboratories
Backofen, Jr., Joseph E.	U. S. Government
Baker, Alan C.	McDonnell Douglas Corporation
Baker, Carolyn S.	Rockwell International Missile Systems Division
Baker, Jay R.	Naval Research Laboratory
Baldini, Luigi F.	U. S. Army ARDEC
Banks, Norman E.	Science Applications International Corporation
Battaglia, LTCDR. Silvano	Italian Navy
Behrmann, L. A.	Schlumberger Well Services
Bentley, Bedford T.	U. S. Army Ballistic Research Laboratory
Berg, Julie A.	U. S. Government
Bergmann, Ernest P.	Southwest Research Institute
Bergmann, O. R.	E. I. DuPont de Nemours & Co.
Berry, Ray A.	Idaho National Engineering Laboratory
Birnbaum, Naury K.	Century Dynamics, Inc.
Bjork, Robert L.	Pacifica Technology
Bjorkman, Michael D.	Boeing Aerospace Company
Blanks, James R.	Calspan Field Services, Inc.
Blaylock, Jr., Neil W.	Southwest Research Institute
Bless, Stephan J.	University of Dayton Research Institute
Boecker, Juergen	Rheinmetall GmbH
Boslough, Mark B.	Sandia National Laboratories
Bowden, Andrew V. B.	Royal Armament Research & Development Establishment
Bowen, Barry E.	E. I. DuPont de Nemours & Co.
Boyd, Price W.	U. S. Army Strategic Defense Command
Brassard, Daniel J.	NSRG Center for Advance Concepts
Brockway, Stanley J.	U. S. Army Strategic Defense Command
Bronstein, Martin P.	TRW, Inc.
Brooks, James E.	Schlumberger Well Services
Brown, Joe H.	Battelle Columbus Division
Brown, Joseph M.	Potomac Research, Inc.
Bryan, Jon B.	Lawrence Livermore National Laboratory
Burke, Wayne A.	General Dynamics Land Systems Division
Bussell, Margaret E.	Kaman Sciences Corporation
Bustamante, Albino C.	Sandia National Laboratories
Caldwell, Steven G.	Teledyne Firth Sterling
Callens, Jr., E. Eugene	Louisiana Tech University
Calvin, Hugh A.	Westinghouse Electric Corporation
Campbell, C. Warren	BDM Corporation
Carlson, Eric	Geo-Centers, Inc.
Cassatt, Gary G.	Boeing Military Airplane Company
Cerroni, Priscilla	Instituto di Astrofisica Spaziale
Chang, James	Kaman Sciences Corporation
Charters, Alexander C.	General Research Corporation
Chau, Henry H.	Lawrence Livermore National Laboratory
Chen, Yuan-Liang	OBU
Chhabildas, Lalit C.	Sandia National Laboratories
Chou, Pei Chi	Dyna East Corporation
Christiansen, Eric L.	Eagle Engineering, Inc.
Cintala, Mark J.	NASA-Johnson Space Center
Cirillo, Thomas G.	System Planning Corporation
Claude, Bozier J.	Commissariat a l Energie Atomique
Cline, Carl F.	Lawrence Livermore National Laboratory
Coffey, Charles S.	Naval Surface Weapons Center
Colle, Alain A. C.	Defence Research Establishment Valcartier
Combi, Captain Rinaldo	Ministry of Defence-Italian Army

Connolly, John I.	Science Applications International Corporation
Coradini, Marcello	Instituto di Astrofisica Spaziale
Cour-Palais, Burton G.	NASA-Johnson Space Center
Cowan, Jr., Maynard (Bill)	Sandia National Laboratories
Crews, Jeanne L.	NASA-Johnson Space Center
Cuadros, Jaime H.	General Dynamics Valley Systems Division
Dandekar, Dattatraya P.	U. S. Army Materials Technology Laboratory
Davies, Frank W.	Physics International Co.
Davis, Cecil G.	Los Alamos National Laboratory
Davis, Jo O.	Sandia National Laboratories
Dehn, James T.	U. S. Army Ballistic Research Laboratory
DeIanni, Richard	Cordin
Der Kacy, James A.	General Dynamics Land Systems Division
DeWitt, Jack R.	Calspan Corporation
Dickinson, David L.	Naval Surface Weapons Center
Doerr, Stephen E.	Sandia National Laboratories
Doyle, James C.	Los Alamos National Laboratory
Draper, Keith J.	Lockheed Engineering & Management Services Co.
Dreitlein, Kenneth C.	United Technologies Research Center
Dressler, Burkhard B. O.	Ontario Geological Survey
Drumheller, Douglas S.	Sandia National Laboratories
Dunbar, Lyle E.	Sparta, Inc.
Dunn, William P.	U. S. Army ARDEC
Dunphy, James R.	United Technologies Research Center
Economou, Thanasis	University of Chicago
Eldred, Philip N.	AAI Corporation
Elfer, Norman C.	Martin Marietta Michoud Division
Fair, Harry D.	Defense Advanced Research Projects Agency
Fanget, Alain	Centre d'Etudes de GRAMAT
Farenthald, Eric P.	University of Texas Center for Electromechanics
Finnegan, Lorraine C.	BDM Corporation
Finnegan, Stephen A.	Naval Weapons Center
Fishburn, Barry D.	U. S. Army ARDEC
Flatau, Abraham	Aerojet Ordnance Company
Flis, William J.	Dyna East Corporation
Foster, G. J.	Hadland Photonics
Fowler, Joe R.	Stress Engineering Services Inc.
Frank, Konrad	U. S. Army Ballistic Research Laboratory
Fry, Mark A.	Science Applications International Corporation
Garcia, John R.	Sandia National Laboratories
Garibotti, Joseph F.	AMETEK, Composite Materials and Appl. Division
Gathers, G. Roger	Lawrence Livermore National Laboratory
Genta, Philippe	Commissariat a l'Energie Atomique
Gilleland, John R.	GA Technologies, Inc.
Ginsberg, Michael J.	Los Alamos National Laboratory
Glaser, Robert J.	Jet Propulsion Laboratory
Glenn, Donald C.	TRW, Inc.
Glenn, Lewis A.	Lawrence Livermore National Laboratory
Godwin, Robert P.	Los Alamos National Laboratory
Gogolewski, Raymond P.	Consultant
Gora, Thaddeus	U. S. Army ARDEC
Gottlieb, James J.	Institute for Aerospace Studies, Ontario
Gould, Robert L.	R & D Associates
Grady, Dennis E.	Sandia National Laboratories
Grossman, Robert E.	Honeywell, DSD
Grover, Richard	Lawrence Livermore National Laboratory
Groves, Kenneth O.	Dow Chemical Company
Gualco, Massimo	Oto Melara S.p.A.
Guy, David J.	B. S. C. Stainless Cyclops
Gyllenspetz, Ingrid	Dynamec Research AB
Ham, Colin J.	Hunting Engineering Limited
Harmon Jr., Darell B.	BDM Corporation
Harris, J. L. (IO)	U. S. Army Missile Command
Harrison, John T.	U. S. Army Ballistic Research Laboratory
Hazelden, Peter S.	United Kingdom Laison Officer, U.S. Army ARDEC
Heatherly, Ewell R.	U. S. Army Strategic Defense Command
Heimdahl, Olaf E. R.	Naval Weapons Center
Hendrich, William R.	Martin Marietta Energy Systems, Inc.
Hereil, Pierre L.	Centre d'Etudes de GRAMAT
Herrmann, Walter	Sandia National Laboratories
Hinckley, William M.	Naval Surface Weapons Center
Hoffman, Samuel J.	Martin Marietta Orlando Aerospace

Hogan, Billy M.	Los Alamos National Laboratory
Hohler, Volker	Ernst-Mach-Institut
Holian, Brad L.	Los Alamos National Laboratory
Holian, Kathleen S.	Los Alamos National Laboratory
Holsapple, Keith A.	University of Washington
Holt, Alan C.	NASA-Johnson Space Center
Holt, William H.	Naval Surface Weapons Center
Hopper, V. D.	University of Melbourne
Hopson, John W.	Los Alamos National Laboratory
Horak, Henry L.	Los Alamos National Laboratory
Horz, Friedrich	NASA-Johnson Space Center
Hough, Gary R.	LTV Aerospace and Defense Company
Hüdepohl, Axel	Technische Universitaet Muenchen
Hughes, Christopher P.	Croton Institute
Igenbergs, Eduardo	Technische Universitaet Muenchen
Ilsen, Roland R.	The Aerospace Corporation
Isbell, William M.	General Research Corporation
Iyer, Kailasam R.	U. S. Army Research Office
Johnson, Gordon R.	Honeywell, Inc.
Johnson, Wallace E.	Computer Code Consultants
Jordan, Timothy A.	BDM Corporation
Kamegai, Minao	Lawrence Livermore National Laboratory
Karlsson, Hans E. V.	FFV, Sweden
Keller, Donald V.	Ktech Corporation
Kemmey, Peter J.	Defense Advanced Research Projects Agency
Kendall, Donald R.	TRW, Inc.
Kennedy, Donald R.	D. R. Kennedy & Associates, Inc.
Kennedy, Neil M.	FMC Corporation
Kerley, Gerald I.	Sandia National Laboratories
Ketchum, Donald E.	Southwest Research Institute
Kidron, Naftali	Rafael Armament Development Authority
Kineke, John H.	U. S. Army Ballistic Research Laboratory
Klein, Morris M.	Los Alamos National Laboratory
Knapp, Donald T.	The Aerospace Corporation
Knoth, Artur M.	Dornier GmbH
Kolkert, Willem J.	Prins Maurits Laboratory TNO
Konradi, Andrei	NASA-Johnson Space Center
Kramer, David	Rockwell International Corp., Rocketdyne Division
Kreyenhagen, Kenneth N.	California Research & Technology, Inc.
Krotiuk, William J.	Battelle Pacific Northwest Laboratory
Kusper, Richard L.	Xontech, Inc.
Kustas, Frank M.	Martin Marietta Denver Aerospace
Kusubov, Andre S.	Lawrence Livermore National Laboratory
Landwehr, George W.	Teledyne Brown Engineering
Laurent, Patrick	Commissariat a l'Energie Atomique
Lawrence, R. Jeffery	Sandia National Laboratories
Lawrence, William	U. S. Army Ballistic Research Laboratory
Leake, Gordon J.	Lawrence Livermore National Laboratory
Leiper, Graham A.	ICI PLC, Scotland
Lin, Jerry	Northwestern University
Linder, LTC David L.	Army Space Institute
Louie, Nikolai A.	McDonnell Douglas Corporation
Lowman Jr., Charles J.	System Planning Corporation
Lundberg, Lynn B.	Los Alamos National Laboratory
Lundberg, Randy L.	Defense Advanced Research Projects Agency
Maini, Tidu	Principia Mechanica
Majerus, Mark E.	California Research & Technology, Inc.
Malley, Michael M.	JAYCOR
Maruo, Toshiaki	The Japan Steel Works, Ltd.
Mathews, Floyd H.	Sandia National Laboratories
McCutchen, Don K.	NASA-Johnson Space Center
McKay, Michael W.	Science Applications International Corporation
McKinnon, William B.	Washington University
Meier, John K.	Los Alamos National Laboratory
Meier, Robert W.	Los Alamos National Laboratory
Melosh, Henry J.	University of Arizona
Meltz, Gerald	United Technologies Research Center
Meni, James	Science and Technology Associates, Inc.
Mescall, John F.	U. S. Army Materials Technology Laboratory
Meyers, Marc	U. S. Army Research Office
Miller, Clayton W.	Avco Systems Textron
Miller, Lewis D.	Army Foreign Science & Technology Center

Miller, Warner A.	Air Force Weapons Laboratory
Milleur, Mac B.	Martin Marietta Denver Aerospace
Mills Jr., Myron E.	LTV Aerospace and Defense Company
Minnich, Michael A.	General Motors, Military Vehicles Operation
Mixon, Larry C.	Test Track Division, Holloman AFB
Morgan, Bruce R.	Lawrence Livermore National Laboratory
Morgan, Thomas H.	National Research Council
Mott, Richard M.	General Research Corporation
Mullin, Scott A.	Southwest Research Institute
Mullins, John R.	New Technology, Inc.
Murphy, John A.	British Aerospace
Murty, Susarla S.	Lawrence Livermore National Laboratory
Nash, Phillip T.	Southwest Research Institute
Naz, Pierre	French-German Research Institute I.S.L.
Nellis, William J.	Lawrence Livermore National Laboratory
Nemat-Nasser, S.	University of California, San Diego
Nieder, Raymond L.	NASA-Johnson Space Center
Nuber, Edward	Geo-Centers, Inc.
Ogata, K.	Kobe Development Corporation
Ohlsson. Hans G.	BOFORS, Nobel Industries
Okajima, Keisuke	Mitsubishi Heavy Industries, Ltd.
Orphal, Dennis L.	California Research & Technology, Inc.
Ortell, Alex R.	Teledyne Brown Engineering
Osher, John E.	Lawrence Livermore National Laboratory
Otto, Jr., William L.	Aluminum Company of America
Ouzilleau, Alain	Canada Arsenals, Ltd.
O'Donoghue, Padraic E.	Southwest Research Institute
Parker, Jerald V.	Los Alamos National Laboratory
Pasman, Hans J.	Prins Maurits Laboratory TNO
Patton, Walter B.	Martin Marietta Orlando Aerospace
Pavel, Wieland W. P.	Industrieanlagen-Betriebsgelsellschaft mbH
Pearson, James C.	U. S. Army ARDEC
Peng, Steven T. J.	Jet Propulsion Laboratory
Perez, Elio	MATRA
Perie, Frederic	Engineering System International
Persechino, Mario A.	Potomac Research, Inc.
Persson, Ake	Dynamec Research AB
Pfanshauser, Wilhelm	Marco Scientific
Phillips, Gary	Science Applications International Corporation
Phinney, Ralph E.	Naval Surface Weapons Center
Picchi, CDR. Paolo	Italian Navy
Piekutowski, Andrew J.	University of Dayton Research Institute
Pitts, David E.	NASA-Johnson Space Center
Place, James R.	Los Alamos Technical Associates
Pomykal, Glenn W.	Lawrence Livermore National Laboratory
Ponsford, Henry T.	Pacifica Technology
Porter, James C.	Los Alamos National Laboratory
Principe, Frank S.	E. I. DuPont de Nemours & Co.
Puckett, Lawrence J.	U. S. Army Ballistic Research Laboratory
Radkowski, Peter P.	Radkowski Associates
Radkowski III, Peter P. F.	Radkowski Associates
Ragan, Charles E.	Los Alamos National Laboratory
Rambaut, Michel	Commissariat a l Energie Atomique
Rand, James L.	Winzen International, Inc.
Randles, Philip W.	Naval Research Laboratory
Rashid, Yusef R.	ANATECH
Recht, Rodney F.	Denver Research Institute
Reece, Eric W.	Sandia National Laboratories
Reed, Len	S-Cubed
Reid, P. W.	Royal Armament Research & Development Establishment
Rice, Thomas G.	Martin Marietta Michoud Division
Rice Jr., Thomas J.	Naval Surface Weapons Center
Riegel III, John P.	Southwest Research Institute
Rish, Frederick	Rockwell International
Roach, Jeffrey M.	New Mexico Tech/TERA Group
Roark, Glenn L.	Kaman Sciences Corporation
Robbins, Donald E.	NASA-Johnson Space Center
Robertson, William R.	Kaman Sciences Corporation
Rodriguez, David A.	Ford Aerospace & Comm. Corp., Aeronautic Division
Roschke, Paul	University of Texas at El Paso
Roth, George J.	University of Dayton Research Institute
Rott, Martin	Technische Universitaet Muenchen
Rottler, J. Steve	Sandia National Laboratories

Rucker, Ken	Hewlett Packard Co.
Rudolf, Karl P.	Diehl GmbH
Sabourin, Brian M.	Naval Surface Weapons Center
Schaffer, Marvin B.	The Rand Corporation
Scharff, Morris F.	Science Applications International Corporation
Schiesser, Ricardo	ENGESA, Brazil
Schmidt, Robert M.	Boeing Aerospace Company
Schneider, Eberhard	Ernst-Mach-Institut
Schriempf, J. T.	Physical Sciences, Inc.
Schultz, Peter	Brown University
Schumacher, Bruce J.	Lawrence Livermore National Laboratory
Scott, Stephen H.	U. S. Army Engineer Waterways Experiment Station
Seaman, Lynn	SRI International
Sears, Gerard A.	The Rand Corporation
Sedgwick, Robert T.	S-Cubed
See, Thomas H.	Lockheed Engineering & Management Services Co.
Sharma, Jagadish	Naval Surface Weapons Center
Shaw, Donald E.	System Planning Corporation
Shepard, William K.	New Mexico Tech/TERA Group
Sherohman, John W.	Lawrence Livermore National Laboratory
Shockey, Donald A.	SRI International
Simonds, Charles H.	Eagle Engineering, Inc.
Simpson, James S.	McDonnell Douglas Corporation
Smith, Jack E.	Oak Ridge National Laboratory
Smith, Vernon M.	Kaman Sciences Corporation
Snow, Peter B.	Kaman Sciences Corporation
Snowden, William E.	Office of Undersecretary of Defense for Res. & Eng.
Sobolik, Steven R.	Sandia National Laboratories
Sommers, Jr., William J.	Nichols Research Corporation
Sparks, Warren M.	Los Alamos National Laboratory
Speyer, Brian A.	Hadland Photonics, Ltd.
Spillman, George R.	Los Alamos National Laboratory
Stanley, Patrick M.	Los Alamos National Laboratory
Starostecki, Julian Z.	U. S. Army ARDEC
Starrett, John E.	University of California, San Diego
Stastny, Theodore G.	AAI Corporation
Stein, Robert A.	Battelle Columbus Division
Steinberg, Daniel J.	Lawrence Livermore National Laboratory
Sternberg, Joseph	Naval Postgraduate School
Stilp, Alois J.	Ernst-Mach-Institut
Stoesser, Paul R.	Dow Chemical Company
Sucov, Eugene W.	Westinghouse Electric Corporation
Summers, David A.	University of Missouri, Rolla
Swenson, Peter A.	Honeywell, Inc.
Swift, Hallock F.	Physics Applications, Inc.
Tabuchi, Tokiharu	The Japan Steel Works, Ltd.
Tackett, Gregory B.	Naval Surface Weapons Center
Tadjbakhsh, Iradj G.	U. S. Army Research Office
Tan, Tai-Ho	Los Alamos National Laboratory
Tasdemiroglu, Server	U. S. Army Tank & Automotive Command
Taylor, Raymond M.	Ford Aerospace & Comm. Corp., Aeronutronic Division
Taylor, Robert J.	LTV Aerospace and Defense Company
Thompson, Tommy	Lockheed Engineering & Management Services Co.
Thompson, Bruce M.	Intera Technologies, Inc.
Toeroek, Erwin I.	Institute Stravmann
Tosello, Robert	Direction des Armements Terrestres
Tower, Michael M.	LTV Aerospace and Defense Company
Trucano, Timothy G.	Sandia National Laboratories
Trzaska, Thomas J.	General Electric Co., Electromagnetic Launcher Opn.
Tucker, Thomas R.	Physical Sciences, Inc.
Tullos, Randy J.	Southwest Research Institute
Vanderbeek, F. James	U. S. Army Ballistic Research Laboratory
van der Hoek, Melis J.	Pisces International B.V.
Vander Matt, Paul	Los Alamos National Laboratory
Vanpoperynghe, Jehan	Commissariat a l Energie Atomique
Vickery, Ann M.	University of Arizona
von Laar, Klaus	Diehl GmbH
Waggener, Samuel S.	Naval Surface Weapons Center
Walsh, J. M.	Los Alamos National Laboratory
Warren, Truman W.	Mason & Hanger-Silas Mason Co., Inc.
Wasmund, Thomas L.	Naval Surface Weapons Center
Watt, J. William	Naval Surface Weapons Center
Weber, Adolf	Diehl GmbH

Weintraub, Herbert S.	Adelman Associates
Welch, Charles R.	U. S. Army Engineer Waterways Experiment Station
Weldon, William F.	University of Texas Center for Electromechanics
Wenzel, Alex B.	Southwest Research Institute
Westine, Peter S.	Southwest Research Institute
Wilbeck, James S.	Winzen International, Inc.
Wilkens, Mark L.	Lawrence Livermore National Laboratory
Williams, Andrew E.	Naval Research Laboratory
Wilson, Leonard L.	Air Force Armament Laboratory
Wright, Valri A.	Naval Surface Weapons Center
Yatteau, Jerome D.	Denver Research Institute
Yew, Ching H.	University of Texas
Yoon, Barbara L.	R & D Associates
Zernow, Louis	Zernow Technical Services, Inc.
Zook, John A.	U. S. Army Ballistic Research Laboratory
Zowarka, Raymond C.	University of Texas Center for Electromechanics
Zuieback, Joel M.	Sparta, Inc.
Zwissler, James G.	Jet Propulsion Laboratory

AUTHOR INDEX

SUBJECT INDEX

CONTENTS BY SUBJECT CATEGORY

NUMERICAL TECHNIQUES

MATERIAL RESPONSE MODELING

ANALYTICAL TECHNIQUES

PHENOMENOLOGY STUDIES

Space and Planetary Science Studies

Defense Related Studies

PHYSICOCHEMICAL HYDRODYNAMICS

Editor-in-Chief: D B SPALDING,
Bakery House, 40 High Street,
Wimbledon,
London SW19 5AU, UK

This international journal actively promotes the growth and recognition of physicochemical hydrodynamics as a coherent and structured interdisciplinary science. The scope of the journal includes: (i) Reporting the results of new research and the development of useful new techniques. (ii) Reviewing the state of knowledge in specific branches of physicochemical hydrodynamics in a way which will be comprehensible to experts in other fields. (iii) Providing useful computer programs for the prediction or analysis of phenomena. (iv) Announcing or reporting on relevant conferences. (v) Book reviews.

Subscription Information

Published 8 times per annum (Volumes 8 & 9)

Annual subscription (1987)	**DM614.00**
Two-year rate (1987/88)	**DM1166.60**

A selection of papers

Experimental and theoretical studies of submicron particle thermophoresis in combustion gases, A D EISNER & D E ROSNER.
Fragmentation of tin in water in a vapour explosion study, C P TSO *et al.*
Natural convection in a rectangular enclosure with differentially heated end walls and a free liquid surface, R W STICKLES *et al.*
Turbulence modelling for incompressible flows, W RODI.
Numerical studies of propagating flames exhibiting the Landau and Rayleigh-Taylore instabilities, D B SPALDING & J Z Y WU.
Waves in vertical annular flow, C J MARTIN & B J AZZOPARDI.
Tracer analysis of developing two-phase annular flow, G W LEMAN *et al.*
Developments in multidirectional holographic interferometry applied to crystal growth from solution, F BEDARIDA *et al.*
Prediction of flame spread hydrodynamics over liquid fuel, J A C HUMPHREY *et al.*
Limitations of the classical LDA formula for velocity measurements for large particles in two-phase suspension flows, A CARTELLIER & J L ACHARD.

FREE SAMPLE COPIES AVAILABLE ON REQUEST
Advertising rate card available on request. Back issues and current subscriptions are also available in microform.
The Deutsche Mark prices shown include postage and insurance. For subscription rates in Africa, Asia, Australasia, UK and Eire and the Americas apply to your nearest Pergamon office. Prices are subject to change without notice.

Pergamon

Headington Hill Hall, Oxford OX3 0BW, UK
Fairview Park, Elmsford, New York 10523, USA

5C/06/08/86

CORROSION SCIENCE

Official Journal of the Institution of Corrosion Science & Technology

Editor-in-Chief: J C SCULLY, *Department of Metallurgy, Houldsworth School of Applied Science, University of Leeds, Leeds LS2 9JT, UK*

Occurrence of corrosion and its practical control is an area of study covering a wide range of scientific knowledge. *Corrosion Science* provides a medium for the communication of ideas, developments and research in all aspects of the field.

The scope of this international journal is very wide. Published papers range from the highly theoretical to the essentially practical and cover such areas as high temperature oxidation, passivity anodic oxidation, biochemical corrosion, stress corrosion cracking and mechanisms and methods of corrosion control.

Corrosion Science is a well-established journal which forms an appropriate balance between corrosion science and technology and will serve as an authoritative reference work for anyone involved in engineering and materials sciences.

Subscription Information
Published monthly (Volume 27)

Annual subscription (1987)	**DM940.00**
Two-year rate (1987/88)	**DM1786.00**

A selection of papers

The segregation of alloy components in scales and sub-scales formed by binary alloys of manganese, iron, cobalt and nickel, B D BASTOW *et al.*

Some characteristic corrosion of stainless surface alloys laser processed on a mild steel, S CHIBA *et al.*

The effect of alloying elements on the grain boundary segregation of phosphorous in iron and the intergranular corrosion of the Fe-P System, W WEI & H J GRABKE.

A critical review of the protection of metals by paints, G W WALTER.

The prediction of corrosion by statistical analysis of corrosion profiles, J E STRUTT *et al.*

A review of environment-sensitive fracture in water reactor materials, P M SCOTT.

FREE SAMPLE COPIES AVAILABLE ON REQUEST
Advertising rate card available on request. Back issues and current subscriptions are also available in microform.
The Deutsche Mark prices shown include postage and insurance. For subscription rates in Africa, Asia, Australasia, UK and Eire and the Americas apply to your nearest Pergamon office. Prices are subject to change without notice.

Pergamon

Headington Hill Hall, Oxford OX3 0BW, UK
Fairview Park, Elmsford, New York 10523, USA

3D/01/08/86

INTERNATIONAL JOURNAL OF IMPACT ENGINEERING

AIMS AND SCOPE

The *International Journal of Impact Engineering* exists for the publication of original research work concerned with the response of structures and bodies to dynamic loads arising from exposure to blast, collision or other impact events. The topics encompassed by the journal include the elastic and plastic response of structures and bodies to impact and blast loading, terminal ballistics, vehicle crashworthiness, containment and other processes and phenomena in which effects due to impact predominate, such as explosive welding. In addition, related matters such as research into fire hazards in association with impact loads will be included. Every paper accepted for publication is strictly refereed and assessed for its contribution to the understanding of impact phenomena or the insight for engineering design in areas of high rate loading.

Papers whose concerns are only in the fields of mathematics, materials science or computation are not requested. Similarly, papers whose contents are purely descriptive will not be accepted.

The international character of the journal will, it is hoped, be reflected in contributions from all parts of the world; it will restrict itself, however, to publishing in English.

NOTES AND INSTRUCTIONS TO CONTRIBUTORS

1. The journal is devoted to original papers of high standard which contribute to the understanding of impact in various branches of engineering, and papers offered for publication are refereed and assessed by reference to these aims.

2. The language of the journal is English; however, contributions from all parts of the world are encouraged.

3. Two copies of the manuscript and figures, etc. of any paper which is offered for publication should be sent to:

Professor Norman Jones
Department of Mechanical Engineering
The University of Liverpool
P.O. Box 147, Liverpool L69 3BX, U.K.

If there are special considerations which an author wishes to raise they may be sent direct to the Editor-in-Chief.

4. Only papers not previously published will be accepted and authors must agree not to publish elsewhere a paper submitted to and accepted by the journal.

5. Papers of a length exceeding thirty A4 pages (manuscript and figures) will not be acceptable unless a prior arrangement has been made with the Editors.

Authors should examine previous issues of the journal as a general guide to the preparation of papers.

Papers not submitted in accordance with the following requirements will be returned.

(i) Manuscripts should be typed with double spacing and ample margins. Pages should be numbered consecutively.

(ii) Manuscripts should begin with the title of the article, the author's name, and the address from which the communication comes. In the case of co-authors, respective addresses should be clearly indicated, as well as which author is to receive correspondence and proofs for correction.

(continued on outside back cover)